Rational (Reciprocal) Function

$f(x) = \dfrac{1}{x}$

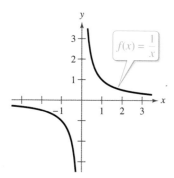

Domain: $(-\infty, 0) \cup (0, \infty)$
Range: $(-\infty, 0) \cup (0, \infty)$
No intercepts
Decreasing on $(-\infty, 0)$ and $(0, \infty)$
Odd function
Origin symmetry
Vertical asymptote: y-axis
Horizontal asymptote: x-axis

Exponential Function

$f(x) = a^x, \ a > 1$

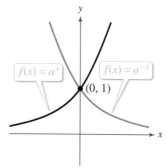

Domain: $(-\infty, \infty)$
Range: $(0, \infty)$
Intercept: $(0, 1)$
Increasing on $(-\infty, \infty)$
 for $f(x) = a^x$
Decreasing on $(-\infty, \infty)$
 for $f(x) = a^{-x}$
Horizontal asymptote: x-axis
Continuous

Logarithmic Function

$f(x) = \log_a x, \ a > 1$

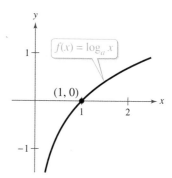

Domain: $(0, \infty)$
Range: $(-\infty, \infty)$
Intercept: $(1, 0)$
Increasing on $(0, \infty)$
Vertical asymptote: y-axis
Continuous
Reflection of graph of $f(x) = a^x$
 in the line $y = x$

Sine Function

$f(x) = \sin x$

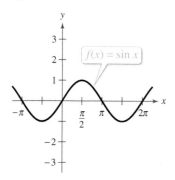

Domain: $(-\infty, \infty)$
Range: $[-1, 1]$
Period: 2π
x-intercepts: $(n\pi, 0)$
y-intercept: $(0, 0)$
Odd function
Origin symmetry

Cosine Function

$f(x) = \cos x$

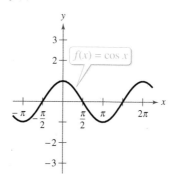

Domain: $(-\infty, \infty)$
Range: $[-1, 1]$
Period: 2π
x-intercepts: $\left(\dfrac{\pi}{2} + n\pi, 0\right)$
y-intercept: $(0, 1)$
Even function
y-axis symmetry

Tangent Function

$f(x) = \tan x$

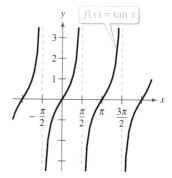

Domain: all $x \neq \dfrac{\pi}{2} + n\pi$

Range: $(-\infty, \infty)$
Period: π
x-intercepts: $(n\pi, 0)$
y-intercept: $(0, 0)$
Vertical asymptotes:

 $x = \dfrac{\pi}{2} + n\pi$

Odd function
Origin symmetry

Cosecant Function

$f(x) = \csc x$

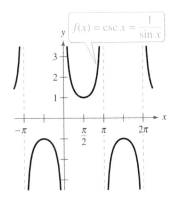

Domain: all $x \neq n\pi$
Range: $(-\infty, -1] \cup [1, \infty)$
Period: 2π
No intercepts
Vertical asymptotes: $x = n\pi$
Odd function
Origin symmetry

Secant Function

$f(x) = \sec x$

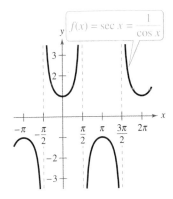

Domain: all $x \neq \dfrac{\pi}{2} + n\pi$

Range: $(-\infty, -1] \cup [1, \infty)$
Period: 2π
y-intercept: $(0, 1)$
Vertical asymptotes:

$$x = \frac{\pi}{2} + n\pi$$

Even function
y-axis symmetry

Cotangent Function

$f(x) = \cot x$

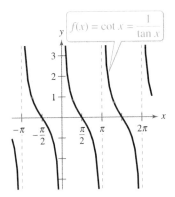

Domain: all $x \neq n\pi$
Range: $(-\infty, \infty)$
Period: π

x-intercepts: $\left(\dfrac{\pi}{2} + n\pi, 0\right)$

Vertical asymptotes: $x = n\pi$
Odd function
Origin symmetry

Inverse Sine Function

$f(x) = \arcsin x$

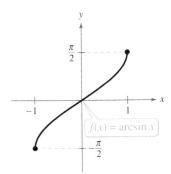

Domain: $[-1, 1]$

Range: $\left[-\dfrac{\pi}{2}, \dfrac{\pi}{2}\right]$

Intercept: $(0, 0)$
Odd function
Origin symmetry

Inverse Cosine Function

$f(x) = \arccos x$

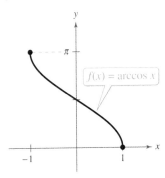

Domain: $[-1, 1]$
Range: $[0, \pi]$

y-intercept: $\left(0, \dfrac{\pi}{2}\right)$

Inverse Tangent Function

$f(x) = \arctan x$

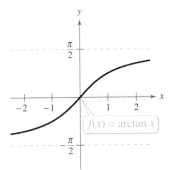

Domain: $(-\infty, \infty)$

Range: $\left(-\dfrac{\pi}{2}, \dfrac{\pi}{2}\right)$

Intercept: $(0, 0)$
Horizontal asymptotes:

$$y = \pm\frac{\pi}{2}$$

Odd function
Origin symmetry

ALGEBRA & TRIG

CalcChat® and CalcView®

11e

Ron Larson

The Pennsylvania State University
The Behrend College

CENGAGE

Australia • Brazil • Canada • Mexico • Singapore • United Kingdom • United States

Algebra and Trig
with CalcChat® and CalcView®
Eleventh Edition
Ron Larson

Product Director: Mark Santee

Senior Product Manager: Gary Whalen

Product Assistant: Tim Rogers

Senior Learning Designer: Laura Gallus

Executive Marketing Manager: Tom Ziolkowski

Content Manager: Rachel Pancare

Digital Delivery Lead: Nikkita Kendrick

IP Analyst: Ashley Maynard

IP Project Manager: Nick Barrows

Manufacturing Planner: Ron Montgomery

Production Service: Larson Texts, Inc.

Compositor: Larson Texts, Inc.

Illustrator: Larson Texts, Inc.

Text and Cover Designer: Larson Texts, Inc.

Cover Image: Romolo Tavani/Shutterstock.com

For product information and technology assistance, contact us at
Cengage Customer & Sales Support, 1-800-354-9706 or support.cengage.com.

For permission to use material from this text or product, submit all requests online at **www.cengage.com/permissions.**

Student Edition
ISBN: 978-0-357-45208-0

Loose-leaf Edition
ISBN: 978-0-357-45244-8

Cengage
200 Pier 4 Boulevard
Boston, MA 02210
USA

Cengage is a leading provider of customized learning solutions with employees residing in nearly 40 different countries and sales in more than 125 countries around the world. Find your local representative at **www.cengage.com.**

To learn more about Cengage platforms and services, register or access your online learning solution, or purchase materials for your course, visit **www.cengage.com.**

QR Code is a registered trademark of Denso Wave Incorporated.

Printed in the United States of America
Print Number: 02   Print Year: 2021

Contents

*Available at the text companion website *LarsonPrecalculus.com*

Preface

Welcome to *Algebra & Trig* with CalcChat® & CalcView®, Eleventh Edition. I am excited to offer you a new edition with more resources than ever that will help you understand and master algebra and trigonometry. This text includes features and resources that continue to make *Algebra & Trig* a valuable learning tool for students and a trustworthy teaching tool for instructors.

Algebra & Trig provides the clear instruction, precise mathematics, and thorough coverage that you expect for your course. Additionally, this new edition provides you with **free** access to a variety of digital resources:

- **GO DIGITAL**—direct access to digital content on your mobile device or computer

- **CalcView.com**—video solutions to selected exercises

- **CalcChat.com**—worked-out solutions to odd-numbered exercises and access to online tutors

- **LarsonPrecalculus.com**—companion website with resources to supplement your learning

These digital resources will help enhance and reinforce your understanding of the material presented in this text and prepare you for future mathematics courses. CalcView® and CalcChat® are also available as free mobile apps.

Features

NEW **GO DIGITAL**

Scan the QR codes ▦ on the pages of this text to *GO DIGITAL* on your mobile device. This will give you easy access from anywhere to instructional videos, solutions to exercises and Checkpoint problems, Skills Refresher videos, Interactive Activities, and many other resources.

UPDATED **CalcView** ®

The website *CalcView.com* provides video solutions of selected exercises. Watch instructors progress step-by-step through solutions, providing guidance to help you solve the exercises. The CalcView mobile app is available for free at the Apple® App Store® or Google Play™ store. You can access the video solutions by scanning the QR Code® at the beginning of the Section exercises, or visiting the *CalcView.com* website.

UPDATED **CalcChat** ®

Solutions to all odd-numbered exercises and tests are provided for free at *CalcChat.com*. Additionally, you can chat with a tutor, at no charge, during the hours posted at the site. For many years, millions of students have visited my site for help. The CalcChat mobile app is also available as a free download at the Apple® App Store® or Google Play™ store.

REVISED LarsonPrecalculus.com

All companion website features have been updated based on this revision, including two new features: Skills Refresher and Review & Refresh. Access to these features is free. You can view and listen to worked-out solutions of Checkpoint problems in English or Spanish, explore examples, download data sets, watch lesson videos, and much more.

NEW Skills Refresher

This feature directs you to an instructional video where you can review algebra skills needed to master the current topic. Scan the on-page code or go to *LarsonPrecalculus.com* to access the video.

▶▶ SKILLS REFRESHER

For a refresher on finding the sum, difference, product, or quotient of two polynomials, watch the video at *LarsonPrecalculus.com.*

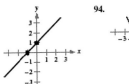

Review & Refresh ▶ Video solutions at *LarsonPrecalculus.com*

Evaluating an Expression In Exercises 89–92, evaluate the expression. (If not possible, state the reason.)

89. $\dfrac{5-7}{12-18}$ 90. $\dfrac{16-6}{6-11}$

91. $\dfrac{3-3}{4-0}$ 92. $\dfrac{1-(-1)}{9-9}$

Identifying x- and y-Intercepts In Exercises 93 and 94, identify x- and y-intercepts of the graph.

93.   94.

Sketching the Graph of an Equation In Exercises 95–98, test for symmetry and graph the equation. Then identify any intercepts.

95. $2x + y = 1$ 96. $3x - y = 7$
97. $y = x^2 + 2$ 98. $y = 2 - x^2$

NEW Review and Refresh

These exercises will help you to reinforce previously learned skills and concepts and to prepare for the next section. View and listen to worked-out solutions of the Review & Refresh exercises in English or Spanish by scanning the code on the first page of the section exercises or go to *LarsonPrecalculus.com*.

NEW Vocabulary and Concept Check

The Vocabulary and Concept Check appears at the beginning of the exercise set for each section. It includes fill-in-the-blank, matching, or non-computational questions designed to help you learn mathematical terminology and to test basic understanding of the concepts of the section.

NEW Summary and Study Strategies

The "What Did You Learn?" feature is a section-by-section overview that ties the learning objectives from the chapter to the Review Exercises for extra practice. The Study Strategies give concrete ways that you can use to help yourself with your study of mathematics.

REVISED Algebra Help

These notes reinforce or expand upon concepts, help you learn how to study mathematics, address special cases, or show alternative or additional steps to a solution of an example.

REVISED Exercise Sets

The exercise sets have been carefully and extensively examined to ensure they are rigorous and relevant, and include topics our users have suggested. The exercises have been reorganized and titled so you can better see the connections between examples and exercises. Multi-step, real-life exercises reinforce problem-solving skills and mastery of concepts by giving you the opportunity to apply the concepts in real-life situations. Two new sets of exercises, Vocabulary and Concept Check and Review & Refresh, have been added to help you develop and maintain your skills.

Section Objectives

A bulleted list of learning objectives provides you the opportunity to preview what will be presented in the upcoming section.

Side-By-Side Examples

Throughout the text, we present solutions to many examples from multiple perspectives—algebraically, graphically, and numerically. The side-by-side format of this pedagogical feature helps you to see that a problem can be solved in more than one way and to see that different methods yield the same result. The side-by-side format also addresses many different learning styles.

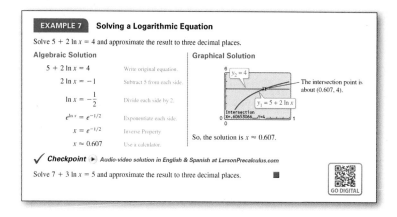

Checkpoints

Accompanying every example, the Checkpoint problems encourage immediate practice and check your understanding of the concepts presented in the example. View and listen to worked-out solutions of the Checkpoint problems in English or Spanish at *LarsonPrecalculus.com*. Scan the on-page code to access the solutions.

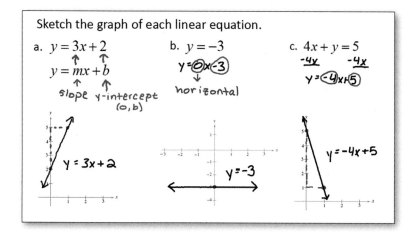

Technology

The technology feature gives suggestions for effectively using tools such as calculators, graphing utilities, and spreadsheet programs to help deepen your understanding of concepts, ease lengthy calculations, and provide alternate solution methods for verifying answers obtained by hand.

Historical Notes

These notes provide helpful information regarding famous mathematicians and their work.

Summarize (Section 3.2)

1. Explain how to use transformations to sketch graphs of polynomial functions (*page 252*). For an example of sketching transformations of monomial functions, see Example 1.
2. Explain how to apply the Leading Coefficient Test (*page 253*). For an example of applying the Leading Coefficient Test, see Example 2.
3. Explain how to find real zeros of polynomial functions and use them as sketching aids (*page 255*). For examples involving finding real zeros of polynomial functions, see Examples 3–5.
4. Explain how to use the Intermediate Value Theorem to help locate real zeros of polynomial functions (*page 258*). For an example of using the Intermediate Value Theorem, see Example 6.

Summarize
The Summarize feature at the end of each section helps you organize the lesson's key concepts into a concise summary, providing you with a valuable study tool. Use this feature to prepare for a homework assignment, to help you study for an exam, or as a review for previously covered sections.

Algebra of Calculus
Throughout the text, special emphasis is given to the algebraic techniques used in calculus. Algebra of Calculus examples and exercises are integrated throughout the text and are identified by the symbol $\int$.

Error Analysis
This exercise presents a sample solution that contains a common error which you are asked to identify.

How Do You See It?
The How Do You See It? feature in each section presents a real-life exercise that you will solve by visual inspection using the concepts learned in the lesson. This exercise is excellent for classroom discussion or test preparation.

Project
The projects at the end of selected sections involve in-depth applied exercises in which you will work with large, real-life data sets, often creating or analyzing models. These projects are offered online at *LarsonPrecalculus.com*.

Collaborative Project
You can find these extended group projects at *LarsonPrecalculus.com*. Check your understanding of the chapter concepts by solving in-depth, real-life problems. These collaborative projects provide an interesting and engaging way for you and other students to work together and investigate ideas.

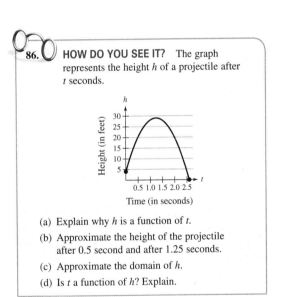

86. **HOW DO YOU SEE IT?**   The graph represents the height h of a projectile after t seconds.

(a) Explain why h is a function of t.
(b) Approximate the height of the projectile after 0.5 second and after 1.25 seconds.
(c) Approximate the domain of h.
(d) Is t a function of h? Explain.

CENGAGE | WEBASSIGN

Built by educators, WebAssign from Cengage is a fully customizable online solution for STEM disciplines. WebAssign includes the flexibility, tools, and content you need to create engaging learning experiences for your students. The patented grading engine provides unparalleled answer evaluation, giving students instant feedback, and insightful analytics highlight exactly where students are struggling. For more information, visit *cengage.com/webassign*.

Complete Solutions Manual

This manual contains solutions to all exercises from the text, including Chapter Review Exercises and Chapter Tests, and Practice Tests with solutions. The Complete Solutions Manual is available on the Instructor Companion Site.

Cengage Testing Powered by Cognero®

Cengage Testing, Powered by Cognero®, is a flexible online system that allows you to author, edit, and manage test bank content online. You can create multiple versions of your test in an instant and deliver tests from your LMS or exportable PDF or Word docs you print for in-class assessment. Cengage Testing is available online via *cengage.com*.

Instructor Companion Site

Everything you need for your course in one place! Access and download PowerPoint® presentations, test banks, the solutions manual, and more. This collection of book-specific lecture and class tools is available online via *cengage.com*.

Test Bank

The test bank contains text-specific multiple-choice and free response test forms and is available online at the Instructor Companion Site.

LarsonPrecalculus.com

In addition to its student resources, *LarsonPrecalculus.com* also has resources to help instructors. If you wish to challenge your students with multi-step and group projects, you can assign the Section Projects and Collaborative Projects. You can assess the knowledge of your students before and after each chapter using the pre- and post-tests. You can also give your students experience using an online graphing calculator with the Interactive Activities. You can access these features by going to *LarsonPrecalculus.com* or by scanning the on-page code ▦.

MathGraphs.com

For exercises that ask students to draw on the graph, I have provided **free,** printable graphs at *MathGraphs.com*. You can access these features by going to *MathGraphs.com* or by scanning the on-page code ▦ at the beginning of the section exercises, review exercises, or tests.

CENGAGE | WEBASSIGN

Prepare for class with confidence using WebAssign from Cengage. This online learning platform, which includes an interactive eBook, fuels practice, so that you truly absorb what you learn and prepare better for tests. Videos and tutorials walk you through concepts and deliver instant feedback and grading, so you always know where you stand in class. Focus your study time and get extra practice where you need it most. Study smarter with WebAssign! Ask your instructor today how you can get access to WebAssign, or learn about self-study options at *cengage.com/webassign*.

Student Study Guide and Solutions Manual
This guide offers step-by-step solutions for all odd-numbered text exercises, Chapter Tests, and Cumulative Tests. It also contains Practice Tests. For more information on how to access this digital resource, go to cengage.com

Note-Taking Guide
This is an innovative study aid, in the form of a notebook organizer, that helps students develop a section-by-section summary of key concepts. For more information on how to access this digital resource, go to cengage.com

LarsonPrecalculus.com
Of the many features at this website, students have told me that the videos are the most helpful. You can watch lesson videos by Dana Mosely as he explains various mathematical concepts. Other helpful features are the data downloads (editable spreadsheets so you do not have to enter the data), video solutions of the Checkpoint problems in English or Spanish, and the Student Success Organizer. The Student Success Organizer will help you organize the important concepts of each section using chapter outlines. You can access these features by going to *LarsonPrecalculus.com* or by scanning the on-page code.

CalcChat.com
This website provides free step-by-step solutions to all odd-numbered exercises and tests. Additionally, you can chat with a tutor, at no charge, during the hours posted at the site. You can access the solutions by going to *CalcChat.com* or by scanning the on-page code on the first page of the section exercises, review exercises, or tests.

CalcView.com
This website has video solutions of selected exercises. Watch instructors progress step-by-step through solutions, providing guidance to help you solve the exercises. You can access the videos by going to *CalcView.com* or by scanning the on-page code on the first page of the section exercises, review exercises, or tests.

MathGraphs.com
For exercises that ask you to draw on the graph, I have provided **free,** printable graphs at *MathGraphs.com.* You can access the printable graphs by going to *MathGraphs.com* or by scanning the on-page code on the first page of the section exercises, review exercises, or tests.

Acknowledgments

I would like to thank the many people who have helped me prepare the text and the supplements package. Their encouragement, criticisms, and suggestions have been invaluable.

Thank you to all of the instructors who took the time to review the changes in this edition and to provide suggestions for improving it. Without your help, this book would not be possible.

Reviewers of the Eleventh Edition

Ivette Chuca, *El Paso Community College*
Russell Murray, *St. Louis Community College-Meramec*
My Linh Nguyen, *The University of Texas at Dallas*
Michael Wallace, *Northern Virginia Community College*

Reviewers of the Previous Editions

Gurdial Arora, *Xavier University of Louisiana;* Darin Bauguess, *Surry Community College;* Timothy Andrew Brown, *South Georgia College;* Blair E. Caboot, *Keystone College;* Russell C. Chappell, *Twinsburg High School, Ohio;* Shannon Cornell, *Amarillo College;* Gayla Dance, *Millsaps College;* John Elias, *Glenda Dawson High School;* John Fellers, *North Allegheny School District;* Paul Finster, *El Paso Community College;* Paul A. Flasch, *Pima Community College West Campus;* Vadas Gintautas, *Chatham University;* Lorraine A. Hughes, *Mississippi State University;* Shu-Jen Huang, *University of Florida;* Renyetta Johnson, *East Mississippi Community College;* George Keihany, *Fort Valley State University;* Brianna Kurtz, *Daytona State College;* Mulatu Lemma, *Savannah State University;* Darlene Martin, *Lawson State Community College;* William Mays Jr., *Salem Community College;* Marcella Melby, *University of Minnesota;* Jonathan Prewett, *University of Wyoming;* Denise Reid, *Valdosta State University;* Professor Steven Sikes, *Collin College;* Ann Slate, *Surry Community College;* David L. Sonnier, *Lyon College;* David H. Tseng, *Miami Dade College—Kendall Campus;* Kimberly Walters, *Mississippi State University;* Richard Weil, *Brown College;* Solomon Willis, *Cleveland Community College;* Kathy Wood, *Lansing Catholic High School;* Bradley R. Young, *Darton College*

My thanks to Robert Hostetler, The Behrend College, The Pennsylvania State University, David Heyd, The Behrend College, The Pennsylvania State University, and David C. Falvo, The Behrend College, The Pennsylvania State University, for their significant contributions to previous editions of this text.

I would also like to thank the staff at Larson Texts, Inc. who assisted with proofreading the manuscript, preparing and proofreading the art package, checking and typesetting the supplements, and developing the websites *LarsonPrecalculus.com*, *CalcView.com*, *CalcChat.com*, and *MathGraphs.com*.

On a personal level, I am grateful to my spouse, Deanna Gilbert Larson, for her love, patience, and support. Also, a special thanks goes to R. Scott O'Neil. If you have suggestions for improving this text, please feel free to write to me. Over the past two decades, I have received many useful comments from both instructors and students, and I value these comments very highly.

Ron Larson, Ph.D.
Professor of Mathematics
Penn State University
www.RonLarson.com

P Prerequisites

GO DIGITAL

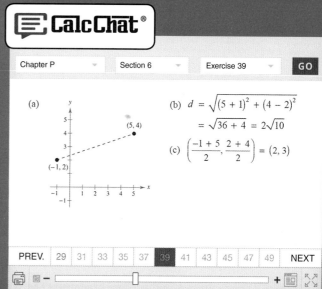

P.1 Federal Deficit *(Exercises 47–50, p. 13)*

P.6 Flying Distance *(Exercise 44, p. 58)*

1

P.1 Review of Real Numbers and Their Properties

Real numbers can represent many real-life quantities. For example, in Exercises 47–50 on page 13, you will use real numbers to represent the federal surplus or deficit.

> ● Represent and classify real numbers.
> ● Order real numbers and use inequalities.
> ● Find the absolute values of real numbers and find the distance between two real numbers.
> ● Evaluate algebraic expressions.
> ● Use the basic rules and properties of algebra.

Real Numbers

Real numbers can describe quantities in everyday life such as age, miles per gallon, and population. Real numbers are represented by symbols such as

$$-5, 9, 0, \frac{4}{3}, 0.666\ldots, 28.21, \sqrt{2}, \pi, \text{ and } \sqrt[3]{-32}.$$

Three commonly used **subsets** of real numbers are listed below. Each member in these subsets is also a member of the set of real numbers. (The three dots, called an *ellipsis*, indicate that the pattern continues indefinitely.)

$$\{1, 2, 3, 4, \ldots\} \qquad \text{Set of natural numbers}$$
$$\{0, 1, 2, 3, 4, \ldots\} \qquad \text{Set of whole numbers}$$
$$\{\ldots, -3, -2, -1, 0, 1, 2, 3, \ldots\} \qquad \text{Set of integers}$$

A real number is **rational** when it can be written as the ratio p/q of two integers, where $q \neq 0$. For example, the numbers

$$\frac{1}{3} = 0.3333\ldots = 0.\overline{3}, \quad \frac{1}{8} = 0.125, \quad \text{and} \quad \frac{125}{111} = 1.126126\ldots = 1.\overline{126}$$

are rational. The decimal form of a rational number either repeats $\left(\text{as in } \frac{173}{55} = 3.1\overline{45}\right)$ or terminates $\left(\text{as in } \frac{1}{2} = 0.5\right)$. A real number that cannot be written as the ratio of two integers is **irrational.** The decimal form of an irrational number neither terminates nor repeats. For example, the numbers

$$\sqrt{2} = 1.4142135\ldots \approx 1.41 \quad \text{and} \quad \pi = 3.1415926\ldots \approx 3.14$$

are irrational. (The symbol $\approx$ means "is approximately equal to.")

Several common subsets of the real numbers and their relationships to each other are shown in Figure P.1.

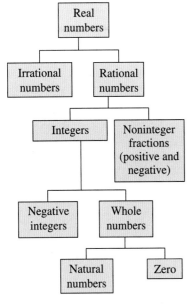

Common subsets of the real numbers
Figure P.1

EXAMPLE 1   Classifying Real Numbers

Determine which numbers in the set $\left\{-13, -\sqrt{5}, -1, -\frac{1}{3}, 0, \frac{5}{8}, \sqrt{2}, \pi, 7\right\}$ are (a) natural numbers, (b) whole numbers, (c) integers, (d) rational numbers, and (e) irrational numbers.

Solution

a. Natural numbers: $\{7\}$ **b.** Whole numbers: $\{0, 7\}$

c. Integers: $\{-13, -1, 0, 7\}$ **d.** Rational numbers: $\left\{-13, -1, -\frac{1}{3}, 0, \frac{5}{8}, 7\right\}$

e. Irrational numbers: $\left\{-\sqrt{5}, \sqrt{2}, \pi\right\}$

✓ *Checkpoint* ▶ *Audio-video solution in English & Spanish at LarsonPrecalculus.com*

Repeat Example 1 for the set $\left\{-\pi, -\frac{1}{4}, \frac{6}{3}, \frac{1}{2}\sqrt{2}, -7.5, -1, 8, -22\right\}$.   ■

Scan the [image] to access digital content available for this page.

GO DIGITAL

Real numbers are represented graphically on the **real number line.** When you draw a point on the real number line that corresponds to a real number, you are **plotting** the real number. The point representing 0 on the real number line is the **origin.** Numbers to the right of 0 are positive, and numbers to the left of 0 are negative, as shown in Figure P.2. The term **nonnegative** describes a number that is either positive or zero.

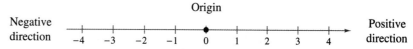

The real number line
Figure P.2

As illustrated in Figure P.3, there is a *one-to-one correspondence* between real numbers and points on the real number line.

Every real number corresponds to exactly one point on the real number line.
Figure P.3

Every point on the real number line corresponds to exactly one real number.

EXAMPLE 2   **Plotting Points on the Real Number Line**

Plot the real numbers on the real number line.

a. $-\dfrac{7}{4}$ **b.** 2.3   **c.** $\dfrac{2}{3}$ **d.** -1.8

Solution   The figure below shows all four points.

a. The point representing the real number

$$-\frac{7}{4} = -1.75 \qquad \text{Write in decimal form.}$$

lies between -2 and -1, but closer to -2, on the real number line.

b. The point representing the real number 2.3 lies between 2 and 3, but closer to 2, on the real number line.

c. The point representing the real number

$$\frac{2}{3} = 0.666\ldots \qquad \text{Write in decimal form.}$$

lies between 0 and 1, but closer to 1, on the real number line.

d. The point representing the real number -1.8 lies between -2 and -1, but closer to -2, on the real number line. Note that the point representing -1.8 lies slightly to the left of the point representing $-\frac{7}{4}$.

✓ *Checkpoint* ▶ *Audio-video solution in English & Spanish at LarsonPrecalculus.com*

Plot the real numbers on the real number line.

a. $\dfrac{5}{2}$ **b.** -1.6 **c.** $-\dfrac{3}{4}$ **d.** 0.7

Ordering Real Numbers

One important property of real numbers is that they are *ordered*. If a and b are real numbers, then a is *less than* b when $b - a$ is positive. The **inequality** $a < b$ denotes the **order** of a and b. This relationship can also be described by saying that b is *greater than a* and writing $b > a$. The inequality $a \leq b$ means that a is *less than or equal to b*, and the inequality $b \geq a$ means that b is *greater than or equal to a*. The symbols $<$, $>$, $\leq$, and $\geq$ are *inequality symbols*.

Geometrically, this implies that $a < b$ if and only if a lies to the *left* of b on the real number line, as shown in Figure P.4.

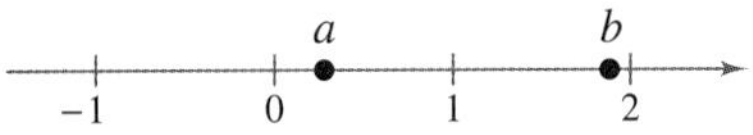

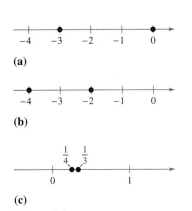

$a < b$ if and only if a lies to the left of b.

Figure P.4

EXAMPLE 3   Ordering Real Numbers

Place the appropriate inequality symbol ($<$ or $>$) between the pair of real numbers.

a. $-3, 0$ **b.** $-2, -4$ **c.** $\frac{1}{4}, \frac{1}{3}$

Solution

a. On the real number line, -3 lies to the left of 0, as shown in Figure P.5(a). So, you can say that -3 is *less than* 0, and write $-3 < 0$.

b. On the real number line, -2 lies to the right of -4, as shown in Figure P.5(b). So, you can say that -2 is *greater than* -4, and write $-2 > -4$.

c. On the real number line, $\frac{1}{4}$ lies to the left of $\frac{1}{3}$, as shown in Figure P.5(c). So, you can say that $\frac{1}{4}$ is *less than* $\frac{1}{3}$, and write $\frac{1}{4} < \frac{1}{3}$.

✓ *Checkpoint* ▶ Audio-video solution in English & Spanish at LarsonPrecalculus.com

Place the appropriate inequality symbol ($<$ or $>$) between the pair of real numbers.

a. $1, -5$ **b.** $\frac{3}{2}, 7$ **c.** $-\frac{2}{3}, -\frac{3}{4}$

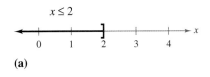

(a)

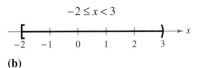

(b)

$\frac{1}{4}$ $\frac{1}{3}$

(c)

Figure P.5

EXAMPLE 4   Interpreting Inequalities

▶▶▶ *See LarsonPrecalculus.com for an interactive version of this type of example.*

Describe the subset of real numbers that the inequality represents.

a. $x \leq 2$ **b.** $-2 \leq x < 3$

Solution

a. The inequality $x \leq 2$ denotes all real numbers less than or equal to 2, as shown in Figure P.6(a). In the figure, the bracket at 2 indicates 2 is *included* in the interval.

b. The inequality $-2 \leq x < 3$ means that $x \geq -2$ *and* $x < 3$. This "double inequality" denotes all real numbers between -2 and 3, including -2 but not including 3, as shown in Figure P.6(b). In the figure, the bracket at -2 indicates -2 is *included* in the interval, and the parenthesis at 3 indicates that 3 is *not* included in the interval.

✓ *Checkpoint* ▶ Audio-video solution in English & Spanish at LarsonPrecalculus.com

Describe the subset of real numbers that the inequality represents.

a. $x > -3$ **b.** $0 < x \leq 4$

$x \leq 2$

(a)

$-2 \leq x < 3$

(b)

Figure P.6

Inequalities can describe subsets of real numbers called **intervals.** In the bounded intervals on the next page, the real numbers a and b are the **endpoints** of each interval. The endpoints of a closed interval are included in the interval, whereas the endpoints of an open interval are not included in the interval.

GO DIGITAL

Bounded Intervals on the Real Number Line

Let a and b be real numbers such that $a < b$.

Notation	Interval Type	Inequality	Graph
$[a, b]$	Closed	$a \le x \le b$	
(a, b)	Open	$a < x < b$	
$[a, b)$		$a \le x < b$	
$(a, b]$		$a < x \le b$	

The reason that the four types of intervals above are called **bounded** is that each has a finite length. An interval that does not have a finite length is **unbounded.** Note in the unbounded intervals below that the symbols ∞, **positive infinity,** and $-\infty$, **negative infinity,** do not represent real numbers. They are convenient symbols used to describe the unboundedness of intervals such as $(1, \infty)$ or $(-\infty, 3]$.

Unbounded Intervals on the Real Number Line

Let a and b be real numbers.

Notation	Interval Type	Inequality	Graph
$[a, \infty)$		$x \ge a$	
(a, ∞)	Open	$x > a$	
$(-\infty, b]$		$x \le b$	
$(-\infty, b)$	Open	$x < b$	
$(-\infty, \infty)$	Entire real line	$-\infty < x < \infty$	

GO DIGITAL

EXAMPLE 5   **Representing Intervals**

Verbal	Algebraic	Graphical
a. All real numbers greater than -1 and less than 3	$(-1, 3)$ or $-1 < x < 3$	See Figure P.7(a).
b. All real numbers greater than or equal to 2	$[2, \infty)$ or $x \ge 2$	See Figure P.7(b).
c. All real numbers less than or equal to 2	$(-\infty, 2]$ or $x \le 2$	See Figure P.7(c).
d. All real numbers greater than -3 and less than or equal to 5	$(-3, 5]$ or $-3 < x \le 5$	See Figure P.7(d).

✓ *Checkpoint* ▶ Audio-video solution in English & Spanish at LarsonPrecalculus.com

a. Represent the interval $[-2, 5)$ verbally, as an inequality, and as a graph.

b. Represent the statement "x is less than 4 and at least -2" as an interval, an inequality, and a graph.

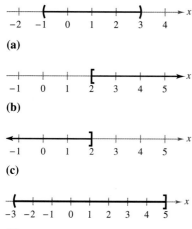

(a)

(b)

(c)

(d)

Figure P.7

Absolute Value and Distance

The **absolute value** of a real number is its *magnitude,* or the distance between the origin and the point representing the real number on the real number line.

> **Definition of Absolute Value**
>
> If a is a real number, then the **absolute value** of a is
>
> $$|a| = \begin{cases} a, & a \geq 0 \\ -a, & a < 0 \end{cases}.$$

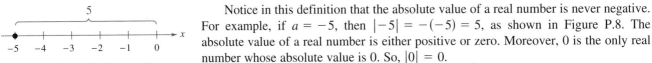

Absolute value as the distance from the origin

Figure P.8

Notice in this definition that the absolute value of a real number is never negative. For example, if $a = -5$, then $|-5| = -(-5) = 5$, as shown in Figure P.8. The absolute value of a real number is either positive or zero. Moreover, 0 is the only real number whose absolute value is 0. So, $|0| = 0$.

> **Properties of Absolute Values**
>
> **1.** $|a| \geq 0$ **2.** $|-a| = |a|$
>
> **3.** $|ab| = |a||b|$ **4.** $\left|\dfrac{a}{b}\right| = \dfrac{|a|}{|b|}, \quad b \neq 0$

EXAMPLE 6 **Finding Absolute Values**

a. $|-15| = 15$ **b.** $\left|\dfrac{2}{3}\right| = \dfrac{2}{3}$

c. $|-4.3| = 4.3$ **d.** $-|-6| = -(6) = -6$

✓ *Checkpoint*  Audio-video solution in English & Spanish at LarsonPrecalculus.com

Evaluate each expression.

a. $|1|$ **b.** $-\left|\dfrac{3}{4}\right|$ **c.** $\dfrac{2}{|-3|}$ **d.** $-|0.7|$

EXAMPLE 7 **Evaluating an Absolute Value Expression**

Evaluate $\dfrac{|x|}{x}$ for (a) $x > 0$ and (b) $x < 0$.

Solution

a. If $x > 0$, then x is positive and $|x| = x$. So, $\dfrac{|x|}{x} = \dfrac{x}{x} = 1$.

b. If $x < 0$, then x is negative and $|x| = -x$. So, $\dfrac{|x|}{x} = \dfrac{-x}{x} = -1$.

✓ *Checkpoint*  Audio-video solution in English & Spanish at LarsonPrecalculus.com

Evaluate $\dfrac{|x + 3|}{x + 3}$ for (a) $x > -3$ and (b) $x < -3$.

GO DIGITAL

The **Law of Trichotomy** states that for any two real numbers a and b, *precisely* one of three relationships is possible:

$$a = b, \quad a < b, \quad \text{or} \quad a > b. \qquad \text{Law of Trichotomy}$$

In words, this property tells you that if a and b are any two real numbers, then a is equal to b, a is less than b, or a is greater than b.

EXAMPLE 8   Comparing Real Numbers

Place the appropriate symbol ($<$, $>$, or $=$) between the pair of real numbers.

a. $|-4|$ ▨ $|3|$ **b.** $|-10|$ ▨ $|10|$ **c.** $-|-7|$ ▨ $|-7|$

Solution

a. $|-4| > |3|$ because $|-4| = 4$ and $|3| = 3$, and 4 is greater than 3.
b. $|-10| = |10|$ because $|-10| = 10$ and $|10| = 10$.
c. $-|-7| < |-7|$ because $-|-7| = -7$ and $|-7| = 7$, and -7 is less than 7.

✓ *Checkpoint* ▶ Audio-video solution in English & Spanish at LarsonPrecalculus.com

Place the appropriate symbol ($<$, $>$, or $=$) between the pair of real numbers.

a. $|-3|$ ▨ $|4|$ **b.** $-|-4|$ ▨ $-|4|$ **c.** $|-3|$ ▨ $-|-3|$ ■

Absolute value can be used to find the distance between two points on the real number line. For example, the distance between -3 and 4 is

$$\begin{aligned} |-3 - 4| &= |-7| \\ &= 7 \qquad \text{Distance between } -3 \text{ and } 4 \end{aligned}$$

as shown in Figure P.9.

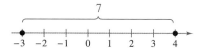

The distance between -3 and 4 is 7.
Figure P.9

Distance Between Two Points on the Real Number Line

Let a and b be real numbers. The **distance between a and b** is

$$d(a, b) = |b - a| = |a - b|.$$

EXAMPLE 9   Finding a Distance

Find the distance between -25 and 13.

Solution

The distance between -25 and 13 is

$$|-25 - 13| = |-38| = 38. \qquad \text{Distance between } -25 \text{ and } 13$$

The distance can also be found as follows.

$$|13 - (-25)| = |38| = 38 \qquad \text{Distance between } -25 \text{ and } 13$$

✓ *Checkpoint* ▶ Audio-video solution in English & Spanish at LarsonPrecalculus.com

Find the distance between each pair of real numbers.

a. 35 and -23 **b.** -35 and -23 **c.** 35 and 23 ■

One application of finding the distance between two points on the real number line is finding a change in temperature.

GO DIGITAL

Algebraic Expressions

One characteristic of algebra is the use of letters to represent numbers. The letters are **variables,** and combinations of letters and numbers are **algebraic expressions.** Here are a few examples of algebraic expressions.

$$5x, \qquad 2x - 3, \qquad \frac{4}{x^2 + 2}, \qquad 7x + y$$

> **Definition of an Algebraic Expression**
>
> An **algebraic expression** is a collection of letters (**variables**) and real numbers (**constants**) combined using the operations of addition, subtraction, multiplication, division, and exponentiation.

The **terms** of an algebraic expression are those parts that are separated by *addition.* For example, $x^2 - 5x + 8 = x^2 + (-5x) + 8$ has three terms: x^2 and $-5x$ are the **variable terms** and 8 is the **constant term.** For terms such as x^2, $-5x$, and 8, the numerical factor is the **coefficient.** Here, the coefficients are 1, -5, and 8.

EXAMPLE 10   **Identifying Terms and Coefficients**

Algebraic Expression	Terms	Coefficients
a. $5x - \dfrac{1}{7}$	$5x, -\dfrac{1}{7}$	$5, -\dfrac{1}{7}$
b. $2x^2 - 6x + 9$	$2x^2, -6x, 9$	$2, -6, 9$
c. $\dfrac{3}{x} + \dfrac{1}{2}x^4 - y$	$\dfrac{3}{x}, \dfrac{1}{2}x^4, -y$	$3, \dfrac{1}{2}, -1$

✓ *Checkpoint* ▶ *Audio-video solution in English & Spanish at LarsonPrecalculus.com*

Identify the terms and coefficients of $-2x + 4$. ■

The **Substitution Principle** states, "If $a = b$, then b can replace a in any expression involving a." Use the Substitution Principle to **evaluate** an algebraic expression by substituting values for each of the variables in the expression. The next example illustrates this.

EXAMPLE 11   **Evaluating Algebraic Expressions**

Expression	Value of Variable	Substitution	Value of Expression
a. $-3x + 5$	$x = 3$	$-3(3) + 5$	$-9 + 5 = -4$
b. $3x^2 + 2x - 1$	$x = -1$	$3(-1)^2 + 2(-1) - 1$	$3 - 2 - 1 = 0$
c. $\dfrac{2x}{x + 1}$	$x = -3$	$\dfrac{2(-3)}{-3 + 1}$	$\dfrac{-6}{-2} = 3$

Note that you must substitute the value for *each* occurrence of the variable.

✓ *Checkpoint* ▶ *Audio-video solution in English & Spanish at LarsonPrecalculus.com*

Evaluate $4x - 5$ when $x = 0$. ■

Basic Rules of Algebra

There are four arithmetic operations with real numbers: *addition, multiplication, subtraction,* and *division,* denoted by the symbols $+$, $\times$ or $\cdot$, $-$, and $\div$ or $/$, respectively. Of these, addition and multiplication are the two primary operations. Subtraction and division are the inverse operations of addition and multiplication, respectively.

Definitions of Subtraction and Division

Subtraction: Add the opposite.   **Division:** Multiply by the reciprocal.

$$a - b = a + (-b) \qquad\qquad \text{If } b \neq 0, \text{ then } a/b = a\!\left(\frac{1}{b}\right) = \frac{a}{b}.$$

In these definitions, $-b$ is the **additive inverse** (or opposite) of b, and $1/b$ is the **multiplicative inverse** (or reciprocal) of b. In the fractional form a/b, a is the **numerator** of the fraction and b is the **denominator.**

The properties of real numbers below are true for variables and algebraic expressions as well as for real numbers, so they are often called the **Basic Rules of Algebra.** Formulate a verbal description of each of these properties. For example, the first property states that *the order in which two real numbers are added does not affect their sum.*

Basic Rules of Algebra

Let a, b, and c be real numbers, variables, or algebraic expressions.

Property		Example
Commutative Property of Addition:	$a + b = b + a$	$4x + x^2 = x^2 + 4x$
Commutative Property of Multiplication:	$ab = ba$	$(4 - x)x^2 = x^2(4 - x)$
Associative Property of Addition:	$(a + b) + c = a + (b + c)$	$(x + 5) + x^2 = x + (5 + x^2)$
Associative Property of Multiplication:	$(ab)c = a(bc)$	$(2x \cdot 3y)(8) = (2x)(3y \cdot 8)$
Distributive Properties:	$a(b + c) = ab + ac$	$3x(5 + 2x) = 3x \cdot 5 + 3x \cdot 2x$
	$(a + b)c = ac + bc$	$(y + 8)y = y \cdot y + 8 \cdot y$
Additive Identity Property:	$a + 0 = a$	$5y^2 + 0 = 5y^2$
Multiplicative Identity Property:	$a \cdot 1 = a$	$(4x^2)(1) = 4x^2$
Additive Inverse Property:	$a + (-a) = 0$	$5x^3 + (-5x^3) = 0$
Multiplicative Inverse Property:	$a \cdot \dfrac{1}{a} = 1, \quad a \neq 0$	$(x^2 + 4)\!\left(\dfrac{1}{x^2 + 4}\right) = 1$

Subtraction is defined as "adding the opposite," so the Distributive Properties are also true for subtraction. For example, the "subtraction form" of $a(b + c) = ab + ac$ is $a(b - c) = ab - ac$. Note that the operations of subtraction and division are neither commutative nor associative. The examples

$$7 - 3 \neq 3 - 7 \quad \text{and} \quad 20 \div 4 \neq 4 \div 20$$

show that subtraction and division are not commutative. Similarly

$$5 - (3 - 2) \neq (5 - 3) - 2 \quad \text{and} \quad 16 \div (4 \div 2) \neq (16 \div 4) \div 2$$

demonstrate that subtraction and division are not associative.

GO DIGITAL

EXAMPLE 12   **Identifying Rules of Algebra**

Identify the rule of algebra illustrated by the statement.

a. $\left(5x^3\right)2 = 2\left(5x^3\right)$ **b.** $(4x + 3) - (4x + 3) = 0$

c. $7x \cdot \dfrac{1}{7x} = 1, \quad x \neq 0$ **d.** $\left(2 + 5x^2\right) + x^2 = 2 + \left(5x^2 + x^2\right)$

Solution

a. This statement illustrates the Commutative Property of Multiplication. In other words, you obtain the same result whether you multiply $5x^3$ by 2, or 2 by $5x^3$.

b. This statement illustrates the Additive Inverse Property. In terms of subtraction, this property states that when any expression is subtracted from itself, the result is 0.

c. This statement illustrates the Multiplicative Inverse Property. Note that x must be a nonzero number. The reciprocal of x is undefined when x is 0.

d. This statement illustrates the Associative Property of Addition. In other words, to form the sum $2 + 5x^2 + x^2$, it does not matter whether 2 and $5x^2$, or $5x^2$ and x^2 are added first.

✓ *Checkpoint* **Audio-video solution in English & Spanish at LarsonPrecalculus.com**

Identify the rule of algebra illustrated by the statement.

a. $x + 9 = 9 + x$ **b.** $5(x^3 \cdot 2) = (5x^3)2$ **c.** $(2 + 5x^2)y^2 = 2 \cdot y^2 + 5x^2 \cdot y^2$

GO DIGITAL

Properties of Negation and Equality

Let a, b, and c be real numbers, variables, or algebraic expressions.

Property	Example
1. $(-1)a = -a$	$(-1)7 = -7$
2. $-(-a) = a$	$-(-6) = 6$
3. $(-a)b = -(ab) = a(-b)$	$(-5)3 = -(5 \cdot 3) = 5(-3)$
4. $(-a)(-b) = ab$	$(-2)(-x) = 2x$
5. $-(a + b) = (-a) + (-b)$	$-(x + 8) = (-x) + (-8)$
	$\qquad\qquad = -x - 8$
6. If $a = b$, then $a \pm c = b \pm c$.	$\frac{1}{2} + 3 = 0.5 + 3$
7. If $a = b$, then $ac = bc$.	$4^2 \cdot 2 = 16 \cdot 2$
8. If $a \pm c = b \pm c$, then $a = b$.	$1.4 - 1 = \frac{7}{5} - 1 \implies 1.4 = \frac{7}{5}$
9. If $ac = bc$ and $c \neq 0$, then $a = b$.	$3x = 3 \cdot 4 \implies x = 4$

Properties of Zero

Let a and b be real numbers, variables, or algebraic expressions.

1. $a + 0 = a$ and $a - 0 = a$ **2.** $a \cdot 0 = 0$

3. $\dfrac{0}{a} = 0, \quad a \neq 0$ **4.** $\dfrac{a}{0}$ is undefined.

5. Zero-Factor Property: If $ab = 0$, then $a = 0$ or $b = 0$.

Properties and Operations of Fractions

Let a, b, c, and d be real numbers, variables, or algebraic expressions such that $b \neq 0$ and $d \neq 0$.

1. **Equivalent Fractions:** $\dfrac{a}{b} = \dfrac{c}{d}$ if and only if $ad = bc$.

2. **Rules of Signs:** $-\dfrac{a}{b} = \dfrac{-a}{b} = \dfrac{a}{-b}$ and $\dfrac{-a}{-b} = \dfrac{a}{b}$

3. **Generate Equivalent Fractions:** $\dfrac{a}{b} = \dfrac{ac}{bc}, \quad c \neq 0$

4. **Add or Subtract with Like Denominators:** $\dfrac{a}{b} \pm \dfrac{c}{b} = \dfrac{a \pm c}{b}$

5. **Add or Subtract with Unlike Denominators:** $\dfrac{a}{b} \pm \dfrac{c}{d} = \dfrac{ad \pm bc}{bd}$

6. **Multiply Fractions:** $\dfrac{a}{b} \cdot \dfrac{c}{d} = \dfrac{ac}{bd}$

7. **Divide Fractions:** $\dfrac{a}{b} \div \dfrac{c}{d} = \dfrac{a}{b} \cdot \dfrac{d}{c} = \dfrac{ad}{bc}, \quad c \neq 0$

EXAMPLE 13 **Properties and Operations of Fractions**

a. $\dfrac{x}{5} = \dfrac{3 \cdot x}{3 \cdot 5} = \dfrac{3x}{15}$ Property 3   b. $\dfrac{7}{x} \div \dfrac{3}{2} = \dfrac{7}{x} \cdot \dfrac{2}{3} = \dfrac{14}{3x}$ Property 7

✓ *Checkpoint* 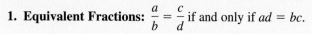 Audio-video solution in English & Spanish at LarsonPrecalculus.com

a. Multiply fractions: $\dfrac{3}{5} \cdot \dfrac{x}{6}$ b. Add fractions: $\dfrac{x}{10} + \dfrac{2x}{5}$ ∎

If a, b, and c are integers such that $ab = c$, then a and b are **factors** or **divisors** of c. A **prime number** is an integer that has exactly two positive factors—itself and 1—such as 2, 3, 5, 7, and 11. The numbers 4, 6, 8, 9, and 10 are **composite** because each can be written as the product of two or more prime numbers. The number 1 is neither prime nor composite. The **Fundamental Theorem of Arithmetic** states that every positive integer greater than 1 is prime or can be written as the product of prime numbers in precisely one way (disregarding order). For example, the **prime factorization** of 24 is $24 = 2 \cdot 2 \cdot 2 \cdot 3$.

Summarize (Section P.1)

1. Explain how to represent and classify real numbers (*pages 2 and 3*). For examples of representing and classifying real numbers, see Examples 1 and 2.

2. Explain how to order real numbers and use inequalities (*pages 4 and 5*). For examples of ordering real numbers and using inequalities, see Examples 3–5.

3. State the definition of the absolute value of a real number (*page 6*). For examples of using absolute value, see Examples 6–9.

4. Explain how to evaluate an algebraic expression (*page 8*). For examples involving algebraic expressions, see Examples 10 and 11.

5. State the basic rules and properties of algebra (*pages 9–11*). For examples involving the basic rules and properties of algebra, see Examples 12 and 13.

GO DIGITAL

P.1 Exercises

See CalcChat.com for tutorial help and worked-out solutions to odd-numbered exercises.

GO DIGITAL

Vocabulary and Concept Check

In Exercises 1–4, fill in the blanks.

1. The decimal form of an _____ number neither terminates nor repeats.

2. The point representing 0 on the real number line is the _____.

3. The _____ of an algebraic expression are those parts that are separated by addition.

4. The _____ _____ states that if $ab = 0$, then $a = 0$ or $b = 0$.

5. Is $|3 - 10|$ equal to $|10 - 3|$? Explain.

6. Match each property with its name.

   (a) Commutative Property of Addition

   (b) Additive Inverse Property

   (c) Distributive Property

   (d) Associative Property of Addition

   (e) Multiplicative Identity Property

   (i) $a \cdot 1 = a$

   (ii) $a(b + c) = ab + ac$

   (iii) $a + b = b + a$

   (iv) $a + (-a) = 0$

   (v) $(a + b) + c = a + (b + c)$

Skills and Applications

Classifying Real Numbers In Exercises 7–10, determine which numbers in the set are (a) natural numbers, (b) whole numbers, (c) integers, (d) rational numbers, and (e) irrational numbers.

7. $\left\{ -9, -\frac{7}{2}, 5, \frac{2}{3}, \sqrt{3}, 0, 8, -4, 2, -11 \right\}$

8. $\left\{ \sqrt{5}, -7, -\frac{7}{3}, 0, 3.14, \frac{5}{4}, -3, 12, 5 \right\}$

9. $\{ 2.01, 0.\overline{6}, -13, 0.010110111\ldots, 1, -6 \}$

10. $\left\{ 25, -17, -\frac{12}{5}, \sqrt{9}, 3.12, \frac{1}{2}\pi, 18, -11.1, 13 \right\}$

Plotting and Ordering Real Numbers In Exercises 11–16, plot the two real numbers on the real number line. Then place the appropriate inequality symbol ($<$ or $>$) between them.

11. $-4, -8$

12. $1, \frac{16}{3}$

13. $\frac{5}{6}, \frac{2}{3}$

14. $-\frac{8}{7}, -\frac{3}{7}$

15. $-5.2, -8.5$

16. $-\frac{4}{3}, -4.75$

Interpreting an Inequality In Exercises 17–20, describe the subset of real numbers that the inequality represents.

17. $x \leq 5$

18. $x < 0$

19. $-2 < x < 2$

20. $0 < x \leq 6$

Representing an Interval In Exercises 21–24, represent the interval verbally, as an inequality, and as a graph.

21. $[4, \infty)$

22. $(-\infty, 2)$

23. $[-5, 2)$

24. $(-1, 2]$

Representing an Interval In Exercises 25–28, represent the statement as an interval, an inequality, and a graph.

25. y is nonpositive.

26. y is no more than 25.

27. t is at least 10 and at most 22.

28. k is less than 5 but no less than -3.

Evaluating an Absolute Value Expression In Exercises 29–38, evaluate the expression.

29. $|-10|$

30. $|0|$

31. $|3 - 8|$

32. $|6 - 2|$

33. $|-1| - |-2|$

34. $-3 - |-3|$

35. $5|-5|$

36. $-4|-4|$

37. $\dfrac{|x + 2|}{x + 2}, \quad x < -2$

38. $\dfrac{|x - 1|}{x - 1}, \quad x > 1$

Comparing Real Numbers In Exercises 39–42, place the appropriate symbol ($<$, $>$, or $=$) between the pair of real numbers.

39. $|-4|$ ▨ $|4|$

40. -5 ▨ $-|5|$

41. $-|-6|$ ▨ $|-6|$

42. $-|-2|$ ▨ $-|2|$

Finding a Distance In Exercises 43–46, find the distance between a and b.

43. $a = 126, b = 75$

44. $a = -20, b = 30$

45. $a = -\frac{5}{2}, b = 0$

46. $a = -\frac{1}{4}, b = -\frac{11}{4}$

A blue exercise number indicates that a video solution can be seen at *CalcView.com*.

Federal Deficit

In Exercises 47–50, use the bar graph, which shows the receipts of the federal government (in billions of dollars) for selected years from 2012 through 2018. In each exercise, you are given the expenditures of the federal government. Find the magnitude of the surplus or deficit for the year. *(Source: U.S. Office of Management and Budget)*

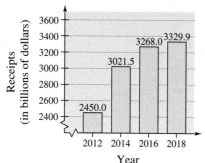

Year	Receipts, R	Expenditures, E	$|R - E|$
47. 2012		\$3526.6 billion	
48. 2014		\$3506.3 billion	
49. 2016		\$3852.6 billion	
50. 2018		\$4109.0 billion	

Identifying Terms and Coefficients In Exercises 51–54, identify the terms. Then identify the coefficients of the variable terms of the expression.

51. $7x + 4$

52. $6x^3 - 5x$

53. $4x^3 + 0.5x - 5$

54. $3\sqrt{3}x^2 + 1$

Evaluating an Algebraic Expression In Exercises 55 and 56, evaluate the expression for each value of x. (If not possible, state the reason.)

55. $x^2 - 3x + 2$ (a) $x = 0$ (b) $x = -1$

56. $\dfrac{x - 2}{x + 2}$ (a) $x = 2$ (b) $x = -2$

Operations with Fractions In Exercises 57–60, perform the operation. (Write fractional answers in simplest form.)

57. $\dfrac{2x}{3} - \dfrac{x}{4}$

58. $\dfrac{3x}{4} + \dfrac{x}{5}$

59. $\dfrac{3x}{10} \cdot \dfrac{5}{6}$

60. $\dfrac{2x}{3} \div \dfrac{6}{7}$

Exploring the Concepts

True or False? In Exercises 61 and 62, determine whether the statement is true or false. Justify your answer.

61. Every nonnegative number is positive.

62. If $a < 0$ and $b < 0$, then $ab > 0$.

63. Error Analysis Describe the error.

$$5(2x + 3) = 5 \cdot 2x + 3 = 10x + 3 \quad ✗$$

64. HOW DO YOU SEE IT? Match each description with its graph. Explain.

(i)
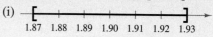

(ii)

(a) The price of an item is within \$0.03 of \$1.90.

(b) The distance between the prongs of an electric plug may not differ from 1.9 centimeters by more than 0.03 centimeter.

65. Conjecture Make a conjecture about the value of the expression $5/n$ as n approaches 0. Explain.

66. Conjecture Make a conjecture about the value of the expression $5/n$ as n increases without bound. Explain.

Review & Refresh ▶ *Video solutions at LarsonPrecalculus.com*

Finding Greatest Common Factor and Least Common Multiple In Exercises 67–70, find (a) the greatest common factor and (b) the least common multiple of the numbers.

67. 6, 8

68. 10, 25

69. 27, 36, 54

70. 49, 98, 112

Evaluating an Expression In Exercises 71–74, evaluate the expression.

71. $-(3 \cdot 3)$

72. $(-3)(-3)$

73. $\dfrac{(-5)(-5)}{-5}$

74. $\dfrac{-(5 \cdot 5)}{(-5)(-5)(-5)}$

Finding a Prime Factorization In Exercises 75–78, find the prime factorization of the number.

75. 48

76. 250

77. 792

78. 4802

Evaluating an Expression In Exercises 79–82, evaluate the expression.

79. $3.785(10,000)$

80. $1.42(1,000,000)$

81. $6.09/1000$

82. $8.603/100,000$

P.2 Exponents and Radicals

Real numbers and algebraic expressions are often written with exponents and radicals. For example, in Exercises 85 and 86 on page 25, you will use an expression involving rational exponents to find the times required for a funnel to empty for different water heights.

❯ Use properties of exponents.
❯ Use scientific notation to represent real numbers.
❯ Use properties of radicals.
❯ Simplify and combine radical expressions.
❯ Use properties of rational exponents.

Integer Exponents and Their Properties

Repeated *multiplication* can be written in **exponential form.**

Repeated Multiplication	Exponential Form
$a \cdot a \cdot a \cdot a \cdot a$	a^5
$(-4)(-4)(-4)$	$(-4)^3$
$(2x)(2x)(2x)(2x)$	$(2x)^4$

Exponential Notation

If a is a real number and n is a positive integer, then

$$a^n = \underbrace{a \cdot a \cdot a \cdots a}_{n \text{ factors}}$$

where n is the **exponent** and a is the **base.** You read a^n as "a to the nth **power.**"

An exponent can also be negative or zero. Properties 3 and 4 below show how to use negative and zero exponents.

Properties of Exponents

Let a and b be real numbers, variables, or algebraic expressions, and let m and n be integers. (All denominators and bases are nonzero.)

Property	Example
1. $a^m a^n = a^{m+n}$	$3^2 \cdot 3^4 = 3^{2+4} = 3^6 = 729$
2. $\dfrac{a^m}{a^n} = a^{m-n}$	$\dfrac{x^7}{x^4} = x^{7-4} = x^3$
3. $a^{-n} = \dfrac{1}{a^n} = \left(\dfrac{1}{a}\right)^n$	$y^{-4} = \dfrac{1}{y^4} = \left(\dfrac{1}{y}\right)^4$
4. $a^0 = 1$	$(x^2 + 1)^0 = 1$
5. $(ab)^m = a^m b^m$	$(5x)^3 = 5^3 x^3 = 125x^3$
6. $(a^m)^n = a^{mn}$	$(y^3)^{-4} = y^{3(-4)} = y^{-12} = \dfrac{1}{y^{12}}$
7. $\left(\dfrac{a}{b}\right)^m = \dfrac{a^m}{b^m}$	$\left(\dfrac{2}{x}\right)^3 = \dfrac{2^3}{x^3} = \dfrac{8}{x^3}$
8. $\lvert a^2 \rvert = \lvert a \rvert^2 = a^2$	$\lvert (-2)^2 \rvert = \lvert -2 \rvert^2 = 2^2 = 4 = (-2)^2$

GO DIGITAL

© iStockPhoto.com/micropic

The properties of exponents listed on the preceding page apply to *all* integers *m* and *n*, not just to positive integers. For instance, by Property 2, you can write

$$\frac{2^4}{2^{-5}} = 2^{4-(-5)} = 2^{4+5} = 2^9.$$

Note how the properties of exponents are used in Examples 1–4.

EXAMPLE 1   Evaluating Exponential Expressions

Evaluate each expression.

a. $(-5)^2$ **b.** -5^2 **c.** $2 \cdot 2^4$ **d.** $\dfrac{4^4}{4^6}$ **e.** $\left(\dfrac{7}{2}\right)^2$

Solution

a. $(-5)^2 = (-5)(-5) = 25$ Negative sign is part of the base.

b. $-5^2 = -(5)(5) = -25$ Negative sign is *not* part of the base.

c. $2 \cdot 2^4 = 2^{1+4} = 2^5 = 32$ Property 1

d. $\dfrac{4^4}{4^6} = 4^{4-6} = 4^{-2} = \dfrac{1}{4^2} = \dfrac{1}{16}$ Properties 2 and 3

e. $\left(\dfrac{7}{2}\right)^2 = \dfrac{7^2}{2^2} = \dfrac{49}{4}$ Property 7

✓ **Checkpoint** 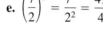 Audio-video solution in English & Spanish at *LarsonPrecalculus.com*

Evaluate each expression.

a. -3^4 **b.** $(-3)^4$

c. $3^2 \cdot 3$ **d.** $\dfrac{3^5}{3^8}$

EXAMPLE 2   Evaluating Algebraic Expressions

Evaluate each algebraic expression when $x = 3$.

a. $5x^{-2}$ **b.** $\dfrac{1}{3}(-x)^3$

Solution

a. When $x = 3$, the expression $5x^{-2}$ has a value of

$$5x^{-2} = 5(3)^{-2} = \frac{5}{3^2} = \frac{5}{9}.$$

b. When $x = 3$, the expression $\dfrac{1}{3}(-x)^3$ has a value of

$$\frac{1}{3}(-x)^3 = \frac{1}{3}(-3)^3 = \frac{1}{3}(-27) = -9.$$

✓ **Checkpoint**  Audio-video solution in English & Spanish at *LarsonPrecalculus.com*

Evaluate each algebraic expression when $x = 4$.

a. $-x^{-2}$ **b.** $\dfrac{1}{4}(-x)^4$

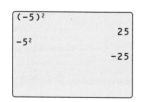

GO DIGITAL

EXAMPLE 3 **Using Properties of Exponents**

Use the properties of exponents to simplify each expression.

a. $(-3ab^4)(4ab^{-3})$ b. $(2xy^2)^3$ c. $3a(-4a^2)^0$ d. $\left(\dfrac{5x^3}{y}\right)^2$

Solution

a. $(-3ab^4)(4ab^{-3}) = (-3)(4)(a)(a)(b^4)(b^{-3}) = -12a^2b$

b. $(2xy^2)^3 = 2^3(x)^3(y^2)^3 = 8x^3y^6$

c. $3a(-4a^2)^0 = 3a(1) = 3a$

d. $\left(\dfrac{5x^3}{y}\right)^2 = \dfrac{5^2(x^3)^2}{y^2} = \dfrac{25x^6}{y^2}$

✓ *Checkpoint* ▶ Audio-video solution in English & Spanish at LarsonPrecalculus.com

Use the properties of exponents to simplify each expression.

a. $(2x^{-2}y^3)(-x^4y)$ b. $(4a^2b^3)^0$ c. $(-5z)^3(z^2)$ d. $\left(\dfrac{3x^4}{x^2y^2}\right)^2$

EXAMPLE 4 **Rewriting with Positive Exponents**

ALGEBRA HELP

Rarely in algebra is there only one way to solve a problem. Do not be concerned when the steps you use to solve a problem are not exactly the same as the steps presented in this text. It is important to use steps that you understand and, of course, steps that are justified by the rules of algebra. For example, the fractional form of Property 3 is

$$\left(\frac{a}{b}\right)^{-m} = \left(\frac{b}{a}\right)^{m}.$$

So, you might prefer the steps below for Example 4(d).

$$\left(\frac{3x^2}{y}\right)^{-2} = \left(\frac{y}{3x^2}\right)^{2} = \frac{y^2}{9x^4}$$

▶▶▶▶

a. $x^{-1} = \dfrac{1}{x}$ Property 3

b. $\dfrac{1}{3x^{-2}} = \dfrac{1(x^2)}{3}$ Property 3 (The exponent -2 does not apply to 3.)

     $= \dfrac{x^2}{3}$ Simplify.

c. $\dfrac{12a^3b^{-4}}{4a^{-2}b} = \dfrac{12a^3 \cdot a^2}{4b \cdot b^4}$ Property 3

     $= \dfrac{3a^5}{b^5}$ Property 1

d. $\left(\dfrac{3x^2}{y}\right)^{-2} = \dfrac{3^{-2}(x^2)^{-2}}{y^{-2}}$ Properties 5 and 7

     $= \dfrac{3^{-2}x^{-4}}{y^{-2}}$ Property 6

     $= \dfrac{y^2}{3^2x^4}$ Property 3

     $= \dfrac{y^2}{9x^4}$ Simplify.

✓ *Checkpoint* ▶ Audio-video solution in English & Spanish at LarsonPrecalculus.com

Rewrite each expression with positive exponents. Simplify, if possible.

a. $2a^{-2}$ b. $\dfrac{3a^{-3}b^4}{15ab^{-1}}$

c. $\left(\dfrac{x}{10}\right)^{-1}$ d. $(-2x^2)^3(4x^3)^{-1}$

Scientific Notation

Exponents provide an efficient way of writing and computing with very large (or very small) numbers. For example, there are about 366 billion billion gallons of water on Earth—that is, 366 followed by 18 zeros.

$$366,000,000,000,000,000,000 \qquad \text{Decimal form}$$

It is convenient to write such numbers in **scientific notation.** This notation has the form $\pm c \times 10^{n}$, where $1 \leq c < 10$ and n is an integer. So, the number of gallons of water on Earth, written in scientific notation, is

$$3.66 \times 100,000,000,000,000,000,000 = 3.66 \times 10^{20}.$$

The *positive* exponent 20 tells you that the number is *large* (10 or greater) and that the decimal point has been moved 20 places. A *negative* exponent tells you that the number is *small* (less than 1). For example, the mass (in grams) of one electron is approximately

$$9.1 \times 10^{-28} = 0.00000000000000000000000000091.$$

28 decimal places

EXAMPLE 5   **Scientific Notation**

a. $0.0000782 = 7.82 \times 10^{-5}$ Small number ⟹ negative exponent

b. $836,100,000 = 8.361 \times 10^{8}$ Large number ⟹ positive exponent

✓ *Checkpoint* ▶ *Audio-video solution in English & Spanish at LarsonPrecalculus.com*

Write 45,850 in scientific notation.

EXAMPLE 6   **Decimal Form**

a. $-9.36 \times 10^{-6} = -0.00000936$ Negative exponent ⟹ small number

b. $1.345 \times 10^{2} = 134.5$ Positive exponent ⟹ large number

✓ *Checkpoint* ▶ *Audio-video solution in English & Spanish at LarsonPrecalculus.com*

Write -2.718×10^{-3} in decimal form.

EXAMPLE 7   **Using Scientific Notation**

Evaluate $\dfrac{(2,400,000,000)(0.0000045)}{(0.00003)(1500)}$.

Solution   Begin by rewriting each number in scientific notation. Then simplify.

$$\frac{(2,400,000,000)(0.0000045)}{(0.00003)(1500)} = \frac{(2.4 \times 10^{9})(4.5 \times 10^{-6})}{(3.0 \times 10^{-5})(1.5 \times 10^{3})}$$

$$= \frac{(2.4)(4.5)(10^{3})}{(4.5)(10^{-2})}$$

$$= (2.4)(10^{5})$$

$$= 240,000$$

✓ *Checkpoint* ▶ *Audio-video solution in English & Spanish at LarsonPrecalculus.com*

Evaluate $(24,000,000,000)(0.00000012)(300,000)$.   ■

GO DIGITAL

Radicals and Their Properties

A **square root** of a number is one of its two equal factors. For example, 5 is a square root of 25 because 5 is one of the two equal factors of 25. In a similar way, a **cube root** of a number is one of its three equal factors, as in $125 = 5^3$.

Definition of *n*th Root of a Number

Let a and b be real numbers, and let n be a positive integer, where $n \geq 2$. If

$$a = b^n$$

then b is an ***n*th root of *a*.** If $n = 2$, then the root is a **square root.** If $n = 3$, then the root is a **cube root.**

Some numbers have more than one nth root. For example, both 5 and -5 are square roots of 25. The *principal square* root of 25, written as $\sqrt{25}$, is the positive root, 5.

Principal *n*th Root of a Number

Let a be a real number that has at least one nth root. The **principal *n*th root of *a*** is the nth root that has the same sign as a. It is denoted by a **radical symbol**

$$\sqrt[n]{a}. \qquad \text{Principal } n\text{th root}$$

The number n is the **index** of the radical, and the number a is the **radicand.** When $n = 2$, omit the index and write $\sqrt{a}$ rather than $\sqrt[2]{a}$. (The plural of index is *indices*.)

A common misunderstanding is that the square root sign implies both negative and positive roots. This is not correct. The square root sign implies only a positive root. When a negative root is needed, you must use the negative sign with the square root sign.

$$\text{Incorrect: } \sqrt{4} = \pm 2 \qquad \bigtimes \qquad \text{Correct: } -\sqrt{4} = -2 \quad \text{and} \quad \sqrt{4} = 2$$

EXAMPLE 8 Evaluating Radical Expressions

a. $\sqrt{36} = 6$ because $6^2 = 36$.

b. $-\sqrt{36} = -6$ because $-\left(\sqrt{36}\right) = -\left(\sqrt{6^2}\right) = -(6) = -6$.

c. $\sqrt[3]{\dfrac{125}{64}} = \dfrac{5}{4}$ because $\left(\dfrac{5}{4}\right)^3 = \dfrac{5^3}{4^3} = \dfrac{125}{64}$.

d. $\sqrt[5]{-32} = -2$ because $(-2)^5 = -32$.

e. $\sqrt[4]{-81}$ is not a real number because no real number raised to the fourth power produces -81.

✓ *Checkpoint* ▶ *Audio-video solution in English & Spanish at LarsonPrecalculus.com*

Evaluate each expression, if possible.

a. $-\sqrt{144}$ **b.** $\sqrt{-144}$

c. $\sqrt{\dfrac{25}{64}}$ **d.** $-\sqrt[3]{\dfrac{8}{27}}$

Here are some generalizations about the *n*th roots of real numbers.

Generalizations About *n*th Roots of Real Numbers			
Real Number *a*	Index *n*	Root(s) of *a*	Example
$a > 0$	*n* is even.	$\sqrt[n]{a}, \ -\sqrt[n]{a}$	$\sqrt[4]{81} = 3, \ -\sqrt[4]{81} = -3$
$a > 0$ or $a < 0$	*n* is odd.	$\sqrt[n]{a}$	$\sqrt[3]{-8} = -2$
$a < 0$	*n* is even.	No real roots	$\sqrt{-4}$ is not a real number.
$a = 0$	*n* is even or odd.	$\sqrt[n]{0} = 0$	$\sqrt[5]{0} = 0$

Integers such as 1, 4, 9, 16, 25, and 36 are **perfect squares** because they have integer square roots. Similarly, integers such as 1, 8, 27, 64, and 125 are **perfect cubes** because they have integer cube roots.

Properties of Radicals

Let *a* and *b* be real numbers, variables, or algebraic expressions such that the indicated roots are real numbers, and let *m* and *n* be positive integers.

Property	Example
1. $\sqrt[n]{a^m} = \left(\sqrt[n]{a}\right)^m$	$\sqrt[3]{8^2} = \left(\sqrt[3]{8}\right)^2 = (2)^2 = 4$
2. $\sqrt[n]{a} \cdot \sqrt[n]{b} = \sqrt[n]{ab}$	$\sqrt{5} \cdot \sqrt{7} = \sqrt{5 \cdot 7} = \sqrt{35}$
3. $\dfrac{\sqrt[n]{a}}{\sqrt[n]{b}} = \sqrt[n]{\dfrac{a}{b}}, \quad b \neq 0$	$\dfrac{\sqrt[4]{27}}{\sqrt[4]{9}} = \sqrt[4]{\dfrac{27}{9}} = \sqrt[4]{3}$
4. $\sqrt[m]{\sqrt[n]{a}} = \sqrt[mn]{a}$	$\sqrt[3]{\sqrt{10}} = \sqrt[6]{10}$
5. $\left(\sqrt[n]{a}\right)^n = a$	$\left(\sqrt{3}\right)^2 = 3$
6. For *n* even, $\sqrt[n]{a^n} = |a|$.	$\sqrt{(-12)^2} = |-12| = 12$
For *n* odd, $\sqrt[n]{a^n} = a$.	$\sqrt[3]{(-12)^3} = -12$

ALGEBRA HELP

A common special case of Property 6 is

$$\sqrt{a^2} = |a|.$$

EXAMPLE 9   **Using Properties of Radicals**

Use the properties of radicals to simplify each expression.

a. $\sqrt{8} \cdot \sqrt{2}$ **b.** $\left(\sqrt[3]{5}\right)^3$ **c.** $\sqrt[3]{x^3}$ **d.** $\sqrt[6]{y^6}$

Solution

a. $\sqrt{8} \cdot \sqrt{2} = \sqrt{8 \cdot 2} = \sqrt{16} = 4$ Property 2

b. $\left(\sqrt[3]{5}\right)^3 = 5$ Property 5

c. $\sqrt[3]{x^3} = x$ Property 6

d. $\sqrt[6]{y^6} = |y|$ Property 6

✓ *Checkpoint* ▶ Audio-video solution in English & Spanish at LarsonPrecalculus.com

Use the properties of radicals to simplify each expression.

a. $\dfrac{\sqrt{125}}{\sqrt{5}}$ **b.** $\sqrt[3]{125^2}$ **c.** $\sqrt[3]{x^2} \cdot \sqrt[3]{x}$ **d.** $\sqrt{\sqrt{x}}$ ∎

Simplifying Radical Expressions

An expression involving radicals is in **simplest form** when the three conditions below are satisfied.

1. All possible factors are removed from the radical.
2. All fractions have radical-free denominators (a process called *rationalizing the denominator* accomplishes this).
3. The index of the radical is reduced.

To simplify a radical, factor the radicand into factors whose exponents are multiples of the index. Write the roots of these factors outside the radical. The "leftover" factors make up the new radicand.

EXAMPLE 10   **Simplifying Radical Expressions**

a. $\overset{\text{Perfect cube}}{\downarrow}\quad\overset{\text{Leftover factor}}{\downarrow}$

$\sqrt[3]{24} = \sqrt[3]{8 \cdot 3} = \sqrt[3]{2^3 \cdot 3} = 2\sqrt[3]{3}$

b. $\overset{\substack{\text{Perfect}\\\text{4th power}}}{\downarrow}\quad\overset{\substack{\text{Leftover}\\\text{factor}}}{\downarrow}$

$\sqrt[4]{48} = \sqrt[4]{16 \cdot 3} = \sqrt[4]{2^4 \cdot 3} = 2\sqrt[4]{3}$

c. $\sqrt{75x^3} = \sqrt{25x^2 \cdot 3x} = \sqrt{(5x)^2 \cdot 3x} = 5x\sqrt{3x}$

d. $\sqrt[3]{24a^4} = \sqrt[3]{8a^3 \cdot 3a} = \sqrt[3]{(2a)^3 \cdot 3a} = 2a\sqrt[3]{3a}$

e. $\sqrt[4]{(5x)^4} = |5x| = 5|x|$

✓ **Checkpoint** ▶ Audio-video solution in English & Spanish at LarsonPrecalculus.com

Simplify each radical expression.

a. $\sqrt{32}$ b. $\sqrt[3]{250}$ c. $\sqrt{24a^5}$ d. $\sqrt[3]{-135x^3}$ ■

Radical expressions can be combined (added or subtracted) when they are **like radicals**—that is, when they have the same index and radicand. For example, $\sqrt{2}$, $3\sqrt{2}$, and $\frac{1}{2}\sqrt{2}$ are like radicals, but $\sqrt{3}$ and $\sqrt{2}$ are unlike radicals. To determine whether two radicals can be combined, first simplify each radical.

EXAMPLE 11   **Combining Radical Expressions**

a. $2\sqrt{48} - 3\sqrt{27} = 2\sqrt{16 \cdot 3} - 3\sqrt{9 \cdot 3}$ Find square factors.

$\qquad\qquad\qquad = 8\sqrt{3} - 9\sqrt{3}$ Find square roots and multiply by coefficients.

$\qquad\qquad\qquad = (8 - 9)\sqrt{3}$ Combine like radicals.

$\qquad\qquad\qquad = -\sqrt{3}$ Simplify.

b. $\sqrt[3]{16x} - \sqrt[3]{54x^4} = \sqrt[3]{8 \cdot 2x} - \sqrt[3]{27x^3 \cdot 2x}$ Find cube factors.

$\qquad\qquad\qquad = 2\sqrt[3]{2x} - 3x\sqrt[3]{2x}$ Find cube roots.

$\qquad\qquad\qquad = (2 - 3x)\sqrt[3]{2x}$ Combine like radicals.

✓ **Checkpoint** ▶ Audio-video solution in English & Spanish at LarsonPrecalculus.com

Simplify each radical expression.

a. $3\sqrt{8} + \sqrt{18}$ b. $\sqrt[3]{81x^5} - \sqrt[3]{24x^2}$ ■

To remove radicals from a denominator or a numerator, use a process called **rationalizing the denominator** or **rationalizing the numerator.** This involves multiplying by an appropriate form of 1 to obtain a perfect nth power (see Example 12). Note that pairs of expressions of the form $a\sqrt{b} + c\sqrt{d}$ and $a\sqrt{b} - c\sqrt{d}$ are **conjugates.** The product of these two expressions contains no radicals. You can use this fact to rationalize a denominator or a numerator (see Examples 13 and 14).

EXAMPLE 12   Rationalizing Single-Term Denominators

a. $\dfrac{5}{2\sqrt{3}} = \dfrac{5}{2\sqrt{3}} \cdot \dfrac{\sqrt{3}}{\sqrt{3}}$ $\sqrt{3}$ is rationalizing factor.

$\qquad = \dfrac{5\sqrt{3}}{2(3)}$ Multiply.

$\qquad = \dfrac{5\sqrt{3}}{6}$ Simplify.

b. $\dfrac{2}{\sqrt[3]{5}} = \dfrac{2}{\sqrt[3]{5}} \cdot \dfrac{\sqrt[3]{5^2}}{\sqrt[3]{5^2}}$ $\sqrt[3]{5^2}$ is rationalizing factor.

$\qquad = \dfrac{2\sqrt[3]{5^2}}{\sqrt[3]{5^3}}$ Multiply.

$\qquad = \dfrac{2\sqrt[3]{25}}{5}$ Simplify.

✓ *Checkpoint* ▶ Audio-video solution in English & Spanish at *LarsonPrecalculus.com*

Rationalize the denominators of (a) $\dfrac{5}{3\sqrt{2}}$ and (b) $\dfrac{1}{\sqrt[3]{25}}$.

EXAMPLE 13   Rationalizing a Denominator with Two Terms

$\dfrac{2}{3 + \sqrt{7}} = \dfrac{2}{3 + \sqrt{7}} \cdot \dfrac{3 - \sqrt{7}}{3 - \sqrt{7}}$ Multiply numerator and denominator by conjugate of denominator.

$\qquad = \dfrac{2(3 - \sqrt{7})}{3(3 - \sqrt{7}) + \sqrt{7}(3 - \sqrt{7})}$ Distributive Property

$\qquad = \dfrac{2(3 - \sqrt{7})}{3(3) - 3(\sqrt{7}) + \sqrt{7}(3) - \sqrt{7}(\sqrt{7})}$ Distributive Property

$\qquad = \dfrac{2(3 - \sqrt{7})}{(3)^2 - (\sqrt{7})^2}$ Simplify.

$\qquad = \dfrac{2(3 - \sqrt{7})}{2}$ Simplify.

$\qquad = 3 - \sqrt{7}$ Divide out common factor.

✓ *Checkpoint* ▶ Audio-video solution in English & Spanish at *LarsonPrecalculus.com*

Rationalize the denominator: $\dfrac{8}{\sqrt{6} - \sqrt{2}}$. ■

Sometimes it is necessary to rationalize the numerator of an expression. For instance, in Section P.5 you will use the technique shown in Example 14 on the next page to rationalize the numerator of an expression from calculus.

GO DIGITAL

EXAMPLE 14   **Rationalizing a Numerator**

$$\frac{\sqrt{5} - \sqrt{7}}{2} = \frac{\sqrt{5} - \sqrt{7}}{2} \cdot \frac{\sqrt{5} + \sqrt{7}}{\sqrt{5} + \sqrt{7}} \qquad \text{Multiply numerator and denominator by conjugate of numerator.}$$

$$= \frac{(\sqrt{5})^2 - (\sqrt{7})^2}{2(\sqrt{5} + \sqrt{7})} \qquad \text{Simplify.}$$

$$= \frac{5 - 7}{2(\sqrt{5} + \sqrt{7})} \qquad \text{Property 5 of radicals}$$

$$= \frac{-2}{2(\sqrt{5} + \sqrt{7})} \qquad \text{Simplify.}$$

$$= \frac{-1}{\sqrt{5} + \sqrt{7}} \qquad \text{Divide out common factor.}$$

✓ *Checkpoint* ▶ *Audio-video solution in English & Spanish at LarsonPrecalculus.com*

Rationalize the numerator: $\dfrac{2 - \sqrt{2}}{3}$. ∎

Rational Exponents and Their Properties

> **Definition of Rational Exponents**
>
> If a is a real number and n is a positive integer ($n \geq 2$) such that the principal nth root of a exists, then $a^{1/n}$ is defined as
>
> $$a^{1/n} = \sqrt[n]{a}.$$
>
> Moreover, if m is a positive integer that has no common factor with n, then
>
> $$a^{m/n} = (a^{1/n})^m = (\sqrt[n]{a})^m \quad \text{and} \quad a^{m/n} = (a^m)^{1/n} = \sqrt[n]{a^m}.$$

The numerator of a rational exponent denotes the *power* to which the base is raised, and the denominator denotes the *index* or the *root* to be taken.

$$b^{m/n} = (\sqrt[n]{b})^m = \sqrt[n]{b^m}$$

(Power, Index)

When you are working with rational exponents, the properties of integer exponents still apply. For example, $2^{1/2}2^{1/3} = 2^{(1/2) + (1/3)} = 2^{5/6}$.

EXAMPLE 15   **Changing From Radical to Exponential Form**

a. $\sqrt{3} = 3^{1/2}$

b. $\sqrt{(3xy)^5} = \sqrt[2]{(3xy)^5} = (3xy)^{5/2}$

c. $2x\sqrt[4]{x^3} = (2x)(x^{3/4}) = 2x^{1 + (3/4)} = 2x^{7/4}$

✓ *Checkpoint* ▶ *Audio-video solution in English & Spanish at LarsonPrecalculus.com*

Write (a) $\sqrt[3]{27}$, (b) $\sqrt{x^3 y^5 z}$, and (c) $3x\sqrt[3]{x^2}$ in exponential form. ∎

There are several ways to use a graphing utility to evaluate radicals and rational exponents, as shown below. Consult the user's guide for your graphing utility for specific keystrokes.

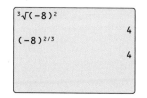

$^3\sqrt{(-8)^2}$

4

$(-8)^{2/3}$

4

EXAMPLE 16 **Changing From Exponential to Radical Form**

▶▶▶ *See LarsonPrecalculus.com for an interactive version of this type of example.*

a. $(x^2 + y^2)^{3/2} = \left(\sqrt{x^2 + y^2}\right)^3 = \sqrt{(x^2 + y^2)^3}$

b. $2y^{3/4}z^{1/4} = 2(y^3z)^{1/4} = 2\sqrt[4]{y^3z}$

c. $a^{-3/2} = \dfrac{1}{a^{3/2}} = \dfrac{1}{\sqrt{a^3}}$

d. $x^{0.2} = x^{1/5} = \sqrt[5]{x}$

✓ *Checkpoint* **Audio-video solution in English & Spanish at LarsonPrecalculus.com**

Write each expression in radical form.

a. $(x^2 - 7)^{-1/2}$ **b.** $-3b^{1/3}c^{2/3}$

c. $a^{0.75}$ **d.** $(x^2)^{2/5}$

Rational exponents are useful for evaluating roots of numbers on a calculator, for reducing the index of a radical, and for simplifying expressions in calculus.

EXAMPLE 17 **Simplifying with Rational Exponents**

a. $(-32)^{-4/5} = \left(\sqrt[5]{-32}\right)^{-4} = (-2)^{-4} = \dfrac{1}{(-2)^4} = \dfrac{1}{16}$

b. $(-5x^{5/3})(3x^{-3/4}) = -15x^{(5/3)-(3/4)} = -15x^{11/12}, \quad x \neq 0$

c. $\sqrt[9]{a^3} = a^{3/9} = a^{1/3} = \sqrt[3]{a}$ Reduce index.

d. $\sqrt[3]{\sqrt{125}} = \sqrt[6]{125} = \sqrt[6]{(5)^3} = 5^{3/6} = 5^{1/2} = \sqrt{5}$

e. $(2x - 1)^{4/3}(2x - 1)^{-1/3} = (2x - 1)^{(4/3)-(1/3)} = 2x - 1, \quad x \neq \dfrac{1}{2}$

✓ *Checkpoint* **Audio-video solution in English & Spanish at LarsonPrecalculus.com**

Simplify each expression.

a. $(-125)^{-2/3}$ **b.** $(4x^2y^{3/2})(-3x^{-1/3}y^{-3/5})$

c. $\sqrt[3]{\sqrt[4]{27}}$ **d.** $(3x + 2)^{5/2}(3x + 2)^{-1/2}$

▶▶▶

ALGEBRA HELP

The expression in Example 17(b) is not defined when $x = 0$ because $0^{-3/4}$ is not a real number. Similarly, the expression in Example 17(e) is not defined when $x = \frac{1}{2}$ because

$$\left(2 \cdot \tfrac{1}{2} - 1\right)^{-1/3} = (0)^{-1/3}$$

is not a real number.

Summarize (Section P.2)

1. Make a list of the properties of exponents *(page 14)*. For examples that use these properties, see Examples 1–4.

2. Explain how to write a number in scientific notation *(page 17)*. For examples involving scientific notation, see Examples 5–7.

3. Make a list of the properties of radicals *(page 19)*. For examples involving radicals, see Examples 8 and 9.

4. Explain how to simplify a radical expression *(page 20)*. For examples of simplifying radical expressions, see Examples 10 and 11.

5. Explain how to rationalize a denominator or a numerator *(page 21)*. For examples of rationalizing denominators and numerators, see Examples 12–14.

6. State the definition of a rational exponent *(page 22)*. For examples involving rational exponents, see Examples 15–17.

GO DIGITAL

Vocabulary and Concept Check

In Exercises 1 and 2, fill in the blanks.

1. In the exponential form a^n, n is the _____ and a is the _____.

2. In the radical form $\sqrt[n]{a}$, the number n is the _____ of the radical and the number a is the _____.

3. When is an expression involving radicals in simplest form?

4. Is 64 a perfect square, a perfect cube, or both?

Skills and Applications

Evaluating an Exponential Expression In Exercises 5–16, evaluate the expression.

5. $5 \cdot 5^3$

6. $(2^3 \cdot 3^2)^2$

7. $(3^3)^2$

8. $(-2)^0$

9. $\dfrac{5^2}{5^4}$

10. $\left(-\dfrac{3}{5}\right)^3 \left(\dfrac{5}{3}\right)^2$

11. -3^2

12. $(-4)^{-3}$

13. $\dfrac{4 \cdot 3^{-2}}{2^{-2} \cdot 3^{-1}}$

14. $\dfrac{3}{3^{-4}}$

15. $3^2 + 2^3$

16. $(3^{-2})^2$

Evaluating an Algebraic Expression In Exercises 17–20, evaluate the expression for the given value of x.

17. $-3x^3$, $\quad x = 2$

18. $7x^{-2}$, $\quad x = 4$

19. $6x^2$, $\quad x = 0.1$

20. $12(-x)^3$, $\quad x = -\dfrac{1}{3}$

Using Properties of Exponents In Exercises 21–30, simplify the expression.

21. $(5z)^3$

22. $(4x^3)^0$

23. $6y^2(2y^0)^2$

24. $(-z)^3(3z^4)$

25. $\dfrac{7x^2}{x^3}$

26. $\dfrac{12(x+y)^3}{9(x+y)}$

27. $\left(\dfrac{4}{y}\right)^3 \left(\dfrac{3}{y}\right)^4$

28. $\left(\dfrac{b^{-2}}{a^{-2}}\right)\left(\dfrac{b}{a}\right)^2$

29. $[(x^2y^{-2})^{-1}]^{-1}$

30. $(5x^2z^6)^3(5x^2z^6)^{-3}$

Rewriting with Positive Exponents In Exercises 31–36, rewrite the expression with positive exponents. Simplify, if possible.

31. $(2x^2)^{-2}$

32. $(4y^{-2})(8y^{-4})$

33. $\left(\dfrac{x^{-3}y^4}{5}\right)^{-3}$

34. $\left(\dfrac{a^{-2}}{b^{-2}}\right)\left(\dfrac{b}{a}\right)^{-3}$

35. $\dfrac{3^n \cdot 3^{2n}}{3^{3n} \cdot 3^2}$

36. $\dfrac{x^2 \cdot x^n}{x^3 \cdot x^n}$

Scientific Notation In Exercises 37 and 38, write the number in scientific notation.

37. $10{,}250.4$

38. -0.000125

Decimal Form In Exercises 39 and 40, write the number in decimal form.

39. 3.14×10^{-4}

40. -2.058×10^6

Using Scientific Notation In Exercises 41–44, evaluate the expression without using a calculator.

41. $(2.0 \times 10^9)(3.4 \times 10^{-4})$

42. $(1.2 \times 10^7)(5.0 \times 10^{-3})$

43. $\dfrac{6.0 \times 10^8}{3.0 \times 10^{-3}}$

44. $\dfrac{2.5 \times 10^{-3}}{5.0 \times 10^2}$

Evaluating Radical Expressions In Exercises 45 and 46, evaluate each expression without using a calculator.

45. (a) $\sqrt{9}$ (b) $\sqrt[3]{\dfrac{27}{8}}$

46. (a) $\sqrt[3]{27}$ (b) $\left(\sqrt{36}\right)^3$

Using Properties of Radicals In Exercises 47 and 48, use the properties of radicals to simplify each expression.

47. (a) $\left(\sqrt[5]{2}\right)^5$ (b) $\sqrt[5]{32x^5}$

48. (a) $\sqrt{12} \cdot \sqrt{3}$ (b) $\sqrt[4]{(3x^2)^4}$

Simplifying a Radical Expression In Exercises 49–62, simplify the radical expression.

49. $\sqrt{20}$

50. $\sqrt[3]{128}$

51. $\sqrt[3]{\dfrac{16}{27}}$

52. $\sqrt{\dfrac{75}{4}}$

53. $\sqrt{72x^3}$

54. $\sqrt{54xy^4}$

55. $\sqrt{\dfrac{18^2}{z^4}}$

56. $\sqrt{\dfrac{32a^4}{b^2}}$

57. $\sqrt{75x^2y^{-4}}$

58. $\sqrt[4]{3x^4y^2}$

59. $2\sqrt{20x^2} + 5\sqrt{125x^2}$

60. $8\sqrt{147x} - 3\sqrt{48x}$

61. $3\sqrt[3]{54x^3} + \sqrt[3]{16x^3}$

62. $\sqrt[3]{64x} - \sqrt[3]{27x^4}$

Rationalizing a Denominator In Exercises 63–66, rationalize the denominator of the expression. Then simplify your answer.

63. $\dfrac{1}{\sqrt{3}}$

64. $\dfrac{8}{\sqrt[3]{2}}$

65. $\dfrac{5}{\sqrt{14}-2}$

66. $\dfrac{3}{\sqrt{5}+\sqrt{6}}$

∫ **Rationalizing a Numerator** In Exercises 67 and 68, rationalize the numerator of the expression. Then simplify your answer.

67. $\dfrac{\sqrt{5}+\sqrt{3}}{3}$

68. $\dfrac{\sqrt{7}-3}{4}$

Writing Exponential and Radical Forms In Exercises 69–72, fill in the missing form of the expression.

Radical Form	Rational Exponent Form
69. $\sqrt[3]{64}$	
70. $x^2\sqrt{x}$	
71.	$3x^{-2/3}$
72.	$a^{0.4}$

Simplifying an Expression In Exercises 73–84, simplify the expression.

73. $32^{-3/5}$

74. $\left(\dfrac{16}{81}\right)^{-3/4}$

75. $\left(\dfrac{9}{4}\right)^{-1/2}$

76. $100^{-3/2}$

77. $\sqrt[4]{3^2}$

78. $\sqrt[4]{(3x^2)^4}$

79. $\sqrt{\sqrt{32}}$

80. $\sqrt{\sqrt[4]{2x}}$

81. $(x-1)^{1/3}(x-1)^{2/3}$

82. $(x-1)^{1/3}(x-1)^{-4/3}$

83. $(4x+3)^{5/2}(4x+3)^{-5/3}$

84. $(4x+3)^{-5/2}(4x+3)^{2/3}$

Mathematical Modeling

In Exercises 85 and 86, use the following information. A funnel is filled with water to a height of h centimeters. The formula $t = 0.03\left[12^{5/2} - (12 - h)^{5/2}\right]$, $0 \le h \le 12$, represents the amount of time t (in seconds) that it will take for the funnel to empty.

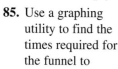

85. Use a graphing utility to find the times required for the funnel to empty for integer-valued water heights from 0 to 12 centimeters.

86. Use the graphing utility to find the water height corresponding to an emptying time of 10 seconds.

Exploring the Concepts

True or False? In Exercises 87 and 88, determine whether the statement is true or false. Justify your answer.

87. $\dfrac{x^{k+1}}{x} = x^k$

88. $\dfrac{a}{\sqrt{b}} = \dfrac{a^2}{\left(\sqrt{b}\right)^2} = \dfrac{a^2}{b}$

89. Error Analysis Describe the error.

$$\left(\dfrac{a^4}{a^6}\right)^{-3} = (a^{-2})^{-3} = \left(\dfrac{1}{a^2}\right)^{-3} = \dfrac{-1}{a^{-6}} = -a^6 \quad ✗$$

90. HOW DO YOU SEE IT?
Package A is a cube with a volume of 500 cubic inches. Package B is a cube with a volume of 250 cubic inches. Is the length x of a side of package A greater than, less than, or equal to twice the length of a side of package B? Explain.

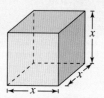

91. Think About It Verify that $a^0 = 1$, $a \ne 0$. (*Hint:* Use the property of exponents $a^m/a^n = a^{m-n}$.)

92. Exploration List all possible digits that occur in the units place of the square of a positive integer. Use that list to determine whether $\sqrt{5233}$ is an integer.

Review & Refresh ▶ Video solutions at LarsonPrecalculus.com

Finding Surface Area and Volume In Exercises 93–96, find the (a) surface area and (b) volume of the rectangular solid.

93.
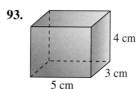
4 cm, 3 cm, 5 cm

94.
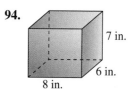
7 in., 6 in., 8 in.

95.
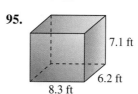
7.1 ft, 6.2 ft, 8.3 ft

96.
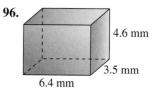
4.6 mm, 3.5 mm, 6.4 mm

Evaluating an Algebraic Expression In Exercises 97 and 98, evaluate the algebraic expression when (a) $x = 2$ and (b) $x = -3$.

97. $(x+4)(x-4)$

98. $(9x+5)(3x-1)$

Identifying Terms and Coefficients In Exercises 99 and 100, identify the terms. Then identify the coefficients of the variable terms of the expression.

99. $2x - 3$

100. $4x^3 - x^2 + 5x + 1$

P.3 Polynomials and Special Products

Polynomials have many real-life applications. For example, in Exercise 75 on page 32, you will work with polynomials that model uniformly distributed safe loads for steel beams.

> ❯ Write polynomials in standard form.
> ❯ Add, subtract, and multiply polynomials, and use special products.
> ❯ Use polynomials to solve real-life problems.

Polynomials

One of the most common types of algebraic expressions is the **polynomial.** Some examples are $2x + 5$, $3x^4 - 7x^2 + 2x + 4$, and $5x^2y^2 - xy + 3$. The first two are *polynomials in x* and the third is a *polynomial in x and y*. The terms of a polynomial in x have the form ax^k, where a is the **coefficient** and k is the **degree** of the term. For example, the polynomial $2x^3 - 5x^2 + 1 = 2x^3 + (-5)x^2 + (0)x + 1$ has coefficients 2, -5, 0, and 1.

Definition of a Polynomial in x

Let $a_0, a_1, a_2, \ldots, a_n$ be real numbers and let n be a nonnegative integer. A **polynomial in x** is an expression of the form

$$a_n x^n + a_{n-1} x^{n-1} + \cdots + a_1 x + a_0$$

where $a_n \neq 0$. The polynomial is of **degree** n, a_n is the **leading coefficient,** and a_0 is the **constant term.**

In **standard form**, a polynomial in x is written with descending powers of x. Polynomials with one, two, and three terms are **monomials, binomials,** and **trinomials,** respectively.

EXAMPLE 1   Writing Polynomials in Standard Form

Polynomial	Standard Form	Degree	Leading Coefficient
a. $4x^2 - 5x^7 - 2 + 3x$	$-5x^7 + 4x^2 + 3x - 2$	7	-5
b. $4 - 9x^2$	$-9x^2 + 4$	2	-9
c. 8	8 or $8x^0$	0	8

✓ *Checkpoint* ▶ *Audio-video solution in English & Spanish at LarsonPrecalculus.com*

Write the polynomial $6 - 7x^3 + 2x$ in standard form. Then identify the degree and leading coefficient of the polynomial.   ∎

A polynomial that has all zero coefficients is called the **zero polynomial,** denoted by 0. No degree is assigned to the zero polynomial. For polynomials in more than one variable, the degree of a *term* is the sum of the exponents of the variables in the term. The degree of the *polynomial* is the highest degree of its terms. For example, the degree of the polynomial $-2x^3y^6 + 4xy - x^7y^4$ is 11 because the sum of the exponents in the last term is the greatest. The leading coefficient of the polynomial is the coefficient of the highest-degree term. Expressions are not polynomials when a variable is underneath a radical or when a polynomial expression (with degree greater than 0) is in the denominator of a term. For example, the expressions $x^3 - \sqrt{3x} = x^3 - (3x)^{1/2}$ and $x^2 + (5/x) = x^2 + 5x^{-1}$ are not polynomials.

Operations with Polynomials and Special Products

You can add and subtract polynomials in much the same way you add and subtract real numbers. Add or subtract the *like terms* (terms having the same variables to the same powers) by adding or subtracting their coefficients. For example, $-3xy^2$ and $5xy^2$ are like terms and their sum is

$$-3xy^2 + 5xy^2 = (-3 + 5)xy^2 = 2xy^2.$$

EXAMPLE 2   **Adding or Subtracting Polynomials**

a. $(5x^3 - 7x^2 - 3) + (x^3 + 2x^2 - x + 8)$

$\quad = (5x^3 + x^3) + (-7x^2 + 2x^2) + (-x) + (-3 + 8)$ Group like terms.

$\quad = 6x^3 - 5x^2 - x + 5$ Combine like terms.

b. $(7x^4 - x^2 - 4x + 2) - (3x^4 - 4x^2 + 3x)$

$\quad = 7x^4 - x^2 - 4x + 2 - 3x^4 + 4x^2 - 3x$ Distributive Property

$\quad = (7x^4 - 3x^4) + (-x^2 + 4x^2) + (-4x - 3x) + 2$ Group like terms.

$\quad = 4x^4 + 3x^2 - 7x + 2$ Combine like terms.

 Checkpoint ▶ *Audio-video solution in English & Spanish at LarsonPrecalculus.com*

Find the difference $(2x^3 - x + 3) - (x^2 - 2x - 3)$ and write the resulting polynomial in standard form. ■

>
>
> **ALGEBRA HELP**
>
> When a negative sign precedes an expression inside parentheses, remember to distribute the negative sign to each term inside the parentheses. In other words, multiply each term by -1.
>
> $-(3x^4 - 4x^2 + 3x)$
>
> $\quad = -3x^4 + 4x^2 - 3x$

To find the *product* of two polynomials, use the right and left Distributive Properties. For example, you can find the product of $3x - 2$ and $5x + 7$ by first treating $5x + 7$ as a single quantity.

$$(3x - 2)(5x + 7) = 3x(5x + 7) - 2(5x + 7)$$

$$= (3x)(5x) + (3x)(7) - (2)(5x) - (2)(7)$$

$$= 15x^2 + 21x - 10x - 14$$

Product of **First terms**	Product of **Outer terms**	Product of **Inner terms**	Product of **Last terms**

$$= 15x^2 + 11x - 14$$

Note that when using the **FOIL Method** to multiply two binomials, some of the terms in the product may be like terms that can be combined into one term.

EXAMPLE 3   **Finding a Product by the FOIL Method**

Use the FOIL Method to find the product of $2x - 4$ and $x + 5$.

Solution

$$\overset{\text{F}\quad\;\text{O}\quad\;\;\text{I}\quad\;\;\text{L}}{(2x - 4)(x + 5) = 2x^2 + 10x - 4x - 20 = 2x^2 + 6x - 20}$$

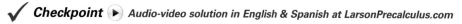 ***Checkpoint*** ▶ *Audio-video solution in English & Spanish at LarsonPrecalculus.com*

Use the FOIL Method to find the product of $3x - 1$ and $x - 5$. ■

When multiplying two polynomials, be sure to multiply *each* term of one polynomial by *each* term of the other. A vertical arrangement can be helpful, as shown in the next example.

EXAMPLE 4   **A Vertical Arrangement for Multiplication**

Multiply $-2x + 2 + x^2$ by $x^2 + 2x + 2$ using a vertical arrangement.

Solution   First, write $-2x + 2 + x^2$ in standard form, $x^2 - 2x + 2$.

$x^2 - 2x + 2$	Write in standard form.
$\times\ x^2 + 2x + 2$	Write in standard form.
$2x^2 - 4x + 4$	$\Longleftarrow$ $2(x^2 - 2x + 2)$
$2x^3 - 4x^2 + 4x$	$\Longleftarrow$ $2x(x^2 - 2x + 2)$
$x^4 - 2x^3 + 2x^2$	$\Longleftarrow$ $x^2(x^2 - 2x + 2)$
$x^4 + 0x^3 + 0x^2 + 0x + 4 = x^4 + 4$	Combine like terms.

So, $(x^2 - 2x + 2)(x^2 + 2x + 2) = x^4 + 4.$

✓ **Checkpoint** ▶ *Audio-video solution in English & Spanish at LarsonPrecalculus.com*

Multiply $x^2 + 2x + 3$ by $x^2 - 2x + 3$ using a vertical arrangement. ■

Some binomial products have special forms that occur frequently in algebra. You do not need to memorize these formulas because you can use the Distributive Property to multiply. However, becoming familiar with these formulas will enable you to manipulate the algebra more quickly.

Special Products

Let u and v be real numbers, variables, or algebraic expressions.

Special Product	**Example**
Sum and Difference of Same Terms	
$(u + v)(u - v) = u^2 - v^2$	$(x + 4)(x - 4) = x^2 - 4^2$
	$= x^2 - 16$
Square of a Binomial	
$(u + v)^2 = u^2 + 2uv + v^2$	$(x + 3)^2 = x^2 + 2(x)(3) + 3^2$
	$= x^2 + 6x + 9$
$(u - v)^2 = u^2 - 2uv + v^2$	$(3x - 2)^2 = (3x)^2 - 2(3x)(2) + 2^2$
	$= 9x^2 - 12x + 4$
Cube of a Binomial	
$(u + v)^3 = u^3 + 3u^2v + 3uv^2 + v^3$	$(x + 2)^3 = x^3 + 3x^2(2) + 3x(2^2) + 2^3$
	$= x^3 + 6x^2 + 12x + 8$
$(u - v)^3 = u^3 - 3u^2v + 3uv^2 - v^3$	$(x - 1)^3 = x^3 - 3x^2(1) + 3x(1^2) - 1^3$
	$= x^3 - 3x^2 + 3x - 1$

ALGEBRA HELP

When multiplying two polynomials, it is best to write each in standard form before using either the horizontal or the vertical format.

ALGEBRA HELP

Note that $u + v$ and $u - v$ are conjugates. In words, you can say that the product of conjugates equals the square of the first term minus the square of the second term.

GO DIGITAL

EXAMPLE 5 Sum and Difference of Same Terms

Find the product of $5x + 9$ and $5x - 9$.

Solution

The product of a sum and a difference of the *same* two terms has no middle term and takes the form $(u + v)(u - v) = u^2 - v^2$.

$$(5x + 9)(5x - 9) = (5x)^2 - 9^2 = 25x^2 - 81$$

✓ **Checkpoint** ▶ ***Audio-video solution in English & Spanish at LarsonPrecalculus.com***

Find the product of $3x - 2$ and $3x + 2$.

ALGEBRA HELP

When squaring a binomial, note that the resulting middle term is always *twice* the product of the two terms of the binomial.

EXAMPLE 6 Square of a Binomial

Find $(6x - 5)^2$.

Solution

The square of the binomial $u - v$ is $(u - v)^2 = u^2 - 2uv + v^2$.

$$(6x - 5)^2 = (6x)^2 - 2(6x)(5) + 5^2 = 36x^2 - 60x + 25$$

✓ **Checkpoint** ▶ ***Audio-video solution in English & Spanish at LarsonPrecalculus.com***

Find $(x + 10)^2$.

EXAMPLE 7 Cube of a Binomial

Find $(3x + 2)^3$.

Solution

The cube of the binomial $u + v$ is $(u + v)^3 = u^3 + 3u^2v + 3uv^2 + v^3$. Note the *decreasing* powers of u and the *increasing* powers of v. Letting $u = 3x$ and $v = 2$,

$$(3x + 2)^3 = (3x)^3 + 3(3x)^2(2) + 3(3x)(2^2) + 2^3 = 27x^3 + 54x^2 + 36x + 8.$$

✓ **Checkpoint** ▶ ***Audio-video solution in English & Spanish at LarsonPrecalculus.com***

Find $(4x - 1)^3$.

EXAMPLE 8 Multiplying Two Trinomials

See LarsonPrecalculus.com for an interactive version of this type of example.

Find the product of $x + y - 2$ and $x + y + 2$.

Solution

One way to find this product is to group $x + y$ and form a special product.

$$\begin{aligned}(x + y - 2)(x + y + 2) &= [\underbrace{(x + y) - 2}_{\text{Difference}}][\underbrace{(x + y) + 2}_{\text{Sum}}] \\ &= (x + y)^2 - 2^2 && \text{Sum and difference of same terms} \\ &= x^2 + 2xy + y^2 - 4 && \text{Square of a binomial}\end{aligned}$$

✓ **Checkpoint** ▶ ***Audio-video solution in English & Spanish at LarsonPrecalculus.com***

Find the product of $x - 2 + 3y$ and $x - 2 - 3y$. ■

GO DIGITAL

Application

EXAMPLE 9   **Finding the Volume of a Box**

An open box is made by cutting squares from the corners of a piece of metal that is 20 inches by 16 inches, as shown in the figure. The edge of each cut-out square is x inches. Find the volume of the box in terms of x. Then find the volume of the box when $x = 1$, $x = 2$, and $x = 3$.

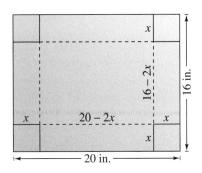

Solution

The volume of a rectangular box is equal to the product of its length, width, and height. From the figure, the length is $20 - 2x$, the width is $16 - 2x$, and the height is x. So, the volume of the box is

$$\text{Volume} = (20 - 2x)(16 - 2x)(x)$$

$$= (320 - 72x + 4x^2)(x)$$

$$= 320x - 72x^2 + 4x^3.$$

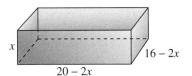

When $x = 1$ inch, the volume of the box is

$$\text{Volume} = 320(1) - 72(1)^2 + 4(1)^3$$

$$= 252 \text{ cubic inches.}$$

When $x = 2$ inches, the volume of the box is

$$\text{Volume} = 320(2) - 72(2)^2 + 4(2)^3$$

$$= 384 \text{ cubic inches.}$$

When $x = 3$ inches, the volume of the box is

$$\text{Volume} = 320(3) - 72(3)^2 + 4(3)^3$$

$$= 420 \text{ cubic inches.}$$

✓ *Checkpoint* ▶ *Audio-video solution in English & Spanish at LarsonPrecalculus.com*

In Example 9, find the volume of the box in terms of x when the piece of metal is 12 inches by 10 inches. Then find the volume when $x = 2$ and $x = 3$. ■

Summarize   (Section P.3)

1. State the definition of a polynomial in x and explain what is meant by the standard form of a polynomial *(page 26)*. For an example of writing polynomials in standard form, see Example 1.

2. Explain how to add and subtract polynomials *(page 27)*. For an example of adding and subtracting polynomials, see Example 2.

3. Explain the FOIL Method *(page 27)*. For an example of finding a product using the FOIL Method, see Example 3.

4. Explain how to find binomial products that have special forms *(page 28)*. For examples of binomial products that have special forms, see Examples 5–8.

5. Describe an example of how to use polynomials to model and solve a real-life problem *(page 30, Example 9)*.

GO DIGITAL

P.3 Exercises

See CalcChat.com for tutorial help and worked-out solutions to odd-numbered exercises.

GO DIGITAL

Vocabulary and Concept Check

In Exercises 1 and 2, fill in the blanks.

1. For the polynomial $a_n x^n + a_{n-1} x^{n-1} + \cdots + a_1 x + a_0$, $a_n \neq 0$, the degree is _____, the leading coefficient is _____, and the constant term is _____.

2. The letters in "FOIL" stand for F _____, O _____, I _____, and L _____.

3. Is it possible for a binomial and a trinomial to have the same degree? If so, give examples. If not, explain why.

4. Match each special product with its equivalent form.

   (a) $(u + v)(u - v)$ (i) $u^3 - 3u^2v + 3uv^2 - v^3$

   (b) $(u + v)^2$ (ii) $u^3 + 3u^2v + 3uv^2 + v^3$

   (c) $(u - v)^2$ (iii) $u^2 + 2uv + v^2$

   (d) $(u + v)^3$ (iv) $u^2 - v^2$

   (e) $(u - v)^3$ (v) $u^2 - 2uv + v^2$

Skills and Applications

Writing a Polynomial in Standard Form In Exercises 5–10, (a) write the polynomial in standard form, (b) identify the degree and leading coefficient of the polynomial, and (c) state whether the polynomial is a monomial, a binomial, or a trinomial.

5. $7x$

6. 3

7. $14x - \frac{1}{2}x^5$

8. $3 + 2x$

9. $1 + 6x^4 - 4x^5$

10. $-y + 25y^2 + 1$

Identifying Polynomials In Exercises 11–16, determine whether the expression is a polynomial. If so, write the polynomial in standard form.

11. $2x - 3x^3 + 8$

12. $5x^4 - 2x^2 + x^{-2}$

13. $\dfrac{3x + 4}{x}$

14. $\dfrac{x^2 + 2x - 3}{2}$

15. $y^2 - y^4 + y^3$

16. $y^4 - \sqrt{y}$

Adding or Subtracting Polynomials In Exercises 17–24, add or subtract and write the result in standard form.

17. $(6x + 5) - (8x + 15)$

18. $(t^3 - 1) + (6t^3 - 5t)$

19. $(4y^2 - 3) + (-7y^2 + 9)$

20. $(2x^2 + 1) - (x^2 - 2x + 1)$

21. $(15x^2 - 6) + (-8.3x^3 - 14.7x^2 - 17)$

22. $(15.6w^4 - 14w - 17.4) + (16.9w^4 - 9.2w + 13)$

23. $5z - [3z - (10z + 8)]$

24. $(y^3 + 1) - [(y^2 + 1) + (3y - 7)]$

Multiplying Polynomials In Exercises 25–36, multiply the polynomials.

25. $3x(x^2 - 2x + 1)$

26. $y^2(4y^2 + 2y - 3)$

27. $-5z(3z - 1)$

28. $-3x(5x + 2)$

29. $(1.5t^2 + 5)(-3t)$

30. $(2 - 3.5y)(2y^3)$

31. $(3x - 5)(2x + 1)$

32. $(7x - 2)(4x - 3)$

33. $(x + 7)(x^2 + 2x + 5)$

34. $(x - 8)(2x^2 + x + 4)$

35. $(x^2 - x + 2)(x^2 + x + 1)$

36. $(2x^2 - x + 4)(x^2 + 3x + 2)$

Finding Special Products In Exercises 37–60, find the special product.

37. $(x + 10)(x - 10)$

38. $(2x + 3)(2x - 3)$

39. $(x + 2y)(x - 2y)$

40. $(4a + 5b)(4a - 5b)$

41. $(2x + 3)^2$

42. $(5 - 8x)^2$

43. $(4x^3 - 3)^2$

44. $(8x + 3)^2$

45. $(x + 3)^3$

46. $(x - 2)^3$

47. $(2x - y)^3$

48. $(3x + 2y)^3$

49. $\left(\frac{1}{5}x - 3\right)\left(\frac{1}{5}x + 3\right)$

50. $(1.5x - 4)(1.5x + 4)$

51. $\left(\frac{1}{4}x - 5\right)^2$

52. $(2.4x + 3)^2$

53. $[(x - 3) + y]^2$

54. $[(x + 1) - y]^2$

55. $(3y - 6x)(-3y - 6x)$

56. $(3a^3 - 4b^2)(3a^3 + 4b^2)$

57. $[(m - 3) + n][(m - 3) - n]$

58. $[(x - 3y) + z][(x - 3y) - z]$

59. $(u + 2)(u - 2)(u^2 + 4)$

60. $(x + y)(x - y)(x^2 + y^2)$

Operations with Polynomials In Exercises 61–64, perform the operation.

61. Subtract $4x^2 - 5$ from $-3x^3 + x^2 + 9$.

62. Subtract $-7t^4 + 5t^2 - 1$ from $2t^4 - 10t^3 - 4t$.

63. Multiply $y^2 + 3y - 5$ by $y^2 - 6y + 4$.

64. Multiply $x^2 + 4x - 1$ by $x^2 - x + 3$.

Finding a Product In Exercises 65–68, find the product. (The expressions are not polynomials, but the special products formulas can still be used.)

65. $\left(\sqrt{x} + \sqrt{y}\right)\left(\sqrt{x} - \sqrt{y}\right)$

66. $\left(5 + \sqrt{x}\right)\left(5 - \sqrt{x}\right)$

67. $\left(x - \sqrt{y}\right)^2$

68. $\left(x + \sqrt{y}\right)^2$

69. **Genetics** In deer, the gene N is for normal coloring and the gene a is for albino. Any gene combination with an N results in normal coloring. The Punnett square shows the possible gene combinations of an offspring and the resulting colors when both parents have the gene combination Na.

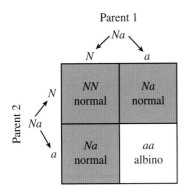

Parent 1

(a) What percent of the possible gene combinations result in albino coloring?

(b) Each parent's gene combination is represented by the polynomial $0.5N + 0.5a$. The product $(0.5N + 0.5a)^2$ represents the possible gene combinations of an offspring. Find this product.

(c) The coefficient of each term of the polynomial you wrote in part (b) is the probability (in decimal form) of the offspring having that gene combination. Use this polynomial to confirm your answer in part (a). Explain.

70. **Construction Management** A square-shaped foundation for a building with 100-foot sides is reduced by x feet on one side and extended by x feet on an adjacent side.

(a) The area of the new foundation is represented by $(100 - x)(100 + x)$. Find this product.

(b) Does the area of the foundation increase, decrease, or stay the same? Explain.

(c) Use the polynomial in part (a) to find the area of the new foundation when $x = 21$.

Geometry In Exercises 71 and 72, find the area of the shaded region in terms of x. Write your result as a polynomial in standard form.

71.
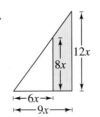

72.
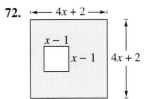

73. **Volume of a Box** A take-out fast-food restaurant is constructing an open box by cutting squares from the corners of the piece of cardboard shown in the figure. The edge of each cut-out square is x centimeters.

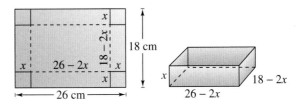

(a) Find the volume of the box in terms of x.

(b) Find the volume when $x = 1$, $x = 2$, and $x = 3$.

74. **Volume of a Box** An overnight shipping company designs a closed box by cutting along the solid lines and folding along the broken lines on the rectangular piece of corrugated cardboard shown in the figure.

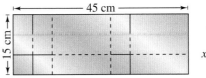

(a) Find the volume of the shipping box in terms of x.

(b) Find the volume when $x = 3$, $x = 5$, and $x = 7$.

75. **Engineering**

A one-inch-wide steel beam has a uniformly distributed load. When the span of the beam is x feet and its depth is 6 inches, the safe load S (in pounds) is approximately $S_6 = (0.06x^2 - 2.42x + 38.71)^2$. When the depth is 8 inches, the safe load is approximately $S_8 = (0.08x^2 - 3.30x + 51.93)^2$.

(a) Approximate the difference of the safe loads for these two beams when the span is 12 feet.

(b) How does the difference of the safe loads change as the span increases?

76. Stopping Distance The stopping distance of an automobile is the distance traveled during the driver's reaction time plus the distance traveled after the driver applies the brakes. In an experiment, researchers measured these distances (in feet) when the automobile was traveling at a speed of x miles per hour on dry, level pavement, as shown in the bar graph. The distance traveled during the reaction time R was

$$R = 1.1x$$

and the braking distance B was

$$B = 0.0475x^2 - 0.001x + 0.23.$$

(a) Determine the polynomial that represents the total stopping distance T.

(b) Use the result of part (a) to estimate the total stopping distance when $x = 30$, $x = 40$, and $x = 55$ miles per hour.

(c) Use the bar graph to make a statement about the total stopping distance required for increasing speeds.

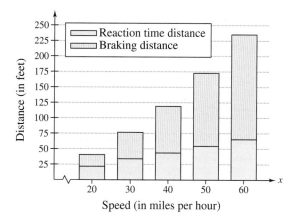

Exploring the Concepts

True or False? In Exercises 77–80, determine whether the statement is true or false. Justify your answer.

77. The product of two binomials is always a second-degree polynomial.

78. The sum of two second-degree polynomials is always a second-degree polynomial.

79. The sum of two binomials is always a binomial.

80. The leading coefficient of the product of two polynomials is always the product of the leading coefficients of the two polynomials.

81. Degree of a Product Find the degree of the product of two polynomials of degrees m and n.

82. Degree of a Sum Find the degree of the sum of two polynomials of degrees m and n, where $m < n$.

83. Error Analysis Describe the error.

$(x - 3)^2 = x^2 + 9$

84. **HOW DO YOU SEE IT?** An open box has a length of $(52 - 2x)$ inches, a width of $(42 - 2x)$ inches, and a height of x inches, as shown.

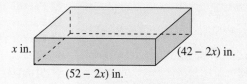

x in. $(42 - 2x)$ in.
$(52 - 2x)$ in.

(a) Describe a way that you could make the box from a rectangular piece of cardboard. Give the original dimensions of the cardboard.

(b) What degree is the polynomial that represents the volume of the box? Explain.

(c) Describe a procedure for finding the value of x (to the nearest tenth of an inch) that yields the maximum possible volume of the box.

85. Think About It When the polynomial $2x - 1$ is subtracted from an unknown polynomial, the difference is $5x^2 + 8$. Find the unknown polynomial.

86. Reasoning Verify that $(x + y)^2$ is not equal to $x^2 + y^2$ by letting $x = 3$ and $y = 4$ and evaluating both expressions. Are there any values of x and y for which $(x + y)^2$ and $x^2 + y^2$ are equal? Explain.

Review & Refresh ▶ *Video solutions at LarsonPrecalculus.com*

Multiplying Radicals In Exercises 87–90, find each product.

87. $\sqrt{3}\left(\sqrt{3}\right)$

88. $\sqrt{6}\left(-\sqrt{6}\right)$

89. $-\sqrt{15}\left(\sqrt{15}\right)$

90. $-\sqrt{21}\left(-\sqrt{21}\right)$

Finding a Greatest Common Factor In Exercises 91–94, find the greatest common factor of the expressions.

91. $2x^2, x^3, 4x$

92. $3x^3, 12x^2, 42x^3$

93. x^{10}, x^{20}, x^{30}

94. $45x^5, 9x^3, 15x^2$

Using Properties of Exponents In Exercises 95 and 96, simplify the expression.

95. $(3x^2)^3$

96. $(4x^4)^2$

Identifying Rules of Algebra In Exercises 97–102, identify the rule(s) of algebra illustrated by the statement.

97. $\dfrac{1}{h + 6}(h + 6) = 1, \; h \neq -6$

98. $(x + 3) - (x + 3) = 0$

99. $x(3y) = (x \cdot 3)y = (3x)y$

100. $\frac{1}{7}(7 \cdot 12) = \left(\frac{1}{7} \cdot 7\right)12 = 1 \cdot 12 = 12$

101. $x(2x^2 + 3x - 1) = 2x^3 + 3x^2 - x$

102. $(x - 2)(2x + 3) = (x - 2)(2x) + (x - 2)(3)$

P.4 Factoring Polynomials

- ❯ **Factor out common factors from polynomials.**
- ❯ **Factor special polynomial forms.**
- ❯ **Factor trinomials as the product of two binomials.**
- ❯ **Factor polynomials by grouping.**

Polynomials with Common Factors

Polynomial factoring has many real-life applications. For example, in Exercise 80 on page 40, you will use polynomial factoring to write an alternative form of an expression that models the rate of change of an autocatalytic chemical reaction.

The process of writing a polynomial as a product is called **factoring.** It is an important tool for solving equations and for simplifying rational expressions.

Unless noted otherwise, when you are asked to factor a polynomial, assume that you are looking for factors that have integer coefficients. If a polynomial does not factor using integer coefficients, then it is **prime** or **irreducible over the integers.** For example, the polynomial $x^2 - 3$ is irreducible over the integers. Over the *real numbers,* this polynomial factors as

$$x^2 - 3 = \left(x + \sqrt{3}\right)\left(x - \sqrt{3}\right).$$

A polynomial is **completely factored** when each of its factors is prime. For example,

$$x^3 - x^2 + 4x - 4 = (x - 1)(x^2 + 4) \qquad \text{Completely factored}$$

is completely factored, but

$$x^3 - x^2 - 4x + 4 = (x - 1)(x^2 - 4) \qquad \text{Not completely factored}$$

is not completely factored. Its complete factorization is

$$x^3 - x^2 - 4x + 4 = (x - 1)(x + 2)(x - 2).$$

The simplest type of factoring involves a polynomial that can be written as the product of a monomial and another polynomial. The technique used here is the Distributive Property, $a(b + c) = ab + ac$, in the *reverse* direction.

$$ab + ac = a(b + c) \qquad a \text{ is a common factor.}$$

Factoring out any common factors is the first step in completely factoring a polynomial.

EXAMPLE 1   Factoring Out Common Factors

Factor each expression.

a. $6x^3 - 4x$ **b.** $-4x^2 + 12x - 16$ **c.** $(x - 2)(2x) + (x - 2)(3)$

Solution

a. $6x^3 - 4x = 2x(3x^2) - 2x(2)$ $2x$ is a common factor.

$\qquad\qquad = 2x(3x^2 - 2)$

b. $-4x^2 + 12x - 16 = -4(x^2) + (-4)(-3x) + (-4)4$ -4 is a common factor.

$\qquad\qquad\qquad\qquad = -4(x^2 - 3x + 4)$

c. $(x - 2)(2x) + (x - 2)(3) = (x - 2)(2x + 3)$ $(x - 2)$ is a common factor.

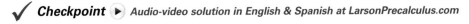 ✔ ***Checkpoint*** ▶ *Audio-video solution in English & Spanish at LarsonPrecalculus.com*

Factor each expression.

a. $5x^3 - 15x^2$ **b.** $-3 + 6x - 12x^3$ **c.** $(x + 1)(x^2) - (x + 1)(2)$ ◼

Factoring Special Polynomial Forms

Some polynomials have special forms that arise from the special product forms on page 28. You should learn to recognize these forms so that you can factor such polynomials efficiently.

Factoring Special Polynomial Forms

Factored Form	Example
Difference of Two Squares	
$u^2 - v^2 = (u + v)(u - v)$	$9x^2 - 4 = (3x)^2 - 2^2 = (3x + 2)(3x - 2)$
Perfect Square Trinomial	
$u^2 + 2uv + v^2 = (u + v)^2$	$x^2 + 6x + 9 = x^2 + 2(x)(3) + 3^2 = (x + 3)^2$
$u^2 - 2uv + v^2 = (u - v)^2$	$x^2 - 6x + 9 = x^2 - 2(x)(3) + 3^2 = (x - 3)^2$
Sum or Difference of Two Cubes	
$u^3 + v^3 = (u + v)(u^2 - uv + v^2)$	$x^3 + 8 = x^3 + 2^3 = (x + 2)(x^2 - 2x + 4)$
$u^3 - v^3 = (u - v)(u^2 + uv + v^2)$	$27x^3 - 1 = (3x)^3 - 1^3 = (3x - 1)(9x^2 + 3x + 1)$

For the difference of two squares, you can think of this form as

$$u^2 - v^2 = (u + v)(u - v).$$

Factors are conjugates.

Difference   Opposite signs

To recognize perfect square terms, look for coefficients that are squares of integers and variables raised to *even powers.*

EXAMPLE 2 Factoring Out a Common Factor First

$$3 - 12x^2 = 3(1 - 4x^2)$$ 3 is a common factor.

$$= 3[1^2 - (2x)^2]$$ Rewrite $1 - 4x^2$ as the difference of two squares.

$$= 3(1 + 2x)(1 - 2x)$$ Factor.

✓ **Checkpoint** ▶ Audio-video solution in English & Spanish at LarsonPrecalculus.com

Factor $100 - 4y^2$.

EXAMPLE 3 Factoring the Difference of Two Squares

a. $(x + 2)^2 - y^2 = [(x + 2) + y][(x + 2) - y]$

$$= (x + 2 + y)(x + 2 - y)$$

b. $16x^4 - 81 = (4x^2)^2 - 9^2$ Rewrite as the difference of two squares.

$$= (4x^2 + 9)(4x^2 - 9)$$ Factor.

$$= (4x^2 + 9)[(2x)^2 - 3^2]$$ Rewrite $4x^2 - 9$ as the difference of two squares.

$$= (4x^2 + 9)(2x + 3)(2x - 3)$$ Factor.

✓ **Checkpoint** ▶ Audio-video solution in English & Spanish at LarsonPrecalculus.com

Factor $(x - 1)^2 - 9y^4$.

ALGEBRA HELP

In Example 2, note that the first step in factoring a polynomial is to check for any common factors. Once you have removed any common factors, it is often possible to recognize patterns that were not immediately obvious.

GO DIGITAL

A perfect square trinomial is the square of a binomial, and it has the form

$$u^2 + 2uv + v^2 = (u + v)^2 \quad \text{or} \quad u^2 - 2uv + v^2 = (u - v)^2.$$

Like signs   Like signs

Note that the first and last terms are squares and the middle term is twice the product of u and v.

EXAMPLE 4   Factoring Perfect Square Trinomials

Factor each trinomial.

a. $x^2 - 10x + 25$ **b.** $16x^2 + 24x + 9$

Solution

a. $x^2 - 10x + 25 = x^2 - 2(x)(5) + 5^2 = (x - 5)^2$

b. $16x^2 + 24x + 9 = (4x)^2 + 2(4x)(3) + 3^2 = (4x + 3)^2$

✓ *Checkpoint* ▶ *Audio-video solution in English & Spanish at LarsonPrecalculus.com*

Factor $9x^2 - 30x + 25$. ∎

The next two formulas show the sum and difference of two cubes. Pay special attention to the signs of the terms.

Like signs   Like signs

$$u^3 + v^3 = (u + v)(u^2 - uv + v^2) \quad u^3 - v^3 = (u - v)(u^2 + uv + v^2)$$

Unlike signs   Unlike signs

EXAMPLE 5   Factoring the Difference of Two Cubes

$$x^3 - 27 = x^3 - 3^3 \qquad \text{Rewrite 27 as } 3^3.$$

$$= (x - 3)(x^2 + 3x + 9) \qquad \text{Factor.}$$

✓ *Checkpoint* ▶ *Audio-video solution in English & Spanish at LarsonPrecalculus.com*

Factor $64x^3 - 1$.

EXAMPLE 6   Factoring the Sum of Two Cubes

a. $y^3 + 8 = y^3 + 2^3$ Rewrite 8 as 2^3.

      $= (y + 2)(y^2 - 2y + 4)$ Factor.

b. $3x^3 + 192 = 3(x^3 + 64)$ 3 is a common factor.

      $= 3(x^3 + 4^3)$ Rewrite 64 as 4^3.

      $= 3(x + 4)(x^2 - 4x + 16)$ Factor.

✓ *Checkpoint* ▶ *Audio-video solution in English & Spanish at LarsonPrecalculus.com*

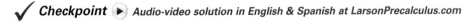

Factor each expression.

a. $x^3 + 216$ **b.** $5y^3 + 135$ ∎

Trinomials with Binomial Factors

To factor a trinomial of the form $ax^2 + bx + c$, use the pattern below.

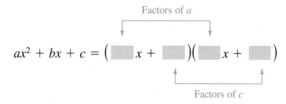

The goal is to find a combination of factors of a and c such that the sum of the outer and inner products is the middle term bx. For example, for the trinomial $6x^2 + 17x + 5$, you can write all possible factorizations and determine which one has outer and inner products whose sum is $17x$.

$$(6x + 5)(x + 1), (6x + 1)(x + 5), (2x + 1)(3x + 5), (2x + 5)(3x + 1)$$

The correct factorization is $(2x + 5)(3x + 1)$ because the sum of the outer (O) and inner (I) products is $17x$.

$$
\begin{array}{ccccc}
\text{F} & \text{O} & \text{I} & \text{L} & \text{O} + \text{I} \\
\downarrow & \downarrow & \downarrow & \downarrow & \downarrow
\end{array}
$$

$$(2x + 5)(3x + 1) = 6x^2 + 2x + 15x + 5 = 6x^2 + 17x + 5$$

ALGEBRA HELP

Factoring a trinomial can involve trial and error. However, you can check your answer by multiplying the factors. The product should be the original trinomial. For instance, in Example 7, verify that $(x - 3)(x - 4) = x^2 - 7x + 12$.

EXAMPLE 7 **Factoring a Trinomial: Leading Coefficient Is 1**

Factor $x^2 - 7x + 12$.

Solution For this trinomial, $a = 1$, $b = -7$, and $c = 12$. Because b is negative and c is positive, both factors of 12 must be negative. So, the possible factorizations of $x^2 - 7x + 12$ are

$$(x - 1)(x - 12), \quad (x - 2)(x - 6), \quad \text{and} \quad (x - 3)(x - 4).$$

Testing the middle term, you will find the correct factorization to be

$$x^2 - 7x + 12 = (x - 3)(x - 4). \qquad \text{O} + \text{I} = -4x + (-3x) = -7x$$

✓ **Checkpoint** Audio-video solution in English & Spanish at LarsonPrecalculus.com

Factor $x^2 + x - 6$.

EXAMPLE 8 **Factoring a Trinomial: Leading Coefficient Is Not 1**

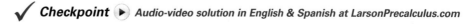 See LarsonPrecalculus.com for an interactive version of this type of example.

Factor $2x^2 + x - 15$.

Solution For this trinomial, $a = 2$, $b = 1$, and $c = -15$. Because c is negative, its factors must have unlike signs. The eight possible factorizations are below.

$$(2x - 1)(x + 15) \quad (2x + 1)(x - 15) \quad (2x - 3)(x + 5) \quad (2x + 3)(x - 5)$$

$$(2x - 5)(x + 3) \quad (2x + 5)(x - 3) \quad (2x - 15)(x + 1) \quad (2x + 15)(x - 1)$$

Testing the middle term, you will find the correct factorization to be

$$2x^2 + x - 15 = (2x - 5)(x + 3). \qquad \text{O} + \text{I} = 6x + (-5x) = x$$

✓ **Checkpoint** ▶ Audio-video solution in English & Spanish at LarsonPrecalculus.com

Factor $2x^2 - 5x + 3$.

Factoring by Grouping

Sometimes, polynomials with more than three terms can be **factored by grouping.**

EXAMPLE 9 Factoring by Grouping

$$x^3 - 2x^2 - 3x + 6 = (x^3 - 2x^2) - (3x - 6) \qquad \text{Group terms.}$$
$$= x^2(x - 2) - 3(x - 2) \qquad \text{Factor each group.}$$
$$= (x - 2)(x^2 - 3) \qquad (x - 2) \text{ is a common factor.}$$

✓ **Checkpoint** ▶ Audio-video solution in English & Spanish at LarsonPrecalculus.com

Factor $x^3 + x^2 - 5x - 5$.

> **ALGEBRA HELP**
>
> Sometimes, more than one grouping will work. For instance, another way to factor the polynomial in Example 9 is
>
> $x^3 - 2x^2 - 3x + 6$
> $= (x^3 - 3x) - (2x^2 - 6)$
> $= x(x^2 - 3) - 2(x^2 - 3)$
> $= (x^2 - 3)(x - 2).$
>
> Notice that this is the same result as in Example 9.

Factoring by grouping can eliminate some of the trial and error involved in factoring a trinomial. To factor a trinomial of the form $ax^2 + bx + c$ by grouping, choose factors of the product ac that sum to b and use these factors to rewrite the middle term. Example 10 illustrates this technique.

EXAMPLE 10 Factoring a Trinomial by Grouping

In the trinomial $2x^2 + 5x - 3$, $a = 2$ and $c = -3$, so the product ac is -6. Now, -6 factors as $(6)(-1)$ and $6 + (-1) = 5 = b$. So, rewrite the middle term as $5x = 6x - x$ and factor by grouping.

$$2x^2 + 5x - 3 = 2x^2 + 6x - x - 3 \qquad \text{Rewrite middle term.}$$
$$= (2x^2 + 6x) - (x + 3) \qquad \text{Group terms.}$$
$$= 2x(x + 3) - (x + 3) \qquad \text{Factor } 2x^2 + 6x.$$
$$= (x + 3)(2x - 1) \qquad (x + 3) \text{ is a common factor.}$$

✓ **Checkpoint** ▶ Audio-video solution in English & Spanish at LarsonPrecalculus.com

Use factoring by grouping to factor $2x^2 + 5x - 12$.

> **Guidelines for Factoring Polynomials**
>
> 1. Factor out any common factors using the Distributive Property.
> 2. Factor according to one of the special polynomial forms.
> 3. Factor as $ax^2 + bx + c = (mx + r)(nx + s)$.
> 4. Factor by grouping.

Summarize (Section P.4)

1. Explain what it means to completely factor a polynomial *(page 34)*. For an example of factoring out common factors, see Example 1.
2. Make a list of the special polynomial forms of factoring *(page 35)*. For examples of factoring these special forms, see Examples 2–6.
3. Explain how to factor a trinomial of the form $ax^2 + bx + c$ *(page 37)*. For examples of factoring trinomials of this form, see Examples 7 and 8.
4. Explain how to factor a polynomial by grouping *(page 38)*. For examples of factoring by grouping, see Examples 9 and 10.

GO DIGITAL

GO DIGITAL

Vocabulary and Concept Check

In Exercises 1 and 2, fill in the blanks.

1. The process of writing a polynomial as a product is called _____.

2. A _____ _____ _____ is the square of a binomial, and it has the form $u^2 + 2uv + v^2$ or $u^2 - 2uv + v^2$.

3. When is a polynomial completely factored?

4. List four guidelines for factoring polynomials.

Skills and Applications

Factoring Out a Common Factor In Exercises 5–8, factor out the common factor.

5. $2x^3 - 6x$

6. $3z^3 - 6z^2 + 9z$

7. $3x(x - 5) + 8(x - 5)$

8. $(x + 3)^2 - 4(x + 3)$

Factoring the Difference of Two Squares In Exercises 9–18, completely factor the difference of two squares.

9. $x^2 - 81$

10. $x^2 - 64$

11. $25y^2 - 4$

12. $4y^2 - 49$

13. $64 - 9z^2$

14. $81 - 36z^2$

15. $(x - 1)^2 - 4$

16. $25 - (z + 5)^2$

17. $81u^4 - 1$

18. $x^4 - 16y^4$

Factoring a Perfect Square Trinomial In Exercises 19–24, factor the perfect square trinomial.

19. $x^2 - 4x + 4$

20. $4t^2 + 4t + 1$

21. $25z^2 - 30z + 9$

22. $36y^2 + 84y + 49$

23. $4y^2 - 12y + 9$

24. $9u^2 + 24uv + 16v^2$

Factoring the Sum or Difference of Two Cubes In Exercises 25–30, factor the sum or difference of two cubes.

25. $x^3 - 8$

26. $x^3 + 125$

27. $8t^3 - 1$

28. $27z^3 + 1$

29. $27x^3 + 8$

30. $64y^3 - 125$

Factoring a Trinomial In Exercises 31–40, factor the trinomial.

31. $x^2 + x - 2$

32. $x^2 + 5x + 6$

33. $s^2 - 5s + 6$

34. $t^2 - t - 6$

35. $3x^2 + 10x - 8$

36. $2x^2 - 3x - 27$

37. $5x^2 + 31x + 6$

38. $8x^2 + 51x + 18$

39. $-5y^2 - 8y + 4$

40. $-6z^2 + 17z + 3$

Factoring by Grouping In Exercises 41–48, factor by grouping.

41. $x^3 - x^2 + 2x - 2$

42. $x^3 + 5x^2 - 5x - 25$

43. $2x^3 - x^2 - 6x + 3$

44. $3x^3 + x^2 - 15x - 5$

45. $6 + 2x - 3x^3 - x^4$

46. $x^5 + 2x^3 + x^2 + 2$

47. $3x^5 + 6x^3 - 2x^2 - 4$

48. $8x^5 - 6x^2 + 12x^3 - 9$

Factoring a Trinomial by Grouping In Exercises 49–52, factor the trinomial by grouping.

49. $2x^2 + 9x + 9$

50. $6x^2 + x - 2$

51. $6x^2 - x - 15$

52. $12x^2 - 13x + 1$

Factoring Completely In Exercises 53–70, completely factor the expression.

53. $6x^2 - 54$

54. $12x^2 - 48$

55. $x^3 - x^2$

56. $x^3 - 16x$

57. $1 - 4x + 4x^2$

58. $-9x^2 + 6x - 1$

59. $2x^2 + 4x - 2x^3$

60. $9x^2 + 12x - 3x^3$

61. $(x^2 + 3)^2 - 16x^2$

62. $(x^2 + 8)^2 - 36x^2$

63. $2x^3 + x^2 - 8x - 4$

64. $3x^3 + x^2 - 27x - 9$

65. $2x(3x + 1) + (3x + 1)^2$

66. $4x(2x - 1) + (2x - 1)^2$

67. $2(x - 2)(x + 1)^2 - 3(x - 2)^2(x + 1)$

68. $2(x + 1)(x - 3)^2 - 3(x + 1)^2(x - 3)$

69. $5(2x + 1)^2(x + 1)^2 + (2x + 1)(x + 1)^3$

70. $7(3x + 2)^2(1 - x)^2 + (3x + 2)(1 - x)^3$

Fractional Coefficients In Exercises 71–76, completely factor the expression. (*Hint:* The factors will contain fractional coefficients.)

71. $16x^2 - \frac{1}{9}$

72. $\frac{4}{25}y^2 - 64$

73. $z^2 + z + \frac{1}{4}$

74. $9y^2 - \frac{3}{2}y + \frac{1}{16}$

75. $y^3 + \frac{8}{27}$

76. $x^3 - \frac{27}{64}$

Geometric Modeling In Exercises 77 and 78, draw a "geometric factoring model" to represent the factorization. For example, a factoring model for $2x^2 + 3x + 1 = (2x + 1)(x + 1)$ is shown below.

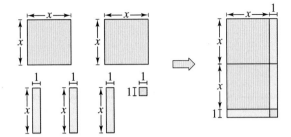

77. $x^2 + 3x + 2 = (x + 2)(x + 1)$

78. $3x^2 + 7x + 2 = (3x + 1)(x + 2)$

79. Geometry The volume V of the cylindrical shell shown in the figure is given by $V = \pi R^2 h - \pi r^2 h$. Factor the expression for the volume. Then use the result to show that

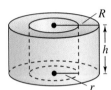

$V = 2\pi(\text{average radius})(\text{thickness of the shell})h.$

80. Chemistry

The rate of change of an autocatalytic chemical reaction is given by $kQx - kx^2$, where Q is the amount of the original substance, x is the amount of substance formed, and k is a constant of proportionality. Factor the expression.

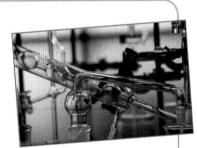

Factoring a Trinomial In Exercises 81 and 82, find all values of b for which the trinomial is factorable.

81. $x^2 + bx - 15$ **82.** $x^2 + bx + 24$

Factoring a Trinomial In Exercises 83 and 84, find two integer values of c such that the trinomial is factorable. (There are many correct answers.)

83. $2x^2 + 5x + c$ **84.** $3x^2 - x + c$

Exploring the Concepts

True or False? In Exercises 85 and 86, determine whether the statement is true or false. Justify your answer.

85. The difference of two squares can be factored as a product of conjugates.

86. A perfect square trinomial can always be factored as the square of a binomial.

87. Think About It Is $(3x - 6)(x + 1)$ completely factored? Explain.

88. HOW DO YOU SEE IT? The figure shows a large square with an area of a^2 that contains a smaller square with an area of b^2.

(a) Describe the regions that represent $a^2 - b^2$. How can you rearrange these regions to show that $a^2 - b^2 = (a - b)(a + b)$?

(b) How can you use the figure to show that $(a - b)^2 = a^2 - 2ab + b^2$?

(c) Draw another figure to show that $(a + b)^2 = a^2 + 2ab + b^2$. Explain how the figure shows this.

89. Error Analysis Describe the error.

$$9x^2 - 9x - 54 = (3x + 6)(3x - 9)$$
$$= 3(x + 2)(x - 3) \quad \textbf{✗}$$

90. Difference of Two Sixth Powers Rewrite $u^6 - v^6$ as the difference of two squares. Then find a formula for completely factoring $u^6 - v^6$. Use your formula to completely factor $x^6 - 1$ and $x^6 - 64$.

Review & Refresh ▶ *Video solutions at LarsonPrecalculus.com*

Operations with Fractions In Exercises 91–94, perform the operation. (Write fractional answers in simplest form.)

91. $\dfrac{4x}{5} - \dfrac{5x}{6}$ **92.** $\dfrac{3x}{8} + \dfrac{x}{6}$

93. $\dfrac{2x}{15} \cdot \dfrac{9}{8}$ **94.** $\dfrac{18x}{11} \div \dfrac{14x^2}{33}$

Evaluating an Expression In Exercises 95 and 96, evaluate the expression for each value of x. (If not possible, state the reason.)

95. $\dfrac{x + 1}{x - 1}$ (a) $x = 1$ (b) $x = -1$

96. $\dfrac{1 - x}{x}$ (a) $x = 1$ (b) $x = 0$

Finding a Least Common Multiple In Exercises 97–100, find the least common multiple of the expressions.

97. x, x^2, x^3 **98.** $2x^2, 12x, 42x^3$

99. $x, x + 1, x^2 - 1$

100. $x + 2, x^2 - 4, x^2 - 2x - 8$

P.5 Rational Expressions

Rational expressions have many real-life applications. For example, in Exercise 73 on page 49, you will work with a rational expression that models the temperature of food in a refrigerator.

- ❯ Find the domains of algebraic expressions.
- ❯ Simplify rational expressions.
- ❯ Add, subtract, multiply, and divide rational expressions.
- ❯ Simplify complex fractions.
- ❯ Simplify expressions from calculus.

The Domain of an Algebraic Expression

The set of real numbers for which an algebraic expression is defined is the **domain** of the expression. Two algebraic expressions are **equivalent** when they have the same domain and yield the same values for all numbers in their domain. For example,

$$(x + 1) + (x + 2) \quad \text{and} \quad 2x + 3$$

are equivalent because

$$(x + 1) + (x + 2) = x + 1 + x + 2 \qquad \text{Remove parentheses.}$$
$$= x + x + 1 + 2 \qquad \text{Commutative Property of Addition}$$
$$= 2x + 3. \qquad \text{Combine like terms.}$$

EXAMPLE 1 Finding the Domains of Algebraic Expressions

Find the domain of each expression.

a. $2x^3 + 3x + 4$ **b.** $\sqrt{x - 2}$ **c.** $\dfrac{x + 2}{x - 3}$

Solution

a. The domain of the polynomial $2x^3 + 3x + 4$ is the set of all real numbers. In fact, the domain of any polynomial is the set of all real numbers, unless the domain is specifically restricted.

b. The domain of the radical expression $\sqrt{x - 2}$ is the set of real numbers greater than or equal to 2, because the square root of a negative number is not a real number.

c. The domain of the expression

$$\frac{x + 2}{x - 3}$$

is the set of all real numbers except $x = 3$, which would result in division by zero, which is undefined.

✓ *Checkpoint* ▶ Audio-video solution in English & Spanish at LarsonPrecalculus.com

Find the domain of each expression.

a. $4x^3 + 3, \ x \geq 0$ **b.** $\sqrt{x + 7}$ **c.** $\dfrac{1 - x}{x}$ ■

The quotient of two algebraic expressions is a *fractional expression*. Moreover, the quotient of two *polynomials* such as

$$\frac{1}{x}, \quad \frac{2x - 1}{x + 1}, \quad \text{or} \quad \frac{x^2 - 1}{x^2 + 1}$$

is a **rational expression.**

GO DIGITAL

Simplifying Rational Expressions

Recall that a fraction is in simplest form when its numerator and denominator have no factors in common other than ± 1. To write a fraction in simplest form, divide out common factors.

$$\frac{ac}{bc} = \frac{a \cdot \cancel{c}}{b \cdot \cancel{c}} = \frac{a}{b}, \quad c \neq 0$$

The key to success in simplifying rational expressions lies in your ability to *factor* polynomials. When simplifying rational expressions, factor each polynomial completely to determine whether the numerator and denominator have factors in common.

EXAMPLE 2   Simplifying a Rational Expression

$$\frac{x^2 + 4x - 12}{3x - 6} = \frac{(x + 6)(x - 2)}{3(x - 2)} \qquad \text{Factor completely.}$$

$$= \frac{x + 6}{3}, \quad x \neq 2 \qquad \text{Divide out common factor.}$$

Note that the original expression is undefined when $x = 2$ (because division by zero is undefined). To make the simplified expression *equivalent* to the original expression, you must restrict the domain of the simplified expression by excluding the value $x = 2$.

✓ **Checkpoint** ▶ *Audio-video solution in English & Spanish at LarsonPrecalculus.com*

Write $\dfrac{4x + 12}{x^2 - 3x - 18}$ in simplest form.   ■

Sometimes it may be necessary to change the sign of a factor by factoring out (-1) to simplify a rational expression, as shown in Example 3.

EXAMPLE 3   Simplifying a Rational Expression

$$\frac{12 + x - x^2}{2x^2 - 9x + 4} = \frac{(4 - x)(3 + x)}{(2x - 1)(x - 4)} \qquad \text{Factor completely.}$$

$$= \frac{-(x - 4)(3 + x)}{(2x - 1)(x - 4)} \qquad (4 - x) = -(x - 4)$$

$$= -\frac{3 + x}{2x - 1}, \quad x \neq 4 \qquad \text{Divide out common factor.}$$

✓ **Checkpoint** ▶ *Audio-video solution in English & Spanish at LarsonPrecalculus.com*

Write $\dfrac{3x^2 - x - 2}{5 - 4x - x^2}$ in simplest form.   ■

In this text, the domain is usually not listed with a rational expression. It is *implied* that the real numbers that make the denominator zero are excluded from the domain. Also, when performing operations with rational expressions, this text follows the convention of listing *by the simplified expression* all values of x that must be specifically excluded from the domain to make the domains of the simplified and original expressions agree. Example 3, for instance, lists the restriction $x \neq 4$ with the simplified expression to make the two domains agree. Note that the value $x = \frac{1}{2}$ is excluded from *both* domains, so it is not necessary to list this value.

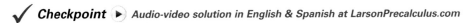

▶▶▶▶

ALGEBRA HELP

In Example 2, do not make the mistake of trying to simplify further by dividing out terms.

$$\frac{x + 6}{3} = \frac{x + \cancel{6}}{\cancel{3}}$$

$$= x + 2 \quad ✗$$

To simplify fractions, divide out common *factors*, not terms. To learn about other common errors, see Appendix A.

GO DIGITAL

Operations with Rational Expressions

To multiply or divide rational expressions, use the properties of fractions discussed in Section P.1. Recall that to divide fractions, you invert the divisor and multiply.

EXAMPLE 4 Multiplying Rational Expressions

$$\frac{2x^2 + x - 6}{x^2 + 4x - 5} \cdot \frac{x^3 - 3x^2 + 2x}{4x^2 - 6x} = \frac{(2x - 3)(x + 2)}{(x + 5)(x - 1)} \cdot \frac{x(x - 2)(x - 1)}{2x(2x - 3)}$$

$$= \frac{(x + 2)(x - 2)}{2(x + 5)}, \quad x \neq 0, x \neq 1, x \neq \frac{3}{2}$$

✓ **Checkpoint** Audio-video solution in English & Spanish at LarsonPrecalculus.com

Multiply and simplify: $\dfrac{15x^2 + 5x}{x^3 - 3x^2 - 18x} \cdot \dfrac{x^2 - 2x - 15}{3x^2 - 8x - 3}.$

EXAMPLE 5 Dividing Rational Expressions

$$\frac{x^3 - 8}{x^2 - 4} \div \frac{x^2 + 2x + 4}{x^3 + 8} = \frac{x^3 - 8}{x^2 - 4} \cdot \frac{x^3 + 8}{x^2 + 2x + 4} \qquad \text{Invert and multiply.}$$

$$= \frac{(x - 2)(x^2 + 2x + 4)}{(x + 2)(x - 2)} \cdot \frac{(x + 2)(x^2 - 2x + 4)}{(x^2 + 2x + 4)}$$

$$= x^2 - 2x + 4, \quad x \neq \pm 2 \qquad \text{Divide out common factors.}$$

✓ **Checkpoint**  Audio-video solution in English & Spanish at LarsonPrecalculus.com

Divide and simplify: $\dfrac{x^3 - 1}{x^2 - 1} \div \dfrac{x^2 + x + 1}{x^2 + 2x + 1}.$ ▪

To add or subtract rational expressions, use the LCD (least common denominator) method or the *basic definition*

$$\frac{a}{b} \pm \frac{c}{d} = \frac{ad \pm bc}{bd}, \quad b \neq 0, d \neq 0. \qquad \text{Basic definition}$$

The basic definition provides an efficient way of adding or subtracting two fractions that have no common factors in their denominators.

EXAMPLE 6 Subtracting Rational Expressions

$$\frac{x}{x - 3} - \frac{2}{3x + 4} = \frac{x(3x + 4) - 2(x - 3)}{(x - 3)(3x + 4)} \qquad \text{Basic definition}$$

$$= \frac{3x^2 + 4x - 2x + 6}{(x - 3)(3x + 4)} \qquad \text{Distributive Property}$$

$$= \frac{3x^2 + 2x + 6}{(x - 3)(3x + 4)} \qquad \text{Combine like terms.}$$

✓ **Checkpoint**  Audio-video solution in English & Spanish at LarsonPrecalculus.com

Subtract and simplify: $\dfrac{x}{2x - 1} - \dfrac{1}{x + 2}.$ ▪

ALGEBRA HELP

Note that Example 4 lists the restrictions $x \neq 0$, $x \neq 1$, and $x \neq \frac{3}{2}$ with the simplified expression to make the two domains agree. Also note that the value $x = -5$ is excluded from both domains, so it is not necessary to list this value.

ALGEBRA HELP

When subtracting rational expressions, remember to distribute the negative sign to *all* the terms in the quantity that is being subtracted.

For three or more fractions, or for fractions with a repeated factor in the denominators, the LCD method works well. Recall that the least common denominator of several fractions consists of the product of all prime factors in the denominators, with each factor given the highest power of its occurrence in any denominator. Here is a numerical example.

$$\frac{1}{6} + \frac{3}{4} - \frac{2}{3} = \frac{1 \cdot 2}{6 \cdot 2} + \frac{3 \cdot 3}{4 \cdot 3} - \frac{2 \cdot 4}{3 \cdot 4}$$

The LCD is 12.

$$= \frac{2}{12} + \frac{9}{12} - \frac{8}{12}$$

$$= \frac{3}{12}$$

$$= \frac{1}{4}$$

Sometimes, the numerator of the answer has a factor in common with the denominator. In such cases, simplify the answer, as shown in the example above.

EXAMPLE 7 Combining Rational Expressions: The LCD Method

⏩ *See LarsonPrecalculus.com for an interactive version of this type of example.*

Perform the operations and simplify.

$$\frac{3}{x - 1} - \frac{2}{x} + \frac{x + 3}{x^2 - 1}$$

Solution Use the factored denominators $x - 1$, x, and $(x + 1)(x - 1)$ to determine that the LCD is $x(x + 1)(x - 1)$.

$$\frac{3}{x - 1} - \frac{2}{x} + \frac{x + 3}{x^2 - 1}$$

Write original expression.

$$= \frac{3}{x - 1} - \frac{2}{x} + \frac{x + 3}{(x + 1)(x - 1)}$$

Factor denominator.

$$= \frac{3(x)(x + 1)}{x(x + 1)(x - 1)} - \frac{2(x + 1)(x - 1)}{x(x + 1)(x - 1)} + \frac{(x + 3)(x)}{x(x + 1)(x - 1)}$$

$$= \frac{3(x)(x + 1) - 2(x + 1)(x - 1) + (x + 3)(x)}{x(x + 1)(x - 1)}$$

$$= \frac{3x^2 + 3x - 2x^2 + 2 + x^2 + 3x}{x(x + 1)(x - 1)}$$

Multiply.

$$= \frac{(3x^2 - 2x^2 + x^2) + (3x + 3x) + 2}{x(x + 1)(x - 1)}$$

Group like terms.

$$= \frac{2x^2 + 6x + 2}{x(x + 1)(x - 1)}$$

Combine like terms.

$$= \frac{2(x^2 + 3x + 1)}{x(x + 1)(x - 1)}$$

Factor.

✓ *Checkpoint* ▶ **Audio-video solution in English & Spanish at LarsonPrecalculus.com**

Perform the operations and simplify.

$$\frac{4}{x} - \frac{x + 5}{x^2 - 4} + \frac{4}{x + 2}$$

Complex Fractions

Complex fractions are fractional expressions with separate fractions in the numerator, denominator, or both. For instance,

$$\frac{\left(\dfrac{1}{x}\right)}{x^2 + 1} \quad \text{and} \quad \frac{\left(\dfrac{1}{x}\right)}{\left(\dfrac{1}{x^2 + 1}\right)}$$

are complex fractions.

One way to simplify a complex fraction is to combine the fractions in the numerator into a single fraction and then combine the fractions in the denominator into a single fraction. Then invert the denominator and multiply. Example 8 shows this method.

EXAMPLE 8 **Simplifying a Complex Fraction**

$$\frac{\left(\dfrac{2}{x} - 3\right)}{\left(1 - \dfrac{1}{x - 1}\right)} = \frac{\left[\dfrac{2 - 3(x)}{x}\right]}{\left[\dfrac{1(x - 1) - 1}{x - 1}\right]} \qquad \text{Combine fractions.}$$

$$= \frac{\left(\dfrac{2 - 3x}{x}\right)}{\left(\dfrac{x - 2}{x - 1}\right)} \qquad \text{Simplify.}$$

$$= \frac{2 - 3x}{x} \cdot \frac{x - 1}{x - 2} \qquad \text{Invert and multiply.}$$

$$= \frac{(2 - 3x)(x - 1)}{x(x - 2)}, \quad x \neq 1$$

✓ **Checkpoint** ▶ Audio-video solution in English & Spanish at LarsonPrecalculus.com

Simplify the complex fraction $\dfrac{\left(\dfrac{1}{x + 2} + 1\right)}{\left(\dfrac{x}{3} - 1\right)}$. ■

Another way to simplify a complex fraction is to multiply its numerator and denominator by the LCD of all fractions in its numerator and denominator. This method is demonstrated below using the original fraction in Example 8. In this case, the LCD is $x(x - 1)$.

$$\frac{\left(\dfrac{2}{x} - 3\right)}{\left(1 - \dfrac{1}{x - 1}\right)} \cdot \frac{x(x - 1)}{x(x - 1)} = \frac{\left(\dfrac{2}{x}\right)(x)(x - 1) - (3)(x)(x - 1)}{(1)(x)(x - 1) - \left(\dfrac{1}{x - 1}\right)(x)(x - 1)}$$

$$= \frac{2(x - 1) - 3x(x - 1)}{x(x - 1) - x} \qquad \text{Simplify.}$$

$$= \frac{(2 - 3x)(x - 1)}{x(x - 2)}, \quad x \neq 1 \qquad \text{Factor.}$$

Notice that this is the same result as the one in Example 8.

Simplifying Expressions from Calculus

The next three examples illustrate some methods for simplifying expressions involving negative exponents and radicals. These types of expressions occur frequently in calculus.

To simplify an expression with negative exponents, one method is to begin by factoring out the common factor with the *lesser* exponent. Remember that when factoring, you *subtract* exponents. For example, in $3x^{-5/2} + 2x^{-3/2}$, the lesser exponent is $-5/2$ and the common factor is $x^{-5/2}$.

$$3x^{-5/2} + 2x^{-3/2} = x^{-5/2}[3 + 2x^{-3/2-(-5/2)}]$$
$$= x^{-5/2}(3 + 2x^1)$$
$$= \frac{3 + 2x}{x^{5/2}}$$

EXAMPLE 9 **Simplifying an Expression**

Simplify $x(1 - 2x)^{-3/2} + (1 - 2x)^{-1/2}$.

Solution Begin by factoring out the common factor with the lesser exponent.

$$x(1 - 2x)^{-3/2} + (1 - 2x)^{-1/2} = (1 - 2x)^{-3/2}[x + (1 - 2x)^{(-1/2)-(-3/2)}]$$
$$= (1 - 2x)^{-3/2}[x + (1 - 2x)^1]$$
$$= \frac{1 - x}{(1 - 2x)^{3/2}}$$

 Checkpoint ▶ *Audio-video solution in English & Spanish at LarsonPrecalculus.com*

Simplify $(x - 1)^{-1/3} - x(x - 1)^{-4/3}$. ■

Another method for simplifying an expression with negative exponents involves multiplying the numerator and denominator by another expression to eliminate the negative exponent, as shown in the next example.

EXAMPLE 10 **Simplifying an Expression**

Simplify $\dfrac{(4 - x^2)^{1/2} + x^2(4 - x^2)^{-1/2}}{4 - x^2}$.

Solution To eliminate the negative exponent $-1/2$, multiply the expression by
$(4 - x^2)^{1/2}/(4 - x^2)^{1/2}$.

$$\frac{(4 - x^2)^{1/2} + x^2(4 - x^2)^{-1/2}}{4 - x^2} = \frac{(4 - x^2)^{1/2} + x^2(4 - x^2)^{-1/2}}{4 - x^2} \cdot \frac{(4 - x^2)^{1/2}}{(4 - x^2)^{1/2}}$$
$$= \frac{(4 - x^2)^1 + x^2(4 - x^2)^0}{(4 - x^2)^{3/2}}$$
$$= \frac{4 - x^2 + x^2}{(4 - x^2)^{3/2}}$$
$$= \frac{4}{(4 - x^2)^{3/2}}$$

 Checkpoint ▶ *Audio-video solution in English & Spanish at LarsonPrecalculus.com*

Simplify $\dfrac{x^2(x^2 - 2)^{-1/2} + (x^2 - 2)^{1/2}}{x^2 - 2}$. ■

GO DIGITAL

The last example involves an important expression in calculus called a *difference quotient*. One form of a difference quotient is the expression

$$\frac{\sqrt{x + h} - \sqrt{x}}{h}$$ A difference quotient

Often, a difference quotient needs to be rewritten in an equivalent form, as shown in Example 11.

EXAMPLE 11   Rewriting a Difference Quotient

Rewrite the difference quotient

$$\frac{\sqrt{x + h} - \sqrt{x}}{h}$$

by rationalizing its numerator.

Solution

$$\frac{\sqrt{x + h} - \sqrt{x}}{h} = \frac{\sqrt{x + h} - \sqrt{x}}{h} \cdot \frac{\sqrt{x + h} + \sqrt{x}}{\sqrt{x + h} + \sqrt{x}}$$

$$= \frac{\left(\sqrt{x + h}\right)^2 - \left(\sqrt{x}\right)^2}{h\left(\sqrt{x + h} + \sqrt{x}\right)}$$

$$= \frac{x + h - x}{h\left(\sqrt{x + h} + \sqrt{x}\right)}$$

$$= \frac{h}{h\left(\sqrt{x + h} + \sqrt{x}\right)}$$

$$= \frac{1}{\sqrt{x + h} + \sqrt{x}}, \quad h \neq 0$$

✓ *Checkpoint* **Audio-video solution in English & Spanish at LarsonPrecalculus.com**

Rewrite the difference quotient

$$\frac{\sqrt{9 + h} - 3}{h}$$

by rationalizing its numerator.

Summarize   (Section P.5)

1. State the definition of the domain of an algebraic expression *(page 41)*. For an example of finding the domains of algebraic expressions, see Example 1.

2. State the definition of a rational expression and explain how to simplify a rational expression *(pages 41 and 42)*. For examples of simplifying rational expressions, see Examples 2 and 3.

3. Explain how to multiply, divide, add, and subtract rational expressions *(page 43)*. For examples of operations with rational expressions, see Examples 4–7.

4. State the definition of a complex fraction *(page 45)*. For an example of simplifying a complex fraction, see Example 8.

5. Explain how to rewrite a difference quotient *(page 47)*. For an example of rewriting a difference quotient, see Example 11.

GO DIGITAL

P.5 Exercises

See CalcChat.com for tutorial help and worked-out solutions to odd-numbered exercises.

Vocabulary and Concept Check

In Exercises 1–4, fill in the blanks.

1. The quotient of two algebraic expressions is a fractional expression, and the quotient of two polynomials is a _____ _____.

2. Fractional expressions with separate fractions in the numerator, denominator, or both are _____ fractions.

3. Two algebraic expressions that have the same domain and yield the same values for all numbers in their domains are _____.

4. To simplify an expression with negative exponents, it is possible to begin by factoring out the common factor with the _____ exponent.

5. When is a rational expression in simplest form?

6. What values are excluded from the domain of a rational expression?

Skills and Applications

Finding the Domain of an Algebraic Expression **In Exercises 7–18, find the domain of the expression.**

7. $3x^2 - 4x + 7$

8. $6x^2 - 9, \quad x > 0$

9. $\dfrac{1}{3 - x}$

10. $\dfrac{1}{x + 5}$

11. $\dfrac{x + 6}{3x + 2}$

12. $\dfrac{x - 4}{1 - 2x}$

13. $\dfrac{x^2 - 5x + 6}{x^2 + 6x + 8}$

14. $\dfrac{x^2 - 1}{x^2 + 3x - 10}$

15. $\sqrt{x - 7}$

16. $\sqrt{2x - 5}$

17. $\dfrac{1}{\sqrt{x - 3}}$

18. $\dfrac{1}{\sqrt{x + 2}}$

Simplifying a Rational Expression **In Exercises 19–32, write the rational expression in simplest form.**

19. $\dfrac{15x^2}{10x}$

20. $\dfrac{18y^2}{60y^5}$

21. $\dfrac{x - 5}{10 - 2x}$

22. $\dfrac{12 - 4x}{x - 3}$

23. $\dfrac{y^2 - 16}{y + 4}$

24. $\dfrac{x^2 - 25}{5 - x}$

25. $\dfrac{6y + 9y^2}{12y + 8}$

26. $\dfrac{4y - 8y^2}{10y - 5}$

27. $\dfrac{x^2 + 4x - 5}{x^2 + 8x + 15}$

28. $\dfrac{x^2 + 8x - 20}{x^2 + 11x + 10}$

29. $\dfrac{x^2 - x - 2}{10 - 3x - x^2}$

30. $\dfrac{4 + 3x - x^2}{2x^2 - 7x - 4}$

31. $\dfrac{x^2 - 16}{x^3 + x^2 - 16x - 16}$

32. $\dfrac{x^2 - 1}{x^3 + x^2 + 9x + 9}$

Multiplying or Dividing Rational Expressions **In Exercises 33–40, multiply or divide and simplify.**

33. $\dfrac{5}{x - 1} \cdot \dfrac{x - 1}{25(x - 2)}$

34. $\dfrac{x + 13}{x^3(3 - x)} \cdot \dfrac{x(x - 3)}{5}$

35. $\dfrac{x^2 - 4}{12} \div \dfrac{2 - x}{2x + 4}$

36. $\dfrac{r}{r - 1} \div \dfrac{r^2}{r^2 - 1}$

37. $\dfrac{4y - 16}{5y + 15} \div \dfrac{4 - y}{2y + 6}$

38. $\dfrac{t^2 - t - 6}{t^2 + 6t + 9} \cdot \dfrac{t + 3}{t^2 - 4}$

39. $\dfrac{x^2 + xy - 2y^2}{x^3 + x^2y} \cdot \dfrac{x}{x^2 + 3xy + 2y^2}$

40. $\dfrac{x^2 - 14x + 49}{x^2 - 49} \div \dfrac{3x - 21}{x + 7}$

Adding or Subtracting Rational Expressions **In Exercises 41–50, add or subtract and simplify.**

41. $\dfrac{x - 1}{x + 2} - \dfrac{x - 4}{x + 2}$

42. $\dfrac{2x - 1}{x + 3} + \dfrac{1 - x}{x + 3}$

43. $\dfrac{1}{3x + 2} + \dfrac{x}{x + 1}$

44. $\dfrac{x}{x + 4} - \dfrac{6}{x - 1}$

45. $\dfrac{3}{2x + 4} - \dfrac{x}{x + 2}$

46. $\dfrac{2}{x^2 - 9} + \dfrac{4}{x + 3}$

47. $\dfrac{1}{x^2 - x - 2} - \dfrac{x}{x^2 - 5x + 6}$

48. $\dfrac{2}{x^2 - x - 2} + \dfrac{10}{x^2 + 2x - 8}$

49. $-\dfrac{1}{x} + \dfrac{2}{x^2 + 1} + \dfrac{1}{x^3 + x}$

50. $\dfrac{2}{x + 1} + \dfrac{2}{x - 1} + \dfrac{1}{x^2 - 1}$

Simplifying a Complex Fraction In Exercises 51–56, simplify the complex fraction.

51. $\dfrac{\left(\dfrac{x}{2} - 1\right)}{x - 2}$

52. $\dfrac{\left(\dfrac{x}{5} - 5\right)}{x + 5}$

53. $\dfrac{\left[\dfrac{x^2}{(x+1)^2}\right]}{\left[\dfrac{x}{(x+1)^3}\right]}$

54. $\dfrac{\left(\dfrac{x^2-1}{x}\right)}{\left[\dfrac{(x-1)^2}{x}\right]}$

55. $\dfrac{\left(\sqrt{x} - \dfrac{1}{2\sqrt{x}}\right)}{\sqrt{x}}$

56. $\dfrac{\left(\dfrac{t^2}{\sqrt{t^2+1}} - \sqrt{t^2+1}\right)}{t^2}$

Factoring an Expression In Exercises 57–60, factor the expression by factoring out the common factor with the lesser exponent.

57. $x^2(x^2 + 3)^{-4} + (x^2 + 3)^3$

58. $2x(x - 5)^{-3} - 4x^2(x - 5)^{-4}$

59. $2x^2(x - 1)^{1/2} - 5(x - 1)^{-1/2}$

60. $4x^3(x + 1)^{-3/2} - x(x + 1)^{-1/2}$

Simplifying an Expression In Exercises 61 and 62, simplify the expression.

61. $\dfrac{3x^{1/3} - x^{-2/3}}{3x^{-2/3}}$

62. $\dfrac{-x^3(1 - x^2)^{-1/2} - 2x(1 - x^2)^{1/2}}{x^4}$

Simplifying a Difference Quotient In Exercises 63–66, simplify the difference quotient.

63. $\dfrac{\left(\dfrac{1}{x+h} - \dfrac{1}{x}\right)}{h}$

64. $\dfrac{\left[\dfrac{1}{(x+h)^2} - \dfrac{1}{x^2}\right]}{h}$

65. $\dfrac{\left(\dfrac{1}{x+h-4} - \dfrac{1}{x-4}\right)}{h}$

66. $\dfrac{\left(\dfrac{x+h}{x+h+1} - \dfrac{x}{x+1}\right)}{h}$

Rewriting a Difference Quotient In Exercises 67–72, rewrite the difference quotient by rationalizing the numerator.

67. $\dfrac{\sqrt{x+2} - \sqrt{x}}{2}$

68. $\dfrac{\sqrt{z-3} - \sqrt{z}}{-3}$

69. $\dfrac{\sqrt{t+3} - \sqrt{3}}{t}$

70. $\dfrac{\sqrt{x+5} - \sqrt{5}}{x}$

71. $\dfrac{\sqrt{x+h+1} - \sqrt{x+1}}{h}$

72. $\dfrac{\sqrt{x+h-2} - \sqrt{x-2}}{h}$

73. Refrigeration

After placing food (at room temperature) in a refrigerator, the time required for the food to cool depends on the amount of food, the air circulation in the refrigerator, the original temperature of the food, and the temperature of the refrigerator. The model that gives the temperature of food that has an original temperature of 75°F and is placed in a 40°F refrigerator is

$$T = 10\left(\dfrac{4t^2 + 16t + 75}{t^2 + 4t + 10}\right)$$

where T is the temperature (in degrees Fahrenheit) and t is the time (in hours). Use a graphing utility to create a table that shows even integer values of t, $0 \le t \le 22$, and corresponding values of T. What value does T appear to be approaching?

74. Television and Mobile Devices The table shows the average time (in hours) spent per person per day in the United States watching television and using mobile devices from 2016 through 2019. (*Source: eMarketer*)

Year	Television	Mobile Devices
2016	4.1	3.1
2017	3.9	3.4
2018	3.7	3.6
2019	3.6	3.7

Mathematical models for the data are

$$\text{Hours watching television} = \dfrac{0.846t + 51.92}{t}$$

$$\text{Hours using mobile devices} = \dfrac{4.52 - 0.324t}{1 - 0.076t}$$

where t represents the year, with $t = 16$ corresponding to 2016.

(a) Using the models, create a table showing the hours watching television and the hours using mobile devices for the given years.

(b) Compare the values from the models with the actual data.

(c) Determine a model for the ratio of the hours watching television to the hours using mobile devices.

(d) Use the model from part (c) to find the ratios for the given years. Interpret your results.

Probability In Exercises 75 and 76, consider an experiment in which a marble is tossed into a box whose base is shown in the figure. The probability that the marble will come to rest in the shaded portion of the base is equal to the ratio of the shaded area to the total area of the figure. Find the probability.

75.
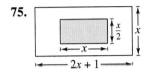

76.
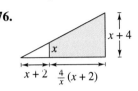

77. Finance The formula that approximates the annual interest rate r of a monthly installment loan is

$$r = \frac{24(NM - P)}{N} \div \left(P + \frac{NM}{12}\right)$$

where N is the total number of payments, M is the monthly payment, and P is the amount financed.

(a) Approximate the annual interest rate for a five-year car loan of $28,000 that has monthly payments of $525.

(b) Simplify the expression for the annual interest rate r, and then rework part (a).

78. Electrical Engineering The formula for the total resistance R_T (in ohms) of two resistors connected in parallel is

$$R_T = \frac{1}{\left(\dfrac{1}{R_1} + \dfrac{1}{R_2}\right)}$$

where R_1 and R_2 are the resistance values of the first and second resistors, respectively. Simplify the expression for the total resistance R_T.

Exploring the Concepts

True or False? In Exercises 79 and 80, determine whether the statement is true or false. Justify your answer.

79. $\dfrac{x^{2n} - 1^{2n}}{x^n - 1^n} = x^n + 1^n$

80. $\dfrac{x^2 - 3x + 2}{x - 1} = x - 2$, for all values of x

81. Evaluating a Rational Expression Complete the table. What can you conclude?

x	0	1	2	3	4	5	6
$\dfrac{x-3}{x^2-x-6}$							
$\dfrac{1}{x+2}$							

82. **HOW DO YOU SEE IT?** The mathematical model

$$P = 100\left(\frac{t^2 - t + 1}{t^2 + 1}\right), \quad t \ge 0$$

gives the percent P of the normal level of oxygen in a pond, where t is the time (in weeks) after organic waste is dumped into the pond. The bar graph shows the situation. What conclusions can you draw from the bar graph?

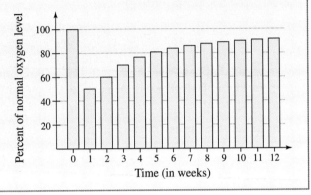

Error Analysis In Exercises 83 and 84, describe the error.

83. $\dfrac{5x^3}{2x^3 + 4} = \dfrac{5x^{\cancel{3}}}{2x^{\cancel{3}} + 4} = \dfrac{5}{2 + 4} = \dfrac{5}{6}$

84. $\dfrac{x + 3}{x - 2} - \dfrac{x - 1}{x - 2} = \dfrac{x + 3 - x - 1}{x - 2} = \dfrac{2}{x - 2}$ ✗

Review & Refresh ▶ *Video solutions at LarsonPrecalculus.com*

Plotting Points on the Real Number Line In Exercises 85 and 86, plot the real numbers on the real number line.

85. (a) 3　　(b) $\frac{7}{2}$　　(c) $-\frac{5}{2}$　　(d) -5.2

86. (a) 8.5　　(b) $\frac{4}{3}$　　(c) -4.75　　(d) $-\frac{8}{3}$

Using Absolute Value Notation In Exercises 87 and 88, use absolute value notation to represent the situation.

87. The distance between x and 5 is no more than 3.

88. The distance between x and -10 is at least 6.

Simplifying Radical Expressions In Exercises 89–92, simplify the radical expression.

89. $(\sqrt{605})^2$　　　　**90.** $(\sqrt{45})^2$

91. $\sqrt{25 + 20}$　　　　**92.** $\sqrt{284 + 321}$

Determining Whether a Triangle is a Right Triangle In Exercises 93 and 94, determine whether the triangle is a right triangle.

93.

94.
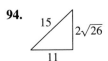

P.6 The Rectangular Coordinate System and Graphs

The Cartesian plane can help you visualize relationships between two variables. For example, in Exercise 44 on page 58, given how far north and west one city is from another, plotting points to represent the cities can help you visualize these distances and determine the flying distance between the cities.

- Plot points in the Cartesian plane.
- Use the Distance Formula to find the distance between two points.
- Use the Midpoint Formula to find the midpoint of a line segment.
- Translate points in the plane.

The Cartesian Plane

Just as you can represent real numbers by points on a real number line, you can represent ordered pairs of real numbers by points in a plane called the **rectangular coordinate system,** or the **Cartesian plane,** named after the French mathematician René Descartes (1596–1650).

Two real number lines intersecting at right angles form the Cartesian plane, as shown in Figure P.10. The horizontal real number line is usually called the **x-axis,** and the vertical real number line is usually called the **y-axis.** The point of intersection of these two axes is the **origin,** and the two axes divide the plane into four **quadrants.**

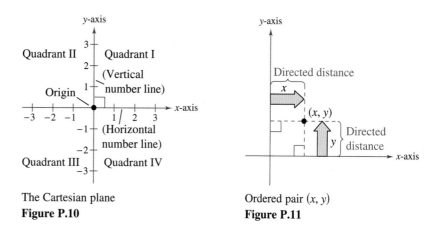

The Cartesian plane
Figure P.10

Ordered pair (x, y)
Figure P.11

Each point in the plane corresponds to an **ordered pair** (x, y) of real numbers x and y, called **coordinates** of the point. The **x-coordinate** represents the directed distance from the y-axis to the point, and the **y-coordinate** represents the directed distance from the x-axis to the point, as shown in Figure P.11.

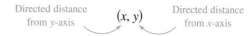

The notation (x, y) denotes both a point in the plane and an open interval on the real number line. The context will tell you which meaning is intended.

EXAMPLE 1 Plotting Points in the Cartesian Plane

Plot the points $(-1, 2)$, $(3, 4)$, $(0, 0)$, $(3, 0)$, and $(-2, -3)$.

Solution To plot the point $(-1, 2)$, imagine a vertical line through -1 on the x-axis and a horizontal line through 2 on the y-axis. The intersection of these two lines is the point $(-1, 2)$. Plot the other four points in a similar way, as shown in Figure P.12.

✓ **Checkpoint** ▶ *Audio-video solution in English & Spanish at LarsonPrecalculus.com*

Plot the points $(-3, 2)$, $(4, -2)$, $(3, 1)$, $(0, -2)$, and $(-1, -2)$.

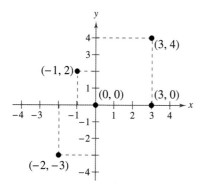

Figure P.12

The beauty of a rectangular coordinate system is that it allows you to *see* relationships between two variables. It would be difficult to overestimate the importance of Descartes's introduction of coordinates in the plane. Today, his ideas are in common use in virtually every scientific and business-related field.

In the next example, data are represented graphically by points plotted in a rectangular coordinate system. This type of graph is called a **scatter plot.**

EXAMPLE 2   Sketching a Scatter Plot

The table shows the numbers N (in millions) of AT&T wireless subscribers from 2013 through 2018, where t represents the year. Sketch a scatter plot of the data.   *(Source: AT&T Inc.)*

DATA	Year, t	2013	2014	2015	2016	2017	2018
	Subscribers, N	110	121	129	134	141	153

Spreadsheet at LarsonPrecalculus.com

Solution

Before sketching the scatter plot, represent each pair of values in the table by an ordered pair (t, N), as shown below.

(2013, 110), (2014, 121), (2015, 129), (2016, 134), (2017, 141), (2018, 153)

To sketch the scatter plot, first draw a vertical axis to represent the number of subscribers (in millions) and a horizontal axis to represent the year. Then plot a point for each ordered pair, as shown in the figure below. In the scatter plot, the break in the t-axis indicates omission of the numbers less than 2013, and the break in the N-axis indicates omission of the numbers less than 100 million. Also, the scatter plot shows that the number of subscribers has increased each year since 2013.

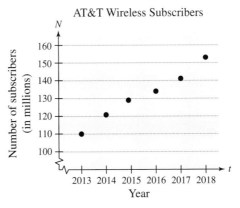

Checkpoint ▶ *Audio-video solution in English & Spanish at LarsonPrecalculus.com*

The table shows the numbers N of Costco stores from 2014 through 2019, where t represents the year. Sketch a scatter plot of the data.   *(Source: Costco Wholesale Corp.)*

DATA	Year, t	2014	2015	2016	2017	2018	2019
	Stores, N	663	686	715	741	762	782

Spreadsheet at LarsonPrecalculus.com

Another way to make the scatter plot in Example 2 is to let $t = 1$ represent the year 2013. In this scatter plot, the horizontal axis does not have a break, and the labels for the tick marks are 1 through 6 (instead of 2013 through 2018).

GO DIGITAL

≫ TECHNOLOGY

The scatter plot in Example 2 is only one way to represent the data graphically. You could also represent the data using a bar graph or a line graph. Use a graphing utility to represent the data given in Example 2 graphically.

The Distance Formula

Before developing the Distance Formula, recall from the **Pythagorean Theorem** that, for a right triangle with hypotenuse of length c and sides of lengths a and b, you have

$$a^2 + b^2 = c^2 \qquad \text{Pythagorean Theorem}$$

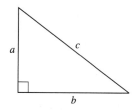

The Pythagorean Theorem:
$a^2 + b^2 = c^2$
Figure P.13

as shown in Figure P.13. (The converse is also true. That is, if $a^2 + b^2 = c^2$, then the triangle is a right triangle.)

Consider two points (x_1, y_1) and (x_2, y_2) that do not lie on the same horizontal or vertical line in the plane. With these two points, you can form a right triangle (see Figure P.14). To determine the distance d between these two points, note that the length of the vertical side of the triangle is $|y_2 - y_1|$ and the length of the horizontal side is $|x_2 - x_1|$. By the Pythagorean Theorem,

$$d^2 = |x_2 - x_1|^2 + |y_2 - y_1|^2 \qquad \text{Pythagorean Theorem}$$

$$d = \sqrt{|x_2 - x_1|^2 + |y_2 - y_1|^2} \qquad \text{Distance } d \text{ must be positive.}$$

$$d = \sqrt{(x_2 - x_1)^2 + (y_2 - y_1)^2}. \qquad \text{Property of exponents}$$

This result is the **Distance Formula.** Note that for the special case in which the two points lie on the same horizontal or vertical line, the Distance Formula still works. (See Exercise 62.)

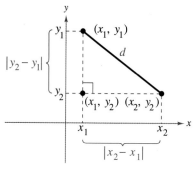

Figure P.14

The Distance Formula

The distance d between the points (x_1, y_1) and (x_2, y_2) in the plane is

$$d = \sqrt{(x_2 - x_1)^2 + (y_2 - y_1)^2}.$$

EXAMPLE 3 **Finding a Distance**

GO DIGITAL

Find the distance between the points $(-2, 1)$ and $(3, 4)$.

Algebraic Solution

Let $(x_1, y_1) = (-2, 1)$ and $(x_2, y_2) = (3, 4)$. Then apply the Distance Formula.

$$d = \sqrt{(x_2 - x_1)^2 + (y_2 - y_1)^2} \qquad \text{Distance Formula}$$

$$= \sqrt{[3 - (-2)]^2 + (4 - 1)^2} \qquad \text{Substitute for } x_1, y_1, x_2, \text{ and } y_2.$$

$$= \sqrt{(5)^2 + (3)^2} \qquad \text{Simplify.}$$

$$= \sqrt{34} \qquad \text{Simplify.}$$

$$\approx 5.83 \qquad \text{Use a calculator.}$$

So, the distance between the points is about 5.83 units.

Check

$$d^2 \overset{?}{=} 5^2 + 3^2 \qquad \text{Pythagorean Theorem}$$

$$\left(\sqrt{34}\right)^2 \overset{?}{=} 5^2 + 3^2 \qquad \text{Substitute for } d.$$

$$34 = 34 \qquad \text{Distance checks. ✓}$$

Graphical Solution

Use centimeter graph paper to plot the points $A(-2, 1)$ and $B(3, 4)$. Carefully sketch the line segment from A to B. Then use a centimeter ruler to measure the length of the segment.

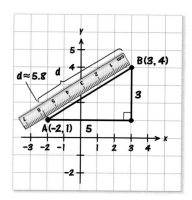

The line segment measures about 5.8 centimeters. So, the distance between the points is about 5.8 units.

✓ *Checkpoint* ▶ *Audio-video solution in English & Spanish at LarsonPrecalculus.com*

Find the distance between the points $(3, 1)$ and $(-3, 0)$. ■

When the Distance Formula is used, it does not matter which point is (x_1, y_1) and which is (x_2, y_2), because the result will be the same. For instance, in Example 3, let $(x_1, y_1) = (3, 4)$ and $(x_2, y_2) = (-2, 1)$. Then

$$d = \sqrt{(-2 - 3)^2 + (1 - 4)^2} = \sqrt{(-5)^2 + (-3)^2} = \sqrt{34} \approx 5.83.$$

EXAMPLE 4   Verifying a Right Triangle

Show that the points $(2, 1)$, $(4, 0)$, and $(5, 7)$ are vertices of a right triangle.

Solution

The three points are plotted in Figure P.15. Use the Distance Formula to find the lengths of the three sides.

$$d_1 = \sqrt{(5 - 2)^2 + (7 - 1)^2} = \sqrt{9 + 36} = \sqrt{45}$$
$$d_2 = \sqrt{(4 - 2)^2 + (0 - 1)^2} = \sqrt{4 + 1} = \sqrt{5}$$
$$d_3 = \sqrt{(5 - 4)^2 + (7 - 0)^2} = \sqrt{1 + 49} = \sqrt{50}$$

Using the converse of the Pythagorean Theorem and the fact that

$$(d_1)^2 + (d_2)^2 = 45 + 5 = 50 = (d_3)^2$$

you can conclude that the triangle is a right triangle.

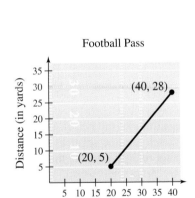

Figure P.15

✓ *Checkpoint* ▶ *Audio-video solution in English & Spanish at LarsonPrecalculus.com*

Show that the points $(2, -1)$, $(5, 5)$, and $(6, -3)$ are vertices of a right triangle.

EXAMPLE 5   Finding the Length of a Pass

A football quarterback throws a pass from the 28-yard line, 40 yards from the sideline. A wide receiver catches the pass on the 5-yard line, 20 yards from the same sideline, as shown in Figure P.16. How long is the pass?

Solution

The length of the pass is the distance between the points $(40, 28)$ and $(20, 5)$.

$$d = \sqrt{(x_2 - x_1)^2 + (y_2 - y_1)^2}$$ Distance Formula
$$= \sqrt{(40 - 20)^2 + (28 - 5)^2}$$ Substitute for x_1, y_1, x_2, and y_2.
$$= \sqrt{20^2 + 23^2}$$ Simplify.
$$= \sqrt{400 + 529}$$ Simplify.
$$= \sqrt{929}$$ Simplify.
$$\approx 30$$ Use a calculator.

So, the pass is about 30 yards long.

Football Pass

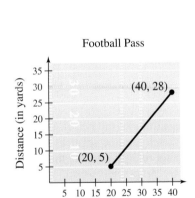

Figure P.16

✓ *Checkpoint* ▶ *Audio-video solution in English & Spanish at LarsonPrecalculus.com*

A football quarterback throws a pass from the 10-yard line, 10 yards from the sideline. A wide receiver catches the pass on the 32-yard line, 25 yards from the same sideline. How long is the pass? ∎

In Example 5, the horizontal and vertical scales do not normally appear on a football field. However, when you use coordinate geometry to solve real-life problems, you may place the coordinate system in any way that helps you solve the problem.

The Midpoint Formula

To find the **midpoint** of the line segment that joins two points in a coordinate plane, find the average values of the respective coordinates of the two endpoints using the **Midpoint Formula.**

The Midpoint Formula

The midpoint of the line segment joining the points (x_1, y_1) and (x_2, y_2) is

$$\text{Midpoint} = \left(\frac{x_1 + x_2}{2}, \frac{y_1 + y_2}{2}\right).$$

For a proof of the Midpoint Formula, see Proofs in Mathematics on page 66.

EXAMPLE 6   Finding the Midpoint of a Line Segment

Find the midpoint of the line segment joining the points $(-5, -3)$ and $(9, 3)$.

Solution   Let $(x_1, y_1) = (-5, -3)$ and $(x_2, y_2) = (9, 3)$.

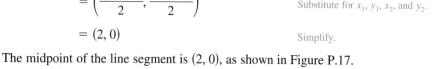

$$\text{Midpoint} = \left(\frac{x_1 + x_2}{2}, \frac{y_1 + y_2}{2}\right) \qquad \text{Midpoint Formula}$$

$$= \left(\frac{-5 + 9}{2}, \frac{-3 + 3}{2}\right) \qquad \text{Substitute for } x_1, y_1, x_2, \text{ and } y_2.$$

$$= (2, 0) \qquad \text{Simplify.}$$

The midpoint of the line segment is $(2, 0)$, as shown in Figure P.17.

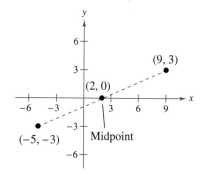

Figure P.17

✓ *Checkpoint* ▶ Audio-video solution in English & Spanish at LarsonPrecalculus.com

Find the midpoint of the line segment joining the points $(-2, 8)$ and $(4, -10)$. ■

EXAMPLE 7   Estimating Annual Revenues

Microsoft Corp. had annual revenues of about $96.7 billion in 2017 and about $125.8 billion in 2019. Estimate the revenues in 2018.   *(Source: Microsoft Corp.)*

Solution   One way to solve this problem is to assume that the revenues followed a *linear* pattern. Then, to estimate the 2018 revenues, find the midpoint of the line segment connecting the points $(2017, 96.7)$ and $(2019, 125.8)$.

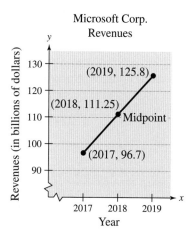

Figure P.18

$$\text{Midpoint} = \left(\frac{x_1 + x_2}{2}, \frac{y_1 + y_2}{2}\right) \qquad \text{Midpoint Formula}$$

$$= \left(\frac{2017 + 2019}{2}, \frac{96.7 + 125.8}{2}\right) \qquad \text{Substitute for } x_1, x_2, y_1, \text{ and } y_2.$$

$$= (2018, 111.25) \qquad \text{Simplify.}$$

So, you would estimate the 2018 revenues to have been about $111.25 billion, as shown in Figure P.18. (The actual 2018 revenues were about $110.36 billion.)

✓ *Checkpoint* ▶ Audio-video solution in English & Spanish at LarsonPrecalculus.com

The Proctor & Gamble Co. had annual sales of about $65.1 billion in 2017 and about $67.7 billion in 2019. Estimate the sales in 2018.   *(Source: Proctor & Gamble Co.)* ■

GO DIGITAL

Application

Much of computer graphics, including the computer-generated tessellation shown at the left, consists of transformations of points in a coordinate plane. One type of transformation, a translation, is illustrated in Example 8. Other types of transformations include reflections, rotations, and stretches.

EXAMPLE 8 Translating Points in the Plane

▶▶▶ *See LarsonPrecalculus.com for an interactive version of this type of example.*

The triangle in Figure P.19 has vertices at $(-1, 2)$, $(1, -2)$, and $(2, 3)$. Shift the triangle three units to the right and two units up. What are the coordinates of the vertices of the shifted triangle (see Figure P.20)?

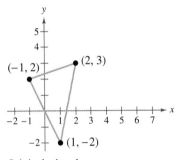

Original triangle
Figure P.19

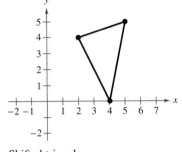

Shifted triangle
Figure P.20

Solution To shift the vertices three units to the right, add 3 to each of the x-coordinates. To shift the vertices two units up, add 2 to each of the y-coordinates.

Original Point	Translated Point
$(-1, 2)$	$(-1 + 3, 2 + 2) = (2, 4)$
$(1, -2)$	$(1 + 3, -2 + 2) = (4, 0)$
$(2, 3)$	$(2 + 3, 3 + 2) = (5, 5)$

✓ **Checkpoint** ▶ **Audio-video solution in English & Spanish at LarsonPrecalculus.com**

The parallelogram in Figure P.21 has vertices at $(1, 4)$, $(1, 0)$, $(3, 2)$, and $(3, 6)$. Shift the parallelogram two units to the left and four units down. What are the coordinates of the vertices of the shifted parallelogram?

The figures in Example 8 were not really essential to the solution. Nevertheless, you should develop the habit of including sketches with your solutions because they serve as useful problem-solving tools.

Figure P.21

Summarize (Section P.6)

1. Describe the Cartesian plane *(page 51)*. For examples of plotting points in the Cartesian plane, see Examples 1 and 2.

2. State the Distance Formula *(page 53)*. For examples of using the Distance Formula to find the distance between two points, see Examples 3–5.

3. State the Midpoint Formula *(page 55)*. For examples of using the Midpoint Formula to find the midpoint of a line segment, see Examples 6 and 7.

4. Describe how to translate points in the plane *(page 56)*. For an example of translating points in the plane, see Example 8.

GO DIGITAL

P.6 Exercises

See CalcChat.com for tutorial help and worked-out solutions to odd-numbered exercises.

GO DIGITAL

Vocabulary and Concept Check

In Exercises 1 and 2, fill in the blanks.

1. An ordered pair of real numbers can be represented in a plane called the rectangular coordinate system or the _____ plane.

2. Finding the average values of the respective coordinates of the two endpoints of a line segment in a coordinate plane is also known as using the _____ _____.

In Exercises 3–8, match each term with its definition.

3. x-axis
4. y-axis
5. origin
6. quadrants
7. x-coordinate
8. y-coordinate

(a) point of intersection of vertical axis and horizontal axis
(b) directed distance from the x-axis
(c) horizontal real number line
(d) four regions of the coordinate plane
(e) directed distance from the y-axis
(f) vertical real number line

Skills and Applications

Approximating Coordinates of Points In Exercises 9 and 10, approximate the coordinates of the points.

9.

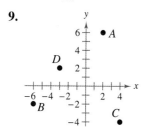

10.

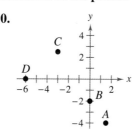

Plotting Points in the Cartesian Plane In Exercises 11 and 12, plot the points.

11. $(2, 4), (3, -1), (-6, 2), (-4, 0), (-1, -8), (1.5, -3.5)$
12. $(1, -5), (-2, -7), (3, 3), (-2, 4), (0, 5), \left(\frac{2}{3}, \frac{5}{2}\right)$

Finding the Coordinates of a Point In Exercises 13 and 14, find the coordinates of the point.

13. The point is three units to the left of the y-axis and four units above the x-axis.

14. The point is on the x-axis and 12 units to the left of the y-axis.

Determining Quadrant(s) for a Point In Exercises 15–20, determine the quadrant(s) in which (x, y) could be located.

15. $x > 0$ and $y < 0$
16. $x < 0$ and $y < 0$
17. $x = -4$ and $y > 0$
18. $x < 0$ and $y = 7$
19. $x + y = 0, x \neq 0, y \neq 0$
20. $xy > 0$

Sketching a Scatter Plot In Exercises 21 and 22, sketch a scatter plot of the data shown in the table.

21. The table shows the number y of Dollar General stores for each year x from 2012 through 2018. *(Source: Dollar General Corporation)*

DATA	Year, x	Number of Stores, y
	2012	10,506
	2013	11,132
	2014	11,789
	2015	12,483
	2016	13,320
	2017	14,609
	2018	15,472

Spreadsheet at LarsonPrecalculus.com

22. The table shows the annual revenues y (in billions of dollars) for Amazon.com for each year x from 2011 through 2018. *(Source: Amazon.com)*

DATA	Month, x	Annual Revenue, y
	2011	48.1
	2012	61.1
	2013	74.5
	2014	89.0
	2015	107.0
	2016	136.0
	2017	177.9
	2018	232.9

Spreadsheet at LarsonPrecalculus.com

Finding a Distance In Exercises 23–28, find the distance between the points.

23. $(-2, 6)$, $(3, -6)$
24. $(8, 5)$, $(0, 20)$
25. $(1, 4)$, $(-5, -1)$
26. $(1, 3)$, $(3, -2)$
27. $\left(\frac{1}{2}, \frac{4}{3}\right)$, $(2, -1)$
28. $(9.5, -2.6)$, $(-3.9, 8.2)$

Verifying a Right Triangle In Exercises 29 and 30, (a) find the length of each side of the right triangle and (b) show that these lengths satisfy the Pythagorean Theorem.

29.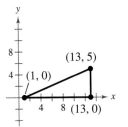

30.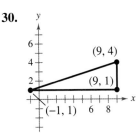

Verifying a Polygon In Exercises 31–34, show that the points form the vertices of the polygon.

31. Right triangle: $(4, 0)$, $(2, 1)$, $(-1, -5)$
32. Right triangle: $(-1, 3)$, $(3, 5)$, $(5, 1)$
33. Isosceles triangle: $(1, -3)$, $(3, 2)$, $(-2, 4)$
34. Isosceles triangle: $(2, 3)$, $(4, 9)$, $(-2, 7)$

Plotting, Distance, and Midpoint In Exercises 35–42, (a) plot the points, (b) find the distance between the points, and (c) find the midpoint of the line segment joining the points.

35. $(6, -3)$, $(6, 5)$
36. $(1, 4)$, $(8, 4)$
37. $(1, 1)$, $(9, 7)$
38. $(1, 12)$, $(6, 0)$
39. $(-1, 2)$, $(5, 4)$
40. $(2, 10)$, $(10, 2)$
41. $(-16.8, 12.3)$, $(5.6, 4.9)$
42. $\left(\frac{1}{2}, 1\right)$, $\left(-\frac{5}{2}, \frac{4}{3}\right)$

43. **Sports** A soccer player passes the ball from a point that is 18 yards from the endline and 12 yards from the sideline. A teammate who is 42 yards from the same endline and 50 yards from the same sideline receives the pass. (See figure.) How long is the pass?

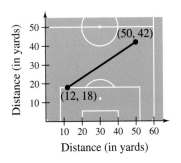

44. **Flying Distance**

An airplane flies from Naples, Italy, in a straight line to Rome, Italy, which is about 120 kilometers north and 150 kilometers west of Naples. How far does the plane fly?

45. **Sales** Walmart had sales of $485.9 billion in 2016 and $514.4 billion in 2018. Use the Midpoint Formula to estimate the sales in 2017. Assume that the sales followed a linear pattern. *(Source: Walmart, Inc.)*

46. **Earnings per Share** The earnings per share for Facebook, Inc. were $6.16 in 2017 and $7.57 in 2018. Use the Midpoint Formula to estimate the earnings per share in 2019. Assume that the earnings per share followed a linear pattern. *(Source: Facebook, Inc.)*

Translating Points in the Plane In Exercises 47–50, find the coordinates of the vertices of the polygon after the given translation to a new position in the plane.

47.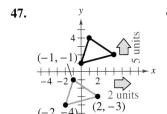

48.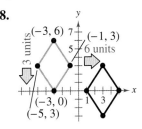

49. Original coordinates of vertices: $(-7, -2)$, $(-2, 2)$, $(-2, -4)$, $(-7, -4)$

Shift: eight units up, four units to the right

50. Original coordinates of vertices: $(5, 8)$, $(3, 6)$, $(7, 6)$

Shift: 6 units down, 10 units to the left

Exploring the Concepts

True or False? In Exercises 51–54, determine whether the statement is true or false. Justify your answer.

51. If the point (x, y) is in Quadrant II, then the point $(2x, -3y)$ is in Quadrant III.

52. To divide a line segment into 16 equal parts, you have to use the Midpoint Formula 16 times.

53. The points $(-8, 4)$, $(2, 11)$, and $(-5, 1)$ represent the vertices of an isosceles triangle.

54. If four points represent the vertices of a polygon, and the four side lengths are equal, then the polygon must be a square.

55. **Think About It** What is the y-coordinate of any point on the x-axis? What is the x-coordinate of any point on the y-axis?

56. Error Analysis Describe the error.

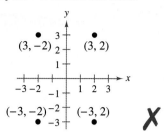

57. Using the Midpoint Formula A line segment has (x_1, y_1) as one endpoint and (x_m, y_m) as its midpoint. Find the other endpoint (x_2, y_2) of the line segment in terms of $x_1, y_1, x_m,$ and y_m.

58. Using the Midpoint Formula Use the Midpoint Formula three times to find the three points that divide the line segment joining (x_1, y_1) and (x_2, y_2) into four equal parts.

59. Proof Prove that the diagonals of the parallelogram in the figure intersect at their midpoints.

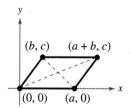

60. HOW DO YOU SEE IT?

Use the plot of the point (x_0, y_0) in the figure. Match the transformation of the point with the correct plot. Explain. [The plots are labeled (i), (ii), (iii), and (iv).]

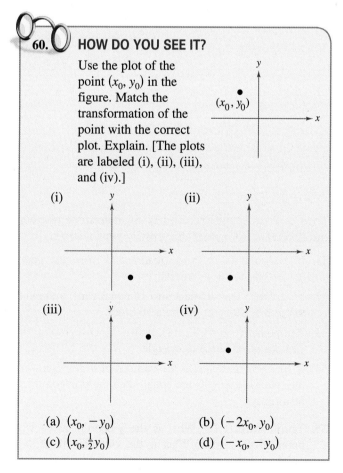

(a) $(x_0, -y_0)$ (b) $(-2x_0, y_0)$
(c) $\left(x_0, \frac{1}{2}y_0\right)$ (d) $(-x_0, -y_0)$

61. Collinear Points Three or more points are collinear when they all lie on the same line. Use the steps below to determine whether the set of points $\{A(2, 3), B(2, 6), C(6, 3)\}$ and the set of points $\{A(8, 3), B(5, 2), C(2, 1)\}$ are collinear.

(a) For each set of points, use the Distance Formula to find the distances from A to B, from B to C, and from A to C. What relationship exists among these distances for each set of points?

(b) Plot each set of points in the Cartesian plane. Do all the points of either set appear to lie on the same line?

(c) Compare your conclusions from part (a) with the conclusions you made from the graphs in part (b). Make a general statement about how to use the Distance Formula to determine collinearity.

62. Points on Vertical and Horizontal Lines Use the Distance Formula to find the distance between each pair of points. Are the results what you expected? Explain.

(a) *On the same vertical line:* (x_1, y_1) and (x_1, y_2)

(b) *On the same horizontal line:* (x_1, y_1) and (x_2, y_1)

Review & Refresh ▶ *Video solutions at LarsonPrecalculus.com*

Evaluating an Expression In Exercises 63–66, evaluate the expression for each value of x. (If not possible, state the reason.)

63. $4x - 6$ (a) $x = -1$ (b) $x = 0$
64. $9 - 7x$ (a) $x = -3$ (b) $x = 3$
65. $2x^3$ (a) $x = -3$ (b) $x = 0$
66. $-3x^{-4}$ (a) $x = 0$ (b) $x = -2$

Simplifying an Expression In Exercises 67–70, simplify the expression.

67. $(-x)^2 - 2$ **68.** $(-x)^3 - (-x)^2 + 2$
69. $-x^2 + (-x)^2$
70. $-x^3 + (-x)^3 + (-x)^4 - x^4$

71. Cost, Revenue, and Profit A manufacturer can produce and sell x electronic devices per week. The total cost C (in dollars) of producing x electronic devices is $C = 93x + 35,000$, and the total revenue R (in dollars) is $R = 135x$.

(a) Find the profit P in terms of x.

(b) Find the profit obtained by selling 5000 electronic devices per week.

72. Rate A copier copies at a rate of 50 pages per minute.

(a) Find the time required to copy one page.

(b) Find the time required to copy x pages.

(c) Find the time required to copy 120 pages.

(d) Find the time required to copy 10,000 pages.

Summary and Study Strategies

GO DIGITAL

What Did You Learn?

The list below reviews the skills covered in the chapter and correlates each one to the Review Exercises (see page 62) that practice the skill.

Section P.1	Review Exercises
■ Represent and classify real numbers *(p. 2)*.	*1, 2*
■ Order real numbers and use inequalities *(p. 4)*.	*3–8*
■ Find the absolute values of real numbers and find the distance between two real numbers *(p. 6)*.	*9–16*
■ Evaluate algebraic expressions *(p. 8)*.	*17–20*
■ Use the basic rules and properties of algebra *(p. 9)*.	*21–34*

Section P.2

■ Use properties of exponents *(p. 14)*.
35–50

$$a^m a^n = a^{m+n} \qquad \frac{a^m}{a^n} = a^{m-n} \qquad a^{-n} = \frac{1}{a^n} = \left(\frac{1}{a}\right)^n \qquad a^0 = 1$$

$$(ab)^m = a^m b^m \qquad (a^m)^n = a^{mn} \qquad \left(\frac{a}{b}\right)^m = \frac{a^m}{b^m} \qquad |a^2| = |a|^2 = a^2$$

■ Use scientific notation to represent real numbers *(p. 17)*.
51–54

■ Use properties of radicals *(p. 18)*.
55–60

$$\sqrt[n]{a^m} = \left(\sqrt[n]{a}\right)^m \qquad \sqrt[n]{a} \cdot \sqrt[n]{b} = \sqrt[n]{ab} \qquad \frac{\sqrt[n]{a}}{\sqrt[n]{b}} = \sqrt[n]{\frac{a}{b}}, \, b \neq 0$$

$$\sqrt[m]{\sqrt[n]{a}} = \sqrt[mn]{a} \qquad \left(\sqrt[n]{a}\right)^n = a$$

For *n* even, $\sqrt[n]{a^n} = |a|$.   For *n* odd, $\sqrt[n]{a^n} = a$.

■ Simplify and combine radical expressions *(p. 20)*.
61–74

■ Use properties of rational exponents *(p. 22)*.
75–78

$$a^{1/n} = \sqrt[n]{a} \qquad a^{m/n} = (a^{1/n})^m = \left(\sqrt[n]{a}\right)^m \qquad a^{m/n} = (a^m)^{1/n} = \sqrt[n]{a^m}$$

Section P.3

■ Write polynomials in standard form *(p. 26)*.
79–82

■ Add, subtract, and multiply polynomials, and use special products *(p. 27)*.
83–94

$$(u + v)(u - v) = u^2 - v^2 \qquad (u + v)^2 = u^2 + 2uv + v^2$$

$$(u - v)^2 = u^2 - 2uv + v^2$$

$$(u + v)^3 = u^3 + 3u^2 v + 3uv^2 + v^3 \qquad (u - v)^3 = u^3 - 3u^2 v + 3uv^2 - v^3$$

■ Use polynomials to solve real-life problems *(p. 30)*.
95–98

Section P.4	**Review Exercises**
■ Factor out common factors from polynomials *(p. 34)*.	*99, 100*
■ Factor special polynomial forms *(p. 35)*.	*101–104*
■ Factor trinomials as the product of two binomials *(p. 37)*.	*105–108*
■ Factor polynomials by grouping *(p. 38)*.	*109, 110*

Section P.5	
■ Find the domains of algebraic expressions *(p. 41)*.	*111–114*
■ Simplify rational expressions *(p. 42)*.	*115, 116*
■ Add, subtract, multiply, and divide rational expressions *(p. 43)*.	*117–120*
■ Simplify complex fractions *(p. 45)*.	*121, 122*
■ Simplify expressions from calculus *(p. 46)*.	*123, 124*

Section P.6	
■ Plot points in the Cartesian plane *(p. 51)*.	*125–128, 136*
■ Use the Distance Formula to find the distance between two points *(p. 53)*. $$d = \sqrt{(x_2 - x_1)^2 + (y_2 - y_1)^2}$$	*129–132*
■ Use the Midpoint Formula to find the midpoint of a line segment *(p. 55)*. $$\text{Midpoint} = \left(\frac{x_1 + x_2}{2}, \frac{y_1 + y_2}{2}\right)$$	*129–132, 135*
■ Translate points in the plane *(p. 56)*.	*133, 134*

Study Strategies

Keeping a Positive Attitude   Your experiences during the first three weeks in a math course often determine whether you stick with it or not. A positive attitude and good study habits will help set you up for success. Here are some strategies to get you started.

1. After the first math class, set aside time for reviewing your notes and the textbook, reworking your notes, and completing homework. One aid that can help you to organize the important concepts of each section is the free *Student Success Organizer* at *LarsonPrecalculus.com*.

2. Find a productive study environment on campus. Most colleges have a tutoring center where students can study and receive assistance as needed.

3. Set up a place for studying at home that is comfortable, but not too comfortable. It needs to be away from all potential distractions.

4. Make at least two other *collegial friends* in class. Collegial friends are students who study well together, help each other out when someone gets sick, and keep each other's attitudes positive.

5. Meet with your instructor at least once during the first two weeks. Ask the instructor what he or she advises for study strategies in the class. This will help you and let the instructor know that you really want to do well.

Review Exercises

See CalcChat.com for tutorial help and worked-out solutions to odd-numbered exercises.

GO DIGITAL

P.1 Classifying Real Numbers In Exercises 1 and 2, determine which numbers in the set are (a) natural numbers, (b) whole numbers, (c) integers, (d) rational numbers, and (e) irrational numbers.

1. $\left\{ 11, -14, -\frac{8}{9}, \frac{5}{2}, \sqrt{6}, 0.4 \right\}$

2. $\left\{ \sqrt{15}, -22, -\frac{10}{3}, 0, 5.2, \frac{3}{7} \right\}$

Plotting and Ordering Real Numbers In Exercises 3 and 4, plot the two real numbers on the real number line. Then place the appropriate inequality symbol (< or >) between them.

3. $\frac{5}{4}, \frac{7}{8}$

4. $-\frac{9}{25}, -\frac{5}{7}$

Interpreting an Inequality In Exercises 5 and 6, describe the subset of real numbers that the inequality represents.

5. $x \geq 6$

6. $-4 < x < 4$

Representing an Interval In Exercises 7 and 8, represent the interval verbally, as an inequality, and as a graph.

7. $[-3, 4)$

8. $[3, \infty)$

Evaluating an Absolute Value Expression In Exercises 9–14, evaluate the expression.

9. $|5|$

10. $|-16|$

11. $|8 - 23|$

12. $|14 - 7|$

13. $-|9| - |-9|$

14. $|-10| - |-11|$

Finding a Distance In Exercises 15 and 16, find the distance between a and b.

15. $a = -74, b = 48$

16. $a = -112, b = -6$

Evaluating an Algebraic Expression In Exercises 17–20, evaluate the expression for each value of x. (If not possible, state the reason.)

17. $12x - 7$ (a) $x = 0$ (b) $x = -1$

18. $x^2 - 6x + 5$ (a) $x = -2$ (b) $x = 2$

19. $-x^2 + x - 1$ (a) $x = 1$ (b) $x = -1$

20. $\dfrac{x}{x - 3}$ (a) $x = -3$ (b) $x = 3$

Identifying Rules of Algebra In Exercises 21–26, identify the rule(s) of algebra illustrated by the statement.

21. $0 + (a - 5) = a - 5$

22. $1 \cdot (3x + 4) = 3x + 4$

23. $2x + (3x - 10) = (2x + 3x) - 10$

24. $4(t + 2) = 4 \cdot t + 4 \cdot 2$

25. $(t^2 + 1) + 3 = 3 + (t^2 + 1)$

26. $\dfrac{2}{y + 4} \cdot \dfrac{y + 4}{2} = 1, \quad y \neq -4$

Performing Operations In Exercises 27–34, perform the operation(s). (Write fractional answers in simplest form.)

27. $-6 + 6$

28. $2 - (-3)$

29. $(-8)(-4)$

30. $5(20 + 7)$

31. $\dfrac{x}{5} + \dfrac{7x}{12}$

32. $\dfrac{x}{2} - \dfrac{2x}{5}$

33. $\dfrac{3x}{10} \cdot \dfrac{5}{3}$

34. $\dfrac{9}{x} \div \dfrac{1}{6}$

P.2 Using Properties of Exponents In Exercises 35–42, simplify each expression.

35. $3x^2(4x^3)^3$

36. $(3a)^2(6a^3)$

37. $\dfrac{5y^6}{10y}$

38. $\dfrac{36x^5}{9x^{10}}$

39. $(-2z)^3$

40. $[(x + 2)^2]^3$

41. $\dfrac{(8y)^0}{y^2}$

42. $\dfrac{40(b - 3)^5}{75(b - 3)^2}$

Rewriting with Positive Exponents In Exercises 43–50, rewrite each expression with positive exponents. Simplify, if possible.

43. $\dfrac{a^2}{b^{-2}}$

44. $(a^2b^4)(3ab^{-2})$

45. $\dfrac{6^2u^3v^{-3}}{12u^{-2}v}$

46. $\dfrac{3^{-4}m^{-1}n^{-3}}{9^{-2}mn^{-3}}$

47. $\dfrac{(5a)^{-2}}{(5a)^2}$

48. $\dfrac{4(x^{-1})^{-3}}{4^{-2}(x^{-1})^{-1}}$

49. $(x + y^{-1})^{-1}$

50. $\left(\dfrac{x^{-3}}{y} \right)\left(\dfrac{x}{y} \right)^{-1}$

Scientific Notation In Exercises 51 and 52, write the number in scientific notation.

51. $274,400,000$

52. 0.3048

Decimal Form In Exercises 53 and 54, write the number in decimal form.

53. 4.84×10^8

54. 2.74×10^{-3}

Simplifying a Radical Expression In Exercises 55–66, simplify the radical expression.

55. $\sqrt[3]{27^2}$

56. $\sqrt{49^3}$

57. $\sqrt[3]{\frac{64}{125}}$

58. $\sqrt{\frac{81}{100}}$

59. $\left(\sqrt[3]{216}\right)^3$

60. $\sqrt[4]{32^4}$

61. $\sqrt[3]{\frac{2x^3}{27}}$

62. $\sqrt[5]{64x^6}$

63. $\sqrt{12x^3} + \sqrt{3x}$

64. $\sqrt{27x^3} - \sqrt{3x^3}$

65. $\sqrt{8x^3} + \sqrt{2x}$

66. $\sqrt{18x^5} - \sqrt{8x^3}$

67. Writing Explain why $\sqrt{5u} + \sqrt{3u} \neq 2\sqrt{2u}$.

68. Engineering The rectangular cross section of a wooden beam cut from a log of diameter 24 inches (see figure) will have a maximum strength when its width w and height h are

$$w = 8\sqrt{3} \quad \text{and} \quad h = \sqrt{24^2 - \left(8\sqrt{3}\right)^2}.$$

Find the area of the rectangular cross section and write the answer in simplest form.

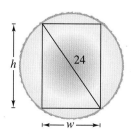

Rationalizing a Denominator In Exercises 69–72, rationalize the denominator of the expression. Then simplify your answer.

69. $\dfrac{3}{4\sqrt{3}}$

70. $\dfrac{12}{\sqrt[3]{4}}$

71. $\dfrac{1}{2 - \sqrt{3}}$

72. $\dfrac{1}{\sqrt{5} + 1}$

Rationalizing a Numerator In Exercises 73 and 74, rationalize the numerator of the expression. Then simplify your answer.

73. $\dfrac{\sqrt{7} + 1}{2}$

74. $\dfrac{\sqrt{2} - \sqrt{11}}{3}$

Simplifying an Expression In Exercises 75–78, simplify the expression.

75. $16^{3/2}$

76. $\left(\frac{64}{125}\right)^{-2/3}$

77. $\sqrt{(6x^2)^2}$

78. $\sqrt{\sqrt{128}}$

P.3 **Writing a Polynomial in Standard Form** In Exercises 79–82, write the polynomial in standard form. Identify the degree and leading coefficient.

79. $3 - 11x^2$

80. $3x^3 - 5x^5 + x - 4$

81. $-4 - 12x^2$

82. $12x - 7x^2 + 6$

Adding or Subtracting Polynomials In Exercises 83 and 84, perform the operation and write the result in standard form.

83. $-(3x^2 + 2x) + (1 - 5x)$

84. $8y - [2y^2 - (3y - 8)]$

Multiplying Polynomials In Exercises 85–92, find the product.

85. $2x(x^2 - 5x + 6)$

86. $(3x^3 - 1.5x^2 + 4)(-3x)$

87. $(3x - 6)(5x + 1)$

88. $(x + 2)(x^2 - 2)$

89. $(6x + 5)(6x - 5)$

90. $\left(3\sqrt{5} + 2x\right)\left(3\sqrt{5} - 2x\right)$

91. $(2x - 3)^2$

92. $(x - 4)^3$

Operations with Polynomials In Exercises 93 and 94, perform the operation.

93. Multiply $x^2 + x + 5$ and $x^2 - 7x - 2$.

94. Subtract $9x^4 - 11x^2 + 16$ from $6x^4 - 20x^2 - x + 3$.

95. Compound Interest An investment of \$2500 compounded annually for 2 years at an interest rate r (in decimal form) yields an amount of $2500(1 + r)^2$. Write this polynomial in standard form.

96. Surface Area The surface area S of a right circular cylinder is $S = 2\pi r^2 + 2\pi rh$.

(a) Draw a right circular cylinder of radius r and height h. Use your drawing to explain how to obtain the surface area formula.

(b) Find the surface area when the radius is 6 inches and the height is 8 inches.

97. Geometry Find a polynomial that represents the total number of square feet for the floor plan shown in the figure.

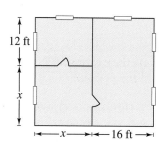

98. Geometry Use the figure below to write two different expressions for the area. Then equate the two expressions and name the algebraic property illustrated.

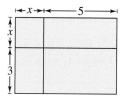

P.4 Factoring Completely In Exercises 99–110, completely factor the expression.

99. $x^3 - x$

100. $x(x - 3) + 4(x - 3)$

101. $25x^2 - 49$

102. $36x^2 - 81$

103. $x^3 - 64$

104. $8x^3 + 27$

105. $2x^2 + 21x + 10$

106. $3x^2 + 14x + 8$

107. $x^3 - x^2 - 12x$

108. $4x^4 - 6x^3 - 10x^2$

109. $x^3 + x^2 - 2x - 2$

110. $x^3 - 4x^2 + 2x - 8$

P.5 Finding the Domain of an Algebraic Expression In Exercises 111–114, find the domain of the expression.

111. $\dfrac{1}{x + 1}$

112. $\dfrac{1}{x^2 + 3x - 18}$

113. $\sqrt{x + 2}$

114. $\sqrt{5x - 4}$

Simplifying a Rational Expression In Exercises 115 and 116, write the rational expression in simplest form.

115. $\dfrac{x^2 - 64}{5(3x + 24)}$

116. $\dfrac{x^3 + 27}{x^2 + x - 6}$

Operations with Rational Expressions In Exercises 117–120, perform the operation and simplify.

117. $\dfrac{x^2 - 7x + 12}{x^2 + 8x + 16} \cdot \dfrac{x + 4}{x^2 - 9}$

118. $\dfrac{2x - 6}{3x + 6} \div \dfrac{3 - x}{2x + 4}$

119. $\dfrac{2}{x - 4} + \dfrac{6}{x^2 - 16}$

120. $\dfrac{3x}{x^2 - 25} - \dfrac{2}{x^2 + x - 20}$

Simplifying a Complex Fraction In Exercises 121 and 122, simplify the complex fraction.

121. $\dfrac{\left[\dfrac{3a}{(a^2/x) - 1} \right]}{\left(\dfrac{a}{x} - 1 \right)}$

122. $\dfrac{\left(\dfrac{1}{2x - 3} - \dfrac{1}{2x + 3} \right)}{\left(\dfrac{1}{2x} - \dfrac{1}{2x + 3} \right)}$

Simplifying a Difference Quotient In Exercises 123 and 124, simplify the difference quotient.

123. $\dfrac{\left[\dfrac{1}{2(x + h)} - \dfrac{1}{2x} \right]}{h}$

124. $\dfrac{\left(\dfrac{1}{x + h - 3} - \dfrac{1}{x - 3} \right)}{h}$

P.6 Plotting Points in the Cartesian Plane In Exercises 125 and 126, plot the points.

125. $(5, 5), (-2, 0), (-3, 6), (-1, -7)$

126. $(0, 6), (8, 1), (5, -4), (-3, -3)$

Determining Quadrant(s) for a Point In Exercises 127 and 128, determine the quadrant(s) in which (x, y) could be located.

127. $x > 0$ and $y = -2$

128. $xy = 4$

Plotting, Distance, and Midpoint In Exercises 129–132, (a) plot the points, (b) find the distance between the points, and (c) find the midpoint of the line segment joining the points.

129. $(-3, 8), (1, 5)$

130. $(-2, 6), (4, -3)$

131. $(5.6, 0), (0, 8.2)$

132. $(1.8, 7.4), (-0.6, -14.5)$

Translating Points in the Plane In Exercises 133 and 134, find the coordinates of the vertices of the polygon after the given translation to a new position in the plane.

133. Original coordinates of vertices:

$(4, 8), (6, 8), (4, 3), (6, 3)$

Shift: eight units down, four units to the left

134. Original coordinates of vertices:

$(0, 1), (3, 3), (0, 5), (-3, 3)$

Shift: three units up, two units to the left

135. **Number of Stores** Target Corporation had 1802 stores in 2016 and 1844 stores in 2018. Use the Midpoint Formula to estimate the number of stores in 2017. Assume that the number of stores followed a linear pattern. *(Source: Target Corporation)*

136. **Meteorology** The apparent temperature is a measure of relative discomfort to a person from heat and high humidity. The table shows the actual temperatures x (in degrees Fahrenheit) versus the apparent temperatures y (in degrees Fahrenheit) for a relative humidity of 75%.

x	70	75	80	85	90	95	100
y	70	77	85	95	109	130	150

(a) Sketch a scatter plot of the data shown in the table.

(b) Find the change in the apparent temperature when the actual temperature changes from 70°F to 100°F.

Exploring the Concepts

True or False? In Exercises 137 and 138, determine whether the statement is true or false. Justify your answer.

137. A binomial sum squared is equal to the sum of the terms squared.

138. $x^n - y^n$ factors as conjugates for all values of n.

Chapter Test

See CalcChat.com for tutorial help and worked-out solutions to odd-numbered exercises.

GO DIGITAL

Take this test as you would take a test in class. When you are finished, check your work against the answers given in the back of the book.

1. Place the appropriate inequality symbol ($<$ or $>$) between the real numbers $-\frac{10}{3}$ and $-\frac{5}{3}$. *(Section P.1)*

2. Find the distance between the real numbers $-\frac{7}{4}$ and $\frac{5}{4}$. *(Section P.1)*

3. Identify the rule of algebra illustrated by $(5 - x) + 0 = 5 - x$. *(Section P.1)*

In Exercises 4 and 5, evaluate each expression without using a calculator. *(Section P.2)*

4. (a) $\left(-\frac{3}{5}\right)^3$ (b) $\left(\frac{3^2}{2}\right)^{-3}$ (c) $\dfrac{5^3 \cdot 7^{-1}}{5^2 \cdot 7}$ (d) $(2^3)^{-2}$

5. (a) $\sqrt{5} \cdot \sqrt{125}$ (b) $\dfrac{\sqrt{27}}{\sqrt{2}}$

    (c) $\dfrac{5.4 \times 10^8}{3 \times 10^3}$ (d) $(4.0 \times 10^8)(2.4 \times 10^{-3})$

In Exercises 6 and 7, simplify each expression. *(Section P.2)*

6. (a) $3z^2(2z^3)^2$ (b) $(u - 2)^{-4}(u - 2)^{-3}$ (c) $\left(\dfrac{x^{-2}y^2}{3}\right)^{-1}$

7. (a) $9z\sqrt{8z} - 3\sqrt{2z^3}$ (b) $(4x^{3/5})(x^{1/3})$ (c) $\sqrt[3]{\dfrac{16}{v^5}}$

8. Write the polynomial $3 - 2x^5 + 3x^3 - x^4$ in standard form. Identify the degree and leading coefficient. *(Section P.3)*

In Exercises 9–14, perform the operation and simplify. *(Sections P.3 and P.5)*

9. $(x^2 + 3) - [3x + (8 - x^2)]$ 10. $\left(x + \sqrt{5}\right)\left(x - \sqrt{5}\right)$

11. $(x + 1)(2x^2 - 3x + 3)$ 12. $\dfrac{y^2 + 8y + 16}{2y - 4} \cdot \dfrac{8y - 16}{(y + 4)^3}$

13. $\dfrac{\left[\dfrac{x^3}{(x - 1)^3}\right]}{\left[\dfrac{x}{(x - 1)^4}\right]}$ 14. $\dfrac{5x}{x - 4} - \dfrac{20}{16 - x^2}$

15. Completely factor (a) $2x^4 - 3x^3 - 2x^2$ and (b) $x^3 + 2x^2 - 4x - 8$. *(Section P.4)*

16. Rationalize each denominator and simplify. *(Section P.2)*

    (a) $\dfrac{16}{\sqrt[3]{16}}$ (b) $\dfrac{4}{1 - \sqrt{2}}$

17. Find the domain of $\dfrac{6 - x}{2 - x - x^2}$. *(Section P.5)*

18. A T-shirt company can produce and sell x T-shirts per day. The total cost C (in dollars) for producing x T-shirts is $C = 1480 + 6x$, and the total revenue R (in dollars) is $R = 15x$. Find the profit obtained by selling 225 T-shirts per day. *(Section P.3)*

19. Plot the points $(-2, 5)$ and $(6, 0)$. Then find the distance between the points and the midpoint of the line segment joining the points. *(Section P.6)*

20. Write an expression for the area of the shaded region in the figure at the left, and simplify the result. *(Section P.3)*

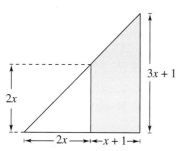

Figure for 20

What does the word *proof* mean to you? In mathematics, the word *proof* means a valid argument. When you prove a statement or theorem, you must use facts, definitions, and accepted properties in a logical order. You can also use previously proved theorems in your proof. For example, the proof of the Midpoint Formula below uses the Distance Formula. There are several different proof methods, which you will see in later chapters.

The Midpoint Formula *(p. 55)*

The midpoint of the line segment joining the points (x_1, y_1) and (x_2, y_2) is

$$\text{Midpoint} = \left(\frac{x_1 + x_2}{2}, \frac{y_1 + y_2}{2}\right).$$

Proof

Using the figure, you must show that $d_1 = d_2$ and $d_1 + d_2 = d_3$.

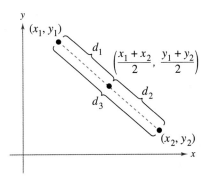

Use the Distance Formula to find d_1, d_2, and d_3. For d_1, you obtain

$$d_1 = \sqrt{\left(\frac{x_1 + x_2}{2} - x_1\right)^2 + \left(\frac{y_1 + y_2}{2} - y_1\right)^2} \qquad \text{Find } d_1.$$

$$= \sqrt{\left(\frac{x_2 - x_1}{2}\right)^2 + \left(\frac{y_2 - y_1}{2}\right)^2}$$

$$= \frac{1}{2}\sqrt{(x_2 - x_1)^2 + (y_2 - y_1)^2}.$$

For d_2, you obtain

$$d_2 = \sqrt{\left(x_2 - \frac{x_1 + x_2}{2}\right)^2 + \left(y_2 - \frac{y_1 + y_2}{2}\right)^2} \qquad \text{Find } d_2.$$

$$= \sqrt{\left(\frac{x_2 - x_1}{2}\right)^2 + \left(\frac{y_2 - y_1}{2}\right)^2}$$

$$= \frac{1}{2}\sqrt{(x_2 - x_1)^2 + (y_2 - y_1)^2}.$$

For d_3, you obtain

$$d_3 = \sqrt{(x_2 - x_1)^2 + (y_2 - y_1)^2}. \qquad \text{Find } d_3.$$

So, it follows that $d_1 = d_2$ and $d_1 + d_2 = d_3$.

P.S. Problem Solving

See CalcChat.com for tutorial help and worked-out solutions to odd-numbered exercises.

GO DIGITAL

1. **Volume and Density** According to USA Track & Field regulations, the men's and women's shots for track and field competition must comply with the specifications below. *(Source: USA Track & Field)*

	Men's	Women's
Weight	7.26 kg	4.0 kg
Diameter (minimum)	110 mm	95 mm
Diameter (maximum)	130 mm	110 mm

   (a) Find the maximum and minimum volumes of both the men's and women's shots.

   (b) To find the *density* of an object, divide its mass (weight) by its volume. Find the maximum and minimum densities of both the men's and women's shots.

   (c) A shot is sometimes made out of iron. When a ball of cork has the same volume as an iron shot, do you think they would have the same density? Explain.

2. **Proof** Find an example for which

   $$|a - b| > |a| - |b|$$

   and an example for which

   $$|a - b| = |a| - |b|.$$

   Then prove that

   $$|a - b| \geq |a| - |b|$$

   for all a, b.

3. **Significant Digits** The accuracy of an approximation to a number is related to how many significant digits there are in the approximation. Write a definition of significant digits and illustrate the concept with examples.

4. **Stained Glass Window** A stained glass window is in the shape of a rectangle with a semicircular arch (see figure). The width of the window is 2 feet and the perimeter is approximately 13.14 feet. Find the least amount of glass required to construct the window.

   ⊢— 2 ft —⊣

5. **Heartbeats** The life expectancies at birth in 2017 for men and women were 76.1 years and 81.1 years, respectively. Assuming an average healthy heart rate of 70 beats per minute, find the numbers of beats in a lifetime for a man and for a woman. *(Source: National Center for Health Statistics)*

6. **Population** The table shows the population y (in millions) of the United States for selected years x from 1960 through 2020. *(Source: U.S. Census Bureau)*

DATA	Year, x	Population, y
	1960	186.72
	1970	209.51
	1980	229.48
	1990	252.12
	2000	281.71
	2010	309.01
	2020	331.00

Spreadsheet at LarsonPrecalculus.com

   (a) Sketch a scatter plot of the data. Describe any trends in the data.

   (b) Find the increase in population from each given year to the next.

   (c) Over which decade did the population increase the most? the least?

   (d) Find the percent increase in population from each given year to the next.

   (e) Over which decade was the percent increase the greatest? the least?

7. **Declining Balances Method** Find the annual depreciation rate r (in decimal form) from the bar graph below. To find r by the declining balances method, use the formula

   $$r = 1 - \left(\frac{S}{C}\right)^{1/n}$$

   where n is the useful life of the item (in years), S is the salvage value (in dollars), and C is the original cost (in dollars).

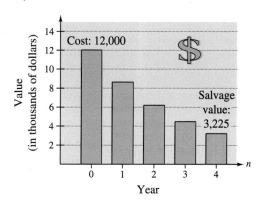

Cost: 12,000

Salvage value: 3,225

8. Planetary Distance and Period Johannes Kepler (1571–1630), a well-known German astronomer, discovered a relationship between the average distance of a planet from the sun and the time (or period) it takes the planet to orbit the sun. People then knew that planets that are closer to the sun take less time to complete an orbit than planets that are farther from the sun. Kepler discovered that the distance and period are related by an exact mathematical formula.

The table shows the average distances x (in astronomical units) and periods y (in years) for the five planets that are closest to the sun. By completing the table, can you rediscover Kepler's relationship? Write a paragraph that summarizes your conclusions.

Planet	Mercury	Venus	Earth	Mars	Jupiter
x	0.387	0.723	1.000	1.524	5.203
$\sqrt{x}$					
y	0.241	0.615	1.000	1.881	11.862
$\sqrt[3]{y}$					

9. Surface Area of a Box A mathematical model for the volume V (in cubic inches) of the box shown below is

$$V = 2x^3 + x^2 - 8x - 4$$

where x is in inches. Find an expression for the surface area of the box. Then find the surface area when $x = 6$ inches.

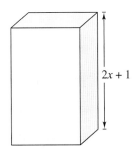

$2x + 1$

10. Proof

(a) Prove that

$$\left(\frac{2x_1 + x_2}{3}, \frac{2y_1 + y_2}{3}\right)$$

is one of the points of trisection of the line segment joining (x_1, y_1) and (x_2, y_2). Find the midpoint of the line segment joining

$$\left(\frac{2x_1 + x_2}{3}, \frac{2y_1 + y_2}{3}\right)$$

and (x_2, y_2) to find the second point of trisection.

(b) Find the points of trisection of the line segment joining each pair of points.

(i) $(1, -2), (4, 1)$

(ii) $(-2, -3), (0, 0)$

11. Nonequivalent and Equivalent Equations Verify that $y_1 \neq y_2$ by letting $x = 0$ and evaluating y_1 and y_2.

$$y_1 = 2x\sqrt{1 - x^2} - \frac{x^3}{\sqrt{1 - x^2}}$$

$$y_2 = \frac{2 - 3x^2}{\sqrt{1 - x^2}}$$

Change y_2 so that $y_1 = y_2$.

12. Weight of a Golf Ball A major feature of Epcot Center at Disney World is Spaceship Earth. The building is spherically shaped and weighs 1.6×10^7 pounds, which is equal in weight to 1.58×10^8 golf balls. Use these values to find the approximate weight (in pounds) of one golf ball. Then convert the weight to ounces. *(Source: Disney.com)*

13. Misleading Graphs Although graphs can help visualize relationships between two variables, they can also be misleading. The graphs shown below represent the same data points.

(a) Which of the two graphs is misleading, and why? Discuss other ways in which graphs can be misleading.

(b) Why would it be beneficial for someone to use a misleading graph?

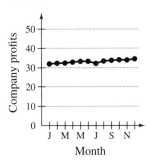

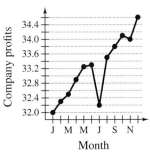

1 Equations, Inequalities, and Mathematical Modeling

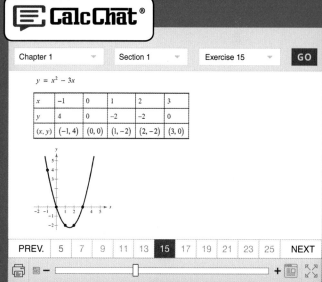

1.1 Population Statistics *(Exercise 74, p. 80)*

1.6 Saturated Steam *(Exercise 97, p. 129)*

69

1.1 Graphs of Equations

The graph of an equation can help you visualize relationships between real-life quantities. For example, in Exercise 74 on page 80, you will use a graph to analyze life expectancy.

- ⬦ Sketch graphs of equations.
- ⬦ Identify *x*- and *y*-intercepts of graphs of equations.
- ⬦ Use symmetry to sketch graphs of equations.
- ⬦ Write equations of circles.
- ⬦ Use graphs of equations to solve real-life problems.

The Graph of an Equation

In Section P.6, you used a coordinate system to graphically represent the relationship between two quantities as points in a coordinate plane. In this section, you will review some basic procedures for sketching the graph of an *equation in two variables*.

Frequently, a relationship between two quantities is expressed as an **equation in two variables.** For example, $y = 7 - 3x$ is an equation in x and y. An ordered pair (a, b) is a **solution** or **solution point** of an equation in x and y when the substitutions $x = a$ and $y = b$ result in a true statement. For example, $(1, 4)$ is a solution of $y = 7 - 3x$ because $4 = 7 - 3(1)$ is a true statement.

EXAMPLE 1	**Determining Solution Points**

Determine whether (a) $(2, 13)$ and (b) $(-1, -3)$ lie on the graph of $y = 10x - 7$.

Solution

a. $y = 10x - 7$ Write original equation.

$13 \overset{?}{=} 10(2) - 7$ Substitute 2 for x and 13 for y.

$13 = 13$ $(2, 13)$ is a solution. ✓

The point $(2, 13)$ *does* lie on the graph of $y = 10x - 7$ because it is a solution point of the equation.

b. $y = 10x - 7$ Write original equation.

$-3 \overset{?}{=} 10(-1) - 7$ Substitute -1 for x and -3 for y.

$-3 \neq -17$ $(-1, -3)$ is not a solution.

The point $(-1, -3)$ *does not* lie on the graph of $y = 10x - 7$ because it is *not* a solution point of the equation.

✓ *Checkpoint* ▶ Audio-video solution in English & Spanish at LarsonPrecalculus.com

Determine whether (a) $(3, -5)$ and (b) $(-2, 26)$ lie on the graph of $y = 14 - 6x$. ■

The set of all solution points of an equation is the **graph of the equation.** The basic technique used for sketching the graph of an equation is the **point-plotting method.**

The Point-Plotting Method of Sketching a Graph

1. If possible, rewrite the equation so that one of the variables is isolated on one side of the equation.
2. Construct a table of values showing several solution points.
3. Plot these points in a rectangular coordinate system.
4. Connect the points with a smooth curve or line.

▶▶ **SKILLS REFRESHER**

For a refresher on how to use the Basic Rules of Algebra to evaluate an expression or an equation, watch the video at *LarsonPrecalculus.com.*

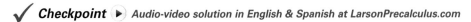

Scan the ▦ to access digital content available for this page.

GO DIGITAL

It is important to use negative values, zero, and positive values for *x* (if possible) when constructing a table. The choice of values to use in the table is somewhat arbitrary. The more values you choose, however, the easier it will be to recognize a pattern.

EXAMPLE 2   **Sketching the Graph of an Equation**

Sketch the graph of $3x + y = 7$.

Solution

Rewrite the equation so that *y* is isolated on the left.

$3x + y = 7$ Write original equation.

$y = -3x + 7$ Subtract $3x$ from each side.

Next, construct a table of values that consists of several solution points of the equation. For instance, when $x = -2$,

$y = -3(-2) + 7 = 13$

which implies that $(-2, 13)$ is a solution point of the equation.

x	−2	−1	0	1	2	3
$y = -3x + 7$	13	10	7	4	1	−2
(x, y)	$(-2, 13)$	$(-1, 10)$	$(0, 7)$	$(1, 4)$	$(2, 1)$	$(3, -2)$

From the table, it follows that

$(-2, 13),\quad (-1, 10),\quad (0, 7),\quad (1, 4),\quad (2, 1),\quad$ and $\quad (3, -2)$

are solution points of the equation. Plot these points, as shown in Figure 1.1(a). It appears that all six points lie on a line, so complete the sketch by drawing a line through the points, as shown in Figure 1.1(b).

ALGEBRA HELP

Example 2 shows three common ways to represent the relationship between two variables. The equation $y = -3x + 7$ is the *algebraic* representation, the table of values is the *numerical* representation, and the graph in Figure 1.1(b) is the *graphical* representation. You will see and use algebraic, numerical, and graphical representations throughout this course.

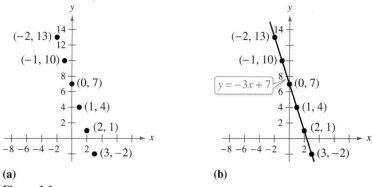

(a)

(b)

Figure 1.1

✓ *Checkpoint* ▶ Audio-video solution in English & Spanish at LarsonPrecalculus.com

Sketch the graph of each equation.

a. $3x + y = 2$ **b.** $-2x + y = 1$ ■

One of your goals in this course is to learn to classify the basic shape of a graph from its equation. For instance, you will learn that the *linear equation* in Example 2 can be written in the form $y = mx + b$ and its graph is a line. Similarly, the *quadratic equation* in Example 3 on the next page has the form $y = ax^2 + bx + c$ and its graph is a *parabola*.

GO DIGITAL

> **EXAMPLE 3**   **Sketching the Graph of an Equation**

▶️🔽 *See LarsonPrecalculus.com for an interactive version of this type of example.*

Sketch the graph of

$$y = x^2 - 2.$$

Solution

The equation is already solved for y, so begin by constructing a table of values.

x	-2	-1	0	1	2	3
$y = x^2 - 2$	2	-1	-2	-1	2	7
(x, y)	$(-2, 2)$	$(-1, -1)$	$(0, -2)$	$(1, -1)$	$(2, 2)$	$(3, 7)$

Next, plot the points given in the table, as shown in Figure 1.2(a). Finally, connect the points with a smooth curve, as shown in Figure 1.2(b).

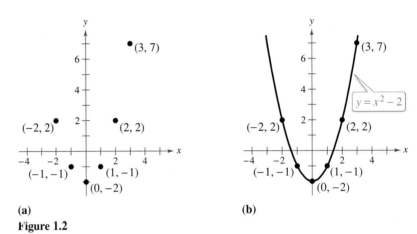

(a)

(b)

Figure 1.2

✔️ ***Checkpoint*** ▶️ *Audio-video solution in English & Spanish at LarsonPrecalculus.com*

Sketch the graph of each equation.

a. $y = x^2 + 3$ **b.** $y = 1 - x^2$ ■

The point-plotting method demonstrated in Examples 2 and 3 is straightforward, but it has shortcomings. For instance, with too few solution points, it is possible to misrepresent the graph of an equation. To illustrate, when you only plot the four points

$$(-2, 2), \quad (-1, -1), \quad (1, -1), \quad \text{and} \quad (2, 2)$$

in Example 3, any one of the three graphs below is reasonable.

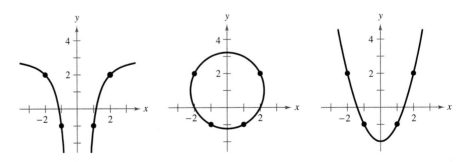

▶️🔽 **TECHNOLOGY**

To graph an equation involving x and y on a graphing utility, use the procedure below.

1. If necessary, rewrite the equation so that y is isolated on the left side.

2. Enter the equation in the graphing utility.

3. Determine a *viewing window* that shows all important features of the graph.

4. Graph the equation.

GO DIGITAL

Intercepts of a Graph

Solution points of an equation that have zero as either the *x*-coordinate or the *y*-coordinate are called **intercepts.** They are the points at which the graph intersects or touches the *x*- or *y*-axis. It is possible for a graph to have no intercepts, one intercept, or several intercepts, as shown in the graphs below.

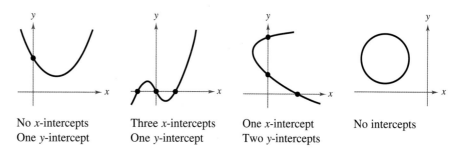

No *x*-intercepts
One *y*-intercept

Three *x*-intercepts
One *y*-intercept

One *x*-intercept
Two *y*-intercepts

No intercepts

Note that an *x*-intercept can be written as the ordered pair $(a, 0)$ and a *y*-intercept can be written as the ordered pair $(0, b)$. Sometimes it is convenient to denote the *x*-intercept as the *x*-coordinate a of the point $(a, 0)$, or the *y*-intercept as the *y*-coordinate b of the point $(0, b)$. Unless it is necessary to make a distinction, the term *intercept* will refer to either the point or the coordinate.

EXAMPLE 4   **Identifying *x*- and *y*-Intercepts**

Identify the *x*- and *y*-intercepts of each graph.

a. 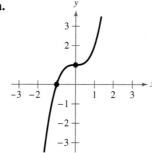   b.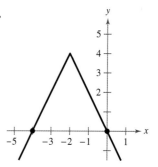

Solution

a. The graph intersects the *x*-axis at $(-1, 0)$, so the *x*-intercept is $(-1, 0)$. The graph intersects the *y*-axis at $(0, 1)$, so the *y*-intercept is $(0, 1)$.

b. The graph intersects the *x*-axis at $(-4, 0)$ and $(0, 0)$, so the *x*-intercepts are $(-4, 0)$ and $(0, 0)$. The graph intersects the *y*-axis at $(0, 0)$, so the *y*-intercept is $(0, 0)$.

✓ *Checkpoint* ▶ *Audio-video solution in English & Spanish at LarsonPrecalculus.com*

Identify the *x*- and *y*-intercepts of each graph.

a. 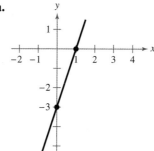   b.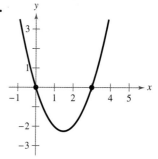

GO DIGITAL

Symmetry

Graphs of equations can have **symmetry** with respect to one of the coordinate axes or with respect to the origin. Symmetry with respect to the *x*-axis means that when you fold the Cartesian plane along the *x*-axis, the portion of the graph above the *x*-axis coincides with the portion below the *x*-axis. Symmetry with respect to the *y*-axis or the origin can be described in a similar manner. The graphs below show these three types of symmetry.

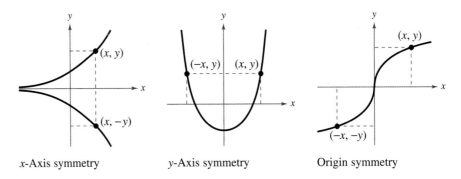

x-Axis symmetry   *y*-Axis symmetry   Origin symmetry

Knowing the symmetry of a graph *before* attempting to sketch it is helpful because you need only half as many solution points to sketch the graph. Graphical and algebraic tests for these three basic types of symmetry are described below.

Tests for Symmetry

Graphical

1. A graph is **symmetric with respect to the *x*-axis** if, whenever (x, y) is on the graph, $(x, -y)$ is also on the graph.

2. A graph is **symmetric with respect to the *y*-axis** if, whenever (x, y) is on the graph, $(-x, y)$ is also on the graph.

3. A graph is **symmetric with respect to the origin** if, whenever (x, y) is on the graph, $(-x, -y)$ is also on the graph.

Algebraic

1. The graph of an equation is **symmetric with respect to the *x*-axis** when replacing *y* with $-y$ yields an equivalent equation.

2. The graph of an equation is **symmetric with respect to the *y*-axis** when replacing *x* with $-x$ yields an equivalent equation.

3. The graph of an equation is **symmetric with respect to the origin** when replacing *x* with $-x$ and *y* with $-y$ yields an equivalent equation.

GO DIGITAL

Using the graphical tests for symmetry, the graph of $y = x^2 - 2$ is symmetric with respect to the *y*-axis because (x, y) and $(-x, y)$ are on its graph, as shown in Figure 1.3. To verify this algebraically, replace *x* with $-x$ in $y = x^2 - 2$

$$y = (-x)^2 - 2 = x^2 - 2 \qquad \text{Replace } x \text{ with } -x \text{ in } y = x^2 - 2.$$

and note that the result is an equivalent equation. To support this result numerically, create a table of values (see below).

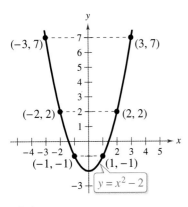

y-Axis symmetry
Figure 1.3

x	-3	-2	-1	1	2	3
y	7	2	-1	-1	2	7
(x, y)	$(-3, 7)$	$(-2, 2)$	$(-1, -1)$	$(1, -1)$	$(2, 2)$	$(3, 7)$

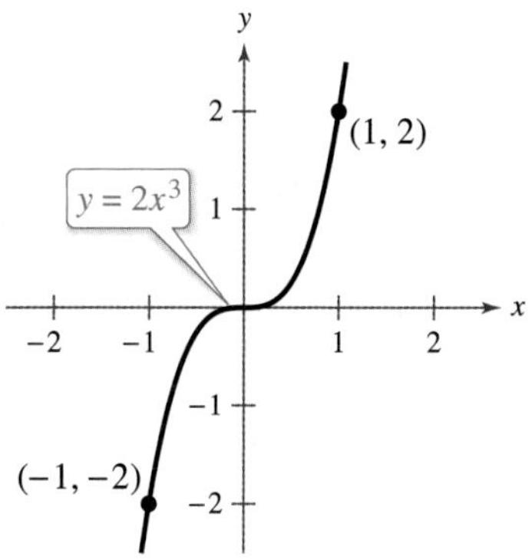

EXAMPLE 5 **Testing for Symmetry**

Test $y = 2x^3$ for symmetry with respect to both axes and the origin.

Solution

x-Axis:
$$y = 2x^3$$ Write original equation.

$$-y = 2x^3$$ Replace y with $-y$. The result is *not* an equivalent equation.

y-Axis:
$$y = 2x^3$$ Write original equation.

$$y = 2(-x)^3$$ Replace x with $-x$.

$$y = -2x^3$$ Simplify. The result is *not* an equivalent equation.

Origin:
$$y = 2x^3$$ Write original equation.

$$-y = 2(-x)^3$$ Replace y with $-y$ and x with $-x$.

$$-y = -2x^3$$ Simplify.

$$y = 2x^3$$ Simplify. The result is an equivalent equation.

Of the three tests for symmetry, the test for origin symmetry is the only one satisfied. So, the graph of $y = 2x^3$ is symmetric with respect to the origin (see Figure 1.4).

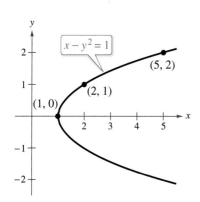

$y = 2x^3$

Origin symmetry
Figure 1.4

✔ *Checkpoint* ▶ Audio-video solution in English & Spanish at LarsonPrecalculus.com

Test $y^2 = 6 - x$ for symmetry with respect to both axes and the origin.

EXAMPLE 6 **Using Symmetry as a Sketching Aid**

Use symmetry to sketch the graph of $x - y^2 = 1$.

Solution Of the three tests for symmetry, the test for x-axis symmetry is the only one satisfied, because $x - (-y)^2 = 1$ is equivalent to $x - y^2 = 1$. So, the graph is symmetric with respect to the x-axis. Find solution points above (or below) the x-axis and then use symmetry to obtain the graph, as shown in Figure 1.5.

$x - y^2 = 1$

x-Axis symmetry
Figure 1.5

✔ *Checkpoint* ▶ Audio-video solution in English & Spanish at LarsonPrecalculus.com

Use symmetry to sketch the graph of $y = x^2 - 4$.

EXAMPLE 7 **Sketching the Graph of an Equation**

Sketch the graph of $y = |x - 1|$.

Solution This equation fails all three tests for symmetry, so its graph is not symmetric with respect to either axis or to the origin. The absolute value bars tell you that y is always nonnegative. Construct a table of values. Then plot and connect the points, as shown in Figure 1.6. Notice from the table that $x = 0$ when $y = 1$. So, the y-intercept is $(0, 1)$. Similarly, $y = 0$ when $x = 1$. So, the x-intercept is $(1, 0)$.

GO DIGITAL

x	-2	-1	0	1	2	3	4
$y = \|x - 1\|$	3	2	1	0	1	2	3
(x, y)	$(-2, 3)$	$(-1, 2)$	$(0, 1)$	$(1, 0)$	$(2, 1)$	$(3, 2)$	$(4, 3)$

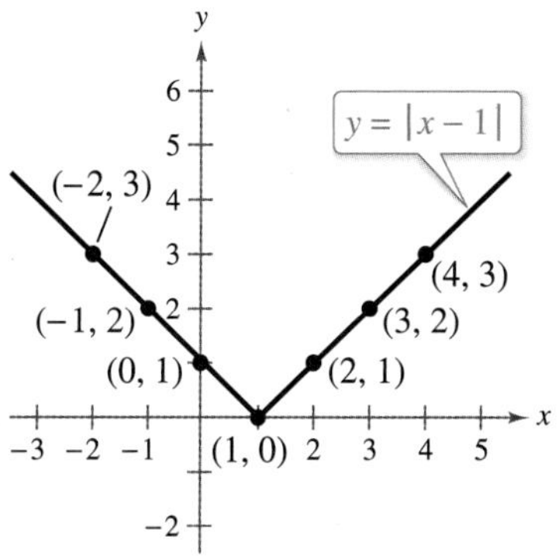

$y = |x - 1|$

Figure 1.6

✔ *Checkpoint* ▶ Audio-video solution in English & Spanish at LarsonPrecalculus.com

Sketch the graph of $y = |x - 2|$.

Circles

A **circle** is the set of all points in a plane that are the same distance from a fixed point. The fixed point is the **center** of the circle, and the distance between the center and a point on the circle is the **radius,** as shown in the figure.

You can use the Distance Formula to write an equation for the circle with center (h, k) and radius r. Let (x, y) be any point on the circle. Then the distance between (x, y) and the center (h, k) is

$$\sqrt{(x - h)^2 + (y - k)^2} = r.$$

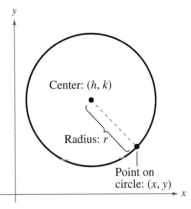

Definition of a circle

By squaring each side of this equation, you obtain the **standard form of the equation of a circle.**

Standard Form of the Equation of a Circle

A point (x, y) lies on the circle of **radius** r and **center** (h, k) if and only if

$$(x - h)^2 + (y - k)^2 = r^2.$$

From this result, the standard form of the equation of a circle with radius r *and center at the origin,* $(h, k) = (0, 0)$, is

$$x^2 + y^2 = r^2.$$ Circle with radius r and center at origin

When $r = 1$, the circle is called the **unit circle.**

EXAMPLE 8   Writing the Equation of a Circle

The point $(3, 4)$ lies on a circle whose center is at $(-1, 2)$, as shown in Figure 1.7. Write the standard form of the equation of this circle.

Solution

The radius of the circle is the distance between $(-1, 2)$ and $(3, 4)$.

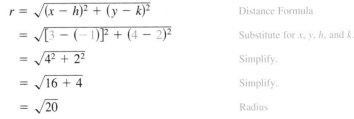

$$r = \sqrt{(x - h)^2 + (y - k)^2}$$ Distance Formula

$$= \sqrt{[3 - (-1)]^2 + (4 - 2)^2}$$ Substitute for x, y, h, and k.

$$= \sqrt{4^2 + 2^2}$$ Simplify.

$$= \sqrt{16 + 4}$$ Simplify.

$$= \sqrt{20}$$ Radius

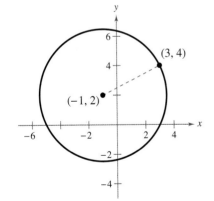

Figure 1.7

Using $(h, k) = (-1, 2)$ and $r = \sqrt{20}$, the equation of the circle is

$$(x - h)^2 + (y - k)^2 = r^2$$ Equation of circle

$$[x - (-1)]^2 + (y - 2)^2 = \left(\sqrt{20}\right)^2$$ Substitute for h, k, and r.

$$(x + 1)^2 + (y - 2)^2 = 20.$$ Standard form

✓ *Checkpoint* ▶ *Audio-video solution in English & Spanish at LarsonPrecalculus.com*

The point $(1, -2)$ lies on a circle whose center is at $(-3, -5)$. Write the standard form of the equation of this circle. ■

Application

In this course, you will learn that there are many ways to approach a problem. Three common approaches are listed below.

A numerical approach: Construct and use a table.

A graphical approach: Draw and use a graph.

An algebraic approach: Use the rules of algebra.

Note how Example 9 uses these approaches.

EXAMPLE 9   **Maximum Weight**

The maximum allowable weight y (in pounds) for a male in the United States Marine Corps can be approximated by the mathematical model

$$y = 0.040x^2 - 0.11x + 3.9, \quad 58 \le x \le 80$$

where x is the male's height (in inches).   *(Source: U.S. Department of Defense)*

a. Construct a table of values that shows the maximum allowable weights for males with heights of 62, 64, 66, 68, 70, 72, 74, and 76 inches.

b. Use the table of values to sketch a graph of the model. Then use the graph to estimate *graphically* the maximum allowable weight for a male whose height is 71 inches.

c. Use the model to estimate the weight in part (b) *algebraically*.

Solution

a. Use a calculator to construct a table, as shown at the left.

b. Use the table of values to sketch the graph of the equation, as shown in Figure 1.8. From the graph, you can estimate that a height of 71 inches corresponds to a maximum allowable weight of about 198 pounds.

c. To estimate the weight in part (b) *algebraically*, substitute 71 for x in the model.

$$y = 0.040(71)^2 - 0.11(71) + 3.9 \approx 197.7$$

The estimate is about 197.7 pounds, which is similar to the estimate in part (b).

✓ *Checkpoint*    **Audio-video solution in English & Spanish at LarsonPrecalculus.com**

Use Figure 1.8 to estimate *graphically* the maximum allowable weight for a man whose height is 75 inches. Then estimate the weight *algebraically*. ∎

ALGEBRA HELP

You should develop the habit of using at least two approaches to solve every problem. This helps build your intuition and helps you check that your answers are reasonable.

DATA	Height, x	Weight, y
	62	150.8
	64	160.7
	66	170.9
	68	181.4
	70	192.2
	72	203.3
	74	214.8
	76	226.6

Spreadsheet at LarsonPrecalculus.com

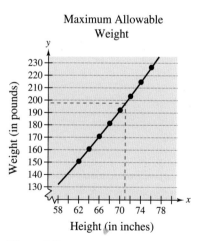

Figure 1.8

GO DIGITAL

Summarize (Section 1.1)

1. Explain how to sketch the graph of an equation *(page 70)*. For examples of sketching graphs of equations, see Examples 2 and 3.

2. Explain how to identify the *x*- and *y*-intercepts of a graph *(page 73)*. For an example of identifying *x*- and *y*-intercepts, see Example 4.

3. Explain how to use symmetry to graph an equation *(page 74)*. For an example of using symmetry to graph an equation, see Example 6.

4. State the standard form of the equation of a circle *(page 76)*. For an example of writing the standard form of the equation of a circle, see Example 8.

5. Describe an example of how to use the graph of an equation to solve a real-life problem *(page 77, Example 9)*.

1.1 Exercises

See CalcChat.com for tutorial help and worked-out solutions to odd-numbered exercises.

GO DIGITAL

Vocabulary and Concept Check

In Exercises 1–6, fill in the blanks.

1. An ordered pair (a, b) is a _____ of an equation in x and y when the substitutions $x = a$ and $y = b$ result in a true statement.
2. The set of all solution points of an equation is the _____ of the equation.
3. The points at which a graph intersects or touches an axis are the _____ of the graph.
4. A graph is symmetric with respect to the _____ if, whenever (x, y) is on the graph, $(-x, y)$ is also on the graph.
5. A graph is symmetric with respect to the _____ if, whenever (x, y) is on the graph, $(-x, -y)$ is also on the graph.
6. When you construct and use a table to solve a problem, you are using a _____ approach.
7. Besides your answer for Exercise 6, name two other approaches you can use to solve problems mathematically.
8. Explain how to use the Distance Formula to write an equation for the circle with center (h, k) and radius r.

Skills and Applications

Determining Solution Points   In Exercises 9–12, determine whether each point lies on the graph of the equation.

Equation	Points	
9. $y = x^2 - 3x + 2$	(a) $(2, 0)$	(b) $(-2, 8)$
10. $y = \sqrt{x + 4}$	(a) $(0, 2)$	(b) $(5, 3)$
11. $y = 4 - |x - 2|$	(a) $(1, 5)$	(b) $(6, 0)$
12. $2x^2 + 5y^2 = 8$	(a) $(6, 0)$	(b) $(0, 4)$

Sketching the Graph of an Equation   In Exercises 13–16, complete the table. Use the resulting solution points to sketch the graph of the equation.

13. $y = -2x + 5$

x	-1	0	1	2	$\frac{5}{2}$
y					
(x, y)					

14. $y + 1 = \frac{3}{4}x$

x	-2	0	1	$\frac{4}{3}$	2
y					
(x, y)					

15. $y + 3x = x^2$

x	-1	0	1	2	3
y					
(x, y)					

16. $y = 5 - x^2$

x	-2	-1	0	1	2
y					
(x, y)					

Identifying x- and y-Intercepts   In Exercises 17–22, identify the x- and y-intercepts of the graph.

17.
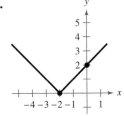

18.
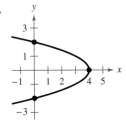

19.
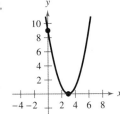

20.
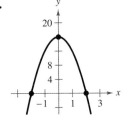

21.
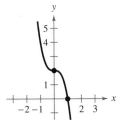

22.
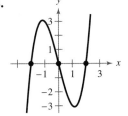

A blue exercise number indicates that a video solution can be seen at *CalcView.com*.

Testing for Symmetry In Exercises 23–30, use the algebraic tests to check for symmetry with respect to both axes and the origin.

23. $x^2 - y = 0$

24. $x - y^2 = 0$

25. $y = x^3$

26. $y = x^4 - x^2 + 3$

27. $y = \dfrac{x}{x^2 + 1}$

28. $y = \dfrac{1}{x^2 + 1}$

29. $xy^2 + 10 = 0$

30. $xy = 4$

Using Symmetry as a Sketching Aid In Exercises 31–34, assume that the graph has the given type of symmetry. Complete the graph of the equation. To print an enlarged copy of the graph, go to *MathGraphs.com*.

31. *y*-Axis symmetry

32. *x*-Axis symmetry

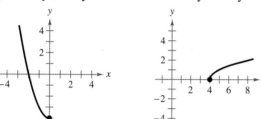

33. Origin symmetry

34. *y*-Axis symmetry

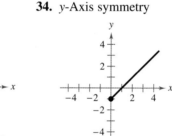

Sketching the Graph of an Equation In Exercises 35–46, test for symmetry and graph the equation. Then identify any intercepts.

35. $y = -3x + 1$

36. $y = 2x - 3$

37. $y = x^2 - 2x$

38. $y = -x^2 - 2x$

39. $y = x^3 + 3$

40. $y = x^3 - 1$

41. $y = \sqrt{x - 3}$

42. $y = \sqrt{1 - x}$

43. $y = |x - 6|$

44. $y = 1 - |x|$

45. $x = y^2 - 1$

46. $x = y^2 - 5$

Approximating Intercepts In Exercises 47–56, use a graphing utility to graph the equation. Use a standard setting. Approximate any intercepts.

47. $y = 5 - \frac{1}{2}x$

48. $y = \frac{2}{3}x - 1$

49. $y = x^2 - 4x + 3$

50. $y = x^2 + x - 2$

51. $y = \dfrac{2x}{x - 1}$

52. $y = \dfrac{4}{x^2 + 1}$

53. $y = \sqrt[3]{x + 1}$

54. $y = x\sqrt{x + 6}$

55. $y = |x + 3|$

56. $y = 2 - |x|$

Writing the Equation of a Circle In Exercises 57–64, write the standard form of the equation of the circle with the given characteristics.

57. Center: $(0, 0)$; Radius: 3

58. Center: $(0, 0)$; Radius: 7

59. Center: $(-4, 5)$; Radius: 2

60. Center: $(1, -3)$; Radius: $\sqrt{11}$

61. Center: $(3, 8)$; Solution point: $(-9, 13)$

62. Center: $(-2, -6)$; Solution point: $(1, -10)$

63. Endpoints of a diameter: $(3, 2), (-9, -8)$

64. Endpoints of a diameter: $(11, -5), (3, 15)$

Sketching a Circle In Exercises 65–70, find the center and radius of the circle with the given equation. Then sketch the circle.

65. $x^2 + y^2 = 25$

66. $x^2 + y^2 = 36$

67. $(x - 1)^2 + (y + 3)^2 = 9$

68. $x^2 + (y - 1)^2 = 1$

69. $\left(x - \frac{1}{2}\right)^2 + \left(y - \frac{1}{2}\right)^2 = \frac{9}{4}$

70. $(x - 2)^2 + (y + 3)^2 = \frac{16}{9}$

71. **Depreciation** A hospital purchases a new magnetic resonance imaging (MRI) machine for $1.2 million. The depreciated value y (reduced value) after t years is given by $y = 1,200,000 - 80,000t$, $0 \le t \le 10$. Sketch the graph of the equation.

72. **Depreciation** You purchase an all-terrain vehicle (ATV) for $9500. The depreciated value y (reduced value) after t years is given by $y = 9500 - 1000t$, $0 \le t \le 6$. Sketch the graph of the equation.

73. **Geometry** A regulation NFL playing field of length x and width y has a perimeter of $346\frac{2}{3}$ or $\frac{1040}{3}$ yards.

(a) Draw a rectangle that gives a visual representation of the problem. Use the specified variables to label the sides of the rectangle.

(b) Show that the width of the rectangle is $y = \frac{520}{3} - x$ and its area is $A = x\left(\frac{520}{3} - x\right)$.

(c) Use a graphing utility to graph the area equation. Be sure to adjust your window settings.

(d) From the graph in part (c), estimate the dimensions of the rectangle that yield a maximum area.

(e) Use an appropriate research source to determine the actual dimensions and area of a regulation NFL playing field and compare your findings with the results of part (d).

The symbol ⚡ indicates an exercise or a part of an exercise in which you are instructed to use a graphing utility.

74. Population Statistics

The table shows the life expectancies of a child (at birth) in the United States for selected years from 1950 through 2020. *(Source: Macrotrends LLC)*

Year	Life Expectancy, y
1950	68.14
1960	69.84
1970	70.78
1980	73.70
1990	75.19
2000	76.75
2010	78.49
2020	78.93

Spreadsheet at LarsonPrecalculus.com

A model for the life expectancy during this period is

$$y = \frac{68.0 + 0.33t}{1 + 0.002t}, \quad 0 \le t \le 70$$

where y represents the life expectancy and t is the time in years, with $t = 0$ corresponding to 1950.

(a) Use a graphing utility to graph the data from the table and the model in the same viewing window. How well does the model fit the data? Explain.

(b) Determine the life expectancy in 1990 both graphically and algebraically.

(c) Use the graph to determine the year when life expectancy was approximately 70.1. Verify your answer algebraically.

(d) Identify the y-intercept of the graph of the model. What does it represent in the context of the problem?

(e) Do you think this model can be used to predict the life expectancy of a child 50 years from now? Explain.

Exploring the Concepts

True or False? In Exercises 75 and 76, determine whether the statement is true or false. Justify your answer.

75. The graph of a linear equation cannot be symmetric with respect to the origin.

76. The graph of a linear equation can have either no x-intercepts or only one x-intercept.

77. Error Analysis Describe the error.

The graph of $x = 3y^2$ is symmetric with respect to the y-axis because

$$\begin{aligned} x &= 3(-y)^2 \\ &= 3y^2. \end{aligned} \quad ✗$$

78. HOW DO YOU SEE IT?

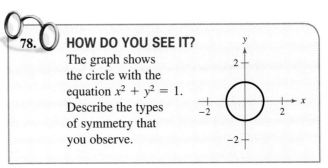

The graph shows the circle with the equation $x^2 + y^2 = 1$. Describe the types of symmetry that you observe.

79. **Think About It** Find a and b when the graph of $y = ax^2 + bx^3$ is symmetric with respect to (a) the y-axis and (b) the origin. (There are many correct answers.)

80. **Graph of a Linear Equation** When the graph of a linear equation in two variables has a negative x-intercept and a positive y-intercept, does the line rise or fall from left to right? Through which quadrant(s) does the line pass? Use a graph to illustrate your answer.

Review & Refresh ▶ Video solutions at LarsonPrecalculus.com

Using the Distributive Property In Exercises 81–84, use the Distributive Property to rewrite the expression. Simplify your results.

81. $3(7x + 1)$

82. $5(x - 6)$

83. $6(x - 1) + 4$

84. $4(x + 2) - 12$

Finding the Least Common Denominator In Exercises 85–88, find the least common denominator.

85. $\dfrac{x}{3}, \dfrac{3x}{4}$

86. $\dfrac{4x}{9}, \dfrac{1}{3}, x, \dfrac{5}{3}$

87. $\dfrac{3x}{x - 4}, 5, \dfrac{12}{x - 4}$

88. $\dfrac{1}{x - 2}, \dfrac{3}{x + 2}, \dfrac{6x}{x^2 - 4}$

Simplifying an Expression In Exercises 89–92, simplify the expression.

89. $7\sqrt{72} - 5\sqrt{18}$

90. $-10\sqrt{25y} - \sqrt{y}$

91. $7^{3/2} \cdot 7^{11/2}$

92. $\dfrac{10^{17/4}}{10^{5/4}}$

Operations with Polynomials In Exercises 93–96, perform the operation and write the result in standard form.

93. $(9x - 4) + (2x^2 - x + 15)$

94. $4x(11 - x + 3x^2)$

95. $(2x + 9)(x - 7)$

96. $(3x^2 - 5)(-x^2 + 1)$

1.2 Linear Equations in One Variable

Linear equations have many real-life applications, such as in forensics. For example, in Exercises 65 and 66 on page 88, you will use linear equations to determine height from femur length.

- ❯ Identify different types of equations.
- ❯ Solve linear equations in one variable.
- ❯ Solve rational equations that lead to linear equations.
- ❯ Find x- and y-intercepts of graphs of equations algebraically.
- ❯ Use linear equations to model and solve real-life problems.

Equations and Solutions of Equations

An **equation** in x is a statement that two algebraic expressions are equal. For example,

$$3x - 5 = 7, \quad x^2 - x - 6 = 0, \quad \text{and} \quad \sqrt{2x} = 4$$

are equations in x. To **solve** an equation in x means to find all values of x for which the equation is true. Such values are **solutions.** For example, to determine whether $x = 4$ is a solution of the equation $3x - 5 = 7$, substitute 4 for x. Then use the rules of algebra to simplify each side of the equation to see whether you obtain a true statement.

$3x - 5 = 7$	Original equation
$3(4) - 5 \overset{?}{=} 7$	Substitute 4 for x.
$12 - 5 \overset{?}{=} 7$	Multiply.
$7 = 7$	Subtract. Solution checks. ✓

Because $7 = 7$ is a true statement, you can conclude that $x = 4$ is a solution of the equation $3x - 5 = 7$.

The solutions of an equation depend on the kinds of numbers being considered. For example, in the set of rational numbers, $x^2 = 10$ has no solution because there is no rational number whose square is 10. However, in the set of real numbers, the equation has the two solutions

$$x = \sqrt{10} \quad \text{and} \quad x = -\sqrt{10}.$$

An equation that is true for *every* real number in the domain of the variable is an **identity.** For example,

$$x^2 - 9 = (x + 3)(x - 3) \qquad \text{Identity}$$

is an identity because it is a true statement for any real value of x. The equation

$$\frac{x}{3x^2} = \frac{1}{3x} \qquad \text{Identity}$$

is an identity because it is true for any nonzero real value of x.

An equation that is true for just *some* (but not all) of the real numbers in the domain of the variable is a **conditional equation.** For example, the equation

$$x^2 - 9 = 0 \qquad \text{Conditional equation}$$

is conditional because $x = 3$ and $x = -3$ are the only values in the domain that satisfy the equation. The equation $2x + 4 = 6$ is conditional because $x = 1$ is the only value in the domain that satisfies the equation.

A **contradiction** is an equation that is false for *every* real number in the domain of the variable. For example, the equation

$$2x - 4 = 2x + 1 \qquad \text{Contradiction}$$

is a contradiction because there are no real values of x for which the equation is true.

GO DIGITAL

Linear Equations in One Variable

A common type of equation in one variable is a *linear equation*.

Definition of a Linear Equation in One Variable

A **linear equation in one variable** x is an equation that can be written in the standard form

$$ax + b = 0 \qquad \text{Standard form}$$

where a and b are real numbers with $a \neq 0$.

This ancient Egyptian papyrus, discovered in 1858, contains one of the earliest examples of mathematical writing in existence. The papyrus itself dates back to around 1650 B.C., but it is actually a copy of writings from two centuries earlier. The algebraic equations on the papyrus were written in words. Diophantus, a Greek who lived around A.D. 250, is often called the Father of Algebra. He was the first to use abbreviated word forms in equations.

Some examples of linear equations in one variable that are written in the standard form $ax + b = 0$ are $3x + 2 = 0$ and $5x - 9 = 0$.

A linear equation in one variable has exactly one solution. To see this, consider the steps below. (Remember that $a \neq 0$.)

$$ax + b = 0 \qquad \text{Original equation}$$

$$ax = -b \qquad \text{Subtract } b \text{ from each side.}$$

$$x = -\frac{b}{a} \qquad \text{Divide each side by } a.$$

It is clear that the last equation has only one solution, $x = -b/a$, and that this equation is equivalent to the original equation. So, you can conclude that every linear equation in one variable, written in standard form, has exactly one solution.

To solve a conditional equation in x, isolate x on one side of the equation in the form

$$x = \boxed{\text{a number}} \, . \qquad \text{Isolate } x \text{ on one side of the equation.}$$

To accomplish this, use a sequence of **equivalent equations,** each having the same solution as the original equation. The operations that yield equivalent equations come from the properties of equality reviewed in Section P.1.

Generating Equivalent Equations

An equation can be transformed into an *equivalent equation* by one or more of the steps listed below.

	Given Equation	**Equivalent Equation**
1. Remove symbols of grouping, combine like terms, or simplify fractions on one or both sides of the equation.	$2x - x = 4$	$x = 4$
2. Add (or subtract) the same quantity to (or from) *each* side of the equation.	$x + 1 = 6$	$x = 5$
3. Multiply (or divide) *each* side of the equation by the same *nonzero* quantity.	$2x = 6$	$x = 3$
4. Interchange the two sides of the equation.	$2 = x$	$x = 2$

EXAMPLE 1 **Solving Linear Equations**

a. $3x - 6 = 0$ — Original equation

$3x = 6$ — Add 6 to each side.

$x = 2$ — Divide each side by 3.

b. $5x + 4 = 3x - 8$ — Original equation

$2x + 4 = -8$ — Subtract $3x$ from each side.

$2x = -12$ — Subtract 4 from each side.

$x = -6$ — Divide each side by 2.

✓ *Checkpoint* ▶ *Audio-video solution in English & Spanish at LarsonPrecalculus.com*

Solve each equation.

a. $7 - 2x = 15$ **b.** $7x - 9 = 5x + 7$

After solving an equation, you should check each solution in the original equation. For instance, to check the solution to Example 1(a), substitute 2 for x in the original equation and simplify.

$3x - 6 = 0$ — Write original equation.

$3(2) - 6 \overset{?}{=} 0$ — Substitute 2 for x.

$0 = 0$ — Solution checks. ✓

Check the solution to Example 1(b) on your own.

EXAMPLE 2 **Solving a Linear Equation**

Solve $6(x - 1) + 4 = 3(7x + 1)$.

Solution

$6(x - 1) + 4 = 3(7x + 1)$ — Write original equation.

$6x - 6 + 4 = 21x + 3$ — Distributive Property

$6x - 2 = 21x + 3$ — Simplify.

$-5 = 15x$ — Simplify.

$x = -\frac{1}{3}$ — Interchange sides and divide each side by 15.

Check

$6(x - 1) + 4 = 3(7x + 1)$ — Write original equation.

$6\left(-\frac{1}{3} - 1\right) + 4 \overset{?}{=} 3\left[7\left(-\frac{1}{3}\right) + 1\right]$ — Substitute $-\frac{1}{3}$ for x.

$6\left(-\frac{4}{3}\right) + 4 \overset{?}{=} 3\left[-\frac{7}{3} + 1\right]$ — Simplify.

$-8 + 4 \overset{?}{=} 3\left(-\frac{4}{3}\right)$ — Simplify.

$-4 = -4$ — Solution checks. ✓

✓ *Checkpoint* ▶ *Audio-video solution in English & Spanish at LarsonPrecalculus.com*

Solve $4(x + 2) - 12 = 5(x - 6)$.

▶▶▶ **TECHNOLOGY**

To graphically check the solution of an equation, use a graphing utility to graph each side of the original equation. Then find the x-coordinate of the intersection point of the graphs. So, to check the solution to Example 2, graph $y_1 = 6(x - 1) + 4$ and $y_2 = 3(7x + 1)$ in the same viewing window (see figure below). Use the *intersect* feature to determine that $x \approx -0.3333333 \approx -1/3$.

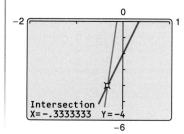

GO DIGITAL

Rational Equations That Lead to Linear Equations

SKILLS REFRESHER

For a refresher on how to find the least common denominator (LCD) of two or more rational expressions, watch the video at *LarsonPrecalculus.com*.

A **rational equation** involves one or more rational expressions. To solve a rational equation, multiply every term by the least common denominator (LCD) of all the terms. This clears the original equation of fractions and produces a simpler equation.

EXAMPLE 3 **Solving a Rational Equation**

Solve $\dfrac{x}{3} + \dfrac{3x}{4} = 2$.

Solution The LCD is 12, so multiply each term by 12.

$$\frac{x}{3} + \frac{3x}{4} = 2 \qquad \text{Write original equation.}$$

$$(12)\frac{x}{3} + (12)\frac{3x}{4} = (12)2 \qquad \text{Multiply each term by the LCD.}$$

$$4x + 9x = 24 \qquad \text{Simplify.}$$

$$13x = 24 \qquad \text{Combine like terms.}$$

$$x = \frac{24}{13} \qquad \text{Divide each side by 13.}$$

The solution is $x = \frac{24}{13}$. Check this in the original equation.

✓ *Checkpoint* ▶ Audio-video solution in English & Spanish at LarsonPrecalculus.com

Solve $\dfrac{4x}{9} - \dfrac{1}{3} = x + \dfrac{5}{3}$.

When multiplying or dividing an equation by a *variable expression,* it is possible to introduce an **extraneous solution,** which is a solution that does not satisfy the original equation. So, it is essential to check your solutions.

EXAMPLE 4 **An Equation with an Extraneous Solution**

 See LarsonPrecalculus.com for an interactive version of this type of example.

Solve $\dfrac{1}{x - 2} = \dfrac{3}{x + 2} - \dfrac{6x}{x^2 - 4}$.

Solution The LCD is $(x + 2)(x - 2)$. Multiply each term by the LCD.

$$\frac{1}{x-2}(x+2)(x-2) = \frac{3}{x+2}(x+2)(x-2) - \frac{6x}{x^2-4}(x+2)(x-2)$$

$$x + 2 = 3(x - 2) - 6x, \quad x \neq \pm 2$$

$$x + 2 = 3x - 6 - 6x$$

$$x + 2 = -3x - 6$$

$$4x = -8 \implies x = -2 \qquad \text{Extraneous solution}$$

ALGEBRA HELP

In Example 4, the factored forms of the denominators are $x - 2$, $x + 2$, and $(x + 2)(x - 2)$. The factors $x - 2$ and $x + 2$ each appear once, so the LCD is $(x + 2)(x - 2)$.

In the original equation, $x = -2$ yields a denominator of zero. So, $x = -2$ is an extraneous solution, and the original equation has *no solution.*

✓ *Checkpoint* ▶ Audio-video solution in English & Spanish at LarsonPrecalculus.com

Solve $\dfrac{3x}{x - 4} = 5 + \dfrac{12}{x - 4}$.

Finding Intercepts Algebraically

In Section 1.1, you learned to find x- and y-intercepts using a graphical approach. All points on the x-axis have a y-coordinate equal to zero, and all points on the y-axis have an x-coordinate equal to zero. This suggests an algebraic approach to finding x- and y-intercepts.

> **Finding Intercepts Algebraically**
>
> 1. To find x-intercepts, set y equal to zero and solve the equation for x.
> 2. To find y-intercepts, set x equal to zero and solve the equation for y.

EXAMPLE 5 **Finding Intercepts Algebraically**

Find the x- and y-intercepts of the graph of each equation algebraically.

a. $y = 4x + 1$ **b.** $3x + 2y = 6$

Solution

a. To find the x-intercept, set y equal to zero and solve for x.

$y = 4x + 1$	Write original equation.
$0 = 4x + 1$	Substitute 0 for y.
$-1 = 4x$	Subtract 1 from each side.
$-\frac{1}{4} = x$	Divide each side by 4.

So, the x-intercept is $\left(-\frac{1}{4}, 0\right)$. To find the y-intercept, set x equal to zero and solve for y.

$y = 4x + 1$	Write original equation.
$y = 4(0) + 1$	Substitute 0 for x.
$y = 1$	Simplify.

So, the y-intercept is $(0, 1)$. Check this by sketching a graph [see Figure 1.9(a)].

b. To find the x-intercept, set y equal to zero and solve for x.

$3x + 2y = 6$	Write original equation.
$3x + 2(0) = 6$	Substitute 0 for y.
$3x = 6$	Simplify.
$x = 2$	Divide each side by 3.

So, the x-intercept is $(2, 0)$. To find the y-intercept, set x equal to zero and solve for y.

$3x + 2y = 6$	Write original equation.
$3(0) + 2y = 6$	Substitute 0 for x.
$2y = 6$	Simplify.
$y = 3$	Divide each side by 2.

So, the y-intercept is $(0, 3)$. Check this by sketching a graph [see Figure 1.9(b)].

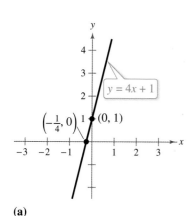

(a)

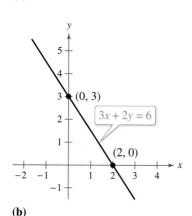

(b)

Figure 1.9

✔ *Checkpoint* *Audio-video solution in English & Spanish at LarsonPrecalculus.com*

Find the x- and y-intercepts of the graph of each equation algebraically.

a. $y = -3x - 2$ **b.** $5x + 3y = 15$

Application

Female Participants in Soccer

The number y (in thousands) of female participants in high school soccer in the United States from 2008 through 2019 can be approximated by the linear model

$$y = 4.4730t + 355.675, \quad -2 \leq t \leq 9$$

where t represents the year, with $t = 0$ corresponding to 2010. (a) Find algebraically and interpret the y-intercept of the graph of the linear model shown in Figure 1.10. (b) According to the model, in which year did participation reach 387,000? *(Source: National Federation of State High School Associations)*

Solution

a. To find the y-intercept, let $t = 0$ and solve for y.

$$y = 4.4730t + 355.675 \qquad \text{Write original equation.}$$
$$= 4.4730(0) + 355.675 \qquad \text{Substitute 0 for } t.$$
$$= 355.675 \qquad \text{Simplify.}$$

So, the y-intercept is $(0, 355.675)$. This means that, according to the model, there were about 355,675 female participants in 2010.

b. Let $y = 387$ and solve for t.

$$y = 4.4730t + 355.675 \qquad \text{Write original equation.}$$
$$387 = 4.4730t + 355.675 \qquad \text{Substitute 387 for } y.$$
$$31.325 = 4.4730t \qquad \text{Subtract 355.675 from each side.}$$
$$7 \approx t \qquad \text{Divide each side by 4.4730.}$$

Because $t = 0$ represents 2010, $t = 7$ must represent 2017. This means that, according to the model, the number of participants reached 387,000 in 2017.

Female Participants in Soccer

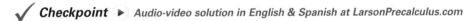

Number (in thousands)

Year (0 ↔ 2010)

Figure 1.10

✓ *Checkpoint* ▶ *Audio-video solution in English & Spanish at LarsonPrecalculus.com*

The number y (in thousands) of male participants in high school lacrosse in the United States from 2009 through 2019 can be approximated by the linear model

$$y = 3.4175t + 68.676, \quad -1 \leq t \leq 9$$

where t represents the year, with $t = 0$ corresponding to 2010. (a) Find algebraically and interpret the y-intercept of the graph of the linear model. (b) According to the model, in which year did participation reach 96,200? *(Source: National Federation of State High School Associations)*

Summarize *(Section 1.2)*

1. State the definitions of an identity, a conditional equation, and a contradiction *(page 81)*.

2. State the definition of a linear equation in one variable and list the four steps that can be used to form an equivalent equation *(page 82)*. For examples of solving linear equations, see Examples 1 and 2.

3. Explain how to solve a rational equation *(page 84)*. For examples of solving rational equations, see Examples 3 and 4.

4. Explain how to find intercepts algebraically *(page 85)*. For an example of finding intercepts algebraically, see Example 5.

5. Describe a real-life application involving a linear equation *(page 86, Example 6)*.

GO DIGITAL

1.2   Exercises

See CalcChat.com for tutorial help and worked-out solutions to odd-numbered exercises.

Vocabulary and Concept Check

In Exercises 1–6, fill in the blanks.

1. An _____ is a statement that equates two algebraic expressions.

2. There are three types of equations: _____, _____ equations, and _____.

3. A linear equation in one variable x is an equation that can be written in the standard form _____.

4. An _____ equation has the same solution(s) as the original equation.

5. A _____ equation is an equation that involves one or more rational expressions.

6. An _____ solution is a solution that does not satisfy the original equation.

7. Are $8 = x - 3$ and $x = 11$ equivalent equations?

8. How can you clear the equation $\dfrac{x}{2} + 1 = \dfrac{1}{4}$ of fractions?

Skills and Applications

Classifying an Equation   In Exercises 9–16, determine whether the equation is an identity, a conditional equation, or a contradiction.

9. $3(x - 1) = 3x - 3$

10. $2(x + 1) = 2x - 1$

11. $2(x - 1) = 3x + 1$

12. $4(x + 2) = 2x + 2$

13. $3(x + 2) = 3x + 2$

14. $5(x + 2) = 5x + 10$

15. $2(x + 3) - 5 = 2x + 1$

16. $3(x - 1) + 2 = 4x - 2$

Solving a Linear Equation   In Exercises 17–28, solve the equation and check your solution. (If not possible, explain why.)

17. $2x + 11 = 15$

18. $7x + 2 = 23$

19. $7 - 2x = 25$

20. $7 - x = 19$

21. $3x - 5 = 2x + 7$

22. $5x + 3 = 6 - 2x$

23. $4y + 2 - 5y = 7 - 6y$

24. $5y + 1 = 8y - 5 + 6y$

25. $x - 3(2x + 3) = 8 - 5x$

26. $9x - 10 = 5x + 2(2x - 5)$

27. $0.25x + 0.75(10 - x) = 3$

28. $0.60x + 0.40(100 - x) = 50$

Solving a Rational Equation   In Exercises 29–42, solve the equation and check your solution. (If not possible, explain why.)

29. $\dfrac{3x}{8} - \dfrac{4x}{3} = 4$

30. $\dfrac{x}{5} - \dfrac{x}{2} = 3 + \dfrac{3x}{10}$

31. $\dfrac{5x - 4}{5x + 4} = \dfrac{2}{3}$

32. $\dfrac{10x + 3}{5x + 6} = \dfrac{1}{2}$

33. $10 - \dfrac{13}{x} = 4 + \dfrac{5}{x}$

34. $\dfrac{15}{x} - 4 = \dfrac{6}{x} + 3$

35. $3 = 2 + \dfrac{2}{z + 2}$

36. $\dfrac{1}{x} + \dfrac{2}{x - 5} = 0$

37. $\dfrac{x}{x + 4} + \dfrac{4}{x + 4} + 2 = 0$

38. $\dfrac{7}{2x + 1} - \dfrac{8x}{2x - 1} = -4$

39. $\dfrac{2}{(x - 4)(x - 2)} = \dfrac{1}{x - 4} + \dfrac{2}{x - 2}$

40. $\dfrac{12}{(x - 1)(x + 3)} = \dfrac{3}{x - 1} + \dfrac{2}{x + 3}$

41. $\dfrac{1}{x - 3} + \dfrac{1}{x + 3} = \dfrac{10}{x^2 - 9}$

42. $\dfrac{1}{x - 2} + \dfrac{3}{x + 3} = \dfrac{4}{x^2 + x - 6}$

Finding Intercepts Algebraically   In Exercises 43–52, find the x- and y-intercepts of the graph of the equation algebraically.

43. $y = 12 - 5x$

44. $y = 16 - 3x$

45. $y = -3(2x + 1)$

46. $y = 5 - (6 - x)$

47. $2x + 3y = 10$

48. $4x - 5y = 12$

49. $4y - 0.75x + 1.2 = 0$

50. $3y + 2.5x - 3.4 = 0$

51. $\dfrac{2x}{5} + 8 - 3y = 0$

52. $\dfrac{8x}{3} + 5 - 2y = 0$

Approximating Intercepts   In Exercises 53–58, use a graphing utility to graph the equation and approximate any x-intercepts. Then set $y = 0$ and solve the resulting equation. Compare the result with the graph's x-intercept.

53. $y = 2(x - 1) - 4$

54. $y = \frac{4}{3}x + 2$

55. $y = 20 - (3x - 10)$

56. $y = 10 + 2(x - 2)$

57. $y = -38 + 5(9 - x)$

58. $y = 6x - 6\left(\frac{16}{11} + x\right)$

Solving an Equation   In Exercises 59–62, solve the equation. (Round your solution to three decimal places.)

59. $\dfrac{2}{7.398} - \dfrac{4.405}{x} = \dfrac{1}{x}$ **60.** $\dfrac{3}{6.350} - \dfrac{6}{x} = 18$

61. $0.275x + 0.725(500 - x) = 300$

62. $2.763 - 4.5(2.1x - 5.1432) = 6.32x + 5$

63. Geometry   The surface area S of the circular cylinder shown in the figure is given by

$$S = 2\pi(25) + 2\pi(5h).$$

Find the height h of the cylinder when the surface area is 471 square feet. Use 3.14 for π.

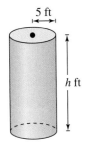

5 ft

h ft

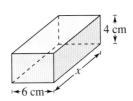

4 cm

x

6 cm

Figure for 63   Figure for 64

64. Geometry   The surface area S of the rectangular solid shown in the figure is given by $S = 2(24) + 2(4x) + 2(6x)$. Find the length x of the solid when the surface area is 248 square centimeters.

Forensics

In Exercises 65 and 66, use the following information. The relationship between the length of an adult's femur (thigh bone) and the height of the adult can be approximated by the linear equations

$y = 0.514x - 14.75$ Female

$y = 0.532x - 17.03$ Male

where y is the length of the femur in inches and x is the height of the adult in inches (see figure).

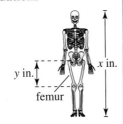

y in.

x in.

femur

65. A crime scene investigator discovers a femur belonging to an adult human female. The bone is 18 inches long. Estimate the height of the female.

66. Officials search a forest for a missing man who is 6 feet 3 inches tall. They find an adult male femur that is 23 inches long. Is it possible that the femur belongs to the missing man?

67. Population   The population y (in thousands) of Raleigh, North Carolina, from 2010 to 2018 can be approximated by the model

$$y = 8.28t + 406.6, \quad 0 \le t \le 8$$

where t represents the year, with $t = 0$ corresponding to 2010 (see figure).   *(Source: U.S. Census Bureau)*

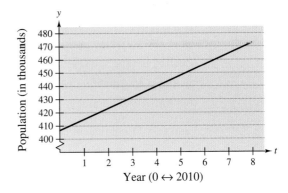

Year (0 ↔ 2010)

(a) Graphically estimate the y-intercept of the graph.

(b) Find algebraically and interpret the y-intercept of the graph.

(c) According to the model, in which year did the population reach 448,000?

68. Population   The population y (in thousands) of Flint, Michigan, from 2010 to 2018 can be approximated by the model $y = -0.77t + 101.8$, $0 \le t \le 8$, where t represents the year, with $t = 0$ corresponding to 2010 (see figure).   *(Source: U.S. Census Bureau)*

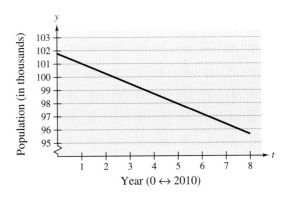

Year (0 ↔ 2010)

(a) Graphically estimate the y-intercept of the graph.

(b) Find algebraically and interpret the y-intercept of the graph.

(c) According to the model, in which year was the population about 99,500?

69. Operating Cost   A delivery company has a fleet of vans. The annual operating cost C (in dollars) per van is given by $C = 0.37m + 2600$, where m is the number of miles traveled by a van in a year. What number of miles yields an annual operating cost of $10,000?

70. Flood Control A river is 8 feet above its flood stage. The water is receding at a rate of 3 inches per hour. Write a mathematical model that shows the number of feet above flood stage after t hours. Assuming the water continually recedes at this rate, when will the river be 1 foot above its flood stage?

Exploring the Concepts

True or False? In Exercises 71–74, determine whether the statement is true or false. Justify your answer.

71. The equation $x(3 - x) = 10$ is a linear equation.

72. The equation $2(x + 3) = 3x + 3$ has no solution.

73. The equation $3(x - 1) - 2 = 3x - 6$ is an identity.

74. The equation

$$2 - \frac{1}{x - 2} = \frac{3}{x - 2}$$

has no solution because $x = 2$ is an extraneous solution.

75. Think About It Are $\dfrac{3x + 2}{5} = 7$ and $x + 9 = 20$ equivalent equations? Explain.

76. HOW DO YOU SEE IT? Use the information below about a possible tax credit for a family consisting of two adults and two children (see figure).

Earned income: E

Subsidy (a grant of money):

$S = 10{,}000 - \frac{1}{2}E, \quad 0 \le E \le 20{,}000$

Total income:

$T = E + S$

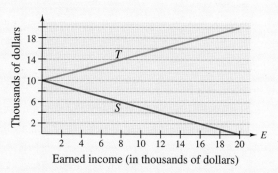

Earned income (in thousands of dollars)

(a) Graphically estimate the intercepts of the graph of the subsidy equation. Then interpret the meaning of each intercept.

(b) Explain how to solve part (a) algebraically.

(c) Graphically estimate the earned income for which the total income is $14,000.

(d) Explain how to solve part (c) algebraically.

77. Graphical Reasoning

(a) Use a graphing utility to graph the equation $y = 3x - 6$.

(b) Use the result of part (a) to estimate the x-intercept.

(c) Explain how the x-intercept is related to the solution of $3x - 6 = 0$.

78. Finding Intercepts Consider the linear equation

$ax + by = c$

where a, b, and c are real numbers.

(a) What is the x-intercept of the graph of the equation when $a \ne 0$?

(b) What is the y-intercept of the graph of the equation when $b \ne 0$?

(c) Use your results from parts (a) and (b) to find the x- and y-intercepts of the graph of $2x + 7y = 11$.

Review & Refresh ▶ Video solutions at LarsonPrecalculus.com

Geometry In Exercises 79 and 80, find the area of the shaded region in terms of x. Write your result as a polynomial in standard form.

79.
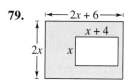

80.
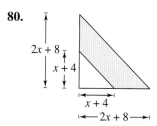

Solving a Proportion In Exercises 81–86, solve the proportion.

81. $\dfrac{144}{x} = \dfrac{36}{55}$

82. $\dfrac{19}{2} = \dfrac{x}{19}$

83. $\dfrac{1}{9} = \dfrac{7}{x}$

84. $\dfrac{x}{72} = \dfrac{18}{5}$

85. $\dfrac{x}{5} = \dfrac{5}{x}$

86. $\dfrac{14}{x} = \dfrac{15}{x + 1}$

Solving a Percent Problem In Exercises 87–90, solve the percent problem.

87. What number is 24% of 600?

88. Fifty-five percent of what number is 110?

89. What percent of 176 is 66.88?

90. What percent of $16x$ is $9x$? (Assume x is nonzero.)

Simplifying a Radical Expression In Exercises 91–94, simplify the radical expression.

91. $\sqrt[3]{16x^5}$

92. $\sqrt[5]{\dfrac{x^8 z^4}{32}}$

93. $\sqrt[6]{x^3}$

94. $\sqrt[6]{(x + 1)^4}$

1.3 Modeling with Linear Equations

Linear equations can model many real-life situations. For example, in Exercise 55 on page 99, you will use a linear model to determine how many gallons of gasoline to add to a gasoline-oil mixture to bring the mixture to the desired concentration for a chainsaw engine.

❯ Write and use mathematical models to solve real-life problems.
❯ Solve mixture problems.
❯ Use common formulas to solve real-life problems.

Using Mathematical Models

In this section, you will use algebra to solve problems that occur in real-life situations. The process of translating phrases or sentences into algebraic expressions or equations is called **mathematical modeling.**

A good approach to mathematical modeling is to use two stages. Begin by using the verbal description of the problem to form a *verbal model.* Then, after assigning labels to the quantities in the verbal model, form a *mathematical model* or *algebraic equation.*

$$\text{Verbal description} \Rightarrow \text{Verbal model} \Rightarrow \text{Assign labels} \Rightarrow \text{Algebraic equation}$$

When you are constructing a verbal model, it is helpful to look for a *hidden equality—* a statement that two algebraic expressions are equal.

EXAMPLE 1   Using a Verbal Model

You accept a job with an annual income of $32,300. This includes your salary and a year-end bonus of $500. You are paid twice a month. What is your gross pay (pay before taxes) for each paycheck?

Solution   There are 12 months in a year and you are paid twice a month, so you receive 24 paychecks during the year.

Verbal model: $\boxed{\text{Annual income}} = \boxed{24 \text{ paychecks}} \cdot \boxed{\text{Amount of each paycheck}} + \boxed{\text{Bonus}}$

Labels: Annual income = 32,300   (dollars)
        Amount of each paycheck = x (dollars)
        Bonus = 500   (dollars)

Equation: $32,300 = 24x + 500$

The algebraic equation for this problem is a linear equation in one variable x. Use the methods discussed in Section 1.2 to solve the equation, as shown below.

$32,300 = 24x + 500$ Write original equation.

$31,800 = 24x$ Subtract 500 from each side.

$1325 = x$ Divide each side by 24.

So, your gross pay for each paycheck is $1325.

✓ *Checkpoint* ▶ Audio-video solution in English & Spanish at LarsonPrecalculus.com

You accept a job with an annual income of $58,400. This includes your salary and a $1200 year-end bonus. You are paid weekly. What is your salary per pay period? ■

GO DIGITAL

© Olivkairishka/Shutterstock.com

A fundamental step in writing a mathematical model to represent a real-life problem is translating key words and phrases into algebraic expressions and equations. The table below gives several examples.

Translating Key Words and Phrases

Key Words and Phrases	Verbal Description	Algebraic Expression or Equation
Addition:		
Sum, plus, increased by, more than, total of	• The sum of 5 and x • Seven more than y	$5 + x$ or $x + 5$ $7 + y$ or $y + 7$
Subtraction:		
Difference, minus, less than, decreased by, subtracted from, reduced by	• The difference of 4 and b • Three less than z	$4 - b$ $z - 3$
Multiplication:		
Product, multiplied by, twice, times, percent of	• Two times x • Three percent of t	$2x$ $0.03t$
Division:		
Quotient, divided by, ratio, per	• The ratio of x to 8	$\dfrac{x}{8}$
Equality:		
Equals, equal to, is, are, was, will be, represents	• The sale price S is $10 less than the list price L.	$S = L - 10$

EXAMPLE 2   **Finding a Percent Raise**

You accept a job that pays $20 per hour. After a two-month probationary period, your hourly wage will increase to $21 per hour. What percent raise will you receive after the two-month period?

Solution

Verbal model: Raise = Percent · Old wage

Labels: Old wage = 20   (dollars per hour)

Raise = 21 − 20 = 1   (dollars per hour)

Percent = r (in decimal form)

Equation: $1 = r \cdot 20$ Write equation.

$\dfrac{1}{20} = r$ Divide each side by 20.

$0.05 = r$ Rewrite the fraction as a decimal.

You will receive a raise of 0.05 or 5%.

✓ *Checkpoint*  *Audio-video solution in English & Spanish at LarsonPrecalculus.com*

You buy stock at $15 per share. You sell the stock at $18 per share. What is the percent increase in the stock's value?

GO DIGITAL

EXAMPLE 3 **Finding a Percent of Annual Income**

Your family has an annual income of $57,000 and these monthly expenses: mortgage ($1100), car payment ($375), food ($900), utilities ($240), and credit cards ($220). What percent of your family's annual income does the total amount of the monthly expenses represent?

Solution The total amount of your family's monthly expenses is $2835. The total monthly expenses for 1 year are $34,020.

Verbal model:

$$\boxed{\text{Monthly expenses}} = \boxed{\text{Percent}} \cdot \boxed{\text{Income}}$$

Labels: Income = 57,000   (dollars)
          Monthly expenses = 34,020   (dollars)
          Percent = r   (in decimal form)

Equation: $34,020 = r \cdot 57,000$ Write equation.

$\dfrac{34,020}{57,000} = r$ Divide each side by 57,000.

$0.597 \approx r$ Use a calculator.

Your family's monthly expenses are approximately 0.597 or 59.7% of your family's annual income.

✓ *Checkpoint* ▶ Audio-video solution in English & Spanish at LarsonPrecalculus.com

Your family has annual loan payments equal to 28% of its annual income. During the year, the loan payments total $17,920. What is your family's annual income?

EXAMPLE 4 **Finding the Dimensions of a Room**

A rectangular kitchen is twice as long as it is wide, and its perimeter is 84 feet. Find the dimensions of the kitchen.

Solution For this problem, it helps to draw a diagram, as shown in Figure 1.11.

Verbal model:

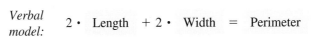

$$2 \cdot \boxed{\text{Length}} + 2 \cdot \boxed{\text{Width}} = \boxed{\text{Perimeter}}$$

Labels: Perimeter = 84   (feet)
          Width = w   (feet)
          Length = l = 2w   (feet)

Equation: $2(2w) + 2w = 84$ Write equation.

$6w = 84$ Combine like terms.

$w = 14$ Divide each side by 6.

The length is twice the width, so

$l = 2w$ Length is twice width.

$= 2(14)$ Substitute 14 for w.

$= 28.$ Simplify.

The dimensions of the kitchen are 14 feet by 28 feet.

✓ *Checkpoint* ▶ Audio-video solution in English & Spanish at LarsonPrecalculus.com

A rectangular family room is 3 times as long as it is wide, and its perimeter is 112 feet. Find the dimensions of the family room.

Figure 1.11

ALGEBRA HELP

Writing units for each of the labels in a real-life problem helps you determine the units for the answer. This is called *unit analysis*. When the same unit of measure occurs in the numerator and denominator of an expression, divide out the unit. For example, unit analysis verifies that time in the formula below is in hours.

$$\text{Time} = \frac{\text{distance}}{\text{rate}}$$

$$= \frac{\text{miles}}{\dfrac{\text{miles}}{\text{hour}}}$$

$$= \text{miles} \cdot \frac{\text{hours}}{\text{miles}}$$

$$= \text{hours}$$

EXAMPLE 5   Estimating Travel Time

A plane flies nonstop from Portland, Oregon, to Atlanta, Georgia, a distance of about 2170 miles. After 3 hours in the air, the plane flies over Topeka, Kansas (a distance of about 1440 miles from Portland). Assuming the plane flies at a constant speed, how long does the entire trip take?

Solution

Verbal model: Distance = Rate · Time

Labels: Distance = 2170   (miles)
Time = t (hours)
Rate = $\dfrac{\text{distance to Topeka}}{\text{time to Topeka}} = \dfrac{1440}{3}$ (miles per hour)

Equation:

$2170 = \dfrac{1440}{3}t$	Write equation.
$2170 = 480t$	Simplify.
$\dfrac{2170}{480} = t$	Divide each side by 480.
$4.52 \approx t$	Use a calculator.

The entire trip takes about 4.52 hours, or about 4 hours and 31 minutes.

✓ **Checkpoint** ▶ Audio-video solution in English & Spanish at LarsonPrecalculus.com

A boat travels at a constant speed to an island 14 miles away. It takes 0.5 hour to travel the first 5 miles. How long does the entire trip take?

EXAMPLE 6   Estimating the Height of a Building

To estimate the height of a building, you measure the shadow cast by the building and find it to be 142 feet long (see Figure 1.12). Then you measure the shadow cast by a four-foot post and find it to be 6 inches long. Estimate the building's height.

Solution   To solve this problem, use the result from geometry that the ratios of corresponding sides of similar triangles are equal.

Verbal model: $\dfrac{\text{Height of building}}{\text{Length of building's shadow}} = \dfrac{\text{Height of post}}{\text{Length of post's shadow}}$

Labels: Height of building = x (feet)
Length of building's shadow = 142   (feet)
Height of post = 4 feet = 48 inches   (inches)
Length of post's shadow = 6   (inches)

Equation: $\dfrac{x}{142} = \dfrac{48}{6}$ ⟹ $x = 1136$

So, the height of the building is 1136 feet.

✓ **Checkpoint** ▶ Audio-video solution in English & Spanish at LarsonPrecalculus.com

You measure the shadow cast by a building and find that it is 55 feet long. Then you measure the shadow cast by a nearby four-foot post and find that it is 1.8 feet long. Determine the building's height. ■

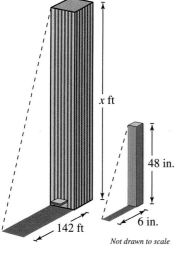

x ft

48 in.

142 ft   6 in.

Not drawn to scale

Figure 1.12

Mixture Problems

Problems that involve two or more rates are called **mixture problems.**

EXAMPLE 7　**A Simple Interest Problem**

You invested a total of $10,000 at $4\frac{1}{2}\%$ and $5\frac{1}{2}\%$ simple interest. During one year, the two accounts earned $508.75. How much did you invest in each account?

Solution　Let x represent the amount invested at $4\frac{1}{2}\%$. Then the amount invested at $5\frac{1}{2}\%$ is $10{,}000 - x$.

| *Verbal model:* | Interest from $4\frac{1}{2}\%$ $+$ Interest from $5\frac{1}{2}\%$ $=$ Total interest |

Labels:　Interest from $4\frac{1}{2}\% = Prt = (x)(0.045)(1)$ 　　　(dollars)
　　　　　 Interest from $5\frac{1}{2}\% = Prt = (10{,}000 - x)(0.055)(1)$　(dollars)
　　　　　 Total interest $= 508.75$ 　　　　　　　　　　　　　　(dollars)

Equation:　$0.045x + 0.055(10{,}000 - x) = 508.75$
　　　　　　　　　　　　　　$-0.01x = -41.25$
　　　　　　　　　　　　　　　　　$x = 4125$

So, you invested $4125 at $4\frac{1}{2}\%$ and $10{,}000 - x = \$5875$ at $5\frac{1}{2}\%$.

 ✓ *Checkpoint*　▶ *Audio-video solution in English & Spanish at LarsonPrecalculus.com*

You invested a total of $5000 at $2\frac{1}{2}\%$ and $3\frac{1}{2}\%$ simple interest. During one year, the two accounts earned $151.25. How much did you invest in each account?

EXAMPLE 8　**An Inventory Problem**

A store has $30,000 of inventory in 24-inch and 50-inch televisions. The profit on a 24-inch television is 22% and the profit on a 50-inch television is 40%. The profit on the entire stock is 35%. How much was invested in each type of television?

Solution　Let x represent the amount invested in 24-inch televisions. Then the amount invested in 50-inch televisions is $30{,}000 - x$.

| *Verbal model:* | Profit from 24-inch televisions $+$ Profit from 50-inch televisions $=$ Total profit |

Labels:　Inventory of 24-inch televisions $= x$ 　　　　　(dollars)
　　　　　 Inventory of 50-inch televisions $= 30{,}000 - x$　(dollars)
　　　　　 Profit from 24-inch televisions $= 0.22x$ 　　　(dollars)
　　　　　 Profit from 50-inch televisions $= 0.40(30{,}000 - x)$　(dollars)
　　　　　 Total profit $= 0.35(30{,}000) = 10{,}500$　　　　(dollars)

Equation:　$0.22x + 0.40(30{,}000 - x) = 10{,}500$

　　　　　　　　　　$-0.18x = -1500$　⟹　$x \approx 8333.33$

So, $8333.33 was invested in 24-inch televisions and $30{,}000 - x = \$21{,}666.67$ was invested in 50-inch televisions.

✓ *Checkpoint*　▶ *Audio-video solution in English & Spanish at LarsonPrecalculus.com*

In Example 8, the profit on a 24-inch television is 24% and the profit on a 50-inch television is 42%. The profit on the entire stock is 36%. How much was invested in each type of television? ◼

ALGEBRA HELP

Example 7 uses the simple interest formula $I = Prt$, where I is the interest, P is the principal (original deposit), r is the annual interest rate (in decimal form), and t is the time in years. Notice that in this example the amount invested, $10,000, is separated into two parts, x and $10{,}000 - x$.

GO DIGITAL

GO DIGITAL

Common Formulas

A **literal equation** is an equation that contains more than one variable. Many common types of geometric, scientific, and investment problems use ready-made literal equations, or **formulas.** Knowing these formulas will help you translate and solve a wide variety of real-life applications.

Common Formulas for Area *A*, Perimeter *P*, Circumference *C*, and Volume *V*

Square

$A = s^2$

$P = 4s$

Rectangle

$A = lw$

$P = 2l + 2w$

Circle

$A = \pi r^2$

$C = 2\pi r$

Triangle

$A = \dfrac{1}{2}bh$

$P = a + b + c$

Cube

$V = s^3$

Rectangular Solid

$V = lwh$

Circular Cylinder

$V = \pi r^2 h$

Sphere

$V = \dfrac{4}{3}\pi r^3$

Miscellaneous Common Formulas

Temperature:

$$F = \frac{9}{5}C + 32$$

$$C = \frac{5}{9}(F - 32)$$

F = degrees Fahrenheit, C = degrees Celsius

Simple Interest:

$$I = Prt$$

I = interest, P = principal (original deposit),
r = annual interest rate (in decimal form), t = time in years

Compound Interest:

$$A = P\left(1 + \frac{r}{n}\right)^{nt}$$

A = balance, P = principal (original deposit), r = annual interest rate (in decimal form),
n = compoundings (number of times interest is calculated) per year, t = time in years

Distance:

$$d = rt$$

d = distance traveled, r = rate, t = time

When solving an applied problem, you may find it helpful to rewrite a common formula. For example, the formula for the perimeter of a rectangle, $P = 2l + 2w$, can be solved for w as $w = \frac{1}{2}(P - 2l)$.

EXAMPLE 9 **Using a Formula**

▶▷▷▷ *See LarsonPrecalculus.com for an interactive version of this type of example.*

The cylindrical can shown below has a volume of 200 cubic centimeters and a radius of 4 centimeters. Find the height of the can.

Solution

The formula for the *volume of a cylinder* is $V = \pi r^2 h$. To find the height of the can, solve for h.

$$h = \frac{V}{\pi r^2}$$

Then, using $V = 200$ and $r = 4$, find the height.

$$h = \frac{200}{\pi(4)^2} \qquad \text{Substitute 200 for } V \text{ and 4 for } r.$$

$$= \frac{200}{16\pi} \qquad \text{Simplify denominator.}$$

$$\approx 3.98 \qquad \text{Use a calculator.}$$

Use unit analysis to determine the units of measure for the height. (Note that the constant π has no units.)

$$\text{Height} = \frac{\text{volume}}{(\text{radius})^2} = \frac{\text{centimeters}^3}{(\text{centimeters})^2} = \frac{\text{centimeters}^3}{(\text{centimeters})^2} = \text{centimeters}$$

So, the height of the can is about 3.98 centimeters.

Check

$$V = \pi r^2 h \approx \pi(4)^2(3.98) \approx 200 \text{ cubic centimeters}$$

✓ *Checkpoint* ▶ *Audio-video solution in English & Spanish at LarsonPrecalculus.com*

A cylindrical container has a volume of 84 cubic inches and a radius of 3 inches. Find the height of the container. ■

Summarize (Section 1.3)

1. Describe the process of mathematical modeling *(page 90)*. For examples of writing and using mathematical models, see Examples 1–6.

2. Explain how to solve a mixture problem *(page 94)*. For examples of solving mixture problems, see Examples 7 and 8.

3. State some common formulas used to solve real-life problems *(page 95)*. For an example that uses a volume formula, see Example 9.

1.3 Exercises

See CalcChat.com for tutorial help and worked-out solutions to odd-numbered exercises.

GO DIGITAL

Vocabulary and Concept Check

In Exercises 1 and 2, fill in the blanks.

1. The process of translating phrases or sentences into algebraic expressions or equations is called _____ _____.

2. A good approach to mathematical modeling is a two-stage approach, using a verbal description to form a _____ _____, and then, after assigning labels to the quantities, forming an _____ _____.

3. What is a hidden equality?

4. When writing a mathematical model to represent a real-life problem, what are some key words and phrases to look for to indicate addition, subtraction, multiplication, division, and equality?

Skills and Applications

Writing a Verbal Description In Exercises 5–12, write a verbal description of the algebraic expression without using the variable.

5. $y + 2$

6. $x - 8$

7. $\dfrac{t}{6}$

8. $\dfrac{1}{3}u$

9. $\dfrac{z - 2}{3}$

10. $\dfrac{x + 9}{5}$

11. $-2(d + 5)$

12. $10y(y - 3)$

Writing an Algebraic Expression In Exercises 13–22, write an algebraic expression for the verbal description.

13. The product of two consecutive odd integers, the first of which is $2n - 1$

14. The sum of the squares of two consecutive even integers, the first of which is $2n$

15. The distance a car travels in t hours at a rate of 55 miles per hour

16. The travel time for a plane traveling at a rate of r kilometers per hour for 900 kilometers

17. The amount of acid in x liters of a 20% acid solution

18. The sale price of an item with a 33% discount on its list price L

19. The perimeter of a rectangle with a width x and a length that is twice the width

20. The area of a triangle with a base that is 16 inches and a height that is h inches

21. The total cost of producing x units for which the fixed costs are $2500 and the cost per unit is $40

22. The total revenue obtained by selling x units at $12.99 per unit

Translating a Statement In Exercises 23–26, translate the statement into an algebraic equation.

23. The discount d is 30% of the list price L.

24. The amount A of water in q quarts of a liquid is 72% of the liquid.

25. The number N represents p percent of 672.

26. The sales for this month S_2 are 20% greater than the sales from last month S_1.

Writing an Expression In Exercises 27 and 28, write an expression for the area of the figure.

27.
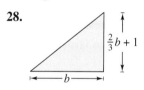

28.

Number Problems In Exercises 29–34, write a mathematical model for the problem and solve.

29. The sum of two consecutive natural numbers is 525. Find the numbers.

30. The sum of three consecutive natural numbers is 804. Find the numbers.

31. One positive number is 5 times another number. The difference between the numbers is 148. Find the numbers.

32. One positive number is $\frac{1}{5}$ of another number. The difference between the numbers is 76. Find the numbers.

33. Find two consecutive integers whose product is 5 less than the square of the smaller number.

34. Find two consecutive natural numbers such that the difference of their reciprocals is $\frac{1}{4}$ the reciprocal of the smaller number.

35. Finance A salesperson's weekly paycheck is 15% less than a second salesperson's paycheck. The two paychecks total $1125. Find the amount of each paycheck.

36. Discount The price of a train ticket after a 16.5% discount is $116.90. Find the original list price of the ticket.

37. Finance A family has annual loan payments equal to 32% of their annual income. During the year, the loan payments total $15,680. What is the family's annual income?

38. Finance A family has a monthly mortgage payment of $760, which is 16% of their monthly income. What is the family's monthly income?

39. Dimensions A rectangular room is 1.5 times as long as it is wide, and its perimeter is 25 meters.

(a) Draw a diagram that gives a visual representation of the problem. Let l represent the length and let w represent the width.

(b) Write l in terms of w and write an equation for the perimeter in terms of w.

(c) Find the dimensions of the room.

40. Dimensions A rectangular soccer field has a perimeter of 400 yards. The width of the field is $\frac{2}{3}$ times its length. Find the dimensions of the soccer field.

41. Course Grade To get an A in a course, you must have an average of at least 90% on four tests worth 100 points each. Your scores so far are 87, 92, and 84. What must you score on the fourth test to get an A in the course?

42. Course Grade You are taking a course that has four tests. The first three tests are worth 100 points each and the fourth test is worth 200 points. To get an A in the course, you must have an average of at least 90% on the four tests. Your scores so far are 87, 92, and 84. What must you score on the fourth test to get an A in the course?

43. Travel Time You are driving on a freeway to a town that is 500 kilometers from your home. After 30 minutes, you pass a freeway exit that you know is 50 kilometers from your home. Assuming that you continue at the same constant speed, how long does the entire trip take?

44. Average Speed A truck driver travels at an average speed of 55 miles per hour on a 200-mile trip to pick up a load of freight. On the return trip (with the truck fully loaded), the average speed is 40 miles per hour. What is the average speed for the round trip?

45. Physics Light travels at the speed of approximately 3.0×10^8 meters per second. Find the time in minutes required for light to travel from the sun to Earth (an approximate distance of 1.5×10^{11} meters).

46. Physics Radio waves travel at the same speed as light, approximately 3.0×10^8 meters per second. Find the time required for a radio wave to travel from Mission Control in Houston to NASA astronauts on the surface of the moon 3.84×10^8 meters away.

47. Height of a Building You measure the shadow cast by One Liberty Place in Philadelphia, Pennsylvania, and find that it is 105 feet long. Then you measure the shadow cast by a nearby three-foot post and find that it is 4 inches long. Determine the building's height.

48. Height of a Tree You measure a tree's shadow and find that it is 8 meters long. Then you measure the shadow of a nearby two-meter lamppost and find that it is 75 centimeters long. (See figure.) How tall is the tree?

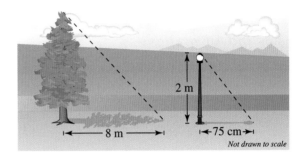

Not drawn to scale

49. Flagpole Height A person who is 6 feet tall walks away from a flagpole toward the tip of the shadow of the flagpole. When the person is 30 feet from the flagpole, the tips of the person's shadow and the shadow cast by the flagpole coincide at a point 5 feet in front of the person.

(a) Draw a diagram that gives a visual representation of the problem. Let h represent the height of the flagpole.

(b) Find the height of the flagpole.

50. Shadow Length A person who is 6 feet tall walks away from a 50-foot tower toward the tip of the tower's shadow. At a distance of 32 feet from the tower, the person's shadow begins to emerge beyond the tower's shadow. How much farther must the person walk to be completely out of the tower's shadow?

51. Simple Interest A business invests a total of $12,000 at $4\frac{1}{2}$% and 5% simple interest. During one year, the two accounts earn $580. How much did the business invest in each account?

52. Simple Interest A business invests a total of $25,000 at 3% and $4\frac{1}{2}$% simple interest. During one year, the two accounts earn $900. How much did the business invest in each account?

53. Inventory A nursery has $40,000 of inventory in dogwood trees and red maple trees. The profit on a dogwood tree is 25% and the profit on a red maple tree is 17%. The profit for the entire inventory is 20%. How much was invested in each type of tree?

54. Inventory An automobile dealer has $600,000 of inventory in all-electric and hybrid vehicles. The profit on an all-electric vehicle is 24% and the profit on a hybrid vehicle is 28%. The profit for the entire stock is 25%. How much was invested in each type of vehicle?

55. Mixture Problem

A forester is making a gasoline-oil mixture for a chainsaw engine. The forester has 2 gallons of a mixture that is 32 parts gasoline and 1 part oil. How many gallons of gasoline should the forester add to bring the mixture to 50 parts gasoline and 1 part oil?

56. Mixture Problem A grocer mixes peanuts that cost $1.49 per pound and walnuts that cost $2.69 per pound to make 100 pounds of a mixture that costs $2.21 per pound. How much of each nut is in the mixture?

57. Area of a Triangle Solve for h: $A = \frac{1}{2}bh$.

58. Volume of a Rectangular Prism Solve for l: $V = lwh$.

59. Markup Solve for C: $S = C + RC$.

60. Discount Solve for L: $S = L - RL$.

61. Investment at Simple Interest Solve for r: $A = P + Prt$.

62. Area of a Trapezoid Solve for b: $A = \frac{1}{2}(a + b)h$.

63. Body Temperature The average body temperature of a person is 98.6°F. What is this temperature in degrees Celsius?

64. Chemistry The melting point of francium is 27°C. What is this temperature in degrees Fahrenheit?

65. Volume of a Billiard Ball A billiard ball has a volume of 5.96 cubic inches. Find the radius of the billiard ball.

66. Length of a Tank The diameter of a cylindrical propane gas tank is 4 feet. The total volume of the tank is 603.2 cubic feet. Find the length of the tank.

Exploring the Concepts

True or False? In Exercises 67 and 68, determine whether the statement is true or false. Justify your answer.

67. The expression $x^3/(x - 4)^2$ can be described as "x cubed divided by the square of the difference of x and 4."

68. The area of a circle with a radius of 2 inches is less than the area of a square with a side length of 4 inches.

69. Writing Give two interpretations of "the quotient of 5 and a number times 3." Explain why $(3n)/5$ is not a possible interpretation.

70. HOW DO YOU SEE IT? To determine a building's height, you measure the shadows cast by the building and a nearby four-foot post (see figure).

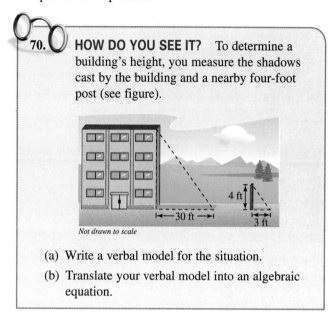

Not drawn to scale

(a) Write a verbal model for the situation.

(b) Translate your verbal model into an algebraic equation.

Error Analysis In Exercises 71 and 72, describe the error.

71. $d = rt \implies t = r/d$ ✗

72. $P = 2l + 2w \implies w = (P - 2l)/2 = P - l$ ✗

Review & Refresh ▶ *Video solutions at LarsonPrecalculus.com*

Finding a Product In Exercises 73–76, find the product.

73. $(x + \sqrt{3})(x - \sqrt{3})$ **74.** $(x + 3\sqrt{2})(x - 3\sqrt{2})$

75. $(x - 3 + \sqrt{7})(x - 3 - \sqrt{7})$

76. $(x + \sqrt{2} + 2)(x + \sqrt{2} - 2)$

Factoring Completely In Exercises 77–82, completely factor the expression.

77. $4x^2 + 4x + 1$ **78.** $x^2 - 22x + 121$

79. $u^3 + 27v^3$ **80.** $(x + 2)^3 - y^3$

81. $2x^2 + 9x + 4$ **82.** $2x^2 - 3x - 5$

Simplifying a Radical Expression In Exercises 83–86, simplify the radical expression.

83. $\dfrac{-3 + \sqrt{3^2 - 4(-9)}}{2}$ **84.** $\dfrac{-2 - \sqrt{2^2 - 4(3)(-10)}}{2(3)}$

85. $\left[3(2 - \sqrt{2}) - 6\right]^2 - 18$

86. $(-1 + \sqrt{7})^2 + 2(-1 + \sqrt{7}) - 6$

Decimal Form In Exercises 87–90, write the number in decimal form.

87. 9.46×10^{12} **88.** 9.02×10^{-6}

89. -3.75×10^{-4} **90.** 1.83×10^8

1.4 Quadratic Equations and Applications

❯ Solve quadratic equations by factoring.
❯ Solve quadratic equations by extracting square roots.
❯ Solve quadratic equations by completing the square.
❯ Use the Quadratic Formula to solve quadratic equations.
❯ Use quadratic equations to model and solve real-life problems.

Solving Quadratic Equations by Factoring

A **quadratic equation** in x is an equation that can be written in the general form

$$ax^2 + bx + c = 0 \qquad \text{General form}$$

where a, b, and c are real numbers with $a \neq 0$. A quadratic equation in x is also called a **second-degree polynomial equation** in x.

In this section, you will study four methods for solving quadratic equations: *factoring, extracting square roots, completing the square*, and the *Quadratic Formula*. The method of factoring is based on the Zero-Factor Property from Section P.1.

$$\text{If } ab = 0, \text{ then } a = 0 \text{ or } b = 0. \qquad \text{Zero-Factor Property}$$

To use this property to solve a quadratic equation in general form, first rewrite the left side of the equation as the product of two linear factors. Then set each linear factor equal to zero and solve, as shown in Example 1.

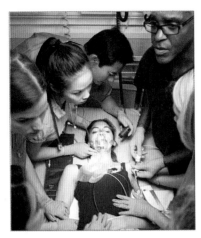

Quadratic equations have many real-life applications. For example, in Exercise 113 on page 112, you will use a quadratic equation to analyze a patient's blood oxygen level.

❯❯ **SKILLS REFRESHER**

For a refresher on factoring quadratic polynomials, watch the video at *LarsonPrecalculus.com*.

EXAMPLE 1 **Solving Quadratic Equations by Factoring**

a. $2x^2 + 9x + 7 = 3$ Original equation

$\quad\quad 2x^2 + 9x + 4 = 0$ Write in general form.

$\quad (2x + 1)(x + 4) = 0$ Factor.

$\quad\quad\quad 2x + 1 = 0 \implies x = -\frac{1}{2}$ Set 1st factor equal to 0.

$\quad\quad\quad\quad x + 4 = 0 \implies x = -4$ Set 2nd factor equal to 0.

The solutions are $x = -\frac{1}{2}$ and $x = -4$. Check these in the original equation.

b. $6x^2 - 3x = 0$ Original equation

$\quad 3x(2x - 1) = 0$ Factor.

$\quad\quad\quad 3x = 0 \implies x = 0$ Set 1st factor equal to 0.

$\quad\quad 2x - 1 = 0 \implies x = \frac{1}{2}$ Set 2nd factor equal to 0.

The solutions are $x = 0$ and $x = \frac{1}{2}$. Check these in the original equation.

✓ *Checkpoint* ▶ *Audio-video solution in English & Spanish at LarsonPrecalculus.com*

Solve $2x^2 - 3x + 1 = 6$ by factoring. ■

The Zero-Factor Property applies *only* to equations written in general form (in which the right side of the equation is zero). So, collect all terms on one side *before* factoring. For example, in the equation $(x - 5)(x + 2) = 8$, it is *incorrect* to set each factor equal to 8. To solve this equation, first multiply the binomials on the left side of the equation. Then subtract 8 from each side. After simplifying the equation, factor the left side and use the Zero-Factor Property to find the solutions. Solve this equation correctly on your own. Then check the solutions in the original equation.

Extracting Square Roots

SKILLS REFRESHER

For a refresher on how to simplify radical expressions with real numbers, watch the video at *LarsonPrecalculus.com*.

Consider a quadratic equation of the form $u^2 = d$, where $d > 0$ and u is an algebraic expression. Factoring shows that this equation has two solutions.

$u^2 = d$	Write original equation.
$u^2 - d = 0$	Write in general form.
$(u + \sqrt{d})(u - \sqrt{d}) = 0$	Factor.
$u + \sqrt{d} = 0 \implies u = -\sqrt{d}$	Set 1st factor equal to 0.
$u - \sqrt{d} = 0 \implies u = \sqrt{d}$	Set 2nd factor equal to 0.

The two solutions differ only in sign, so you can write the solutions together, using a "plus or minus sign," as $u = \pm\sqrt{d}$. This form of the solution is read as "u is equal to plus or minus the square root of d." Solving an equation of the form $u^2 = d$ without going through the steps of factoring is called **extracting square roots.**

Extracting Square Roots

The equation $u^2 = d$, where $d > 0$, has exactly two solutions:

$$u = \sqrt{d} \quad \text{and} \quad u = -\sqrt{d}.$$

These solutions can also be written as $u = \pm\sqrt{d}$.

EXAMPLE 2   **Extracting Square Roots**

a.

$4x^2 = 12$	Original equation
$x^2 = 3$	Divide each side by 4.
$x = \pm\sqrt{3}$	Extract square roots.

The solutions are $x = \sqrt{3}$ and $x = -\sqrt{3}$. Check these in the original equation.

b.

$(x - 3)^2 = 7$	Original equation
$x - 3 = \pm\sqrt{7}$	Extract square roots.
$x = 3 \pm \sqrt{7}$	Add 3 to each side.

The solutions are $x = 3 \pm \sqrt{7}$. Check these in the original equation.

c.

$(3x - 6)^2 - 18 = 0$	Original equation
$(3x - 6)^2 = 18$	Add 18 to each side.
$3x - 6 = \pm 3\sqrt{2}$	Extract square roots.
$3x = 6 \pm 3\sqrt{2}$	Add 6 to each side.
$x = 2 \pm \sqrt{2}$	Divide each side by 3.

ALGEBRA HELP

When extracting square roots in Example 2(c), note that $\pm\sqrt{18} = \pm\sqrt{3^2 \cdot 2} = \pm 3\sqrt{2}$.

The solutions are $x = 2 \pm \sqrt{2}$. Check these in the original equation.

✓ *Checkpoint* Audio-video solution in English & Spanish at LarsonPrecalculus.com

Solve each equation by extracting square roots.

a. $3x^2 = 36$

b. $(x - 1)^2 = 10$

Completing the Square

The equation $(x - 3)^2 = 7$ in Example 2(b) was given in the form $u^2 = d$ so that you could find the solutions by extracting square roots. Suppose you were given the equation $(x - 3)^2 = 7$ in its general form,

$$x^2 - 6x + 2 = 0. \qquad \text{General form}$$

How would you solve this form of the equation? You could try factoring, but the left side of the equation $x^2 - 6x + 2 = 0$ is not factorable using integer coefficients. One way to overcome this problem is to rewrite the equation by **completing the square,** and then solve the rewritten equation by extracting square roots.

> ### Completing the Square
>
> To **complete the square** for the expression $x^2 + bx$, add $(b/2)^2$, which is the square of half the coefficient of x. Consequently,
>
> $$x^2 + bx + \left(\frac{b}{2}\right)^2 = \left(x + \frac{b}{2}\right)^2.$$

When solving quadratic equations by completing the square, you must add $(b/2)^2$ to *each side* to maintain equality.

EXAMPLE 3 Completing the Square: Leading Coefficient Is 1

Solve $x^2 + 2x - 6 = 0$ by completing the square.

Solution

$$x^2 + 2x - 6 = 0 \qquad \text{Write original equation.}$$
$$x^2 + 2x = 6 \qquad \text{Add 6 to each side.}$$
$$x^2 + 2x + 1^2 = 6 + 1^2 \qquad \text{Add } 1^2 \text{ to each side.}$$

(Half of 2)^2

$$(x + 1)^2 = 7 \qquad \text{Simplify.}$$
$$x + 1 = \pm\sqrt{7} \qquad \text{Extract square roots.}$$
$$x = -1 \pm \sqrt{7} \qquad \text{Subtract 1 from each side.}$$

The solutions are $x = -1 \pm \sqrt{7}$. Check the first solution, $x = -1 + \sqrt{7}$, as shown below.

Check $x = -1 + \sqrt{7}$

$$x^2 + 2x - 6 = 0 \qquad \text{Write original equation.}$$
$$\left(-1 + \sqrt{7}\right)^2 + 2\left(-1 + \sqrt{7}\right) - 6 \stackrel{?}{=} 0 \qquad \text{Substitute } -1 + \sqrt{7} \text{ for } x.$$
$$8 - 2\sqrt{7} - 2 + 2\sqrt{7} - 6 \stackrel{?}{=} 0 \qquad \text{Multiply.}$$
$$8 - 2 - 6 = 0 \qquad \text{Solution checks. } \checkmark$$

Check $x = -1 - \sqrt{7}$ on your own.

✓ **Checkpoint** ▶ Audio-video solution in English & Spanish at LarsonPrecalculus.com

Solve $x^2 - 4x - 1 = 0$ by completing the square. ■

When the leading coefficient of a quadratic equation is *not* 1, divide each side of the equation by the leading coefficient *before* completing the square. This process is shown in Examples 4 and 5 on the next page.

ALGEBRA HELP

Note that when you complete the square to solve a quadratic equation, you are rewriting the equation so it can be solved by extracting square roots, as shown in Example 3.

EXAMPLE 4   **Completing the Square: Leading Coefficient Is Not 1**

Solve $2x^2 + 8x + 3 = 0$ by completing the square.

Solution   Note that the leading coefficient is 2.

$$2x^2 + 8x + 3 = 0 \qquad \text{Write original equation.}$$

$$2x^2 + 8x = -3 \qquad \text{Subtract 3 from each side.}$$

$$x^2 + 4x = -\frac{3}{2} \qquad \text{Divide each side by 2.}$$

$$x^2 + 4x + 2^2 = -\frac{3}{2} + 2^2 \qquad \text{Add } 2^2 \text{ to each side.}$$

$$\underbrace{\qquad}_{\text{(Half of 4)}^2}$$

$$(x + 2)^2 = \frac{5}{2} \qquad \text{Simplify.}$$

$$x + 2 = \pm\sqrt{\frac{5}{2}} \qquad \text{Extract square roots.}$$

$$x + 2 = \pm\frac{\sqrt{10}}{2} \qquad \text{Rationalize denominator.}$$

$$x = -2 \pm \frac{\sqrt{10}}{2} \qquad \text{Subtract 2 from each side.}$$

Check these in the original equation.

 Checkpoint ▶ *Audio-video solution in English & Spanish at LarsonPrecalculus.com*

Solve $2x^2 - 4x + 1 = 0$ by completing the square.

ALGEBRA HELP

When rationalizing the denominator of the radical in Example 4, note that

$$\sqrt{\frac{5}{2}} = \frac{\sqrt{5}}{\sqrt{2}} = \frac{\sqrt{5}}{\sqrt{2}} \cdot \frac{\sqrt{2}}{\sqrt{2}}$$

$$= \frac{\sqrt{10}}{2}.$$

EXAMPLE 5   **Completing the Square: Leading Coefficient Is Not 1**

Solve $3x^2 - 4x - 5 = 0$ by completing the square.

Solution   Note that the leading coefficient is 3.

$$3x^2 - 4x - 5 = 0 \qquad \text{Write original equation.}$$

$$3x^2 - 4x = 5 \qquad \text{Add 5 to each side.}$$

$$x^2 - \frac{4}{3}x = \frac{5}{3} \qquad \text{Divide each side by 3.}$$

$$x^2 - \frac{4}{3}x + \left(-\frac{2}{3}\right)^2 = \frac{5}{3} + \left(-\frac{2}{3}\right)^2 \qquad \text{Add } \left(-\frac{2}{3}\right)^2 \text{ to each side.}$$

$$\left(x - \frac{2}{3}\right)^2 = \frac{19}{9} \qquad \text{Simplify.}$$

$$x - \frac{2}{3} = \pm\frac{\sqrt{19}}{3} \qquad \text{Extract square roots.}$$

$$x = \frac{2}{3} \pm \frac{\sqrt{19}}{3} \qquad \text{Add } \tfrac{2}{3} \text{ to each side.}$$

Check these in the original equation.

 Checkpoint ▶ *Audio-video solution in English & Spanish at LarsonPrecalculus.com*

Solve $3x^2 - 10x - 2 = 0$ by completing the square. ■

The Quadratic Formula

Often in mathematics you learn the long way of solving a problem first. Then, you use the longer method to develop shorter techniques. The long way stresses understanding and the short way stresses efficiency.

For example, completing the square is a "long way" of solving a quadratic equation. When you use completing the square to solve quadratic equations, you must complete the square for *each* equation separately. In the derivation below, you complete the square *once* for the general form of a quadratic equation

$$ax^2 + bx + c = 0$$

to obtain the **Quadratic Formula**—a shortcut for solving quadratic equations.

$$ax^2 + bx + c = 0 \qquad \text{General form, } a \neq 0$$

$$ax^2 + bx = -c \qquad \text{Subtract } c \text{ from each side.}$$

$$x^2 + \frac{b}{a}x = -\frac{c}{a} \qquad \text{Divide each side by } a.$$

$$x^2 + \frac{b}{a}x + \left(\frac{b}{2a}\right)^2 = -\frac{c}{a} + \left(\frac{b}{2a}\right)^2 \qquad \text{Add } \left(\frac{b}{2a}\right)^2 \text{ to each side.}$$

$$\left(\text{Half of } \frac{b}{a}\right)^2$$

$$\left(x + \frac{b}{2a}\right)^2 = \frac{b^2 - 4ac}{4a^2} \qquad \text{Simplify.}$$

$$x + \frac{b}{2a} = \pm\sqrt{\frac{b^2 - 4ac}{4a^2}} \qquad \text{Extract square roots.}$$

$$x = -\frac{b}{2a} \pm \frac{\sqrt{b^2 - 4ac}}{2|a|} \qquad \text{Subtract } \frac{b}{2a} \text{ from each side.}$$

Note that $\pm 2|a|$ represents the same numbers as $\pm 2a$, so the formula simplifies to

$$x = \frac{-b \pm \sqrt{b^2 - 4ac}}{2a}. \qquad \begin{array}{l}\text{Solutions of a quadratic equation}\\\text{given in general form}\end{array}$$

The Quadratic Formula

The solutions of a quadratic equation in the general form

$$ax^2 + bx + c = 0, \quad a \neq 0$$

are given by the **Quadratic Formula**

$$x = \frac{-b \pm \sqrt{b^2 - 4ac}}{2a}.$$

The Quadratic Formula is one of the most important formulas in algebra. It is possible to solve every quadratic equation by completing the square or using the Quadratic Formula. To help you to remember this important formula, learn a verbal statement of the rule, such as

"Negative b, plus or minus the square root of b squared minus $4ac$, all divided by $2a$."

In the Quadratic Formula, the quantity under the radical sign, $b^2 - 4ac$, is the **discriminant** of the quadratic equation $ax^2 + bx + c = 0$. It can be used to determine the number of real solutions of a quadratic equation.

GO DIGITAL

Solutions of a Quadratic Equation

The solutions of a quadratic equation

$$ax^2 + bx + c = 0, \ a \neq 0$$

can be classified in three ways.

1. If the discriminant $b^2 - 4ac$ is *positive*, then the quadratic equation has *two* distinct real solutions and its graph has *two* x-intercepts.

2. If the discriminant $b^2 - 4ac$ is *zero*, then the quadratic equation has *one* repeated real solution and its graph has *one* x-intercept.

3. If the discriminant $b^2 - 4ac$ is *negative*, then the quadratic equation has *no* real solutions and its graph has *no* x-intercepts.

If the discriminant of a quadratic equation is negative, then its square root is not a real number and the Quadratic Formula yields two complex solutions. You will study complex solutions in Section 1.5.

When you use the Quadratic Formula, remember that *before* applying the formula, you must first write the quadratic equation in general form.

EXAMPLE 6 **Using the Quadratic Formula**

▶▶▶ *See LarsonPrecalculus.com for an interactive version of this type of example.*

Use the Quadratic Formula to solve $x^2 + 3x = 9$.

Solution Before using the Quadratic Formula, write the quadratic equation in general form.

$$x^2 + 3x = 9 \qquad \text{Write original equation.}$$
$$x^2 + 3x - 9 = 0 \qquad \text{Write in general form.}$$

From the general form, $a = 1$, $b = 3$, and $c = -9$. The discriminant is

$$b^2 - 4ac = 3^2 - 4(1)(-9) = 9 + 36 = 45.$$

Because the discriminant is positive, the equation has two real solutions. Now, use the Quadratic Formula to find the solutions.

$$x^2 + 3x - 9 = 0 \qquad \text{Write in general form.}$$
$$x = \frac{-b \pm \sqrt{b^2 - 4ac}}{2a} \qquad \text{Quadratic Formula}$$
$$x = \frac{-3 \pm \sqrt{(3)^2 - 4(1)(-9)}}{2(1)} \qquad \text{Substitute 1 for } a, \text{ 3 for } b, \text{ and } -9 \text{ for } c.$$
$$x = \frac{-3 \pm \sqrt{45}}{2} \qquad \text{Simplify.}$$
$$x = \frac{-3 \pm 3\sqrt{5}}{2} \qquad \text{Simplify.}$$

The two solutions are

$$x = \frac{-3 + 3\sqrt{5}}{2} \quad \text{and} \quad x = \frac{-3 - 3\sqrt{5}}{2}.$$

Check these in the original equation.

✓ *Checkpoint* ▶ *Audio-video solution in English & Spanish at LarsonPrecalculus.com*

Use the Quadratic Formula to solve $3x^2 + 2x = 10$.

▶▶▶ **TECHNOLOGY**

To use a graphing utility to check the real solutions of a quadratic equation, begin by writing the equation in general form. Then set y equal to the left side and graph the resulting equation. The x-intercepts of the graph represent the real solutions of the original equation. Use the *zero* or *root* feature of the graphing utility to approximate the x-intercepts of the graph. For instance, to check the solutions to Example 6, graph $y = x^2 + 3x - 9$, and find that the x-intercepts are about $(1.85, 0)$ and $(-4.85, 0)$, as shown in the figure below. These x-intercepts are the decimal approximations of the solutions found in Example 6.

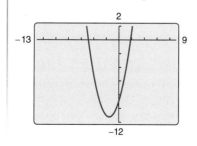

GO DIGITAL

Applications

Quadratic equations are often used in problems dealing with area. Here is a relatively simple example.

A square room has an area of 144 square feet. Find the dimensions of the room.

To solve this problem, let x represent the length of each side of the room. Then use the formula for the area of a square to write and solve the equation

$$x^2 = 144$$

and conclude that each side of the room is 12 feet long. Note that although the equation $x^2 = 144$ has two solutions, $x = -12$ and $x = 12$, the negative solution does not make sense in the context of the problem, so choose the positive solution.

EXAMPLE 7 **Finding the Dimensions of a Room**

A rectangular sunroom is 3 feet longer than it is wide (see figure) and has an area of 154 square feet. Find the dimensions of the room.

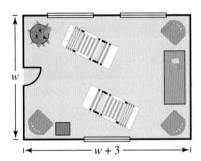

Solution

Verbal model: $\quad \dfrac{\text{Width}}{\text{of room}} \cdot \dfrac{\text{Length}}{\text{of room}} = \dfrac{\text{Area}}{\text{of room}}$

Labels: Width of room $= w$ (feet)
         Length of room $= w + 3$ (feet)
         Area of room $= 154$ (square feet)

Equation:
$$w(w + 3) = 154$$
$$w^2 + 3w - 154 = 0 \qquad \text{Write in general form.}$$
$$(w - 11)(w + 14) = 0 \qquad \text{Factor.}$$
$$w - 11 = 0 \implies w = 11 \qquad \text{Set 1st factor equal to 0.}$$
$$w + 14 = 0 \implies w = -14 \qquad \text{Set 2nd factor equal to 0.}$$

Choosing the positive value, the width is 11 feet and the length is $11 + 3 = 14$ feet.

Check

The length of 14 feet is 3 more than the width of 11 feet. ✓
The area of the sunroom is $11(14) = 154$ square feet. ✓

✓ *Checkpoint* ▶ *Audio-video solution in English & Spanish at LarsonPrecalculus.com*

A rectangular kitchen is 6 feet longer than it is wide and has an area of 112 square feet. Find the dimensions of the kitchen. ∎

GO DIGITAL

Another common application of quadratic equations involves an object that is falling (or vertically projected into the air). The general equation that gives the height of such an object is a **position equation,** and on Earth's surface it has the form

$$s = -16t^2 + v_0t + s_0.$$

In this equation, s represents the height of the object (in feet), v_0 represents the initial velocity of the object (in feet per second), s_0 represents the initial height of the object (in feet), and t represents the time (in seconds).

ALGEBRA HELP

The position equation described here ignores air resistance.

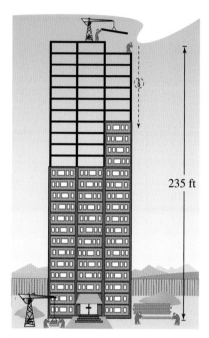

Figure 1.13

EXAMPLE 8   A Falling Object

A construction worker accidentally drops a wrench from a height of 235 feet (see Figure 1.13) and immediately yells "Look out below!" Could a person at ground level hear this warning in time to get out of the way? (*Note:* The speed of sound is about 1100 feet per second.)

Solution

You are given that the distance to the ground is 235 feet, and that the speed of sound is about 1100 feet per second. So, a person at ground level hears the warning in about

$$\frac{235}{1100} \approx 0.21 \text{ second.}$$

To set up a mathematical model for the height of the wrench, use the position equation

$$s = -16t^2 + v_0t + s_0.$$

The object is dropped rather than thrown, so the initial velocity is $v_0 = 0$ feet per second. Moreover, the initial height is $s_0 = 235$ feet, so you have the model

$$s = -16t^2 + (0)t + 235 = -16t^2 + 235.$$

After 1 second, the wrench's height is

$$-16(1)^2 + 235 = 219 \text{ feet.}$$

After 2 seconds, the wrench's height is

$$-16(2)^2 + 235 = 171 \text{ feet.}$$

When the wrench hits the ground, its height is 0 feet. So, to find the number of seconds it takes the wrench to hit the ground, let the height s be zero and solve the equation for t.

$s = -16t^2 + 235$	Write position equation.
$0 = -16t^2 + 235$	Substitute 0 for height.
$16t^2 = 235$	Add $16t^2$ to each side.
$t^2 = \dfrac{235}{16}$	Divide each side by 16.
$t = \dfrac{\sqrt{235}}{4}$	Extract positive square root.
$t \approx 3.83$	Use a calculator.

The wrench will take about 3.83 seconds to hit the ground. So, a person who hears the warning 0.21 second after the wrench is dropped has more than 3 seconds to get out of the way.

✓ *Checkpoint* 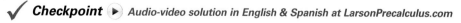 ▶ *Audio-video solution in English & Spanish at LarsonPrecalculus.com*

You drop a rock from a height of 196 feet. How long does it take the rock to hit the ground? ∎

GO DIGITAL

A third application of quadratic equations is modeling change over time.

EXAMPLE 9   **Quadratic Modeling: Social Media**

From 2012 through 2019, the number of monthly active Facebook users F (in billions) worldwide can be approximated by the quadratic equation

$$F = 0.00429t^2 + 0.0802t - 0.588, \quad 12 \le t \le 19$$

where t represents the year, with $t = 12$ corresponding to 2012. According to the model, in which year did the number of users reach 1.8 billion?   *(Source: Facebook, Inc.)*

Algebraic Solution

To find the year in which the number of users reached 1.8 billion, solve the equation

$$0.00429t^2 + 0.0802t - 0.588 = 1.8.$$

To begin, write the equation in general form.

$$0.00429t^2 + 0.0802t - 2.388 = 0$$

Then apply the Quadratic Formula.

$$t = \frac{-b \pm \sqrt{b^2 - 4ac}}{2a}$$

$$= \frac{-0.0802 \pm \sqrt{(0.0802)^2 - 4(0.00429)(-2.388)}}{2(0.00429)}$$

$$= \frac{-0.0802 \pm \sqrt{0.04741012}}{0.00858}$$

$$\approx 16 \text{ or } -35$$

Choose $t \approx 16$ because it is in the domain of F. Because $t = 12$ corresponds to 2012, it follows that $t \approx 16$ must correspond to 2016. This means that the number of users reached 1.8 billion during the year 2016.

Numerical Solution

Use a table to estimate the year in which the number of users reached 1.8 billion.

Year	t	F
2012	12	1.0
2013	13	1.2
2014	14	1.4
2015	15	1.6
2016	16	1.8
2017	17	2.0
2018	18	2.2
2019	19	2.5

From the table, you can estimate that the number of users reached 1.8 billion during the year 2016.

 Checkpoint ▶ *Audio-video solution in English & Spanish at LarsonPrecalculus.com*

According to the model in Example 9, in which year did the number of users reach 2.25 billion?

⟫⟫ TECHNOLOGY

You can also use a graphical approach to solve Example 9. Use a graphing utility to graph

$$y_1 = 0.00429t^2 + 0.0802t - 0.588 \quad \text{and} \quad y_2 = 1.8$$

in the same viewing window. Then use the *intersect* feature to find that the graphs intersect when $t \approx 16$ and when $t \approx -35$. Choose $t \approx 16$ because it is in the domain of F.

A fourth application of quadratic equations involves the Pythagorean Theorem. Recall from Section P.6 that the theorem states that

$$a^2 + b^2 = c^2 \qquad \text{Pythagorean Theorem}$$

where a and b are the lengths of the legs of a right triangle and c is the length of the hypotenuse.

GO DIGITAL

EXAMPLE 10 **An Application Involving the Pythagorean Theorem**

The figure below shows an L-shaped sidewalk from the athletic center to the library on a college campus. The length of one sidewalk forming the L is twice as long as the other. The length of the diagonal sidewalk that cuts across the grounds between the two buildings is 102 feet. How many feet does a person save by walking on the diagonal sidewalk?

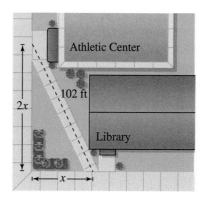

Solution Use the Pythagorean Theorem.

$a^2 + b^2 = c^2$	Pythagorean Theorem
$x^2 + (2x)^2 = 102^2$	Substitute for a, b, and c.
$5x^2 = 10{,}404$	Simplify.
$x^2 = 2080.8$	Divide each side by 5.
$x = \sqrt{2080.8}$	Extract positive square root.

The total distance covered by walking on the L-shaped sidewalk is

$$x + 2x = 3x = 3\sqrt{2080.8} \approx 136.85 \text{ feet.}$$

Walking on the diagonal sidewalk saves a person about $136.85 - 102 = 34.85$ feet.

✓ **Checkpoint** ▶ *Audio-video solution in English & Spanish at LarsonPrecalculus.com*

In Example 10, how many feet does a person save by walking on the diagonal sidewalk when the length of one sidewalk forming the L is three times as long as the other? ∎

Summarize (Section 1.4)

1. Explain how to solve a quadratic equation by factoring *(page 100)*. For an example of solving quadratic equations by factoring, see Example 1.

2. Explain how to solve a quadratic equation by extracting square roots *(page 101)*. For an example of solving quadratic equations by extracting square roots, see Example 2.

3. Explain how to solve a quadratic equation by completing the square *(page 102)*. For examples of solving quadratic equations by completing the square, see Examples 3–5.

4. Explain how to solve a quadratic equation using the Quadratic Formula *(page 104)*. For an example of solving a quadratic equation using the Quadratic Formula, see Example 6.

5. Describe real-life applications of quadratic equations *(pages 106–109, Examples 7–10)*.

1.4   Exercises

See CalcChat.com for tutorial help and worked-out solutions to odd-numbered exercises.

GO DIGITAL

Vocabulary and Concept Check

In Exercises 1–4, fill in the blanks.

1. A _____ _____ in x is an equation that can be written in the general form $ax^2 + bx + c = 0$, where a, b, and c are real numbers with $a \neq 0$.

2. A quadratic equation in x is also called a _____ _____ equation in x.

3. The part of the Quadratic Formula, $b^2 - 4ac$, known as the _____, determines the number of real solutions of a quadratic equation.

4. An important theorem that is sometimes used in applications that require solving quadratic equations is the _____ _____.

5. List four methods that can be used to solve a quadratic equation.

6. What does the equation $s = -16t^2 + v_0 t + s_0$ represent? What do v_0 and s_0 represent?

Skills and Applications

Solving a Quadratic Equation by Factoring In Exercises 7–18, solve the quadratic equation by factoring.

7. $6x^2 + 3x = 0$
8. $8x^2 - 2x = 0$
9. $3 + 5x - 2x^2 = 0$
10. $x^2 + 6x + 9 = 0$
11. $x^2 + 10x + 25 = 0$
12. $4x^2 + 12x + 9 = 0$
13. $16x^2 - 9 = 0$
14. $x^2 - 2x - 8 = 0$
15. $2x^2 = 19x + 33$
16. $-x^2 + 4x = 3$
17. $\frac{3}{4}x^2 + 8x + 20 = 0$
18. $\frac{1}{8}x^2 - x - 16 = 0$

Extracting Square Roots In Exercises 19–32, solve the equation by extracting square roots. When a solution is irrational, list both the exact solution *and* its approximation rounded to two decimal places.

19. $x^2 = 49$
20. $x^2 = 144$
21. $x^2 = 19$
22. $x^2 = 43$
23. $3x^2 = 81$
24. $9x^2 = 36$
25. $(x - 4)^2 = 49$
26. $(x - 5)^2 = 25$
27. $(x + 2)^2 = 14$
28. $(x + 9)^2 = 24$
29. $(2x - 1)^2 = 18$
30. $(4x + 7)^2 = 44$
31. $(x - 7)^2 = (x + 3)^2$
32. $(x + 5)^2 = (x + 4)^2$

Completing the Square In Exercises 33–42, solve the quadratic equation by completing the square.

33. $x^2 + 4x - 32 = 0$
34. $x^2 - 2x - 3 = 0$
35. $x^2 + 4x + 2 = 0$
36. $x^2 + 8x + 14 = 0$
37. $6x^2 - 12x = -3$
38. $4x^2 - 4x = 1$
39. $7 + 2x - x^2 = 0$
40. $-x^2 + x - 1 = 0$
41. $2x^2 + 5x - 8 = 0$
42. $3x^2 - 4x - 7 = 0$

Approximating Intercepts In Exercises 43–50, (a) use a graphing utility to graph the equation, (b) use the graph to approximate any x-intercepts, (c) set $y = 0$ and solve the resulting equation, and (d) compare the result of part (c) with the x-intercepts of the graph.

43. $y = (x + 3)^2 - 4$
44. $y = (x - 5)^2 - 1$
45. $y = 1 - (x - 2)^2$
46. $y = 9 - (x - 8)^2$
47. $y = -4x^2 + 4x + 3$
48. $y = 4x^2 - 1$
49. $y = x^2 + 3x - 4$
50. $y = x^2 - 5x - 24$

Using the Discriminant In Exercises 51–60, use the discriminant to determine the number of real solutions of the quadratic equation.

51. $9x^2 + 12x + 4 = 0$
52. $x^2 + 2x + 4 = 0$
53. $2x^2 - 5x + 5 = 0$
54. $-5x^2 - 4x + 1 = 0$
55. $2x^2 - x - 1 = 0$
56. $x^2 - 4x + 4 = 0$
57. $\frac{1}{3}x^2 - 5x + 25 = 0$
58. $\frac{4}{7}x^2 - 8x + 28 = 0$
59. $0.2x^2 + 1.2x - 8 = 0$
60. $9 + 2.4x - 8.3x^2 = 0$

Using the Quadratic Formula In Exercises 61–78, use the Quadratic Formula to solve the equation.

61. $2x^2 + x - 1 = 0$
62. $2x^2 - x - 1 = 0$
63. $16x^2 + 8x - 3 = 0$
64. $25x^2 - 20x + 3 = 0$
65. $x^2 + 8x - 4 = 0$
66. $9x^2 + 30x + 25 = 0$
67. $2x^2 - 7x + 1 = 0$
68. $36x^2 + 24x - 7 = 0$
69. $2 + 2x - x^2 = 0$
70. $x^2 + 10 + 8x = 0$
71. $x^2 + 16 = -12x$
72. $4x = 8 - x^2$
73. $4x^2 + 6x = 8$
74. $16x^2 + 5 = 40x$
75. $28x - 49x^2 = 4$
76. $3x + x^2 - 1 = 0$
77. $8t = 5 + 2t^2$
78. $25h^2 + 80h = -61$

116. Boating A winch tows a boat to a dock. The rope is attached to the boat at a point 15 feet below the level of the winch (see figure).

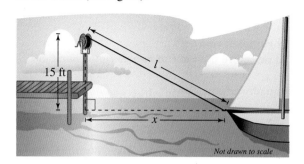

Not drawn to scale

(a) Use the Pythagorean Theorem to write an equation giving the relationship between l and x.

(b) Find the distance from the boat to the dock when the length l is 75 feet.

Exploring the Concepts

True or False? In Exercises 117 and 118, determine whether the statement is true or false. Justify your answer.

117. The quadratic equation $-3x^2 + x = -5$ has two real solutions.

118. If $(2x - 3)(x + 5) = 8$, then either $2x - 3 = 8$ or $x + 5 = 8$.

119. Think About It Is it possible for the graph of a quadratic equation to have more than two x-intercepts? Explain.

120. HOW DO YOU SEE IT? Use the graph to determine whether the discriminant of each equation is positive, zero, or negative. Explain.

(a) $x^2 - 2x = 0$ (b) $x^2 - 2x + 1 = 0$

(c) $x^2 - 2x + 2 = 0$

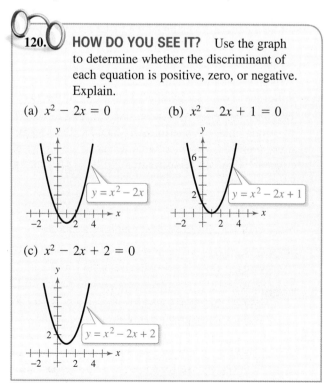

121. Think About It To solve the equation $3x^2 = 3 - x$ using the Quadratic Formula, what are the values of a, b, and c?

122. Error Analysis Describe the error.

A quadratic equation that has solutions $x = 2$ and $x = 4$ is
$(x + 2)(x + 4) = 0 \implies x^2 + 6x + 8 = 0.$

Think About It In Exercises 123–128, write a quadratic equation that has the given solutions. (There are many correct answers.)

123. 0 and 4   **124.** -2 and -8

125. 8 and 14   **126.** $\frac{1}{6}$ and $-\frac{2}{5}$

127. $1 + \sqrt{2}$ and $1 - \sqrt{2}$

128. $-3 + \sqrt{5}$ and $-3 - \sqrt{5}$

Review & Refresh ▶ *Video solutions at LarsonPrecalculus.com*

Adding or Subtracting Polynomials In Exercises 129–132, add or subtract and write the result in standard form.

129. $(x^2 + 5x + 11) + (2x^2 - 13x + 16)$

130. $(7x^2 - 8x + 4) + (9x^3 + 3x^2 + x)$

131. $(12x^2 - 15) - (x^2 - 19x - 5)$

132. $(x^2 - 3x - 2) - (x^2 - 2) - (x - 3)$

Multiplying Polynomials In Exercises 133–138, multiply the polynomials.

133. $(x + 6)(3x - 5)$ **134.** $(3x + 13)(4x - 7)$

135. $(2x - 9)(2x + 9)$ **136.** $(4x + 1)^2$

137. $(2x^2 - y)(3x^2 + 4y)$ **138.** $(4x^3 + 7y^2)(3x^3 - y^2)$

Multiplying Radical Expressions In Exercises 139–142, find each product and simplify.

139. $(\sqrt{3} + \sqrt{2})(\sqrt{3} - \sqrt{2})$

140. $(\sqrt{5} - 1)(\sqrt{5} + 1)$

141. $(7\sqrt{2} - 4 + \sqrt{5})(7\sqrt{2} + 4 - \sqrt{5})$

142. $(2\sqrt{3} + 3\sqrt{2})(2\sqrt{3} - 3\sqrt{2})$

Rationalizing a Denominator In Exercises 143–146, rationalize the denominator of the expression. Then simplify your answer.

143. $\dfrac{12}{5\sqrt{3}}$ **144.** $\dfrac{4}{\sqrt{10} - 2}$

145. $\dfrac{3}{8 + \sqrt{11}}$ **146.** $\dfrac{14}{3\sqrt{10} - 1}$

Project: Population To work an extended application analyzing the population of the United States, visit this text's website at *LarsonPrecalculus.com*. *(Source: U.S. Census Bureau)*

1.5 Complex Numbers

Complex numbers are often used in electrical engineering. For example, in Exercises 59 and 60 on page 120, you will use complex numbers to find the impedance of an electrical circuit.

- ❯ **Use the imaginary unit *i* to write complex numbers.**
- ❯ **Add, subtract, and multiply complex numbers.**
- ❯ **Use complex conjugates to write the quotient of two complex numbers in standard form.**
- ❯ **Find complex solutions of quadratic equations.**

The Imaginary Unit *i*

In Section 1.4, you learned that some quadratic equations have no real solutions. For example, the quadratic equation $x^2 + 1 = 0$ has no real solution because there is no real number x that can be squared to produce -1. To overcome this deficiency, mathematicians created an expanded system of numbers using the **imaginary unit *i*,** defined as

$$i = \sqrt{-1} \qquad \text{Imaginary unit}$$

where $i^2 = -1$. By adding real numbers to real multiples of this imaginary unit, you obtain the set of **complex numbers.** Each complex number can be written in the **standard form $a + bi$.** For example, the standard form of the complex number $-5 + \sqrt{-9}$ is $-5 + 3i$ because

$$-5 + \sqrt{-9} = -5 + \sqrt{3^2(-1)} = -5 + 3\sqrt{-1} = -5 + 3i.$$

Definition of a Complex Number

Let a and b be real numbers. The number $a + bi$ is a **complex number** written in **standard form.** The real number a is the **real part** and the number bi (where b is a real number) is the **imaginary part** of the complex number.

When $b = 0$, the number $a + bi$ is a real number. When $b \neq 0$, the number $a + bi$ is an **imaginary number.** A number of the form bi, where $b \neq 0$, is a **pure imaginary number.**

Every real number a can be written as a complex number using $b = 0$. That is, for every real number a, $a = a + 0i$. So, the set of real numbers is a subset of the set of complex numbers, as shown in the figure below.

Real numbers ($a + 0i$)
Examples: $-1, \sqrt{2}, \frac{300}{97}, \pi$

Imaginary numbers
($a + bi, b \neq 0$)
Examples: $8 - i, 5i, 1 - \sqrt{7}i$

Complex numbers ($a + bi$)
Examples: $-1, 8 - i, \sqrt{2}, 5i,$ $\frac{300}{97}, 1 - \sqrt{7}i, \pi$

Equality of Complex Numbers

Two complex numbers $a + bi$ and $c + di$, written in standard form, are equal to each other

$$a + bi = c + di \qquad \text{Equality of two complex numbers}$$

if and only if $a = c$ and $b = d$.

GO DIGITAL

Operations with Complex Numbers

To add (or subtract) two complex numbers, add (or subtract) the real and imaginary parts of the numbers separately.

> ### Addition and Subtraction of Complex Numbers
>
> For two complex numbers $a + bi$ and $c + di$ written in standard form, their sum is
>
> $$(a + bi) + (c + di) = (a + c) + (b + d)i \qquad \text{Sum}$$
>
> and their difference is
>
> $$(a + bi) - (c + di) = (a - c) + (b - d)i. \qquad \text{Difference}$$

The **additive identity** in the complex number system is zero (the same as in the real number system). Furthermore, the **additive inverse** of the complex number $a + bi$ is

$$-(a + bi) = -a - bi. \qquad \text{Additive inverse}$$

So, you have

$$(a + bi) + (-a - bi) = 0 + 0i = 0.$$

EXAMPLE 1   **Adding and Subtracting Complex Numbers**

a. $(4 + 7i) + (1 - 6i) = 4 + 7i + 1 - 6i$ — Remove parentheses.

$\qquad\qquad\qquad\quad = (4 + 1) + (7 - 6)i$ — Definition of complex addition

$\qquad\qquad\qquad\quad = 5 + i$ — Write in standard form.

b. $(1 + 2i) + (3 - 2i) = 1 + 2i + 3 - 2i$ — Remove parentheses.

$\qquad\qquad\qquad\quad = (1 + 3) + (2 - 2)i$ — Definition of complex addition

$\qquad\qquad\qquad\quad = 4 + 0i$ — Simplify.

$\qquad\qquad\qquad\quad = 4$ — Write in standard form.

c. $3i - (-2 + 3i) - (2 + 5i) = 3i + 2 - 3i - 2 - 5i$

$\qquad\qquad\qquad\qquad\qquad = (2 - 2) + (3 - 3 - 5)i$

$\qquad\qquad\qquad\qquad\qquad = 0 - 5i$

$\qquad\qquad\qquad\qquad\qquad = -5i$

d. $(3 + 2i) + (4 - i) - (7 + i) = 3 + 2i + 4 - i - 7 - i$

$\qquad\qquad\qquad\qquad\qquad = (3 + 4 - 7) + (2 - 1 - 1)i$

$\qquad\qquad\qquad\qquad\qquad = 0 + 0i$

$\qquad\qquad\qquad\qquad\qquad = 0$

ALGEBRA HELP

Note that the sum of two complex numbers can be a real number.

✓ *Checkpoint* *Audio-video solution in English & Spanish at LarsonPrecalculus.com*

Perform each operation and write the result in standard form.

a. $(7 + 3i) + (5 - 4i)$ 　　　　 **b.** $(3 + 4i) - (5 - 3i)$

c. $2i + (-3 - 4i) - (-3 - 3i)$ 　　 **d.** $(5 - 3i) + (3 + 5i) - (8 + 2i)$ ■

GO DIGITAL

Many of the properties of real numbers are valid for complex numbers as well. Here are some examples.

Associative Properties of Addition and Multiplication

Commutative Properties of Addition and Multiplication

Distributive Property of Multiplication over Addition

▶▶▶   SKILLS REFRESHER

For a refresher on the FOIL Method, watch the video at *LarsonPrecalculus.com*.

Note the use of these properties when multiplying two complex numbers.

$$(a + bi)(c + di) = a(c + di) + bi(c + di) \qquad \text{Distributive Property}$$
$$= ac + (ad)i + (bc)i + (bd)i^2 \qquad \text{Distributive Property}$$
$$= ac + (ad)i + (bc)i + (bd)(-1) \qquad i^2 = -1$$
$$= ac - bd + (ad)i + (bc)i \qquad \text{Commutative Property}$$
$$= (ac - bd) + (ad + bc)i \qquad \text{Associative Property}$$

The procedure shown above is similar to multiplying two binomials and combining like terms, as in the FOIL Method discussed in Section P.3. So, you do not need to memorize this procedure.

EXAMPLE 2   Multiplying Complex Numbers

▶▶▶ *See LarsonPrecalculus.com for an interactive version of this type of example.*

Perform each operation and write the result in standard form.

a. $4(-2 + 3i)$ **b.** $(2 - i)(4 + 3i)$ **c.** $(3 + 2i)(3 - 2i)$ **d.** $(3 + 2i)^2$

Solution

a. $4(-2 + 3i) = 4(-2) + 4(3i)$ Distributive Property

$\qquad\qquad\qquad = -8 + 12i$ Write in standard form.

b. $(2 - i)(4 + 3i) = 8 + 6i - 4i - 3i^2$ FOIL Method

$\qquad\qquad\qquad = 8 + 6i - 4i - 3(-1)$ $i^2 = -1$

$\qquad\qquad\qquad = (8 + 3) + (6 - 4)i$ Group like terms.

$\qquad\qquad\qquad = 11 + 2i$ Write in standard form.

c. $(3 + 2i)(3 - 2i) = 9 - 6i + 6i - 4i^2$ FOIL Method

$\qquad\qquad\qquad = 9 - 6i + 6i - 4(-1)$ $i^2 = -1$

$\qquad\qquad\qquad = 9 + 4$ Simplify.

$\qquad\qquad\qquad = 13$ Write in standard form.

d. $(3 + 2i)^2 = (3 + 2i)(3 + 2i)$ Property of exponents

$\qquad\qquad = 9 + 6i + 6i + 4i^2$ FOIL Method

$\qquad\qquad = 9 + 6i + 6i + 4(-1)$ $i^2 = -1$

$\qquad\qquad = 9 + 12i - 4$ Simplify.

$\qquad\qquad = 5 + 12i$ Write in standard form.

▶▶▶   TECHNOLOGY

Some graphing utilities can perform operations with complex numbers (see below). For specific keystrokes, consult the user's guide for your graphing utility.

```
4(-2+3i)
              -8+12i
(2-i)(4+3i)
               11+2i
(3+2i)(3-2i)
                  13
```

✓ *Checkpoint* ▶ *Audio-video solution in English & Spanish at LarsonPrecalculus.com*

Perform each operation and write the result in standard form.

a. $-5(3 - 2i)$ **b.** $(2 - 4i)(3 + 3i)$

c. $(4 + 5i)(4 - 5i)$ **d.** $(4 + 2i)^2$

Complex Conjugates

Notice in Example 2(c) that the product of two complex numbers can be a real number. This occurs with pairs of complex numbers of the form $a + bi$ and $a - bi$, called **complex conjugates.**

$$(a + bi)(a - bi) = a^2 - abi + abi - b^2i^2 = a^2 - b^2(-1) = a^2 + b^2$$

EXAMPLE 3   Multiplying Conjugates

Multiply each complex number by its complex conjugate.

a. $1 + i$ **b.** $4 - 3i$

Solution

a. The complex conjugate of $1 + i$ is $1 - i$.

$$(1 + i)(1 - i) = 1^2 - i^2 = 1 - (-1) = 2$$

b. The complex conjugate of $4 - 3i$ is $4 + 3i$.

$$(4 - 3i)(4 + 3i) = 4^2 - (3i)^2 = 16 - 9i^2 = 16 - 9(-1) = 25$$

✓ **Checkpoint** ▶ Audio-video solution in English & Spanish at LarsonPrecalculus.com

Multiply each complex number by its complex conjugate.

a. $3 + 6i$ **b.** $2 - 5i$

To write the quotient of $a + bi$ and $c + di$ in standard form, where c and d are not both zero, multiply the numerator and denominator by the complex conjugate of the *denominator* to obtain

$$\frac{a + bi}{c + di} = \frac{a + bi}{c + di}\left(\frac{c - di}{c - di}\right) = \frac{(ac + bd) + (bc - ad)i}{c^2 + d^2} = \frac{ac + bd}{c^2 + d^2} + \left(\frac{bc - ad}{c^2 + d^2}\right)i.$$

ALGEBRA HELP

Note that when you multiply a quotient of complex numbers by

$$\frac{c - di}{c - di}$$

you are multiplying the quotient by a form of 1. So, you are not changing the original expression. You are only writing an equivalent expression.

EXAMPLE 4   Writing a Quotient in Standard Form

Write $\dfrac{2 + 3i}{4 - 2i}$ in standard form.

Solution

$$\frac{2 + 3i}{4 - 2i} = \frac{2 + 3i}{4 - 2i}\left(\frac{4 + 2i}{4 + 2i}\right) \qquad \text{Multiply numerator and denominator by complex conjugate of denominator.}$$

$$= \frac{8 + 4i + 12i + 6i^2}{16 - 4i^2} \qquad \text{Expand.}$$

$$= \frac{8 + 4i + 12i - 6}{16 + 4} \qquad i^2 = -1$$

$$= \frac{2 + 16i}{20} \qquad \text{Simplify.}$$

$$= \frac{1}{10} + \frac{4}{5}i \qquad \text{Write in standard form.}$$

✓ **Checkpoint** ▶ Audio-video solution in English & Spanish at LarsonPrecalculus.com

Write $\dfrac{2 + i}{2 - i}$ in standard form.

GO DIGITAL

Complex Solutions of Quadratic Equations

The standard form of the complex number $\sqrt{-3}$ is $\sqrt{3}i$. The number $\sqrt{3}i$ is the *principal square root* of -3.

> **Principal Square Root of a Negative Number**
>
> Let a be a positive real number. The **principal square root** of $-a$ is defined as
> $$\sqrt{-a} = \sqrt{a}i.$$

ALGEBRA HELP

The definition of principal square root uses the rule

$$\sqrt{ab} = \sqrt{a}\sqrt{b}$$

for $a > 0$ and $b < 0$. This rule is not valid when *both a and b* are negative. For example,

$$\sqrt{-5}\sqrt{-5} = \sqrt{5(-1)}\sqrt{5(-1)}$$
$$= \sqrt{5}i\sqrt{5}i$$
$$= \sqrt{25}i^2$$
$$= 5i^2$$
$$= -5$$

whereas

$$\sqrt{(-5)(-5)} = \sqrt{25} = 5.$$

Be sure to convert complex numbers to standard form *before* performing any operations.

EXAMPLE 5 **Writing Complex Numbers in Standard Form**

a. $\sqrt{-3}\sqrt{-12} = \sqrt{3}i\sqrt{12}i = \sqrt{36}i^2 = 6(-1) = -6$

b. $\sqrt{-48} - \sqrt{-27} = \sqrt{48}i - \sqrt{27}i = 4\sqrt{3}i - 3\sqrt{3}i = \sqrt{3}i$

c. $\left(-1 + \sqrt{-3}\right)^2 = \left(-1 + \sqrt{3}i\right)^2$
$$= (-1)^2 - 2\sqrt{3}i + \left(\sqrt{3}\right)^2(i^2)$$
$$= 1 - 2\sqrt{3}i + 3(-1)$$
$$= -2 - 2\sqrt{3}i$$

✓ **Checkpoint** ▶ Audio-video solution in English & Spanish at LarsonPrecalculus.com

Write $\sqrt{-14}\sqrt{-2}$ in standard form.

EXAMPLE 6 **Complex Solutions of a Quadratic Equation**

Solve $3x^2 - 2x + 5 = 0$.

Solution

$$x = \frac{-(-2) \pm \sqrt{(-2)^2 - 4(3)(5)}}{2(3)} \qquad \text{Quadratic Formula}$$

$$= \frac{2 \pm \sqrt{-56}}{6} \qquad \text{Simplify.}$$

$$= \frac{2 \pm 2\sqrt{14}i}{6} \qquad \text{Write } \sqrt{-56} \text{ in standard form.}$$

$$= \frac{1}{3} \pm \frac{\sqrt{14}}{3}i \qquad \text{Write solution in standard form.}$$

✓ **Checkpoint** ▶ Audio-video solution in English & Spanish at LarsonPrecalculus.com

Solve $8x^2 + 14x + 9 = 0$. ■

Summarize (Section 1.5)

1. Explain how to write complex numbers using the imaginary unit i *(page 114)*.

2. Explain how to add, subtract, and multiply complex numbers *(pages 115 and 116, Examples 1 and 2)*.

3. Explain how to use complex conjugates to write the quotient of two complex numbers in standard form *(page 117, Example 4)*.

4. Explain how to find complex solutions of a quadratic equation *(page 118, Example 6)*.

GO DIGITAL

1.5 Exercises

See CalcChat.com for tutorial help and worked-out solutions to odd-numbered exercises.

GO DIGITAL

Vocabulary and Concept Check

In Exercises 1 and 2, fill in the blanks.

1. The imaginary unit i is defined as $i =$ _____, where $i^2 =$ _____.

2. Let a be a positive real number. The _____ _____ root of $-a$ is defined as $\sqrt{-a} = \sqrt{a}i$.

3. Match the type of complex number with its definition.
   (a) real number
   (b) imaginary number
   (c) pure imaginary number
   (i) $a + bi$, $a = 0$, $b \neq 0$
   (ii) $a + bi$, $b = 0$
   (iii) $a + bi$, $a \neq 0$, $b \neq 0$

4. What method for multiplying two polynomials can you use when multiplying two complex numbers in standard form?

5. What is the additive inverse of the complex number $2 - 4i$?

6. What is the complex conjugate of the complex number $2 - 4i$?

Skills and Applications

Equality of Complex Numbers In Exercises 7 and 8, find real numbers a and b such that the equation is true.

7. $a + bi = 9 + 8i$
8. $a + bi = b + (2a - 1)i$

Adding or Subtracting Complex Numbers In Exercises 9–16, perform the operation and write the result in standard form.

9. $(5 + i) + (2 + 3i)$
10. $(13 - 2i) + (-5 + 6i)$
11. $(9 - i) - (8 - i)$
12. $(3 + 2i) - (6 + 13i)$
13. $\left(-2 + \sqrt{-8}\right) + \left(5 - \sqrt{-50}\right)$
14. $\left(8 + \sqrt{-18}\right) - \left(4 + 3\sqrt{2}i\right)$
15. $13i - (14 - 7i) + (2 - 11i)$
16. $(25 + 6i) + (-10 + 11i) - (17 - 15i)$

Multiplying Complex Numbers In Exercises 17–24, perform the operation and write the result in standard form.

17. $(1 + i)(3 - 2i)$
18. $(7 - 2i)(3 - 5i)$
19. $12i(1 - 9i)$
20. $-8i(9 + 4i)$
21. $\left(\sqrt{2} + 3i\right)\left(\sqrt{2} - 3i\right)$
22. $\left(4 + \sqrt{7}i\right)\left(4 - \sqrt{7}i\right)$
23. $(6 + 7i)^2$
24. $(5 - 4i)^2$

Multiplying Conjugates In Exercises 25–32, multiply the complex number by its complex conjugate.

25. $9 + 2i$
26. $8 - 10i$
27. $-1 - \sqrt{5}i$
28. $-3 + \sqrt{2}i$
29. $\sqrt{-20}$
30. $\sqrt{-15}$
31. $1 - \sqrt{-6}$
32. $1 + \sqrt{-8}$

Writing a Quotient in Standard Form In Exercises 33–40, write the quotient in standard form.

33. $\dfrac{2}{4 - 5i}$
34. $\dfrac{13}{1 - i}$
35. $\dfrac{5 + i}{5 - i}$
36. $\dfrac{6 - 7i}{1 - 2i}$
37. $\dfrac{9 - 4i}{i}$
38. $\dfrac{8 + 16i}{2i}$
39. $\dfrac{3i}{(4 - 5i)^2}$
40. $\dfrac{5i}{(2 + 3i)^2}$

Writing a Complex Number in Standard Form In Exercises 41–48, write the complex number in standard form.

41. $\sqrt{-6}\sqrt{-2}$
42. $\sqrt{-5}\sqrt{-10}$
43. $\left(\sqrt{-15}\right)^2$
44. $\left(\sqrt{-75}\right)^2$
45. $\sqrt{-8} + \sqrt{-50}$
46. $\sqrt{-45} - \sqrt{-5}$
47. $\left(3 + \sqrt{-5}\right)\left(7 - \sqrt{-10}\right)$
48. $\left(2 - \sqrt{-6}\right)^2$

Complex Solutions of a Quadratic Equation In Exercises 49–58, use the Quadratic Formula to solve the quadratic equation.

49. $x^2 - 2x + 2 = 0$
50. $x^2 + 6x + 10 = 0$
51. $4x^2 + 16x + 17 = 0$
52. $9x^2 - 6x + 37 = 0$
53. $4x^2 + 16x + 21 = 0$
54. $16t^2 - 4t + 3 = 0$
55. $\frac{3}{2}x^2 - 6x + 9 = 0$
56. $\frac{7}{8}x^2 - \frac{3}{4}x + \frac{5}{16} = 0$
57. $1.4x^2 - 2x + 10 = 0$
58. $4.5x^2 - 3x + 12 = 0$

Impedance of a Circuit

The opposition to current in an electrical circuit is called its *impedance*. The impedance z in a parallel circuit with two pathways satisfies the equation

$$\frac{1}{z} = \frac{1}{z_1} + \frac{1}{z_2}$$

where z_1 is the impedance (in ohms) of pathway 1 and z_2 is the impedance (in ohms) of pathway 2. The impedance of each pathway in a parallel circuit is found by adding the impedances of all components in the pathway (see table). In Exercises 59 and 60, find z_1, z_2, and z for the circuit shown in the figure.

	Resistor	Inductor	Capacitor
Symbol	—⟋⟍⟋⟍— $a\ \Omega$	—⟋⟍⟋⟍— $b\ \Omega$	—‖— $c\ \Omega$
Impedance	a	bi	$-ci$

59.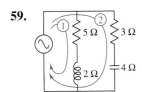

60.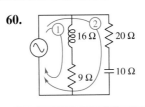

Exploring the Concepts

True or False? In Exercises 61–64, determine whether the statement is true or false. Justify your answer.

61. The sum of two complex numbers is always a real number.

62. There is no complex number that is equal to its complex conjugate.

63. $-i\sqrt{6}$ is a solution of $x^4 - x^2 + 14 = 56$.

64. $i^{44} + i^{150} - i^{74} - i^{109} + i^{61} = -1$

65. Pattern Recognition Find the missing values.

$i^1 = i$ $i^2 = -1$ $i^3 = -i$ $i^4 = 1$

$i^5 = $ ▢ $i^6 = $ ▢ $i^7 = $ ▢ $i^8 = $ ▢

$i^9 = $ ▢ $i^{10} = $ ▢ $i^{11} = $ ▢ $i^{12} = $ ▢

What pattern do you see? Write a brief description of how you would find i raised to any positive integer power.

66. Cube of a Complex Number Cube each complex number.

(a) $-1 + \sqrt{3}i$ (b) $-1 - \sqrt{3}i$

67. Error Analysis Describe the error.

$$\sqrt{-6}\sqrt{-6} = \sqrt{(-6)(-6)} = \sqrt{36} = 6 \quad ✗$$

68. 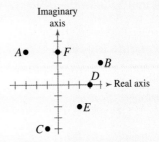 **HOW DO YOU SEE IT?** The coordinate system shown below is called the complex plane. In the complex plane, the point (a, b) corresponds to the complex number $a + bi$.

Match each complex number with its corresponding point.

(i) 3 (ii) $3i$ (iii) $4 + 2i$

(iv) $2 - 2i$ (v) $-3 + 3i$ (vi) $-1 - 4i$

69. Proof Prove that the complex conjugate of the product of two complex numbers $a_1 + b_1i$ and $a_2 + b_2i$ is the product of their complex conjugates.

70. Proof Prove that the complex conjugate of the sum of two complex numbers $a_1 + b_1i$ and $a_2 + b_2i$ is the sum of their complex conjugates.

Review & Refresh ▶ *Video solutions at LarsonPrecalculus.com*

Factoring Completely In Exercises 71–80, completely factor the expression.

71. $3x^4 - 48x^2$

72. $9x^4 - 12x^2$

73. $x^3 - 3x^2 + 3x - 9$

74. $x^3 - 5x^2 - 2x + 10$

75. $6x^3 - 27x^2 - 54x$

76. $12x^3 - 16x^2 - 60x$

77. $x^4 - 3x^2 + 2$

78. $x^4 - 7x^2 + 12$

79. $9x^4 - 37x^2 + 4$

80. $4x^4 - 37x^2 + 9$

Evaluating an Expression In Exercises 81–90, evaluate the expression for each value of x.

81. $\sqrt{2x + 7} - x$ (a) $x = -3$ (b) $x = 1$

82. $x + \sqrt{40 - 9x}$ (a) $x = 4$ (b) $x = -9$

83. $\sqrt{2x - 5} - \sqrt{x - 3} - 1$ (a) $x = 3$ (b) $x = 7$

84. $\sqrt{5x - 4} + \sqrt{x} - 1$ (a) $x = 4$ (b) $x = 1$

85. $(x - 4)^{2/3}$ (a) $x = 129$ (b) $x = -121$

86. $(x - 5)^{3/2}$ (a) $x = 69$ (b) $x = 14$

87. $\dfrac{2}{x} - \dfrac{3}{x - 2} + 1$ (a) $x = 4$ (b) $x = -1$

88. $\dfrac{4}{x} + \dfrac{2}{x + 3} + 3$ (a) $x = -4$ (b) $x = -1$

89. $|x^2 - 3x| + 4x - 6$ (a) $x = -3$ (b) $x = 2$

90. $|x^2 + 4x| - 7x - 18$ (a) $x = -3$ (b) $x = -9$

1.6 Other Types of Equations

Polynomial equations, radical equations, rational equations, and absolute value equations have many real-life applications. For example, in Exercise 97 on page 129, you will use a radical equation to analyze the relationship between the pressure and temperature of saturated steam.

> ❯ **Solve polynomial equations of degree three or greater.**
> ❯ **Solve radical equations.**
> ❯ **Solve rational equations and absolute value equations.**
> ❯ **Use nonlinear and nonquadratic models to solve real-life problems.**

Polynomial Equations

In this section, you will extend the techniques for solving equations to nonlinear and nonquadratic equations. At this point in the text, you have only four basic methods for solving nonlinear equations—*factoring, extracting square roots, completing the square*, and the *Quadratic Formula*. So, the main goal of this section is to learn to *rewrite* nonlinear equations in a form that enables you to apply one of these methods.

Example 1 shows how to use factoring to solve a **polynomial equation,** which is an equation that can be written in the general form

$$a_n x^n + a_{n-1} x^{n-1} + \cdots + a_2 x^2 + a_1 x + a_0 = 0.$$

EXAMPLE 1 Solving a Polynomial Equation by Factoring

Solve $3x^4 = 48x^2$.

Solution First write the polynomial equation in general form. Then factor the polynomial, set each factor equal to zero, and solve.

$3x^4 = 48x^2$	Write original equation.
$3x^4 - 48x^2 = 0$	Write in general form.
$3x^2(x^2 - 16) = 0$	Factor out common factor.
$3x^2(x + 4)(x - 4) = 0$	Factor difference of two squares.
$3x^2 = 0 \implies x = 0$	Set 1st factor equal to 0.
$x + 4 = 0 \implies x = -4$	Set 2nd factor equal to 0.
$x - 4 = 0 \implies x = 4$	Set 3rd factor equal to 0.

Check these solutions by substituting in the original equation.

Check

$3(0)^4 \overset{?}{=} 48(0)^2 \implies$	$0 = 0$	0 checks. ✓
$3(-4)^4 \overset{?}{=} 48(-4)^2 \implies$	$768 = 768$	−4 checks. ✓
$3(4)^4 \overset{?}{=} 48(4)^2 \implies$	$768 = 768$	4 checks. ✓

So, the solutions are $x = 0$, $x = -4$, and $x = 4$.

 Checkpoint ▶ Audio-video solution in English & Spanish at LarsonPrecalculus.com

Solve $9x^4 - 12x^2 = 0$. ■

A common mistake when solving an equation such as that in Example 1 is to divide each side of the equation by the variable factor x^2. This loses the solution $x = 0$. When solving a polynomial equation, always write the equation in general form, then factor the polynomial and set each factor equal to zero. Do not divide each side of an equation by a variable factor in an attempt to simplify the equation.

▶❯❯ **SKILLS REFRESHER**

For a refresher on factoring polynomials of degree three or higher, watch the video at *LarsonPrecalculus.com*.

GO DIGITAL

EXAMPLE 2   **Solving a Polynomial Equation by Factoring**

Solve $x^3 - 3x^2 + 3x - 9 = 0$.

Solution

$$x^3 - 3x^2 + 3x - 9 = 0 \qquad \text{Write original equation.}$$
$$x^2(x - 3) + 3(x - 3) = 0 \qquad \text{Group terms and factor.}$$
$$(x - 3)(x^2 + 3) = 0 \qquad (x - 3) \text{ is a common factor.}$$
$$x - 3 = 0 \implies x = 3 \qquad \text{Set 1st factor equal to 0.}$$
$$x^2 + 3 = 0 \implies x = \pm\sqrt{3}i \qquad \text{Set 2nd factor equal to 0.}$$

The solutions are $x = 3$, $x = \sqrt{3}i$, and $x = -\sqrt{3}i$. Check these in the original equation.

✓ *Checkpoint* ▶ Audio-video solution in English & Spanish at LarsonPrecalculus.com

Solve each equation.

a. $x^3 - 5x^2 - 2x + 10 = 0$

b. $6x^3 - 27x^2 - 54x = 0$

Occasionally, mathematical models involve equations that are of **quadratic type.** In general, an equation is of quadratic type when it can be written in the form

$$au^2 + bu + c = 0$$

where $a \neq 0$ and u is an algebraic expression.

EXAMPLE 3   **Solving an Equation of Quadratic Type**

Solve $x^4 - 3x^2 + 2 = 0$.

Solution

This equation is of quadratic type with $u = x^2$.

$$x^4 - 3x^2 + 2 = 0 \qquad \text{Write original equation.}$$
$$(x^2)^2 - 3(x^2) + 2 = 0 \qquad \text{Quadratic form}$$
$$u^2 - 3u + 2 = 0 \qquad u = x^2$$
$$(u - 1)(u - 2) = 0 \qquad \text{Factor.}$$
$$u - 1 = 0 \implies u = 1 \qquad \text{Set 1st factor equal to 0.}$$
$$u - 2 = 0 \implies u = 2 \qquad \text{Set 2nd factor equal to 0.}$$

Next, replace u with x^2 and solve for x in each equation.

$$\begin{array}{ll} u = 1 & u = 2 \\ x^2 = 1 & x^2 = 2 \\ x = \pm 1 & x = \pm\sqrt{2} \end{array}$$

The solutions are $x = 1$, $x = -1$, $x = \sqrt{2}$, and $x = -\sqrt{2}$. Check these in the original equation.

✓ *Checkpoint* ▶ Audio-video solution in English & Spanish at LarsonPrecalculus.com

Solve each equation.

a. $x^4 - 7x^2 + 12 = 0$

b. $9x^4 - 37x^2 + 4 = 0$

▶▶▶ TECHNOLOGY

To graphically check the *real* solution to Example 2, use a graphing utility to graph the equation $y = x^3 - 3x^2 + 3x - 9$. Then use the *zero* or *root* feature to approximate any *x*-intercepts, as shown below. The *x*-intercept is 3, confirming the real solution $x = 3$.

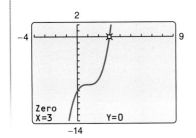

Radical Equations

A **radical equation** is an equation that involves one or more radical expressions. A radical equation can often be cleared of radicals by raising each side of the equation to an appropriate power. This procedure may introduce extraneous solutions, so checking your solutions is crucial.

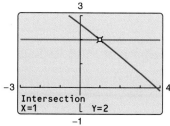

The solution of $\sqrt{2x + 7} - x = 2$ is $x = 1$.
Figure 1.14

> **EXAMPLE 4**   **Solving Radical Equations**

a.

$\sqrt{2x + 7} - x = 2$	Original equation
$\sqrt{2x + 7} = x + 2$	Isolate radical.
$2x + 7 = x^2 + 4x + 4$	Square each side.
$0 = x^2 + 2x - 3$	Write in general form.
$0 = (x + 3)(x - 1)$	Factor.
$x + 3 = 0 \implies x = -3$	Set 1st factor equal to 0.
$x - 1 = 0 \implies x = 1$	Set 2nd factor equal to 0.

Checking these values shows that the only solution is $x = 1$. Figure 1.14 shows the solution graphically.

b.

$\sqrt{2x - 5} - \sqrt{x - 3} = 1$	Original equation
$\sqrt{2x - 5} = \sqrt{x - 3} + 1$	Isolate $\sqrt{2x - 5}$.
$2x - 5 = x - 3 + 2\sqrt{x - 3} + 1$	Square each side.
$x - 3 = 2\sqrt{x - 3}$	Isolate $2\sqrt{x - 3}$.
$x^2 - 6x + 9 = 4(x - 3)$	Square each side.
$x^2 - 10x + 21 = 0$	Write in general form.
$(x - 3)(x - 7) = 0$	Factor.
$x - 3 = 0 \implies x = 3$	Set 1st factor equal to 0.
$x - 7 = 0 \implies x = 7$	Set 2nd factor equal to 0.

The solutions are $x = 3$ and $x = 7$. Check these in the original equation.

✓ **Checkpoint** 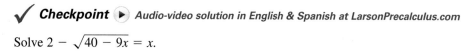 Audio-video solution in English & Spanish at *LarsonPrecalculus.com*

Solve $2 - \sqrt{40 - 9x} = x$.

> **EXAMPLE 5**   **Solving an Equation Involving a Rational Exponent**

$(x - 4)^{2/3} = 25$	Original equation
$\sqrt[3]{(x - 4)^2} = 25$	Rewrite in radical form.
$(x - 4)^2 = 15{,}625$	Cube each side.
$x - 4 = \pm 125$	Extract square roots.
$x = 129, \ x = -121$	Add 4 to each side.

The solutions are $x = 129$ and $x = -121$. Check these in the original equation.

✓ **Checkpoint** 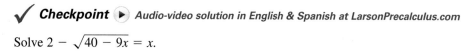 Audio-video solution in English & Spanish at *LarsonPrecalculus.com*

Solve $(x - 5)^{2/3} = 16$. ∎

ALGEBRA HELP

When an equation contains two radical expressions, it may not be possible to isolate both of them in the first step. In such cases, you may have to isolate radical expressions at *two* different stages in the solution, as shown in Example 4(b).

▶▶▶▶ **SKILLS REFRESHER**

For a refresher on rewriting expressions with rational exponents, watch the video at *LarsonPrecalculus.com*.

GO DIGITAL

Rational Equations and Absolute Value Equations

In Section 1.2, you learned how to solve rational equations. Recall that the first step is to multiply each term of the equation by the least common denominator (LCD).

EXAMPLE 6   Solving a Rational Equation

Solve $\dfrac{2}{x} = \dfrac{3}{x-2} - 1$.

Solution   For this equation, the LCD of the three terms is $x(x-2)$, so begin by multiplying each term of the equation by this expression.

$$\frac{2}{x} = \frac{3}{x-2} - 1 \qquad \text{Write original equation.}$$

$$x(x-2)\frac{2}{x} = x(x-2)\frac{3}{x-2} - x(x-2)(1) \qquad \text{Multiply each term by the LCD.}$$

$$2(x-2) = 3x - x(x-2), \quad x \neq 0, 2 \qquad \text{Simplify.}$$

$$2x - 4 = -x^2 + 5x \qquad \text{Simplify.}$$

$$x^2 - 3x - 4 = 0 \qquad \text{Write in general form.}$$

$$(x-4)(x+1) = 0 \qquad \text{Factor.}$$

$$x - 4 = 0 \implies x = 4 \qquad \text{Set 1st factor equal to 0.}$$

$$x + 1 = 0 \implies x = -1 \qquad \text{Set 2nd factor equal to 0.}$$

Both $x = 4$ and $x = -1$ are possible solutions. Multiplying each side of an equation by a variable expression can introduce extraneous solutions, so it is important to check your solutions.

Check $x = 4$

$$\frac{2}{x} = \frac{3}{x-2} - 1 \qquad \text{Write original equation.}$$

$$\frac{2}{4} \overset{?}{=} \frac{3}{4-2} - 1 \qquad \text{Substitute 4 for } x.$$

$$\frac{1}{2} \overset{?}{=} \frac{3}{2} - 1 \qquad \text{Simplify.}$$

$$\frac{1}{2} = \frac{1}{2} \qquad \text{4 checks. } \checkmark$$

Check $x = -1$

$$\frac{2}{x} = \frac{3}{x-2} - 1 \qquad \text{Write original equation.}$$

$$\frac{2}{-1} \overset{?}{=} \frac{3}{-1-2} - 1 \qquad \text{Substitute } -1 \text{ for } x.$$

$$-2 \overset{?}{=} -1 - 1 \qquad \text{Simplify.}$$

$$-2 = -2 \qquad -1 \text{ checks. } \checkmark$$

So, the solutions are $x = 4$ and $x = -1$.

ALGEBRA HELP

Notice that the values $x = 0$ and $x = 2$ are excluded because they result in division by zero in the original equation.

✓ **Checkpoint** ▶ *Audio-video solution in English & Spanish at LarsonPrecalculus.com*

Solve $\dfrac{4}{x} + \dfrac{2}{x+3} = -3$.

An **absolute value equation** is an equation that involves one or more absolute value expressions. To solve an absolute value equation, remember that the expression inside the absolute value bars can be positive or negative. This results in *two* separate equations, each of which must be solved. For example, the equation $|x - 2| = 3$ results in the two equations

$$x - 2 = 3 \quad \text{and} \quad -(x - 2) = 3.$$

Both equations are linear, so use the techniques of Section 1.2 to determine that the solutions are $x = 5$ and $x = -1$.

EXAMPLE 7   Solving an Absolute Value Equation

▶▶▶ *See LarsonPrecalculus.com for an interactive version of this type of example.*

Solve $|x^2 - 3x| = -4x + 6$.

Solution   The variable expression inside the absolute value signs can be positive or negative, so you must solve the two *quadratic* equations

$$x^2 - 3x = -4x + 6 \quad \text{and} \quad -(x^2 - 3x) = -4x + 6.$$

First Equation

$x^2 - 3x = -4x + 6$	Use positive expression.
$x^2 + x - 6 = 0$	Write in general form.
$(x + 3)(x - 2) = 0$	Factor.
$x + 3 = 0 \implies x = -3$	Set 1st factor equal to 0.
$x - 2 = 0 \implies x = 2$	Set 2nd factor equal to 0.

Second Equation

$-(x^2 - 3x) = -4x + 6$	Use negative expression.
$x^2 - 7x + 6 = 0$	Write in general form.
$(x - 1)(x - 6) = 0$	Factor.
$x - 1 = 0 \implies x = 1$	Set 1st factor equal to 0.
$x - 6 = 0 \implies x = 6$	Set 2nd factor equal to 0.

Check

$|(-3)^2 - 3(-3)| \stackrel{?}{=} -4(-3) + 6$	Substitute -3 for x.
$18 = 18$	-3 checks. ✓
$|(2)^2 - 3(2)| \stackrel{?}{=} -4(2) + 6$	Substitute 2 for x.
$2 \neq -2$	2 does not check.
$|(1)^2 - 3(1)| \stackrel{?}{=} -4(1) + 6$	Substitute 1 for x.
$2 = 2$	1 checks. ✓
$|(6)^2 - 3(6)| \stackrel{?}{=} -4(6) + 6$	Substitute 6 for x.
$18 \neq -18$	6 does not check.

Only two of the four values satisfy the original equation. So, the solutions are $x = -3$ and $x = 1$. To check the solutions graphically, note that the original equation can be rewritten as $|x^2 - 3x| + 4x - 6 = 0$. Figure 1.15 shows that the graph of $y = |x^2 - 3x| + 4x - 6$ has x-intercepts at $(-3, 0)$ and $(1, 0)$, confirming that the solutions are $x = -3$ and $x = 1$.

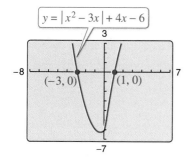

The solutions of $|x^2 - 3x| = -4x + 6$ are $x = -3$ and $x = 1$.
Figure 1.15

✓ *Checkpoint* ▶ Audio-video solution in English & Spanish at LarsonPrecalculus.com

Solve $|x^2 + 4x| = 7x + 18$.

GO DIGITAL

Applications

It would be virtually impossible to categorize all of the different types of applications that involve nonlinear and nonquadratic models. However, you will see a variety of applications in the next two examples and in the exercises.

EXAMPLE 8 **Reduced Rates**

A ski club charters a bus for a ski trip at a cost of $480. To lower the bus fare per skier, the club invites nonmembers to go on the trip. After 5 nonmembers join the trip, the fare per skier decreases by $4.80. How many club members are going on the trip?

Solution Begin the solution by creating a verbal model and assigning labels.

Verbal model: Cost per skier $\cdot$ Number of skiers $=$ Cost of trip

Labels:

Cost of trip $= 480$	(dollars)
Number of ski club members $= x$	(people)
Number of skiers $= x + 5$	(people)
Original cost per member $= \dfrac{480}{x}$	(dollars per person)
Cost per skier $= \dfrac{480}{x} - 4.80$	(dollars per person)

Equation:

$$\left(\frac{480}{x} - 4.80\right)(x + 5) = 480$$

$$\left(\frac{480 - 4.8x}{x}\right)(x + 5) = 480 \qquad \text{Rewrite first factor.}$$

$$(480 - 4.8x)(x + 5) = 480x, \; x \neq 0 \qquad \text{Multiply each side by } x.$$

$$480x + 2400 - 4.8x^2 - 24x = 480x \qquad \text{Multiply.}$$

$$-4.8x^2 - 24x + 2400 = 0 \qquad \text{Write in general form.}$$

$$x^2 + 5x - 500 = 0 \qquad \text{Divide each side by } -4.8.$$

$$(x + 25)(x - 20) = 0 \qquad \text{Factor.}$$

$$x + 25 = 0 \implies x = -25 \qquad \text{Set 1st factor equal to 0.}$$

$$x - 20 = 0 \implies x = 20 \qquad \text{Set 2nd factor equal to 0.}$$

Only the positive value of x makes sense in the context of the problem, so 20 ski club members are going on the trip. Check this in the original statement of the problem.

Check

$$\left(\frac{480}{20} - 4.80\right)(20 + 5) \overset{?}{=} 480 \qquad \text{Substitute 20 for } x.$$

$$(24 - 4.80)25 \overset{?}{=} 480 \qquad \text{Simplify.}$$

$$480 = 480 \qquad \text{20 checks.} \checkmark$$

✓ *Checkpoint*  *Audio-video solution in English & Spanish at LarsonPrecalculus.com*

A high school charters a bus for $560 to take a group of students to an observatory. When 8 more students join the trip, the cost per student decreases by $3.50. How many students were in the original group?

GO DIGITAL

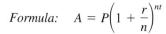

EXAMPLE 9   **Compound Interest**

When you were born, your grandparents deposited $5000 in a long-term investment in which the interest was compounded quarterly. On your 25th birthday, the value of the investment is $25,062.59. What is the annual interest rate for this investment?

Solution

ALGEBRA HELP

Recall from Section 1.3 that the formula for interest that is compounded n times per year is

$$A = P\left(1 + \frac{r}{n}\right)^{nt}.$$

In this formula, A is the balance in the account, P is the principal (or original deposit), r is the annual interest rate (in decimal form), n is the number of compoundings per year, and t is the time in years.

Formula: $A = P\left(1 + \dfrac{r}{n}\right)^{nt}$

Labels: Balance $= A = 25{,}062.59$ (dollars)

Principal $= P = 5000$ (dollars)

Time $= t = 25$ (years)

Compoundings per year $= n = 4$ (compoundings per year)

Annual interest rate $= r$ (percent in decimal form)

Equation: $25{,}062.59 = 5000\left(1 + \dfrac{r}{4}\right)^{4(25)}$ Substitute.

$\dfrac{25{,}062.59}{5000} = \left(1 + \dfrac{r}{4}\right)^{100}$ Divide each side by 5000.

$5.0125 \approx \left(1 + \dfrac{r}{4}\right)^{100}$ Use a calculator.

$(5.0125)^{1/100} \approx 1 + \dfrac{r}{4}$ Raise each side to reciprocal power.

$1.01625 \approx 1 + \dfrac{r}{4}$ Use a calculator.

$0.01625 \approx \dfrac{r}{4}$ Subtract 1 from each side.

$0.065 \approx r$ Multiply each side by 4.

The annual interest rate is about 0.065, or 6.5%. Check this in the formula using $r = 0.065$ and the values of A, P, t, and n listed above.

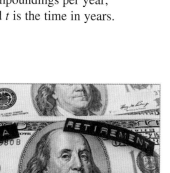

With compound interest, it is beneficial to begin saving for retirement as soon as possible. Use the formula above to verify that investing $10,000 at 5.5% compounded quarterly yields $123,389.89 after 46 years, but only $60,656.10 after 33 years.

✓ **Checkpoint** ▶ *Audio-video solution in English & Spanish at LarsonPrecalculus.com*

You deposit $2500 in a long-term investment in which the interest is compounded monthly. After 5 years, the balance is $3544.06. What is the annual interest rate?   ■

Summarize **(Section 1.6)**

1. Explain how to solve a polynomial equation of degree three or greater by factoring *(page 121)*. For examples of solving polynomial equations by factoring, see Examples 1–3.

2. Explain how to solve a radical equation *(page 123)*. For an example of solving radical equations, see Example 4.

3. Explain how to solve rational equations and absolute value equations *(pages 124 and 125)*. For examples of solving these types of equations, see Examples 6 and 7.

4. Describe real-life applications that use nonlinear and nonquadratic models *(pages 126 and 127, Examples 8 and 9)*.

GO DIGITAL

1.6   Exercises

See CalcChat.com for tutorial help and worked-out solutions to odd-numbered exercises.

GO DIGITAL

Vocabulary and Concept Check

In Exercises 1 and 2, fill in the blanks.

1. The general form of a _____ equation in x is $a_nx^n + a_{n-1}x^{n-1} + \cdots + a_2x^2 + a_1x + a_0 = 0$.
2. To clear the equation $4/x + 5 = 6/(x - 3)$ of fractions, multiply each side of the equation by the least common denominator _____.

3. Describe the step needed to remove the radical from the equation $\sqrt{x + 2} = x$.
4. Is the equation $x^4 - 2x + 4 = 0$ of quadratic type?

Skills and Applications

Solving a Polynomial Equation **In Exercises 5–14, solve the equation. Check your solutions.**

5. $6x^4 - 54x^2 = 0$
6. $36x^3 - 100x = 0$
7. $5x^3 + 30x^2 + 45x = 0$
8. $9x^4 - 24x^3 + 16x^2 = 0$
9. $x^4 - 81 = 0$
10. $x^6 - 64 = 0$
11. $x^3 + 512 = 0$
12. $27x^3 - 343 = 0$
13. $x^3 + 2x^2 + 3x = -6$
14. $x^4 + 2x^3 - 8x = 16$

Solving an Equation of Quadratic Type **In Exercises 15–26, solve the equation. Check your solutions.**

15. $x^4 - 4x^2 + 3 = 0$
16. $x^4 - 13x^2 + 36 = 0$
17. $4x^4 - 65x^2 + 16 = 0$
18. $36t^4 + 29t^2 - 7 = 0$
19. $2x + 9\sqrt{x} = 5$
20. $6x - 7\sqrt{x} - 3 = 0$
21. $9t^{2/3} + 24t^{1/3} = -16$
22. $3x^{1/3} + 2x^{2/3} = 5$
23. $\dfrac{1}{x^2} + \dfrac{8}{x} + 15 = 0$
24. $1 + \dfrac{3}{x} = -\dfrac{2}{x^2}$
25. $2\left(\dfrac{x}{x + 2}\right)^2 - 3\left(\dfrac{x}{x + 2}\right) - 2 = 0$
26. $6\left(\dfrac{x}{x + 1}\right)^2 + 5\left(\dfrac{x}{x + 1}\right) - 6 = 0$

Solving a Radical Equation **In Exercises 27–38, solve the equation, if possible. Check your solutions.**

27. $\sqrt{5x} - 10 = 0$
28. $\sqrt{3x + 1} = 7$
29. $4 + \sqrt[3]{2x - 9} = 0$
30. $\sqrt[3]{12 - x} - 3 = 0$
31. $\sqrt{x + 8} = 2 + x$
32. $2x = \sqrt{-5x + 24} - 3$
33. $\sqrt{x - 3} + 1 = \sqrt{x}$
34. $\sqrt{x} + \sqrt{x - 24} = 2$
35. $2\sqrt{x + 1} - \sqrt{2x + 3} = 1$
36. $4\sqrt{x - 3} - \sqrt{6x - 17} = 3$
37. $\sqrt{4\sqrt{4x + 9}} = \sqrt{8x + 2}$
38. $\sqrt{16 + 9\sqrt{x}} = 4 + \sqrt{x}$

Solving an Equation Involving a Rational Exponent **In Exercises 39–44, solve the equation. Check your solutions.**

39. $(x - 5)^{3/2} = 8$
40. $(x + 2)^{2/3} = 9$
41. $(x^2 - 5)^{3/2} = 27$
42. $(x^2 - x - 22)^{3/2} = 27$
43. $3x(x - 1)^{1/2} + 2(x - 1)^{3/2} = 0$
44. $4x^2(x - 1)^{1/3} + 6x(x - 1)^{4/3} = 0$

Solving a Rational Equation **In Exercises 45–50, solve the equation. Check your solutions.**

45. $\dfrac{1}{x} - \dfrac{1}{x + 1} = 3$
46. $\dfrac{4}{x + 1} - \dfrac{3}{x + 2} = 1$
47. $3 - \dfrac{14}{x} - \dfrac{5}{x^2} = 0$
48. $5 = \dfrac{18}{x} + \dfrac{8}{x^2}$
49. $\dfrac{x + 1}{3} - \dfrac{x + 1}{x + 2} = 0$
50. $\dfrac{x}{x^2 - 4} + \dfrac{1}{x + 2} = 3$

Solving an Absolute Value Equation **In Exercises 51–56, solve the equation. Check your solutions.**

51. $|2x - 5| = 11$
52. $|3x + 2| = 7$
53. $|x| = x^2 + x - 24$
54. $|x^2 + 6x| = 3x + 18$
55. $|x + 1| = x^2 - 5$
56. $|x - 15| = x^2 - 15x$

Approximating Intercepts **In Exercises 57–64, (a) use a graphing utility to graph the equation, (b) use the graph to approximate any x-intercepts, (c) set $y = 0$ and solve the resulting equation, and (d) compare the result of part (c) with the x-intercept(s) of the graph.**

57. $y = x^3 - 2x^2 - 3x$
58. $y = x^4 - 29x^2 + 100$
59. $y = \sqrt{11x - 30} - x$
60. $y = 2x - \sqrt{15 - 4x}$
61. $y = \dfrac{1}{x} - \dfrac{4}{x - 1} - 1$
62. $y = x + \dfrac{9}{x + 1} - 5$
63. $y = |x + 1| - 2$
64. $y = |x - 2| - 3$

Solving an Equation In Exercises 65–78, find the real solution(s) of the equation. (Round your answer(s) to three decimal places, if necessary.)

65. $x^3 - 3x^2 - 1.21x = -3.63$

66. $x^4 - 1.7x^3 + x = 1.7$

67. $3.2x^4 - 1.5x^2 = 2.1$

68. $0.1x^4 - 2.4x^2 = 3.6$

69. $7.08x^6 + 4.15x^3 = 9.6$

70. $5.25x^6 - 0.2x^3 = 1.55$

71. $11.5 - 5.6\sqrt{x} = 0$

72. $\sqrt{x + 8.2} - 5.55 = 0$

73. $1.8x - 6\sqrt{x} = 5.6$

74. $5.3x + 3.1 = 9.8\sqrt{x}$

75. $4x^{2/3} + 8x^{1/3} = -3.6$

76. $8.4x^{2/3} - 1.2x^{1/3} = 24$

77. $x = \dfrac{3.3}{x} + \dfrac{1}{2.2}$

78. $\dfrac{4.4}{x} - \dfrac{5.5}{3.3} = \dfrac{x}{6.6}$

Writing an Equation In Exercises 79–88, write an equation that has the given solutions. (There are many correct answers.)

79. $-4, 7$

80. $0, 2, 9$

81. $-\frac{7}{3}, \frac{6}{7}$

82. $-\frac{1}{8}, -\frac{4}{5}$

83. $\sqrt{3}, -\sqrt{3}, 4$

84. $2\sqrt{7}, -\sqrt{7}$

85. $i, -i$

86. $2i, -2i$

87. $-1, 1, i, -i$

88. $4i, -4i, 6, -6$

89. Reduced Rates A college charters a bus for $1700 to take a group of students to see a Broadway production. When 6 more students join the trip, the cost per student decreases by $7.50. How many students were in the original group?

90. Reduced Rates Three students plan to divide the rent for an apartment equally. When they add a fourth student, the cost per student decreases by $150 per month. What is the total monthly rent for the apartment?

91. Average Speed An airline runs a commuter flight between Portland, Oregon, and Seattle, Washington, which are 145 miles apart. An increase of 40 miles per hour in the average speed of the plane decreases the travel time by 12 minutes. What initial average speed results in this decrease in travel time?

92. Average Speed A family drives 1080 miles to a vacation lodge. On the return trip, it takes the family $2\frac{1}{2}$ hours longer, traveling at an average speed that is 6 miles per hour slower. Determine the average speed on the way to the lodge.

93. Compound Interest You deposit $2500 in a long-term investment fund in which the interest is compounded monthly. After 5 years, the balance is $2694.58. What is the annual interest rate?

94. Compound Interest You deposit $6000 in a long-term investment fund in which the interest is compounded quarterly. After 5 years, the balance is $7734.27. What is the annual interest rate?

95. Airline Passengers An airline offers daily flights between Chicago and Denver. The total monthly cost C (in millions of dollars) of these flights can be modeled by $C = \sqrt{0.2x + 1}$, where x is the number of passengers (in thousands). The total cost of the flights for June is 2.5 million dollars. How many passengers flew in June?

96. Nursing The number N (in thousands) of nursing graduates who passed the National Council Licensure Examination for Registered Nurses in the United States from 2013 through 2018 can be modeled by $N = \sqrt{4241.855 + 1404.727t}$, $13 \le t \le 18$, where t represents the year, with $t = 13$ corresponding to 2013. *(Source: National Council of State Boards of Nursing)*

(a) In which year did the number of nursing graduates who passed the examination reach 160,000?

(b) In which year did the number of nursing graduates who passed the examination reach 170,000?

97. Saturated Steam

The temperature T (in degrees Fahrenheit) of saturated steam increases as pressure increases. This relationship can be approximated by the model $T = 75.82 - 2.11x + 43.51\sqrt{x}$, $5 \le x \le 40$, where x is the absolute pressure (in pounds per square inch).

(a) The temperature of saturated steam at sea level is 212°F. Find the absolute pressure at this temperature.

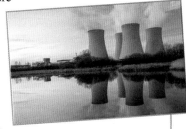

(b) Use a graphing utility to verify your solution for part (a).

98. Voting-Age Population The total voting-age population P (in millions) in the United States from 2010 through 2018 can be modeled by

$$P = \frac{218.22 + 0.9052t^2}{1 + 0.0031t^2}, \quad 10 \le t \le 18$$

where t represents the year, with $t = 10$ corresponding to 2010. *(Source: U.S. Census Bureau)*

(a) In which year did the total voting-age population reach 240 million?

(b) In which year did the total voting-age population reach 250 million?

99. Demand The demand for a video game can be modeled by $p = 40 - \sqrt{0.01x + 1}$, where x is the number of units demanded per day and p is the price per unit. Find the demand when the price is $37.55.

100. Power Line A power station is on one side of a river that is $\frac{3}{4}$ mile wide, and a factory is 8 miles downstream on the other side of the river. It costs $24 per foot to run power lines over land and $30 per foot to run them under water. The project's cost is $1,098,662.40. Find the length x labeled in the figure.

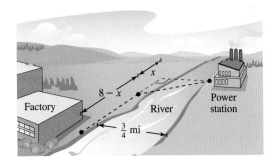

101. Tiling a Floor Working alone, you can tile a floor in t hours. When you work with a friend, the time y (in hours) it takes to tile the floor satisfies the equation

$$\frac{1}{t} + \frac{1}{t+3} = \frac{1}{y}.$$

Find the time it takes you to tile the floor working alone when you and your friend can tile the floor in 2 hours working together.

102. Painting a Fence Working alone, you can paint a fence in t hours. Working with a friend, the time y (in hours) it takes to paint the fence satisfies the equation

$$\frac{1}{t} + \frac{1}{t+2} = \frac{1}{y}.$$

Find the time it takes you to paint the fence working alone when you and your friend can paint the fence in 3 hours working together.

103. Potential Energy of a Spring The distance d a spring is stretched can be calculated using the formula

$$d = \sqrt{\frac{2U}{k}}$$

where U is the potential energy and k is the spring constant. Solve the formula for U.

104. Circumference of an Ellipse The circumference C of an ellipse with major axis length $2a$ and minor axis length $2b$ can be approximated using the formula below. Solve the formula for a.

$$C \approx 2\pi \sqrt{\frac{a^2 + b^2}{2}}$$

Exploring the Concepts

True or False? In Exercises 105 and 106, determine whether the statement is true or false. Justify your answer.

105. An equation can never have more than one extraneous solution.

106. The equation $\sqrt{x+10} - \sqrt{x-10} = 0$ has no solution.

107. Error Analysis Describe the error(s).

$$\sqrt{3x} = \sqrt{7x+4}$$
$$3x^2 = 7x + 4$$
$$x = \frac{-7 \pm \sqrt{7^2 - 4(3)(4)}}{2(3)}$$
$$x = -1 \text{ and } x = -\frac{4}{3} \quad \textbf{✗}$$

108. 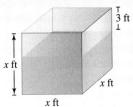 **HOW DO YOU SEE IT?** The figure shows a glass cube partially filled with water.

(a) What does the expression $x^2(x-3)$ represent?

(b) Given $x^2(x-3) = 320$, explain how to find the volume of the cube.

Solving an Absolute Value Equation In Exercises 109 and 110, solve the equation. Check your solutions.

109. $|x-3| = |4x+9|$ **110.** $|x^2+1| = |2x^2-1|$

Review & Refresh ▶ *Video solutions at LarsonPrecalculus.com*

Plotting and Ordering Real Numbers In Exercises 111–116, plot the numbers on the real number line. Then place the appropriate inequality symbol (< or >) between them.

111. $-5, -12$ **112.** $1, \frac{4}{3}$

113. $\frac{3}{4}, \frac{4}{5}$ **114.** $-\frac{9}{8}, -\frac{3}{8}$

115. $-6.5, -9.2$ **116.** $-\frac{7}{2}, -3.4$

Interpreting an Inequality In Exercises 117–120, describe the subset of real numbers that the inequality represents.

117. $x \le 3$ **118.** $x > 0$

119. $-5 < x < 5$ **120.** $0 < x \le 12$

Representing an Interval In Exercises 121–124, represent the interval verbally, as an inequality, and as a graph.

121. $(2, \infty)$ **122.** $(-\infty, 4]$

123. $[-10, 0)$ **124.** $(-2, 3]$

Solving a Linear Equation In Exercises 125–128, solve the equation and check your solution.

125. $5x - 7 = 3x + 9$ **126.** $7x - 3 = 2x + 7$

127. $1 - \frac{3}{2}x = x - 4$ **128.** $2 - \frac{5}{3}x = x - 6$

1.7   Linear Inequalities in One Variable

❯ Represent solutions of linear inequalities in one variable.
❯ Use properties of inequalities to write equivalent inequalities.
❯ Solve linear inequalities in one variable.
❯ Solve absolute value inequalities.
❯ Use linear inequalities to model and solve real-life problems.

Linear inequalities have many real-life applications. For example, in Exercise 92 on page 139, you will use an absolute value inequality to describe the distance between two locations.

Introduction

Simple inequalities were discussed in Section P.1. There, the inequality symbols $<$, $\leq$, $>$, and $\geq$ were used to compare two numbers and to denote subsets of real numbers. For example, the simple inequality $x \geq 3$ denotes all real numbers x that are greater than or equal to 3.

Now, you will expand your work with inequalities to include more involved statements such as $5x - 7 < 3x + 9$ and $-3 \leq 6x - 1 < 3$. As with an equation, you **solve an inequality** in the variable x by finding all values of x for which the inequality is true. Such values are **solutions** that **satisfy** the inequality. The set of all real numbers that are solutions of an inequality is the **solution set** of the inequality. For example, the solution set of $x + 1 < 4$ is all real numbers that are less than 3.

The set of all points on the real number line that represents the solution set is the **graph of the inequality.** Graphs of many types of inequalities consist of intervals on the real number line. See Section P.1 to review the nine basic types of intervals on the real number line. Note that each type of interval can be classified as *bounded* or *unbounded*.

EXAMPLE 1   Intervals and Inequalities

State whether each interval is bounded or unbounded, represent the interval with an inequality, and sketch its graph.

a. $(-3, 5]$ **b.** $(4, \infty)$ **c.** $(-\infty, 2]$

Solution

a. The interval $(-3, 5]$ is bounded and corresponds to the inequality $-3 < x \leq 5$.

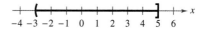

b. The interval $(4, \infty)$ is unbounded and corresponds to the inequality $4 < x < \infty$.

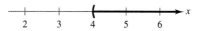

c. The interval $(-\infty, 2]$ is unbounded and corresponds to the inequality $-\infty < x \leq 2$.

 Checkpoint ▶ *Audio-video solution in English & Spanish at LarsonPrecalculus.com*

State whether each interval is bounded or unbounded, represent the interval with an inequality, and sketch its graph.

a. $[-1, 3]$ **b.** $(-1, 6)$ **c.** $(-\infty, 4)$ **d.** $[0, \infty)$ ■

ALGEBRA HELP

The intervals in Example 1 are represented algebraically and graphically. The intervals can also be represented verbally. For instance, in words, the interval in Example 1(a) is "all real numbers greater than -3 and less than or equal to 5." Represent the other intervals in Example 1 verbally.

GO DIGITAL

Properties of Inequalities

The procedures for solving linear inequalities in one variable are similar to those for solving linear equations. To isolate the variable, use the **properties of inequalities.** These properties are similar to the properties of equality, but there are two important exceptions. When you multiply or divide each side of an inequality by a negative number, you must *reverse the direction of the inequality symbol* to maintain a true statement. Here is an example.

$$-2 < 5 \qquad \text{Original inequality}$$
$$(-3)(-2) > (-3)(5) \qquad \text{Multiply each side by } -3 \text{ and reverse the inequality symbol.}$$
$$6 > -15 \qquad \text{Simplify.}$$

Notice that when you do not reverse the inequality symbol in the example above, you obtain the false statement

$$6 < -15. \qquad \text{False statement}$$

Two inequalities that have the same solution set are **equivalent.** For example, the inequalities

$$x + 2 < 5 \quad \text{and} \quad x < 3$$

are equivalent, as shown in the figure.

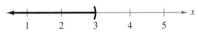

The inequalities $x + 2 < 5$ and $x < 3$
have the same solution set.

To obtain the second inequality from the first, subtract 2 from each side of the inequality. The properties listed below describe operations used to write equivalent inequalities.

Properties of Inequalities

Let a, b, c, and d be real numbers.

1. Transitive Property

$$a < b \text{ and } b < c \implies a < c$$

2. Addition of Inequalities

$$a < b \text{ and } c < d \implies a + c < b + d$$

3. Addition of a Constant

$$a < b \implies a + c < b + c$$

4. Multiplication by a Constant

$$\text{For } c > 0, a < b \implies ac < bc$$
$$\text{For } c < 0, a < b \implies ac > bc \qquad \text{Reverse the inequality symbol.}$$

Each of the properties above is true when the symbol $<$ is replaced with $\leq$ and $>$ is replaced with $\geq$. For example, another form of the multiplication property is shown below.

$$\text{For } c > 0, a \leq b \implies ac \leq bc$$
$$\text{For } c < 0, a \leq b \implies ac \geq bc$$

Solving Linear Inequalities in One Variable

The simplest type of inequality to solve is a **linear inequality** in one variable. For example, $2x + 3 > 4$ is a linear inequality in x.

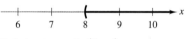

EXAMPLE 2 Solving a Linear Inequality

Solve $5x - 7 > 3x + 9$. Then graph the solution set.

Solution

$5x - 7 > 3x + 9$	Write original inequality.
$2x - 7 > 9$	Subtract $3x$ from each side.
$2x > 16$	Add 7 to each side.
$x > 8$	Divide each side by 2.

The solution set is all real numbers that are greater than 8. The interval notation for this solution set is $(8, \infty)$. The graph of this solution set is shown below. Note that a parenthesis at 8 on the real number line indicates that 8 *is not* part of the solution set.

Solution interval: $(8, \infty)$

✓ **Checkpoint** ▶ Audio-video solution in English & Spanish at LarsonPrecalculus.com

Solve $7x - 3 \le 2x + 7$. Then graph the solution set.

EXAMPLE 3 Solving a Linear Inequality

▶▶▶ *See LarsonPrecalculus.com for an interactive version of this type of example.*

Solve $1 - \frac{3}{2}x \ge x - 4$.

Algebraic Solution

$1 - \frac{3}{2}x \ge x - 4$	Write original inequality.
$2 - 3x \ge 2x - 8$	Multiply each side by 2.
$2 - 5x \ge -8$	Subtract $2x$ from each side.
$-5x \ge -10$	Subtract 2 from each side.
$x \le 2$	Divide each side by -5 and reverse the inequality symbol.

The solution set is all real numbers that are less than or equal to 2. The interval notation for this solution set is $(-\infty, 2]$. The graph of this solution set is shown below. Note that a bracket at 2 on the real number line indicates that 2 *is* part of the solution set.

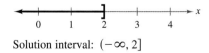

Solution interval: $(-\infty, 2]$

Graphical Solution

Use a graphing utility to graph $y_1 = 1 - \frac{3}{2}x$ and $y_2 = x - 4$ in the same viewing window.

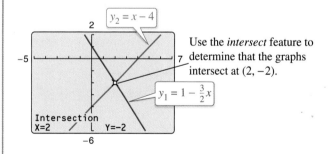

Use the *intersect* feature to determine that the graphs intersect at $(2, -2)$.

The graph of y_1 lies above the graph of y_2 to the left of their point of intersection, which implies that $y_1 \ge y_2$ for all $x \le 2$.

✓ **Checkpoint** ▶ Audio-video solution in English & Spanish at LarsonPrecalculus.com

Solve $2 - \frac{5}{3}x > x - 6$ (a) algebraically and (b) graphically.

Sometimes it is possible to write two inequalities as a **double inequality.** For example, you can write the two inequalities

$$-4 \le 5x - 2 \quad \text{and} \quad 5x - 2 < 7$$

as the double inequality

$$-4 \le 5x - 2 < 7. \qquad \text{Double inequality}$$

This form allows you to solve the two inequalities together, as demonstrated in Example 4.

EXAMPLE 4   **Solving a Double Inequality**

Solve $-3 \le 6x - 1 < 3$. Then graph the solution set.

Solution   One way to solve this double inequality is to isolate x as the middle term.

$$-3 \le 6x - 1 < 3 \qquad \text{Write original inequality.}$$

$$-3 + 1 \le 6x - 1 + 1 < 3 + 1 \qquad \text{Add 1 to each part.}$$

$$-2 \le 6x < 4 \qquad \text{Simplify.}$$

$$\frac{-2}{6} \le \frac{6x}{6} < \frac{4}{6} \qquad \text{Divide each part by 6.}$$

$$-\frac{1}{3} \le x < \frac{2}{3} \qquad \text{Simplify.}$$

The solution set is all real numbers that are greater than or equal to $-\frac{1}{3}$ and less than $\frac{2}{3}$. The interval notation for this solution set is

$$\left[-\frac{1}{3}, \frac{2}{3}\right). \qquad \text{Solution interval}$$

The graph of this solution set is shown below.

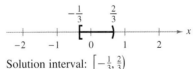

Solution interval: $\left[-\frac{1}{3}, \frac{2}{3}\right)$

✓ *Checkpoint* ▶ *Audio-video solution in English & Spanish at LarsonPrecalculus.com*

Solve $1 < 2x + 7 < 11$. Then graph the solution set.

Another way to solve the double inequality in Example 4 is to solve it in two parts.

$$-3 \le 6x - 1 \quad \text{and} \quad 6x - 1 < 3$$

$$-2 \le 6x \qquad\qquad\quad 6x < 4$$

$$-\frac{1}{3} \le x \qquad\qquad\quad x < \frac{2}{3}$$

The solution set consists of all real numbers that satisfy *both* inequalities. In other words, the solution set is the set of all values of x for which

$$-\frac{1}{3} \le x < \frac{2}{3}.$$

When combining two inequalities to form a double inequality, be sure that the inequalities satisfy the Transitive Property. For example, it is *incorrect* to combine the inequalities $3 < x$ and $x \le -1$ as $3 < x \le -1$. This "inequality" is wrong because 3 is not less than -1.

Absolute Value Inequalities

>>> TECHNOLOGY

A graphing utility can help you identify the solution set of an inequality. For instance, to find the solution set of $|x - 5| < 2$ in Example 5(a), first rewrite the inequality as $|x - 5| - 2 < 0$. Then graph $y = |x - 5| - 2$, as shown below. The graph of y is below the x-axis on the interval $(3, 7)$, which is the solution set of the original inequality.

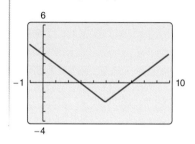

Solving an Absolute Value Inequality

Let x be a variable or an algebraic expression, and let a be a real number such that $a > 0$.

1. The solutions of $|x| < a$ are all values of x that lie between $-a$ and a. That is,

$$|x| < a \quad \text{if and only if} \quad -a < x < a.$$

2. The solutions of $|x| > a$ are all values of x that are less than $-a$ or greater than a. That is,

$$|x| > a \quad \text{if and only if} \quad x < -a \quad \text{or} \quad x > a.$$

These rules are also valid when $<$ is replaced by $\leq$ and $>$ is replaced by $\geq$.

EXAMPLE 5   **Solving Absolute Value Inequalities**

a. $|x - 5| < 2$ Original inequality

$\qquad -2 < x - 5 < 2$ Equivalent double inequality

$\quad -2 + 5 < x - 5 + 5 < 2 + 5$ Add 5 to each part.

$\qquad\qquad 3 < x < 7$ Simplify.

The solution set is all real numbers that are greater than 3 and less than 7, which in interval notation is $(3, 7)$. The graph of this solution set is shown below. Note that the graph of the inequality can be described as all real numbers less than two units from 5.

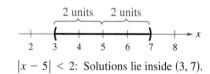

$|x - 5| < 2$: Solutions lie inside $(3, 7)$.

b. $|x + 3| \geq 7$ Original inequality

$\quad x + 3 \leq -7 \qquad \text{or} \qquad x + 3 \geq 7$ Equivalent inequalities

$x + 3 - 3 \leq -7 - 3 \qquad x + 3 - 3 \geq 7 - 3$ Subtract 3 from each side.

$\qquad x \leq -10 \qquad\qquad\qquad x \geq 4$ Simplify.

The solution set is all real numbers that are less than or equal to -10 *or* greater than or equal to 4. The interval notation for this solution set is $(-\infty, -10] \cup [4, \infty)$. The symbol $\cup$ is the *union* symbol, which denotes the combining of two sets. The graph of this solution set is shown below. Note that the graph of the inequality can be described as all real numbers at least seven units from -3.

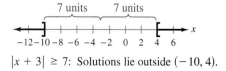

$|x + 3| \geq 7$: Solutions lie outside $(-10, 4)$.

✓ **Checkpoint** ▶ *Audio-video solution in English & Spanish at LarsonPrecalculus.com*

Solve $|x - 20| \leq 4$. Then graph the solution set.   ■

GO DIGITAL

Application

A problem-solving plan like the one used in Section 1.3 can be used to model and solve real-life problems that involve inequalities, as illustrated in Example 6.

> **EXAMPLE 6** **Comparative Shopping**

You are choosing between two different car-sharing plans. Plan A has a membership fee of $12.00 per month plus $9.25 per hour. Plan B does not have a membership fee and charges $10.75 per hour. How many hours must you use the car in one month for plan B to cost more than plan A?

Solution

Verbal model: | Monthly cost for plan B | > | Monthly cost for plan A |

Labels: Hours used in one month = h (hours)

Monthly cost for Plan A = $9.25h + 12$ (dollars)

Monthly cost for Plan B = $10.75h$ (dollars)

Inequality: $10.75h > 9.25h + 12$ Write inequality.

$1.5h > 12$ Subtract 9.25h from each side.

$h > 8$ Divide each side by 1.5.

So, Plan B costs more when you use the car for more than 8 hours in one month. The table below helps confirm this conclusion.

Hours	4	5	6	7	8	9
Plan A	$49.00	$58.25	$67.50	$76.75	$86.00	$95.25
Plan B	$43.00	$53.75	$64.50	$75.25	$86.00	$96.75

✓ *Checkpoint*  Audio-video solution in English & Spanish at LarsonPrecalculus.com

You are considering two job offers. The first job pays $14.50 per hour. The second job pays $10.00 per hour plus $0.75 per unit produced per hour. Write an inequality for the number of units that must be produced per hour so that the second job yields the greater hourly wage. Solve the inequality. ■

> **Summarize** (Section 1.7)
>
> 1. Explain how to use inequalities to represent intervals *(page 131)*. For an example of writing inequalities that represent intervals, see Example 1.
> 2. State the properties of inequalities *(page 132)*.
> 3. Explain how to solve a linear inequality in one variable *(page 133)*. For examples of solving linear inequalities in one variable, see Examples 2–4.
> 4. Explain how to solve an absolute value inequality *(page 135)*. For an example of solving absolute value inequalities, see Example 5.
> 5. Describe a real-life application of a linear inequality in one variable *(page 136, Example 6)*.

GO DIGITAL

1.7 Exercises

See CalcChat.com for tutorial help and worked-out solutions to odd-numbered exercises.

GO DIGITAL

Vocabulary and Concept Check

In Exercises 1–4, fill in the blanks.

1. The set of all real numbers that are solutions of an inequality is the _____ _____ of the inequality.

2. The set of all points on the real number line that represents the solution set of an inequality is the _____ of the inequality.

3. It is sometimes possible to write two inequalities as a _____ inequality.

4. The symbol $\cup$ is the _____ symbol, which denotes the combining of two sets.

5. Are the inequalities $x - 4 < 5$ and $x > 9$ equivalent?

6. Which property of inequalities is shown below?

$a < b$ and $b < c \implies a < c$

Skills and Applications

Intervals and Inequalities In Exercises 7–14, state whether the interval is bounded or unbounded, represent the interval with an inequality, and sketch its graph.

7. $[-2, 6)$ 8. $(-7, 4)$

9. $[-1, 5]$ 10. $(2, 10]$

11. $(11, \infty)$ 12. $[-5, \infty)$

13. $(-\infty, 7]$ 14. $(-\infty, -2)$

Solving a Linear Inequality In Exercises 15–32, solve the inequality. Then graph the solution set.

15. $4x < 12$ 16. $10x < -40$

17. $-2x > -3$ 18. $-6x > 15$

19. $2x - 5 \geq 7$ 20. $5x + 7 \leq 12$

21. $2x + 7 < 3 + 4x$ 22. $3x + 1 \geq 2 + x$

23. $3x - 4 \geq 4 - 5x$ 24. $6x - 4 \leq 2 + 8x$

25. $4 - 2x < 3(3 - x)$ 26. $4(x + 1) < 2x + 3$

27. $\frac{3}{4}x - 6 \leq x - 7$ 28. $3 + \frac{2}{7}x > x - 2$

29. $\frac{1}{2}(8x + 1) \geq 3x + \frac{5}{2}$ 30. $9x - 1 < \frac{3}{4}(16x - 2)$

31. $3.6x + 11 \geq -3.4$ 32. $15.6 - 1.3x < -5.2$

Solving a Double Inequality In Exercises 33–40, solve the inequality. Then graph the solution set.

33. $1 < 2x + 3 < 9$ 34. $-9 \leq -2x - 7 < 5$

35. $-1 \leq -(x - 4) < 7$ 36. $0 < 3(x + 7) \leq 20$

37. $-4 < \dfrac{2x - 3}{3} < 4$ 38. $0 \leq \dfrac{x + 3}{2} < 5$

39. $-1 < \dfrac{-x - 2}{3} \leq 1$

40. $-1 \leq \dfrac{-3x + 5}{7} \leq 2$

Solving an Absolute Value Inequality In Exercises 41–50, solve the inequality. Then graph the solution set. (Some inequalities have no solution.)

41. $|x| < 5$ 42. $|x| \geq 8$

43. $\left|\dfrac{x}{2}\right| > 1$ 44. $\left|\dfrac{x}{3}\right| < 2$

45. $|x - 5| < -1$ 46. $|x - 7| < -5$

47. $|7 - 2x| \geq 9$ 48. $|1 - 2x| < 5$

49. $\left|\dfrac{x - 3}{2}\right| \geq 4$ 50. $\left|1 - \dfrac{2x}{3}\right| < 1$

Identifying a Solution Set In Exercises 51–58, use a graphing utility to graph the inequality and identify the solution set.

51. $8 - 3x \geq 2$ 52. $20 < 6x - 1$

53. $4(x - 3) \leq 8 - x$ 54. $3(x + 1) < x + 7$

55. $|x - 8| \leq 14$ 56. $|2x + 9| > 13$

57. $2|x + 7| \geq 13$ 58. $\frac{1}{2}|x + 1| \leq 3$

Approximating Solution Sets In Exercises 59–64, use a graphing utility to graph the equation. Use the graph to approximate the values of x that satisfy each inequality.

Equation	Inequalities	
59. $y = 3x - 1$	(a) $y \geq 2$	(b) $y \leq 0$
60. $y = \frac{2}{3}x + 1$	(a) $y \leq 5$	(b) $y \geq 0$
61. $y = -\frac{1}{2}x + 2$	(a) $0 \leq y \leq 3$	(b) $y \geq 0$
62. $y = -3x + 8$	(a) $-1 \leq y \leq 3$	(b) $y \leq 0$
63. $y = |x - 3|$	(a) $y \leq 2$	(b) $y \geq 4$
64. $y = \left|\frac{1}{2}x + 1\right|$	(a) $y \leq 4$	(b) $y \geq 1$

Using Absolute Value **In Exercises 65–72, use absolute value notation to define the interval (or pair of intervals) on the real number line.**

65.

66.

67.

68.

69. All real numbers at least three units from 7

70. All real numbers more than five units from 8

71. All real numbers less than four units from -3

72. All real numbers no more than seven units from -6

Writing an Inequality **In Exercises 73–76, write an inequality to describe the situation.**

73. During a trading day, the price P of a stock is no less than \$7.25 and no more than \$7.75.

74. During a month, a person's weight w is greater than 180 pounds but less than 185.5 pounds.

75. The expected return r on an investment is no more than 8%.

76. The expected net income I of a company is no less than \$239 million.

Physiology **One formula that relates a person's maximum heart rate r (in beats per minute) to the person's age A (in years) is $r = 220 - A$. In Exercises 77 and 78, determine the interval in which the person's heart rate is from 50% to 85% of the maximum heart rate.** *(Source: American Heart Association)*

77. a 20-year-old

78. a 40-year-old

79. **Job Offers** You are considering two job offers. The first job pays \$13.50 per hour. The second job pays \$9.00 per hour plus \$0.75 per unit produced per hour. How many units must you produce per hour for the second job to pay more per hour than the first job?

80. **Job Offers** You are considering two job offers. The first job pays \$13.75 per hour. The second job pays \$10.00 per hour plus \$1.25 per unit produced per hour. How many units must you produce per hour for the second job to pay more than the first job?

81. **Investment** What annual simple interest rates yield a balance of more than \$2000 on a 10-year investment of \$1000?

82. **Investment** What annual simple interest rates yield a balance of more than \$750 on a 5-year investment of \$500?

83. **Cost, Revenue, and Profit** The revenue from selling x units of a product is $R = 115.95x$. The cost of producing x units is $C = 95x + 750$. To obtain a profit, the revenue must be greater than the cost. For what values of x does this product return a profit?

84. **Cost, Revenue, and Profit** The revenue from selling x units of a product is $R = 24.55x$. The cost of producing x units is $C = 15.4x + 150,000$. To obtain a profit, the revenue must be greater than the cost. For what values of x does this product return a profit?

85. **Daily Sales** A doughnut shop sells a dozen doughnuts for \$7.95. Beyond the fixed costs (rent, utilities, and insurance) of \$165 per day, it costs \$1.45 for enough materials and labor to produce a dozen doughnuts. The daily profit from doughnut sales varies between \$400 and \$1200. Between what levels (in dozens of doughnuts) do the daily sales vary?

86. **Weight Loss Program** A person enrolls in a diet and exercise program that guarantees a loss of at least $1\frac{1}{2}$ pounds per week. The person's weight at the beginning of the program is 164 pounds. Find the maximum number of weeks before the person attains a goal weight of 128 pounds.

87. **GPA** An equation that relates the college grade-point averages y and high school grade-point averages x of the students at a college is $y = 0.692x + 0.988$.

   (a) Use a graphing utility to graph the model.

   (b) Use the graph to estimate the values of x that predict a college grade-point average of at least 3.0.

   (c) Verify your estimate from part (b) algebraically.

88. **Weightlifting** The 6RM load for a weightlifting exercise is the maximum weight at which a person can perform six repetitions. An equation that relates an athlete's 6RM bench press load x (in kilograms) and the athlete's 6RM barbell curl load y (in kilograms) is $y = 0.33x + 6.20$. *(Source: Journal of Sports Science & Medicine)*

   (a) Use a graphing utility to graph the model.

   (b) Use the graph to estimate the values of x that predict a 6RM barbell curl load of no more than 80 kilograms.

   (c) Verify your estimate from part (b) algebraically.

89. **Civil Engineers' Wages** The mean hourly wage W (in dollars) of civil engineers in the United States from 2010 through 2018 can be modeled by

$$W = 0.693t + 32.23, \quad 10 \le t \le 18$$

where t represents the year, with $t = 10$ corresponding to 2010. *(Source: U.S. Bureau of Labor Statistics)*

   (a) According to the model, when was the mean hourly wage at least \$40, but no more than \$42?

   (b) Use the model to determine when the mean hourly wage exceeded \$44.

90. Milk Production Milk production M (in billions of pounds) in the United States from 2010 through 2018 can be modeled by $M = 3.125t + 161.93$, $10 \le t \le 18$, where t represents the year, with $t = 10$ corresponding to 2010. *(Source: U.S. Department of Agriculture)*

(a) According to the model, when was the annual milk production greater than 200 billion pounds, but no more than 210 billion pounds?

(b) Use the model to determine when milk production exceeded 212 billion pounds.

91. Time Study The times required to perform a task in a manufacturing process by approximately two-thirds of the employees in a study satisfy the inequality $|t - 15.6| \le 1.9$, where t is time in minutes. Determine the interval in which these times lie.

92. Geography

A geographic information system reports that the distance between two locations is 206 meters. The system is accurate to within 3 meters.

(a) Write an absolute value inequality for the possible distances between the locations.

(b) Graph the solution set.

93. Accuracy of Measurement You buy 6 T-bone steaks that cost \$8.99 per pound. The weight that is listed on the package is 5.72 pounds. The scale that weighed the package is accurate to within $\frac{1}{32}$ pound. How much might you be undercharged or overcharged?

94. Geometry The side length of a square machined automobile part is 24.2 centimeters with a possible error of 0.25 centimeter. Determine the interval containing the possible areas of the part.

Exploring the Concepts

True or False? **In Exercises 95–98, determine whether the statement is true or false. Justify your answer.**

95. If a, b, and c are real numbers, and $a < b$, then $a + c < b + c$.

96. If a, b, and c are real numbers, and $a \le b$, then $ac \le bc$.

97. If $-10 \le x \le 8$, then $-10 \ge -x$ and $-x \ge -8$.

98. If $|2x - 5| \le 0$, then $x = \frac{5}{2}$.

99. Think About It Give an example of an inequality whose solution set is $(-\infty, \infty)$.

100. **HOW DO YOU SEE IT?** The graph shows the relationship between volume and mass for aluminum bronze.

(a) Estimate the mass when the volume is 2 cubic centimeters.

(b) Approximate the interval for the mass when the volume is greater than or equal to 0 cubic centimeters and less than 4 cubic centimeters.

101. Error Analysis Describe the error.

$$|3x - 4| \ge -5$$
$$3x - 4 \le -5 \quad \text{or} \quad 3x - 4 \ge 5$$
$$3x \le -1 \qquad\qquad 3x \ge 9$$
$$x \le -\tfrac{1}{3} \qquad\qquad x \ge 3 \quad \boldsymbol{\times}$$

Review & Refresh ▶ Video solutions at LarsonPrecalculus.com

102. Compound Interest An investment of \$500 compounded annually for 2 years at interest rate r (in decimal form) yields an amount of $500(1 + r)^2$.

(a) Write this polynomial in standard form.

(b) Use a calculator to evaluate the polynomial for the values of r given in the table.

r	$2\frac{1}{2}\%$	3%	4%	$4\frac{1}{2}\%$	5%
$500(1 + r)^2$					

Solving an Equation **In Exercises 103–110, solve the equation. Check your solutions.**

103. $x^2 - x - 6 = 0$

104. $x^2 - x - 20 = 0$

105. $4x^2 - 5x = 6$

106. $2x^2 + 3x = 5$

107. $2x^3 - 3x^2 = 32x - 48$

108. $3x^3 - x^2 = 12x - 4$

109. $\dfrac{2x - 7}{x - 5} = 3$

110. $\dfrac{1}{x - 3} = \dfrac{3}{2x + 1}$

Sketching the Graph of an Equation **In Exercises 111–114, test for symmetry and graph the equation. Then identify any intercepts.**

111. $y = x^2 - 2x - 3$

112. $y = x^2 + 2x + 1$

113. $y = 64 - 4x^2$

114. $y = 2x^2 - 18$

1.8 Other Types of Inequalities

Nonlinear inequalities have many real-life applications. For example, in Exercises 63 and 64 on page 148, you will use a polynomial inequality to model the height of a projectile.

◉ Solve polynomial inequalities.
◉ Solve rational inequalities.
◉ Use nonlinear inequalities to model and solve real-life problems.

Polynomial Inequalities

To solve a polynomial inequality such as

$$x^2 - 2x - 3 < 0 \qquad \text{Polynomial inequality}$$

use the fact that a polynomial can change signs only at its *zeros* (the x-values that make the polynomial equal to zero). Between two consecutive zeros, a polynomial must be entirely positive or entirely negative. This means that when the real zeros of a polynomial are put in order, they divide the real number line into intervals in which the polynomial has no sign changes. These zeros are the **key numbers** of the inequality, and the resulting open intervals are the **test intervals** for the inequality.

For example, the polynomial $x^2 - 2x - 3$ factors as

$$x^2 - 2x - 3 = (x + 1)(x - 3)$$

so it has two zeros: $x = -1$ and $x = 3$. These zeros divide the real number line into three test intervals:

$$(-\infty, -1), \quad (-1, 3), \quad \text{and} \quad (3, \infty)$$

as shown in the figure below.

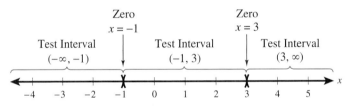

Three test intervals for $x^2 - 2x - 3$

To solve the inequality $x^2 - 2x - 3 < 0$, you need to test only one value in each test interval. When a value from a test interval satisfies the original inequality, you can conclude that the interval is a solution of the inequality. This approach, summarized below, can be used to determine the test intervals for any polynomial.

Finding Test Intervals for a Polynomial

To determine the intervals on which the values of a polynomial are entirely negative or entirely positive, use the steps below.

1. Find all real zeros of the polynomial, and arrange the zeros in increasing order. These zeros are the key numbers of the inequality.

2. Use the key numbers of the inequality to determine the test intervals.

3. Choose one representative x-value in each test interval and evaluate the polynomial at that value. When the value of the polynomial is negative, the polynomial has negative values for every x-value in the interval. When the value of the polynomial is positive, the polynomial has positive values for every x-value in the interval.

GO DIGITAL

© Digidreamgrafix/Shutterstock.com

EXAMPLE 1 **Solving a Polynomial Inequality**

Solve $x^2 - x - 6 < 0$. Then graph the solution set.

Solution Factoring the polynomial

$$x^2 - x - 6 = (x + 2)(x - 3)$$

shows that the key numbers are $x = -2$ and $x = 3$. So, the test intervals are

$$(-\infty, -2), \quad (-2, 3), \quad \text{and} \quad (3, \infty). \qquad \text{Test interval}$$

In each test interval, choose a representative x-value and evaluate the polynomial.

Test Interval	x-Value	Polynomial Value	Conclusion
$(-\infty, -2)$	$x = -3$	$(-3)^2 - (-3) - 6 = 6$	Positive
$(-2, 3)$	$x = 0$	$(0)^2 - (0) - 6 = -6$	Negative
$(3, \infty)$	$x = 4$	$(4)^2 - (4) - 6 = 6$	Positive

The inequality is satisfied for all x-values in $(-2, 3)$. This implies that the solution set of the inequality

$$x^2 - x - 6 < 0$$

is the interval $(-2, 3)$, as shown on the number line below. Note that the original inequality contains a "less than" symbol. This means that the solution set does not contain the endpoints of the test interval $(-2, 3)$.

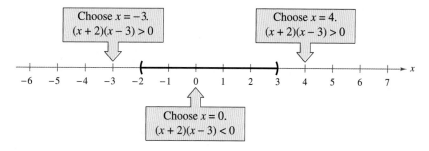

$$\begin{array}{l}\text{Choose } x = -3. \\ (x + 2)(x - 3) > 0\end{array}$$

$$\begin{array}{l}\text{Choose } x = 4. \\ (x + 2)(x - 3) > 0\end{array}$$

$$\begin{array}{l}\text{Choose } x = 0. \\ (x + 2)(x - 3) < 0\end{array}$$

 Checkpoint Audio-video solution in English & Spanish at LarsonPrecalculus.com

Solve $x^2 - x - 20 < 0$. Then graph the solution set. ∎

As with linear inequalities, you can check the reasonableness of a solution by substituting x-values into the original inequality. For instance, to check the solution to Example 1, substitute several x-values from the interval $(-2, 3)$ into the inequality

$$x^2 - x - 6 < 0.$$

Regardless of which x-values you choose, the inequality should be satisfied.

You can also use a graph to check the result of Example 1. Sketch the graph of

$$y = x^2 - x - 6$$

as shown in Figure 1.16. Notice that the graph is below the x-axis on the interval $(-2, 3)$, which supports the solution to Example 1 that $x^2 - x - 6 < 0$ on the interval $(-2, 3)$.

In Example 1, the polynomial inequality is in general form (with the polynomial on one side and zero on the other). Whenever this is not the case, you should begin by writing the inequality in general form.

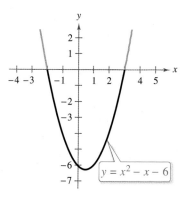

Figure 1.16

EXAMPLE 2 **Solving a Polynomial Inequality**

$$2x^3 - 3x^2 - 32x + 48 > 0 \qquad \text{Original inequality}$$
$$x^2(2x - 3) - 16(2x - 3) > 0 \qquad \text{Factor by grouping.}$$
$$(x^2 - 16)(2x - 3) > 0 \qquad \text{Distributive Property}$$
$$(x - 4)(x + 4)(2x - 3) > 0 \qquad \text{Factor difference of two squares.}$$

The key numbers are $x = -4$, $x = \frac{3}{2}$, and $x = 4$, and the test intervals are $(-\infty, -4)$, $\left(-4, \frac{3}{2}\right)$, $\left(\frac{3}{2}, 4\right)$, and $(4, \infty)$.

Test Interval	x-Value	Polynomial Value	Conclusion
$(-\infty, -4)$	$x = -5$	$2(-5)^3 - 3(-5)^2 - 32(-5) + 48 = -117$	Negative
$\left(-4, \frac{3}{2}\right)$	$x = 0$	$2(0)^3 - 3(0)^2 - 32(0) + 48 = 48$	Positive
$\left(\frac{3}{2}, 4\right)$	$x = 2$	$2(2)^3 - 3(2)^2 - 32(2) + 48 = -12$	Negative
$(4, \infty)$	$x = 5$	$2(5)^3 - 3(5)^2 - 32(5) + 48 = 63$	Positive

The inequality is satisfied on the open intervals $\left(-4, \frac{3}{2}\right)$ and $(4, \infty)$. So, the solution set is $\left(-4, \frac{3}{2}\right) \cup (4, \infty)$, as shown on the number line below.

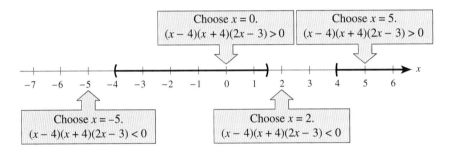

Choose $x = 0$.
$(x - 4)(x + 4)(2x - 3) > 0$

Choose $x = 5$.
$(x - 4)(x + 4)(2x - 3) > 0$

Choose $x = -5$.
$(x - 4)(x + 4)(2x - 3) < 0$

Choose $x = 2$.
$(x - 4)(x + 4)(2x - 3) < 0$

GO DIGITAL

✓ *Checkpoint* ▶ Audio-video solution in English & Spanish at *LarsonPrecalculus.com*

Solve $3x^3 - x^2 - 12x > -4$. Then graph the solution set.

EXAMPLE 3 **Solving a Polynomial Inequality**

▶▶▶ *See LarsonPrecalculus.com for an interactive version of this type of example.*

Solve $4x^2 - 5x > 6$.

Algebraic Solution

$$4x^2 - 5x - 6 > 0 \qquad \text{Write in general form.}$$
$$(x - 2)(4x + 3) > 0 \qquad \text{Factor.}$$

Key numbers: $x = -\frac{3}{4}$, $x = 2$

Test intervals: $\left(-\infty, -\frac{3}{4}\right)$, $\left(-\frac{3}{4}, 2\right)$, $(2, \infty)$

Test: Is $(x - 2)(4x + 3) > 0$?

Testing these intervals shows that the polynomial $4x^2 - 5x - 6$ is positive on the open intervals $\left(-\infty, -\frac{3}{4}\right)$ and $(2, \infty)$. So, the solution set of the inequality is $\left(-\infty, -\frac{3}{4}\right) \cup (2, \infty)$.

Graphical Solution

First write the polynomial inequality $4x^2 - 5x > 6$ as $4x^2 - 5x - 6 > 0$. Then use a graphing utility to graph $y = 4x^2 - 5x - 6$.

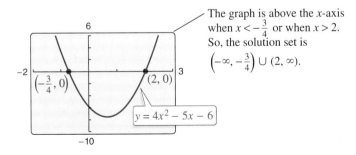

The graph is above the x-axis when $x < -\frac{3}{4}$ or when $x > 2$. So, the solution set is $\left(-\infty, -\frac{3}{4}\right) \cup (2, \infty)$.

$y = 4x^2 - 5x - 6$

✓ *Checkpoint* ▶ Audio-video solution in English & Spanish at *LarsonPrecalculus.com*

Solve $2x^2 + 3x < 5$ (a) algebraically and (b) graphically.

You may find it easier to determine the sign of a polynomial from its *factored* form. For instance, in Example 3, when you substitute the test value $x = 1$ into the factored form

$$(x - 2)(4x + 3)$$

the sign pattern of the factors is $(-)(+)$, which yields a negative result. Rework the other examples in this section using the factored forms of the polynomials to determine the signs of the polynomials in the test intervals. Do you get the same results?

When solving a polynomial inequality, be sure to account for the inequality symbol. For instance, in Example 3, note that the original inequality symbol is "greater than" and the solution consists of two open intervals. If the original inequality had been

$$4x^2 - 5x \geq 6$$

then the solution set would have been $\left(-\infty, -\frac{3}{4}\right] \cup [2, \infty)$.

Each of the polynomial inequalities in Examples 1, 2, and 3 has a solution set that consists of a single interval or the union of two intervals. When solving the exercises for this section, watch for unusual solution sets, as illustrated in Example 4.

EXAMPLE 4   Unusual Solution Sets

a. The solution set of $x^2 + 2x + 4 > 0$ consists of the entire set of real numbers, $(-\infty, \infty)$. In other words, the value of the quadratic polynomial $x^2 + 2x + 4$ is positive for every real value of x, as shown in Figure 1.17(a).

b. The solution set of $x^2 + 2x + 1 \leq 0$ consists of the single real number $\{-1\}$, because the inequality has only one key number, $x = -1$, and it is the only value that satisfies the inequality, as shown in Figure 1.17(b).

c. The solution set of $x^2 + 3x + 5 < 0$ is empty. In other words, $x^2 + 3x + 5$ is not less than zero for any value of x, as shown in Figure 1.17(c).

d. The solution set of $x^2 - 4x + 4 > 0$ consists of all real numbers except $x = 2$. This solution set can be written in interval notation as $(-\infty, 2) \cup (2, \infty)$, as shown in Figure 1.17(d).

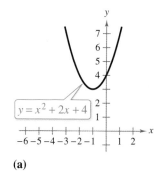 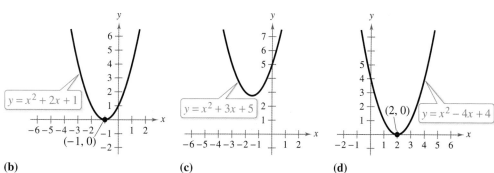

(a)   (b)   (c)   (d)

Figure 1.17

✓ *Checkpoint* ▶ Audio-video solution in English & Spanish at LarsonPrecalculus.com

What is unusual about the solution set of each inequality?

a. $x^2 + 6x + 9 < 0$

b. $x^2 + 4x + 4 \leq 0$

c. $x^2 - 6x + 9 > 0$

d. $x^2 - 2x + 1 \geq 0$

GO DIGITAL

Rational Inequalities

The concepts of key numbers and test intervals can be extended to rational inequalities. Use the fact that the value of a rational expression can change sign at its *zeros* (the *x*-values for which its numerator is zero) and at its *undefined values* (the *x*-values for which its denominator is zero). These two types of numbers make up the *key numbers* of a rational inequality. When solving a rational inequality, begin by writing the inequality in general form, that is, with zero on the right side of the inequality.

EXAMPLE 5   **Solving a Rational Inequality**

Solve $\dfrac{2x - 7}{x - 5} \le 3$. Then graph the solution set.

Solution

$$\frac{2x - 7}{x - 5} \le 3 \qquad \text{Write original inequality.}$$

$$\frac{2x - 7}{x - 5} - 3 \le 0 \qquad \text{Write in general form.}$$

$$\frac{2x - 7 - 3x + 15}{x - 5} \le 0 \qquad \text{Find the LCD and subtract fractions.}$$

$$\frac{-x + 8}{x - 5} \le 0 \qquad \text{Simplify.}$$

Key numbers: $x = 5, x = 8$ Zeros and undefined values of rational expression

Test intervals: $(-\infty, 5), (5, 8), (8, \infty)$

Test: Is $\dfrac{-x + 8}{x - 5} \le 0$?

Testing these intervals, as shown in the figure below, the inequality is satisfied on the open intervals $(-\infty, 5)$ and $(8, \infty)$. Moreover,

$$\frac{-x + 8}{x - 5} = 0$$

when $x = 8$, so the solution set is $(-\infty, 5) \cup [8, \infty)$. (Be sure to use a bracket to signify that $x = 8$ is included in the solution set.)

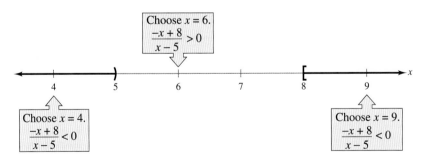

✓ *Checkpoint* ▶ Audio-video solution in English & Spanish at LarsonPrecalculus.com

Solve each inequality. Then graph the solution set.

a. $\dfrac{x - 2}{x - 3} \ge -3$

b. $\dfrac{4x - 1}{x - 6} > 3$

GO DIGITAL

Applications

One common application of inequalities comes from business and involves profit, revenue, and cost. The formula that relates these three quantities is

Profit = Revenue − Cost

$$P = R - C.$$

EXAMPLE 6   Profit from a Product

The marketing department of a calculator manufacturer determines that the demand for a new model of calculator is

$$p = 100 - 0.00001x, \quad 0 \le x \le 10{,}000{,}000 \qquad \text{Demand equation}$$

where p is the price per calculator (in dollars) and x represents the number of calculators sold. (According to this model, no one would be willing to pay \$100 for the calculator. At the other extreme, the company could not *give* away more than 10 million calculators.) The revenue for selling x calculators is

$$R = xp = x(100 - 0.00001x). \qquad \text{Revenue equation}$$

The total cost of producing x calculators is \$10 per calculator plus a one-time development cost of \$2,500,000. So, the total cost is

$$C = 10x + 2{,}500{,}000. \qquad \text{Cost equation}$$

What prices can the company charge per calculator to obtain a profit of at least \$190,000,000?

Solution

Verbal model: Profit = Revenue − Cost

Equation: $P = R - C$

$$P = 100x - 0.00001x^2 - (10x + 2{,}500{,}000)$$

$$P = -0.00001x^2 + 90x - 2{,}500{,}000$$

To answer the question, solve the inequality

$$P \ge 190{,}000{,}000$$

$$-0.00001x^2 + 90x - 2{,}500{,}000 \ge 190{,}000{,}000.$$

Write the inequality in general form, find the key numbers and the test intervals, and then test a value in each test interval to find that the solution is

$$3{,}500{,}000 \le x \le 5{,}500{,}000$$

as shown in Figure 1.18. Substituting the x-values in the original demand equation shows that prices of

$$\$45.00 \le p \le \$65.00$$

yield a profit of at least \$190,000,000.

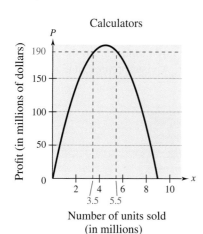

Calculators

Profit (in millions of dollars)

Number of units sold
(in millions)

Figure 1.18

✓ *Checkpoint* ▶ *Audio-video solution in English & Spanish at LarsonPrecalculus.com*

The revenue and cost equations for a product are

$$R = x(60 - 0.0001x) \quad \text{and} \quad C = 12x + 1{,}800{,}000$$

where R and C are measured in dollars and x represents the number of units sold. How many units must be sold to obtain a profit of at least \$3,600,000?

GO DIGITAL

Another common application of inequalities is finding the domain of an expression that involves a square root, as shown in Example 7.

EXAMPLE 7 Finding the Domain of an Expression

Find the domain of $\sqrt{64 - 4x^2}$.

Algebraic Solution

Recall that the domain of an expression is the set of all x-values for which the expression is defined. The expression $\sqrt{64 - 4x^2}$ is defined only when $64 - 4x^2$ is nonnegative, so the inequality $64 - 4x^2 \geq 0$ gives the domain.

$$64 - 4x^2 \geq 0 \qquad \text{Write in general form.}$$

$$16 - x^2 \geq 0 \qquad \text{Divide each side by 4.}$$

$$(4 - x)(4 + x) \geq 0 \qquad \text{Write in factored form.}$$

The inequality has two key numbers: $x = -4$ and $x = 4$. Use these two numbers to test the inequality.

Key numbers: $x = -4, x = 4$

Test intervals: $(-\infty, -4), (-4, 4), (4, \infty)$

Test: Is $(4 - x)(4 + x) \geq 0$?

A test shows that the inequality is satisfied in the *closed interval* $[-4, 4]$. So, the domain of the expression $\sqrt{64 - 4x^2}$ is the closed interval $[-4, 4]$.

Graphical Solution

Begin by sketching the graph of the equation $y = \sqrt{64 - 4x^2}$, as shown below. The graph shows that the x-values extend from -4 to 4 (including -4 and 4). So, the domain of the expression $\sqrt{64 - 4x^2}$ is the closed interval $[-4, 4]$.

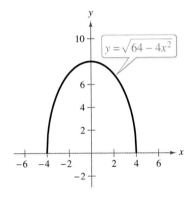

$y = \sqrt{64 - 4x^2}$

✓ *Checkpoint* ▶ **Audio-video solution in English & Spanish at LarsonPrecalculus.com**

Find the domain of $\sqrt{x^2 - 7x + 10}$.

You can check the reasonableness of the solution to Example 7 by choosing a representative x-value in the interval and evaluating the radical expression at that value. When you substitute any number from the closed interval $[-4, 4]$ into the expression $\sqrt{64 - 4x^2}$, you obtain a nonnegative number under the radical symbol that simplifies to a real number. When you substitute any number from the intervals $(-\infty, -4)$ and $(4, \infty)$, you obtain a complex number. A visual representation of the intervals is shown below.

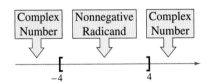

Summarize (Section 1.8)

1. Explain how to solve a polynomial inequality *(page 140)*. For examples of solving polynomial inequalities, see Examples 1–4.

2. Explain how to solve a rational inequality *(page 144)*. For an example of solving a rational inequality, see Example 5.

3. Describe applications of polynomial inequalities *(pages 145 and 146, Examples 6 and 7)*.

GO DIGITAL

1.8 Exercises

See CalcChat.com for tutorial help and worked-out solutions to odd-numbered exercises.

Vocabulary and Concept Check

In Exercises 1 and 2, fill in the blanks.

1. Between two consecutive zeros, a polynomial must be entirely _____ or entirely _____.
2. A rational expression can change sign at its _____ and its _____ _____.

3. The test intervals of a polynomial inequality are $(-\infty, -2)$, $(-2, 5)$, and $(5, \infty)$. What are the key numbers of the inequality?
4. Is $x = 4$ a solution of the inequality $x(x - 4) < 0$? Explain.

Skills and Applications

Checking Solutions In Exercises 5–8, determine whether each value of x is a solution of the inequality.

Inequality	Values
5. $x^2 - 3 < 0$	(a) $x = 3$ (b) $x = 0$
	(c) $x = \frac{3}{2}$ (d) $x = -5$
6. $x^2 - 2x - 8 \geq 0$	(a) $x = -2$ (b) $x = 0$
	(c) $x = -4$ (d) $x = 1$
7. $\dfrac{x + 2}{x - 4} \geq 3$	(a) $x = 5$ (b) $x = 4$
	(c) $x = -\frac{9}{2}$ (d) $x = \frac{9}{2}$
8. $\dfrac{3x^2}{x^2 + 4} < 1$	(a) $x = -2$ (b) $x = -1$
	(c) $x = 0$ (d) $x = 3$

Finding Zeros and Undefined Values In Exercises 9–12, find the zeros and the undefined values (if any) of the expression.

9. $x^2 - 3x - 18$
10. $9x^3 - 25x^2$
11. $\dfrac{1}{x - 5} + 1$
12. $\dfrac{x}{x + 2} - \dfrac{2}{x - 1}$

Solving a Polynomial Inequality In Exercises 13–34, solve the inequality. Then graph the solution set.

13. $2x^2 + 4x < 0$
14. $3x^2 - 9x \geq 0$
15. $x^2 < 9$
16. $x^2 \leq 25$
17. $(x + 2)^2 \leq 25$
18. $(x - 3)^2 \geq 1$
19. $x^2 + 6x + 1 \geq -7$
20. $x^2 - 8x + 2 < 11$
21. $x^2 + x < 6$
22. $x^2 + 2x > 3$
23. $x^2 < 3 - 2x$
24. $x^2 > 2x + 8$
25. $3x^2 - 11x > 20$
26. $-2x^2 + 6x \leq -15$
27. $x^3 - 3x^2 - x + 3 > 0$
28. $x^3 + 2x^2 - 4x \leq 8$
29. $-x^3 + 7x^2 + 9x > 63$
30. $2x^3 + 13x^2 - 8x \geq 52$
31. $4x^3 - 6x^2 < 0$
32. $x^3 - 4x \geq 0$
33. $(x - 1)^2(x + 2)^3 \geq 0$
34. $x^4(x - 3) \leq 0$

Unusual Solution Sets In Exercises 35–38, explain what is unusual about the solution set of the inequality.

35. $4x^2 - 4x + 1 \leq 0$
36. $x^2 + 3x + 8 > 0$
37. $x^2 - 6x + 12 \leq 0$
38. $x^2 - 8x + 16 > 0$

Solving a Rational Inequality In Exercises 39–48, solve the inequality. Then graph the solution set.

39. $\dfrac{4x - 1}{x} > 0$
40. $\dfrac{x - 1}{x} < 0$
41. $\dfrac{3x + 5}{x - 1} < 2$
42. $\dfrac{x + 12}{x + 2} \geq 3$
43. $\dfrac{2}{x + 5} > \dfrac{1}{x - 3}$
44. $\dfrac{5}{x - 6} > \dfrac{3}{x + 2}$
45. $\dfrac{x^2 + 2x}{x^2 - 9} \leq 0$
46. $\dfrac{x^2 + x - 6}{x^2 - 4x} \geq 0$
47. $\dfrac{3}{x - 1} + \dfrac{2x}{x + 1} > -1$
48. $\dfrac{3x}{x - 1} \leq \dfrac{x}{x + 4} + 3$

Approximating Solution Sets In Exercises 49–56, use a graphing utility to graph the equation. Use the graph to approximate the values of x that satisfy each inequality.

Equation	Inequalities
49. $y = -x^2 + 2x + 3$	(a) $y \leq 0$ (b) $y \geq 3$
50. $y = \frac{1}{2}x^2 - 2x + 1$	(a) $y \leq 0$ (b) $y \geq 7$
51. $y = \frac{1}{8}x^3 - \frac{1}{2}x$	(a) $y \geq 0$ (b) $y \leq 6$
52. $y = x^3 - x^2 - 16x + 16$	(a) $y \leq 0$ (b) $y \geq 36$
53. $y = \dfrac{3x}{x - 2}$	(a) $y \leq 0$ (b) $y \geq 6$
54. $y = \dfrac{2(x - 2)}{x + 1}$	(a) $y \leq 0$ (b) $y \geq 8$
55. $y = \dfrac{2x^2}{x^2 + 4}$	(a) $y \geq 1$ (b) $y \leq 2$
56. $y = \dfrac{5x}{x^2 + 4}$	(a) $y \geq 1$ (b) $y \leq 0$

Solving an Inequality   In Exercises 57–62, solve the inequality. (**Round your answers to two decimal places.**)

57. $0.3x^2 + 6.26 < 10.8$

58. $-1.3x^2 + 3.78 > 2.12$

59. $12.5x + 1.6 > 0.5x^2$

60. $1.2x^2 + 4.8x + 3.1 < 5.3$

61. $\dfrac{1}{2.3x - 5.2} > 3.4$

62. $\dfrac{2}{3.1x - 3.7} > 5.8$

Height of a Projectile

In Exercises 63 and 64, use the position equation

$$s = -16t^2 + v_0 t + s_0$$

where s represents the height of an object (in feet), v_0 represents the initial velocity of the object (in feet per second), s_0 represents the initial height of the object (in feet), and t represents the time (in seconds).

63. A projectile is fired straight upward from ground level ($s_0 = 0$) with an initial velocity of 160 feet per second.

   (a) At what instant will it be back at ground level?

   (b) When will the height exceed 384 feet?

64. A projectile is fired straight upward from ground level ($s_0 = 0$) with an initial velocity of 128 feet per second.

   (a) At what instant will it be back at ground level?

   (b) When will the height be less than 128 feet?

65. **Cost, Revenue, and Profit**   The revenue and cost equations for a product are $R = x(75 - 0.0005x)$ and $C = 30x + 250,000$, where R and C are measured in dollars and x represents the number of units sold. How many units must be sold to obtain a profit of at least $750,000? What is the price per unit?

66. **Cost, Revenue, and Profit**   The revenue and cost equations for a product are $R = x(50 - 0.0002x)$ and $C = 12x + 150,000$, where R and C are measured in dollars and x represents the number of units sold. How many units must be sold to obtain a profit of at least $1,650,000? What is the price per unit?

Finding the Domain of an Expression   In Exercises 67–72, find the domain of the expression. Use a graphing utility to verify your result.

67. $\sqrt{4 - x^2}$

68. $\sqrt{x^2 - 9}$

69. $\sqrt{x^2 - 9x + 20}$

70. $\sqrt{49 - x^2}$

71. $\sqrt{\dfrac{x}{x^2 - 2x - 35}}$

72. $\sqrt{\dfrac{x}{x^2 - 9}}$

73. **Teachers' Salaries**   The average annual salary S (in thousands of dollars) of public school teachers in the United States from 2010 through 2018 can be approximated by the model

$$S = \frac{52.88 - 1.89t}{1 - 0.038t}, \quad 10 \le t \le 18$$

where t represents the year, with $t = 10$ corresponding to 2010. (*Source: National Center for Education Statistics*)

   (a) Use a graphing utility to graph the model.

   (b) Use the model to determine when the mean salary was less than $57,000.

   (c) Is the model valid for long-term predictions of public school teacher salaries? Explain.

74. **School Enrollment**   The table shows the projected numbers N (in millions) of students enrolled in public elementary and secondary schools in the United States from 2021 through 2028. (*Source: National Center for Education Statistics*)

DATA	Year	Number, N
	2021	50.89
	2022	51.01
	2023	51.10
	2024	51.12
	2025	51.12
	2026	51.12
	2027	51.23
	2028	51.42

Spreadsheet at LarsonPrecalculus.com

   (a) Use a graphing utility to create a scatter plot of the data. Let t represent the year, with $t = 21$ corresponding to 2021.

   (b) Use the *regression* feature of the graphing utility to find a *cubic* model for the data. (A cubic model has the form $at^3 + bt^2 + ct + d$, where a, b, c, and d are constant and t is variable.)

   (c) Graph the model and the scatter plot in the same viewing window. How well does the model fit the data?

   (d) According to the model, when will the number of students enrolled in public elementary and secondary schools exceed 51.20 million?

75. **Safe Load**   The maximum safe load uniformly distributed over a one-foot section of a two-inch-wide wooden beam can be approximated by the model

$$\text{Load} = 168.5d^2 - 472.1$$

where d is the depth of the beam.

   (a) Evaluate the model for $d = 4$, $d = 6$, $d = 8$, $d = 10$, and $d = 12$. Use the results to create a bar graph.

   (b) Determine the minimum depth of the beam that will safely support a load of 2000 pounds.

76. Resistors When two resistors of resistances R_1 and R_2 are connected in parallel (see figure), the total resistance R satisfies the equation

$$\frac{1}{R} = \frac{1}{R_1} + \frac{1}{R_2}.$$

Find R_1 for a parallel circuit in which $R_2 = 2$ ohms and R must be at least 1 ohm.

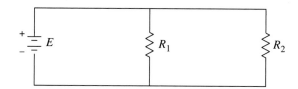

77. Geometry A rectangular playing field with a perimeter of 100 meters is to have an area of at least 500 square meters. Within what bounds must the length of the rectangle lie?

78. Geometry A rectangular parking lot with a perimeter of 440 feet is to have an area of at least 8000 square feet. Within what bounds must the length of the rectangle lie?

Exploring the Concepts

True or False? In Exercises 79 and 80, determine whether the statement is true or false. Justify your answer.

79. The zeros of the polynomial

$$x^3 - 2x^2 - 11x + 12 = (x + 3)(x - 1)(x - 4)$$

divide the real number line into three test intervals.

80. The solution set of the inequality $\frac{3}{2}x^2 + 3x + 6 \geq 0$ is the entire set of real numbers.

81. Graphical Reasoning Use a graphing utility to verify the results in Example 4. For instance, the graph of $y = x^2 + 2x + 4$ is shown below. Notice that the y-values are greater than 0 for all values of x, as stated in Example 4(a). Use the graphing utility to graph $y = x^2 + 2x + 1$, $y = x^2 + 3x + 5$, and $y = x^2 - 4x + 4$. Explain how you can use the graphs to verify the results of parts (b), (c), and (d) of Example 4.

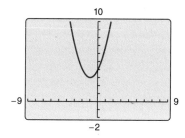

82. Writing Explain how the procedure for finding the key numbers of a polynomial inequality is different from the procedure for finding the key numbers of a rational inequality.

83. Error Analysis Describe the error.

The solution set of the inequality $1/x \geq 0$ is $[0, \infty)$.

84. **HOW DO YOU SEE IT?** Consider the polynomial

$$(x - a)(x - b)$$

and the real number line shown below.

(a) Identify the points on the line at which the polynomial is zero.

(b) For each of the three subintervals of the real number line, write the sign of each factor and the sign of the product.

(c) At what x-values does the polynomial change signs?

Conjecture In Exercises 85–88, (a) find the interval(s) for b such that the equation has at least one real solution and (b) write a conjecture about the interval(s) based on the values of the coefficients.

85. $x^2 + bx + 9 = 0$ **86.** $x^2 + bx - 9 = 0$

87. $3x^2 + bx + 10 = 0$ **88.** $2x^2 + bx + 5 = 0$

Review & Refresh ▶ *Video solutions at LarsonPrecalculus.com*

Evaluating an Expression In Exercises 89–92, evaluate the expression. (If not possible, state the reason.)

89. $\dfrac{5 - 7}{12 - 18}$ **90.** $\dfrac{16 - 6}{6 - 11}$

91. $\dfrac{3 - 3}{4 - 0}$ **92.** $\dfrac{1 - (-1)}{9 - 9}$

Identifying x- and y-Intercepts In Exercises 93 and 94, identify x- and y-intercepts of the graph.

93. 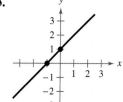   **94.**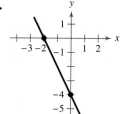

Sketching the Graph of an Equation In Exercises 95–98, test for symmetry and graph the equation. Then identify any intercepts.

95. $2x + y = 1$ **96.** $3x - y = 7$

97. $y = x^2 + 2$ **98.** $y = 2 - x^2$

Summary and Study Strategies

GO DIGITAL

What Did You Learn?

The list below reviews the skills covered in the chapter and correlates each one to the Review Exercises (see page 152) that practice the skill.

Section 1.1	**Review Exercises**
■ Sketch graphs of equations *(p. 70)*.	*1, 2*
■ Identify *x*- and *y*-intercepts of graphs of equations *(p. 73)*.	*3, 4*
■ Use symmetry to sketch graphs of equations *(p. 74)*.	*5–12*
■ Write equations of circles *(p. 76)*. Circle with radius r and center at (h, k): $(x - h)^2 + (y - k)^2 = r^2$	*13–18*
■ Use graphs of equations to solve real-life problems *(p. 77)*.	*19, 20*

Section 1.2	
■ Identify different types of equations *(p. 81)*.	*21–24*
■ Solve linear equations in one variable *(p. 82)*. Standard form of a linear equation in x: $ax + b = 0$	*25–28*
■ Solve rational equations that lead to linear equations *(p. 84)*.	*29–32*
■ Find *x*- and *y*-intercepts of graphs of equations algebraically *(p. 85)*. To find *x*-intercepts, set *y* equal to zero and solve the equation for *x*. To find *y*-intercepts, set *x* equal to zero and solve the equation for *y*.	*33–38*
■ Use linear equations to model and solve real-life problems *(p. 86)*.	*39, 40*

Section 1.3	
■ Write and use mathematical models to solve real-life problems *(p. 90)*.	*41–44*
■ Solve mixture problems *(p. 94)*.	*45, 46*
■ Use common formulas to solve real-life problems *(p. 95)*.	*47, 48*

Section 1.4	
■ Solve quadratic equations by factoring *(p. 100)*. General form of a quadratic equation in x: $ax^2 + bx + c = 0$	*49, 50*
■ Solve quadratic equations by extracting square roots *(p. 101)*.	*51–54*
■ Solve quadratic equations by completing the square *(p. 102)*.	*55, 56*
■ Use the Quadratic Formula to solve quadratic equations *(p. 104)*. Quadratic Formula: $x = \dfrac{-b \pm \sqrt{b^2 - 4ac}}{2a}$	*57, 58*
■ Use quadratic equations to model and solve real-life problems *(p. 106)*.	*59, 60*

Section 1.5	**Review Exercises**
■ Use the imaginary unit i to write complex numbers *(p. 114)*.	*61–64*
■ Add, subtract, and multiply complex numbers *(p. 115)*.	*65–70*
■ Use complex conjugates to write the quotient of two complex numbers in standard form *(p. 117)*.	*71–76*
■ Find complex solutions of quadratic equations *(p. 118)*.	*77–80*

Section 1.6	
■ Solve polynomial equations of degree three or greater *(p. 121)*.	*81–86*
■ Solve radical equations *(p. 123)*.	*87–90*
■ Solve rational equations and absolute value equations *(p. 124)*.	*91–96*
■ Use nonlinear and nonquadratic models to solve real-life problems *(p. 126)*.	*97, 98*

Section 1.7	
■ Represent solutions of linear inequalities in one variable *(p. 131)*.	*99–102*
■ Use properties of inequalities to write equivalent inequalities *(p. 132)*.	*103–108*
■ Solve linear inequalities in one variable *(p. 133)*.	*103–108*
■ Solve absolute value inequalities *(p. 135)*. $\|x\| < a$ if and only if $-a < x < a$ $\|x\| > a$ if and only if $x < -a$ or $x > a$	*109, 110*
■ Use linear inequalities to model and solve real-life problems *(p. 136)*.	*111, 112*

Section 1.8	
■ Solve polynomial inequalities *(p. 140)*.	*113–116*
■ Solve rational inequalities *(p. 144)*.	*117, 118*
■ Use nonlinear inequalities to model and solve real-life problems *(p. 145)*.	*119, 120*

Study Strategies

Absorbing Details Sequentially   Math is a sequential subject. Learning new math concepts successfully depends on how well you understand the previous concepts. So, it is important to learn and remember concepts as they are encountered. One way to work through a section sequentially is by following the steps listed below.

1. Work through an example. If you have trouble, consult your notes or seek help from a classmate or instructor.
2. Complete the Checkpoint problem following the example.
3. View and listen to the worked-out solution for the Checkpoint problem at *LarsonPrecalculus.com*. If you get the Checkpoint problem correct, move on to the next example. If not, make sure you understand why you got the problem wrong before moving on.
4. When you have finished working through the examples in the section, take a short break of 5 to 10 minutes. This will give your brain time to process everything.
5. Start the homework exercises. Check the answers to odd-numbered exercises at *CalcChat.com*.

Review Exercises

See CalcChat.com for tutorial help and worked-out solutions to odd-numbered exercises.

GO DIGITAL

1.1 **Sketching the Graph of an Equation** In Exercises 1 and 2, construct a table of values that consists of several solution points of the equation. Use the resulting solution points to sketch the graph of the equation.

1. $y = -4x + 1$

2. $y = x^2 + 2x$

Identifying *x*- and *y*-Intercepts In Exercises 3 and 4, identify the *x*- and *y*-intercepts of the graph.

3.

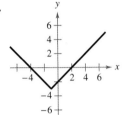

4.

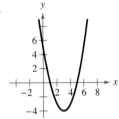

Testing for Symmetry In Exercises 5–12, use the algebraic tests to check for symmetry with respect to both axes and the origin. Then sketch the graph of the equation.

5. $y = -3x + 7$

6. $x = -8$

7. $x = y^2 - 5$

8. $y = 3x^3$

9. $y = -x^4 + 6x^2$

10. $y = x^4 + x^3 - 1$

11. $y = \dfrac{3}{x}$

12. $y = |x| - 4$

Sketching a Circle In Exercises 13–16, find the center and radius of the circle with the given equation. Then sketch the circle.

13. $x^2 + y^2 = 9$

14. $x^2 + y^2 = 4$

15. $(x + 2)^2 + y^2 = 16$

16. $x^2 + (y - 8)^2 = 81$

Writing the Equation of a Circle In Exercises 17 and 18, write the standard form of the equation of the circle for which the endpoints of a diameter are the given points.

17. $(0, 0), (4, -6)$

18. $(-2, -3), (4, -10)$

19. Sales The sales S (in millions of dollars) for Jazz Pharmaceuticals for the years 2010 through 2018 can be approximated by the model

$$S = -6.876t^2 + 411.94t - 3329.1, \quad 10 \le t \le 18$$

where t represents the year, with $t = 10$ corresponding to 2010. *(Source: Jazz Pharmaceuticals)*

(a) Use a graphing utility to graph the model.

(b) Use the graph to estimate the year in which the sales were $1 billion.

20. Physics The force F (in pounds) required to stretch a spring x inches from its natural length (see figure) is

$$F = \frac{5}{4}x, \quad 0 \le x \le 20.$$

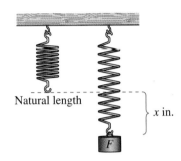

Natural length

x in.

F

(a) Use the model to complete the table.

x	0	4	8	12	16	20
Force, F						

(b) Sketch a graph of the model.

(c) Use the graph to estimate the force necessary to stretch the spring 10 inches.

1.2 **Classifying an Equation** In Exercises 21–24, determine whether the equation is an identity, a conditional equation, or a contradiction.

21. $2(x - 2) = 2x - 4$

22. $2(x + 3) = 2x - 2$

23. $3(x - 2) + 2x = 2(x + 3)$

24. $5(x - 1) - 2x = 3x - 5$

Solving an Equation In Exercises 25–32, solve the equation and check your solution.

25. $8x - 5 = 3x + 20$

26. $7x + 3 = 3x - 17$

27. $2(x + 5) - 7 = x + 9$

28. $7(x - 4) = 1 - (x + 9)$

29. $\dfrac{x}{5} - 3 = \dfrac{x}{3} + 1$

30. $\dfrac{4x - 3}{6} + \dfrac{x}{4} = x - 2$

31. $3 + \dfrac{2}{x - 5} = \dfrac{2x}{x - 5}$

32. $\dfrac{1}{x^2 + 3x - 18} - \dfrac{3}{x + 6} = \dfrac{4}{x - 3}$

Finding Intercepts Algebraically In Exercises 33–38, find the *x*- and *y*-intercepts of the graph of the equation algebraically.

33. $y = 3x - 1$

34. $y = -5x + 6$

35. $y = 2(x - 4)$

36. $y = 4(7x + 1)$

37. $y = -\frac{1}{2}x + \frac{2}{3}$

38. $y = \frac{3}{4}x - \frac{1}{4}$

39. Geometry The surface area S of the cylinder shown in the figure is approximated by the equation $S = 2(3.14)(3)^2 + 2(3.14)(3)h$. The surface area is 244.92 square inches. Find the height h of the cylinder.

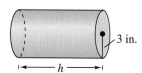

3 in.

h

40. Temperature The Fahrenheit and Celsius temperature scales are related by the equation $C = \frac{5}{9}F - \frac{160}{9}$. Find the Fahrenheit temperature that corresponds to $100°$ Celsius.

1.3

41. Revenue In 2019, LinkedIn's total revenue was 28.3% more than it was in 2018. The total revenue for the two years was $12.1 billion. Find the revenue for each year. *(Source: LinkedIn Corp.)*

42. Discount The sale price of a digital camera after a 20% discount is $340. Find the original price.

43. Business Venture You are starting a small business. You have nine investors who are willing to share equally in the venture. When you add three more investors, each person's share decreases by $2500. What is the total investment required to start the business?

44. Average Speed You commute 56 miles one way to work. The trip to work takes 10 minutes longer than the trip home. Your average speed on the trip home is 8 miles per hour faster. What is your average speed on the trip home?

45. Mixture Problem A car radiator contains 10 liters of a 30% antifreeze solution. How many liters should you replace with pure antifreeze to get a 50% antifreeze solution?

46. Simple Interest A business invests a total of $6000 at $4\frac{1}{2}$% and $5\frac{1}{2}$% simple interest. During the first year, the two accounts earn $305. Find the initial investment for each account.

47. Volume of a Cone The volume V of a cone can be calculated by the formula $V = \frac{1}{3}\pi r^2 h$, where r is the radius and h is the height. Solve for h.

48. Kinetic Energy The kinetic energy E of an object can be calculated by the formula $E = \frac{1}{2}mv^2$, where m is the mass and v is the velocity. Solve for m.

1.4 Choosing a Method In Exercises 49–58, solve the equation using any convenient method.

49. $15 + x - 2x^2 = 0$

50. $2x^2 - x - 28 = 0$

51. $6 = 3x^2$

52. $16x^2 = 25$

53. $(x + 13)^2 = 25$

54. $(x - 5)^2 = 30$

55. $x^2 + 12x = -25$

56. $9x^2 - 12x = 14$

57. $-2x^2 - 5x + 27 = 0$

58. $-20 - 3x + 3x^2 = 0$

59. Simply Supported Beam A simply supported 20-foot beam supports a uniformly distributed load of 1000 pounds per foot. The bending moment M (in foot-pounds) x feet from one end of the beam is given by $M = 500x(20 - x)$.

(a) Where is the bending moment zero?

(b) Use a graphing utility to graph the equation.

(c) Use the graph to determine the point on the beam where the bending moment is the greatest.

60. Sports You throw a softball straight up into the air at a velocity of 30 feet per second. You release the softball at a height of 5.8 feet and catch it when it falls back to a height of 6.2 feet.

(a) Use the position equation to write a mathematical model for the height of the softball.

(b) What is the height of the softball after 1 second?

(c) How many seconds is the softball in the air?

1.5 Writing a Complex Number in Standard Form In Exercises 61–64, write the complex number in standard form.

61. $\sqrt{-18}\sqrt{-6}$

62. $\sqrt{-27} + \sqrt{-3}$

63. $\left(5 + \sqrt{-10}\right)\left(10 - \sqrt{-5}\right)$

64. $\left(6 - \sqrt{-2}\right)^2$

Performing Operations with Complex Numbers In Exercises 65–70, perform the operation and write the result in standard form.

65. $(6 - 4i) + (-9 + i)$

66. $(7 - 2i) - (3 - 8i)$

67. $-3i(-2 + 5i)$

68. $(4 + i)(3 - 10i)$

69. $(1 + 7i)(1 - 7i)$

70. $(5 - 9i)^2$

Writing a Quotient in Standard Form In Exercises 71–74, write the quotient in standard form.

71. $\dfrac{4}{1 - 2i}$

72. $\dfrac{6 - 5i}{i}$

73. $\dfrac{3 + 2i}{5 + i}$

74. $\dfrac{7i}{(3 + 2i)^2}$

Performing Operations with Complex Numbers In Exercises 75 and 76, perform the operation and write the result in standard form.

75. $\dfrac{4}{2 - 3i} + \dfrac{2}{1 + i}$

76. $\dfrac{1}{2 + i} - \dfrac{5}{1 + 4i}$

Complex Solutions of a Quadratic Equation In Exercises 77–80, use the Quadratic Formula to solve the quadratic equation.

77. $x^2 - 2x + 10 = 0$

78. $x^2 + 6x + 34 = 0$

79. $4x^2 + 4x + 7 = 0$

80. $6x^2 + 3x + 27 = 0$

1.6 **Solving an Equation** In Exercises 81–96, solve the equation. Check your solutions.

81. $5x^4 - 12x^3 = 0$ **82.** $4x^3 - 6x^2 = 0$

83. $x^3 - 7x^2 + 4x = 28$

84. $9x^4 + 27x^3 - 4x^2 = 12x$

85. $x^6 - 7x^3 - 8 = 0$ **86.** $x^4 - 13x^2 - 48 = 0$

87. $\sqrt{2x + 3} = 2 + x$ **88.** $5\sqrt{x} - \sqrt{x - 1} = 6$

89. $(x - 1)^{2/3} - 25 = 0$ **90.** $(x + 2)^{3/4} = 27$

91. $\dfrac{5}{x} = 1 + \dfrac{3}{x + 2}$ **92.** $\dfrac{6}{x} + \dfrac{8}{x + 5} = 3$

93. $|x - 5| = 10$ **94.** $|2x + 3| = 7$

95. $|x^2 - 3| = 2x$ **96.** $|x^2 - 6| = x$

97. Demand The demand equation for a hair dryer is $p = 42 - \sqrt{0.001x + 2}$, where x is the number of units demanded per day and p is the price per unit. Find the demand when the price is set at $29.95.

 98. Newspapers The paid circulation C (in thousands) of daily newspapers in the United States from 2011 through 2018 can be approximated by the model $C = 51{,}663 - 16.772t^{5/2}$, $11 \le t \le 18$, where t represents the year, with $t = 11$ corresponding to 2011. The table shows the paid circulation of daily newspapers for each year during this time period. (Source: Statista)

DATA	Year	Paid Circulation, C
	2011	44,421
	2012	43,433
	2013	40,712
	2014	40,420
	2015	37,711
	2016	34,657
	2017	30,948
	2018	28,554

Spreadsheet at LarsonPrecalculus.com

(a) Use a graphing utility to plot the data and graph the model in the same viewing window. How well does the model fit the data?

(b) Use the graph in part (a) to estimate the year in which there was a paid circulation of about 37 million daily newspapers.

(c) Use the model to verify algebraically the estimate from part (b).

1.7 **Intervals and Inequalities** In Exercises 99–102, state whether the interval is bounded or unbounded, represent the interval with an inequality, and sketch its graph.

99. $(-7, 2]$ **100.** $(3, \infty)$

101. $(-\infty, -10]$ **102.** $[-2, 2]$

Solving an Inequality In Exercises 103–110, solve the inequality. Then graph the solution set.

103. $3(x + 2) < 2x - 12$ **104.** $2(x + 5) \ge 5(x - 3)$

105. $4(5 - 2x) \le \frac{1}{2}(8 - x)$ **106.** $\frac{1}{2}(3 - x) > \frac{1}{3}(2 - 3x)$

107. $3.2 \le 0.4x - 1 \le 4.4$ **108.** $1.6 < 0.3x + 1 < 2.8$

109. $|x + 6| < 5$ **110.** $\frac{2}{3}|3 - x| \ge 4$

111. Cost, Revenue, and Profit The revenue for selling x units of a product is $R = 125.33x$. The cost of producing x units is $C = 92x + 1200$. To obtain a profit, the revenue must be greater than the cost. For what values of x does this product return a profit?

112. Geometry The side length of a square machined automobile part is 19.3 centimeters with a possible error of 0.5 centimeter. Determine the interval containing the possible areas of the part.

1.8 **Solving an Inequality** In Exercises 113–118, solve the inequality. Then graph the solution set.

113. $x^2 - 6x - 27 < 0$ **114.** $x^2 - 2x \ge 3$

115. $5x^3 - 45x < 0$ **116.** $2x^3 - 5x^2 - 3x \ge 0$

117. $\dfrac{2}{x + 1} \le \dfrac{3}{x - 1}$ **118.** $\dfrac{x - 5}{3 - x} < 0$

119. Investment An investment of P dollars at interest rate r (in decimal form) compounded annually increases to an amount $A = P(1 + r)^2$ in 2 years. An investment of $5000 increases to an amount greater than $5500 in 2 years. The interest rate must be greater than what percent?

120. Biology A biologist introduces 200 ladybugs into a crop field. The population P of the ladybugs can be approximated by the model $P = [1000(1 + 3t)]/(5 + t)$, where t is the time in days. Find the time required for the population to increase to at least 2000 ladybugs.

Review & Refresh ▶ *Video solutions at LarsonPrecalculus.com*

True or False? In Exercises 121 and 122, determine whether the statement is true or false. Justify your answer.

121. $\sqrt{-18}\sqrt{-2} = \sqrt{(-18)(-2)}$

122. The equation $325x^2 - 717x + 398 = 0$ has no solution.

123. Writing Explain why it is essential to check your solutions to radical, absolute value, and rational equations.

124. Error Analysis Describe the error.

$$|11x + 4| \ge 26$$

$$11x + 4 \le 26 \quad \text{or} \quad 11x + 4 \ge 26$$

$$11x \le 22 \qquad\qquad 11x \ge 22$$

$$x \le 2 \qquad\qquad\qquad x \ge 2 \quad ✗$$

See CalcChat.com for tutorial help and worked-out solutions
to odd-numbered exercises.

GO DIGITAL

Take this test as you would take a test in class. When you are finished, check your work against the answers given in the back of the book.

In Exercises 1–6, use the algebraic tests to check for symmetry with respect to both axes and the origin. Then sketch the graph of the equation. Identify any x- and y-intercepts. *(Section 1.1)*

1. $y = 4 - \frac{3}{4}x$

2. $y = 4 - \frac{3}{4}|x|$

3. $y = 4 - (x - 2)^2$

4. $y = x - x^3$

5. $y = \sqrt{5 - x}$

6. $(x - 3)^2 + y^2 = 9$

In Exercises 7–12, solve the equation and check your solution. (If not possible, explain why.) *(Sections 1.2, 1.4, 1.5, and 1.6)*

7. $\frac{2}{3}(x - 1) + \frac{1}{4}x = 10$

8. $(x - 4)(x + 2) = 7$

9. $\dfrac{x - 2}{x + 2} + \dfrac{4}{x + 2} + 4 = 0$

10. $x^4 + x^2 - 6 = 0$

11. $2\sqrt{x} - \sqrt{2x + 1} = 1$

12. $|3x - 1| = 7$

In Exercises 13–16, solve the inequality. Then graph the solution set. *(Sections 1.7 and 1.8)*

13. $-3 \le 2(x + 4) < 14$

14. $\dfrac{2}{x} > \dfrac{5}{x + 6}$

15. $2x^2 + 5x > 12$

16. $|3x + 5| \ge 10$

17. Perform each operation and write the result in standard form. *(Section 1.5)*

   (a) $\sqrt{-16} - 2(7 + 2i)$

   (b) $(5 - i)(3 + 4i)$

18. Write the quotient in standard form: $\dfrac{8}{1 + 2i}$. *(Section 1.5)*

19. Solve $2x^2 - 6x + 5 = 0$. *(Section 1.5)*

20. The sales S (in millions of dollars) for Paycom Software, Inc. from 2012 through 2018 can be approximated by the model

   $$S = 10.437t^2 - 231.05t + 1345.5, \quad 12 \le t \le 18$$

   where t represents the year, with $t = 12$ corresponding to 2012. *(Source: Paycom Software, Inc.)* *(Section 1.1)*

   (a) Use a graphing utility to graph the model.

   (b) Use the graph in part (a) to estimate the sales in 2017.

   (c) Use the model to verify algebraically your estimate from part (b).

21. A basketball has a volume of about 455.9 cubic inches. Find the radius of the basketball (accurate to three decimal places). *(Section 1.3)*

22. On the first part of a 350-kilometer trip, a salesperson travels 2 hours and 15 minutes at an average speed of 100 kilometers per hour. The salesperson needs to arrive at the destination in another hour and 20 minutes. Find the average speed required for the remainder of the trip. *(Section 1.6)*

23. The area of the ellipse in the figure is $A = \pi ab$. Find a and b such that the area of the ellipse equals the area of the circle when a and b satisfy the constraint $a + b = 100$. *(Section 1.4)*

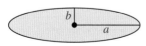

Figure for 23

Proofs in Mathematics

Conditional Statements

Many theorems are written in the **if-then** form "if p, then q," which is denoted by

$p \rightarrow q$ Conditional statement

where p is the **hypothesis** and q is the **conclusion.** Here are some other ways to express the conditional statement $p \rightarrow q$.

p implies q. p only if q. p is sufficient for q.

Conditional statements can be either true or false. The conditional statement $p \rightarrow q$ is false only when p is true and q is false. To show that a conditional statement is true, you must prove that the conclusion follows for all cases that fulfill the hypothesis. To show that a conditional statement is false, you need to describe only a single **counterexample** that shows that the statement is not always true.

For instance, $x = -4$ is a counterexample that shows that the statement below is false.

If $x^2 = 16$, then $x = 4$.

The hypothesis "$x^2 = 16$" is true because $(-4)^2 = 16$. However, the conclusion "$x = 4$" is false. This implies that the given conditional statement is false.

For the conditional statement $p \rightarrow q$, there are three important associated conditional statements.

1. The **converse** of $p \rightarrow q$: $q \rightarrow p$

2. The **inverse** of $p \rightarrow q$: $\sim p \rightarrow \sim q$

3. The **contrapositive** of $p \rightarrow q$: $\sim q \rightarrow \sim p$

The symbol $\sim$ means the **negation** of a statement. For example, the negation of "The engine is running" is "The engine is not running."

EXAMPLE   **Writing the Converse, Inverse, and Contrapositive**

Write the converse, inverse, and contrapositive of the conditional statement "If I get a B on my test, then I will pass the course."

Solution

Converse: If I pass the course, then I got a B on my test.

Inverse: If I do not get a B on my test, then I will not pass the course.

Contrapositive: If I do not pass the course, then I did not get a B on my test.   ■

In the example above, notice that neither the converse nor the inverse is logically equivalent to the original conditional statement. On the other hand, the contrapositive *is* logically equivalent to the original conditional statement.

P.S. Problem Solving

See CalcChat.com for tutorial help and worked-out solutions to odd-numbered exercises.

1. **Time and Distance** Let x represent the time (in seconds), and let y represent the distance (in feet) between you and a tree. Sketch a possible graph that shows how x and y are related when you are walking toward the tree.

2. **Sum of the First n Natural Numbers**

   (a) Find each sum.

   $$1 + 2 + 3 + 4 + 5 = \rule{1cm}{0.3cm}$$

   $$1 + 2 + 3 + 4 + 5 + 6 + 7 + 8 = \rule{1cm}{0.3cm}$$

   $$1 + 2 + 3 + 4 + 5 + 6$$
   $$+ 7 + 8 + 9 + 10 = \rule{1cm}{0.3cm}$$

   (b) Use the formula below for the sum of the first n natural numbers to verify your answers to part (a).

   $$1 + 2 + 3 + \cdots + n = \frac{1}{2}n(n + 1)$$

   (c) Use the formula in part (b) to find n such that the sum of the first n natural numbers is 210.

3. **Area of an Ellipse** The area of an ellipse is equal to $A = \pi ab$ (see figure). For the ellipse below, $a + b = 20$.

   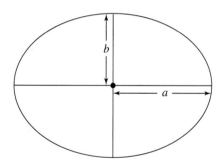

   (a) Show that

   $$A = \pi a(20 - a).$$

   (b) Complete the table.

a	4	7	10	13	16
A					

   (c) Find two values of a such that $A = 300$.

   (d) Use a graphing utility to graph the area equation.

   (e) Find the a-intercepts of the graph of the area equation. What do these values represent?

   (f) What is the maximum area? What values of a and b yield the maximum area?

4. **Using a Graph to Solve an Inequality** Use the graph of

   $$y = x^4 - x^3 - 6x^2 + 4x + 8$$

   to solve the inequality

   $$x^4 - x^3 - 6x^2 + 4x + 8 > 0.$$

5. **Wind Pressure** A building code requires that a building be able to withstand a specific amount of wind pressure. The pressure P (in pounds per square foot) from wind blowing at s miles per hour is given by

   $$P = 0.00256s^2.$$

   (a) A structural engineer is designing a library. The building is required to withstand wind pressure of 20 pounds per square foot. Under this requirement, how fast must the wind blow to produce excessive stress on the building?

   (b) To be safe, the engineer designs the library so that it can withstand wind pressure of 40 pounds per square foot. Does this mean that the library can survive wind blowing at twice the speed you found in part (a)? Justify your answer.

   (c) Use the pressure formula to explain why even a relatively small increase in the wind speed could have potentially serious effects on a building.

6. **Water Height** For a bathtub with a rectangular base, Torricelli's Law implies that the height h of water in the tub t seconds after it begins draining is given by

   $$h = \left(\sqrt{h_0} - \frac{2\pi d^2 \sqrt{3}}{lw} t \right)^2$$

   where l and w are the tub's length and width, d is the diameter of the drain, and h_0 is the water's initial height. (All measurements are in inches.) You completely fill a tub with water. The tub is 60 inches long by 30 inches wide by 25 inches high and has a drain with a two-inch diameter.

   (a) Find the time it takes for the tub to go from being full to half-full.

   (b) Find the time it takes for the tub to go from being half-full to empty.

   (c) Based on your results in parts (a) and (b), what general statement can you make about the speed at which the water drains?

7. **Sum of Squares**

   (a) Consider the sum of squares $x^2 + 9$. If the sum can be factored, then there are integers m and n such that $x^2 + 9 = (x + m)(x + n)$. Write two equations relating the sum and the product of m and n to the coefficients in $x^2 + 9$.

   (b) Show that there are no integers m and n that satisfy both equations you wrote in part (a). What can you conclude?

8. **Finding Values** Use the equation $4\sqrt{x} = 2x + k$ to find three different values of k such that the equation has two solutions, one solution, and no solution. Describe the process you used to find the values.

9. **Pythagorean Triples** A Pythagorean Triple is a group of three integers, such as 3, 4, and 5, that could be the lengths of the sides of a right triangle.

   (a) Find two other Pythagorean Triples.

   (b) Notice that $3 \cdot 4 \cdot 5 = 60$. Is the product of the three numbers in each Pythagorean Triple evenly divisible by 3? by 4? by 5?

   (c) Write a conjecture involving Pythagorean Triples and divisibility by 60.

10. **Sums and Products of Solutions** Determine the solutions x_1 and x_2 of each quadratic equation. Use the values of x_1 and x_2 to fill in the boxes.

Equation	x_1, x_2	$x_1 + x_2$	$x_1 \cdot x_2$
(a) $x^2 - x - 12 = 0$			
(b) $2x^2 + 5x - 3 = 0$			
(c) $4x^2 - 9 = 0$			
(d) $x^2 - 10x + 34 = 0$			

11. **Proof** The solutions of a quadratic equation are

$$x = \frac{-b \pm \sqrt{b^2 - 4ac}}{2a}.$$

   (a) Prove that the sum of the solutions is

   $$S = -\frac{b}{a}.$$

   (b) Prove that the product of the solutions is

   $$P = \frac{c}{a}.$$

12. **Principal Cube Root**

   (a) The principal cube root of 125, $\sqrt[3]{125}$, is 5. Evaluate the expression x^3 for each value of x.

   (i) $x = \dfrac{-5 + 5\sqrt{3}i}{2}$

   (ii) $x = \dfrac{-5 - 5\sqrt{3}i}{2}$

   (b) The principal cube root of 27, $\sqrt[3]{27}$, is 3. Evaluate the expression x^3 for each value of x.

   (i) $x = \dfrac{-3 + 3\sqrt{3}i}{2}$

   (ii) $x = \dfrac{-3 - 3\sqrt{3}i}{2}$

   (c) Use the results of parts (a) and (b) to list possible cube roots of (i) 1, (ii) 8, and (iii) 64. Verify your results algebraically.

13. **Multiplicative Inverse of a Complex Number** The multiplicative inverse of a complex number z is a complex number z_m such that $z \cdot z_m = 1$. Find the multiplicative inverse of each complex number.

   (a) $z = 1 + i$

   (b) $z = 3 - i$

   (c) $z = -2 + 8i$

14. **Proof** Prove that the product of a complex number $a + bi$ and its complex conjugate is a real number.

15. **The Mandelbrot Set** A **fractal** is a geometric figure that consists of a pattern that is repeated infinitely on a smaller and smaller scale. The most famous fractal is the **Mandelbrot Set,** named after the Polish-born mathematician Benoit Mandelbrot (1924–2010). To draw the Mandelbrot Set, consider the sequence of numbers below.

$$c, \; c^2 + c, \; (c^2 + c)^2 + c, \; [(c^2 + c)^2 + c]^2 + c, \ldots$$

The behavior of this sequence depends on the value of the complex number c. If the sequence is bounded (the absolute value of each number in the sequence,

$$|a + bi| = \sqrt{a^2 + b^2}$$

is less than some fixed number N), then the complex number c is in the Mandelbrot Set, and if the sequence is unbounded (the absolute value of the terms of the sequence become infinitely large), then the complex number c is not in the Mandelbrot Set. Determine whether the complex number c is in the Mandelbrot Set.

   (a) $c = i$

   (b) $c = 1 + i$

   (c) $c = -2$

The figure below shows a graph of the Mandelbrot Set, where the horizontal and vertical axes represent the real and imaginary parts of c, respectively.

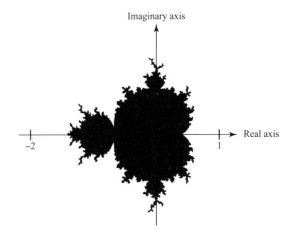

2 Functions and Their Graphs

GO DIGITAL

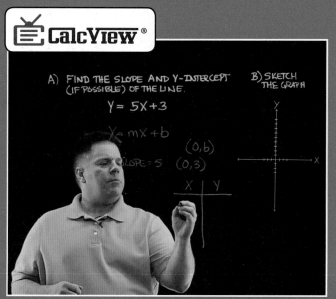

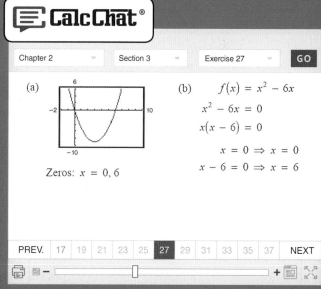

2.3 Temperature (*Exercise 87, p. 196*)

2.7 Diesel Mechanics (*Exercise 70, p. 229*)

159

left, © iStockPhoto.com/Dsafanda; right, © ShaunI/Getty Images

2.1 Linear Equations in Two Variables

Linear equations in two variables can help you model and solve real-life problems. For example, in Exercise 90 on page 171, you will use a surveyor's measurements to find a linear equation that models a mountain road.

❯ Use slope to graph linear equations in two variables.
❯ Find the slope of a line given two points on the line.
❯ Write linear equations in two variables.
❯ Use slope to identify parallel and perpendicular lines.
❯ Use slope and linear equations in two variables to model and solve real-life problems.

Using Slope

The simplest mathematical model relating two variables x and y is the **linear equation**

$$y = mx + b \qquad \text{Linear equation in two variables } x \text{ and } y$$

where m and b are constants. The equation is called *linear* because its graph is a line. (In mathematics, the term *line* means *straight line*.) By letting $x = 0$, you obtain

$$y = m(0) + b = b.$$

So, the line crosses the y-axis at $y = b$, as shown in Figure 2.1. In other words, the y-intercept is $(0, b)$. The steepness, or *slope,* of the line is m.

$$y = mx + b$$

Slope ⟶ | ⟵ y-Intercept

The **slope** of a nonvertical line is the number of units the line rises (or falls) vertically for each unit of horizontal change from left to right. When the line rises from left to right, the slope is positive (see Figure 2.1). When the line falls from left to right, the slope is negative (see Figure 2.2).

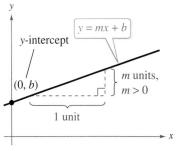

Positive slope, line rises
Figure 2.1

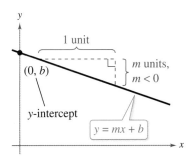

Negative slope, line falls
Figure 2.2

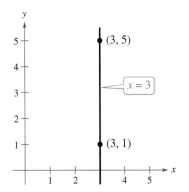

Slope is undefined.
Figure 2.3

The Slope-Intercept Form of the Equation of a Line

The linear equation $y = mx + b$ is in **slope-intercept form.** The graph of the equation $y = mx + b$ is a line whose slope is m and whose y-intercept is $(0, b)$.

Once you determine the slope and the y-intercept of a line, it is relatively simple to sketch its graph. In the next example, note that none of the lines is vertical. A vertical line has an equation of the form

$$x = a \qquad \text{Vertical line}$$

where a is a real number. The equation of a vertical line cannot be written in the form $y = mx + b$ because the slope of a vertical line is undefined (see Figure 2.3).

GO DIGITAL

EXAMPLE 1 **Graphing Linear Equations**

▶▶▶ *See LarsonPrecalculus.com for an interactive version of this type of example.*

Sketch the graph of each linear equation.

a. $y = 2x + 1$

b. $y = 2$

c. $x + y = 2$

Solution

a. Because $b = 1$, the y-intercept is $(0, 1)$. Moreover, the slope is $m = 2$, so the line *rises* two units for each unit the line moves to the right, as shown in Figure 2.4(a).

b. By writing this equation in the form $y = (0)x + 2$, you find that the y-intercept is $(0, 2)$ and the slope is $m = 0$. A slope of 0 implies that the line is horizontal—that is, it does not rise *or* fall, as shown in Figure 2.4(b).

c. By writing this equation in slope-intercept form

$$x + y = 2 \qquad \text{Write original equation.}$$
$$y = -x + 2 \qquad \text{Subtract } x \text{ from each side.}$$
$$y = (-1)x + 2 \qquad \text{Write in slope-intercept form.}$$

you find that the y-intercept is $(0, 2)$. Moreover, the slope is $m = -1$, so the line *falls* one unit for each unit the line moves to the right, as shown in Figure 2.4(c).

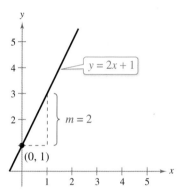

(a) When m is positive, the line rises.
Figure 2.4

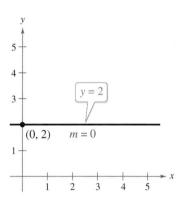

(b) When m is 0, the line is horizontal.

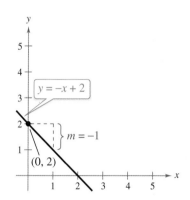

(c) When m is negative, the line falls.

✓ *Checkpoint* ▶ *Audio-video solution in English & Spanish at LarsonPrecalculus.com*

Sketch the graph of each linear equation.

a. $y = 3x + 2$ **b.** $y = -3$ **c.** $4x + y = 5$ ∎

From the lines shown in Figures 2.3 and 2.4, you can make several generalizations about the slope of a line.

1. A line with positive slope ($m > 0$) *rises* from left to right. [See Figure 2.4(a).]

2. A line with negative slope ($m < 0$) *falls* from left to right. [See Figure 2.4(c).]

3. A line with zero slope ($m = 0$) is *horizontal*. [See Figure 2.4(b).]

4. A line with undefined slope is *vertical*. (See Figure 2.3.)

From the slope-intercept form of the equation of a line, you can see that a horizontal line ($m = 0$) has an equation of the form

$$y = (0)x + b \quad \text{or} \quad y = b. \qquad \text{Horizontal line}$$

GO DIGITAL

Finding the Slope of a Line

Given an equation of a line, you can find its slope by writing the equation in slope-intercept form. When you are not given an equation, you can still find the slope by using two points on the line. For example, consider the line passing through the points (x_1, y_1) and (x_2, y_2) in the figure below.

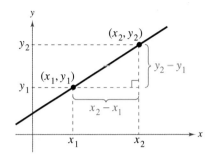

As you move from left to right along this line, a change of $(y_2 - y_1)$ units in the vertical direction corresponds to a change of $(x_2 - x_1)$ units in the horizontal direction. That is,

$$y_2 - y_1 = \text{change in } y = \text{rise}$$

and

$$x_2 - x_1 = \text{change in } x = \text{run}.$$

The ratio of $(y_2 - y_1)$ to $(x_2 - x_1)$ represents the slope of the line that passes through the points (x_1, y_1) and (x_2, y_2).

$$\text{Slope} = \frac{\text{change in } y}{\text{change in } x} = \frac{\text{rise}}{\text{run}} = \frac{y_2 - y_1}{x_2 - x_1}$$

ALGEBRA HELP

Be sure you understand that the definition of slope does not apply to *vertical* lines. For instance, consider the points $(3, 5)$ and $(3, 1)$ on the vertical line shown in Figure 2.3. Applying the formula for slope, you obtain

$$m = \frac{5 - 1}{3 - 3} = \frac{4}{0}. \quad \text{✗}$$

Because division by zero is undefined, the slope of a vertical line is undefined.

> **Definition of the Slope of a Line**
>
> The **slope** m of the nonvertical line through (x_1, y_1) and (x_2, y_2) is
>
> $$m = \frac{y_2 - y_1}{x_2 - x_1}$$
>
> where $x_1 \neq x_2$.

When using the formula for slope, the *order of subtraction* is important. Given two points on a line, you are free to label either one of them as (x_1, y_1) and the other as (x_2, y_2). However, once you do this, you must form the numerator and denominator using the same order of subtraction.

$$m = \frac{y_2 - y_1}{x_2 - x_1} \qquad m = \frac{y_1 - y_2}{x_1 - x_2} \qquad m = \frac{y_2 - y_1}{x_1 - x_2} \quad \text{✗}$$

$\qquad\quad$ Correct $\qquad\qquad\quad$ Correct $\qquad\qquad\quad$ Incorrect

For example, the slope of the line passing through the points $(3, 4)$ and $(5, 7)$ is

$$m = \frac{y_2 - y_1}{x_2 - x_1} = \frac{7 - 4}{5 - 3} = \frac{3}{2}$$

or

$$m = \frac{y_1 - y_2}{x_1 - x_2} = \frac{4 - 7}{3 - 5} = \frac{-3}{-2} = \frac{3}{2}.$$

GO DIGITAL

EXAMPLE 2 **Finding the Slope of a Line Through Two Points**

Find the slope of the line passing through each pair of points.

a. $(-2, 0)$ and $(3, 1)$ **b.** $(-1, 2)$ and $(2, 2)$

c. $(0, 4)$ and $(1, -1)$ **d.** $(3, 4)$ and $(3, 1)$

Solution

a. Letting $(x_1, y_1) = (-2, 0)$ and $(x_2, y_2) = (3, 1)$, you find that the slope is

$$m = \frac{y_2 - y_1}{x_2 - x_1} = \frac{1 - 0}{3 - (-2)} = \frac{1}{5}.$$ See Figure 2.5(a).

b. The slope of the line passing through $(-1, 2)$ and $(2, 2)$ is

$$m = \frac{2 - 2}{2 - (-1)} = \frac{0}{3} = 0.$$ See Figure 2.5(b).

c. The slope of the line passing through $(0, 4)$ and $(1, -1)$ is

$$m = \frac{-1 - 4}{1 - 0} = \frac{-5}{1} = -5.$$ See Figure 2.5(c).

d. The slope of the line passing through $(3, 4)$ and $(3, 1)$ is

$$m = \frac{1 - 4}{3 - 3} = \frac{-3}{0}.$$ ✗   See Figure 2.5(d).

Division by 0 is undefined, so the slope is undefined and the line is vertical.

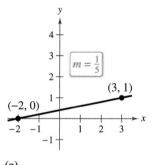

(a)

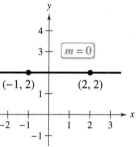

(b)

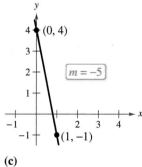

(c)

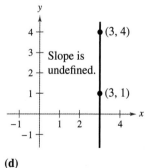

(d)

Figure 2.5

✓ *Checkpoint* ▶ *Audio-video solution in English & Spanish at LarsonPrecalculus.com*

Find the slope of the line passing through each pair of points.

a. $(-5, -6)$ and $(2, 8)$ **b.** $(4, 2)$ and $(2, 5)$

c. $(0, 0)$ and $(0, -6)$ **d.** $(0, -1)$ and $(3, -1)$

Writing Linear Equations in Two Variables

When you know the slope of a line *and* you also know the coordinates of one point on the line, you can find an equation of the line. For example, in the figure at the right, let (x_1, y_1) be a point on the line whose slope is m. When (x, y) is any *other* point on the line, it follows that

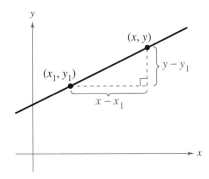

$$\frac{y - y_1}{x - x_1} = m.$$

This equation in the variables x and y can be rewritten in the **point-slope form** of the equation of a line.

Point-Slope Form of the Equation of a Line

The equation of the line with slope m passing through the point (x_1, y_1) is

$$y - y_1 = m(x - x_1). \qquad \text{Point-slope form}$$

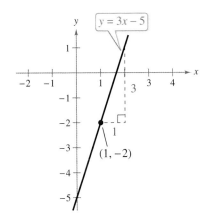

Figure 2.6

EXAMPLE 3 **Using the Point-Slope Form**

Find the slope-intercept form of the equation of the line that has a slope of 3 and passes through the point $(1, -2)$.

Solution Use the point-slope form with $m = 3$ and $(x_1, y_1) = (1, -2)$.

$y - y_1 = m(x - x_1)$	Point-slope form
$y - (-2) = 3(x - 1)$	Substitute for m, x_1, and y_1.
$y + 2 = 3x - 3$	Simplify.
$y = 3x - 5$	Write in slope-intercept form.

The slope-intercept form of the equation of the line is $y = 3x - 5$. Figure 2.6 shows the graph of this equation.

✓ *Checkpoint* ▶ *Audio-video solution in English & Spanish at LarsonPrecalculus.com*

Find the slope-intercept form of the equation of the line that has the given slope and passes through the given point.

a. $m = 2$, $(3, -7)$ **b.** $m = -\dfrac{2}{3}$, $(1, 1)$ **c.** $m = 0$, $(1, 1)$ ■

ALGEBRA HELP

When you find an equation of the line that passes through two given points, you only need to substitute the coordinates of one of the points in the point-slope form. It does not matter which point you choose because both points will yield the same result.

The point-slope form can be used to find an equation of the line passing through two points (x_1, y_1) and (x_2, y_2). To do this, first find the slope of the line.

$$m = \frac{y_2 - y_1}{x_2 - x_1}, \quad x_1 \neq x_2$$

Then use the point-slope form to obtain the equation.

$$y - y_1 = \frac{y_2 - y_1}{x_2 - x_1}(x - x_1) \qquad \text{Two-point form}$$

This is sometimes called the **two-point form** of the equation of a line.

GO DIGITAL

Parallel and Perpendicular Lines

The slope of a line is a convenient way for determining whether two lines are parallel or perpendicular. Specifically, nonvertical lines with the same slope are parallel, and nonvertical lines whose slopes are negative reciprocals are perpendicular.

Parallel and Perpendicular Lines

1. Two distinct nonvertical lines are **parallel** if and only if their slopes are equal. That is,

$$m_1 = m_2.$$ Parallel ⟺ Slopes are equal.

2. Two nonvertical lines are **perpendicular** if and only if their slopes are negative reciprocals of each other. That is,

$$m_1 = -\frac{1}{m_2}.$$ Perpendicular ⟺ Slopes are negative reciprocals.

Note that $m_1 m_2 = -1$.

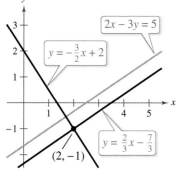

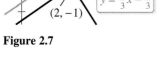

Figure 2.7

>>> **TECHNOLOGY**

The standard viewing window on some graphing utilities does not give a true geometric perspective because the screen is rectangular, which distorts the image. So, perpendicular lines will not appear to be perpendicular, and circles will not appear to be circular. To overcome this, use a square setting, in which the horizontal and vertical tick marks have equal spacing. On many graphing utilities, a square setting can be obtained when the ratio of the range of y to the range of x is 2 to 3.

EXAMPLE 4 **Finding Parallel and Perpendicular Lines**

Find the slope-intercept form of the equations of the lines that pass through the point $(2, -1)$ and are (a) parallel to and (b) perpendicular to the line $2x - 3y = 5$.

Solution Write the equation $2x - 3y = 5$ in slope-intercept form.

$2x - 3y = 5$	Write original equation.
$-3y = -2x + 5$	Subtract $2x$ from each side.
$y = \frac{2}{3}x - \frac{5}{3}$	Write in slope-intercept form.

Notice that the line has a slope of $m = \frac{2}{3}$.

a. Any line parallel to the given line must also have a slope of $\frac{2}{3}$. Use the point-slope form with $m = \frac{2}{3}$ and $(x_1, y_1) = (2, -1)$.

$y - (-1) = \frac{2}{3}(x - 2)$	Write in point-slope form.
$y + 1 = \frac{2}{3}x - \frac{4}{3}$	Simplify.
$y = \frac{2}{3}x - \frac{7}{3}$	Write in slope-intercept form.

Notice the similarity between the slope-intercept form of this equation and the slope-intercept form of the given equation, $y = \frac{2}{3}x - \frac{5}{3}$.

b. Any line perpendicular to the given line must have a slope of $-\frac{3}{2}$ (because $-\frac{3}{2}$ is the negative reciprocal of $\frac{2}{3}$). Use the point-slope form with $m = -\frac{3}{2}$ and $(x_1, y_1) = (2, -1)$.

$y - (-1) = -\frac{3}{2}(x - 2)$	Write in point-slope form.
$y + 1 = -\frac{3}{2}x + 3$	Simplify.
$y = -\frac{3}{2}x + 2$	Write in slope-intercept form.

The graphs of all three equations are shown in Figure 2.7.

✓ *Checkpoint* ▶ *Audio-video solution in English & Spanish at LarsonPrecalculus.com*

Find the slope-intercept form of the equations of the lines that pass through the point $(-4, 1)$ and are (a) parallel to and (b) perpendicular to the line $5x - 3y = 8$. ■

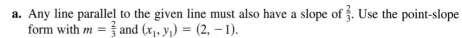

Applications

In real-life problems, the slope of a line can be interpreted as either a *ratio* or a *rate*. When the *x*-axis and *y*-axis have the same unit of measure, the slope has no units and is a **ratio.** When the *x*-axis and *y*-axis have different units of measure, the slope is a **rate** or **rate of change.**

EXAMPLE 5 **Using Slope as a Ratio**

The maximum recommended slope of a wheelchair ramp is $\frac{1}{12}$. A business installs a wheelchair ramp that rises 22 inches over a horizontal length of 24 feet (see figure). Is the ramp steeper than recommended? *(Source: ADA Standards for Accessible Design)*

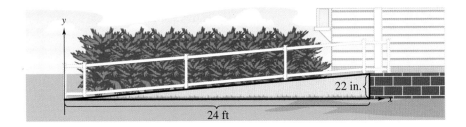

Solution The horizontal length of the ramp is 24 feet or $12(24) = 288$ inches. The slope of the ramp is the ratio of its height (the rise) to its length (the run).

$$\text{Slope} = \frac{\text{rise}}{\text{run}} = \frac{22 \text{ in.}}{288 \text{ in.}} \approx 0.076$$

The slope of the ramp is about 0.076, which is less than $\frac{1}{12} \approx 0.083$. So, the ramp is not steeper than recommended. Note that the slope of the ramp is a ratio and has no units.

 Checkpoint ▶ *Audio-video solution in English & Spanish at LarsonPrecalculus.com*

The business in Example 5 installs a second ramp that rises 36 inches over a horizontal length of 32 feet. Is the ramp steeper than recommended?

EXAMPLE 6 **Using Slope as a Rate of Change**

A kitchen appliance manufacturing company determines that the total cost *C* (in dollars) of producing *x* units of a blender is given by

$$C = 25x + 3500. \qquad \text{Cost equation}$$

Interpret the *y*-intercept and slope of this line.

Solution The *y*-intercept $(0, 3500)$ tells you that the cost of producing 0 units is $3500. This is the *fixed cost* of production—it includes costs that must be paid regardless of the number of units produced. The slope of $m = 25$ tells you that the cost of producing each unit is $25, as shown in Figure 2.8. Economists call the cost per unit the *marginal cost*. When the production increases by one unit, the "margin," or extra amount of cost, is $25. So, the cost increases at a rate of $25 per unit.

 Checkpoint ▶ *Audio-video solution in English & Spanish at LarsonPrecalculus.com*

An accounting firm determines that the value *V* (in dollars) of a copier *t* years after its purchase is given by

$$V = -300t + 1500.$$

Interpret the *y*-intercept and slope of this line.

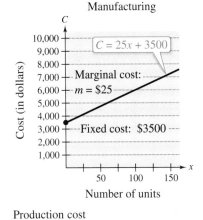

Manufacturing

Production cost
Figure 2.8

Businesses can deduct most of their expenses in the same year they occur. One exception is the cost of property that has a useful life of more than 1 year. Such costs must be *depreciated* (decreased in value) over the useful life of the property. Depreciating the *same amount* each year is called *linear* or *straight-line depreciation*. The *book value* is the difference between the original value and the total amount of depreciation accumulated to date.

EXAMPLE 7   Straight-Line Depreciation

A college purchases exercise equipment worth $12,000 for the new campus fitness center. The equipment has a useful life of 8 years. The salvage value at the end of 8 years is $2000. Write a linear equation that describes the book value of the equipment each year.

Solution   Let V represent the value of the equipment at the end of year t. Represent the initial value of the equipment by the data point $(0, 12{,}000)$ and the salvage value of the equipment by the data point $(8, 2000)$. The slope of the line is

$$m = \frac{2000 - 12{,}000}{8 - 0} = -\$1250 \text{ per year}$$

which represents the annual depreciation in *dollars per year*. So, the value of the equipment decreases $1250 per year. Using the point-slope form, write an equation of the line.

$$V - 12{,}000 = -1250(t - 0) \qquad \text{Write in point-slope form.}$$
$$V = -1250t + 12{,}000 \qquad \text{Write in slope-intercept form.}$$

Note that the domain of the equation is $0 \le t \le 8$. The table shows the book value at the end of each year, and Figure 2.9 shows the graph of the equation.

Year, t	Value, V
0	12,000
1	10,750
2	9500
3	8250
4	7000
5	5750
6	4500
7	3250
8	2000

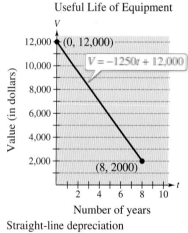

Useful Life of Equipment

Straight-line depreciation
Figure 2.9

✓ *Checkpoint* ▶ *Audio-video solution in English & Spanish at LarsonPrecalculus.com*

A manufacturing firm purchases a machine worth $24,750. The machine has a useful life of 6 years. After 6 years, the machine will have to be discarded and replaced, because it will have no salvage value. Write a linear equation that describes the book value of the machine each year.   ■

In many real-life applications, the two data points that determine the line are often given in a disguised form. Note how the data points are described in Example 7.

GO DIGITAL

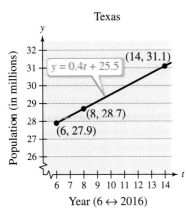

Texas

Figure 2.10

EXAMPLE 8 Predicting a Population

The population of Texas was about 27.9 million in 2016 and about 28.7 million in 2018. Use this information to write a linear equation that gives the population (in millions) in terms of the year. Predict the population of Texas in 2024. *(Source: U.S. Census Bureau)*

Solution Let $t = 6$ represent 2016. Then the two given values are represented by the data points $(6, 27.9)$ and $(8, 28.7)$. The slope of the line through these points is

$$m = \frac{28.7 - 27.9}{8 - 6} = \frac{0.8}{2} = 0.4 \text{ million people per year.}$$

Use the point-slope form to write an equation that relates the population y and the year t.

$$y - 27.9 = 0.4(t - 6) \implies y = 0.4t + 25.5$$

According to this equation, the population in 2024 will be

$$y = 0.4(14) + 25.5 = 5.6 + 25.5 = 31.1 \text{ million people.} \qquad \text{See Figure 2.10.}$$

✓ *Checkpoint* ▶ *Audio-video solution in English & Spanish at LarsonPrecalculus.com*

The population of Oregon was about 3.9 million in 2013 and about 4.2 million in 2018. Repeat Example 8 using this information. *(Source: U.S. Census Bureau)* ∎

The prediction method illustrated in Example 8 is called **linear extrapolation.** Note in Figure 2.11 that an extrapolated point does not lie between the given points. When the estimated point lies between two given points, as shown in Figure 2.12, the procedure is called **linear interpolation.**

The slope of a vertical line is undefined, so its equation cannot be written in slope-intercept form. However, every line has an equation that can be written in the **general form** $Ax + By + C = 0$, where A and B are not both zero.

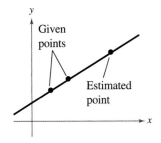

Linear extrapolation
Figure 2.11

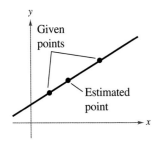

Linear interpolation
Figure 2.12

Summary of Equations of Lines

1. General form: $Ax + By + C = 0$
2. Vertical line: $x = a$
3. Horizontal line: $y = b$
4. Slope-intercept form: $y = mx + b$
5. Point-slope form: $y - y_1 = m(x - x_1)$
6. Two-point form: $y - y_1 = \dfrac{y_2 - y_1}{x_2 - x_1}(x - x_1)$

Summarize (Section 2.1)

1. Explain how to use slope to graph a linear equation in two variables *(page 160)* and how to find the slope of a line passing through two points *(page 162)*. For examples of using and finding slopes, see Examples 1 and 2.

2. State the point-slope form of the equation of a line *(page 164)*. For an example of using point-slope form, see Example 3.

3. Explain how to use slope to identify parallel and perpendicular lines *(page 165)*. For an example of finding parallel and perpendicular lines, see Example 4.

4. Describe examples of how to use slope and linear equations in two variables to model and solve real-life problems *(pages 166–168, Examples 5–8)*.

GO DIGITAL

2.1 Exercises

See CalcChat.com for tutorial help and worked-out solutions to odd-numbered exercises.

Vocabulary and Concept Check

In Exercises 1–6, fill in the blanks.

1. The simplest mathematical model for relating two variables is the _____ equation in two variables $y = mx + b$.

2. For a line, the ratio of the change in y to the change in x is the _____ of the line.

3. The _____-_____ form of the equation of a line with slope m passing through the point (x_1, y_1) is $y - y_1 = m(x - x_1)$.

4. Two distinct nonvertical lines are _____ if and only if their slopes are equal.

5. When the x-axis and y-axis have different units of measure, the slope can be interpreted as a _____.

6. _____ _____ is the prediction method used to estimate a point on a line when the point does not lie between the given points.

7. What is the relationship between two lines whose slopes are -3 and $\frac{1}{3}$?

8. Write the point-slope form equation $y - y_1 = m(x - x_1)$ in general form.

Skills and Applications

Identifying Lines In Exercises 9 and 10, identify the line that has each slope.

9. (a) $m = \frac{2}{3}$
   (b) m is undefined.
   (c) $m = -2$

10. (a) $m = 0$
    (b) $m = -\frac{3}{4}$
    (c) $m = 1$

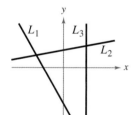

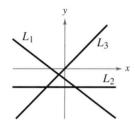

Sketching Lines In Exercises 11 and 12, sketch the lines through the point with the given slopes on the same set of coordinate axes.

	Point	Slopes
11.	$(2, 3)$	(a) 0 (b) 1 (c) 2 (d) -3
12.	$(-4, 1)$	(a) 3 (b) -3 (c) $\frac{1}{2}$ (d) Undefined

Estimating the Slope of a Line In Exercises 13 and 14, estimate the slope of the line.

13.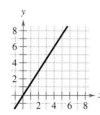

14.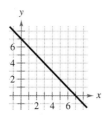

Graphing a Linear Equation In Exercises 15–24, find the slope and y-intercept (if possible) of the line. Sketch the line.

15. $y = 5x + 3$
16. $y = -x - 10$
17. $y = -\frac{3}{4}x - 1$
18. $y = \frac{2}{3}x + 2$
19. $y - 5 = 0$
20. $x + 4 = 0$
21. $5x - 2 = 0$
22. $3y + 5 = 0$
23. $7x - 6y = 30$
24. $2x + 3y = 9$

Finding the Slope of a Line Through Two Points In Exercises 25–34, find the slope of the line passing through the pair of points.

25. $(0, 9), (6, 0)$
26. $(10, 0), (0, -5)$
27. $(-3, -2), (1, 6)$
28. $(2, -1), (-2, 1)$
29. $(5, -7), (8, -7)$
30. $(-2, 1), (-4, -5)$
31. $(-6, -1), (-6, 4)$
32. $(0, -10), (-4, 0)$
33. $(4.8, 3.1), (-5.2, 1.6)$
34. $\left(\frac{11}{2}, -\frac{4}{3}\right), \left(-\frac{3}{2}, -\frac{1}{3}\right)$

Using the Slope and a Point In Exercises 35–42, use the slope of the line and the point on the line to find three additional points through which the line passes. (There are many correct answers.)

35. $m = 0$, $(5, 7)$
36. $m = 0$, $(3, -2)$
37. $m = 2$, $(-5, 4)$
38. $m = -2$, $(0, -9)$
39. $m = -\frac{1}{3}$, $(4, 5)$
40. $m = \frac{1}{4}$, $(3, -4)$
41. m is undefined, $(-4, 3)$
42. m is undefined, $(2, 14)$

Using the Point-Slope Form   In Exercises 43–54, find the slope-intercept form of the equation of the line that has the given slope and passes through the given point. Sketch the line.

43. $m = 3$, $(0, -2)$
44. $m = -1$, $(0, 10)$
45. $m = -2$, $(-3, 6)$
46. $m = 4$, $(0, 0)$
47. $m = -\frac{1}{3}$, $(4, 0)$
48. $m = \frac{1}{4}$, $(8, 2)$
49. $m = -\frac{1}{2}$, $(2, -3)$
50. $m = \frac{3}{4}$, $(-2, -5)$
51. $m = 0$, $\left(4, \frac{5}{7}\right)$
52. $m = 6$, $\left(2, \frac{3}{2}\right)$
53. $m = 5$, $(-5.1, 1.8)$
54. $m = 0$, $(-2.5, 3.25)$

Finding an Equation of a Line   In Exercises 55–64, find an equation of the line passing through the pair of points. Sketch the line.

55. $(5, -1), (-5, 5)$
56. $(4, 3), (-4, -4)$
57. $(-7, 2), (-7, 5)$
58. $(-6, -3), (2, -3)$
59. $\left(2, \frac{1}{2}\right), \left(\frac{1}{2}, \frac{5}{4}\right)$
60. $(1, 1), \left(6, -\frac{2}{3}\right)$
61. $(1, 0.6), (-2, -0.6)$
62. $(-8, 0.6), (2, -2.4)$
63. $(2, -1), \left(\frac{1}{3}, -1\right)$
64. $\left(\frac{7}{3}, -8\right), \left(\frac{7}{3}, 1\right)$

Parallel and Perpendicular Lines   In Exercises 65–68, determine whether the lines are parallel, perpendicular, or neither.

65. L_1: $y = -\frac{2}{3}x - 3$
    L_2: $y = -\frac{2}{3}x + 4$
66. L_1: $y = \frac{1}{4}x - 1$
    L_2: $y = 4x + 7$
67. L_1: $y = \frac{1}{2}x - 3$
    L_2: $y = -\frac{1}{2}x + 1$
68. L_1: $y = -\frac{4}{5}x - 5$
    L_2: $y = \frac{5}{4}x + 1$

Parallel and Perpendicular Lines   In Exercises 69–72, determine whether the lines L_1 and L_2 passing through the pairs of points are parallel, perpendicular, or neither.

69. L_1: $(0, -1), (5, 9)$
    L_2: $(0, 3), (4, 1)$
70. L_1: $(-2, -1), (1, 5)$
    L_2: $(1, 3), (5, -5)$
71. L_1: $(-6, -3), (2, -3)$
    L_2: $\left(3, -\frac{1}{2}\right), \left(6, -\frac{1}{2}\right)$
72. L_1: $(4, 8), (-4, 2)$
    L_2: $(3, -5), \left(-1, \frac{1}{3}\right)$

Finding Parallel and Perpendicular Lines   In Exercises 73–80, find equations of the lines that pass through the given point and are (a) parallel to and (b) perpendicular to the given line.

73. $4x - 2y = 3$, $(2, 1)$
74. $x + y = 7$, $(-3, 2)$
75. $3x + 4y = 7$, $\left(-\frac{2}{3}, \frac{7}{8}\right)$
76. $5x + 3y = 0$, $\left(\frac{7}{8}, \frac{3}{4}\right)$
77. $y + 5 = 0$, $(-2, 4)$
78. $x - 4 = 0$, $(3, -2)$
79. $x - y = 4$, $(2.5, 6.8)$
80. $6x + 2y = 9$, $(-3.9, -1.4)$

Using Intercept Form   In Exercises 81–86, use the *intercept form* to find the general form of the equation of the line with the given intercepts. The intercept form of the equation of a line with intercepts $(a, 0)$ and $(0, b)$ is

$$\frac{x}{a} + \frac{y}{b} = 1, \quad a \neq 0, \quad b \neq 0.$$

81. x-intercept: $(3, 0)$; y-intercept: $(0, 5)$
82. x-intercept: $(-3, 0)$; y-intercept: $(0, 4)$
83. x-intercept: $\left(-\frac{1}{6}, 0\right)$; y-intercept: $\left(0, -\frac{2}{3}\right)$
84. x-intercept: $\left(\frac{2}{3}, 0\right)$; y-intercept: $(0, -2)$
85. Point on line: $(1, 2)$
    x-intercept: $(c, 0)$, $c \neq 0$
    y-intercept: $(0, c)$, $c \neq 0$
86. Point on line: $(-3, 4)$
    x-intercept: $(d, 0)$, $d \neq 0$
    y-intercept: $(0, d)$, $d \neq 0$

87. **Sales**   The slopes of lines representing annual sales y in terms of time x in years are given below. Use the slopes to interpret any change in annual sales for a one-year increase in time.

    (a) The line has a slope of $m = 135$.
    (b) The line has a slope of $m = 0$.
    (c) The line has a slope of $m = -40$.

88. **Sales**   The graph shows the sales (in billions of dollars) for Apple Inc. in the years 2013 through 2019. *(Source: Apple Inc.)*

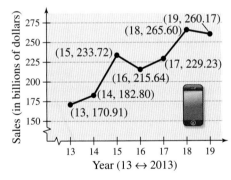

    (a) Use the slopes of the line segments to determine the years in which the sales showed the greatest increase and the greatest decrease.

    (b) Find the slope of the line segment connecting the points for the years 2013 and 2019.

    (c) Interpret the meaning of the slope in part (b) in the context of the problem.

89. **Road Grade**   You are driving on a road that has a 6% uphill grade. This means that the slope of the road is $\frac{6}{100}$. Approximate the amount of vertical change in your position when you drive 200 feet.

90. Road Grade

From the top of a mountain road, a surveyor takes several horizontal measurements x and several vertical measurements y, as shown in the table (x and y are measured in feet).

x	300	600	900	1200
y	-25	-50	-75	-100

x	1500	1800	2100
y	-125	-150	-175

(a) Sketch a scatter plot of the data.

(b) Use a straightedge to sketch the line that you think best fits the data.

(c) Find an equation for the line you sketched in part (b).

(d) Interpret the meaning of the slope of the line in part (c) in the context of the problem.

(e) The surveyor needs to put up a road sign that indicates the steepness of the road. For example, a surveyor would put up a sign that states "8% grade" on a road with a downhill grade that has a slope of $-\frac{8}{100}$. What should the sign state for the road in this problem?

91. Temperature Conversion   Write a linear equation that expresses the relationship between the temperature in degrees Celsius C and degrees Fahrenheit F. Use the fact that water freezes at 0°C (32°F) and boils at 100°C (212°F).

92. Neurology   The average weight of a male child's brain is 970 grams at age 1 and 1270 grams at age 3. *(Source: American Neurological Association)*

(a) Assuming that the relationship between brain weight y and age t is linear, write a linear model for the data.

(b) What is the slope and what does it tell you about brain weight?

(c) Use your model to estimate the average brain weight at age 2.

(d) Use your school's library, the Internet, or some other reference source to find the actual average brain weight at age 2. How close was your estimate?

(e) Do you think your model could be used to determine the average brain weight of an adult? Explain.

93. Depreciation   A sandwich shop purchases a used pizza oven for $830. After 5 years, the oven will have to be discarded and replaced. Write a linear equation giving the value V of the equipment during the 5 years it will be in use.

94. Depreciation   A school district purchases a high-volume printer, copier, and scanner for $24,000. After 10 years, the equipment will have to be replaced. Its value at that time is expected to be $2000. Write a linear equation giving the value V of the equipment during the 10 years it will be in use.

95. Cost, Revenue, and Profit   A roofing contractor purchases a shingle delivery truck with a shingle elevator for $42,000. The vehicle requires an average expenditure of $9.50 per hour for fuel and maintenance, and the operator is paid $11.50 per hour.

(a) Write a linear equation giving the total cost C of operating this equipment for t hours. (Include the purchase cost of the equipment.)

(b) Assuming that customers are charged $45 per hour of machine use, write an equation for the revenue R obtained from t hours of use.

(c) Use the formula for profit $P = R - C$ to write an equation for the profit obtained from t hours of use.

(d) Use the result of part (c) to find the break-even point—that is, the number of hours this equipment must be used to yield a profit of 0 dollars.

96. Geometry   The length and width of a rectangular garden are 15 meters and 10 meters, respectively. A walkway of width x surrounds the garden.

(a) Draw a diagram that gives a visual representation of the problem.

(b) Write the equation for the perimeter y of the walkway in terms of x. Explain what the slope of the equation represents.

(c) Use a graphing utility to graph the equation for the perimeter.

Exploring the Concepts

True or False?   In Exercises 97 and 98, determine whether the statement is true or false. Justify your answer.

97. A line with a slope of $-\frac{5}{7}$ is steeper than a line with a slope of $-\frac{6}{7}$.

98. The line through $(-8, 2)$ and $(-1, 4)$ and the line through $(0, -4)$ and $(-7, 7)$ are parallel.

99. Right Triangle   Explain how you can use slope to show that the points $A(-1, 5)$, $B(3, 7)$, and $C(5, 3)$ are the vertices of a right triangle.

100. Perpendicular Segments Find d_1 and d_2 in terms of m_1 and m_2, respectively (see figure). Then use the Pythagorean Theorem to find a relationship between m_1 and m_2.

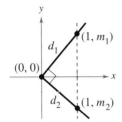

101. Error Analysis Describe the error in finding the equation of each graph.

(a) $y = 2x - 1$ ✗

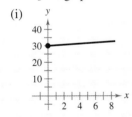

(b) $y = \frac{3}{4}x + 4$ ✗

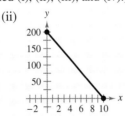

102. **HOW DO YOU SEE IT?** Match the description of the situation with its graph. Also determine the slope and y-intercept of each graph and interpret the slope and y-intercept in the context of the situation. [The graphs are labeled (i), (ii), (iii), and (iv).]

(i)

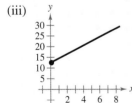

(ii)

(iii)

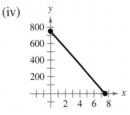

(iv)

(a) A person is paying $20 per week to a friend to repay a $200 loan.

(b) An employee receives $12.50 per hour plus $2 for each unit produced per hour.

(c) A sales representative receives $30 per day for food plus $0.32 for each mile traveled.

(d) A computer that was purchased for $750 depreciates $100 per year.

103. Comparing Slopes Use a graphing utility to compare the slopes of the lines $y = mx$, where $m = 0.5, 1, 2,$ and 4. Which line rises most quickly? Now, let $m = -0.5, -1, -2,$ and -4. Which line falls most quickly? Use a square setting to obtain a true geometric perspective. What can you conclude about the slope and the "rate" at which the line rises or falls?

104. Slope and Steepness The slopes of two lines are -4 and $\frac{5}{2}$. Which is steeper? Explain.

Review & Refresh ▶ Video solutions at LarsonPrecalculus.com

Solving for a Variable In Exercises 105–108, evaluate the equation when $x = -2, 0, 3,$ and 6.

105. $y = x^2 - x$

106. $y = x + 5y$

107. $2f = \dfrac{7 - x^3}{5}$

108. $\dfrac{g}{4} = \dfrac{\sqrt{2 + x}}{3}$

Solving an Equation In Exercises 109–118, solve the equation, if possible. Check your solutions.

109. $2x^3 - 5x^2 - 2x = -5$

110. $x^4 - 3x^3 - x - 2 = 1 - 2x$

111. $x^6 + 4x^3 - 5 = 0$

112. $x^4 - 5x^2 + 4 = 0$

113. $3\sqrt{x} - 5\sqrt{18} = 0$

114. $(x + 1)^{3/5} - 8 = 0$

115. $x = \dfrac{2}{x} + 1$

116. $\dfrac{3}{x} - \dfrac{1}{2} = \dfrac{x}{4}$

117. $x + |x - 9| = 7$

118. $|3x - 6| = x^2 - 4$

Simplifying an Expression In Exercises 119–122, simplify the expression.

119. $\dfrac{[(x - 1)^2 + 1] - (x^2 + 1)}{x}$

120. $\dfrac{[(t + 1)^2 - (t + 1) - 2] - (t^2 - t - 2)}{t}$

121. $\dfrac{4 - x^2 - 3}{x - 1} + x^2 + x + 4$

122. $\dfrac{x^3 + x}{(x^2 - 4)} - \dfrac{5}{2(x + 2)} + \dfrac{5}{2(x - 2)} + 2$

Project: Bachelor's Degrees To work an extended application analyzing the numbers of bachelor's degrees earned by women in the United States from 2006 through 2017, visit this text's website at *LarsonPrecalculus.com*. (*Source: National Center for Education Statistics*)

2.2 Functions

Functions are used to model and solve real-life problems. For example, in Exercise 64 on page 184, you will use a function that models the force of water against the face of a dam.

⊘ Determine whether relations between two variables are functions, and use function notation.
⊘ Find the domains of functions.
⊘ Use functions to model and solve real-life problems, and evaluate difference quotients.

Introduction to Functions and Function Notation

Many everyday phenomena involve two quantities that are related to each other by some rule of correspondence. The mathematical term for such a rule of correspondence is a **relation.** In mathematics, equations and formulas often represent relations. For example, the simple interest I earned on $1000 for 1 year is related to the annual interest rate r by the formula $I = 1000r$.

The formula $I = 1000r$ represents a special kind of relation that matches each item from one set with *exactly one* item from a different set. Such a relation is a **function.**

Definition of Function

A **function** f from a set A to a set B is a relation that assigns to each element x in the set A exactly one element y in the set B. The set A is the **domain** (or set of inputs) of the function f, and the set B contains the **range** (or set of outputs).

To help understand this definition, look at the figure below, which shows a function that relates the time of day to the temperature.

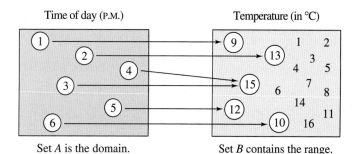

Set A is the domain.
Inputs: 1, 2, 3, 4, 5, 6

Set B contains the range.
Outputs: 9, 10, 12, 13, 15

This function can be represented by the ordered pairs

$$\{(1, 9), (2, 13), (3, 15), (4, 15), (5, 12), (6, 10)\}.$$

In each ordered pair, the first coordinate (x-value) is the **input** and the second coordinate (y-value) is the **output**.

Characteristics of a Function from Set A to Set B

1. Each element in A must be matched with an element in B.
2. Some elements in B may not be matched with any element in A.
3. Two or more elements in A may be matched with the same element in B.
4. An element in A (the domain) cannot be matched with two different elements in B.

GO DIGITAL

There are four common ways to represent a function—verbally, numerically, graphically, and algebraically.

> ### Four Ways to Represent a Function
>
> 1. *Verbally* by a sentence that describes how the input variable is related to the output variable
> 2. *Numerically* by a table or a list of ordered pairs that matches input values with output values
> 3. *Graphically* by points in a coordinate plane in which the horizontal positions represent the input values and the vertical positions represent the output values
> 4. *Algebraically* by an equation in two variables

To determine whether a relation is a function, you must decide whether each input value is matched with exactly one output value. When any input value is matched with two or more output values, the relation is not a function.

EXAMPLE 1 Testing for Functions

Determine whether the relation represents *y* as a function of *x*.

a. The input value *x* is the number of representatives from a state, and the output value *y* is the number of senators.

b.

Input, *x*	Output, *y*
2	11
2	10
3	8
4	5
5	1

c.

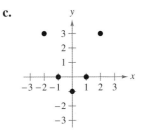

Solution

a. This verbal description *does* describe *y* as a function of *x*. Regardless of the value of *x*, the value of *y* is always 2. This is an example of a *constant function*.

b. This table *does not* describe *y* as a function of *x*. The input value 2 is matched with two different *y*-values.

c. The graph *does* describe *y* as a function of *x*. Each input value is matched with exactly one output value.

 Checkpoint ▶ *Audio-video solution in English & Spanish at LarsonPrecalculus.com*

Determine whether the relation represents *y* as a function of *x*.

a. *Domain, x*   *Range, y*

b.

Input, *x*	0	1	2	3	4
Output, *y*	−4	−2	0	2	4

HISTORICAL NOTE

Many consider Leonhard Euler (1707–1783), a Swiss mathematician, to be the most prolific and productive mathematician in history. One of his greatest influences on mathematics was his use of symbols, or notation. Euler introduced the function notation $y = f(x)$.

Representing functions by sets of ordered pairs is common in *discrete mathematics.* In algebra, however, it is more common to represent functions by equations or formulas involving two variables. For example, the equation

$$y = x^2 \qquad\qquad \textit{y is a function of x.}$$

represents the variable y as a function of the variable x. In this equation, x is the **independent variable** (the input) and y is the **dependent variable** (the output). The domain of the function is the set of all values taken on by the independent variable x, and the range of the function is the set of all values taken on by the dependent variable y.

EXAMPLE 2   Testing for Functions Represented Algebraically

▶▶▶ *See LarsonPrecalculus.com for an interactive version of this type of example.*

Determine whether each equation represents y as a function of x.

a. $x^2 + y = 1$ **b.** $-x + y^2 = 1$

Solution   To determine whether y is a function of x, solve for y in terms of x.

a. Solving for y yields

$$x^2 + y = 1 \qquad\qquad \text{Write original equation.}$$
$$y = 1 - x^2. \qquad\qquad \text{Solve for } y.$$

To each value of x there corresponds exactly one value of y. So, y is a function of x.

b. Solving for y yields

$$-x + y^2 = 1 \qquad\qquad \text{Write original equation.}$$
$$y^2 = 1 + x \qquad\qquad \text{Add } x \text{ to each side.}$$
$$y = \pm\sqrt{1 + x}. \qquad\qquad \text{Solve for } y.$$

The $\pm$ indicates that to a given value of x there correspond two values of y. For instance, when $x = 3$, $y = 2$ or $y = -2$. So, y is not a function of x.

✓ *Checkpoint* ▶ **Audio-video solution in English & Spanish at LarsonPrecalculus.com**

Determine whether each equation represents y as a function of x.

a. $x^2 + y^2 = 8$ **b.** $y - 4x^2 = 36$ ▪

When using an equation to represent a function, it is convenient to name the function for easy reference. For instance, you know from Example 2(a) that the equation $y = 1 - x^2$ describes y as a function of x. By naming this function "f," you can write the input, output, and equation using **function notation,** as shown below.

Input	Output	Equation
x	$f(x)$	$f(x) = 1 - x^2$

The symbol $f(x)$ is read as *the value of f at x* or simply *f of x.* The symbol $f(x)$ corresponds to the y-value for a given x. So, $y = f(x)$. Keep in mind that f is the *name* of the function, whereas $f(x)$ is the *output value* of the function at the *input value x.* For example, the function $f(x) = 3 - 2x$ has *function values* denoted by $f(-1)$, $f(0)$, $f(2)$, and so on. To find these values, substitute the specified input values into f.

For $x = -1$, $f(-1) = 3 - 2(-1) = 3 + 2 = 5$.

For $x = 0$, $f(0) = 3 - 2(0) = 3 - 0 = 3$.

For $x = 2$, $f(2) = 3 - 2(2) = 3 - 4 = -1$.

GO DIGITAL

Although it is often convenient to use f as a function name and x as the independent variable, other letters may be used as well. For example,

$$f(x) = x^2 - 4x + 7, \quad f(t) = t^2 - 4t + 7, \quad \text{and} \quad g(s) = s^2 - 4s + 7$$

all define the same function. In fact, the role of the independent variable is that of a "placeholder." Consequently, the function can be described by

$$f\left(\boxed{}\right) = \left(\boxed{}\right)^2 - 4\left(\boxed{}\right) + 7.$$

EXAMPLE 3 **Evaluating a Function**

Let $g(x) = -x^2 + 4x + 1$. Find $g(2)$, $g(t)$, and $g(x + 2)$.

Solution

To find $g(2)$, replace x with 2 in $g(x) = -x^2 + 4x + 1$ and simplify.

$$\begin{aligned} g(2) &= -(2)^2 + 4(2) + 1 \\ &= -4 + 8 + 1 \\ &= 5 \end{aligned}$$

To find $g(t)$, replace x with t and simplify.

$$\begin{aligned} g(t) &= -(t)^2 + 4(t) + 1 \\ &= -t^2 + 4t + 1 \end{aligned}$$

To find $g(x + 2)$, replace x with $x + 2$ and simplify.

$$\begin{aligned} g(x + 2) &= -(x + 2)^2 + 4(x + 2) + 1 & \text{Substitute } x + 2 \text{ for } x. \\ &= -(x^2 + 4x + 4) + 4x + 8 + 1 & \text{Multiply.} \\ &= -x^2 - 4x - 4 + 4x + 8 + 1 & \text{Distributive Property} \\ &= -x^2 + 5 & \text{Simplify.} \end{aligned}$$

ALGEBRA HELP

In Example 3, note that $g(x + 2)$ is not equal to $g(x) + g(2)$. In general, $g(u + v) \neq g(u) + g(v)$.

✓ *Checkpoint* ▶ *Audio-video solution in English & Spanish at LarsonPrecalculus.com*

Let $f(x) = 10 - 3x^2$. Find $f(2)$, $f(-4)$, and $f(x - 1)$. ■

A function defined by two or more equations over a specified domain is called a **piecewise-defined function.**

EXAMPLE 4 **A Piecewise-Defined Function**

Evaluate the function f when $x = -1$, 0, and 1.

$$f(x) = \begin{cases} x^2 + 1, & x < 0 \\ x - 1, & x \geq 0 \end{cases}$$

Solution Because $x = -1$ is less than 0, use $f(x) = x^2 + 1$ to obtain

$$f(-1) = (-1)^2 + 1 = 2.$$

For $x = 0$, use $f(x) = x - 1$ to obtain

$$f(0) = (0) - 1 = -1.$$

For $x = 1$, use $f(x) = x - 1$ to obtain

$$f(1) = (1) - 1 = 0.$$

✓ *Checkpoint* ▶ *Audio-video solution in English & Spanish at LarsonPrecalculus.com*

Evaluate the function f given in Example 4 when $x = -2$, 2, and 3. ■

GO DIGITAL

EXAMPLE 5   **Finding Values for Which f(x) = 0**

Find all real values of x for which $f(x) = 0$.

a. $f(x) = -2x + 10$ **b.** $f(x) = x^2 - 5x + 6$

Solution   For each function, set $f(x) = 0$ and solve for x.

a. $-2x + 10 = 0$ Set $f(x)$ equal to 0.

$\qquad -2x = -10$ Subtract 10 from each side.

$\qquad\qquad x = 5$ Divide each side by -2.

So, $f(x) = 0$ when $x = 5$.

b. $x^2 - 5x + 6 = 0$ Set $f(x)$ equal to 0.

$(x - 2)(x - 3) = 0$ Factor.

$\qquad x - 2 = 0 \implies x = 2$ Set 1st factor equal to 0.

$\qquad x - 3 = 0 \implies x = 3$ Set 2nd factor equal to 0.

So, $f(x) = 0$ when $x = 2$ or $x = 3$.

✓ **Checkpoint** Audio-video solution in English & Spanish at *LarsonPrecalculus.com*

Find all real values of x for which $f(x) = 0$, where $f(x) = x^2 - 16$.

EXAMPLE 6   **Finding Values for Which f(x) = g(x)**

Find the values of x for which $f(x) = g(x)$.

a. $f(x) = x^2 + 1$ and $g(x) = 3x - x^2$

b. $f(x) = x^2 - 1$ and $g(x) = -x^2 + x + 2$

Solution

a. $\qquad x^2 + 1 = 3x - x^2$ Set $f(x)$ equal to $g(x)$.

$\quad 2x^2 - 3x + 1 = 0$ Write in general form.

$(2x - 1)(x - 1) = 0$ Factor.

$\qquad 2x - 1 = 0 \implies x = \tfrac{1}{2}$ Set 1st factor equal to 0.

$\qquad x - 1 = 0 \implies x = 1$ Set 2nd factor equal to 0.

So, $f(x) = g(x)$ when $x = 1/2$ or $x = 1$.

b. $\qquad x^2 - 1 = -x^2 + x + 2$ Set $f(x)$ equal to $g(x)$.

$\quad 2x^2 - x - 3 = 0$ Write in general form.

$(2x - 3)(x + 1) = 0$ Factor.

$\qquad 2x - 3 = 0 \implies x = \tfrac{3}{2}$ Set 1st factor equal to 0.

$\qquad x + 1 = 0 \implies x = -1$ Set 2nd factor equal to 0.

So, $f(x) = g(x)$ when $x = 3/2$ or $x = -1$.

✓ **Checkpoint** Audio-video solution in English & Spanish at *LarsonPrecalculus.com*

Find the values of x for which $f(x) = g(x)$, where $f(x) = x^2 + 6x - 24$ and $g(x) = 4x - x^2$.   ∎

GO DIGITAL

The Domain of a Function

The domain of a function can be described explicitly or it can be *implied* by the expression used to define the function. The **implied domain** is the set of all real numbers for which the expression is defined. For example, the function

$$f(x) = \frac{1}{x^2 - 4} \qquad \text{Domain excludes } x\text{-values that result in division by zero.}$$

has an implied domain consisting of all real numbers x other than $x = \pm 2$. These two values are excluded from the domain because division by zero is undefined. Another common type of implied domain is that used to avoid even roots of negative numbers. For example, the function

$$f(x) = \sqrt{x} \qquad \text{Domain excludes } x\text{-values that result in even roots of negative numbers.}$$

is defined only for $x \geq 0$. So, its implied domain is the interval $[0, \infty)$. In general, the domain of a function *excludes* values that cause division by zero *or* that result in the even root of a negative number.

EXAMPLE 7   Finding the Domains of Functions

Find the domain of each function.

a. f: $\{(-3, 0), (-1, 4), (0, 2), (2, 2), (4, -1)\}$ **b.** $g(x) = \dfrac{1}{x + 5}$

c. Volume of a sphere: $V = \frac{4}{3}\pi r^3$ **d.** $h(x) = \sqrt{4 - 3x}$

Solution

a. The domain of f consists of all first coordinates in the set of ordered pairs.

$$\text{Domain} = \{-3, -1, 0, 2, 4\}$$

b. Excluding x-values that yield zero in the denominator, the domain of g is the set of all real numbers x except $x = -5$.

c. This function represents the volume of a sphere, so the values of the radius r must be positive. The domain is the set of all real numbers r such that $r > 0$.

d. This function is defined only for x-values for which

$$4 - 3x \geq 0.$$

Using the methods described in Section 1.7, you can conclude that $x \leq \frac{4}{3}$. So, the domain is the interval $\left(-\infty, \frac{4}{3}\right]$.

✓ *Checkpoint* ▶ *Audio-video solution in English & Spanish at LarsonPrecalculus.com*

Find the domain of each function.

a. f: $\{(-2, 2), (-1, 1), (0, 3), (1, 1), (2, 2)\}$ **b.** $g(x) = \dfrac{1}{3 - x}$

c. Circumference of a circle: $C = 2\pi r$ **d.** $h(x) = \sqrt{x - 16}$ ■

In Example 7(c), note that the domain of a function may be implied by the physical context. For example, from the equation

$$V = \frac{4}{3}\pi r^3$$

you have no reason to restrict r to positive values, but the physical context implies that a sphere cannot have a negative or zero radius.

GO DIGITAL

Applications

EXAMPLE 8 **The Dimensions of a Container**

You work in the marketing department of a soft-drink company and are experimenting with a new can for iced tea that is slightly narrower and taller than a standard can. For your experimental can, the ratio of the height to the radius is 4.

a. Write the volume of the can as a function of the radius r.

b. Write the volume of the can as a function of the height h.

Solution

a. $V(r) = \pi r^2 h = \pi r^2(4r) = 4\pi r^3$ Write V as a function of r.

b. $V(h) = \pi r^2 h = \pi\left(\dfrac{h}{4}\right)^2 h = \dfrac{\pi h^3}{16}$ Write V as a function of h.

✓ *Checkpoint* ▶ Audio-video solution in English & Spanish at LarsonPrecalculus.com

For the experimental can described in Example 8, write the *surface area* as a function of (a) the radius r and (b) the height h.

EXAMPLE 9 **The Path of a Baseball**

A batter hits a baseball at a point 3 feet above ground at a velocity of 100 feet per second and an angle of 45°. The path of the baseball is given by the function

$$f(x) = -0.0032x^2 + x + 3$$

where $f(x)$ is the height of the baseball (in feet) and x is the horizontal distance from home plate (in feet). Will the baseball clear a 10-foot fence located 300 feet from home plate?

Algebraic Solution

Find the height of the baseball when $x = 300$.

$f(x) = -0.0032x^2 + x + 3$ Write original function.

$f(300) = -0.0032(300)^2 + 300 + 3$ Substitute 300 for x.

$\quad = 15$ Simplify.

When $x = 300$, the height of the baseball is 15 feet. So, the baseball will clear a 10-foot fence.

Graphical Solution

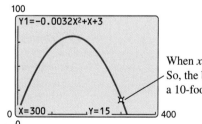

When $x = 300$, $y = 15$. So, the ball will clear a 10-foot fence.

✓ *Checkpoint* ▶ Audio-video solution in English & Spanish at LarsonPrecalculus.com

A second baseman throws a baseball toward the first baseman 60 feet away. The path of the baseball is given by the function

$$f(x) = -0.004x^2 + 0.3x + 6$$

where $f(x)$ is the height of the baseball (in feet) and x is the horizontal distance from the second baseman (in feet). The first baseman can reach 8 feet high. Can the first baseman catch the baseball without jumping? ■

GO DIGITAL

EXAMPLE 10   **Mortgage Debt**

The total mortgage debt M (in trillions of dollars) for all U.S. consumers decreased in a linear pattern from 2009 through 2013, and then increased in a quadratic pattern from 2014 through 2019, as shown in the bar graph. These two patterns can be approximated by the function

$$M(t) = \begin{cases} -0.268t + 11.47, & 9 \le t < 14 \\ 0.0376t^2 - 1.003t + 14.77, & 14 \le t \le 19 \end{cases}$$

where t represents the year, with $t = 9$ corresponding to 2009. Use this function to approximate the mortgage debt for all U.S. consumers in 2010, 2014, and 2016. *(Source: New York Fed Consumer Credit Panel/Equifax)*

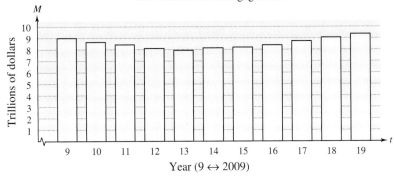

U.S. Consumers: Mortgage Debt

Year (9 ↔ 2009)

Solution   The year 2010 corresponds to $t = 10$, so use $M(t) = -0.268t + 11.47$.

$M(10) = -0.268(10) + 11.47$ Substitute 10 for t.

$\quad\quad\quad = \$8.79$ trillion   Simplify.

The year 2014 corresponds to $t = 14$, so use $M(t) = 0.0376t^2 - 1.003t + 14.77$.

$M(14) = 0.0376(14)^2 - 1.003(14) + 14.77$ Substitute 14 for t.

$\quad\quad\quad = \$8.0976$ trillion   Simplify.

The year 2016 corresponds to $t = 16$, so use $M(t) = 0.0376t^2 - 1.003t + 14.77$.

$M(16) = 0.0376(16)^2 - 1.003(16) + 14.77$ Substitute 16 for t.

$\quad\quad\quad = \$8.3476$ trillion   Simplify.

✓ *Checkpoint* ▶ *Audio-video solution in English & Spanish at LarsonPrecalculus.com*

The total credit card debt C (in trillions of dollars) for all U.S. consumers from 2009 through 2019 can be approximated by the function

$$C(t) = \begin{cases} 0.0135t^2 - 0.339t + 2.81, & 9 \le t < 13 \\ 0.037t + 0.16, & 13 \le t \le 19 \end{cases}$$

where t represents the year, with $t = 9$ corresponding to 2009. Use this function to approximate the credit card debt for all U.S. consumers in 2010, 2013, and 2018. *(Source: New York Fed Consumer Credit Panel/Equifax)*

One of the basic definitions in calculus uses the ratio

$$\frac{f(x + h) - f(x)}{h}, \quad h \ne 0.$$

This ratio is a **difference quotient,** as illustrated in Example 11.

GO DIGITAL

EXAMPLE 11   **Evaluating a Difference Quotient**

For $f(x) = x^2 - 4x + 7$, find $\dfrac{f(x + h) - f(x)}{h}$.

Solution

$$\frac{f(x + h) - f(x)}{h} = \frac{[(x + h)^2 - 4(x + h) + 7] - (x^2 - 4x + 7)}{h}$$

$$= \frac{x^2 + 2xh + h^2 - 4x - 4h + 7 - x^2 + 4x - 7}{h}$$

$$= \frac{2xh + h^2 - 4h}{h}$$

$$= \frac{h(2x + h - 4)}{h}$$

$$= 2x + h - 4, \quad h \neq 0$$

✓ *Checkpoint* ▶ Audio-video solution in English & Spanish at LarsonPrecalculus.com

For $f(x) = x^2 + 2x - 3$, find $\dfrac{f(x + h) - f(x)}{h}$. ■

ALGEBRA HELP

You may find it easier to calculate the difference quotient in Example 11 by first finding $f(x + h)$, and then substituting the resulting expression into the difference quotient

$$\frac{f(x + h) - f(x)}{h}.$$

Summary of Function Terminology

Function: A **function** is a relation between two variables such that to each value of the independent variable there corresponds exactly one value of the dependent variable.

Function notation: For the function $y = f(x)$, f is the *name* of the function, y is the **dependent variable,** or output value, x is the **independent variable,** or input value, and $f(x)$ is the *value of the function at x.*

Domain: The **domain** of a function is the set of all values (inputs) of the independent variable for which the function is defined. If x is in the domain of f, then f is *defined* at x. If x is not in the domain of f, then f is *undefined* at x.

Range: The **range** of a function is the set of all values (outputs) taken on by the dependent variable (that is, the set of all function values).

Implied domain: If f is defined by an algebraic expression and the domain is not specified, then the **implied domain** consists of all real numbers for which the expression is defined.

Summarize (Section 2.2)

1. State the definition of a function and describe function notation *(pages 173–177)*. For examples of determining functions and using function notation, see Examples 1–6.

2. State the definition of the implied domain of a function *(page 178)*. For an example of finding the domains of functions, see Example 7.

3. Describe examples of how functions can model real-life problems *(pages 179 and 180, Examples 8–10)*.

4. State the definition of a difference quotient *(page 180)*. For an example of evaluating a difference quotient, see Example 11.

GO DIGITAL

2.2 Exercises

See CalcChat.com for tutorial help and worked-out solutions to odd-numbered exercises.

GO DIGITAL

Vocabulary and Concept Check

In Exercises 1 and 2, fill in the blanks.

1. For an equation that represents y as a function of x, the set of all values taken on by the _____ variable x is the domain, and the set of all values taken on by the _____ variable y is the range.

2. One of the basic definitions in calculus uses the ratio $\dfrac{f(x + h) - f(x)}{h}$, $h \neq 0$. This ratio is a _____ _____.

3. Explain the difference between a relation and a function.

4. Let $g(x) = 4x - 5$. Explain how to determine $g(x + 2)$.

5. Is the domain of a piecewise-defined function *implied* or *explicitly described*?

6. Explain how the domain of $y = \sqrt{4 - x^2}$ can be determined.

Skills and Applications

Testing for Functions In Exercises 7–10, determine whether the relation represents y as a function of x.

7. Domain, x   Range, y

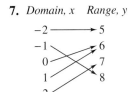

8. Domain, x   Range, y

9.
Input, x	10	7	4	7	10
Output, y	3	6	9	12	15

10.
Input, x	-2	0	2	4	6
Output, y	1	1	1	1	1

Testing for Functions In Exercises 11 and 12, which sets of ordered pairs represent functions from A to B? Explain.

11. $A = \{0, 1, 2, 3\}$ and $B = \{-2, -1, 0, 1, 2\}$
   (a) $\{(0, 1), (1, -2), (2, 0), (3, 2)\}$
   (b) $\{(0, 0), (1, 0), (2, 0), (3, 0)\}$
   (c) $\{(0, 2), (3, 0), (1, 1)\}$

12. $A = \{a, b, c\}$ and $B = \{0, 1, 2, 3\}$
   (a) $\{(a, 1), (c, 2), (c, 3), (b, 3)\}$
   (b) $\{(a, 1), (b, 2), (c, 3)\}$
   (c) $\{(1, a), (0, a), (2, c), (3, b)\}$

Testing for Functions Represented Algebraically In Exercises 13–20, determine whether the equation represents y as a function of x.

13. $x^2 + y^2 = 4$

14. $x^2 - y = 9$

15. $y = \sqrt{16 - x^2}$

16. $y = \sqrt{x + 5}$

17. $y = 4 - |x|$

18. $|y| = 4 - x$

19. $y = -75$

20. $x - 1 = 0$

Evaluating a Function In Exercises 21–30, find each function value, if possible.

21. $g(t) = 4t^2 - 3t + 5$
   (a) $g(2)$ (b) $g(-1)$ (c) $g(t + 2)$

22. $V(r) = \frac{4}{3}\pi r^3$
   (a) $V(3)$ (b) $V\left(\frac{3}{2}\right)$ (c) $V(2r)$

23. $f(y) = 3 - \sqrt{y}$
   (a) $f(4)$ (b) $f(0.25)$ (c) $f(4x^2)$

24. $f(x) = \sqrt{x + 8} + 2$
   (a) $f(-8)$ (b) $f(1)$ (c) $f(x - 8)$

25. $q(x) = 1/(x^2 - 9)$
   (a) $q(0)$ (b) $q(3)$ (c) $q(y + 3)$

26. $q(t) = (2t^2 + 3)/t^2$
   (a) $q(2)$ (b) $q(0)$ (c) $q(-x)$

27. $f(x) = |x|/x$
   (a) $f(2)$ (b) $f(-2)$ (c) $f(x - 1)$

28. $f(x) = |x| + 4$
   (a) $f(2)$ (b) $f(-2)$ (c) $f(x^2)$

29. $f(x) = \begin{cases} 2x + 1, & x < 0 \\ 2x + 2, & x \geq 0 \end{cases}$
   (a) $f(-1)$ (b) $f(0)$ (c) $f(2)$

30. $f(x) = \begin{cases} -3x - 3, & x < -1 \\ x^2 + 2x - 1, & x \geq -1 \end{cases}$
   (a) $f(-2)$ (b) $f(-1)$ (c) $f(1)$

Evaluating a Function In Exercises 31–34, complete the table.

31. $f(x) = -x^2 + 5$

x	-2	-1	0	1	2
$f(x)$					

32. $h(t) = \frac{1}{2}|t + 3|$

t	-5	-4	-3	-2	-1
$h(t)$					

33. $f(x) = \begin{cases} -\frac{1}{2}x + 4, & x \le 0 \\ (x-2)^2, & x > 0 \end{cases}$

x	-2	-1	0	1	2
$f(x)$					

34. $f(x) = \begin{cases} 9 - x^2, & x < 3 \\ x - 3, & x \ge 3 \end{cases}$

x	1	2	3	4	5
$f(x)$					

Finding Values for Which $f(x) = 0$ In Exercises 35–42, find all real values of x for which $f(x) = 0$.

35. $f(x) = 15 - 3x$

36. $f(x) = 4x + 6$

37. $f(x) = \frac{3x - 4}{5}$

38. $f(x) = \frac{12 - x^2}{8}$

39. $f(x) = x^2 - 81$

40. $f(x) = x^2 - 6x - 16$

41. $f(x) = x^3 - x$

42. $f(x) = x^3 - x^2 - 3x + 3$

Finding Values for Which $f(x) = g(x)$ In Exercises 43–46, find the value(s) of x for which $f(x) = g(x)$.

43. $f(x) = x^2$, $g(x) = x + 2$

44. $f(x) = x^2 + 2x + 1$, $g(x) = 5x + 19$

45. $f(x) = x^4 - 2x^2$, $g(x) = 2x^2$

46. $f(x) = \sqrt{x} - 4$, $g(x) = 2 - x$

Finding the Domain of a Function In Exercises 47–54, find the domain of the function.

47. $f(x) = 5x^2 + x - 1$

48. $g(x) = 1 - 2x^2$

49. $g(y) = \sqrt{y + 6}$

50. $f(t) = \sqrt[3]{t + 4}$

51. $g(x) = \frac{1}{x} - \frac{3}{x + 2}$

52. $h(x) = \frac{6}{x^2 - 4x}$

53. $f(s) = \frac{\sqrt{s - 1}}{s - 4}$

54. $f(x) = \frac{x + 2}{\sqrt{x - 10}}$

55. Maximum Volume An open box of maximum volume is made from a square piece of material 24 centimeters on a side by cutting equal squares from the corners and turning up the sides (see figure).

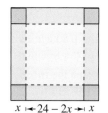 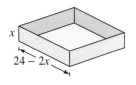

(a) The table shows the volumes V (in cubic centimeters) of the box for various heights x (in centimeters). Use the table to estimate the maximum volume.

Height, x	1	2	3	4	5	6
Volume, V	484	800	972	1024	980	864

(b) Plot the points (x, V) from the table in part (a). Does the relation defined by the ordered pairs represent V as a function of x? If it does, write the function and determine its domain.

56. Maximum Profit The cost per unit in the production of an MP3 player is $60. The manufacturer charges $90 per unit for orders of 100 or less. To encourage large orders, the manufacturer reduces the charge by $0.15 per MP3 player for each unit ordered in excess of 100 (for example, the charge is reduced to $87 per MP3 player for an order size of 120).

(a) The table shows the profits P (in dollars) for various numbers of units ordered, x. Use the table to estimate the maximum profit.

Units, x	130	140	150	160	170
Profit, P	3315	3360	3375	3360	3315

(b) Plot the points (x, P) from the table in part (a). Does the relation defined by the ordered pairs represent P as a function of x? If it does, write the function and determine its domain. (*Note:* $P = R - C$, where R is revenue and C is cost.)

57. Path of a Ball You throw a baseball to a child 25 feet away. The height y (in feet) of the baseball is given by

$$y = -\frac{1}{10}x^2 + 3x + 6$$

where x is the horizontal distance (in feet) from where you threw the ball. Can the child catch the baseball while holding a baseball glove at a height of 5 feet?

58. Postal Regulations A rectangular package has a combined length and girth (perimeter of a cross section) of 108 inches (see figure).

(a) Write the volume V of the package as a function of x. What is the domain of the function?

(b) Use a graphing utility to graph the function. Be sure to use an appropriate window setting.

(c) What dimensions will maximize the volume of the package? Explain.

59. Geometry A right triangle is formed in the first quadrant by the x- and y-axes and a line through the point $(2, 1)$ (see figure). Write the area A of the triangle as a function of a, and determine the domain of the function.

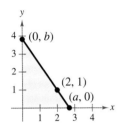

Figure for 59

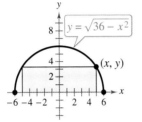

Figure for 60

60. Geometry A rectangle is bounded by the x-axis and the semicircle $y = \sqrt{36 - x^2}$ (see figure). Write the area A of the rectangle as a function of x, and graphically determine the domain of the function.

61. Pharmacology The percent p of prescriptions filled with generic drugs at CVS Pharmacies from 2012 through 2018 (see figure) can be approximated by the model

$$p(t) = \begin{cases} 1.76t + 58.3, & 12 \le t < 16 \\ 0.90t + 71.5, & 16 \le t \le 18 \end{cases}$$

where t represents the year, with $t = 12$ corresponding to 2012. Use this model to find the percent of prescriptions filled with generic drugs in each year from 2012 through 2018. (*Source: CVS Health*)

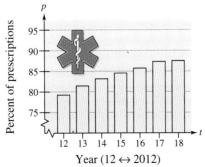

Year (12 ↔ 2012)

62. Median Sale Price The median sale price p (in thousands of dollars) of houses sold in the United States from 2007 through 2019 (see figure) can be approximated by the model

$$p(t) = \begin{cases} 4.011t^2 - 76.89t + 586.7, & 7 \le t < 12 \\ 14.94t + 70.0, & 12 \le t < 17 \\ -4.263t^2 + 152.04t - 1030.4, & 17 \le t \le 19 \end{cases}$$

where t represents the year, with $t = 7$ corresponding to 2007. Use this model to find the median sale price of houses sold in each year from 2007 through 2019. (*Source: Federal Reserve Bank of St. Louis*)

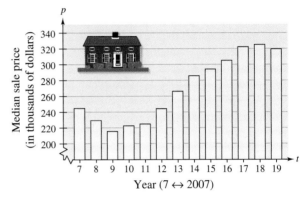

Year (7 ↔ 2007)

63. Height of a Balloon A balloon carrying a transmitter ascends vertically from a point 3000 feet from the receiving station.

(a) Draw a diagram that gives a visual representation of the problem. Let h represent the height of the balloon and let d represent the distance between the balloon and the receiving station.

(b) Write the height of the balloon as a function of d. What is the domain of the function?

64. Physics

The function $F(y) = 149.76\sqrt{10}\,y^{5/2}$ estimates the force F (in tons) of water against the face of a dam, where y is the depth of the water (in feet).

(a) Complete the table. What can you conclude from the table?

y	5	10	20	30	40
$F(y)$					

(b) Use the table to approximate the depth at which the force against the dam is 1,000,000 tons.

(c) Find the depth at which the force against the dam is 1,000,000 tons algebraically.

65. Cost, Revenue, and Profit A company produces a product for which the variable cost is $12.30 per unit and the fixed costs are $98,000. The product sells for $17.98. Let x be the number of units produced and sold.

(a) The total cost for a business is the sum of the variable cost and the fixed costs. Write the total cost C as a function of the number of units produced.

(b) Write the revenue R as a function of the number of units sold.

(c) Write the profit P as a function of the number of units sold. (*Note:* $P = R - C$)

66. Average Cost The inventor of a new game believes that the variable cost for producing the game is $0.95 per unit and the fixed costs are $6000. The inventor sells each game for $1.69. Let x be the number of games produced.

(a) The total cost for a business is the sum of the variable cost and the fixed costs. Write the total cost C as a function of the number of games produced.

(b) Write the average cost per unit $\overline{C} = \dfrac{C}{x}$ as a function of x.

67. Transportation For groups of 80 or more people, a charter bus company determines the rate per person according to the formula

Rate $= 8 - 0.05(n - 80)$, $n \geq 80$

where the rate is given in dollars and n is the number of people.

(a) Write the revenue R for the bus company as a function of n.

(b) Use the function in part (a) to complete the table. What can you conclude?

n	90	100	110	120	130	140	150
$R(n)$							

68. E-Filing The table shows the numbers of tax returns (in millions) made through e-file from 2011 through 2018. Let $f(t)$ represent the number of tax returns made through e-file in the year t. (*Source: eFile*)

DATA	Year	Number of Tax Returns Made Through E-File
	2011	112.2
	2012	112.1
	2013	114.4
	2014	125.8
	2015	128.8
	2016	131.9
	2017	135.5
	2018	137.9

Spreadsheet at LarsonPrecalculus.com

(a) Find $\dfrac{f(2018) - f(2011)}{2018 - 2011}$ and interpret the result in the context of the problem.

(b) Make a scatter plot of the data.

(c) Find a linear model for the data algebraically. Let N represent the number of tax returns made through e-file and let $t = 11$ correspond to 2011.

(d) Use the model found in part (c) to complete the table.

t	11	12	13	14	15	16	17	18
N								

(e) Compare your results from part (d) with the actual data.

(f) Use a graphing utility to find a linear model for the data. Let $x = 11$ correspond to 2011. How does the model you found in part (c) compare with the model given by the graphing utility?

Evaluating a Difference Quotient In Exercises 69–76, find the difference quotient and simplify your answer.

69. $f(x) = x^2 - 2x + 4$, $\dfrac{f(2 + h) - f(2)}{h}$, $h \neq 0$

70. $f(x) = 5x - x^2$, $\dfrac{f(5 + h) - f(5)}{h}$, $h \neq 0$

71. $f(x) = x^3 + 3x$, $\dfrac{f(x + h) - f(x)}{h}$, $h \neq 0$

72. $f(x) = 4x^3 - 2x$, $\dfrac{f(x + h) - f(x)}{h}$, $h \neq 0$

73. $g(x) = \dfrac{1}{x^2}$, $\dfrac{g(x) - g(3)}{x - 3}$, $x \neq 3$

74. $f(t) = \dfrac{1}{t - 2}$, $\dfrac{f(t) - f(1)}{t - 1}$, $t \neq 1$

75. $f(x) = \sqrt{5x}$, $\dfrac{f(x) - f(5)}{x - 5}$, $x \neq 5$

76. $f(x) = x^{2/3} + 1$, $\dfrac{f(x) - f(8)}{x - 8}$, $x \neq 8$

Exploring the Concepts

True or False? In Exercises 77–80, determine whether the statement is true or false. Justify your answer.

77. Every relation is a function.

78. Every function is a relation.

79. For the function

$f(x) = x^4 - 1$

the domain is $(-\infty, \infty)$ and the range is $(0, \infty)$.

80. The set of ordered pairs $\{(-8, -2), (-6, 0), (-4, 0), (-2, 2), (0, 4), (2, -2)\}$ represents a function.

Modeling Data In Exercises 81–84, determine which of the following functions

$$f(x) = cx, \ g(x) = cx^2, \ h(x) = c\sqrt{|x|}, \ \text{and} \ r(x) = \frac{c}{x}$$

can be used to model the data and determine the value of the constant c that will make the function fit the data in the table.

81.

x	-4	-1	0	1	4
y	-32	-2	0	-2	-32

82.

x	-4	-1	0	1	4
y	-1	$-\frac{1}{4}$	0	$\frac{1}{4}$	1

83.

x	-4	-1	0	1	4
y	-8	-32	Undefined	32	8

84.

x	-4	-1	0	1	4
y	6	3	0	3	6

85. Error Analysis Describe the error.

The functions

$$f(x) = \sqrt{x-1} \quad \text{and} \quad g(x) = \frac{1}{\sqrt{x-1}}$$

have the same domain, which is the set of all real numbers x such that $x \geq 1$.

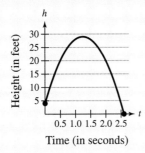

86. HOW DO YOU SEE IT? The graph represents the height h of a projectile after t seconds.

(a) Explain why h is a function of t.

(b) Approximate the height of the projectile after 0.5 second and after 1.25 seconds.

(c) Approximate the domain of h.

(d) Is t a function of h? Explain.

87. Think About It Given $f(x) = x^2$, is f the independent variable? Why or why not?

88. Think About It Consider

$$f(x) = \sqrt{x-2} \quad \text{and} \quad g(x) = \sqrt[3]{x-2}.$$

Why are the domains of f and g different?

Review & Refresh 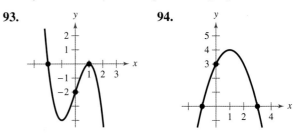 Video solutions at LarsonPrecalculus.com

Solving an Equation In Exercises 89–92, solve the equation. Check your solutions.

89. $x^3 - 3x^2 - x + 3 = 0$

90. $x^3 + 2x^2 - 4x - 8 = 0$

91. $0 = -2x^2 - 7x + 15$

92. $0 = 2x^2 - 13x + 20$

Identifying Intercepts In Exercises 93 and 94, identify the x- and y-intercepts of the graph.

93.

94.

Approximating Intercepts In Exercises 95–98, use a graphing utility to approximate any x-intercepts.

95. $y = 2x^4 - 15x^3 + 18x^2$

96. $y = x^4 - 10x^2 + 9$

97. $y = x^2 - 3.61x + 2.86$

98. $y = x^3 + 1.27x^2 + 5.49$

Evaluating an Expression In Exercises 99–104, evaluate the expression.

99. $\dfrac{\left((3)^2 + 4\right) - (1^2 + 4)}{3 - 1}$

100. $\dfrac{(9^{3/2}) - (4^{3/2})}{9 - 4}$

101. $\dfrac{\frac{1}{3} - \frac{1}{2}}{6 - 4}$

102. $\dfrac{-\sqrt{\frac{1}{9}} + \sqrt{\frac{1}{4}}}{9 - 4}$

103. $\dfrac{\sqrt{3^2 + 4^2}}{\frac{3}{4} - \frac{1}{3}}$

104. $\dfrac{\sqrt{4^3 - (3^3 + 1)}}{\left(1 + \sqrt{13}\right)\left(1 - \sqrt{13}\right)}$

2.3 Analyzing Graphs of Functions

Graphs of functions can help you visualize relationships between variables in real life. For example, in Exercise 87 on page 196, you will use the graph of a function to visually represent the temperature in a city over a 24-hour period.

❯ **Use the Vertical Line Test for functions.**
❯ **Find the zeros of functions.**
❯ **Determine intervals on which functions are increasing or decreasing.**
❯ **Determine relative minimum and relative maximum values of functions.**
❯ **Determine the average rate of change of a function.**
❯ **Identify even and odd functions.**

The Graph of a Function

In Section 2.2, you studied functions from an algebraic point of view. In this section, you will study functions from a graphical perspective.

The **graph of a function** f is the collection of ordered pairs $(x, f(x))$ such that x is in the domain of f. As you study this section, remember that x is the directed distance from the y-axis, and $y = f(x)$ is the directed distance from the x-axis (see figure at the right).

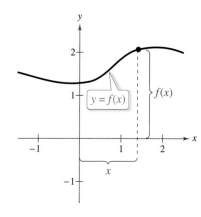

EXAMPLE 1 **Finding the Domain and Range of a Function**

Use the graph of the function f, shown in Figure 2.13, to find (a) the domain of f, (b) the function values $f(-1)$ and $f(2)$, and (c) the range of f.

Solution

a. The closed dot at $(-1, 1)$ indicates that $x = -1$ is in the domain of f, whereas the open dot at $(5, 2)$ indicates that $x = 5$ is not in the domain. So, the domain of f is all x in the interval $[-1, 5)$.

b. One point on the graph of f is $(-1, 1)$, so $f(-1) = 1$. Another point on the graph of f is $(2, -3)$, so $f(2) = -3$.

c. The graph does not extend below $f(2) = -3$ or above $f(0) = 3$, so the range of f is the interval $[-3, 3]$.

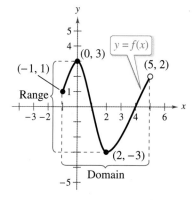

Figure 2.13

✓ **Checkpoint** ▶ *Audio-video solution in English & Spanish at LarsonPrecalculus.com*

Use the graph of the function f, shown in Figure 2.14, to find (a) the domain of f, (b) the function values $f(0)$ and $f(3)$, and (c) the range of f. ∎

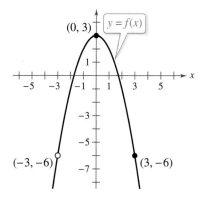

Figure 2.14

In Figures 2.13 and 2.14, note the use of dots (open or closed). A closed dot is a point on the graph whereas an open dot is not. Also, when a curve stops at a dot, such as in Figure 2.13 at $(-1, 1)$, the graph does not extend beyond the dot. Note that a curve does not have to stop at a dot. The curve in Figure 2.14 does not stop at any of the dots but continues past them.

GO DIGITAL

By the definition of a function, at most one y-value corresponds to a given x-value. So, no two points on the graph of a function have the same x-coordinate, or lie on the same vertical line. It follows, then, that a vertical line can intersect the graph of a function at most once. This observation provides a convenient visual test called the **Vertical Line Test** for functions.

> **Vertical Line Test for Functions**
>
> A set of points in a coordinate plane is the graph of y as a function of x if and only if no *vertical* line intersects the graph at more than one point.

EXAMPLE 2   Vertical Line Test for Functions

Use the Vertical Line Test to determine whether each graph represents y as a function of x.

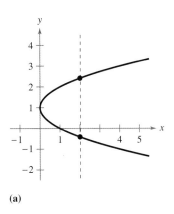

(a)

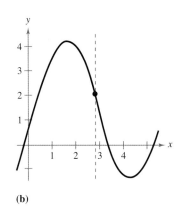

(b)

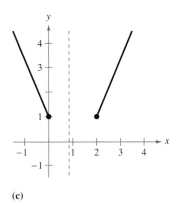

(c)

Solution

a. This *is not* a graph of y as a function of x, because there are vertical lines that intersect the graph twice. That is, for a particular input x, there is more than one output y.

b. This *is* a graph of y as a function of x, because every vertical line intersects the graph at most once. That is, for a particular input x, there is at most one output y.

c. This *is* a graph of y as a function of x, because every vertical line intersects the graph at most once. That is, for a particular input x, there is at most one output y. (Note that when a vertical line does not intersect a graph, it simply means that the function is undefined for that particular value of x.)

✓ *Checkpoint* ▶ *Audio-video solution in English & Spanish at LarsonPrecalculus.com*

Use the Vertical Line Test to determine whether the graph represents y as a function of x.

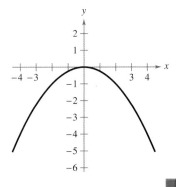

 TECHNOLOGY

Most graphing utilities graph functions of x more easily than other types of equations. For example, the graph shown in Example 2(a) represents the equation $x - (y - 1)^2 = 0$. To duplicate this graph using a graphing utility, you may have to solve the equation for y to obtain $y = 1 \pm \sqrt{x}$, and then graph the two equations $y_1 = 1 + \sqrt{x}$ and $y_2 = 1 - \sqrt{x}$ in the same viewing window.

GO DIGITAL

Zeros of a Function

If the graph of a function of x has an x-intercept at $(a, 0)$, then a is a **zero** of the function.

> **Zeros of a Function**
>
> The **zeros of a function** $y = f(x)$ are the x-values for which $f(x) = 0$.

EXAMPLE 3   **Finding the Zeros of Functions**

Find the zeros of each function algebraically.

a. $f(x) = 3x^2 + x - 10$

b. $g(x) = \sqrt{10 - x^2}$

c. $h(t) = \dfrac{2t - 3}{t + 5}$

Solution   To find the zeros of a function, set the function equal to zero and solve for the independent variable.

a. $3x^2 + x - 10 = 0$ Set $f(x)$ equal to 0.

$(3x - 5)(x + 2) = 0$ Factor.

$3x - 5 = 0 \implies x = \frac{5}{3}$ Set 1st factor equal to 0.

$x + 2 = 0 \implies x = -2$ Set 2nd factor equal to 0.

The zeros of f are $x = \frac{5}{3}$ and $x = -2$. In Figure 2.15(a), note that the graph of f has x-intercepts at $\left(\frac{5}{3}, 0\right)$ and $(-2, 0)$.

b. $\sqrt{10 - x^2} = 0$ Set $g(x)$ equal to 0.

$10 - x^2 = 0$ Square each side.

$10 = x^2$ Add x^2 to each side.

$\pm\sqrt{10} = x$ Extract square roots.

The zeros of g are $x = -\sqrt{10}$ and $x = \sqrt{10}$. In Figure 2.15(b), note that the graph of g has x-intercepts at $\left(-\sqrt{10}, 0\right)$ and $\left(\sqrt{10}, 0\right)$.

c. $\dfrac{2t - 3}{t + 5} = 0$ Set $h(t)$ equal to 0.

$2t - 3 = 0$ Multiply each side by $t + 5$.

$2t = 3$ Add 3 to each side.

$t = \dfrac{3}{2}$ Divide each side by 2.

The zero of h is $t = \frac{3}{2}$. In Figure 2.15(c), note that the graph of h has an x-intercept at $\left(\frac{3}{2}, 0\right)$.

✔ *Checkpoint* ▶ Audio-video solution in English & Spanish at LarsonPrecalculus.com

Find the zeros of each function algebraically.

a. $f(x) = 2x^2 + 13x - 24$ **b.** $g(t) = \sqrt{t - 25}$ **c.** $h(x) = \dfrac{x^2 - 2}{x - 1}$ ∎

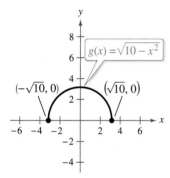

(a) Zeros of f: $x = -2$, $x = \frac{5}{3}$

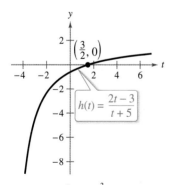

(b) Zeros of g: $x = \pm\sqrt{10}$

(c) Zero of h: $t = \frac{3}{2}$

Figure 2.15

GO DIGITAL

Increasing and Decreasing Functions

The more you know about the graph of a function, the more you know about the function itself. Consider the graph shown in Figure 2.16. As you move from *left to right,* this graph falls (decreases) from $x = -2$ to $x = 0$, is constant from $x = 0$ to $x = 2$, and rises (increases) from $x = 2$ to $x = 4$.

The next definition summarizes when a function is increasing, decreasing, or constant on an open interval.

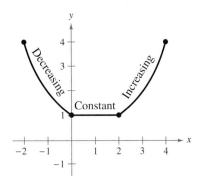

Figure 2.16

Increasing, Decreasing, and Constant Functions

A function f is **increasing** on an interval when, for any x_1 and x_2 in the interval,

$$x_1 < x_2 \quad \text{implies} \quad f(x_1) < f(x_2).$$

A function f is **decreasing** on an interval when, for any x_1 and x_2 in the interval,

$$x_1 < x_2 \quad \text{implies} \quad f(x_1) > f(x_2).$$

A function f is **constant** on an interval when, for any x_1 and x_2 in the interval,

$$f(x_1) = f(x_2).$$

EXAMPLE 4   **Describing Function Behavior**

Determine the open intervals on which each function is increasing, decreasing, or constant.

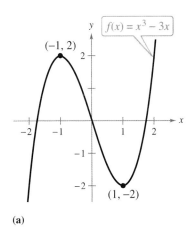

(a)

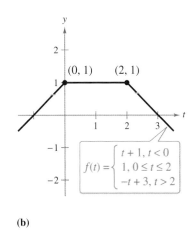

(b)

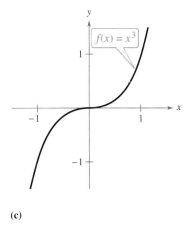

(c)

Solution

a. This function is increasing on the interval $(-\infty, -1)$, decreasing on the interval $(-1, 1)$, and increasing on the interval $(1, \infty)$.

b. This function is increasing on the interval $(-\infty, 0)$, constant on the interval $(0, 2)$, and decreasing on the interval $(2, \infty)$.

c. This function may appear to be constant on an interval near $x = 0$, but for all real values of x_1 and x_2, if $x_1 < x_2$, then $(x_1)^3 < (x_2)^3$. So, the function is increasing on the interval $(-\infty, \infty)$.

✓ *Checkpoint*  *Audio-video solution in English & Spanish at LarsonPrecalculus.com*

Graph the function $f(x) = x^3 + 3x^2 - 1$. Then determine the open intervals on which the function is increasing, decreasing, or constant.

Relative Minimum and Relative Maximum Values

The points at which a function changes its increasing, decreasing, or constant behavior are helpful in determining the **relative minimum** or **relative maximum** values of the function.

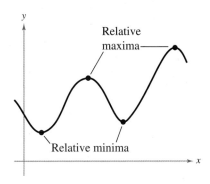

Figure 2.17

> ### Definitions of Relative Minimum and Relative Maximum
>
> A function value $f(a)$ is a **relative minimum** of f when there exists an interval (x_1, x_2) that contains a such that
>
> $$x_1 < x < x_2 \quad \text{implies} \quad f(a) \le f(x).$$
>
> A function value $f(a)$ is a **relative maximum** of f when there exists an interval (x_1, x_2) that contains a such that
>
> $$x_1 < x < x_2 \quad \text{implies} \quad f(a) \ge f(x).$$

Figure 2.17 shows several different examples of relative minima and relative maxima. In Section 3.1, you will study a technique for finding the *exact point* at which a second-degree polynomial function has a relative minimum or relative maximum. For the time being, however, you can use a graphing utility to find reasonable approximations of these points.

EXAMPLE 5   **Approximating a Relative Minimum**

Use a graphing utility to approximate the relative minimum of the function

$$f(x) = 3x^2 - 4x - 2.$$

Solution   The graph of f is shown in Figure 2.18. By using the *zoom* and *trace* features or the *minimum* feature of a graphing utility, you can approximate that the relative minimum of the function occurs at the point $(0.67, -3.33)$. So, the relative minimum is approximately -3.33. Later, in Section 3.1, you will learn how to determine that the exact point at which the relative minimum occurs is $\left(\frac{2}{3}, -\frac{10}{3}\right)$ and the exact relative minimum is $-\frac{10}{3}$.

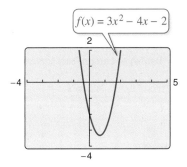

Figure 2.18

$f(x) = 3x^2 - 4x - 2$

✓ *Checkpoint* ▶ Audio-video solution in English & Spanish at LarsonPrecalculus.com

Use a graphing utility to approximate the relative maximum of the function

$$f(x) = -4x^2 - 7x + 3. \qquad ■$$

You can also use the *table* feature of a graphing utility to numerically approximate the relative minimum of the function in Example 5. Using a table that begins at 0.6 and increments the value of x by 0.01, you can approximate that the minimum of

$$f(x) = 3x^2 - 4x - 2$$

occurs at the point $(0.67, -3.33)$.

▶▶▶ **TECHNOLOGY**

When you use a graphing utility to approximate the x- and y-values of the point where a relative minimum or relative maximum occurs, the *zoom* feature will often produce graphs that are nearly flat. To overcome this problem, manually change the vertical setting of the viewing window. The graph will stretch vertically when the values of Ymin and Ymax are closer together.

GO DIGITAL

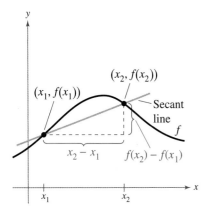

Figure 2.19

Average Rate of Change

In Section 2.1, you learned that the slope of a line can be interpreted as a *rate of change*. For a nonlinear graph, the **average rate of change** between any two points $(x_1, f(x_1))$ and $(x_2, f(x_2))$ is the slope of the line through the two points (see Figure 2.19). The line through the two points is called a **secant line,** and the slope of this line is denoted as m_{sec}.

$$\text{Average rate of change of } f \text{ from } x_1 \text{ to } x_2 = \frac{f(x_2) - f(x_1)}{x_2 - x_1}$$

$$= \frac{\text{change in } y}{\text{change in } x}$$

$$= m_{\text{sec}}$$

EXAMPLE 6   **Average Rate of Change of a Function**

Find the average rates of change of $f(x) = x^3 - 3x$ (a) from $x_1 = -2$ to $x_2 = -1$ and (b) from $x_1 = 0$ to $x_2 = 1$ (see Figure 2.20).

Solution

a. The average rate of change of f from $x_1 = -2$ to $x_2 = -1$ is

$$\frac{f(x_2) - f(x_1)}{x_2 - x_1} = \frac{f(-1) - f(-2)}{-1 - (-2)} = \frac{2 - (-2)}{1} = 4. \qquad \text{Secant line has positive slope.}$$

b. The average rate of change of f from $x_1 = 0$ to $x_2 = 1$ is

$$\frac{f(x_2) - f(x_1)}{x_2 - x_1} = \frac{f(1) - f(0)}{1 - 0} = \frac{-2 - 0}{1} = -2. \qquad \text{Secant line has negative slope.}$$

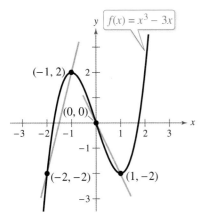

Figure 2.20

✓ *Checkpoint* ▶ *Audio-video solution in English & Spanish at LarsonPrecalculus.com*

Find the average rates of change of $f(x) = x^2 + 2x$ (a) from $x_1 = -3$ to $x_2 = -2$ and (b) from $x_1 = -2$ to $x_2 = 0$.

EXAMPLE 7   **Finding Average Speed**

The distance s (in feet) a moving car is from a traffic signal is given by the function

$$s(t) = 20t^{3/2}$$

where t is the time (in seconds). Find the average speed of the car (a) from $t_1 = 0$ to $t_2 = 4$ seconds and (b) from $t_1 = 4$ to $t_2 = 9$ seconds.

Solution

a. The average speed of the car from $t_1 = 0$ to $t_2 = 4$ seconds is

$$\frac{s(t_2) - s(t_1)}{t_2 - t_1} = \frac{s(4) - s(0)}{4 - 0} = \frac{160 - 0}{4} = 40 \text{ feet per second.}$$

b. The average speed of the car from $t_1 = 4$ to $t_2 = 9$ seconds is

$$\frac{s(t_2) - s(t_1)}{t_2 - t_1} = \frac{s(9) - s(4)}{9 - 4} = \frac{540 - 160}{5} = 76 \text{ feet per second.}$$

✓ *Checkpoint* ▶ *Audio-video solution in English & Spanish at LarsonPrecalculus.com*

In Example 7, find the average speed of the car (a) from $t_1 = 0$ to $t_2 = 1$ second and (b) from $t_1 = 1$ second to $t_2 = 4$ seconds.   ■

Even and Odd Functions

In Section 1.1, you studied different types of symmetry of a graph. In the terminology of functions, a function is said to be **even** when its graph is symmetric with respect to the *y*-axis and **odd** when its graph is symmetric with respect to the origin. The symmetry tests in Section 1.1 yield the tests for even and odd functions below.

Tests for Even and Odd Functions

A function $y = f(x)$ is **even** when, for each x in the domain of f, $f(-x) = f(x)$.

A function $y = f(x)$ is **odd** when, for each x in the domain of f, $f(-x) = -f(x)$.

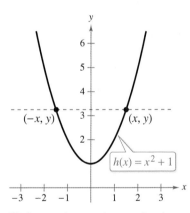

(a) Symmetric to origin: odd function

(b) Symmetric to *y*-axis: even function

Figure 2.21

EXAMPLE 8 Even and Odd Functions

▶▶▶ *See LarsonPrecalculus.com for an interactive version of this type of example.*

a. The function $g(x) = x^3 - x$ is odd because $g(-x) = -g(x)$, as follows.

$$g(-x) = (-x)^3 - (-x) \qquad \text{Substitute } -x \text{ for } x.$$
$$= -x^3 + x \qquad \text{Simplify.}$$
$$= -(x^3 - x) \qquad \text{Distributive Property}$$
$$= -g(x) \qquad \text{Test for odd function}$$

b. The function $h(x) = x^2 + 1$ is even because $h(-x) = h(x)$, as follows.

$$h(-x) = (-x)^2 + 1 \qquad \text{Substitute } -x \text{ for } x.$$
$$= x^2 + 1 \qquad \text{Simplify.}$$
$$= h(x) \qquad \text{Test for even function}$$

Figure 2.21 shows the graphs and symmetry of these two functions.

✓ *Checkpoint*  ▶ **Audio-video solution in English & Spanish at LarsonPrecalculus.com**

Determine whether each function is even, odd, or neither. Then describe the symmetry.

a. $f(x) = 5 - 3x$ **b.** $g(x) = x^4 - x^2 - 1$ **c.** $h(x) = 2x^3 + 3x$ ∎

Summarize (Section 2.3)

1. State the Vertical Line Test for functions *(page 188)*. For an example of using the Vertical Line Test, see Example 2.

2. Explain how to find the zeros of a function *(page 189)*. For an example of finding the zeros of functions, see Example 3.

3. Explain how to determine intervals on which functions are increasing or decreasing *(page 190)*. For an example of describing function behavior, see Example 4.

4. Explain how to determine relative minimum and relative maximum values of functions *(page 191)*. For an example of approximating a relative minimum, see Example 5.

5. Explain how to determine the average rate of change of a function *(page 192)*. For examples of determining average rates of change, see Examples 6 and 7.

6. State the definitions of an even function and an odd function *(page 193)*. For an example of identifying even and odd functions, see Example 8.

GO DIGITAL

2.3 **Exercises**

See CalcChat.com for tutorial help and worked-out solutions to odd-numbered exercises.

GO DIGITAL

Vocabulary and Concept Check

In Exercises 1–4, fill in the blanks.

1. The _____ of a function $y = f(x)$ are the values of x for which $f(x) = 0$.

2. A function f is _____ on an interval when, for any x_1 and x_2 in the interval, $x_1 < x_2$ implies $f(x_1) > f(x_2)$.

3. The _____ _____ _____ _____ between any two points $(x_1, f(x_1))$ and $(x_2, f(x_2))$ is the slope of the line through the two points, and this line is called the _____ line.

4. A function f is _____ when, for each x in the domain of f, $f(-x) = -f(x)$.

5. A vertical line intersects a graph twice. Does the graph represent a function?

6. Let f be a function such that $f(2) \geq f(x)$ for all values of x in the interval $(0, 3)$. Does $f(2)$ represent a relative minimum or a relative maximum?

Skills and Applications

Domain, Range, and Values of a Function
In Exercises 7–10, use the graph of the function to find the domain and range of f and each function value.

7. (a) $f(-1)$ (b) $f(0)$
   (c) $f(1)$ (d) $f(2)$

8. (a) $f(-1)$ (b) $f(0)$
   (c) $f(1)$ (d) $f(3)$

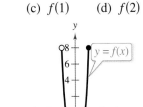

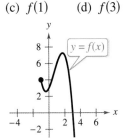

9. (a) $f(2)$ (b) $f(1)$
   (c) $f(3)$ (d) $f(-1)$

10. (a) $f(-2)$ (b) $f(1)$
    (c) $f(0)$ (d) $f(2)$

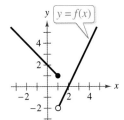

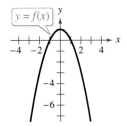

Vertical Line Test for Functions **In Exercises 11–14, use the Vertical Line Test to determine whether the graph represents y as a function of x.**

11.
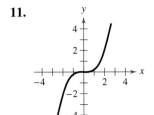

12.
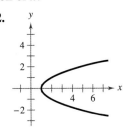

13.
14.

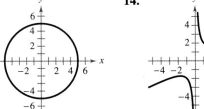

Finding the Zeros of a Function **In Exercises 15–26, find the zeros of the function algebraically.**

15. $f(x) = 2x^2 - 7x - 30$ 16. $f(x) = 3x^2 + 22x - 16$

17. $f(x) = \frac{1}{3}x^3 - 2x$ 18. $f(x) = -25x^4 + 9x^2$

19. $f(x) = x^3 - 4x^2 - 9x + 36$

20. $f(x) = 4x^3 - 24x^2 - x + 6$

21. $f(x) = \sqrt{2x - 1}$

22. $f(x) = \sqrt{3x + 2}$

23. $f(x) = \sqrt{x^2 - 1}$

24. $f(x) = \sqrt{x^2 + 2x + 1}$

25. $f(x) = \dfrac{x + 3}{2x^2 - 6}$

26. $f(x) = \dfrac{x^2 - 9x + 14}{4x}$

Graphing and Finding Zeros **In Exercises 27–32, (a) use a graphing utility to graph the function and find the zeros of the function and (b) verify your results from part (a) algebraically.**

27. $f(x) = x^2 - 6x$ 28. $f(x) = 2x^2 - 13x - 7$

29. $f(x) = \sqrt{2x + 11}$ 30. $f(x) = \sqrt{3x - 14} - 8$

31. $f(x) = \dfrac{3x - 1}{x - 6}$ 32. $f(x) = \dfrac{2x^2 - 9}{3 - x}$

Describing Function Behavior   In Exercises 33–40, determine the open intervals on which the function is increasing, decreasing, or constant.

33. $f(x) = -\frac{1}{2}x^3$

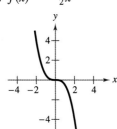

34. $f(x) = x^2 - 4x$

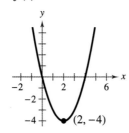

35. $f(x) = \sqrt{x^2 - 1}$

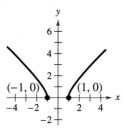

36. $f(x) = x^3 - 3x^2 + 2$

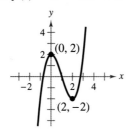

37. $f(x) = |x + 1| + |x - 1|$ **38.** $f(x) = \dfrac{x^2 + x + 1}{x + 1}$

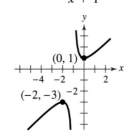

39. $f(x) = \begin{cases} 2x + 1, & x \leq -1 \\ x^2 - 2, & x > -1 \end{cases}$

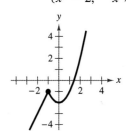

40. $f(x) = \begin{cases} x + 3, & x \leq 0 \\ 3, & 0 < x \leq 2 \\ 2x + 1, & x > 2 \end{cases}$

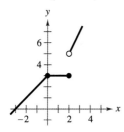

Describing Function Behavior   In Exercises 41–48, use a graphing utility to graph the function and visually determine the open intervals on which the function is increasing, decreasing, or constant. Use a table of values to verify your results.

41. $f(x) = 3$

42. $g(x) = x$

43. $g(x) = \frac{1}{2}x^2 - 3$

44. $f(x) = 3x^4 - 6x^2$

45. $f(x) = \sqrt{1 - x}$

46. $f(x) = x\sqrt{x + 3}$

47. $f(x) = x^{3/2}$

48. $f(x) = x^{2/3}$

Approximating Relative Minima or Maxima   In Exercises 49–54, use a graphing utility to approximate (to two decimal places) any relative minima or maxima of the function.

49. $f(x) = x(x + 3)$

50. $f(x) = -x^2 + 3x - 2$

51. $h(x) = x^3 - 6x^2 + 15$

52. $f(x) = x^3 - 3x^2 - x + 1$

53. $h(x) = (x - 1)\sqrt{x}$

54. $g(x) = x\sqrt{4 - x}$

Graphical Reasoning   In Exercises 55–60, graph the function and determine the interval(s) for which $f(x) \geq 0$.

55. $f(x) = 4 - x$

56. $f(x) = 4x + 2$

57. $f(x) = 9 - x^2$

58. $f(x) = x^2 - 4x$

59. $f(x) = \sqrt{x - 1}$

60. $f(x) = |x + 5|$

Average Rate of Change of a Function   In Exercises 61–64, find the average rate of change of the function from x_1 to x_2.

Function	x-Values
61. $f(x) = -2x + 15$	$x_1 = 0, x_2 = 3$
62. $f(x) = x^2 - 2x + 8$	$x_1 = 1, x_2 = 5$
63. $f(x) = x^3 - 3x^2 - x$	$x_1 = -1, x_2 = 2$
64. $f(x) = -x^3 + 6x^2 + x$	$x_1 = 1, x_2 = 6$

65. Applied Research   The amounts (in billions of dollars) the U.S. federal government spent on applied research from 2013 through 2018 can be approximated by the model

$$y = 0.0729t^2 - 0.526t + 24.34$$

where t represents the year, with $t = 13$ corresponding to 2013. *(Source: National Center for Science and Engineering Statistics)*

(a) Use a graphing utility to graph the model.

(b) Find the average rate of change of the model from 2013 to 2018. Interpret your answer in the context of the problem.

66. Finding Average Speed Use the information in Example 7 to find the average speed of the car from $t_1 = 0$ to $t_2 = 9$ seconds. Explain why the result is less than the value obtained in part (b) of Example 7.

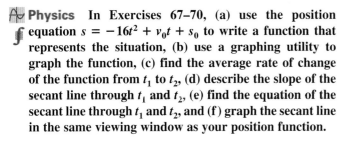 **Physics** In Exercises 67–70, (a) use the position equation $s = -16t^2 + v_0 t + s_0$ to write a function that represents the situation, (b) use a graphing utility to graph the function, (c) find the average rate of change of the function from t_1 to t_2, (d) describe the slope of the secant line through t_1 and t_2, (e) find the equation of the secant line through t_1 and t_2, and (f) graph the secant line in the same viewing window as your position function.

67. An object is thrown upward from a height of 6 feet at a velocity of 64 feet per second.

$t_1 = 0, t_2 = 3$

68. An object is thrown upward from a height of 6.5 feet at a velocity of 72 feet per second.

$t_1 = 0, t_2 = 4$

69. An object is thrown upward from ground level at a velocity of 120 feet per second.

$t_1 = 3, t_2 = 5$

70. An object is dropped from a height of 80 feet.

$t_1 = 1, t_2 = 2$

Even, Odd, or Neither? In Exercises 71–76, determine whether the function is even, odd, or neither. Then describe the symmetry.

71. $f(x) = x^6 - 2x^2 + 3$ **72.** $g(x) = x^3 - 5x$
73. $h(x) = x\sqrt{x+5}$ **74.** $f(x) = x\sqrt{1-x^2}$
75. $f(s) = 4s^{3/2}$ **76.** $g(s) = 4s^{2/3}$

Even, Odd, or Neither? In Exercises 77–82, sketch a graph of the function and determine whether it is even, odd, or neither. Verify your answer algebraically.

77. $f(x) = -9$ **78.** $f(x) = 5 - 3x$
79. $f(x) = -|x - 5|$ **80.** $h(x) = x^2 - 4$
81. $f(x) = \sqrt[3]{4x}$ **82.** $f(x) = \sqrt[3]{x-4}$

Height of a Rectangle In Exercises 83 and 84, write the height h of the rectangle as a function of x.

83.
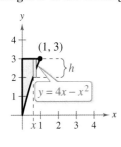

84.
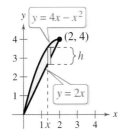

Length of a Rectangle In Exercises 85 and 86, write the length L of the rectangle as a function of y.

85.
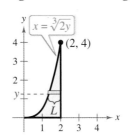

86.
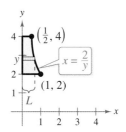

87. Temperature
The table shows the temperatures y (in degrees Fahrenheit) in a city over a 24-hour period. Let x represent the time of day, where $x = 0$ corresponds to 6 A.M.

Time, x	Temperature, y
0	34
2	50
4	60
6	64
8	63
10	59
12	53
14	46
16	40
18	36
20	34
22	37
24	45

These data can be approximated by the model

$y = 0.026x^3 - 1.03x^2 + 10.2x + 34, \quad 0 \le x \le 24.$

(a) Use a graphing utility to create a scatter plot of the data. Then graph the model in the same viewing window.

(b) How well does the model fit the data?

(c) Use the graph to approximate the times when the temperature was increasing and decreasing.

(d) Use the graph to approximate the maximum and minimum temperatures during this 24-hour period.

(e) Could this model predict the temperatures in the city during the next 24-hour period? Why or why not?

88. Geometry Corners of equal size are cut from a square with sides of length 8 meters (see figure).

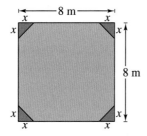

(a) Write the area A of the resulting figure as a function of x. Determine the domain of the function.

(b) Use a graphing utility to graph the area function over its domain. Use the graph to find the range of the function.

(c) Identify the figure that results when x is the maximum value in the domain of the function. What would be the length of each side of the figure?

Exploring the Concepts

True or False? In Exercises 89 and 90, determine whether the statement is true or false. Justify your answer.

89. A function with a square root cannot have a domain that is the set of real numbers.

90. It is possible for an odd function to have the interval $[0, \infty)$ as its domain.

91. Error Analysis Describe the error.

The function $f(x) = 2x^3 - 5$ is odd because $f(-x) = -f(x)$, as follows.

$$f(-x) = 2(-x)^3 - 5$$
$$= -2x^3 - 5$$
$$= -(2x^3 - 5)$$
$$= -f(x) \qquad \times$$

92. **HOW DO YOU SEE IT?** Use the graph of the function to answer parts (a)–(e).

(a) Find the domain and range of f.

(b) Find the zero(s) of f.

(c) Determine the open intervals on which f is increasing, decreasing, or constant.

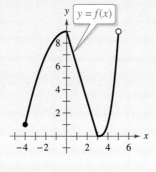

(d) Approximate any relative minimum or relative maximum values of f.

(e) Is f even, odd, or neither?

Think About It In Exercises 93 and 94, find the coordinates of a second point on the graph of a function f when the given point is on the graph and the function is (a) even and (b) odd.

93. $\left(-\frac{5}{3}, -7\right)$ **94.** $(2a, 2c)$

95. Writing Use a graphing utility to graph each function. Write a paragraph describing any similarities and differences you observe among the graphs.

(a) $y = x$ (b) $y = x^2$ (c) $y = x^3$

(d) $y = x^4$ (e) $y = x^5$ (f) $y = x^6$

96. Reasoning Determine whether each function is even, odd, or neither.

$$f(x) = x^2 - x^4 \qquad\qquad g(x) = 2x^3 + 1$$
$$h(x) = x^5 - 2x^3 + x \qquad j(x) = 2 - x^6 - x^8$$
$$k(x) = x^5 - 2x^4 + x - 2 \quad p(x) = x^9 + 3x^5 - x^3 + x$$

What do you notice about the equations of functions that are odd? What do you notice about the equations of functions that are even? Can you describe a way to identify a function as odd or even by inspecting the equation? Can you describe a way to identify a function as neither odd nor even by inspecting the equation?

Review & Refresh ▶ *Video solutions at LarsonPrecalculus.com*

Finding an Equation of a Line In Exercises 97–100, find an equation of the line passing through the pair of points. Sketch the line.

97. $(1, 3), (4, 0)$ **98.** $(-2, 6), (4, -9)$

99. $(5, 0), (5, 1)$ **100.** $(6, -1), (-6, -1)$

Evaluating a Function In Exercises 101–104, find each function value.

101. $f(x) = 5x - 3$

(a) $f(-3)$ (b) $f(3)$ (c) $f(x + 3)$

102. $f(x) = x^2 + 3x - 1$

(a) $f(-2)$ (b) $f(4)$ (c) $f(x - 1)$

103. $f(x) = \begin{cases} 2x + 3, & x \le 1 \\ -x + 4, & x > 1 \end{cases}$

(a) $f(0)$ (b) $f(1)$ (c) $f(2)$

104. $f(x) = \begin{cases} -\frac{1}{2}x - 6, & x \le -4 \\ x + 5, & x > -4 \end{cases}$

(a) $f(-8)$ (b) $f(-4)$ (c) $f(0)$

Writing an Algebraic Expression In Exercises 105 and 106, write an algebraic expression for the verbal description.

105. The sum of two consecutive natural numbers

106. The product of two consecutive natural numbers

2.4 A Library of Parent Functions

Piecewise-defined functions model many real-life situations. For example, in Exercise 45 on page 204, you will write a piecewise-defined function to model the depth of snow during a snowstorm.

❷ Identify and graph linear and squaring functions.
❷ Identify and graph cubic, square root, and reciprocal functions.
❷ Identify and graph step and other piecewise-defined functions.
❷ Recognize graphs of commonly used parent functions.

Linear and Squaring Functions

One of the goals of this text is to enable you to recognize the basic shapes of the graphs of different types of functions. For example, you know that the graph of the **linear function** $f(x) = ax + b$ is a line with slope $m = a$ and y-intercept at $(0, b)$. The graph of a linear function has the characteristics below.

- The domain of the function is the set of all real numbers.
- When $m \neq 0$, the range of the function is the set of all real numbers.
- The graph has an x-intercept at $(-b/m, 0)$ and a y-intercept at $(0, b)$.
- The graph is increasing when $m > 0$, decreasing when $m < 0$, and constant when $m = 0$.

EXAMPLE 1 **Writing a Linear Function**

Write the linear function f for which $f(1) = 3$ and $f(4) = 0$.

Solution To find the equation of the line that passes through $(x_1, y_1) = (1, 3)$ and $(x_2, y_2) = (4, 0)$, first find the slope of the line.

$$m = \frac{y_2 - y_1}{x_2 - x_1} = \frac{0 - 3}{4 - 1} = \frac{-3}{3} = -1$$

Next, use the point-slope form of the equation of a line.

$$y - y_1 = m(x - x_1) \qquad \text{Point-slope form}$$
$$y - 3 = -1(x - 1) \qquad \text{Substitute for } x_1, y_1, \text{ and } m.$$
$$y = -x + 4 \qquad \text{Simplify.}$$
$$f(x) = -x + 4 \qquad \text{Function notation}$$

The figure below shows the graph of f. Note that the points $(1, 3)$ and $(4, 0)$ lie on the graph, corresponding to the given function values $f(1) = 3$ and $f(4) = 0$, respectively.

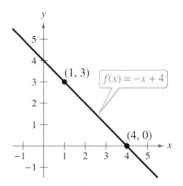

✓ **Checkpoint** ▶ *Audio-video solution in English & Spanish at LarsonPrecalculus.com*

Write the linear function f for which $f(-2) = 6$ and $f(4) = -9$. ■

GO DIGITAL

© DenisTangneyJr/Getty Images

There are two special types of linear functions, the **constant function** and the **identity function.** A constant function has the form

$$f(x) = c \qquad \text{Constant function}$$

and has a domain of all real numbers with a range consisting of a single real number c. The graph of a constant function is a horizontal line, as shown in Figure 2.22. The identity function has the form

$$f(x) = x. \qquad \text{Identity function}$$

Its domain and range are the set of all real numbers. The identity function has a slope of $m = 1$ and a y-intercept at $(0, 0)$. The graph of the identity function is a line for which each x-coordinate equals the corresponding y-coordinate. The graph is always increasing, as shown in Figure 2.23.

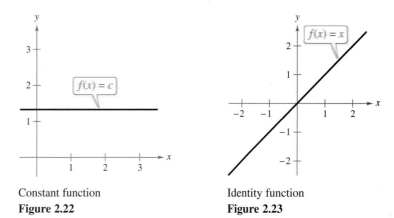

Constant function
Figure 2.22

Identity function
Figure 2.23

The graph of the **squaring function**

$$f(x) = x^2$$

is a U-shaped curve with the characteristics below.

- The domain of the function is the set of all real numbers.
- The range of the function is the set of all nonnegative real numbers.
- The function is even.
- The graph has an intercept at $(0, 0)$.
- The graph is decreasing on the interval $(-\infty, 0)$ and increasing on the interval $(0, \infty)$.
- The graph is symmetric with respect to the y-axis.
- The graph has a relative minimum at $(0, 0)$.

The figure below shows the graph of the squaring function.

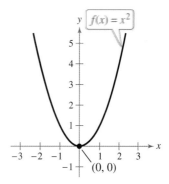

Squaring function

Cubic, Square Root, and Reciprocal Functions

Here are the basic characteristics of the graphs of the **cubic, square root,** and **reciprocal functions.**

1. The graph of the *cubic* function

   $$f(x) = x^3$$

   has the characteristics below.

   - The domain of the function is the set of all real numbers.
   - The range of the function is the set of all real numbers.
   - The function is odd.
   - The graph has an intercept at $(0, 0)$.
   - The graph is increasing on the interval $(-\infty, \infty)$.
   - The graph is symmetric with respect to the origin.

   The figure shows the graph of the cubic function.

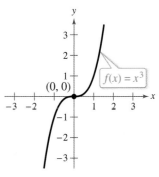

Cubic function

2. The graph of the *square root* function

   $$f(x) = \sqrt{x}$$

   has the characteristics below.

   - The domain of the function is the set of all nonnegative real numbers.
   - The range of the function is the set of all nonnegative real numbers.
   - The graph has an intercept at $(0, 0)$.
   - The graph is increasing on the interval $(0, \infty)$.

   The figure shows the graph of the square root function.

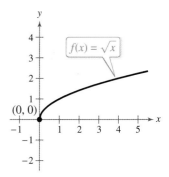

Square root function

3. The graph of the *reciprocal* function

   $$f(x) = \frac{1}{x}$$

   has the characteristics below.

   - The domain of the function is $(-\infty, 0) \cup (0, \infty)$.
   - The range of the function is $(-\infty, 0) \cup (0, \infty)$.
   - The function is odd.
   - The graph does not have any intercepts.
   - The graph is decreasing on the intervals $(-\infty, 0)$ and $(0, \infty)$.
   - The graph is symmetric with respect to the origin.

   The figure shows the graph of the reciprocal function.

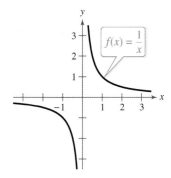

Reciprocal function

GO DIGITAL

Step and Piecewise-Defined Functions

Functions whose graphs resemble sets of stairsteps are known as **step functions.** One common type of step function is the **greatest integer function,** denoted by $[\![x]\!]$ and defined as

$f(x) = [\![x]\!] = $ *the greatest integer less than or equal to* x.

Here are several examples of evaluating the greatest integer function.

$$[\![-1]\!] = (\text{greatest integer} \le -1) = -1$$

$$\left[\!\left[-\frac{1}{2}\right]\!\right] = \left(\text{greatest integer} \le -\frac{1}{2}\right) = -1$$

$$\left[\!\left[\frac{1}{10}\right]\!\right] = \left(\text{greatest integer} \le \frac{1}{10}\right) = 0$$

$$[\![1.9]\!] = (\text{greatest integer} \le 1.9) = 1$$

The graph of the greatest integer function $f(x) = [\![x]\!]$ has the characteristics below, as shown in Figure 2.24.

- The domain of the function is the set of all real numbers.
- The range of the function is the set of all integers.
- The graph has a y-intercept at $(0, 0)$ and x-intercepts in the interval $[0, 1)$.
- The graph is constant between each pair of consecutive integer values of x.
- The graph jumps vertically one unit at each integer value of x.

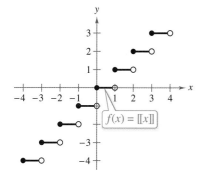

Greatest integer function
Figure 2.24

>>> TECHNOLOGY

Most graphing utilities display graphs in *connected* mode, which works well for graphs that do not have breaks. For graphs that do have breaks, such as the graph of the greatest integer function, it may be better to use *dot* mode. Graph the greatest integer function [often called Int(x)] in *connected* and *dot* modes, and compare the two results.

EXAMPLE 2   **Evaluating a Step Function**

Evaluate the function $f(x) = [\![x]\!] + 1$ when $x = -1$, 2, and $\frac{3}{2}$.

Solution   For $x = -1$, the greatest integer ≤ -1 is -1, so

$$f(-1) = [\![-1]\!] + 1 = -1 + 1 = 0.$$

For $x = 2$, the greatest integer ≤ 2 is 2, so

$$f(2) = [\![2]\!] + 1 = 2 + 1 = 3.$$

For $x = \frac{3}{2}$, the greatest integer $\le \frac{3}{2}$ is 1, so

$$f\left(\frac{3}{2}\right) = \left[\!\left[\frac{3}{2}\right]\!\right] + 1 = 1 + 1 = 2.$$

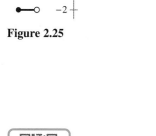

Figure 2.25

Verify your answers by examining the graph of $f(x) = [\![x]\!] + 1$ shown in Figure 2.25.

✓ *Checkpoint* 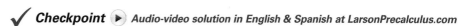 Audio-video solution in English & Spanish at LarsonPrecalculus.com

Evaluate the function $f(x) = [\![x + 2]\!]$ when $x = -\frac{3}{2}$, 1, and $-\frac{5}{2}$.   ■

Recall from Section 2.2 that a piecewise-defined function is defined by two or more equations over a specified domain. To graph a piecewise-defined function, graph each equation separately over the specified domain, as shown in Example 3.

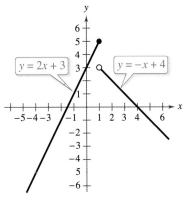

Figure 2.26

EXAMPLE 3 **Graphing a Piecewise-Defined Function**

>>> *See LarsonPrecalculus.com for an interactive version of this type of example.*

Sketch the graph of $f(x) = \begin{cases} 2x + 3, & x \le 1 \\ -x + 4, & x > 1 \end{cases}$.

Solution This piecewise-defined function consists of two linear functions. At $x = 1$ and to the left of $x = 1$, the graph is the line $y = 2x + 3$, and to the right of $x = 1$, the graph is the line $y = -x + 4$, as shown in Figure 2.26. Notice that the point $(1, 5)$ is a solid dot and the point $(1, 3)$ is an open dot. This is because $f(1) = 2(1) + 3 = 5$.

✓ *Checkpoint* ▶ Audio-video solution in English & Spanish at LarsonPrecalculus.com

Sketch the graph of $f(x) = \begin{cases} -\frac{1}{2}x - 6, & x \le -4 \\ x + 5, & x > -4 \end{cases}$.

Commonly Used Parent Functions

The graphs below represent the most commonly used functions in algebra. Familiarity with the characteristics of these graphs will help you analyze more complicated graphs obtained from these graphs by the transformations studied in the next section.

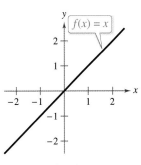

(a) Identity function

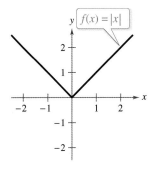

(b) Absolute value function

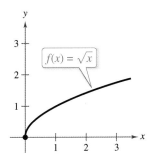

(c) Square root function

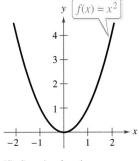

(d) Squaring function

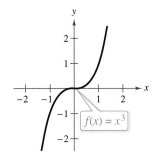

(e) Cubic function

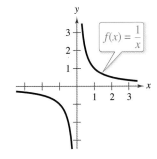

(f) Reciprocal function

Summarize (Section 2.4)

1. Explain how to identify and graph linear and squaring functions (*pages 198 and 199*). For an example involving a linear function, see Example 1.
2. Explain how to identify and graph cubic, square root, and reciprocal functions (*page 200*).
3. Explain how to identify and graph step and other piecewise-defined functions (*page 201*). For examples involving these functions, see Examples 2 and 3.
4. Identify and sketch the graphs of parent functions (*page 202*).

GO DIGITAL

2.4 Exercises

See CalcChat.com for tutorial help and worked-out solutions to odd-numbered exercises.

GO DIGITAL

Vocabulary and Concept Check

In Exercises 1–9, write the most specific name of the function.

1. $f(x) = [\![x]\!]$

2. $f(x) = x$

3. $f(x) = 1/x$

4. $f(x) = x^2$

5. $f(x) = \sqrt{x}$

6. $f(x) = c$

7. $f(x) = |x|$

8. $f(x) = x^3$

9. $f(x) = ax + b$

10. Fill in the blank: The constant function and the identity function are two special types of _____ functions.

Skills and Applications

Writing a Linear Function **In Exercises 11–14, (a) write the linear function f that has the given function values and (b) sketch the graph of the function.**

11. $f(1) = 4$, $f(0) = 6$ **12.** $f(-3) = -8$, $f(1) = 2$

13. $f\left(\frac{1}{2}\right) = -\frac{5}{3}$, $f(6) = 2$

14. $f\left(\frac{3}{5}\right) = \frac{1}{2}$, $f(4) = 9$

Graphing a Function **In Exercises 15–26, use a graphing utility to graph the function. Be sure to choose an appropriate viewing window.**

15. $f(x) = 2.5x - 4.25$ **16.** $f(x) = \frac{5}{6} - \frac{2}{3}x$

17. $g(x) = x^2 + 3$ **18.** $f(x) = -2x^2 - 1$

19. $f(x) = x^3 - 1$ **20.** $f(x) = (x - 1)^3 + 2$

21. $f(x) = \sqrt{x} + 4$ **22.** $h(x) = \sqrt{x + 2} + 3$

23. $f(x) = \dfrac{1}{x - 2}$ **24.** $k(x) = 3 + \dfrac{1}{x + 3}$

25. $g(x) = |x| - 5$ **26.** $f(x) = |x - 1|$

Evaluating a Step Function **In Exercises 27–30, evaluate the function for the given values.**

27. $f(x) = [\![x]\!]$

(a) $f(2.1)$ (b) $f(2.9)$ (c) $f(-3.1)$ (d) $f\left(\frac{7}{2}\right)$

28. $h(x) = [\![x + 3]\!]$

(a) $h(-2)$ (b) $h\left(\frac{1}{2}\right)$ (c) $h(4.2)$ (d) $h(-21.6)$

29. $k(x) = [\![2x + 1]\!]$

(a) $k\left(\frac{1}{3}\right)$ (b) $k(-2.1)$ (c) $k(1.1)$ (d) $k\left(\frac{2}{3}\right)$

30. $g(x) = -7[\![x + 4]\!] + 6$

(a) $g\left(\frac{1}{8}\right)$ (b) $g(9)$ (c) $g(-4)$ (d) $g\left(\frac{3}{2}\right)$

Graphing a Step Function **In Exercises 31–34, sketch the graph of the function.**

31. $g(x) = -[\![x]\!]$ **32.** $g(x) = 4[\![x]\!]$

33. $g(x) = [\![x]\!] - 1$ **34.** $g(x) = [\![x - 3]\!]$

Graphing a Piecewise-Defined Function **In Exercises 35–40, sketch the graph of the function.**

35. $g(x) = \begin{cases} x + 6, & x \le -4 \\ \frac{1}{2}x - 4, & x > -4 \end{cases}$

36. $f(x) = \begin{cases} 4 + x, & x \le 2 \\ x^2 + 2, & x > 2 \end{cases}$

37. $f(x) = \begin{cases} 1 - (x - 1)^2, & x \le 2 \\ \sqrt{x - 2}, & x > 2 \end{cases}$

38. $f(x) = \begin{cases} \sqrt{4 + x}, & x < 0 \\ \sqrt{4 - x}, & x \ge 0 \end{cases}$

39. $h(x) = \begin{cases} 4 - x^2, & x < -2 \\ 3 + x, & -2 \le x < 0 \\ x^2 + 1, & x \ge 0 \end{cases}$

40. $k(x) = \begin{cases} 2x + 1, & x \le -1 \\ 2x^2 - 1, & -1 < x \le 1 \\ 1 - x^2, & x > 1 \end{cases}$

Graphing a Function **In Exercises 41 and 42, (a) use a graphing utility to graph the function and (b) state the domain and range of the function.**

41. $s(x) = 2\left(\frac{1}{4}x - \left[\!\left[\frac{1}{4}x\right]\!\right]\right)$ **42.** $k(x) = 4\left(\frac{1}{2}x - \left[\!\left[\frac{1}{2}x\right]\!\right]\right)^2$

43. Wages A mechanic's pay is \$14 per hour for regular time and time-and-a-half for overtime. The weekly wage function is

$$W(h) = \begin{cases} 14h, & 0 < h \le 40 \\ 21(h - 40) + 560, & h > 40 \end{cases}$$

where h is the number of hours worked in a week.

(a) Evaluate $W(30)$, $W(40)$, $W(45)$, and $W(50)$.

(b) The company decreases the regular work week to 36 hours. What is the new weekly wage function?

(c) The company increases the mechanic's pay to \$16 per hour. What is the new weekly wage function? Use a regular work week of 40 hours.

44. Delivery Charges The cost of mailing a package weighing up to, but not including, 1 pound is $2.80. Each additional pound or portion of a pound costs $0.53.

(a) Use the greatest integer function to create a model for the cost C of mailing a package weighing x pounds, where $x > 0$.

(b) Sketch the graph of the function.

45. Snowstorm

During a nine-hour snowstorm, it snows at a rate of 1 inch per hour for the first 2 hours, at a rate of 2 inches per hour for the next 6 hours, and at a rate of 0.5 inch per hour for the final hour. Write and graph a piecewise-defined function that gives the depth of the snow during the snowstorm. How many inches of snow accumulated from the storm?

46. HOW DO YOU SEE IT? For each graph of f shown below, answer parts (a)–(d).

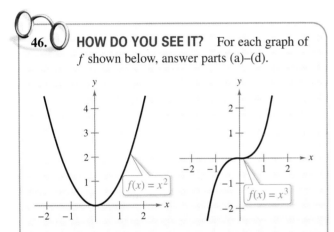

(a) Find the domain and range of f.

(b) Find the x- and y-intercepts of the graph of f.

(c) Determine the open intervals on which f is increasing, decreasing, or constant.

(d) Determine whether f is even, odd, or neither. Then describe the symmetry.

Exploring the Concepts

True or False? **In Exercises 47 and 48, determine whether the statement is true or false. Justify your answer.**

47. A piecewise-defined function will always have at least one x-intercept or at least one y-intercept.

48. A linear equation will always have an x-intercept and a y-intercept.

49. Error Analysis Describe the error.

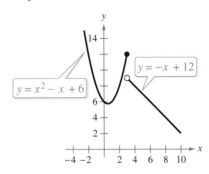

The piecewise-defined function represented in the graph is

$$f(x) = \begin{cases} x^2 - x + 6, & x < 3 \\ -x + 12, & x \geq 3 \end{cases}. \quad \text{✗}$$

50. Error Analysis Describe the error in evaluating $f(x) = [\![x]\!] + 5$ when $x = -\frac{4}{3}$.

$$f(x) = \left[\!\!\left[-\frac{4}{3} \right]\!\!\right] + 5 = -1 + 5 = 4 \quad \text{✗}$$

Review & Refresh ▶ Video solutions at LarsonPrecalculus.com

Testing for Functions **In Exercises 51 and 52, determine whether the statements use the word *function* in ways that are mathematically correct. Explain.**

51. (a) The sales tax on a purchased item is a function of the selling price.

(b) Your score on the next algebra exam is a function of the number of hours you study the night before the exam.

52. (a) The amount in your savings account is a function of your salary.

(b) The speed at which a free-falling acorn strikes the ground is a function of the height from which it was dropped.

Sketching the Graph of a Function **In Exercises 53–56, sketch a graph of the function and identify any intercepts.**

53. $f(x) = x^2 - 2x - 8$ **54.** $g(x) = -x^2 + 6x - 9$

55. $p(x) = \sqrt{8 - x^2} - 2$ **56.** $q(x) = -5|x + 2| + 3$

Simplifying an Expression **In Exercises 57–60, simplify the expression.**

57. $\dfrac{\sqrt[3]{(-x)^2} + \sqrt[3]{-x}}{-x}$

58. $-\dfrac{\sqrt{(2 - x)^3} + \sqrt{2 - x}}{x - 2}$

59. $(x - 3)^2 - 3(x - 3) + 2$

60. $3(x - 1)^3 - 2(x - 1)^2 - 3(x - 1) + 2$

2.5 Transformations of Functions

Transformations of functions model many real-life applications. For example, in Exercise 67 on page 212, you will use a transformation of a function to model the number of horsepower required to overcome wind drag on an automobile.

❯ **Use vertical and horizontal shifts to sketch graphs of functions.**
❯ **Use reflections to sketch graphs of functions.**
❯ **Use nonrigid transformations to sketch graphs of functions.**

Shifting Graphs

Many functions have graphs that are transformations of the graphs of parent functions summarized in Section 2.4. For example, to obtain the graph of $h(x) = x^2 + 2$, shift the graph of $f(x) = x^2$ *up* two units, as shown in Figure 2.27. In function notation, h and f are related as follows.

$$h(x) = x^2 + 2 = f(x) + 2 \qquad \text{Upward shift of two units}$$

Similarly, to obtain the graph of

$$g(x) = (x - 2)^2$$

shift the graph of $f(x) = x^2$ to the *right* two units, as shown in Figure 2.28. In this case, the functions g and f have the following relationship.

$$g(x) = (x - 2)^2 = f(x - 2) \qquad \text{Right shift of two units}$$

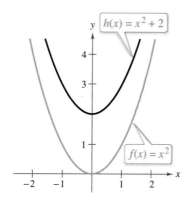

Vertical shift upward: two units
Figure 2.27

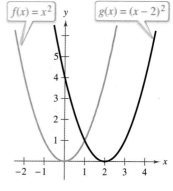

Horizontal shift to the right: two units
Figure 2.28

ALGEBRA HELP

For horizontal shifts, be sure you see that $h(x) = f(x - c)$ corresponds to a *right* shift and $h(x) = f(x + c)$ corresponds to a *left* shift for $c > 0$.

Vertical and Horizontal Shifts

Let c be a positive real number. **Vertical and horizontal shifts** in the graph of $y = f(x)$ are represented as follows.

1. Vertical shift c units *up*: $\qquad\qquad h(x) = f(x) + c$
2. Vertical shift c units *down*: $\qquad\quad\; h(x) = f(x) - c$
3. Horizontal shift c units to the *right*: $h(x) = f(x - c)$
4. Horizontal shift c units to the *left*: $\;\; h(x) = f(x + c)$

Some graphs are obtained from combinations of vertical and horizontal shifts, as demonstrated in Example 1(b) on the next page. Vertical and horizontal shifts generate a *family of functions*, each with a graph that has the same shape but at a different location in the plane.

GO DIGITAL

EXAMPLE 1 **Shifting the Graph of a Function**

a. To sketch the graph of $g(x) = x^3 - 1$, shift the graph of $f(x) = x^3$ one unit down [see Figure 2.29(a)].

b. To sketch the graph of $h(x) = (x + 2)^3 + 1$, shift the graph of $f(x) = x^3$ two units to the left and then one unit up [see Figure 2.29(b)]. Note that you obtain the same graph whether the vertical shift precedes the horizontal shift or the horizontal shift precedes the vertical shift.

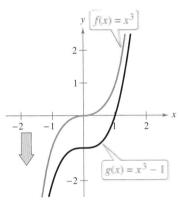
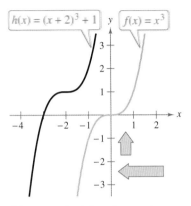

(a) Vertical shift: one unit down   **(b)** Two units left and one unit up

Figure 2.29

✓ *Checkpoint* ▶ *Audio-video solution in English & Spanish at LarsonPrecalculus.com*

Use the graph of $f(x) = x^3$ to sketch the graph of each function.

a. $h(x) = x^3 + 5$ **b.** $g(x) = (x - 3)^3 + 2$

EXAMPLE 2 **Writing Equations from Graphs**

Each graph is a transformation of the graph of $f(x) = x^2$. Write an equation for the function represented by each graph.

a.

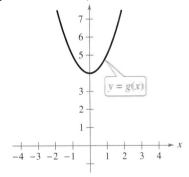

b.

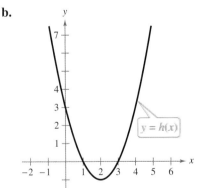

Solution

a. The graph of g is an upward shift of four units of the graph of $f(x) = x^2$. So, an equation for g is $g(x) = x^2 + 4$.

b. The graph of h is a right shift of two units and a downward shift of one unit of the graph of $f(x) = x^2$. So, an equation for h is $h(x) = (x - 2)^2 - 1$.

✓ *Checkpoint* ▶ *Audio-video solution in English & Spanish at LarsonPrecalculus.com*

The graph in Figure 2.30 is a transformation of the graph of $f(x) = x^2$. Write an equation for the function represented by the graph. ■

Figure 2.30

GO DIGITAL

Reflecting Graphs

Another common type of transformation is a **reflection.** For example, if you consider the *x*-axis to be a mirror, then the graph of $h(x) = -x^2$ is the mirror image (or reflection) of the graph of $f(x) = x^2$, as shown in Figure 2.31.

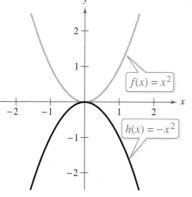

Figure 2.31

> ### Reflections in the Coordinate Axes
>
> **Reflections** in the coordinate axes of the graph of $y = f(x)$ are represented as follows.
>
> **1.** Reflection in the *x*-axis: $h(x) = -f(x)$
>
> **2.** Reflection in the *y*-axis: $h(x) = f(-x)$

EXAMPLE 3   **Writing Equations from Graphs**

Each graph is a transformation of the graph of $f(x) = x^2$. Write an equation for the function represented by each graph.

a.

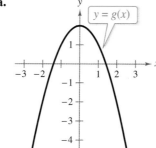

b.

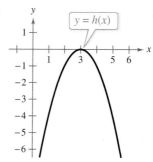

Solution

a. The graph of *g* is a reflection in the *x*-axis *followed by* an upward shift of two units of the graph of $f(x) = x^2$. So, an equation for *g* is

$$g(x) = -x^2 + 2.$$

b. The graph of *h* is a right shift of three units *followed by* a reflection in the *x*-axis of the graph of $f(x) = x^2$. So, an equation for *h* is

$$h(x) = -(x - 3)^2.$$

 Checkpoint ▶ *Audio-video solution in English & Spanish at LarsonPrecalculus.com*

The graph below is a transformation of the graph of $f(x) = x^2$. Write an equation for the function represented by the graph.

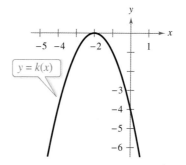

EXAMPLE 4   **Reflections and Shifts**

Compare the graph of each function with the graph of $f(x) = \sqrt{x}$.

a. $g(x) = -\sqrt{x}$ **b.** $h(x) = \sqrt{-x}$ **c.** $k(x) = -\sqrt{x+2}$

Algebraic Solution

a. The graph of g is a reflection of the graph of f in the x-axis because

$$g(x) = -\sqrt{x}$$
$$= -f(x).$$

b. The graph of h is a reflection of the graph of f in the y-axis because

$$h(x) = \sqrt{-x}$$
$$= f(-x).$$

c. The graph of k is a left shift of two units followed by a reflection in the x-axis because

$$k(x) = -\sqrt{x+2}$$
$$= -f(x+2).$$

Graphical Solution

a. Graph f and g on the same set of coordinate axes. The graph of g is a reflection of the graph of f in the x-axis.

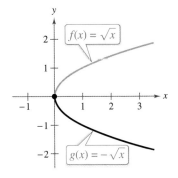

b. Graph f and h on the same set of coordinate axes. The graph of h is a reflection of the graph of f in the y-axis.

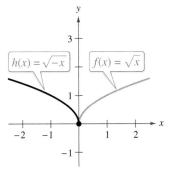

c. Graph f and k on the same set of coordinate axes. The graph of k is a left shift of two units followed by a reflection in the x-axis of the graph of f.

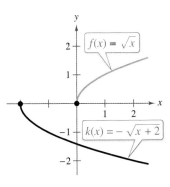

✓ *Checkpoint* ▶ *Audio-video solution in English & Spanish at LarsonPrecalculus.com*

Compare the graph of each function with the graph of

$$f(x) = \sqrt{x-1}.$$

a. $g(x) = -\sqrt{x-1}$ **b.** $h(x) = \sqrt{-x-1}$

When sketching the graphs of functions involving square roots, remember that you must restrict the domain to exclude negative numbers inside the radical. For instance, here are the domains of the functions in Example 4.

Domain of $g(x) = -\sqrt{x}$: $x \geq 0$

Domain of $h(x) = \sqrt{-x}$: $x \leq 0$

Domain of $k(x) = -\sqrt{x+2}$: $x \geq -2$

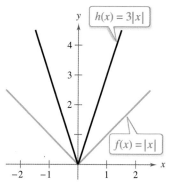

Figure 2.32

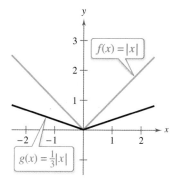

Figure 2.33

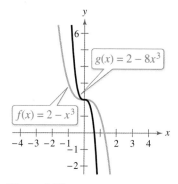

Figure 2.34

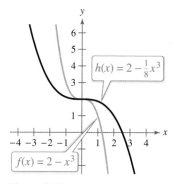

Figure 2.35

Nonrigid Transformations

Horizontal shifts, vertical shifts, and reflections are **rigid transformations** because the basic shape of the graph is unchanged. These transformations change only the *position* of the graph in the coordinate plane. **Nonrigid transformations** are those that cause a *distortion*—a change in the shape of the original graph. For example, a nonrigid transformation of the graph of $y = f(x)$ is represented by $g(x) = cf(x)$, where the transformation is a **vertical stretch** when $c > 1$ and a **vertical shrink** when $0 < c < 1$. Another nonrigid transformation of the graph of $y = f(x)$ is represented by $h(x) = f(cx)$, where the transformation is a **horizontal shrink** when $c > 1$ and a **horizontal stretch** when $0 < c < 1$.

EXAMPLE 5 **Nonrigid Transformations**

Compare the graph of each function with the graph of $f(x) = |x|$.

a. $h(x) = 3|x|$ **b.** $g(x) = \frac{1}{3}|x|$

Solution

a. Relative to the graph of $f(x) = |x|$, the graph of $h(x) = 3|x| = 3f(x)$ is a vertical stretch (each y-value is multiplied by 3). (See Figure 2.32.)

b. Similarly, the graph of $g(x) = \frac{1}{3}|x| = \frac{1}{3}f(x)$ is a vertical shrink $\left(\text{each } y\text{-value is multiplied by } \frac{1}{3}\right)$ of the graph of f. (See Figure 2.33.)

✔ *Checkpoint* ▶ *Audio-video solution in English & Spanish at LarsonPrecalculus.com*

Compare the graph of each function with the graph of $f(x) = x^2$.

a. $g(x) = 4x^2$ **b.** $h(x) = \frac{1}{4}x^2$

EXAMPLE 6 **Nonrigid Transformations**

▶▶▶ *See LarsonPrecalculus.com for an interactive version of this type of example.*

Compare the graph of each function with the graph of $f(x) = 2 - x^3$.

a. $g(x) = f(2x)$ **b.** $h(x) = f\left(\frac{1}{2}x\right)$

Solution

a. Relative to the graph of $f(x) = 2 - x^3$, the graph of $g(x) = f(2x) = 2 - (2x)^3 = 2 - 8x^3$ is a horizontal shrink ($c > 1$). (See Figure 2.34.)

b. Similarly, the graph of $h(x) = f\left(\frac{1}{2}x\right) = 2 - \left(\frac{1}{2}x\right)^3 = 2 - \frac{1}{8}x^3$ is a horizontal stretch ($0 < c < 1$) of the graph of f. (See Figure 2.35.)

✔ *Checkpoint* ▶ *Audio-video solution in English & Spanish at LarsonPrecalculus.com*

Compare the graph of each function with the graph of $f(x) = x^2 + 3$.

a. $g(x) = f(2x)$ **b.** $h(x) = f\left(\frac{1}{2}x\right)$ ■

Summarize (Section 2.5)

1. Explain how to shift the graph of a function vertically and horizontally *(page 205)*. For examples of shifting the graphs of functions, see Examples 1 and 2.

2. Explain how to reflect the graph of a function in the x-axis and in the y-axis *(page 207)*. For examples of reflecting graphs of functions, see Examples 3 and 4.

3. Describe nonrigid transformations of the graph of a function *(page 209)*. For examples of nonrigid transformations, see Examples 5 and 6.

2.5 Exercises

GO DIGITAL

Vocabulary and Concept Check

In Exercises 1 and 2, fill in the blanks.

1. A reflection in the x-axis of the graph of $y = f(x)$ is represented by $h(x) = $ _____,
   while a reflection in the y-axis of the graph of $y = f(x)$ is represented by $h(x) = $ _____.

2. A nonrigid transformation of the graph of $y = f(x)$ represented by $g(x) = cf(x)$ is
   a _____ _____ when $c > 1$ and a _____ _____ when $0 < c < 1$.

3. Name three types of rigid transformations.

4. Match each function h with the transformation it represents, where $c > 0$.
   (a) $h(x) = f(x) + c$ (i) A horizontal shift of f, c units to the right
   (b) $h(x) = f(x) - c$ (ii) A vertical shift of f, c units down
   (c) $h(x) = f(x + c)$ (iii) A horizontal shift of f, c units to the left
   (d) $h(x) = f(x - c)$ (iv) A vertical shift of f, c units up

Skills and Applications

5. **Shifting the Graph of a Function** For each function, sketch the graphs of the function when $c = -2, -1, 1$, and 2 on the same set of coordinate axes.
   (a) $f(x) = |x| + c$ (b) $f(x) = |x - c|$

6. **Shifting the Graph of a Function** For each function, sketch the graphs of the function when $c = -3, -2, 2$, and 3 on the same set of coordinate axes.
   (a) $f(x) = \sqrt{x} + c$ (b) $f(x) = \sqrt{x - c}$

7. **Shifting the Graph of a Function** For each function, sketch the graphs of the function when $c = -4, -1, 2$, and 5 on the same set of coordinate axes.
   (a) $f(x) = [\![x]\!] + c$ (b) $f(x) = [\![x + c]\!]$

8. **Shifting the Graph of a Function** For each function, sketch the graphs of the function when $c = -3, -2, 1$, and 2 on the same set of coordinate axes.
   (a) $f(x) = \begin{cases} x^2 + c, & x < 0 \\ -x^2 + c, & x \geq 0 \end{cases}$

   (b) $f(x) = \begin{cases} (x + c)^2, & x < 0 \\ -(x + c)^2, & x \geq 0 \end{cases}$

Sketching Transformations In Exercises 9 and 10, use the graph of f to sketch each graph. To print an enlarged copy of the graph, go to *MathGraphs.com*.

9. (a) $y = f(-x)$
   (b) $y = f(x) + 4$
   (c) $y = 2f(x)$
   (d) $y = -f(x - 4)$
   (e) $y = f(x) - 3$
   (f) $y = -f(x) - 1$
   (g) $y = f(2x)$

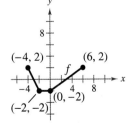

10. (a) $y = f(x - 5)$
    (b) $y = -f(x) + 3$
    (c) $y = \frac{1}{3}f(x)$
    (d) $y = -f(x + 1)$
    (e) $y = f(-x)$
    (f) $y = f(x) - 10$
    (g) $y = f(\frac{1}{3}x)$
    (h) $y = 2f(x) - 2$

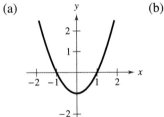

11. **Writing Equations from Graphs** Use the graph of $f(x) = x^2$ to write an equation for the function represented by each graph.
    (a)   (b)

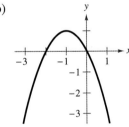

12. **Writing Equations from Graphs** Use the graph of $f(x) = x^3$ to write an equation for the function represented by each graph.
    (a)   (b)

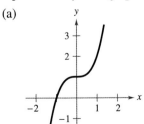

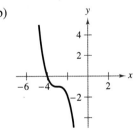

13. Writing Equations from Graphs Use the graph of $f(x) = |x|$ to write an equation for the function represented by each graph.

(a)   (b)

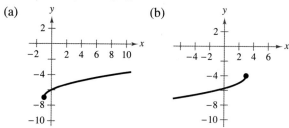

14. Writing Equations from Graphs Use the graph of $f(x) = \sqrt{x}$ to write an equation for the function represented by each graph.

(a)   (b)

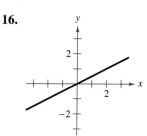

Writing an Equation from a Graph In Exercises 15–20, identify the parent function and the transformation represented by the graph. Write an equation for the function represented by the graph.

15.   16.

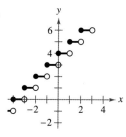

17.   18.

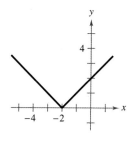

19.   20.

Describing Transformations In Exercises 21–44, g is related to one of the parent functions described in Section 2.4. (a) Identify the parent function f. (b) Describe the sequence of transformations from f to g. (c) Sketch the graph of g. (d) Use function notation to write g in terms of f.

21. $g(x) = x^2 + 6$ 22. $g(x) = x^2 - 2$

23. $g(x) = -(x - 2)^3$ 24. $g(x) = -(x + 1)^3$

25. $g(x) = -3 - (x + 1)^2$ 26. $g(x) = 4 - (x - 2)^2$

27. $g(x) = -2x^2 + 1$ 28. $g(x) = \frac{1}{2}x^2 - 2$

29. $g(x) = |x - 1| + 2$ 30. $g(x) = |x + 3| - 2$

31. $g(x) = |2x|$ 32. $g(x) = \left|\frac{1}{2}x\right|$

33. $g(x) = 3|x - 1| + 2$ 34. $g(x) = -2|x + 1| - 3$

35. $g(x) = 2\sqrt{x}$ 36. $g(x) = \frac{1}{2}\sqrt{x}$

37. $g(x) = \sqrt{x - 9}$ 38. $g(x) = \sqrt{3x} + 1$

39. $g(x) = \sqrt{7 - x} - 2$ 40. $g(x) = \sqrt{x + 4} + 8$

41. $g(x) = 2[\![x]\!] - 1$ 42. $g(x) = -[\![x]\!] + 1$

43. $g(x) = 3 - [\![x]\!]$ 44. $g(x) = 2[\![x + 5]\!]$

Writing an Equation from a Description In Exercises 45–52, write an equation for the function whose graph is described.

45. The shape of $f(x) = x^2$, but shifted three units to the right and seven units down

46. The shape of $f(x) = x^2$, but shifted two units to the left, nine units up, and then reflected in the x-axis

47. The shape of $f(x) = x^3$, but shifted 13 units to the right

48. The shape of $f(x) = x^3$, but shifted six units to the left, six units down, and then reflected in the y-axis

49. The shape of $f(x) = |x|$, but shifted 12 units up and then reflected in the x-axis

50. The shape of $f(x) = |x|$, but shifted four units to the left and eight units down

51. The shape of $f(x) = \sqrt{x}$, but shifted six units to the left and then reflected in both the x-axis and the y-axis

52. The shape of $f(x) = \sqrt{x}$, but shifted nine units down and then reflected in both the x-axis and the y-axis

53. Writing Equations from Graphs Use the graph of $f(x) = x^2$ to write an equation for the function represented by each graph.

(a)   (b)

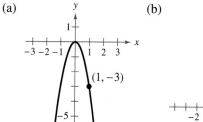
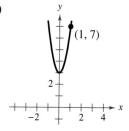

54. Writing Equations from Graphs Use the graph of

$f(x) = x^3$

to write an equation for the function represented by each graph.

(a)
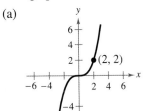

(b)
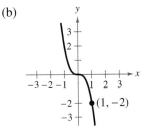

55. Writing Equations from Graphs Use the graph of

$f(x) = |x|$

to write an equation for the function represented by each graph.

(a)
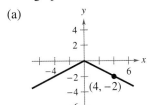

(b)
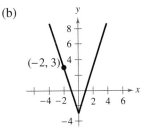

56. Writing Equations from Graphs Use the graph of

$f(x) = \sqrt{x}$

to write an equation for the function represented by each graph.

(a)
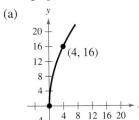

(b)
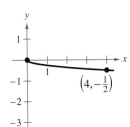

Writing an Equation from a Graph In Exercises 57–62, identify the parent function and the transformation represented by the graph. Write an equation for the function represented by the graph. Then use a graphing utility to verify your answer.

57.
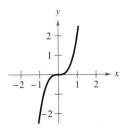

58.
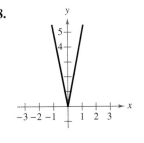

59.
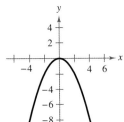

60.
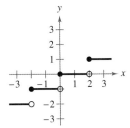

61.
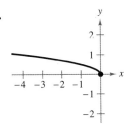

62.
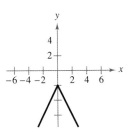

Writing an Equation from a Graph In Exercises 63–66, write an equation for the transformation of the parent function.

63.
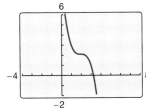

64.
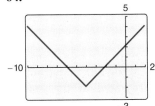

65.
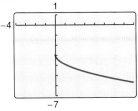

66.
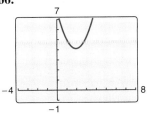

67. Automobile Aerodynamics

The horsepower H required to overcome wind drag on a particular automobile is given by

$H(x) = 0.00004636x^3$

where x is the speed of the car (in miles per hour).

(a) Use a graphing utility to graph the function.

(b) Rewrite the horsepower function so that x represents the speed in kilometers per hour. [Find $H(x/1.6)$.] Identify the type of transformation applied to the graph of the horsepower function.

68. Households The number N (in millions) of households in the United States from 2005 through 2019 can be approximated by

$$N(x) = 0.0022(x + 239.16)^2 - 17.8, \quad 5 \le t \le 19$$

where t represents the year, with $t = 5$ corresponding to 2005. *(Source: U.S. Census Bureau)*

(a) Describe the transformation of the parent function $f(x) = x^2$. Then use a graphing utility to graph the function over the specified domain.

(b) Find the average rate of change of the function from 2005 to 2019. Interpret your answer in the context of the problem.

(c) Use the model to determine the number of households in the United States in 2015.

Exploring the Concepts

True or False? In Exercises 69–72, determine whether the statement is true or false. Justify your answer.

69. The graph of $y = f(-x)$ is a reflection of the graph of $y = f(x)$ in the x-axis.

70. The graph of $y = -f(x)$ is a reflection of the graph of $y = f(x)$ in the y-axis.

71. The graphs of $f(x) = |x| + 6$ and $f(x) = |-x| + 6$ are identical.

72. If the graph of the parent function $f(x) = x^2$ is shifted six units to the right, three units up, and reflected in the x-axis, then the point $(-2, 19)$ will lie on the graph of the transformation.

73. Finding Points on a Graph The graph of $y = f(x)$ passes through the points $(0, 1)$, $(1, 2)$, and $(2, 3)$. Find the corresponding points on the graph of $y = f(x + 2) - 1$.

74. Think About It Two methods of graphing a function are plotting points and translating a parent function as shown in this section. Which method of graphing do you prefer to use for each function? Explain.

(a) $f(x) = 3x^2 - 4x + 1$ (b) $f(x) = 2(x - 1)^2 - 6$

75. Error Analysis Describe the error.

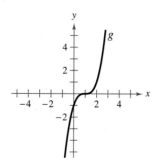

The graph of g is a right shift of one unit of the graph of $f(x) = x^3$. So, an equation for g is $g(x) = (x + 1)^3$.

76. 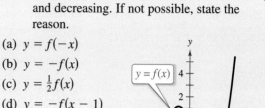 **HOW DO YOU SEE IT?** Use the graph of $y = f(x)$ to find the open intervals on which the graph of each transformation is increasing and decreasing. If not possible, state the reason.

(a) $y = f(-x)$
(b) $y = -f(x)$
(c) $y = \frac{1}{2}f(x)$
(d) $y = -f(x - 1)$
(e) $y = f(x - 2) + 1$

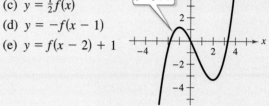

77. Reversing the Order of Transformations Reverse the order of transformations in Example 3(a). Do you obtain the same graph? Do the same for Example 3(b). Do you obtain the same graph? Explain.

Review & Refresh ▶ Video solutions at LarsonPrecalculus.com

78. Exam Scores The table shows the mathematics entrance test scores x and the final examination scores y in an algebra course for a sample of 10 students.

x	22	29	35	40	44	48	53	58	65	76
y	53	74	57	66	79	90	76	93	83	99

(a) Sketch a scatter plot of the data.

(b) Find the entrance test score of any student with a final exam score in the 80s.

(c) Does a higher entrance test score imply a higher final exam score? Explain.

Adding or Subtracting Polynomials In Exercises 79–82, add or subtract and write the result in standard form.

79. $(2x + 1) + (x^2 + 2x - 1)$

80. $(2x + 1) - (x^2 + 2x - 1)$

81. $(3x^2 + x - 1) - (1 - x)$

82. $(3x^2 + x - 1) + (1 - x)$

Multiplying Polynomials In Exercises 83–86, multiply and write the result in standard form.

83. $x^2(x - 3)$ **84.** $x^2(1 - x)$

85. $-2x(0.1x + 17)$ **86.** $6y\left(5 - \frac{3}{8}y\right)$

Dividing Polynomials In Exercises 87 and 88, divide and write the result in simplest form.

87. $(3x + 5) \div (6x^2 + 10x)$ **88.** $(20x^2 - 43x) \div x$

2.6 Combinations of Functions: Composite Functions

Arithmetic combinations of functions are used to model and solve real-life problems. For example, in Exercise 56 on page 220, you will use arithmetic combinations of functions to analyze numbers of pets in the United States.

- ❯ Add, subtract, multiply, and divide functions.
- ❯ Find the composition of one function with another function.
- ❯ Use combinations and compositions of functions to model and solve real-life problems.

Arithmetic Combinations of Functions

Just as two real numbers can be combined by the operations of addition, subtraction, multiplication, and division to form other real numbers, two *functions* can be combined to create new functions. For example, the functions $f(x) = 2x - 3$ and $g(x) = x^2 - 1$ can be combined to form the sum, difference, product, and quotient of f and g.

$$f(x) + g(x) = (2x - 3) + (x^2 - 1) = x^2 + 2x - 4 \qquad \text{Sum}$$

$$f(x) - g(x) = (2x - 3) - (x^2 - 1) = -x^2 + 2x - 2 \qquad \text{Difference}$$

$$f(x)g(x) = (2x - 3)(x^2 - 1) = 2x^3 - 3x^2 - 2x + 3 \qquad \text{Product}$$

$$\frac{f(x)}{g(x)} = \frac{2x - 3}{x^2 - 1}, \quad x \neq \pm 1 \qquad \text{Quotient}$$

The domain of an **arithmetic combination** of functions f and g consists of all real numbers that are common to the domains of f and g. In the case of the quotient $f(x)/g(x)$, there is the further restriction that $g(x) \neq 0$.

Sum, Difference, Product, and Quotient of Functions

Let f and g be two functions with overlapping domains. Then, for all x common to both domains, the *sum, difference, product,* and *quotient* of f and g are defined as follows.

1. Sum: $\quad (f + g)(x) = f(x) + g(x)$

2. Difference: $(f - g)(x) = f(x) - g(x)$

3. Product: $\quad (fg)(x) = f(x) \cdot g(x)$

4. Quotient: $\left(\dfrac{f}{g}\right)(x) = \dfrac{f(x)}{g(x)}, \quad g(x) \neq 0$

❯❯❯ **SKILLS REFRESHER**

For a refresher on finding the sum, difference, product, or quotient of two polynomials, watch the video at *LarsonPrecalculus.com*.

EXAMPLE 1 **Finding the Sum of Two Functions**

Given $f(x) = 2x + 1$ and $g(x) = x^2 + 2x - 1$, find $(f + g)(x)$. Then evaluate the sum when $x = 3$.

Solution The sum of f and g is

$$(f + g)(x) = f(x) + g(x) = (2x + 1) + (x^2 + 2x - 1) = x^2 + 4x.$$

When $x = 3$, the value of this sum is

$$(f + g)(3) = 3^2 + 4(3) = 21.$$

✓ *Checkpoint* ▶ *Audio-video solution in English & Spanish at LarsonPrecalculus.com*

Given $f(x) = x^2$ and $g(x) = 1 - x$, find $(f + g)(x)$. Then evaluate the sum when $x = 2$.

EXAMPLE 2 Finding the Difference of Two Functions

Given $f(x) = 2x + 1$ and $g(x) = x^2 + 2x - 1$, find $(f - g)(x)$. Then evaluate the difference when $x = 2$.

Solution The difference of f and g is

$$(f - g)(x) = f(x) - g(x) \qquad \text{Difference of } f \text{ and } g.$$
$$= (2x + 1) - (x^2 + 2x - 1) \qquad \text{Substitute.}$$
$$= 2x + 1 - x^2 - 2x + 1 \qquad \text{Distributive Property}$$
$$= -x^2 + 2. \qquad \text{Simplify.}$$

When $x = 2$, the value of this difference is $(f - g)(2) = -(2)^2 + 2 = -2$.

✓ **Checkpoint** ▶ Audio-video solution in English & Spanish at LarsonPrecalculus.com

Given $f(x) = x^2$ and $g(x) = 1 - x$, find $(f - g)(x)$. Then evaluate the difference when $x = 3$.

EXAMPLE 3 Finding the Product of Two Functions

Given $f(x) = x^2$ and $g(x) = x - 3$, find $(fg)(x)$. Then evaluate the product when $x = 4$.

Solution The product of f and g is

$$(fg)(x) = f(x)g(x) = (x^2)(x - 3) = x^3 - 3x^2.$$

When $x = 4$, the value of this product is $(fg)(4) = 4^3 - 3(4)^2 = 16$.

✓ **Checkpoint** ▶ Audio-video solution in English & Spanish at LarsonPrecalculus.com

Given $f(x) = x^2$ and $g(x) = 1 - x$, find $(fg)(x)$. Then evaluate the product when $x = 3$. ∎

In Examples 1–3, both f and g have domains that consist of all real numbers. So, the domains of $f + g$, $f - g$, and fg are also the set of all real numbers. Remember to consider any restrictions on the domains of f and g when forming the sum, difference, product, or quotient of f and g.

EXAMPLE 4 Finding the Quotient of Two Functions

Find $(f/g)(x)$ for the functions $f(x) = \sqrt{x}$ and $g(x) = \sqrt{4 - x}$. Then find the domain of f/g.

Solution The quotient of f and g is

$$\left(\frac{f}{g}\right)(x) = \frac{f(x)}{g(x)} = \frac{\sqrt{x}}{\sqrt{4 - x}}.$$

The function f is only defined when $x \geq 0$, and g is only defined when $4 - x \geq 0$, or $x \leq 4$. So, the domain of f is $[0, \infty)$, and the domain of g is $(-\infty, 4]$. The intersection of these two domains is $[0, 4]$. When $x = 4$, however, $(f/g)(4)$ is

$$\left(\frac{f}{g}\right)(4) = \frac{\sqrt{4}}{\sqrt{4 - 4}} = \frac{2}{0} \qquad \text{Division by zero is undefined.}$$

which is undefined. So, the domain of f/g is $[0, 4)$.

✓ **Checkpoint** ▶ Audio-video solution in English & Spanish at LarsonPrecalculus.com

Find $(f/g)(x)$ for the functions $f(x) = \sqrt{x - 3}$ and $g(x) = \sqrt{16 - x}$. Then find the domain of f/g. ∎

GO DIGITAL

Compositions of Functions

Another way of combining two functions is to form the **composition** of one with the other. For example, if $f(x) = x^2$ and $g(x) = x + 1$, then the composition of f with g is

$$f(g(x)) = f(x + 1)$$
$$= (x + 1)^2.$$

This composition is denoted as $f \circ g$ and reads as "f composed with g."

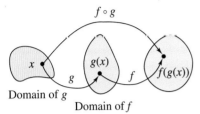

$f \circ g$

Domain of g

Domain of f

Figure 2.36

Definition of Composition of Two Functions

The **composition** of the function f with the function g is

$$(f \circ g)(x) = f(g(x)).$$

The domain of $f \circ g$ is the set of all x in the domain of g such that $g(x)$ is in the domain of f. (See Figure 2.36.)

EXAMPLE 5 **Compositions of Functions**

▶▶▶ *See LarsonPrecalculus.com for an interactive version of this type of example.*

Given $f(x) = x + 2$ and $g(x) = 4 - x^2$, find the following.

a. $(f \circ g)(x)$ **b.** $(g \circ f)(x)$ **c.** $(g \circ f)(-2)$

Solution

a. The composition of f with g is

$$\begin{aligned}(f \circ g)(x) &= f(g(x)) && \text{Definition of } f \circ g \\ &= f(4 - x^2) && \text{Definition of } g(x) \\ &= (4 - x^2) + 2 && \text{Definition of } f(x) \\ &= -x^2 + 6. && \text{Simplify.}\end{aligned}$$

b. The composition of g with f is

$$\begin{aligned}(g \circ f)(x) &= g(f(x)) && \text{Definition of } g \circ f \\ &= g(x + 2) && \text{Definition of } f(x) \\ &= 4 - (x + 2)^2 && \text{Definition of } g(x) \\ &= 4 - (x^2 + 4x + 4) && \text{Square of a binomial} \\ &= 4 - x^2 - 4x - 4 && \text{Distributive Property} \\ &= -x^2 - 4x. && \text{Simplify.}\end{aligned}$$

Note that, in this case, $(f \circ g)(x) \neq (g \circ f)(x)$.

c. Evaluate the result of part (b) when $x = -2$.

$$\begin{aligned}(g \circ f)(-2) &= -(-2)^2 - 4(-2) && \text{Substitute.} \\ &= -4 + 8 && \text{Simplify.} \\ &= 4 && \text{Simplify.}\end{aligned}$$

✓ *Checkpoint* ▶ *Audio-video solution in English & Spanish at LarsonPrecalculus.com*

Given $f(x) = 2x + 5$ and $g(x) = 4x^2 + 1$, find the following.

a. $(f \circ g)(x)$ **b.** $(g \circ f)(x)$ **c.** $(f \circ g)\left(-\frac{1}{2}\right)$

GO DIGITAL

▶▶▶▶

ALGEBRA HELP

The tables of values below help illustrate the composition $(f \circ g)(x)$ in Example 5(a).

x	0	1	2	3
$g(x)$	4	3	0	-5

$g(x)$	4	3	0	-5
$f(g(x))$	6	5	2	-3

x	0	1	2	3
$f(g(x))$	6	5	2	-3

Note that the first two tables are combined (or "composed") to produce the values in the third table.

EXAMPLE 6 **Finding the Domain of a Composite Function**

Find the domain of $f \circ g$ for the functions

$$f(x) = x^2 - 9 \quad \text{and} \quad g(x) = \sqrt{9 - x^2}.$$

Algebraic Solution

Find the composition of f with g.

$$
\begin{aligned}
(f \circ g)(x) &= f(g(x)) && \text{Definition of } f \circ g \\
&= f\left(\sqrt{9 - x^2}\right) && \text{Definition of } g(x) \\
&= \left(\sqrt{9 - x^2}\right)^2 - 9 && \text{Definition of } f(x) \\
&= 9 - x^2 - 9 && \left(\sqrt[n]{a}\right)^n = a \\
&= -x^2 && \text{Simplify.}
\end{aligned}
$$

The domain of $f \circ g$ is restricted to the x-values in the domain of g for which $g(x)$ is in the domain of f. The domain of $f(x) = x^2 - 9$ is the set of all real numbers, which includes all real values of g. So, the domain of $f \circ g$ is the entire domain of $g(x) = \sqrt{9 - x^2}$, which is $[-3, 3]$.

Graphical Solution

The x-coordinates of points on the graph extend from -3 to 3. So, the domain of $f \circ g$ is $[-3, 3]$.

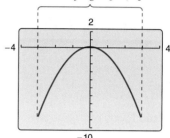

✓ **Checkpoint** ▶ Audio-video solution in English & Spanish at LarsonPrecalculus.com

Find the domain of $f \circ g$ for the functions $f(x) = \sqrt{x}$ and $g(x) = x^2 + 4$. ■

In Examples 5 and 6, you formed the composition of two given functions. In calculus, it is also important to be able to identify two functions that make up a given composite function. For example, the function $h(x) = (3x - 5)^3$ is the composition of $f(x) = x^3$ and $g(x) = 3x - 5$. That is,

$$h(x) = (3x - 5)^3 = [g(x)]^3 = f(g(x)).$$

Basically, to "decompose" a composite function, look for an "inner" function and an "outer" function. In the function h above, $g(x) = 3x - 5$ is the inner function and $f(x) = x^3$ is the outer function.

EXAMPLE 7 **Writing a Composite Function**

Write the function $h(x) = \dfrac{1}{(x - 2)^2}$ as a composition of two functions.

Solution One way to write h as a composition of two functions is to let $g(x) = x - 2$ be the inner function and let

$$f(x) = \frac{1}{x^2} = x^{-2} \qquad \frac{1}{a^n} = a^{-n}$$

be the outer function. Then you can write h as

$$
\begin{aligned}
h(x) &= \frac{1}{(x - 2)^2} \\
&= (x - 2)^{-2} \\
&= f(x - 2) \\
&= f(g(x)).
\end{aligned}
$$

✓ **Checkpoint** ▶ Audio-video solution in English & Spanish at LarsonPrecalculus.com

Write the function $h(x) = \dfrac{\sqrt[3]{8 - x}}{5}$ as a composition of two functions. ■

Application

EXAMPLE 8   **Bacteria Count**

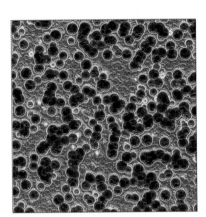

Refrigerated foods can have two types of bacteria: pathogenic bacteria, which can cause foodborne illness, and spoilage bacteria, which give foods an unpleasant look, smell, taste, or texture.

The number N of bacteria in a refrigerated food is given by

$$N(T) = 20T^2 - 80T + 500, \quad 2 \le T \le 14$$

where T is the temperature of the food in degrees Celsius. When the food is removed from refrigeration, the temperature of the food is given by

$$T(t) = 4t + 2, \quad 0 \le t \le 3$$

where t is the time in hours. (a) Find and interpret $(N \circ T)(t)$. (b) Find the time when the bacteria count reaches 2000.

Solution

a.
$$
\begin{aligned}
(N \circ T)(t) &= N(T(t)) \\
&= 20(4t + 2)^2 - 80(4t + 2) + 500 \\
&= 20(16t^2 + 16t + 4) - 320t - 160 + 500 \\
&= 320t^2 + 320t + 80 - 320t - 160 + 500 \\
&= 320t^2 + 420
\end{aligned}
$$

The composite function $N \circ T$ represents the number of bacteria in the food as a function of the amount of time the food has been out of refrigeration.

b. The bacteria count will reach $N = 2000$ when $320t^2 + 420 = 2000$. You can solve this equation for t algebraically as shown.

$$320t^2 + 420 = 2000 \implies 320t^2 = 1580 \implies t^2 = \frac{79}{16}$$

Extract square roots to find that $t \approx \pm 2.22$. Reject the negative value because $0 \le t \le 3$. So, the count will reach 2000 when $t \approx 2.22$ hours.

✓ *Checkpoint* ▶ *Audio-video solution in English & Spanish at LarsonPrecalculus.com*

The number N of bacteria in a refrigerated food is given by

$$N(T) = 8T^2 - 14T + 200, \quad 2 \le T \le 12$$

where T is the temperature of the food in degrees Celsius. When the food is removed from refrigeration, the temperature of the food is given by

$$T(t) = 2t + 2, \quad 0 \le t \le 5$$

where t is the time in hours. Find (a) $(N \circ T)(t)$ and (b) the time when the bacteria count reaches 1000. ■

Summarize (Section 2.6)

1. Explain how to add, subtract, multiply, and divide functions *(page 214)*. For examples of finding arithmetic combinations of functions, see Examples 1–4.

2. Explain how to find the composition of one function with another function *(page 216)*. For examples that use compositions of functions, see Examples 5–7.

3. Describe a real-life example that uses a composition of functions *(page 218, Example 8)*.

2.6 Exercises

Vocabulary and Concept Check

In Exercises 1 and 2, fill in the blanks.

1. Two functions f and g can be combined by the arithmetic operations of
_____, _____, _____, and _____ to create new functions.

2. The _____ of the function f with the function g is $(f \circ g)(x) = f(g(x))$.

3. If $f(x) = x^2 + 1$ and $(fg)(x) = 2x(x^2 + 1)$, then what is $g(x)$?

4. If $(f \circ g)(x) = f(x^2 + 1)$, then what is $g(x)$?

Skills and Applications

Graphing the Sum of Two Functions   In Exercises 5 and 6, use the graphs of f and g to graph $h(x) = (f + g)(x)$. To print an enlarged copy of the graph, go to *MathGraphs.com*.

5.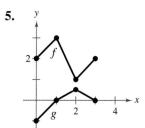

6.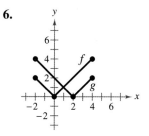

Finding Arithmetic Combinations of Functions   In Exercises 7–14, find (a) $(f + g)(x)$, (b) $(f - g)(x)$, (c) $(fg)(x)$, and (d) $(f/g)(x)$. What is the domain of f/g?

7. $f(x) = x + 2,\quad g(x) = x - 2$

8. $f(x) = 2x - 5,\quad g(x) = 2 - x$

9. $f(x) = x^2,\quad g(x) = 4x - 5$

10. $f(x) = 3x + 1,\quad g(x) = x^2 - 16$

11. $f(x) = x^2 + 6,\quad g(x) = \sqrt{1 - x}$

12. $f(x) = \sqrt{x^2 - 4},\quad g(x) = \sqrt{x + 2}$

13. $f(x) = \dfrac{x}{x + 1},\quad g(x) = x^3$

14. $f(x) = \dfrac{2}{x},\quad g(x) = \dfrac{1}{x^2 - 1}$

Evaluating an Arithmetic Combination of Functions   In Exercises 15–22, evaluate the function for $f(x) = x + 3$ and $g(x) = x^2 - 2$.

15. $(f + g)(2)$

16. $(f - g)(1)$

17. $(f - g)(3t)$

18. $(f + g)(t - 2)$

19. $(fg)(6)$

20. $(fg)(-6)$

21. $(f/g)(5)$

22. $(f/g)(0)$

Graphical Reasoning   In Exercises 23–26, use a graphing utility to graph f, g, and $f + g$ in the same viewing window. Which function contributes most to the magnitude of the sum when $0 \le x \le 2$? Which function contributes most to the magnitude of the sum when $x > 6$?

23. $f(x) = 3x,\quad g(x) = -\dfrac{x^3}{10}$

24. $f(x) = \dfrac{x}{2},\quad g(x) = \sqrt{x}$

25. $f(x) = 3x + 2,\quad g(x) = -\sqrt{x + 5}$

26. $f(x) = x^2 - \frac{1}{2},\quad g(x) = -3x^2 - 1$

Finding Compositions of Functions   In Exercises 27–30, find (a) $f \circ g$, (b) $g \circ f$, and (c) $g \circ g$.

27. $f(x) = x + 8,\quad g(x) = x - 3$

28. $f(x) = 3x,\quad g(x) = x^4$

29. $f(x) = \sqrt[3]{x - 1},\quad g(x) = x^3 + 1$

30. $f(x) = x^3,\quad g(x) = \dfrac{1}{x}$

Finding Domains of Functions and Composite Functions   In Exercises 31–36, find (a) $f \circ g$ and (b) $g \circ f$. Find the domain of each function and of each composite function.

31. $f(x) = \sqrt{x + 4},\quad g(x) = x^2$

32. $f(x) = \sqrt[3]{x - 5},\quad g(x) = x^3 + 1$

33. $f(x) = |x|,\quad g(x) = x + 6$

34. $f(x) = |x - 4|,\quad g(x) = 3 - x$

35. $f(x) = \dfrac{1}{x},\quad g(x) = x + 3$

36. $f(x) = \dfrac{3}{x^2 - 1},\quad g(x) = x + 1$

Graphing Combinations of Functions In Exercises 37 and 38, on the same set of coordinate axes, (a) graph the functions f, g, and $f + g$ and (b) graph the functions f, g, and $f \circ g$.

37. $f(x) = \frac{1}{2}x$, $g(x) = x - 4$

38. $f(x) = x + 3$, $g(x) = x^2$

Evaluating Combinations of Functions In Exercises 39–42, use the graphs of f and g to evaluate the functions.

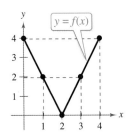
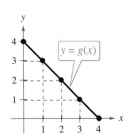

39. (a) $(f + g)(3)$ (b) $(f/g)(2)$

40. (a) $(f - g)(1)$ (b) $(fg)(4)$

41. (a) $(f \circ g)(2)$ (b) $(g \circ f)(2)$

42. (a) $(f \circ g)(1)$ (b) $(g \circ f)(3)$

Decomposing a Composite Function In Exercises 43–52, find two functions f and g such that $(f \circ g)(x) = h(x)$. (There are many correct answers.)

43. $h(x) = (2x + 1)^2$

44. $h(x) = (1 - x)^3$

45. $h(x) = \sqrt[3]{x^2 - 4}$ **46.** $h(x) = \sqrt{9 - x}$

47. $h(x) = \dfrac{1}{x + 2}$ **48.** $h(x) = \dfrac{4}{(5x + 2)^2}$

49. $h(x) = \dfrac{-x^2 + 3}{4 - x^2}$ **50.** $h(x) = \dfrac{27x^3 + 6x}{10 - 27x^3}$

51. $h(x) = \sqrt{\dfrac{1}{x^2 + 1}}$ **52.** $h(x) = \sqrt{\dfrac{x^2 - 4}{x^2 + 16}}$

53. Stopping Distance The research and development department of an automobile manufacturer determines that when a driver is required to stop quickly to avoid an accident, the distance (in feet) the car travels during the driver's reaction time is given by $R(x) = \frac{3}{4}x$, where x is the speed of the car in miles per hour. The distance (in feet) the car travels while the driver is braking is given by $B(x) = \frac{1}{15}x^2$.

(a) Find the function that represents the total stopping distance T.

(b) Graph the functions R, B, and T on the same set of coordinate axes for $0 \le x \le 60$.

(c) Which function contributes most to the magnitude of the sum at higher speeds? Explain.

54. Business The annual cost C (in thousands of dollars) and revenue R (in thousands of dollars) for a company each year from 2014 through 2020 can be approximated by the models

$$C = 145 - 9t + 1.1t^2 \quad \text{and} \quad R = 341 + 3.2t$$

where t is the year, with $t = 14$ corresponding to 2014.

(a) Write a function P that represents the annual profit of the company.

(b) Use a graphing utility to graph C, R, and P in the same viewing window.

55. Vital Statistics Let $b(t)$ be the number of births in the United States in year t, and let $d(t)$ represent the number of deaths in the United States in year t, where $t = 14$ corresponds to 2014.

(a) If $p(t)$ is the population of the United States in year t, find the function $c(t)$ that represents the percent change in the population of the United States.

(b) Interpret $c(20)$.

56. Pets

Let $d(t)$ be the number of dogs in the United States in year t, and let $c(t)$ be the number of cats in the United States in year t, where $t = 14$ corresponds to 2014.

(a) Find the function $p(t)$ that represents the total number of dogs and cats in the United States.

(b) Interpret $p(20)$.

(c) Let $n(t)$ represent the population of the United States in year t, where $t = 14$ corresponds to 2014. Find and interpret

$$h(t) = p(t)/n(t).$$

57. Geometry A square concrete foundation is a base for a cylindrical tank (see figure).

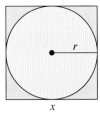

(a) Write the radius r of the tank as a function of the length x of the sides of the square.

(b) Write the area A of the circular base of the tank as a function of the radius r.

(c) Find and interpret $(A \circ r)(x)$.

58. Biology The number N of bacteria in a refrigerated food is given by

$$N(T) = 10T^2 - 20T + 600, \quad 2 \le T \le 20$$

where T is the temperature of the food in degrees Celsius. When the food is removed from refrigeration, the temperature of the food is given by

$$T(t) = 3t + 2, \quad 0 \le t \le 6$$

where t is the time in hours.

(a) Find and interpret $(N \circ T)(t)$.

(b) Find the bacteria count after 0.5 hour.

(c) Find the time when the bacteria count reaches 1500.

Exploring the Concepts

True or False? In Exercises 59 and 60, determine whether the statement is true or false. Justify your answer.

59. If $f(x) = x + 1$ and $g(x) = 6x$, then

$$(f \circ g)(x) = (g \circ f)(x).$$

60. When you are given two functions f and g and a constant c, you can find $(f \circ g)(c)$ if and only if $g(c)$ is in the domain of f.

61. Writing Functions Write two unique functions f and g such that $(f \circ g)(x) = (g \circ f)(x)$ and f and g are (a) linear functions and (b) polynomial functions with degrees greater than one.

62. 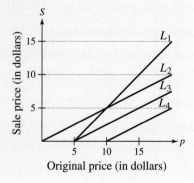 **HOW DO YOU SEE IT?** The graphs labeled L_1, L_2, L_3, and L_4 represent four different pricing discounts, where p is the original price (in dollars) and S is the sale price (in dollars). Match each function with its graph. Describe the situations in parts (c) and (d).

(a) $f(p)$: A 50% discount is applied.

(b) $g(p)$: A \$5 discount is applied.

(c) $(g \circ f)(p)$

(d) $(f \circ g)(p)$

63. Proof Prove that the product of two odd functions is an even function, and that the product of two even functions is an even function.

64. Proof

(a) Given a function f, prove that g is even and h is odd, where $g(x) = \frac{1}{2}[f(x) + f(-x)]$ and

$$h(x) = \frac{1}{2}[f(x) - f(-x)].$$

(b) Use the result of part (a) to prove that any function can be written as a sum of even and odd functions. [*Hint:* Add the two equations in part (a).]

(c) Use the result of part (b) to write each function as a sum of even and odd functions.

$$f(x) = x^2 - 2x + 1, \quad k(x) = \frac{1}{x + 1}$$

65. Conjecture Use examples to hypothesize whether the product of an odd function and an even function is even or odd. Then prove your hypothesis.

66. Error Analysis Describe the error.

Given $f(x) = 2x + 1$ and $g(x) = 3x + 2$, $(f \circ g)(x) = 3(2x + 1) + 2 = 6x + 5.$

Review & Refresh ▶ Video solutions at LarsonPrecalculus.com

Testing for Symmetry In Exercises 67–70, use the algebraic tests to check for symmetry with respect to both axes and the origin.

67. $y^2 = x - 5$ **68.** $2x + 5y = 17$

69. $y = x^2 + 1$ **70.** $x^2 + y^2 = 72$

Even, Odd, or Neither? In Exercises 71–74, sketch a graph of the function and determine whether it is even, odd, or neither. Verify your answer algebraically.

71. $y = 5x$ **72.** $y = 3x^2$

73. $y = x - 5$ **74.** $y = x^3 + 4x$

Solving for y In Exercises 75–78, solve the equation for y.

75. $2x + 3y = 5$ **76.** $xy - 1 = 3y + x$

77. $x = \sqrt{y + 1}$ **78.** $x = \dfrac{5 - y}{3y + 2}$

Geometry In Exercises 79 and 80, write an expression in factored form for the area of the shaded portion of the figure.

79.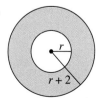

80.

2.7 Inverse Functions

❯ Find inverse functions informally and verify that two functions are inverse functions of each other.
❯ Use graphs to verify that two functions are inverse functions of each other.
❯ Use the Horizontal Line Test to determine whether functions are one-to-one.
❯ Find inverse functions algebraically.

Inverse functions can help you model and solve real-life problems. For example, in Exercise 70 on page 229, you will write an inverse function and use it to determine the percent load interval for a diesel engine.

Inverse Functions

Recall from Section 2.2 that a function can be represented by a set of ordered pairs. For example, the function $f(x) = x + 4$ from the set $A = \{1, 2, 3, 4\}$ to the set $B = \{5, 6, 7, 8\}$ can be written as

$$f(x) = x + 4 \colon \{(1, 5), (2, 6), (3, 7), (4, 8)\}.$$

In this case, by interchanging the first and second coordinates of each ordered pair, you form the **inverse function** of f, which is denoted by f^{-1}. It is a function from the set B to the set A and can be written as

$$f^{-1}(x) = x - 4 \colon \{(5, 1), (6, 2), (7, 3), (8, 4)\}.$$

Note that the domain of f is equal to the range of f^{-1}, and vice versa, as shown in the figure below. Also note that the functions f and f^{-1} have the effect of "undoing" each other. In other words, when you form the composition of f with f^{-1} or the composition of f^{-1} with f, you obtain the identity function.

$$f(f^{-1}(x)) = f(x - 4) = (x - 4) + 4 = x$$
$$f^{-1}(f(x)) = f^{-1}(x + 4) = (x + 4) - 4 = x$$

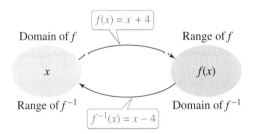

Domain of f		Range of f

$f(x) = x + 4$

x $\quad\quad$ $f(x)$

Range of f^{-1} $\quad\quad$ Domain of f^{-1}

$f^{-1}(x) = x - 4$

EXAMPLE 1   **Finding an Inverse Function Informally**

Find the inverse function of $f(x) = 4x$. Then verify that both $f(f^{-1}(x))$ and $f^{-1}(f(x))$ are equal to the identity function.

Solution   The function f *multiplies* each input by 4. To "undo" this function, you need to *divide* each input by 4. So, the inverse function of $f(x) = 4x$ is

$$f^{-1}(x) = \frac{x}{4}.$$

Verify that $f(f^{-1}(x)) = x$ and $f^{-1}(f(x)) = x$.

$$f(f^{-1}(x)) = f\left(\frac{x}{4}\right) = 4\left(\frac{x}{4}\right) = x \qquad f^{-1}(f(x)) = f^{-1}(4x) = \frac{4x}{4} = x$$

✓ *Checkpoint* ▶ *Audio-video solution in English & Spanish at LarsonPrecalculus.com*

Find the inverse function of $f(x) = \frac{1}{5}x$. Then verify that both $f(f^{-1}(x))$ and $f^{-1}(f(x))$ are equal to the identity function.

GO DIGITAL

> ### Definition of Inverse Function
>
> Let f and g be two functions such that
>
> $$f(g(x)) = x \quad \text{for every } x \text{ in the domain of } g$$
>
> and
>
> $$g(f(x)) = x \quad \text{for every } x \text{ in the domain of } f.$$
>
> Under these conditions, the function g is the **inverse function** of the function f. The function g is denoted by f^{-1} (read "f-inverse"). So,
>
> $$f(f^{-1}(x)) = x \quad \text{and} \quad f^{-1}(f(x)) = x.$$
>
> The domain of f must be equal to the range of f^{-1}, and the range of f must be equal to the domain of f^{-1}.

Do not be confused by the use of -1 to denote the inverse function f^{-1}. In this text, whenever f^{-1} is written, it *always* refers to the inverse function of the function f and *not* to the reciprocal of $f(x)$.

If the function g is the inverse function of the function f, then it must also be true that the function f is the inverse function of the function g. So, it is correct to say that the functions f and g are *inverse functions of each other*.

EXAMPLE 2   Verifying Inverse Functions

Which of the functions is the inverse function of $f(x) = \dfrac{5}{x - 2}$?

$$g(x) = \frac{x - 2}{5} \qquad h(x) = \frac{5}{x} + 2$$

Solution   By forming the composition of f with g, you have

$$f(g(x)) = f\left(\frac{x - 2}{5}\right) = \frac{5}{\left(\dfrac{x - 2}{5}\right) - 2} = \frac{25}{x - 12} \neq x.$$

This composition is not equal to the identity function x, so g *is not* the inverse function of f. By forming the composition of f with h, you have

$$f(h(x)) = f\left(\frac{5}{x} + 2\right) = \frac{5}{\left(\dfrac{5}{x} + 2\right) - 2} = \frac{5}{\left(\dfrac{5}{x}\right)} = x.$$

So, it appears that h *is* the inverse function of f. Confirm this by showing that the composition of h with f is also equal to the identity function.

$$h(f(x)) = h\left(\frac{5}{x - 2}\right) = \frac{5}{\left(\dfrac{5}{x - 2}\right)} + 2 = x - 2 + 2 = x$$

Check to see that the domain of f is the same as the range of h and vice versa.

✓ *Checkpoint* ▶ *Audio-video solution in English & Spanish at LarsonPrecalculus.com*

Which of the functions is the inverse function of $f(x) = \dfrac{x - 4}{7}$?

$$g(x) = 7x + 4 \qquad h(x) = \frac{7}{x - 4}$$

GO DIGITAL

The Graph of an Inverse Function

The graphs of a function f and its inverse function f^{-1} are related to each other in this way: If the point (a, b) lies on the graph of f, then the point (b, a) must lie on the graph of f^{-1}, and vice versa. This means that the graph of f^{-1} is a *reflection* of the graph of f in the line $y = x$, as shown in Figure 2.37.

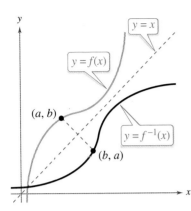

Figure 2.37

EXAMPLE 3   Verifying Inverse Functions Graphically

Verify graphically that the functions $f(x) = 2x - 3$ and $g(x) = \frac{1}{2}(x + 3)$ are inverse functions of each other.

Solution   Sketch the graphs of f and g on the same rectangular coordinate system, as shown in Figure 2.38. It appears that the graphs are reflections of each other in the line $y = x$. Further verify this reflective property by testing a few points on each graph. Note that for each point (a, b) on the graph of f, the point (b, a) is on the graph of g.

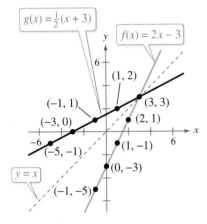

Figure 2.38

Graph of $f(x) = 2x - 3$	Graph of $g(x) = \frac{1}{2}(x + 3)$
$(-1, -5)$	$(-5, -1)$
$(0, -3)$	$(-3, 0)$
$(1, -1)$	$(-1, 1)$
$(2, 1)$	$(1, 2)$
$(3, 3)$	$(3, 3)$

The graphs of f and g are reflections of each other in the line $y = x$. So, f and g are inverse functions of each other.

✓ *Checkpoint* ▶ Audio-video solution in English & Spanish at LarsonPrecalculus.com

Verify graphically that the functions $f(x) = 4x - 1$ and $g(x) = \frac{1}{4}(x + 1)$ are inverse functions of each other.

EXAMPLE 4   Verifying Inverse Functions Graphically

Verify graphically that the functions $f(x) = x^2$ $(x \geq 0)$ and $g(x) = \sqrt{x}$ are inverse functions of each other.

Solution   Sketch the graphs of f and g on the same rectangular coordinate system, as shown in Figure 2.39. It appears that the graphs are reflections of each other in the line $y = x$. Test a few points on each graph.

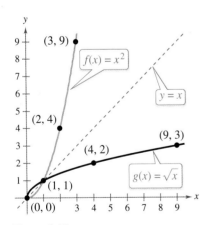

Figure 2.39

Graph of $f(x) = x^2$, $x \geq 0$	Graph of $g(x) = \sqrt{x}$
$(0, 0)$	$(0, 0)$
$(1, 1)$	$(1, 1)$
$(2, 4)$	$(4, 2)$
$(3, 9)$	$(9, 3)$

The graphs of f and g are reflections of each other in the line $y = x$. So, f and g are inverse functions of each other.

✓ *Checkpoint* ▶ Audio-video solution in English & Spanish at LarsonPrecalculus.com

Verify graphically that the functions $f(x) = x^2 + 1$ $(x \geq 0)$ and $g(x) = \sqrt{x - 1}$ are inverse functions of each other. ■

GO DIGITAL

One-to-One Functions

The reflective property of the graphs of inverse functions gives you a graphical test for determining whether a function has an inverse function. This test is the **Horizontal Line Test** for inverse functions.

> ### Horizontal Line Test for Inverse Functions
>
> A function f has an inverse function if and only if no *horizontal* line intersects the graph of f at more than one point.

If no horizontal line intersects the graph of f at more than one point, then no y-value corresponds to more than one x-value. This is the essential characteristic of **one-to-one functions.**

> ### One-to-One Functions
>
> A function f is **one-to-one** when each value of the dependent variable corresponds to exactly one value of the independent variable. A function f has an inverse function if and only if f is one-to-one.

Consider the table of values for the function $f(x) = x^2$ on the left. The output $f(x) = 4$ corresponds to two inputs, $x = -2$ and $x = 2$, so f is not one-to-one. In the table on the right, x and y are interchanged. Here $x = 4$ corresponds to both $y = -2$ and $y = 2$, so this table does not represent a function. So, $f(x) = x^2$ is not one-to-one and does not have an inverse function.

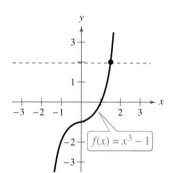

Figure 2.40

x	$f(x) = x^2$
-2	4
-1	1
0	0
1	1
2	4
3	9

x	y
4	-2
1	-1
0	0
1	1
4	2
9	3

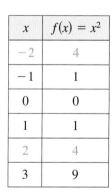

Figure 2.41

EXAMPLE 5   Applying the Horizontal Line Test

▶▶▶ *See LarsonPrecalculus.com for an interactive version of this type of example.*

a. The graph of the function $f(x) = x^3 - 1$ is shown in Figure 2.40. No horizontal line intersects the graph of f at more than one point, so f *is* a one-to-one function and *does* have an inverse function.

b. The graph of the function $f(x) = x^2 - 1$ is shown in Figure 2.41. It is possible to find a horizontal line that intersects the graph of f at more than one point, so f *is not* a one-to-one function and *does not* have an inverse function.

✓ *Checkpoint* ▶ *Audio-video solution in English & Spanish at LarsonPrecalculus.com*

Use the graph of f to determine whether the function has an inverse function.

a. $f(x) = \frac{1}{2}(3 - x)$ b. $f(x) = |x|$

GO DIGITAL

Note what happens when you try to find the inverse function of a function that is not one-to-one.

$$f(x) = x^2 + 1 \qquad \text{Original function}$$

$$y = x^2 + 1 \qquad \text{Replace } f(x) \text{ with } y.$$

$$x = y^2 + 1 \qquad \text{Interchange } x \text{ and } y.$$

$$x - 1 = y^2 \qquad \text{Isolate } y\text{-term.}$$

$$y = \pm\sqrt{x - 1} \qquad \text{Solve for } y.$$

You obtain two y-values for each x.

Finding Inverse Functions Algebraically

For relatively simple functions (such as the one in Example 1), you can find inverse functions by inspection. For more complicated functions, however, it is best to use the guidelines below. The key step in these guidelines is Step 3—interchanging the roles of x and y. This step corresponds to the fact that inverse functions have ordered pairs with the coordinates reversed.

Finding an Inverse Function

1. Use the Horizontal Line Test to decide whether f has an inverse function.
2. In the equation for $f(x)$, replace $f(x)$ with y.
3. Interchange the roles of x and y, and solve for y.
4. Replace y with $f^{-1}(x)$ in the new equation.
5. Verify that f and f^{-1} are inverse functions of each other by showing that the domain of f is equal to the range of f^{-1}, the range of f is equal to the domain of f^{-1}, and $f(f^{-1}(x)) = x$ and $f^{-1}(f(x)) = x$.

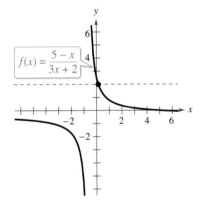

$$f(x) = \dfrac{5 - x}{3x + 2}$$

Figure 2.42

EXAMPLE 6   **Finding an Inverse Function Algebraically**

Find the inverse function of

$$f(x) = \frac{5 - x}{3x + 2}.$$

Solution   The graph of f is shown in Figure 2.42. This graph passes the Horizontal Line Test. So, you know that f is one-to-one and has an inverse function.

$$f(x) = \frac{5 - x}{3x + 2} \qquad \text{Write original function.}$$

$$y = \frac{5 - x}{3x + 2} \qquad \text{Replace } f(x) \text{ with } y.$$

$$x = \frac{5 - y}{3y + 2} \qquad \text{Interchange } x \text{ and } y.$$

$$x(3y + 2) = 5 - y \qquad \text{Multiply each side by } 3y + 2.$$

$$3xy + 2x = 5 - y \qquad \text{Distributive Property}$$

$$3xy + y = 5 - 2x \qquad \text{Collect terms with } y.$$

$$y(3x + 1) = 5 - 2x \qquad \text{Factor.}$$

$$y = \frac{5 - 2x}{3x + 1} \qquad \text{Solve for } y.$$

$$f^{-1}(x) = \frac{5 - 2x}{3x + 1} \qquad \text{Replace } y \text{ with } f^{-1}(x).$$

Verify that $f(f^{-1}(x)) = x$ and $f^{-1}(f(x)) = x$.

✓ *Checkpoint* ▶ *Audio-video solution in English & Spanish at LarsonPrecalculus.com*

Find the inverse function of

$$f(x) = \frac{5 - 3x}{x + 2}.$$

GO DIGITAL

EXAMPLE 7 **Finding an Inverse Function Algebraically**

Find the inverse function of

$$f(x) = \sqrt{2x - 3}.$$

Solution The graph of f is shown in the figure below. This graph passes the Horizontal Line Test. So, you know that f is one-to-one and has an inverse function.

$f(x) = \sqrt{2x - 3}$	Write original function.
$y = \sqrt{2x - 3}$	Replace $f(x)$ with y.
$x = \sqrt{2y - 3}$	Interchange x and y.
$x^2 = 2y - 3$	Square each side.
$2y = x^2 + 3$	Isolate y-term.
$y = \dfrac{x^2 + 3}{2}$	Solve for y.
$f^{-1}(x) = \dfrac{x^2 + 3}{2}, \ x \geq 0$	Replace y with $f^{-1}(x)$.

The graph of f^{-1} in the figure is the reflection of the graph of f in the line $y = x$. Note that the range of f is the interval $[0, \infty)$, which implies that the domain of f^{-1} is the interval

$$[0, \infty). \qquad \text{Domain of } f^{-1}$$

Moreover, the domain of f is the interval $\left[\frac{3}{2}, \infty\right)$, which implies that the range of f^{-1} is the interval

$$\left[\frac{3}{2}, \infty\right). \qquad \text{Range of } f^{-1}$$

Verify that $f(f^{-1}(x)) = x$ and $f^{-1}(f(x)) = x$.

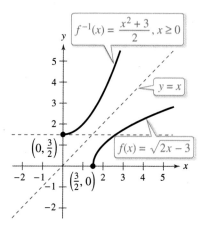

✓ **Checkpoint** ▶ *Audio-video solution in English & Spanish at LarsonPrecalculus.com*

Find the inverse function of

$$f(x) = \sqrt[3]{10 + x}.$$

Summarize (Section 2.7)

1. State the definition of an inverse function *(page 223)*. For examples of finding inverse functions informally and verifying inverse functions, see Examples 1 and 2.

2. Explain how to use graphs to verify that two functions are inverse functions of each other *(page 224)*. For examples of verifying inverse functions graphically, see Examples 3 and 4.

3. Explain how to use the Horizontal Line Test to determine whether a function is one-to-one *(page 225)*. For an example of applying the Horizontal Line Test, see Example 5.

4. Explain how to find an inverse function algebraically *(page 226)*. For examples of finding inverse functions algebraically, see Examples 6 and 7.

GO DIGITAL

2.7 Exercises

See CalcChat.com for tutorial help and worked-out solutions to odd-numbered exercises.

Vocabulary and Concept Check

In Exercises 1–4, fill in the blanks.

1. If $f(g(x))$ and $g(f(x))$ both equal x, then the function g is the _____ function of the function f.
2. The inverse function of f is denoted by _____.
3. The domain of f is the _____ of f^{-1}, and the _____ of f^{-1} is the range of f.
4. The graphs of f and f^{-1} are reflections of each other in the line _____.

5. To show that two functions f and g are inverse functions, you must show that both $f(g(x))$ and $g(f(x))$ are equal to what?
6. Can $(1, 4)$ and $(2, 4)$ be two ordered pairs of a one-to-one function?
7. How many times can a horizontal line intersect the graph of a function that is one-to-one?
8. Give an example of a function that does not pass the Horizontal Line Test.

Skills and Applications

Finding an Inverse Function Informally In Exercises 9–16, find the inverse function of f informally. Verify that $f(f^{-1}(x)) = x$ and $f^{-1}(f(x)) = x$.

9. $f(x) = 6x$

10. $f(x) = \dfrac{1}{3}x$

11. $f(x) = 3x + 1$

12. $f(x) = \dfrac{x - 3}{2}$

13. $f(x) = x^3 + 1$

14. $f(x) = \dfrac{x^5}{4}$

15. $f(x) = x^2 - 4, \ x \ge 0$

16. $f(x) = x^2 + 2, \ x \ge 0$

Verifying Inverse Functions In Exercises 17–20, verify that f and g are inverse functions algebraically.

17. $f(x) = \dfrac{x - 9}{4}, \quad g(x) = 4x + 9$

18. $f(x) = -\dfrac{3}{2}x - 4, \quad g(x) = -\dfrac{2x + 8}{3}$

19. $f(x) = \dfrac{x^3}{4}, \quad g(x) = \sqrt[3]{4x}$

20. $f(x) = x^3 + 5, \quad g(x) = \sqrt[3]{x - 5}$

Sketching the Graph of an Inverse Function In Exercises 21 and 22, use the graph of the function to sketch the graph of its inverse function $y = f^{-1}(x)$.

21.

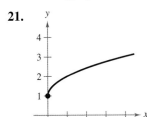

22.

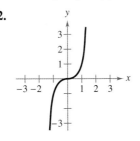

Verifying Inverse Functions In Exercises 23–32, verify that f and g are inverse functions (a) algebraically and (b) graphically.

23. $f(x) = x - 5, \quad g(x) = x + 5$

24. $f(x) = 2x, \quad g(x) = \dfrac{x}{2}$

25. $f(x) = 7x + 1, \quad g(x) = \dfrac{x - 1}{7}$

26. $f(x) = 3 - 4x, \quad g(x) = \dfrac{3 - x}{4}$

27. $f(x) = x^3, \quad g(x) = \sqrt[3]{x}$

28. $f(x) = \dfrac{x^3}{3}, \quad g(x) = \sqrt[3]{3x}$

29. $f(x) = \sqrt{x + 5}, \quad g(x) = x^2 - 5, \quad x \ge 0$

30. $f(x) = 1 - x^3, \quad g(x) = \sqrt[3]{1 - x}$

31. $f(x) = \dfrac{x - 1}{x + 5}, \quad g(x) = -\dfrac{5x + 1}{x - 1}$

32. $f(x) = \dfrac{x + 3}{x - 2}, \quad g(x) = \dfrac{2x + 3}{x - 1}$

Using a Table to Determine an Inverse Function In Exercises 33 and 34, does the function have an inverse function?

33.

x	-1	0	1	2	3	4
$f(x)$	-2	1	2	1	-2	-6

34.

x	-3	-2	-1	0	2	3
$f(x)$	10	6	4	1	-3	-10

Using a Table to Find an Inverse Function In Exercises 35 and 36, use the table of values for $y = f(x)$ to complete a table for $y = f^{-1}(x)$.

35.

x	-1	0	1	2	3	4
$f(x)$	3	5	7	9	11	13

36.

x	-3	-2	-1	0	1	2
$f(x)$	10	5	0	-5	-10	-15

Applying the Horizontal Line Test In Exercises 37–40, does the function have an inverse function?

37.

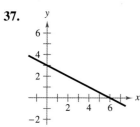

38.

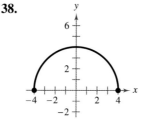

39.

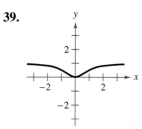

40.

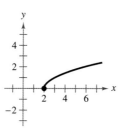

Applying the Horizontal Line Test In Exercises 41–44, use a graphing utility to graph the function, and use the Horizontal Line Test to determine whether the function has an inverse function.

41. $g(x) = (x + 3)^2 + 2$ **42.** $f(x) = \frac{1}{5}(x + 2)^3$

43. $f(x) = x\sqrt{9 - x^2}$

44. $h(x) = |x| - |x - 4|$

Finding and Analyzing Inverse Functions In Exercises 45–54, (a) find the inverse function of f, (b) graph both f and f^{-1} on the same set of coordinate axes, (c) describe the relationship between the graphs of f and f^{-1}, and (d) state the domains and ranges of f and f^{-1}.

45. $f(x) = x^5 - 2$

46. $f(x) = x^3 + 8$

47. $f(x) = \sqrt{4 - x^2}$, $0 \le x \le 2$

48. $f(x) = x^2 - 2$, $x \le 0$

49. $f(x) = \dfrac{4}{x}$ **50.** $f(x) = -\dfrac{2}{x}$

51. $f(x) = \dfrac{x + 1}{x - 2}$ **52.** $f(x) = \dfrac{x - 2}{3x + 5}$

53. $f(x) = \sqrt[3]{x - 1}$ **54.** $f(x) = x^{3/5}$

Finding an Inverse Function In Exercises 55–68, determine whether the function has an inverse function. If it does, find the inverse function.

55. $f(x) = x^4$ **56.** $f(x) = \dfrac{1}{x^2}$

57. $g(x) = \dfrac{x + 1}{6}$ **58.** $f(x) = 3x + 5$

59. $p(x) = -4$ **60.** $f(x) = 0$

61. $f(x) = \sqrt{2x + 3}$ **62.** $f(x) = \sqrt{x - 2}$

63. $f(x) = \dfrac{6x + 4}{4x + 5}$ **64.** $f(x) = \dfrac{5x - 3}{2x + 5}$

65. $f(x) = (x + 3)^2$, $x \ge -3$

66. $f(x) = |x - 2|$, $x \le 2$

67. $f(x) = \begin{cases} x + 3, & x < 0 \\ 6 - x, & x \ge 0 \end{cases}$

68. $f(x) = \begin{cases} -x, & x \le 0 \\ x^2 - 3x, & x > 0 \end{cases}$

69. Hourly Wage Your wage is \$10.00 per hour plus \$0.75 for each unit produced per hour. So, your hourly wage y in terms of the number of units produced x is $y = 10 + 0.75x$.

(a) Find the inverse function. What does each variable represent in the inverse function?

(b) Determine the number of units produced when your hourly wage is \$24.25.

70. Diesel Mechanics

The function

$$y = 0.03x^2 + 245.50, \quad 0 < x < 100$$

approximates the exhaust temperature y in degrees Fahrenheit, where x is the percent load for a diesel engine.

(a) Find the inverse function. What does each variable represent in the inverse function?

(b) Use a graphing utility to graph the inverse function.

(c) The exhaust temperature of the engine must not exceed 500 degrees Fahrenheit. What is the percent load interval?

Composition with Inverses In Exercises 71–74, use the functions $f(x) = x + 4$ and $g(x) = 2x - 5$ to find the function.

71. $g^{-1} \circ f^{-1}$ **72.** $f^{-1} \circ g^{-1}$

73. $(f \circ g)^{-1}$ **74.** $(g \circ f)^{-1}$

Composition with Inverses In Exercises 75–80, use the functions $f(x) = \frac{1}{8}x - 3$ and $g(x) = x^3$ to find the value or function.

75. $(f^{-1} \circ g^{-1})(1)$

76. $(g^{-1} \circ f^{-1})(-3)$

77. $(f^{-1} \circ f^{-1})(4)$

78. $(g^{-1} \circ g^{-1})(-1)$

79. $(f \circ g)^{-1}$

80. $g^{-1} \circ f^{-1}$

Exploring the Concepts

True or False? In Exercises 81 and 82, determine whether the statement is true or false. Justify your answer.

81. If f is an even function, then f^{-1} exists.

82. If the inverse function of f exists and the graph of f has a y-intercept, then the y-intercept of f is an x-intercept of f^{-1}.

83. **Creating a Table** Use the graph of each function f to create a table of values for the given points. Then create a second table that can be used to find f^{-1}, and sketch the graph of f^{-1}, if possible.

(a)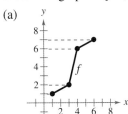

(b)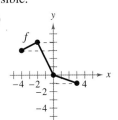

84. **HOW DO YOU SEE IT?** The cost C for a business to make personalized T-shirts is given by

$$C(x) = 7.50x + 1500$$

where x represents the number of T-shirts.

(a) The graphs of C and C^{-1} are shown below. Match each function with its graph.

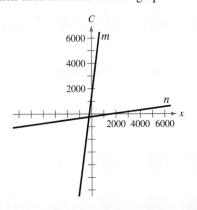

(b) Explain what $C(x)$ and $C^{-1}(x)$ represent in the context of the problem.

85. **Proof** Prove that if f and g are one-to-one functions, then $(f \circ g)^{-1}(x) = (g^{-1} \circ f^{-1})(x)$.

86. **Proof** Prove that if f is a one-to-one odd function, then f^{-1} is an odd function.

87. **Think About It** Restrict the domain of $f(x) = x^2 + 1$ to $x \geq 0$. Use a graphing utility to graph the function. Does the restricted function have an inverse function? Explain.

88. **Think About It** Consider the functions $f(x) = x + 2$ and $f^{-1}(x) = x - 2$. Evaluate $f(f^{-1}(x))$ and $f^{-1}(f(x))$ for the given values of x. What can you conclude about the functions?

x	-10	0	7	45
$f(f^{-1}(x))$				
$f^{-1}(f(x))$				

Review & Refresh ▶ Video solutions at LarsonPrecalculus.com

Factoring a Quadratic Function In Exercises 89–92, factor the quadratic function. Check your solutions.

89. $y = -(x - 5)^2 + 1$

90. $y = 3(x - 4)^2 - 12$

91. $y = -\left(x - \frac{13}{2}\right)^2 + \frac{25}{4}$

92. $y = 3\left(x + \frac{1}{4}\right)^2 - \frac{363}{16}$

Describing Function Behavior In Exercises 93–96, determine the open intervals on which the function is increasing, decreasing, or constant.

93.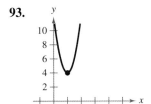

94.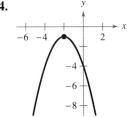

95.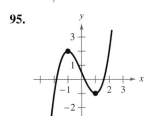

96.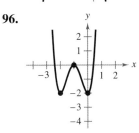

Approximating Relative Minima or Maxima In Exercises 97 and 98, approximate any relative minima or maxima of the function.

97.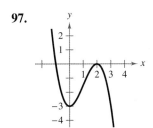

98.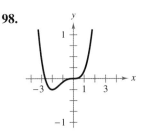

Summary and Study Strategies

GO DIGITAL

What Did You Learn?

The list below reviews the skills covered in the chapter and correlates each one to the Review Exercises (see page 233) that practice the skill.

Section 2.1	Review Exercises
■ Use slope to graph linear equations in two variables *(p. 160)*. Slope-intercept form: $y = mx + b$	*1–4*
■ Find the slope of a line given two points on the line *(p. 162)*.	*5, 6*
■ Write linear equations in two variables *(p. 164)*. Point-slope form: $y - y_1 = m(x - x_1)$	*7–10*
■ Use slope to identify parallel and perpendicular lines *(p. 165)*. Parallel lines: $m_1 = m_2$ Perpendicular lines: $m_1 = -1/m_2$	*11, 12*
■ Use slope and linear equations in two variables to model and solve real-life problems *(p. 166)*.	*13, 14*

Section 2.2	
■ Determine whether relations between two variables are functions, and use function notation *(p. 173)*.	*15–20*
■ Find the domains of functions *(p. 178)*.	*21, 22*
■ Use functions to model and solve real-life problems, and evaluate difference quotients *(p. 179)*.	*23–26*

Section 2.3	
■ Use the Vertical Line Test for functions *(p. 187)*. A set of points in a coordinate plane is the graph of y as a function of x if and only if no *vertical* line intersects the graph at more than one point.	*27, 28*
■ Find the zeros of functions *(p. 189)*. The zeros of a function $y = f(x)$ are the x-values for which $f(x) = 0$.	*29, 30*
■ Determine intervals on which functions are increasing or decreasing *(p. 190)*.	*31, 32*
■ Determine relative minimum and relative maximum values of functions *(p. 191)*.	*33, 34*
■ Determine the average rate of change of a function *(p. 192)*.	*35, 36*
■ Identify even and odd functions *(p. 193)*. In an even function, $f(-x) = f(x)$. In an odd function, $f(-x) = -f(x)$.	*37–40*

Section 2.4	
■ Identify and graph linear and squaring functions *(p. 198)*.	*41–43*
■ Identify and graph cubic, square root, and reciprocal functions *(p. 200)*.	*44–46*
■ Identify and graph step and other piecewise-defined functions *(p. 201)*.	*47–50*
■ Recognize graphs of commonly used parent functions *(p. 202)*.	*43–46*

Section 2.5	**Review Exercises**
■ Use vertical and horizontal shifts to sketch graphs of functions *(p. 205)*. Vertical shifts: $h(x) = f(x) \pm c$ Horizontal shifts: $h(x) = f(x \pm c)$	*51–59*
■ Use reflections to sketch graphs of functions *(p. 207)*. Reflection in the x-axis: $h(x) = -f(x)$ Reflection in the y-axis: $h(x) = f(-x)$	*54, 55, 57, 59, 60*
■ Use nonrigid transformations to sketch graphs of functions *(p. 209)*. $g(x) = cf(x)$: vertical stretch when $c > 1$, vertical shrink when $0 < c < 1$. $h(x) = f(cx)$: horizontal shrink when $c > 1$, horizontal stretch when $0 < c < 1$.	*58, 60*

Section 2.6	
■ Add, subtract, multiply, and divide functions *(p. 214)*.	*61, 62*
■ Find the composition of one function with another function *(p. 216)*.	*63, 64*
■ Use combinations and compositions of functions to model and solve real-life problems *(p. 218)*.	*65, 66*

Section 2.7	
■ Find inverse functions informally and verify that two functions are inverse functions of each other *(p. 222)*.	*67, 68*
■ Use graphs to verify that two functions are inverse functions of each other *(p. 224)*.	*69–72*
■ Use the Horizontal Line Test to determine whether functions are one-to-one *(p. 225)*.	*69, 70*
■ Find inverse functions algebraically *(p. 226)*.	*71–74*

Study Strategies

Knowing and Using Your Preferred Learning Modality   Math is a specific system of rules, properties, and calculations used to solve problems. However, you can take different approaches to learning this specific system based on *learning modalities*. A learning modality is a preferred way of taking in information that is then transferred into the brain for processing. Three basic modalities are visual, auditory, and kinesthetic. A brief description of each modality and several ways to use it are listed below. You may find that one approach, or multiple approaches, works best for you.

- **Visual**   You take in information more productively when you see the information.

  1. Draw a diagram of a word problem before writing a verbal model.

  2. When making a review card for a word problem, include a picture. This will help you recall the information while taking a test.

- **Auditory**   You take in information more productively when you listen to an explanation and talk about it.

  1. Explain how to do a word problem to another student. This is a form of thinking out loud. Write the instructions down on a review card.

  2. Teach the material to an imaginary person when studying alone.

- **Kinesthetic**   You take in information more productively when you experience it or use physical activity in studying.

  1. Act out a word problem as much as possible. Use props when you can.

  2. Solve a problem on a large whiteboard—writing is more kinesthetic when the writing is larger and you move around while doing it.

  3. Make a review card.

Review Exercises

See CalcChat.com for tutorial help and worked-out solutions to odd-numbered exercises.

GO DIGITAL

2.1 Graphing a Linear Equation In Exercises 1–4, find the slope and y-intercept (if possible) of the line. Sketch the line.

1. $y = -\frac{1}{2}x + 1$

2. $2x - 3y = 6$

3. $y = 1$

4. $x = -6$

Finding the Slope of a Line Through Two Points In Exercises 5 and 6, find the slope of the line passing through the pair of points.

5. $(5, -2), (-1, 4)$

6. $(-1, 6), (3, -2)$

Using the Point-Slope Form In Exercises 7 and 8, find the slope-intercept form of the equation of the line that has the given slope and passes through the given point. Sketch the line.

7. $m = \frac{1}{3}$, $(6, -5)$

8. $m = -\frac{3}{4}$, $(-4, -2)$

Finding an Equation of a Line In Exercises 9 and 10, find an equation of the line passing through the pair of points. Sketch the line.

9. $(-6, 4), (4, 9)$

10. $(-9, -3), (-3, -5)$

Finding Parallel and Perpendicular Lines In Exercises 11 and 12, find equations of the lines that pass through the given point and are (a) parallel to and (b) perpendicular to the given line.

11. $5x - 4y = 8$, $(3, -2)$ 12. $2x + 3y = 5$, $(-8, 3)$

13. **Sales** A discount outlet offers a 20% discount on all items. Write a linear equation giving the sale price S for an item with a list price L.

14. **Wage** A manuscript translator charges a starting fee of \$50 plus \$2.50 per page translated. Write a linear equation for the amount A earned for translating p pages.

2.2 Testing for Functions Represented Algebraically In Exercises 15–18, determine whether the equation represents y as a function of x.

15. $16x - y^4 = 0$

16. $2x - y - 3 = 0$

17. $y = \sqrt{1 - x}$

18. $|y| = x + 2$

Evaluating a Function In Exercises 19 and 20, find each function value.

19. $g(x) = x^{4/3}$

   (a) $g(8)$ (b) $g(t + 1)$ (c) $g(-27)$ (d) $g(-x)$

20. $h(x) = |x - 2|$

   (a) $h(-4)$ (b) $h(-2)$ (c) $h(0)$ (d) $h(-x + 2)$

Finding the Domain of a Function In Exercises 21 and 22, find the domain of the function.

21. $f(x) = \sqrt{25 - x^2}$

22. $h(x) = \dfrac{x}{x^2 - x - 6}$

Physics In Exercises 23 and 24, the velocity of a ball projected upward from ground level is given by $v(t) = -32t + 48$, where t is the time in seconds and v is the velocity in feet per second.

23. Find the velocity when $t = 1$.

24. Find the time when the ball reaches its maximum height. [*Hint:* Find the time when $v(t) = 0$.]

Evaluating a Difference Quotient In Exercises 25 and 26, find the difference quotient and simplify your answer.

25. $f(x) = 2x^2 + 3x - 1$, $\dfrac{f(x + h) - f(x)}{h}$, $h \neq 0$

26. $f(x) = x^3 - 5x^2 + x$, $\dfrac{f(x + h) - f(x)}{h}$, $h \neq 0$

2.3 Vertical Line Test for Functions In Exercises 27 and 28, use the Vertical Line Test to determine whether the graph represents y as a function of x.

27.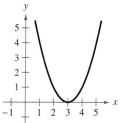

28.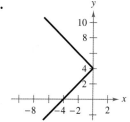

Finding the Zeros of a Function In Exercises 29 and 30, find the zeros of the function algebraically.

29. $f(x) = 5x^2 + 4x - 1$ 30. $f(x) = x^3 - x^2$

Describing Function Behavior In Exercises 31 and 32, use a graphing utility to graph the function and visually determine the open intervals on which the function is increasing, decreasing, or constant. Use a table of values to verify your results.

31. $f(x) = |x| + |x + 1|$ 32. $f(x) = (x^2 - 4)^2$

Approximating Relative Minima or Maxima In Exercises 33 and 34, use a graphing utility to approximate (to two decimal places) any relative minima or maxima of the function.

33. $f(x) = -x^2 + 2x + 1$ 34. $f(x) = x^3 - 4x^2 - 1$

Average Rate of Change of a Function In Exercises 35 and 36, find the average rate of change of the function from x_1 to x_2.

Function	x-Values
35. $f(x) = -x^2 + 8x - 4$	$x_1 = 0, x_2 = 4$
36. $f(x) = x^3 + 2x + 1$	$x_1 = 1, x_2 = 3$

Even, Odd, or Neither? In Exercises 37–40, determine whether the function is even, odd, or neither. Then describe the symmetry.

37. $f(x) = x^5 + 4x - 7$ **38.** $f(x) = x^4 - 20x^2$

39. $f(x) = 2x\sqrt{x^2 + 3}$ **40.** $f(x) = \sqrt[5]{6x^2}$

2.4 **Writing a Linear Function** In Exercises 41 and 42, (a) write the linear function f that has the given function values and (b) sketch the graph of the function.

41. $f(2) = -6$, $f(-1) = 3$

42. $f(0) = -5$, $f(4) = -8$

Graphing a Function In Exercises 43–50, sketch the graph of the function.

43. $f(x) = x^2$ **44.** $f(x) = x^3$

45. $f(x) = \sqrt{x}$ **46.** $f(x) = \dfrac{1}{x}$

47. $g(x) = [\![x]\!] - 2$

48. $g(x) = [\![x + 4]\!]$

49. $f(x) = \begin{cases} 5x - 3, & x \geq -1 \\ -4x + 5, & x < -1 \end{cases}$

50. $f(x) = \begin{cases} 2x + 1, & x \leq 2 \\ x^2 + 1, & x > 2 \end{cases}$

2.5 **Describing Transformations** In Exercises 51–60, h is related to one of the parent functions described in this chapter. (a) Identify the parent function f. (b) Describe the sequence of transformations from f to h. (c) Sketch the graph of h. (d) Use function notation to write h in terms of f.

51. $h(x) = x^2 - 9$ **52.** $h(x) = (x - 2)^3 + 2$

53. $h(x) = |x + 3| - 5$ **54.** $h(x) = -\sqrt{x} + 4$

55. $h(x) = -(x + 2)^2 + 3$ **56.** $h(x) = \frac{1}{2}(x - 1)^2 - 2$

57. $h(x) = -[\![x]\!] + 6$ **58.** $h(x) = 5[\![x - 9]\!]$

59. $h(x) = -\sqrt{x + 1} + 9$ **60.** $h(x) = -\frac{1}{3}x^3$

2.6 **Finding Arithmetic Combinations of Functions** In Exercises 61 and 62, find (a) $(f + g)(x)$, (b) $(f - g)(x)$, (c) $(fg)(x)$, and (d) $(f/g)(x)$. What is the domain of f/g?

61. $f(x) = x^2 + 3$, $g(x) = 2x - 1$

62. $f(x) = x^2 - 4$, $g(x) = \sqrt{3 - x}$

Finding Domains of Functions and Composite Functions In Exercises 63 and 64, find (a) $f \circ g$ and (b) $g \circ f$. Find the domain of each function and of each composite function.

63. $f(x) = \frac{1}{3}x - 3$, $g(x) = 3x + 1$

64. $f(x) = \sqrt{x + 1}$, $g(x) = x^2$

Retail In Exercises 65 and 66, the price of a washing machine is x dollars. The function $f(x) = x - 100$ gives the price of the washing machine after a \$100 rebate. The function $g(x) = 0.95x$ gives the price of the washing machine after a 5% discount.

65. Find and interpret $(f \circ g)(x)$.

66. Find and interpret $(g \circ f)(x)$.

2.7 **Finding an Inverse Function Informally** In Exercises 67 and 68, find the inverse function of f informally. Verify that $f(f^{-1}(x)) = x$ and $f^{-1}(f(x)) = x$.

67. $f(x) = \dfrac{x - 4}{5}$ **68.** $f(x) = x^3 - 1$

Applying the Horizontal Line Test In Exercises 69 and 70, use a graphing utility to graph the function, and use the Horizontal Line Test to determine whether the function has an inverse function.

69. $f(x) = (x - 1)^2$ **70.** $h(t) = \dfrac{2}{t - 3}$

Finding and Analyzing Inverse Functions In Exercises 71 and 72, (a) find the inverse function of f, (b) graph both f and f^{-1} on the same set of coordinate axes, (c) describe the relationship between the graphs of f and f^{-1}, and (d) state the domains and ranges of f and f^{-1}.

71. $f(x) = \frac{1}{2}x - 3$ **72.** $f(x) = \sqrt{x + 1}$

Restricting the Domain In Exercises 73 and 74, restrict the domain of the function f to an interval on which the function is increasing, and find f^{-1} on that interval.

73. $f(x) = 2(x - 4)^2$ **74.** $f(x) = |x - 2|$

Exploring the Concepts

True or False? In Exercises 75 and 76, determine whether the statement is true or false. Justify your answer.

75. Relative to the graph of $f(x) = \sqrt{x}$, the graph of the function $h(x) = -\sqrt{x + 9} - 13$ is shifted 9 units to the left and 13 units down, then reflected in the x-axis.

76. If f and g are two inverse functions, then the domain of g is equal to the range of f.

Chapter Test

See CalcChat.com for tutorial help and worked-out solutions to odd-numbered exercises.

Take this test as you would take a test in class. When you are finished, check your work against the answers given in the back of the book.

In Exercises 1 and 2, find an equation of the line passing through the pair of points. Sketch the line. *(Section 2.1)*

1. $(-2, 5), (1, -7)$ **2.** $(-4, -7), \left(1, \frac{4}{3}\right)$

3. Find equations of the lines that pass through the point $(0, 4)$ and are (a) parallel to and (b) perpendicular to the line $5x + 2y = 3$. *(Section 2.1)*

In Exercises 4 and 5, find each function value. *(Section 2.2)*

4. $f(x) = |x + 2| - 15$
     (a) $f(-8)$ (b) $f(14)$ (c) $f(x - 6)$

5. $f(x) = \dfrac{\sqrt{x + 9}}{x^2 - 81}$
     (a) $f(7)$ (b) $f(-5)$ (c) $f(x - 9)$

In Exercises 6 and 7, find (a) the domain and (b) the zeros of the function. *(Sections 2.2 and 2.3)*

6. $f(x) = \dfrac{x - 5}{2x^2 - x}$ **7.** $f(x) = 10 - \sqrt{3 - x}$

In Exercises 8–10, (a) use a graphing utility to graph the function, (b) approximate the open intervals on which the function is increasing, decreasing, or constant, and (c) determine whether the function is even, odd, or neither. *(Section 2.3)*

8. $f(x) = 2x^6 + 5x^4 - x^2$ **9.** $f(x) = 4x\sqrt{3 - x}$ **10.** $f(x) = |x + 5|$

11. Use a graphing utility to approximate (to two decimal places) any relative minima or maxima of $f(x) = -x^3 + 2x - 1$. *(Section 2.3)*

12. Find the average rate of change of $f(x) = -2x^2 + 5x - 3$ from $x_1 = 1$ to $x_2 = 3$. *(Section 2.3)*

13. Sketch the graph of $f(x) = \begin{cases} 3x + 7, & x \le -3 \\ 4x^2 - 1, & x > -3 \end{cases}$. *(Section 2.4)*

In Exercises 14–16, (a) identify the parent function f in the transformation, (b) describe the sequence of transformations from f to h, (c) sketch the graph of h, and (d) use function notation to write h in terms of f. *(Section 2.5)*

14. $h(x) = 4[\![x]\!]$ **15.** $h(x) = -\sqrt{x + 5} + 8$ **16.** $h(x) = -2(x - 5)^3 + 3$

In Exercises 17 and 18, find (a) $(f + g)(x)$, (b) $(f - g)(x)$, (c) $(fg)(x)$, (d) $(f/g)(x)$, (e) $(f \circ g)(x)$, and (f) $(g \circ f)(x)$. What is the domain of f/g? *(Section 2.6)*

17. $f(x) = 3x^2 - 7, \quad g(x) = -x^2 - 4x + 5$ **18.** $f(x) = \dfrac{1}{x}, \quad g(x) = 2\sqrt{x}$

In Exercises 19–21, determine whether the function has an inverse function. If it does, find the inverse function. *(Section 2.7)*

19. $f(x) = x^3 + 9$ **20.** $f(x) = |x^2 - 3| + 6$ **21.** $f(x) = 3x\sqrt{x}$

22. It costs a company $58 to produce 6 units of a product and $78 to produce 10 units. Assuming that the cost function is linear, how much does it cost to produce 25 units? *(Section 2.1)*

Cumulative Test for Chapters P–2 See CalcChat.com for tutorial help and worked-out solutions to odd-numbered exercises.

Take this test as you would take a test in class. When you are finished, check your work against the answers given in the back of the book.

In Exercises 1 and 2, simplify the expression. *(Section P.2)*

1. $\dfrac{8x^2y^{-3}}{30x^{-1}y^2}$

2. $\sqrt{20x^2y^3}$

In Exercises 3–5, perform the operation(s) and simplify the result. *(Sections P.3 and P.5)*

3. $4x - [2x + 3(2 - x)]$ 4. $(x - 2)(x^2 + x - 3)$ 5. $\dfrac{3}{s + 5} - \dfrac{2}{s - 3}$

In Exercises 6–8, completely factor the expression. *(Section P.4)*

6. $36 - (x + 1)^2$

7. $x - 5x^2 - 6x^3$

8. $54x^3 + 16$

In Exercises 9 and 10, write an expression for the area of the figure as a polynomial in standard form. *(Section P.4)*

9.

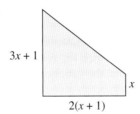

10.

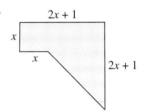

In Exercises 11–13, sketch the graph of the equation. *(Section 1.1)*

11. $x - 3y + 12 = 0$ 12. $y = x^2 - 9$ 13. $y = \sqrt{4 - x}$

In Exercises 14–16, solve the equation and check your solution. *(Section 1.2)*

14. $3x - 5 = 6x + 8$

15. $-(x + 3) = 14(x - 6)$

16. $\dfrac{1}{x - 2} = \dfrac{10}{4x + 3}$

In Exercises 17–22, solve the equation using any convenient method. *(Section 1.4)*

17. $x^2 - 4x + 3 = 0$ 18. $-2x^2 + 4x + 6 = 0$

19. $3x^2 + 9x + 1 = 0$ 20. $3x^2 + 5x - 6 = 0$

21. $\frac{2}{3}x^2 = 24$ 22. $\frac{1}{2}x^2 - 7 = 25$

In Exercises 23–28, solve the equation, if possible. Check your solutions. *(Section 1.6)*

23. $x^4 + 12x^3 + 4x^2 + 48x = 0$ 24. $8x^3 - 48x^2 + 72x = 0$

25. $x^{3/2} + 21 = 13$ 26. $\sqrt{x + 10} = x - 2$

27. $|2(x - 1)| = 8$ 28. $|x - 12| = -2$

In Exercises 29–32, solve the inequality. Then graph the solution set.
(Sections 1.7 and 1.8)

29. $|x + 1| \leq 6$ **30.** $|5 + 6x| > 3$

31. $5x^2 + 12x + 7 \geq 0$

32. $\dfrac{-1}{8x^2 - 2x - 3} > 0$

33. Find the slope-intercept form of the equation of the line passing through $\left(-\frac{1}{2}, 1\right)$ and $(3, 8)$. Sketch the line.   *(Section 2.1)*

34. Let $f(x) = \dfrac{x}{x - 2}$. Find each function value, if possible.   *(Section 2.2)*

    (a) $f(6)$ (b) $f(2)$ (c) $f(s + 2)$

35. Explain why the graph at the left does not represent y as a function of x. *(Section 2.3)*

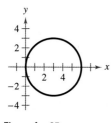

Figure for 35

In Exercises 36–38, determine whether the function is even, odd, or neither. Then describe the symmetry.   *(Section 2.3)*

36. $f(x) = 5 + \sqrt{4 - x}$ **37.** $f(x) = 2x^3 - 4x$ **38.** $f(x) = x^4 + 1$

39. Compare the graph of each function with the graph of $y = \sqrt[3]{x}$. (*Note:* It is not necessary to sketch the graphs.)   *(Section 2.5)*

    (a) $r(x) = \frac{1}{2}\sqrt[3]{x}$ (b) $h(x) = \sqrt[3]{x} + 2$ (c) $g(x) = \sqrt[3]{x + 2}$

In Exercises 40 and 41, find (a) $(f + g)(x)$, (b) $(f - g)(x)$, (c) $(fg)(x)$, and (d) $(f/g)(x)$. What is the domain of f/g?   *(Section 2.6)*

40. $f(x) = x - 4, \quad g(x) = 3x + 1$

41. $f(x) = \sqrt{x - 1}, \quad g(x) = x^2 + 1$

In Exercises 42 and 43, find (a) $f \circ g$ and (b) $g \circ f$. Find the domain of each function and of each composite function.   *(Section 2.6)*

42. $f(x) = 2x^2, \quad g(x) = \sqrt{x + 6}$

43. $f(x) = x - 2, \quad g(x) = |x|$

44. Determine whether $h(x) = 3x - 4$ has an inverse function. If it does, find the inverse function.   *(Section 2.7)*

45. A group of n people decide to buy a \$36,000 single-engine plane. Each person will pay an equal share of the cost. When three additional people join the group, the cost per person will decrease by \$1000. Find n.   *(Section 1.6)*

46. For groups of 60 or more people, a charter bus company determines the rate per person according to the formula

    Rate $= 10 - 0.05(n - 60), \quad n \geq 60$

    where the rate is given in dollars and n is the number of people.   *(Section 2.2)*

    (a) Write the revenue R as a function of n.

    (b) Use a graphing utility to graph the revenue function. Use the graph to approximate the number of people that will maximize the revenue.

47. The height of an object thrown upward from a height of 8 feet at a velocity of 36 feet per second can be modeled by $s(t) = -16t^2 + 36t + 8$, where s is the height (in feet) and t is the time (in seconds). Find the average rate of change of the function from $t_1 = 0$ to $t_2 = 2$. Interpret your answer in the context of the problem.   *(Section 2.3)*

Proofs in Mathematics

Biconditional Statements

Recall from the Proofs in Mathematics in Chapter 1 that a conditional statement is a statement of the form "if p, then q." A statement of the form "p if and only if q" is a **biconditional statement.** A biconditional statement, denoted by

$$p \leftrightarrow q \qquad \text{Biconditional statement}$$

is the conjunction of the conditional statement $p \to q$ and its converse $q \to p$.

A biconditional statement can be either true or false. To be true, *both* the conditional statement and its converse must be true.

EXAMPLE 1　Analyzing a Biconditional Statement

Consider the statement "$x = 3$ if and only if $x^2 = 9$."

a. Is the statement a biconditional statement?

b. Is the statement true?

Solution

a. The statement is a biconditional statement, because it is of the form "p if and only if q."

b. Rewrite the statement as a conditional statement and its converse.

> *Conditional statement:* If $x = 3$, then $x^2 = 9$.

> *Converse:* If $x^2 = 9$, then $x = 3$.

The conditional statement is true, but the converse is false, because x can also equal -3. So, the biconditional statement is false. ■

Knowing how to use biconditional statements is an important tool for reasoning in mathematics.

EXAMPLE 2　Analyzing a Biconditional Statement

Determine whether the biconditional statement is true or false. If it is false, provide a counterexample.

> A number is divisible by 5 if and only if it ends in 0.

Solution Rewrite the biconditional statement as a conditional statement and its converse.

> *Conditional statement:* If a number is divisible by 5, then it ends in 0.

> *Converse:* If a number ends in 0, then it is divisible by 5.

The conditional statement is false. A counterexample is the number 15, which is divisible by 5 but does not end in 0. So, the biconditional statement is false. ■

P.S. Problem Solving

See CalcChat.com for tutorial help and worked-out solutions to odd-numbered exercises.

1. Monthly Wages As a salesperson, you receive a monthly salary of $2000, plus a commission of 7% of sales. You receive an offer for a new job at $2300 per month, plus a commission of 5% of sales.

(a) Write a linear equation for your current monthly wage W_1 in terms of your monthly sales S.

(b) Write a linear equation for the monthly wage W_2 of your new job offer in terms of the monthly sales S.

(c) Use a graphing utility to graph both equations in the same viewing window. Find the point of intersection. What does the point of intersection represent?

(d) You expect sales of $20,000 per month. Should you change jobs? Explain.

2. Cellphone Keypad For the numbers 2 through 9 on a cellphone keypad (see figure), consider two relations: one mapping numbers onto letters, and the other mapping letters onto numbers. Are both relations functions? Explain.

1	2 ABC	3 DEF
4 GHI	5 JKL	6 MNO
7 PQRS	8 TUV	9 WXYZ
*	0	#

3. Sums and Differences of Functions What can be said about the sum and difference of each pair of functions?

(a) Two even functions

(b) Two odd functions

(c) An odd function and an even function

4. Inverse Functions The functions

$$f(x) = x \quad \text{and} \quad g(x) = -x$$

are their own inverse functions. Graph each function and explain why this is true. Graph other linear functions that are their own inverse functions. Find a formula for a family of linear functions that are their own inverse functions.

5. Proof Prove that a function of the form

$$y = a_{2n}x^{2n} + a_{2n-2}x^{2n-2} + \cdots + a_2x^2 + a_0$$

is an even function.

6. Miniature Golf A golfer is trying to make a hole-in-one on the miniature golf green shown. The golf ball is at the point $(2.5, 2)$ and the hole is at the point $(9.5, 2)$. The golfer wants to bank the ball off the side wall of the green at the point (x, y). Find the coordinates of the point (x, y). Then write an equation for the path of the ball.

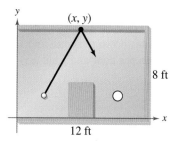

Figure for 6

7. Titanic At 2:00 P.M. on April 11, 1912, the *Titanic* left Cobh, Ireland, on her voyage to New York City. At 11:40 P.M. on April 14, the *Titanic* struck an iceberg and sank, having covered only about 2100 miles of the approximately 3400-mile trip.

(a) What was the total duration of the voyage in hours?

(b) What was the average speed in miles per hour?

(c) Write a function relating the distance of the *Titanic* from New York City and the number of hours traveled. Find the domain and range of the function.

(d) Graph the function in part (c).

8. Average Rate of Change Consider the function $f(x) = -x^2 + 4x - 3$. Find the average rate of change of the function from x_1 to x_2.

(a) $x_1 = 1, x_2 = 2$

(b) $x_1 = 1, x_2 = 1.5$

(c) $x_1 = 1, x_2 = 1.25$

(d) $x_1 = 1, x_2 = 1.125$

(e) $x_1 = 1, x_2 = 1.0625$

(f) Does the average rate of change seem to be approaching one value? If so, state the value.

(g) Find the equations of the secant lines through the points $(x_1, f(x_1))$ and $(x_2, f(x_2))$ for parts (a)–(e).

(h) Find the equation of the line through the point $(1, f(1))$ using your answer from part (f) as the slope of the line.

9. Inverse of a Composition Consider the functions $f(x) = 4x$ and $g(x) = x + 6$.

(a) Find $(f \circ g)(x)$.

(b) Find $(f \circ g)^{-1}(x)$.

(c) Find $f^{-1}(x)$ and $g^{-1}(x)$.

(d) Find $(g^{-1} \circ f^{-1})(x)$ and compare the result with that of part (b).

(e) Repeat parts (a) through (d) for $f(x) = x^3 + 1$ and $g(x) = 2x$.

(f) Write two one-to-one functions f and g, and repeat parts (a) through (d) for these functions.

(g) Make a conjecture about $(f \circ g)^{-1}(x)$ and $(g^{-1} \circ f^{-1})(x)$.

10. Trip Time You are in a boat 2 miles from the nearest point on the coast (see figure). You plan to travel to point Q, 3 miles down the coast and 1 mile inland. You row at 2 miles per hour and walk at 4 miles per hour.

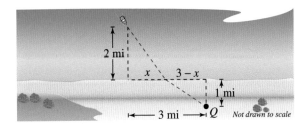

(a) Write the total time T (in hours) of the trip as a function of the distance x (in miles).

(b) Determine the domain of the function.

(c) Use a graphing utility to graph the function. Be sure to choose an appropriate viewing window.

(d) Find the value of x that minimizes T.

(e) Write a brief paragraph interpreting these values.

11. Heaviside Function The **Heaviside function**

$$H(x) = \begin{cases} 1, & x \geq 0 \\ 0, & x < 0 \end{cases}$$

is widely used in engineering applications. (See figure.) To print an enlarged copy of the graph, go to *MathGraphs.com*.

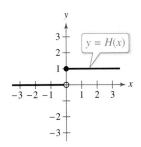

Sketch the graph of each function by hand.

(a) $H(x) - 2$

(b) $H(x - 2)$

(c) $-H(x)$

(d) $H(-x)$

(e) $\frac{1}{2}H(x)$

(f) $-H(x - 2) + 2$

12. Repeated Composition Let $f(x) = \dfrac{1}{1 - x}$.

(a) Find the domain and range of f.

(b) Find $f(f(x))$. What is the domain of this function?

(c) Find $f(f(f(x)))$. Is the graph a line? Why or why not?

13. Associative Property with Compositions Show that the Associative Property holds for compositions of functions—that is,

$$(f \circ (g \circ h))(x) = ((f \circ g) \circ h)(x).$$

14. Graphical Reasoning Use the graph of the function f to sketch the graph of each function. To print an enlarged copy of the graph, go to *MathGraphs.com*.

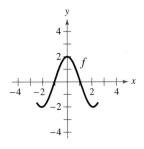

(a) $f(x + 1)$

(b) $f(x) + 1$

(c) $2f(x)$

(d) $f(-x)$

(e) $-f(x)$

(f) $|f(x)|$

(g) $f(|x|)$

15. Graphical Reasoning Use the graphs of f and f^{-1} to complete each table of function values.

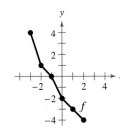 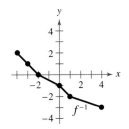

(a)

x	-4	-2	0	4
$(f(f^{-1}(x)))$				

(b)

x	-3	-2	0	1
$(f + f^{-1})(x)$				

(c)

x	-3	-2	0	1
$(f \cdot f^{-1})(x)$				

(d)

x	-4	-3	0	4
$|f^{-1}(x)|$				

3 Polynomial Functions

GO DIGITAL

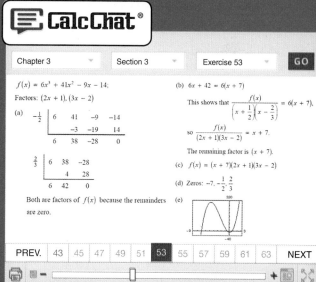

3.5 Ocean Temperatures *(Exercise 65, p. 296)*

3.1 Path of a Diver *(Exercise 61, p. 249)*

241

3.1 Quadratic Functions and Models

> Analyze graphs of quadratic functions.
> Write quadratic functions in standard form and use the results to sketch their graphs.
> Find minimum and maximum values of quadratic functions in real-life applications.

The Graph of a Quadratic Function

In this and the next section, you will study graphs of polynomial functions. Section 2.4 introduced basic functions such as linear, constant, and squaring functions.

$f(x) = ax + b$ Linear function

$f(x) = c$ Constant function

$f(x) = x^2$ Squaring function

These are examples of **polynomial functions.**

Quadratic functions have many real-life applications. For example, in Exercise 61 on page 249, you will use a quadratic function that models the path of a diver.

Definition of a Polynomial Function

Let n be a nonnegative integer and let $a_n, a_{n-1}, \ldots, a_2, a_1, a_0$ be real numbers with $a_n \neq 0$. The function

$$f(x) = a_n x^n + a_{n-1} x^{n-1} + \cdots + a_2 x^2 + a_1 x + a_0$$

is a **polynomial function of x with degree n.**

Polynomial functions are classified by degree. For example, a constant function $f(x) = c$ with $c \neq 0$ has degree 0, and a linear function $f(x) = ax + b$ with $a \neq 0$ has degree 1. In this section, you will study **quadratic functions,** which are second-degree polynomial functions.

For example, each function listed below is a quadratic function.

$f(x) = x^2 + 6x + 2$

$g(x) = 2(x + 1)^2 - 3$

$h(x) = 9 + \frac{1}{4}x^2$

$k(x) = (x - 2)(x + 1)$

Note that the squaring function is a simple quadratic function.

Definition of a Quadratic Function

Let a, b, and c be real numbers with $a \neq 0$. The function

$$f(x) = ax^2 + bx + c$$ Quadratic function

is a **quadratic function.**

Often, quadratic functions can model real-life data. For example, the table at the left shows the heights h (in feet) of a projectile fired from an initial height of 6 feet with an initial velocity of 256 feet per second at selected values of time t (in seconds). A quadratic model for the data in the table is

$$h(t) = -16t^2 + 256t + 6, \quad 0 \leq t \leq 16.$$

Time, t	Height, h
0	6
4	774
8	1030
12	774
16	6

GO DIGITAL

The graph of a quadratic function is a U-shaped curve called a **parabola.** Parabolas occur in many real-life applications—including those that involve reflective properties of satellite dishes and flashlight reflectors. You will study these properties in Section 4.3.

All parabolas are symmetric with respect to a line called the **axis of symmetry,** or simply the **axis** of the parabola. The point where the axis intersects the parabola is the **vertex** of the parabola. When the leading coefficient is positive, the graph of

$$f(x) = ax^2 + bx + c$$

is a parabola that opens upward. When the leading coefficient is negative, the graph is a parabola that opens downward. The next two figures show the axes and vertices of parabolas for cases where $a > 0$ and $a < 0$.

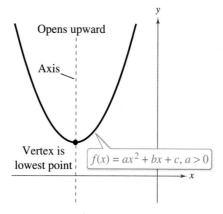

Leading coefficient is positive.

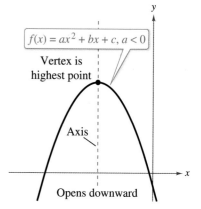

Leading coefficient is negative.

The simplest type of quadratic function is one in which $b = c = 0$. In this case, the function has the form $f(x) = ax^2$. Its graph is a parabola whose vertex is $(0, 0)$. When $a > 0$, the vertex is the point with the *minimum y*-value on the graph, and when $a < 0$, the vertex is the point with the *maximum y*-value on the graph, as shown in the figures below.

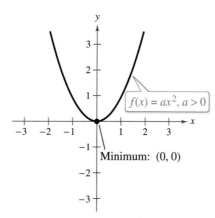

Leading coefficient is positive.

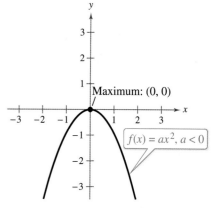

Leading coefficient is negative.

When sketching the graph of $f(x) = ax^2$, it is helpful to use the graph of $y = x^2$ as a reference, as suggested in Section 2.5. There you learned that when $a > 1$, the graph of $y = af(x)$ is a vertical stretch of the graph of $y = f(x)$. When $0 < a < 1$, the graph of $y = af(x)$ is a vertical shrink of the graph of $y = f(x)$. Example 1 demonstrates this again.

EXAMPLE 1   **Sketching Graphs of Quadratic Functions**

▷▷▷ *See LarsonPrecalculus.com for an interactive version of this type of example.*

Sketch the graph of each quadratic function and compare it with the graph of $y = x^2$.

a. $f(x) = \frac{1}{3}x^2$ **b.** $g(x) = 2x^2$

Solution

a. Compared with $y = x^2$, each output of $f(x) = \frac{1}{3}x^2$ "shrinks" by a factor of $\frac{1}{3}$, producing the broader parabola shown in Figure 3.1.

b. Compared with $y = x^2$, each output of $g(x) = 2x^2$ "stretches" by a factor of 2, producing the narrower parabola shown in Figure 3.2.

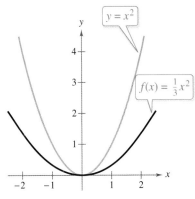

Figure 3.1   Figure 3.2

 Checkpoint ▶ *Audio-video solution in English & Spanish at LarsonPrecalculus.com*

Sketch the graph of each quadratic function and compare it with the graph of $y = x^2$.

a. $f(x) = \frac{1}{4}x^2$ **b.** $g(x) = -\frac{1}{6}x^2$ **c.** $h(x) = \frac{5}{2}x^2$ **d.** $k(x) = -4x^2$ ■

In Example 1, note that the coefficient a determines how wide the parabola $f(x) = ax^2$ opens. The smaller the value of $|a|$, the wider the parabola opens.

Recall from Section 2.5 that the graphs of

$$y = f(x \pm c), \quad y = f(x) \pm c, \quad y = f(-x), \quad \text{and} \quad y = -f(x)$$

are rigid transformations of the graph of $y = f(x)$. For example, in the figures below, notice how transformations of the graph of $y = x^2$ can produce the graphs of

$$f(x) = -x^2 + 1 \quad \text{and} \quad g(x) = (x + 2)^2 - 3.$$

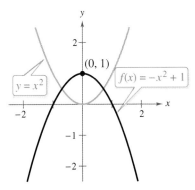

Reflection in x-axis followed by an upward shift of one unit

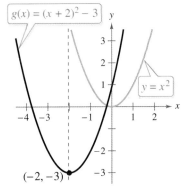

Left shift of two units followed by a downward shift of three units

The Standard Form of a Quadratic Function

The **standard form** of a quadratic function is $f(x) = a(x - h)^2 + k$. This form is especially convenient for sketching a parabola because it identifies the vertex of the parabola as (h, k).

> ### Standard Form of a Quadratic Function
>
> The quadratic function
>
> $$f(x) = a(x - h)^2 + k, \quad a \neq 0$$
>
> is in **standard form**. The graph of f is a parabola whose axis is the vertical line $x = h$ and whose vertex is the point (h, k). When $a > 0$, the parabola opens upward, and when $a < 0$, the parabola opens downward.

To graph a parabola, it is helpful to begin by writing the quadratic function in standard form using the process of completing the square, as illustrated in Example 2. In this example, notice that when completing the square, you *add and subtract* the square of half the coefficient of x within the parentheses instead of adding the value to each side of the equation as is done in Section 1.4.

EXAMPLE 2   Using Standard Form to Graph a Parabola

Sketch the graph of

$$f(x) = 2x^2 + 8x + 7.$$

Identify the vertex and the axis of the parabola.

Solution   Begin by writing the quadratic function in standard form. Notice that the first step in completing the square is to factor out any coefficient of x^2 that is not 1.

$$\begin{aligned}
f(x) &= 2x^2 + 8x + 7 && \text{Write original function.} \\
&= 2(x^2 + 4x) + 7 && \text{Factor 2 out of } x\text{-terms.} \\
&= 2(x^2 + 4x + 4 - 4) + 7 && \text{Add and subtract } (4/2)^2 = 4 \text{ within parentheses.}
\end{aligned}$$

$$\underbrace{\qquad}_{(4/2)^2}$$

After adding and subtracting 4 within the parentheses, you must now regroup the terms to form a perfect square trinomial. To remove the -4 from inside the parentheses, note that, because of the 2 outside of the parentheses, you must multiply by 2, as shown below.

$$\begin{aligned}
f(x) &= 2(x^2 + 4x + 4) - 2(4) + 7 && \text{Regroup terms.} \\
&= 2(x^2 + 4x + 4) - 8 + 7 && \text{Simplify.} \\
&= 2(x + 2)^2 - 1 && \text{Write in standard form.}
\end{aligned}$$

From this form, you can see that the graph of f is a parabola that opens upward and has its vertex at $(-2, -1)$. This corresponds to a left shift of two units and a downward shift of one unit relative to the graph of $y = 2x^2$, as shown in Figure 3.3. In the figure, you can see that the axis of the parabola is the vertical line through the vertex, $x = -2$.

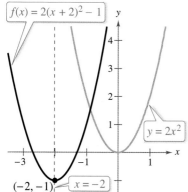

$f(x) = 2(x + 2)^2 - 1$

$y = 2x^2$

$(-2, -1)$ $x = -2$

Figure 3.3

✓ *Checkpoint* ▶ *Audio-video solution in English & Spanish at LarsonPrecalculus.com*

Sketch the graph of

$$f(x) = 3x^2 - 6x + 4.$$

Identify the vertex and the axis of the parabola.

GO DIGITAL

To find the x-intercepts of the graph of $f(x) = ax^2 + bx + c$, you must solve the equation $ax^2 + bx + c = 0$. When $ax^2 + bx + c$ does not factor, use completing the square or the Quadratic Formula to find the x-intercepts. Remember, however, that a parabola may not have x-intercepts.

EXAMPLE 3 **Finding the Vertex and x-Intercepts of a Parabola**

Sketch the graph of $f(x) = -x^2 + 6x - 8$. Identify the vertex and x-intercepts.

Solution

$$\begin{aligned}
f(x) &= -x^2 + 6x - 8 && \text{Write original function.}\\
&= -(x^2 - 6x) - 8 && \text{Factor } -1 \text{ out of } x\text{-terms.}\\
&= -(x^2 - 6x + 9 - 9) - 8 && \text{Add and subtract } (-6/2)^2 = 9 \text{ within parentheses.}
\end{aligned}$$

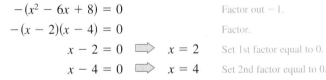

$$\begin{aligned}
&= -(x^2 - 6x + 9) - (-9) - 8 && \text{Regroup terms.}\\
&= -(x - 3)^2 + 1 && \text{Write in standard form.}
\end{aligned}$$

The graph of f is a parabola that opens downward with vertex $(3, 1)$. Next, find the x-intercepts of the graph.

$$\begin{aligned}
-(x^2 - 6x + 8) &= 0 && \text{Factor out } -1.\\
-(x - 2)(x - 4) &= 0 && \text{Factor.}\\
x - 2 = 0 \;\Longrightarrow\; x &= 2 && \text{Set 1st factor equal to 0.}\\
x - 4 = 0 \;\Longrightarrow\; x &= 4 && \text{Set 2nd factor equal to 0.}
\end{aligned}$$

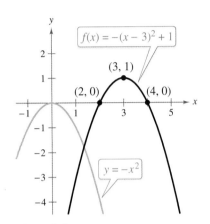

Figure 3.4

So, the x-intercepts are $(2, 0)$ and $(4, 0)$, as shown in Figure 3.4.

✓ **Checkpoint** ▶ *Audio-video solution in English & Spanish at LarsonPrecalculus.com*

Sketch the graph of $f(x) = x^2 - 4x + 3$. Identify the vertex and x-intercepts.

EXAMPLE 4 **Writing a Quadratic Function**

Write the standard form of the quadratic function whose graph is a parabola with vertex $(1, 2)$ and that passes through the point $(3, -6)$.

Solution The vertex is $(h, k) = (1, 2)$, so the equation has the form

$$f(x) = a(x - 1)^2 + 2. \qquad \text{Substitute for } h \text{ and } k \text{ in standard form.}$$

The parabola passes through the point $(3, -6)$, so it follows that $f(3) = -6$.

$$\begin{aligned}
f(x) &= a(x - 1)^2 + 2 && \text{Write in standard form.}\\
-6 &= a(3 - 1)^2 + 2 && \text{Substitute 3 for } x \text{ and } -6 \text{ for } f(x).\\
-6 &= 4a + 2 && \text{Simplify.}\\
-8 &= 4a && \text{Subtract 2 from each side.}\\
-2 &= a && \text{Divide each side by 4.}
\end{aligned}$$

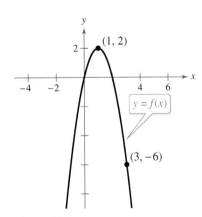

Figure 3.5

The function in standard form is $f(x) = -2(x - 1)^2 + 2$. Figure 3.5 shows the graph of f.

✓ **Checkpoint** ▶ *Audio-video solution in English & Spanish at LarsonPrecalculus.com*

Write the standard form of the quadratic function whose graph is a parabola with vertex $(-4, 11)$ and that passes through the point $(-6, 15)$. ∎

GO DIGITAL

Finding Minimum and Maximum Values

Many applications involve finding the maximum or minimum value of a quadratic function. Note that you can use the completing the square process to rewrite the quadratic function $f(x) = ax^2 + bx + c$ in standard form (see Exercise 71).

$$f(x) = a\left(x + \frac{b}{2a}\right)^2 + \left(c - \frac{b^2}{4a}\right) \qquad \text{Standard form}$$

So, the vertex of the graph of f is $\left(-\dfrac{b}{2a},\ f\left(-\dfrac{b}{2a}\right)\right)$.

Minimum and Maximum Values of Quadratic Functions

Consider the function $f(x) = ax^2 + bx + c$ with vertex $\left(-\dfrac{b}{2a}, f\left(-\dfrac{b}{2a}\right)\right)$.

1. When $a > 0$, f has a *minimum* at $x = -\dfrac{b}{2a}$. The minimum value is $f\left(-\dfrac{b}{2a}\right)$.

2. When $a < 0$, f has a *maximum* at $x = -\dfrac{b}{2a}$. The maximum value is $f\left(-\dfrac{b}{2a}\right)$.

EXAMPLE 5　Maximum Height of a Baseball

The path of a baseball after being hit is modeled by $f(x) = -0.0032x^2 + x + 3$, where $f(x)$ is the height of the baseball (in feet) and x is the horizontal distance from home plate (in feet). What is the maximum height of the baseball?

Algebraic Solution

For this quadratic function, you have

$$f(x) = ax^2 + bx + c = -0.0032x^2 + x + 3$$

which shows that $a = -0.0032$ and $b = 1$. Because $a < 0$, the function has a maximum at $x = -b/(2a)$. So, the baseball reaches its maximum height when it is

$$x = -\frac{b}{2a} = -\frac{1}{2(-0.0032)} = 156.25 \text{ feet}$$

from home plate. At this distance, the maximum height is

$$f(156.25) = -0.0032(156.25)^2 + 156.25 + 3 = 81.125 \text{ feet.}$$

Graphical Solution

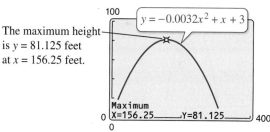

The maximum height is $y = 81.125$ feet at $x = 156.25$ feet.

$y = -0.0032x^2 + x + 3$

Maximum
X=156.25　Y=81.125

✓ *Checkpoint* ▶ Audio-video solution in English & Spanish at LarsonPrecalculus.com

Rework Example 5 when the path of the baseball is modeled by

$$f(x) = -0.007x^2 + x + 4.$$

Summarize　(Section 3.1)

1. State the definition of a quadratic function and describe its graph *(pages 242–244)*. For an example of sketching graphs of quadratic functions, see Example 1.

2. State the standard form of a quadratic function *(page 245)*. For examples that use the standard form of a quadratic function, see Examples 2–4.

3. Explain how to find the minimum or maximum value of a quadratic function *(page 247)*. For a real-life application, see Example 5.

GO DIGITAL

3.1 Exercises

See CalcChat.com for tutorial help and worked-out solutions to odd-numbered exercises.

GO DIGITAL

Vocabulary and Concept Check

In Exercises 1 and 2, fill in the blanks.

1. A polynomial function of x with degree n has the form $f(x) = a_n x^n + a_{n-1} x^{n-1} + \cdots + a_1 x + a_0$ ($a_n \neq 0$), where n is a _____ _____ and $a_n, a_{n-1}, \ldots, a_1, a_0$ are _____ numbers.

2. A _____ function is a second-degree polynomial function, and its graph is called a _____.

3. Is the quadratic function $f(x) = \left(x - \frac{1}{2}\right)^2 + 3$ written in standard form? Identify the vertex of the graph of f.

4. Does the graph of $f(x) = -3x^2 + 5x + 2$ have a minimum value or a maximum value?

Skills and Applications

Matching   In Exercises 5–8, match the quadratic function with its graph. [The graphs are labeled (a), (b), (c), and (d).]

(a)

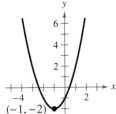

(b)

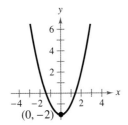

(c)

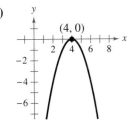

(d)

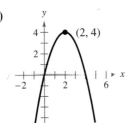

5. $f(x) = x^2 - 2$

6. $f(x) = (x + 1)^2 - 2$

7. $f(x) = -(x - 4)^2$

8. $f(x) = 4 - (x - 2)^2$

Sketching Graphs of Quadratic Functions   In Exercises 9–12, sketch the graph of each quadratic function and compare it with the graph of $y = x^2$.

9. (a) $f(x) = \frac{1}{2}x^2$ (b) $g(x) = -\frac{1}{8}x^2$
   (c) $h(x) = \frac{3}{2}x^2$ (d) $k(x) = -3x^2$

10. (a) $f(x) = x^2 + 1$ (b) $g(x) = x^2 - 1$
    (c) $h(x) = x^2 + 3$ (d) $k(x) = x^2 - 3$

11. (a) $f(x) = (x - 1)^2$ (b) $g(x) = (3x)^2 + 1$
    (c) $h(x) = \left(\frac{1}{3}x\right)^2 - 3$ (d) $k(x) = (x + 3)^2$

12. (a) $f(x) = -\frac{1}{2}(x - 2)^2 + 1$
    (b) $g(x) = \left[\frac{1}{2}(x - 1)\right]^2 - 3$
    (c) $h(x) = -\frac{1}{2}(x + 2)^2 - 1$
    (d) $k(x) = [2(x + 1)]^2 + 4$

Using Standard Form to Graph a Parabola   In Exercises 13–26, write the quadratic function in standard form and sketch its graph. Identify the vertex, axis of symmetry, and x-intercept(s).

13. $f(x) = x^2 - 6x$

14. $g(x) = x^2 - 8x$

15. $h(x) = x^2 - 8x + 16$

16. $g(x) = x^2 + 2x + 1$

17. $f(x) = x^2 - 6x + 2$

18. $f(x) = x^2 + 16x + 61$

19. $f(x) = x^2 + 12x + 40$

20. $f(x) = x^2 - 8x + 21$

21. $f(x) = -x^2 + x - \frac{5}{4}$

22. $f(x) = x^2 + 3x + \frac{1}{4}$

23. $f(x) = -x^2 + 2x + 5$

24. $f(x) = -x^2 - 4x + 1$

25. $h(x) = 4x^2 - 4x + 21$

26. $f(x) = -2x^2 + x - 1$

Graphing a Quadratic Function   In Exercises 27–34, use a graphing utility to graph the quadratic function. Identify the vertex, axis of symmetry, and x-intercept(s). Then check your results algebraically by writing the quadratic function in standard form.

27. $f(x) = -(x^2 + 2x - 3)$

28. $f(x) = -(x^2 + x - 30)$

29. $g(x) = x^2 + 8x + 11$

30. $f(x) = x^2 + 10x + 14$

31. $f(x) = -2x^2 + 12x - 18$

32. $f(x) = -4x^2 + 24x - 41$

33. $g(x) = \frac{1}{2}(x^2 + 4x - 2)$

34. $f(x) = \frac{3}{5}(x^2 + 6x - 5)$

Writing a Quadratic Function   In Exercises 35 and 36, write the standard form of the quadratic function whose graph is the parabola shown.

35.

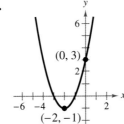

36.

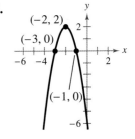

Writing a Quadratic Function In Exercises 37–44, write the standard form of the quadratic function whose graph is a parabola with the given vertex and that passes through the given point.

37. Vertex: $(-2, 5)$; point: $(0, 9)$
38. Vertex: $(-3, -10)$; point: $(0, 8)$
39. Vertex: $(1, -2)$; point: $(-1, 14)$
40. Vertex: $(2, 3)$; point: $(0, 2)$
41. Vertex: $(5, 12)$; point: $(7, 15)$
42. Vertex: $(-2, -2)$; point: $(-1, 0)$
43. Vertex: $\left(-\frac{1}{4}, \frac{3}{2}\right)$; point: $(-2, 0)$
44. Vertex: $(6, 6)$; point: $\left(\frac{61}{10}, \frac{3}{2}\right)$

Graphical Reasoning In Exercises 45–48, determine the x-intercept(s) of the graph visually. Then find the x-intercept(s) algebraically to confirm your results.

45.
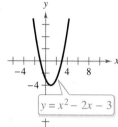
$y = x^2 - 2x - 3$

46.
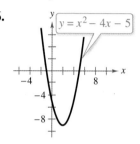
$y = x^2 - 4x - 5$

47.
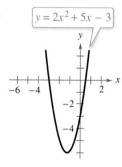
$y = 2x^2 + 5x - 3$

48.
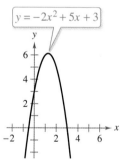
$y = -2x^2 + 5x + 3$

Graphing a Quadratic Function In Exercises 49–52, use a graphing utility to graph the quadratic function. Find the x-intercept(s) of the graph and compare them with the solutions of the corresponding quadratic equation when $f(x) = 0$.

49. $f(x) = x^2 - 4x$
50. $f(x) = -2x^2 + 10x$
51. $f(x) = x^2 - 8x - 20$
52. $f(x) = 2x^2 - 7x - 30$

Finding Quadratic Functions In Exercises 53–56, find a quadratic function whose graph passes through the given points and opens (a) upward and (b) downward. (There are many correct answers.)

53. $(-3, 0), (3, 0)$
54. $(-1, 0), (4, 0)$
55. $(0, 1), (1, 0)$
56. $(0, -4), (2, 0)$

Number Problems In Exercises 57–60, find two positive real numbers whose product is a maximum.

57. The sum is 110.
58. The sum is S.
59. The sum of the first and twice the second is 24.
60. The sum of the first and three times the second is 42.

61. **Path of a Diver**

The path of a diver is modeled by $f(x) = -\frac{4}{9}x^2 + \frac{24}{9}x + 12$, where $f(x)$ is the height (in feet) and x is the horizontal distance (in feet) from the end of the diving board. What is the maximum height of the diver?

62. **Pumpkin-Launching Contest** The path of a pumpkin launched from a compressed air cannon is modeled by

$$f(x) = -\frac{3}{10,000}x^2 + \frac{3}{2}x + 24$$

where $f(x)$ is the height (in feet) and x is the horizontal distance (in feet) from where the pumpkin was launched.

(a) How high is the pumpkin when it is launched?

(b) What is the maximum height of the pumpkin?

(c) The world record distance for a launched pumpkin is 5545 feet. Does this launch break the record? Explain.

63. **Minimum Cost** A manufacturer of lighting fixtures has daily production costs of $C = 800 - 10x + 0.25x^2$, where C is the total cost (in dollars) and x is the number of units produced. What daily production number yields a minimum cost?

64. **Maximum Revenue** The total revenue R earned (in thousands of dollars) from manufacturing handheld video game systems is given by $R(p) = -25p^2 + 1200p$, where p is the price per unit (in dollars). Find the unit price that yields a maximum revenue. What is the maximum revenue?

65. **Maximum Area** A rancher has 200 feet of fencing to enclose two adjacent rectangular corrals (see figure).

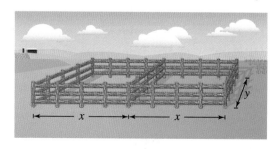

(a) Write the area A of the corrals as a function of x.

(b) What dimensions produce a maximum enclosed area?

66. Maximum Area A Norman window is constructed by adjoining a semicircle to the top of an ordinary rectangular window (see figure). The perimeter of the window is 16 feet.

(a) Write the area A of the window as a function of x.

(b) What dimensions produce a window of maximum area?

Exploring the Concepts

True or False? In Exercises 67 and 68, determine whether the statement is true or false. Justify your answer.

67. The graph of $f(x) = -12x^2 - 1$ has no x-intercepts.

68. The graphs of $f(x) = -4x^2 - 10x + 7$ and $g(x) = 12x^2 + 30x + 1$ have the same axis of symmetry.

69. Think About It Find the value(s) of b such that $f(x) = x^2 + bx - 25$ has a minimum value of -50.

70. HOW DO YOU SEE IT? The graph shows a quadratic function of the form $P(t) = at^2 + bt + c$, which represents the yearly profit for a company, where $P(t)$ is the profit in year t.

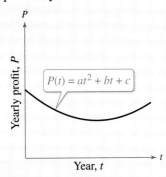

$P(t) = at^2 + bt + c$

(a) Is the value of a positive, negative, or zero? Explain.

(b) Write an expression in terms of a and b that represents the year t when the company made the least profit.

(c) The company made the same yearly profits in 2012 and 2020. In which year did the company make the least profit?

71. Verifying the Vertex Write the quadratic function

$$f(x) = ax^2 + bx + c$$

in standard form to verify that the vertex occurs at

$$\left(-\frac{b}{2a}, f\left(-\frac{b}{2a} \right) \right).$$

72. Proof Assume that the function

$$f(x) = ax^2 + bx + c, \quad a \neq 0$$

has two real zeros. Prove that the x-coordinate of the vertex of the graph is the average of the zeros of f. (*Hint:* Use the Quadratic Formula.)

Review & Refresh ▶ Video solutions at LarsonPrecalculus.com

Writing Polynomials in Standard Form In Exercises 73–78, (a) write the polynomial in standard form, (b) identify the degree and leading coefficient of the polynomial, and (c) state whether the polynomial is a monomial, binomial, or trinomial.

73. $3x^4$

74. -5

75. $x - 4x^2 + 1$

76. $x^5 + x^3$

77. $-29x^3 + \frac{1}{4}x$

78. $x^2 + 8x^4 - 12x^8$

Multiplying Polynomials In Exercises 79–82, multiply the polynomials.

79. $(y - 1)(y - 9)$

80. $(2x + 5)(3x - 6)$

81. $(x^2 + x - 3)(x^2 - 4x - 2)$

82. $(3z^2 - z + 5)(z^2 + 2z + 10)$

Identifying x- and y-intercepts In Exercises 83–86, identify the x- and y-intercepts of the graph.

83. $y = (x + 2)^2$

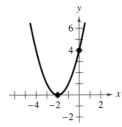

84. $y = 4 - (x - 2)^2$

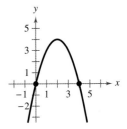

85. $y = |x - 4| - 2$

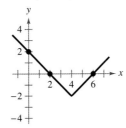

86. $x = 3(1 - y)^3$

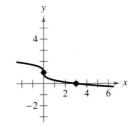

Project: Height of a Basketball To work an extended application analyzing the height of a dropped basketball, visit this text's website at *LarsonPrecalculus.com.*

3.2 Polynomial Functions of Higher Degree

> ❯ **Use transformations to sketch graphs of polynomial functions.**
> ❯ **Use the Leading Coefficient Test to determine the end behaviors of graphs of polynomial functions.**
> ❯ **Find real zeros of polynomial functions and use them as sketching aids.**
> ❯ **Use the Intermediate Value Theorem to help locate real zeros of polynomial functions.**

Polynomial functions have many real-life applications. For example, in Exercise 94 on page 262, you will use a polynomial function to analyze the growth of a red oak tree.

Graphs of Polynomial Functions

In this section, you will study basic features of the graphs of polynomial functions. One feature is that the graph of a polynomial function is **continuous.** Essentially, this means that the graph of a polynomial function has no breaks, holes, or gaps, as shown in Figure 3.6(a). The graph of the piecewise-defined function shown in Figure 3.6(b) has a gap, so this function is *not* continuous.

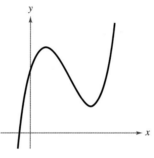

(a) Polynomial functions have continuous graphs.

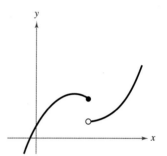

(b) Functions with graphs that are not continuous are not polynomial functions.

Figure 3.6

Another feature of the graph of a polynomial function is that it has only smooth, rounded turns, as shown in Figure 3.7(a). The graph of a polynomial function cannot have a sharp turn, such as the one shown in Figure 3.7(b).

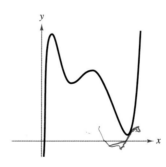

(a) Polynomial functions have graphs with smooth, rounded turns.

Figure 3.7

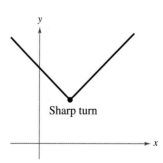

(b) Functions with graphs that have sharp turns are not polynomial functions.

Sketching graphs of polynomial functions of degree greater than 2 is often more involved than sketching graphs of polynomial functions of degree 0, 1, or 2. Using the features presented in this section, however, along with your knowledge of point plotting, intercepts, and symmetry, you should be able to make reasonably accurate sketches by hand.

GO DIGITAL

The polynomial functions that have the simplest graphs are monomial functions of the form $f(x) = x^n$, where n is an integer greater than zero. When n is *even*, the graph is similar to the graph of $f(x) = x^2$, and when n is *odd*, the graph is similar to the graph of $f(x) = x^3$, as shown in Figure 3.8. Moreover, the greater the value of n, the flatter the graph near the origin. Polynomial functions of the form $f(x) = x^n$ are often referred to as **power functions.**

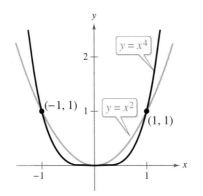

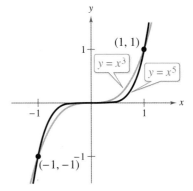

(a) When n is even, the graph of $y = x^n$ touches the x-axis at the x-intercept.

(b) When n is odd, the graph of $y = x^n$ crosses the x-axis at the x-intercept.

Figure 3.8

EXAMPLE 1 Sketching Transformations of Monomial Functions

▶▶▶ *See LarsonPrecalculus.com for an interactive version of this type of example.*

Sketch the graph of each function.

a. $f(x) = -x^5$ **b.** $h(x) = (x + 1)^4$

Solution

a. The degree of $f(x) = -x^5$ is odd, so its graph is similar to the graph of $y = x^3$. In Figure 3.9(a), note that the negative coefficient has the effect of reflecting the graph in the x-axis.

b. The degree of $h(x) = (x + 1)^4$ is even, so its graph is similar to the graph of $y = x^2$. In Figure 3.9(b), note that the graph of h is a left shift by one unit of the graph of $y = x^4$.

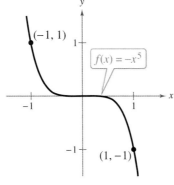

(a)

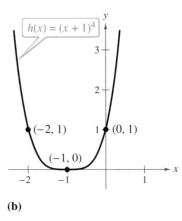

(b)

Figure 3.9

✓ *Checkpoint* ▶ *Audio-video solution in English & Spanish at LarsonPrecalculus.com*

Sketch the graph of each function.

a. $f(x) = (x + 5)^4$ **b.** $g(x) = x^4 - 7$

c. $h(x) = 7 - x^4$ **d.** $k(x) = \frac{1}{4}(x - 3)^4$

The Leading Coefficient Test

In Example 1, note that both graphs eventually rise or fall without bound as x moves to the left or to the right. A nonconstant polynomial function's degree (even or odd) and its leading coefficient (positive or negative) determine whether the graph of the function eventually rises or falls, as described in the **Leading Coefficient Test.**

Leading Coefficient Test

As x moves without bound to the left or to the right, the graph of the nonconstant polynomial function

$$f(x) = a_n x^n + \cdots + a_1 x + a_0, \quad a_n \neq 0$$

eventually rises or falls in the manner described below.

1. When n is *odd:*

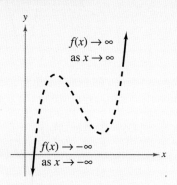 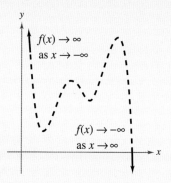

If the leading coefficient is positive $(a_n > 0)$, then the graph falls to the left and rises to the right.

If the leading coefficient is negative $(a_n < 0)$, then the graph rises to the left and falls to the right.

2. When n is *even:*

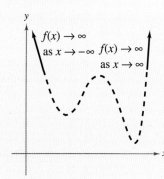 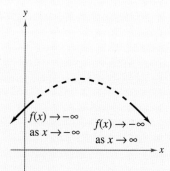

If the leading coefficient is positive $(a_n > 0)$, then the graph rises to the left and to the right.

If the leading coefficient is negative $(a_n < 0)$, then the graph falls to the left and to the right.

The dashed portions of the graphs indicate that the test determines *only* the right-hand and left-hand behavior of the graph.

ALGEBRA HELP

The notation "$f(x) \to -\infty$ as $x \to -\infty$" means that the graph falls to the left, the notation "$f(x) \to \infty$ as $x \to \infty$" means that the graph rises to the right, and so on.

As you continue to study polynomial functions and their graphs, you will notice that the degree of a polynomial plays an important role in determining other characteristics of the polynomial function and its graph.

EXAMPLE 2 **Applying the Leading Coefficient Test**

Describe the left-hand and right-hand behavior of the graph of each function.

a. $f(x) = -x^3 + 4x$ **b.** $f(x) = x^4 - 5x^2 + 4$ **c.** $f(x) = x^5 - x$

Solution

a. The degree is odd and the leading coefficient is negative, so the graph rises to the left and falls to the right, as shown in the figure below.

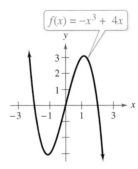

b. The degree is even and the leading coefficient is positive, so the graph rises to the left and to the right, as shown in the figure below.

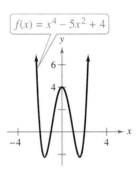

c. The degree is odd and the leading coefficient is positive, so the graph falls to the left and rises to the right, as shown in the figure below.

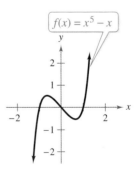

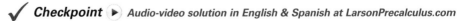

✓ *Checkpoint* ▶ *Audio-video solution in English & Spanish at LarsonPrecalculus.com*

Describe the left-hand and right-hand behavior of the graph of each function.

a. $f(x) = \frac{1}{4}x^3 - 2x$ **b.** $f(x) = -3.6x^5 + 5x^3 - 1$ ■

In Example 2, note that the Leading Coefficient Test tells you only whether the graph *eventually* rises or falls to the left or to the right. You must use other tests to determine other characteristics of the graph, such as intercepts and minimum and maximum points.

Real Zeros of Polynomial Functions

For a nonconstant polynomial function f of degree n, where $n > 0$, it is possible to show that the two statements below are true.

1. The function f has, at most, n real zeros. (You will study this result in detail in the discussion of the Fundamental Theorem of Algebra in Section 3.4.)

2. The graph of f has, at most, $n - 1$ turning points. (Turning points, also called relative minima or relative maxima, are points at which the graph changes from increasing to decreasing or vice versa.)

Finding the zeros of a polynomial function is an important problem in algebra. There is a strong interplay between graphical and algebraic approaches to this problem.

ALGEBRA HELP

Remember that the *zeros* of a function of x are the x-values for which the function is zero.

Real Zeros of Polynomial Functions

When f is a nonconstant polynomial function and a is a real number, the statements listed below are equivalent.

1. $x = a$ is a *zero* of the function f.
2. $x = a$ is a *solution* of the polynomial equation $f(x) = 0$.
3. $(x - a)$ is a *factor* of the polynomial $f(x)$.
4. $(a, 0)$ is an *x-intercept* of the graph of f.

EXAMPLE 3   **Finding Real Zeros of a Polynomial Function**

Find all real zeros of $f(x) = -2x^4 + 2x^2$. Then determine the maximum possible number of turning points of the graph of the function.

Solution

$$-2x^4 + 2x^2 = 0 \qquad \text{Set } f(x) \text{ equal to 0.}$$
$$-2x^2(x^2 - 1) = 0 \qquad \text{Remove common monomial factor.}$$
$$-2x^2(x - 1)(x + 1) = 0 \qquad \text{Factor the difference of two squares.}$$

So, the real zeros are $x = 0$, $x = 1$, and $x = -1$, and the corresponding x-intercepts occur at $(0, 0)$, $(1, 0)$, and $(-1, 0)$. The function is a fourth-degree polynomial, so the graph of f can have at most $4 - 1 = 3$ turning points. In this case, the graph of f has three turning points. Figure 3.10 shows the graph of f.

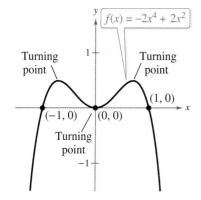

Turning point

Turning point

$f(x) = -2x^4 + 2x^2$

$(1, 0)$

$(-1, 0)$ $(0, 0)$

Turning point

Figure 3.10

 Checkpoint ▶ Audio-video solution in English & Spanish at LarsonPrecalculus.com

Find all real zeros of $f(x) = x^3 - 12x^2 + 36x$. Then determine the maximum possible number of turning points of the graph of the function. ∎

In Example 3, note that the factor $-2x^2$ yields the *repeated* zero $x = 0$, and the graph of f in Figure 3.10 touches (but does not cross) the x-axis at $x = 0$.

Repeated Zeros

If $(x - a)^k$ is a factor of a polynomial function f and $k > 1$, then $(x - a)^k$ yields a **repeated zero** $x = a$ of **multiplicity** k.

1. When k is odd, the graph of f *crosses* the x-axis at $x = a$.
2. When k is even, the graph of f *touches* the x-axis (but does not cross the x-axis) at $x = a$.

GO DIGITAL

A polynomial function is in **standard form** when its terms are in descending order of exponents from left to right. For example, the standard form of the polynomial function $f(x) = 5x^2 - 2x^7 + 4 - 2x$ is

$$f(x) = -2x^7 + 5x^2 - 2x + 4. \qquad \text{Standard form}$$

To avoid making a mistake when applying the Leading Coefficient Test, rewrite the polynomial function in standard form first, if necessary.

EXAMPLE 4 Sketching the Graph of a Polynomial Function

Sketch the graph of $f(x) = -4x^3 + 3x^4$.

Solution

1. *Apply the Leading Coefficient Test.* In standard form, the polynomial function is

$$f(x) = 3x^4 - 4x^3. \qquad \text{Standard form}$$

The leading coefficient is positive and the degree is even, so you know that the graph eventually rises to the left and to the right (see Figure 3.11).

2. *Find the Real Zeros of the Function.* Find the real zeros by factoring $f(x)$.

$$f(x) = 3x^4 - 4x^3 = x^3(3x - 4) \qquad \text{Remove common monomial factor.}$$

The real zeros of f are $x = 0$ and $x = \frac{4}{3}$ (both of odd multiplicity). So, the x-intercepts occur at $(0, 0)$ and $\left(\frac{4}{3}, 0\right)$. Add these points to your graph, as shown in Figure 3.11.

3. *Plot a Few Additional Points.* To sketch the graph, find a few additional points, as shown in the table. Then plot the points (see Figure 3.12).

x	-1	$\dfrac{1}{2}$	1	$\dfrac{3}{2}$
$f(x)$	7	$-\dfrac{5}{16}$	-1	$\dfrac{27}{16}$

4. *Draw the Graph.* Draw a continuous curve through the points, as shown in Figure 3.12. Both zeros are of odd multiplicity, so you know that the graph should cross the x-axis at $x = 0$ and $x = \frac{4}{3}$. If you are unsure of the shape of a portion of a graph, plot additional points.

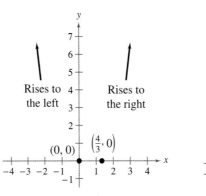

Figure 3.11

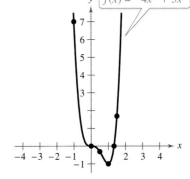

Figure 3.12

✓ **Checkpoint** ▶ Audio-video solution in English & Spanish at LarsonPrecalculus.com

Sketch the graph of $f(x) = 2x^3 - 6x^2$.

>>> TECHNOLOGY

Example 4 uses an *algebraic approach* to describe the graph of the function. A graphing utility can complement this approach. Remember to find a viewing window that shows the significant features of the graph. For instance, the first viewing window below shows the significant features of the function in Example 4, but the second viewing window does not.

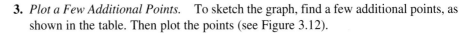

GO DIGITAL

EXAMPLE 5   **Sketching the Graph of a Polynomial Function**

Sketch the graph of

$$f(x) = -2x^3 + 6x^2 - \frac{9}{2}x.$$

Solution

1. *Apply the Leading Coefficient Test.*   The leading coefficient is negative and the degree is odd, so you know that the graph eventually rises to the left and falls to the right (see Figure 3.13).

2. *Find the Real Zeros of the Function.*   Find the real zeros by factoring $f(x)$.

$$f(x) = -2x^3 + 6x^2 - \frac{9}{2}x \qquad \text{Write original function.}$$

$$= -\frac{1}{2}x(4x^2 - 12x + 9) \qquad \text{Remove common monomial factor.}$$

$$= -\frac{1}{2}x(2x - 3)^2 \qquad \text{Factor perfect square trinomial.}$$

The real zeros of f are $x = 0$ (of odd multiplicity) and $x = \frac{3}{2}$ (of even multiplicity). So, the x-intercepts occur at

$$(0, 0) \quad \text{and} \quad \left(\frac{3}{2}, 0\right). \qquad \text{x-intercepts}$$

Add these points to your graph, as shown in Figure 3.13.

3. *Plot a Few Additional Points.*   To sketch the graph, find a few additional points, as shown in the table. Then plot the points (see Figure 3.14).

x	$-\dfrac{1}{2}$	$\dfrac{1}{2}$	1	2
$f(x)$	4	-1	$-\dfrac{1}{2}$	-1

4. *Draw the Graph.*   Draw a continuous curve through the points, as shown in Figure 3.14. From the multiplicities of the zeros, you know that the graph crosses the x-axis at $(0, 0)$ but does not cross the x-axis at $\left(\frac{3}{2}, 0\right)$.

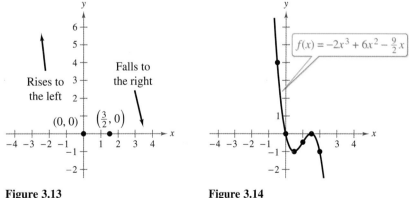

Figure 3.13   **Figure 3.14**

✓ *Checkpoint* ▶ *Audio-video solution in English & Spanish at LarsonPrecalculus.com*

Sketch the graph of $f(x) = -\frac{1}{4}x^4 + \frac{3}{2}x^3 - \frac{9}{4}x^2$.

GO DIGITAL

▶▶▶▶

ALGEBRA HELP

Observe in Example 5 that the sign of $f(x)$ is positive to the left of and negative to the right of the zero $x = 0$. Similarly, the sign of $f(x)$ is negative to the left and to the right of the zero $x = \frac{3}{2}$. This suggests that if the zero of a polynomial function is of *odd* multiplicity, then the sign of $f(x)$ changes from one side to the other side of the zero. If the zero is of *even* multiplicity, then the sign of $f(x)$ does not change from one side to the other side of the zero. The table below helps to illustrate this concept. (This sign analysis may be helpful in graphing polynomial functions.)

x	$-\frac{1}{2}$	0	$\frac{1}{2}$
$f(x)$	4	0	-1
Sign	$+$		$-$

x	1	$\frac{3}{2}$	2
$f(x)$	$-\frac{1}{2}$	0	-1
Sign	$-$		$-$

The Intermediate Value Theorem

The **Intermediate Value Theorem** implies that if

$$(a, f(a)) \quad \text{and} \quad (b, f(b))$$

are two points on the graph of a polynomial function such that $f(a) \neq f(b)$, then for any number d between $f(a)$ and $f(b)$ there must be a number c between a and b such that $f(c) = d$. (See figure below.)

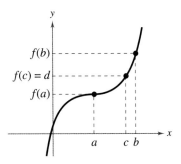

> ### Intermediate Value Theorem
>
> Let a and b be real numbers such that $a < b$. If f is a polynomial function such that $f(a) \neq f(b)$, then, in the interval $[a, b]$, f takes on every value between $f(a)$ and $f(b)$.

ALGEBRA HELP

Note that $f(a)$ and $f(b)$ must be of opposite signs to guarantee that a zero exists between them. If $f(a)$ and $f(b)$ are of the same sign, then it is inconclusive whether a zero exists between them.

One application of the Intermediate Value Theorem is in helping you locate real zeros of a polynomial function. If there exists a value $x = a$ at which a polynomial function is negative, and another value $x = b$ at which it is positive (or if it is positive when $x = a$ and negative when $x = b$), then the function has at least one real zero between these two values. For example, the function

$$f(x) = x^3 + x^2 + 1$$

is negative when $x = -2$ and positive when $x = -1$. So, it follows from the Intermediate Value Theorem that f must have a real zero somewhere between -2 and -1, as shown in the figure below.

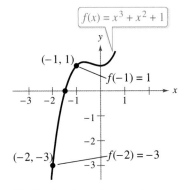

The function f must have a real zero somewhere between -2 and -1.

By continuing this line of reasoning, it is possible to approximate real zeros of a polynomial function to any desired accuracy. Example 6 further demonstrates this concept.

TECHNOLOGY

Using the *table* feature of a graphing utility can help you approximate real zeros of polynomial functions. For instance, in Example 6, construct a table that shows function values for integer values of x. Below, notice that $f(-1)$ and $f(0)$ differ in sign.

X	Y1
-2	-11
-1	-1
0	1
1	1
2	5
3	19
4	49
X=0	

So, by the Intermediate Value Theorem, the function has a real zero between -1 and 0. Adjust your table to show function values for $-1 \le x \le 0$ using increments of 0.1. Below, notice that $f(-0.8)$ and $f(-0.7)$ differ in sign.

X	Y1
-1	-1
-.9	-.539
-.8	-.152
-.7	.167
-.6	.424
-.5	.625
-.4	.776
X=-.7	

So, the function has a real zero between -0.8 and -0.7. Repeat this process to show that the zero is approximately $x \approx -0.755$.

EXAMPLE 6 Using the Intermediate Value Theorem

Use the Intermediate Value Theorem to approximate the real zero of

$$f(x) = x^3 - x^2 + 1.$$

Solution Begin by computing a few function values.

x	-2	-1	0	1
$f(x)$	-11	-1	1	1

The value $f(-1)$ is negative and $f(0)$ is positive, so by the Intermediate Value Theorem, the function has a real zero between -1 and 0. To pinpoint this zero more closely, divide the interval $[-1, 0]$ into tenths and evaluate the function at each point. When you do this, you will find that

$$f(-0.8) = -0.152$$

and

$$f(-0.7) = 0.167.$$

So, f must have a real zero between -0.8 and -0.7, as shown in the figure. For a more accurate approximation, compute function values between $f(-0.8)$ and $f(-0.7)$ and apply the Intermediate Value Theorem again. Continue this process to verify that

$$x \approx -0.755$$

is an approximation (to the nearest thousandth) of the real zero of f.

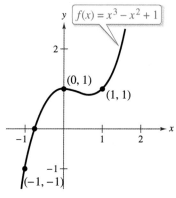

The function f has a real zero between -0.8 and -0.7.

✓ **Checkpoint** ▶ Audio-video solution in English & Spanish at LarsonPrecalculus.com

Use the Intermediate Value Theorem to approximate the real zero of

$$f(x) = x^3 - 3x^2 - 2.$$

Summarize (Section 3.2)

1. Explain how to use transformations to sketch graphs of polynomial functions *(page 252)*. For an example of sketching transformations of monomial functions, see Example 1.

2. Explain how to apply the Leading Coefficient Test *(page 253)*. For an example of applying the Leading Coefficient Test, see Example 2.

3. Explain how to find real zeros of polynomial functions and use them as sketching aids *(page 255)*. For examples involving finding real zeros of polynomial functions, see Examples 3–5.

4. Explain how to use the Intermediate Value Theorem to help locate real zeros of polynomial functions *(page 258)*. For an example of using the Intermediate Value Theorem, see Example 6.

GO DIGITAL

3.2 Exercises

See CalcChat.com for tutorial help and worked-out solutions to odd-numbered exercises.

Vocabulary and Concept Check

In Exercises 1–4, fill in the blanks.

1. The graph of a polynomial function is _____, so it has no breaks, holes, or gaps.

2. A polynomial function of degree n has at most _____ real zeros and at most _____ turning points.

3. When a real zero $x = a$ of a polynomial function f is of even multiplicity, the graph of f _____ the x-axis at $x = a$, and when it is of odd multiplicity, the graph of f _____ the x-axis at $x = a$.

4. A factor $(x - a)^k$, $k > 1$, yields a _____ _____ $x = a$ of _____ k.

In Exercises 5–8, use the graph which shows the right-hand and left-hand behavior of a polynomial function f.

5. Can f be a fourth-degree polynomial function?

6. Can the leading coefficient of f be negative?

7. The graph shows that $f(x_1) < 0$. What other information shown in the graph allows you to apply the Intermediate Value Theorem to guarantee that f has a zero in the interval $[x_1, x_2]$?

8. Is the repeated zero of f in the interval $[x_3, x_4]$ of even or odd multiplicity?

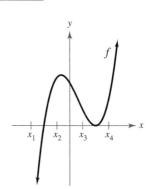

Skills and Applications

Sketching Transformations of Monomial Functions In Exercises 9–12, sketch the graph of $y = x^n$ and each transformation.

9. $y = x^3$
   (a) $f(x) = (x - 4)^3$
   (b) $f(x) = x^3 - 4$
   (c) $f(x) = -\frac{1}{4}x^3$
   (d) $f(x) = (x - 4)^3 - 4$

10. $y = x^5$
   (a) $f(x) = (x + 1)^5$
   (b) $f(x) = x^5 + 1$
   (c) $f(x) = 1 - \frac{1}{2}x^5$
   (d) $f(x) = -\frac{1}{2}(x + 1)^5$

11. $y = x^4$
   (a) $f(x) = (x + 3)^4$
   (b) $f(x) = x^4 - 3$
   (c) $f(x) = 4 - x^4$
   (d) $f(x) = \frac{1}{2}(x - 1)^4$
   (e) $f(x) = (2x)^4 + 1$
   (f) $f(x) = \left(\frac{1}{2}x\right)^4 - 2$

12. $y = x^6$
   (a) $f(x) = (x - 5)^6$
   (b) $f(x) = \frac{1}{8}x^6$
   (c) $f(x) = (x + 3)^6 - 4$
   (d) $f(x) = -\frac{1}{4}x^6 + 1$
   (e) $f(x) = \left(\frac{1}{4}x\right)^6 - 2$
   (f) $f(x) = (2x)^6 - 1$

Matching In Exercises 13–18, match the polynomial function with its graph. [The graphs are labeled (a), (b), (c), (d), (e), and (f).]

13. $f(x) = -2x^2 - 5x$

14. $f(x) = 2x^3 - 3x + 1$

15. $f(x) = -\frac{1}{4}x^4 + 3x^2$

16. $f(x) = -\frac{1}{3}x^3 + x^2 - \frac{4}{3}$

17. $f(x) = x^4 + 2x^3$

18. $f(x) = \frac{1}{5}x^5 - 2x^3 + \frac{9}{5}x$

(a)
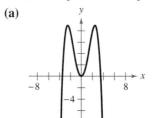

(b)
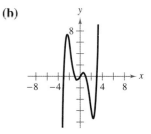

(c)
(d)
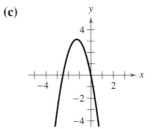

(e)
(f)
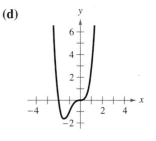

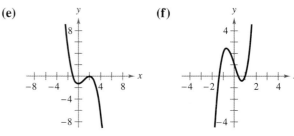

Applying the Leading Coefficient Test   In Exercises 19–28, describe the left-hand and right-hand behavior of the graph of the polynomial function.

19. $f(x) = 12x^3 + 4x$

20. $f(x) = 2x^2 - 3x + 1$

21. $g(x) = 5 - \frac{7}{2}x - 3x^2$

22. $h(x) = 1 - x^6$

23. $h(x) = 6x - 9x^3 + x^2$

24. $g(x) = 8 + \frac{1}{4}x^5 - x^4$

25. $f(x) = 9.8x^6 - 1.2x^3$

26. $h(x) = 1 - 0.5x^5 - 2.7x^3$

27. $f(s) = -\frac{7}{8}(s^3 + 5s^2 - 7s + 1)$

28. $h(t) = -\frac{4}{3}(t - 6t^3 + 2t^4 + 9)$

Comparing Polynomial Functions   In Exercises 29–32, use a graphing utility to graph the functions f and g in the same viewing window. Zoom out sufficiently far to show that the left-hand and right-hand behaviors of f and g appear identical.

29. $f(x) = 3x^3 - 9x + 1$, $g(x) = 3x^3$

30. $f(x) = -\frac{1}{3}(x^3 - 3x + 2)$, $g(x) = -\frac{1}{3}x^3$

31. $f(x) = -(x^4 - 4x^3 + 16x)$, $g(x) = -x^4$

32. $f(x) = 3x^4 - 6x^2$, $g(x) = 3x^4$

Finding Real Zeros of a Polynomial Function   In Exercises 33–48, (a) find all real zeros of the polynomial function, (b) determine whether the multiplicity of each zero is even or odd, (c) determine the maximum possible number of turning points of the graph of the function, and (d) use a graphing utility to graph the function and verify your answers.

33. $f(x) = x^2 - 36$

34. $f(x) = 81 - x^2$

35. $h(t) = t^2 - 6t + 9$

36. $f(x) = x^2 + 10x + 25$

37. $f(x) = \frac{1}{3}x^2 + \frac{1}{3}x - \frac{2}{3}$

38. $f(x) = \frac{1}{2}x^2 + \frac{5}{2}x - \frac{3}{2}$

39. $g(x) = 5x(x^2 - 2x - 1)$

40. $f(t) = t^2(3t^2 - 10t + 7)$

41. $f(x) = -3x^3 + 12x^2 - 3x$

42. $f(x) = x^4 - x^3 - 30x^2$

43. $g(t) = t^5 - 6t^3 + 9t$

44. $f(x) = x^5 + x^3 - 6x$

45. $f(x) = 3x^4 + 9x^2 + 6$

46. $f(t) = 2t^4 - 2t^2 - 40$

47. $g(x) = x^3 + 3x^2 - 4x - 12$

48. $f(x) = x^3 - 4x^2 - 25x + 100$

Graphing a Polynomial Function   In Exercises 49–52, (a) use a graphing utility to graph the function, (b) use the graph to approximate any x-intercepts, (c) find any real zeros of the function algebraically, and (d) compare the results of part (c) with those of part (b).

49. $y = 4x^3 - 20x^2 + 25x$

50. $y = 4x^3 + 4x^2 - 8x - 8$

51. $y = x^5 - 5x^3 + 4x$

52. $y = \frac{1}{5}x^5 - \frac{9}{5}x^3$

Finding a Polynomial Function   In Exercises 53–62, find a polynomial function that has the given zeros. (There are many correct answers.)

53. $0, 7$

54. $-2, 5$

55. $0, -2, -4$

56. $0, 1, 6$

57. $4, -3, 3, 0$

58. $-2, -1, 0, 1, 2$

59. $1 + \sqrt{2}, 1 - \sqrt{2}$

60. $4 + \sqrt{3}, 4 - \sqrt{3}$

61. $2, 2 + \sqrt{5}, 2 - \sqrt{5}$

62. $3, 2 + \sqrt{7}, 2 - \sqrt{7}$

Finding a Polynomial Function   In Exercises 63–68, find a polynomial of degree n that has the given zero(s). (There are many correct answers.)

	Zero(s)	Degree
63.	$x = -3$	$n = 2$
64.	$x = -\sqrt{2}, \sqrt{2}$	$n = 2$
65.	$x = -5, 0, 1$	$n = 3$
66.	$x = -2, 6$	$n = 3$
67.	$x = -5, 1, 2$	$n = 4$
68.	$x = 0, -\sqrt{3}, \sqrt{3}$	$n = 5$

Sketching the Graph of a Polynomial Function   In Exercises 69–82, sketch the graph of the function by (a) applying the Leading Coefficient Test, (b) finding the real zeros of the polynomial, (c) plotting sufficient solution points, and (d) drawing a continuous curve through the points.

69. $f(t) = \frac{1}{4}(t^2 - 2t + 15)$

70. $g(x) = -x^2 + 10x - 16$

71. $f(x) = x^3 - 25x$

72. $g(x) = -9x^2 + x^4$

73. $f(x) = -8 + \frac{1}{2}x^4$

74. $f(x) = 8 - x^3$

75. $f(x) = 3x^3 - 15x^2 + 18x$

76. $f(x) = -4x^3 + 4x^2 + 15x$

77. $f(x) = -5x^2 - x^3$

78. $f(x) = -48x^2 + 3x^4$

79. $f(x) = 9x^2(x + 2)^3$

80. $h(x) = \frac{1}{3}x^3(x - 4)^2$

81. $g(t) = -\frac{1}{4}(t - 2)^2(t + 2)^2$

82. $g(x) = \frac{1}{10}(x + 1)^2(x - 3)^3$

Finding Real Zeros of a Polynomial Function   In Exercises 83–86, use a graphing utility to graph the function. Use the *zero* or *root* feature to approximate the real zeros of the function. Then determine whether the multiplicity of each zero is even or odd.

83. $f(x) = x^3 - 16x$

84. $f(x) = \frac{1}{4}x^4 - 2x^2$

85. $g(x) = \frac{1}{5}(x + 1)^2(x - 3)(2x - 9)$

86. $h(x) = \frac{1}{5}(x + 2)^2(3x - 5)^2$

Using the Intermediate Value Theorem In Exercises 87–90, (a) use the Intermediate Value Theorem and the *table* feature of a graphing utility to find intervals one unit in length in which the polynomial function is guaranteed to have a zero. (b) Adjust the table to approximate the zeros of the function to the nearest thousandth.

87. $f(x) = x^3 - 3x^2 + 3$

88. $f(x) = 0.11x^3 - 2.07x^2 + 9.81x - 6.88$

89. $g(x) = 3x^4 + 4x^3 - 3$ 90. $h(x) = x^4 - 10x^2 + 3$

91. **Maximum Volume** You construct an open box from a square piece of material, 36 inches on a side, by cutting equal squares with sides of length x from the corners and turning up the sides (see figure).

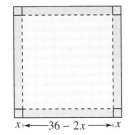
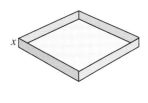

$x \longleftarrow 36 - 2x \longrightarrow x$

(a) Write a function V that represents the volume of the box.

(b) Determine the domain of the function V.

(c) Use a graphing utility to construct a table that shows the box heights x and the corresponding volumes $V(x)$. Use the table to estimate the dimensions that produce a maximum volume.

(d) Use the graphing utility to graph V and use the graph to estimate the value of x for which $V(x)$ is a maximum. Compare your result with that of part (c).

92. **Maximum Volume** You construct an open box with locking tabs from a square piece of material, 24 inches on a side, by cutting equal sections from the corners and folding along the dashed lines (see figure).

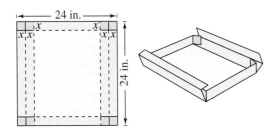

(a) Write a function V that represents the volume of the box.

(b) Determine the domain of the function V.

(c) Sketch a graph of the function and estimate the value of x for which $V(x)$ is a maximum.

93. **Revenue** The revenue R (in millions of dollars) for a beverage company is related to its advertising expense by the function

$$R = \frac{1}{100,000}(-x^3 + 600x^2), \quad 0 \le x \le 400$$

where x is the amount spent on advertising (in tens of thousands of dollars). Use the graph of this function to estimate the point on the graph at which the function is increasing most rapidly. This point is called the *point of diminishing returns* because any expense above this amount will yield less return per dollar invested in advertising.

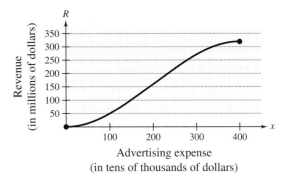

Advertising expense
(in tens of thousands of dollars)

94. **Arboriculture**

The growth of a red oak tree is approximated by the function

$$G = -0.003t^3 + 0.137t^2 + 0.458t - 0.839,$$

$$2 \le t \le 34$$

where G is the height of the tree (in feet) and t is its age (in years).

(a) Use a graphing utility to graph the function.

(b) Estimate the age of the tree when it is growing most rapidly. This point is called the *point of diminishing returns* because the increase in size will be less with each additional year.

(c) Using calculus, the point of diminishing returns can be found by finding the vertex of the parabola

$$y = -0.009t^2 + 0.274t + 0.458.$$

Find the vertex of this parabola.

(d) Compare your results from parts (b) and (c).

Exploring the Concepts

True or False? **In Exercises 95–100, determine whether the statement is true or false. Justify your answer.**

95. It is possible for a fifth-degree polynomial to have no real zeros.

96. It is possible for a sixth-degree polynomial to have only one zero.

97. If the graph of a polynomial function falls to the right, then its leading coefficient is negative.

98. A fifth-degree polynomial function can have five turning points in its graph.

99. It is possible for a polynomial with an even degree to have a range of $(-\infty, \infty)$.

100. If f is a polynomial function of x such that $f(2) = -6$ and $f(6) = 6$, then f has at most one real zero between $x = 2$ and $x = 6$.

101. **Modeling Polynomials** Sketch the graph of a fourth-degree polynomial function that has a zero of multiplicity 2 and a negative leading coefficient. Sketch the graph of another polynomial function with the same characteristics except that the leading coefficient is positive. Compare the graphs.

102. **HOW DO YOU SEE IT?** For each graph, describe a polynomial function that could represent the graph. (Indicate the degree of the function and the sign of its leading coefficient.)

(a)

(b)

(c)

(d)

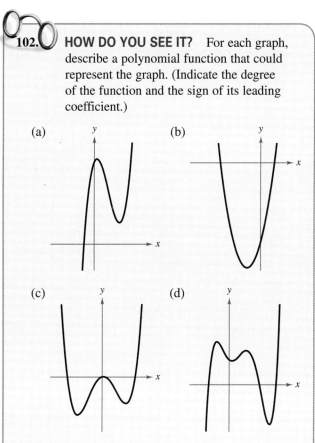

103. **Graphical Reasoning** Sketch the graph of the function $f(x) = x^4$. Explain how the graph of each function g differs (if it does) from the graph of f. Determine whether g is even, odd, or neither.

(a) $g(x) = f(x) + 2$ (b) $g(x) = f(x + 2)$
(c) $g(x) = f(-x)$ (d) $g(x) = -f(x)$
(e) $g(x) = f\left(\frac{1}{2}x\right)$ (f) $g(x) = \frac{1}{2}f(x)$
(g) $g(x) = f(x^{3/4})$ (h) $g(x) = (f \circ f)(x)$

104. **Think About It** Use a graphing utility to graph the functions

$$y_1 = -\frac{1}{3}(x - 2)^5 + 1 \quad \text{and} \quad y_2 = \frac{3}{5}(x + 2)^5 - 3.$$

(a) Determine whether the graphs of y_1 and y_2 are increasing or decreasing. Explain.

(b) Will the graph of

$$g(x) = a(x - h)^5 + k$$

always be only increasing or only decreasing? If so, is this behavior determined by a, h, or k? Explain.

(c) Use a graphing utility to graph

$$f(x) = x^5 - 3x^2 + 2x + 1.$$

Use a graph and the result of part (b) to determine whether f can be written in the form $f(x) = a(x - h)^5 + k$. Explain.

Review & Refresh ▶ Video solutions at LarsonPrecalculus.com

Dividing Integers **In Exercises 105–110, use long division to divide.**

105. $8\overline{)216}$ 106. $13\overline{)1066}$
107. $10\overline{)567}$ 108. $23\overline{)980}$
109. $151\overline{)4336}$ 110. $91\overline{)7571}$

Operations with Polynomials **In Exercises 111–114, perform the operation.**

111. Add $12n - 1$ and $3n^2 - 9n$.
112. Add $-5a + 7a^2$ and $10a^2 - a$.
113. Subtract $2x^2 - 6x$ from $7x^2 + x + 4$.
114. Subtract $6z + 12z^2 - 2$ from $9z^2 + 6z - 3$.

Performing Operations with Complex Numbers **In Exercises 115–122, perform the operation and write the result in standard form.**

115. $(3 + i) - (5 - 4i)$ 116. $(-8 + 3i) + (2 - 3i)$
117. $(1 - 6i)(9 + 2i)$ 118. $(4 - 3i)^2$
119. $(1 - 2i) \div (3 + i)$ 120. $(2 + 12i) \div 4i$
121. $\dfrac{2}{1 + i} - \dfrac{3}{1 - i}$ 122. $\dfrac{2i}{2 + i} + \dfrac{5}{2 - i}$

3.3 Polynomial and Synthetic Division

❯ **Use long division to divide polynomials by other polynomials.**
❯ **Use synthetic division to divide polynomials by binomials of the form $(x - k)$.**
❯ **Use the Remainder Theorem and the Factor Theorem.**

One application of synthetic division is in evaluating polynomial functions. For example, in Exercise 62 on page 271, you will use synthetic division to evaluate a polynomial function that models the number of confirmed cases of Lyme disease in the United States.

Long Division of Polynomials

Consider the graph of

$$f(x) = 6x^3 - 19x^2 + 16x - 4$$

in Figure 3.15. Notice that one of the zeros of f is $x = 2$. This means that $(x - 2)$ is a factor of $f(x)$, and there exists a second-degree polynomial $q(x)$ such that

$$f(x) = (x - 2) \cdot q(x).$$

One way to find $q(x)$ is to use **long division of polynomials,** as illustrated in Example 1.

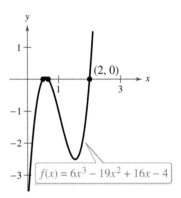

Figure 3.15

Note that in Example 1, the division process requires $-7x^2 + 14x$ to be subtracted from $-7x^2 + 16x$. The difference

$$\begin{array}{r} -7x^2 + 16x \\ -(-7x^2 + 14x) \end{array}$$

is implied and is written as

$$\begin{array}{r} -7x^2 + 16x \\ \underline{-7x^2 + 14x} \\ 2x. \end{array}$$

EXAMPLE 1   **Long Division of Polynomials**

Divide the polynomial $6x^3 - 19x^2 + 16x - 4$ by $x - 2$, and use the result to factor the polynomial completely.

Solution

Think $\dfrac{6x^3}{x} = 6x^2$.

Think $\dfrac{-7x^2}{x} = -7x$.

Think $\dfrac{2x}{x} = 2$.

$$
\begin{array}{r}
6x^2 - 7x + 2 \\
x - 2 \overline{)\, 6x^3 - 19x^2 + 16x - 4} \\
\underline{6x^3 - 12x^2} \\
-7x^2 + 16x \\
\underline{-7x^2 + 14x} \\
2x - 4 \\
\underline{2x - 4} \\
0
\end{array}
$$

Multiply: $6x^2(x - 2)$.
Subtract and bring down $16x$.
Multiply: $-7x(x - 2)$.
Subtract and bring down -4.
Multiply: $2(x - 2)$.
Subtract.

Using the result of the division and then factoring, you can write

$$\begin{aligned} 6x^3 - 19x^2 + 16x - 4 &= (x - 2)(6x^2 - 7x + 2) \\ &= (x - 2)(2x - 1)(3x - 2). \end{aligned}$$

Note that this factorization agrees with the graph of f (see Figure 3.15) in that the three x-intercepts occur at $(2, 0)$, $(1/2, 0)$, and $(2/3, 0)$.

✓ *Checkpoint* ▶ *Audio-video solution in English & Spanish at LarsonPrecalculus.com*

Divide the polynomial $9x^3 + 36x^2 - 49x - 196$ by $x + 4$, and use the result to factor the polynomial completely.

In Example 1, $x - 2$ is a factor of the polynomial

$$6x^3 - 19x^2 + 16x - 4$$

and the long division process produces a remainder of zero. Often, long division will produce a nonzero remainder. For example, when you divide $x^2 + 3x + 5$ by $x + 1$, you obtain a remainder of 3.

$$
\begin{array}{r}
x + 2 \quad \longleftarrow \text{Quotient} \\
\text{Divisor} \longrightarrow \; x + 1 \overline{)\, x^2 + 3x + 5} \quad \longleftarrow \text{Dividend} \\
\underline{x^2 + x} \\
2x + 5 \\
\underline{2x + 2} \\
3 \quad \longleftarrow \text{Remainder}
\end{array}
$$

In fractional form, you can write this result as

$$
\underbrace{\frac{\overbrace{x^3 + 3x + 5}^{\text{Dividend}}}{\underbrace{x + 1}_{\text{Divisor}}}}_{} = \overbrace{x + 2}^{\text{Quotient}} + \frac{\overset{\text{Remainder}}{3}}{\underbrace{x + 1}_{\text{Divisor}}}.
$$

This implies that

$$x^2 + 3x + 5 = (x + 1)(x + 2) + 3 \qquad \text{Multiply each side by } (x + 1).$$

which illustrates the **Division Algorithm.**

The Division Algorithm

If $f(x)$ and $d(x)$ are polynomials such that $d(x) \neq 0$, and the degree of $d(x)$ is less than or equal to the degree of $f(x)$, then there exist unique polynomials $q(x)$ and $r(x)$ such that

$$
\underset{\underset{\text{Dividend}}{\uparrow}}{f(x)} = \underset{\underset{\text{Divisor}}{\uparrow}}{d(x)} \underset{\underset{\text{Quotient}}{\uparrow}}{q(x)} + \underset{\underset{\text{Remainder}}{\uparrow}}{r(x)}
$$

where $r(x) = 0$ or the degree of $r(x)$ is less than the degree of $d(x)$. If the remainder $r(x)$ is zero, then $d(x)$ *divides evenly* into $f(x)$.

Another way to write the Division Algorithm is

$$\frac{f(x)}{d(x)} = q(x) + \frac{r(x)}{d(x)}.$$

In the Division Algorithm, the rational expression $f(x)/d(x)$ is **improper** because the degree of $f(x)$ is greater than or equal to the degree of $d(x)$. On the other hand, the rational expression $r(x)/d(x)$ is **proper** because the degree of $r(x)$ is less than the degree of $d(x)$.

Before applying the Division Algorithm, perform the steps listed below. (Note how Examples 2 and 3 apply these steps.)

1. Make sure the terms of both the dividend and the divisor are written in descending powers of the variable.

2. Insert placeholders with zero coefficients for any missing powers of the variable.

GO DIGITAL

EXAMPLE 2 **Long Division of Polynomials**

Divide $x^3 - 1$ by $x - 1$.

Solution There is no x^2-term or x-term in the dividend $x^3 - 1$, so rewrite the dividend as $x^3 + 0x^2 + 0x - 1$ before applying the Division Algorithm.

ALGEBRA HELP
Note how using zero coefficients for the missing terms helps keep the like terms aligned as you perform the long division.

$$
\begin{array}{r}
x^2 + \ x + 1 \\
x - 1 \overline{)\ x^3 + 0x^2 + 0x - 1} \\
\underline{x^3 - \ x^2} \\
x^2 + 0x \\
\underline{x^2 - \ x} \\
x - 1 \\
\underline{x - 1} \\
0
\end{array}
$$

Multiply: $x^2(x - 1)$.
Subtract and bring down $0x$.
Multiply: $x(x - 1)$.
Subtract and bring down -1.
Multiply: $1(x - 1)$.
Subtract.

So, $x - 1$ divides evenly into $x^3 - 1$, and you can write

$$\frac{x^3 - 1}{x - 1} = x^2 + x + 1, \quad x \neq 1.$$

Check the result by multiplying.

$$(x - 1)(x^2 + x + 1) = x^3 + x^2 + x - x^2 - x - 1$$
$$= x^3 - 1$$

✓ *Checkpoint* ▶ Audio-video solution in English & Spanish at LarsonPrecalculus.com

Divide $x^3 - 2x^2 - 9$ by $x - 3$.

EXAMPLE 3 **Long Division of Polynomials**

▶▶▶ *See LarsonPrecalculus.com for an interactive version of this type of example.*

Divide $-5x^2 - 2 + 3x + 2x^4 + 4x^3$ by $2x - 3 + x^2$.

Solution Write the terms of the dividend and divisor in descending powers of x.

ALGEBRA HELP
In Examples 1 and 2, the divisor was a first-degree polynomial. Note that the Division Algorithm also works with polynomial divisors of degree two or more, as shown in Example 3.

$$
\begin{array}{r}
2x^2 \qquad\ + 1 \\
x^2 + 2x - 3 \overline{)\ 2x^4 + 4x^3 - 5x^2 + 3x - 2} \\
\underline{2x^4 + 4x^3 - 6x^2} \\
x^2 + 3x - 2 \\
\underline{x^2 + 2x - 3} \\
x + 1
\end{array}
$$

Multiply: $2x^2(x^2 + 2x - 3)$.
Subtract and bring down $3x - 2$.
Multiply: $1(x^2 + 2x - 3)$.
Subtract.

Note that the first subtraction eliminated two terms from the dividend. When this happens, the quotient skips a term. You can write the result as

$$\frac{2x^4 + 4x^3 - 5x^2 + 3x - 2}{x^2 + 2x - 3} = 2x^2 + 1 + \frac{x + 1}{x^2 + 2x - 3}.$$

Check the result by multiplying.

$$(x^2 + 2x - 3)(2x^2 + 1) + x + 1 = 2x^4 + x^2 + 4x^3 + 2x - 6x^2 - 3 + x + 1$$
$$= 2x^4 + 4x^3 - 5x^2 + 3x - 2$$

✓ *Checkpoint* ▶ Audio-video solution in English & Spanish at LarsonPrecalculus.com

Divide $-x^3 + 9x + 6x^4 - x^2 - 3$ by $1 + 3x$. ■

Synthetic Division

For long division of polynomials by divisors of the form $x - k$, there is a shortcut called **synthetic division**. The pattern for synthetic division of a cubic polynomial is summarized below. (The pattern for higher-degree polynomials is similar.)

Synthetic Division (for a Cubic Polynomial)

To divide $ax^3 + bx^2 + cx + d$ by $x - k$, use the pattern below.

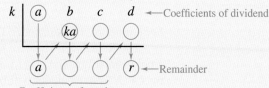

Vertical pattern: Add terms in columns.
Diagonal pattern: Multiply results by k.

This algorithm for synthetic division works only for divisors of the form $x - k$. Remember that $x + k = x - (-k)$.

EXAMPLE 4 **Using Synthetic Division**

Use synthetic division to divide

$$x^4 - 10x^2 - 2x + 4 \quad \text{by} \quad x + 3.$$

Solution Begin by setting up an array, as shown below. Include a zero for the missing x^3-term in the dividend.

$$-3 \; \lvert \; \overset{\frown}{1} \quad 0 \quad -10 \quad -2 \quad 4$$

Then, use the synthetic division pattern by adding terms in columns and multiplying the results by -3.

Divisor: $x + 3$ Dividend: $x^4 - 10x^2 - 2x + 4$

$$
\begin{array}{r|rrrrr}
-3 & 1 & 0 & -10 & -2 & 4 \\
   &   & -3 & 9 & 3 & -3 \\
\hline
   & 1 & -3 & -1 & 1 & 1 \quad \leftarrow \text{Remainder: 1}
\end{array}
$$

Quotient: $x^3 - 3x^2 - x + 1$

So, you have

$$\frac{x^4 - 10x^2 - 2x + 4}{x + 3} = x^3 - 3x^2 - x + 1 + \frac{1}{x + 3}.$$

✓ *Checkpoint* ▶ *Audio-video solution in English & Spanish at LarsonPrecalculus.com*

Use synthetic division to divide $5x^3 + 8x^2 - x + 6$ by $x + 2$. ∎

The Remainder and Factor Theorems

The remainder obtained in the synthetic division process has an important interpretation, as described in the **Remainder Theorem.**

> **The Remainder Theorem**
>
> If a polynomial $f(x)$ is divided by $x - k$, then the remainder is
>
> $$r = f(k).$$

For a proof of the Remainder Theorem, see Proofs in Mathematics on page 305.

The Remainder Theorem tells you that synthetic division can be used to evaluate a polynomial function. That is, to evaluate a polynomial $f(x)$ when $x = k$, divide $f(x)$ by $x - k$. The remainder will be $f(k)$, as illustrated in Example 5.

EXAMPLE 5   **Using the Remainder Theorem**

Use the Remainder Theorem to evaluate

$$f(x) = 3x^3 + 8x^2 + 5x - 7$$

when $x = -2$.

Solution   Using synthetic division gives the result below.

$$
\begin{array}{r|rrrr}
-2 & 3 & 8 & 5 & -7 \\
   &   & -6 & -4 & -2 \\
\hline
   & 3 & 2 & 1 & -9
\end{array}
$$

The remainder is $r = -9$, so by the Remainder Theorem,

$$f(-2) = -9. \qquad {\scriptstyle r = f(k)}$$

This means that $(-2, -9)$ is a point on the graph of f. You can check this result graphically, as shown in Figure 3.16. To check the result algebraically, substitute $x = -2$ in the original function.

Check

$$
\begin{aligned}
f(-2) &= 3(-2)^3 + 8(-2)^2 + 5(-2) - 7 \\
&= 3(-8) + 8(4) - 10 - 7 \\
&= -24 + 32 - 10 - 7 \\
&= -9
\end{aligned}
$$

✓ *Checkpoint* ▶ *Audio-video solution in English & Spanish at LarsonPrecalculus.com*

Use the Remainder Theorem to evaluate

$$f(x) = 4x^3 + 10x^2 - 3x - 8$$

at each function value.

a. $f(-1)$ **b.** $f(4)$ **c.** $f\left(\tfrac{1}{2}\right)$ **d.** $f(-3)$ ■

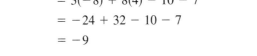

$f(x) = 3x^3 + 8x^2 + 5x - 7$

$(-2, -9)$

Figure 3.16

▷▷▷ TECHNOLOGY

One way to evaluate a function with your graphing utility is to enter the function in the equation editor and use the *table* feature in *ask* mode. When you enter an x-value in the table, the corresponding function value is displayed in the function column.

GO DIGITAL

Another important theorem is the **Factor Theorem.** This theorem states that you can test whether a polynomial has $(x - k)$ as a factor by evaluating the polynomial at $x = k$. If the result is 0, then $(x - k)$ is a factor.

> **The Factor Theorem**
>
> A polynomial $f(x)$ has a factor $(x - k)$ if and only if $f(k) = 0$.

For a proof of the Factor Theorem, see Proofs in Mathematics on page 305.

EXAMPLE 6   **Factoring a Polynomial: Repeated Division**

Show that $(x - 2)$ and $(x + 3)$ are factors of

$$f(x) = 2x^4 + 7x^3 - 4x^2 - 27x - 18.$$

Then find the remaining factors of $f(x)$.

Algebraic Solution

Using synthetic division with the factor $(x - 2)$ gives the result below.

$$
\begin{array}{r|rrrrr}
2 & 2 & 7 & -4 & -27 & -18 \\
  &   & 4 & 22 & 36 & 18 \\
\hline
  & 2 & 11 & 18 & 9 & 0
\end{array}
$$

0 remainder; $(x - 2)$ is a factor.

Take the result of this division and perform synthetic division again using the factor $(x + 3)$.

$$
\begin{array}{r|rrrr}
-3 & 2 & 11 & 18 & 9 \\
   &   & -6 & -15 & -9 \\
\hline
   & 2 & 5 & 3 & 0
\end{array}
$$

0 remainder; $(x + 3)$ is a factor.

$$\underbrace{}_{2x^2 + 5x + 3}$$

The resulting quadratic expression factors as

$$2x^2 + 5x + 3 = (2x + 3)(x + 1)$$

so the complete factorization of $f(x)$ is

$$f(x) = (x - 2)(x + 3)(2x + 3)(x + 1).$$

Graphical Solution

The graph of $f(x) = 2x^4 + 7x^3 - 4x^2 - 27x - 18$ has four x-intercepts (see figure). These occur at $x = -3$, $x = -\frac{3}{2}$, $x = -1$, and $x = 2$. (Check this algebraically.) This implies that $(x + 3)$, $\left(x + \frac{3}{2}\right)$, $(x + 1)$, and $(x - 2)$ are factors of $f(x)$. [Note that $\left(x + \frac{3}{2}\right)$ and $(2x + 3)$ are equivalent factors because they both yield the same zero, $x = -\frac{3}{2}$.]

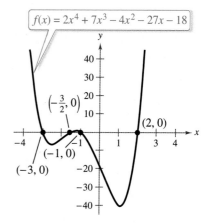

$$f(x) = 2x^4 + 7x^3 - 4x^2 - 27x - 18$$

✓ **Checkpoint** ▶ *Audio-video solution in English & Spanish at LarsonPrecalculus.com*

Show that $(x + 3)$ is a factor of $f(x) = x^3 - 19x - 30$. Then find the remaining factors of $f(x)$. ∎

> **Summarize**   (Section 3.3)
>
> **1.** Explain how to use long division to divide two polynomials *(pages 264 and 265).* For examples of long division of polynomials, see Examples 1–3.
>
> **2.** Describe the algorithm for synthetic division *(page 267).* For an example of synthetic division, see Example 4.
>
> **3.** State the Remainder Theorem and the Factor Theorem *(pages 268 and 269).* For an example of using the Remainder Theorem, see Example 5. For an example of using the Factor Theorem, see Example 6.

GO DIGITAL

3.3 Exercises

See CalcChat.com for tutorial help and worked-out solutions to odd-numbered exercises.

GO DIGITAL

Vocabulary and Concept Check

1. Two forms of the Division Algorithm are shown below. Identify and label each term or function.

$$f(x) = d(x)q(x) + r(x) \qquad \frac{f(x)}{d(x)} = q(x) + \frac{r(x)}{d(x)}$$

2. You divide the polynomial $f(x)$ by $(x - 2)$ and obtain a remainder of 15. What is $f(2)$?

In Exercises 3–6, fill in the blanks.

3. In the Division Algorithm, the rational expression $r(x)/d(x)$ is _____ because the degree of $r(x)$ is less than the degree of $d(x)$.

4. In the Division Algorithm, the rational expression $f(x)/d(x)$ is _____ because the degree of $f(x)$ is greater than or equal to the degree of $d(x)$.

5. A shortcut for long division of polynomials is _____ _____, in which the divisor must be of the form $x - k$.

6. The _____ Theorem states that a polynomial $f(x)$ has a factor $(x - k)$ if and only if $f(k) = 0$.

Skills and Applications

Using the Division Algorithm In Exercises 7 and 8, use long division to verify that $y_1 = y_2$.

7. $y_1 = \dfrac{x^2}{x + 2}, \quad y_2 = x - 2 + \dfrac{4}{x + 2}$

8. $y_1 = \dfrac{x^3 - 3x^2 + 4x - 1}{x + 3}, \quad y_2 = x^2 - 6x + 22 - \dfrac{67}{x + 3}$

Verifying Equivalence In Exercises 9 and 10, (a) use a graphing utility to graph the two equations in the same viewing window, (b) use the graphs to verify that the expressions are equivalent, and (c) use long division to verify the results algebraically.

9. $y_1 = \dfrac{x^2 + 2x - 1}{x + 3}, \quad y_2 = x - 1 + \dfrac{2}{x + 3}$

10. $y_1 = \dfrac{x^4 + x^2 - 1}{x^2 + 1}, \quad y_2 = x^2 - \dfrac{1}{x^2 + 1}$

Long Division of Polynomials In Exercises 11–20, use long division to divide.

11. $(6x + 5) \div (x + 1)$ 12. $(9x - 4) \div (3x + 2)$

13. $(2x^2 + 10x + 12) \div (x + 3)$

14. $(4x^3 - 7x^2 - 11x + 5) \div (4x + 5)$

15. $(x^4 + 5x^3 + 6x^2 - x - 2) \div (x + 2)$

16. $(x^3 + 4x^2 - 3x - 12) \div (x - 3)$

17. $(x^3 - 9) \div (x^2 + 1)$

18. $(x^5 + 7) \div (x^4 - 1)$

19. $(3x + 2x^3 - 9 - 8x^2) \div (x^2 + 1)$

20. $(5x^3 - 16 - 20x + x^4) \div (x^2 - x - 3)$

Using Synthetic Division In Exercises 21–36, use synthetic division to divide.

21. $\dfrac{2x^3 - 10x^2 + 14x - 24}{x - 4}$ 22. $\dfrac{5x^3 + 18x^2 + 7x - 6}{x + 3}$

23. $\dfrac{6x^3 + 7x^2 - x + 26}{x - 3}$ 24. $\dfrac{2x^3 + 12x^2 + 14x - 3}{x + 4}$

25. $\dfrac{4x^3 - 9x + 8x^2 - 18}{x + 2}$ 26. $\dfrac{9x^3 - 16x - 18x^2 + 32}{x - 2}$

27. $(-x^3 + 75x - 250) \div (x + 10)$

28. $(3x^3 - 16x^2 - 72) \div (x - 6)$

29. $(x^3 - 3x^2 + 5) \div (x - 4)$

30. $(5x^3 + 6x + 8) \div (x + 2)$

31. $(10x^4 - 50x^3 - 800) \div (x - 6)$

32. $(x^5 - 13x^4 - 120x + 80) \div (x + 3)$

33. $(x^3 + 512) \div (x + 8)$

34. $(x^3 - 729) \div (x - 9)$

35. $(4x^3 + 16x^2 - 23x - 15) \div \left(x + \frac{1}{2}\right)$

36. $(3x^3 - 4x^2 + 5) \div \left(x - \frac{3}{2}\right)$

Using the Remainder Theorem In Exercises 37–40, write the function in the form $f(x) = (x - k)q(x) + r$ for the given value of k, and demonstrate that $f(k) = r$.

37. $f(x) = x^3 - x^2 - 10x + 7, \quad k = 3$

38. $f(x) = x^3 - 4x^2 - 10x + 8, \quad k = -2$

39. $f(x) = 15x^4 + 10x^3 - 6x^2 + 14, \quad k = -\frac{2}{3}$

40. $f(x) = -3x^3 + 8x^2 + 10x - 8, \quad k = 2 + \sqrt{2}$

Using the Remainder Theorem In Exercises 41–44, use the Remainder Theorem and synthetic division to find each function value. Verify your answers using another method.

41. $f(x) = 2x^3 - 7x + 3$
    (a) $f(1)$ (b) $f(-2)$ (c) $f(3)$ (d) $f(2)$
42. $h(x) = x^3 - 5x^2 - 7x + 4$
    (a) $h(3)$ (b) $h(\frac{1}{2})$ (c) $h(-2)$ (d) $h(-5)$
43. $f(x) = 4x^4 - 16x^3 + 7x^2 + 20$
    (a) $f(1)$ (b) $f(-2)$ (c) $f(5)$ (d) $f(-10)$
44. $g(x) = 2x^6 + 3x^4 - x^2 + 3$
    (a) $g(2)$ (b) $g(1)$ (c) $g(3)$ (d) $g(-1)$

Using the Factor Theorem In Exercises 45–50, use synthetic division to show that x is a solution of the equation, and use the result to factor the polynomial completely. List all real solutions of the equation.

45. $x^3 + 6x^2 + 11x + 6 = 0$, $x = -3$
46. $x^3 - 52x - 96 = 0$, $x = -6$
47. $2x^3 - 15x^2 + 27x - 10 = 0$, $x = \frac{1}{2}$
48. $48x^3 - 80x^2 + 41x - 6 = 0$, $x = \frac{2}{3}$
49. $x^3 + 2x^2 - 3x - 6 = 0$, $x = \sqrt{3}$
50. $x^3 - x^2 - 13x - 3 = 0$, $x = 2 - \sqrt{5}$

Factoring a Polynomial In Exercises 51–54, (a) verify the given factors of $f(x)$, (b) find the remaining factor(s) of $f(x)$, (c) use your results to write the complete factorization of $f(x)$, (d) list all real zeros of f, and (e) confirm your results by using a graphing utility to graph the function.

Function	Factors
51. $f(x) = 2x^3 + x^2 - 5x + 2$	$(x + 2), (x - 1)$
52. $f(x) = x^4 - 8x^3 + 9x^2$ $+ 38x - 40$	$(x - 5), (x + 2)$
53. $f(x) = 6x^3 + 41x^2 - 9x - 14$	$(2x + 1), (3x - 2)$
54. $f(x) = 2x^3 - x^2 - 10x + 5$	$(2x - 1), (x + \sqrt{5})$

Approximating Zeros In Exercises 55–58, (a) use the *zero* or *root* feature of a graphing utility to approximate the zeros of the function accurate to three decimal places, (b) determine the exact value of one of the zeros, and (c) use synthetic division to verify your result from part (b), and then factor the polynomial completely.

55. $f(x) = x^3 - 2x^2 - 5x + 10$
56. $f(s) = s^3 - 12s^2 + 40s - 24$
57. $h(x) = x^5 - 7x^4 + 10x^3 + 14x^2 - 24x$
58. $g(x) = 6x^4 - 11x^3 - 51x^2 + 99x - 27$

Simplifying Rational Expressions In Exercises 59 and 60, simplify the rational expression by using long division or synthetic division.

59. $\dfrac{4x^3 - 8x^2 + x + 3}{2x - 3}$ 60. $\dfrac{x^4 + 6x^3 + 11x^2 + 6x}{x^2 + 3x + 2}$

61. **Profit** A company that produces recycled plastic rugs estimates that the profit P (in dollars) from selling the rugs is given by

$$P = -152x^3 + 7545x^2 - 169{,}625, \quad 0 \le x \le 45$$

where x is the advertising expense (in tens of thousands of dollars). An advertising expense of \$400,000 ($x = 40$) results in a profit of \$2,174,375.

(a) Use a graphing utility to graph the profit function.

(b) Use the graph from part (a) to estimate another amount the company can spend on advertising that results in the same profit.

(c) Use synthetic division to confirm the result of part (b) algebraically.

62. **Lyme Disease**

The numbers N of cases of Lyme disease in the United States from 2014 through 2018 are shown in the table, where t represents the year, with $t = 14$ corresponding to 2014. (*Source: Centers for Disease Control and Prevention*)

Year, t	Number, N
14	25,359
15	28,453
16	26,203
17	29,513
18	33,666

(a) Use a graphing utility to create a scatter plot of the data.

(b) Use the *regression* feature of the graphing utility to find a *quartic* model for the data. (A quartic model has the form $at^4 + bt^3 + ct^2 + dt + e$, where a, b, c, d, and e are constant and t is variable.) Graph the model in the same viewing window as the scatter plot.

(c) Use the model to create a table of estimated values of N. Compare the model with the original data.

(d) Use synthetic division to confirm algebraically your estimated value for the year 2018.

Exploring the Concepts

True or False? In Exercises 63–66, determine whether the statement is true or false. Justify your answer.

63. If $(7x + 4)$ is a factor of some polynomial function $f(x)$, then $\frac{4}{7}$ is a zero of f.

64. The factor $(2x - 1)$ is a factor of the polynomial
$$6x^6 + x^5 - 92x^4 + 45x^3 + 184x^2 + 4x - 48.$$

65. The rational expression $\dfrac{x^3 + 2x^2 - 7x + 4}{x^2 - 4x - 12}$ is improper.

66. The equation $(x^3 - 3x^2 + 4) \div (x + 1) = x^2 - 4x + 4$ is true for all values of x.

67. Error Analysis Describe the error.

Use synthetic division to find the remainder when $x^2 + 3x - 5$ is divided by $x + 1$.

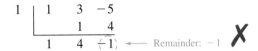

68. HOW DO YOU SEE IT? The graph below shows a company's estimated profits for different advertising expenses. The company's actual profit was $936,660 for an advertising expense of $300,000.

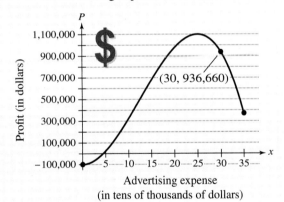

(30, 936,660)

Advertising expense
(in tens of thousands of dollars)

(a) From the graph, it appears that the company could have obtained the same profit for a lesser advertising expense. Use the graph to estimate this expense.

(b) The company's model is
$$P = -140.75x^3 + 5348.3x^2 - 76{,}560,$$
$$0 \le x \le 35$$

where P is the profit (in dollars) and x is the advertising expense (in tens of thousands of dollars). Explain how you could verify the lesser expense from part (a) algebraically.

Think About It In Exercises 69 and 70, perform the division. Assume that n is a positive integer.

69. $\dfrac{x^{3n} + 9x^{2n} + 27x^n + 27}{x^n + 3}$ **70.** $\dfrac{x^{3n} - 3x^{2n} + 5x^n - 6}{x^n - 2}$

Reasoning In Exercises 71 and 72, find the constant c such that the denominator will divide evenly into the numerator.

71. $\dfrac{x^3 + 4x^2 - 3x + c}{x - 5}$ **72.** $\dfrac{x^5 - 2x^2 + x + c}{x + 2}$

73. Think About It Find the value of k such that $x - 4$ is a factor of $x^3 - kx^2 + 2kx - 8$.

74. Writing Perform each polynomial division. Write a brief description of the pattern that you obtain, and use your result to find a formula for the polynomial division $(x^n - 1)/(x - 1)$. Create a numerical example to test your formula.

(a) $\dfrac{x^2 - 1}{x - 1}$ (b) $\dfrac{x^3 - 1}{x - 1}$ (c) $\dfrac{x^4 - 1}{x - 1}$

Review & Refresh ▶ Video solutions at LarsonPrecalculus.com

Finding the Zeros of a Function In Exercises 75–84, find the zeros of the function algebraically.

75. $f(x) = 2x + 5$ **76.** $f(x) = 18 - 3x$

77. $f(x) = x^2 - 8x + 16$ **78.** $f(x) = x^2 - 4x - 45$

79. $f(x) = 6x^2 + x - 15$ **80.** $f(x) = 12x^2 - 65x - 17$

81. $f(x) = x^3 - x^2 - 25x + 25$

82. $f(x) = 12x^3 - 4x^2 - 3x + 1$

83. $f(x) = 6x^4 - 3x^3 - 12x^2 + 6x$

84. $f(x) = 2x^4 + 3x^3 - 34x^2 - 51x$

Writing a Complex Number in Standard Form In Exercises 85–88, write the complex number in standard form.

85. $2 + \sqrt{-25}$ **86.** $4 + \sqrt{-49}$

87. $-6i + i^2$ **88.** $-2i^2 + 4i$

Complex Solutions of a Quadratic Equation In Exercises 89–94, solve the quadratic equation.

89. $x^2 + 4 = 0$ **90.** $x^2 = -9$

91. $x^2 + 2x = -7$ **92.** $x^2 + 4 = 2x$

93. $4x^2 + 8x + 9 = 0$ **94.** $9x^2 - 24x + 34 = 0$

Factoring with Variables in the Exponents In Exercises 95 and 96, factor the expression as completely as possible.

95. $x^{2n} - y^{2n}$ **96.** $x^{3n} + y^{3n}$

97. Think About It Give an example of a polynomial that is prime.

3.4 Zeros of Polynomial Functions

Finding zeros of polynomial functions is an important part of solving many real-life problems. For example, in Exercise 99 on page 285, you will use the zeros of a polynomial function to redesign a storage bin so that it can hold five times as much food.

- Use the Fundamental Theorem of Algebra to determine numbers of zeros of polynomial functions.
- Find rational zeros of polynomial functions.
- Find complex zeros using conjugate pairs.
- Find zeros of polynomials by factoring.
- Use Descartes's Rule of Signs and the Upper and Lower Bound Rules to find zeros of polynomials.
- Find zeros of polynomials in real-life applications.

The Fundamental Theorem of Algebra

In the complex number system, every nth-degree polynomial function has *precisely n* zeros. This important result is derived from the **Fundamental Theorem of Algebra,** first proved by German mathematician Carl Friedrich Gauss (1777–1855).

> **The Fundamental Theorem of Algebra**
>
> If $f(x)$ is a polynomial of degree n, where $n > 0$, then f has at least one zero in the complex number system.

Using the Fundamental Theorem of Algebra and the equivalence of zeros and factors, you obtain the **Linear Factorization Theorem.**

> **Linear Factorization Theorem**
>
> If $f(x)$ is a polynomial of degree n, where $n > 0$, then $f(x)$ has precisely n linear factors, $f(x) = a_n(x - c_1)(x - c_2) \cdots (x - c_n)$, where $c_1, c_2, \ldots, c_n$ are complex numbers.

For a proof of the Linear Factorization Theorem, see Proofs in Mathematics on page 306.

Note that the Fundamental Theorem of Algebra and the Linear Factorization Theorem tell you only that the zeros or factors of a polynomial exist, not how to find them. Such theorems are called **existence theorems.**

ALGEBRA HELP

In Example 1(c), note that when you solve $x^2 + 4 = 0$, you get $x = \pm 2i$. This means that $x^2 + 4 = (x - 2i)(x + 2i)$.

$\ggg$

| **EXAMPLE 1** | **Zeros of Polynomial Functions** |

▶▶▶ *See LarsonPrecalculus.com for an interactive version of this type of example.*

a. The first-degree polynomial function $f(x) = x - 2$ has exactly *one* zero: $x = 2$.

b. The second-degree polynomial function $f(x) = x^2 - 6x + 9 = (x - 3)(x - 3)$ has exactly *two* zeros: $x = 3$ and $x = 3$ (a *repeated zero*).

c. The third-degree polynomial function

$$f(x) = x^3 + 4x = x(x^2 + 4) = x(x - 2i)(x + 2i)$$

has exactly *three* zeros: $x = 0$, $x = 2i$, and $x = -2i$.

✓ *Checkpoint* ▶ *Audio-video solution in English & Spanish at LarsonPrecalculus.com*

Determine the number of zeros of the polynomial function $f(x) = x^4 - 1$. ■

GO DIGITAL

The Rational Zero Test

The **Rational Zero Test** relates the possible rational zeros of a polynomial (having integer coefficients) to the leading coefficient and to the constant term of the polynomial.

The Rational Zero Test

If the polynomial

$$f(x) = a_n x^n + a_{n-1}x^{n-1} + \cdots + a_2 x^2 + a_1 x + a_0 \quad n > 0 \text{ and } a_0 \neq 0$$

has *integer* coefficients, then every rational zero of f has the form

$$\text{Rational zero} = \frac{p}{q}$$

where p and q have no common factors other than 1, p is a factor of the constant term a_0, and q is a factor of the leading coefficient a_n.

HISTORICAL NOTE

Although they were not contemporaries, French mathematician Jean Le Rond d'Alembert (1717–1783) worked independently of Carl Friedrich Gauss in trying to prove the Fundamental Theorem of Algebra. His efforts were such that, in France, the Fundamental Theorem of Algebra is frequently known as d'Alembert's Theorem.

To use the Rational Zero Test, you should first list all rational numbers whose numerators are factors of the constant term and whose denominators are factors of the leading coefficient.

$$\text{Possible rational zeros:} \quad \frac{\text{Factors of constant term}}{\text{Factors of leading coefficient}}$$

Having formed this list of *possible rational zeros,* use a trial-and-error method to determine which, if any, are actual zeros of the polynomial. Note that when the leading coefficient is 1, the possible rational zeros are simply the factors of the constant term. This case is illustrated in Example 2.

EXAMPLE 2   Rational Zero Test with Leading Coefficient of 1

Find (if possible) the rational zeros of

$$f(x) = x^3 + x + 1.$$

Solution   The leading coefficient is 1, so the possible rational zeros are the factors of the constant term, 1.

Possible rational zeros: 1 and -1

Next, determine which of these values, if any, are actual zeros of f. Letting $x = 1$, you can determine that it is *not* a zero of f.

$$f(1) = (1)^3 + 1 + 1 \qquad \text{Substitute 1 for } x.$$
$$= 3 \qquad\qquad\qquad x = 1 \text{ is } not \text{ a zero of } f.$$

Likewise, by letting $x = -1$, you can determine that it is *not* a zero of f.

$$f(-1) = (-1)^3 + (-1) + 1 \qquad \text{Substitute } -1 \text{ for } x.$$
$$= -1 \qquad\qquad\qquad x = -1 \text{ is } not \text{ a zero of } f.$$

So, the given polynomial has no *rational* zeros. Note from the graph of f in Figure 3.17 that f does have one real zero between -1 and 0. By the Rational Zero Test, however, you know that this real zero is *not* a rational number.

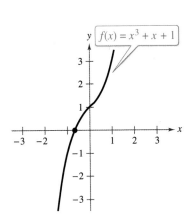

Figure 3.17

✓ *Checkpoint* ▶ *Audio-video solution in English & Spanish at LarsonPrecalculus.com*

Find (if possible) the rational zeros of $f(x) = x^3 + 2x^2 + 6x - 4.$ ∎

GO DIGITAL

EXAMPLE 3 **Rational Zero Test with Leading Coefficient of 1**

Find the rational zeros of $f(x) = x^4 - x^3 + x^2 - 3x - 6$.

Solution The leading coefficient is 1, so the possible rational zeros are the factors of the constant term, -6.

Possible rational zeros: $\pm 1, \pm 2, \pm 3, \pm 6$

By applying synthetic division successively, you find that $x = -1$ and $x = 2$ are the only two rational zeros.

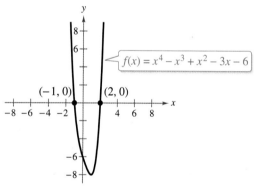

So, $(x + 1)$ and $(x - 2)$ are factors of $f(x)$. Using these factors and the quotient from the last synthetic division above, you can write $f(x)$ as

$$f(x) = (x + 1)(x - 2)(x^2 + 3).$$

The factor $(x^2 + 3)$ produces no real zeros, so $x = -1$ and $x = 2$ are the only *real* zeros of f, as shown in the figure.

✓ *Checkpoint* ▶ Audio-video solution in English & Spanish at LarsonPrecalculus.com

Find the rational zeros of $f(x) = x^3 - 15x^2 + 75x - 125$. ■

When the leading coefficient of a polynomial is not 1, the number of possible rational zeros can increase dramatically. In such cases, the search can be shortened in several ways.

1. Use a graphing utility to speed up the calculations.
2. Use a graph to estimate the locations of any zeros.
3. The Intermediate Value Theorem, along with a table of values, can give approximations of the zeros.
4. The Factor Theorem and synthetic division can be used to test the possible rational zeros.

After finding the first zero, the search becomes simpler by working with the lower-degree polynomial obtained in synthetic division, as shown in Example 3.

ALGEBRA HELP

When there are few possible rational zeros, as in Example 2, it may be quicker to test the zeros by evaluating the function. When there are more possible rational zeros, as in Example 3, it may be quicker to use a different approach to test the zeros, such as using synthetic division or sketching a graph.

GO DIGITAL

Find the rational zeros of $f(x) = 2x^3 + 3x^2 - 8x + 3$.

Solution The leading coefficient is 2, and the constant term is 3.

$$\text{Possible rational zeros: } \frac{\text{Factors of 3}}{\text{Factors of 2}} = \frac{\pm 1, \pm 3}{\pm 1, \pm 2} = \pm 1, \pm 3, \pm \frac{1}{2}, \pm \frac{3}{2}$$

By synthetic division, $x = 1$ is a rational zero.

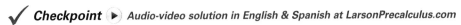

$$\begin{array}{r|rrrr} 1 & 2 & 3 & -8 & 3 \\ & & 2 & 5 & -3 \\ \hline & 2 & 5 & -3 & 0 \end{array} \quad\longrightarrow\quad \text{0 remainder, so } x = 1 \text{ is a zero.}$$

$\underbrace{}_{2x^2 + 5x - 3}$

So, $(x - 1)$ is a factor of $f(x)$. Using this factor and the quotient from the synthetic division above, you can write $f(x)$ as

$$\begin{aligned} f(x) &= (x - 1)(2x^2 + 5x - 3) &\quad \text{Factor out } (x - 1).\\ &= (x - 1)(2x - 1)(x + 3) &\quad \text{Factor trinomial.} \end{aligned}$$

which shows that the rational zeros of f are $x = 1$, $x = \frac{1}{2}$, and $x = -3$.

✓ **Checkpoint** ▶ Audio-video solution in English & Spanish at LarsonPrecalculus.com

Find the rational zeros of

$$f(x) = 2x^3 + x^2 - 13x + 6.$$

Recall from Section 3.2 that if $x = a$ is a zero of the polynomial function f, then $x = a$ is a solution of the polynomial equation $f(x) = 0$.

Find all real solutions of $-10x^3 + 15x^2 + 16x - 12 = 0$.

Solution The leading coefficient is -10 and the constant term is -12.

$$\text{Possible rational solutions: } \frac{\text{Factors of } -12}{\text{Factors of } -10} = \frac{\pm 1, \pm 2, \pm 3, \pm 4, \pm 6, \pm 12}{\pm 1, \pm 2, \pm 5, \pm 10}$$

With so many possibilities (32, in fact), it is worth your time to sketch a graph. In Figure 3.18, three reasonable solutions appear to be $x = -\frac{6}{5}$, $x = \frac{1}{2}$, and $x = 2$. Testing these by synthetic division shows that $x = 2$ is the only rational solution.

$$\begin{array}{r|rrrr} 2 & -10 & 15 & 16 & -12 \\ & & -20 & -10 & 12 \\ \hline & -10 & -5 & 6 & 0 \end{array} \quad\longrightarrow\quad \text{0 remainder, so } x = 2 \text{ is a zero.}$$

$\underbrace{}_{-10x^2 - 5x + 6}$

So, you can write the original equation as $(x - 2)(-10x^2 - 5x + 6) = 0$. Using the Quadratic Formula to solve $-10x^2 - 5x + 6 = 0$, you find that the two additional solutions are irrational numbers.

$$x = \frac{5 + \sqrt{265}}{-20} \approx -1.0639 \quad \text{and} \quad x = \frac{5 - \sqrt{265}}{-20} \approx 0.5639$$

✓ **Checkpoint** ▶ Audio-video solution in English & Spanish at LarsonPrecalculus.com

Find all real solutions of $-2x^3 - 5x^2 + 15x + 18 = 0$.

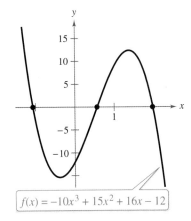

$f(x) = -10x^3 + 15x^2 + 16x - 12$

Figure 3.18

GO DIGITAL

Conjugate Pairs

In Example 1(c), note that the two complex zeros $2i$ and $-2i$ are complex conjugates. That is, they are of the forms $a + bi$ and $a - bi$.

> **Complex Zeros Occur in Conjugate Pairs**
>
> Let f be a polynomial function that has *real coefficients*. If $a + bi$, where $b \neq 0$, is a zero of the function, then the complex conjugate $a - bi$ is also a zero of the function.

Be sure you see that this result is true only when the polynomial function has *real coefficients*. For example, the result applies to the function $f(x) = x^2 + 1$, but not to the function $g(x) = x - i$.

EXAMPLE 6 **Finding a Polynomial Function with Given Zeros**

Find a fourth-degree polynomial function f with real coefficients that has -1, -1, and $3i$ as zeros.

Solution   You are given that $3i$ is a zero of f *and* the polynomial has real coefficients, so you know that the complex conjugate $-3i$ must also be a zero. Using the Linear Factorization Theorem, write $f(x)$ as

$$f(x) = a(x + 1)(x + 1)(x - 3i)(x + 3i).$$

For simplicity, let $a = 1$ to obtain

$$f(x) = (x^2 + 2x + 1)(x^2 + 9) = x^4 + 2x^3 + 10x^2 + 18x + 9.$$

✓ *Checkpoint* ▶ *Audio-video solution in English & Spanish at LarsonPrecalculus.com*

Find a fourth-degree polynomial function f with real coefficients that has 2, -2, and $-7i$ as zeros.

EXAMPLE 7 **Finding a Polynomial Function with Given Zeros**

Find the cubic polynomial function f with real coefficients that has 2 and $1 - i$ as zeros, and $f(1) = 3$.

Solution   You are given that $1 - i$ is a zero of f, so the complex conjugate $1 + i$ is also a zero.

$$\begin{aligned}
f(x) &= a(x - 2)[x - (1 - i)][x - (1 + i)] \\
&= a(x - 2)[(x - 1) + i][(x - 1) - i] \\
&= a(x - 2)[(x - 1)^2 + 1] \\
&= a(x - 2)(x^2 - 2x + 2) \\
&= a(x^3 - 4x^2 + 6x - 4)
\end{aligned}$$

To find the value of a, use the fact that $f(1) = 3$.

$$a[(1)^3 - 4(1)^2 + 6(1) - 4] = 3 \implies a(-1) = 3 \implies a = -3$$

So, $a = -3$ and

$$f(x) = -3(x^3 - 4x^2 + 6x - 4) = -3x^3 + 12x^2 - 18x + 12.$$

✓ *Checkpoint* ▶ *Audio-video solution in English & Spanish at LarsonPrecalculus.com*

Find the quartic polynomial function f with real coefficients that has 1, -2, and $2i$ as zeros, and $f(-1) = 10$.

GO DIGITAL

Factoring a Polynomial

The Linear Factorization Theorem states that you can write any nth-degree polynomial, where $n > 0$, as the product of n linear factors.

$$f(x) = a_n(x - c_1)(x - c_2)(x - c_3) \cdots (x - c_n)$$

This result, however, includes the possibility that some of the values of c_i are imaginary. The theorem below states that you can write $f(x)$ as the product of linear and quadratic factors with real coefficients.

> ### Factors of a Polynomial
>
> Every polynomial of degree $n > 0$ with real coefficients can be written as the product of linear and quadratic factors with real coefficients, where the quadratic factors have no real zeros.

For a proof of this theorem, see Proofs in Mathematics on page 306.

A quadratic factor with no real zeros is *prime* or **irreducible over the reals.** Note that this is not the same as being *irreducible over the rationals.* For example, the quadratic $x^2 + 1 = (x - i)(x + i)$ is irreducible over the reals (and therefore over the rationals). On the other hand, the quadratic $x^2 - 2 = \left(x - \sqrt{2}\right)\left(x + \sqrt{2}\right)$ is irreducible over the rationals but *reducible* over the reals.

>> **TECHNOLOGY**

Another way to find the real zeros of the function in Example 8 is to use a graphing utility to graph the function (see figure).

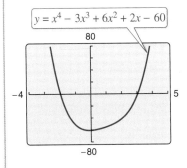

Then use the *zero* or *root* feature of the graphing utility to determine that $x = -2$ and $x = 3$ are the real zeros.

EXAMPLE 8 **Finding the Zeros of a Polynomial Function**

Find all the zeros of $f(x) = x^4 - 3x^3 + 6x^2 + 2x - 60$ given that $1 + 3i$ is a zero of f.

Solution Complex zeros occur in conjugate pairs, so you know that $1 - 3i$ is also a zero of f. This means that both $[x - (1 + 3i)]$ and $[x - (1 - 3i)]$ are factors of $f(x)$. Multiplying these two factors produces

$$[x - (1 + 3i)][x - (1 - 3i)] = [(x - 1) - 3i][(x - 1) + 3i]$$
$$= (x - 1)^2 - 9i^2$$
$$= x^2 - 2x + 10.$$

Using long division, divide $x^2 - 2x + 10$ into $f(x)$.

$$
\begin{array}{r}
x^2 - x - 6 \\
x^2 - 2x + 10 \overline{)\, x^4 - 3x^3 + 6x^2 + 2x - 60} \\
\underline{x^4 - 2x^3 + 10x^2} \\
-x^3 - 4x^2 + 2x \\
\underline{-x^3 + 2x^2 - 10x} \\
-6x^2 + 12x - 60 \\
\underline{-6x^2 + 12x - 60} \\
0
\end{array}
$$

So, you have

$$f(x) = (x^2 - 2x + 10)(x^2 - x - 6) = (x^2 - 2x + 10)(x - 3)(x + 2)$$

and you can conclude that the zeros of f are $x = 1 + 3i$, $x = 1 - 3i$, $x = 3$, and $x = -2$.

✓ *Checkpoint* ▶ Audio-video solution in English & Spanish at LarsonPrecalculus.com

Find all the zeros of $f(x) = 3x^3 - 2x^2 + 48x - 32$ given that $4i$ is a zero of f. ■

In Example 8, without knowing that $1 + 3i$ is a zero of f, it is still possible to find all the zeros of the function. You can first use synthetic division to find the real zeros -2 and 3. Then, factor the polynomial as $(x + 2)(x - 3)(x^2 - 2x + 10)$. Finally, use the Quadratic Formula to solve $x^2 - 2x + 10 = 0$ to obtain the zeros $1 + 3i$ and $1 - 3i$.

In Example 9, you will find all the zeros, including the imaginary zeros, of a fifth-degree polynomial function.

EXAMPLE 9   Finding the Zeros of a Polynomial Function

Write the function $f(x) = x^5 + x^3 + 2x^2 - 12x + 8$ as the product of linear factors and list all the zeros of the function.

Solution   The leading coefficient is 1, so the possible rational zeros are the factors of the constant term.

    Possible rational zeros: $\pm 1, \pm 2, \pm 4,$ and ± 8

By synthetic division, $x = 1$ and $x = -2$ are zeros.

$$
\begin{array}{r|rrrrrr}
1 & 1 & 0 & 1 & 2 & -12 & 8 \\
  &   & 1 & 1 & 2 & 4 & -8 \\
\hline
  & 1 & 1 & 2 & 4 & -8 & 0
\end{array}
$$
    $\longrightarrow$ 0 remainder, so $x = 1$ is a zero.

$$
\begin{array}{r|rrrrr}
-2 & 1 & 1 & 2 & 4 & -8 \\
   &   & -2 & 2 & -8 & 8 \\
\hline
   & 1 & -1 & 4 & -4 & 0
\end{array}
$$
    $\longrightarrow$ 0 remainder, so $x = -2$ is a zero.

$\underbrace{\qquad\qquad\qquad\qquad}_{x^3 - x^2 + 4x - 4}$

So, $(x - 1)$ and $(x + 2)$ are factors of $f(x)$. Using these factors and the quotient from the last synthetic division above, you can write $f(x)$ as

$$
\begin{aligned}
f(x) &= x^5 + x^3 + 2x^2 - 12x + 8 && \text{Write original function.}\\
&= (x - 1)(x + 2)(x^3 - x^2 + 4x - 4). && \text{Factor out } (x - 1) \text{ and } (x + 2).
\end{aligned}
$$

To factor the cubic term $x^3 - x^2 + 4x - 4$, use factoring by grouping.

$$
\begin{aligned}
x^3 - x^2 + 4x - 4 &= (x^3 - x^2) + (4x - 4) && \text{Group terms.}\\
&= x^2(x - 1) + 4(x - 1) && \text{Factor each group.}\\
&= (x - 1)(x^2 + 4) && (x - 1) \text{ is a common factor.}
\end{aligned}
$$

Thus, $f(x) = (x - 1)(x - 1)(x + 2)(x^2 + 4)$. The remaining zeros come from the quadratic factor $(x^2 + 4)$ and can be found by solving $x^2 + 4 = 0$.

$$
x^2 + 4 = 0 \implies x^2 = -4 \implies x = \pm\sqrt{-4} \implies x = \pm 2i
$$

In completely factored form, $f(x) = (x - 1)(x - 1)(x + 2)(x - 2i)(x + 2i)$, which gives all five zeros of f.

$$
x = 1, \quad x = 1, \quad x = -2, \quad x = 2i, \quad \text{and} \quad x = -2i
$$

Figure 3.19 shows the graph of f. Notice that the *real* zeros are the only ones that appear as x-intercepts and that the real zero $x = 1$ is repeated.

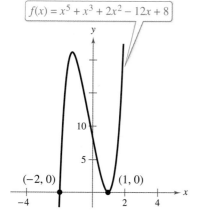

$f(x) = x^5 + x^3 + 2x^2 - 12x + 8$

$(-2, 0)$ $(1, 0)$

Figure 3.19

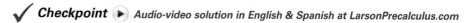

✓ *Checkpoint* ▶ *Audio-video solution in English & Spanish at LarsonPrecalculus.com*

Write the function

$$
f(x) = x^4 + 8x^2 - 9
$$

as the product of linear factors and list all the zeros of the function.   ■

GO DIGITAL

Other Tests for Zeros of Polynomials

You know that an nth-degree polynomial function can have *at most n* real zeros. Of course, many nth-degree polynomial functions do not have that many real zeros. For example, $f(x) = x^2 + 1$ has no real zeros, and $f(x) = x^3 + 1$ has only one real zero. The next theorem, called **Descartes's Rule of Signs,** sheds more light on the number of real zeros of a polynomial.

> ### Descartes's Rule of Signs
>
> Let $f(x) = a_n x^n + a_{n-1} x^{n-1} + \cdots + a_2 x^2 + a_1 x + a_0$ be a polynomial with real coefficients and $a_0 \neq 0$.
>
> 1. The number of *positive real zeros* of f is either equal to the number of variations in sign of $f(x)$ or less than that number by an even integer.
> 2. The number of *negative real zeros* of f is either equal to the number of variations in sign of $f(-x)$ or less than that number by an even integer.

ALGEBRA HELP

For a polynomial in standard form, a *variation in sign* means that two consecutive coefficients have opposite signs. Missing terms (those with zero coefficients) can be ignored.

When using Descartes's Rule of Signs, count a zero of multiplicity k as k zeros. For example, the polynomial

$$+ \text{ to } -$$
$$x^3 - 3x + 2 \qquad \text{The polynomial has two variations in sign.}$$
$$- \text{ to } +$$

has *two* variations in sign, and so it has either two positive or no positive real zeros. This polynomial factors as $x^3 - 3x + 2 = (x - 1)(x - 1)(x + 2)$, so the two positive real zeros are $x = 1$ of multiplicity 2.

EXAMPLE 10 **Using Descartes's Rule of Signs**

Determine the possible numbers of positive and negative real zeros of the function $f(x) = 3x^3 - 5x^2 + 6x - 4$.

Solution The original polynomial has *three* variations in sign.

$$+ \text{ to } - \qquad + \text{ to } -$$
$$f(x) = 3x^3 - 5x^2 + 6x - 4$$
$$- \text{ to } +$$

The polynomial

$$f(-x) = 3(-x)^3 - 5(-x)^2 + 6(-x) - 4$$
$$= -3x^3 - 5x^2 - 6x - 4$$

has no variations in sign. So, from Descartes's Rule of Signs, the polynomial $f(x) = 3x^3 - 5x^2 + 6x - 4$ has either three positive real zeros or one positive real zero, and has no negative real zeros. Figure 3.20 shows that the function has only one real zero, $x = 1$.

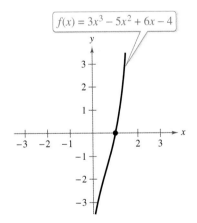

$f(x) = 3x^3 - 5x^2 + 6x - 4$

Figure 3.20

✓ *Checkpoint* ▶ Audio-video solution in English & Spanish at LarsonPrecalculus.com

Determine the possible numbers of positive and negative real zeros of the function $f(x) = 2x^3 + 5x^2 + x + 8$. ■

Another test for real zeros of a polynomial function is related to the sign pattern in the last row of the synthetic division array. This test can give you an upper or lower bound for the real zeros of f, which can help you eliminate possible real zeros. A real number c is an **upper bound** for the real zeros of f when no zeros are greater than c. Similarly, c is a **lower bound** when no real zeros of f are less than c.

Upper and Lower Bound Rules

Let $f(x)$ be a polynomial with real coefficients and a positive leading coefficient. Divide $f(x)$ by $x - c$ using synthetic division.

1. If $c > 0$ and each number in the last row is either positive or zero, then c is an **upper bound** for the real zeros of f.

2. If $c < 0$ and the numbers in the last row are alternately positive and negative (zero entries count as positive or negative), then c is a **lower bound** for the real zeros of f.

EXAMPLE 11   **Finding Real Zeros of a Polynomial Function**

Find all real zeros of

$$f(x) = 6x^3 - 4x^2 + 3x - 2.$$

Solution   List the possible rational zeros of f.

$$\frac{\text{Factors of } -2}{\text{Factors of } 6} = \frac{\pm 1, \pm 2}{\pm 1, \pm 2, \pm 3, \pm 6} = \pm 1, \pm\frac{1}{2}, \pm\frac{1}{3}, \pm\frac{1}{6}, \pm\frac{2}{3}, \pm 2$$

The original polynomial $f(x)$ has three variations in sign. The polynomial

$$f(-x) = 6(-x)^3 - 4(-x)^2 + 3(-x) - 2$$
$$= -6x^3 - 4x^2 - 3x - 2$$

has no variations in sign. So, by Descartes's Rule of Signs, there are three positive real zeros or one positive real zero, and no negative real zeros. Use synthetic division to test $x = 1$.

$$
\begin{array}{r|rrrr}
1 & 6 & -4 & 3 & -2 \\
  &   & 6 & 2 & 5 \\
\hline
  & 6 & 2 & 5 & 3
\end{array}
$$
$\longrightarrow$ Nonzero remainder; $x = 1$ is *not* a zero.

So, $x = 1$ is not a zero. The last row has all positive entries, however, telling you that $x = 1$ is an upper bound for the real zeros. So, restrict the search to zeros between 0 and 1. By trial and error, you can determine that $x = \frac{2}{3}$ is a zero. So, using $\left(x - \frac{2}{3}\right)$ as a factor, you can determine that $f(x)$ factors as

$$f(x) = \left(x - \frac{2}{3}\right)(6x^2 + 3).$$

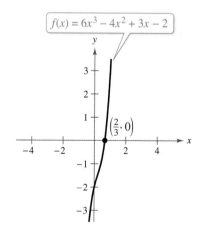

$f(x) = 6x^3 - 4x^2 + 3x - 2$

$\left(\frac{2}{3}, 0\right)$

The factor $6x^2 + 3$ has no real zeros, so it follows that $x = \frac{2}{3}$ is the only real zero of f, as shown in the figure at the right.

✓ *Checkpoint*   ▶ *Audio-video solution in English & Spanish at LarsonPrecalculus.com*

Find all real zeros of $f(x) = 8x^3 - 4x^2 + 6x - 3$.

Application

EXAMPLE 12 **Using a Polynomial Model**

You design candle-making kits. Each kit contains 25 cubic inches of candle wax and a mold for making a pyramid-shaped candle. You want the height of the candle to be 2 inches less than the length of each side of the candle's square base. What should the dimensions of your candle mold be?

Solution The volume of a pyramid is $V = \frac{1}{3}Bh$, where B is the area of the base and h is the height (see Figure 3.21). The area of the base is x^2 and the height is $(x - 2)$. So, the volume of the pyramid is $V = \frac{1}{3}x^2(x - 2)$. Substitute 25 for the volume and solve for x.

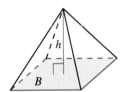

The volume of a pyramid is $V = \frac{1}{3}Bh$.
Figure 3.21

$$25 = \tfrac{1}{3}x^2(x - 2) \qquad \text{Substitute 25 for } V.$$
$$75 = x^3 - 2x^2 \qquad \text{Multiply each side by 3, and distribute } x^2.$$
$$0 = x^3 - 2x^2 - 75 \qquad \text{Write in general form.}$$

The possible rational solutions are $x = \pm 1, \pm 3, \pm 5, \pm 15, \pm 25, \pm 75$. Note that in this case it makes sense to consider only positive x-values. Use synthetic division to test some of the possible solutions and determine that $x = 5$ is a solution.

$$
\begin{array}{r|rrrr}
5 & 1 & -2 & 0 & -75 \\
  &   & 5 & 15 & 75 \\
\hline
  & 1 & 3 & 15 & 0 \\
\end{array}
\quad \longrightarrow \quad \text{0 remainder; } x = 5 \text{ is a solution.}
$$

The other two solutions that satisfy $x^2 + 3x + 15 = 0$ are imaginary, so discard them and conclude that the base of the candle mold should be 5 inches by 5 inches and the height should be $5 - 2 = 3$ inches.

✓ **Checkpoint**  *Audio-video solution in English & Spanish at LarsonPrecalculus.com*

Rework Example 12 when each kit contains 147 cubic inches of candle wax and you want the height of the pyramid-shaped candle to be 2 inches more than the length of each side of the candle's square base. ◼

Before concluding this section, here is an additional hint that can help you find the zeros of a polynomial function. When the terms of $f(x)$ have a common monomial factor, you should factor it out before applying the tests in this section. For example, writing $f(x) = x^4 - 5x^3 + 3x^2 + x = x(x^3 - 5x^2 + 3x + 1)$ shows that $x = 0$ is a zero of f. Obtain the remaining zeros by analyzing the cubic factor.

Summarize (Section 3.4)

1. State the Fundamental Theorem of Algebra and the Linear Factorization Theorem *(page 273, Example 1)*.

2. Explain how to use the Rational Zero Test *(page 274, Examples 2–5)*.

3. Explain how to use complex conjugates when analyzing a polynomial function *(page 277, Examples 6 and 7)*.

4. Explain how to find the zeros of a polynomial function *(page 278, Examples 8 and 9)*.

5. State Descartes's Rule of Signs and the Upper and Lower Bound Rules *(pages 280 and 281, Examples 10 and 11)*.

6. Describe a real-life application of finding the zeros of a polynomial function *(page 282, Example 12)*.

GO DIGITAL

3.4 Exercises

See CalcChat.com for tutorial help and worked-out solutions to odd-numbered exercises.

GO DIGITAL

Vocabulary and Concept Check

In Exercises 1–4, fill in the blanks.

1. The _____ _____ _____ _____ states that if $f(x)$ is a polynomial of degree n $(n > 0)$, then f has at least one zero in the complex number system.

2. The _____ _____ _____ states that if $f(x)$ is a polynomial of degree n $(n > 0)$, then $f(x)$ has precisely n linear factors, $f(x) = a_n(x - c_1)(x - c_2) \cdots (x - c_n)$, where $c_1, c_2, \ldots, c_n$ are complex numbers.

3. The test that gives a list of the possible rational zeros of a polynomial function is the _____ _____ Test.

4. A real number c is a _____ bound for the real zeros of f when no real zeros are less than c, and is a _____ bound when no real zeros are greater than c.

5. How many negative real zeros are possible for a polynomial function f, given that $f(-x)$ has five variations in sign?

6. Let $y = f(x)$ be a quartic (fourth-degree) polynomial with leading coefficient $a = 1$ and $f(i) = f(2i) = 0$. Describe how to write an equation for f.

Skills and Applications

Zeros of Polynomial Functions In Exercises 7–12, determine the number of zeros of the polynomial function.

7. $f(x) = x^3 + 2x^2 + 1$ 8. $f(x) = x^4 - 3x$

9. $g(x) = x^4 - x^5$

10. $f(x) = x^3 - x^6$

11. $f(x) = (x + 5)^2$

12. $h(t) = (t - 1)^2 - (t + 1)^2$

Using the Rational Zero Test In Exercises 13–16, use the Rational Zero Test to list the possible rational zeros of f. Verify that the zeros of f shown in the graph are contained in the list.

13. $f(x) = x^3 + 2x^2 - x - 2$

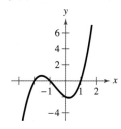

14. $f(x) = x^3 - 4x^2 - 4x + 16$

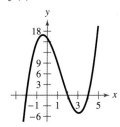

15. $f(x) = 2x^4 - 17x^3 + 35x^2 + 9x - 45$

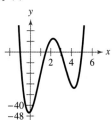

16. $f(x) = 4x^5 - 8x^4 - 5x^3 + 10x^2 + x - 2$

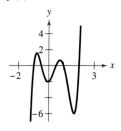

Using the Rational Zero Test In Exercises 17–26, find (if possible) the rational zeros of the function.

17. $f(x) = x^3 - 7x - 6$ 18. $f(x) = x^3 - 13x + 12$

19. $g(t) = t^3 - 4t^2 + 4$

20. $h(x) = x^3 - 19x + 30$

21. $h(t) = t^3 + 8t^2 + 13t + 6$

22. $g(x) = x^3 + 8x^2 + 12x + 18$

23. $C(x) = 2x^3 + 3x^2 - 1$

24. $f(x) = 3x^3 - 19x^2 + 33x - 9$

25. $g(x) = 9x^4 - 9x^3 - 58x^2 + 4x + 24$

26. $f(x) = 2x^4 - 15x^3 + 23x^2 + 15x - 25$

Solving a Polynomial Equation In Exercises 27–30, find all real solutions of the polynomial equation.

27. $-5x^3 + 11x^2 - 4x - 2 = 0$

28. $8x^3 + 10x^2 - 15x - 6 = 0$

29. $x^4 + 6x^3 + 3x^2 - 16x + 6 = 0$

30. $x^4 + 8x^3 + 14x^2 - 17x - 42 = 0$

Using the Rational Zero Test In Exercises 31–34, (a) list the possible rational zeros of f, (b) sketch the graph of f so that some of the possible zeros in part (a) can be disregarded, and then (c) determine all real zeros of f.

31. $f(x) = x^3 + x^2 - 4x - 4$

32. $f(x) = -3x^3 + 20x^2 - 36x + 16$

33. $f(x) = -4x^3 + 15x^2 - 8x - 3$

34. $f(x) = 4x^3 - 12x^2 - x + 15$

Using the Rational Zero Test In Exercises 35–38, (a) list the possible rational zeros of f, (b) use a graphing utility to graph f so that some of the possible zeros in part (a) can be disregarded, and then (c) determine all real zeros of f.

35. $f(x) = -2x^4 + 13x^3 - 21x^2 + 2x + 8$

36. $f(x) = 4x^4 - 17x^2 + 4$

37. $f(x) = 32x^3 - 52x^2 + 17x + 3$

38. $f(x) = 4x^3 + 7x^2 - 11x - 18$

Finding a Polynomial Function with Given Zeros In Exercises 39–44, find a polynomial function with real coefficients that has the given zeros. (There are many correct answers.)

39. $1, 5i$

40. $4, -3i$

41. $2, 2, 1 + i$

42. $-1, 5, 3 - 2i$

43. $\frac{2}{3}, -1, 3 + \sqrt{2}i$

44. $-\frac{5}{2}, -5, 1 + \sqrt{3}i$

Finding a Polynomial Function with Given Zeros In Exercises 45–48, find the polynomial function f with real coefficients that has the given degree, zeros, and solution point.

Degree	Zeros	Solution Point
45. 4	$-2, 1, i$	$f(0) = -4$
46. 4	$-1, 2, \sqrt{2}i$	$f(1) = 12$
47. 3	$-3, 1 + \sqrt{3}i$	$f(-2) = 12$
48. 3	$-2, 1 - \sqrt{2}i$	$f(-1) = -12$

Factoring a Polynomial In Exercises 49–52, write the polynomial (a) as the product of factors that are irreducible over the *rationals*, (b) as the product of linear and quadratic factors that are irreducible over the *reals*, and (c) in completely factored form.

49. $f(x) = x^4 + 2x^2 - 8$

50. $f(x) = x^4 + 6x^2 - 27$

51. $f(x) = x^4 - 2x^3 - 3x^2 + 12x - 18$
   (*Hint:* One factor is $x^2 - 6$.)

52. $f(x) = x^4 - 3x^3 - x^2 - 12x - 20$
   (*Hint:* One factor is $x^2 + 4$.)

Finding the Zeros of a Polynomial Function In Exercises 53–58, use the given zero to find all the zeros of the function.

Function	Zero
53. $f(x) = x^3 - x^2 + 4x - 4$	$2i$
54. $f(x) = 2x^3 + 3x^2 + 18x + 27$	$3i$
55. $g(x) = x^3 - 8x^2 + 25x - 26$	$3 + 2i$
56. $g(x) = x^3 + 9x^2 + 25x + 17$	$-4 + i$
57. $h(x) = x^4 - 6x^3 + 14x^2 - 18x + 9$	$1 - \sqrt{2}i$
58. $h(x) = x^4 + x^3 - 3x^2 - 13x + 14$	$-2 + \sqrt{3}i$

Finding the Zeros of a Polynomial Function In Exercises 59–70, write the polynomial as the product of linear factors and list all the zeros of the function.

59. $f(x) = x^2 + 36$ 60. $f(x) = x^2 + 49$

61. $h(x) = x^2 - 2x + 17$ 62. $g(x) = x^2 + 10x + 17$

63. $f(x) = x^4 - 16$ 64. $f(y) = y^4 - 256$

65. $f(z) = z^2 - 2z + 2$

66. $h(x) = x^3 - 3x^2 + 4x - 2$

67. $g(x) = x^3 - 3x^2 + x + 5$

68. $f(x) = x^3 - x^2 + x + 39$

69. $g(x) = x^4 - 4x^3 + 8x^2 - 16x + 16$

70. $h(x) = x^4 + 6x^3 + 10x^2 + 6x + 9$

Finding the Zeros of a Polynomial Function In Exercises 71–76, find all the zeros of the function. When there is an extended list of possible rational zeros, use a graphing utility to graph the function to disregard any of the possible rational zeros that are obviously not zeros of the function.

71. $f(x) = x^3 + 24x^2 + 214x + 740$

72. $f(s) = 2s^3 - 5s^2 + 12s - 5$

73. $f(x) = 16x^3 - 20x^2 - 4x + 15$

74. $f(x) = 9x^3 - 15x^2 + 11x - 5$

75. $f(x) = 2x^4 + 5x^3 + 4x^2 + 5x + 2$

76. $g(x) = x^5 - 8x^4 + 28x^3 - 56x^2 + 64x - 32$

Using Descartes's Rule of Signs   In Exercises 77–84, use Descartes's Rule of Signs to determine the possible numbers of positive and negative real zeros of the function.

77. $g(x) = 2x^3 - 3x^2 - 3$ **78.** $h(x) = 4x^2 - 8x + 3$

79. $h(x) = 2x^3 + 3x^2 + 1$ **80.** $h(x) = 2x^4 - 3x - 2$

81. $g(x) = 6x^4 + 2x^3 - 3x^2 + 2$

82. $f(x) = 4x^3 - 3x^2 - 2x - 1$

83. $f(x) = 5x^3 + x^2 - x + 5$

84. $f(x) = 3x^3 - 2x^2 - x + 3$

Verifying Upper and Lower Bounds   In Exercises 85–88, use synthetic division to verify the upper and lower bounds of the real zeros of f.

85. $f(x) = x^3 + 3x^2 - 2x + 1$

    (a) Upper: $x = 1$ (b) Lower: $x = -4$

86. $f(x) = x^3 - 4x^2 + 1$

    (a) Upper: $x = 4$ (b) Lower: $x = -1$

87. $f(x) = x^4 - 4x^3 + 16x - 16$

    (a) Upper: $x = 5$ (b) Lower: $x = -3$

88. $f(x) = 2x^4 - 8x + 3$

    (a) Upper: $x = 3$ (b) Lower: $x = -4$

Finding Real Zeros of a Polynomial Function   In Exercises 89–92, find all real zeros of the function.

89. $f(x) = 16x^3 - 12x^2 - 4x + 3$

90. $f(z) = 12z^3 - 4z^2 - 27z + 9$

91. $f(y) = 4y^3 + 3y^2 + 8y + 6$

92. $g(x) = 3x^3 - 2x^2 + 15x - 10$

Finding the Rational Zeros of a Polynomial   In Exercises 93–96, find the rational zeros of the polynomial function.

93. $P(x) = x^4 - \frac{25}{4}x^2 + 9 = \frac{1}{4}(4x^4 - 25x^2 + 36)$

94. $f(x) = x^3 - \frac{3}{2}x^2 - \frac{23}{2}x + 6$
$\quad\quad = \frac{1}{2}(2x^3 - 3x^2 - 23x + 12)$

95. $f(x) = x^3 - \frac{1}{4}x^2 - x + \frac{1}{4} = \frac{1}{4}(4x^3 - x^2 - 4x + 1)$

96. $f(z) = z^3 + \frac{11}{6}z^2 - \frac{1}{2}z - \frac{1}{3} = \frac{1}{6}(6z^3 + 11z^2 - 3z - 2)$

97. Geometry   You want to make an open box from a rectangular piece of material, 15 centimeters by 9 centimeters, by cutting equal squares from the corners and turning up the sides.

    (a) Let x represent the side length of each of the squares removed. Draw a diagram showing the squares removed from the original piece of material and the resulting dimensions of the open box.

    (b) Use the diagram to write the volume V of the box as a function of x. Determine the domain of the function.

    (c) Sketch the graph of the function and approximate the dimensions of the box that yield a maximum volume.

    (d) Find values of x such that $V = 56$. Which of these values is a physical impossibility in the construction of the box? Explain.

98. Geometry   A rectangular package to be sent by a delivery service (see figure) has a combined length and girth (perimeter of a cross section) of 120 inches.

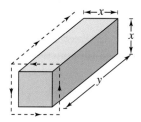

    (a) Use the diagram to write the volume V of the package as a function of x.

    (b) Use a graphing utility to graph the function and approximate the dimensions of the package that yield a maximum volume.

    (c) Find values of x such that $V = 13{,}500$. Which of these values is a physical impossibility in the construction of the package? Explain.

99. Geometry

A bulk food storage bin with dimensions 2 feet by 3 feet by 4 feet needs to be increased in size to hold five times as much food as the current bin.

    (a) Assume each dimension is increased by the same amount. Write a function that represents the volume V of the new bin.

    (b) Find the dimensions of the new bin.

100. Cost   The ordering and transportation cost C (in thousands of dollars) for machine parts is given by

$$C(x) = 100\left(\frac{200}{x^2} + \frac{x}{x + 30}\right), \quad x \ge 1$$

where x is the order size (in hundreds). In calculus, it can be shown that the cost is a minimum when

$$3x^3 - 40x^2 - 2400x - 36{,}000 = 0.$$

Use a graphing utility to approximate the optimal order size to the nearest hundred units.

Exploring the Concepts

True or False? **In Exercises 101 and 102, decide whether the statement is true or false. Justify your answer.**

101. It is possible for a third-degree polynomial function with integer coefficients to have no real zeros.

102. If $x = -i$ is a zero of the function

$$f(x) = x^3 + ix^2 + ix - 1$$

then $x = i$ must also be a zero of f.

Think About It **In Exercises 103–108, determine (if possible) the zeros of the function g when the function f has zeros at $x = r_1$, $x = r_2$, and $x = r_3$.**

103. $g(x) = -f(x)$ **104.** $g(x) = 3f(x)$

105. $g(x) = f(x - 5)$ **106.** $g(x) = f(2x)$

107. $g(x) = 3 + f(x)$ **108.** $g(x) = f(-x)$

109. **Think About It** A cubic polynomial function f has real zeros -2, $\frac{1}{2}$, and 3, and its leading coefficient is negative. Write an equation for f and sketch its graph. How many different polynomial functions are possible for f?

110. **Think About It** Sketch the graph of a fifth-degree polynomial function whose leading coefficient is positive and that has a zero at $x = 3$ of multiplicity 2.

Writing an Equation **In Exercises 111 and 112, the graph of a cubic polynomial function $y = f(x)$ is shown. One of the zeros is $1 + i$. Write an equation for f.**

111. 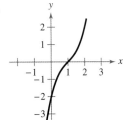   **112.**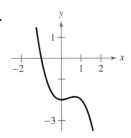

113. **Error Analysis** Describe the error.

The graph of a quartic (fourth-degree) polynomial $y = f(x)$ is shown. One of the zeros is i.

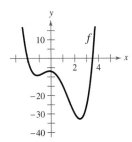

The function is $f(x) = (x + 2)(x - 3.5)(x - i)$.

114. **HOW DO YOU SEE IT?** Use the information in the table to answer each question.

Interval	Value of $f(x)$
$(-\infty, -2)$	Positive
$(-2, 1)$	Negative
$(1, 4)$	Negative
$(4, \infty)$	Positive

(a) What are the three real zeros of the polynomial function f?

(b) What can be said about the behavior of the graph of f at $x = 1$?

(c) What is the least possible degree of f? Explain. Can the degree of f ever be odd? Explain.

(d) Is the leading coefficient of f positive or negative? Explain.

(e) Sketch a graph of a function that exhibits the behavior described in the table.

Review & Refresh ▶ *Video solutions at LarsonPrecalculus.com*

Solving an Inequality **In Exercises 115–120, solve the inequality. Then graph the solution set.**

115. $|x| < 7$ **116.** $|x| > 2$

117. $|4x - 1| \le 11$ **118.** $|3x + 7| \ge 13$

119. $(x - 1)^2 < 4$

120. $(5 - x)^2 + 2 \ge 18$

Finding the Slope of a Line Through Two Points **In Exercises 121–126, find the slope of the line passing through the point $(1, 1)$ and the given point.**

121. $(2, 5)$ **122.** $(1, -3)$

123. $(6, 0.5)$ **124.** $(5, 2.5)$

125. $\left(-\frac{2}{5}, -\frac{3}{5}\right)$ **126.** $\left(-\frac{3}{4}, -\frac{15}{8}\right)$

Solving for a Variable **In Exercises 127–134, evaluate the equation for k when $x = 4$, $y = -2$, and $z = -3$.**

127. $y = \dfrac{2k - 5}{x}$ **128.** $\dfrac{4}{3k} = \dfrac{x}{y}$

129. $z = \dfrac{kx}{y}$ **130.** $kz = \dfrac{x}{2y - 1}$

131. $kz = \dfrac{x}{-5y - 7}$ **132.** $ky = \dfrac{zx}{3y}$

133. $z = kxy$ **134.** $\dfrac{z}{x} = ky^2$

3.5 Mathematical Modeling and Variation

Mathematical models have a wide variety of real-life applications. For example, in Exercise 65 on page 296, you will use variation to model ocean temperatures at various depths.

> ❷ Use mathematical models to approximate sets of data points.
> ❷ Use the *regression* feature of a graphing utility to find equations of least squares regression lines.
> ❷ Write mathematical models for direct variation.
> ❷ Write mathematical models for direct variation as an *n*th power.
> ❷ Write mathematical models for inverse variation.
> ❷ Write mathematical models for combined variation.
> ❷ Write mathematical models for joint variation.

Introduction

In this section, you will study two techniques for fitting models to data: *least squares regression* and *direct and inverse variation*.

EXAMPLE 1   Using a Mathematical Model

The table shows the populations *y* (in millions) of the United States from 2012 through 2019.   *(Source: U.S. Census Bureau)*

Year	2012	2013	2014	2015	2016	2017	2018	2019
Population, *y*	313.8	316.0	318.3	320.6	322.9	325.0	326.7	328.2

Spreadsheet at LarsonPrecalculus.com

A linear model that approximates the data is

$$y = 2.10t + 288.8, \quad 12 \le t \le 19$$

where *t* represents the year, with $t = 12$ corresponding to 2012. Plot the actual data *and* the model on the same graph. How closely does the model represent the data?

Solution   Figure 3.22 shows the actual data and the model plotted on the same graph. From the graph, it appears that the model is a "good fit" for the actual data. To see how well the model fits, compare the actual values of *y* with the values of *y* found using the model. The values found using the model are labeled *y** in the table below.

t	12	13	14	15	16	17	18	19
y	313.8	316.0	318.3	320.6	322.9	325.0	326.7	328.2
y*	314.0	316.1	318.2	320.3	322.4	324.5	326.6	328.7

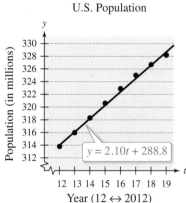

U.S. Population

$y = 2.10t + 288.8$

Year (12 ↔ 2012)

Figure 3.22

✓ *Checkpoint* ▶ Audio-video solution in English & Spanish at *LarsonPrecalculus.com*

The ordered pairs below give the median sales prices *y* (in thousands of dollars) of new homes sold in a neighborhood from 2013 through 2020. *(Spreadsheet at LarsonPrecalculus.com)*

DATA

(2013, 179.4)   (2015, 191.0)   (2017, 202.6)   (2019, 214.9)
(2014, 185.4)   (2016, 196.7)   (2018, 208.7)   (2020, 221.4)

A linear model that approximates the data is $y = 5.96t + 101.7$, $13 \le t \le 20$, where *t* represents the year, with $t = 13$ corresponding to 2013. Plot the actual data *and* the model on the same graph. How closely does the model represent the data?

GO DIGITAL

Least Squares Regression and Graphing Utilities

So far in this text, you have worked with many different types of mathematical models that approximate real-life data. In some instances the model was given (as in Example 1), whereas in other instances you found the model using algebraic techniques or a graphing utility.

To find a model that approximates a set of data most accurately, statisticians use a measure called the **sum of the squared differences,** which is the sum of the squares of the differences between actual data values and model values. The "best-fitting" linear model, called the **least squares regression line,** is the one with the least sum of the squared differences.

Recall that you can approximate this line visually by plotting the data points and drawing the line that appears to best fit the data—or you can enter the data points into a graphing utility and use the *linear regression* feature.

Note that the output of a graphing utility's *linear regression* feature may display an *r*-value called the *correlation coefficient.* The **correlation coefficient *r*** is a measure of the strength of a *linear* relationship between two variables. The closer $|r|$ is to 1, the better the fit.

EXAMPLE 2 Finding a Least Squares Regression Line

▶▶▶ *See LarsonPrecalculus.com for an interactive version of this type of example.*

The table shows the total outstanding student loan debt D (in trillions of dollars) in the United States from 2012 through 2019. Construct a scatter plot that represents the data and find the equation of the least squares regression line for the data.   *(Source: New York Fed Consumer Credit Panel/Equifax)*

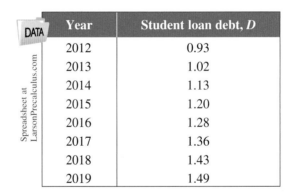

Year	Student loan debt, D
2012	0.93
2013	1.02
2014	1.13
2015	1.20
2016	1.28
2017	1.36
2018	1.43
2019	1.49

Solution Let $t = 12$ represent 2012. Figure 3.23 shows a scatter plot of the data. Using the *regression* feature of a graphing utility, the equation of the least squares regression line is $D = 0.080t - 0.01$. To check this model, compare the actual D-values with the D-values found using the model, which are labeled $D*$ in the table at the left. The correlation coefficient for this model is $r \approx 0.997$, so the model is a good fit.

✓ *Checkpoint* ▶ Audio-video solution in English & Spanish at LarsonPrecalculus.com

The ordered pairs below give the total outstanding auto loan debt D (in trillions of dollars) in the United States from 2012 through 2019. Construct a scatter plot that represents the data and find the equation of the least squares regression line for the data.   *(Source: New York Fed Consumer Credit Panel/Equifax)*

(2012, 0.76) (2014, 0.92) (2016, 1.12) (2018, 1.25)
(2013, 0.83) (2015, 1.02) (2017, 1.20) (2019, 1.30)

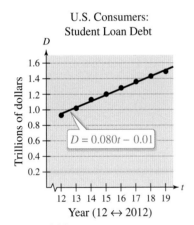

U.S. Consumers:
Student Loan Debt

$D = 0.080t - 0.01$

Year (12 ↔ 2012)

Figure 3.23

t	D	$D*$
12	0.93	0.95
13	1.02	1.03
14	1.13	1.11
15	1.20	1.19
16	1.28	1.27
17	1.36	1.35
18	1.43	1.43
19	1.49	1.51

GO DIGITAL

Direct Variation

There are two basic types of linear models. The more general model has a nonzero *y*-intercept.

$$y = mx + b, \quad b \neq 0$$ Linear model with slope *m* and *y*-intercept at $(0, b)$

The simpler model

$$y = kx$$ Linear model with slope $m = k$ and *y*-intercept at $(0, 0)$

has a *y*-intercept of zero. In the simpler model, *y* **varies directly** as *x*, or is **directly proportional** to *x*.

Direct Variation

The statements below are equivalent.

1. *y* **varies directly** as *x*.
2. *y* is **directly proportional** to *x*.
3. $y = kx$ for some nonzero constant *k*.

The number *k* is the **constant of variation** or the **constant of proportionality**.

EXAMPLE 3   **Direct Variation**

In Pennsylvania, the state income tax is directly proportional to *gross income*. You work in Pennsylvania and your state income tax deduction is $46.05 for a gross monthly income of $1500. Find a mathematical model that gives the Pennsylvania state income tax in terms of gross income.

Solution

Verbal model: State income tax $= k \cdot$ Gross income

Labels: State income tax $= y$ (dollars)
             Gross income $= x$ (dollars)
             Income tax rate $= k$ (percent in decimal form)

Equation: $y = kx$

To find the state income tax rate *k*, substitute the given information into the equation $y = kx$ and solve.

$$y = kx$$ Write direct variation model.
$$46.05 = k(1500)$$ Substitute 46.05 for *y* and 1500 for *x*.
$$0.0307 = k$$ Divide each side by 1500.

So, the equation (or model) for state income tax in Pennsylvania is

$$y = 0.0307x.$$

In other words, Pennsylvania has a state income tax rate of 3.07% of gross income. Figure 3.24 shows the graph of this equation.

 Checkpoint ▶ Audio-video solution in English & Spanish at LarsonPrecalculus.com

The simple interest on an investment is directly proportional to the amount of the investment. For example, an investment of $2500 earns $187.50 after 1 year. Find a mathematical model that gives the interest *I* after 1 year in terms of the amount invested *P*.

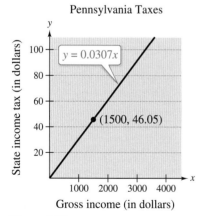

Pennsylvania Taxes

$y = 0.0307x$

(1500, 46.05)

State income tax (in dollars) / Gross income (in dollars)

Figure 3.24

GO DIGITAL

Direct Variation as an *n*th Power

Another type of direct variation relates one variable to a *power* of another variable. For example, in the formula for the area of a circle

$$A = \pi r^2$$ Area of a circle

the area A is directly proportional to the square of the radius r. Note that for this formula, π is the constant of proportionality.

Direct Variation as an *n*th Power

The statements below are equivalent.
1. y **varies directly as the *n*th power** of x.
2. y is **directly proportional to the *n*th power** of x.
3. $y = kx^n$ for some nonzero constant k.

EXAMPLE 4   **Direct Variation as an *n*th Power**

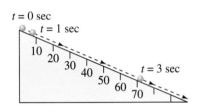

$t = 0$ sec

$t = 1$ sec

$t = 3$ sec

Figure 3.25

The distance a ball rolls down an inclined plane is directly proportional to the square of the time it rolls. During the first second, the ball rolls 8 feet. (See Figure 3.25.)

a. Write an equation relating the distance traveled to the time.

b. How far does the ball roll during the first 3 seconds?

Solution

a. Letting d be the distance (in feet) the ball rolls and letting t be the time (in seconds), you have

$$d = kt^2.$$ The distance d is directly proportional to the 2nd power of time t.

Now, $d = 8$ when $t = 1$, so you have

$$d = kt^2$$ Write direct variation model.
$$8 = k(1)^2$$ Substitute 8 for d and 1 for t.
$$8 = k$$ Simplify.

and, the equation relating distance to time is

$$d = 8t^2.$$

b. When $t = 3$, the distance traveled is

$$d = 8(3)^2$$ Substitute 3 for t.
$$= 8(9)$$ Simplify.
$$= 72 \text{ feet.}$$ Simplify.

So, the ball rolls 72 feet during the first 3 seconds.

✓ *Checkpoint* ▶ Audio-video solution in English & Spanish at LarsonPrecalculus.com

Neglecting air resistance, the distance s an object falls varies directly as the square of the duration t of the fall. An object falls a distance of 144 feet in 3 seconds. How far does it fall in 6 seconds? ∎

In Examples 3 and 4, the direct variations are such that an *increase* in one variable corresponds to an *increase* in the other variable. You should not, however, assume that this always occurs with direct variation. For example, for the model $y = -3x$, an increase in x results in a *decrease* in y, and yet y is said to vary directly as x.

GO DIGITAL

Inverse Variation

Inverse Variation

The statements below are equivalent.

1. y **varies inversely** as x.

2. y is **inversely proportional** to x.

3. $y = \dfrac{k}{x}$ for some nonzero constant k.

If x and y are related by an equation of the form $y = k/x^n$, then y varies inversely as the nth power of x (or y is inversely proportional to the nth power of x).

EXAMPLE 5 Inverse Variation

A company has found that the demand for one of its products varies inversely as the price of the product. When the price is \$6.25, the demand is 400 units. Approximate the demand when the price is \$5.75.

Solution

Let p be the price and let x be the demand. The demand varies inversely as the price, so you have

$$x = \frac{k}{p}.$$ The demand x is inversely proportional to the price p.

Now, $x = 400$ when $p = 6.25$, so you have

$$x = \frac{k}{p}$$ Write inverse variation model.

$$400 = \frac{k}{6.25}$$ Substitute 400 for x and 6.25 for p.

$$(400)(6.25) = k$$ Multiply each side by 6.25.

$$2500 = k$$ Simplify.

and the equation relating price and demand is

$$x = \frac{2500}{p}.$$

When $p = 5.75$, the demand is

$$x = \frac{2500}{p}$$ Write inverse variation model.

$$= \frac{2500}{5.75}$$ Substitute 5.75 for p.

$$\approx 435 \text{ units.}$$ Simplify.

So, the demand for the product is about 435 units when the price is \$5.75.

 Checkpoint ▶ *Audio-video solution in English & Spanish at LarsonPrecalculus.com*

The company in Example 5 has found that the demand for another of its products also varies inversely as the price of the product. When the price is \$2.75, the demand is 600 units. Approximate the demand when the price is \$3.25. ■

Supply and demand are fundamental concepts in economics. The law of demand states that, all other factors remaining equal, the lower the price of the product, the higher the quantity demanded. The law of supply states that the higher the price of the product, the higher the quantity supplied. *Equilibrium* occurs when the demand and the supply are the same.

GO DIGITAL

Combined Variation

Some applications of variation involve problems with *both* direct and inverse variations in the same model. These types of models have **combined variation.**

EXAMPLE 6 **Combined Variation**

A gas law states that the volume of an enclosed gas varies inversely as the pressure (Figure 3.26) *and* directly as the temperature. The pressure of a gas is 0.75 kilogram per square centimeter when the temperature is 294 K and the volume is 8000 cubic centimeters.

a. Write an equation relating pressure, temperature, and volume.

b. Find the pressure when the temperature is 300 K and the volume is 7000 cubic centimeters.

Solution

a. Volume V varies directly as temperature T and inversely as pressure P, so you have $V = (kT)/P$. Now, $P = 0.75$ when $T = 294$ and $V = 8000$, so you have

$$V = \frac{kT}{P}$$ Write combined variation model.

$$8000 = \frac{k(294)}{0.75}$$ Substitute 8000 for V, 294 for T, and 0.75 for P.

$$\frac{6000}{294} = k$$ Simplify.

$$\frac{1000}{49} = k$$ Simplify.

and the equation relating pressure, temperature, and volume is

$$V = \frac{1000}{49}\left(\frac{T}{P}\right).$$

b. To find the pressure, rewrite the equation in part (a) by isolating P. To do this, multiply each side of the equation by P/V.

$$V = \frac{1000}{49}\left(\frac{T}{P}\right) \implies V\left(\frac{P}{V}\right) = \frac{1000}{49}\left(\frac{T}{P}\right)\left(\frac{P}{V}\right) \implies P = \frac{1000}{49}\left(\frac{T}{V}\right)$$

When $T = 300$ and $V = 7000$, the pressure is

$$P = \frac{1000}{49}\left(\frac{T}{V}\right)$$ Combined variation model solved for P.

$$= \frac{1000}{49}\left(\frac{300}{7000}\right)$$ Substitute 300 for T and 7000 for V.

$$= \frac{300}{343}$$ Simplify.

$$\approx 0.87 \text{ kilogram per square centimeter.}$$ Use a calculator.

So, the pressure is about 0.87 kilogram per square centimeter when the temperature is 300 K and the volume is 7000 cubic centimeters.

✓ *Checkpoint* ▶ *Audio-video solution in English & Spanish at LarsonPrecalculus.com*

The resistance of a copper wire carrying an electrical current is directly proportional to its length and inversely proportional to its cross-sectional area. A copper wire with a diameter of 0.0126 inch has a resistance of 64.9 ohms per thousand feet. What length of 0.0201-inch-diameter copper wire will produce a resistance of 33.5 ohms? ∎

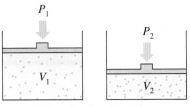

If $P_2 > P_1$, then $V_2 < V_1$.

If the temperature is held constant and pressure increases, then the volume *decreases*.

Figure 3.26

GO DIGITAL

Joint Variation

> ### Joint Variation
>
> The statements below are equivalent.
> 1. z **varies jointly** as x and y.
> 2. z is **jointly proportional** to x and y.
> 3. $z = kxy$ for some nonzero constant k.

If x, y, and z are related by an equation of the form $z = kx^n y^m$, then z varies jointly as the nth power of x and the mth power of y.

EXAMPLE 7 **Joint Variation**

The *simple* interest for an investment is jointly proportional to the time and the principal. After one quarter (3 months), the interest on a principal of \$5000 is \$43.75. (a) Write an equation relating the interest, principal, and time. (b) Find the interest after three quarters.

Solution

a. Interest I (in dollars) is jointly proportional to principal P (in dollars) and time t (in years), so you have

$$I = kPt. \qquad \text{The interest } i \text{ is jointly proportional to the principal } P \text{ and time } t.$$

For $I = 43.75$, $P = 5000$, and $t = \frac{3}{12} = \frac{1}{4}$, you have $43.75 = k(5000)\left(\frac{1}{4}\right)$, which implies that $k = 4(43.75)/5000 = 0.035$. So, the equation relating interest, principal, and time is

$$I = 0.035Pt$$

which is the familiar equation for simple interest where the constant of proportionality, 0.035, represents an annual interest rate of 3.5%.

b. When $P = \$5000$ and $t = \frac{3}{4}$, the interest is $I = (0.035)(5000)\left(\frac{3}{4}\right) = \$131.25$.

✓ *Checkpoint* *Audio-video solution in English & Spanish at LarsonPrecalculus.com*

The kinetic energy E of an object varies jointly with the object's mass m and the square of the object's velocity v. An object with a mass of 50 kilograms traveling at 16 meters per second has a kinetic energy of 6400 joules. What is the kinetic energy of an object with a mass of 70 kilograms traveling at 20 meters per second? ■

Summarize (Section 3.5)

1. Explain how to use a mathematical model to approximate a set of data points *(page 287)*. For an example of using a mathematical model to approximate a set of data points, see Example 1.
2. Explain how to use the *regression* feature of a graphing utility to find the equation of a least squares regression line *(page 288)*. For an example of finding the equation of a least squares regression line, see Example 2.
3. Explain how to write mathematical models for direct variation, direct variation as an nth power, inverse variation, combined variation, and joint variation *(pages 289–293)*. For examples of these types of variation, see Examples 3–7.

GO DIGITAL

3.5 Exercises

See CalcChat.com for tutorial help and worked-out solutions to odd-numbered exercises.

Vocabulary and Concept Check

In Exercises 1–4, fill in the blanks.

1. Statisticians use a measure called the _____ of the _____ _____ to find a model that approximates a set of data most accurately.

2. The linear model with the least sum of the squared differences is called the _____ line.

3. An *r*-value, or _____ _____, of a set of data gives a measure of the strength of a linear correlation between two variables.

4. The direct variation model $y = kx^n$ can be described as "*y* varies directly as the *n*th power of *x*," or "*y* is _____ _____ to the *n*th power of *x*."

5. What are two other ways to describe how *x*, *y*, and *z* are related when $z = kxy$ for some nonzero constant *k*?

6. What type of variation does each mathematical model represent?

   (a) $y = \dfrac{4}{x}$ (b) $z = 12xy$ (c) $y = 3.5x$ (d) $z = \dfrac{6x}{y}$

Skills and Applications

Mathematical Models In Exercises 7 and 8, (a) plot the actual data and the model of the same graph and (b) describe how closely the model represents the data. If the model does not closely represent the data, suggest another type of model that may be a better fit.

7. The ordered pairs below give the civilian noninstitutional U.S. populations *y* (in millions of people) 16 years of age and over in the civilian labor force from 2011 through 2019. *(Spreadsheet at LarsonPrecalculus.com)*

   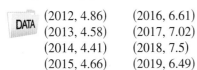

   (2011, 153.6)   (2016, 159.2)
   (2012, 155.0)   (2017, 160.3)
   (2013, 155.4)   (2018, 162.1)
   (2014, 155.9)   (2019, 163.5)
   (2015, 157.1)

   A model for the data is $y = 1.23t + 151.8$, $1 \le t \le 9$, where *t* represents the year, with $t = 1$ corresponding to 2011. *(Source: U.S. Bureau of Labor Statistics)*

8. The ordered pairs below give the revenues *y* (in billions of dollars) for Activision Blizzard, Inc., from 2012 through 2019. *(Spreadsheet at LarsonPrecalculus.com)*

   (2012, 4.86)   (2016, 6.61)
   (2013, 4.58)   (2017, 7.02)
   (2014, 4.41)   (2018, 7.5)
   (2015, 4.66)   (2019, 6.49)

   A model for the data is $y = 0.426t + 3.42$, $2 \le t \le 9$, where *t* represents the year, with $t = 2$ corresponding to 2012. *(Source: Activision Blizzard, Inc.)*

Sketching a Line In Exercises 9–14, sketch the line that you think best approximates the data in the scatter plot. Then find an equation of the line. To print an enlarged copy of the graph, go to *MathGraphs.com*.

9.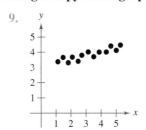

10.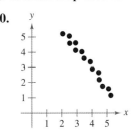

11.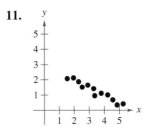

12.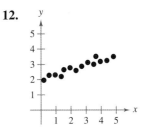

13.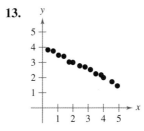

14.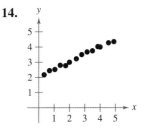

15. Sports The ordered pairs below give the winning times (in seconds) of the women's 100-meter freestyle swimming event in the Olympics from 1988 through 2016. *(Spreadsheet at LarsonPrecalculus.com) (Source: International Olympic Committee)*

DATA
(1988, 54.93)	(2004, 53.84)
(1992, 54.64)	(2008, 53.12)
(1996, 54.50)	(2012, 53.00)
(2000, 53.83)	(2016, 52.70)

(a) Sketch a scatter plot of the data. Let y represent the winning time (in seconds) and let $t = 88$ represent 1988.

(b) Sketch the line that you think best approximates the data and find an equation of the line.

(c) Use the *regression* feature of a graphing utility to find the equation of the least squares regression line that fits the data.

(d) Compare the linear model you found in part (b) with the linear model you found in part (c).

16. Tourism The ordered pairs below give the number of tourists (in millions) that visited New York City during each year from 2001 through 2018. *(Spreadsheet at LarsonPrecalculus.com) (Source: NYC and Company)*

DATA
(2001, 35.2)	(2007, 46.0)	(2013, 53.0)
(2002, 35.3)	(2008, 47.1)	(2014, 54.4)
(2003, 37.8)	(2009, 45.8)	(2015, 55.9)
(2004, 39.9)	(2010, 48.8)	(2016, 60.5)
(2005, 42.6)	(2011, 50.9)	(2017, 62.8)
(2006, 43.8)	(2012, 51.5)	(2018, 65.2)

(a) Use a graphing utility to create a scatter plot of the data. Let $t = 1$ represent 2001.

(b) Use the *regression* feature of the graphing utility to find the equation of the least squares regression line that fits the data.

(c) Use the graphing utility to graph the scatter plot you created in part (a) and the model you found in part (b) in the same viewing window. How closely does the model represent the data?

(d) Use the model to predict the number of tourists that will visit New York City in 2025.

(e) Interpret the meaning of the slope of the linear model in the context of the problem.

Direct Variation   In Exercises 17–22, find a direct variation model that relates y and x.

17. $x = 2, y = 14$ 18. $x = 5, y = 12$

19. $x = -24, y = 3$ 20. $x = 5, y = 1$

21. $x = 4, y = 8\pi$ 22. $x = \pi, y = -1$

Direct Variation as an *n*th Power   In Exercises 23–26, use the given values of k and n to complete the table for the direct variation model $y = kx^n$. Plot the points in a rectangular coordinate system.

x	2	4	6	8	10
$y = kx^n$					

23. $k = 1, n = 2$

24. $k = 2, n = 2$

25. $k = \frac{1}{2}, n = 3$

26. $k = \frac{1}{4}, n = 3$

Inverse Variation as an *n*th Power   In Exercises 27–30, use the given values of k and n to complete the table for the inverse variation model $y = k/x^n$. Plot the points in a rectangular coordinate system.

x	2	4	6	8	10
$y = k/x^n$					

27. $k = 2, n = 1$

28. $k = 5, n = 1$

29. $k = 10, n = 2$

30. $k = 20, n = 2$

Determining Variation   In Exercises 31–34, determine whether the variation model represented by the ordered pairs (x, y) is of the form $y = kx$ or $y = k/x$, and find k. Then write a model that relates y and x.

31. $(5, 1), \left(10, \frac{1}{2}\right), \left(15, \frac{1}{3}\right), \left(20, \frac{1}{4}\right), \left(25, \frac{1}{5}\right)$

32. $(5, 2), (10, 4), (15, 6), (20, 8), (25, 10)$

33. $(5, -3.5), (10, -7), (15, -10.5), (20, -14), (25, -17.5)$

34. $(5, 24), (10, 12), (15, 8), (20, 6), \left(25, \frac{24}{5}\right)$

Finding a Mathematical Model   In Exercises 35–44, find a mathematical model for the verbal statement.

35. A varies directly as the square of r.

36. V varies directly as the cube of l.

37. y varies inversely as the square of x.

38. h varies inversely as the square root of s.

39. F varies directly as g and inversely as r^2.

40. z varies jointly as the square of x and the cube of y.

41. *Newton's Law of Cooling:* The rate of change R of the temperature of an object is directly proportional to the difference between the temperature T of the object and the temperature T_e of the environment.

42. *Boyle's Law:* For a constant temperature, the pressure P of a gas is inversely proportional to the volume V of the gas.

43. *Direct Current:* The electric power P of a direct current circuit is jointly proportional to the voltage V and the electric current I.

44. *Newton's Law of Universal Gravitation:* The gravitational attraction F between two objects of masses m_1 and m_2 is jointly proportional to the masses and inversely proportional to the square of the distance r between the objects.

Describing a Formula In Exercises 45–48, use variation terminology to describe the formula.

45. $y = 2x^2$

46. $t = \dfrac{72}{r}$

47. $A = \dfrac{1}{2}bh$

48. $K = \dfrac{1}{2}mv^2$

Finding a Mathematical Model In Exercises 49–56, find a mathematical model that represents the statement. (Determine the constant of proportionality.)

49. y is directly proportional to x. ($y = 54$ when $x = 3$.)
50. A varies directly as r^2. ($A = 9\pi$ when $r = 3$.)
51. y varies inversely as x. ($y = 3$ when $x = 25$.)
52. y is inversely proportional to x^3. ($y = 7$ when $x = 2$.)
53. z varies jointly as x and y. ($z = 64$ when $x = 4$ and $y = 8$.)
54. F is jointly proportional to r and the third power of s. ($F = 4158$ when $r = 11$ and $s = 3$.)
55. P varies directly as x and inversely as the square of y. ($P = \frac{28}{3}$ when $x = 42$ and $y = 9$.)
56. z varies directly as the square of x and inversely as y. ($z = 6$ when $x = 6$ and $y = 4$.)

57. **Simple Interest** The simple interest on an investment is directly proportional to the amount of the investment. An investment of $3250 earns $113.75 after 1 year. Find a mathematical model that gives the interest I after 1 year in terms of the amount invested P.

58. **Simple Interest** The simple interest on a savings account is directly proportional to the amount of the investment. A savings account with $2500 earns $31.25 after 1 year. Find a mathematical model that gives the interest I after 1 year in terms of the amount invested P.

59. **Measurement** Use the fact that 13 inches is approximately the same length as 33 centimeters to find a mathematical model that relates centimeters y to inches x. Then use the model to find the numbers of centimeters in 10 inches and 20 inches.

60. **Measurement** Use the fact that 14 gallons is approximately the same amount as 53 liters to find a mathematical model that relates liters y to gallons x. Then use the model to find the numbers of liters in 5 gallons and 25 gallons.

Hooke's Law In Exercises 61–64, use Hooke's Law, which states that the distance a spring stretches (or compresses) from its natural, or equilibrium, length varies directly as the applied force on the spring.

61. A force of 220 newtons stretches a spring 0.12 meter. What force stretches the spring 0.16 meter?

62. A force of 265 newtons stretches a spring 0.15 meter.
   (a) What force stretches the spring 0.1 meter?
   (b) How far does a force of 90 newtons stretch the spring?

63. The coiled spring of a toy supports the weight of a child. The weight of a 25-pound child compresses the spring a distance of 1.9 inches. The toy does not work properly when a weight compresses the spring more than 3 inches. What is the maximum weight for which the toy works properly?

64. An overhead garage door has two springs, one on each side of the door. A force of 15 pounds is required to stretch each spring 1 foot. Because of a pulley system, the springs stretch only one-half the distance the door travels. The door moves a total of 8 feet, and the springs are at their natural lengths when the door is open. Find the combined lifting force applied to the door by the springs when the door is closed.

65. **Ocean Temperatures**

The ordered pairs below give the average water temperatures C (in degrees Celsius) at several depths d (in meters) in the Indian Ocean.

(Spreadsheet at LarsonPrecalculus.com)
(Source: NOAA)

DATA
(1000, 4.85)	(2500, 1.888)
(1500, 3.525)	(3000, 1.583)
(2000, 2.468)	(3500, 1.422)

(a) Sketch a scatter plot of the data.
(b) Determine whether a direct variation model or an inverse variation model better fits the data.
(c) Find k for each pair of coordinates. Then find the mean value of k to find the constant of proportionality for the model you chose in part (b).
(d) Use your model to approximate the depth at which the water temperature is $3°C$.

66. Light Intensity The ordered pairs below give the intensities y (in microwatts per square centimeter) of the light measured by a light probe located x centimeters from a light source. (*Spreadsheet at LarsonPrecalculus.com*)

DATA
(30, 0.1881) (38, 0.1172) (46, 0.0775)
(34, 0.1543) (42, 0.0998) (50, 0.0645)

A model that approximates the data is $y = 171.33/x^2$.

(a) Use a graphing utility to plot the data points and the model in the same viewing window.

(b) Use the model to approximate the light intensity 25 centimeters from the light source.

67. Ecology The diameter of the largest particle that a stream can move is approximately directly proportional to the square of the velocity of the stream. When the velocity is $\frac{1}{4}$ mile per hour, the stream can move coarse sand particles about 0.02 inch in diameter. Approximate the velocity required to carry particles 0.12 inch in diameter.

68. Work The work W required to lift an object varies jointly with the object's mass m and the height h that the object is lifted. The work required to lift a 120-kilogram object 1.8 meters is 2116.8 joules. Find the amount of work required to lift a 100-kilogram object 1.5 meters.

69. Music The fundamental frequency (in hertz) of a piano string is directly proportional to the square root of its tension and inversely proportional to its length and the square root of its mass density. A string has a frequency of 100 hertz. Find the frequency of a string with each property.

(a) Four times the tension   (b) Twice the length

(c) Four times the tension and twice the length

70. Beam Load The maximum load that a horizontal beam can safely support varies jointly as the width of the beam and the square of its depth and inversely as the length of the beam. Determine how each change affects the beam's maximum load.

(a) Doubling the width   (b) Doubling the depth

(c) Halving the length

(d) Halving the width and doubling the length

Exploring the Concepts

True or False? In Exercises 71–73, decide whether the statement is true or false. Justify your answer.

71. If y is directly proportional to x and x is directly proportional to z, then y is directly proportional to z.

72. If y is inversely proportional to x and x is inversely proportional to z, then y is inversely proportional to z.

73. In the equation for the surface area S of a sphere, $S = 4\pi r^2$, the surface area S varies jointly with π and the square of the radius r.

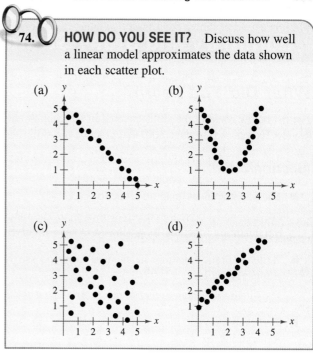

74. HOW DO YOU SEE IT? Discuss how well a linear model approximates the data shown in each scatter plot.

(a)   (b)   (c)   (d)

Review & Refresh ▶ *Video solutions at LarsonPrecalculus.com*

True or False? In Exercises 75 and 76, determine whether the statement is true or false. Justify your answer.

75. If a function $f(x)$ can be factored as $f(x) = (x - 2)p(x)$, where $p(x)$ is a polynomial function, then $f(2) = 0$.

76. If the graph of a polynomial function rises to the left, then its leading coefficient is positive.

Operations with Rational Expressions In Exercises 77–82, perform the operation and simplify.

77. $\dfrac{x - 1}{5x - 4} - \dfrac{2x}{5x - 4}$

78. $\dfrac{3 - x}{x + 2} + \dfrac{x + 8}{x + 2}$

79. $\dfrac{5}{x - 1} + \dfrac{2x}{x + 3}$

80. $\dfrac{3x}{x - 2} - \dfrac{6x}{x^2 - 4}$

81. $\dfrac{x + 3}{2x + 1} \cdot \dfrac{4}{3(x + 3)}$

82. $\dfrac{x^2}{x^2 - 1} \div \dfrac{2x}{x + 1}$

Testing for Functions Represented Algebraically In Exercises 83–88, determine whether the equation represents y as a function of x. If so, find the domain of the function.

83. $x^3 + y = 8$

84. $x = |y|$

85. $1 - y^2 = x$

86. $xy = 3$

87. $y = \sqrt{x + 5}$

88. $8x^2 + 2y^2 = 32$

Project: Fraud and Identity Theft To work an extended application analyzing the numbers of fraud complaints and identity theft victims in the United States in 2018, visit this text's website at *LarsonPrecalculus.com*. (*Source: U.S. Federal Trade Commission*)

Summary and Study Strategies

GO DIGITAL

What Did You Learn?

The list below reviews the skills covered in the chapter and correlates each one to the Review Exercises (see page 300) that practice the skill.

Section 3.1	Review Exercises
■ Analyze graphs of quadratic functions *(p. 242)*.	*1, 2, 15, 16*
Let a, b, and c be real numbers with $a \neq 0$. The function $f(x) = ax^2 + bx + c$ is a quadratic function. Its graph is a U-shaped curve called a parabola.	
■ Write quadratic functions in standard form and use the results to sketch their graphs *(p. 245)*.	*3–20*
The quadratic function $f(x) = a(x - h)^2 + k$, $a \neq 0$, is in standard form.	
■ Find minimum and maximum values of quadratic functions in real-life applications. *(p. 247)*.	*21–26*

Section 3.2	
■ Use transformations to sketch graphs of polynomial functions *(p. 251)*.	*27–32*
■ Use the Leading Coefficient Test to determine the end behaviors of graphs of polynomial functions *(p. 253)*.	*33–36, 43–46*
■ Find real zeros of polynomial functions and use them as sketching aids *(p. 255)*.	*37–46*
■ Use the Intermediate Value Theorem to help locate real zeros of polynomial functions *(p. 258)*.	*47–50*

Section 3.3	
■ Use long division to divide polynomials by other polynomials *(p. 264)*.	*51–56*
■ Use synthetic division to divide polynomials by binomials of the form $(x - k)$ *(p. 267)*.	*57–60, 65–68*
■ Use the Remainder Theorem and the Factor Theorem *(p. 268)*.	*61–68*
The Remainder Theorem states that if a polynomial $f(x)$ is divided by $x - k$, then the remainder is $r = f(k)$. The Factor Theorem states that a polynomial $f(x)$ has a factor $(x - k)$ if and only if $f(k) = 0$.	

Section 3.4	
■ Use the Fundamental Theorem of Algebra to determine numbers of zeros of polynomial functions *(p. 273)*.	*69–74*
If $f(x)$ is a polynomial of degree n, where $n > 0$, then f has at least one zero in the complex number system.	
■ Find rational zeros of polynomial functions *(p. 274)*.	*75–92*
■ Find complex zeros using conjugate pairs *(p. 277)*.	*81–84*
Let f be a polynomial function that has real coefficients. If $a + bi$, where $b \neq 0$, is a zero of the function, then the complex conjugate $a - bi$ is also a zero of the function.	
■ Find zeros of polynomials by factoring *(p. 278)*.	*75–80, 83–92*

Section 3.4 (continued)	**Review Exercises**
■ Use Descartes's Rule of Signs and the Upper and Lower Bound Rules to find zeros of polynomials *(p. 280)*.	*93–96*
■ Find zeros of polynomials in real-life applications *(p. 282)*.	*97, 98*

Section 3.5	
■ Use mathematical models to approximate sets of data points *(p. 287)*.	*99, 100*
■ Use the *regression* feature of a graphing utility to find equations of least squares regression lines *(p. 288)*.	*100*
■ Write mathematical models for direct variation *(p. 289)*. $y = kx$ for some nonzero constant k.	*101*
■ Write mathematical models for direct variation as an nth power *(p. 290)*. $y = kx^n$ for some nonzero constant k.	*102–103*
■ Write mathematical models for inverse variation *(p. 291)*. $y = k/x$ for some nonzero constant k.	*104*
■ Write mathematical models for combined variation *(p. 292)*.	*105*
■ Write mathematical models for joint variation *(p. 293)*. $z = kxy$ for some nonzero constant k.	*106*

Study Strategies

Reading Your Textbook   Many students avoid opening their textbooks due to anxiety and frustration. But not opening your textbook will cause more anxiety and frustration! Your textbook is designed to help you master skills and understand concepts. It contains many features and resources to help you succeed in your course.

1. **Review what you learned**
   - Read the bulleted list at the beginning of the section. If you cannot remember how to perform a skill, review the appropriate example.
   - Read the contents of the concept boxes—these contain important definitions and rules. Use the Summarize feature at the end of the section to organize the lesson's key concepts.
2. **Prepare for homework**
   - Complete the Checkpoint exercises. If you have difficulty with a Checkpoint exercise, reread the example or watch the solution video at *LarsonPrecalculus.com*.
3. **Prepare for quizzes and tests**
   - Make use of the What Did You Learn? and Study Strategies features.
   - Complete the Review Exercises. Then take the Chapter Test or Cumulative Test, as appropriate.

Problem-Solving Strategies   When you get stuck trying to solve a real-life problem, consider the strategies below.

- **Draw a Diagram**   Draw a diagram representing the problem. Label all known values and unknown values on the diagram.
- **Solve a Simpler Problem**   Simplify the problem, or write several simple examples of the problem. For instance, if you are asked to find the dimensions that will produce a maximum area, try calculating the areas of several examples.
- **Rewrite the Problem in Your Own Words**   Rewriting a problem can help you understand it better.
- **Guess and Check**   Try guessing the answer, then check your guess in the statement of the original problem. By refining your guesses, you may be able to think of a general strategy for solving the problem.

Review Exercises

See CalcChat.com for tutorial help and worked-out solutions to odd-numbered exercises.

GO DIGITAL

3.1 **Sketching Graphs of Quadratic Functions** In Exercises 1 and 2, sketch the graph of each quadratic function and compare it with the graph of $y = x^2$.

1. (a) $f(x) = 4x^2$
   (b) $g(x) = -2x^2$
   (c) $h(x) = x^2 + 2$
   (d) $k(x) = (x + 2)^2$

2. (a) $f(x) = x^2 - 4$
   (b) $g(x) = 4 - x^2$
   (c) $h(x) = (x - 5)^2 + 3$
   (d) $k(x) = \frac{1}{2}x^2 - 1$

Using Standard Form to Graph a Parabola In Exercises 3–14, write the quadratic function in standard form and sketch its graph. Identify the vertex, axis of symmetry, and x-intercept(s).

3. $g(x) = x^2 - 2x$

4. $f(x) = 8x - x^2$

5. $f(x) = x^2 - 6x + 1$

6. $h(x) = x^2 + 5x - 4$

7. $f(x) = x^2 + 8x + 10$

8. $f(x) = x^2 - 8x + 12$

9. $h(x) = 3 + 4x - x^2$

10. $f(x) = -2x^2 + 4x + 1$

11. $h(x) = 4x^2 + 4x + 13$

12. $f(x) = 4x^2 + 4x + 5$

13. $f(x) = \frac{1}{3}(x^2 + 5x - 4)$

14. $f(x) = \frac{1}{2}(6x^2 - 24x + 22)$

Writing a Quadratic Function In Exercises 15–20, write the standard form of the quadratic function whose graph is a parabola with the given vertex and that passes through the given point.

15.

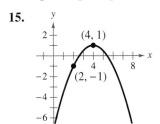

16.

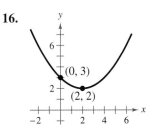

17. Vertex: $(6, 0)$; point: $(3, -9)$

18. Vertex: $(-3, -8)$; point: $(-6, 10)$

19. Vertex: $\left(2, -\frac{5}{2}\right)$; point: $\left(4, \frac{1}{2}\right)$

20. Vertex: $\left(-\frac{1}{4}, 7\right)$; point: $\left(\frac{3}{4}, \frac{49}{8}\right)$

21. **Geometry** A rectangle is inscribed in the region bounded by the x-axis, the y-axis, and the graph of $x + 2y - 8 = 0$, as shown in the figure.

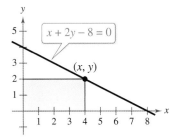

(a) Write the area A of the rectangle as a function of x.

(b) Determine the domain of the function in the context of the problem.

(c) Construct a table showing possible values of x and the corresponding areas of the rectangle. Use the table to estimate the dimensions that produce the maximum area.

(d) Use a graphing utility to graph the area function. Use the graph to approximate the dimensions that produce the maximum area.

(e) Write the area function in standard form to find analytically the dimensions that produce the maximum area.

22. **Geometry** The perimeter of a rectangle is 200 meters.

(a) Draw a diagram that gives a visual representation of the problem. Let x and y represent the length and width of the rectangle, respectively.

(b) Write y as a function of x. Use the result to write the area A as a function of x.

(c) Of all possible rectangles with perimeters of 200 meters, find the dimensions of the one with the maximum area.

23. **Maximum Revenue** The total revenue R earned (in dollars) from producing a gift box of tea is given by

$$R(p) = -10p^2 + 800p$$

where p is the price per box (in dollars).

(a) Find the revenues when the prices per box are $20, $25, and $30.

(b) Find the unit price that yields a maximum revenue. What is the maximum revenue? Explain your results.

24. Maximum Profit A real estate office handles an apartment building that has 50 units. When the rent is $540 per month, all units are occupied. For each $30 increase in rent, however, one unit becomes vacant. Each occupied unit requires an average of $18 per month for service and repairs. What rent should they charge to obtain the maximum profit?

25. Minimum Cost A soft-drink manufacturer has a daily production cost of

$$C = 70{,}000 - 120x + 0.055x^2$$

where C is the total cost (in dollars) and x is the number of units produced. How many units should they produce each day to yield a minimum cost?

26. Maximum Revenue A small theater has a seating capacity of 2000. When the ticket price is $20, attendance is 1500. For each $1 decrease in price, attendance increases by 100.

(a) Write the revenue R of the theater as a function of ticket price x.

(b) What ticket price will yield a maximum revenue? What is the maximum revenue?

3.2 **Sketching a Transformation of a Monomial Function** In Exercises 27–32, sketch the graphs of $y = x^n$ and the transformation.

27. $y = x^3$, $f(x) = (x - 2)^3$

28. $y = x^3$, $f(x) = 4x^3$

29. $y = x^4$, $f(x) = 6 - x^4$

30. $y = x^4$, $f(x) = 2(x - 8)^4$

31. $y = x^5$, $f(x) = (x - 5)^5 + 1$

32. $y = x^5$, $f(x) = \frac{1}{2}x^5 + 3$

Applying the Leading Coefficient Test In Exercises 33–36, describe the left-hand and right-hand behavior of the graph of the polynomial function.

33. $f(x) = -2x^2 - 5x + 12$

34. $f(x) = 4x - \frac{1}{2}x^3$

35. $g(x) = -3.1x^3 - 0.8x^4 + 1.2x^5$

36. $h(x) = 5 + 9x^6 - 6x^5$

Finding Real Zeros of a Polynomial Function In Exercises 37–42, (a) find all real zeros of the polynomial function, (b) determine whether the multiplicity of each zero is even or odd, (c) determine the maximum possible number of turning points of the graph of the function, and (d) use a graphing utility to graph the function and verify your answers.

37. $f(x) = 3x^2 + 20x - 32$

38. $f(x) = x^2 + 12x + 36$

39. $f(t) = t^3 - 3t$

40. $f(x) = x^3 - 8x^2$

41. $f(x) = x^4 - 8x^2 - 9$

42. $g(x) = x^4 + x^3 - 12x^2$

Sketching the Graph of a Polynomial Function In Exercises 43–46, sketch the graph of the function by (a) applying the Leading Coefficient Test, (b) finding the real zeros of the polynomial, (c) plotting sufficient solution points, and (d) drawing a continuous curve through the points.

43. $f(x) = -x^3 + x^2 - 2$

44. $g(x) = 2x^3 + 4x^2$

45. $f(x) = x(x^3 + x^2 - 5x + 3)$

46. $h(x) = 3x^2 - x^4$

Using the Intermediate Value Theorem In Exercises 47–50, (a) use the Intermediate Value Theorem and the *table* feature of a graphing utility to find intervals one unit in length in which the polynomial function is guaranteed to have a zero. (b) Adjust the table to approximate the zeros of the function to the nearest thousandth. Use the *zero* or *root* feature of the graphing utility to verify your results.

47. $f(x) = 4x^3 - x^2 + 2$

48. $f(x) = 0.25x^3 - 3.65x + 6.12$

49. $f(x) = x^4 - 5x - 1$

50. $f(x) = 7x^4 + 3x^3 - 8x^2 + 2$

3.3 **Long Division of Polynomials** In Exercises 51–56, use long division to divide.

51. $\dfrac{30x^2 - 3x + 8}{5x - 3}$

52. $\dfrac{4x + 7}{3x - 2}$

53. $\dfrac{5x^3 - 21x^2 - 25x - 4}{x^2 - 5x - 1}$

54. $\dfrac{3x^4}{x^2 - 1}$

55. $\dfrac{x^4 - 3x^3 + 4x^2 - 6x + 3}{x^2 + 2}$

56. $\dfrac{6x^4 + 10x^3 + 13x^2 - 5x + 2}{2x^2 - 1}$

Using Synthetic Division In Exercises 57–60, use synthetic division to divide.

57. $\dfrac{2x^3 - 25x^2 + 66x + 48}{x - 8}$

58. $\dfrac{5x^3 + 33x^2 + 50x - 8}{x + 4}$

59. $\dfrac{x^4 - 2x^2 + 9x}{x + 3}$

60. $\dfrac{6x^4 - 4x^3 - 27x^2 + 18x}{x - 2}$

Using the Remainder Theorem In Exercises 61 and 62, use the Remainder Theorem and synthetic division to find each function value.

61. $f(x) = x^4 + 10x^3 - 24x^2 + 20x + 44$

(a) $f(-3)$ (b) $f(-1)$

62. $g(t) = 2t^5 - 5t^4 - 8t + 20$

(a) $g(-4)$ (b) $g(\sqrt{2})$

Using the Factor Theorem In Exercises 63 and 64, use synthetic division to determine whether the given values of x are zeros of the function.

63. $f(x) = 20x^4 + 9x^3 - 14x^2 - 3x$

(a) $x = -1$ (b) $x = \frac{3}{4}$ (c) $x = 0$ (d) $x = 1$

64. $f(x) = 3x^3 - 8x^2 - 20x + 16$

(a) $x = 4$ (b) $x = -4$ (c) $x = \frac{2}{3}$ (d) $x = -1$

Factoring a Polynomial In Exercises 65–68, (a) verify the given factor(s) of $f(x)$, (b) find the remaining factors of $f(x)$, (c) use your results to write the complete factorization of $f(x)$, (d) list all real zeros of f, and (e) confirm your results by using a graphing utility to graph the function.

Function	Factor(s)
65. $f(x) = x^3 + 4x^2 - 25x - 28$	$(x - 4)$
66. $f(x) = 2x^3 + 11x^2 - 21x - 90$	$(x + 6)$
67. $f(x) = x^4 - 4x^3 - 7x^2 + 22x + 24$	$(x + 2), (x - 3)$
68. $f(x) = x^4 - 11x^3 + 41x^2 - 61x + 30$	$(x - 2), (x - 5)$

3.4 Zeros of Polynomial Functions In Exercises 69–74, determine the number of zeros of the polynomial function.

69. $f(x) = x - 6$ **70.** $g(x) = x^2 - 2x - 8$

71. $h(t) = t^2 - t^7$ **72.** $f(x) = x^8 + x^9$

73. $f(x) = (x - 8)^3$ **74.** $g(t) = (2t - 1)^2 - t^4$

Using the Rational Zero Test In Exercises 75–80, find the rational zeros of the function.

75. $f(x) = x^3 + 3x^2 - 28x - 60$

76. $f(x) = x^3 - 10x^2 + 17x - 8$

77. $f(x) = 3x^3 + 8x^2 - 4x - 16$

78. $f(x) = 4x^3 - 27x^2 + 11x + 42$

79. $f(x) = x^4 + x^3 - 11x^2 + x - 12$

80. $f(x) = 25x^4 + 25x^3 - 154x^2 - 4x + 24$

Finding a Polynomial Function with Given Zeros In Exercises 81 and 82, find a polynomial function with real coefficients that has the given zeros. (There are many correct answers.)

81. $\frac{2}{3}, 4, \sqrt{3}i$

82. $2, -3, 1 - 2i$

Finding the Zeros of a Polynomial Function In Exercises 83 and 84, use the given zero to find all the zeros of the function.

Function	Zero
83. $h(x) = -x^3 + 2x^2 - 16x + 32$	$-4i$
84. $g(x) = 2x^4 - 3x^3 - 13x^2 + 37x - 15$	$2 + i$

Finding the Zeros of a Polynomial Function In Exercises 85–88, write the polynomial as the product of linear factors and list all the zeros of the function.

85. $f(x) = x^3 + 4x^2 - 5x$

86. $g(x) = x^3 - 7x^2 + 36$

87. $g(x) = x^4 + 4x^3 - 3x^2 + 40x + 208$

88. $f(x) = x^4 + 8x^3 + 8x^2 - 72x - 153$

Finding the Zeros of a Polynomial Function In Exercises 89–92, find all the zeros of the function. When there is an extended list of possible rational zeros, use a graphing utility to graph the function to disregard any of the possible rational zeros that are obviously not zeros of the function.

89. $f(x) = x^3 - 16x^2 + x - 16$

90. $f(x) = 4x^4 - 12x^3 - 71x^2 - 3x - 18$

91. $g(x) = x^4 - 3x^3 - 14x^2 - 12x - 72$

92. $g(x) = 9x^5 - 27x^4 - 86x^3 + 204x^2 - 40x + 96$

Using Descartes's Rule of Signs In Exercises 93 and 94, use Descartes's Rule of Signs to determine the possible numbers of positive and negative real zeros of the function.

93. $g(x) = 5x^3 + 3x^2 - 6x + 9$

94. $h(x) = -2x^5 + 4x^3 - 2x^2 + 5$

Verifying Upper and Lower Bounds In Exercises 95 and 96, use synthetic division to verify the upper and lower bounds of the real zeros of f.

95. $f(x) = 4x^3 - 3x^2 + 4x - 3$

(a) Upper: $x = 1$

(b) Lower: $x = -\frac{1}{4}$

96. $f(x) = 2x^3 - 5x^2 - 14x + 8$

(a) Upper: $x = 8$

(b) Lower: $x = -4$

97. Geometry A right cylindrical water bottle has a volume of 36π cubic inches and a height 9 inches greater than its radius. Find the dimensions of the water bottle.

98. Geometry A kitchen has a volume of 60 cubic meters. The width of the room is 1 meter greater than the length and the height is 1 meter less than the length. Find the dimensions of the room.

3.5

99. Business The table shows the number of restaurants R operated by Chipotle Mexican Grill, Inc. at the end of each year from 2012 through 2019. (*Source: Chipotle Mexican Grill, Inc.*)

Year	Restaurants, R
2012	1410
2013	1595
2014	1783
2015	2010
2016	2250
2017	2408
2018	2491
2019	2622

DATA · Spreadsheet at LarsonPrecalculus.com

A linear model that approximates the data is

$$R = 179.5t - 711, \quad 12 \le t \le 19$$

where t represents the year, with $t = 12$ corresponding to 2012. Plot the actual data and the model on the same graph. How closely does the model represent the data?

100. Agriculture The table shows the number C (in millions) of calves born each year from 2013 through 2019. (*Source: United States Department of Agriculture*)

Year	Calves, C
2013	33.6
2014	33.5
2015	34.1
2016	35.1
2017	35.8
2018	36.3
2019	36.1

DATA · Spreadsheet at LarsonPrecalculus.com

(a) Use a graphing utility to create a scatter plot of the data. Let t represent the year, with $t = 13$ corresponding to 2013.

(b) Use the *regression* feature of the graphing utility to find the equation of the least squares regression line that fits the data. Then graph the model and the scatter plot you found in part (a) in the same viewing window. How closely does the model represent the data?

(c) Use the model to predict the number of calves born in 2025.

(d) Interpret the meaning of the slope of the linear model in the context of the problem.

101. Measurement A billboard says that it is 12.5 miles, or 20 kilometers, to the next gas station. Use this information to find a mathematical model that relates miles x to kilometers y. Then use the model to find the numbers of kilometers in 5 miles and 25 miles.

102. Energy The power P produced by a wind turbine is directly proportional to the cube of the wind speed S. A wind speed of 27 miles per hour produces a power output of 750 kilowatts. Find the output for a wind speed of 40 miles per hour.

103. Frictional Force The frictional force F between the tires and the road required to keep a car on a curved section of a highway is directly proportional to the square of the speed s of the car. If the speed of the car is doubled, the force will change by what factor?

104. Travel Time The travel time between two cities is inversely proportional to the average speed. A train travels between the cities in 3 hours at an average speed of 65 miles per hour. How long does it take to travel between the cities at an average speed of 80 miles per hour?

105. Demand The daily demand d for a company's pastries is directly proportional to the advertising budget x and inversely proportional to the price p. When the price is \$4 and the advertising expenses are \$25,000, the demand is 1250 pastries. Approximate the demand when the price is \$5.50.

106. Cost The cost of constructing a toy box with a square base varies jointly as the height of the box and the square of the width of the box. Constructing a box of height 16 inches and of width 6 inches costs \$28.80. How much does it cost to construct a box of height 14 inches and of width 8 inches?

Exploring the Concepts

True or False? In Exercises 107–109, determine whether the statement is true or false. Justify your answer.

107. The graph of the function

$$f(x) = 2 + x - x^2 + x^3 - x^4 + x^5 + x^6 - x^7$$

rises to the left and falls to the right.

108. A fourth-degree polynomial with real coefficients can have

$$-5, \quad -8i, \quad 4i, \quad \text{and} \quad 5$$

as its zeros.

109. If y is directly proportional to x, then x is directly proportional to y.

110. Writing Explain how to determine the maximum or minimum value of a quadratic function.

111. Writing Explain the connections between factors of a polynomial, zeros of a polynomial function, and solutions of a polynomial equation.

Chapter Test

See CalcChat.com for tutorial help and worked-out solutions to odd-numbered exercises.

GO DIGITAL

y

6
4 (0, 3)
2
x
−4 −2   2 4 6 8
−4
−6   (3, −6)

Figure for 3

Take this test as you would take a test in class. When you are finished, check your work against the answers given in the back of the book.

1. Sketch the graph of each quadratic function and compare it with the graph of $y = x^2$. *(Section 3.1)*
   (a) $g(x) = -x^2 + 4$ (b) $g(x) = \left(x - \frac{3}{2}\right)^2$

2. Identify the vertex and intercepts of the graph of $f(x) = x^2 - 6x + 8$. *(Section 3.1)*

3. Write the standard form of the equation of the parabola shown at the left. *(Section 3.1)*

4. The path of a particle is modeled by the function $f(x) = -\frac{1}{20}x^2 + 3x + 5$, where $f(x)$ is the height (in feet) of the particle and x is the horizontal distance (in feet) from where the particle started moving. *(Section 3.1)*
   (a) What is the maximum height of the particle?
   (b) Which number determines the height at which the particle started moving? Does changing this value change the coordinates of the maximum height of the particle? Explain.

5. Describe the left-hand and right-hand behavior of the graph of the function $h(t) = -\frac{3}{4}t^5 + 2t^2$. Then sketch its graph. *(Section 3.2)*

6. Divide using long division. *(Section 3.3)*

$$\frac{3x^3 + 4x - 1}{x^2 + 1}$$

7. Divide using synthetic division. *(Section 3.3)*

$$\frac{2x^4 - 3x^2 + 4x - 1}{x + 2}$$

8. Use synthetic division to show that $x = \sqrt{3}$ is a zero of the function $f(x) = 2x^3 - 5x^2 - 6x + 15$. Use the result to factor the polynomial function completely and list all the zeros of the function. *(Section 3.3)*

In Exercises 9 and 10, find the rational zeros of the function. *(Section 3.4)*

9. $g(t) = 2t^4 - 3t^3 + 16t - 24$

10. $h(x) = 3x^5 + 2x^4 - 3x - 2$

In Exercises 11 and 12, find a polynomial function with real coefficients that has the given zeros. (There are many correct answers.) *(Section 3.4)*

11. $0, 2, 3i$

12. $1, 1, 2 + \sqrt{3}i$

In Exercises 13 and 14, find all the zeros of the function. *(Section 3.4)*

13. $f(x) = 3x^3 + 14x^2 - 7x - 10$

14. $f(x) = x^4 - 9x^2 - 22x - 24$

In Exercises 15–17, find a mathematical model that represents the statement. (Determine the constant of proportionality.) *(Section 3.5)*

15. v varies directly as the square root of s. ($v = 24$ when $s = 16$.)

16. A varies jointly as x and y. ($A = 500$ when $x = 15$ and $y = 8$.)

17. b varies inversely as a. ($b = 32$ when $a = 1.5$.)

 18. The table at the left shows the numbers y of insured commercial banks in the United States for the years 2013 through 2018, where t represents the year, with $t = 13$ corresponding to 2013. Use the *regression* feature of a graphing utility to find the equation of the least squares regression line that fits the data. How well does the model represent the data? *(Source: Federal Deposit Insurance Corporation)* *(Section 3.5)*

DATA

Spreadsheet at
LarsonPrecalculus.com

Year, t	Insured Commercial Banks, y
13	5851
14	5610
15	5349
16	5116
17	4919
18	4718

Proofs in Mathematics

These two pages contain proofs of four important theorems about polynomial functions. The first two theorems are from Section 3.3, and the second two theorems are from Section 3.4.

The Remainder Theorem *(p. 268)*

If a polynomial $f(x)$ is divided by $x - k$, then the remainder is

$$r = f(k).$$

Proof

Using the Division Algorithm with the divisor $(x - k)$, you have

$$f(x) = (x - k)q(x) + r(x).$$

Either $r(x) = 0$ or the degree of $r(x)$ is less than the degree of $x - k$, so you know that $r(x)$ must be a constant. That is, $r(x) = r$. Now, by evaluating $f(x)$ at $x = k$, you have

$$f(k) = (k - k)q(k) + r$$
$$= (0)q(k) + r$$
$$= r.$$

To be successful in algebra, it is important that you understand the connection among *factors* of a polynomial, *zeros* of a polynomial function, and *solutions* or *roots* of a polynomial equation. The Factor Theorem is the basis for this connection.

The Factor Theorem *(p. 269)*

A polynomial $f(x)$ has a factor $(x - k)$ if and only if $f(k) = 0$.

Proof

Using the Division Algorithm with the factor $(x - k)$, you have

$$f(x) = (x - k)q(x) + r(x).$$

By the Remainder Theorem, $r(x) = r = f(k)$, and you have

$$f(x) = (x - k)q(x) + f(k)$$

where $q(x)$ is a polynomial of lesser degree than $f(x)$. If $f(k) = 0$, then

$$f(x) = (x - k)q(x)$$

and you see that $(x - k)$ is a factor of $f(x)$. Conversely, if $(x - k)$ is a factor of $f(x)$, then division of $f(x)$ by $(x - k)$ yields a remainder of 0. So, by the Remainder Theorem, you have $f(k) = 0$.

The Fundamental Theorem of Algebra, which is closely related to the Linear Factorization Theorem, has a long and interesting history. In the early work with polynomial equations, the Fundamental Theorem of Algebra was thought to have been false, because imaginary solutions were not considered. In fact, in the very early work by mathematicians such as Abu al-Khwarizmi (circa A.D. 800), negative solutions were also not considered.

Once imaginary numbers were considered, several mathematicians attempted to give a general proof of the Fundamental Theorem of Algebra. These included Jean Le Rond d'Alembert (1746), Leonhard Euler (1749), Joseph-Louis Lagrange (1772), and Pierre Simon Laplace (1795). The first substantial proof (although not rigorous by today's standards) of the Fundamental Theorem of Algebra is credited to Carl Friedrich Gauss, who published the proof in his doctoral thesis in 1799.

Linear Factorization Theorem   *(p. 273)*

If $f(x)$ is a polynomial of degree n, where $n > 0$, then $f(x)$ has precisely n linear factors

$$f(x) = a_n(x - c_1)(x - c_2) \cdots (x - c_n)$$

where $c_1, c_2, \ldots, c_n$ are complex numbers.

Proof

Using the Fundamental Theorem of Algebra, you know that f must have at least one zero, c_1. Consequently, $(x - c_1)$ is a factor of $f(x)$, and you have

$$f(x) = (x - c_1)f_1(x).$$

If the degree of $f_1(x)$ is greater than zero, then you again apply the Fundamental Theorem of Algebra to conclude that f_1 must have a zero c_2, which implies that

$$f(x) = (x - c_1)(x - c_2)f_2(x).$$

It is clear that the degree of $f_1(x)$ is $n - 1$, that the degree of $f_2(x)$ is $n - 2$, and that you can repeatedly apply the Fundamental Theorem of Algebra n times until you obtain

$$f(x) = a_n(x - c_1)(x - c_2) \cdots (x - c_n)$$

where a_n is the leading coefficient of the polynomial $f(x)$. ∎

Factors of a Polynomial   *(p. 278)*

Every polynomial of degree $n > 0$ with real coefficients can be written as the product of linear and quadratic factors with real coefficients, where the quadratic factors have no real zeros.

Proof

To begin, use the Linear Factorization Theorem to conclude that $f(x)$ can be *completely* factored in the form

$$f(x) = d(x - c_1)(x - c_2)(x - c_3) \cdots (x - c_n).$$

If each c_i is real, then there is nothing more to prove. If any c_i is imaginary ($c_i = a + bi, b \neq 0$), then you know that the conjugate $c_j = a - bi$ is also a zero, because the coefficients of $f(x)$ are real. By multiplying the corresponding factors, you obtain

$$
\begin{aligned}
(x - c_i)(x - c_j) &= [x - (a + bi)][x - (a - bi)] \\
&= [(x - a) - bi][(x - a) + bi] \\
&= (x - a)^2 + b^2 \\
&= x^2 - 2ax + (a^2 + b^2)
\end{aligned}
$$

where each coefficient is real. ∎

P.S. Problem Solving

See CalcChat.com for tutorial help and worked-out solutions to odd-numbered exercises.

GO DIGITAL

1. Exploring Zeros of Functions

(a) Find the zeros of each quadratic function $g(x)$.

(i) $g(x) = x^2 - 4x - 12$

(ii) $g(x) = x^2 + 5x$

(iii) $g(x) = x^2 + 3x - 10$

(iv) $g(x) = x^2 - 4x + 4$

(v) $g(x) = x^2 - 2x - 6$

(vi) $g(x) = x^2 + 3x + 4$

(b) For each function in part (a), use a graphing utility to graph $f(x) = (x - 2) \cdot g(x)$. Verify that $(2, 0)$ is an x-intercept of the graph of $f(x)$. Describe any similarities or differences in the behaviors of the six functions at this x-intercept.

(c) For each function in part (b), use the graph of $f(x)$ to approximate the other x-intercepts of the graph.

(d) Describe the connections that you find among the results of parts (a), (b), and (c).

2. Exploring Zeros of Functions

(a) Find the zeros of each quadratic function $g(x)$.

(i) $g(x) = 2x^2 + 5x - 3$

(ii) $g(x) = -x^2 - 3x - 2$

(b) For each function in part (a), find the zeros of $f(x) = g\left(\frac{1}{2}x\right)$.

(c) Describe the connection between the results in parts (a) and (b).

3. Building a Quonset Hut Quonset huts were developed during World War II. They were temporary housing structures that could be assembled quickly and easily. A Quonset hut is shaped like a half cylinder. A manufacturer has 600 square feet of material with which to build a Quonset hut.

(a) The formula for the surface area of half a cylinder is $S = \pi r^2 + \pi r l$, where r is the radius and l is the length of the hut. Solve this equation for l when $S = 600$.

(b) The formula for the volume of the hut is $V = \frac{1}{2}\pi r^2 l$. Write the volume V of the Quonset hut as a polynomial function of r.

(c) Use the function you wrote in part (b) to find the maximum volume of a Quonset hut with a surface area of 600 square feet. What are the dimensions of the hut?

4. Verifying the Remainder Theorem Show that if $f(x) = ax^3 + bx^2 + cx + d$, then $f(k) = r$, where $r = ak^3 + bk^2 + ck + d$, using long division. In other words, verify the Remainder Theorem for a third-degree polynomial function.

5. Babylonian Mathematics In 2000 B.C., the Babylonians solved polynomial equations by referring to tables of values. One such table gave the values of $y^3 + y^2$. To be able to use this table, the Babylonians sometimes used the method below to manipulate the equation.

$$ax^3 + bx^2 = c \qquad \text{Original equation}$$

$$\frac{a^3x^3}{b^3} + \frac{a^2x^2}{b^2} = \frac{a^2c}{b^3} \qquad \text{Multiply each side by } \frac{a^2}{b^3}.$$

$$\left(\frac{ax}{b}\right)^3 + \left(\frac{ax}{b}\right)^2 = \frac{a^2c}{b^3} \qquad \text{Rewrite.}$$

Then they would find $(a^2c)/b^3$ in the $y^3 + y^2$ column of the table. They knew that the corresponding y-value was equal to $(ax)/b$, so they could conclude that $x = (by)/a$.

(a) Calculate $y^3 + y^2$ for $y = 1, 2, 3, \ldots, 10$. Record the values in a table.

(b) Use the table from part (a) and the method above to solve each equation.

(i) $x^3 + x^2 = 252$ (ii) $x^3 + 2x^2 = 288$

(iii) $3x^3 + x^2 = 90$ (iv) $2x^3 + 5x^2 = 2500$

(v) $7x^3 + 6x^2 = 1728$ (vi) $10x^3 + 3x^2 = 297$

(c) Using the methods from this chapter, verify your solution of each equation.

6. Zeros of a Cubic Function Can a cubic function with real coefficients have two real zeros and one complex zero? Explain.

7. Sums and Products of Zeros

(a) Complete the table.

Function	Zeros	Sum of Zeros	Product of Zeros
$f_1(x) = x^2 - 5x + 6$			
$f_2(x) = x^3 - 7x + 6$			
$f_3(x) = x^4 + 2x^3 + x^2 + 8x - 12$			
$f_4(x) = x^5 - 3x^4 - 9x^3 + 25x^2 - 6x$			

(b) Use the table to make a conjecture relating the sum of the zeros of a polynomial function to the coefficients of the polynomial function.

(c) Use the table to make a conjecture relating the product of the zeros of a polynomial function to the coefficients of the polynomial function.

307

8. True or False? Determine whether the statement is true or false. If false, provide one or more reasons why the statement is false and correct the statement. Let

$$f(x) = ax^3 + bx^2 + cx + d, \quad a \neq 0$$

and let $f(2) = -1$. Then

$$\frac{f(x)}{x + 1} = q(x) + \frac{2}{x + 1}$$

where $q(x)$ is a second-degree polynomial.

9. Finding the Equation of a Parabola The parabola shown in the figure has an equation of the form $y = ax^2 + bx + c$. Find the equation of this parabola using each method. (a) Find the equation analytically. (b) Use the *regression* feature of a graphing utility to find the equation.

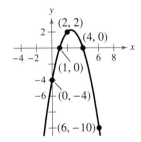

10. Finding the Slope of a Tangent Line One of the fundamental themes of calculus is to find the slope of the tangent line to a curve at a point. To see how this can be done, consider the point $(2, 4)$ on the graph of the quadratic function $f(x) = x^2$, as shown in the figure.

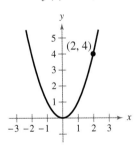

(a) Find the slope m_1 of the line joining $(2, 4)$ and $(3, 9)$.

(b) Find the slope m_2 of the line joining $(2, 4)$ and $(1, 1)$.

(c) Find the slope m_3 of the line joining $(2, 4)$ and $(2.1, 4.41)$.

(d) Find the slope m_h of the line joining $(2, 4)$ and $(2 + h, f(2 + h))$ in terms of the nonzero number h.

(e) Evaluate the slope formula from part (d) for $h = -1, 1$, and 0.1. Compare these values with those in parts (a)–(c).

(f) What can you conclude the slope m_{tan} of the tangent line at $(2, 4)$ to be? Explain.

11. Maximum Area A rancher plans to fence a rectangular pasture adjacent to a river (see figure). The rancher has 100 meters of fencing, and no fencing is needed along the river.

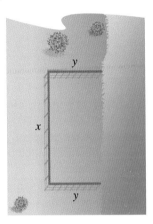

(a) Write the area A of the pasture as a function of x, the length of the side parallel to the river. What is the domain of A?

(b) Graph the function A and estimate the dimensions that yield the maximum area of the pasture.

(c) Find the exact dimensions that yield the maximum area of the pasture by writing the quadratic function in standard form.

12. Maximum and Minimum Area A wire 100 centimeters in length is cut into two pieces. One piece is bent to form a square and the other to form a circle. Let x equal the length of the wire used to form the square.

(a) Write the function that represents the combined area of the two figures.

(b) Determine the domain of the function.

(c) Find the value(s) of x that yield a maximum area and a minimum area. Explain.

13. Finding Dimensions At a glassware factory, molten cobalt glass is poured into molds to make paperweights. Each mold is a rectangular prism whose height is 3 inches greater than the length of each side of the square base. A machine pours 20 cubic inches of liquid glass into each mold. What are the dimensions of the mold?

4 Rational Functions and Conics

4.1 Rational Functions and Asymptotes
4.2 Graphs of Rational Functions
4.3 Conics
4.4 Translations of Conics

GO DIGITAL

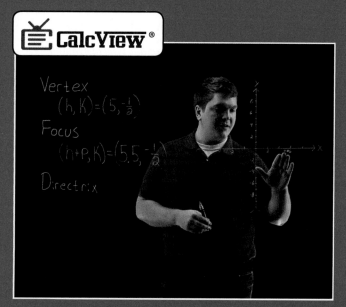

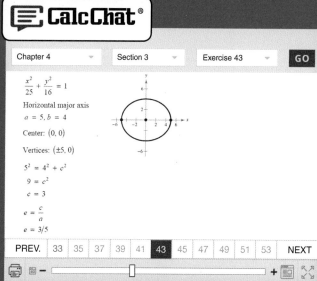

CalcChat®

| Chapter 4 | Section 3 | Exercise 43 | GO |

$\frac{x^2}{25} + \frac{y^2}{16} = 1$

Horizontal major axis
$a = 5, b = 4$

Center: $(0, 0)$

Vertices: $(\pm 5, 0)$

$5^2 = 4^2 + c^2$
$9 = c^2$
$c = 3$

$e = \frac{c}{a}$

$e = 3/5$

PREV. 33 35 37 39 41 **43** 45 47 49 51 53 NEXT

4.1 Recycling *(Exercise 39, p. 316)*

4.4 Satellite Orbit *(Exercise 41, p. 348)*

309

4.1 Rational Functions and Asymptotes

Rational functions have many real-life applications. For example, in Exercise 39 on page 316, you will use a rational function to determine the cost of supplying recycling bins to the population of a rural township.

❯ **Find domains of rational functions.**
❯ **Find vertical and horizontal asymptotes of graphs of rational functions.**
❯ **Use rational functions to model and solve real-life problems.**

Introduction

A **rational function** is a quotient of polynomial functions. It can be written in the form

$$f(x) = \frac{N(x)}{D(x)} \qquad N(x) \text{ and } D(x) \text{ are polynomials, } D(x) \neq 0.$$

where $N(x)$ and $D(x)$ are polynomials and $D(x)$ is not the zero polynomial.

In general, the *domain* of a rational function of x includes all real numbers except x-values that make the denominator zero. Much of the discussion of rational functions will focus on the behavior of their graphs near x-values excluded from the domain.

EXAMPLE 1 **Finding the Domain of a Rational Function**

 See LarsonPrecalculus.com for an interactive version of this type of example.

Find the domain of $f(x) = \dfrac{1}{x}$ and discuss the behavior of f near any excluded x-values.

Solution The denominator is zero when $x = 0$, so the domain of f is all real numbers except $x = 0$. To determine the behavior of f near this excluded value, evaluate $f(x)$ to the left and right of $x = 0$, as shown in the tables below.

x	-1	-0.5	-0.1	-0.01	-0.001	$\to 0$
$f(x)$	-1	-2	-10	-100	-1000	$\to -\infty$

x	$0 \leftarrow$	0.001	0.01	0.1	0.5	1
$f(x)$	$\infty \leftarrow$	1000	100	10	2	1

Note that as x approaches 0 *from the left*, $f(x)$ decreases without bound. In contrast, as x approaches 0 *from the right*, $f(x)$ increases without bound. The graph of f is shown below.

ALGEBRA HELP

In the tables, the notation "$\to 0$" means x "approaches 0 from the left," and "$0 \leftarrow$" means x "approaches 0 from the right." The notation "$\to -\infty$" means $f(x)$ "decreases without bound," and "$\infty \leftarrow$" means $f(x)$ "increases without bound."

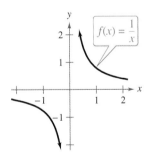

✓ *Checkpoint* ▶ Audio-video solution in English & Spanish at LarsonPrecalculus.com

Find the domain of $f(x) = (3x)/(x - 1)$ and discuss the behavior of f near any excluded x-values.

GO DIGITAL

Vertical and Horizontal Asymptotes

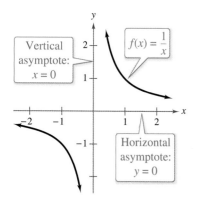

Figure 4.1

In Example 1, the behavior of the graph of $f(x) = 1/x$ as x approaches 0 from the left or from the right is denoted as shown below.

$$f(x) \to -\infty \text{ as } x \to 0^- \qquad f(x) \to \infty \text{ as } x \to 0^+$$

$f(x)$ decreases without bound as x approaches 0 from the left. $f(x)$ increases without bound as x approaches 0 from the right.

The line $x = 0$ is a **vertical asymptote** of the graph of f, as shown in Figure 4.1. From this figure, you can see that the graph of f also has a **horizontal asymptote**—the line $y = 0$. This means $f(x)$ approaches 0 as x increases or decreases without bound. This behavior is denoted as shown below.

$$f(x) \to 0 \text{ as } x \to -\infty \qquad f(x) \to 0 \text{ as } x \to \infty$$

$f(x)$ approaches 0 as x decreases without bound. $f(x)$ approaches 0 as x increases without bound.

Definitions of Vertical and Horizontal Asymptotes

1. The line $x = a$ is a **vertical asymptote** of the graph of f when

$$f(x) \to \infty \quad \text{or} \quad f(x) \to -\infty$$

as $x \to a$, either from the right or from the left.

2. The line $y = b$ is a **horizontal asymptote** of the graph of f when $f(x) \to b$ as $x \to \infty$ or $x \to -\infty$.

Eventually (as $x \to \infty$ or $x \to -\infty$), the distance between the horizontal asymptote and the points on the graph must approach zero. Figure 4.2 shows the vertical and horizontal asymptotes of the graphs of three rational functions.

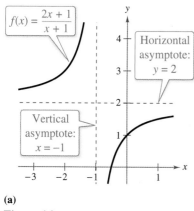

(a)

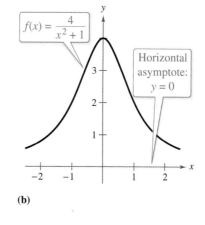

(b)

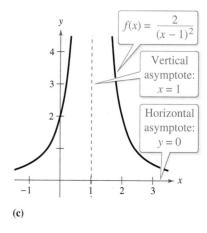

(c)

Figure 4.2

Verify numerically the horizontal asymptotes shown in Figure 4.2. For example, to show that the line $y = 2$ is the horizontal asymptote of the graph of

$$f(x) = \frac{2x + 1}{x + 1} \qquad \text{See Figure 4.2(a).}$$

create a table that shows the value of $f(x)$ as x increases and decreases without bound.

The graphs shown in Figures 4.1 and 4.2(a) are **hyperbolas.** You will study hyperbolas in Sections 4.3 and 4.4.

GO DIGITAL

Vertical and Horizontal Asymptotes

Let f be the rational function

$$f(x) = \frac{N(x)}{D(x)} = \frac{a_n x^n + a_{n-1}x^{n-1} + \cdots + a_1 x + a_0}{b_m x^m + b_{m-1}x^{m-1} + \cdots + b_1 x + b_0}$$

where $N(x)$ and $D(x)$ have no common factors.

1. The graph of f has *vertical* asymptotes at the zeros of $D(x)$.

2. The graph of f has at most one *horizontal* asymptote determined by comparing the degrees of $N(x)$ and $D(x)$.

   a. When $n < m$, the graph of f has the line $y = 0$ (the x-axis) as a horizontal asymptote.

   b. When $n = m$, the graph of f has the line

   $$y = \frac{a_n}{b_m}$$

   as a horizontal asymptote, where a_n is the leading coefficient of the numerator and b_m is the leading coefficient of the denominator.

   c. When $n > m$, the graph of f has no horizontal asymptote.

GO DIGITAL

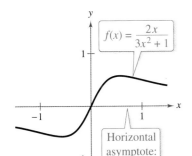

Figure 4.3

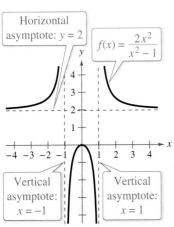

Figure 4.4

| EXAMPLE 2 | **Finding Vertical and Horizontal Asymptotes** |

Find all vertical and horizontal asymptotes of the graph of each rational function.

a. $f(x) = \dfrac{2x}{3x^2 + 1}$ **b.** $f(x) = \dfrac{2x^2}{x^2 - 1}$

Solution

a. For this rational function, the degree of the numerator is *less than* the degree of the denominator, so the graph has the line $y = 0$ as a horizontal asymptote. To find any vertical asymptotes, set the denominator equal to zero and solve the resulting equation for x. The equation

$$3x^2 + 1 = 0 \qquad \text{Set denominator equal to zero.}$$

has no real solution, so the graph has no vertical asymptote (see Figure 4.3).

b. For this rational function, the degree of the numerator is *equal* to the degree of the denominator. The leading coefficient of the numerator is 2 and the leading coefficient of the denominator is 1, so the graph has the line $y = 2/1 = 2$ as a horizontal asymptote. To find any vertical asymptotes, set the denominator equal to zero and solve the resulting equation for x.

$$x^2 - 1 = 0 \qquad \text{Set denominator equal to zero.}$$
$$(x + 1)(x - 1) = 0 \qquad \text{Factor.}$$
$$x + 1 = 0 \implies x = -1 \qquad \text{Set 1st factor equal to 0.}$$
$$x - 1 = 0 \implies x = 1 \qquad \text{Set 2nd factor equal to 0.}$$

This equation has two real solutions, $x = -1$ and $x = 1$, so the graph has the lines $x = -1$ and $x = 1$ as vertical asymptotes, as shown in Figure 4.4.

✓ *Checkpoint* ▶ Audio-video solution in English & Spanish at LarsonPrecalculus.com

Find all vertical and horizontal asymptotes of the graph of $f(x) = \dfrac{5x^2}{x^2 - 1}$. ∎

Values for which a rational function is undefined (the denominator is zero) result in a vertical asymptote or a hole in the graph. When the numerator and the denominator of a rational function have a common factor, the graph of the function has a hole at the zero of the common factor, as shown in Example 3.

EXAMPLE 3 Finding Asymptotes and Holes

Find all asymptotes and holes in the graph of

$$f(x) = \frac{x^2 + x - 2}{x^2 - x - 6}.$$

Solution For this rational function, the degree of the numerator is *equal to* the degree of the denominator. The leading coefficients of the numerator and denominator are both 1, so the graph has the line $y = 1/1 = 1$ as a horizontal asymptote. To find any vertical asymptotes, first factor the numerator and denominator.

$$f(x) = \frac{x^2 + x - 2}{x^2 - x - 6} = \frac{(x - 1)(x + 2)}{(x + 2)(x - 3)} = \frac{x - 1}{x - 3}, \quad x \neq -2$$

Setting the denominator $x - 3$ (of the simplified function) equal to zero shows that the graph has the line $x = 3$ as a vertical asymptote, as shown in Figure 4.5. The numerator and denominator of f have a common factor of $x + 2$, so there is a hole in the graph at $x = -2$. To find the y-coordinate of the hole, substitute $x = -2$ into the simplified form of f.

$$y = \frac{x - 1}{x - 3} = \frac{-2 - 1}{-2 - 3} = \frac{3}{5} \qquad \text{Find } y\text{-coordinate of the hole.}$$

So, the graph of the rational function has a hole at $(-2, 3/5)$. In Figure 4.5, the hole is indicated by an open circle.

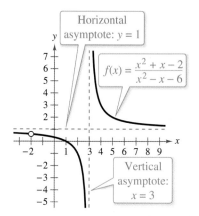

The graph of f has a hole at $(-2, 3/5)$.
Figure 4.5

✓ **Checkpoint** ▶ Audio-video solution in English & Spanish at LarsonPrecalculus.com

Find all asymptotes and holes in the graph of $f(x) = \dfrac{3x^2 + 7x - 6}{x^2 + 4x + 3}.$

EXAMPLE 4 Finding a Function's Domain and Asymptotes

For the function f, find (a) the domain of f, (b) the vertical asymptote of f, and (c) the horizontal asymptote of f.

$$f(x) = \frac{3x^3 + 7x^2 + 2}{-2x^3 + 16}$$

Solution

a. Because the denominator is zero when $-2x^3 + 16 = 0$, solve this equation to determine that the domain of f is all real numbers except $x = 2$.

b. Because the denominator of f has a zero at $x = 2$, and 2 is not a zero of the numerator, the graph of f has the vertical asymptote $x = 2$.

c. Because the degrees of the numerator and denominator are the same, and the leading coefficient of the numerator is 3 and the leading coefficient of the denominator is -2, the horizontal asymptote of f is $y = -\frac{3}{2}$.

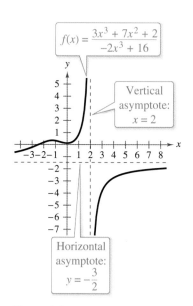

Figure 4.6

Figure 4.6 shows the graph of f.

✓ **Checkpoint** ▶ Audio-video solution in English & Spanish at LarsonPrecalculus.com

Repeat Example 4 using the function $f(x) = \dfrac{3x^2 + x - 5}{x^2 + 1}.$

GO DIGITAL

Application

EXAMPLE 5   **Ultraviolet Radiation Protection**

The percent P of ultraviolet radiation protection provided by a sunscreen can be modeled by

$$P = \frac{100s - 100}{s}, \quad 1 \le s \le 65$$

where s is the Sun Protection Factor, or SPF.   *(Source: U.S. Environmental Protection Agency)*

a. Find the percents of ultraviolet radiation protection provided by sunscreens with SPF values of 15, 30, and 50.

b. If the model were valid for all $s \ge 1$, what would be the horizontal asymptote of the graph of this function, and what would it represent?

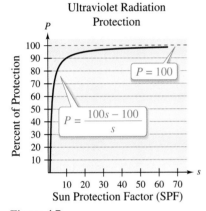

Ultraviolet Radiation Protection

Figure 4.7

Solution

a. When $s = 15$, $P = \dfrac{100(15) - 100}{15}$

$$\approx 93.3\%.$$

When $s = 30$, $P = \dfrac{100(30) - 100}{30}$

$$\approx 96.7\%.$$

When $s = 50$, $P = \dfrac{100(50) - 100}{50}$

$$= 98\%.$$

b. As shown in Figure 4.7, the horizontal asymptote is the line $P = 100$. This line represents 100% protection from ultraviolet radiation.

✓ *Checkpoint* ▶ *Audio-video solution in English & Spanish at LarsonPrecalculus.com*

A business has a cost function of $C = 0.4x + 8000$, where C is measured in dollars and x is the number of units produced. The average cost per unit is given by

$$\overline{C} = \frac{C}{x} = \frac{0.4x + 8000}{x}, \quad x > 0.$$

a. Find the average costs per unit when $x = 1000$, $x = 8000$, $x = 20{,}000$, and $x = 100{,}000$.

b. What is the horizontal asymptote of the graph of this function, and what does it represent? ■

Summarize   (Section 4.1)

1. State the definition of a rational function and explain how to find the domain of a rational function *(page 310)*. For an example of finding the domain of a rational function, see Example 1.

2. Explain how to find the vertical and horizontal asymptotes of the graph of a rational function *(page 311)*. For examples of finding vertical and horizontal asymptotes of graphs of rational functions, see Examples 2–4.

3. Describe a real-life application of a rational function *(page 314, Example 5)*.

GO DIGITAL

4.1 Exercises

See CalcChat.com for tutorial help and worked-out solutions to odd-numbered exercises.

Vocabulary and Concept Check

In Exercises 1–4, fill in the blanks.

1. A _____ _____ is a quotient of polynomial functions.
2. When $f(x) \to \pm\infty$ as $x \to a$ from the left or the right, $x = a$ is a _____ _____ of the graph of f.
3. When $f(x) \to b$ as $x \to \pm\infty$, $y = b$ is a _____ _____ of the graph of f.
4. The graph of $f(x) = 1/x$ is a _____.

5. What feature of the graph of $f(x) = \dfrac{9}{x - 3}$ can you find by solving $x - 3 = 0$?

6. Is $y = 4$ a horizontal asymptote of the function $f(x) = \dfrac{4x}{x^2 - 8}$?

Skills and Applications

Finding the Domain of a Rational Function In Exercises 7–14, find the domain of the function and discuss the behavior of f near any excluded x-values.

7. $f(x) = \dfrac{1}{x - 1}$

8. $f(x) = \dfrac{4}{x + 3}$

9. $f(x) = \dfrac{5x}{x^2 + 7x + 12}$

10. $f(x) = \dfrac{6x}{x^2 - 5x + 6}$

11. $f(x) = \dfrac{3x^2}{x^2 - 1}$

12. $f(x) = \dfrac{2x}{x^2 - 4}$

13. $f(x) = \dfrac{x^2 + 3x + 2}{x^2 - 2x + 1}$

14. $f(x) = \dfrac{x^2 + x - 20}{x^2 + 8x + 16}$

Finding Vertical and Horizontal Asymptotes In Exercises 15–22, find all vertical and horizontal asymptotes of the graph of the rational function.

15. $f(x) = \dfrac{4}{x^2}$

16. $f(x) = \dfrac{1}{(x - 2)^3}$

17. $f(x) = \dfrac{5 + x}{5 - x}$

18. $f(x) = \dfrac{3 - 7x}{3 + 2x}$

19. $f(x) = \dfrac{x^3}{x^2 - 1}$

20. $f(x) = \dfrac{2x^2}{x + 1}$

21. $f(x) = \dfrac{-4x^2 + 1}{x^2 + x + 3}$

22. $f(x) = \dfrac{3x^2 + x - 5}{x^2 + 1}$

Finding Asymptotes and Holes In Exercises 23–28, find all asymptotes and holes in the graph of the rational function.

23. $f(x) = \dfrac{x - 4}{x^2 - 16}$

24. $f(x) = \dfrac{x + 3}{x^2 - 9}$

25. $f(x) = \dfrac{x^2 - 1}{x^2 - 2x - 3}$

26. $f(x) = \dfrac{x^2 - 4}{x^2 - 3x + 2}$

27. $f(x) = \dfrac{x^2 - 3x - 4}{2x^2 + x - 1}$

28. $f(x) = \dfrac{x^2 + x - 2}{2x^2 + 5x + 2}$

Matching In Exercises 29–34, match the rational function with its graph. [The graphs are labeled (a)–(f).]

(a)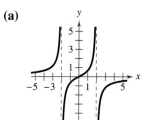

(b)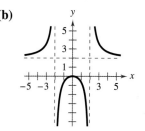

(c)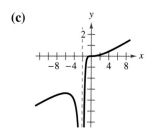

(d)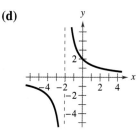

(e)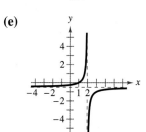

(f)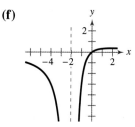

29. $f(x) = \dfrac{4}{x + 2}$

30. $f(x) = -\dfrac{x - 1}{2x - 4}$

31. $f(x) = \dfrac{2x^2}{x^2 - 4}$

32. $f(x) = \dfrac{-2x}{x^2 - 4}$

33. $f(x) = \dfrac{x^3}{4(x + 2)^2}$

34. $f(x) = \dfrac{3x}{(x + 2)^2}$

Analyzing Functions In Exercises 35–38, (a) determine the domains of f and g, (b) simplify f and find any vertical asymptotes of the graph of f, (c) complete the table, and (d) explain how the two functions differ.

35. $f(x) = \dfrac{x^2 - 4}{x + 2}$, $g(x) = x - 2$

x	-4	-3	-2.5	-2	-1.5	-1	0
$f(x)$							
$g(x)$							

36. $f(x) = \dfrac{x^2(x + 3)}{x^2 + 3x}$, $g(x) = x$

x	-3	-2	-1	0	1	2	3
$f(x)$							
$g(x)$							

37. $f(x) = \dfrac{2x - 1}{2x^2 - x}$, $g(x) = \dfrac{1}{x}$

x	-1	-0.5	0	0.5	2	3	4
$f(x)$							
$g(x)$							

38. $f(x) = \dfrac{2x - 8}{x^2 - 9x + 20}$, $g(x) = \dfrac{2}{x - 5}$

x	0	1	2	3	4	5	6
$f(x)$							
$g(x)$							

39. **Recycling**

The cost C (in dollars) of supplying recycling bins to $p\%$ of the population of a rural township is given by

$$C = \frac{25{,}000p}{100 - p}, \quad 0 \le p < 100.$$

(a) Use a graphing utility to graph the cost function.

(b) Find the costs of supplying bins to 15%, 50%, and 90% of the population.

(c) According to the model, is it possible to supply bins to 100% of the population? Explain.

40. **Ecology** The cost C (in millions of dollars) of removing $p\%$ of the industrial and municipal pollutants discharged into a river is given by

$$C = \frac{225p}{100 - p}, \quad 0 \le p < 100.$$

(a) Use a graphing utility to graph the cost function.

(b) Find the costs of removing 10%, 40%, and 75% of the pollutants.

(c) According to the model, is it possible to remove 100% of the pollutants? Explain.

41. **Population Growth** A game commission introduces 100 deer into newly acquired state game lands. The population N of the herd is modeled by

$$N = \frac{20(5 + 3t)}{1 + 0.04t}, \quad t \ge 0$$

where t is the time in years.

(a) Use a graphing utility to graph this model.

(b) Find the populations when $t = 5$, $t = 10$, and $t = 25$.

(c) What is the limiting size of the herd as time increases?

42. **Biology** A science class performs an experiment comparing the quantity of food consumed by a species of moth with the quantity of food supplied. The model for the experimental data is

$$y = \frac{1.568x - 0.001}{6.360x + 1}, \quad x > 0$$

where x is the quantity (in milligrams) of food supplied and y is the quantity (in milligrams) of food consumed.

(a) Use a graphing utility to graph this model.

(b) At what level of consumption will the moth become satiated?

43. **Psychology** Psychologists have developed mathematical models to predict memory performance as a function of the number of trials n of a certain task. Consider the learning curve

$$P = \frac{0.5 + 0.9(n - 1)}{1 + 0.9(n - 1)}, \quad n > 0$$

where P is the fraction of correct responses after n trials.

(a) Complete the table for this model. What does it suggest?

n	1	2	3	4	5	6	7	8	9	10
P										

(b) According to the model, what is the limiting percent of correct responses as n increases?

44. Physics Experiment Consider a physics laboratory experiment designed to determine an unknown mass. A flexible metal meter stick is clamped to a table with 50 centimeters overhanging the edge (see figure). Known masses M ranging from 200 grams to 2000 grams are attached to the end of the meter stick. For each mass, the meter stick is displaced vertically and then allowed to oscillate. The average time t (in seconds) of one oscillation for each mass is recorded in the table.

DATA	Mass, M	Time, t
	200	0.450
	400	0.597
	600	0.712
	800	0.831
	1000	0.906
	1200	1.003
	1400	1.088
	1600	1.168
	1800	1.218
	2000	1.338

Spreadsheet at LarsonPrecalculus.com

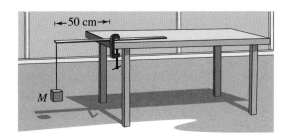

←50 cm→

M

A model for the data that can be used to predict the time of one oscillation is

$$t = \frac{38M + 16{,}965}{10(M + 5000)}.$$

(a) Use this model to create a table showing the predicted time for each of the masses shown in the table above.

(b) Compare the predicted times with the experimental times. What can you conclude?

(c) Use the model to approximate the mass of an object for which $t = 1.056$ seconds.

Exploring the Concepts

True or False? **In Exercises 45 and 46, determine whether the statement is true or false. Justify your answer.**

45. The graph of a rational function can have vertical asymptotes only at the x-values that are the zeros of the denominator of the function.

46. The graph of a rational function never has two horizontal asymptotes.

47. Error Analysis Describe the error.

A real zero of the numerator of a rational function f is $x = c$. So, $x = c$ must also be a zero of f. ✗

48. 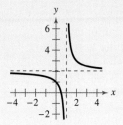 **HOW DO YOU SEE IT?** The graph of a rational function $f(x) = N(x)/D(x)$ is shown below. Determine which of the statements about the function is false. Justify your answer.

(a) When $x = 1$, $D(x) = 0$.

(b) The degree of $N(x)$ is equal to the degree of $D(x)$.

(c) The ratio of the leading coefficients of $N(x)$ and $D(x)$ is 1.

49. Think About It Give an example of a rational function whose domain is the set of all real numbers. Give an example of a rational function whose domain is the set of all real numbers except $x = 15$.

50. Think About It Write a rational function f whose graph has vertical asymptotes $x = -2$ and $x = 1$ and horizontal asymptote $y = 2$. (There are many correct answers.)

Review & Refresh ▶ Video solutions at LarsonPrecalculus.com

Simplifying a Rational Expression **In Exercises 51–54, write the rational expression in simplest form.**

51. $\dfrac{x^2 - 9}{x^2 - 2x - 3}$

52. $\dfrac{x^2 - 4}{x^2 - x - 6}$

53. $\dfrac{x^2 + 11x + 30}{x^2 + x - 30}$

54. $\dfrac{2x^2 - 3x - 9}{2x^2 - 5x - 12}$

Finding Intercepts Algebraically **In Exercises 55–62, find the x- and y-intercepts of the graph of the equation algebraically.**

55. $y = x^2 + 12x + 20$

56. $y = x^2 - 19x - 20$

57. $y = \dfrac{3 + 2x}{1 + x}$

58. $y = \dfrac{3x}{x^2 + 2}$

59. $f(x) = \dfrac{x}{x^2 - x - 2}$

60. $f(x) = \dfrac{3x}{x^2 - 2x - 3}$

61. $f(x) = \dfrac{x^2 - 9}{x^2 - x - 6}$

62. $f(x) = \dfrac{x^2 - x - 2}{x - 1}$

4.2 Graphs of Rational Functions

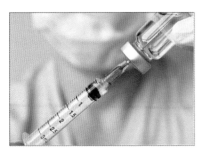

Graphs of rational functions have many real-life applications. For example, in Exercise 72 on page 325, you will use the graph of a rational function to analyze the concentration of a chemical in the bloodstream after injection.

> • Sketch graphs of rational functions.
> • Sketch graphs of rational functions that have slant asymptotes.
> • Use graphs of rational functions to model and solve real-life problems.

Sketching the Graph of a Rational Function

Guidelines for Graphing Rational Functions

Let $f(x) = N(x)/D(x)$, where $N(x)$ and $D(x)$ are polynomials and $D(x)$ is not the zero polynomial.

1. Simplify f, if possible. List any restrictions on the domain of f that are not implied by the simplified function.
2. Find and plot the y-intercept (if any) by evaluating $f(0)$.
3. Find the zeros of the numerator (if any). Then plot the corresponding x-intercepts.
4. Find the zeros of the denominator (if any). Then sketch the corresponding vertical asymptotes and plot the corresponding holes.
5. Find and sketch the horizontal asymptote (if any) by using the rule for finding the horizontal asymptote of a rational function on page 312.
6. Plot at least one point *between* and one point *beyond* each x-intercept and vertical asymptote.
7. Use smooth curves to complete the graph between and beyond the vertical asymptotes.

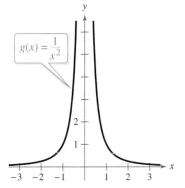

$g(x) = \dfrac{1}{x^2}$

The graph of g is symmetric with respect to the y-axis.
Figure 4.8

When graphing simple rational functions, testing for symmetry may be useful. Recall from Section 2.4 that the graph of $f(x) = 1/x$ is symmetric with respect to the origin. Figure 4.8 shows that the graph of $g(x) = 1/x^2$ is symmetric with respect to the y-axis.

▶▶▶ **TECHNOLOGY**

Some graphing utilities have difficulty graphing rational functions with vertical asymptotes. In *connected* mode, the utility may connect parts of the graph that are not supposed to be connected. For example, the graph in Figure 4.9(a) should consist of two unconnected portions—one to the left of $x = 2$ and the other to the right of $x = 2$. Using the graphing utility's *dot* mode eliminates this problem. In *dot* mode, however, the graph is represented as a collection of dots rather than as a smooth curve [see Figure 4.9(b)].

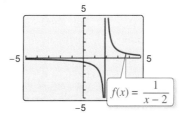

(a) Connected mode

$f(x) = \dfrac{1}{x - 2}$

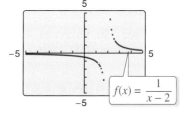

(b) Dot mode

$f(x) = \dfrac{1}{x - 2}$

Figure 4.9

GO DIGITAL

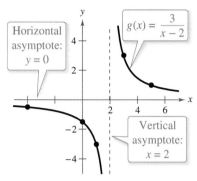

Figure 4.10

EXAMPLE 1 **Sketching the Graph of a Rational Function**

⯈⯈⯈ *See LarsonPrecalculus.com for an interactive version of this type of example.*

State the domain of $g(x) = \dfrac{3}{x-2}$ and sketch its graph.

Solution The domain of g is all real numbers except $x = 2$.

y-intercept:	$\left(0, -\frac{3}{2}\right)$, because $g(0) = -\frac{3}{2}$
x-intercept:	None, because there are no zeros of the numerator
Vertical asymptote:	$x = 2$, zero of denominator
Horizontal asymptote:	$y = 0$, because degree of $N(x) <$ degree of $D(x)$

Additional points:

x	-4	1	2	3	5
$g(x)$	-0.5	-3	Undefined	3	1

By plotting the intercept, asymptotes, and a few additional points, you obtain the graph shown in Figure 4.10.

✓ *Checkpoint* ▶ Audio-video solution in English & Spanish at *LarsonPrecalculus.com*

State the domain of $f(x) = \dfrac{1}{x+3}$ and sketch its graph. ■

Note that the graph of g in Example 1 is a vertical stretch and a right shift of the graph of $f(x) = 1/x$, because

$$g(x) = \frac{3}{x-2} = 3\left(\frac{1}{x-2}\right) = 3f(x-2).$$

EXAMPLE 2 **Sketching the Graph of a Rational Function**

State the domain of $f(x) = \dfrac{2x-1}{x}$ and sketch its graph.

Solution The domain of f is all real numbers except $x = 0$.

y-intercept:	None, because $x = 0$ is not in the domain
x-intercept:	$\left(\frac{1}{2}, 0\right)$, because $2x - 1 = 0$ when $x = \frac{1}{2}$
Vertical asymptote:	$x = 0$, zero of denominator
Horizontal asymptote:	$y = 2$, because degree of $N(x) =$ degree of $D(x)$

Additional points:

x	-4	-1	0	$\frac{1}{4}$	4
$f(x)$	2.25	3	Undefined	-2	1.75

By plotting the intercept, asymptotes, and a few additional points, you obtain the graph shown in Figure 4.11.

✓ *Checkpoint* ▶ Audio-video solution in English & Spanish at *LarsonPrecalculus.com*

State the domain of $g(x) = \dfrac{3+2x}{1+x}$ and sketch its graph. ■

Figure 4.11

EXAMPLE 3 Sketching the Graph of a Rational Function

State the domain of $f(x) = \dfrac{x}{x^2 - x - 2}$ and sketch its graph.

Solution The domain of f is all real numbers except $x = -1$ and $x = 2$. Factor the denominator to more easily determine the zeros of the denominator.

$$f(x) = \frac{x}{x^2 - x - 2} = \frac{x}{(x + 1)(x - 2)} \qquad \text{Factor the denominator.}$$

y-intercept: $(0, 0)$, because $f(0) = 0$

x-intercept: $(0, 0)$

Vertical asymptotes: $x = -1$, $x = 2$, zeros of denominator

Horizontal asymptote: $y = 0$, because degree of $N(x)$ < degree of $D(x)$

Additional points:

x	-3	-1	-0.5	1	2	3
$f(x)$	-0.3	Undefined	0.4	-0.5	Undefined	0.75

Figure 4.12 shows the graph of f.

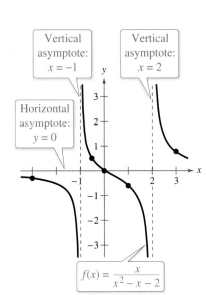

Vertical asymptote: $x = -1$

Vertical asymptote: $x = 2$

Horizontal asymptote: $y = 0$

$f(x) = \dfrac{x}{x^2 - x - 2}$

Figure 4.12

✓ *Checkpoint* ▶ Audio-video solution in English & Spanish at *LarsonPrecalculus.com*

State the domain of $f(x) = \dfrac{3x}{x^2 + x - 2}$ and sketch its graph.

EXAMPLE 4 Sketching the Graph of a Rational Function

State the domain of $f(x) = \dfrac{x^2 - 9}{x^2 - 2x - 3}$ and sketch its graph.

Solution The domain of f is all real numbers except $x = -1$ and $x = 3$. By factoring the numerator and denominator, you have

$$f(x) = \frac{x^2 - 9}{x^2 - 2x - 3} = \frac{(x - 3)(x + 3)}{(x - 3)(x + 1)} = \frac{x + 3}{x + 1}, \quad x \neq 3.$$

y-intercept: $(0, 3)$, because $f(0) = 3$

x-intercept: $(-3, 0)$, because $x + 3 = 0$ when $x = -3$

Vertical asymptote: $x = -1$, zero of (simplified) denominator

Hole: $\left(3, \frac{3}{2}\right)$

Horizontal asymptote: $y = 1$, because degree of $N(x)$ = degree of $D(x)$

Additional points:

x	-5	-2	-1	-0.5	1	3	4
$f(x)$	0.5	-1	Undefined	5	2	Undefined	1.4

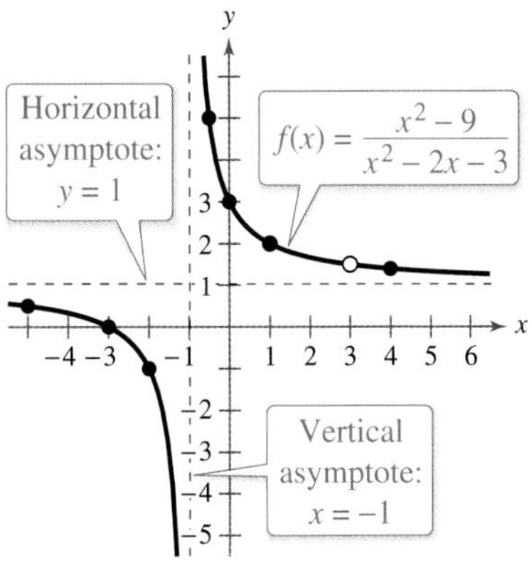

Horizontal asymptote: $y = 1$

$f(x) = \dfrac{x^2 - 9}{x^2 - 2x - 3}$

Vertical asymptote: $x = -1$

Hole at $x = 3$

Figure 4.13

Figure 4.13 shows the graph of f. Notice that there is a hole in the graph at $x = 3$ because the numerator and denominator of f have a common factor of $x - 3$.

✓ *Checkpoint* ▶ Audio-video solution in English & Spanish at *LarsonPrecalculus.com*

State the domain of $f(x) = \dfrac{x^2 - 4}{x^2 - x - 6}$ and sketch its graph.

GO DIGITAL

Slant Asymptotes

Consider a rational function whose denominator is of degree 1 or greater. If the degree of the numerator is exactly *one more* than the degree of the denominator, then the graph of the function has a **slant** (or **oblique**) **asymptote.** For example, the graph of

$$f(x) = \frac{x^2 - x}{x + 1}$$

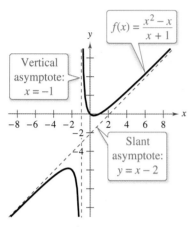

Figure 4.14

has a slant asymptote, as shown in Figure 4.14. To find the equation of a slant asymptote, use long division. For example, by dividing $x + 1$ into $x^2 - x$, you obtain

$$f(x) = \frac{x^2 - x}{x + 1} = \underbrace{x - 2}_{\substack{\text{Slant asymptote} \\ (y = x - 2)}} + \frac{2}{x + 1}. \qquad \text{Rewrite using long division.}$$

As x increases or decreases without bound, the remainder term $2/(x + 1)$ approaches 0, so the graph of f approaches the line $y = x - 2$, as shown in Figure 4.14.

EXAMPLE 5 **A Rational Function with a Slant Asymptote**

State the domain of $f(x) = \dfrac{x^2 - x - 2}{x - 1}$ and sketch its graph.

Solution The domain of f is all real numbers except $x = 1$. Next, write $f(x)$ in two different ways. Factoring the numerator

$$f(x) = \frac{x^2 - x - 2}{x - 1} \qquad \text{Write original function.}$$

$$= \frac{(x - 2)(x + 1)}{x - 1} \qquad \text{Factor the numerator.}$$

enables you to recognize the x-intercepts. Long division

$$f(x) = \frac{x^2 - x - 2}{x - 1} \qquad \text{Write original function.}$$

$$= x - \frac{2}{x - 1} \qquad \text{Rewrite using long division.}$$

enables you to recognize that the line $y = x$ is a slant asymptote of the graph.

y-intercept: $\qquad$ (0, 2), because $f(0) = 2$

x-intercepts: $\qquad$ (2, 0) and (−1, 0), because $x - 2 = 0$ when $x = 2$ and $x + 1 = 0$ when $x = -1$

Vertical asymptote: $x = 1$, zero of denominator

Slant asymptote: $\qquad$ $y = x$

Additional points:

x	-2	0.5	1	1.5	3
$f(x)$	-1.33	4.5	Undefined	-2.5	2

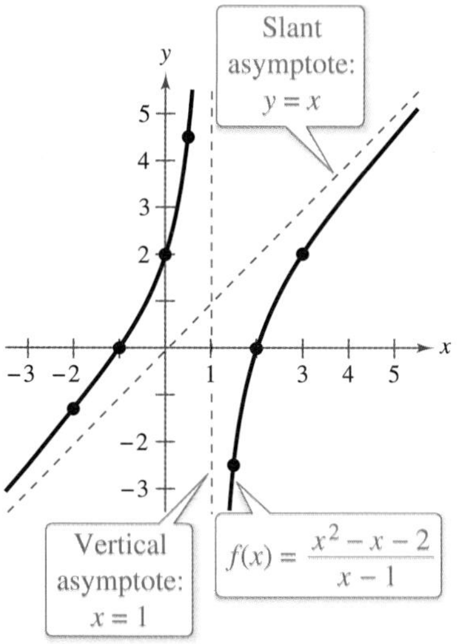

Figure 4.15 shows the graph of f.

✓ *Checkpoint* ▶ *Audio-video solution in English & Spanish at LarsonPrecalculus.com*

State the domain of $f(x) = \dfrac{3x^2 + 1}{x}$ and sketch its graph. ∎

Figure 4.15

Application

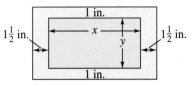

Figure 4.16

EXAMPLE 6 **Finding a Minimum Area**

A rectangular page contains 48 square inches of print. The margins at the top and bottom of the page are each 1 inch deep. The margins on each side are $1\frac{1}{2}$ inches wide. What should the dimensions of the page be to use the least amount of paper?

Graphical Solution

Let A be the area to be minimized. From Figure 4.16, you can write $A = (x + 3)(y + 2)$. The printed area inside the margins is given by $xy = 48$ or $y = 48/x$. To find the minimum area, rewrite the equation for A in terms of just one variable by substituting $48/x$ for y.

$$A = (x + 3)\left(\frac{48}{x} + 2\right)$$
$$= \frac{(x + 3)(48 + 2x)}{x}, \quad x > 0$$

The graph of this rational function is shown below. Because x represents the width of the printed area, you need to consider only the portion of the graph for which x is positive. Use the *minimum* feature of a graphing utility to estimate that the minimum value of A occurs when $x \approx 8.5$ inches. The corresponding value of y is $48/8.5 \approx 5.6$ inches. So, the dimensions should be

$$x + 3 \approx 11.5 \text{ inches} \quad \text{by} \quad y + 2 \approx 7.6 \text{ inches.}$$

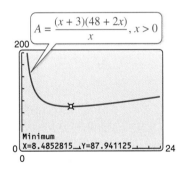

Numerical Solution

Let A be the area to be minimized. From Figure 4.16, you can write $A = (x + 3)(y + 2)$. The printed area inside the margins is given by $xy = 48$ or $y = 48/x$. To find the minimum area, rewrite the equation for A in terms of just one variable by substituting $48/x$ for y.

$$A = (x + 3)\left(\frac{48}{x} + 2\right) = \frac{(x + 3)(48 + 2x)}{x}, \quad x > 0$$

Use the *table* feature of a graphing utility to create a table of values for the function

$$y_1 = \frac{(x + 3)(48 + 2x)}{x}$$

beginning at $x = 1$ and increasing by 1. The minimum value of y_1 occurs when x is somewhere between 8 and 9, as shown in Figure 4.17. To approximate the minimum value of y_1 to one decimal place, change the table to begin at $x = 8$ and increase by 0.1. The minimum value of y_1 occurs when $x \approx 8.5$, as shown in Figure 4.18. The corresponding value of y is $48/8.5 \approx 5.6$ inches. So, the dimensions should be

$$x + 3 \approx 11.5 \text{ inches} \quad \text{by} \quad y + 2 \approx 7.6 \text{ inches.}$$

X	Y₁
6	90
7	88.571
8	88
9	88
10	88.4
11	89.091
12	90
X=8	

Figure 4.17

X	Y₁
8.2	87.961
8.3	87.949
8.4	87.943
8.5	87.941
8.6	87.944
8.7	87.952
8.8	87.964
X=8.5	

Figure 4.18

✓ *Checkpoint* ▶ *Audio-video solution in English & Spanish at LarsonPrecalculus.com*

Rework Example 6 when the margins on each side are 2 inches wide and the page contains 40 square inches of print. ▪

Summarize (Section 4.2)

1. Explain how to sketch the graph of a rational function *(page 318)*. For examples of sketching graphs of rational functions, see Examples 1–4.

2. Explain how to determine whether the graph of a rational function has a slant asymptote *(page 321)*. For an example of sketching the graph of a rational function that has a slant asymptote, see Example 5.

3. Describe an example of how to use the graph of a rational function to model and solve a real-life problem *(page 322, Example 6)*.

GO DIGITAL

4.2 Exercises

See CalcChat.com for tutorial help and worked-out solutions to odd-numbered exercises.

GO DIGITAL

Vocabulary and Concept Check

In Exercises 1 and 2, fill in the blanks.

1. For the rational function $f(x) = N(x)/D(x)$, if the degree of $N(x)$ is exactly one more than the degree of $D(x)$, then the graph of f has a _____ (or oblique) _____.

2. The graph of $g(x) = 3/(x - 2)$ has a _____ asymptote at $x = 2$.

3. Using long division, you find that $f(x) = \dfrac{x^2 + 1}{x + 1} = x - 1 + \dfrac{2}{x + 1}$. What is the slant asymptote of the graph of f?

4. Which features of the graph of a rational function are found at x-values that are zeros of (a) the numerator and (b) the denominator?

Skills and Applications

Sketching a Transformation of a Rational Function
In Exercises 5–8, use the graph of $f(x) = \dfrac{2}{x}$ to sketch the graph of g.

5. $g(x) = \dfrac{2}{x} + 4$

6. $g(x) = \dfrac{2}{x - 4}$

7. $g(x) = -\dfrac{2}{x}$

8. $g(x) = \dfrac{4}{x}$

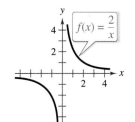

Sketching a Transformation of a Rational Function In Exercises 9–12, use the graph of $f(x) = \dfrac{3}{x^2}$ to sketch the graph of g.

9. $g(x) = \dfrac{3}{x^2} - 1$

10. $g(x) = -\dfrac{3}{x^2}$

11. $g(x) = \dfrac{3}{(x - 1)^2}$

12. $g(x) = \dfrac{1}{x^2} + 2$

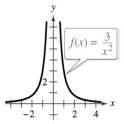

Sketching the Graph of a Rational Function In Exercises 13–42, (a) state the domain of the function, (b) identify all intercepts, (c) find any vertical or horizontal asymptotes or holes, and (d) plot additional solution points as needed to sketch the graph of the rational function.

13. $f(x) = \dfrac{1}{x + 1}$

14. $f(x) = \dfrac{1}{x - 3}$

15. $h(x) = \dfrac{-1}{x + 4}$

16. $g(x) = \dfrac{1}{6 - x}$

17. $C(x) = \dfrac{2x + 3}{x + 2}$

18. $P(x) = \dfrac{1 - 3x}{1 - x}$

19. $g(x) = \dfrac{1}{x + 2} + 2$

20. $f(x) = \dfrac{3}{x - 1} - 5$

21. $f(x) = \dfrac{x^2}{x^2 + 9}$

22. $f(t) = \dfrac{1 - 2t}{t}$

23. $h(x) = \dfrac{x^2}{x^2 - 9}$

24. $g(x) = \dfrac{x}{x^2 - 9}$

25. $g(s) = \dfrac{4s}{s^2 + 4}$

26. $f(x) = -\dfrac{x}{(x - 2)^2}$

27. $g(x) = \dfrac{4(x + 1)}{x(x - 4)}$

28. $h(x) = \dfrac{2}{x^2(x - 2)}$

29. $f(x) = \dfrac{2x}{x^2 - 3x - 4}$

30. $f(x) = \dfrac{3x}{x^2 + 2x - 3}$

31. $f(x) = \dfrac{5(x + 4)}{x^2 + x - 12}$

32. $f(x) = \dfrac{x^2 + 3x}{x^2 + x - 6}$

33. $f(t) = \dfrac{t^2 - 1}{t - 1}$

34. $f(x) = \dfrac{x^2 - 36}{x + 6}$

35. $h(x) = \dfrac{x^2 - 5x + 4}{x^2 - 4}$

36. $g(x) = \dfrac{x^2 - 2x - 8}{x^2 - 9}$

37. $f(x) = \dfrac{x^2 + 4}{x^2 + 3x - 4}$

38. $f(x) = \dfrac{3(x^2 + 1)}{x^2 + 2x - 15}$

39. $f(x) = \dfrac{2x^2 - 5x + 2}{2x^2 - x - 6}$

40. $f(x) = \dfrac{3x^2 - 8x + 4}{2x^2 - 3x - 2}$

41. $f(x) = \dfrac{2x^2 - 5x - 3}{x^3 - 2x^2 - x + 2}$

42. $f(x) = \dfrac{x^2 - x - 2}{x^3 - 2x^2 - 5x + 6}$

Comparing Graphs of Functions In Exercises 43 and 44, (a) state the domains of f and g, (b) use a graphing utility to graph f and g in the same viewing window, and (c) explain why the graphing utility may not show the difference in the domains of f and g.

43. $f(x) = \dfrac{x^2 - 1}{x + 1}, \quad g(x) = x - 1$

44. $f(x) = \dfrac{x - 2}{x^2 - 2x}, \quad g(x) = \dfrac{1}{x}$

A Rational Function with a Slant Asymptote In Exercises 45–60, (a) state the domain of the function, (b) identify all intercepts, (c) find any vertical or slant asymptotes or holes, and (d) plot additional solution points as needed to sketch the graph of the rational function.

45. $h(x) = \dfrac{x^2 - 4}{x}$

46. $g(x) = \dfrac{x^2 + 5}{x}$

47. $f(x) = \dfrac{2x^2 + 1}{x}$

48. $f(x) = \dfrac{-x^2 - 2}{x}$

49. $g(x) = \dfrac{x^2 + 1}{x}$

50. $h(x) = \dfrac{x^2}{x - 1}$

51. $f(t) = -\dfrac{t^2 + 1}{t + 5}$

52. $f(x) = \dfrac{x^2 + 1}{x + 1}$

53. $f(x) = \dfrac{x^3}{x^2 - 4}$

54. $g(x) = \dfrac{x^3}{2x^2 - 8}$

55. $f(x) = \dfrac{x^3 - 1}{x^2 - x}$

56. $f(x) = \dfrac{x^4 + x}{x^3}$

57. $f(x) = \dfrac{x^2 - x + 1}{x - 1}$

58. $f(x) = \dfrac{2x^2 - 5x + 5}{x - 2}$

59. $f(x) = \dfrac{2x^3 - x^2 - 2x + 1}{x^2 + 3x + 2}$

60. $f(x) = \dfrac{2x^3 + x^2 - 8x - 4}{x^2 - 3x + 2}$

Graphical Reasoning In Exercises 61–68, (a) use the graph to determine any x-intercepts of the graph of the rational function and (b) set $y = 0$ and solve the resulting equation to confirm your result in part (a).

61. $y = \dfrac{x + 1}{x - 3}$

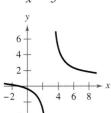

62. $y = \dfrac{2x}{x - 3}$

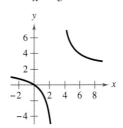

63. $y = \dfrac{3x}{x + 1}$

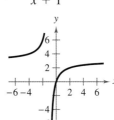

64. $y = \dfrac{x}{x^2 - 4}$

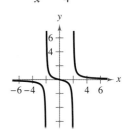

65. $y = \dfrac{1}{x} - x$

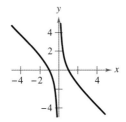

66. $y = x - 3 + \dfrac{2}{x}$

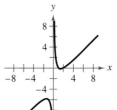

67. $y = x + 2 + \dfrac{1}{x}$

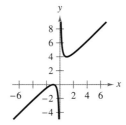

68. $y = x - \dfrac{2}{x + 1}$

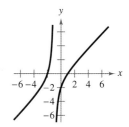

69. Page Design A page that is x inches wide and y inches high contains 30 square inches of print. The top and bottom margins are each 1 inch deep and the margins on each side are 2 inches wide (see figure).

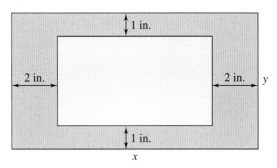

(a) Show that the total area A of the page is

$$A = \dfrac{2x(x + 11)}{x - 4}.$$

(b) Determine the domain of the function based on the physical constraints of the problem.

(c) Use a graphing utility to graph the area function and approximate the dimensions of the page that use the least amount of paper. Verify your answer using the *table* feature of the graphing utility.

70. Page Design A rectangular page contains 64 square inches of print. The margins at the top and bottom of the page are each 1 inch deep. The margins on each side are $1\frac{1}{2}$ inches wide. What should the dimensions of the page be to use the least amount of paper?

71. Concentration of a Mixture A 1000-liter tank contains 50 liters of a 25% brine solution. You add x liters of a 75% brine solution to the tank.

(a) Show that the concentration C, the proportion of brine to total solution, in the final mixture is

$$C = \frac{3x + 50}{4(x + 50)}.$$

(b) Determine the domain of the function based on the physical constraints of the problem.

(c) Sketch the graph of the concentration function.

(d) As the tank is filled, what happens to the rate at which the concentration of brine is increasing? What percent does the concentration of brine appear to approach?

72. Medicine

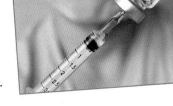

The concentration C of a chemical in the bloodstream t hours after injection into muscle tissue is given by

$$C = \frac{3t^2 + t}{t^3 + 50}, \quad t > 0.$$

(a) Determine the horizontal asymptote of the graph of the function and interpret its meaning in the context of the problem.

(b) Use a graphing utility to graph the function and approximate the time when the bloodstream concentration is greatest.

(c) Use the graphing utility to determine when the concentration is less than 0.345.

Graphical Reasoning In Exercises 73–76, (a) use a graphing utility to graph the function and determine any x-intercepts of the graph and (b) set $y = 0$ and solve the resulting equation to confirm your result in part (a).

73. $y = \dfrac{1}{x + 5} + \dfrac{4}{x}$

74. $y = 20\left(\dfrac{2}{x + 1} - \dfrac{3}{x}\right)$

75. $y = x - \dfrac{6}{x - 1}$

76. $y = x - \dfrac{16}{x}$

© Seasontime/Shutterstock.com

Finding Extrema In Exercises 77–80, use a graphing utility to graph the function and locate any relative maximum or minimum points on the graph.

77. $f(x) = \dfrac{3(x + 1)}{x^2 + x + 1}$

78. $g(x) = \dfrac{6x}{x^2 + x + 1}$

79. $C(x) = x - 2 + \dfrac{32}{x}$

80. $f(x) = x - 4 + \dfrac{16}{x}$

81. Minimum Cost The processing and shipping cost C (in thousands of dollars) for ordering the components used in manufacturing a product is given by

$$C = 100\left(\frac{200}{x^2} + \frac{x}{x + 30}\right), \quad x \geq 1$$

where x is the order size (in hundreds of components). Use a graphing utility to graph the cost function. From the graph, estimate the order size that minimizes cost.

82. Minimum Cost The cost C (in dollars) of producing x units of a product is given by

$$C = 0.2x^2 + 10x + 5$$

and the average cost per unit $\overline{C}$ is given by

$$\overline{C} = \frac{C}{x} = \frac{0.2x^2 + 10x + 5}{x}, \quad x > 0.$$

Sketch the graph of the average cost function and estimate the number of units that should be produced to minimize the average cost per unit.

83. Average Speed A driver's average speed is 50 miles per hour on a round trip between two cities 100 miles apart. The average speeds for going and returning were x and y miles per hour, respectively.

(a) Show that $y = \dfrac{25x}{x - 25}$.

(b) Determine the vertical and horizontal asymptotes of the graph of the function.

(c) Use a graphing utility to graph the function.

(d) Complete the table.

x	30	35	40	45	50	55	60
y							

(e) Are the results in the table what you expected? Explain.

(f) Is it possible to average 20 miles per hour in one direction and still average 50 miles per hour on the round trip? Explain.

Exploring the Concepts

True or False? **In Exercises 84–88, determine whether the statement is true or false. Justify your answer.**

84. The graph of a rational function is continuous only when the denominator is a constant polynomial.

85. The graph of a rational function can never cross one of its asymptotes.

86. The graph of $f(x) = \dfrac{x^2}{x+1}$ has a slant asymptote.

87. The maximum number of vertical asymptotes that the graph of a rational function can have is equal to the degree of the denominator.

88. The graph of every rational function has a horizontal asymptote.

89. **Error Analysis** Describe the error.

The graph of

$$h(x) = \frac{6 - 2x}{3 - x}$$

has the line $x = 3$ as a vertical asymptote because $3 - x = 0$ when $x = 3$.

90. **HOW DO YOU SEE IT?** For each statement, determine which of the graphs show that the statement is incorrect.

(i)

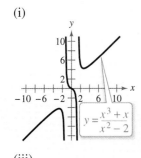

$y = \dfrac{x^3 + x}{x^2 - 2}$

(ii)

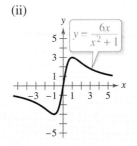

$y = \dfrac{6x}{x^2 + 1}$

(iii)

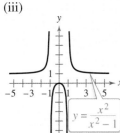

$y = \dfrac{x^2}{x^2 - 1}$

(iv)

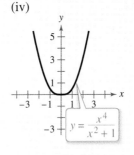

$y = \dfrac{x^4}{x^2 + 1}$

(a) The graph of every rational function has a vertical asymptote.

(b) The graph of every rational function has at least one vertical, horizontal, or slant asymptote.

(c) The graph of a rational function can have at most one vertical asymptote.

91. **Writing**

(a) Given a rational function f, how can you determine whether the graph of f has a slant asymptote?

(b) You determine that the graph of a rational function f has a slant asymptote. How do you find it?

Writing and Graphing a Rational Function **In Exercises 92 and 93, write a rational function that satisfies the given criteria. Then sketch the graph of your function.**

92. Vertical asymptote: $x = 2$

Slant asymptote: $y = x + 1$

Zero of the function: $x = -2$

93. Vertical asymptote: $x = -4$

Slant asymptote: $y = x - 2$

Zero of the function: $x = 3$

94. **Think About It** Is it possible for the graph of a rational function to have all three types of asymptotes? Why or why not?

Review & Refresh ▶ Video solutions at LarsonPrecalculus.com

Using Standard Form to Graph a Parabola **In Exercises 95–98, write the quadratic function in standard form and sketch its graph. Identify the vertex, axis of symmetry, and x-intercept(s).**

95. $f(x) = x^2 + 2x$

96. $f(x) = x^2 - 4x + 4$

97. $f(x) = -x^2 - 16x - 13$

98. $f(x) = -\frac{1}{4}x^2 + 2x$

Writing the Equation of a Circle **In Exercises 99–102, write the standard form of the equation of the circle with the given characteristics.**

99. Center: $(0, 3)$; Radius: 7

100. Center: $(5, -2)$; Radius: 2

101. Center: $(-2, -1)$; Solution point: $(1, 3)$

102. Endpoints of a diameter: $(1, 12)$, $(11, -12)$

One-to-One Function Representation **In Exercises 103 and 104, determine whether the situation can be represented by a one-to-one function. If so, write a statement that best describes the inverse function.**

103. The number of miles n a marathon runner has completed in terms of the time t in hours

104. The depth of the tide d at a beach in terms of the time t over a 24-hour period

Project: NASA To work an extended application analyzing the annual budget of NASA from 1958 to 2020, visit this text's website at *LarsonPrecalculus.com.* (*Source: U.S. Office of Management and Budget*)

4.3 Conics

Conics have many real-life applications and are often used to model and solve engineering problems. For example, in Exercise 31 on page 338, you will use a parabola to model the cables of the Golden Gate Bridge.

- ❯ Describe a conic section.
- ❯ Recognize, graph, and write equations of parabolas (vertex at origin).
- ❯ Recognize, graph, and write equations of ellipses (center at origin).
- ❯ Recognize, graph, and write equations of hyperbolas (center at origin).

Conic Sections

Conic sections were discovered during the classical Greek period. The early Greek studies were largely concerned with the geometric properties of conics. It was not until the early seventeenth century that the broad applicability of conics became apparent and played a prominent role in the early development of calculus.

A **conic section** (or simply **conic**) is the intersection of a plane and a double-napped cone. Notice in Figure 4.19 that in the formation of the four basic conics, the intersecting plane does not pass through the vertex of the cone. When the plane does pass through the vertex, the resulting figure is a **degenerate conic,** as shown in Figure 4.20.

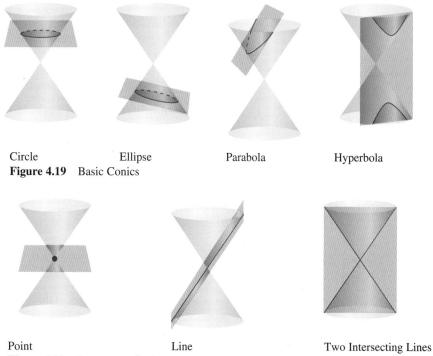

Circle   Ellipse   Parabola   Hyperbola
Figure 4.19   Basic Conics

Point   Line   Two Intersecting Lines
Figure 4.20   Degenerate Conics

There are several ways to approach the study of conics. You could begin by defining conics in terms of the intersections of planes and cones, as the Greeks did, or you could define them algebraically, in terms of the general second-degree equation

$$Ax^2 + Bxy + Cy^2 + Dx + Ey + F = 0.$$

However, you will study a third approach, in which each of the conics is defined as a **locus** (collection) of points satisfying a given geometric property. For example, in Section 1.1 you saw how the definition of a circle as *the collection of all points (x, y) that are equidistant from a fixed point (h, k)* led to the standard form of the equation of a circle

$$(x - h)^2 + (y - k)^2 = r^2. \qquad \text{Equation of a circle}$$

Recall that the center of a circle is at (h, k) and that the radius of the circle is r.

GO DIGITAL

Parabolas

In Section 3.1, you learned that the graph of the quadratic function

$$f(x) = ax^2 + bx + c$$

is a parabola that opens upward or downward. The definition of a parabola below is more general in the sense that it is independent of the orientation of the parabola.

Definition of a Parabola

A **parabola** is the set of all points (x, y) in a plane that are equidistant from a fixed line, the **directrix,** and a fixed point, the **focus,** not on the line. (See figure.) The **vertex** is the midpoint between the focus and the directrix. The **axis** of the parabola is the line passing through the focus and the vertex.

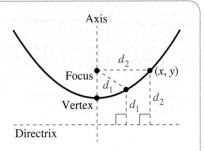

Note in the figure above that a parabola is symmetric with respect to its axis. The definition of a parabola can be used to derive the **standard form of the equation of a parabola** with vertex at $(0, 0)$ and directrix parallel to the x-axis or to the y-axis. This standard form is given in the next definition.

Standard Equation of a Parabola (Vertex at Origin)

The **standard form of the equation of a parabola** with vertex at $(0, 0)$ and directrix $y = -p$ is

$$x^2 = 4py, \quad p \neq 0. \qquad \text{Vertical axis}$$

For directrix $x = -p$, the equation is

$$y^2 = 4px, \quad p \neq 0. \qquad \text{Horizontal axis}$$

The focus lies on the axis p units (directed distance) from the vertex.

For a proof of the standard form of the equation of a parabola, see Proofs in Mathematics on page 356.

Notice that a parabola can have a vertical or a horizontal axis, as shown in the figures below.

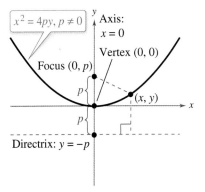

Parabola with vertical axis

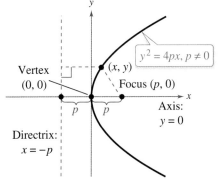

Parabola with horizontal axis

GO DIGITAL

EXAMPLE 1 **Finding the Focus and Directrix of a Parabola**

Find the focus and directrix of the parabola $y = -2x^2$. Then sketch the parabola.

Solution The squared term in the equation involves x, so you know that the axis is vertical, and the equation is of the form

$$x^2 = 4py.$$ Standard form, vertical axis

Next, rewrite the original equation in standard form, as shown below.

$$-2x^2 = y$$ Write original equation.

$$x^2 = -\frac{1}{2}y$$ Divide each side by -2.

Comparing this equation with $x^2 = 4py$, you can conclude that

$$4p = -\frac{1}{2} \quad \Longrightarrow \quad p = -\frac{1}{8}.$$

Because p is negative, the parabola opens downward, as shown in Figure 4.21. The focus of the parabola is

$$(0, p) = \left(0, -\frac{1}{8}\right)$$ Focus

and the directrix of the parabola is

$$y = -p = -\left(-\frac{1}{8}\right) = \frac{1}{8}.$$ Directrix

✔ **Checkpoint** ▶ Audio-video solution in English & Spanish at LarsonPrecalculus.com

Find the focus and directrix of the parabola $y = \frac{1}{4}x^2$. Then sketch the parabola.

EXAMPLE 2 **Finding the Standard Equation of a Parabola**

▶▶▶ *See LarsonPrecalculus.com for an interactive version of this type of example.*

Find the standard form of the equation of the parabola with vertex at the origin and focus at $(2, 0)$.

Solution The axis of the parabola is horizontal, passing through $(0, 0)$ and $(2, 0)$, as shown in Figure 4.22. So, the equation is of the form

$$y^2 = 4px.$$ Standard form, horizontal axis

The focus is $p = 2$ units from the vertex, so the standard form of the equation is

$$y^2 = 4(2)x$$ Substitute 2 for p.

$$y^2 = 8x.$$ Simplify.

✔ **Checkpoint** ▶ Audio-video solution in English & Spanish at LarsonPrecalculus.com

Find the standard form of the equation of the parabola with vertex at the origin and focus at $\left(0, \frac{3}{8}\right)$.

Parabolas occur in a wide variety of applications. For example, a parabolic reflector can be formed by revolving a parabola about its axis. The resulting surface has the property that all incoming rays parallel to the axis reflect through the focus of the parabola. This is the principle behind the construction of the parabolic mirrors used in reflecting telescopes. Conversely, the light rays emanating from the focus of a parabolic reflector used in a flashlight are all parallel to one another, as shown in Figure 4.23.

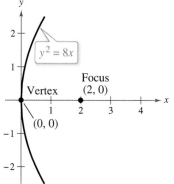

Directrix: $y = \frac{1}{8}$; Focus $\left(0, -\frac{1}{8}\right)$; $y = -2x^2$

Figure 4.21

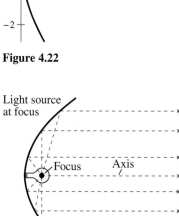

$y^2 = 8x$; Vertex $(0, 0)$; Focus $(2, 0)$

Figure 4.22

Light source at focus; Focus; Axis; Parabolic reflector: light reflects in parallel rays.

Figure 4.23

GO DIGITAL

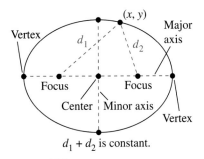

$d_1 + d_2$ is constant.

Figure 4.24

Figure 4.25

Ellipses

Definition of an Ellipse

An **ellipse** is the set of all points (x, y) in a plane, the sum of whose distances from two distinct fixed points, called **foci,** is constant. See Figure 4.24.

The line through the foci intersects the ellipse at two points called **vertices.** The chord joining the vertices is the **major axis,** and its midpoint is the **center** of the ellipse. The chord perpendicular to the major axis at the center is the **minor axis.** (See Figure 4.24.)

To visualize the definition of an ellipse, imagine two thumbtacks placed at the foci (see Figure 4.25). When the ends of a fixed length of string are fastened to the thumbtacks and the string is drawn taut with a pencil, the path traced by the pencil is an ellipse.

The standard form of the equation of an ellipse takes one of two forms, depending on whether the major axis is horizontal or vertical.

Standard Equation of an Ellipse (Center at Origin)

The **standard form of the equation of an ellipse** centered at the origin with major and minor axes of lengths $2a$ and $2b$, respectively, where $0 < b < a$, is

$$\frac{x^2}{a^2} + \frac{y^2}{b^2} = 1 \qquad \text{Major axis is horizontal.}$$

or

$$\frac{x^2}{b^2} + \frac{y^2}{a^2} = 1. \qquad \text{Major axis is vertical.}$$

The vertices and foci lie on the major axis, a and c units, respectively, from the center. Moreover, a, b, and c are related by the equation $c^2 = a^2 - b^2$. The figures below show the horizontal and vertical orientations for an ellipse.

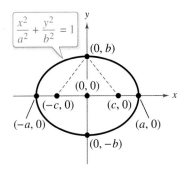

Major axis is horizontal;
minor axis is vertical.

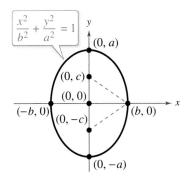

Major axis is vertical;
minor axis is horizontal.

To show that $c^2 = a^2 - b^2$, use either of the figures above and the fact that the sum of the distances from a point on the ellipse to the two foci is constant. For example, using the figure on the left above, you have

Sum of distances from $(0, b)$ to foci $=$ Sum of distances from $(a, 0)$ to foci

$$2\sqrt{b^2 + c^2} = (a + c) + (a - c)$$
$$\sqrt{b^2 + c^2} = a$$
$$c^2 = a^2 - b^2.$$

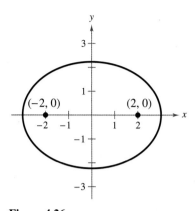

Figure 4.26

EXAMPLE 3   **Finding the Standard Equation of an Ellipse**

Find the standard form of the equation of the ellipse that has a major axis of length 6 and foci at $(-2, 0)$ and $(2, 0)$, as shown in Figure 4.26.

Solution   The center of the ellipse is the midpoint of the line segment joining the foci. So, by the Midpoint Formula, the center of the ellipse is $(0, 0)$. Thus, the distance from the center to one of the foci is $c = 2$. Because $2a = 6$, you know that $a = 3$. Now, from $c^2 = a^2 - b^2$, you have

$$b^2 = a^2 - c^2 = 3^2 - 2^2 = 9 - 4 = 5.$$

The major axis of the ellipse is horizontal, so the standard form of the equation is

$$\frac{x^2}{a^2} + \frac{y^2}{b^2} = 1 \quad \Longrightarrow \quad \frac{x^2}{3^2} + \frac{y^2}{\left(\sqrt{5}\right)^2} = 1.$$

✓ *Checkpoint* ▶  Audio-video solution in English & Spanish at *LarsonPrecalculus.com*

Find the standard form of the equation of the ellipse that has a major axis of length 10 and foci at $(0, -3)$ and $(0, 3)$.

EXAMPLE 4   **Sketching an Ellipse**

Sketch the ellipse $4x^2 + y^2 = 36$ and identify the vertices.

Algebraic Solution

$4x^2 + y^2 = 36$	Write original equation.
$\dfrac{4x^2}{36} + \dfrac{y^2}{36} = \dfrac{36}{36}$	Divide each side by 36.
$\dfrac{x^2}{3^2} + \dfrac{y^2}{6^2} = 1$	Write in standard form.

The center of the ellipse is $(0, 0)$. Because the denominator of the y^2-term is greater than the denominator of the x^2-term, you can conclude that the major axis is vertical. Moreover, because $a = 6$, the vertices are $(0, -6)$ and $(0, 6)$. Finally, because $b = 3$, the endpoints of the minor axis (the *co-vertices*) are $(-3, 0)$ and $(3, 0)$, as shown in the figure.

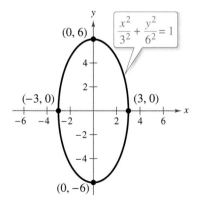

Graphical Solution

Solve the equation of the ellipse for y.

$$4x^2 + y^2 = 36$$
$$y^2 = 36 - 4x^2$$
$$y = \pm\sqrt{36 - 4x^2}$$

Then, use a graphing utility to graph

$$y_1 = \sqrt{36 - 4x^2}$$

and

$$y_2 = -\sqrt{36 - 4x^2}$$

in the same viewing window, as shown in the figure below. Be sure to use a square setting.

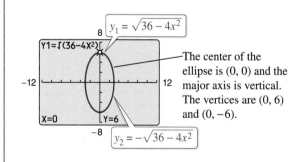

The center of the ellipse is $(0, 0)$ and the major axis is vertical. The vertices are $(0, 6)$ and $(0, -6)$.

✓ *Checkpoint* ▶ Audio-video solution in English & Spanish at *LarsonPrecalculus.com*

Sketch the ellipse $x^2 + 9y^2 = 81$ and identify the vertices.

GO DIGITAL

Ellipses have many practical and aesthetic uses. For example, machine gears, supporting arches, and acoustic designs often involve elliptical shapes. The orbits of satellites and planets are also ellipses.

It was difficult for early astronomers to detect that the orbits of the planets are ellipses because the foci of the planetary orbits are relatively close to their centers, and so the orbits are nearly circular. You can measure the "ovalness" of an ellipse by using the concept of **eccentricity.**

Definition of Eccentricity

The **eccentricity** e of an ellipse is the ratio $e = \dfrac{c}{a}$.

Note that $0 < e < 1$ for *every* ellipse.

To see how this ratio describes the shape of an ellipse, note that because the foci of an ellipse are located along the major axis between the vertices and the center, it follows that $0 < c < a$. For an ellipse that is nearly circular, the foci are close to the center and the ratio c/a is close to 0, as shown in Figure 4.27. On the other hand, for an elongated ellipse, the foci are close to the vertices and the ratio c/a is close to 1, as shown in Figure 4.28.

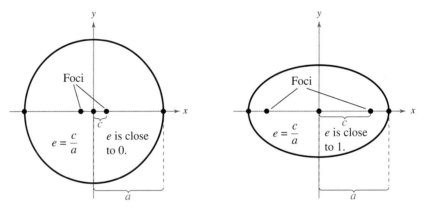

Figure 4.27   **Figure 4.28**

The orbit of the moon has an eccentricity of

$e \approx 0.0549.$ Eccentricity of the moon's orbit

The eccentricities of the eight planetary orbits are listed below.

Planet	Orbit Eccentricity, e
Mercury	0.2056
Venus	0.0067
Earth	0.0167
Mars	0.0935
Jupiter	0.0489
Saturn	0.0565
Uranus	0.0457
Neptune	0.0113

GO DIGITAL

Hyperbolas

The definition of a **hyperbola** is similar to that of an ellipse. For an ellipse, the *sum* of the distances between the foci and a point on the ellipse is constant, whereas for a hyperbola, the absolute value of the *difference* of the distances between the foci and a point on the hyperbola is constant.

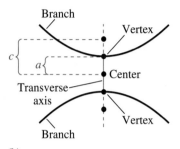

$|d_2 - d_1|$ is constant.

(a)

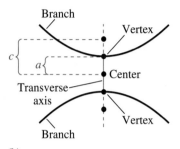

(b)

Figure 4.29

> ### Definition of a Hyperbola
>
> A **hyperbola** is the set of all points (x, y) in a plane for which the absolute value of the difference of the distances from two distinct fixed points, called **foci**, is constant. See Figure 4.29(a).

The graph of a hyperbola has two disconnected parts called **branches.** The line through the foci intersects the hyperbola at two points called **vertices.** The line segment connecting the vertices is the **transverse axis,** and its midpoint is the **center** of the hyperbola. In Figure 4.29(b), notice that the distance between the foci and the center is greater than the distance between the vertices and the center.

As with ellipses, the standard form of the equation of a hyperbola takes one of two forms, depending on whether the major axis is horizontal or vertical.

> ### Standard Equation of a Hyperbola (Center at Origin)
>
> The **standard form of the equation of a hyperbola** with center at the origin is
>
> $$\frac{x^2}{a^2} - \frac{y^2}{b^2} = 1 \qquad \text{Transverse axis is horizontal.}$$
>
> or
>
> $$\frac{y^2}{a^2} - \frac{x^2}{b^2} = 1 \qquad \text{Transverse axis is vertical.}$$
>
> where $a > 0$ and $b > 0$. The vertices and foci are, respectively, a and c units from the center. Moreover, a, b, and c are related by the equation $c^2 = a^2 + b^2$. The figures below show the horizontal and vertical orientations for a hyperbola.

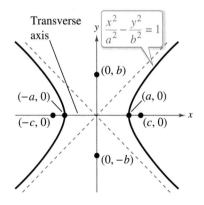

Transverse axis is horizontal.

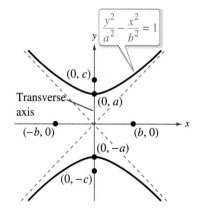

Transverse axis is vertical.

Be careful when finding the foci of ellipses and hyperbolas. Notice that the relationships between a, b, and c differ slightly. When finding the foci of an *ellipse*, use $c^2 = a^2 - b^2$. When finding the foci of a *hyperbola*, use $c^2 = a^2 + b^2$.

GO DIGITAL

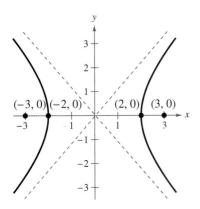

Figure 4.30

EXAMPLE 5   **Finding the Standard Equation of a Hyperbola**

Find the standard form of the equation of the hyperbola with foci at

$$(-3, 0) \quad \text{and} \quad (3, 0)$$

and vertices at

$$(-2, 0) \quad \text{and} \quad (2, 0)$$

as shown in Figure 4.30.

Solution   From the graph, $c = 3$ because the foci are three units from the center $(0, 0)$. Moreover, $a = 2$ because the vertices are two units from the center. So, it follows that

$$b^2 = c^2 - a^2$$
$$= 3^2 - 2^2$$
$$= 9 - 4$$
$$= 5.$$

The transverse axis is horizontal, so the equation is of the form

$$\frac{x^2}{a^2} - \frac{y^2}{b^2} = 1. \qquad \text{Standard form, horizontal transverse axis}$$

Substitute

$$a^2 = 2^2 \quad \text{and} \quad b^2 = 5 = \left(\sqrt{5}\right)^2$$

to obtain

$$\frac{x^2}{2^2} - \frac{y^2}{\left(\sqrt{5}\right)^2} = 1. \qquad \text{Standard form}$$

✓ *Checkpoint*   ▶ *Audio-video solution in English & Spanish at LarsonPrecalculus.com*

Find the standard form of the equation of the hyperbola with foci at $(0, -6)$ and $(0, 6)$ and vertices at $(0, -3)$ and $(0, 3)$. ■

An important aid in sketching the graph of a hyperbola is the determination of its **asymptotes,** as shown in the figures below. Every hyperbola has two asymptotes that intersect at the center of the hyperbola. Furthermore, the asymptotes pass through the corners of a rectangle of dimensions $2a$ by $2b$. The line segment of length $2b$ joining $(0, b)$ and $(0, -b)$ [or $(-b, 0)$ and $(b, 0)$] is the **conjugate axis** of the hyperbola.

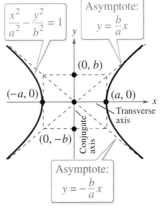

Transverse axis is horizontal; conjugate axis is vertical.

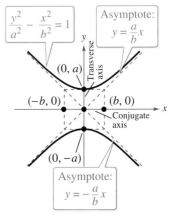

Transverse axis is vertical; conjugate axis is horizontal.

> **Asymptotes of a Hyperbola (Center at Origin)**
>
> The asymptotes of a hyperbola with center at $(0, 0)$ are
>
> $$y = \frac{b}{a}x \quad \text{and} \quad y = -\frac{b}{a}x \qquad \text{Transverse axis is horizontal.}$$
>
> or
>
> $$y = \frac{a}{b}x \quad \text{and} \quad y = -\frac{a}{b}x. \qquad \text{Transverse axis is vertical.}$$

EXAMPLE 6   Sketching a Hyperbola

Sketch the hyperbola $x^2 - \dfrac{y^2}{4} = 4$.

Algebraic Solution

$$x^2 - \frac{y^2}{4} = 4 \qquad \text{Write original equation.}$$

$$\frac{x^2}{4} - \frac{y^2}{16} = 1 \qquad \text{Divide each side by 4.}$$

$$\frac{x^2}{2^2} - \frac{y^2}{4^2} = 1 \qquad \text{Write in standard form.}$$

The x^2-term is positive, so the transverse axis is horizontal and the vertices occur at $(-2, 0)$ and $(2, 0)$. Moreover, the endpoints of the conjugate axis occur at $(0, -4)$ and $(0, 4)$. Use these four points to sketch the rectangle shown in Figure 4.31. Finally, draw the asymptotes

$$y = 2x$$

and

$$y = -2x$$

through the corners of this rectangle and complete the sketch, as shown in Figure 4.32.

Graphical Solution

Solve the equation of the hyperbola for y.

$$x^2 - \frac{y^2}{4} = 4$$

$$x^2 - 4 = \frac{y^2}{4}$$

$$4x^2 - 16 = y^2$$

$$\pm\sqrt{4x^2 - 16} = y$$

Then use a graphing utility to graph

$$y_1 = \sqrt{4x^2 - 16}$$

and

$$y_2 = -\sqrt{4x^2 - 16}$$

in the same viewing window, as shown in the figure below. Be sure to use a square setting.

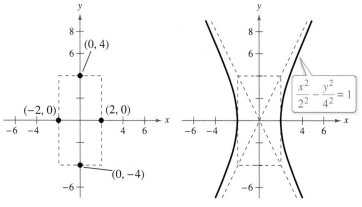

Figure 4.31   **Figure 4.32**

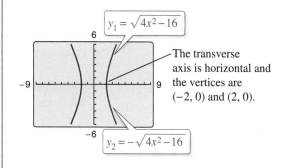

The transverse axis is horizontal and the vertices are $(-2, 0)$ and $(2, 0)$.

✓ *Checkpoint* ▶ Audio-video solution in English & Spanish at LarsonPrecalculus.com

Sketch the hyperbola $2x^2 - \dfrac{y^2}{2} = 2$.

GO DIGITAL

EXAMPLE 7 **Finding the Standard Equation of a Hyperbola**

Find the standard form of the equation of the hyperbola that has vertices at $(0, -3)$ and $(0, 3)$ and asymptotes $y = -2x$ and $y = 2x$, as shown in the figure.

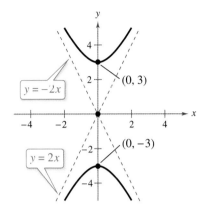

Solution

The transverse axis is vertical, so the asymptotes are of the forms

$$y = \frac{a}{b}x \quad \text{and} \quad y = -\frac{a}{b}x.$$

You are given that the asymptotes are $y = 2x$ and $y = -2x$, so

$$\frac{a}{b} = 2.$$

Because $a = 3$, you can conclude that the value of b is

$$\frac{a}{b} = 2 \quad \Longrightarrow \quad \frac{3}{b} = 2 \quad \Longrightarrow \quad \frac{3}{2} = b.$$

Finally, write the equation of the hyperbola in standard form.

$$\frac{y^2}{3^2} - \frac{x^2}{(3/2)^2} = 1 \qquad \text{Standard form}$$

✓ **Checkpoint** ▶ *Audio-video solution in English & Spanish at LarsonPrecalculus.com*

Find the standard form of the equation of the hyperbola that has vertices at $(-1, 0)$ and $(1, 0)$ and asymptotes $y = -4x$ and $y = 4x$. ∎

Summarize (Section 4.3)

1. List the four basic conic sections *(page 327)*.

2. State the definition of a parabola and the standard form of the equation of a parabola with its vertex at the origin *(page 328)*. For examples involving parabolas, see Examples 1 and 2.

3. State the definition of an ellipse and the standard form of the equation of an ellipse with its center at the origin *(page 330)*. For examples involving ellipses, see Examples 3 and 4.

4. State the definition of a hyperbola and the standard form of the equation of a hyperbola with its center at the origin *(page 333)*. For examples involving hyperbolas, see Examples 5–7.

GO DIGITAL

4.3   Exercises

GO DIGITAL

Vocabulary and Concept Check

In Exercises 1–5, fill in the blanks.

1. A _____ is the intersection of a plane and a double-napped cone.

2. A _____ is the set of all points (x, y) in a plane that are equidistant from a fixed line, called the _____, and a fixed point, called the _____, not on the line.

3. The line that passes through the focus and the vertex of a parabola is the _____ of the parabola.

4. An _____ is the set of all points (x, y) in a plane, the sum of whose distances from two distinct fixed points, called _____, is constant.

5. A _____ is the set of all points (x, y) in a plane for which the absolute value of the difference of the distances from two distinct fixed points, called _____, is constant.

6. What are the names of the two axes of an ellipse? Which one passes through the vertices?

7. How do you find the eccentricity of an ellipse, given the major and minor axes lengths?

8. What are the names of the two axes of a hyperbola? Which one passes through the vertices?

Skills and Applications

Matching   In Exercises 9–14, match the equation with its graph. [The graphs are labeled (a)–(f).]

(a)
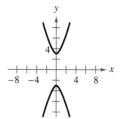

(b)
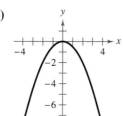

(c)
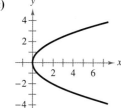

(d)
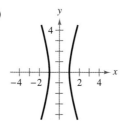

(e)
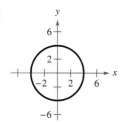

(f)
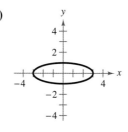

9. $x^2 = -2y$

10. $y^2 = 2x$

11. $\dfrac{x^2}{9} + y^2 = 1$

12. $x^2 - \dfrac{y^2}{9} = 1$

13. $y^2 - 9x^2 = 9$

14. $x^2 + y^2 = 16$

Finding the Focus and Directrix of a Parabola   In Exercises 15–20, find the focus and directrix of the parabola. Then sketch the parabola.

15. $y = \frac{1}{2}x^2$

16. $y = -4x^2$

17. $y^2 = -6x$

18. $y^2 = 3x$

19. $x^2 + 12y = 0$

20. $x + y^2 = 0$

Finding the Standard Equation of a Parabola   In Exercises 21–26, find the standard form of the equation of the parabola with the given characteristic(s) and vertex at the origin.

21. Focus: $(3, 0)$

22. Focus: $\left(0, \frac{1}{2}\right)$

23. Directrix: $y = 2$

24. Directrix: $x = -4$

25. Passes through the point $(-4, 6)$; horizontal axis

26. Passes through the point $\left(2, -\frac{1}{4}\right)$; vertical axis

Finding the Standard Equation of a Parabola   In Exercises 27 and 28, find the standard form of the equation of the parabola and determine the coordinates of the focus.

27.
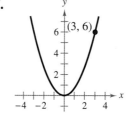

28.
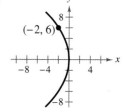

29. Flashlight The light bulb in a flashlight is at the focus of the parabolic reflector, 1.5 centimeters from the vertex of the reflector (see figure). Write an equation for a cross section of the flashlight's reflector with its focus on the positive *x*-axis and its vertex at the origin.

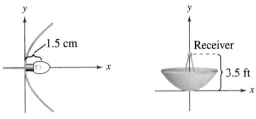

| Figure for 29 | Figure for 30 |

30. Satellite Dish The receiver of a parabolic satellite dish is at the focus of the parabola (see figure). Write an equation for a cross section of the satellite dish.

31. Suspension Bridge

Each cable of the Golden Gate Bridge is suspended (in the shape of a parabola) between two towers that are 1280 meters apart. The top of each tower is 152 meters above the roadway.

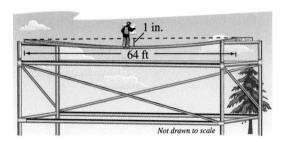

The cables touch the roadway at the midpoint between the towers.

(a) Sketch the bridge on a rectangular coordinate system with the cables touching the roadway at the origin. Label the coordinates of the known points.

(b) Write an equation that models the cables.

32. Beam Deflection A simply supported beam (see figure) is 64 feet long and has a load at the center. The deflection of the beam at its center is 1 inch. The shape of the deflected beam is parabolic.

Not drawn to scale

(a) Write an equation of the parabola. (Assume that the origin is at the center of the beam.)

(b) How far from the center of the beam is the deflection $\frac{1}{2}$ inch?

Finding the Standard Equation of an Ellipse In Exercises 33–42, find the standard form of the equation of the ellipse with the given characteristics and center at the origin.

33. 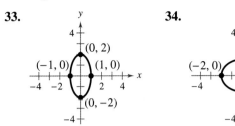 **34.**

35. Vertices: $(\pm 5, 0)$; foci: $(\pm 2, 0)$

36. Vertices: $(0, \pm 8)$; foci: $(0, \pm 4)$

37. Foci: $(\pm 5, 0)$; major axis of length 14

38. Foci: $(\pm 2, 0)$; major axis of length 10

39. Vertices: $(\pm 9, 0)$; minor axis of length 6

40. Vertices: $(0, \pm 10)$; minor axis of length 2

41. Passes through the points $(0, 14)$ and $(-7, 0)$

42. Passes through the points $(0, 4)$ and $(2, 0)$

Sketching an Ellipse In Exercises 43–52, find the vertices and eccentricity of the ellipse. Then sketch the ellipse.

43. $\dfrac{x^2}{25} + \dfrac{y^2}{16} = 1$ **44.** $\dfrac{x^2}{121} + \dfrac{y^2}{144} = 1$

45. $\dfrac{x^2}{25/9} + \dfrac{y^2}{16/9} = 1$ **46.** $\dfrac{x^2}{4} + \dfrac{y^2}{1/4} = 1$

47. $7x^2 + 36y^2 = 252$ **48.** $16x^2 + 7y^2 = 448$

49. $9x^2 + 5y^2 = 45$ **50.** $5x^2 + 4y^2 = 20$

51. $4x^2 + y^2 = 1$ **52.** $x^2 + 9y^2 = 1$

Using Eccentricity In Exercises 53 and 54, find the standard form of the equation of the ellipse with the given vertices, eccentricity *e*, and center at the origin.

53. Vertices: $(\pm 5, 0)$; $e = \frac{4}{5}$ **54.** Vertices: $(0, \pm 8)$; $e = \frac{1}{2}$

55. Architecture A mason is building a semielliptical fireplace arch that has a height of 2 feet at the center and a width of 6 feet along the base (see figure). The mason draws the semiellipse on the wall by the method shown in Figure 4.25 on page 330. Find the positions of the thumbtacks and the length of the string.

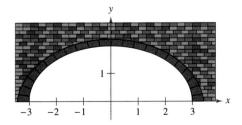

56. Architecture A semielliptical arch over a tunnel for a one-way road through a mountain has a major axis of 50 feet and a height at the center of 10 feet.

(a) Sketch the arch of the tunnel on a rectangular coordinate system with the center of the road entering the tunnel at the origin. Label the coordinates of the known points.

(b) Find an equation of the semielliptical arch over the tunnel.

(c) You are driving a moving truck that has a width of 8 feet and a height of 9 feet. Will the moving truck clear the opening of the arch?

57. Architecture Repeat Exercise 56 for a semielliptical arch with a major axis of 40 feet and a height at the center of 15 feet. The dimensions of the truck are 10 feet wide by 14 feet high.

58. Geometry A line segment through a focus of an ellipse with endpoints on the ellipse and perpendicular to the major axis is called a **latus rectum** of the ellipse. An ellipse has two latera recta. Knowing the length of the latera recta is helpful in sketching an ellipse because it yields other points on the curve (see figure). Show that the length of each latus rectum is $2b^2/a$.

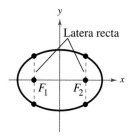

Using Latera Recta   In Exercises 59–62, sketch the ellipse using the latera recta (see Exercise 58).

59. $\dfrac{x^2}{4} + \dfrac{y^2}{1} = 1$ **60.** $\dfrac{x^2}{9} + \dfrac{y^2}{16} = 1$

61. $9x^2 + 4y^2 = 36$

62. $3x^2 + 6y^2 = 30$

Finding the Standard Equation of a Hyperbola   In Exercises 63–70, find the standard form of the equation of the hyperbola with the given characteristics and center at the origin.

63. Vertices: $(0, \pm 2)$; foci: $(0, \pm 6)$

64. Vertices: $(\pm 4, 0)$; foci: $(\pm 5, 0)$

65. Vertices: $(\pm 1, 0)$; asymptotes: $y = \pm 3x$

66. Vertices: $(0, \pm 3)$; asymptotes: $y = \pm 3x$

67. Foci: $(\pm 10, 0)$; asymptotes: $y = \pm \frac{3}{4}x$

68. Foci: $(0, \pm 8)$; asymptotes: $y = \pm 4x$

69. Vertices: $(0, \pm 3)$; passes through the point $(-2, 5)$

70. Vertices: $(\pm 2, 0)$; passes through the point $\left(3, \sqrt{3}\right)$

Sketching a Hyperbola   In Exercises 71–80, find the vertices of the hyperbola. Then sketch the hyperbola using the asymptotes as an aid.

71. $\dfrac{x^2}{25} - \dfrac{y^2}{25} = 1$ **72.** $\dfrac{x^2}{9} - \dfrac{y^2}{16} = 1$

73. $\dfrac{1}{36}y^2 - \dfrac{1}{100}x^2 = 1$ **74.** $\dfrac{1}{144}x^2 - \dfrac{1}{169}y^2 = 1$

75. $2y^2 - \dfrac{x^2}{2} = 2$ **76.** $\dfrac{y^2}{3} - 3x^2 = 3$

77. $25y^2 - 9x^2 = 225$ **78.** $4x^2 - 36y^2 = 144$

79. $9x^2 - y^2 = 1$ **80.** $4y^2 - x^2 = 1$

81. Art A cross section of a sculpture can be modeled by a hyperbola (see figure).

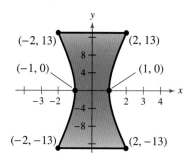

(a) Write an equation that models the curved sides of the sculpture.

(b) Each unit on the coordinate plane represents 1 foot. Find the width of the sculpture at a height of 18 feet.

82. Optics A hyperbolic mirror (used in some telescopes) has the property that a light ray directed at focus A is reflected to focus B. Find the vertex of the mirror when its mount at the top edge of the mirror has coordinates $(24, 24)$.

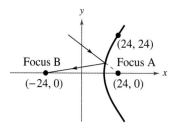

83. Aeronautics When an airplane travels faster than the speed of sound, the sound waves form a cone behind the airplane. If the airplane is flying parallel to the ground, then the sound waves intersect the ground in the shape of one branch of a hyperbola with the airplane directly above its center, and a sonic boom is heard along the hyperbola. You hear a sonic boom that is audible along a hyperbola with the equation $(x^2/100) - (y^2/4) = 1$, where x and y are measured in miles. What is the shortest horizontal distance you could be from the airplane?

84. Navigation Long-distance radio navigation for aircraft and ships uses synchronized pulses transmitted by widely separated transmitting stations. These pulses travel at the speed of light (186,000 miles per second). The difference in the times of arrival of these pulses at an aircraft or ship is constant on a hyperbola having the transmitting stations as foci.

Assume that two stations 300 miles apart are positioned on a rectangular coordinate system with coordinates $(-150, 0)$ and $(150, 0)$ and that a ship is traveling on a path with coordinates $(x, 75)$, as shown in the figure. Find the x-coordinate of the position of the ship when the time difference between the pulses from the transmitting stations is 1000 microseconds (0.001 second).

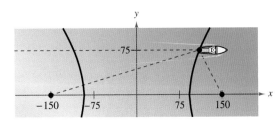

Exploring the Concepts

True or False? In Exercises 85–89, determine whether the statement is true or false. Justify your answer.

85. The equation $x^2 - y^2 = 144$ represents a circle.

86. It is possible for a parabola to intersect its directrix.

87. When the vertex and focus of a parabola are on a horizontal line, the directrix of the parabola is vertical.

88. It is easier to distinguish the graph of an ellipse from the graph of a circle when the eccentricity of the ellipse is large (close to 1).

89. If the asymptotes of the hyperbola

$$\frac{x^2}{a^2} - \frac{y^2}{b^2} = 1, \text{ where } a, b > 0$$

intersect at right angles, then $a = b$.

90. HOW DO YOU SEE IT? In parts (a)–(d), describe how a plane could intersect the double-napped cone to form each conic section (see figure).

(a) Circle

(b) Ellipse

(c) Parabola

(d) Hyperbola

91. Think About It Is the graph of $x^2 - 4y^4 = 4$ a hyperbola? Explain

92. Think About It How can you tell whether an ellipse is a circle from the equation?

93. Writing Explain how to use a graphing utility to check your graph in Exercise 43. What equation(s) would you enter into the graphing utility?

94. Degenerate Conic The graph of $x^2 - y^2 = 0$ is a degenerate conic. Sketch this graph and identify the degenerate conic.

95. Think About It Which part of the graph of the ellipse $4x^2 + 9y^2 = 36$ does each equation represent? Answer without graphing the equations.

(a) $x = -\frac{3}{2}\sqrt{4 - y^2}$ (b) $y = \frac{2}{3}\sqrt{9 - x^2}$

96. Drawing Ellipses Use two thumbtacks, a string, and a pencil to draw an ellipse, as shown in Figure 4.25 on page 330. Vary the length of the string and the distance between the thumbtacks. Explain how to obtain ellipses that are almost circular. Explain how to obtain ellipses that are long and narrow.

97. Using a Definition Use the definition of an ellipse to derive the standard form of the equation of an ellipse. [*Hint:* The sum of the distances from a point (x, y) to the foci is $2a$.]

98. Using a Definition Use the definition of a hyperbola to derive the standard form of the equation of a hyperbola. [*Hint:* The absolute value of the difference of the distances from a point (x, y) to the foci is $2a$.]

Review & Refresh ▶ *Video solutions at LarsonPrecalculus.com*

Completing the Square In Exercises 99–102, solve the quadratic equation by completing the square.

99. $x^2 + 6x - 16 = 0$ **100.** $x^2 - 2x - 4 = 0$

101. $2x^2 - 12x + 3 = 0$ **102.** $-4x^2 + 8x + 5 = 0$

Using Properties of Exponents In Exercises 103–106, simplify each expression.

103. $\dfrac{4}{2x^{-3}}$ **104.** $\dfrac{3c^2d^{-1}}{c^6d^{-3}}$

105. $\left(\dfrac{a^3}{2b^2}\right)^{-4}$ **106.** $\left(x^2yz\right)^{-2}\left(x^2yz^{-3}\right)$

Writing an Equation from a Graph In Exercises 107 and 108, use the graph of $f(x) = x^2$ to write an equation for the function represented by the graph.

107. 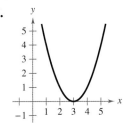   **108.**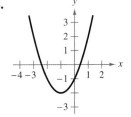

4.4 Translations of Conics

> Recognize equations of conics that are shifted vertically or horizontally in the plane.
> Write and graph equations of conics that are shifted vertically or horizontally in the plane.

Vertical and Horizontal Shifts of Conics

In Section 4.3, you studied conic sections whose graphs are in *standard position*, that is, whose centers or vertices are at the origin. In this section, you will study the equations of conic sections that are shifted vertically or horizontally in the plane.

In some real-life applications, it is convenient to use conics whose centers or vertices are not at the origin. For example, in Exercise 41 on page 348, you will use a parabola whose vertex is not at the origin to model the path of a satellite as it escapes Earth's gravity.

Standard Forms of Equations of Conics

Circle: Center $= (h, k)$; radius $= r$; $(x - h)^2 + (y - k)^2 = r^2$

Parabola: Vertex $= (h, k)$; directed distance from vertex to focus $= p$

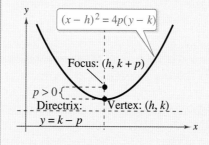

$(x - h)^2 = 4p(y - k)$

Focus: $(h, k + p)$

$p > 0$

Directrix: $y = k - p$

Vertex: (h, k)

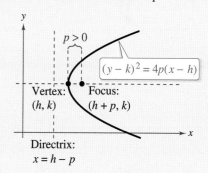

$p > 0$

$(y - k)^2 = 4p(x - h)$

Vertex: (h, k)

Focus: $(h + p, k)$

Directrix: $x = h - p$

Ellipse: Center $= (h, k)$; major axis length $= 2a$; minor axis length $= 2b$

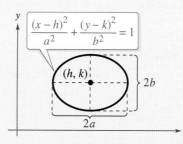

$\dfrac{(x - h)^2}{a^2} + \dfrac{(y - k)^2}{b^2} = 1$

(h, k)

$2b$

$2a$

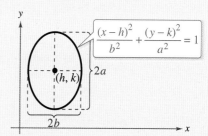

$\dfrac{(x - h)^2}{b^2} + \dfrac{(y - k)^2}{a^2} = 1$

(h, k)

$2a$

$2b$

Hyperbola: Center $= (h, k)$; transverse axis length $= 2a$; conjugate axis length $= 2b$

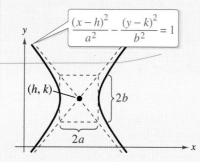

$\dfrac{(x - h)^2}{a^2} - \dfrac{(y - k)^2}{b^2} = 1$

(h, k)

$2b$

$2a$

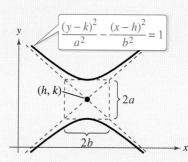

$\dfrac{(y - k)^2}{a^2} - \dfrac{(x - h)^2}{b^2} = 1$

(h, k)

$2a$

$2b$

GO DIGITAL

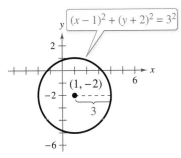

Figure 4.33 Circle

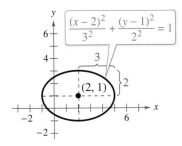

Figure 4.34 Ellipse

EXAMPLE 1 **Equations of Conic Sections**

Identify each conic. Then describe the translation of the conic from standard position.

a. $(x - 1)^2 + (y + 2)^2 = 3^2$ **b.** $\dfrac{(x - 2)^2}{3^2} + \dfrac{(y - 1)^2}{2^2} = 1$

c. $\dfrac{(x - 3)^2}{1^2} - \dfrac{(y - 2)^2}{3^2} = 1$ **d.** $(x - 2)^2 = 4(-1)(y - 3)$

Solution

a. The graph of $(x - 1)^2 + (y + 2)^2 = 3^2$ is a *circle* whose center is the point $(1, -2)$ and whose radius is 3, as shown in Figure 4.33. The circle is shifted one unit to the right and two units down from standard position.

b. The graph of

$$\frac{(x - 2)^2}{3^2} + \frac{(y - 1)^2}{2^2} = 1$$

is an *ellipse* whose center is the point $(2, 1)$. The major axis is horizontal and of length $2(3) = 6$, and the minor axis is vertical and of length $2(2) = 4$, as shown in Figure 4.34. The ellipse is shifted two units to the right and one unit up from standard position.

c. The graph of

$$\frac{(x - 3)^2}{1^2} - \frac{(y - 2)^2}{3^2} = 1$$

is a *hyperbola* whose center is the point $(3, 2)$ The transverse axis is horizontal and of length $2(1) = 2$, and the conjugate axis is vertical and of length $2(3) = 6$, as shown in Figure 4.35. The hyperbola is shifted three units to the right and two units up from standard position.

d. The graph of $(x - 2)^2 = 4(-1)(y - 3)$ is a *parabola* whose vertex is the point $(2, 3)$. The axis of the parabola is vertical. Moreover, $p = -1$, so the focus lies one unit below the vertex, and the directrix is $y = k - p = 3 - (-1) = 3 + 1 = 4$, as shown in Figure 4.36. The parabola is shifted two units to the right and three units up from standard position.

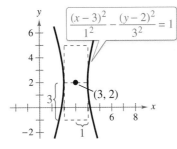

Figure 4.35 Hyperbola

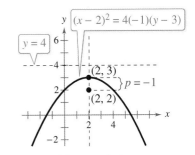

Figure 4.36 Parabola

✓ *Checkpoint* ▶ *Audio-video solution in English & Spanish at LarsonPrecalculus.com*

Identify each conic. Then describe the translation of the conic from standard position.

a. $\dfrac{(x + 1)^2}{3^2} + \dfrac{(y - 2)^2}{5^2} = 1$ **b.** $(x + 1)^2 + (y - 1)^2 = 2^2$

c. $(x + 4)^2 = 4(2)(y - 3)$ **d.** $\dfrac{(x - 3)^2}{2^2} - \dfrac{(y - 1)^2}{1^2} = 1$

GO DIGITAL

>>> **SKILLS REFRESHER**

For a refresher on completing the square in an equation, watch the video at *LarsonPrecalculus.com*.

Equations of Conics in Standard Form

EXAMPLE 2 **Finding Characteristics of a Parabola**

Find the vertex, focus, and directrix of the parabola $x^2 - 2x + 4y - 3 = 0$. Then sketch the parabola.

Solution The squared term in the equation involves x, so you know that the axis is vertical. Begin by isolating the x-terms on the left side of the equation and the rest of the terms on the right side. Then complete the square.

$$x^2 - 2x + 4y - 3 = 0 \qquad \text{Write original equation.}$$
$$x^2 - 2x = -4y + 3 \qquad \text{Isolate the } x\text{-terms on one side of the equation.}$$
$$x^2 - 2x + 1 = -4y + 3 + 1 \qquad \text{Complete the square.}$$
$$(x - 1)^2 = -4y + 4 \qquad \text{Write in completed square form.}$$

To write the equation $(x - 1)^2 = -4y + 4$ in the standard form $(x - h)^2 = 4p(y - k)$, factor out -4 from $-4y + 4$ to obtain

$$(x - 1)^2 = -4(y - 1). \qquad \text{Write in standard form.}$$

From this standard form, you can see that $h = 1$ and $k = 1$, which means the vertex is $(1, 1)$. Because $4p = -4$, you know that $p = -1$. The value of p is negative and the parabola has a vertical axis, so the parabola opens downward. Also, with $p = -1$, the focus is located one unit *below* the vertex and the directrix is one unit *above* the vertex.

Vertex: $(h, k + p)(1, 1 + (-1)) = (1, 0)$

Directrix: $y = k - p = 1 - (-1) = 2$

Next, plot the vertex and focus, draw a dashed line for the directrix, and draw a curve for the parabola, as shown in Figure 4.37.

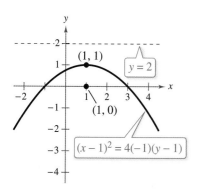

Figure 4.37

✓ *Checkpoint* ▶ Audio-video solution in English & Spanish at *LarsonPrecalculus.com*

Repeat Example 2 for the parabola $y^2 - 6y + 4x + 17 = 0$. ■

Recall that for a parabola, p is the directed distance from the vertex to the focus. In Example 2, the axis of the parabola is vertical and $p = -1$, so the focus is one unit below the vertex.

EXAMPLE 3 **Finding the Standard Equation of a Parabola**

Find the standard form of the equation of the parabola whose vertex is $(2, 1)$ and whose focus is $(4, 1)$, as shown in Figure 4.38.

Solution The vertex and the focus lie on a horizontal line and the focus lies two units to the right of the vertex, so it follows that the axis of the parabola is horizontal and $p = 2$. So the standard form of the parabola is

$$(y - k)^2 = 4p(x - h) \qquad \text{Standard form, horizontal axis}$$
$$(y - 1)^2 = 4(2)(x - 2) \qquad \text{Substitute.}$$
$$(y - 1)^2 = 8(x - 2). \qquad \text{Simplify.}$$

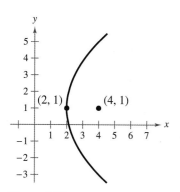

Figure 4.38

✓ *Checkpoint* ▶ Audio-video solution in English & Spanish at *LarsonPrecalculus.com*

Find the standard form of the equation of the parabola whose vertex is $(-1, 1)$ and whose directrix is $y = 0$. ■

GO DIGITAL

> **EXAMPLE 4** **Sketching an Ellipse**

Sketch the ellipse

$$x^2 + 4y^2 + 6x - 8y + 9 = 0.$$

Solution

To write the equation in standard form, begin by completing the square.

$x^2 + 4y^2 + 6x - 8y + 9 = 0$	Write original equation.
$\left(x^2 + 6x + \boxed{}\right) + \left(4y^2 - 8y + \boxed{}\right) = -9$	Group terms.
$\left(x^2 + 6x + \boxed{}\right) + 4\left(y^2 - 2y + \boxed{}\right) = -9$	Factor 4 out of y-terms.
$(x^2 + 6x + 9) + 4(y^2 - 2y + 1) = -9 + 9 + 4(1)$	Complete the squares.
$(x + 3)^2 + 4(y - 1)^2 = 4$	Write in completed square form.
$\dfrac{(x + 3)^2}{4} + \dfrac{(y - 1)^2}{1} = 1$	Divide each side by 4.
$\dfrac{(x + 3)^2}{2^2} + \dfrac{(y - 1)^2}{1^2} = 1$	$\dfrac{(x - h)^2}{a^2} + \dfrac{(y - k)^2}{b^2} = 1$

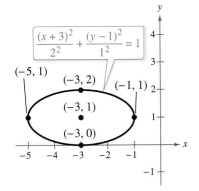

From this standard form, it follows that the center of the ellipse is $(h, k) = (-3, 1)$. The denominator of the x-term is greater than the denominator of the y-term, so the major axis is horizontal and thus $a^2 = 2^2$. Because $a = 2$, the endpoints of the major axis lie two units to the right and left of the center at $(-1, 1)$ and $(-5, 1)$. Similarly, from the denominator of the y-term you know that $b^2 = 1^2$. So, $b = 1$ and the endpoints of the minor axis lie one unit up and down from the center at $(-3, 2)$ and $(-3, 0)$. Use the center and the endpoints of the major and minor axes to sketch the ellipse, as shown in Figure 4.39.

Figure 4.39

✓ *Checkpoint* ▶ *Audio-video solution in English & Spanish at LarsonPrecalculus.com*

Sketch the ellipse $9x^2 + 4y^2 - 36x + 24y + 36 = 0.$

> **EXAMPLE 5** **Finding the Standard Equation of an Ellipse**

▶▶▶ *See LarsonPrecalculus.com for an interactive version of this type of example.*

Find the standard form of the equation of the ellipse whose vertices are $(2, -2)$ and $(2, 4)$, and whose minor axis length is 4, as shown in Figure 4.40.

Solution

The center of the ellipse lies at the midpoint of its vertices. So, the center is

$$(h, k) = \left(\frac{2 + 2}{2}, \frac{4 + (-2)}{2}\right) = (2, 1). \qquad \text{Center}$$

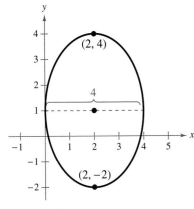

Figure 4.40

The vertices lie on a vertical line and are six units apart, so the major axis is vertical and has a length of $2a = 6$, which implies that $a = 3$. Moreover, the minor axis has a length of 4, so $2b = 4$, which implies that $b = 2$. The standard form of the ellipse is

$\dfrac{(x - h)^2}{b^2} + \dfrac{(y - k)^2}{a^2} = 1$	Standard form, vertical major axis
$\dfrac{(x - 2)^2}{2^2} + \dfrac{(y - 1)^2}{3^2} = 1$	Substitute.

✓ *Checkpoint* ▶ *Audio-video solution in English & Spanish at LarsonPrecalculus.com*

Find the standard form of the equation of the ellipse whose vertices are $(3, 0)$ and $(3, 10)$, and whose minor axis length is 6.

GO DIGITAL

EXAMPLE 6 Sketching a Hyperbola

Sketch the hyperbola

$$y^2 - 4x^2 + 4y + 24x - 41 = 0.$$

Solution

To write the equation in standard form, begin by completing the square.

$$y^2 - 4x^2 + 4y + 24x - 41 = 0 \qquad \text{Write original equation.}$$

$$\left(y^2 + 4y + \boxed{}\right) - \left(4x^2 - 24x + \boxed{}\right) = 41 \qquad \text{Group terms.}$$

$$\left(y^2 + 4y + \boxed{}\right) - 4\left(x^2 - 6x + \boxed{}\right) = 41 \qquad \text{Factor 4 out of } x\text{-terms.}$$

$$(y^2 + 4y + 4) - 4(x^2 - 6x + 9) = 41 + 4 - 4(9) \qquad \text{Complete the squares.}$$

$$(y + 2)^2 - 4(x - 3)^2 = 9 \qquad \text{Write in completed square form.}$$

$$\frac{(y + 2)^2}{9} - \frac{4(x - 3)^2}{9} = 1 \qquad \text{Divide each side by 9.}$$

$$\frac{(y + 2)^2}{9} - \frac{(x - 3)^2}{\dfrac{9}{4}} = 1 \qquad \text{Rewrite } \frac{4}{9} \text{ as } \frac{1}{\frac{9}{4}}.$$

$$\frac{(y + 2)^2}{3^2} - \frac{(x - 3)^2}{\left(\dfrac{3}{2}\right)^2} = 1 \qquad \frac{(y - k)^2}{a^2} - \frac{(x - h)^2}{b^2} = 1$$

From this standard form, it follows that the center of the hyperbola is $(h, k) = (3, -2)$. The x-term is subtracted from the y-term, so the transverse axis is vertical and the hyperbola opens upward and downward. The denominator of the y-term is $a^2 = 3^2$, so $a = 3$. This means the vertices occur three units above and below the center at $(3, 1)$ and $(3, -5)$. The denominator of the x-term is

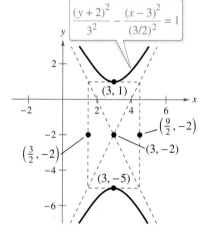

$$\frac{(y + 2)^2}{3^2} - \frac{(x - 3)^2}{(3/2)^2} = 1$$

$$b^2 = \left(\frac{3}{2}\right)^2 \quad \Longrightarrow \quad b = \frac{3}{2}.$$

This means the endpoints of the conjugate axis occur $3/2$ units to the right and left of the center at

$$\left(\frac{9}{2}, -2\right) \quad \text{and} \quad \left(\frac{3}{2}, -2\right).$$

Draw a rectangle through these two points and the vertices. Finally, sketch the asymptotes by drawing lines through the opposite corners of the rectangle. Using these asymptotes, complete the sketch of the hyperbola, as shown in Figure 4.41.

Figure 4.41

✓ **Checkpoint** ▶ *Audio-video solution in English & Spanish at LarsonPrecalculus.com*

Sketch the hyperbola

$$9x^2 - y^2 - 18x - 6y - 9 = 0.$$ ∎

Recall from Section 4.3 that the foci of a hyperbola are located c units from its center, where $c^2 = a^2 + b^2$. So, to find the foci in Example 6, note that $a^2 = 3^2$ and $b^2 = (3/2)^2$. The value of c is

$$c^2 = 3^2 + \left(\frac{3}{2}\right)^2 \quad \Longrightarrow \quad c^2 = 9 + \frac{9}{4} \quad \Longrightarrow \quad c^2 = \frac{45}{4} \quad \Longrightarrow \quad c = \frac{3\sqrt{5}}{2}.$$

The transverse axis is vertical, so the foci lie c units above and below the center.

$$\left(3, -2 + \frac{3}{2}\sqrt{5}\right) \quad \text{and} \quad \left(3, -2 - \frac{3}{2}\sqrt{5}\right) \qquad \text{Foci}$$

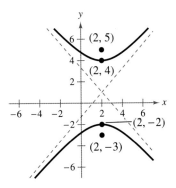

Figure 4.42

EXAMPLE 7   Finding the Standard Equation of a Hyperbola

Find the standard form of the equation of the hyperbola whose vertices are $(2, -2)$ and $(2, 4)$, and whose foci are $(2, -3)$ and $(2, 5)$, as shown in Figure 4.42.

Solution   The center of the hyperbola lies at the midpoint of its vertices. So, the center is $(h, k) = (2, 1)$. The vertices lie on a vertical line and are six units apart, so the transverse axis is vertical and has a length of $2a = 6$, which implies that $a = 3$. Moreover, the foci are four units from the center, so $c = 4$, and

$$
\begin{aligned}
b^2 &= c^2 - a^2 \\
&= 4^2 - 3^2 \\
&= 16 - 9 \\
&= 7 \\
&= \left(\sqrt{7}\right)^2.
\end{aligned}
$$

The transverse axis is vertical, so the standard form of the equation is

$$\frac{(y - k)^2}{a^2} - \frac{(x - h)^2}{b^2} = 1 \qquad \text{Standard form, vertical transverse axis}$$

$$\frac{(y - 1)^2}{3^2} - \frac{(x - 2)^2}{\left(\sqrt{7}\right)^2} = 1. \qquad \text{Substitute.}$$

✓ **Checkpoint** ▶ *Audio-video solution in English & Spanish at LarsonPrecalculus.com*

Find the standard form of the equation of the hyperbola whose vertices are $(3, -1)$ and $(5, -1)$, and whose foci are $(1, -1)$ and $(7, -1)$. ■

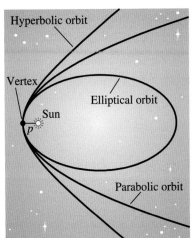

Figure 4.43

An interesting application of conic sections involves the orbits of comets in our solar system. Comets can have elliptical, parabolic, or hyperbolic orbits. The center of the sun is a focus of each of these orbits, and each orbit has a vertex at the point where the comet is closest to the sun, as shown in Figure 4.43. Undoubtedly, many comets with parabolic or hyperbolic orbits have not been identified. You get to see such comets only *once*. Comets with elliptical orbits, such as Halley's comet, are the only ones that remain in our solar system.

If p is the distance between the vertex and the focus (in meters), and v is the velocity of the comet at the vertex (in meters per second), then the type of orbit is determined as follows.

1. Elliptical: $v < \sqrt{2GM/p}$

2. Parabolic: $v = \sqrt{2GM/p}$

3. Hyperbolic: $v > \sqrt{2GM/p}$

In each of the above, $M \approx 1.989 \times 10^{30}$ kilograms (the mass of the sun) and $G \approx 6.67 \times 10^{-11}$ cubic meter per kilogram-second squared (the universal gravitational constant).

Summarize   (Section 4.4)

1. List the standard forms of the equations of conics *(page 341)*. For an example of identifying conics and describing translations of conics, see Example 1.

2. Explain how to write and graph equations of conics that are shifted vertically or horizontally in the plane *(pages 343–346)*. For examples of writing and graphing equations of conics that are shifted vertically or horizontally in the plane, see Examples 2–7.

GO DIGITAL

4.4 Exercises

See CalcChat.com for tutorial help and worked-out solutions to odd-numbered exercises.

Vocabulary and Concept Check

Match the description of the conic with its standard equation. The equations are labeled (a)–(f).

(a) $\dfrac{(x-h)^2}{a^2} + \dfrac{(y-k)^2}{b^2} = 1$ (b) $\dfrac{(x-h)^2}{a^2} - \dfrac{(y-k)^2}{b^2} = 1$ (c) $\dfrac{(y-k)^2}{a^2} - \dfrac{(x-h)^2}{b^2} = 1$

(d) $\dfrac{(x-h)^2}{b^2} + \dfrac{(y-k)^2}{a^2} = 1$ (e) $(x-h)^2 = 4p(y-k)$ (f) $(y-k)^2 = 4p(x-h)$

1. Hyperbola with horizontal transverse axis **2.** Ellipse with vertical major axis **3.** Parabola with vertical axis

4. Hyperbola with vertical transverse axis **5.** Ellipse with horizontal major axis **6.** Parabola with horizontal axis

Skills and Applications

Equations of Conic Sections In Exercises 7–14, identify the conic. Then describe the translation of the conic from standard position.

7. $(x+2)^2 + (y-1)^2 = 4$ **8.** $(y-1)^2 = 4(2)(x+2)$

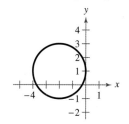   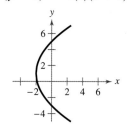

9. $\dfrac{(y+3)^2}{4} - (x-1)^2 = 1$ **10.** $\dfrac{(x-2)^2}{9} + \dfrac{(y+1)^2}{4} = 1$

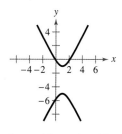   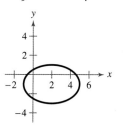

11. $(x+1)^2 = 4(-1)(y-2)$
12. $(x+1)^2 + (y-3)^2 = 6$
13. $\dfrac{(x+4)^2}{9} + \dfrac{(y+2)^2}{16} = 1$
14. $\dfrac{(x+2)^2}{4} - \dfrac{(y-3)^2}{9} = 1$

Finding Characteristics of a Circle In Exercises 15–20, find the center and radius of the circle.

15. $x^2 + y^2 = 49$ **16.** $x^2 + y^2 = 1$
17. $(x-1)^2 + y^2 = 10$ **18.** $x^2 + (y+12)^2 = 24$
19. $(x-4)^2 + (y-5)^2 = 36$
20. $(x+8)^2 + (y+1)^2 = 144$

Writing the Equation of a Circle in Standard Form In Exercises 21–26, write the equation of the circle in standard form, and then find its center and radius.

21. $x^2 + y^2 - 8y = 0$
22. $x^2 + y^2 - 10x + 16 = 0$
23. $x^2 + y^2 - 2x + 6y + 9 = 0$
24. $2x^2 + 2y^2 - 2x - 2y - 7 = 0$
25. $4x^2 + 4y^2 + 12x - 24y + 41 = 0$
26. $9x^2 + 9y^2 + 54x - 36y + 17 = 0$

Finding Characteristics of a Parabola In Exercises 27–34, find the vertex, focus, and directrix of the parabola. Then sketch the parabola.

27. $\left(y + \frac{1}{2}\right)^2 = 2(x-5)$ **28.** $\left(x + \frac{1}{2}\right)^2 = 4(y-3)$
29. $(x-1)^2 + 8(y+2) = 0$
30. $(x+2) + (y-4)^2 = 0$
31. $y = \dfrac{x^2 - 2x + 5}{4}$
32. $\dfrac{y^2 + 14y + 19}{6} = x$
33. $8x + y^2 + 1 = 2y - 4$
34. $-12 - x^2 = 12y + 4x$

Finding the Standard Equation of a Parabola In Exercises 35–40, find the standard form of the equation of the parabola with the given characteristics.

35. Vertex: $(3, 2)$; focus: $(1, 2)$
36. Vertex: $(-1, 2)$; focus: $(-1, 0)$
37. Vertex: $(0, 4)$; directrix: $y = 2$
38. Vertex: $(-2, 1)$; directrix: $x = 1$
39. Focus: $(4, 4)$; directrix: $x = -4$
40. Focus: $(0, 0)$; directrix: $y = 4$

41. Satellite Orbit

A satellite in a 100-mile-high circular orbit around Earth has a velocity of approximately 17,500 miles per hour (see figure). When this velocity is multiplied by $\sqrt{2}$, the satellite has the minimum velocity necessary to escape Earth's gravity and follow a parabolic path with the center of Earth as the focus.

(a) Find the escape velocity.

(b) Find the standard form of the equation that represents the parabolic path. (Assume that the radius of Earth is 4000 miles.)

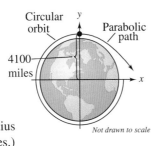

Circular orbit Parabolic path
4100 miles
Not drawn to scale

42. Fluid Flow

Water flowing from a horizontal pipe 48 feet above the ground has the shape of a parabola whose vertex $(0, 48)$ is at the end of the pipe (see figure). The water strikes the ocean at the point $\left(10\sqrt{3}, 0\right)$.

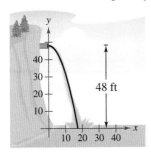

48 ft

(a) Find the standard form of the equation that represents the path of the water.

(b) The pipe is moved up 12 feet. Find an equation that represents the path of the water. Find the coordinates of the point where the water strikes the ocean.

Projectile Motion In Exercises 43 and 44, consider the path of an object projected horizontally with a velocity of v feet per second at a height of s feet, where the model for the path is $x^2 = (-v^2/16)(y - s)$. In this model (in which air resistance is disregarded), y is the height (in feet) of the projectile and x is the horizontal distance (in feet) the projectile travels.

43. A ball is thrown from the top of a 100-foot tower with a velocity of 28 feet per second.

(a) Find the equation that represents the parabolic path.

(b) How far does the ball travel horizontally before it strikes the ground?

44. A cargo plane is flying at an altitude of 500 feet and a speed of 135 miles per hour. A supply crate is dropped from the plane. How many feet will the crate travel horizontally before it hits the ground?

Sketching an Ellipse In Exercises 45–52, find the center, foci, and vertices of the ellipse. Then sketch the ellipse.

45. $\dfrac{(x-1)^2}{9} + \dfrac{(y-5)^2}{25} = 1$

46. $\dfrac{(x-6)^2}{4} + \dfrac{(y+7)^2}{16} = 1$

47. $(x+2)^2 + \dfrac{(y+4)^2}{1/4} = 1$

48. $\dfrac{(x-3)^2}{25/9} + (y-8)^2 = 1$

49. $9x^2 + 25y^2 - 36x - 50y + 52 = 0$

50. $16x^2 + 25y^2 - 32x + 50y + 16 = 0$

51. $25x^2 + 4y^2 + 50x - 75 = 0$

52. $9x^2 + 4y^2 - 36x + 8y + 31 = 0$

Finding the Standard Equation of an Ellipse In Exercises 53–64, find the standard form of the equation of the ellipse with the given characteristics.

53. Vertices: $(3, -3)$, $(3, 3)$; minor axis of length 2

54. Vertices: $(-2, 3)$, $(6, 3)$; minor axis of length 6

55. Foci: $(0, 0)$, $(4, 0)$; major axis of length 8

56. Foci: $(0, 0)$, $(0, 8)$; major axis of length 16

57. Center: $(1, 4)$; $a = 2c$; vertices: $(1, 0)$, $(1, 8)$

58. Center: $(3, 2)$; $a = 3c$; foci: $(1, 2)$, $(5, 2)$

59. Vertices: $(-3, 0)$, $(7, 0)$; foci: $(0, 0)$, $(4, 0)$

60. Vertices: $(2, 0)$, $(2, 4)$; foci: $(2, 1)$, $(2, 3)$

61. Foci: $(-3, -3)$, $(-3, -1)$; eccentricity: $\frac{1}{3}$

62. Foci: $(7, -6)$, $(7, -4)$; eccentricity: $\frac{1}{2}$

63. Vertices: $(2, -1)$, $(2, 3)$; eccentricity: $\sqrt{3}/2$

64. Vertices: $\left(-\frac{7}{2}, 0\right)$, $\left(\frac{3}{2}, 0\right)$; eccentricity: $\frac{3}{5}$

65. **Planetary Motion** The dwarf planet Pluto moves in an elliptical orbit with the sun at one of the foci, as shown in the figure. The length of half of the major axis, a, is 3.67×10^9 miles, and the eccentricity is 0.249. Find the least distance (*perihelion*) and the greatest distance (*aphelion*) of Pluto from the center of the sun.

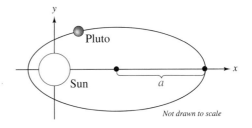

Pluto
Sun a
Not drawn to scale

66. Australian Football In Australia, football by Australian Rules is played on elliptical fields. The field can be a maximum of 155 meters wide and a maximum of 185 meters long. Let the center of a field of maximum size be represented by the point $(0, 77.5)$. Find the standard form of the equation that represents this field. (*Source: United States Australian Football League*)

Sketching a Hyperbola In Exercises 67–74, find the center, foci, and vertices of the hyperbola. Then sketch the hyperbola using the asymptotes as an aid.

67. $\dfrac{(x-2)^2}{16} - \dfrac{(y+1)^2}{9} = 1$ **68.** $\dfrac{(x-1)^2}{144} - \dfrac{(y-4)^2}{25} = 1$

69. $(y+6)^2 - (x-2)^2 = 1$

70. $\dfrac{(y-1)^2}{1/4} - \dfrac{(x+3)^2}{1/9} = 1$

71. $x^2 - 9y^2 + 2x - 54y - 85 = 0$

72. $16y^2 - x^2 + 2x + 64y + 62 = 0$

73. $9y^2 - x^2 - 36y - 6x + 18 = 0$

74. $x^2 - 9y^2 + 36y - 72 = 0$

Finding the Standard Equation of a Hyperbola In Exercises 75–82, find the standard form of the equation of the hyperbola with the given characteristics.

75. Vertices: $(0, 2)$, $(0, 0)$; foci: $(0, 3)$, $(0, -1)$

76. Vertices: $(1, 2)$, $(5, 2)$; foci: $(0, 2)$, $(6, 2)$

77. Foci: $(-8, 5)$, $(0, 5)$; transverse axis of length 4

78. Foci: $(7, -5)$, $(7, 3)$; conjugate axis of length 2

79. Vertices: $(2, 3)$, $(2, -3)$; passes through the point $(0, 5)$

80. Vertices: $(-2, 1)$, $(2, 1)$; passes through the point $(4, 3)$

81. Vertices: $(0, 2)$, $(6, 2)$; asymptotes: $y = \frac{2}{3}x$, $y = 4 - \frac{2}{3}x$

82. Vertices: $(3, 0)$, $(3, 4)$; asymptotes: $y = \frac{2}{3}x$, $y = 4 - \frac{2}{3}x$

Identifying a Conic In Exercises 83–88, identify the conic by writing its equation in standard form. Then sketch its graph and describe the translation from standard position.

83. $y^2 - x^2 + 4y = 0$

84. $4x^2 - y^2 - 4x - 3 = 0$

85. $16y^2 + 128x + 8y - 7 = 0$

86. $x^2 + y^2 - 6x + 4y + 9 = 0$

87. $9x^2 + 16y^2 + 36x + 128y + 148 = 0$

88. $25x^2 - 10x - 200y - 119 = 0$

Exploring the Concepts

True or False? In Exercises 89 and 90, determine whether the statement is true or false. Justify your answer.

89. The conic represented by the equation $3x^2 + 2y^2 - 18x - 16y + 58 = 0$ is a circle.

90. A hyperbola can have vertices $(-9, 4)$ and $(-3, 4)$ and foci $(-6, 9)$ and $(-6, -1)$.

91. Think About It Find an equation of an ellipse such that for any point on the ellipse, the sum of the distances from the point to the points $(2, 2)$ and $(10, 2)$ is 36.

92. 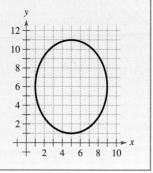 **HOW DO YOU SEE IT?** Consider the ellipse shown.

(a) Find the center, vertices, and foci of the ellipse.

(b) Find the standard form of the equation of the ellipse.

93. Error Analysis Describe the error.

The graph of

$$\dfrac{(x+4)^2}{2^2} - \dfrac{(y-2)^2}{3^2} = 1 \quad \textbf{X}$$

is a hyperbola with center $(4, -2)$, transverse axis of length $2(2) = 4$, and conjugate axis of length $2(3) = 6$.

Review & Refresh ▶ Video solutions at LarsonPrecalculus.com

94. Compound Interest You deposit $1000 in a long-term investment fund in which the interest is compounded quarterly. After 5 years, the balance is $1161.18. What is the annual interest rate?

Finding the Domain of a Function In Exercises 95–100, find the domain of the function.

95. $f(x) = 5x$ **96.** $f(x) = 5 - x$

97. $f(x) = \sqrt{x+1}$ **98.** $f(x) = -\sqrt{x-1}$

99. $f(x) = x^{3/2}$ **100.** $f(x) = x^{2/3}$

Describing Function Behavior In Exercises 101–110, use a graphing utility to graph the function and visually determine the open intervals on which the function is increasing, decreasing, or constant. Use a table of values to verify your results.

101. $f(x) = -x$ **102.** $f(x) = 2x + 1$

103. $f(x) = 3x^2 - 5$ **104.** $f(x) = 1 - x^2$

105. $f(x) = -x^3 + 1$ **106.** $f(x) = x^3 + x$

107. $f(x) = \sqrt{x}$ **108.** $f(x) = \sqrt{-2x}$

109. $f(x) = x^{3/4}$

110. $f(x) = x^{5/3}$

Summary and Study Strategies

GO DIGITAL

What Did You Learn?

The list below reviews the skills covered in the chapter and correlates each one to the Review Exercises (see page 352) that practice the skill.

Section 4.1	**Review Exercises**

■ Find domains of rational functions *(p. 310)*.   *1–4*

■ Find vertical and horizontal asymptotes of graphs of rational functions *(p. 311)*.   *5–10*

Let f be the rational function

$$f(x) = \frac{N(x)}{D(x)} = \frac{a_n x^n + a_{n-1} x^{n-1} + \cdots + a_1 x + a_0}{b_m x^m + b_{m-1} x^{m-1} + \cdots + b_1 x + b_0}$$

where $N(x)$ and $D(x)$ have no common factors.

1. The graph of f has *vertical* asymptotes at the zeros of $D(x)$.

2. The graph of f has at most one *horizontal* asymptote determined by comparing the degrees of $N(x)$ and $D(x)$.

   a. When $n < m$, the graph of f has the line $y = 0$ as a horizontal asymptote.

   b. When $n = m$, the graph of f has the line $y = a_n/b_m$ as a horizontal asymptote, where a_n is the leading coefficient of the numerator and b_m is the leading coefficient of the denominator.

   c. When $n > m$, the graph of f has no horizontal asymptote.

■ Use rational functions to model and solve real-life problems *(p. 314)*.   *11, 12*

Section 4.2

■ Sketch graphs of rational functions *(p. 318)*.   *13–24*

Let $f(x) = N(x)/D(x)$, where $N(x)$ and $D(x)$ are polynomials and $D(x)$ is not the zero polynomial.

1. Simplify f, if possible. List any restrictions on the domain of f that are not implied by the simplified function.

2. Find and plot the y-intercept (if any) by evaluating $f(0)$.

3. Find the zeros of the numerator (if any). Then plot the corresponding x-intercepts.

4. Find the zeros of the denominator (if any). Then sketch the corresponding vertical asymptotes and plot the corresponding holes.

5. Find and sketch the horizontal asymptote (if any).

6. Plot at least one point *between* and one point *beyond* each x-intercept and vertical asymptote.

7. Use smooth curves to complete the graph between and beyond the vertical asymptotes.

■ Sketch graphs of rational functions that have slant asymptotes. *(p. 321)*.   *25–30*

Consider a rational function whose denominator is of degree 1 or greater. If the degree of the numerator is exactly *one more* than the degree of the denominator, then the graph of the function has a slant asymptote.

■ Use graphs of rational functions to model and solve real-life problems *(p. 322)*.   *31–34*

Section 4.3	**Review Exercises**
■ Describe a conic section *(p. 327)*.	*35–42*
■ Recognize, graph, and write equations of parabolas (vertex at origin) *(p. 328)*. Vertical axis: $x^2 = 4py$ Horizontal axis: $y^2 = 4px$	*43–50*
■ Recognize, graph, and write equations of ellipses (center at origin) *(p. 330)*. Horizontal major axis: $\dfrac{x^2}{a^2} + \dfrac{y^2}{b^2} = 1$ Vertical major axis: $\dfrac{x^2}{b^2} + \dfrac{y^2}{a^2} = 1$	*51–58*
■ Recognize, graph, and write equations of hyperbolas (center at origin) *(p. 333)*. Horizontal transverse axis: $\dfrac{x^2}{a^2} - \dfrac{y^2}{b^2} = 1$ Vertical transverse axis: $\dfrac{y^2}{a^2} - \dfrac{x^2}{b^2} = 1$	*59–62*

Section 4.4	
■ Recognize equations of conics that are shifted vertically or horizontally in the plane *(p. 341)*. Parabola: $(x - h)^2 = 4p(y - k)$ or $(y - k)^2 = 4p(x - h)$ Ellipse: $\dfrac{(x - h)^2}{a^2} + \dfrac{(y - k)^2}{b^2} = 1$ or $\dfrac{(x - h)^2}{b^2} + \dfrac{(y - k)^2}{a^2} = 1$ Hyperbola: $\dfrac{(x - h)^2}{a^2} - \dfrac{(y - k)^2}{b^2} = 1$ or $\dfrac{(y - k)^2}{a^2} - \dfrac{(x - h)^2}{b^2} = 1$	*63–70*
■ Write and graph equations of conics that are shifted vertically or horizontally in the plane *(p. 343)*.	*63–86*

Study Strategies

Managing Test Anxiety   Test anxiety is different from the typical nervousness that usually occurs during tests. It interferes with the thinking process. After leaving the classroom, have you suddenly been able to recall what you could not remember during the test? It is likely that this was a result of test anxiety. Test anxiety is a learned reaction or response—no one is born with it. The good news is that most students can learn to manage test anxiety. It is important to get as much information as you can into your long-term memory and to practice retrieving the information before you take a test. The more you practice retrieving information, the easier it will be during the test. One way to get information into your long-term memory is to make note cards.

1. Write down important information on note cards, such as formulas, examples of problems you find difficult, and concepts that always trip you up.

2. Memorize the information on the note cards. Flash through the cards, placing the ones containing information you know in one stack and the ones containing information you do not know in another stack. Keep working on the information you do not know.

3. As soon as you receive your test, turn it over and write down all the information you remember, starting with things you have the greatest difficulty remembering. Having this information available should boost your confidence and free up mental energy for focusing on the test.

Do not wait until the night before the test to make note cards. Make them after you study each section. Then review them two or three times a week.

Review Exercises

See CalcChat.com for tutorial help and worked-out solutions to odd-numbered exercises.

4.1 Finding the Domain of a Rational Function
In Exercises 1–4, find the domain of the function and discuss the behavior of f near any excluded x-values.

1. $f(x) = \dfrac{3x}{x + 10}$ **2.** $f(x) = \dfrac{4x^3}{2 + 5x}$

3. $f(x) = \dfrac{8}{x^2 - 10x + 24}$ **4.** $f(x) = \dfrac{x^2 + x - 2}{x^2 - 4x + 4}$

Finding Vertical and Horizontal Asymptotes In Exercises 5–10, find all vertical and horizontal asymptotes of the graph of the rational function.

5. $f(x) = \dfrac{6x^2}{x + 3}$ **6.** $f(x) = \dfrac{2x^2 + 5x - 3}{x^2 + 2}$

7. $g(x) = \dfrac{x^2}{x^2 - 4}$ **8.** $g(x) = \dfrac{x + 1}{x^2 - 1}$

9. $h(x) = \dfrac{5x + 20}{x^2 - 2x - 24}$ **10.** $h(x) = \dfrac{x^3 - 4x^2}{x^2 + 3x + 2}$

11. Average Cost The cost C (in dollars) of producing x units of a product is given by $C = 0.5x + 500$ and the average cost per unit $\overline{C}$ is given by

$$\overline{C} = \frac{C}{x} = \frac{0.5x + 500}{x}, \quad x > 0.$$

Determine the average cost per unit as x increases without bound.

12. Seizure of Illegal Drugs The cost C (in millions of dollars) for the federal government to seize $p\%$ of an illegal drug as it enters the country is given by

$$C = \frac{528p}{100 - p}, \quad 0 \le p < 100.$$

(a) Use a graphing utility to graph the cost function.

(b) Find the costs of seizing 25%, 50%, and 75% of the drug.

(c) According to the model, is it possible to seize 100% of the drug? Explain.

4.2 Sketching the Graph of a Rational Function
In Exercises 13–24, (a) state the domain of the function, (b) identify all intercepts, (c) find any vertical or horizontal asymptotes or holes, and (d) plot additional solution points as needed to sketch the graph of the rational function.

13. $f(x) = \dfrac{-3}{2x^2}$ **14.** $f(x) = \dfrac{4}{x}$

15. $g(x) = \dfrac{2 + x}{1 - x}$ **16.** $h(x) = \dfrac{x - 4}{x - 7}$

17. $p(x) = \dfrac{5x^2}{4x^2 + 1}$ **18.** $f(x) = \dfrac{-8x}{x^2 + 4}$

19. $f(x) = \dfrac{x}{x^2 - 16}$ **20.** $h(x) = \dfrac{9}{(x - 3)^2}$

21. $f(x) = \dfrac{-6x^2}{x^2 + 1}$ **22.** $y = \dfrac{2x^2}{x^2 - 4}$

23. $f(x) = \dfrac{3x^2 - 10x + 3}{x^2 - 4x + 3}$ **24.** $f(x) = \dfrac{6x^2 - x - 2}{2x^2 - 5x - 3}$

A Rational Function with a Slant Asymptote In Exercises 25–30, (a) state the domain of the function, (b) identify all intercepts, (c) find any vertical or slant asymptotes or holes, and (d) plot additional solution points as needed to sketch the graph of the rational function.

25. $f(x) = \dfrac{2x^3}{x^2 + 1}$ **26.** $f(x) = \dfrac{2x^2 + 1}{x + 1}$

27. $f(x) = \dfrac{x^2 + 3x - 10}{x + 2}$ **28.** $f(x) = \dfrac{x^3}{2x^2 - 18}$

29. $f(x) = \dfrac{3x^3 - 2x^2 - 3x + 2}{3x^2 - x - 4}$

30. $f(x) = \dfrac{3x^3 - 4x^2 - 12x + 16}{3x^2 + 5x - 2}$

31. Average Cost The cost C (in dollars) of producing x units of a product is given by $C = 100{,}000 + 0.9x$ and the average cost per unit $\overline{C}$ is given by

$$\overline{C} = \frac{C}{x} = \frac{100{,}000 + 0.9x}{x}, \quad x > 0.$$

(a) Sketch the graph of the average cost function.

(b) Find the average costs when $x = 1000$, $x = 10{,}000$, and $x = 100{,}000$.

(c) By increasing the level of production, what is the minimum average cost per unit you can obtain? Explain.

32. Page Design A page that is x inches wide and y inches high contains 30 square inches of print. The top and bottom margins are each 2 inches deep and the margins on each side are 2 inches wide.

(a) Show that the total area A of the page is

$$A = \frac{2x(2x + 7)}{x - 4}.$$

(b) Determine the domain of the function based on the physical constraints of the problem.

(c) Use a graphing utility to graph the area function and approximate the dimensions of the page that use the least amount of paper. Verify your answer using the *table* feature of the graphing utility.

33. Biology A parks and wildlife commission releases 80,000 fish into a lake. After t years, the population N of the fish (in thousands) is given by

$$N = \frac{20(3t + 4)}{0.05t + 1}, \quad t \geq 0.$$

(a) Sketch the graph of the function.

(b) Find the population when $t = 5$, $t = 10$, and $t = 25$.

(c) What is the maximum number of fish in the lake as time increases? Explain.

34. Medicine The concentration C of a chemical in the bloodstream t hours after injection into muscle tissue is given by $C(t) = (2t + 1)/(t^2 + 4)$, $t > 0$.

(a) Determine the horizontal asymptote of the graph of the function and interpret its meaning in the context of the problem.

(b) Use a graphing utility to graph the function and approximate the time when the bloodstream concentration is greatest.

4.3 Identifying a Conic In Exercises 35–42, identify the conic.

35. $y^2 = -10x$

36. $x^2 + \dfrac{y^2}{16} = 1$

37. $\dfrac{y^2}{12} - \dfrac{x^2}{9} = 1$

38. $x^2 + y^2 = 20$

39. $4x^2 + 18y^2 = 36$

40. $8x^2 - y^2 = 48$

41. $3x^2 + 3y^2 = 75$

42. $5y = x^2$

Finding the Standard Equation of a Parabola In Exercises 43–48, find the standard form of the equation of the parabola with the given characteristic(s) and vertex at the origin.

43. Focus: $(-6, 0)$

44. Focus: $(0, 7)$

45. Directrix: $y = -3$

46. Directrix: $x = 3$

47. Passes through the point $(3, 6)$; horizontal axis

48. Passes through the point $(4, -2)$; vertical axis

49. Satellite Dish A cross section of a large parabolic satellite dish is modeled by $y = x^2/200$, $-100 \leq x \leq 100$ (see figure). The receiving and transmitting equipment is positioned at the focus. Find the coordinates of the focus.

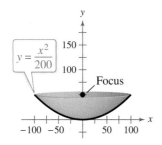

50. Suspension Bridge Each cable of a suspension bridge is suspended (in the shape of a parabola) between two towers (see figure).

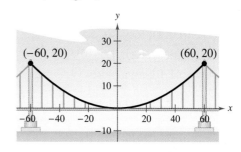

(a) Find the coordinates of the focus.

(b) Write an equation that models the cables.

Finding the Standard Equation of an Ellipse In Exercises 51–56, find the standard form of the equation of the ellipse with the given characteristics and center at the origin.

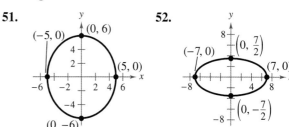

51.

52.

53. Foci: $(\pm 14, 0)$; minor axis of length 10

54. Foci: $(\pm 3, 0)$; major axis of length 12

55. Vertices: $(0, \pm 7)$; foci: $(0, \pm 6)$

56. Passes through the points $(0, -6)$ and $(3, 0)$

57. Architecture A semielliptical archway is formed over the entrance to an estate. The arch is set on pillars that are 10 feet apart and has a height (atop the pillars) of 4 feet (see figure). Describe the location of the foci.

58. Wading Pool You are building a wading pool that is in the shape of an ellipse. An equation for the elliptical shape of the pool, measured in feet, is

$$\frac{x^2}{324} + \frac{y^2}{196} = 1.$$

Find the longest distance across the pool, the shortest distance, and the distance between the foci.

Finding the Standard Equation of a Hyperbola In Exercises 59–62, find the standard form of the equation of the hyperbola with the given characteristics and center at the origin.

59. Vertices: $(0, \pm 1)$; foci: $(0, \pm 5)$

60. Vertices: $(\pm 4, 0)$; foci: $(\pm 6, 0)$

61. Vertices: $(\pm 1, 0)$; asymptotes: $y = \pm 2x$

62. Vertices: $(0, \pm 2)$; asymptotes: $y = \pm \dfrac{2}{\sqrt{5}}x$

4.4 **Identifying a Conic** In Exercises 63–70, identify the conic by writing its equation in standard form. Then sketch its graph and describe the translation from standard position.

63. $x^2 - 6x + 2y + 9 = 0$

64. $y^2 - 12y - 8x + 20 = 0$

65. $x^2 + y^2 - 2x - 4y + 1 = 0$

66. $16x^2 + 16y^2 - 16x + 24y - 3 = 0$

67. $x^2 + 9y^2 + 10x - 18y + 25 = 0$

68. $4x^2 + y^2 - 16x + 15 = 0$

69. $9x^2 - y^2 - 72x + 8y + 119 = 0$

70. $y^2 - 9x^2 + 10y + 18x + 7 = 0$

Finding the Standard Equation of a Parabola In Exercises 71–74, find the standard form of the equation of the parabola with the given characteristics.

71. Vertex: $(4, 2)$; focus: $(4, 0)$

72. Vertex: $(2, 0)$; focus: $(0, 0)$

73. Vertex: $(8, -8)$; directrix: $x = 1$

74. Focus: $(5, 6)$; directrix: $y = 0$

Finding the Standard Equation of an Ellipse In Exercises 75–80, find the standard form of the equation of the ellipse with the given characteristics.

75. Vertices: $(0, 2), (4, 2)$; minor axis of length 2

76. Vertices: $(5, 0), (5, 12)$; minor axis of length 10

77. Vertices: $(2, -2), (2, 8)$; foci: $(2, 0), (2, 6)$

78. Vertices: $(-9, -4), (11, -4)$; foci: $(-5, -4), (7, -4)$

79. Foci: $(0, -4), (0, 0)$; eccentricity: $\frac{2}{3}$

80. Foci: $(5, 1), (11, 1)$; eccentricity: $\frac{3}{5}$

Finding the Standard Equation of a Hyperbola In Exercises 81–84, find the standard form of the equation of the hyperbola with the given characteristics.

81. Vertices: $(-10, 3), (6, 3)$; foci: $(-12, 3), (8, 3)$

82. Vertices: $(2, -2), (2, 2)$; foci: $(2, -4), (2, 4)$

83. Vertices: $(3, -4), (3, 4)$; passes through the point $(4, 6)$

84. Vertices: $(\pm 6, 7)$;
asymptotes: $y = -\frac{1}{2}x + 7, y = \frac{1}{2}x + 7$

85. Architecture A parabolic archway is 12 meters high at the vertex. At a height of 10 meters, the width of the archway is 8 meters (see figure). How wide is the archway at ground level?

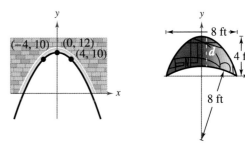

Figure for 85   Figure for 86

86. Architecture A stained-glass window is bounded above by a parabola and below by the arc of a circle (see figure).

(a) Find equations for the parabola and the circle.

(b) Complete the table by filling in the vertical distance d between the circle and the parabola for each given value of x.

x	0	1	2	3	4
d					

Exploring the Concepts

True or False? In Exercises 87 and 88, determine whether the statement is true or false. Justify your answer.

87. The graph of the equation

$$Ax^2 + Bxy + Cy^2 + Dx + Ey + F = 0$$

can be a single point.

88. If two ellipses have the same major axis length and same eccentricity, then they have the same minor axis length.

89. Alternate Form of the Equation of an Ellipse Consider the ellipse $(x - h)^2/a^2 + (y - k)^2/b^2 = 1$.

(a) Show that the equation of the ellipse can be written as

$$\frac{(x - h)^2}{a^2} + \frac{(y - k)^2}{a^2(1 - e^2)} = 1$$

where e is the eccentricity.

(b) Use a graphing utility to graph the ellipse

$$\frac{(x - 2)^2}{4} + \frac{(y - 3)^2}{4(1 - e^2)} = 1$$

for $e = 0.95, 0.75, 0.5, 0.25,$ and 0. Make a conjecture about the change in the shape of the ellipse as e approaches 0.

Chapter Test

See CalcChat.com for tutorial help and worked-out solutions to odd-numbered exercises.

Take this test as you would take a test in class. When you are finished, check your work against the answers given in the back of the book.

In Exercises 1–3, find the domain of the function and identify any asymptotes of the graph of the function. *(Section 4.1)*

1. $y = \dfrac{3x}{x + 1}$

2. $f(x) = \dfrac{3 - x^2}{3 + x^2}$

3. $g(x) = \dfrac{x - 4}{x^2 - 9x + 20}$

In Exercises 4–9, identify any intercepts and asymptotes of the graph of the function. Then sketch the graph of the function. *(Section 4.2)*

4. $h(x) = \dfrac{3}{x^2} - 1$

5. $g(x) = \dfrac{x^2 + 2}{x - 1}$

6. $f(x) = \dfrac{x + 1}{x^2 + x - 12}$

7. $f(x) = \dfrac{2x^2 - 5x - 12}{x^2 - 16}$

8. $f(x) = \dfrac{2x^2 + 9}{5x^2 + 9}$

9. $g(x) = \dfrac{2x^3 + 3x^2 - 8x - 12}{x^2 - x - 2}$

10. A rectangular page contains 36 square inches of print. The margins at the top and bottom of the page are each 2 inches deep. The margins on each side are 1 inch wide. What should the dimensions of the page be to use the least amount of paper? *(Section 4.2)*

11. A triangle is formed by the coordinate axes and a line through the point (2, 1), as shown in the figure. The value of y is given by

$$y = 1 + \frac{2}{x - 2}. \quad \textit{(Section 4.2)}$$

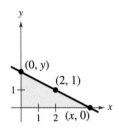

Figure for 11

(a) Write the area A of the triangle as a function of x. Determine the domain of the function in the context of the problem.

(b) Sketch the graph of the area function. Estimate the minimum area of the triangle from the graph.

In Exercises 12–17, sketch the conic and identify the center, vertices, and foci, if applicable. *(Sections 4.3 and 4.4)*

12. $4x^2 - y^2 = 4$

13. $4y^2 - 5x^2 = 80$

14. $y^2 - 4x = 0$

15. $x^2 + y^2 - 10x + 4y + 4 = 0$

16. $x^2 - 10x - 2y + 19 = 0$

17. $x^2 + 3y^2 - 2x + 36y + 100 = 0$

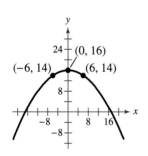

Figure for 20

18. Find the standard form of the equation of the ellipse with vertices (0, 2) and (8, 2) and minor axis of length 4. *(Sections 4.3 and 4.4)*

19. Find the standard form of the equation of the hyperbola with vertices (0, ±3) and asymptotes $y = \pm\frac{3}{2}x$. *(Section 4.3)*

20. A parabolic archway is 16 meters high at the vertex. At a height of 14 meters, the width of the archway is 12 meters, as shown in the figure. How wide is the archway at ground level? *(Section 4.4)*

21. The moon orbits Earth in an elliptical path with the center of Earth at one focus, as shown in the figure. The major and minor axes of the orbit have lengths of 768,800 kilometers and 767,641 kilometers, respectively. Find the least distance (perigee) and the greatest distance (apogee) from the center of the moon to the center of Earth. *(Section 4.4)*

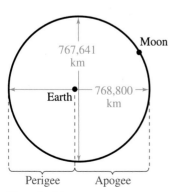

Figure for 21

Proofs in Mathematics

HISTORICAL NOTE

There are many natural occurrences of parabolas in real life. For example, Italian astronomer and mathematician Galileo Galilei discovered in the seventeenth century that an object projected upward and obliquely to the pull of gravity travels in a parabolic path. Examples of this include the path of a jumping dolphin and the path of water molecules from a drinking water fountain.

The definition of a parabola can be used to derive the standard form of the equation of a parabola whose vertex is at $(0, 0)$, and whose directrix is parallel to the x-axis or to the y-axis.

> ### Standard Equation of a Parabola (Vertex at Origin) *(p. 328)*
>
> The **standard form of the equation of a parabola** with vertex at $(0, 0)$ and directrix $y = -p$ is
> $$x^2 = 4py, \quad p \neq 0. \qquad \text{Vertical axis}$$
> For directrix $x = -p$, the equation is
> $$y^2 = 4px, \quad p \neq 0. \qquad \text{Horizontal axis}$$
> The focus lies on the axis p units (directed distance) from the vertex.

Proof

For the first case, assume that the directrix is parallel to the x-axis and is given by $y = -p$. In Figure 1, $p > 0$, and p is the directed distance from the vertex to the focus, so the focus must lie above the vertex. By the definition of a parabola, the point (x, y) is equidistant from $(0, p)$ and $y = -p$. Apply the Distance Formula to obtain

$$\sqrt{(x - 0)^2 + (y - p)^2} = y - (-p) \qquad \text{Distance Formula}$$
$$\sqrt{x^2 + (y - p)^2} = y + p \qquad \text{Simplify.}$$
$$x^2 + (y - p)^2 = (y + p)^2 \qquad \text{Square each side.}$$
$$x^2 + y^2 - 2py + p^2 = y^2 + 2py + p^2 \qquad \text{Expand.}$$
$$x^2 = 4py. \qquad \text{Simplify.}$$

A proof of the second case is similar to the proof of the first case. Assume that the directrix is parallel to the y-axis and is given by $x = -p$. In Figure 2, $p > 0$, and p is the directed distance from the vertex to the focus, so the focus must lie to the right of the vertex. By the definition of a parabola, the point (x, y) is equidistant from $(p, 0)$ and $x = -p$. Apply the Distance Formula to obtain

$$\sqrt{(x - p)^2 + (y - 0)^2} = x - (-p) \qquad \text{Distance Formula}$$
$$\sqrt{(x - p)^2 + y^2} = x + p \qquad \text{Simplify.}$$
$$(x - p)^2 + y^2 = (x + p)^2 \qquad \text{Square each side.}$$
$$x^2 - 2px + p^2 + y^2 = x^2 + 2px + p^2 \qquad \text{Expand.}$$
$$y^2 = 4px. \qquad \text{Simplify.}$$

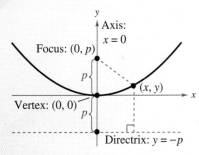

Parabola with vertical axis
Figure 1

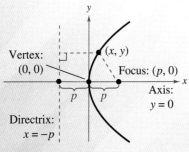

Parabola with horizontal axis
Figure 2

P.S. Problem Solving

See CalcChat.com for tutorial help and worked-out solutions to odd-numbered exercises.

1. Matching Match the graph of the rational function

$$f(x) = \frac{ax + b}{cx + d}$$

with the given conditions.

(a)

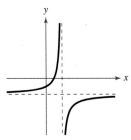

(b)

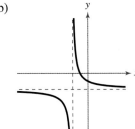

(c)

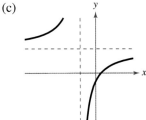

(d)

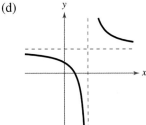

(i) $a > 0$	(ii) $a > 0$	(iii) $a < 0$	(iv) $a > 0$
$b < 0$	$b > 0$	$b > 0$	$b < 0$
$c > 0$	$c < 0$	$c > 0$	$c > 0$
$d < 0$	$d < 0$	$d < 0$	$d > 0$

2. Effects of Values on a Graph Consider the function $f(x) = (ax)/(x - b)^2$.

(a) Determine the effect on the graph of f when $b \neq 0$ and a is varied. Consider cases in which a is positive and a is negative.

(b) Determine the effect on the graph of f when $a \neq 0$ and b is varied.

3. Distinct Vision The endpoints of the interval over which distinct vision is possible are called the *near point* and *far point* of the eye (see figure). With increasing age, these points normally change. The table shows the approximate near points y (in inches) for various ages x (in years).

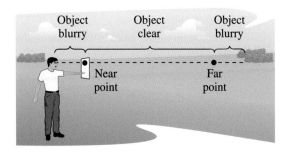

Object blurry — Object clear — Object blurry

Near point   Far point

Age, x	Near Point, y
16	3.0
32	4.7
44	9.8
50	19.7
60	39.4

(a) Use the *regression* feature of a graphing utility to find a quadratic model for the data. Use the graphing utility to plot the data and graph the model in the same viewing window.

(b) Find a rational model for the data. Take the reciprocals of the near points to generate the points $(x, 1/y)$. Use the *regression* feature of the graphing utility to find a linear model for the data. The resulting line has the form

$$\frac{1}{y} = ax + b.$$

Solve for y. Use the graphing utility to plot the original data and graph the model in the same viewing window.

(c) Use the *table* feature of the graphing utility to construct a table showing the predicted near point based on each model for each of the ages in the original table. How well do the models fit the original data?

(d) Use both models to estimate the near point for a person who is 25 years old. Which model is a better fit?

(e) Do you think either model can be used to predict the near point for a person who is 70 years old? Explain.

4. Statuary Hall Statuary Hall is an elliptical room in the United States Capitol in Washington, D.C. The room is also called the Whispering Gallery because a person standing at one focus of the room can hear even a whisper spoken by a person standing at the other focus. This occurs because any sound emitted from one focus of an ellipse reflects off the side of the ellipse to the other focus. Statuary Hall is 46 feet wide and 97 feet long.

(a) Find an equation that models the shape of the room.

(b) How far apart are the two foci?

(c) What is the area of the floor of the room? (The area of an ellipse is $A = \pi ab$.)

5. Property of a Hyperbola Use the figure to show that

$$|d_2 - d_1| = 2a.$$

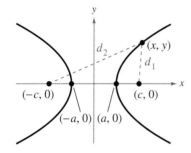

6. Finding the Equation of a Hyperbola Find an equation of a hyperbola such that for any point on the hyperbola, the difference between its distances from the points $(2, 2)$ and $(10, 2)$ is 6.

7. Tour Boat A tour boat travels between two islands that are 12 miles apart (see figure). For each trip between the islands, there is enough fuel for a 20-mile trip.

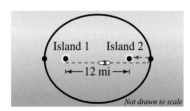

Not drawn to scale

(a) Explain why the region in which the boat can travel is bounded by an ellipse.

(b) Let $(0, 0)$ represent the center of the ellipse. Find the coordinates of the center of each island.

(c) The boat travels from Island 1, past Island 2 to one vertex of the ellipse, and then to Island 2 (see figure). How many miles does the boat travel? Use your answer to find the coordinates of the vertex.

(d) Use the results of parts (b) and (c) to write an equation of the ellipse that bounds the region in which the boat can travel.

8. Car Headlight The filament of a light bulb is a thin wire that glows when electricity passes through it. The filament of a car headlight is at the focus of a parabolic reflector, which sends light out in a straight beam. Given that the filament is 1.5 inches from the vertex, find an equation for the cross section of the reflector. The reflector is 7 inches wide. How deep is it?

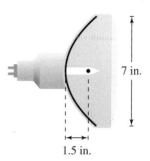

7 in.

1.5 in.

9. Analyzing Parabolas Consider the parabola $x^2 = 4py$.

(a) Use a graphing utility to graph the parabola for $p = 1$, $p = 2$, $p = 3$, and $p = 4$. Describe the effect on the graph when p increases.

(b) Locate the focus for each parabola in part (a).

(c) For each parabola in part (a), find the length of the chord passing through the focus and parallel to the directrix. How can you determine the length of this chord directly from the standard form of the equation of the parabola?

(d) Explain how the result of part (c) can be used as an aid when sketching parabolas.

10. Tangent Line Let (x_1, y_1) be the coordinates of a point on the parabola $x^2 = 4py$. The equation of the line that just touches the parabola at the point (x_1, y_1), called a *tangent line*, is

$$y - y_1 = \frac{x_1}{2p}(x - x_1).$$

(a) What is the slope of the tangent line?

(b) For each parabola in Exercise 9, find the equations of the tangent lines at the endpoints of the chord. Use a graphing utility to graph the parabola and tangent lines.

11. Proof Prove that the graph of the equation

$$Ax^2 + Cy^2 + Dx + Ey + F = 0$$

is one of the following (except in degenerate cases).

Conic	Condition
(a) Circle	$A = C, A \neq 0$
(b) Parabola	$A = 0$ or $C = 0$, but not both
(c) Ellipse	$AC > 0, A \neq C$
(d) Hyperbola	$AC < 0$

5 Exponential and Logarithmic Functions

GO DIGITAL

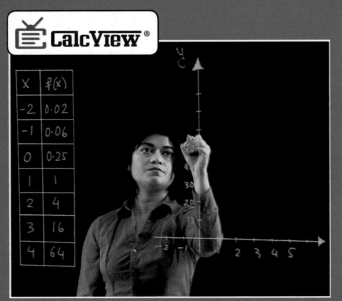

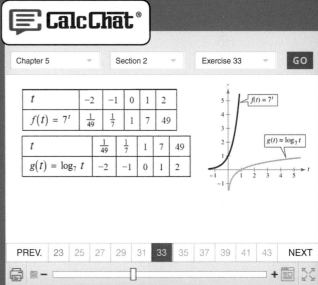

5.1 Chemistry *(Exercise 60, p. 369)*

5.2 Human Memory Model *(Exercise 81, p. 380)*

359

5.1 Exponential Functions and Their Graphs

❯ Recognize and evaluate exponential functions with base *a*.
❯ Graph exponential functions and use the One-to-One Property.
❯ Recognize, evaluate, and graph exponential functions with base e.
❯ Use exponential functions to model and solve real-life problems.

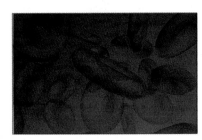

Exponential functions can help you model and solve real-life problems. For example, in Exercise 60 on page 369, you will use an exponential function to model the concentration of a drug in the bloodstream.

Exponential Functions

So far, this text has dealt mainly with **algebraic functions,** which include polynomial functions and rational functions. In this chapter, you will study two types of nonalgebraic functions—*exponential functions* and *logarithmic functions*. These functions are examples of **transcendental functions.** This section will focus on exponential functions.

Definition of Exponential Function

The **exponential function** *f* **with base** *a* is denoted by

$$f(x) = a^x$$

where $a > 0$, $a \neq 1$, and x is any real number.

The base *a* of an exponential function cannot be 1 because $a = 1$ yields $f(x) = 1^x = 1$. This is a constant function, not an exponential function.

You have evaluated a^x for integer and rational values of *x*. For example, you know that $4^3 = 64$ and $4^{1/2} = 2$. However, to evaluate 4^x for any real number *x*, you need to interpret forms with *irrational* exponents. For the purposes of this text, it is sufficient to think of $a^{\sqrt{2}}$, where $\sqrt{2} \approx 1.41421356$, as the number that has the successively closer approximations

$$a^{1.4},\ a^{1.41},\ a^{1.414},\ a^{1.4142},\ a^{1.41421},\ \ldots\ .$$

❯❯❯ **SKILLS REFRESHER**

For a refresher on using the rules of exponents, watch the video at *LarsonPrecalculus.com.*

EXAMPLE 1   **Evaluating Exponential Functions**

Use a calculator to evaluate each function at the given value of *x*.

Function	Value
a. $f(x) = 2^x$	$x = -3.1$
b. $f(x) = 2^{-x}$	$x = \pi$
c. $f(x) = 0.6^x$	$x = \frac{3}{2}$

Solution

Function Value	Calculator Keystrokes	Display
a. $f(-3.1) = 2^{-3.1}$	2 〔∧〕〔(−)〕 3.1 〔ENTER〕	0.1166291
b. $f(\pi) = 2^{-\pi}$	2 〔∧〕〔(−)〕 π 〔ENTER〕	0.1133147
c. $f\left(\frac{3}{2}\right) = (0.6)^{3/2}$	.6 〔∧〕〔(〕 3 〔÷〕 2 〔)〕 〔ENTER〕	0.4647580

✓ *Checkpoint* ▶ Audio-video solution in English & Spanish at *LarsonPrecalculus.com*

Use a calculator to evaluate $f(x) = 8^{-x}$ at $x = \sqrt{2}$. ∎

On some calculators, you may need to enclose exponents in parentheses to obtain the correct result.

© Kagrafi/Shutterstock.com

Graphs of Exponential Functions

The graphs of all exponential functions have similar characteristics, as shown in Examples 2, 3, and 5.

EXAMPLE 2 Graphs of $y = a^x$

In the same coordinate plane, sketch the graphs of $f(x) = 2^x$ and $g(x) = 4^x$.

Solution Begin by constructing a table of values.

x	-3	-2	-1	0	1	2
2^x	$\frac{1}{8}$	$\frac{1}{4}$	$\frac{1}{2}$	1	2	4
4^x	$\frac{1}{64}$	$\frac{1}{16}$	$\frac{1}{4}$	1	4	16

To sketch the graph of each function, plot the points from the table and connect them with a smooth curve, as shown in Figure 5.1. Note that both graphs are increasing. Moreover, the graph of $g(x) = 4^x$ is increasing more rapidly than the graph of $f(x) = 2^x$.

✓ **Checkpoint** ▶ Audio-video solution in English & Spanish at LarsonPrecalculus.com

In the same coordinate plane, sketch the graphs of $f(x) = 3^x$ and $g(x) = 9^x$. ■

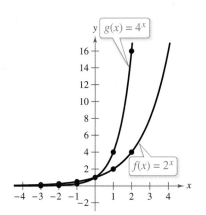

Figure 5.1

The table in Example 2 was evaluated by hand for integer values of x. You can also evaluate $f(x)$ and $g(x)$ for noninteger values of x by using a calculator.

EXAMPLE 3 Graphs of $y = a^{-x}$

In the same coordinate plane, sketch the graphs of $F(x) = 2^{-x}$ and $G(x) = 4^{-x}$.

Solution Begin by constructing a table of values.

x	-2	-1	0	1	2	3
2^{-x}	4	2	1	$\frac{1}{2}$	$\frac{1}{4}$	$\frac{1}{8}$
4^{-x}	16	4	1	$\frac{1}{4}$	$\frac{1}{16}$	$\frac{1}{64}$

To sketch the graph of each function, plot the points from the table and connect them with a smooth curve, as shown in Figure 5.2. Note that both graphs are decreasing. Moreover, the graph of $G(x) = 4^{-x}$ is decreasing more rapidly than the graph of $F(x) = 2^{-x}$.

✓ **Checkpoint** ▶ Audio-video solution in English & Spanish at LarsonPrecalculus.com

In the same coordinate plane, sketch the graphs of $F(x) = 3^{-x}$ and $G(x) = 9^{-x}$. ■

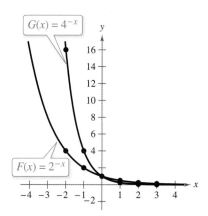

Figure 5.2

Note that the functions in Example 3 can be rewritten with positive exponents using the property

$$a^{-n} = \frac{1}{a^n} = \left(\frac{1}{a}\right)^n.$$ Property of exponents (See Section P.2.)

So, the functions $F(x) = 2^{-x}$ and $G(x) = 4^{-x}$ can be rewritten as

$$F(x) = 2^{-x} = \frac{1}{2^x} = \left(\frac{1}{2}\right)^x \quad \text{and} \quad G(x) = 4^{-x} = \frac{1}{4^x} = \left(\frac{1}{4}\right)^x.$$

GO DIGITAL

Comparing the functions in Examples 2 and 3, observe that $F(x) = 2^{-x} = f(-x)$ and $G(x) = 4^{-x} = g(-x)$. Consequently, the graph of F is a reflection (in the y-axis) of the graph of f. The graphs of G and g have the same relationship. The graphs in Figures 5.1 and 5.2 are typical of the exponential functions $y = a^x$ and $y = a^{-x}$. They have one y-intercept and one horizontal asymptote (the x-axis), and they are continuous. Here is a summary of the basic characteristics of the graphs of these exponential functions.

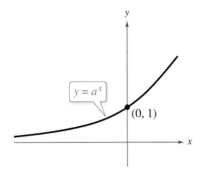

Graph of $y = a^x$, $a > 1$
- Domain: $(-\infty, \infty)$
- Range: $(0, \infty)$
- y-intercept: $(0, 1)$
- Increasing on $(-\infty, \infty)$
- x-axis is a horizontal asymptote $(a^x \to 0$ as $x \to -\infty)$.
- Continuous

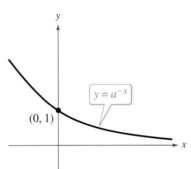

Graph of $y = a^{-x}$, $a > 1$
- Domain: $(-\infty, \infty)$
- Range: $(0, \infty)$
- y-intercept: $(0, 1)$
- Decreasing on $(-\infty, \infty)$
- x-axis is a horizontal asymptote $(a^{-x} \to 0$ as $x \to \infty)$.
- Continuous

Notice that the graph of an exponential function is always increasing or always decreasing, so the graph passes the Horizontal Line Test. Therefore, an exponential function is a one-to-one function. You can use the following **One-to-One Property** to solve simple exponential equations.

For $a > 0$ and $a \neq 1$, $a^x = a^y$ if and only if $x = y$. One-to-One Property

EXAMPLE 4 **Using the One-to-One Property**

a. $9 = 3^{x+1}$ Original equation

$\quad 3^2 = 3^{x+1}$ $9 = 3^2$

$\quad\quad 2 = x + 1$ One-to-One Property

$\quad\quad 1 = x$ Solve for x.

b. $\left(\dfrac{1}{2}\right)^x = 8$ Original equation

$\quad 2^{-x} = 2^3$ $\left(\dfrac{1}{2}\right)^x = 2^{-x}, 8 = 2^3$

$\quad\quad x = -3$ One-to-One Property

✓ *Checkpoint* ▶ *Audio-video solution in English & Spanish at LarsonPrecalculus.com*

Use the One-to-One Property to solve (a) $8 = 2^{2x-1}$ and (b) $\left(\dfrac{1}{3}\right)^{-x} = 27$ for x. ∎

GO DIGITAL

In Example 5, notice how the graph of $y = a^x$ can be used to sketch the graphs of functions of the form $f(x) = b \pm a^{x+c}$.

EXAMPLE 5 Transformations of Graphs of $f(x) = a^x$

▶▶▶ *See LarsonPrecalculus.com for an interactive version of this type of example.*

Describe the transformation of the graph of $f(x) = 3^x$ that yields each graph.

a.
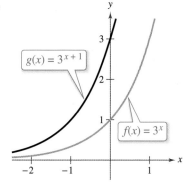
$g(x) = 3^{x+1}$
$f(x) = 3^x$

b.
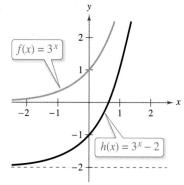
$f(x) = 3^x$
$h(x) = 3^x - 2$

c.
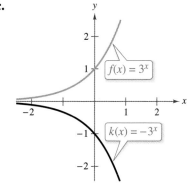
$f(x) = 3^x$
$k(x) = -3^x$

d.
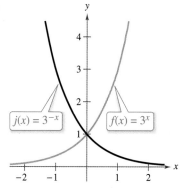
$j(x) = 3^{-x}$
$f(x) = 3^x$

Solution

a. Because $g(x) = 3^{x+1} = f(x + 1)$, the graph of g is obtained by shifting the graph of f one unit to the *left*.

b. Because $h(x) = 3^x - 2 = f(x) - 2$, the graph of h is obtained by shifting the graph of f *down* two units.

c. Because $k(x) = -3^x = -f(x)$, the graph of k is obtained by *reflecting* the graph of f in the x-axis.

d. Because $j(x) = 3^{-x} = f(-x)$, the graph of j is obtained by *reflecting* the graph of f in the y-axis.

✓ *Checkpoint* ▶ *Audio-video solution in English & Spanish at LarsonPrecalculus.com*

Describe the transformation of the graph of $f(x) = 4^x$ that yields the graph of each function.

a. $g(x) = 4^{x-2}$ b. $h(x) = 4^x + 3$ c. $k(x) = 4^{-x} - 3$ ■

Note how each transformation in Example 5 affects the y-intercept and the horizontal asymptote.

GO DIGITAL

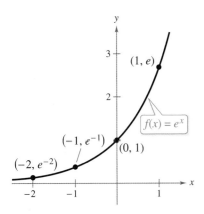

Figure 5.3

The Natural Base e

In many applications, the most convenient choice for a base is the irrational number

$$e \approx 2.718281828 \ldots .$$

This number is called the **natural base.** The function $f(x) = e^x$ is called the **natural exponential function,** and its graph is shown in Figure 5.3. The graph of the natural exponential function has the same basic characteristic as the graph of the function $f(x) = a^x$. Be sure you see that for the exponential function $f(x) = e^x$, e is the constant $2.718281828 \ldots$, whereas x is the variable.

EXAMPLE 6 **Evaluating the Natural Exponential Function**

Use a calculator to evaluate the function $f(x) = e^x$ at each value of x.

a. $x = -2$ **b.** $x = -1$

c. $x = 0.25$ **d.** $x = -0.3$

Solution

Function Value	Calculator Keystrokes	Display
a. $f(-2) = e^{-2}$	e^x $(-)$ 2 ENTER	0.1353353
b. $f(-1) = e^{-1}$	e^x $(-)$ 1 ENTER	0.3678794
c. $f(0.25) = e^{0.25}$	e^x 0.25 ENTER	1.2840254
d. $f(-0.3) = e^{-0.3}$	e^x $(-)$ 0.3 ENTER	0.7408182

✓ **Checkpoint** ▶ Audio-video solution in English & Spanish at LarsonPrecalculus.com

Use a calculator to evaluate the function $f(x) = e^x$ at each value of x.

a. $x = 0.3$

b. $x = -1.2$

c. $x = 6.2$

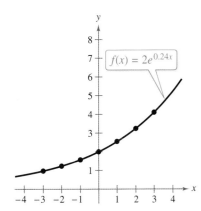

Figure 5.4

EXAMPLE 7 **Graphing Natural Exponential Functions**

Sketch the graphs of the natural exponential functions

$$f(x) = 2e^{0.24x} \quad \text{and} \quad g(x) = \frac{1}{2}e^{-0.58x}.$$

Solution Begin by using a graphing utility to construct a table of values.

x	-3	-2	-1	0	1	2	3
$f(x)$	0.974	1.238	1.573	2.000	2.542	3.232	4.109
$g(x)$	2.849	1.595	0.893	0.500	0.280	0.157	0.088

To graph each function, plot the points from the table and connect them with a smooth curve, as shown in Figures 5.4 and 5.5. Note that the graph in Figure 5.4 is increasing, whereas the graph in Figure 5.5 is decreasing.

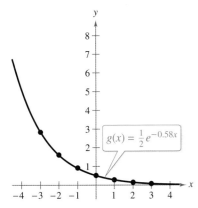

Figure 5.5

✓ **Checkpoint** ▶ Audio-video solution in English & Spanish at LarsonPrecalculus.com

Sketch the graph of $f(x) = 5e^{0.17x}$.

GO DIGITAL

Applications

One of the most familiar examples of exponential growth is an investment earning *continuously compounded interest*. Recall from page 95 in Section 1.3 that the formula for *interest compounded n times per year* is

$$A = P\left(1 + \frac{r}{n}\right)^{nt}.$$

In this formula, A is the balance in the account, P is the principal (or original deposit), r is the annual interest rate (in decimal form), n is the number of compoundings per year, and t is the time in years. Exponential functions can be used to *develop* this formula and show how it leads to continuous compounding.

Consider a principal P invested at an annual interest rate r, compounded once per year. When the interest is added to the principal at the end of the first year, the new balance P_1 is

$$P_1 = P + Pr$$
$$= P(1 + r).$$

This pattern of multiplying the balance by $1 + r$ repeats each successive year, as shown here.

Year	Balance After Each Compounding
0	$P = P$
1	$P_1 = P(1 + r)$
2	$P_2 = P_1(1 + r) = P(1 + r)(1 + r) = P(1 + r)^2$
3	$P_3 = P_2(1 + r) = P(1 + r)^2(1 + r) = P(1 + r)^3$
$\vdots$	$\vdots$
t	$P_t = P(1 + r)^t$

To accommodate more frequent compounding of interest, such as quarterly, monthly, or daily, let n be the number of compoundings per year and let t be the number of years. Then the rate per compounding is r/n, and the account balance after t years is

$$A = P\left(1 + \frac{r}{n}\right)^{nt}.$$ Amount (balance) with n compoundings per year

When the number of compoundings n increases without bound, the process approaches what is called **continuous compounding.** In the formula for n compoundings per year, let $m = n/r$. This yields a new expression.

$$A = P\left(1 + \frac{r}{n}\right)^{nt}$$ Amount with n compoundings per year

$$= P\left(1 + \frac{r}{mr}\right)^{mrt}$$ Substitute mr for n.

$$= P\left(1 + \frac{1}{m}\right)^{mrt}$$ Simplify.

$$= P\left[\left(1 + \frac{1}{m}\right)^m\right]^{rt}$$ Property of exponents

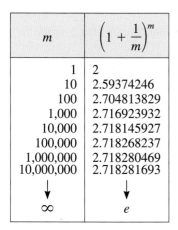

m	$\left(1 + \dfrac{1}{m}\right)^m$
1	2
10	2.59374246
100	2.704813829
1,000	2.716923932
10,000	2.718145927
100,000	2.718268237
1,000,000	2.718280469
10,000,000	2.718281693
$\downarrow$	$\downarrow$
∞	e

As m increases without bound (that is, as $m \to \infty$), the table at the left shows that $[1 + (1/m)]^m$ approaches e. This allows you to conclude that the formula for continuous compounding is

$$A = Pe^{rt}.$$ Substitute e for $[1 + (1/m)]^m$.

GO DIGITAL

ALGEBRA HELP

When using the formulas for compound interest, you must write the annual interest rate in decimal form. For example, you must write 2.5% as 0.025.

Formulas for Compound Interest

After t years, the balance A in an account with principal P and annual interest rate r (in decimal form) is given by one of these two formulas.

1. For n compoundings per year: $A = P\left(1 + \dfrac{r}{n}\right)^{nt}$

2. For continuous compounding: $A = Pe^{rt}$

EXAMPLE 8 ▶ **Compound Interest**

You invest $12,000 at an annual interest rate of 3%. Find the balance after 5 years for each type of compounding.

a. Quarterly   **b.** Monthly   **c.** Continuously

Solution The investment is $12,000, so $P = 12{,}000$. The annual interest rate of 3% is 0.03 in decimal form, so $r = 0.03$. To find the balance after 5 years, let $t = 5$.

a. For quarterly compounding, use $n = 4$ to find the balance after 5 years.

$A = P\left(1 + \dfrac{r}{n}\right)^{nt}$ Formula for compound interest

$= 12{,}000\left(1 + \dfrac{0.03}{4}\right)^{4(5)}$ Substitute for P, r, n, and t.

$\approx 13{,}934.21$ Use a calculator.

With quarterly compounding, the balance in the account after 5 years is $13,934.21.

b. For monthly compounding, use $n = 12$ to find the balance after 5 years.

$A = P\left(1 + \dfrac{r}{n}\right)^{nt}$ Formula for compound interest

$= 12{,}000\left(1 + \dfrac{0.03}{12}\right)^{12(5)}$ Substitute for P, r, n, and t.

$\approx 13{,}939.40$ Use a calculator.

With monthly compounding, the balance in the account after 5 years is $13,939.40.

c. Use the formula for continuous compounding to find the balance after 5 years.

$A = Pe^{rt}$ Formula for continuous compounding

$= 12{,}000e^{0.03(5)}$ Substitute for P, r, and t.

$\approx 13{,}942.01$ Use a calculator.

With continuous compounding, the balance in the account after 5 years is $13,942.01.

✓ **Checkpoint** ▶ Audio-video solution in English & Spanish at LarsonPrecalculus.com

You invest $6000 at an annual rate of 4%. Find the balance after 7 years for each type of compounding.

a. Quarterly   **b.** Monthly   **c.** Continuously   ■

In Example 8, note that continuous compounding yields more than quarterly and monthly compounding. This is typical of the two types of compounding. That is, for a given principal, interest rate, and time, continuous compounding will always yield a larger balance than compounding n times per year.

GO DIGITAL

EXAMPLE 9 **Radioactive Decay**

In 1986, a nuclear reactor accident occurred in Chernobyl in what was then the Soviet Union. The explosion spread highly toxic radioactive chemicals, such as plutonium $\left(^{239}\text{Pu}\right)$, over hundreds of square miles, and the government evacuated the city and the surrounding area. To see why the city is now uninhabited, consider the model

$$P = 10\left(\frac{1}{2}\right)^{t/24,100}$$

which represents the amount of plutonium P that remains (from an initial amount of 10 pounds) after t years. Sketch the graph of this function over the interval from $t = 0$ to $t = 100{,}000$, where $t = 0$ represents 1986. How much of the 10 pounds will remain in the year 2030? How much of the 10 pounds will remain after 100,000 years?

Solution The graph of this function is shown in the figure at the right. Note from this graph that plutonium has a *half-life* of about 24,100 years. That is, after 24,100 years, *half* of the original amount will remain. After another 24,100 years, one-quarter of the original amount will remain, and so on. In the year 2030 ($t = 44$), there will still be

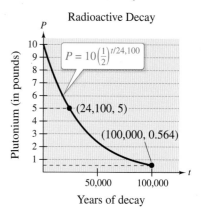

Radioactive Decay

$$P = 10\left(\frac{1}{2}\right)^{44/24,100}$$

$$\approx 10\left(\frac{1}{2}\right)^{0.0018257}$$

$$\approx 9.987 \text{ pounds}$$

of plutonium remaining. After 100,000 years, there will still be

$$P = 10\left(\frac{1}{2}\right)^{100,000/24,100}$$

$$\approx 0.564 \text{ pound}$$

of plutonium remaining.

✓ *Checkpoint* ▶ *Audio-video solution in English & Spanish at LarsonPrecalculus.com*

In Example 9, how much of the 10 pounds will remain in the year 2089? How much of the 10 pounds will remain after 125,000 years? ▪

Summarize *(Section 5.1)*

1. State the definition of the exponential function f with base a *(page 360)*. For an example of evaluating exponential functions, see Example 1.

2. Describe the basic characteristics of the graphs of the exponential functions $y = a^x$ and $y = a^{-x}$, $a > 1$ *(page 362)*. For examples of graphing exponential functions, see Examples 2, 3, and 5.

3. State the definitions of the natural base and the natural exponential function *(page 364)*. For examples of evaluating and graphing natural exponential functions, see Examples 6 and 7.

4. Describe real-life applications involving exponential functions *(pages 366 and 367, Examples 8 and 9)*.

GO DIGITAL

5.1 Exercises

GO DIGITAL

Vocabulary and Concept Check

In Exercises 1 and 2, fill in the blanks.

1. Exponential and logarithmic functions are examples of nonalgebraic functions, also called _____ functions.

2. The exponential function $f(x) = e^x$ is called the _____ _____ function, and the base e is called the _____ base.

3. What type of transformation of the graph of $f(x) = 5^x$ is the graph of $f(x + 1)$?

4. The formula $A = Pe^{rt}$ gives the balance A of an account earning what type of interest?

Skills and Applications

Evaluating an Exponential Function   In Exercises 5–8, evaluate the function at the given value of x. Round your result to three decimal places.

Function	Value
5. $f(x) = 0.9^x$	$x = 1.4$
6. $f(x) = 3^x$	$x = \frac{2}{5}$
7. $f(x) = 5000(2^x)$	$x = -1.5$
8. $f(x) = 200(1.2)^{12x}$	$x = 24$

Matching an Exponential Function with Its Graph   In Exercises 9–12, match the exponential function with its graph. [The graphs are labeled (a), (b), (c), and (d).]

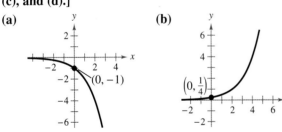

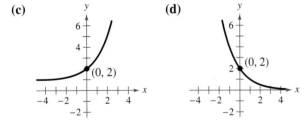

9. $f(x) = 2^{1-x}$

10. $f(x) = 2^x + 1$

11. $f(x) = -2^x$

12. $f(x) = 2^{x-2}$

Graphing an Exponential Function   In Exercises 13–20, construct a table of values for the function. Then sketch the graph of the function.

13. $f(x) = 7^x$

14. $f(x) = 7^{-x}$

15. $f(x) = \left(\frac{1}{4}\right)^{-x}$

16. $f(x) = \left(\frac{1}{4}\right)^x$

17. $f(x) = 4^{x-1}$

18. $f(x) = 4^{x+1}$

19. $f(x) = 2^{x+1} + 3$

20. $f(x) = 3^{x-2} + 1$

Using the One-to-One Property   In Exercises 21–24, use the One-to-One Property to solve the equation for x.

21. $3^{x+1} = 27$

22. $243 = 9^x$

23. $\left(\frac{1}{2}\right)^x = 32$

24. $5^{x-2} = \frac{1}{125}$

Transformations of the Graph of $f(x) = a^x$ In Exercises 25–28, describe any transformations of the graph of f that yield the graph of g.

25. $f(x) = 3^x$, $g(x) = 3^x + 1$

26. $f(x) = \left(\frac{7}{2}\right)^x$, $g(x) = -\left(\frac{7}{2}\right)^{-x}$

27. $f(x) = 10^x$, $g(x) = 10^{-x+3}$

28. $f(x) = 0.3^x$, $g(x) = -0.3^x + 5$

Evaluating a Natural Exponential Function   In Exercises 29–32, evaluate the function at the given value of x. Round your result to three decimal places.

Function	Value
29. $f(x) = e^x$	$x = 1.9$
30. $f(x) = 1.5e^{x/2}$	$x = 240$
31. $f(x) = 5000e^{0.06x}$	$x = 6$
32. $f(x) = 250e^{0.05x}$	$x = 20$

Graphing a Natural Exponential Function   In Exercises 33–36, construct a table of values for the function. Then sketch the graph of the function.

33. $f(x) = 3e^{x+4}$

34. $f(x) = 2e^{-1.5x}$

35. $f(x) = 2e^{x-2} + 4$

36. $f(x) = 2 + e^{x-5}$

Graphing a Natural Exponential Function In Exercises 37–40, use a graphing utility to graph the exponential function.

37. $s(t) = 2e^{0.5t}$ 38. $s(t) = 3e^{-0.2t}$

39. $g(x) = 1 + e^{-x}$ 40. $h(x) = e^{x-2}$

Using the One-to-One Property In Exercises 41–44, use the One-to-One Property to solve the equation for x.

41. $e^{3x+2} = e^3$ 42. $e^{2x-1} = e^4$

43. $e^{x^2-3} = e^{2x}$ 44. $e^{x^2+6} = e^{5x}$

Compound Interest In Exercises 45–48, complete the table by finding the balance A when P dollars is invested at rate r for t years and compounded n times per year.

n	1	2	4	12	365	Continuous
A						

45. $P = \$1500$, $r = 2\%$, $t = 10$ years

46. $P = \$2500$, $r = 3.5\%$, $t = 10$ years

47. $P = \$2500$, $r = 4\%$, $t = 20$ years

48. $P = \$1000$, $r = 6\%$, $t = 40$ years

Compound Interest In Exercises 49–52, complete the table by finding the balance A when $12,000$ is invested at rate r for t years, compounded continuously.

t	10	20	30	40	50
A					

49. $r = 4\%$

50. $r = 6\%$

51. $r = 6.5\%$

52. $r = 3.5\%$

53. **Investment** A business deposits $30,000 in a fund that pays 5% interest, compounded continuously. Determine the balance in this account after 25 years.

54. **College Savings Fund** A deposit of $5000 is made in a college savings fund that pays 7.5% interest, compounded continuously. The balance will be given to a student after the money has earned interest for 18 years. How much will the student receive?

55. **Inflation** Assuming that the annual rate of inflation averages 4% over the next 10 years, the approximate costs C of goods or services during any year in that decade can be modeled by $C(t) = P(1.04)^t$, where t is the time in years and P is the present cost. The price of an oil change for your car is presently $29.88. Estimate the price 10 years from now.

56. **Population Growth** The projected population of the United States for the years 2025 through 2055 can be modeled by $P = 313.39e^{0.0043t}$, where P is the population (in millions) and t is the time (in years), with $t = 25$ corresponding to 2025. (Source: U.S. Census Bureau)

(a) Use a graphing utility to graph the function for the years 2025 through 2055.

(b) Use the *table* feature of the graphing utility to create a table of values for the same time period as in part (a).

(c) According to the model, in what year will the population of the United States be about 355 million?

57. **Radioactive Decay** Let Q represent a mass (in grams) of radioactive plutonium (^{239}Pu), whose half-life is 24,100 years. The quantity of plutonium present after t years is $Q = 16\left(\frac{1}{2}\right)^{t/24,100}$.

(a) Determine the initial quantity (when $t = 0$).

(b) Determine the quantity present after 75,000 years.

(c) Use a graphing utility to graph the function over the interval $t = 0$ to $t = 150,000$.

58. **Radioactive Decay** Let Q represent a mass (in grams) of carbon (^{14}C), whose half-life is 5715 years. The quantity of carbon-14 present after t years is $Q = 10\left(\frac{1}{2}\right)^{t/5715}$.

(a) Determine the initial quantity (when $t = 0$).

(b) Determine the quantity present after 2000 years.

(c) Sketch the graph of the function over the interval $t = 0$ to $t = 10,000$.

59. **Coronavirus Infections** A country reports 52,490 coronavirus infections in its first month of testing. Each month thereafter, the number of infections reduces to $\frac{7}{8}$ the number of the previous month.

(a) Find a model for $V(t)$, the number of coronavirus infections after t months.

(b) Determine the number of infections after 4 months.

60. **Chemistry**

Immediately following an injection, the concentration of a drug in the bloodstream is 300 milligrams per milliliter. After t hours, the concentration is 75% of the level of the previous hour.

(a) Find a model for $C(t)$, the concentration of the drug after t hours.

(b) Determine the concentration of the drug after 8 hours.

Exploring the Concepts

True or False? In Exercises 61 and 62, determine whether the statement is true or false. Justify your answer.

61. The line $y = -2$ is an asymptote for the graph of $f(x) = 10^x - 2$.

62. $e = \dfrac{271{,}801}{99{,}990}$

Think About It In Exercises 63–66, use properties of exponents to determine which functions (if any) are the same.

63. $f(x) = 3^{x-2}$
$g(x) = 3^x - 9$
$h(x) = \frac{1}{9}(3^x)$

64. $f(x) = 4^x + 12$
$g(x) = 2^{2x+6}$
$h(x) = 64(4^x)$

65. $f(x) = 16(4^{-x})$
$g(x) = \left(\frac{1}{4}\right)^{x-2}$
$h(x) = 16(2^{-2x})$

66. $f(x) = e^{-x} + 3$
$g(x) = e^{3-x}$
$h(x) = -e^{x-3}$

 67. Graphical Reasoning Use a graphing utility to graph $y_1 = [1 + (1/x)]^x$ and $y_2 = e$ in the same viewing window. Using the *trace* feature, explain what happens to the graph of y_1 as x increases.

68. Graphical Reasoning Use a graphing utility to graph

$$f(x) = \left(1 + \frac{0.5}{x}\right)^x \quad \text{and} \quad g(x) = e^{0.5}$$

in the same viewing window. What is the relationship between f and g as x increases and decreases without bound?

69. Comparing Graphs Use a graphing utility to graph each pair of functions in the same viewing window. Describe any similarities and differences in the graphs.

(a) $y_1 = 2^x,\ y_2 = x^2$ (b) $y_1 = 3^x,\ y_2 = x^3$

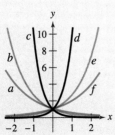

70. HOW DO YOU SEE IT? The figure shows the graphs of $y = 2^x$, $y = e^x$, $y = 10^x$, $y = 2^{-x}$, $y = e^{-x}$, and $y = 10^{-x}$. Match each function with its graph. [The graphs are labeled (a) through (f).] Explain your reasoning.

71. Graphical Reasoning Use a graphing utility to graph each function. Use the graph to find where the function is increasing and decreasing, and approximate any relative maximum or minimum values.

(a) $f(x) = x^2 e^{-x}$ (b) $g(x) = x2^{3-x}$

72. Error Analysis Describe the error.

For the function $f(x) = 2^x - 1$, the domain is $(-\infty, \infty)$ and the range is $(0, \infty)$.

73. Think About It Which functions are exponential?

(a) $f(x) - 3x$ (b) $g(x) = 3x^2$
(c) $h(x) = 3^x$ (d) $k(x) = 2^{-x}$

74. Compound Interest Use the formula

$$A = P\left(1 + \frac{r}{n}\right)^{nt}$$

to calculate the balance A of an investment when $P = \$3000$, $r = 6\%$, and $t = 10$ years, and compounding is done (a) by the day, (b) by the hour, (c) by the minute, and (d) by the second. Does increasing the number of compoundings per year result in unlimited growth of the balance? Explain.

Review & Refresh ▶ Video solutions at LarsonPrecalculus.com

Verifying Inverse Functions In Exercises 75–80, verify that f and g are inverse functions (a) algebraically and (b) graphically.

75. $f(x) = x,\ g(x) = x$ **76.** $f(x) = 4x,\ g(x) = \frac{1}{4}x$

77. $f(x) = \dfrac{1}{x},\ g(x) = \dfrac{1}{x}$

78. $f(x) = \dfrac{x+1}{x-1},\ g(x) = \dfrac{x+1}{x-1}$

79. $f(x) = \dfrac{2x+1}{x+1},\ g(x) = \dfrac{1-x}{x-2}$

80. $f(x) = \dfrac{1}{1+x},\ x \ge 0,\ g(x) = \dfrac{1-x}{x},\ 0 < x \le 1$

Restricting the Domain In Exercises 81–88, restrict the domain of the function f so that the function is one-to-one and has an inverse function. Then find the inverse function f^{-1}. State the domains and ranges of f and f^{-1}. Explain your results. (There are many correct answers.)

81. $f(x) = |x + 2|$ **82.** $f(x) = |x - 5|$
83. $f(x) = (x + 6)^2$ **84.** $f(x) = (x - 4)^2$
85. $f(x) = -2x^2 + 5$ **86.** $f(x) = \frac{1}{2}x^2 - 1$
87. $f(x) = |x - 4| + 1$ **88.** $f(x) = -|x - 1| - 2$

Project: Population per Square Mile To work an extended application analyzing the population per square mile of the United States, visit this text's website at *LarsonPrecalculus.com*. (*Source: U.S. Census Bureau*)

5.2 Logarithmic Functions and Their Graphs

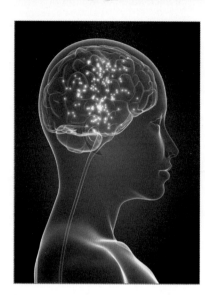

⊘ Recognize and evaluate logarithmic functions with base *a*.
⊘ Graph logarithmic functions.
⊘ Recognize, evaluate, and graph natural logarithmic functions.
⊘ Use logarithmic functions to model and solve real-life problems.

Logarithmic Functions

Recall from Section 2.7 that a function f has an inverse function if and only if f is one-to-one. In the preceding section, you learned that $f(x) = a^x$ is one-to-one, so f must have an inverse function. This inverse function is the **logarithmic function with base *a*** and is denoted by the symbol $\log_a x$, read as "log base *a* of *x*."

> **Definition of Logarithmic Function with Base *a***
>
> For $x > 0$, $a > 0$, and $a \neq 1$, $y = \log_a x$ if and only if $x = a^y$. The function $f(x) = \log_a x$ is called the **logarithmic function with base *a*.**

Logarithmic functions can often model scientific observations. For example, in Exercise 81 on page 380, you will use a logarithmic function that models human memory.

The equations $y = \log_a x$ and $x = a^y$ are equivalent, where the first equation is written in logarithmic form and the second equation is written in exponential form. For example, $2 = \log_3 9$ is equivalent to $9 = 3^2$, and $5^3 = 125$ is equivalent to $\log_5 125 = 3$.

When evaluating logarithms, remember that *a logarithm is an exponent*. This means that $\log_a x$ is the exponent to which a must be raised to obtain x. For example, $\log_2 8 = 3$ because 2 raised to the third power is 8.

EXAMPLE 1   Evaluating Logarithms

Evaluate each logarithm at the given value of x.

a. $f(x) = \log_2 x, \ x = 32$ **b.** $f(x) = \log_3 x, \ x = 1$

c. $f(x) = \log_4 x, \ x = 2$ **d.** $f(x) = \log_{10} x, \ x = \frac{1}{100}$

Solution

a. $f(32) = \log_2 32 = 5$ because $2^5 = 32$.

b. $f(1) = \log_3 1 = 0$ because $3^0 = 1$.

c. $f(2) = \log_4 2 = \frac{1}{2}$ because $4^{1/2} = \sqrt{4} = 2$.

d. $f\left(\dfrac{1}{100}\right) = \log_{10} \dfrac{1}{100} = -2$ because $10^{-2} = \dfrac{1}{10^2} = \dfrac{1}{100}$.

✓ *Checkpoint* ▶ *Audio-video solution in English & Spanish at LarsonPrecalculus.com*

Evaluate each logarithm at the given value of x.

a. $f(x) = \log_6 x, x = 1$ **b.** $f(x) = \log_5 x, x = \frac{1}{125}$ **c.** $f(x) = \log_7 x, x = 343$ ■

The logarithmic function with base 10 is called the **common logarithmic function.** It is denoted by $\log_{10}$ or simply log. On most calculators, it is denoted by ⌐LOG⌐. Example 2 on the next page shows how to use a calculator to evaluate common logarithmic functions. You will learn how to use a calculator to evaluate logarithms with any base in the next section.

GO DIGITAL

© SciePro/Shutterstock.com

EXAMPLE 2   **Evaluating Common Logarithms on a Calculator**

Use a calculator to evaluate the function $f(x) = \log x$ at each value of x.

a. $x = 10$ **b.** $x = \frac{1}{3}$ **c.** $x = -2$

Solution

Function Value	Calculator Keystrokes	Display
a. $f(10) = \log 10$	LOG 10 ENTER	1
b. $f\left(\frac{1}{3}\right) = \log \frac{1}{3}$	LOG (1 ÷ 3) ENTER	-0.4771213
c. $f(-2) = \log(-2)$	LOG (−) 2 ENTER	ERROR

Note that the calculator displays an error message when you try to evaluate $\log(-2)$ because there is no real number power to which 10 can be raised to obtain -2.

✓ **Checkpoint**  Audio-video solution in English & Spanish at LarsonPrecalculus.com

Use a calculator to evaluate the function $f(x) = \log x$ at each value of x.

a. $x = 275$ **b.** $x = -\frac{1}{2}$ **c.** $x = \frac{1}{2}$

>>> **TECHNOLOGY**

In Example 2(c), note that some graphing utilities do not give an error message for $\log(-2)$. Instead the graphing utility will display an imaginary number. In this text, the domain of a logarithmic function is the set of *positive real numbers*.

The definition of the logarithmic function with base a leads to several properties.

Properties of Logarithms

1. $\log_a 1 = 0$ because $a^0 = 1$.

2. $\log_a a = 1$ because $a^1 = a$.

3. $\log_a a^x = x$ and $a^{\log_a x} = x$ Inverse Properties

4. If $\log_a x = \log_a y$, then $x = y$.   One-to-One Property

---

EXAMPLE 3   **Using Properties of Logarithms**

a. To simplify $\log_4 1$, use Property 1 to write $\log_4 1 = 0$.

b. To simplify $\log_{\sqrt{7}} \sqrt{7}$, use Property 2 to write $\log_{\sqrt{7}} \sqrt{7} = 1$.

c. To simplify $6^{\log_6 20}$, use Property 3 to write $6^{\log_6 20} = 20$.

d. To solve $\log_3 x = \log_3 12$ for x, use the One-to-One Property.

$\log_3 x = \log_3 12$ Write original equation.

$x = 12$ One-to-One Property

e. To solve $\log(2x + 1) = \log 3x$ for x, use the One-to-One Property.

$\log(2x + 1) = \log 3x$ Write original equation.

$2x + 1 = 3x$ One-to-One Property

$1 = x$ Subtract $2x$ from each side.

✓ **Checkpoint** 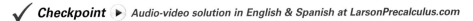 Audio-video solution in English & Spanish at LarsonPrecalculus.com

a. Simplify $\log_9 9$.

b. Simplify $20^{\log_{20} 3}$.

c. Simplify $\log_{\sqrt{3}} 1$.

d. Solve $\log(x - 5) = \log 20$ for x.

GO DIGITAL

Graphs of Logarithmic Functions

To sketch the graph of $y = \log_a x$, use the fact that the graphs of inverse functions are reflections of each other in the line $y = x$.

Figure 5.6

EXAMPLE 4   Graphing Exponential and Logarithmic Functions

In the same coordinate plane, sketch the graph of each function.

a. $f(x) = 2^x$ **b.** $g(x) = \log_2 x$

Solution

a. For $f(x) = 2^x$, construct a table of values. By plotting these points and connecting them with a smooth curve, you obtain the graph shown in Figure 5.6.

x	-2	-1	0	1	2	3
$f(x) = 2^x$	$\frac{1}{4}$	$\frac{1}{2}$	1	2	4	8

b. Because $g(x) = \log_2 x$ is the inverse function of $f(x) = 2^x$, the graph of g is obtained by plotting the points $(f(x), x)$ and connecting them with a smooth curve. The graph of g is a reflection of the graph of f in the line $y = x$, as shown in Figure 5.6.

✓ **Checkpoint** ▶ *Audio-video solution in English & Spanish at LarsonPrecalculus.com*

In the same coordinate plane, sketch the graphs of (a) $f(x) = 8^x$ and (b) $g(x) = \log_8 x$.

EXAMPLE 5   Sketching the Graph of a Logarithmic Function

Sketch the graph of $f(x) = \log x$. Identify the vertical asymptote.

Solution   Begin by constructing a table of values. Note that some of the values can be obtained without a calculator by using the properties of logarithms. Others require a calculator.

	Without calculator				With calculator		
x	$\frac{1}{100}$	$\frac{1}{10}$	1	10	2	5	8
$f(x) = \log x$	-2	-1	0	1	0.301	0.699	0.903

Next, plot the points and connect them with a smooth curve, as shown in the figure below. The vertical asymptote is $x = 0$ (y-axis).

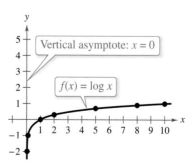

✓ **Checkpoint** ▶ *Audio-video solution in English & Spanish at LarsonPrecalculus.com*

Sketch the graph of $f(x) = \log_3 x$ by constructing a table of values without using a calculator. Identify the vertical asymptote.

GO DIGITAL

The graph in Example 5 is typical for functions of the form $f(x) = \log_a x, a > 1$. They have one x-intercept and one vertical asymptote. Notice how slowly the graph rises for $x > 1$. Here are the basic characteristics of logarithmic graphs.

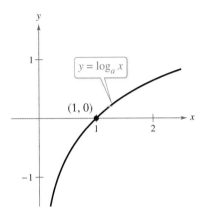

Graph of $y = \log_a x, a > 1$

- Domain: $(0, \infty)$
- Range: $(-\infty, \infty)$
- x-intercept: $(1, 0)$
- Increasing on $(0, \infty)$
- One-to-one, therefore has an inverse function
- y-axis is a vertical asymptote
  ($\log_a x \to -\infty$ as $x \to 0^+$).
- Continuous
- Reflection of graph of $y = a^x$ in the line $y = x$

Some basic characteristics of the graph of $f(x) = a^x$ are listed below to illustrate the inverse relation between $f(x) = a^x$ and $g(x) = \log_a x$.

- Domain: $(-\infty, \infty)$ • Range: $(0, \infty)$
- y-intercept: $(0, 1)$ • x-axis is a horizontal asymptote ($a^x \to 0$ as $x \to -\infty$).

The next example uses the graph of $y = \log_a x$ to sketch the graphs of functions of the form $f(x) = b \pm \log_a(x + c)$.

EXAMPLE 6   Shifting Graphs of Logarithmic Functions

▶▶▶   *See LarsonPrecalculus.com for an interactive version of this type of example.*

Use the graph of $f(x) = \log x$ to sketch the graph of each function.

a. $g(x) = \log(x - 1)$ **b.** $h(x) = 2 + \log x$

Solution

a. Because $g(x) = \log(x - 1) = f(x - 1)$, the graph of g can be obtained by shifting the graph of f one unit to the right, as shown in Figure 5.7.

b. Because $h(x) = 2 + \log x = 2 + f(x)$, the graph of h can be obtained by shifting the graph of f two units up, as shown in Figure 5.8.

ALGEBRA HELP

Notice that the vertical transformation in Figure 5.8 keeps the y-axis as the vertical asymptote, but the horizontal transformation in Figure 5.7 yields a new vertical asymptote of $x = 1$.

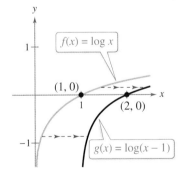

Figure 5.7

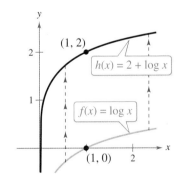

Figure 5.8

✓ *Checkpoint* ▶ *Audio-video solution in English & Spanish at LarsonPrecalculus.com*

Use the graph of $f(x) = \log_3 x$ to sketch the graph of each function.

a. $g(x) = -1 + \log_3 x$ **b.** $h(x) = \log_3(x + 3)$

The Natural Logarithmic Function

By looking back at the graph of the natural exponential function introduced in the preceding section, you will see that $f(x) = e^x$ is one-to-one and so has an inverse function. This inverse function is called the **natural logarithmic function** and is denoted by the special symbol ln x, read as "the natural log of x" or "el en of x."

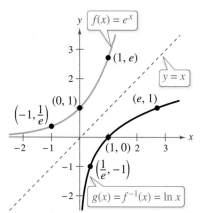

Reflection of graph of $f(x) = e^x$ in the line $y = x$

Figure 5.9

> **The Natural Logarithmic Function**
>
> The function $f(x) = \log_e x = \ln x$, $x > 0$ is called the **natural logarithmic function.**

The equations $y = \ln x$ and $x = e^y$ are equivalent. Note that the natural logarithm ln x is written without a base. The base is understood to be e.

Because the functions $f(x) = e^x$ and $g(x) = \ln x$ are inverse functions of each other, their graphs are reflections of each other in the line $y = x$, as shown in Figure 5.9.

On most calculators, the natural logarithm is denoted by ⟨LN⟩, as illustrated in the next example.

EXAMPLE 7 **Evaluating the Natural Logarithmic Function**

Use a calculator to evaluate the function $f(x) = \ln x$ at each value of x.

a. $x = 2$ **b.** $x = 0.3$ **c.** $x = 1 + \sqrt{2}$ **d.** $x = -1$

Solution

Function Value	Calculator Keystrokes	Display
a. $f(2) = \ln 2$	⟨LN⟩ 2 ⟨ENTER⟩	0.6931472
b. $f(0.3) = \ln 0.3$	⟨LN⟩ .3 ⟨ENTER⟩	−1.2039728
c. $f\left(1 + \sqrt{2}\right) = \ln\left(1 + \sqrt{2}\right)$	⟨LN⟩ ⟨(⟩ 1 ⟨+⟩ ⟨√⟩ 2 ⟨)⟩ ⟨ENTER⟩	0.8813736
d. $f(-1) = \ln(-1)$	⟨LN⟩ ⟨(−)⟩ 1 ⟨ENTER⟩	ERROR

Note that the calculator displays an error message when you try to evaluate $\ln(-1)$. This occurs because the domain of $\ln x$ is the set of *positive real numbers* (see Figure 5.9). So, $\ln(-1)$ is undefined.

✓ *Checkpoint*  *Audio-video solution in English & Spanish at LarsonPrecalculus.com*

Use a calculator to evaluate the function $f(x) = \ln x$ at each value of x.

a. $x = 0.01$ **b.** $x = 4$ **c.** $x = \sqrt{3} + 2$ **d.** $x = \sqrt{3} - 2$ ∎

The properties of logarithms on page 372 are also valid for natural logarithms.

> **Properties of Natural Logarithms**
>
> **1.** $\ln 1 = 0$ because $e^0 = 1$.
>
> **2.** $\ln e = 1$ because $e^1 = e$.
>
> **3.** $\ln e^x = x$ and $e^{\ln x} = x$ Inverse Properties
>
> **4.** If $\ln x = \ln y$, then $x = y$.   One-to-One Property

GO DIGITAL

EXAMPLE 8 **Using Properties of Natural Logarithms**

Use the properties of natural logarithms to simplify each expression.

a. $\ln \dfrac{1}{e}$ **b.** $e^{\ln 5}$ **c.** $\dfrac{\ln 1}{3}$ **d.** $2 \ln e$

Solution

a. $\ln \dfrac{1}{e} = \ln e^{-1} = -1$ Property 3 (Inverse Property)

b. $e^{\ln 5} = 5$ Property 3 (Inverse Property)

c. $\dfrac{\ln 1}{3} = \dfrac{0}{3} = 0$ Property 1

d. $2 \ln e = 2(1) = 2$ Property 2

✓ **Checkpoint** ▶ *Audio-video solution in English & Spanish at LarsonPrecalculus.com*

Use the properties of natural logarithms to simplify each expression.

a. $\ln e^{1/3}$ **b.** $5 \ln 1$ **c.** $\frac{3}{4} \ln e$ **d.** $e^{\ln 7}$

EXAMPLE 9 **Finding the Domains of Logarithmic Functions**

Find the domain of each function.

a. $f(x) = \ln(x - 2)$ **b.** $g(x) = \ln(2 - x)$ **c.** $h(x) = \ln x^2$

Solution

a. Because $\ln(x - 2)$ is defined only when

$$x - 2 > 0$$

it follows that the domain of f is $(2, \infty)$, as shown in Figure 5.10.

b. Because $\ln(2 - x)$ is defined only when

$$2 - x > 0$$

it follows that the domain of g is $(-\infty, 2)$, as shown in Figure 5.11.

c. Because $\ln x^2$ is defined only when

$$x^2 > 0$$

it follows that the domain of h is all real numbers except $x = 0$, as shown in Figure 5.12.

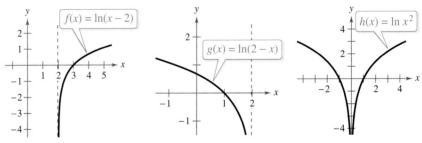

Figure 5.10   Figure 5.11   Figure 5.12

✓ **Checkpoint** ▶ *Audio-video solution in English & Spanish at LarsonPrecalculus.com*

Find the domain of $f(x) = \ln(x + 3)$.

Application

EXAMPLE 10 **Human Memory Model**

Students participating in a psychology experiment attended several lectures on a subject and took an exam. Every month for a year after the exam, the students took a retest to see how much of the material they remembered. The average scores for the group are given by the *human memory model* $f(t) = 75 - 6 \ln(t + 1)$, $0 \le t \le 12$, where t is the time in months.

a. What was the average score on the original exam ($t = 0$)?

b. What was the average score at the end of $t = 2$ months?

c. What was the average score at the end of $t = 6$ months?

Algebraic Solution	**Graphical Solution**
a. The original average score was	**a.** 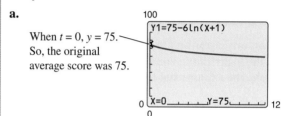
$f(0) = 75 - 6 \ln(0 + 1)$ Substitute 0 for *t*.	When $t = 0$, $y = 75$. So, the original average score was 75.
$= 75 - 6 \ln 1$ Simplify.	
$= 75 - 6(0)$ Property of natural logarithms	
$= 75$. Solution	
b. After 2 months, the average score was	**b.** 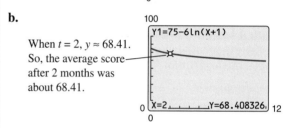
$f(2) = 75 - 6 \ln(2 + 1)$ Substitute 2 for *t*.	When $t = 2$, $y \approx 68.41$. So, the average score after 2 months was about 68.41.
$= 75 - 6 \ln 3$ Simplify.	
$\approx 75 - 6(1.0986)$ Use a calculator.	
≈ 68.41. Solution	
c. After 6 months, the average score was	**c.** 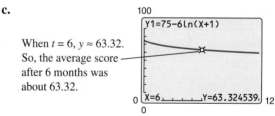
$f(6) = 75 - 6 \ln(6 + 1)$ Substitute 6 for *t*.	When $t = 6$, $y \approx 63.32$. So, the average score after 6 months was about 63.32.
$= 75 - 6 \ln 7$ Simplify.	
$\approx 75 - 6(1.9459)$ Use a calculator.	
≈ 63.32. Solution	

✓ *Checkpoint* ▶ *Audio-video solution in English & Spanish at LarsonPrecalculus.com*

In Example 10, find the average score at the end of (a) $t = 1$ month, (b) $t = 9$ months, and (c) $t = 12$ months. ■

Summarize (Section 5.2)

1. State the definition of the logarithmic function with base *a (page 371)* and make a list of the properties of logarithms *(page 372)*. For examples of evaluating logarithmic functions and using the properties of logarithms, see Examples 1–3.

2. Explain how to graph a logarithmic function *(pages 373 and 374)*. For examples of graphing logarithmic functions, see Examples 4–6.

3. State the definition of the natural logarithmic function and make a list of the properties of natural logarithms *(page 375)*. For examples of evaluating natural logarithmic functions and using the properties of natural logarithms, see Examples 7 and 8.

4. Describe a real-life application that uses a logarithmic function to model and solve a problem *(page 377, Example 10)*.

GO DIGITAL

5.2 Exercises

See CalcChat.com for tutorial help and worked-out solutions to odd-numbered exercises.

GO DIGITAL

Vocabulary and Concept Check

In Exercises 1–3, fill in the blanks.

1. The inverse function of the exponential function $f(x) = a^x$ is the _____ function with base a.

2. The common logarithmic function has base _____.

3. The logarithmic function $f(x) = \ln x$ is the _____ logarithmic function and has base _____.

4. What exponential equation is equivalent to the logarithmic equation $\log_a b = c$?

5. Use the Inverse Properties of logarithms to simplify (a) $\log_a a^x$ and (b) $b^{\log_b y^2}$.

6. State the domain of the natural logarithmic function.

Skills and Applications

Writing an Exponential Equation **In Exercises 7–10,** write the logarithmic equation in exponential form. For example, the exponential form of $\log_5 25 = 2$ is $5^2 = 25$.

7. $\log_4 16 = 2$

8. $\log 1000 = 3$

9. $\log_9 \frac{1}{81} = -2$

10. $\log_{32} 4 = \frac{2}{5}$

Writing a Logarithmic Equation **In Exercises 11–14,** write the exponential equation in logarithmic form. For example, the logarithmic form of $2^3 = 8$ is $\log_2 8 = 3$.

11. $5^3 = 125$

12. $9^{3/2} = 27$

13. $4^{-3} = \frac{1}{64}$

14. $24^0 = 1$

Evaluating a Logarithm **In Exercises 15–20,** evaluate the logarithm at the given value of x without using a calculator.

Function	Value
15. $f(x) = \log_2 x$	$x = 64$
16. $f(x) = \log_{25} x$	$x = 5$
17. $f(x) = \log x$	$x = 10$
18. $f(x) = \log_8 x$	$x = 1$
19. $g(x) = \log_b x$	$x = \sqrt{b}$
20. $g(x) = \log_a x$	$x = a^{-2}$

Evaluating a Common Logarithm on a Calculator **In Exercises 21–24,** use a calculator to evaluate $f(x) = \log x$ at the given value of x. Round your result to three decimal places.

21. $x = 12.5$

22. $x = 96.75$

23. $x = \frac{7}{8}$

24. $x = \frac{1}{500}$

Using Properties of Logarithms **In Exercises 25–28,** use the properties of logarithms to simplify the expression.

25. $\log_8 8$

26. $\log_{7.5} 1$

27. $\log_\pi \pi^2$

28. $5^{\log_5 3}$

Using the One-to-One Property **In Exercises 29–32,** use the One-to-One Property to solve the equation for x.

29. $\log_5(x + 1) = \log_5 6$

30. $\log_2(x - 3) = \log_2 9$

31. $\log 11 = \log(x^2 + 7)$

32. $\log(x^2 + 6x) = \log 27$

Graphing Exponential and Logarithmic Functions **In Exercises 33–36,** sketch the graphs of f and g in the same coordinate plane.

33. $f(t) = 7^t$, $g(t) = \log_7 t$

34. $f(t) = 5^t$, $g(t) = \log_5 t$

35. $f(t) = 6^t$, $g(t) = \log_6 t$

36. $f(t) = 10^t$, $g(t) = \log t$

Matching a Logarithmic Function with Its Graph **In Exercises 37–40,** use the graph of $g(x) = \log_3 x$ to match the given function with its graph. Then describe the relationship between the graphs of f and g. [The graphs are labeled (a), (b), (c), and (d).]

(a)
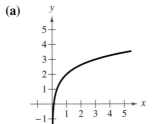

(b)
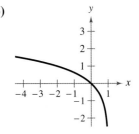

(c)
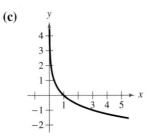

(d)
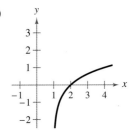

37. $f(x) = \log_3 x + 2$

38. $f(x) = \log_3(x - 1)$

39. $f(x) = \log_3(1 - x)$

40. $f(x) = -\log_3 x$

Sketching the Graph of a Logarithmic Function In Exercises 41–48, find the domain, x-intercept, and vertical asymptote of the logarithmic function and sketch its graph.

41. $f(x) = \log_4 x$
42. $g(x) = \log_6 x$
43. $y = \log_3 x + 1$
44. $h(x) = \log_4(x - 3)$
45. $f(x) = -\log_6(x + 2)$
46. $y = \log_5(x - 1) + 4$
47. $y = \log(x/7)$
48. $y = \log(-2x)$

Writing a Natural Exponential Equation In Exercises 49–52, write the logarithmic equation in exponential form.

49. $\ln \frac{1}{2} = -0.693\ldots$
50. $\ln 7 = 1.945\ldots$
51. $\ln 250 = 5.521\ldots$
52. $\ln 1 = 0$

Writing a Natural Logarithmic Equation In Exercises 53–56, write the exponential equation in logarithmic form.

53. $e^2 = 7.3890\ldots$
54. $e^{-3/4} = 0.4723\ldots$
55. $e^{-4x} = \frac{1}{2}$
56. $e^{2x} = 3$

Evaluating a Logarithmic Function In Exercises 57–60, use a calculator to evaluate $f(x) = 8 \ln x$ at the given value of x. Round your result to three decimal places.

57. $x = 18$
58. $x = 0.74$
59. $x = \sqrt{5}$
60. $x = 2\sqrt{3} - 1$

Using Properties of Natural Logarithms In Exercises 61–66, use the properties of natural logarithms to simplify the expression.

61. $e^{\ln 4}$
62. $\ln(1/e^2)$
63. $2.5 \ln 1$
64. $(\ln e)/\pi$
65. $\ln e^{\ln e}$
66. $e^{\ln(1/e)}$

Graphing a Natural Logarithmic Function In Exercises 67–70, find the domain, x-intercept, and vertical asymptote of the logarithmic function and sketch its graph.

67. $f(x) = \ln(x - 4)$
68. $h(x) = \ln(x + 5)$
69. $g(x) = \ln(-x)$
70. $f(x) = \ln(3 - x)$

Graphing a Natural Logarithmic Function In Exercises 71 and 72, use a graphing utility to graph the function. Use an appropriate viewing window.

71. $f(x) = \ln(x - 1)$
72. $f(x) = 3 \ln x - 1$

Using the One-to-One Property In Exercises 73–76, use the One-to-One Property to solve the equation for x.

73. $\ln(x + 4) = \ln 12$
74. $\ln(x - 7) = \ln 7$
75. $\ln(x^2 - x) = \ln 6$
76. $\ln(x^2 - 2) = \ln 23$

77. Monthly Payment The model
$$t = 16.708 \ln \frac{x}{x - 750}$$
approximates the term of a home mortgage of \$150,000 at a 6% interest rate, where t is the term of the mortgage in years and x is the monthly payment in dollars.

(a) Approximate the terms of a \$150,000 mortgage at 6% when the monthly payment is \$897.72 and when the monthly payment is \$1659.24.

(b) Approximate the total amounts paid over the term of the mortgage with a monthly payment of \$897.72 and with a monthly payment of \$1659.24. What amount of the total is interest costs in each case?

(c) What is the vertical asymptote for the model? Interpret its meaning in the context of the problem.

78. Aeronautics The time t (in minutes) for a commercial jet to climb to an altitude of h feet is given by
$$t = 10 \ln\left(\frac{35,000}{35,000 - h}\right)$$
where 35,000 feet is the cruising altitude.

(a) Determine the domain of the function appropriate for the context of the problem.

(b) Use a graphing utility to graph the function and identify any asymptotes.

(c) Find the amount of time it will take for the jet to reach an altitude of 10,000 feet.

(d) Find the altitude of the jet after 10 minutes.

79. Population The time t (in years) for the world population to double when it is increasing at a continuous rate r (in decimal form) is given by
$$t = \frac{\ln 2}{r}.$$

(a) Complete the table and interpret your results.

r	0.005	0.010	0.015	0.020	0.025	0.030
t						

(b) Use a graphing utility to graph the function.

80. Compound Interest A principal P, invested at $5\frac{1}{2}\%$ and compounded continuously, increases to an amount K times the original principal after t years, where $t = (\ln K)/0.055$.

(a) Complete the table and interpret your results.

K	1	2	4	6	8	10	12
t							

(b) Sketch a graph of the function.

81. Human Memory Model

Students in a mathematics class took an exam and then took a retest monthly with an equivalent exam. The average scores for the class are given by the human memory model

$$f(t) = 80 - 17 \log(t + 1), \quad 0 \le t \le 12$$

where t is the time in months.

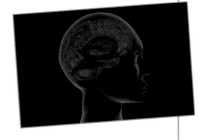

(a) Use a graphing utility to graph the model over the specified domain.

(b) What was the average score on the original exam ($t = 0$)?

(c) What was the average score after 4 months?

(d) What was the average score after 10 months?

Exploring the Concepts

True or False? In Exercises 82 and 83, determine whether the statement is true or false. Justify your answer.

82. You can determine the graph of $f(x) = \log_6 x$ by graphing $g(x) = 6^x$ and reflecting it about the x-axis.

83. The graph of $f(x) = \ln(-x)$ is a reflection of the graph of $h(x) = e^{-x}$ in the line $y = -x$.

84. HOW DO YOU SEE IT? The figure shows the graphs of $f(x) = 3^x$ and $g(x) = \log_3 x$. [The graphs are labeled m and n.]

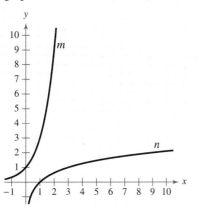

(a) Match each function with its graph.

(b) Given that $f(a) = b$, what is $g(b)$? Explain.

85. Think About It Find the value of the base b so that the graph of $f(x) = \log_b x$ contains the point $\left(\frac{1}{81}, 2\right)$.

86. Writing Explain why $\log_a x$ is defined only for $0 < a < 1$ and $a > 1$.

Error Analysis In Exercises 87 and 88, describe the error.

87. $(1, 0), (2, 1), (8, 3)$

From the ordered pairs (x, y), you can conclude that y is an exponential function of x.

88.

x	1	2	5
y	2	4	32

From the table, you can conclude that y is a logarithmic function of x.

89. Exploration Let $f(x) = \ln x$ and $g(x) = x^{1/n}$.

(a) Use a graphing utility to graph g (for $n = 2$) and f in the same viewing window.

(b) Determine which function is increasing at a greater rate as x approaches infinity.

(c) Repeat parts (a) and (b) for $n = 3, 4$, and 5. What do you notice?

90. Numerical Analysis

(a) Complete the table for the function $f(x) = (\ln x)/x$.

x	1	5	10	10^2	10^4	10^6
$f(x)$						

(b) Use the table in part (a) to determine what value $f(x)$ approaches as x increases without bound.

(c) Use a graphing utility to confirm the result of part (b).

Review & Refresh ▶ Video solutions at LarsonPrecalculus.com

Using Properties of Exponents In Exercises 91–94, simplify each expression.

91. $(x^{-1}y^2)(x^4y^3)$

92. $(a^2b^{-5})^2$

93. $\dfrac{m^0n^2}{m^{-3}n^3}$

94. $\left(\dfrac{(xy)^{-3}}{x^{-4}y^2}\right)^2$

Dividing Polynomials In Exercises 95–98, use long division or synthetic division to divide.

95. $(5x^2 - 17x - 12) \div (x - 4)$

96. $(6x^3 - 16x^2 + 17x - 6) \div (3x - 2)$

97. $(180x - x^4) \div (x - 6)$

98. $(5 - 3x + 2x^2 - x^3) \div (x + 1)$

Finding Compositions of Functions In Exercises 99–102, find (a) $f \circ g$, (b) $g \circ f$, and (c) $g \circ g$.

99. $f(x) = x - 3, g(x) = 4x + 1$

100. $f(x) = x + 2, g(x) = x^2 - 4$

101. $f(x) = \sqrt{x^2 - 9}, \ g(x) = -1/x$

102. $f(x) = \sqrt[4]{x + 2}, \ g(x) = x^4 - 2$

5.3 Properties of Logarithms

Logarithmic functions have many real-life applications. For example, in Exercise 79 on page 386, you will find a logarithmic equation that relates an animal's weight and its lowest stride frequency while galloping.

❯ Use the change-of-base formula to rewrite and evaluate logarithmic expressions.
❯ Use properties of logarithms to evaluate or rewrite logarithmic expressions.
❯ Use properties of logarithms to expand or condense logarithmic expressions.
❯ Use logarithmic functions to model and solve real-life problems.

Change of Base

Most calculators have only two types of log keys, ⌧LOG⌧ for common logarithms (base 10) and ⌧LN⌧ for natural logarithms (base e). Although common logarithms and natural logarithms are the most frequently used, you may occasionally need to evaluate logarithms with other bases. To do this, use the **change-of-base formula.**

> ### Change-of-Base Formula
>
> Let a, b, and x be positive real numbers such that $a \neq 1$ and $b \neq 1$. To convert $\log_a x$ to a different base, use one of the formulas listed below.
>
> **Base b**
> $$\log_a x = \frac{\log_b x}{\log_b a}$$
>
> **Base 10**
> $$\log_a x = \frac{\log x}{\log a}$$
>
> **Base e**
> $$\log_a x = \frac{\ln x}{\ln a}$$

One way to look at the change-of-base formula is that logarithms with base a are *constant multiples* of logarithms with base b. The constant multiplier is

$$\frac{1}{\log_b a}.$$

EXAMPLE 1   Changing Bases Using Common Logarithms

$$\log_4 25 = \frac{\log 25}{\log 4} \qquad \log_a x = \frac{\log x}{\log a}$$

$$\approx \frac{1.39794}{0.60206} \qquad \text{Use a calculator.}$$

$$\approx 2.3219 \qquad \text{Simplify.}$$

✓ *Checkpoint* ▶ Audio-video solution in English & Spanish at LarsonPrecalculus.com

Evaluate $\log_2 12$ using the change-of-base formula and common logarithms.

EXAMPLE 2   Changing Bases Using Natural Logarithms

$$\log_4 25 = \frac{\ln 25}{\ln 4} \qquad \log_a x = \frac{\ln x}{\ln a}$$

$$\approx \frac{3.21888}{1.38629} \qquad \text{Use a calculator.}$$

$$\approx 2.3219 \qquad \text{Simplify.}$$

✓ *Checkpoint* ▶ Audio-video solution in English & Spanish at LarsonPrecalculus.com

Evaluate $\log_2 12$ using the change-of-base formula and natural logarithms.

GO DIGITAL

Properties of Logarithms

You know from the preceding section that the logarithmic function with base a is the *inverse function* of the exponential function with base a. So, it makes sense that the properties of exponents have corresponding properties involving logarithms. For example, the exponential property $a^m a^n = a^{m+n}$ has the corresponding logarithmic property $\log_a(uv) = \log_a u + \log_a v$.

Properties of Logarithms

Let a be a positive number such that $a \neq 1$, let n be a real number, and let u and v be positive real numbers.

	Logarithm with Base a	Natural Logarithm
1. Product Property:	$\log_a(uv) = \log_a u + \log_a v$	$\ln(uv) = \ln u + \ln v$
2. Quotient Property:	$\log_a \dfrac{u}{v} = \log_a u - \log_a v$	$\ln \dfrac{u}{v} = \ln u - \ln v$
3. Power Property:	$\log_a u^n = n \log_a u$	$\ln u^n = n \ln u$

For proofs of the properties listed above, see Proofs in Mathematics on page 418.

EXAMPLE 3 **Using Properties of Logarithms**

Write each logarithm in terms of $\ln 2$ and $\ln 3$.

a. $\ln 6$ **b.** $\ln \dfrac{2}{27}$

Solution

a. $\ln 6 = \ln(2 \cdot 3)$ 　　　　Rewrite 6 as $2 \cdot 3$.

　　　$= \ln 2 + \ln 3$ 　　　Product Property

b. $\ln \dfrac{2}{27} = \ln 2 - \ln 27$ 　　　Quotient Property

　　　　$= \ln 2 - \ln 3^3$ 　　　Rewrite 27 as 3^3.

　　　　$= \ln 2 - 3 \ln 3$ 　　　Power Property

✓ *Checkpoint* ▶ *Audio-video solution in English & Spanish at LarsonPrecalculus.com*

Write each logarithm in terms of $\log 3$ and $\log 5$.

a. $\log 75$ **b.** $\log \dfrac{9}{125}$

EXAMPLE 4 **Using Properties of Logarithms**

Find the exact value of $\log_5 \sqrt[3]{5}$ without using a calculator.

Solution

$\log_5 \sqrt[3]{5} = \log_5 5^{1/3} = \tfrac{1}{3} \log_5 5 = \tfrac{1}{3}(1) = \tfrac{1}{3}$

✓ *Checkpoint* ▶ *Audio-video solution in English & Spanish at LarsonPrecalculus.com*

Find the exact value of $\ln e^6 - \ln e^2$ without using a calculator. ■

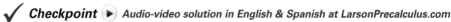

Rewriting Logarithmic Expressions

The properties of logarithms are useful for rewriting logarithmic expressions in forms that simplify the operations of algebra. This is true because these properties convert complicated products, quotients, and exponential forms into simpler sums, differences, and products, respectively.

SKILLS REFRESHER

For a refresher on rewriting expressions with rational exponents, watch the video at *LarsonPrecalculus.com*.

EXAMPLE 5 **Expanding Logarithmic Expressions**

Expand each logarithmic expression.

a. $\log_4 5x^3 y$ **b.** $\ln \dfrac{\sqrt{3x - 5}}{7}$

Solution

a. $\log_4 5x^3 y = \log_4 5 + \log_4 x^3 + \log_4 y$ Product Property

$\qquad\qquad = \log_4 5 + 3 \log_4 x + \log_4 y$ Power Property

b. $\ln \dfrac{\sqrt{3x - 5}}{7} = \ln \dfrac{(3x - 5)^{1/2}}{7}$ Rewrite using rational exponent.

$\qquad\qquad = \ln(3x - 5)^{1/2} - \ln 7$ Quotient Property

$\qquad\qquad = \dfrac{1}{2} \ln(3x - 5) - \ln 7$ Power Property

✓ *Checkpoint* Audio-video solution in English & Spanish at LarsonPrecalculus.com

Expand the expression $\log_3 \dfrac{4x^2}{\sqrt{y}}$. ■

Example 5 uses the properties of logarithms to *expand* logarithmic expressions. Example 6 reverses this procedure and uses the properties of logarithms to *condense* logarithmic expressions.

EXAMPLE 6 **Condensing Logarithmic Expressions**

▶▶▶ *See LarsonPrecalculus.com for an interactive version of this type of example.*

Condense each logarithmic expression.

a. $\frac{1}{2} \log x + 3 \log(x + 1)$ **b.** $2 \ln(x + 2) - \ln x$ **c.** $\frac{1}{3}[\log_2 x + \log_2(x + 1)]$

Solution

a. $\frac{1}{2} \log x + 3 \log(x + 1) = \log x^{1/2} + \log(x + 1)^3$ Power Property

$\qquad\qquad = \log\left[\sqrt{x}\,(x + 1)^3\right]$ Product Property

b. $2 \ln(x + 2) - \ln x = \ln(x + 2)^2 - \ln x$ Power Property

$\qquad\qquad = \ln \dfrac{(x + 2)^2}{x}$ Quotient Property

c. $\frac{1}{3}[\log_2 x + \log_2(x + 1)] = \frac{1}{3} \log_2[x(x + 1)]$ Product Property

$\qquad\qquad = \log_2[x(x + 1)]^{1/3}$ Power Property

$\qquad\qquad = \log_2 \sqrt[3]{x(x + 1)}$ Rewrite with a radical.

✓ *Checkpoint* Audio-video solution in English & Spanish at LarsonPrecalculus.com

Condense the expression $2[\log(x + 3) - 2 \log(x - 2)]$. ■

GO DIGITAL

Application

One way to determine a possible relationship between the x- and y-values of a set of nonlinear data is to take the natural logarithm of each x-value and each y-value. If the plotted points $(\ln x, \ln y)$ lie on a line, then x and y are related by the equation $\ln y = m \ln x$, where m is the slope of the line.

EXAMPLE 7 Finding a Mathematical Model

The table shows the mean distance x from the sun and the period y (the time it takes a planet to orbit the sun, in years) for each of the six planets that are closest to the sun. In the table, the mean distance is given in astronomical units (where one astronomical unit is defined as Earth's mean distance from the sun). The points from the table are plotted in Figure 5.13. Find an equation that relates y and x.

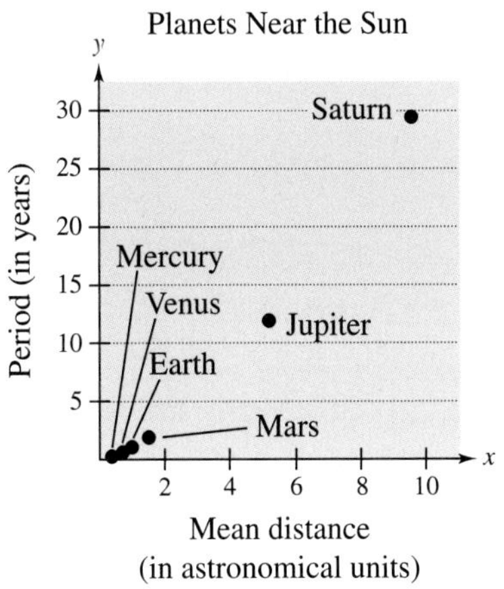

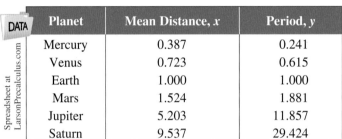

DATA	Planet	Mean Distance, x	Period, y
	Mercury	0.387	0.241
	Venus	0.723	0.615
	Earth	1.000	1.000
	Mars	1.524	1.881
	Jupiter	5.203	11.857
	Saturn	9.537	29.424

Spreadsheet at LarsonPrecalculus.com

Planets Near the Sun

Period (in years)

Mean distance (in astronomical units)

Figure 5.13

Planet	ln x	ln y
Mercury	−0.949	−1.423
Venus	−0.324	−0.486
Earth	0.000	0.000
Mars	0.421	0.632
Jupiter	1.649	2.473
Saturn	2.255	3.382

Solution From Figure 5.13, it is not clear how to find an equation that relates y and x. To solve this problem, make a table of values giving the natural logarithms of all x- and y-values of the data (see the table at the left). Plot each point $(\ln x, \ln y)$. These points appear to lie on a line (see Figure 5.14). Choose two points to determine the slope of the line. Using the points $(0.421, 0.632)$ and $(0, 0)$, the slope of the line is

$$m = \frac{0.632 - 0}{0.421 - 0} \approx 1.5 = \frac{3}{2}.$$

By the point-slope form, the equation of the line is $Y = \frac{3}{2} X$, where $Y = \ln y$ and $X = \ln x$. So, an equation that relates y and x is $\ln y = \frac{3}{2} \ln x$.

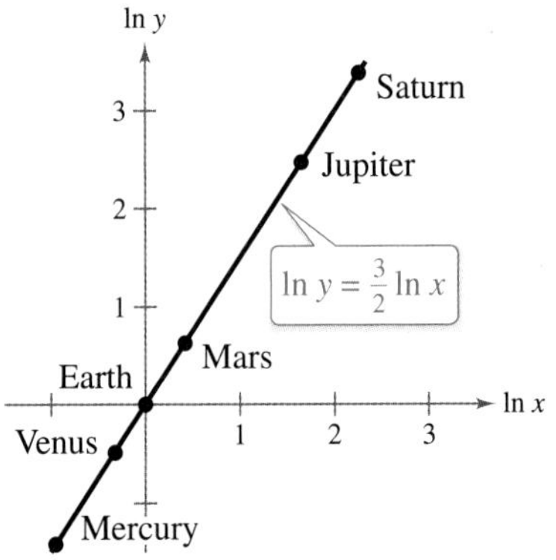

✓ *Checkpoint* ▶ Audio-video solution in English & Spanish at LarsonPrecalculus.com

Find a logarithmic equation that relates y and x for the following ordered pairs.

$(0.37, 0.51)$, $(1.00, 1.00)$, $(2.72, 1.95)$, $(7.39, 3.79)$, $(20.09, 7.39)$ ■

ln y

$$\ln y = \frac{3}{2} \ln x$$

ln x

Figure 5.14

Summarize (Section 5.3)

1. State the change-of-base formula *(page 381)*. For examples of using the change-of-base formula to rewrite and evaluate logarithmic expressions, see Examples 1 and 2.

2. Make a list of the properties of logarithms *(page 382)*. For examples of using the properties of logarithms to evaluate or rewrite logarithmic expressions, see Examples 3 and 4.

3. Explain how to use the properties of logarithms to expand or condense logarithmic expressions *(page 383)*. For examples of expanding and condensing logarithmic expressions, see Examples 5 and 6.

4. Describe an example of how to use a logarithmic function to model and solve a real-life problem *(page 384, Example 7)*.

GO DIGITAL

5.3 Exercises

See CalcChat.com for tutorial help and worked-out solutions to odd-numbered exercises.

GO DIGITAL

Vocabulary and Concept Check

In Exercises 1 and 2, fill in the blanks.

1. To evaluate a logarithm to any base, use the _____ formula.
2. When you consider $\log_a x$ to be a constant multiple of $\log_b x$, the constant multiplier is _____.
3. Is $\log_3 24 = \dfrac{\ln 3}{\ln 24}$ or $\log_3 24 = \dfrac{\ln 24}{\ln 3}$ correct?
4. Name the property of logarithms illustrated by each statement.

    (a) $\ln(uv) = \ln u + \ln v$ (b) $\log_a u^n = n \log_a u$ (c) $\ln \dfrac{u}{v} = \ln u - \ln v$

Skills and Applications

Changing Bases In Exercises 5–8, rewrite the logarithm as a ratio of (a) common logarithms and (b) natural logarithms.

5. $\log_5 16$
6. $\log_{1/5} 4$
7. $\log_x \frac{3}{10}$
8. $\log_{2.6} x$

Using the Change-of-Base Formula In Exercises 9–12, evaluate the logarithm using the change-of-base formula. Round your result to three decimal places.

9. $\log_3 17$
10. $\log_{0.4} 12$
11. $\log_\pi 0.5$
12. $\log_{2/3} 0.125$

Using Properties of Logarithms In Exercises 13–18, use the properties of logarithms to write the logarithm in terms of $\log_3 5$ and $\log_3 7$.

13. $\log_3 35$
14. $\log_3 \frac{5}{7}$
15. $\log_3 \frac{7}{25}$
16. $\log_3 175$
17. $\log_3 \frac{21}{5}$
18. $\log_3 \frac{45}{49}$

Using Properties of Logarithms In Exercises 19–32, find the exact value of the logarithmic expression without using a calculator. (If this is not possible, state the reason.)

19. $\log_3 9$
20. $\log_5 \frac{1}{125}$
21. $\log_6 \sqrt[3]{\frac{1}{6}}$
22. $\log_2 \sqrt[4]{8}$
23. $\log_2(-2)$
24. $\log_3(-27)$
25. $\ln \sqrt[4]{e^3}$
26. $\ln(1/\sqrt{e})$
27. $\ln e^2 + \ln e^5$
28. $2 \ln e^6 - \ln e^5$
29. $\log_5 75 - \log_5 3$
30. $\log_4 2 + \log_4 32$
31. $\log_4 8$
32. $\log_8 16$

Using Properties of Logarithms In Exercises 33–40, approximate the logarithm using the properties of logarithms, given $\log_b 2 \approx 0.3562$, $\log_b 3 \approx 0.5646$, and $\log_b 5 \approx 0.8271$.

33. $\log_b 10$
34. $\log_b \frac{2}{3}$
35. $\log_b 0.04$
36. $\log_b \sqrt{2}$
37. $\log_b 45$
38. $\log_b(3b^2)$
39. $\log_b(2b)^{-2}$
40. $\log_b \sqrt[3]{3b}$

Expanding a Logarithmic Expression In Exercises 41–62, use the properties of logarithms to expand the expression as a sum, difference, and/or constant multiple of logarithms. (Assume all variables are positive.)

41. $\ln 7x$
42. $\log_3 13z$
43. $\log_8 x^4$
44. $\ln(xy)^3$
45. $\log_5 \dfrac{5}{x}$
46. $\log_6 \dfrac{w^2}{v}$
47. $\ln \sqrt{z}$
48. $\ln \sqrt[3]{t}$
49. $\ln xyz^2$
50. $\log_4 11b^2c$
51. $\ln z(z-1)^2, \quad z > 1$
52. $\ln \dfrac{x^2 - 1}{x^3}, \quad x > 1$
53. $\log_2 \dfrac{\sqrt{a^2 - 4}}{7}, \quad a > 2$
54. $\ln \dfrac{3}{\sqrt{x^2 + 1}}$
55. $\log_5 \dfrac{x^2}{y^2 z^3}$
56. $\log_{10} \dfrac{xy^4}{z^5}$
57. $\ln \sqrt{\dfrac{x^2}{y^3}}$
58. $\ln \sqrt[3]{\dfrac{yz}{x^2}}$
59. $\ln x^2 \sqrt{\dfrac{y}{z}}$
60. $\log_2 x^4 \sqrt{\dfrac{y}{z^3}}$
61. $\ln \sqrt[4]{x^3(x^2 + 3)}$
62. $\ln \sqrt{x^2(x + 2)}$

Condensing a Logarithmic Expression **In Exercises 63–76, condense the expression to the logarithm of a single quantity.**

63. $\ln 3 + \ln x$

64. $\log_5 8 - \log_5 t$

65. $\frac{2}{3} \log_7(z - 2)$

66. $-4 \ln 3x$

67. $\log_3 5x - 4 \log_3 x$

68. $2 \log_2 x + 4 \log_2 y$

69. $\log x + 2 \log(x + 1)$

70. $2 \ln 8 - 5 \ln(z - 4)$

71. $\log x - 2 \log y + 3 \log z$

72. $3 \log_3 x + \frac{1}{4} \log_3 y - 4 \log_3 z$

73. $\ln x - [\ln(x + 1) + \ln(x - 1)]$

74. $4[\ln z + \ln(z + 5)] - 2 \ln(z - 5)$

75. $\frac{1}{2}[2 \ln(x + 3) + \ln x - \ln(x^2 - 1)]$

76. $\frac{1}{2}[\log_4(x + 1) + 2 \log_4(x - 1)] + 6 \log_4 x$

Curve Fitting **In Exercises 77 and 78, find a logarithmic equation that relates y and x.**

77.

x	1	2	3	4	5	6
y	2.5	2.102	1.9	1.768	1.672	1.597

78.

x	1	2	3	4	5	6
y	1	1.189	1.316	1.414	1.495	1.565

79. Stride Frequency of Animals

Four-legged animals run with two different types of motion: trotting and galloping. An animal that is trotting has at least one foot on the ground at all times, whereas an animal that is galloping

has all four feet off the ground at some point in its stride. The number of strides per minute at which an animal breaks from a trot to a gallop depends on the weight of the animal. Use the table to find a logarithmic equation that relates an animal's weight x (in pounds) and its lowest stride frequency while galloping y (in strides per minute).

DATA	Weight, x	Stride Frequency, y
	25	191.5
	35	182.7
	50	173.8
	75	164.2
	500	125.9
	1000	114.2

Spreadsheet at LarsonPrecalculus.com

80. Nail Length The approximate lengths and diameters (in inches) of bright common wire nails are shown in the table. Find a logarithmic equation that relates the diameter y of a bright common wire nail to its length x.

Length, x	Diameter, y
2	0.113
3	0.148
4	0.192
5	0.225
6	0.262

81. Comparing Models A cup of water at an initial temperature of 78°C is placed in a room at a constant temperature of 21°C. The temperature of the water is measured every 5 minutes during a half-hour period. The results are recorded as ordered pairs of the form (t, T), where t is the time (in minutes) and T is the temperature (in degrees Celsius).

$(0, 78.0°)$, $(5, 66.0°)$, $(10, 57.5°)$, $(15, 51.2°)$, $(20, 46.3°)$, $(25, 42.4°)$, $(30, 39.6°)$

(a) Subtract the room temperature from each of the temperatures in the ordered pairs. Use a graphing utility to plot the data points (t, T) and $(t, T - 21)$.

(b) An exponential model for the data $(t, T - 21)$ is $T - 21 = 54.4(0.964)^t$. Solve for T and graph the model. Compare the result with the plot of the original data.

(c) Use the graphing utility to plot the points $(t, \ln(T - 21))$ and observe that the points appear to be linear. Use the *regression* feature of the graphing utility to fit a line to these data. This resulting line has the form $\ln(T - 21) = at + b$, which is equivalent to $e^{\ln(T-21)} = e^{at+b}$. Solve for T, and verify that the result is equivalent to the model in part (b).

(d) Fit a rational model to the data. Take the reciprocals of the y-coordinates of the revised data points to generate the points

$$\left(t, \frac{1}{T - 21}\right).$$

Use the graphing utility to graph these points and observe that they appear to be linear. Use the *regression* feature of the graphing utility to fit a line to these data. The resulting line has the form

$$\frac{1}{T - 21} = at + b.$$

Solve for T, and use the graphing utility to graph the rational function and the original data points.

82. Writing Write a short paragraph explaining why the transformations of the data in Exercise 81 were necessary to obtain the models. Why did taking the logarithms of the temperatures lead to a linear scatter plot? Why did taking the reciprocals of the temperatures lead to a linear scatter plot?

Exploring the Concepts

True or False? In Exercises 83–88, determine whether the statement is true or false given that $f(x) = \ln x$. Justify your answer.

83. $f(0) = 0$

84. $f(ax) = f(a) + f(x), \quad a > 0, \quad x > 0$

85. $f(x - 2) = f(x) - f(2), \quad x > 2$

86. $\sqrt{f(x)} = \frac{1}{2}f(x)$

87. If $f(u) = 2f(v)$, then $v = u^2$.

88. If $f(x) < 0$, then $0 < x < 1$.

Error Analysis In Exercises 89 and 90, describe the error.

89. $(\ln e)^2 = 2(\ln e)$
$$= 2(1)$$
$$= 2 \quad ✗$$

90. $\log_2 8 = \log_2(4 + 4)$
$$= \log_2 4 + \log_2 4$$
$$= \log_2 2^2 + \log_2 2^2$$
$$= 2 + 2$$
$$= 4 \quad ✗$$

91. Graphical Reasoning Use a graphing utility to graph the functions $y_1 = \ln x - \ln(x - 3)$ and $y_2 = \ln \dfrac{x}{x - 3}$ in the same viewing window. Does the graphing utility show the functions with the same domain? If not, explain why some numbers are in the domain of one function but not the other.

92. HOW DO YOU SEE IT? The figure shows the graphs of $y = \ln x$, $y = \ln x^2$, $y = \ln 2x$, and $y = \ln 2$. Match each function with its graph. (The graphs are labeled A through D.) Explain.

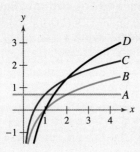

Comparing Logarithmic Quantities In Exercises 93 and 94, determine which (if any) of the logarithmic expressions are equal. Justify your answer.

93. $\dfrac{\log_2 32}{\log_2 4}, \quad \log_2 \dfrac{32}{4}, \quad \log_2 32 - \log_2 4$

94. $\log_7 \sqrt{70}, \quad \log_7 35, \quad \frac{1}{2} + \log_7 \sqrt{10}$

Using the Change-of-Base Formula In Exercises 95–98, use the change-of-base formula to rewrite the logarithm as a ratio of logarithms. Then use a graphing utility to graph the ratio.

95. $f(x) = \log_2 x$

96. $f(x) = \log_{1/2} x$

97. $f(x) = \log_{1/4} x$

98. $f(x) = \log_{11.8} x$

99. Think About It For which integers between 1 and 20 can you approximate natural logarithms, given the values $\ln 2 \approx 0.6931$, $\ln 3 \approx 1.0986$, and $\ln 5 \approx 1.6094$? Approximate these logarithms. (Do not use a calculator.)

100. Proof Prove that $\dfrac{\log_a x}{\log_{a/b} x} = 1 + \log_a \dfrac{1}{b}$.

Review & Refresh ▶ Video solutions at LarsonPrecalculus.com

Solving an Equation In Exercises 101–108, solve the equation. Check your answers.

101. $x(x + 5) = 24$

102. $x(x - 4) = 12$

103. $x - 4\sqrt{x} + 3 = 0$

104. $2x - 3\sqrt{x} - 2 = 0$

105. $5^x = 625$

106. $\left(\frac{1}{4}\right)^x = 8$

107. $e^{x+5} = e^2$

108. $e^{x^2} = e$

Solving a Quadratic Equation by Factoring In Exercises 109–114, solve the quadratic equation by factoring.

109. $x^2 - 5x = 0$

110. $4x^2 + x = 0$

111. $x^2 = 3x + 4$

112. $x^2 - 5x = 50$

113. $x^2 + 8x + 16 = 0$

114. $4x^2 + 9 = 12x$

Determining Solution Points In Exercises 115–120, determine whether each point lies on the graph of the equation.

	Equation	Points	
115.	$y = \sqrt{5 - x}$	(a) $(1, 2)$	(b) $(5, 0)$
116.	$y = 3 - 2x^2$	(a) $(-1, 1)$	(b) $(-2, 11)$
117.	$y = |x - 1| + 2$	(a) $(2, 3)$	(b) $(-1, 0)$
118.	$x^2 + y^2 = 20$	(a) $(3, -2)$	(b) $(-4, 2)$
119.	$y = x^{3/2} + 2$	(a) $(4, 10)$	(b) $(-4, -6)$
120.	$y = \ln(x + 5)$	(a) $(e - 5, 1)$	(b) $(e^3 - 5, 3)$

5.4 Exponential and Logarithmic Equations

Exponential and logarithmic equations have many life science applications. For example, Exercise 79 on page 396 uses an exponential function to model the beaver population in a given area.

❯ Solve simple exponential and logarithmic equations.
❯ Solve more complicated exponential equations.
❯ Solve more complicated logarithmic equations.
❯ Use exponential and logarithmic equations to model and solve real-life problems.

Introduction

So far in this chapter, you have studied the definitions, graphs, and properties of exponential and logarithmic functions. In this section, you will study procedures for *solving equations* involving exponential and logarithmic expressions.

There are two basic strategies for solving exponential or logarithmic equations. The first is based on the One-to-One Properties and was used to solve simple exponential and logarithmic equations in Sections 5.1 and 5.2. The second is based on the Inverse Properties. For $a > 0$ and $a \neq 1$, the properties below are true for all x and y for which $\log_a x$ and $\log_a y$ are defined.

One-to-One Properties

$a^x = a^y$ if and only if $x = y$.

$\log_a x = \log_a y$ if and only if $x = y$.

Inverse Properties

$a^{\log_a x} = x$

$\log_a a^x = x$

 EXAMPLE 1   **Solving Simple Equations**

Original Equation	Rewritten Equation	Solution	Property
a. $2^x = 32$	$2^x = 2^5$	$x = 5$	One-to-One
b. $\ln x - \ln 3 = 0$	$\ln x = \ln 3$	$x = 3$	One-to-One
c. $\left(\dfrac{1}{3}\right)^x = 9$	$3^{-x} = 3^2$	$x = -2$	One-to-One
d. $e^x = 7$	$\ln e^x = \ln 7$	$x = \ln 7$	Inverse
e. $\ln x = -3$	$e^{\ln x} = e^{-3}$	$x = e^{-3}$	Inverse
f. $\log x = -1$	$10^{\log x} = 10^{-1}$	$x = 10^{-1} = \dfrac{1}{10}$	Inverse
g. $\log_3 x = 4$	$3^{\log_3 x} = 3^4$	$x = 81$	Inverse

✓ *Checkpoint* *Audio-video solution in English & Spanish at LarsonPrecalculus.com*

Solve each equation for x.

a. $2^x = 512$ **b.** $\log_6 x = 3$ **c.** $5 - e^x = 0$ **d.** $9^x = \dfrac{1}{3}$ ∎

ALGEBRA HELP

Remember that the natural logarithmic function has a base of e.

Strategies for Solving Exponential and Logarithmic Equations

1. Rewrite the original equation in a form that allows the use of the One-to-One Properties of exponential or logarithmic functions.

2. Rewrite an *exponential* equation in logarithmic form and apply the Inverse Property of logarithmic functions.

3. Rewrite a *logarithmic* equation in exponential form and apply the Inverse Property of exponential functions.

GO DIGITAL

Solving Exponential Equations

EXAMPLE 2 **Solving an Exponential Equation**

$$e^{-x^2} = e^{-3x-4}$$ Original equation.

$$-x^2 = -3x - 4$$ One-to-One Property

$$x^2 - 3x - 4 = 0$$ Write in general form.

$$(x + 1)(x - 4) = 0$$ Factor.

$$x + 1 = 0 \implies x = -1$$ Set 1st factor equal to 0.

$$x - 4 = 0 \implies x = 4$$ Set 2nd factor equal to 0.

The solutions are $x = -1$ and $x = 4$. Check these in the original equation.

✓ *Checkpoint* ▶ *Audio-video solution in English & Spanish at LarsonPrecalculus.com*

Solve $e^{2x} = e^{x^2-8}$.

EXAMPLE 3 **Solving Exponential Equations**

Solve each equation and approximate the result to three decimal places.

a. $3(2^x) = 42$ **b.** $e^x + 5 = 60$

Solution

a. $3(2^x) = 42$ Write original equation.

$$2^x = 14$$ Divide each side by 3.

$$\log_2 2^x = \log_2 14$$ Take log (base 2) of each side.

$$x = \log_2 14$$ Inverse Property

$$x = \frac{\ln 14}{\ln 2}$$ Change-of-base formula

$$x \approx 3.807$$ Use a calculator.

The solution is $x = \log_2 14 \approx 3.807$. Check this in the original equation.

b. $e^x + 5 = 60$ Write original equation.

$$e^x = 55$$ Subtract 5 from each side.

$$\ln e^x = \ln 55$$ Take natural log of each side.

$$x = \ln 55$$ Inverse Property

$$x \approx 4.007$$ Use a calculator.

The solution is $x = \ln 55 \approx 4.007$. Check this in the original equation.

✓ *Checkpoint* ▶ *Audio-video solution in English & Spanish at LarsonPrecalculus.com*

Solve each equation and approximate the result to three decimal places.

a. $2(5^x) = 32$ **b.** $e^x - 7 = 23$ ■

In Example 3(a), the exact solution is $x = \log_2 14$, and the approximate solution is $x \approx 3.807$. An exact solution is preferred when it is an intermediate step in a larger problem. For a final answer, an approximate solution is more practical.

ALGEBRA HELP

Another way to solve Example 3(a) is by taking the natural log of each side and then applying the Power Property.

$$3(2^x) = 42$$

$$2^x = 14$$

$$\ln 2^x = \ln 14$$

$$x \ln 2 = \ln 14$$

$$x = \frac{\ln 14}{\ln 2}$$

$$x \approx 3.807$$

EXAMPLE 4   **Solving an Exponential Equation**

$2(3^{2t-5}) - 4 = 8$	Original equation.
$2(3^{2t-5}) = 12$	Add 4 to each side.
$3^{2t-5} = 6$	Divide each side by 2.
$\log_3 3^{2t-5} = \log_3 6$	Take log (base 3) of each side.
$2t - 5 = \log_3 6$	Inverse Property
$2t = 5 + \log_3 6$	Add 5 to each side.
$t = \dfrac{5}{2} + \dfrac{1}{2}\log_3 6$	Divide each side by 2.
$t = \dfrac{5}{2} + \dfrac{1}{2}\left(\dfrac{\ln 6}{\ln 3}\right)$	Change-of-base formula
$t \approx 3.315$	Use a calculator.

The solution is $t = \dfrac{5}{2} + \dfrac{1}{2}\log_3 6 \approx 3.315$. Check this in the original equation.

✓ *Checkpoint* ▶ *Audio-video solution in English & Spanish at LarsonPrecalculus.com*

Solve $6(2^{t+5}) + 4 = 22$ and approximate the result to three decimal places. ■

When an equation involves two or more exponential expressions, you can still use a procedure similar to that demonstrated in Examples 2, 3, and 4. However, it may include additional algebraic techniques.

EXAMPLE 5   **Solving an Exponential Equation of Quadratic Type**

Solve $e^{2x} - 3e^x + 2 = 0$.

Algebraic Solution

$e^{2x} - 3e^x + 2 = 0$	Write original equation.
$(e^x)^2 - 3e^x + 2 = 0$	Write in quadratic form.
$(e^x - 2)(e^x - 1) = 0$	Factor.
$e^x - 2 = 0$	Set 1st factor equal to 0.
$e^x = 2$	Add 2 to each side.
$x = \ln 2$	Inverse Property
$e^x - 1 = 0$	Set 2nd factor equal to 0.
$e^x = 1$	Add 1 to each side.
$x = \ln 1$	Inverse Property
$x = 0$	Property of natural logarithms

The solutions are $x = \ln 2 \approx 0.693$ and $x = 0$. Check these in the original equation.

Graphical Solution

Use a graphing utility to graph $y = e^{2x} - 3e^x + 2$ and then find the zeros.

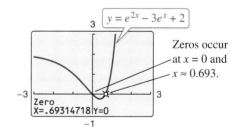

$y = e^{2x} - 3e^x + 2$

Zeros occur at $x = 0$ and $x \approx 0.693$.

Zero
X=.69314718 Y=0

So, the solutions are $x = 0$ and $x \approx 0.693$.

✓ *Checkpoint* ▶ *Audio-video solution in English & Spanish at LarsonPrecalculus.com*

Solve $e^{2x} - 7e^x + 12 = 0$. ■

GO DIGITAL

Solving Logarithmic Equations

To solve a logarithmic equation, write it in exponential form. This procedure is called *exponentiating* each side of an equation.

$$\ln x = 3 \qquad \text{Logarithmic form}$$

$$e^{\ln x} = e^3 \qquad \text{Exponentiate each side.}$$

$$x = e^3 \qquad \text{Exponential form}$$

EXAMPLE 6 **Solving Logarithmic Equations**

a. $\ln x = 2$ Original equation

$\quad e^{\ln x} = e^2$ Exponentiate each side.

$\quad\quad x = e^2$ Inverse Property

b. $\log_3(5x - 1) = \log_3(x + 7)$ Original equation

$\quad\quad\quad 5x - 1 = x + 7$ One-to-One Property

$\quad\quad\quad\quad\quad x = 2$ Solve for x.

c. $\log_6(3x + 14) - \log_6 5 = \log_6 2x$ Original equation

$$\log_6\left(\frac{3x + 14}{5}\right) = \log_6 2x \qquad \text{Quotient Property of Logarithms}$$

$$\frac{3x + 14}{5} = 2x \qquad \text{One-to-One Property}$$

$$3x + 14 = 10x \qquad \text{Multiply each side by 5.}$$

$$x = 2 \qquad \text{Solve for } x.$$

✓ *Checkpoint* ▶ Audio-video solution in English & Spanish at *LarsonPrecalculus.com*

Solve each equation.

a. $\ln x = \dfrac{2}{3}$ **b.** $\log_2(2x - 3) = \log_2(x + 4)$ **c.** $\log 4x - \log(12 + x) = \log 2$

EXAMPLE 7 **Solving a Logarithmic Equation**

Solve $5 + 2 \ln x = 4$ and approximate the result to three decimal places.

Algebraic Solution

$$5 + 2 \ln x = 4 \qquad \text{Write original equation.}$$

$$2 \ln x = -1 \qquad \text{Subtract 5 from each side.}$$

$$\ln x = -\frac{1}{2} \qquad \text{Divide each side by 2.}$$

$$e^{\ln x} = e^{-1/2} \qquad \text{Exponentiate each side.}$$

$$x = e^{-1/2} \qquad \text{Inverse Property}$$

$$x \approx 0.607 \qquad \text{Use a calculator.}$$

Graphical Solution

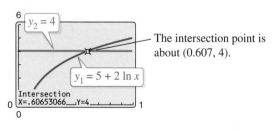

The intersection point is about $(0.607, 4)$.

So, the solution is $x \approx 0.607$.

✓ *Checkpoint* ▶ Audio-video solution in English & Spanish at *LarsonPrecalculus.com*

Solve $7 + 3 \ln x = 5$ and approximate the result to three decimal places. ■

> **EXAMPLE 8** **Solving a Logarithmic Equation**

Solve $2 \log_5 3x = 4$.

Solution

$$2 \log_5 3x = 4 \qquad \text{Write original equation.}$$

$$\log_5 3x = 2 \qquad \text{Divide each side by 2.}$$

$$5^{\log_5 3x} = 5^2 \qquad \text{Exponentiate each side (base 5).}$$

$$3x = 25 \qquad \text{Inverse Property}$$

$$x = \frac{25}{3} \qquad \text{Divide each side by 3.}$$

The solution is $x = 25/3$. Check this in the original equation.

✓ *Checkpoint* ▶ Audio-video solution in English & Spanish at LarsonPrecalculus.com

Solve $3 \log_4 6x = 9$. ■

The domain of a logarithmic function generally does not include all real numbers, so you should be sure to check for extraneous solutions of logarithmic equations, as shown in the next example.

> **EXAMPLE 9** **Checking for Extraneous Solutions**

Solve $\log 5x + \log(x - 1) = 2$.

Algebraic Solution

$$\log 5x + \log(x - 1) = 2 \qquad \text{Write original equation.}$$

$$\log[5x(x - 1)] = 2 \qquad \text{Product Property of Logarithms}$$

$$10^{\log(5x^2 - 5x)} = 10^2 \qquad \text{Exponentiate each side (base 10).}$$

$$5x^2 - 5x = 100 \qquad \text{Inverse Property}$$

$$x^2 - x - 20 = 0 \qquad \text{Write in general form.}$$

$$(x - 5)(x + 4) = 0 \qquad \text{Factor.}$$

$$x - 5 = 0 \qquad \text{Set 1st factor equal to 0.}$$

$$x = 5 \qquad \text{Solve for } x.$$

$$x + 4 = 0 \qquad \text{Set 2nd factor equal to 0.}$$

$$x = -4 \qquad \text{Solve for } x.$$

The solutions appear to be $x = 5$ and $x = -4$. However, when you check these in the original equation, you can see that $x = 5$ is the only solution.

Graphical Solution

First, rewrite the original equation as

$$\log 5x + \log(x - 1) - 2 = 0.$$

Then use a graphing utility to graph the equation

$$y = \log 5x + \log(x - 1) - 2$$

and find the zero(s).

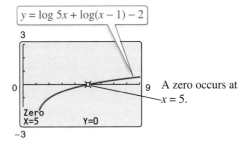

A zero occurs at $x = 5$.

So, the solution is $x = 5$.

✓ *Checkpoint* ▶ Audio-video solution in English & Spanish at LarsonPrecalculus.com

Solve $\log x + \log(x - 9) = 1$. ■

In Example 9, the domain of $\log 5x$ is $x > 0$ and the domain of $\log(x - 1)$ is $x > 1$, so the domain of the original equation is $x > 1$. This means that the solution $x = -4$ is extraneous. The graphical solution supports this conclusion.

Applications

EXAMPLE 10   **Doubling an Investment**

▶▶▶ *See LarsonPrecalculus.com for an interactive version of this type of example.*

You invest \$500 at an annual interest rate of 6.75%, compounded continuously. How long will it take your money to double?

Solution   Using the formula for continuous compounding, the balance is

$$A = Pe^{rt}$$

$$A = 500e^{0.0675t}.$$

To find the time required for the balance to double, let $A = 1000$ and solve the resulting equation for t.

$$500e^{0.0675t} = 1000 \qquad \text{Let } A = 1000.$$

$$e^{0.0675t} = 2 \qquad \text{Divide each side by 500.}$$

$$\ln e^{0.0675t} = \ln 2 \qquad \text{Take natural log of each side.}$$

$$0.0675t = \ln 2 \qquad \text{Inverse Property}$$

$$t = \frac{\ln 2}{0.0675} \qquad \text{Divide each side by 0.0675.}$$

$$t \approx 10.27 \qquad \text{Use a calculator.}$$

The balance in the account will double after approximately 10.27 years. This result is demonstrated graphically below.

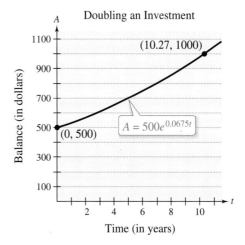

✓ Checkpoint ▶ *Audio-video solution in English & Spanish at LarsonPrecalculus.com*

You invest \$500 at an annual interest rate of 5.25%, compounded continuously. How long will it take your money to double? Compare your result with that of Example 10. ■

In Example 10, an approximate answer of 10.27 years is given. Within the context of the problem, the exact solution

$$t = \frac{\ln 2}{0.0675}$$

does not make sense as an answer.

EXAMPLE 11   **Movie Ticket Sales**

The total ticket sales y (in millions of dollars) of a movie t days after its release can be modeled by

$$y = 36.5 + 141.14 \ln t, \quad t \geq 1.$$

After how many days will the total sales reach \$330 million? Estimate the answer (a) numerically, (b) graphically, and (c) algebraically.

Solution

a. To estimate the number of days numerically, construct a table.

t	1	2	3	4	5	6	7	8
y	37	134	192	232	264	289	311	330

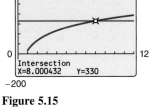

600

0

−200

Figure 5.15

12

Intersection
X=8.000432   Y=330

From the table, the total sales will reach \$330 million after 8 days.

b. Use a graphing utility to graph $y_1 = 36.5 + 141.14 \ln t$ and $y_2 = 330$ in the same viewing window, as shown in Figure 5.15. Using the *intersect* feature, the point of intersection is about $(8, 330)$. So, the total sales will reach \$330 million after 8 days.

c. Let $y = 330$ in the original equation and solve for t.

$36.5 + 141.14 \ln t = y$	Write original equation.
$36.5 + 141.14 \ln t = 330$	Substitute 330 for y.
$141.14 \ln t = 293.5$	Subtract 36.5 from each side.
$\ln t = \dfrac{293.5}{141.14}$	Divide each side by 141.14.
$t = e^{293.5/141.14}$	Inverse Property
$t \approx 8$	Use a calculator.

The solution is $t \approx 8$. So, the total sales will reach \$330 million after 8 days.

✓ *Checkpoint* *Audio-video solution in English & Spanish at LarsonPrecalculus.com*

The total ticket sales y (in millions of dollars) of a movie t days after its release can be modeled by $y = 17 + 46.1 \ln t$, where $t \geq 1$. After how many days will the total sales reach \$100 million? Estimate the answer (a) numerically, (b) graphically, and (c) algebraically. ∎

Summarize   (Section 5.4)

1. State the One-to-One Properties and the Inverse Properties that are used to solve simple exponential and logarithmic equations *(page 388)*. For an example of solving simple exponential and logarithmic equations, see Example 1.

2. Describe strategies for solving exponential equations *(pages 389 and 390)*. For examples of solving exponential equations, see Examples 2–5.

3. Describe strategies for solving logarithmic equations *(pages 391 and 392)*. For examples of solving logarithmic equations, see Examples 6–9.

4. Describe examples of how to use exponential and logarithmic equations to model and solve real-life problems *(pages 393 and 394, Examples 10 and 11)*.

GO DIGITAL

5.4 Exercises

See CalcChat.com for tutorial help and worked-out solutions to odd-numbered exercises.

GO DIGITAL

Vocabulary and Concept Check

In Exercises 1 and 2, fill in the blanks.

1. To solve exponential and logarithmic equations, you can use the One-to-One and Inverse Properties below.

   (a) $a^x = a^y$ if and only if _____.   (b) $\log_a x = \log_a y$ if and only if _____.

   (c) $a^{\log_a x} =$ _____   (d) $\log_a a^x =$ _____

2. An _____ solution does not satisfy the original equation.

3. Describe how a One-to-One Property can be used to solve $5^x = 125$.

4. Do you solve $\log_4 x = 2$ by using a One-to-One Property or an Inverse Property?

Skills and Applications

Determining Solutions In Exercises 5–8, determine whether each x-value is a solution (or an approximate solution) of the equation.

5. $4^{2x-7} = 64$
   (a) $x = 2$
   (b) $x = \frac{1}{2}(\log_4 64 + 7)$

6. $4e^{x-1} = 60$
   (a) $x = 1 + \ln 15$
   (b) $x = \ln 16$

7. $\log_2(x + 3) = 10$
   (a) $x = 1021$
   (b) $x = 10^2 - 3$

8. $\ln(2x + 3) = 5.8$
   (a) $x = \frac{1}{2}(-3 + \ln 5.8)$
   (b) $x = \frac{1}{2}(-3 + e^{5.8})$

Solving a Simple Equation In Exercises 9–18, solve for x.

9. $4^x = 16$
10. $\left(\frac{1}{2}\right)^x = 32$
11. $\ln x - \ln 2 = 0$
12. $\log x - \log 10 = 0$
13. $e^x = 2$
14. $e^x = \frac{1}{3}$
15. $\ln x = -1$
16. $\log x = -2$
17. $\log_4 x = 3$
18. $\log_5 x = \frac{1}{2}$

Approximating a Point of Intersection In Exercises 19 and 20, approximate the point of intersection of the graphs of f and g. Then solve the equation $f(x) = g(x)$ algebraically to verify your approximation.

19. $f(x) = 2^x$, $g(x) = 8$

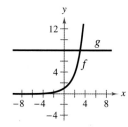

20. $f(x) = \log_3 x$, $g(x) = 2$

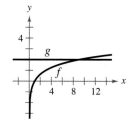

Solving an Exponential Equation In Exercises 21–44, solve the exponential equation algebraically. Approximate the result to three decimal places, if necessary.

21. $e^x = e^{x^2-2}$
22. $e^{x^2-3} = e^{x-2}$
23. $4(3^x) = 20$
24. $4e^x = 91$
25. $e^x - 8 = 31$
26. $5^x + 8 = 26$
27. $3^{2-x} = 400$
28. $7^{-3-x} = 242$
29. $8(10^{3x}) = 12$
30. $8(3^{6-x}) = 40$
31. $e^{3x} = 12$
32. $500e^{-2x} = 125$
33. $7 - 2e^x = 5$
34. $-14 + 3e^x = 11$
35. $6(2^{3x-1}) - 7 = 9$
36. $8(4^{6-2x}) + 13 = 41$
37. $4^x = 5^{x^2}$
38. $3^{x^2} = 7^{6-x}$
39. $e^{2x} - 4e^x - 5 = 0$
40. $e^{2x} - 5e^x + 6 = 0$
41. $\dfrac{1}{1 - e^x} = 5$
42. $\dfrac{100}{1 + e^{2x}} = 1$
43. $\left(1 + \dfrac{0.065}{365}\right)^{365t} = 4$
44. $\left(1 + \dfrac{0.10}{12}\right)^{12t} = 2$

Solving a Logarithmic Equation In Exercises 45–58, solve the logarithmic equation algebraically, if possible. Approximate the result to three decimal places, if necessary.

45. $\ln x = -3$
46. $\ln x - 7 = 0$
47. $2.1 = \ln 6x$
48. $\log 3z = 2$
49. $3 - 4\ln x = 11$
50. $3 + 8\ln x = 7$
51. $6\log_3 0.5x = 11$
52. $4\log(x - 6) = 11$
53. $\ln(x + 5) = \ln(x - 1) - \ln(x + 1)$
54. $\ln(x + 1) - \ln(x - 2) = \ln x$
55. $\log(3x + 4) = \log(x - 10)$
56. $\log_2 x + \log_2(x + 2) = \log_2(x + 6)$
57. $\log_4 x - \log_4(x - 1) = \frac{1}{2}$
58. $\log 8x - \log\left(1 + \sqrt{x}\right) = 2$

Solving an Exponential or Logarithmic Equation In Exercises 59–66, use a graphing utility to graphically solve the equation. Approximate the result to three decimal places. Verify your result algebraically.

59. $5^x = 212$

60. $6e^{1-x} = 25$

61. $8e^{-2x/3} = 11$

62. $e^{0.09t} = 3$

63. $3 - \ln x = 0$

64. $10 - 4\ln(x - 2) = 0$

65. $2\ln(x + 3) = 3$

66. $\ln(x + 1) = 2 - \ln x$

Compound Interest In Exercises 67 and 68, you invest \$2500 in an account at interest rate r, compounded continuously. Find the time required for the amount to (a) double and (b) triple.

67. $r = 0.025$

68. $r = 0.0375$

Algebra of Calculus In Exercises 69–76, solve the equation algebraically. Round your result to three decimal places, if necessary. Verify your answer using a graphing utility.

69. $2x^2 e^{2x} + 2xe^{2x} = 0$

70. $-x^2 e^{-x} + 2xe^{-x} = 0$

71. $-xe^{-x} + e^{-x} = 0$

72. $e^{-2x} - 2xe^{-2x} = 0$

73. $\dfrac{1 + \ln x}{2} = 0$

74. $\dfrac{1 - \ln x}{x^2} = 0$

75. $2x \ln x + x = 0$

76. $2x \ln\left(\dfrac{1}{x}\right) - x = 0$

77. Average Heights The percent m of American males between the ages of 20 and 29 who are under x inches tall is modeled by

$$m(x) = \frac{100}{1 + e^{-0.5536(x - 69.51)}}$$

and the percent f of American females between the ages of 20 and 29 who are under x inches tall is modeled by

$$f(x) = \frac{100}{1 + e^{-0.5834(x - 64.49)}}.$$

(Source: U.S. National Center for Health Statistics)

(a) Use the graph to determine any horizontal asymptotes of the graphs of the functions. Interpret the meaning in the context of the problem.

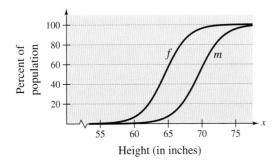

(b) What is the average height of each sex?

78. Effective Yield The *effective yield* of an investment plan is the percent increase in the balance after 1 year. Find the effective yield for each investment plan. Which investment plan has the greatest effective yield? Which investment plan will have the highest balance after 5 years?

(a) The annual interest rate is 7% and is compounded annually.

(b) The annual interest rate is 7% and is compounded continuously.

(c) The annual interest rate is 7% and is compounded quarterly.

(d) The annual interest rate is 7.25% and is compounded quarterly.

79. Ecology

The number N of beavers in a given area after x years can be approximated by

$$N = 5.5(10^{0.23t}), \quad 0 \le t \le 9.$$

Use the model to approximate how many years it will take for the beaver population to reach 78.

80. Ecology The number N of trees of a given species per acre is approximated by the model

$$N = 3500(10^{-0.12x}), \quad 3 \le x \le 30$$

where x is the average diameter of the trees (in inches) 4.5 feet above the ground. Use the model to approximate the average diameter of the trees in a test plot when $N = 22$.

81. Population The population P (in thousands) of the District of Columbia in the years 2009 through 2019 can be modeled by

$$P = 161 \ln t + 236, \quad 9 \le t \le 19$$

where t represents the year, with $t = 9$ corresponding to 2009. During which year did the population of the District of Columbia exceed 660 thousand? *(Source: U.S. Census Bureau)*

82. Population The population P (in thousands) of Montana in the years 2009 through 2019 can be modeled by

$$P = 116 \ln t + 720, \quad 9 \le t \le 19$$

where t represents the year, with $t = 9$ corresponding to 2009. During which year did the population of Montana exceed 995 thousand? *(Source: U.S. Census Bureau)*

83. Finance You are investing P dollars at an annual interest rate of r, compounded continuously, for t years. Which change below results in the highest value of the investment? Explain.

(a) Double the amount you invest.

(b) Double your interest rate.

(c) Double the number of years.

84. Temperature An object at a temperature of 160°C was removed from a furnace and placed in a room at 20°C. The temperature T of the object was measured each hour h and recorded in the table. A model for the data is

$$T = 20 + 140e^{-0.68h}.$$

Hour, h	Temperature, T
0	160°
1	90°
2	56°
3	38°
4	29°
5	24°

Spreadsheet at LarsonPrecalculus.com

(a) The figure below shows the graph of the model. Use the graph to identify the horizontal asymptote of the model and interpret the asymptote in the context of the problem.

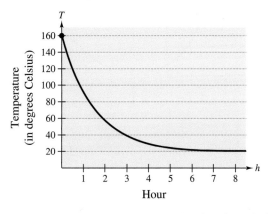

(b) Use the model to approximate the time it took for the object to reach a temperature of 100°C.

Exploring the Concepts

True or False? In Exercises 85–88, rewrite each verbal statement as an equation. Then decide whether the statement is true or false. Justify your answer.

85. The logarithm of the product of two numbers is equal to the sum of the logarithms of the numbers.

86. The logarithm of the sum of two numbers is equal to the product of the logarithms of the numbers.

87. The logarithm of the difference of two numbers is equal to the difference of the logarithms of the numbers.

88. The logarithm of the quotient of two numbers is equal to the difference of the logarithms of the numbers.

89. Think About It Is it possible for a logarithmic equation to have more than one extraneous solution? Explain.

90. HOW DO YOU SEE IT? Solving $\log_3 x + \log_3(x - 8) = 2$ algebraically, the solutions appear to be $x = 9$ and $x = -1$. Use the graph of

$$y = \log_3 x + \log_3(x - 8) - 2$$

to determine whether each value is an actual solution of the equation. Explain.

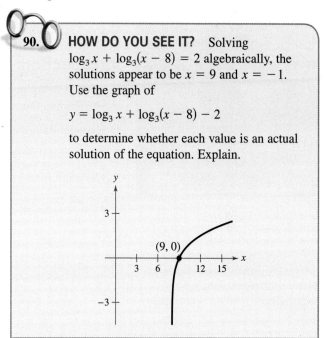

Review & Refresh Video solutions at LarsonPrecalculus.com

Graphical Reasoning In Exercises 91 and 92, find and approximate any vertical and horizontal asymptotes.

91. **92.**

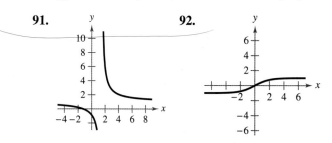

Solving for a Variable In Exercises 93–96, solve the equation for the given variable.

93. $(2e^{-3})^2 = 4^x$

94. $(3x^2)^3 = (9x)^2$

95. $(2k)^4 = k^2(8 + \ln 3)$

96. $5(1 - 3^{x/2}) = 2$

Rewriting with Positive Exponents In Exercises 97 and 98, rewrite the expression with positive exponents. Simplify, if possible.

97. $\dfrac{(5x^{-3})^2}{2^{-2}}$

98. $\left(\dfrac{x^{-2}}{2^{-5}}\right)\left(\dfrac{4}{x}\right)^{-3}$

5.5 Exponential and Logarithmic Models

Exponential growth and decay models can often represent populations. For example, in Exercise 27 on page 406, you will use exponential growth and decay models to compare the populations of several countries.

❯ Recognize the five most common types of models involving exponential and logarithmic functions.
❯ Use exponential growth and decay functions to model and solve real-life problems.
❯ Use Gaussian functions to model and solve real-life problems.
❯ Use logistic growth functions to model and solve real-life problems.
❯ Use logarithmic functions to model and solve real-life problems.

Introduction

The five most common types of mathematical models involving exponential functions and logarithmic functions are listed below.

1. **Exponential growth model:** $y = ae^{bx}, \quad b > 0$

2. **Exponential decay model:** $y = ae^{-bx}, \quad b > 0$

3. **Gaussian model:** $y = ae^{-(x-b)^2/c}$

4. **Logistic growth model:** $y = \dfrac{a}{1 + be^{-rx}}$

5. **Logarithmic models:** $y = a + b \ln x, \quad y = a + b \log x$

The basic shapes of the graphs of these functions are shown below.

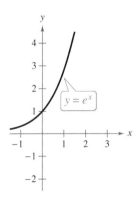

Exponential growth model

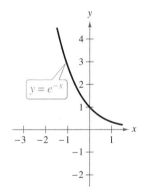

Exponential decay model

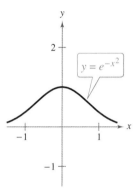

Gaussian model

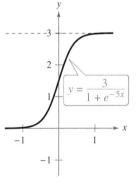

Logistic growth model

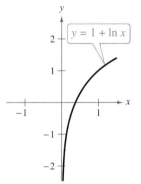

Natural logarithmic model

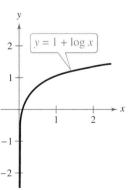

Common logarithmic model

You often gain insight into a situation modeled by an exponential or logarithmic function by identifying and interpreting the asymptotes of the graph of the function. Identify any asymptotes of the graph of each function shown above.

GO DIGITAL

Exponential Growth and Decay

| EXAMPLE 1 | **Retail Sales** |

The U.S. retail e-commerce sales (in billions of dollars) from 2012 through 2018 are shown in the table. A scatter plot of the data is shown at the right.   *(Source: U.S. Census Bureau)*

U.S. Retail e-Commerce

Year	2012	2013	2014	2015	2016	2017	2018
Sales	232	264	303	347	397	461	524

An exponential growth model for the data is $S = 44.73e^{0.1368t}$, where S is the amount of sales (in billions of dollars) and t represents the year, with $t = 12$ corresponding to 2012. Compare the values found using the model with the amounts shown in the table. According to the model, in what year were the sales approximately $690 billion?

Algebraic Solution

The table below compares the actual amounts with the values found using the model, rounded to the nearest billion.

Year	2012	2013	2014	2015	2016	2017	2018
Sales	232	264	303	347	397	461	524
Model	231	265	304	348	399	458	525

To find when the sales are about $690 billion, let $S = 690$ in the model and solve for t.

$$44.73e^{0.1368t} = S \qquad \text{Write original model.}$$

$$44.73e^{0.1368t} = 690 \qquad \text{Substitute 690 for } S.$$

$$e^{0.1368t} = \frac{690}{44.73} \qquad \text{Divide each side by 44.73.}$$

$$\ln e^{0.1368t} = \ln \frac{690}{44.73} \qquad \text{Take natural log of each side.}$$

$$0.1368t = \ln \frac{690}{44.73} \qquad \text{Inverse Property}$$

$$t \approx 20 \qquad \text{Divide each side by 0.1368.}$$

According to the model, the sales were about $690 billion in 2020.

Graphical Solution

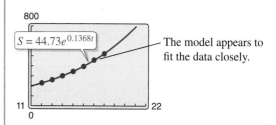

The model appears to fit the data closely.

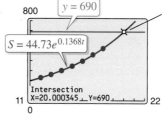

The intersection point of the model and the line $y = 690$ is about $(20, 690)$. So, according to the model, the sales were $690 billion in 2020.

 *Checkpoint* *Audio-video solution in English & Spanish at LarsonPrecalculus.com*

In Example 1, in what year were the sales approximately $795 billion? ■

$\ggg$ TECHNOLOGY

Some graphing utilities have an *exponential regression* feature that can help you find exponential models to represent data. If you have such a graphing utility, use it to find an exponential model for the data given in Example 1. How does your model compare with the model given in Example 1?

GO DIGITAL

In Example 1, the exponential growth model is given. Sometimes you must find such a model. One technique for doing this is shown in Example 2.

EXAMPLE 2 **Modeling Population Growth**

In a research experiment, a population of fruit flies is increasing according to the law of exponential growth. After 2 days there are 100 flies, and after 4 days there are 300 flies. How many flies will there be after 5 days?

Solution Let y be the number of flies at time t (in days). From the given information, you know that $y = 100$ when $t = 2$ and $y = 300$ when $t = 4$. Substituting this information into the model $y = ae^{bt}$ produces

$$100 = ae^{2b} \quad \text{and} \quad 300 = ae^{4b}.$$

To solve for b, solve for a in the first equation.

$$100 = ae^{2b} \qquad \text{Write first equation.}$$

$$\frac{100}{e^{2b}} = a \qquad \text{Solve for } a.$$

Then substitute the result into the second equation.

$$300 = ae^{4b} \qquad \text{Write second equation.}$$

$$300 = \left(\frac{100}{e^{2b}}\right)e^{4b} \qquad \text{Substitute } \tfrac{100}{e^{2b}} \text{ for } a.$$

$$300 = 100e^{2b} \qquad \text{Simplify.}$$

$$3 = e^{2b} \qquad \text{Divide each side by 100.}$$

$$\ln 3 = \ln e^{2b} \qquad \text{Take natural log of each side.}$$

$$\ln 3 = 2b \qquad \text{Inverse Property}$$

$$\frac{1}{2}\ln 3 = b \qquad \text{Solve for } b.$$

Now substitute $\frac{1}{2}\ln 3$ for b in the expression you found for a.

$$a = \frac{100}{e^{2[(1/2)\ln 3]}} \qquad \text{Substitute } \tfrac{1}{2}\ln 3 \text{ for } b.$$

$$= \frac{100}{e^{\ln 3}} \qquad \text{Simplify.}$$

$$= \frac{100}{3} \qquad \text{Inverse Property}$$

$$\approx 33.33 \qquad \text{Divide.}$$

So, with $a \approx 33.33$ and $b = \frac{1}{2}\ln 3 \approx 0.5493$, the exponential growth model is

$$y = 33.33e^{0.5493t}$$

as shown in Figure 5.16. After 5 days, the population will be

$$y = 33.33e^{0.5493(5)} \approx 520 \text{ flies.}$$

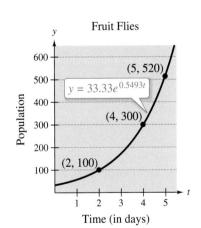

Fruit Flies

$y = 33.33e^{0.5493t}$

(5, 520)

(4, 300)

(2, 100)

Figure 5.16

✓ *Checkpoint* ▶ *Audio-video solution in English & Spanish at LarsonPrecalculus.com*

The number of bacteria in a culture is increasing according to the law of exponential growth. After 1 hour there are 100 bacteria, and after 2 hours there are 200 bacteria. How many bacteria will there be after 3 hours? ■

GO DIGITAL

In living organic material, the ratio of the number of radioactive carbon isotopes (carbon-14) to the number of nonradioactive carbon isotopes (carbon-12) is about 1 to 10^{12}. When organic material dies, its carbon-12 content remains fixed, whereas its radioactive carbon-14 begins to decay with a half-life of about 5700 years. To estimate the age (the number of years since death) of organic material, scientists use the carbon dating model

$$R = \frac{1}{10^{12}} e^{-t/8223}$$ Carbon dating model

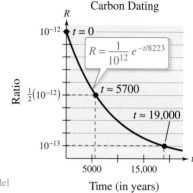

Carbon Dating

where R represents the ratio of carbon-14 to carbon-12 of organic material t years after death. The graph of R is shown above. Note that R decreases as t increases.

EXAMPLE 3   Carbon Dating

The ratio of carbon-14 to carbon-12 in a newly discovered fossil is

$$R = \frac{1}{10^{13}}.$$

Estimate the age of the fossil.

Algebraic Solution

Substitute the given value of R in the carbon dating model. Then solve for t.

$\dfrac{1}{10^{12}} e^{-t/8223} = R$	Write original model.
$\dfrac{e^{-t/8223}}{10^{12}} = \dfrac{1}{10^{13}}$	Substitute $\frac{1}{10^{13}}$ for R.
$e^{-t/8223} = \dfrac{1}{10}$	Multiply each side by 10^{12}.
$\ln e^{-t/8223} = \ln \dfrac{1}{10}$	Take natural log of each side.
$-\dfrac{t}{8223} = \ln \dfrac{1}{10}$	Inverse Property
$t = -8223 \ln \dfrac{1}{10}$	Multiply each side by -8223.
$t \approx 18,934$	Use a calculator.

So, to the nearest thousand years, the age of the fossil is about 19,000 years.

Graphical Solution

Use a graphing utility to graph

$$y_1 = \frac{1}{10^{12}} e^{-x/8223} \quad \text{and} \quad y_2 = \frac{1}{10^{13}}$$

in the same viewing window.

Use the *intersect* feature to estimate that $x \approx 18,934$ when $y = 1/10^{13}$.

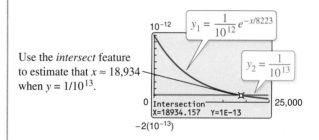

So, to the nearest thousand years, the age of the fossil is about 19,000 years.

✓ *Checkpoint* ▶ Audio-video solution in English & Spanish at LarsonPrecalculus.com

Estimate the age of a newly discovered fossil for which the ratio of carbon-14 to carbon-12 is $R = 1/10^{14}$. ■

GO DIGITAL

The carbon dating model in Example 3 assumed that the carbon-14 to carbon-12 ratio of the fossil was one part in 10,000,000,000,000. Suppose an error in measurement occurred and the actual ratio was only one part in 8,000,000,000,000. The fossil age corresponding to the actual ratio would then be approximately 17,000 years. Try checking this result.

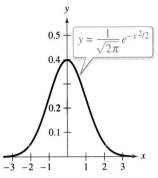

The standard normal curve, like all Gaussian models, is bell shaped.

Figure 5.17

Gaussian Models

As mentioned at the beginning of this section, Gaussian models are of the form

$$y = ae^{-(x-b)^2/c}.$$

This type of model is commonly used in probability and statistics to represent populations that are **normally distributed.** For the *standard normal distribution*, the model takes the form

$$y = \frac{1}{\sqrt{2\pi}}e^{-x^2/2}.$$

The graph of a Gaussian model is called a **bell-shaped curve.** For example, the graph of the standard normal curve is shown in Figure 5.17. Can you see why it is called a bell-shaped curve?

The **average value** of a population can be found from the bell-shaped curve by observing where the maximum *y*-value of the function occurs. The *x*-value corresponding to the maximum *y*-value of the function represents the average value of the independent variable—in this case, *x*.

EXAMPLE 4 **Test Scores**

▶▶▶ *See LarsonPrecalculus.com for an interactive version of this type of example.*

The Scholastic Aptitude Test (SAT) mathematics scores for U.S. high school graduates in 2019 roughly followed the normal distribution.

$$y = 0.0034e^{-(x-528)^2/27,378}, \quad 200 \le x \le 800$$

where *x* is the SAT score for mathematics. Use a graphing utility to graph this function and estimate the average SAT mathematics score. *(Source: The College Board)*

Solution The graph of the function is shown in Figure 5.18(a). On this bell-shaped curve, the average score is the *x*-value corresponding to the maximum *y*-value of the function. Using the *maximum* feature of the graphing utility, you can see that the average mathematics score for high school graduates in 2019 was about 528. Note that 50% of the graduates who took the test earned scores greater than 528, as shown in Figure 5.18(b).

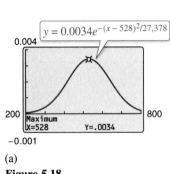

(a)

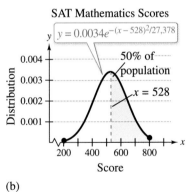

(b)

Figure 5.18

✓ **Checkpoint** ▶ *Audio-video solution in English & Spanish at LarsonPrecalculus.com*

The SAT reading and writing scores for high school graduates in 2019 roughly followed the normal distribution

$$y = 0.0038e^{-(x-531)^2/21,632}, \quad 200 \le x \le 800$$

where *x* is the SAT score for reading and writing. Use a graphing utility to graph this function and estimate the average SAT reading and writing score. *(Source: The College Board)*

GO DIGITAL

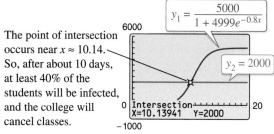

Figure 5.19

Logistic Growth Models

Some populations initially have rapid growth, followed by a declining rate of growth, as illustrated by the graph in Figure 5.19. One model for describing this type of growth pattern is the **logistic curve** given by the function

$$y = \frac{a}{1 + be^{-rx}}$$

where y is the population size and x is the time. An example is a bacteria culture that is initially allowed to grow under ideal conditions and then under less favorable conditions that inhibit growth. A logistic growth curve is also called a **sigmoidal curve.**

EXAMPLE 5   **Spread of a Virus**

On a college campus of 5000 students, one student returns from vacation with a contagious and long-lasting flu virus. The spread of the virus is modeled by

$$y = \frac{5000}{1 + 4999e^{-0.8t}}, \quad t \geq 0$$

where y is the total number of students infected after t days. The college will cancel classes when 40% or more of the students are infected.

a. How many students are infected after 5 days?

b. After how many days will the college cancel classes?

Algebraic Solution

a. After 5 days, the number of students infected is

$$y = \frac{5000}{1 + 4999e^{-0.8(5)}} = \frac{5000}{1 + 4999e^{-4}} \approx 54.$$

b. The college will cancel classes when the number of infected students is $(0.40)(5000) = 2000$.

$$2000 = \frac{5000}{1 + 4999e^{-0.8t}}$$

$$1 + 4999e^{-0.8t} = 2.5$$

$$e^{-0.8t} = \frac{1.5}{4999}$$

$$\ln e^{-0.8t} = \ln \frac{1.5}{4999}$$

$$-0.8t = \ln \frac{1.5}{4999}$$

$$t = -\frac{1}{0.8} \ln \frac{1.5}{4999}$$

$$t \approx 10.14$$

So, after about 10 days, at least 40% of the students will be infected, and the college will cancel classes.

Graphical Solution

a.

Use the *value* feature to estimate that $y \approx 54$ when $x = 5$. So, after 5 days, about 54 students are infected.

6000

Y1=5000/(1+4999e^(−.8X))

$$y = \frac{5000}{1 + 4999e^{-0.8x}}$$

0   20

X=5   Y=54.019085

−1000

b. The college will cancel classes when the number of infected students is $(0.40)(5000) = 2000$. Use a graphing utility to graph

$$y_1 = \frac{5000}{1 + 4999e^{-0.8x}} \quad \text{and} \quad y_2 = 2000$$

in the same viewing window. Use the *intersect* feature of the graphing utility to find the point of intersection of the graphs.

The point of intersection occurs near $x \approx 10.14$. So, after about 10 days, at least 40% of the students will be infected, and the college will cancel classes.

$$y_1 = \frac{5000}{1 + 4999e^{-0.8x}}$$

6000

$$y_2 = 2000$$

0   Intersection   20
X=10.13941   Y=2000

−1000

 Checkpoint ▶ *Audio-video solution in English & Spanish at LarsonPrecalculus.com*

In Example 5, after how many days are 250 students infected?

GO DIGITAL

Logarithmic Models

ALGEBRA HELP

The symbol β is the lowercase Greek letter beta.

The level of sound β (in decibels) with an intensity of I (in watts per square meter) is

$$\beta = 10 \log \frac{I}{I_0}$$

where I_0 represents the faintest sound that can be heard by the human ear and is approximately equal to 10^{-12} watt per square meter.

EXAMPLE 6   **Sound Intensity**

You and your roommate are playing music on your soundbars at the same time and at the same intensity. How much louder is the music when both soundbars are playing than when just one soundbar is playing?

Solution   Let β_1 represent the level of sound when one soundbar is playing and let β_2 represent the level of sound when both soundbars are playing. Using the formula for level of sound, you can express β_1 as

$$\beta_1 = 10 \log \frac{I}{10^{-12}}.$$

Because β_2 represents the level of sound when *two* soundbars are playing at the same intensity, multiply I by 2 in the formula for level of sound and write β_2 as

$$\beta_2 = 10 \log \frac{2I}{10^{-12}}.$$

Next, to determine the increase in loudness, subtract β_1 from β_2.

$$\beta_2 - \beta_1 = 10 \log \frac{2I}{10^{-12}} - 10 \log \frac{1}{10^{-12}}$$

$$= 10 \left(\log \frac{2I}{10^{-12}} - \log \frac{I}{10^{-12}} \right)$$

$$= 10 \left(\log 2 + \log \frac{I}{10^{-12}} - \log \frac{I}{10^{-12}} \right)$$

$$= 10 \log 2$$

$$\approx 3 \text{ decibels}$$

So, the music is about 3 decibels louder when both soundbars are playing than when just one soundbar is playing. Notice that the variable I "drops out" of the equation when it is simplified. This means that the loudness increases by about 3 decibels when both soundbars are playing at the same intensity, regardless of their individual intensities.

✓ *Checkpoint* *Audio-video solution in English & Spanish at LarsonPrecalculus.com*

Two sounds have intensities of $I_1 = 10^{-6}$ watt per square meter and $I_2 = 10^{-9}$ watt per square meter. Use the formula for the level of sound to find the difference in loudness between the two sounds. ■

Summarize   (Section 5.5)

1. State the five most common types of models involving exponential and logarithmic functions *(page 398)*.

2. Describe examples of real-life applications that use exponential growth and decay functions *(pages 399–401, Examples 1–3)*, a Gaussian function *(page 402, Example 4)*, a logistic growth function *(page 403, Example 5)*, and a logarithmic function *(page 404, Example 6)*.

GO DIGITAL

5.5 Exercises

See CalcChat.com for tutorial help and worked-out solutions to odd-numbered exercises.

Vocabulary and Concept Check

In Exercises 1 and 2, fill in the blanks.

1. A logarithmic model has the form _____ or _____.
2. In probability and statistics, Gaussian models commonly represent populations that are _____ _____.

3. What is the difference between an exponential growth model and an exponential decay model?
4. Describe the shape of the graph of a logistic growth model.

Skills and Applications

Solving for a Variable In Exercises 5 and 6, solve for P and solve for t in the compound interest formula (discussed in Section 5.1).

5. $A = Pe^{rt}$

6. $A = P\left(1 + \dfrac{r}{n}\right)^{nt}$

Compound Interest In Exercises 7–12, find the missing values assuming continuously compounded interest.

	Initial Investment	Annual % Rate	Time to Double	Amount After 10 Years
7.	$1000	3.5%		
8.	$750	$10\frac{1}{2}\%$		
9.	$750		$7\frac{3}{4}$ yr	
10.	$500			$1505.00
11.		4.5%		$10,000.00
12.			12 yr	$2000.00

Compound Interest In Exercises 13 and 14, determine the time necessary for P dollars to double when it is invested at interest rate r compounded (a) annually, (b) monthly, (c) daily, and (d) continuously.

13. $r = 10\%$
14. $r = 6.5\%$

15. **Compound Interest** Complete the table for the time t (in years) necessary for P dollars to triple when it is invested at an interest rate r compounded (a) continuously and (b) annually.

r	2%	4%	6%	8%	10%	12%
t						

16. **Modeling Data** Draw scatter plots of the data in Exercise 15. Use the *regression* feature of a graphing utility to find models for the data.

17. **Website Growth** The number of hits a website receives in its first five months of operation are shown in the table.

Month	April	May	June	July	August
Hits	5611	7402	9978	13,454	18,105

An exponential growth model for the data is $y = 4147e^{0.294t}$, where y represents the number of hits and t represents the month, with $t = 1$ corresponding to April. Compare the values found using the model with the amounts shown in the table. Predict the first month in which the website will receive over 50,000 hits.

18. **Population** The populations P (in thousands) of Horry County, South Carolina, from 1971 through 2018 can be modeled by $P = 75.9e^{0.0316t}$ where t represents the year, with $t = 1$ corresponding to 1971. *(Source: U.S. Census Bureau)*

    (a) Use the model to find the population in 1980, 1990, 2000, and 2010.

    (b) According to the model, when will the population of Horry County reach 380,000?

    (c) Do you think the model is valid for long-term predictions of the population? Explain.

Finding an Exponential Model In Exercises 19–22, find the exponential model that fits the points shown in the graph or table.

19.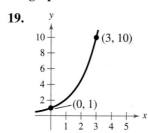

20.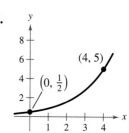

21.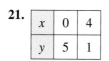

22.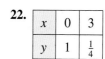

Radioactive Decay In Exercises 23–26, find the missing value for the radioactive isotope.

Isotope	Half-life (years)	Initial Quantity	Amount After 1000 Years
23. ^{226}Ra	1599	10 g	
24. ^{14}C	5715	6.5 g	
25. ^{14}C	5715		2 g
26. ^{239}Pu	24,100		0.4 g

27. **Population**

The table shows the mid-year estimated populations (in millions) of five countries in 2019 and the projected populations (in millions) for the year 2029. *(Source: U.S. Census Bureau)*

Country	2019	2029
Bulgaria	7.0	6.5
Canada	37.4	40.0
China	1389.6	1405.7
United Kingdom	65.4	68.2
United States	330.3	353.0

(a) Find the exponential growth or decay model $y = ae^{bt}$ or $y = ae^{-bt}$ for the population of each country by letting $t = 19$ correspond to 2019. Use the model to predict the population of each country in 2039.

(b) You can see that the populations of the United States and the United Kingdom are growing at different rates. What constant in the equation $y = ae^{bt}$ gives the growth rate? Discuss the relationship between the different growth rates and the magnitude of the constant.

28. **Population** The population P (in thousands) of Tallahassee, Florida, from 2000 through 2018 can be modeled by $P = 152.7e^{kt}$, where t represents the year, with $t = 0$ corresponding to 2000. In 2008, the population of Tallahassee was about 171,847. *(Source: U.S. Census Bureau)*

(a) Estimate the value of k. Is the population increasing or decreasing? Explain.

(b) Use the model to predict the populations of Tallahassee in 2025 and 2030. Are the results reasonable? Explain.

29. **Bacteria Growth** The number of bacteria in a culture is increasing according to the law of exponential growth. After 3 hours there are 100 bacteria, and after 5 hours there are 400 bacteria. How many bacteria will there be after 6 hours?

30. **Bacteria Growth** The number of bacteria in a culture is increasing according to the law of exponential growth. The initial population is 250 bacteria, and the population after 10 hours is double the population after 1 hour. How many bacteria will there be after 6 hours?

31. **Depreciation** A laptop computer that costs $575 new has a book value of $275 after 2 years.

(a) Find the linear model $V = mt + b$.

(b) Find the exponential model $V = ae^{kt}$.

(c) Use a graphing utility to graph the two models in the same viewing window. Which model depreciates faster in the first 2 years?

(d) Find the book values of the computer after 1 year and after 3 years using each model.

(e) Explain the advantages and disadvantages of using each model to a buyer and a seller.

32. **Learning Curve** The management at a plastics factory has found that the maximum number of units a worker can produce in a day is 30. The learning curve for the number N of units produced per day after a new employee has worked t days is modeled by $N = 30(1 - e^{kt})$. After 20 days on the job, a new employee produces 19 units.

(a) Find the learning curve for this employee.

(b) How many days does the model predict will pass before this employee is producing 25 units per day?

33. **Carbon Dating** The ratio of carbon-14 to carbon-12 in a piece of wood discovered in a cave is $R = 1/8^{14}$. Estimate the age of the piece of wood.

34. **Carbon Dating** The ratio of carbon-14 to carbon-12 in a piece of paper buried in a tomb is $R = 1/13^{11}$. Estimate the age of the piece of paper.

35. **IQ Scores** The IQ scores for a sample of students at a small college roughly follow the normal distribution

$$y = 0.0266e^{-(x - 100)^2/450}, \quad 70 \le x \le 115$$

where x is the IQ score. Use a graphing utility to graph the function and estimate the average IQ score of a student.

36. **Education** The amount of time (in hours per week) a student utilizes a math-tutoring center roughly follows the normal distribution

$$y = 0.7979e^{-(x - 5.4)^2/0.5}, \quad 4 \le x \le 7$$

where x is the number of hours. Use a graphing utility to graph the function and estimate the average number of hours per week a student uses the tutoring center.

37. Cell Sites A cell site is a site where electronic communications equipment is placed in a cellular network for the use of mobile phones. The numbers y of cell sites from 1985 through 2018 can be modeled by

$$y = \frac{336{,}011}{1 + 293e^{-0.236t}}$$

where t represents the year, with $t = 5$ corresponding to 1985. *(Source: CTIA-The Wireless Association)*

(a) Use the model to find the numbers of cell sites in the years 1998, 2008, and 2015.

(b) Use a graphing utility to graph the function.

(c) Use the graph to determine the year in which the number of cell sites reached 240,000.

(d) Confirm your answer to part (c) algebraically.

38. Population The population P (in thousands) of a city from 2000 through 2020 can be modeled by

$$P = \frac{2632}{1 + 0.083e^{0.050t}}$$

where t represents the year, with $t = 0$ corresponding to 2000.

(a) Use the model to find the populations of the city in the years 2000, 2005, 2010, 2015, and 2020.

(b) Use a graphing utility to graph the function.

(c) Use the graph to determine the year in which the population reached 2.2 million.

(d) Confirm your answer to part (c) algebraically.

39. Population Growth A conservation organization released 100 animals of an endangered species into a game preserve. The preserve has a carrying capacity of 1000 animals. The growth of the pack is modeled by the logistic curve $p(t) = 1000/(1 + 9e^{-0.1656t})$ where t is measured in months (see figure).

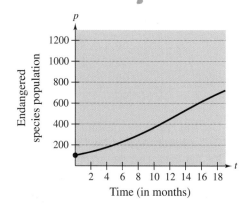

Time (in months)

(a) Estimate the population after 5 months.

(b) After how many months is the population 500?

(c) Use a graphing utility to graph the function. Use the graph to determine the horizontal asymptotes, and interpret the meaning of the asymptotes in the context of the problem.

40. Sales After discontinuing all advertising for a tool kit in 2010, the manufacturer noted that sales began to drop according to the model

$$S = \frac{500{,}000}{1 + 0.1e^{kt}}$$

where S represents the number of units sold and t represents the year, with $t = 0$ corresponding to 2010 (see figure). In 2014, 300,000 units were sold.

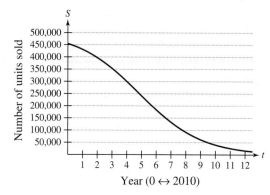

Year ($0 \leftrightarrow 2010$)

(a) Use the graph to estimate sales in 2020.

(b) Complete the model by solving for k.

(c) Use the model to estimate sales in 2020. Compare your results with that of part (a).

Intensity of Sound In Exercises 41–44, use the following information for determining sound intensity. The number of decibels β of a sound with an intensity of I watts per square meter is given by

$$\beta = 10 \log(I/I_0)$$

where I_0 is an intensity of 10^{-12} watt per square meter. In Exercises 41 and 42, find the number of decibels β of the sound.

41. (a) $I = 10^{-10}$ watt per m² (quiet room)

(b) $I = 10^{-5}$ watt per m² (busy street corner)

(c) $I = 10^{-8}$ watt per m² (quiet radio)

(d) $I = 10^{-3}$ watt per m² (loud car horn)

42. (a) $I = 10^{-11}$ watt per m² (rustle of leaves)

(b) $I = 10^2$ watt per m² (jet at 30 meters)

(c) $I = 10^{-4}$ watt per m² (door slamming)

(d) $I = 10^{-6}$ watt per m² (normal conversation)

43. Due to the installation of noise-suppression materials, the noise level in an auditorium decreased from 93 to 80 decibels. Find the percent decrease in the intensity of the noise as a result of the installation of these materials.

44. Due to the installation of a muffler, the noise level of an engine decreased from 88 to 72 decibels. Find the percent decrease in the intensity of the noise as a result of the installation of the muffler.

Geology In Exercises 45 and 46, use the Richter scale

$$R = \log \frac{I}{I_0}$$

where I_0 is the minimum intensity used for comparison, for measuring the magnitude R of an earthquake.

45. Find the intensity I of an earthquake measuring R on the Richter scale (let $I_0 = 1$).

    (a) Peru in 2015: $R = 7.6$

    (b) Pakistan in 2015: $R = 5.6$

    (c) Indonesia in 2015: $R = 6.6$

46. Find the magnitude R of each earthquake of intensity I (let $I_0 = 1$).

    (a) $I = 199,500,000$

    (b) $I = 48,275,000$

    (c) $I = 17,000$

pH Levels In Exercises 47–52, use the acidity model pH = $-\log[H^+]$, where acidity (pH) is a measure of the hydrogen ion concentration $[H^+]$ (measured in moles of hydrogen per liter) of a solution.

47. Find the pH when $[H^+] = 2.3 \times 10^{-5}$.

48. Find the pH when $[H^+] = 1.13 \times 10^{-5}$.

49. Compute $[H^+]$ for a solution in which pH = 5.8.

50. Compute $[H^+]$ for a solution in which pH = 3.2.

51. Apple juice has a pH of 2.9 and drinking water has a pH of 8.0. The hydrogen ion concentration of the apple juice is how many times the concentration of drinking water?

52. The pH of a solution decreases by one unit. By what factor does the hydrogen ion concentration increase?

53. **Forensics** At 8:30 A.M., a coroner went to the home of a person who had died during the night. To estimate the time of death, the coroner took the person's temperature twice. At 9:00 A.M. the temperature was 85.7°F, and at 11:00 A.M. the temperature was 82.8°F. From these two temperatures, the coroner was able to determine that the time elapsed since death and the body temperature were related by the formula

$$t = -10 \ln \frac{T - 70}{98.6 - 70}$$

where t is the time in hours elapsed since the person died and T is the temperature (in degrees Fahrenheit) of the person's body. (This formula comes from a general cooling principle called *Newton's Law of Cooling*. It uses the assumptions that the person had a normal body temperature of 98.6°F at death and that the room temperature was a constant 70°F.) Use the formula to estimate the time of death of the person.

54. **Home Mortgage** A $120,000 home mortgage for 30 years at $7\frac{1}{2}\%$ has a monthly payment of $839.06. Part of the monthly payment covers the interest charge on the unpaid balance, and the remainder of the payment reduces the principal. The amount paid toward the interest is

$$u = M - \left(M - \frac{Pr}{12}\right)\left(1 + \frac{r}{12}\right)^{12t}$$

and the amount paid toward the reduction of the principal is

$$v = \left(M - \frac{Pr}{12}\right)\left(1 + \frac{r}{12}\right)^{12t}.$$

In these formulas, P is the amount of the mortgage, r is the interest rate (in decimal form), M is the monthly payment, and t is the time in years.

    (a) Use a graphing utility to graph each function in the same viewing window. (The viewing window should show all 30 years of mortgage payments.)

    (b) In the early years of the mortgage, is the greater part of the monthly payment paid toward the interest or the principal? Approximate the time when the monthly payment is evenly divided between interest and principal reduction.

    (c) Repeat parts (a) and (b) for a repayment period of 20 years ($M = \$966.71$). What can you conclude?

55. **Car Speed** The table shows the time t (in seconds) required for a car to attain a speed of s miles per hour from a standing start. Two models for these data are $t_1 = 40.757 + 0.556s - 15.817 \ln s$ and $t_2 = 1.2259 + 0.0023s^2$.

DATA	Speed, s	Time, t
	30	3.4
	40	5.0
	50	7.0
	60	9.3
	70	12.0
	80	15.8
	90	20.0

Spreadsheet at LarsonPrecalculus.com

    (a) Use the *regression* feature of a graphing utility to find a linear model t_3 and an exponential model t_4 for the data. Use the graphing utility to graph the data and each model in the same viewing window.

    (b) Create a table comparing the data with estimates obtained from each model.

    (c) Use the results of part (c) to find the sum of the absolute values of the differences between the data and the estimated values found using each model. Based on the four sums, which model do you think best fits the data? Explain.

56. Home Mortgage The total interest u paid on a home mortgage of P dollars at interest rate r for t years is

$$u = P\left[\dfrac{rt}{1 - \left(\dfrac{1}{1 + r/12}\right)^{12t}} - 1\right].$$

Consider a \$120,000 home mortgage at $7\frac{1}{2}\%$ $(r = 0.075)$.

(a) Use a graphing utility to graph $u(t)$.

(b) Approximate the length of the mortgage for which the total interest paid is the same as the size of the mortgage. Is it possible to pay twice as much in interest charges as the size of the mortgage?

Exploring the Concepts

True or False? **In Exercises 57–59, determine whether the statement is true or false. Justify your answer.**

57. The domain of a logistic growth function cannot be the set of real numbers.

58. A logistic growth function will always have a zero.

59. A Gaussian model will never have a zero.

60. HOW DO YOU SEE IT? Identify each model as exponential growth, exponential decay, Gaussian, logarithmic, logistic growth, quadratic, or none of the above. Explain your reasoning.

(a)

(b)

(c)

(d)

(e)

(f)

Identifying Models **In Exercises 61–64, identify the type of mathematical model you studied in this section that has the given characteristic.**

61. The maximum value of the function occurs at the average value of the independent variable.

62. A horizontal asymptote of its graph represents the limiting value of a population.

63. Its graph shows a steadily increasing rate of growth.

64. The only asymptote of its graph is a vertical asymptote.

Review & Refresh ▶ *Video solutions at LarsonPrecalculus.com*

Operations with Rational Expressions **In Exercises 65–68, perform the operation.**

65. $\dfrac{x}{x - 3} + \dfrac{3}{x + 1}$

66. $\dfrac{x + 2}{x} - \dfrac{5}{x - 2}$

67. $\dfrac{x^2 + x}{2x} \cdot \dfrac{x - 1}{x + 1}$

68. $\dfrac{x^2 + x - 12}{x^2 - 16} \div \dfrac{x - 3}{4 + x}$

Determining the Quadrant **In Exercises 69–72, determine the quadrant in which (x, y) could be located.**

69. $x < 0$ and $y < 0$

70. $x = 2$ and $y > 0$

71. $x + y = 0$ and $x - 5 > 0$

72. $xy > 0$ and $x + y < 0$

Sketching the Graph of a Quadratic Function **In Exercises 73–76, sketch the graph of each quadratic function and compare it with the graph of $y = x^2$.**

73. $h(x) = -\frac{2}{3}x^2$

74. $f(x) = x^2 - 6$

75. $g(x) = (2x)^2 - 1$

76. $p(x) = \frac{1}{4}(x + 3)^2 + 4$

Finding the Domain of a Function **In Exercises 77–80, find the domain of the function.**

77. $f(x) = 1 + 2x^3$

78. $f(x) = \sqrt{x - 5}$

79. $f(x) = 1/(2x^2 + 6x)$

80. $f(x) = \sqrt{x}/(\sqrt[3]{x - 3})$

Using Symmetry as a Sketching Aid **In Exercises 81 and 82, assume that the graph has the given type of symmetry. Complete the graph of the equation.**

81.

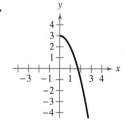

y-axis symmetry

82.

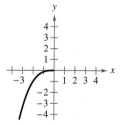

origin symmetry

Project: Sales per Share To work an extended application analyzing the sales per share for Kohl's Corporation from 2001 through 2018, visit this text's website at *LarsonPrecalculus.com*. *(Source: Kohl's Corporation)*

Summary and Study Strategies

GO DIGITAL

What Did You Learn?

The list below reviews the skills covered in the chapter and correlates each one to the Review Exercises (see page 412) that practice the skill.

Section 5.1	**Review Exercises**
■ Recognize and evaluate exponential functions with base a *(p. 360)*.	*1–6*
The exponential function f with base a is denoted by $f(x) = a^x$, where $a > 0$, $a \neq 1$, and x is any real number.	
■ Graph exponential functions and use the One-to-One Property *(p. 361)*.	*7–20*
One-to-One Property: For $a > 0$ and $a \neq 1$, $a^x = a^y$ if and only if $x = y$.	
■ Recognize, evaluate, and graph exponential functions with base e *(p. 364)*.	*21–28*
The function $f(x) = e^x$ is called the natural exponential function.	
■ Use exponential functions to model and solve real-life problems *(p. 365)*.	*29–32*

Section 5.2	
■ Recognize and evaluate logarithmic functions with base a *(p. 371)*.	*33–44*
For $x > 0$, $a > 0$, and $a \neq 1$, $y = \log_a x$ if and only if $x = a^y$. The function $f(x) = \log_a x$ is the logarithmic function with base a. The logarithmic function with base 10, the common logarithmic function, is denoted by $\log_{10}$ or log.	
■ Graph logarithmic functions *(p. 373)*.	*45–48*
■ Recognize, evaluate, and graph natural logarithmic functions *(p. 375)*.	*49–56*
The function $f(x) = \ln x$ is called the natural logarithmic function.	
■ Use logarithmic functions to model and solve real-life problems *(p. 377)*.	*57, 58*

Section 5.3	
■ Use the change-of-base formula to rewrite and evaluate logarithmic expressions *(p. 381)*.	*59–62*

Base b	**Base 10**	**Base e**
$\log_a x = \dfrac{\log_b x}{\log_b a}$	$\log_a x = \dfrac{\log x}{\log a}$	$\log_a x = \dfrac{\ln x}{\ln a}$

■ Use properties of logarithms to evaluate or rewrite logarithmic expressions *(p. 382)*.	*63–66*

Logarithm with Base a	**Natural Logarithm**
Product Property: $\log_a(uv) = \log_a u + \log_a v$	$\ln(uv) = \ln u + \ln v$
Quotient Property: $\log_a(u/v) = \log_a u - \log_a v$	$\ln(u/v) = \ln u - \ln v$
Power Property: $\log_a u^n = n \log_a u$	$\ln u^n = n \ln u$

■ Use properties of logarithms to expand or condense logarithmic expressions *(p. 383)*.	*67–78*
■ Use logarithmic functions to model and solve real-life problems *(p. 384)*.	*79, 80*

Section 5.4	Review Exercises
■ Solve simple exponential and logarithmic equations *(p. 388)*.	*81–86*
■ Solve more complicated exponential equations *(p. 389)*.	*87–90, 99, 100*
■ Solve more complicated logarithmic equations *(p. 391)*.	*91–98, 101, 102*
■ Use exponential and logarithmic equations to model and solve real-life problems *(p. 393)*.	*103, 104*

Section 5.5	
■ Recognize the five most common types of models involving exponential and logarithmic functions *(p. 398)*.	*105–110*
■ Use exponential growth and decay functions to model and solve real-life problems *(p. 399)*. Exponential growth model: $y = ae^{bx}$, $b > 0$ Exponential decay model: $y = ae^{-bx}$, $b > 0$	*111, 112*
■ Use Gaussian functions to model and solve real-life problems *(p. 402)*. Gaussian model: $y = ae^{-(x-b)^2/c}$	*113*
■ Use logistic growth functions to model and solve real-life problems *(p. 403)*. Logistic growth model: $y = \dfrac{a}{1 + be^{-rx}}$	*114*
■ Use logarithmic functions to model and solve real-life problems *(p. 404)*. Logarithmic models: $y = a + b \ln x$, $y = a + b \log x$	*115*

Study Strategies

Build a Support System

1. **Surround yourself with positive classmates.** Find another student in class with whom to study. Make sure this person is not anxious about math because you do not want another student's anxiety to increase your own. Arrange to meet on campus and compare notes, homework, and so on at least two times per week.

2. **Find a place on campus to study where other students are also studying.** Libraries, learning centers, and tutoring centers are great places to study. While studying in such places, you will be able to ask for assistance when you have questions. You do not want to study alone if you typically get down on yourself with lots of negative self-talk.

3. **Establish a relationship with a learning assistant.** Get to know someone who can help you find assistance for any type of academic issue. Learning assistants, tutors, and instructors are excellent resources.

4. **Seek out assistance before you are overwhelmed.** Visit your instructor when you need help. Instructors are more than willing to help their students, particularly during office hours. Go with a friend if you are nervous about visiting your instructor. Also, during the hours posted at the website *CalcChat.com*, you can chat with a tutor about any odd-numbered exercise in the text.

5. **Be your own support.** Listen to what you tell yourself when frustrated with studying math. Replace any negative self-talk dialogue with more positive statements.

Review Exercises

See CalcChat.com for tutorial help and worked-out solutions to odd-numbered exercises.

5.1 **Evaluating an Exponential Function** In Exercises 1–6, evaluate the function at the given value of x. Round your result to three decimal places.

1. $f(x) = 0.3^x$, $x = 1.5$
2. $f(x) = 30^x$, $x = \sqrt{3}$
3. $f(x) = 2^x$, $x = \frac{2}{3}$
4. $f(x) = \left(\frac{1}{2}\right)^{2x}$, $x = \pi$
5. $f(x) = 7(0.2^x)$, $x = -\sqrt{11}$
6. $f(x) = -14(5^x)$, $x = -0.8$

Graphing an Exponential Function In Exercises 7–12, construct a table of values for the function. Then sketch the graph of the function.

7. $f(x) = 4^{-x} + 4$
8. $f(x) = 2.65^{x-1}$
9. $f(x) = 5^{x-2} + 4$
10. $f(x) = 2^{x-6} - 5$
11. $f(x) = \left(\frac{1}{2}\right)^{-x} + 3$
12. $f(x) = \left(\frac{1}{8}\right)^{x+2} - 5$

Using the One-to-One Property In Exercises 13–16, use a One-to-One Property to solve the equation for x.

13. $\left(\frac{1}{3}\right)^{x-3} = 9$
14. $3^{x+3} = \frac{1}{81}$
15. $e^{3x-5} = e^7$
16. $e^{8-2x} = e^{-3}$

Transformations of the Graph of $f(x) = a^x$ In Exercises 17–20, describe any transformations of the graph of f that yield the graph of g.

17. $f(x) = 5^x$, $g(x) = 5^x + 1$
18. $f(x) = 6^x$, $g(x) = 6^{x+1}$
19. $f(x) = 3^x$, $g(x) = 1 - 3^x$
20. $f(x) = \left(\frac{1}{2}\right)^x$, $g(x) = -\left(\frac{1}{2}\right)^{x+2}$

Evaluating the Natural Exponential Function In Exercises 21–24, evaluate $f(x) = e^x$ at the given value of x. Round your result to three decimal places.

21. $x = 3.4$
22. $x = -2.5$
23. $x = \frac{3}{5}$
24. $x = \frac{2}{7}$

Graphing a Natural Exponential Function In Exercises 25–28, construct a table of values for the function. Then sketch the graph of the function.

25. $h(x) = e^{-x/2}$
26. $h(x) = 2 - e^{-x/2}$
27. $f(x) = e^{x+2}$
28. $s(t) = 4e^{t-1}$

29. **Waiting Times** The average time between new posts on a message board is 3 minutes. The probability F of waiting less than t minutes until the next post is approximated by the model $F(t) = 1 - e^{-t/3}$. A message has just been posted. Find the probability that the next post will be within (a) 1 minute, (b) 2 minutes, and (c) 5 minutes.

30. **Depreciation** After t years, the value V of a car that originally cost \$23,970 is given by $V(t) = 23,970\left(\frac{3}{4}\right)^t$.

   (a) Use a graphing utility to graph the function.
   (b) Find the value of the car 2 years after it was purchased.
   (c) According to the model, when does the car depreciate most rapidly? Is this realistic? Explain.
   (d) According to the model, when will the car have no value?

Compound Interest In Exercises 31 and 32, complete the table by finding the balance A when P dollars is invested at rate r for t years and compounded n times per year.

n	1	2	4	12	365	Continuous
A						

31. $P = \$5000$, $r = 3\%$, $t = 10$ years
32. $P = \$4500$, $r = 2.5\%$, $t = 30$ years

5.2 **Writing a Logarithmic Equation** In Exercises 33–36, write the exponential equation in logarithmic form. For example, the logarithmic form of $2^3 = 8$ is $\log_2 8 = 3$.

33. $3^3 = 27$
34. $25^{3/2} = 125$
35. $e^{0.8} = 2.2255\ldots$
36. $e^0 = 1$

Evaluating a Logarithm In Exercises 37–40, evaluate the logarithm at the given value of x without using a calculator.

37. $f(x) = \log x$, $x = 1000$
38. $g(x) = \log_9 x$, $x = 3$
39. $g(x) = \log_2 x$, $x = \frac{1}{4}$
40. $f(x) = \log_3 x$, $x = \frac{1}{81}$

Using the One-to-One Property In Exercises 41–44, use a One-to-One Property to solve the equation for x.

41. $\log_4(x + 7) = \log_4 14$
42. $\log_8(3x - 10) = \log_8 5$
43. $\ln(x + 9) = \ln 4$
44. $\log(3x - 2) = \log 7$

Sketching the Graph of a Logarithmic Function In Exercises 45–48, find the domain, x-intercept, and vertical asymptote of the logarithmic function and sketch its graph.

45. $g(x) = \log_7 x$
46. $f(x) = \log \dfrac{x}{3}$
47. $f(x) = 4 - \log(x + 5)$
48. $f(x) = \log(x - 3) + 1$

Evaluating a Logarithmic Function In Exercises 49–52, use a calculator to evaluate the function at the given value of x. Round your result to three decimal places, if necessary.

49. $f(x) = \ln x$, $x = 22.6$

50. $f(x) = \ln x$, $x = e^{-12}$

51. $f(x) = \frac{1}{2}\ln x$, $x = \sqrt{e}$

52. $f(x) = 5\ln x$, $x = 0.98$

Graphing a Natural Logarithmic Function In Exercises 53–56, find the domain, x-intercept, and vertical asymptote of the logarithmic function and sketch its graph.

53. $f(x) = \ln x + 6$ **54.** $f(x) = \ln x - 5$

55. $h(x) = \ln(x - 6)$ **56.** $f(x) = \ln(x + 4)$

57. Astronomy The formula $M = m - 5\log(d/10)$ gives the distance d (in parsecs) from Earth to a star with apparent magnitude m and absolute magnitude M. The star Rasalhague has an apparent magnitude of 2.08 and an absolute magnitude of 1.3. Find the distance from Earth to Rasalhague.

58. Snow Removal The number of miles s of roads cleared of snow is approximated by the model

$$s = 25 - \frac{13\ln(h/12)}{\ln 3}, \quad 2 \le h \le 15$$

where h is the depth (in inches) of the snow. Use this model to find s when $h = 10$ inches.

5.3 **Using the Change-of-Base Formula** In Exercises 59–62, evaluate the logarithm using the change-of-base formula (a) with common logarithms and (b) with natural logarithms. Round your results to three decimal places.

59. $\log_2 6$ **60.** $\log_{12} 200$

61. $\log_{1/2} 5$ **62.** $\log_4 0.75$

Using Properties of Logarithms In Exercises 63–66, use the properties of logarithms to write the logarithm in terms of $\log_2 3$ and $\log_2 5$.

63. $\log_2 \frac{5}{3}$ **64.** $\log_2 45$

65. $\log_2 \frac{9}{5}$ **66.** $\log_2 \frac{20}{9}$

Expanding a Logarithmic Expression In Exercises 67–72, use the properties of logarithms to expand the expression as a sum, difference, and/or constant multiple of logarithms. (Assume all variables are positive.)

67. $\log 7x^2$ **68.** $\log 11x^3$

69. $\log_3 \frac{9}{\sqrt{x}}$ **70.** $\log_7 \frac{\sqrt[3]{x}}{19}$

71. $\ln x^2y^2z$ **72.** $\ln\left(\frac{y-1}{3}\right)^2$, $\quad y > 1$

Condensing a Logarithmic Expression In Exercises 73–78, condense the expression to the logarithm of a single quantity.

73. $\ln 7 + \ln x$

74. $\log_2 y - \log_2 3$

75. $\log x - \frac{1}{2}\log y$

76. $3\ln x + 2\ln(x + 1)$

77. $\frac{1}{2}\log_3 x - 2\log_3(y + 8)$

78. $5\ln(x - 2) - \ln(x + 2) - 3\ln x$

79. Climb Rate The time t (in minutes) for a small plane to climb to an altitude of h feet is modeled by

$$t = 50\log[18{,}000/(18{,}000 - h)]$$

where 18,000 feet is the plane's absolute ceiling.

  (a) Determine the domain of the function in the context of the problem.

  (b) Use a graphing utility to graph the function and identify any asymptotes.

  (c) As the plane approaches its absolute ceiling, what can be said about the time required to increase its altitude?

  (d) Find the time it takes for the plane to climb to an altitude of 4000 feet.

80. Human Memory Model Students in a learning theory study took an exam and then retested monthly for 6 months with an equivalent exam. The data obtained in the study are given by the ordered pairs (t, s), where t is the time (in months) after the initial exam and s is the average score for the class. Use the data to find a logarithmic equation that relates t and s.

$(1, 84.2)$, $(2, 78.4)$, $(3, 72.1)$,
$(4, 68.5)$, $(5, 67.1)$, $(6, 65.3)$

5.4 **Solving a Simple Equation** In Exercises 81–86, solve for x.

81. $5^x = 125$

82. $6^x = \frac{1}{216}$

83. $e^x = 3$

84. $\log x - \log 5 = 0$

85. $\ln x = 4$

86. $\ln x = -1.6$

Solving an Exponential Equation In Exercises 87–90, solve the exponential equation algebraically. Approximate the result to three decimal places, if necessary.

87. $e^{4x} = e^{x^2 + 3}$

88. $e^{3x} = 25$

89. $2^x - 3 = 29$

90. $e^{2x} - 6e^x + 8 = 0$

Solving a Logarithmic Equation In Exercises 91–98, solve the logarithmic equation algebraically, if possible. Approximate the result to three decimal places, if necessary.

91. $\ln 3x = 8.2$ **92.** $4 \ln 3x = 15$

93. $\ln x + \ln(x - 3) = 1$

94. $\ln(x + 2) - \ln x = 2$

95. $\log_8(x - 1) = \log_8(x - 2) - \log_8(x + 2)$

96. $\log_6(x + 2) - \log_6 x = \log_6(x + 5)$

97. $\log(1 - x) = -1$

98. $\log(-x - 4) = 2$

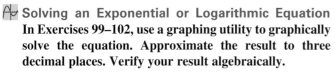 **Solving an Exponential or Logarithmic Equation** In Exercises 99–102, use a graphing utility to graphically solve the equation. Approximate the result to three decimal places. Verify your result algebraically.

99. $25e^{-0.3x} = 12$

100. $2 = 5 - e^{x+7}$

101. $2 \ln(x + 3) - 5 = 0$

102. $2 \ln x - \ln(3x - 1) = 0$

103. Compound Interest A business deposits $8500 in an account that pays 1.5% interest, compounded continuously. How long will it take for the money to triple?

104. Meteorology The speed of the wind S (in miles per hour) near the center of a tornado and the distance d (in miles) the tornado travels are related by the model $S = 93 \log d + 65$. On March 18, 1925, a large tornado struck portions of Missouri, Illinois, and Indiana with a wind speed at the center of about 283 miles per hour. Approximate the distance traveled by this tornado.

5.5 **Matching a Function with Its Graph** In Exercises 105–110, match the function with its graph. [The graphs are labeled (a), (b), (c), (d), (e), and (f).]

(a)
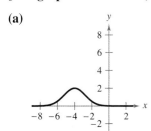

(b)
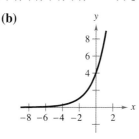

(c)
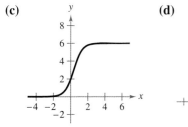

(d)
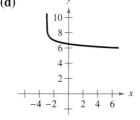

(e)
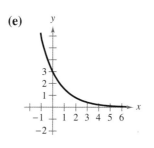

(f)
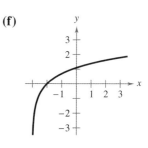

105. $y = 3e^{-2x/3}$ **106.** $y = 4e^{2x/3}$

107. $y = \ln(x + 3)$ **108.** $y = 7 - \log(x + 3)$

109. $y = 2e^{-(x+4)^2/3}$ **110.** $y = \dfrac{6}{1 + 2e^{-2x}}$

111. Finding an Exponential Model Find the exponential model $y = ae^{bx}$ that fits the points $(0, 2)$ and $(4, 3)$.

112. Wildlife Population A species of bat is in danger of becoming extinct. Five years ago, the total population of the species was 2000. Two years ago, the total population of the species was 1400. What was the total population of the species one year ago?

113. Test Scores The test scores for a biology test follow the normal distribution

$$y = 0.0499e^{-(x-71)^2/128}, \quad 40 \le x \le 100$$

where x is the test score. Use a graphing utility to graph the equation and estimate the average test score.

114. Typing Speed In a typing class, the average number N of words per minute typed after t weeks of lessons is

$$N = 157/(1 + 5.4e^{-0.12t}).$$

Find the time necessary to type (a) 50 words per minute and (b) 75 words per minute.

115. Sound Intensity The relationship between the number of decibels β and the intensity of a sound I (in watts per square meter) is $\beta = 10 \log(I/10^{-12})$. Find the intensity I for each decibel level β.

(a) $\beta = 60$ (b) $\beta = 135$ (c) $\beta = 1$

Exploring the Concepts

116. Graph of an Exponential Function Consider the graph of $y = e^{kt}$. Describe the characteristics of the graph when k is positive and when k is negative.

True or False? In Exercises 117 and 118, determine whether the equation is true or false. Justify your answer.

117. $\log_b b^{2x} = 2x$

118. $\ln(x + y) = \ln x + \ln y$

119. Writing Explain why $\log_a x$ is defined only for $0 < a < 1$ and $a > 1$.

Chapter Test

See CalcChat.com for tutorial help and worked-out solutions to odd-numbered exercises.

GO DIGITAL

Take this test as you would take a test in class. When you are finished, check your work against the answers given in the back of the book.

In Exercises 1–4, evaluate the expression. Round your result to three decimal places. *(Section 5.1)*

1. $0.7^{2.5}$ **2.** $3^{-\pi}$ **3.** $e^{-7/10}$ **4.** $e^{3.1}$

In Exercises 5–7, construct a table of values for the function. Then sketch the graph of the function. *(Section 5.1)*

5. $f(x) = 10^{-x}$ **6.** $f(x) = -6^{x-2}$ **7.** $f(x) = 1 - e^{2x}$

8. Evaluate (a) $\log_7 7^{-0.89}$ and (b) $4.6 \ln e^2$. *(Section 5.2)*

In Exercises 9–11, find the domain, x-intercept, and vertical asymptote of the logarithmic function and sketch its graph. *(Section 5.2)*

9. $f(x) = 4 + \log x$ **10.** $f(x) = \ln(x - 5)$ **11.** $f(x) = 1 + \ln(x + 6)$

In Exercises 12–14, evaluate the logarithm using the change-of-base formula. Round your result to three decimal places. *(Section 5.3)*

12. $\log_5 35$ **13.** $\log_{16} 0.63$ **14.** $\log_{3/4} 24$

In Exercises 15–17, use the properties of logarithms to expand the expression as a sum, difference, and/or constant multiple of logarithms. (Assume all variables are positive.) *(Section 5.3)*

15. $\log_2 3a^4$ **16.** $\ln \dfrac{\sqrt{x}}{7}$ **17.** $\log \dfrac{10x^2}{y^3}$

In Exercises 18–20, condense the expression to the logarithm of a single quantity. *(Section 5.3)*

18. $\log_3 13 + \log_3 y$ **19.** $4 \ln x - 4 \ln y$

20. $3 \ln x - \ln(x + 3) + 2 \ln y$

In Exercises 21–26, solve the equation algebraically. Approximate the result to three decimal places, if necessary. *(Section 5.4)*

21. $5^{x^2} = 5$ **22.** $3e^{-5x} = 132$

23. $\dfrac{1025}{8 + e^{4x}} = 5$ **24.** $\ln x - \ln x^2 = \dfrac{1}{2}$

25. $\log x + \log(x - 15) = 2$ **26.** $\log(2x - 1) = \log(4x - 3)$

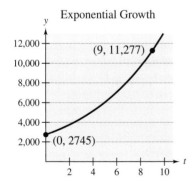

Exponential Growth

(9, 11,277)

(0, 2745)

Figure for 27

27. Find the exponential growth model that fits the points shown in the graph. *(Section 5.5)*

28. The half-life of radioactive actinium (^{227}Ac) is 21.77 years. What percent of a present amount of radioactive actinium will remain after 19 years? *(Section 5.1)*

29. A model that can predict a child's height H (in centimeters) based on the child's age is $H = 70.228 + 5.104x + 9.222 \ln x$, $\frac{1}{4} \le x \le 6$, where x is the child's age in years. *(Source: Snapshots of Applications in Mathematics)* *(Section 5.2)*

(a) Construct a table of values for the model. Then sketch the graph of the model.

(b) Use the graph from part (a) to predict the height of a four-year-old child. Then confirm your prediction algebraically.

Cumulative Test for Chapters 3–5   See CalcChat.com for tutorial help and worked-out solutions to odd-numbered exercises.

Take this test as you would take a test in class. When you are finished, check your work against the answers given in the back of the book.

1. Write the standard form of the quadratic function whose graph is a parabola with vertex at $(-8, 5)$ and that passes through the point $(-4, -7)$. *(Section 3.1)*

In Exercises 2–4, sketch the graph of the function. *(Sections 3.1 and 3.2)*

2. $h(x) = -x^2 + 10x - 21$ 3. $f(t) = -\frac{1}{2}(t-1)^2(t+2)^2$ 4. $g(s) - s^3 - 3s^2$

In Exercises 5 and 6, find all the zeros of the function. *(Section 3.4)*

5. $f(x) = x^3 + 2x^2 + 4x + 8$ 6. $f(x) = x^4 + 4x^3 - 21x^2$

7. Use long division to divide: $\dfrac{6x^3 - 4x^2}{2x^2 + 1}$. *(Section 3.3)*

8. Use synthetic division to divide $3x^4 + 2x^2 - 5x + 3$ by $x - 2$. *(Section 3.3)*

 9. Use a graphing utility to approximate (to three decimal places) the real zero of the function $g(x) = x^3 + 3x^2 - 6$. *(Section 3.2)*

10. Find a polynomial function with real coefficients that has -5, -2, and $2 + \sqrt{3}i$ as its zeros. (There are many correct answers.) *(Section 3.4)*

11. Find a direct variation model that relates x and y, given that $x = 12$ when $y = 16$. *(Section 3.5)*

In Exercises 12–16, state the domain of the rational function. Sketch the graph of the function, identify all intercepts, and find any asymptotes or holes. *(Sections 4.1 and 4.2)*

12. $f(x) = \dfrac{2x}{x - 3}$

13. $f(x) = \dfrac{4x^2}{x - 5}$

14. $f(x) = \dfrac{2x}{x^2 + 2x - 3}$

15. $f(x) = \dfrac{x^2 - 4}{x^2 + x - 2}$

16. $f(x) = \dfrac{x^3 - 2x^2 - 9x + 18}{x^2 + 4x + 3}$

In Exercises 17–19, sketch the conic. *(Sections 4.3 and 4.4)*

17. $x^2 = -8y$

18. $\dfrac{(x+3)^2}{16} - \dfrac{(y+4)^2}{25} = 1$

19. $\dfrac{(x-2)^2}{4} + \dfrac{(y+1)^2}{9} = 1$

20. Find the standard form of the equation of the parabola shown in the figure below. *(Section 3.1)*

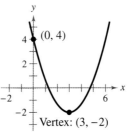

21. Find the standard form of the equation of the hyperbola whose vertices are $(-1, -5)$ and $(-1, 1)$, and whose foci are $(-1, -7)$ and $(-1, 3)$. *(Section 4.3)*

In Exercises 22 and 23, describe any transformations of the graph of f that yield the graph of g. *(Section 5.1)*

22. $f(x) = \left(\frac{2}{5}\right)^x$, $g(x) = -\left(\frac{2}{5}\right)^{-x+3}$

23. $f(x) = 2.2^x$, $g(x) = -2.2^x + 4$

In Exercises 24–27, use a calculator to evaluate the expression. Round your result to three decimal places. *(Section 5.2)*

24. $\log 98$

25. $\log \frac{6}{7}$

26. $\ln \sqrt{31}$

27. $\ln\left(\sqrt{30} - 4\right)$

In Exercises 28–30, evaluate the logarithm using the change-of-base formula. Round your result to three decimal places. *(Section 5.3)*

28. $\log_5 4.3$

29. $\log_3 0.149$

30. $\log_{1/2} 17$

31. Use the properties of logarithms to expand $\ln\left(\dfrac{x^2 - 25}{x^4}\right)$, where $x > 5$. *(Section 5.3)*

32. Condense $2 \ln x - \frac{1}{2} \ln(x + 5)$ to the logarithm of a single quantity. *(Section 5.3)*

In Exercises 33–38, solve the equation algebraically. Approximate the result to three decimal places, if necessary. *(Section 5.4)*

33. $6e^{2x} = 72$

34. $4^{x-5} + 21 = 30$

35. $e^{2x} - 13e^x + 42 = 0$

36. $\log_2 x + \log_2 5 = 6$

37. $\ln 4x - \ln 2 = 8$

38. $\ln \sqrt{x + 2} = 3$

39. The profit P that a company makes depends on the amount x the company spends on advertising according to the model $P = 230 + 20x - 0.5x^2$, where P and x are in thousands of dollars. What expenditure for advertising yields a maximum profit? *(Section 3.1)*

40. A rectangular region of length x and width y has an area of 600 square meters. *(Section 4.2)*

   (a) Write y as a function of x.

   (b) Determine the domain of the function based on the physical constraints of the problem.

   (c) Sketch the graph of the function and determine the width of the rectangle when $x = 35$ meters.

41. On the day a grandchild is born, a grandparent deposits $2500 in a fund earning 7.5% interest, compounded continuously. Determine the balance in the account on the grandchild's 25th birthday. *(Section 5.1)*

42. The number N of bacteria in a culture is given by the model $N = 175e^{kt}$, where t is the time in hours. Given that $N = 420$ when $t = 8$, estimate the time required for the population to double in size. *(Section 5.5)*

43. The population P (in thousands) of Austin, Texas from 2010 through 2018 can be approximated by the model $P = 647.965e^{0.0229t}$, where t represents the year, with $t = 10$ corresponding to 2010. According to this model, when will the population reach 1.2 million? *(Source: U.S. Census Bureau)* *(Section 5.5)*

44. The population p of a species of bird t years after it is introduced into a new habitat is given by

$$p = \frac{1200}{1 + 3e^{-t/5}}. \quad \textit{(Section 5.5)}$$

   (a) Determine the population size that was introduced into the habitat.

   (b) Determine the population after 5 years.

   (c) After how many years will the population be 800?

Proofs in Mathematics

Each of the three properties of logarithms listed below can be proved by using properties of exponential functions.

HISTORICAL NOTE

William Oughtred (1574–1660) is credited with inventing the slide rule. The slide rule is a computational device with a sliding portion and a fixed portion. A slide rule enables you to perform multiplication by using the Product Property of Logarithms. There are other slide rules that allow for the calculation of roots and trigonometric functions. Mathematicians and engineers used slide rules until the handheld calculator came into widespread use in the 1970s.

Properties of Logarithms (p. 382)

Let a be a positive number such that $a \neq 1$, let n be a real number, and let u and v be positive real numbers.

	Logarithm with Base a	Natural Logarithm
1. Product Property:	$\log_a(uv) = \log_a u + \log_a v$	$\ln(uv) = \ln u + \ln v$
2. Quotient Property:	$\log_a \dfrac{u}{v} = \log_a u - \log_a v$	$\ln \dfrac{u}{v} = \ln u - \ln v$
3. Power Property:	$\log_a u^n = n \log_a u$	$\ln u^n = n \ln u$

Proof

Let

$$x = \log_a u \quad \text{and} \quad y = \log_a v.$$

The corresponding exponential forms of these two equations are

$$a^x = u \quad \text{and} \quad a^y = v.$$

To prove the Product Property, multiply u and v to obtain

$$uv = a^x a^y$$
$$= a^{x+y}.$$

The corresponding logarithmic form of $uv = a^{x+y}$ is $\log_a(uv) = x + y$. So,

$$\log_a(uv) = \log_a u + \log_a v.$$

To prove the Quotient Property, divide u by v to obtain

$$\frac{u}{v} = \frac{a^x}{a^y}$$
$$= a^{x-y}.$$

The corresponding logarithmic form of $\dfrac{u}{v} = a^{x-y}$ is $\log_a \dfrac{u}{v} = x - y$. So,

$$\log_a \frac{u}{v} = \log_a u - \log_a v.$$

To prove the Power Property, substitute a^x for u in the expression $\log_a u^n$.

$$\log_a u^n = \log_a (a^x)^n \qquad \text{Substitute } a^x \text{ for } u.$$
$$= \log_a a^{nx} \qquad \text{Property of exponents}$$
$$= nx \qquad \text{Inverse Property}$$
$$= n \log_a u \qquad \text{Substitute } \log_a u \text{ for } x.$$

So, $\log_a u^n = n \log_a u.$

P.S. Problem Solving

See CalcChat.com for tutorial help and worked-out solutions to odd-numbered exercises.

GO DIGITAL

1. **Graphical Reasoning** Graph the exponential function $y = a^x$ for $a = 0.5, 1.2,$ and 2.0. Which of these curves intersects the line $y = x$? Determine all positive numbers a for which the curve $y = a^x$ intersects the line $y = x$.

2. **Graphical Reasoning** Use a graphing utility to graph each of the functions $y_1 = e^x$, $y_2 = x^2$, $y_3 = x^3$, $y_4 = \sqrt{x}$, and $y_5 = |x|$. Which function increases at the greatest rate as x approaches ∞?

3. **Conjecture** Use the result of Exercise 2 to make a conjecture about the rate of growth of $y_1 = e^x$ and $y = x^n$, where n is a natural number and x approaches ∞.

4. **Implication of "Growing Exponentially"** Use the results of Exercises 2 and 3 to describe what is implied when it is stated that a quantity is growing exponentially.

5. **Exponential Function** Given the exponential function

    $$f(x) = a^x$$

    show that

    (a) $f(u + v) = f(u) \cdot f(v)$ and (b) $f(2x) = [f(x)]^2$.

6. **Difference of Squares of Two Functions** Given that

    $$f(x) = \frac{e^x + e^{-x}}{2} \quad \text{and} \quad g(x) = \frac{e^x - e^{-x}}{2}$$

    show that

    $$[f(x)]^2 - [g(x)]^2 = 1.$$

7. **Graphical Reasoning** Use a graphing utility to compare the graph of the function $y = e^x$ with the graph of each function. [$n!$ (read "n factorial") is defined as $n! = 1 \cdot 2 \cdot 3 \cdots (n - 1) \cdot n$.]

    (a) $y_1 = 1 + \dfrac{x}{1!}$

    (b) $y_2 = 1 + \dfrac{x}{1!} + \dfrac{x^2}{2!}$

    (c) $y_3 = 1 + \dfrac{x}{1!} + \dfrac{x^2}{2!} + \dfrac{x^3}{3!}$

8. **Identifying a Pattern** Identify the pattern of successive polynomials given in Exercise 7. Extend the pattern one more term and compare the graph of the resulting polynomial function with the graph of $y = e^x$. What do you think this pattern implies?

9. **Finding an Inverse Function** Graph the function

    $$f(x) = e^x - e^{-x}.$$

    From the graph, the function appears to be one-to-one. Assume that f has an inverse function and find $f^{-1}(x)$.

10. **Finding a Pattern for an Inverse Function** Find a pattern for $f^{-1}(x)$ when

    $$f(x) = \frac{a^x + 1}{a^x - 1}$$

    where $a > 0$, $a \neq 1$.

11. **Determining the Equation of a Graph** Determine whether the graph represents equation (a), (b), or (c). Explain your reasoning.

    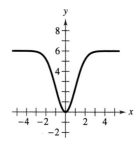

    (a) $y = 6e^{-x^2/2}$

    (b) $y = \dfrac{6}{1 + e^{-x/2}}$

    (c) $y = 6(1 - e^{-x^2/2})$

12. **Simple and Compound Interest** You have two options for investing \$500. The first earns 7% interest compounded annually, and the second earns 7% simple interest. The figure shows the growth of each investment over a 30-year period.

    (a) Determine which graph represents each type of investment. Explain your reasoning.

    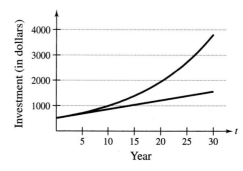

    (b) Verify your answer in part (a) by finding the equations that model the investment growth and by graphing the models.

    (c) Which option would you choose? Explain.

13. **Radioactive Decay** Two different samples of radioactive isotopes are decaying. The isotopes have initial amounts of c_1 and c_2 and half-lives of k_1 and k_2, respectively. Find an expression for the time t required for the samples to decay to equal amounts.

14. Bacteria Decay A lab culture initially contains 500 bacteria. Two hours later, the number of bacteria decreases to 200. Find the exponential decay model of the form

$$B = B_0 a^{kt}$$

that approximates the number of bacteria B in the culture after t hours.

15. Colonial Population The table shows the colonial population estimates of the American colonies for each decade from 1700 through 1780. *(Source: U.S. Census Bureau)*

Year	Population
1700	250,900
1710	331,700
1720	466,200
1730	629,400
1740	905,600
1750	1,170,800
1760	1,593,600
1770	2,148,100
1780	2,780,400

DATA Spreadsheet at LarsonPrecalculus.com

Let y represent the population in the year t, with $t = 0$ corresponding to 1700.

(a) Use the *regression* feature of a graphing utility to find an exponential model for the data.

(b) Use the *regression* feature of the graphing utility to find a quadratic model for the data.

(c) Use the graphing utility to plot the data and the models from parts (a) and (b) in the same viewing window.

(d) Which model is a better fit for the data? Would you use this model to predict the population of the United States in 2025? Explain your reasoning.

16. Ratio of Logarithms Show that

$$\frac{\log_a x}{\log_{a/b} x} = 1 + \log_a \frac{1}{b}.$$

17. Solving a Logarithmic Equation Solve

$$(\ln x)^2 = \ln x^2.$$

18. Graphical Reasoning Use a graphing utility to compare the graph of each function with the graph of $y = \ln x$.

(a) $y_1 = x - 1$

(b) $y_2 = (x - 1) - \frac{1}{2}(x - 1)^2$

(c) $y_3 = (x - 1) - \frac{1}{2}(x - 1)^2 + \frac{1}{3}(x - 1)^3$

19. Identifying a Pattern Identify the pattern of successive polynomials given in Exercise 18. Extend the pattern one more term and compare the graph of the resulting polynomial function with the graph of $y = \ln x$. What do you think the pattern implies?

20. Finding Slope and y-Intercept Take the natural log of each side of each equation below.

$$y = ab^x, \quad y = ax^b$$

(a) What are the slope and y-intercept of the line relating x and $\ln y$ for $y = ab^x$?

(b) What are the slope and y-intercept of the line relating $\ln x$ and $\ln y$ for $y = ax^b$?

Ventilation Rate In Exercises 21 and 22, use the model

$$y = 80.4 - 11 \ln x, \quad 100 \leq x \leq 1500$$

which approximates the minimum required ventilation rate in terms of the air space per child in a public school classroom. In the model, x is the air space (in cubic feet) per child and y is the ventilation rate (in cubic feet per minute) per child.

21. Use a graphing utility to graph the model and approximate the required ventilation rate when there are 300 cubic feet of air space per child.

22. In a classroom designed for 30 students, the air conditioning system can move 450 cubic feet of air per minute.

(a) Determine the ventilation rate per child in a full classroom.

(b) Estimate the air space required per child.

(c) Determine the minimum number of square feet of floor space required for the room when the ceiling height is 30 feet.

Using Technology In Exercises 23–26, (a) use a graphing utility to create a scatter plot of the data, (b) decide whether the data could best be modeled by a linear model, an exponential model, or a logarithmic model, (c) explain why you chose the model you did in part (b), (d) use the *regression* feature of the graphing utility to find the model you chose in part (b) for the data and graph the model with the scatter plot, and (e) determine how well the model you chose fits the data.

23. $(1, 2.0), (1.5, 3.5), (2, 4.0), (4, 5.8), (6, 7.0), (8, 7.8)$

24. $(1, 4.4), (1.5, 4.7), (2, 5.5), (4, 9.9), (6, 18.1), (8, 33.0)$

25. $(1, 7.5), (1.5, 7.0), (2, 6.8), (4, 5.0), (6, 3.5), (8, 2.0)$

26. $(1, 5.0), (1.5, 6.0), (2, 6.4), (4, 7.8), (6, 8.6), (8, 9.0)$

6 Trigonometry

GO DIGITAL

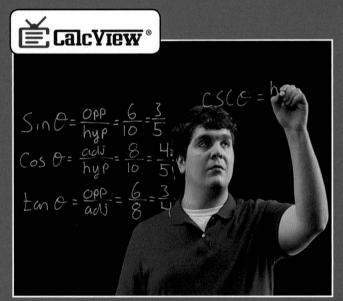

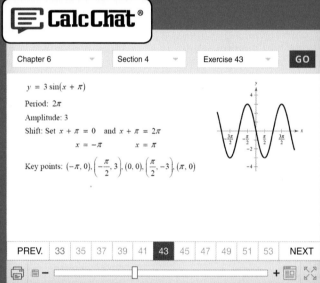

6.1 Speed of a Bicycle (*Exercise 105, p. 433*)

6.4 Respiratory Cycle (*Exercise 80, p. 466*)

6.1 Angles and Their Measure

Angles and their measure have a wide variety of real-life applications. For example, in Exercise 105 on page 433, you will use angles and their measure to model the distance a cyclist travels.

❯ **Describe angles.**
❯ **Use degree measure.**
❯ **Use radian measure.**
❯ **Convert between degrees and radians.**
❯ **Use angles and their measure to model and solve real-life problems.**

Angles

As derived from the Greek language, the word **trigonometry** means "measurement of triangles." Originally, trigonometry dealt with relationships among the sides and angles of triangles and was instrumental in the development of astronomy, navigation, and surveying. With the development of calculus and the physical sciences in the seventeenth century, a different perspective arose—one that viewed the classic trigonometric relationships as *functions* with the set of real numbers as their domains. Consequently, the applications of trigonometry expanded to include a vast number of physical phenomena, such as sound waves, planetary orbits, vibrating strings, pendulums, and orbits of atomic particles. This text incorporates *both* perspectives, starting with angles and their measure.

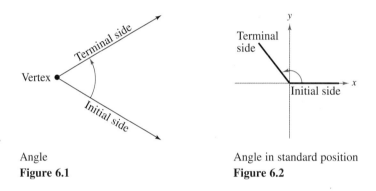

Angle
Figure 6.1

Angle in standard position
Figure 6.2

Rotating a ray (half-line) about its endpoint determines an **angle.** The starting position of the ray is the **initial side** of the angle, and the position after rotation is the **terminal side,** as shown in Figure 6.1. The endpoint of the ray is the **vertex** of the angle. This perception of an angle fits a coordinate system in which the origin is the vertex and the initial side coincides with the positive *x*-axis. Such an angle is in **standard position,** as shown in Figure 6.2. Counterclockwise rotation generates **positive angles,** and clockwise rotation generates **negative angles,** as shown in Figure 6.3. Labels for angles can be Greek letters such as α (alpha), β (beta), and θ (theta), or uppercase letters such as *A*, *B*, and *C*. In Figure 6.4, note that angles α and β have the same initial and terminal sides. Such angles are **coterminal.**

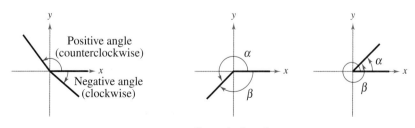

Figure 6.3

Coterminal angles
Figure 6.4

GO DIGITAL

Degree Measure

The amount of rotation from the initial side to the terminal side determines the **measure of an angle.** The most common unit of angle measure is the **degree,** denoted by the symbol °. A measure of one degree (1°) is equivalent to a rotation of $\frac{1}{360}$ of a complete revolution about the vertex. To measure angles, it is convenient to mark degrees on the circumference of a circle, as shown in Figure 6.5. So, a full revolution (counterclockwise) corresponds to 360°, a half revolution corresponds to 180°, a quarter revolution corresponds to 90°, and so on.

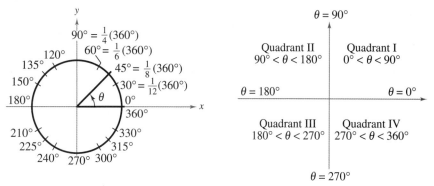

Figure 6.5   **Figure 6.6**

Recall that the four quadrants in a coordinate system are numbered I, II, III, and IV. Figure 6.6 shows which angles between 0° and 360° lie in each of the four quadrants. An angle whose terminal side lies on the *x*- or *y*-axis, such as 0°, 90°, 180°, 270°, and so on, is a **quadrantal angle.** The terminal sides of the quadrantal angles do not lie within quadrants.

Figure 6.7 shows several common angles with their degree measures. Note that angles between 0° and 90° are **acute** and angles between 90° and 180° are **obtuse.**

> **ALGEBRA HELP**
>
> The phrase "θ lies in a quadrant" is another way of saying "the terminal side of θ lies in a quadrant."

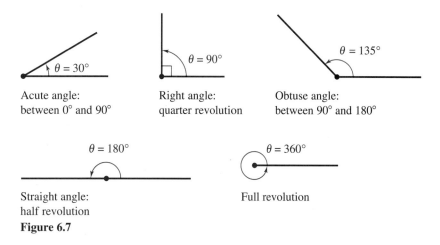

Acute angle:
between 0° and 90°

Right angle:
quarter revolution

Obtuse angle:
between 90° and 180°

Straight angle:
half revolution

Full revolution

Figure 6.7

Two angles are coterminal when they have the same initial and terminal sides. For example, the angles 30° and 390° are coterminal, as shown in the figure at the right.

To find an angle that is coterminal to a given angle θ, add or subtract 360° (one revolution), as demonstrated in Example 1 on the next page. A given angle θ has infinitely many coterminal angles. For example, $\theta = 30°$ is coterminal with $30° + n(360°)$, where n is an integer.

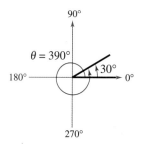

GO DIGITAL

With calculators, it is convenient to use *decimal* degrees to denote fractional parts of degrees. Historically, however, fractional parts of degrees were expressed in *minutes* and *seconds*, using the prime (′) and double prime (″) notations, respectively. That is,

$$1' = \text{one minute} = \tfrac{1}{60}(1°)$$

and

$$1'' = \text{one second} = \tfrac{1}{3600}(1°).$$

For example, you would write an angle θ of 64 degrees, 32 minutes, and 47 seconds as $\theta = 64° \, 32' \, 47''$.

Many calculators have special keys for converting an angle in degrees, minutes, and seconds ($D° \, M' \, S''$) to decimal degree form and vice versa.

GO DIGITAL

▶▶▶

ALGEBRA HELP

Consider a positive angle θ.
- When θ is less than 90°, the complement of θ is $90° - \theta$.
- When θ is less than 180°, the supplement of θ is $180° - \theta$.

EXAMPLE 1 **Finding Coterminal Angles**

▶▶▶ *See LarsonPrecalculus.com for an interactive version of this type of example.*

a. For the positive angle 405°, subtract 360° to obtain a positive coterminal angle.

$$405° - 360° = 45° \qquad \text{See Figure 6.8(a).}$$

Subtract $2(360°) = 720°$ to obtain a negative coterminal angle.

$$405° - 720° = -315°$$

b. For the negative angle $-120°$, add 360° to obtain a positive coterminal angle.

$$-120° + 360° = 240° \qquad \text{See Figure 6.8(b).}$$

Subtract 360° to obtain a negative coterminal angle.

$$-120° - 360° = -480°$$

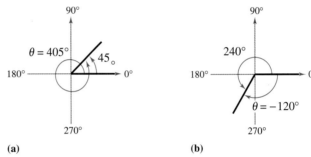

(a)   **(b)**
Figure 6.8

✓ *Checkpoint* ▶ Audio-video solution in English & Spanish at LarsonPrecalculus.com

Determine two coterminal angles (one positive and one negative) for each angle.

a. $\theta = 55°$ **b.** $\theta = -28°$

Two positive angles α and β are **complementary** (complements of each other) when their sum is 90°. Two positive angles are **supplementary** (supplements of each other) when their sum is 180°. (See figures below.)

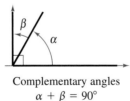

Complementary angles   Supplementary angles
$\alpha + \beta = 90°$ $\alpha + \beta = 180°$

EXAMPLE 2 **Complementary and Supplementary Angles**

a. For the angle 72°, the complement is $90° - 72° = 18°$, and the supplement is $180° - 72° = 108°$.

b. The angle 148° does not have a complement because $148° > 90°$. (Remember that complements are *positive* angles.) The supplement of 148° is $180° - 148° = 32°$.

✓ *Checkpoint* ▶ Audio-video solution in English & Spanish at LarsonPrecalculus.com

Find (if possible) the complement and supplement of (a) $\theta = 23°$ and (b) $\theta = -28°$.

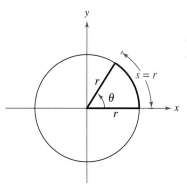

Arc length = radius when $\theta = 1$ radian
Figure 6.9

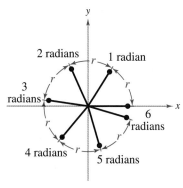

Figure 6.10

ALGEBRA HELP

One revolution around a circle
of radius r corresponds to an
angle of 2π radians because

$$\theta = \frac{s}{r} = \frac{2\pi r}{r} = 2\pi \text{ radians.}$$

Radian Measure

Another way to measure angles is in radians. This type of measure is especially useful
in calculus. To define a radian, use a **central angle** of a circle, which is an angle whose
vertex is the center of the circle, as shown in Figure 6.9.

Definition of a Radian

One **radian** (rad) is the measure of a central angle θ that intercepts an arc s
equal in length to the radius r of the circle. (See Figure 6.9.) Algebraically,
this means that

$$\theta = \frac{s}{r}$$

where θ is measured in radians. (Note that $\theta = 1$ when $s = r$.)

The circumference of a circle is $2\pi r$ units, so it follows that a central angle of one
full revolution (counterclockwise) corresponds to an arc length of $s = 2\pi r$. Moreover,
$2\pi \approx 6.28$, so there are just over six radius lengths in a full circle, as shown in
Figure 6.10. The units of measure for s and r are the same, so the ratio s/r has no
units—it is simply a real number.

EXAMPLE 3   **Finding Angles**

Find (a) an angle that is coterminal to $\theta = 17\pi/6$, (b) the complement of $\theta = \pi/12$,
and (c) the supplement of $\theta = 5\pi/6$.

Solution

a. To find a coterminal angle in radian measure, add or subtract 2π, which is equivalent
to 360°. For $\theta = 17\pi/6$, subtract 2π to obtain a coterminal angle.

$(17\pi/6) - 2\pi = (17\pi/6) - (12\pi/6) = 5\pi/6$ See Figure 6.11.

b. To find the complement of an angle in radian measure, subtract the angle from $\pi/2$,
which is equivalent to 90°. So, the complement of $\theta = \pi/12$ is

$(\pi/2) - (\pi/12) = (6\pi/12) - (\pi/12) = 5\pi/12.$ See Figure 6.12.

c. To find the supplement of an angle in radian measure, subtract the angle from π,
which is equivalent to 180°. So, the supplement of $\theta = 5\pi/6$ is

$\pi - (5\pi/6) = (6\pi/6) - (5\pi/6) = \pi/6.$ See Figure 6.13.

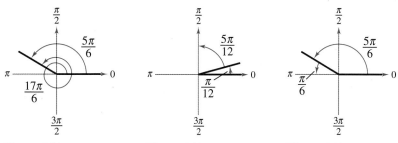

Figure 6.11   **Figure 6.12**   **Figure 6.13**

✓ *Checkpoint* ▶ *Audio-video solution in English & Spanish at LarsonPrecalculus.com*

Find (a) an angle that is coterminal to $\theta = -4\pi/3$, (b) the complement of $\theta = 3\pi/16$,
and (c) the supplement of $\theta = 5\pi/12$. ∎

GO DIGITAL

Conversion of Angle Measure

One complete revolution corresponds to 2π radians, so degrees and radians are related by the equations

$$360° = 2\pi \text{ rad} \quad \text{and} \quad 180° = \pi \text{ rad}.$$

From these equations, you obtain

$$1° = \frac{\pi}{180} \text{ rad} \quad \text{and} \quad 1 \text{ rad} = \left(\frac{180}{\pi}\right)°$$

which lead to the conversion rules below.

Figure 6.14

Conversions Between Degrees and Radians

1. To convert degrees to radians, multiply degrees by $\dfrac{\pi \text{ rad}}{180°}$.

2. To convert radians to degrees, multiply radians by $\dfrac{180°}{\pi \text{ rad}}$.

To apply these two conversion rules, use the basic relationship π rad $= 180°$. (See Figure 6.14.)

When no units of angle measure are specified, *radian measure is implied*. For example, $\theta = 2$ implies that $\theta = 2$ radians.

EXAMPLE 4 **Converting from Degrees to Radians**

a. $135° = (135 \text{ deg})\left(\dfrac{\pi \text{ rad}}{180 \text{ deg}}\right) = \dfrac{3\pi}{4}$ radians   *Multiply by $\frac{\pi \text{ rad}}{180°}$.*

b. $540° = (540 \text{ deg})\left(\dfrac{\pi \text{ rad}}{180 \text{ deg}}\right) = 3\pi$ radians   *Multiply by $\frac{\pi \text{ rad}}{180°}$.*

✓ *Checkpoint* ▶ Audio-video solution in English & Spanish at LarsonPrecalculus.com

Convert each degree measure to radian measure as a multiple of π. Do not use a calculator.

a. $60°$ **b.** $320°$

EXAMPLE 5 **Converting from Radians to Degrees**

a. $-\dfrac{\pi}{2} \text{ rad} = \left(-\dfrac{\pi}{2} \text{ rad}\right)\left(\dfrac{180 \text{ deg}}{\pi \text{ rad}}\right) = -90°$ *Multiply by $\frac{180°}{\pi \text{ rad}}$.*

b. $2 \text{ rad} = (2 \text{ rad})\left(\dfrac{180 \text{ deg}}{\pi \text{ rad}}\right) = \dfrac{360°}{\pi} \approx 114.59°$ *Multiply by $\frac{180°}{\pi \text{ rad}}$.*

✓ *Checkpoint* ▶ Audio-video solution in English & Spanish at LarsonPrecalculus.com

Convert each radian measure to degree measure. Do not use a calculator.

a. $\pi/6$ **b.** $5\pi/3$

Use a calculator with a "radian-to-degree" conversion key to verify the result shown in part (b) of Example 5.

GO DIGITAL

Applications

To measure arc length along a circle, use the radian measure formula, $\theta = s/r$.

Arc Length

For a circle of radius r, a central angle θ intercepts an arc of length s given by

$$s = r\theta \qquad \text{Length of circular arc}$$

where θ is measured in radians. Note that if $r = 1$, then $s = \theta$, and the radian measure of θ equals the arc length.

EXAMPLE 6   Finding Arc Length

A circle has a radius of 4 inches. Find the length of the arc intercepted by a central angle of 240°, as shown in Figure 6.15.

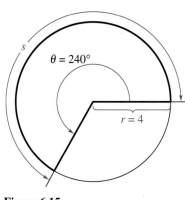

$\theta = 240°$

$r = 4$

Figure 6.15

Solution   To use the formula $s = r\theta$, first convert 240° to radian measure.

$$240° = (240 \ \text{deg})\left(\frac{\pi \ \text{rad}}{180 \ \text{deg}}\right) \qquad \text{Multiply by } \tfrac{\pi \ \text{rad}}{180°}.$$

$$= \frac{4\pi}{3} \ \text{radians} \qquad \text{Simplify.}$$

Then, using a radius of $r = 4$ inches, find the arc length.

$$s = r\theta \qquad \text{Length of circular arc}$$

$$= 4\left(\frac{4\pi}{3}\right) \qquad \text{Substitute for } r \text{ and } \theta.$$

$$\approx 16.76 \ \text{inches} \qquad \text{Use a calculator.}$$

Note that the units for r determine the units for $r\theta$ because θ is in radian measure, which has no units.

✓ **Checkpoint** Audio-video solution in English & Spanish at *LarsonPrecalculus.com*

A circle has a radius of 27 inches. Find the length of the arc intercepted by a central angle of 160°. ∎

The formula for the length of a circular arc can be used to analyze the motion of a particle moving at a *constant speed* along a circular path.

Linear and Angular Speeds

Consider a particle moving at a constant speed along a circular arc of radius r. If s is the length of the arc traveled in time t, then the **linear speed** v of the particle is

$$\text{Linear speed } v = \frac{\text{arc length}}{\text{time}} = \frac{s}{t}.$$

Moreover, if θ is the angle (in radian measure) corresponding to the arc length s, then the **angular speed** ω (the lowercase Greek letter omega) of the particle is

$$\text{Angular speed } \omega = \frac{\text{central angle}}{\text{time}} = \frac{\theta}{t}.$$

Linear speed measures how fast the particle moves, and angular speed measures how fast the angle changes. To establish a relationship between linear speed v and angular speed ω, divide each side of the formula for arc length by t, as shown below.

$$s = r\theta \implies \frac{s}{t} = \frac{r\theta}{t} \implies v = r\omega$$

EXAMPLE 7   Finding Linear Speed

The second hand of a clock is 10.2 centimeters long, as shown at the right. Find the linear speed of the tip of the second hand as it passes around the clock face.

Solution   In one revolution, the arc length traveled is

$$s = 2\pi r = 2\pi(10.2) = 20.4\pi \text{ centimeters.}$$

The time required for the second hand to travel this distance is $t = 1$ minute $= 60$ seconds. So, the linear speed of the tip of the second hand is

$$v = \frac{s}{t}$$
$$= \frac{20.4\pi \text{ centimeters}}{60 \text{ seconds}}$$
$$\approx 1.07 \text{ centimeters per second.}$$

 Checkpoint ▶ *Audio-video solution in English & Spanish at LarsonPrecalculus.com*

The second hand of a clock is 8 centimeters long. Find the linear speed of the tip of the second hand as it passes around the clock face.

EXAMPLE 8   Finding Angular and Linear Speeds

The blades of a wind turbine are 116 feet long (see Figure 6.16). The propeller rotates at 15 revolutions per minute.

a. Find the angular speed of the propeller in radians per minute.

b. Find the linear speed of the tips of the blades.

Solution

a. Each revolution corresponds to 2π radians, so the propeller turns $15(2\pi) = 30\pi$ radians per minute. In other words, the angular speed is

$$\omega = \frac{\theta}{t} = \frac{30\pi \text{ radians}}{1 \text{ minute}} = 30\pi \text{ radians per minute.}$$

b. The linear speed is

$$v = \frac{s}{t} = \frac{r\theta}{t} = \frac{116(30\pi) \text{ feet}}{1 \text{ minute}} \approx 10{,}933 \text{ feet per minute.}$$

Figure 6.16

 Checkpoint ▶ *Audio-video solution in English & Spanish at LarsonPrecalculus.com*

The circular blade on a saw has a radius of 4 inches and it rotates at 2400 revolutions per minute.

a. Find the angular speed of the blade in radians per minute.

b. Find the linear speed of the edge of the blade.

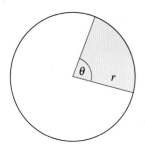

Figure 6.17

A **sector** of a circle is the region bounded by two radii of the circle and their intercepted arc (see Figure 6.17).

Area of a Sector of a Circle

For a circle of radius r, the area A of a sector of the circle with central angle θ is

$$A = \frac{1}{2}r^2\theta$$

where θ is measured in radians.

EXAMPLE 9   **Area of a Sector of a Circle**

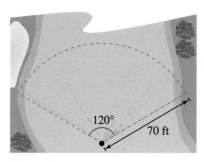

Figure 6.18

A sprinkler on a golf course fairway sprays water over a distance of 70 feet and rotates through an angle of 120°, as shown in Figure 6.18. Find the area of the fairway watered by the sprinkler.

Solution   First convert 120° to radian measure.

$$\theta = 120°$$

$$= (120 \; \text{deg})\left(\frac{\pi \; \text{rad}}{180 \; \text{deg}}\right) \qquad \text{Multiply by } \frac{\pi \; \text{rad}}{180°}.$$

$$= \frac{2\pi}{3} \; \text{radians}$$

Then, using $\theta = 2\pi/3$ and $r = 70$, the area is

$$A = \frac{1}{2}r^2\theta \qquad\qquad \text{Formula for the area of a sector of a circle}$$

$$= \frac{1}{2}(70)^2\left(\frac{2\pi}{3}\right) \qquad \text{Substitute for } r \text{ and } \theta.$$

$$= \frac{4900\pi}{3} \qquad\qquad \text{Multiply.}$$

$$\approx 5131 \; \text{square feet.} \qquad \text{Use a calculator.}$$

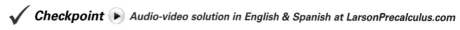 **Checkpoint** ▶ *Audio-video solution in English & Spanish at LarsonPrecalculus.com*

A sprinkler sprays water over a distance of 40 feet and rotates through an angle of 80°. Find the area watered by the sprinkler. ∎

Summarize   (Section 6.1)

1. Describe an angle *(page 422)*.

2. Explain how to use degree measure *(page 423)*. For examples involving degree measure, see Examples 1 and 2.

3. Explain how to use radian measure *(page 425)*. For an example involving radian measure, see Example 3.

4. Explain how to convert between degrees and radians *(page 426)*. For examples of converting between degrees and radians, see Examples 4 and 5.

5. Describe real-life applications involving angles and their measure *(pages 427–429, Examples 6–9)*.

GO DIGITAL

6.1 Exercises

See CalcChat.com for tutorial help and worked-out solutions to odd-numbered exercises.

Vocabulary and Concept Check

In Exercises 1–5, fill in the blanks.

1. _____ means "measurement of triangles."

2. Rotating a ray about its endpoint determines an _____.

3. The angle measure that is equivalent to a rotation of $\frac{1}{360}$ of a complete revolution about an angle's vertex is one _____.

4. One _____ is the measure of a central angle that intercepts an arc equal in length to the radius of the circle.

5. The area A of a sector of a circle with radius r and central angle θ, where θ is measured in radians, is given by the formula _____.

6. What is the sum (in radians) of two complementary angles? two supplementary angles?

7. How can you find an angle that is coterminal with a $5\pi/4$ angle?

8. Is the angle $2\pi/3$ acute or obtuse?

9. How do you convert an angle measured in degrees to radians? radians to degrees?

10. Compare the linear speed and angular speed of a particle.

Skills and Applications

Estimating an Angle **In Exercises 11–14, estimate the number of degrees in the angle.**

11.

12.

13.

14.

Sketching Angles **In Exercises 15–18, sketch each angle in standard position and tell whether the angle is a quadrantal angle. If not, determine the quadrant in which the angle lies.**

15. (a) $30°$ (b) $150°$

16. (a) $-270°$ (b) $-120°$

17. (a) $405°$ (b) $-540°$

18. (a) $-750°$ (b) $630°$

Finding Coterminal Angles **In Exercises 19–22, determine two coterminal angles (one positive and one negative) for each angle. Give your answers in degrees.**

19. (a)   (b)

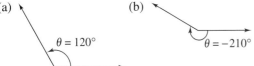

20. (a)   (b)

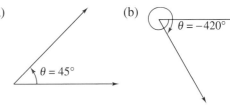

21. (a) $\theta = -300°$ (b) $\theta = 740°$

22. (a) $\theta = -520°$ (b) $\theta = 230°$

Converting to Decimal Degree Form **In Exercises 23–26, convert each angle measure to decimal degree form.**

23. $135° \, 36''$ 24. $-330° \, 25''$

25. $-408° \, 16' \, 20''$

26. $85° \, 18' \, 30''$

Converting to D° M′ S″ Form **In Exercises 27–30, convert each angle measure to D° M′ S″ form.**

27. $2.5°$ 28. $0.45°$

29. $-345.12°$ 30. $-3.58°$

Complementary and Supplementary Angles **In Exercises 31–36, find (if possible) the complement and supplement of each angle.**

31. $18°$ 32. $85°$

33. $93°$ 34. $46°$

35. $24°$ 36. $126°$

Estimating an Angle In Exercises 37–40, estimate the angle to the nearest one-half radian.

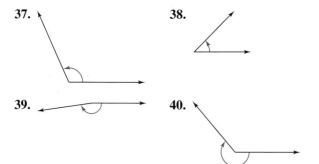

37.

38.

39.

40.

Determining Quadrants In Exercises 41–46, determine the quadrant in which each angle lies.

41. (a) $\dfrac{\pi}{4}$ (b) $\dfrac{3\pi}{4}$ **42.** (a) $\dfrac{11\pi}{8}$ (b) $\dfrac{13\pi}{8}$

43. (a) $-\dfrac{\pi}{5}$ (b) $\dfrac{7\pi}{5}$ **44.** (a) $\dfrac{\pi}{12}$ (b) $-\dfrac{11\pi}{9}$

45. (a) -1 (b) 8 **46.** (a) -7 (b) 1.5

Sketching Angles In Exercises 47–50, sketch each angle in standard position.

47. (a) $\dfrac{\pi}{3}$ (b) $-\dfrac{2\pi}{3}$ **48.** (a) $-\dfrac{7\pi}{4}$ (b) $\dfrac{5\pi}{2}$

49. (a) $\dfrac{17\pi}{6}$ (b) -3 **50.** (a) 4 (b) 7π

Finding Coterminal Angles In Exercises 51–56, determine two coterminal angles (one positive and one negative) for each angle. Give your answers in radians.

51. (a)   (b)

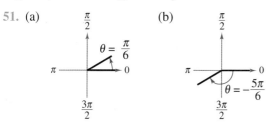

52. (a)   (b)

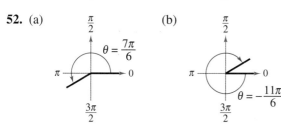

53. (a) $\theta = \dfrac{2\pi}{3}$ (b) $\theta = -\dfrac{\pi}{12}$

54. (a) $\theta = -\dfrac{3\pi}{4}$ (b) $\theta = -\dfrac{7\pi}{4}$

55. (a) $\theta = -\dfrac{9\pi}{4}$ (b) $\theta = -\dfrac{2\pi}{15}$

56. (a) $\theta = \dfrac{8\pi}{3}$ (b) $\theta = \dfrac{8\pi}{45}$

Complementary and Supplementary Angles In Exercises 57–60, find (if possible) the complement and supplement of each angle.

57. (a) $\dfrac{\pi}{12}$ (b) $\dfrac{11\pi}{12}$ **58.** (a) $\dfrac{\pi}{6}$ (b) $\dfrac{3\pi}{4}$

59. (a) $\dfrac{12\pi}{5}$ (b) $\dfrac{\pi}{4}$ **60.** (a) $\dfrac{2\pi}{3}$ (b) $\dfrac{3\pi}{2}$

Converting from Degrees to Radians In Exercises 61–64, convert each degree measure to radian measure as a multiple of π. Do not use a calculator.

61. (a) $30°$ (b) $45°$
62. (a) $315°$ (b) $120°$
63. (a) $20°$ (b) $-60°$
64. (a) $-270°$ (b) $144°$

Converting from Radians to Degrees In Exercises 65–68, convert each radian measure to degree measure. Do not use a calculator.

65. (a) $\dfrac{3\pi}{2}$ (b) $\dfrac{7\pi}{6}$ **66.** (a) $-\dfrac{7\pi}{12}$ (b) $\dfrac{\pi}{9}$

67. (a) $\dfrac{5\pi}{12}$ (b) $-\dfrac{7\pi}{3}$ **68.** (a) $\dfrac{11\pi}{6}$ (b) $\dfrac{34\pi}{15}$

Converting from Degrees to Radians In Exercises 69–74, convert the degree measure to radian measure. Round to three decimal places.

69. $45°$ **70.** $400°$
71. $532.76°$
72. $-216.35°$
73. $-0.83°$
74. $0.54°$

Converting from Radians to Degrees In Exercises 75–78, convert the radian measure to degree measure. Round to three decimal places.

75. $\dfrac{\pi}{7}$ **76.** $\dfrac{5\pi}{11}$

77. -2 **78.** -0.57

Finding Arc Length In Exercises 79–82, find the length of the arc on a circle of radius r intercepted by a central angle θ.

79. $r = 15$ inches, $\theta = 120°$
80. $r = 9$ feet, $\theta = 60°$
81. $r = 3$ meters, $\theta = 2$ radians
82. $r = 20$ centimeters, $\theta = \pi/4$ radian

Finding the Central Angle In Exercises 83–88, find the radian measure of the central angle of a circle of radius *r* that intercepts an arc of length *s*.

83. $r = 4$ inches, $s = 18$ inches
84. $r = 14$ feet, $s = 8$ feet
85. $r = 14.5$ centimeters, $s = 0.2$ meter
86. $r = 700$ meters, $s = 3.5$ kilometers
87.   88.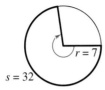

Area of a Sector of a Circle In Exercises 89–92, find the area of the sector of a circle of radius *r* and central angle θ.

89. $r = 6$ inches, $\theta = \pi/3$ radians
90. $r = 12$ millimeters, $\theta = \pi/4$ radian
91. $r = 2.5$ kilometers, $\theta = 225°$
92. $r = 1.4$ miles, $\theta = 330°$

Earth-Space Science In Exercises 93 and 94, find the distance between the cities. Assume that Earth is a sphere of radius 4000 miles and that the cities are on the same longitude (one city is due north of the other).

	City	Latitude
93.	Dallas, Texas	32° 47′ 9″ N
	Omaha, Nebraska	41° 15′ 50″ N
94.	San Francisco, California	37° 47′ 36″ N
	Seattle, Washington	47° 37′ 18″ N

95. **Instrumentation** The pointer on a voltmeter is 6 centimeters in length (see figure). Find the number of degrees through which the pointer rotates when it moves 2.5 centimeters on the scale.

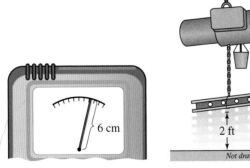

Figure for 95   Figure for 96

96. **Electric Hoist** The diameter of the drum on an electric hoist is 10 inches (see figure). Find the number of degrees through which the drum must rotate to lift a beam 2 feet.

97. **Linear Speed** A satellite in circular orbit 1250 kilometers above Earth makes one complete revolution every 110 minutes. Assuming that Earth is a sphere of radius 6400 kilometers, what is the linear speed (in kilometers per minute) of the satellite?

98. **Angular Speed** A car moves at a rate of 65 miles per hour. The diameter of its wheels is 2.5 feet.
    (a) Find the number of revolutions per minute the wheels are rotating.
    (b) Find the angular speed of the wheels in radians per minute.

99. **Linear and Angular Speed** A $7\frac{1}{4}$-inch circular power saw blade rotates at 5200 revolutions per minute.
    (a) Find the angular speed of the saw blade in radians per minute.
    (b) Find the linear speed (in feet per minute) of the saw teeth as they contact the wood being cut.

100. **Linear and Angular Speed** A carousel with a 50-foot diameter makes 4 revolutions per minute.
    (a) Find the angular speed of the carousel in radians per minute.
    (b) Find the linear speed (in feet per minute) of the platform rim of the carousel.

101. **Linear and Angular Speed** A Blu-ray disc is approximately 12 centimeters in diameter. The drive motor of a Blu-ray player is able to rotate up to 10,000 revolutions per minute.
    (a) Find the maximum angular speed (in radians per second) of a Blu-ray disc as it rotates.
    (b) Find the maximum linear speed (in meters per second) of a point on the outermost track as the disc rotates.

102. **Linear and Angular Speed** A computerized spin balance machine rotates a 25-inch-diameter tire at 480 revolutions per minute.
    (a) Find the road speed (in miles per hour) at which the tire is being balanced.
    (b) At what rate should the spin balance machine be set so that the tire is being tested for 55 miles per hour?

103. **Area** A sprinkler on a golf green is set to spray water over a distance of 15 meters and to rotate through an angle of 150°. Draw a diagram that shows the region that can be irrigated with the sprinkler. Find the area of the region.

104. **Area** A sprinkler system on a farm is set to spray water over a distance of 35 meters and to rotate through an angle of 140°. Draw a diagram that shows the region that can be irrigated with the sprinkler. Find the area of the region.

105. Speed of a Bicycle

The radii of the pedal sprocket, the wheel sprocket, and the wheel of the bicycle in the figure are 4 inches, 2 inches, and 14 inches, respectively. A cyclist pedals at a rate of 1 revolution per second.

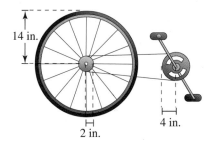

14 in.

4 in.

2 in.

(a) Find the speed of the bicycle in feet per second and miles per hour.

(b) Use your result from part (a) to write a function for the distance d (in miles) a cyclist travels in terms of the number n of revolutions of the pedal sprocket.

(c) Write a function for the distance d (in miles) a cyclist travels in terms of the time t (in seconds). Compare this function with the function from part (b).

(d) Classify the types of functions you found in parts (b) and (c). Explain.

106. Area
A car's rear windshield wiper rotates $125°$. The total length of the wiper mechanism is 25 inches and the length of the wiper blade is 14 inches.

(a) Find the area wiped by the wiper blade.

(b) The wiper takes 1 second to rotate and return to its starting position. Find the angular speed of the tip of the wiper blade. Does the entire wiper blade move at the same angular speed? Explain.

(c) Find the linear speed of the tip of the wiper blade. Does the entire wiper blade move at the same linear speed? Explain.

Exploring the Concepts

True or False? In Exercises 107–109, determine whether the statement is true or false. Justify your answer.

107. An angle measure containing π must be in radian measure.

108. A measurement of 4 radians corresponds to two complete revolutions from the initial side to the terminal side of an angle.

109. A degree is a larger unit of measure than a radian.

110. HOW DO YOU SEE IT? Determine which angles in the figure are coterminal angles with angle A. Explain.

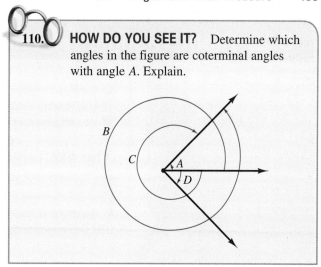

111. Think About It A fan motor turns at a given angular speed. How does the speed of the tips of the blades change when a fan of greater diameter is installed on the motor? Explain.

112. Proof Prove that the area of a circular sector of radius r with central angle θ is $A = \frac{1}{2}\theta r^2$, where θ is measured in radians.

Review & Refresh ▶ Video solutions at LarsonPrecalculus.com

Using the Pythagorean Theorem In Exercises 113–116, solve for x in the right triangle.

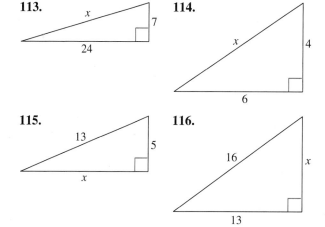

113.

x

7

24

114.

x

4

6

115.

13

5

x

116.

16

x

13

Finding the Standard Equation of a Conic In Exercises 117–120, find the standard equation of the given conic with the given characteristics.

117. Parabola; vertex: $(0, 0)$; focus: $(-2, 0)$

118. Ellipse; center: $(0, 0)$; vertices: $(\pm 4, 0)$; foci: $(\pm 3, 0)$

119. Hyperbola; center: $(0, 0)$; vertices: $(\pm 5, 0)$; foci: $(\pm 6, 0)$

120. Hyperbola; center: $(0, 0)$; vertices: $(0, \pm 3)$; asymptotes: $y = \pm x$

6.2 Right Triangle Trigonometry

Right triangle trigonometry has many real-life applications. For example, in Exercise 68 on page 444, you will use right triangle trigonometry to analyze the height of a helium-filled balloon.

❯ Evaluate trigonometric functions of acute angles.
❯ Use fundamental trigonometric identities.
❯ Use trigonometric functions to model and solve real-life problems.

The Six Trigonometric Functions

This section introduces the trigonometric functions from a *right triangle* perspective. Consider the right triangle shown below, in which one acute angle is labeled θ. Relative to the angle θ, the three sides of the triangle are the **hypotenuse,** the **opposite side** (the side opposite the angle θ), and the **adjacent side** (the side adjacent to the angle θ).

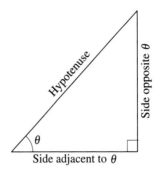

Side adjacent to θ

Using the lengths of these three sides, you can form six ratios that define the six trigonometric functions of the acute angle θ.

sine cosecant cosine secant tangent cotangent

Abbreviations for these six functions are sin, csc, cos, sec, tan, and cot, respectively. In the definitions below, $0° < \theta < 90°$ (θ lies in the first quadrant). For such angles, the value of each trigonometric function is *positive*.

HISTORICAL NOTE

Georg Joachim Rheticus (1514–1576) was the leading Teutonic mathematical astronomer of the sixteenth century. He was the first to define the trigonometric functions as ratios of the sides of a right triangle.

Right Triangle Definitions of Trigonometric Functions

Let θ be an *acute* angle of a right triangle. The six trigonometric functions of the angle θ are defined below. (Note that the functions in the second column are the *reciprocals* of the corresponding functions in the first column.)

$$\sin \theta = \frac{\text{opp}}{\text{hyp}} \qquad \csc \theta = \frac{\text{hyp}}{\text{opp}}$$

$$\cos \theta = \frac{\text{adj}}{\text{hyp}} \qquad \sec \theta = \frac{\text{hyp}}{\text{adj}}$$

$$\tan \theta = \frac{\text{opp}}{\text{adj}} \qquad \cot \theta = \frac{\text{adj}}{\text{opp}}$$

The abbreviations opp, adj, and hyp represent the lengths of the three sides of a right triangle.

opp = the length of the side *opposite* θ

adj = the length of the side *adjacent* to θ

hyp = the length of the *hypotenuse*

GO DIGITAL

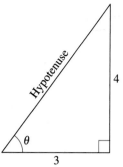

Figure 6.19

▶︎▶︎▶︎ **SKILLS REFRESHER**

For a refresher on the properties of triangles and finding unknown dimensions, watch the video at *LarsonPrecalculus.com*.

EXAMPLE 1 **Evaluating Trigonometric Functions**

▶▶▶ *See LarsonPrecalculus.com for an interactive version of this type of example.*

Use the triangle in Figure 6.19 to find the values of the six trigonometric functions of θ.

Solution In Figure 6.19, opp = 4 and adj = 3, but hyp is not given. To find hyp, note that by the Pythagorean Theorem, $(\text{hyp})^2 = (\text{opp})^2 + (\text{adj})^2$. So, it follows that

$$\text{hyp} = \sqrt{4^2 + 3^2} = \sqrt{25} = 5.$$

So, the six trigonometric functions of θ are

$$\sin\theta = \frac{\text{opp}}{\text{hyp}} = \frac{4}{5} \qquad \csc\theta = \frac{\text{hyp}}{\text{opp}} = \frac{5}{4}$$

$$\cos\theta = \frac{\text{adj}}{\text{hyp}} = \frac{3}{5} \qquad \sec\theta = \frac{\text{hyp}}{\text{adj}} = \frac{5}{3}$$

$$\tan\theta = \frac{\text{opp}}{\text{adj}} = \frac{4}{3} \qquad \cot\theta = \frac{\text{adj}}{\text{opp}} = \frac{3}{4}.$$

✓ *Checkpoint* ▶ Audio-video solution in English & Spanish at *LarsonPrecalculus.com*

Use the triangle below to find the values of the six trigonometric functions of θ.

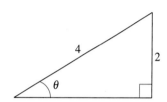

In Example 1, you were given the lengths of two sides of the right triangle but not the angle θ. Often, you will be asked to find the trigonometric functions of a *given* acute angle θ. To do this, construct a right triangle having θ as one of its angles.

EXAMPLE 2 **Evaluating Trigonometric Functions of 45°**

Find the values of $\sin 45°$, $\cos 45°$, and $\tan 45°$.

Solution Construct a right triangle having 45° as one of its acute angles, as shown in Figure 6.20. For convenience, choose 1 as the length of the adjacent side. From geometry, you know that the other acute angle is also 45°. So, the triangle is isosceles and the length of the opposite side is also 1. By the Pythagorean Theorem, the length of the hypotenuse is $\sqrt{2}$.

$$\sin 45° = \frac{\text{opp}}{\text{hyp}} = \frac{1}{\sqrt{2}} = \frac{\sqrt{2}}{2}$$

$$\cos 45° = \frac{\text{adj}}{\text{hyp}} = \frac{1}{\sqrt{2}} = \frac{\sqrt{2}}{2}$$

$$\tan 45° = \frac{\text{opp}}{\text{adj}} = \frac{1}{1} = 1$$

✓ *Checkpoint* ▶ Audio-video solution in English & Spanish at *LarsonPrecalculus.com*

Find the values of $\cot 45°$, $\sec 45°$, and $\csc 45°$.

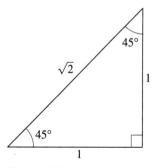

Figure 6.20

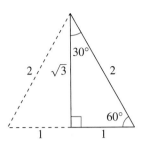

Figure 6.21

ALGEBRA HELP

These special angles occur frequently in trigonometry, so you should learn to construct the triangles shown in Figures 6.20 and 6.21.

TECHNOLOGY

When evaluating trigonometric functions with a calculator, remember to enclose all fractional angle measures in parentheses. For example, to evaluate $\sin \theta$ for $\theta = \pi/6$, enter

$\boxed{\text{SIN}}\ \boxed{(\!(}\ \boxed{\pi}\ \boxed{\div}\ 6\ \boxed{)\!)}\ \boxed{\text{ENTER}}$.

The keystrokes yield the correct value of 0.5. Note that some calculators automatically place a left parenthesis after trigonometric functions.

GO DIGITAL

EXAMPLE 3   **Evaluating Trigonometric Functions of 30° and 60°**

Use the equilateral triangle shown in Figure 6.21 to find the values of sin 60°, cos 60°, sin 30°, and cos 30°.

Solution   For $\theta = 60°$, you have adj $= 1$, opp $= \sqrt{3}$, and hyp $= 2$. So,

$$\sin 60° = \frac{\text{opp}}{\text{hyp}} = \frac{\sqrt{3}}{2} \quad \text{and} \quad \cos 60° = \frac{\text{adj}}{\text{hyp}} = \frac{1}{2}.$$

For $\theta = 30°$, adj $= \sqrt{3}$, opp $= 1$, and hyp $= 2$. So,

$$\sin 30° = \frac{\text{opp}}{\text{hyp}} = \frac{1}{2} \quad \text{and} \quad \cos 30° = \frac{\text{adj}}{\text{hyp}} = \frac{\sqrt{3}}{2}.$$

 ✓ *Checkpoint* ▶ *Audio-video solution in English & Spanish at LarsonPrecalculus.com*

Use the equilateral triangle shown in Figure 6.21 to find the values of tan 60° and tan 30°.

Sines, Cosines, and Tangents of Special Angles

$$\sin 30° = \sin \frac{\pi}{6} = \frac{1}{2} \qquad \cos 30° = \cos \frac{\pi}{6} = \frac{\sqrt{3}}{2} \qquad \tan 30° = \tan \frac{\pi}{6} = \frac{\sqrt{3}}{3}$$

$$\sin 45° = \sin \frac{\pi}{4} = \frac{\sqrt{2}}{2} \qquad \cos 45° = \cos \frac{\pi}{4} = \frac{\sqrt{2}}{2} \qquad \tan 45° = \tan \frac{\pi}{4} = 1$$

$$\sin 60° = \sin \frac{\pi}{3} = \frac{\sqrt{3}}{2} \qquad \cos 60° = \cos \frac{\pi}{3} = \frac{1}{2} \qquad \tan 60° = \tan \frac{\pi}{3} = \sqrt{3}$$

Note that $\sin 30° = \frac{1}{2} = \cos 60°$. This occurs because 30° and 60° are complementary angles. In general, it can be shown from the right triangle definitions that *cofunctions of complementary angles are equal*. That is, if θ is an acute angle, then the relationships below are true.

$$\sin(90° - \theta) = \cos \theta \qquad\qquad \cos(90° - \theta) = \sin \theta$$
$$\tan(90° - \theta) = \cot \theta \qquad\qquad \cot(90° - \theta) = \tan \theta$$
$$\sec(90° - \theta) = \csc \theta \qquad\qquad \csc(90° - \theta) = \sec \theta$$

When evaluating a trigonometric function with a calculator, set the calculator to the desired *mode* of measurement (degree or radian). Most calculators do not have keys for the cosecant, secant, and cotangent functions. To evaluate these functions, you can use the $\boxed{x^{-1}}$ key with their respective reciprocal functions, sine, cosine, and tangent. For example, to evaluate $\csc(\pi/8)$, use the fact that $\csc(\pi/8) = 1/\sin(\pi/8)$ and enter the keystroke sequence below in *radian* mode.

$\boxed{(\!(}\ \boxed{\text{SIN}}\ \boxed{(\!(}\ \boxed{\pi}\ \boxed{\div}\ 8\ \boxed{)\!)}\ \boxed{)\!)}\ \boxed{x^{-1}}\ \boxed{\text{ENTER}}$ Display 2.6131259

EXAMPLE 4   **Using a Calculator**

Function	Mode	Calculator Keystrokes	Display
a. $\sin 76.4°$	Degree	$\boxed{\text{SIN}}\ \boxed{(\!(}\ 76.4\ \boxed{)\!)}\ \boxed{\text{ENTER}}$	0.9719610
b. $\cot 1.5$	Radian	$\boxed{(\!(}\ \boxed{\text{TAN}}\ \boxed{(\!(}\ 1.5\ \boxed{)\!)}\ \boxed{)\!)}\ \boxed{x^{-1}}\ \boxed{\text{ENTER}}$	0.0709148

✓ *Checkpoint* ▶ *Audio-video solution in English & Spanish at LarsonPrecalculus.com*

Use a calculator to evaluate (a) sin 32.8° and (b) csc 1.2.

Trigonometric Identities

Trigonometric identities are relationships between trigonometric functions.

Fundamental Trigonometric Identities

Reciprocal Identities

$$\sin \theta = \frac{1}{\csc \theta} \qquad \cos \theta = \frac{1}{\sec \theta} \qquad \tan \theta = \frac{1}{\cot \theta}$$

$$\csc \theta = \frac{1}{\sin \theta} \qquad \sec \theta = \frac{1}{\cos \theta} \qquad \cot \theta = \frac{1}{\tan \theta}$$

Quotient Identities

$$\tan \theta = \frac{\sin \theta}{\cos \theta} \qquad \cot \theta = \frac{\cos \theta}{\sin \theta}$$

Pythagorean Identities

$$\sin^2 \theta + \cos^2 \theta = 1$$
$$1 + \tan^2 \theta = \sec^2 \theta$$
$$1 + \cot^2 \theta = \csc^2 \theta$$

ALGEBRA HELP

Note that $\sin^2 \theta$ represents $(\sin \theta)^2$, $\cos^2 \theta$ represents $(\cos \theta)^2$, and so on.

EXAMPLE 5   Applying Trigonometric Identities

Let θ be an acute angle such that $\sin \theta = 0.6$. Use trigonometric identities to find (a) $\cos \theta$ and (b) $\tan \theta$.

Solution

a. To find the value of $\cos \theta$, use the Pythagorean identity $\sin^2 \theta + \cos^2 \theta = 1$ with $\sin \theta = 0.6$, and solve for $\cos \theta$.

$$(0.6)^2 + \cos^2 \theta = 1 \qquad \text{Substitute 0.6 for } \sin \theta.$$

$$0.36 + \cos^2 \theta = 1 \qquad \text{Evaluate power.}$$

$$\cos^2 \theta = 0.64 \qquad \text{Subtract 0.36 from each side.}$$

$$\cos \theta = \sqrt{0.64} \qquad \text{Extract positive square root.}$$

$$\cos \theta = 0.8 \qquad \text{Simplify.}$$

b. Now, knowing the sine and cosine of θ, you can find the tangent of θ.

$$\tan \theta = \frac{\sin \theta}{\cos \theta} \qquad \text{Quotient identity}$$

$$= \frac{0.6}{0.8} \qquad \text{Substitute for } \sin \theta \text{ and } \cos \theta.$$

$$= 0.75 \qquad \text{Simplify.}$$

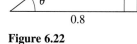

Figure 6.22

Use the definitions of $\cos \theta$ and $\tan \theta$ and the triangle shown in Figure 6.22 to check these results.

 ✓ **Checkpoint** ▶ *Audio-video solution in English & Spanish at LarsonPrecalculus.com*

Let θ be an acute angle such that $\cos \theta = 0.96$. Use trigonometric identities to find (a) $\sin \theta$ and (b) $\tan \theta$. ∎

GO DIGITAL

SKILLS REFRESHER

For a refresher on operations with fractions, watch the video at *LarsonPrecalculus.com*.

EXAMPLE 6   **Applying Trigonometric Identities**

Let θ be an acute angle such that $\tan \theta = \frac{1}{3}$. Use trigonometric identities to find (a) $\cot \theta$ and (b) $\sec \theta$.

Solution

a. $\cot \theta = \dfrac{1}{\tan \theta}$ Reciprocal identity

$\qquad = \dfrac{1}{1/3}$ Substitute $\frac{1}{3}$ for $\tan \theta$.

$\qquad = 3$ Simplify.

b. $\sec^2 \theta = 1 + \tan^2 \theta$ Pythagorean identity

$\quad \sec^2 \theta = 1 + \left(\dfrac{1}{3}\right)^2$ Substitute $\frac{1}{3}$ for $\tan \theta$.

$\quad \sec^2 \theta = \dfrac{10}{9}$ Simplify.

$\quad \sec \theta = \dfrac{\sqrt{10}}{3}$ Extract positive square root and simplify.

Use the definitions of $\cot \theta$ and $\sec \theta$ and the triangle below to check these results.

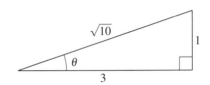

✓ *Checkpoint* ▶ Audio-video solution in English & Spanish at LarsonPrecalculus.com

Let θ be an acute angle such that $\tan \theta = 2$. Use trigonometric identities to find (a) $\cot \theta$ and (b) $\sec \theta$.

EXAMPLE 7   **Using Trigonometric Identities**

Use trigonometric identities to transform the left side of the equation into the right side $(0 < \theta < \pi/2)$.

a. $\sin \theta \csc \theta = 1$ **b.** $(\csc \theta + \cot \theta)(\csc \theta - \cot \theta) = 1$

Solution

a. $\sin \theta \csc \theta = \left(\dfrac{1}{\csc \theta}\right) \csc \theta = 1$ Use a reciprocal identity and simplify.

b. $(\csc \theta + \cot \theta)(\csc \theta - \cot \theta)$

$\qquad = \csc^2 \theta - \cot^2 \theta$ Sum and difference of same terms

$\qquad = 1$ Pythagorean identity

✓ *Checkpoint* ▶ Audio-video solution in English & Spanish at LarsonPrecalculus.com

Use trigonometric identities to transform the left side of the equation into the right side $(0 < \theta < \pi/2)$.

a. $\tan \theta \csc \theta = \sec \theta$ **b.** $(\csc \theta + 1)(\csc \theta - 1) = \cot^2 \theta$

GO DIGITAL

Applications Involving Right Triangles

Many applications of trigonometry involve **solving right triangles.** In this type of application, you are usually given one side of a right triangle and one of the acute angles and are asked to find one of the other sides, *or* you are given two sides and are asked to find one of the acute angles.

In Example 8, you are given the **angle of elevation,** which represents the angle from the horizontal upward to an object. In other applications you may be given the **angle of depression,** which represents the angle from the horizontal downward to an object. (See Figure 6.23.)

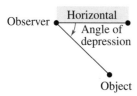

Figure 6.23

EXAMPLE 8 **Solving a Right Triangle**

A surveyor stands 115 feet from the base of the Washington Monument, as shown in Figure 6.24. The surveyor measures the angle of elevation to the top of the monument to be 78.3°. How tall is the Washington Monument? (Round your answer to the nearest foot.)

Solution From Figure 6.24,

$$\tan 78.3° = \frac{\text{opp}}{\text{adj}} = \frac{y}{115}$$

where y is the height of the monument. So, to the nearest foot, the height of the Washington Monument is

$$y = 115 \tan 78.3° \approx 555 \text{ feet.}$$

✓ *Checkpoint* ▶ *Audio-video solution in English & Spanish at LarsonPrecalculus.com*

The angle of elevation to the top of a flagpole at a distance of 19 feet from its base is 64.6°. How tall is the flagpole?

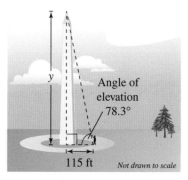

Figure 6.24

EXAMPLE 9 **Solving a Right Triangle**

A lighthouse is 200 yards from a bike path along the edge of a lake. A walkway to the lighthouse is 400 yards long. (See Figure 6.25.) Find the acute angle θ between the bike path and the walkway.

Solution From Figure 6.25, the sine of the angle θ is

$$\sin \theta = \frac{\text{opp}}{\text{hyp}} = \frac{200}{400} = \frac{1}{2}.$$

You should recognize that $\theta = 30°$.

✓ *Checkpoint* ▶ *Audio-video solution in English & Spanish at LarsonPrecalculus.com*

Find the acute angle θ between the two paths shown below.

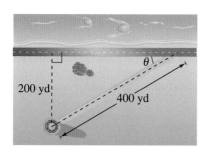

Figure 6.25

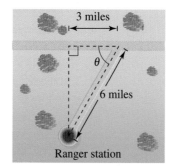

GO DIGITAL

In Example 9, you were able to recognize that the special angle $\theta = 30°$ satisfies the equation $\sin \theta = \frac{1}{2}$. However, when θ is not a special angle, you can *estimate* its value. For example, to estimate the acute angle θ in the equation $\sin \theta = 0.6$, you could reason that $\sin 30° = \frac{1}{2} = 0.5000$ and $\sin 45° = 1/\sqrt{2} \approx 0.7071$, so θ lies somewhere between 30° and 45°. In a later section, you will study a method of determining a more precise value of θ.

> **EXAMPLE 10**   **Solving a Right Triangle**

Find the length c and the height b of the skateboard ramp below.

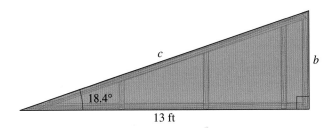

Solution   From the figure,

$$\cos 18.4° = \frac{\text{adj}}{\text{hyp}} = \frac{13}{c}.$$

So, the length of the skateboard ramp is

$$c = \frac{13}{\cos 18.4°} \approx 13.7 \text{ feet.}$$

Also from the figure,

$$\tan 18.4° = \frac{\text{opp}}{\text{adj}} = \frac{b}{13}.$$

So, the height is

$$b = 13 \tan 18.4° \approx 4.3 \text{ feet.}$$

✓ *Checkpoint* ▶ *Audio-video solution in English & Spanish at LarsonPrecalculus.com*

Find the length c and the horizontal length a of the loading ramp below.

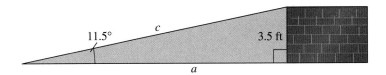

> **Summarize**   (Section 6.2)
>
> **1.** State the right triangle definitions of the six trigonometric functions *(page 434)*. For examples of evaluating trigonometric functions of acute angles, see Examples 1–4.
>
> **2.** List the reciprocal, quotient, and Pythagorean identities *(page 437)*. For examples of using these identities, see Examples 5–7.
>
> **3.** Describe real-life applications of trigonometric functions *(pages 439 and 440, Examples 8–10)*.

GO DIGITAL

6.2 Exercises

See CalcChat.com for tutorial help and worked-out solutions to odd-numbered exercises.

GO DIGITAL

Vocabulary and Concept Check

1. Use the figure to answer each question.

   (a) What is the length of the side opposite the angle θ?

   (b) What is the length of the side adjacent to the angle θ?

   (c) What is the length of the hypotenuse?

2. Use the figure to determine each function value.

   (a) $\sin \alpha$ (b) $\cos \beta$

   (c) $\tan \beta$ (d) $\cot \alpha$

   (e) $\sec \alpha$ (f) $\csc \beta$

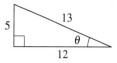

Figure for 1

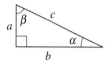

Figure for 2

In Exercises 3 and 4, fill in the blanks.

3. Cofunctions of _____ angles are equal.

4. An angle of _____ represents the angle from the horizontal upward to an object, whereas an angle of _____ represents the angle from the horizontal downward to an object.

Skills and Applications

Evaluating Trigonometric Functions In Exercises 5–10, find the exact values of the six trigonometric functions of the angle θ.

5.

6.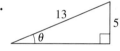

7.

8.

9.

10.

Evaluating Trigonometric Functions In Exercises 11–14, find the exact values of the six trigonometric functions of the angle θ for each of the two triangles. Explain why the function values are the same.

11.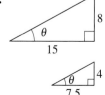

12.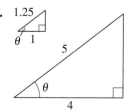

13.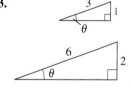

14.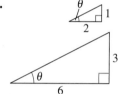

Evaluating Trigonometric Functions In Exercises 15–22, sketch a right triangle corresponding to the trigonometric function of the acute angle θ. Then find the exact values of the other five trigonometric functions of θ.

15. $\cos \theta = \frac{15}{17}$ 16. $\sin \theta = \frac{3}{5}$

17. $\sec \theta = \frac{6}{5}$ 18. $\tan \theta = \frac{4}{5}$

19. $\sin \theta = \frac{1}{5}$ 20. $\sec \theta = \frac{17}{7}$

21. $\cot \theta = 3$ 22. $\csc \theta = 9$

Evaluating Trigonometric Functions of 30°, 45°, and 60° In Exercises 23–26, construct an appropriate triangle to find the missing values. ($0° \le \theta \le 90°, 0 \le \theta \le \pi/2$)

Function	θ (deg)	θ (rad)	Function Value
23. tan	30°		
24. cos	45°		
25. sec		$\frac{\pi}{4}$	
26. csc		$\frac{\pi}{6}$	

Using a Calculator In Exercises 27–32, use a calculator to evaluate each function. Round your answers to four decimal places. (Be sure the calculator is in the correct mode.)

27. (a) $\sin 20.2°$ (b) $\csc 69.8°$

28. (a) $\tan 12.75°$ (b) $\cot 12.75°$

29. (a) $\cot \dfrac{\pi}{16}$ (b) $\tan \dfrac{\pi}{16}$

30. (a) $\sec \dfrac{\pi}{9}$ (b) $\cos \dfrac{\pi}{9}$

31. (a) $\csc 1$ (b) $\tan \dfrac{1}{2}$

32. (a) $\sec\left(\dfrac{\pi}{2} - 1\right)$ (b) $\cot\left(\dfrac{\pi}{2} - \dfrac{1}{2}\right)$

Applying Trigonometric Identities In Exercises 33–38, use the given function value(s) and the trigonometric identities to find the exact value of each indicated trigonometric function.

33. $\sin \theta = \dfrac{\sqrt{3}}{2}$

   (a) $\csc \theta$ (b) $\cos \theta$

   (c) $\tan \theta$ (d) $\cot \theta$

34. $\tan \theta = \dfrac{\sqrt{3}}{3}$

   (a) $\csc \theta$ (b) $\cot \theta$

   (c) $\cos \theta$ (d) $\sin \theta$

35. $\cos \theta = \dfrac{1}{3}$

   (a) $\sin \theta$ (b) $\tan \theta$

   (c) $\sec \theta$ (d) $\csc(90° - \theta)$

36. $\sec \theta = 5$

   (a) $\cos \theta$ (b) $\cot \theta$

   (c) $\csc \theta$ (d) $\sin \theta$

37. $\cot \alpha = 3$

   (a) $\tan \alpha$ (b) $\csc \alpha$

   (c) $\sec \alpha$ (d) $\sin \alpha$

38. $\cos \beta = \dfrac{\sqrt{7}}{4}$

   (a) $\sec \beta$ (b) $\sin \beta$

   (c) $\cot \beta$ (d) $\sin(90° - \beta)$

Using Trigonometric Identities In Exercises 39–48, use trigonometric identities to transform the left side of the equation into the right side $(0 < \theta < \pi/2)$.

39. $\tan \theta \cot \theta = 1$

40. $\cos \theta \sec \theta = 1$

41. $\tan \alpha \cos \alpha = \sin \alpha$

42. $\cot \alpha \sin \alpha = \cos \alpha$

43. $(1 + \sin \theta)(1 - \sin \theta) = \cos^2 \theta$

44. $(1 + \cos \theta)(1 - \cos \theta) = \sin^2 \theta$

45. $(\sec \theta + \tan \theta)(\sec \theta - \tan \theta) = 1$

46. $\sin^2 \theta - \cos^2 \theta = 2 \sin^2 \theta - 1$

47. $\dfrac{\sin \theta}{\cos \theta} + \dfrac{\cos \theta}{\sin \theta} = \csc \theta \sec \theta$

48. $\dfrac{\tan \beta + \cot \beta}{\tan \beta} = \csc^2 \beta$

Finding Special Angles of a Triangle In Exercises 49–54, find each value of θ in degrees $(0° < \theta < 90°)$ and radians $(0 < \theta < \pi/2)$ without using a calculator.

49. (a) $\sin \theta = \dfrac{1}{2}$ (b) $\csc \theta = 2$

50. (a) $\cos \theta = \dfrac{\sqrt{2}}{2}$ (b) $\tan \theta = 1$

51. (a) $\sec \theta = 2$ (b) $\cot \theta = 1$

52. (a) $\tan \theta = \sqrt{3}$ (b) $\csc \theta = \sqrt{2}$

53. (a) $\csc \theta = \dfrac{2\sqrt{3}}{3}$ (b) $\sin \theta = \dfrac{\sqrt{2}}{2}$

54. (a) $\cot \theta = \dfrac{\sqrt{3}}{3}$ (b) $\sec \theta = \sqrt{2}$

Finding Side Lengths of a Triangle In Exercises 55–58, find the exact values of the indicated variables.

55. Find x and y.

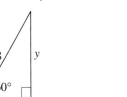

56. Find x and r.

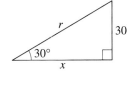

57. Find x and r.

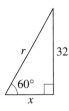

58. Find x and r.

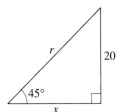

59. **Empire State Building** You are standing 45 meters from the base of the Empire State Building. You estimate that the angle of elevation to the top of the 86th floor (the observatory) is $82°$. The total height of the building is another 123 meters above the 86th floor. What is the approximate height of the building? One of your friends is on the 86th floor. What is the distance between you and your friend?

60. Distance   A passenger in an airplane at an altitude of 10 kilometers sees a town to the east of the plane. The angle of depression to the town is 22°.

(a) Draw a right triangle that gives a visual representation of the problem. Label the known quantities of the triangle and use variables to represent the number of kilometers east the town is from the plane and the straight-line distance from the town to the plane.

(b) Use trigonometric functions to write equations involving the unknown quantities.

(c) How many kilometers east of the plane is the town?

(d) What is the straight-line distance from the town to the plane?

61. Angle of Elevation   You are skiing down a mountain with a vertical height of 1250 feet. The distance from the top of the mountain to the base is 2500 feet. What is the angle of elevation from the base to the top of the mountain?

62. Biology   A biologist wants to know the width w of a river. From point A, the biologist walks downstream 100 feet and sights to point C (see figure). From this sighting, it is determined that $\theta = 54°$. How wide is the river?

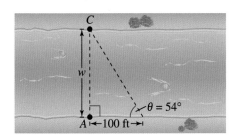

63. Guy Wire   A guy wire runs from the ground to a cell tower. The wire is attached to the cell tower 150 feet above the ground. The angle formed between the wire and the ground is 43° (see figure).

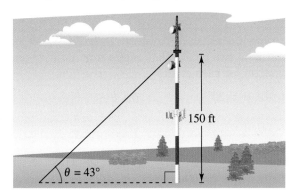

(a) How long is the guy wire?

(b) How far from the base of the tower is the guy wire anchored to the ground?

64. Height of a Mountain   In traveling across flat land, you see a mountain directly in front of you. Its angle of elevation (to the peak) is 3.5°. After you drive 13 miles closer to the mountain, the angle of elevation is 9° (see figure). Approximate the height of the mountain.

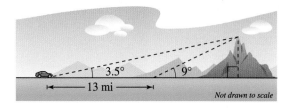

65. Machine Shop Calculations   A steel plate has the form of one-fourth of a circle with a radius of 60 centimeters. Two two-centimeter holes are drilled in the plate, positioned as shown in the figure. Find the coordinates of the center of each hole.

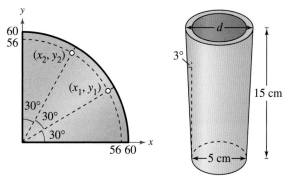

Figure for 65   Figure for 66

66. Machine Shop Calculations   A tapered shaft has a diameter of 5 centimeters at the small end and is 15 centimeters long (see figure). The taper is 3°. Find the diameter d of the large end of the shaft.

67. Geometry   Use a compass to sketch a quarter of a circle of radius 10 centimeters. Using a protractor, construct an angle of 20° in standard position (see figure). Construct a line perpendicular to the x-axis from the point of intersection of the terminal side of the angle and the arc of the circle. By actual measurement, calculate the coordinates (x, y) of the point of intersection and use these measurements to approximate the six trigonometric functions of a 20° angle. Use a calculator to check the accuracy of your results.

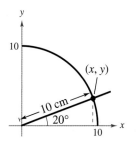

68. Helium-Filled Balloon

A 20-meter line is used to tether a helium-filled balloon. The line makes an angle of approximately 85° with the ground because of a breeze.

(a) Draw a right triangle that gives a visual representation of the problem. Label the known quantities of the triangle and use a variable to represent the height of the balloon.

(b) Use a trigonometric function to write and solve an equation for the height of the balloon.

(c) The breeze becomes stronger and the angle the line makes with the ground decreases. How does this affect the triangle you drew in part (a)?

(d) Complete a table that shows the heights of the balloon for $\theta = 80°, 70°, 60°, \ldots, 10°$.

(e) As θ approaches 0°, how does this affect the height of the balloon? Draw a right triangle to explain your reasoning.

69. Johnstown Inclined Plane The Johnstown Inclined Plane in Pennsylvania is one of the longest and steepest hoists in the world. The railway cars travel a distance of 896.5 feet at an angle of approximately 35.4°, rising to a height of 1693.5 feet above sea level.

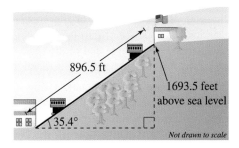

(a) Find the vertical rise.

(b) Find the elevation of the lower end.

70. Johnstown Inclined Plane In Exercise 69, the cars move up the mountain at a rate of 300 feet per minute. Find the rate at which they rise vertically.

Exploring the Concepts

True or False? In Exercises 71–76, determine whether the statement is true or false. Justify your answer.

71. $\sin 60° \csc 60° = 1$ **72.** $\sec 30° = \csc 30°$

73. $\sin 45° + \cos 45° = 1$ **74.** $\cos 60° - \sin 30° = 0$

75. $\dfrac{\sin 60°}{\sin 30°} = \sin 2°$ **76.** $\tan[(5°)^2] = \tan^2 5°$

77. Think About It You are given the value of $\tan \theta$. Is it possible to find the value of $\sec \theta$ without finding the measure of θ? Explain.

78. **HOW DO YOU SEE IT?** Use the figure below.

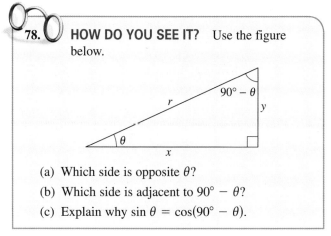

(a) Which side is opposite θ?

(b) Which side is adjacent to $90° - \theta$?

(c) Explain why $\sin \theta = \cos(90° - \theta)$.

79. Error Analysis Describe the error.

$$\cos 60° = \frac{\text{opp}}{\text{hyp}} = \frac{1}{2} \quad \textbf{✗}$$

80. Writing In right triangle trigonometry, explain why $\sin 30° = \frac{1}{2}$ regardless of the size of the triangle.

Review & Refresh ▶ Video solutions at LarsonPrecalculus.com

Determining the Quadrant for a Point In Exercises 81–84, determine the quadrant in which the point is located.

81. $(1, 3)$ **82.** $(9, -2)$

83. $(-5, 4)$ **84.** $(-6, -7)$

Determining the Quadrant for a Line In Exercises 85–88, determine the quadrants in which the graph of the line is located. Give examples of points that lie on the line in each of these quadrants.

85. $y = -x$ **86.** $y = \frac{1}{3}x$

87. $2x - y = 0$ **88.** $4x + 3y = 0$

Finding Coterminal Angles In Exercises 89 and 90, determine two coterminal angles (one positive and one negative) for each angle.

89. (a) $\theta = 20°$ (b) $\theta = -30°$

90. (a) $\theta = \dfrac{\pi}{12}$ (b) $\theta = -\dfrac{\pi}{7}$

Even, Odd, or Neither? In Exercises 91–96, determine whether the function is even, odd, or neither. Then describe the symmetry.

91. $f(x) = \dfrac{1}{x^3 - x}$ **92.** $f(x) = \dfrac{x^2}{4x^2 + 1}$

93. $f(x) = e^{x^2}$ **94.** $f(x) = e^x + e^{x^3}$

95. $f(x) = -\log(-x)$ **96.** $f(x) = x \ln 5 + x^3 \ln 4$

6.3 Trigonometric Functions of Any Angle

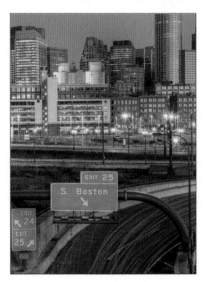

> Evaluate trigonometric functions of any angle.
> Use reference angles to evaluate trigonometric functions.
> Evaluate trigonometric functions of real numbers.

Introduction

In the preceding section, the definitions of trigonometric functions were restricted to acute angles. In this section, the definitions are extended to cover *any* angle. When θ is an *acute* angle, the definitions here coincide with those in the preceding section.

Definitions of Trigonometric Functions of Any Angle

Let θ be an angle in standard position with (x, y) a point on the terminal side of θ and $r = \sqrt{x^2 + y^2} \neq 0$.

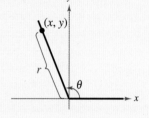

$$\sin \theta = \frac{y}{r} \qquad \cos \theta = \frac{x}{r}$$

$$\tan \theta = \frac{y}{x}, \quad x \neq 0 \qquad \cot \theta = \frac{x}{y}, \quad y \neq 0$$

$$\sec \theta = \frac{r}{x}, \quad x \neq 0 \qquad \csc \theta = \frac{r}{y}, \quad y \neq 0$$

Because $r = \sqrt{x^2 + y^2}$ *cannot* be zero, it follows that the sine and cosine functions are defined for any real value of θ. However, when $x = 0$, the tangent and secant of θ are undefined. For example, the tangent of $90°$ is undefined. Similarly, when $y = 0$, the cotangent and cosecant of θ are undefined.

Trigonometric functions have a wide variety of real-life applications. For example, in Exercise 107 on page 455, you will use trigonometric functions to model the average high temperatures in two cities.

EXAMPLE 1 **Evaluating Trigonometric Functions**

Let $(-3, 4)$ be a point on the terminal side of θ. Find the sine, cosine, and tangent of θ.

Solution   Referring to Figure 6.26, $x = -3$, $y = 4$, and

$$\begin{aligned} r &= \sqrt{x^2 + y^2} \\ &= \sqrt{(-3)^2 + 4^2} \\ &= 5. \end{aligned}$$

So, you have

$$\sin \theta = \frac{y}{r} = \frac{4}{5} \qquad \text{Definition of sine of any angle}$$

$$\cos \theta = \frac{x}{r} = -\frac{3}{5} \qquad \text{Definition of cosine of any angle}$$

and

$$\tan \theta = \frac{y}{x} = -\frac{4}{3}. \qquad \text{Definition of tangent of any angle}$$

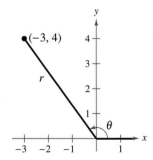

Figure 6.26

✓ **Checkpoint** ▶ *Audio-video solution in English & Spanish at LarsonPrecalculus.com*

Let $(-2, 3)$ be a point on the terminal side of θ. Find the sine, cosine, and tangent of θ.

GO DIGITAL

© iStockphoto.com/DenisTangneyJr

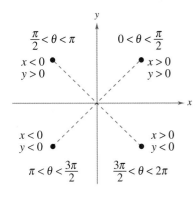

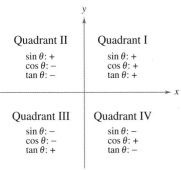

Figure 6.27

The *signs* of the trigonometric functions in the four quadrants can be determined from the definitions of the functions. For example, $\cos \theta = x/r$, so $\cos \theta$ is positive wherever $x > 0$, which is in Quadrants I and IV, as shown in Figure 6.27. (Remember, r is always positive.) In a similar manner, you can verify the other results shown in Figure 6.27.

EXAMPLE 2   **Evaluating Trigonometric Functions**

Given $\tan \theta = -\frac{5}{4}$ and $\cos \theta > 0$, find $\sin \theta$ and $\sec \theta$.

Solution   Note that θ lies in Quadrant IV because that is the only quadrant in which the tangent is negative and the cosine is positive. Moreover, using

$$\tan \theta = \frac{y}{x} = -\frac{5}{4}$$

and the fact that y is negative in Quadrant IV, let $y = -5$ and $x = 4$. So, $r = \sqrt{16 + 25} = \sqrt{41}$ and you have the results below.

$$\sin \theta = \frac{y}{r}$$

$$= \frac{-5}{\sqrt{41}} \qquad \text{Exact value}$$

$$\approx -0.7809 \qquad \text{Approximate value}$$

$$\sec \theta = \frac{r}{x}$$

$$= \frac{\sqrt{41}}{4} \qquad \text{Exact value}$$

$$\approx 1.6008 \qquad \text{Approximate value}$$

✓ *Checkpoint* ▶ *Audio-video solution in English & Spanish at LarsonPrecalculus.com*

Given $\sin \theta = \frac{4}{5}$ and $\tan \theta < 0$, find $\cos \theta$ and $\tan \theta$.

EXAMPLE 3   **Trigonometric Functions of Quadrantal Angles**

Evaluate the cosine and tangent functions at the quadrantal angles 0, $\frac{\pi}{2}$, π, and $\frac{3\pi}{2}$.

Solution   To begin, choose a point on the terminal side of each angle, as shown in Figure 6.28. For each of the four points, $r = 1$ and you have the results below.

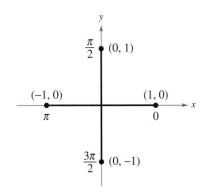

Figure 6.28

$$\cos 0 = \frac{x}{r} = \frac{1}{1} = 1 \qquad \tan 0 = \frac{y}{x} = \frac{0}{1} = 0 \qquad (x, y) = (1, 0)$$

$$\cos \frac{\pi}{2} = \frac{x}{r} = \frac{0}{1} = 0 \qquad \tan \frac{\pi}{2} = \frac{y}{x} = \frac{1}{0} \implies \text{undefined} \qquad (x, y) = (0, 1)$$

$$\cos \pi = \frac{x}{r} = \frac{-1}{1} = -1 \qquad \tan \pi = \frac{y}{x} = \frac{0}{-1} = 0 \qquad (x, y) = (-1, 0)$$

$$\cos \frac{3\pi}{2} = \frac{x}{r} = \frac{0}{1} = 0 \qquad \tan \frac{3\pi}{2} = \frac{y}{x} = \frac{-1}{0} \implies \text{undefined} \qquad (x, y) = (0, -1)$$

✓ *Checkpoint* ▶ *Audio-video solution in English & Spanish at LarsonPrecalculus.com*

Evaluate the sine and cotangent functions at the quadrantal angle $\frac{3\pi}{2}$.   ■

Reference Angles

The values of the trigonometric functions of angles greater than 90° (or less than 0°) can be determined from their values at corresponding acute angles called **reference angles.**

> **Definition of a Reference Angle**
>
> Let θ be an angle in standard position. Its **reference angle** is the acute angle θ' formed by the terminal side of θ and the horizontal axis.

The three figures below show the reference angles for θ in Quadrants II, III, and IV.

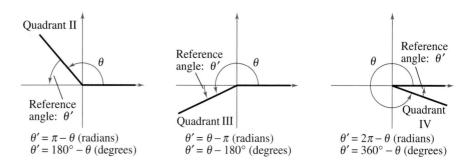

$\theta' = \pi - \theta$ (radians)
$\theta' = 180° - \theta$ (degrees)

$\theta' = \theta - \pi$ (radians)
$\theta' = \theta - 180°$ (degrees)

$\theta' = 2\pi - \theta$ (radians)
$\theta' = 360° - \theta$ (degrees)

EXAMPLE 4 **Finding Reference Angles**

Find the reference angle θ'.

a. $\theta = 300°$ **b.** $\theta = 2.3$ **c.** $\theta = -135°$

Solution

a. Because 300° lies in Quadrant IV, the angle it makes with the x-axis is

$$\theta' = 360° - 300°$$

$$= 60°. \qquad \text{Degrees}$$

Figure 6.29 shows the angle $\theta = 300°$ and its reference angle $\theta' = 60°$.

b. Because 2.3 lies between $\pi/2 \approx 1.5708$ and $\pi \approx 3.1416$, it follows that it is in Quadrant II and its reference angle is

$$\theta' = \pi - 2.3$$

$$\approx 0.8416. \qquad \text{Radians}$$

Figure 6.30 shows the angle $\theta = 2.3$ and its reference angle $\theta' = \pi - 2.3$.

c. First, determine that $-135°$ is coterminal with 225°, which lies in Quadrant III. So, the reference angle is

$$\theta' = 225° - 180°$$

$$= 45°. \qquad \text{Degrees}$$

Figure 6.31 shows the angle $\theta = -135°$ and its reference angle $\theta' = 45°$.

✓ **Checkpoint** ▶ *Audio-video solution in English & Spanish at LarsonPrecalculus.com*

Find the reference angle θ'.

a. $\theta = 213°$ **b.** $\theta = \dfrac{14\pi}{9}$ **c.** $\theta = \dfrac{4\pi}{5}$

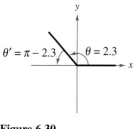

Figure 6.29

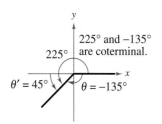

Figure 6.30

GO DIGITAL

To see how to use a reference angle to evaluate a trigonometric function, consider the point (x, y) on the terminal side of the angle θ, as shown at the right. You know that

$$\sin \theta = \frac{y}{r}$$

and

$$\tan \theta = \frac{y}{x}.$$

For the right triangle with acute angle θ' and sides of lengths $|x|$ and $|y|$, you have

$$\sin \theta' = \frac{\text{opp}}{\text{hyp}} = \frac{|y|}{r}$$

and

$$\tan \theta' = \frac{\text{opp}}{\text{adj}} = \frac{|y|}{|x|}.$$

opp $= |y|$, adj $= |x|$

So, it follows that $\sin \theta$ and $\sin \theta'$ are equal, *except possibly in sign*. The same is true for $\tan \theta$ and $\tan \theta'$ and for the other four trigonometric functions. In all cases, the quadrant in which θ lies determines the sign of the function value.

Evaluating Trigonometric Functions of Any Angle

To find the value of a trigonometric function of any angle θ:

1. Determine the function value of the associated reference angle θ'.

2. Depending on the quadrant in which θ lies, affix the appropriate sign to the function value.

Using reference angles and the special angles discussed in the preceding section enables you to greatly extend the scope of *exact* trigonometric function values. For example, knowing the function values of 30° means that you know the function values of all angles for which 30° is a reference angle. For convenience, the table below shows the exact values of the sine, cosine, and tangent functions of special angles and quadrantal angles.

ALGEBRA HELP

Learning the table of values at the right is worth the effort because doing so will increase both your efficiency and your confidence when working in trigonometry. Below is a pattern for the sine function that may help you remember the values.

θ	0°	30°	45°	60°	90°
$\sin \theta$	$\dfrac{\sqrt{0}}{2}$	$\dfrac{\sqrt{1}}{2}$	$\dfrac{\sqrt{2}}{2}$	$\dfrac{\sqrt{3}}{2}$	$\dfrac{\sqrt{4}}{2}$

Reverse the order to get cosine values of the same angles.

Trigonometric Values of Common Angles

θ (degrees)	0°	30°	45°	60°	90°	180°	270°
θ (radians)	0	$\dfrac{\pi}{6}$	$\dfrac{\pi}{4}$	$\dfrac{\pi}{3}$	$\dfrac{\pi}{2}$	π	$\dfrac{3\pi}{2}$
$\sin \theta$	0	$\dfrac{1}{2}$	$\dfrac{\sqrt{2}}{2}$	$\dfrac{\sqrt{3}}{2}$	1	0	-1
$\cos \theta$	1	$\dfrac{\sqrt{3}}{2}$	$\dfrac{\sqrt{2}}{2}$	$\dfrac{1}{2}$	0	-1	0
$\tan \theta$	0	$\dfrac{\sqrt{3}}{3}$	1	$\sqrt{3}$	Undef.	0	Undef.

GO DIGITAL

| EXAMPLE 5 | **Using Reference Angles** |

▶▶▶ *See LarsonPrecalculus.com for an interactive version of this type of example.*

Evaluate each trigonometric function.

a. $\cos \dfrac{4\pi}{3}$ **b.** $\tan(-210°)$ **c.** $\csc \dfrac{11\pi}{4}$

Solution

a. Because $\theta = 4\pi/3$ lies in Quadrant III, the reference angle is

$$\theta' = \dfrac{4\pi}{3} - \pi = \dfrac{\pi}{3}$$

as shown at the right. The cosine is negative in Quadrant III, so

$$\cos \dfrac{4\pi}{3} = (-)\cos \dfrac{\pi}{3}$$

$$= -\dfrac{1}{2}.$$

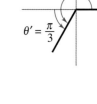

b. Because $-210° + 360° = 150°$, it follows that $-210°$ is coterminal with the second-quadrant angle $150°$. So, the reference angle is

$$\theta' = 180° - 150° = 30°$$

as shown at the right. The tangent is negative in Quadrant II, so

$$\tan(-210°) = (-)\tan 30°$$

$$= -\dfrac{\sqrt{3}}{3}.$$

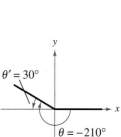

c. Because $(11\pi/4) - 2\pi = 3\pi/4$, it follows that $11\pi/4$ is coterminal with the second-quadrant angle $3\pi/4$. So, the reference angle is

$$\theta' = \pi - \dfrac{3\pi}{4} = \dfrac{\pi}{4}$$

as shown at the right. The cosecant is positive in Quadrant II, so

$$\csc \dfrac{11\pi}{4} = (+)\csc \dfrac{\pi}{4}$$

$$= \dfrac{1}{\sin(\pi/4)}$$

$$= \sqrt{2}.$$

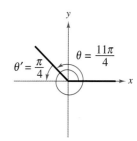

✓ *Checkpoint* ▶ *Audio-video solution in English & Spanish at LarsonPrecalculus.com*

Evaluate each trigonometric function.

a. $\sin \dfrac{7\pi}{4}$ **b.** $\cos(-120°)$ **c.** $\tan \dfrac{11\pi}{6}$

The fundamental trigonometric identities listed in the preceding section (for an acute angle θ) are also valid when θ is any angle in the domain of the function.

EXAMPLE 6   Using Trigonometric Identities

Let θ be an angle in Quadrant II such that $\sin \theta = \frac{1}{3}$. Find (a) $\cos \theta$ and (b) $\tan \theta$ by using trigonometric identities.

Solution

a. Using the Pythagorean identity $\sin^2 \theta + \cos^2 \theta = 1$, you obtain

$$\left(\frac{1}{3}\right)^2 + \cos^2 \theta = 1 \qquad \text{Substitute } \tfrac{1}{3} \text{ for } \sin \theta.$$

$$\cos^2 \theta = 1 - \frac{1}{9} \qquad \text{Evaluate power and subtract.}$$

$$\cos^2 \theta = \frac{8}{9}. \qquad \text{Simplify.}$$

You know that $\cos \theta < 0$ in Quadrant II, so use the negative root to obtain

$$\cos \theta = -\frac{\sqrt{8}}{\sqrt{9}} = -\frac{2\sqrt{2}}{3}.$$

b. Using the trigonometric identity $\tan \theta = (\sin \theta)/(\cos \theta)$, you obtain

$$\tan \theta = \frac{1/3}{-2\sqrt{2}/3} \qquad \text{Substitute for } \sin \theta \text{ and } \cos \theta.$$

$$= -\frac{1}{2\sqrt{2}} \qquad \text{Simplify.}$$

$$= -\frac{\sqrt{2}}{4}. \qquad \text{Rationalize denominator.}$$

✓ **Checkpoint** ▶ Audio-video solution in English & Spanish at LarsonPrecalculus.com

Let θ be an angle in Quadrant III such that $\sin \theta = -\frac{4}{5}$. Find (a) $\cos \theta$ and (b) $\tan \theta$ by using trigonometric identities.

EXAMPLE 7   Using a Calculator

Use a calculator to evaluate (a) $\cot 410°$, (b) $\sin(-7)$, and (c) $\sec(\pi/9)$.

Solution

	Function	Mode	Calculator Keystrokes	Display
a.	$\cot 410°$	Degree	(TAN (410)) x⁻¹ ENTER	0.8390996
b.	$\sin(-7)$	Radian	SIN ((−) 7) ENTER	−0.6569866
c.	$\sec \dfrac{\pi}{9}$	Radian	(COS (π ÷ 9)) x⁻¹ ENTER	1.0641778

✓ **Checkpoint** ▶ Audio-video solution in English & Spanish at LarsonPrecalculus.com

Use a calculator to evaluate each trigonometric function.

a. $\tan 119°$ **b.** $\csc 5$ **c.** $\cos \dfrac{\pi}{5}$

Trigonometric Functions of Real Numbers

To define a trigonometric function of a real number (rather than an angle), let t represent any real number. Then imagine that the real number line is wrapped around a *unit circle*, as shown in the figures below. (Recall from Section 1.1 that a unit circle has a radius of 1.) Note that positive numbers correspond to a counterclockwise wrapping and negative numbers correspond to a clockwise wrapping.

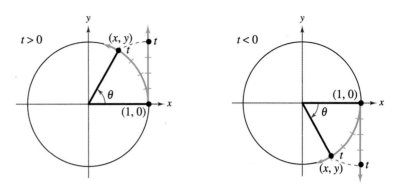

As the real number line wraps around the unit circle, each real number t corresponds to a central angle θ (in standard position). Moreover, the circle has a radius of 1, so the arc intercepted by the angle θ will have a (directional) length of t. This means that *if θ is measured in radians, then $t = \theta$*. So, you can define $\sin t$ as $\sin t = \sin(t \text{ radians})$, $\cos t = \cos(t \text{ radians})$, $\tan t = \tan(t \text{ radians})$, and so on. Furthermore, each t-value corresponds to a point (x, y) on the unit circle, so

$$\sin t = \frac{y}{1} = y, \quad \cos t = \frac{x}{1}, \quad \text{and} \quad \tan t = \frac{y}{x}.$$

EXAMPLE 8   Evaluating Trigonometric Functions

a. Evaluate $f(t) = \sin t$ for $t = 1$ and $t = 7\pi/2$.

b. Evaluate $f(t) = \cos t$ for $t = -2\pi/3$, which corresponds to the point $\left(-1/2, -\sqrt{3}/2\right)$ on the unit circle.

Solution

a. Using a calculator in *radian* mode, $f(1) = \sin 1 \approx 0.8415$. The angles

$$t = \frac{7\pi}{2} \quad \text{and} \quad t = \frac{3\pi}{2}$$

are coterminal, so

$$f\left(\frac{7\pi}{2}\right) = \sin \frac{7\pi}{2} = \sin \frac{3\pi}{2} = -1.$$

b. Using the point $\left(-1/2, -\sqrt{3}/2\right)$, it follows that

$$f\left(-\frac{2\pi}{3}\right) = \cos\left(-\frac{2\pi}{3}\right) = x = -\frac{1}{2}.$$

✓ *Checkpoint* ▶ *Audio-video solution in English & Spanish at LarsonPrecalculus.com*

a. Evaluate $f(t) = \cos t$ for $t = 2$ and $t = 5\pi/2$.

b. Evaluate $f(t) = \tan t$ for $t = 7\pi/3$, which corresponds to the point $(1/2, \sqrt{3}/2)$ on the unit circle. ∎

GO DIGITAL

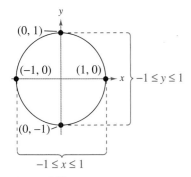

Figure 6.32

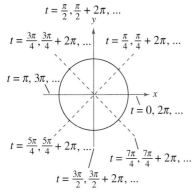

Figure 6.33

The *domain* of the sine and cosine functions is the set of all real numbers. To determine the *range* of these two functions, consider the unit circle shown in Figure 6.32. You know that $\sin t = y$ and $\cos t = x$. Moreover, (x, y) is on the unit circle, so you also know that $-1 \le y \le 1$ and $-1 \le x \le 1$. This means that the values of sine and cosine also range between -1 and 1.

$$\begin{array}{ccc} -1 \le & y & \le 1 \\ -1 \le & \sin t & \le 1 \end{array} \quad \text{and} \quad \begin{array}{ccc} -1 \le & x & \le 1 \\ -1 \le & \cos t & \le 1 \end{array}$$

Adding 2π to each value of t in the interval $[0, 2\pi]$ results in a revolution around the unit circle, as shown in Figure 6.33. The values of $\sin(t + 2\pi)$ and $\cos(t + 2\pi)$ correspond to those of $\sin t$ and $\cos t$. Repeated revolutions (positive or negative) on the unit circle yield similar results. This leads to the general result

$$\sin(t + 2\pi n) = \sin t \quad \text{and} \quad \cos(t + 2\pi n) = \cos t$$

for any integer n and real number t. Functions that behave in such a repetitive (or cyclic) manner are **periodic.**

Definition of Periodic Function

A function f is **periodic** when there exists a positive real number c such that

$$f(t + c) = f(t)$$

for all t in the domain of f. The smallest number c for which f is periodic is the **period** of f.

From this definition, it follows that the sine and cosine functions are periodic and have a period of 2π. The other four trigonometric functions are also periodic and will be discussed further in Section 6.5.

Recall from Section 2.3 that a function f is *even* when $f(-t) = f(t)$ and is *odd* when $f(-t) = -f(t)$.

Even and Odd Trigonometric Functions

The cosine and secant functions are *even.*

$$\cos(-t) = \cos t \qquad \sec(-t) = \sec t$$

The sine, cosecant, tangent, and cotangent functions are *odd.*

$$\sin(-t) = -\sin t \qquad \csc(-t) = -\csc t$$
$$\tan(-t) = -\tan t \qquad \cot(-t) = -\cot t$$

Summarize   (Section 6.3)

1. State the definitions of the trigonometric functions of any angle *(page 445)*. For examples of evaluating trigonometric functions, see Examples 1–3.

2. Explain how to use a reference angle to evaluate a trigonometric function *(pages 447 and 448)*. For examples of finding and using reference angles, see Examples 4 and 5.

3. Explain how to evaluate a trigonometric function of a real number *(page 451)*. For an example of evaluating trigonometric functions of real numbers, see Example 8.

GO DIGITAL

6.3 Exercises

See CalcChat.com for tutorial help and worked-out solutions to odd-numbered exercises.

Vocabulary and Concept Check

In Exercises 1 and 2, fill in the blanks.

1. A function f is _____ when there exists a positive real number c such that $f(t + c) = f(t)$ for all t in the domain of f.

2. The cosine and secant functions are _____ functions, and the sine, cosecant, tangent, and cotangent functions are _____ functions.

3. What do you call the acute angle formed by the terminal side of an angle θ in standard position and the horizontal axis?

4. Is the value of $\cos 170°$ equal to the value of $\cos 10°$?

Skills and Applications

Evaluating Trigonometric Functions In Exercises 5–8, find the exact values of the six trigonometric functions of each angle θ.

5. (a)   (b)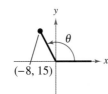

6. (a) 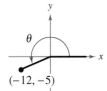 (b)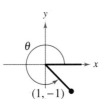

7. (a) 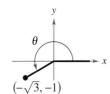 (b)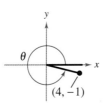

8. (a) 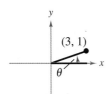 (b)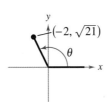

Evaluating Trigonometric Functions In Exercises 9–14, the point is on the terminal side of an angle in standard position. Find the exact values of the six trigonometric functions of the angle.

9. $(5, 12)$

10. $(8, 15)$

11. $(-5, -2)$

12. $(-4, 10)$

13. $(-5.4, 7.2)$

14. $\left(3\frac{1}{2}, -2\sqrt{15}\right)$

Determining a Quadrant In Exercises 15–18, determine the quadrant in which θ lies.

15. $\sin \theta > 0$, $\cos \theta > 0$

16. $\sin \theta < 0$, $\cos \theta < 0$

17. $\csc \theta > 0$, $\tan \theta < 0$

18. $\sec \theta > 0$, $\cot \theta < 0$

Evaluating Trigonometric Functions In Exercises 19–28, find the exact values of the remaining trigonometric functions of θ satisfying the given conditions.

19. $\tan \theta = \dfrac{15}{8}$, $\sin \theta > 0$

20. $\cos \theta = \dfrac{8}{17}$, $\tan \theta < 0$

21. $\sin \theta = 0.6$, θ lies in Quadrant II.

22. $\cos \theta = -0.8$, θ lies in Quadrant III.

23. $\cot \theta = -3$, $\cos \theta > 0$

24. $\csc \theta = 4$, $\cot \theta < 0$

25. $\cos \theta = 0$, $\csc \theta = 1$

26. $\sin \theta = 0$, $\sec \theta = -1$

27. $\cot \theta$ is undefined, $\dfrac{\pi}{2} \le \theta \le \dfrac{3\pi}{2}$

28. $\tan \theta$ is undefined, $\pi \le \theta \le 2\pi$

An Angle Formed by a Line Through the Origin In Exercises 29–32, the terminal side of θ lies on the given line in the specified quadrant. Find the exact values of the six trigonometric functions of θ by finding a point on the line.

Line	Quadrant
29. $y = -x$	II
30. $y = \frac{1}{3}x$	III
31. $2x - y = 0$	I
32. $4x + 3y = 0$	IV

Trigonometric Function of a Quadrantal Angle In Exercises 33–42, evaluate the trigonometric function of the quadrantal angle, if possible.

33. $\sin 0$

34. $\csc \dfrac{3\pi}{2}$

35. $\sec \dfrac{3\pi}{2}$

36. $\sec \pi$

37. $\sin \dfrac{\pi}{2}$

38. $\cot 0$

39. $\csc \pi$

40. $\cot \dfrac{\pi}{2}$

41. $\cos \dfrac{9\pi}{2}$

42. $\tan\left(-\dfrac{\pi}{2}\right)$

Finding a Reference Angle In Exercises 43–50, find the reference angle θ'. Then sketch θ in standard position and label θ'.

43. $\theta = 160°$

44. $\theta = 309°$

45. $\theta = -125°$

46. $\theta = -215°$

47. $\theta = \dfrac{2\pi}{3}$

48. $\theta = \dfrac{7\pi}{6}$

49. $\theta = 4.8$

50. $\theta = 12.9$

Using a Reference Angle In Exercises 51–64, evaluate the sine, cosine, and tangent of the angle without using a calculator.

51. $225°$

52. $300°$

53. $750°$

54. $675°$

55. $-120°$

56. $-570°$

57. $\dfrac{2\pi}{3}$

58. $\dfrac{3\pi}{4}$

59. $-\dfrac{\pi}{6}$

60. $-\dfrac{2\pi}{3}$

61. $\dfrac{11\pi}{4}$

62. $\dfrac{13\pi}{6}$

63. $-\dfrac{17\pi}{6}$

64. $-\dfrac{23\pi}{4}$

Using a Trigonometric Identity In Exercises 65–70, use the function value to find the indicated trigonometric value in the specified quadrant.

Function Value	Quadrant	Trigonometric Value
65. $\sin \theta = -\frac{3}{5}$	IV	$\cos \theta$
66. $\cot \theta = -3$	II	$\csc \theta$
67. $\tan \theta = \frac{3}{2}$	III	$\sec \theta$
68. $\csc \theta = -2$	IV	$\cot \theta$
69. $\cos \theta = \frac{5}{8}$	I	$\csc \theta$
70. $\sec \theta = -\frac{9}{4}$	III	$\cot \theta$

Using a Calculator In Exercises 71–86, use a calculator to evaluate the trigonometric function. Round your answer to four decimal places. (Be sure the calculator is in the correct mode.)

71. $\sin 10°$

72. $\tan 304°$

73. $\cos(-110°)$

74. $\sin(-330°)$

75. $\cot 178°$

76. $\sec 72°$

77. $\csc 405°$

78. $\cot(-560°)$

79. $\tan \dfrac{\pi}{9}$

80. $\cos \dfrac{2\pi}{7}$

81. $\sec \dfrac{11\pi}{8}$

82. $\csc \dfrac{15\pi}{4}$

83. $\sin(-0.65)$

84. $\cos 1.35$

85. $\csc(-10)$

86. $\sec(-4.6)$

Solving for θ In Exercises 87–92, find two solutions of each equation. Give your answers in degrees ($0° \le \theta < 360°$) and in radians ($0 \le \theta < 2\pi$). Do not use a calculator.

87. (a) $\sin \theta = \frac{1}{2}$
    (b) $\sin \theta = -\frac{1}{2}$

88. (a) $\cos \theta = \dfrac{\sqrt{2}}{2}$

    (b) $\cos \theta = -\dfrac{\sqrt{2}}{2}$

89. (a) $\cos \theta = \frac{1}{2}$
    (b) $\sec \theta = 2$

90. (a) $\sin \theta = \dfrac{\sqrt{3}}{2}$

    (b) $\csc \theta = \dfrac{2\sqrt{3}}{3}$

91. (a) $\tan \theta = 1$
    (b) $\cot \theta = -\sqrt{3}$

92. (a) $\cot \theta = 0$
    (b) $\sec \theta = -\sqrt{2}$

Evaluating Trigonometric Functions In Exercises 93–100, find the point (x, y) on the unit circle that corresponds to the real number t. Use the result to evaluate $\sin t$, $\cos t$, and $\tan t$.

93. $t = \dfrac{\pi}{4}$

94. $t = \dfrac{\pi}{3}$

95. $t = \dfrac{5\pi}{6}$

96. $t = \dfrac{3\pi}{4}$

97. $t = \dfrac{4\pi}{3}$

98. $t = \dfrac{5\pi}{3}$

99. $t = \dfrac{\pi}{2}$

100. $t = \pi$

Estimation In Exercises 101 and 102, use the figure below and a straightedge to approximate the value of each trigonometric function to the nearest tenth.

101. (a) sin 5
   (b) cos 2

102. (a) sin 0.75
   (b) cos 2.5

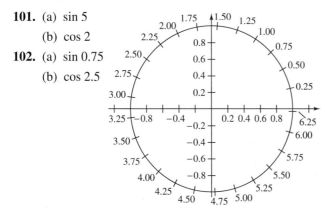

Estimation In Exercises 103 and 104, use the figure above and a straightedge to approximate two solutions of each equation to the nearest tenth, where $0 \le t \le 2\pi$.

103. (a) sin t = 0.25   (b) cos t = −0.25
104. (a) sin t = −0.75   (b) cos t = 0.75

105. **Harmonic Motion** The displacement from equilibrium of an oscillating weight suspended by a spring is given by $y(t) = 2 \cos 6t$, where y is the displacement in centimeters and t is the time in seconds (see figure). Find the displacement for each time.

   (a) $t = 0$ (b) $t = \dfrac{1}{4}$ (c) $t = \dfrac{1}{2}$

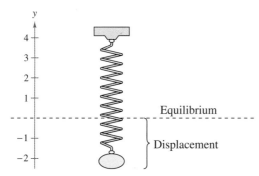

Figure for 105 and 106

106. **Harmonic Motion** The displacement from equilibrium of an oscillating weight suspended by a spring and subject to the damping effect of friction is given by

$y(t) = 2e^{-t} \cos 6t$

where y is the displacement in centimeters and t is the time in seconds (see figure). Find the displacement for each time.

   (a) $t = 0$ (b) $t = \dfrac{1}{4}$ (c) $t = \dfrac{1}{2}$

107. **Temperature**

The table shows the average high temperatures (in degrees Fahrenheit) in Boston, Massachusetts (B), and Fairbanks, Alaska (F), for selected months in 2019. (*Source: U.S. Climate Data*)

DATA	Month	Boston, B	Fairbanks, F
	January	39	2
	March	47	40
	June	76	75
	August	82	64
	November	50	21

Spreadsheet at LarsonPrecalculus.com

(a) Use the *regression* feature of a graphing utility to find a model of the form

$y = a \sin(bt + c) + d$

for each city. Let t represent the month, with $t = 1$ corresponding to January.

(b) Use the models from part (a) to estimate the monthly average high temperatures for the two cities in February, April, May, July, September, October, and December.

(c) Use a graphing utility to graph both models in the same viewing window. Compare the temperatures for the two cities.

108. **Sales** A company that produces snowboards forecasts monthly sales over the next 2 years to be

$$S = 23.1 + 0.442t + 4.3 \cos \frac{\pi t}{6}$$

where S is measured in thousands of units and t is the time in months, with $t = 1$ corresponding to January 2022. Predict the sales for each of the following months.

(a) February 2022
(b) February 2023
(c) June 2022
(d) June 2023

109. **Electric Circuits** The current I (in amperes) when 100 volts is applied to a circuit is given by

$I = 5e^{-2t} \sin t$

where t is the time (in seconds) after the voltage is applied. Approximate the current at $t = 0.7$ second after the voltage is applied.

110. Distance An airplane, flying at an altitude of 6 miles, is on a flight path that passes directly over an observer (see figure). Let θ be the angle of elevation from the observer to the plane. Find the distance d from the observer to the plane when (a) $\theta = 30°$, (b) $\theta = 90°$, and (c) $\theta = 120°$.

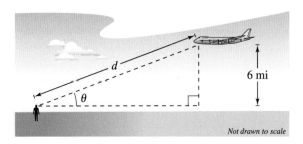

6 mi

Not drawn to scale

Exploring the Concepts

True or False? In Exercises 111 and 112, determine whether the statement is true or false. Justify your answer.

111. In each of the four quadrants, the signs of the secant function and the sine function are the same.

112. The reference angle for an angle θ (in degrees) is the angle $\theta' = 360°n - \theta$, where n is an integer and $0° \le \theta' \le 360°$.

113. Think About It When $f(t) = \sin t$ and $g(t) = \cos t$, is $h(t) = f(t)g(t)$ even, odd, or neither? Explain.

114. **HOW DO YOU SEE IT?** Consider an angle in standard position with $r = 12$ centimeters, as shown in the figure. Describe the changes in the values of x, y, $\sin \theta$, $\cos \theta$, and $\tan \theta$ as θ increases continuously from $0°$ to $90°$.

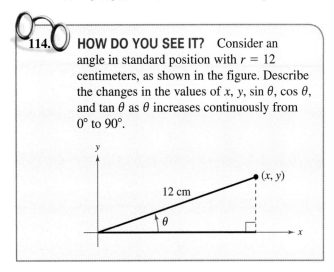

115. Error Analysis Describe the error.

Your classmate uses a calculator to evaluate $\tan(\pi/2)$ and gets a result of 0.0274224385.

116. Writing Use a graphing utility to graph each of the six trigonometric functions. Determine the domain, range, period, and zeros of each function. Identify, and write a short paragraph describing, any inherent patterns in the trigonometric functions. What can you conclude?

117. Using Technology With a graphing utility in *radian* and *parametric* modes, enter the equations $X_{1T} = \cos T$ and $Y_{1T} = \sin T$ and use the settings below.

Tmin = 0, Tmax = 6.3, Tstep = 0.1

Xmin = −1.5, Xmax = 1.5, Xscl = 1

Ymin = −1, Ymax = 1, Yscl = 1

(a) Graph the entered equations and describe the graph.

(b) Use the *trace* feature to move the cursor around the graph. What do the *t*-values represent? What do the *x*- and *y*-values represent?

(c) What are the least and greatest values of x and y?

118. Think About It Let (x_1, y_1) and (x_2, y_2) be points on the unit circle corresponding to $t = t_1$ and $t = \pi - t_1$, respectively.

(a) Identify the symmetry of the points (x_1, y_1) and (x_2, y_2).

(b) Make a conjecture about any relationship between $\sin t_1$ and $\sin(\pi - t_1)$.

Review & Refresh ▶ *Video solutions at LarsonPrecalculus.com*

Describing Transformations In Exercises 119 and 120, g is related to one of the parent functions. (a) Identify the parent function f. (b) Describe the sequence of transformations from f to g.

119. $g(x) = -\sqrt{x + 12} + 9$

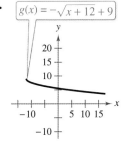

120. $g(x) = 15(x + 2)^3 - 4$

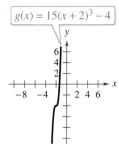

Finding Relative Minima or Maxima In Exercises 121 and 122, use the graph to find any relative minima or maxima of the function.

121.

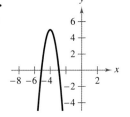

122.

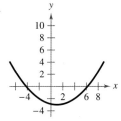

Finding the *x*-intercepts of a Function In Exercises 123–126, find the *x*-intercepts of the function.

123. $y = x^2 + x - 132$

124. $y = 2x^7 - x^5 - x^3$

125. $y = 2e^{x-5} - 4$

126. $y = \frac{1}{5}\ln(x - 3) + \frac{3}{10}$

6.4 Graphs of Sine and Cosine Functions

Graphs of sine and cosine functions have many scientific applications. For example, in Exercise 80 on page 466, you will use the graph of a sine function to analyze airflow during a respiratory cycle.

❯ Sketch the graphs of basic sine and cosine functions.
❯ Use amplitude and period to help sketch the graphs of sine and cosine functions.
❯ Sketch translations of the graphs of sine and cosine functions.
❯ Use sine and cosine functions to model real-life data.

Basic Sine and Cosine Curves

In this section, you will study techniques for sketching the graphs of the sine and cosine functions. The graph of the sine function, shown in Figure 6.34, is a **sine curve.** In the figure, the black portion of the graph from $x = 0$ to $x = 2\pi$ represents one period of the function and is **one cycle** of the sine curve. Note that the basic sine curve repeats indefinitely to the left and right. Figure 6.35 shows the graph of the cosine function.

Recall from Section 6.3 that the domain of the sine and cosine functions is the set of all real numbers. Moreover, the range of each function is the interval $[-1, 1]$, and each function has a period of 2π. This information is consistent with the basic graphs shown in Figures 6.34 and 6.35.

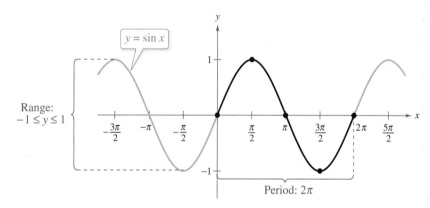

Figure 6.34

 TECHNOLOGY

To use a graphing utility to produce the graphs shown in Figures 6.34 and 6.35, make sure you set the graphing utility to *radian* mode.

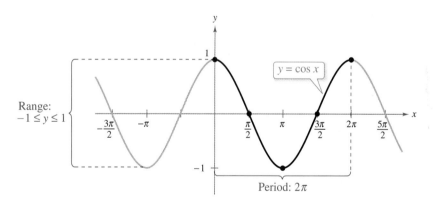

Figure 6.35

Note in Figures 6.34 and 6.35 that the sine curve is symmetric with respect to the *origin*, whereas the cosine curve is symmetric with respect to the *y-axis*. These properties of symmetry follow from the fact that the sine function is odd and the cosine function is even.

GO DIGITAL

To sketch the graphs of the basic sine and cosine functions, it helps to note five **key points** in one period of each graph: the *intercepts, maximum points,* and *minimum points.* The figures below show the five key points on the graphs of $y = \sin x$ and $y = \cos x$.

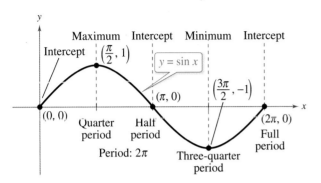
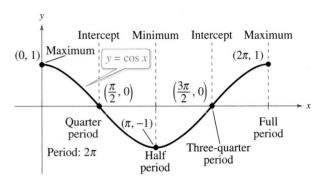

EXAMPLE 1 **Using Key Points to Sketch a Sine Curve**

▶▶▶ *See LarsonPrecalculus.com for an interactive version of this type of example.*

Sketch the graph of

$$y = 2 \sin x$$

on the interval $[-\pi, 4\pi]$.

Solution Note that

$$y = 2 \sin x$$
$$= 2(\sin x).$$

So, the y-values for the key points have twice the magnitude of those on the graph of $y = \sin x$. Divide the period 2π into four equal parts to obtain the key points

Intercept	Maximum	Intercept	Minimum	Intercept
$(0, 0)$,	$\left(\dfrac{\pi}{2}, 2\right)$,	$(\pi, 0)$,	$\left(\dfrac{3\pi}{2}, -2\right)$, and	$(2\pi, 0)$.

By connecting these key points with a smooth curve and extending the curve in both directions over the interval $[-\pi, 4\pi]$, you obtain the graph below.

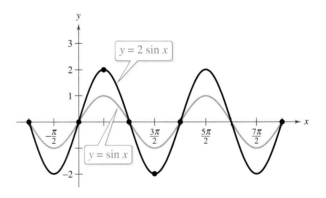

✓ *Checkpoint* ▶ *Audio-video solution in English & Spanish at LarsonPrecalculus.com*

Sketch the graph of

$$y = 2 \cos x$$

on the interval $[-\pi/2, 9\pi/2]$.

▶▶▶ **TECHNOLOGY**

When using a graphing utility to graph trigonometric functions, pay special attention to the viewing window you use. For example, graph

$$y = \frac{\sin 10x}{10}$$

in the standard viewing window in *radian* mode. What do you observe? Use the *zoom* feature to find a viewing window that displays a good view of the graph.

GO DIGITAL

Amplitude and Period

In the rest of this section, you will study the effect of each of the constants a, b, c, and d on the graphs of equations of the forms

$$y = d + a\sin(bx - c)$$

and

$$y = d + a\cos(bx - c).$$

A quick review of the transformations you studied in Section 2.5 will help in this investigation.

The constant factor a in $y = a\sin x$ and $y = a\cos x$ acts as a *scaling factor*—a *vertical stretch* or *vertical shrink* of the basic curve. When $|a| > 1$, the basic curve is stretched, and when $0 < |a| < 1$, the basic curve is shrunk. The result is that the graphs of $y = a\sin x$ and $y = a\cos x$ range between $-a$ and a instead of between -1 and 1. The absolute value of a is the **amplitude** of the function. The range of the function for $a > 0$ is $-a \leq y \leq a$.

Definition of the Amplitude of Sine and Cosine Curves

The **amplitude** of $y = a\sin x$ and $y = a\cos x$ represents half the distance between the maximum and minimum values of the function and is given by

Amplitude $= |a|$.

EXAMPLE 2   **Scaling: Vertical Shrinking and Stretching**

In the same coordinate plane, sketch the graph of each function.

a. $y = \dfrac{1}{2}\cos x$

b. $y = 3\cos x$

Solution

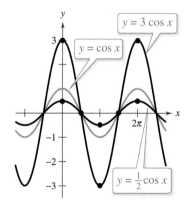

Figure 6.36

a. The amplitude of $y = \frac{1}{2}\cos x$ is $\frac{1}{2}$, so the maximum value is $\frac{1}{2}$ and the minimum value is $-\frac{1}{2}$. Divide one cycle, $0 \leq x \leq 2\pi$, into four equal parts to obtain the key points

Maximum	Intercept	Minimum	Intercept	Maximum
$\left(0, \dfrac{1}{2}\right),$	$\left(\dfrac{\pi}{2}, 0\right),$	$\left(\pi, -\dfrac{1}{2}\right),$	$\left(\dfrac{3\pi}{2}, 0\right),$	and $\left(2\pi, \dfrac{1}{2}\right).$

b. A similar analysis shows that the amplitude of $y = 3\cos x$ is 3, and the key points are

Maximum	Intercept	Minimum	Intercept	Maximum
$(0, 3),$	$\left(\dfrac{\pi}{2}, 0\right),$	$(\pi, -3),$	$\left(\dfrac{3\pi}{2}, 0\right),$	and $(2\pi, 3).$

Figure 6.36 shows the graphs of these two functions. Notice that the graph of $y = \frac{1}{2}\cos x$ is a vertical *shrink* of the graph of $y = \cos x$, and the graph of $y = 3\cos x$ is a vertical *stretch* of the graph of $y = \cos x$.

✓ **Checkpoint**  *Audio-video solution in English & Spanish at LarsonPrecalculus.com*

In the same coordinate plane, sketch the graph of each function.

a. $y = \dfrac{1}{3}\sin x$

b. $y = 3\sin x$

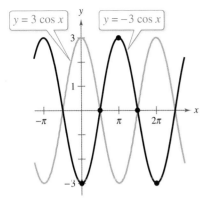

Figure 6.37

You know from Section 2.5 that the graph of $y = -f(x)$ is a **reflection** in the x-axis of the graph of $y = f(x)$. For example, the graph of $y = -3 \cos x$ is a reflection of the graph of $y = 3 \cos x$, as shown in Figure 6.37.

Next, consider the effect of the positive real number b on the graphs of $y = a \sin bx$ and $y = a \cos bx$. For example, compare the graphs of $y = a \sin x$ and $y = a \sin bx$. The graph of $y = a \sin x$ completes one cycle from $x = 0$ to $x = 2\pi$, so it follows that the graph of $y = a \sin bx$ completes one cycle from $x = 0$ to $x = 2\pi/b$.

Period of Sine and Cosine Functions

Let b be a positive real number. The **period** of $y = a \sin bx$ and $y = a \cos bx$ is given by

$$\text{Period} = \frac{2\pi}{b}.$$

Note that when $0 < b < 1$, the period of $y = a \sin bx$ is greater than 2π and represents a *horizontal stretch* of the basic curve. Similarly, when $b > 1$, the period of $y = a \sin bx$ is less than 2π and represents a *horizontal shrink* of the basic curve. These two statements are also true for $y = a \cos bx$. When b is negative, rewrite the function using the identity $\sin(-x) = -\sin x$ or $\cos(-x) = \cos x$.

EXAMPLE 3 **Scaling: Horizontal Stretching**

Sketch the graph of

$$y = \sin \frac{x}{2}.$$

Solution The amplitude is 1. Moreover, $b = \frac{1}{2}$, so the period is

$$\frac{2\pi}{b} = \frac{2\pi}{1/2} = 4\pi. \qquad \text{Substitute for } b.$$

Now, divide the period-interval $[0, 4\pi]$ into four equal parts using the values π, 2π, and 3π to obtain the key points

Intercept	Maximum	Intercept	Minimum		Intercept
$(0, 0),$	$(\pi, 1),$	$(2\pi, 0),$	$(3\pi, -1),$	and	$(4\pi, 0).$

The graph is shown below.

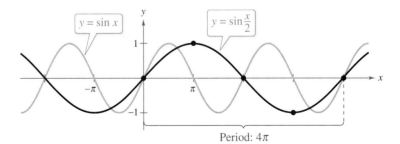

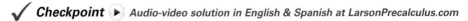

Period: 4π

ALGEBRA HELP

In general, to divide a period-interval into four equal parts, successively add "period/4," starting with the left endpoint of the interval. For example, for the period-interval $[-\pi/6, \pi/2]$ of length $2\pi/3$, you would successively add

$$\frac{2\pi/3}{4} = \frac{\pi}{6}$$

to obtain $-\pi/6$, 0, $\pi/6$, $\pi/3$, and $\pi/2$ as the x-values for the key points on the graph.

✓ *Checkpoint* ▶ Audio-video solution in English & Spanish at LarsonPrecalculus.com

Sketch the graph of

$$y = \cos \frac{x}{3}.$$

Translations of Sine and Cosine Curves

The constant c in the equations

$$y = a\sin(bx - c) \quad \text{and} \quad y = a\cos(bx - c)$$

results in *horizontal translations* (shifts) of the basic curves. For example, compare the graphs of $y = a\sin bx$ and $y = a\sin(bx - c)$. The graph of $y = a\sin(bx - c)$ completes one cycle from $bx - c = 0$ to $bx - c = 2\pi$. Solve for x to find that the interval for one cycle is

Left endpoint   Right endpoint

$$\frac{c}{b} \le x \le \frac{c}{b} + \frac{2\pi}{b}.$$

Period

This implies that the period of $y = a\sin(bx - c)$ is $2\pi/b$, and the graph of $y = a\sin bx$ is shifted by an amount c/b. The number c/b is the **phase shift.**

> ### Graphs of Sine and Cosine Functions
>
> The graphs of $y = a\sin(bx - c)$ and $y = a\cos(bx - c)$ have the characteristics below. (Assume $b > 0$.)
>
> $$\text{Amplitude} = |a| \qquad \text{Period} = \frac{2\pi}{b}$$
>
> The left and right endpoints of a one-cycle interval can be determined by solving the equations $bx - c = 0$ and $bx - c = 2\pi$.

EXAMPLE 4   **Horizontal Translation**

Analyze the graph of $y = \dfrac{1}{2}\sin\left(x - \dfrac{\pi}{3}\right)$.

Algebraic Solution

The amplitude is $\frac{1}{2}$ and the period is $2\pi/1 = 2\pi$. Solving the equations

$$x - \frac{\pi}{3} = 0 \quad \Longrightarrow \quad x = \frac{\pi}{3}$$

and

$$x - \frac{\pi}{3} = 2\pi \quad \Longrightarrow \quad x = \frac{7\pi}{3}$$

shows that the interval $[\pi/3, 7\pi/3]$ corresponds to one cycle of the graph. Dividing this interval into four equal parts produces the key points

Intercept	Maximum	Intercept	Minimum	Intercept
$\left(\dfrac{\pi}{3}, 0\right)$,	$\left(\dfrac{5\pi}{6}, \dfrac{1}{2}\right)$,	$\left(\dfrac{4\pi}{3}, 0\right)$,	$\left(\dfrac{11\pi}{6}, -\dfrac{1}{2}\right)$, and	$\left(\dfrac{7\pi}{3}, 0\right)$.

Graphical Solution

Use a graphing utility set in *radian* mode to graph $y = (1/2)\sin[x - (\pi/3)]$, as shown in the figure below. Use the *minimum, maximum,* and *zero* or *root* features of the graphing utility to approximate the key points $(1.05, 0)$, $(2.62, 0.5)$, $(4.19, 0)$, $(5.76, -0.5)$, and $(7.33, 0)$.

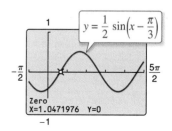

✓ *Checkpoint* ▶ Audio-video solution in English & Spanish at LarsonPrecalculus.com

Analyze the graph of $y = 2\cos\left(x - \dfrac{\pi}{2}\right)$.

■

GO DIGITAL

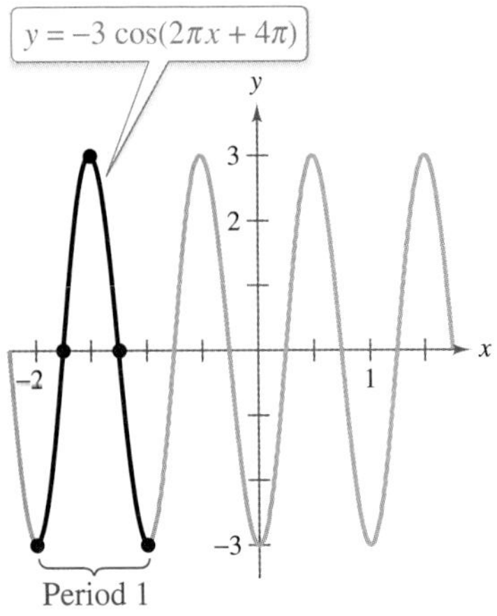

$y = -3\cos(2\pi x + 4\pi)$

Figure 6.38

Period 1

EXAMPLE 5 **Horizontal Translation**

Sketch the graph of

$$y = -3\cos(2\pi x + 4\pi).$$

Solution   The amplitude is 3 and the period is $2\pi/2\pi = 1$. Solving the equations

$$2\pi x + 4\pi = 0$$
$$2\pi x = -4\pi$$
$$x = -2$$

and

$$2\pi x + 4\pi = 2\pi$$
$$2\pi x = -2\pi$$
$$x = -1$$

shows that the interval $[-2, -1]$ corresponds to one cycle of the graph. Dividing this interval into four equal parts produces the key points

Minimum	Intercept	Maximum	Intercept	Minimum
$(-2, -3)$,	$\left(-\dfrac{7}{4}, 0\right)$,	$\left(-\dfrac{3}{2}, 3\right)$,	$\left(-\dfrac{5}{4}, 0\right)$, and	$(-1, -3)$.

Figure 6.38 shows the graph.

✓ *Checkpoint* ▶ *Audio-video solution in English & Spanish at LarsonPrecalculus.com*

Sketch the graph of

$$y = -\frac{1}{2}\sin(\pi x + \pi).$$

The constant d in the equations

$$y = d + a\sin(bx - c) \quad \text{and} \quad y = d + a\cos(bx - c)$$

results in *vertical translations* of the basic curves. The shift is d units up for $d > 0$ and d units down for $d < 0$. In other words, the graph oscillates about the horizontal line $y = d$ instead of about the x-axis.

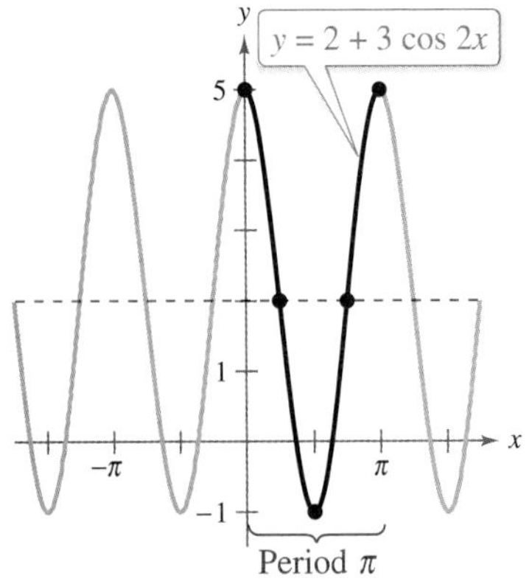

$y = 2 + 3\cos 2x$

Period π

Figure 6.39

EXAMPLE 6 **Vertical Translation**

Sketch the graph of

$$y = 2 + 3\cos 2x.$$

Solution   The amplitude is 3 and the period is $2\pi/2 = \pi$. The key points over the interval $[0, \pi]$ are

$$(0, 5), \quad \left(\frac{\pi}{4}, 2\right), \quad \left(\frac{\pi}{2}, -1\right), \quad \left(\frac{3\pi}{2}, 2\right), \quad \text{and} \quad (\pi, 5).$$

Figure 6.39 shows the graph. Compared with the graph of $f(x) = 3\cos 2x$, the graph of $y = 2 + 3\cos 2x$ is shifted up two units.

✓ *Checkpoint* ▶ *Audio-video solution in English & Spanish at LarsonPrecalculus.com*

Sketch the graph of

$$y = 2\cos x - 5.$$

GO DIGITAL

Mathematical Modeling

DATA	Time, t	Depth, y
	0	3.4
	2	8.7
	4	11.3
	6	9.1
	8	3.8
	10	0.1
	12	1.2

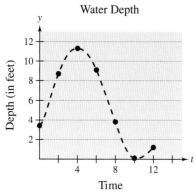

Figure 6.40

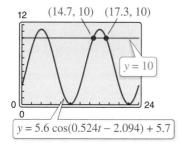

$y = 5.6 \cos(0.524t - 2.094) + 5.7$

Figure 6.41

EXAMPLE 7 **Finding a Trigonometric Model**

The table shows the depths (in feet) of the water at the end of a dock every two hours from midnight to noon, where $t = 0$ corresponds to midnight. (a) Use a trigonometric function to model the data. (b) A boat needs at least 10 feet of water to moor at the dock. Use a graphing utility to determine the times in the afternoon that the boat can safely dock.

Solution

a. Begin by graphing the data, as shown in Figure 6.40. Use either a sine or cosine model. For instance, a cosine model has the form $y = a \cos(bt - c) + d$. The difference between the maximum value and the minimum value is twice the amplitude of the function. So, the amplitude is

$$a = \tfrac{1}{2}[(\text{maximum depth}) - (\text{minimum depth})] = \tfrac{1}{2}(11.3 - 0.1) = 5.6.$$

The cosine function completes one half of a cycle between the times at which the maximum and minimum depths occur. So, the period p is

$$p = 2[(\text{time of min. depth}) - (\text{time of max. depth})] = 2(10 - 4) = 12.$$

This implies that the value of b is

$$b = \frac{2\pi}{p} = \frac{2\pi}{12} = \frac{\pi}{6}.$$

The maximum depth occurs 4 hours after midnight, so the left endpoint of one cycle is $c/b = 4$. So, the value of c is

$$\frac{c}{b} = 4 \implies \frac{c}{\pi/6} = 4 \implies c = 4\left(\frac{\pi}{6}\right) \implies c \approx 2.094.$$

Moreover, the average depth is $\tfrac{1}{2}(11.3 + 0.1) = 5.7$, so it follows that $d = 5.7$. So, with $a = 5.6$, $b = \pi/6 \approx 0.524$, $c \approx 2.094$, and $d = 5.7$, a model for the depth is

$$y = 5.6 \cos(0.524t - 2.094) + 5.7.$$

b. Using a graphing utility, graph the model with the line $y = 10$. Using the *intersect* feature, determine that the depth is at least 10 feet between 2:42 P.M. ($t \approx 14.7$) and 5:18 P.M. ($t \approx 17.3$), as shown in Figure 6.41.

✓ *Checkpoint* ▶ *Audio-video solution in English & Spanish at LarsonPrecalculus.com*

Find a sine model for the data in Example 7. ◼

Summarize (Section 6.4)

1. Explain how to sketch the graphs of basic sine and cosine functions *(page 457)*. For an example of sketching the graph of a sine function, see Example 1.

2. Explain how to use amplitude and period to help sketch the graphs of sine and cosine functions *(pages 459 and 460)*. For examples of using amplitude and period to sketch graphs of sine and cosine functions, see Examples 2 and 3.

3. Explain how to sketch translations of the graphs of sine and cosine functions *(page 461)*. For examples of translating the graphs of sine and cosine functions, see Examples 4–6.

4. Give an example of using a sine or cosine function to model real-life data *(page 463, Example 7)*.

GO DIGITAL

See CalcChat.com for tutorial help and worked-out solutions to odd-numbered exercises.

6.4 Exercises

GO DIGITAL

Vocabulary and Concept Check

In Exercises 1 and 2, fill in the blanks.

1. One period of a sine or cosine function is one _____ of the sine or cosine curve.

2. The _____ of a sine or cosine curve represents half the distance between the maximum and minimum values of the function.

3. Describe the effect of the constant d on the graph of $y = \sin x + d$.

4. Find an interval that corresponds to one cycle of the graph of $y = 3\cos(2x - \pi/4)$.

Skills and Applications

Finding the Amplitude and Period In Exercises 5–12, find the amplitude and period.

5. $y = 2\sin 5x$

6. $y = 3\cos 2x$

7. $y = \dfrac{3}{4}\cos\dfrac{\pi x}{2}$

8. $y = -5\sin\dfrac{\pi x}{3}$

9. $y = -\dfrac{1}{2}\sin\dfrac{5x}{4}$

10. $y = \dfrac{1}{4}\sin\dfrac{x}{6}$

11. $y = -\dfrac{5}{3}\cos\dfrac{\pi x}{12}$

12. $y = -\dfrac{2}{5}\cos 10\pi x$

Describing the Relationship Between Graphs In Exercises 13–24, describe the relationship between the graphs of f and g. Consider amplitude, period, and shifts.

13. $f(x) = \cos x$
    $g(x) = \cos 5x$

14. $f(x) = \sin x$
    $g(x) = 2\sin x$

15. $f(x) = \cos 2x$
    $g(x) = -\cos 2x$

16. $f(x) = \sin 3x$
    $g(x) = \sin(-3x)$

17. $f(x) = \sin x$
    $g(x) = \sin(x - \pi)$

18. $f(x) = \cos x$
    $g(x) = \cos(x + \pi)$

19. $f(x) = \sin 2x$
    $g(x) = 3 + \sin 2x$

20. $f(x) = \cos 4x$
    $g(x) = -2 + \cos 4x$

21.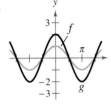

22.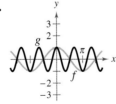

23.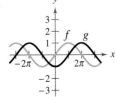

24.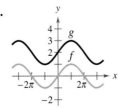

Sketching Graphs of Sine or Cosine Functions In Exercises 25–30, sketch the graphs of f and g in the same coordinate plane. (Include two full periods.)

25. $f(x) = \sin x$
    $g(x) = 4\sin x$

26. $f(x) = -\sin x$
    $g(x) = -3\sin x$

27. $f(x) = \cos x$
    $g(x) = 2 + \cos x$

28. $f(x) = \cos x$
    $g(x) = \cos\left(x + \dfrac{\pi}{2}\right)$

29. $f(x) = -\cos x$
    $g(x) = -\cos(x - \pi)$

30. $f(x) = \sin x$
    $g(x) = \sin\dfrac{x}{3}$

Sketching the Graph of a Sine or Cosine Function In Exercises 31–52, sketch the graph of the function. (Include two full periods.)

31. $y = 5\sin x$

32. $y = \frac{1}{4}\sin x$

33. $y = \frac{1}{3}\cos x$

34. $y = 4\cos x$

35. $y = \cos\dfrac{x}{2}$

36. $y = \sin 4x$

37. $y = \cos 2\pi x$

38. $y = \sin\dfrac{\pi x}{4}$

39. $y = -\sin\dfrac{2\pi x}{3}$

40. $y = 10\cos\dfrac{\pi x}{6}$

41. $y = \cos\left(x - \dfrac{\pi}{2}\right)$

42. $y = \sin(x - 2\pi)$

43. $y = 3\sin(x + \pi)$

44. $y = -4\cos\left(x + \dfrac{\pi}{4}\right)$

45. $y = 2 - \sin\dfrac{2\pi x}{3}$

46. $y = -3 + 5\cos\dfrac{\pi t}{12}$

47. $y = 2 + 5\cos 6\pi x$

48. $y = 2\sin 3x + 5$

49. $y = 3\sin(x + \pi) - 3$

50. $y = -3\sin(6x + \pi)$

51. $y = \dfrac{2}{3}\cos\left(\dfrac{x}{2} - \dfrac{\pi}{4}\right)$

52. $y = 4\cos\left(\pi x + \dfrac{\pi}{2}\right) - 1$

Describing a Transformation   In Exercises 53–58, g is related to a parent function $f(x) = \sin(x)$ or $f(x) = \cos(x)$. (a) Describe the sequence of transformations from f to g. (b) Sketch the graph of g. (c) Use function notation to write g in terms of f.

53. $g(x) = \sin(4x - \pi)$

54. $g(x) = \sin(2x + \pi)$

55. $g(x) = \cos\left(x - \dfrac{\pi}{2}\right) + 2$

56. $g(x) = 1 + \cos(x + \pi)$

57. $g(x) = 2 \sin(4x - \pi) - 3$

58. $g(x) = 4 - \sin\left(2x + \dfrac{\pi}{2}\right)$

 Graphing a Sine or Cosine Function   In Exercises 59–64, use a graphing utility to graph the function. (Include two full periods.) Be sure to choose an appropriate viewing window.

59. $y = -2 \sin(4x + \pi)$

60. $y = -4 \sin\left(\dfrac{2}{3}x - \dfrac{\pi}{3}\right)$

61. $y = \cos\left(2\pi x - \dfrac{\pi}{2}\right) + 1$

62. $y = 3 \cos\left(\dfrac{\pi x}{2} + \dfrac{\pi}{2}\right) - 2$

63. $y = -0.1 \sin\left(\dfrac{\pi x}{10} + \pi\right)$

64. $y = \dfrac{1}{100} \cos 120\pi t$

Graphical Reasoning   In Exercises 65–68, find a and d for the function $f(x) = a \cos x + d$ such that the graph of f matches the figure.

65.
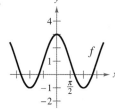

66.
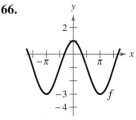

67.
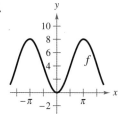

68.
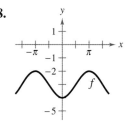

Graphical Reasoning   In Exercises 69–72, find a, b, and c for the function $f(x) = a \sin(bx - c)$ such that the graph of f matches the figure.

69.
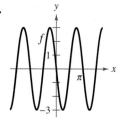

70.
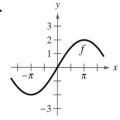

71.
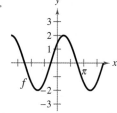

72.
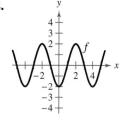

 Finding Real Numbers   In Exercises 73 and 74, use a graphing utility to graph y_1 and y_2 in the interval $[-2\pi, 2\pi]$. Use the graphs to find real numbers x such that $y_1 = y_2$.

73. $y_1 = \sin x$, $y_2 = -\dfrac{1}{2}$

74. $y_1 = \cos x$, $y_2 = -1$

Writing an Equation   In Exercises 75–78, write an equation for a function with the given characteristics.

75. A sine curve with a period of π, an amplitude of 2, a right phase shift of $\pi/2$, and a vertical translation up 1 unit

76. A sine curve with a period of 4π, an amplitude of 3, a left phase shift of $\pi/4$, and a vertical translation down 1 unit

77. A cosine curve with a period of π, an amplitude of 1, a left phase shift of π, and a vertical translation down $\dfrac{3}{2}$ units

78. A cosine curve with a period of 4π, an amplitude of 3, a right phase shift of $\pi/2$, and a vertical translation up 2 units

79. **Respiratory Cycle**   For a person exercising, the velocity v (in liters per second) of airflow during a respiratory cycle (the time from the beginning of one breath to the beginning of the next) is modeled by

    $$v = 1.75 \sin(\pi t / 2)$$

    where t is the time (in seconds). (Inhalation occurs when $v > 0$, and exhalation occurs when $v < 0$.)

    (a) Find the time for one full respiratory cycle.

    (b) Find the number of cycles per minute.

    (c) Sketch the graph of the velocity function.

80. Respiratory Cycle

For a person at rest, the velocity v (in liters per second) of airflow during a respiratory cycle (the time from the beginning of one breath to the beginning of the next) is modeled by $v = 0.85 \sin(\pi t/3)$, where t is the time (in seconds).

(a) Find the time for one full respiratory cycle.

(b) Find the number of cycles per minute.

(c) Sketch the graph of the velocity function. Use the graph to confirm your answer in part (a) by finding two times when new breaths begin. (Inhalation occurs when $v > 0$, and exhalation occurs when $v < 0$.)

81. Blood Pressure The function $P = 100 - 20 \cos(5\pi t/3)$ approximates the blood pressure P (in millimeters of mercury) at time t (in seconds) for a person at rest.

(a) Find the period of the function.

(b) Find the number of heartbeats per minute.

82. Piano Tuning When tuning a piano, a technician strikes a tuning fork for the A above middle C and sets up a wave motion that can be approximated by $y = 0.001 \sin 880\pi t$, where t is the time (in seconds).

(a) What is the period p of the function?

(b) The frequency f is given by $f = 1/p$. What is the frequency of the note?

83. Astronomy The table shows the percent y (in decimal form) of the moon's face illuminated on day x in the year 2021, where $x = 1$ corresponds to January 1. *(Source: U.S. Naval Observatory)*

DATA	x	y
	28	1.00
	35	0.50
	42	0.00
	50	0.49
	58	1.00
	64	0.54

(a) Create a scatter plot of the data.

(b) Find a trigonometric model for the data.

(c) Add the graph of your model in part (b) to the scatter plot. How well does the model fit the data?

(d) What is the period of the model?

(e) Estimate the percent of the moon's face illuminated on March 28, 2021 ($x = 87$).

84. Meteorology The table shows the normal daily maximum temperatures (in degrees Fahrenheit) in Las Vegas L and International Falls I for month t, where $t = 1$ corresponds to January. *(Source: NOAA)*

DATA Month, t	Las Vegas, L	International Falls, I
1	58.0	15.4
2	62.5	22.0
3	70.3	34.7
4	78.3	51.5
5	88.9	64.8
6	98.7	73.2
7	104.2	77.8
8	102.0	75.9
9	94.0	65.4
10	80.6	51.1
11	66.3	33.7
12	56.6	19.0

(a) A model for the temperatures in Las Vegas is

$$L(t) = 80.4 + 23.8 \cos\left(\frac{\pi t}{6} - \frac{7\pi}{6}\right).$$

Find a trigonometric model for the temperatures in International Falls.

(b) Use a graphing utility to graph the data points and the model for the temperatures in Las Vegas. How well does the model fit the data?

(c) Use the graphing utility to graph the data points and the model for the temperatures in International Falls. How well does the model fit the data?

(d) Use the models to estimate the average maximum temperature in each city. Which value in each model did you use? Explain.

(e) What is the period of each model? Are the periods what you expected? Explain.

(f) Which city has the greater variability in temperature throughout the year? Which value in each model determines this variability? Explain.

85. Ferris Wheel The height h (in feet) above ground of a seat on a Ferris wheel at time t (in seconds) is modeled by

$$h(t) = 53 + 50 \sin\left(\frac{\pi}{10}t - \frac{\pi}{2}\right).$$

(a) Find the period of the model. What does the period tell you about the ride?

(b) Find the amplitude of the model. What does the amplitude tell you about the ride?

(c) Use a graphing utility to graph one cycle of the model.

86. Fuel Consumption The daily consumption C (in gallons) of diesel fuel on a farm is modeled by

$$C = 30.3 + 21.6 \sin\left(\frac{2\pi t}{365} + 10.9\right)$$

where t is the time (in days), with $t = 1$ corresponding to January 1.

(a) What is the period of the model? Is it what you expected? Explain.

(b) What is the average daily fuel consumption? Which value in the model did you use? Explain.

(c) Use a graphing utility to graph the model. Use the graph to approximate the time of the year when consumption exceeds 40 gallons per day.

Exploring the Concepts

True or False? In Exercises 87 and 88, determine whether the statement is true or false. Justify your answer.

87. The function $y = \frac{1}{2}\cos 2x$ has an amplitude that is twice that of the function $y = \cos x$.

88. The graph of $y = -\cos x$ is a reflection of the graph of $y = \sin[x + (\pi/2)]$ in the x-axis.

89. Error Analysis Describe the error.

The key points in the graph of $y = \sin x$ on the interval $[0, 2\pi]$ are

$(0, 1)$, $\left(\frac{\pi}{2}, 0\right)$, $(\pi, -1)$, $\left(\frac{3\pi}{2}, 0\right)$, and $(2\pi, 1)$.

90. HOW DO YOU SEE IT? The figure below shows the graph of $y = \sin(x - c)$ for

$$c = -\frac{\pi}{4},\quad 0,\quad \text{and}\quad \frac{\pi}{4}.$$

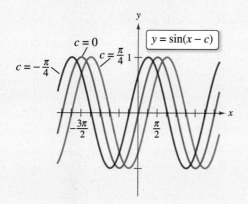

(a) How does the value of c affect the graph?

(b) Which graph is equivalent to that of

$$y = -\cos\left(x + \frac{\pi}{4}\right)?$$

91. Writing Sketch the graph of $y = \cos bx$ for $b = \frac{1}{2}$, 2, and 3. How does the value of b affect the graph? How many complete cycles of the graph occur between 0 and 2π for each value of b?

92. Polynomial Approximations Using calculus, it can be shown that the sine and cosine functions can be approximated by the polynomials

$$\sin x \approx x - \frac{x^3}{3!} + \frac{x^5}{5!} \quad \text{and} \quad \cos x \approx 1 - \frac{x^2}{2!} + \frac{x^4}{4!}$$

where x is in radians.

(a) Use a graphing utility to graph the sine function and its polynomial approximation in the same viewing window. How do the graphs compare?

(b) Use the graphing utility to graph the cosine function and its polynomial approximation in the same viewing window. How do the graphs compare?

(c) Predict the next term in the polynomial approximations of the sine and cosine functions. Then repeat parts (a) and (b). How does the accuracy of the approximations change when an additional term is added?

Review & Refresh ▶ *Video solutions at LarsonPrecalculus.com*

Sketching the Graph of an Equation In Exercises 93–96, (a) complete the table, (b) use the algebraic tests to check for symmetry with respect to both axes and the origin, and (c) use the results of parts (a) and (b) to sketch the graph of the equation.

93. $y = x^2 + 4$

x	0	1	2	3	4
y					

94. $y^2 = 1 - x$

x					
y	0	1	2	3	4

95. $y = x^3 - x$

x	0	$\frac{1}{2}$	1	$\frac{3}{2}$	2
y					

96. $y = 2/x^3$

x	-2	$-\frac{3}{2}$	-1	$-\frac{1}{2}$	$-\frac{1}{4}$
y					

97. Solving Inequalities Graph the functions $y = 3^x$ and $y = 4^x$ and use the graphs to solve each inequality.

(a) $4^x < 3^x$ (b) $4^x > 3^x$

Project: Meteorology To work an extended application analyzing the normal daily mean temperatures and normal precipitation for Honolulu, Hawaii, visit this text's website at *LarsonPrecalculus.com*. (*Source: NOAA*)

6.5 Graphs of Other Trigonometric Functions

Graphs of trigonometric functions have many real-life applications, such as in modeling the distance from a television camera to a unit in a parade, as in Exercise 81 on page 477.

❯ Sketch the graphs of tangent functions.
❯ Sketch the graphs of cotangent functions.
❯ Sketch the graphs of secant and cosecant functions.
❯ Sketch the graphs of damped trigonometric functions.

Graph of the Tangent Function

Recall that the tangent function is odd. That is, $\tan(-x) = -\tan x$. Consequently, the graph of $y = \tan x$ is symmetric with respect to the origin. You also know from the identity $\tan x = (\sin x)/(\cos x)$ that the tangent function is undefined for values at which $\cos x = 0$. Two such values are $x = \pm\pi/2 \approx \pm1.5708$. As shown in the table below, $\tan x$ increases without bound as x approaches $\pi/2$ from the left and decreases without bound as x approaches $-\pi/2$ from the right.

x	$-\dfrac{\pi}{2}$	-1.57	-1.5	$-\dfrac{\pi}{4}$	0	$\dfrac{\pi}{4}$	1.5	1.57	$\dfrac{\pi}{2}$
$\tan x$	Undef.	-1255.8	-14.1	-1	0	1	14.1	1255.8	Undef.

So, the graph of $y = \tan x$ has *vertical asymptotes* at $x = \pi/2$ and $x = -\pi/2$, as shown in the figure below. Moreover, the period of the tangent function is π, so vertical asymptotes also occur at $x = (\pi/2) + n\pi$, where n is an integer. The domain of the tangent function is the set of all real numbers other than $x = (\pi/2) + n\pi$, and the range is the set of all real numbers.

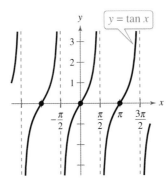

Period: π
Domain: all $x \neq \dfrac{\pi}{2} + n\pi$
Range: $(-\infty, \infty)$
Vertical asymptotes: $x = \dfrac{\pi}{2} + n\pi$
x-intercepts: $(n\pi, 0)$
y-intercept: $(0, 0)$
Symmetry: origin
Odd function

Sketching the graph of $y = a\tan(bx - c)$ is similar to sketching the graph of $y = a\sin(bx - c)$ in that you locate key points of the graph. When sketching the graph of $y = a\tan(bx - c)$, the key points identify the intercepts and asymptotes. Two consecutive vertical asymptotes can be found by solving the equations

$$bx - c = -\frac{\pi}{2} \quad \text{and} \quad bx - c = \frac{\pi}{2}.$$

On the x-axis, the point halfway between two consecutive vertical asymptotes is an x-intercept of the graph. The period of the function $y = a\tan(bx - c)$ is the distance between two consecutive vertical asymptotes. The amplitude of a tangent function is not defined. After plotting two consecutive asymptotes and the x-intercept between them, plot additional points between the asymptotes and sketch one cycle. Finally, sketch one or two additional cycles to the left and right.

GO DIGITAL

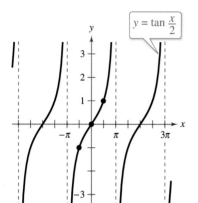

Figure 6.42

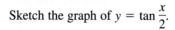

EXAMPLE 1   **Sketching the Graph of a Tangent Function**

Sketch the graph of $y = \tan \dfrac{x}{2}$.

Solution

Solving the equations

$$\frac{x}{2} = -\frac{\pi}{2} \quad \text{and} \quad \frac{x}{2} = \frac{\pi}{2}$$

shows that two consecutive vertical asymptotes occur at $x = -\pi$ and $x = \pi$. Between these two asymptotes, find a few points, including the x-intercept, as shown in the table. Figure 6.42 shows three cycles of the graph.

x	$-\pi$	$-\dfrac{\pi}{2}$	0	$\dfrac{\pi}{2}$	π
$\tan \dfrac{x}{2}$	Undef.	-1	0	1	Undef.

 Checkpoint ▶ *Audio-video solution in English & Spanish at LarsonPrecalculus.com*

Sketch the graph of $y = \tan \dfrac{x}{4}$.

EXAMPLE 2   **Sketching the Graph of a Tangent Function**

Sketch the graph of $y = -3 \tan 2x$.

Solution

Solving the equations

$$2x = -\frac{\pi}{2} \quad \text{and} \quad 2x = \frac{\pi}{2}$$

shows that two consecutive vertical asymptotes occur at $x = -\pi/4$ and $x = \pi/4$. Between these two asymptotes, find a few points, including the x-intercept, as shown in the table. Figure 6.43 shows three cycles of the graph.

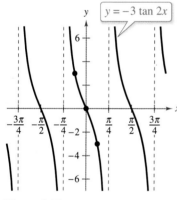

Figure 6.43

x	$-\dfrac{\pi}{4}$	$-\dfrac{\pi}{8}$	0	$\dfrac{\pi}{8}$	$\dfrac{\pi}{4}$
$-3 \tan 2x$	Undef.	3	0	-3	Undef.

✓ **Checkpoint** ▶ *Audio-video solution in English & Spanish at LarsonPrecalculus.com*

Sketch the graph of $y = \tan 2x$. ■

Compare the graphs in Examples 1 and 2. The graph of $y = a \tan(bx - c)$ increases between consecutive vertical asymptotes when $a > 0$ and decreases between consecutive vertical asymptotes when $a < 0$. In other words, the graph for $a < 0$ is a reflection in the x-axis of the graph for $a > 0$. Also, the period is greater when $0 < b < 1$ than when $b > 1$. In other words, compared with the case where $b = 1$, the period represents a horizontal stretch when $0 < b < 1$ and a horizontal shrink when $b > 1$.

GO DIGITAL

Graph of the Cotangent Function

The graph of the cotangent function is similar to the graph of the tangent function. It also has a period of π. However, the identity

$$y = \cot x = \frac{\cos x}{\sin x}$$

shows that the cotangent function has vertical asymptotes when $\sin x$ is zero, which occurs at $x = n\pi$, where n is an integer. The graph of the cotangent function is shown below. Note that two consecutive vertical asymptotes of the graph of $y = a \cot(bx - c)$ can be found by solving the equations $bx - c = 0$ and $bx - c = \pi$.

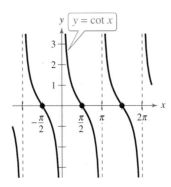

Period: π
Domain: all $x \neq n\pi$
Range: $(-\infty, \infty)$
Vertical asymptotes: $x = n\pi$
x-intercepts: $\left(\dfrac{\pi}{2} + n\pi, 0\right)$
Symmetry: origin
Odd function

EXAMPLE 3 **Sketching the Graph of a Cotangent Function**

Sketch the graph of $y = 2 \cot \dfrac{x}{3}$.

Solution Solving the equations

$$\frac{x}{3} = 0 \quad \text{and} \quad \frac{x}{3} = \pi$$

shows that two consecutive vertical asymptotes occur at $x = 0$ and $x = 3\pi$. Between these two asymptotes, find a few points, including the x-intercept, as shown in the table. Figure 6.44 shows three cycles of the graph. Note that the period is 3π, the distance between consecutive asymptotes.

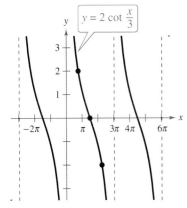

Figure 6.44

x	0	$\dfrac{3\pi}{4}$	$\dfrac{3\pi}{2}$	$\dfrac{9\pi}{4}$	3π
$2 \cot \dfrac{x}{3}$	Undef.	2	0	-2	Undef.

 Checkpoint ▶ Audio-video solution in English & Spanish at LarsonPrecalculus.com

Sketch the graph of $y = \cot \dfrac{x}{4}$. ■

⟫⟫ TECHNOLOGY

Some graphing utilities do not have a key for the cotangent function. To overcome this, use the identity $\cot x = (\cos x)/(\sin x)$ to graph the function. For instance, to graph the function in Example 3, rewrite it as

$$y = 2 \cot \frac{x}{3} = 2\left(\frac{\cos (x/3)}{\sin (x/3)}\right).$$

GO DIGITAL

Graphs of the Reciprocal Functions

You can obtain the graphs of the cosecant and secant functions from the graphs of the sine and cosine functions, respectively, using the reciprocal identities

$$\csc x = \frac{1}{\sin x} \quad \text{and} \quad \sec x = \frac{1}{\cos x}.$$

▶▶▶ **TECHNOLOGY**

Some graphing utilities have difficulty graphing trigonometric functions that have vertical asymptotes. In *connected* mode, your graphing utility may connect parts of the graphs of tangent, cotangent, secant, and cosecant functions that are not supposed to be connected. In *dot* mode, the graphs are represented as collections of dots, so the graphs do not resemble solid curves.

For example, at a given value of x, the y-coordinate of $\sec x$ is the reciprocal of the y-coordinate of $\cos x$. Of course, when $\cos x = 0$, the reciprocal does not exist. Near such values of x, the behavior of the secant function is similar to that of the tangent function. In other words, the graphs of

$$\tan x = \frac{\sin x}{\cos x} \quad \text{and} \quad \sec x = \frac{1}{\cos x}$$

have vertical asymptotes where $\cos x = 0$, that is, at $x = (\pi/2) + n\pi$, where n is an integer. Similarly,

$$\cot x = \frac{\cos x}{\sin x} \quad \text{and} \quad \csc x = \frac{1}{\sin x}$$

have vertical asymptotes where $\sin x = 0$, that is, at $x = n\pi$, where n is an integer.

To sketch the graph of a secant or cosecant function, first make a sketch of its reciprocal function. For example, to sketch the graph of $y = \csc x$, first sketch the graph of $y = \sin x$. Then find reciprocals of the y-coordinates to obtain points on the graph of $y = \csc x$. You can use this procedure to obtain the graphs below.

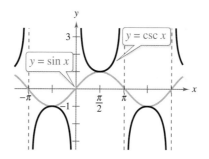

Period: 2π
Domain: all $x \neq n\pi$
Range: $(-\infty, -1] \cup [1, \infty)$
Vertical asymptotes: $x = n\pi$
No intercepts
Symmetry: origin
Odd function

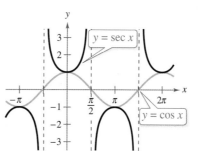

Period: 2π
Domain: all $x \neq \dfrac{\pi}{2} + n\pi$
Range: $(-\infty, -1] \cup [1, \infty)$
Vertical asymptotes: $x = \dfrac{\pi}{2} + n\pi$
y-intercept: $(0, 1)$
Symmetry: y-axis
Even function

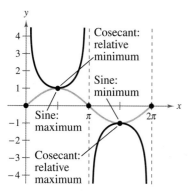

Figure 6.45

GO DIGITAL

In comparing the graphs of the cosecant and secant functions with those of the sine and cosine functions, respectively, note that the "hills" and "valleys" are interchanged. For example, a hill (or maximum point) on the sine curve corresponds to a valley (a relative minimum) on the cosecant curve, and a valley (or minimum point) on the sine curve corresponds to a hill (a relative maximum) on the cosecant curve, as shown in Figure 6.45. Additionally, x-intercepts of the sine and cosine functions become vertical asymptotes of the cosecant and secant functions, respectively (see Figure 6.45).

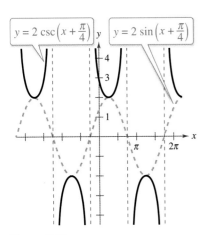

Figure 6.46

EXAMPLE 4   **Sketching the Graph of a Cosecant Function**

Sketch the graph of $y = 2 \csc\left(x + \dfrac{\pi}{4}\right)$.

Solution

Begin by sketching the graph of

$$y = 2 \sin\left(x + \frac{\pi}{4}\right).$$

For this function, the amplitude is 2 and the period is 2π. Solving the equations

$$x + \frac{\pi}{4} = 0 \quad \text{and} \quad x + \frac{\pi}{4} = 2\pi$$

shows that one cycle of the sine function corresponds to the interval from $x = -\pi/4$ to $x = 7\pi/4$. The dashed gray curve in Figure 6.46 represents the graph of the sine function. At the midpoint and endpoints of this interval, the sine function is zero. So, the corresponding cosecant function

$$y = 2 \csc\left(x + \frac{\pi}{4}\right)$$

$$= 2\left(\frac{1}{\sin[x + (\pi/4)]}\right)$$

has vertical asymptotes at $x = -\pi/4$, $x = 3\pi/4$, $x = 7\pi/4$, and so on. The solid black curve in Figure 6.46 represents the graph of the cosecant function.

✓ *Checkpoint* ▶ *Audio-video solution in English & Spanish at LarsonPrecalculus.com*

Sketch the graph of $y = 2 \csc\left(x + \dfrac{\pi}{2}\right)$.

EXAMPLE 5   **Sketching the Graph of a Secant Function**

▶▶▶ *See LarsonPrecalculus.com for an interactive version of this type of example.*

Sketch the graph of $y = \sec 2x$.

Solution

Begin by sketching the graph of $y = \cos 2x$, shown as the dashed gray curve in Figure 6.47. Then, form the graph of $y = \sec 2x$, shown as the solid black curve in the figure. Note that the x-intercepts of $y = \cos 2x$

$$\left(-\frac{\pi}{4}, 0\right), \quad \left(\frac{\pi}{4}, 0\right), \quad \left(\frac{3\pi}{4}, 0\right), \ldots$$

correspond to the vertical asymptotes

$$x = -\frac{\pi}{4}, \quad x = \frac{\pi}{4}, \quad x = \frac{3\pi}{4}, \ldots$$

of the graph of $y = \sec 2x$. Moreover, notice that the period of $y = \cos 2x$ and $y = \sec 2x$ is π.

✓ *Checkpoint* ▶ *Audio-video solution in English & Spanish at LarsonPrecalculus.com*

Sketch the graph of $y = \sec \dfrac{x}{2}$.

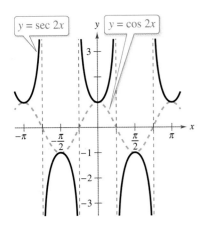

Figure 6.47

GO DIGITAL

Damped Trigonometric Graphs

You can graph a *product* of two functions using properties of the individual functions. For example, consider the function

$$f(x) = x \sin x$$

as the product of the functions $y = x$ and $y = \sin x$. Using properties of absolute value and the fact that $|\sin x| \le 1$, you have

$$0 \le |x||\sin x| \le |x|.$$

Consequently,

$$-|x| \le x \sin x \le |x|$$

which means that the graph of $f(x) = x \sin x$ lies between the lines $y = -x$ and $y = x$. Furthermore,

$$f(x) = x \sin x = \pm x \quad \text{at} \quad x = \frac{\pi}{2} + n\pi$$

and

$$f(x) = x \sin x = 0 \quad \text{at} \quad x = n\pi$$

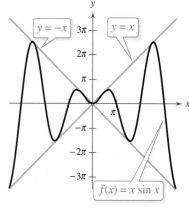

where n is an integer, so the graph of f touches the line $y = x$ or the line $y = -x$ at $x = (\pi/2) + n\pi$ and has x-intercepts at $x = n\pi$. A sketch of f is shown at the right. In the function $f(x) = x \sin x$, the factor x is called the **damping factor.**

ALGEBRA HELP

Do you see why the graph of $f(x) = x \sin x$ touches the lines $y = \pm x$ at $x = (\pi/2) + n\pi$ and why the graph has x-intercepts at $x = n\pi$? Recall that the sine function is equal to ± 1 at odd multiples of $\pi/2$ and is equal to 0 at multiples of π.

EXAMPLE 6 **Damped Sine Curve**

Sketch the graph of $f(x) = e^{-x} \sin 3x$.

Solution

Consider f as the product of the two functions $y = e^{-x}$ and $y = \sin 3x$, each of which has the set of real numbers as its domain. For any real number x, you know that $e^{-x} > 0$ and $|\sin 3x| \le 1$. So,

$$e^{-x}|\sin 3x| \le e^{-x}$$

which means that

$$-e^{-x} \le e^{-x} \sin 3x \le e^{-x}.$$

Furthermore,

$$f(x) = e^{-x} \sin 3x = \pm e^{-x} \quad \text{at} \quad x = \frac{\pi}{6} + \frac{n\pi}{3}$$

and

$$f(x) = e^{-x} \sin 3x = 0 \quad \text{at} \quad x = \frac{n\pi}{3}$$

so the graph of f touches the curve $y = e^{-x}$ or the curve $y = -e^{-x}$ at $x = (\pi/6) + (n\pi/3)$ and has intercepts at $x = n\pi/3$. Figure 6.48 shows a sketch of f.

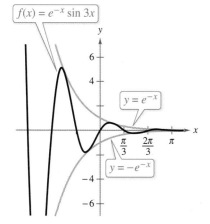

Figure 6.48

✓ *Checkpoint* ▶ *Audio-video solution in English & Spanish at LarsonPrecalculus.com*

Sketch the graph of $f(x) = e^x \sin 4x$.

Below is a summary of the characteristics of the six basic trigonometric functions.

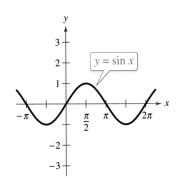

Domain: $(-\infty, \infty)$
Range: $[-1, 1]$
Period: 2π
x-intercepts: $(n\pi, 0)$
y-intercept: $(0, 0)$
Odd function
Origin symmetry

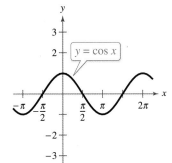

Domain: $(-\infty, \infty)$
Range: $[-1, 1]$
Period: 2π
x-intercepts: $\left(\dfrac{\pi}{2} + n\pi, 0\right)$
y-intercept: $(0, 1)$
Even function
y-axis symmetry

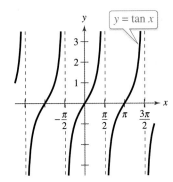

Domain: all $x \neq \dfrac{\pi}{2} + n\pi$
Range: $(-\infty, \infty)$
Period: π
x-intercepts: $(n\pi, 0)$
y-intercept: $(0, 0)$
Vertical asymptotes:
$$x = \dfrac{\pi}{2} + n\pi$$
Odd function
Origin symmetry

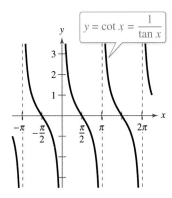

Domain: all $x \neq n\pi$
Range: $(-\infty, \infty)$
Period: π
x-intercepts: $\left(\dfrac{\pi}{2} + n\pi, 0\right)$
Vertical asymptotes: $x = n\pi$
Odd function
Origin symmetry

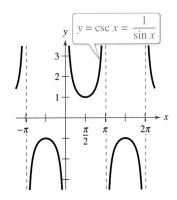

Domain: all $x \neq n\pi$
Range: $(-\infty, -1] \cup [1, \infty)$
Period: 2π
No intercepts
Vertical asymptotes: $x = n\pi$
Odd function
Origin symmetry

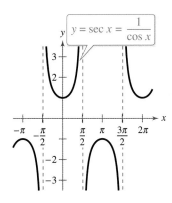

Domain: all $x \neq \dfrac{\pi}{2} + n\pi$
Range: $(-\infty, -1] \cup [1, \infty)$
Period: 2π
y-intercept: $(0, 1)$
Vertical asymptotes:
$$x = \dfrac{\pi}{2} + n\pi$$
Even function
y-axis symmetry

Summarize (Section 6.5)

1. Explain how to sketch the graph of $y = a \tan(bx - c)$ *(page 468)*. For examples of sketching graphs of tangent functions, see Examples 1 and 2.

2. Explain how to sketch the graph of $y = a \cot(bx - c)$ *(page 470)*. For an example of sketching the graph of a cotangent function, see Example 3.

3. Explain how to sketch the graphs of $y = a \csc(bx - c)$ and $y = a \sec(bx - c)$ *(page 471)*. For examples of sketching graphs of cosecant and secant functions, see Examples 4 and 5.

4. Explain how to sketch the graph of a damped trigonometric function *(page 473)*. For an example of sketching the graph of a damped trigonometric function, see Example 6.

GO DIGITAL

6.5 Exercises

See CalcChat.com for tutorial help and worked-out solutions to odd-numbered exercises.

GO DIGITAL

Vocabulary and Concept Check

In Exercises 1–4, fill in the blanks.

1. The tangent, cotangent, and cosecant functions are _____, so the graphs of these functions have symmetry with respect to the _____.

2. To sketch the graph of a secant or cosecant function, first make a sketch of its _____ function.

3. For the function $f(x) = g(x) \cdot \sin x$, $g(x)$ is called the _____ factor.

4. The domain of $y = \cot x$ is all real numbers such that _____.

5. What is the range of $y = \csc x$ and $y = \sec x$?

6. Which four parent trigonometric functions have a period of 2π?

Skills and Applications

Matching In Exercises 7–12, match the function with its graph. State the period of the function. [The graphs are labeled (a), (b), (c), (d), (e), and (f).]

(a)

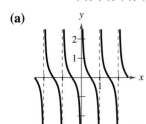

(b)

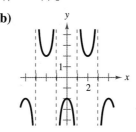

(c)

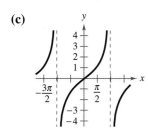

(d)

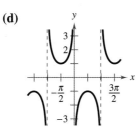

(e)

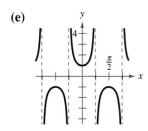

(f)

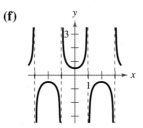

7. $y = \sec 2x$

8. $y = \tan(x/2)$

9. $y = \frac{1}{2} \cot \pi x$

10. $y = -\csc x$

11. $y = \frac{1}{2} \sec(\pi x/2)$

12. $y = -2 \sec(\pi x/2)$

Sketching the Graph of a Trigonometric Function In Exercises 13–36, sketch the graph of the function. (Include two full periods.)

13. $y = \frac{1}{3} \tan x$

14. $y = -\frac{1}{2} \tan x$

15. $y = \tan(\pi x/4)$

16. $y = \tan 4x$

17. $y = -2 \tan 3x$

18. $y = -3 \tan \pi x$

19. $y = 3 \cot 2x$

20. $y = 3 \cot(\pi x/2)$

21. $y = -\frac{1}{2} \sec x$

22. $y = \frac{1}{4} \sec x$

23. $y = \csc(x/2)$

24. $y = \csc(x/3)$

25. $y = \frac{1}{2} \sec \pi x$

26. $y = 2 \sec 3x$

27. $y = \csc \pi x - 2$

28. $y = 3 \csc 4x + 1$

29. $y = -\sec \pi x + 1$

30. $y = -2 \sec 4x + 2$

31. $y = \tan(x + \pi)$

32. $y = 2 \cot\left(x + \dfrac{\pi}{2}\right)$

33. $y = 2 \sec(x + \pi)$

34. $y = 2 \csc(x - \pi)$

35. $y = \dfrac{1}{4} \csc\left(x + \dfrac{\pi}{4}\right)$

36. $y = \dfrac{3}{2} \sec(2x - \pi)$

Graphing a Trigonometric Function In Exercises 37–46, use a graphing utility to graph the function. (Include two full periods.)

37. $y = \cot(x/3)$

38. $y = -\tan 2x$

39. $y = -2 \sec 4x$

40. $y = \csc \pi x$

41. $y = \tan\left(x - \dfrac{\pi}{4}\right)$

42. $y = \dfrac{1}{4} \cot\left(x - \dfrac{\pi}{2}\right)$

43. $y = -\csc(4x - \pi)$

44. $y = 2 \sec(2x - \pi)$

45. $y = 0.1 \tan\left(\dfrac{\pi x}{4} + \dfrac{\pi}{4}\right)$

46. $y = \dfrac{1}{3} \sec\left(\dfrac{\pi x}{2} + \dfrac{\pi}{2}\right)$

Solving a Trigonometric Equation In Exercises 47–52, find the solutions of the equation in the interval $[-2\pi, 2\pi]$. Use a graphing utility to verify your results.

47. $\tan x = 1$

48. $\tan x = \sqrt{3}$

49. $\cot x = -\sqrt{3}$

50. $\cot x = 1$

51. $\sec x = 2$

52. $\csc x = \sqrt{2}$

Even and Odd Trigonometric Functions In Exercises 53–60, use the graph of the function to determine whether the function is even, odd, or neither. Verify your answer algebraically.

53. $f(x) = \sec x$

54. $f(x) = \tan x$

55. $f(x) = \csc 2x$

56. $f(x) = \cot 2x$

57. $f(x) = x + \tan x$

58. $f(x) = x^2 - \sec x$

59. $g(x) = x \csc x$

60. $g(x) = x^2 \cot x$

Identifying Damped Trigonometric Functions In Exercises 61–64, match the function with its graph. Describe the behavior of the function as x approaches zero. [The graphs are labeled (a), (b), (c), and (d).]

(a)
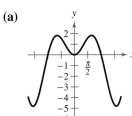

(b)
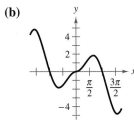

(c)
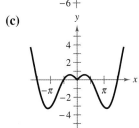

(d)
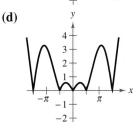

61. $f(x) = |x \cos x|$

62. $f(x) = x \sin x$

63. $g(x) = |x| \sin x$

64. $g(x) = |x| \cos x$

Conjecture In Exercises 65–68, graph the functions f and g. Use the graphs to make a conjecture about the relationship between the functions.

65. $f(x) = \sin x + \cos\left(x + \dfrac{\pi}{2}\right)$, $g(x) = 0$

66. $f(x) = \sin x - \cos\left(x + \dfrac{\pi}{2}\right)$, $g(x) = 2 \sin x$

67. $f(x) = \sin^2 x$, $g(x) = \dfrac{1}{2}(1 - \cos 2x)$

68. $f(x) = \cos^2 \dfrac{\pi x}{2}$, $g(x) = \dfrac{1}{2}(1 + \cos \pi x)$

Analyzing a Damped Trigonometric Graph In Exercises 69–72, use a graphing utility to graph the function and the damping factor of the function in the same viewing window. Describe the behavior of the function as x increases without bound.

69. $g(x) = e^{-x^2/2} \sin x$

70. $f(x) = e^{-x} \cos x$

71. $f(x) = 2^{-x/4} \cos \pi x$

72. $h(x) = 2^{-x^2/4} \sin x$

Analyzing a Trigonometric Graph In Exercises 73–78, use a graphing utility to graph the function. Describe the behavior of the function as x approaches zero.

73. $y = \dfrac{6}{x} + \cos x$, $x > 0$

74. $y = \dfrac{4}{x} + \sin 2x$, $x > 0$

75. $g(x) = \dfrac{\sin x}{x}$

76. $f(x) = \dfrac{1 - \cos x}{x}$

77. $f(x) = \sin \dfrac{1}{x}$

78. $h(x) = x \sin \dfrac{1}{x}$

79. **Meteorology** The normal monthly high temperatures H (in degrees Fahrenheit) in Erie, Pennsylvania, are approximated by

$$H(t) = 57.54 - 18.53 \cos \dfrac{\pi t}{6} - 14.03 \sin \dfrac{\pi t}{6}$$

and the normal monthly low temperatures L are approximated by

$$L(t) = 42.03 - 15.99 \cos \dfrac{\pi t}{6} - 14.32 \sin \dfrac{\pi t}{6}$$

where t is the time (in months), with $t = 1$ corresponding to January (see figure). *(Source: NOAA)*

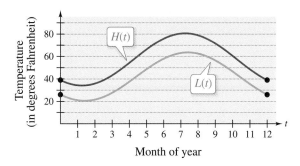

(a) What is the period of each function?

(b) During what part of the year is the difference between the normal high and normal low temperatures greatest? When is it least?

(c) The sun is northernmost in the sky around June 21, but the graph shows the warmest temperatures at a later date. Approximate the lag time of the temperatures relative to the position of the sun.

80. **Sales** The projected monthly sales S (in thousands of units) of lawn mowers are modeled by

$$S = 74 + 3t - 40 \cos \dfrac{\pi t}{6}$$

where t is the time (in months), with $t = 1$ corresponding to January.

(a) Graph the sales function over 1 year.

(b) What are the projected sales for June?

81. Television Coverage

A television camera is on a reviewing platform 27 meters from the street on which a parade passes from left to right (see figure). Write the distance d from the camera to a unit in the parade as a function of the angle x, and graph the function over the interval

$$-\pi/2 < x < \pi/2.$$

(Consider x as negative when a unit in the parade approaches from the left.)

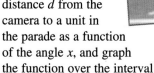

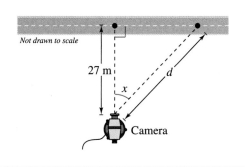

Not drawn to scale

27 m

x

d

Camera

82. Distance
A plane flying at an altitude of 7 miles above a radar antenna passes directly over the radar antenna (see figure). Let d be the ground distance from the antenna to the point directly under the plane and let x be the angle of elevation to the plane from the antenna. (d is positive as the plane approaches the antenna.) Write d as a function of x and graph the function over the interval $0 < x < \pi$.

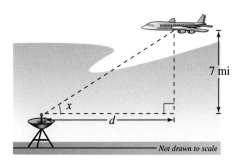

7 mi

x

d

Not drawn to scale

Exploring the Concepts

True or False? In Exercises 83 and 84, determine whether the statement is true or false. Justify your answer.

83. You can obtain the graph of $y = \csc x$ on a calculator by graphing the reciprocal of $y = \sin x$.

84. You can obtain the graph of $y = \sec x$ on a calculator by graphing a translation of the reciprocal of $y = \sin x$.

85. Think About It   Let $f(x) = x - \cos x$.

(a) Use a graphing utility to graph the function and verify that there exists a zero between 0 and 1. Use the graph to approximate the zero.

(b) Starting with $x_0 = 1$, generate a sequence x_1, x_2, x_3, . . . , where $x_n = \cos(x_{n-1})$. For example, $x_0 = 1, x_1 = \cos(x_0), x_2 = \cos(x_1), x_3 = \cos(x_2), \ldots$. What value does the sequence approach?

86. HOW DO YOU SEE IT?   Determine which function each graph represents. Do not use a calculator. Explain.

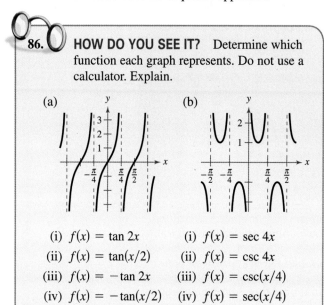

(a)

(b)

(i) $f(x) = \tan 2x$ (i) $f(x) = \sec 4x$

(ii) $f(x) = \tan(x/2)$ (ii) $f(x) = \csc 4x$

(iii) $f(x) = -\tan 2x$ (iii) $f(x) = \csc(x/4)$

(iv) $f(x) = -\tan(x/2)$ (iv) $f(x) = \sec(x/4)$

Review & Refresh ▶ *Video solutions at LarsonPrecalculus.com*

Finding Domains of Functions and Composite Functions In Exercises 87–90, find (a) $f \circ g$ and (b) $g \circ f$. Find the domain of each function and of each composite function.

87. $f(x) = \sqrt[3]{x}, g(x) = x^2$

88. $f(x) = x + 4, g(x) = 1/x$

89. $f(x) = x^4, g(x) = \sqrt{x - 1}$

90. $f(x) = x/(x^2 - 4), g(x) = 2x$

Finding an Inverse Function In Exercises 91–96, determine whether the function has an inverse function. If it does, find the inverse function.

91. $f(x) = 9 - x^2$ **92.** $g(x) = 4x + 1$

93. $h(x) = 1/x^3$ **94.** $h(x) = |x - 3|$

95. $f(x) = \sqrt{3x - 6}$

96. $f(x) = (x - 5)^2, x \le 5$

Solving an Exponential Equation In Exercises 97 and 98, solve the exponential equation algebraically. Approximate the result to three decimal places.

97. $10e^x + 2 = 7$

98. $3^{2x} = 8^{x^2}$

6.6 Inverse Trigonometric Functions

Inverse trigonometric functions have many applications in real life. For example, in Exercise 94 on page 486, you will use an inverse trigonometric function to model the angle of elevation from a television camera to a space shuttle.

> ❯ Evaluate and graph the inverse sine function.
> ❯ Evaluate and graph other inverse trigonometric functions.
> ❯ Evaluate composite functions involving inverse trigonometric functions.

Inverse Sine Function

Recall from Section 2.7 that for a function to have an inverse function, it must be one-to-one—that is, it must pass the Horizontal Line Test. Notice in Figure 6.49 that $y = \sin x$ does not pass the test because different values of x yield the same y-value.

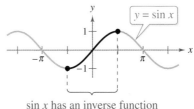

$\sin x$ has an inverse function
on this interval.

Figure 6.49

However, when you restrict the domain to the interval $-\pi/2 \le x \le \pi/2$ (corresponding to the black portion of the graph in Figure 6.49), the properties listed below hold.

1. On the interval $[-\pi/2, \pi/2]$, the function $y = \sin x$ is increasing.
2. On the interval $[-\pi/2, \pi/2]$, $y = \sin x$ takes on its full range of values, $-1 \le \sin x \le 1$.
3. On the interval $[-\pi/2, \pi/2]$, $y = \sin x$ is one-to-one.

So, on the restricted domain $-\pi/2 \le x \le \pi/2$, $y = \sin x$ has a unique inverse function called the **inverse sine function.** It is denoted by

$$y = \arcsin x \quad \text{or} \quad y = \sin^{-1} x.$$

The notation $\sin^{-1} x$ is consistent with the inverse function notation $f^{-1}(x)$. The arcsin x notation (read as "the arcsine of x") comes from the association of a central angle with its intercepted *arc length* on a unit circle. So, arcsin x means the angle (or arc) whose sine is x. Both notations, arcsin x and $\sin^{-1} x$, are commonly used in mathematics. You must remember that $\sin^{-1} x$ denotes the *inverse* sine function, *not* $1/\sin x$. The values of arcsin x lie in the interval

$$-\frac{\pi}{2} \le \arcsin x \le \frac{\pi}{2}.$$

The graph of $y = \arcsin x$ is shown in Figure 6.50 on the next page.

GO DIGITAL

❯❯❯❯

ALGEBRA HELP

When evaluating the inverse sine function, it helps to remember the phrase "the arcsine of x is the angle (or number) whose sine is x."

Definition of Inverse Sine Function

The **inverse sine function** is defined by

$$y = \arcsin x \quad \text{if and only if} \quad \sin y = x$$

where $-1 \le x \le 1$ and $-\pi/2 \le y \le \pi/2$. The domain of $y = \arcsin x$ is $[-1, 1]$, and the range is $[-\pi/2, \pi/2]$.

ALGEBRA HELP

As with trigonometric functions, some of the work with inverse trigonometric functions can be done by *exact* calculations rather than by calculator approximations. Exact calculations help to increase your understanding of inverse functions by relating them to the right triangle definitions of trigonometric functions.

EXAMPLE 1 Evaluating the Inverse Sine Function

If possible, find the exact value of each expression.

a. $\arcsin\left(-\frac{1}{2}\right)$ **b.** $\sin^{-1}\frac{\sqrt{3}}{2}$ **c.** $\sin^{-1} 2$

Solution

a. You know that $\sin\left(-\frac{\pi}{6}\right) = -\frac{1}{2}$ and $-\frac{\pi}{6}$ lies in $\left[-\frac{\pi}{2}, \frac{\pi}{2}\right]$, so

$$\arcsin\left(-\frac{1}{2}\right) = -\frac{\pi}{6}. \qquad \text{Angle whose sine is } -\tfrac{1}{2}$$

b. You know that $\sin\frac{\pi}{3} = \frac{\sqrt{3}}{2}$ and $\frac{\pi}{3}$ lies in $\left[-\frac{\pi}{2}, \frac{\pi}{2}\right]$, so

$$\sin^{-1}\frac{\sqrt{3}}{2} = \frac{\pi}{3}. \qquad \text{Angle whose sine is } \sqrt{3}/2$$

c. It is not possible to evaluate $y = \sin^{-1} x$ when $x = 2$ because there is no angle whose sine is 2. Remember that the domain of the inverse sine function is $[-1, 1]$.

✓ ***Checkpoint*** ▶ ***Audio-video solution in English & Spanish at LarsonPrecalculus.com***

If possible, find the exact value of each expression.

a. $\arcsin 1$ **b.** $\sin^{-1}(-2)$

EXAMPLE 2 Graphing the Arcsine Function

See LarsonPrecalculus.com for an interactive version of this type of example.

Sketch the graph of $y = \arcsin x$.

Solution

By definition, the equations $y = \arcsin x$ and $\sin y = x$ are equivalent for $-\pi/2 \le y \le \pi/2$. So, their graphs are the same. From the interval $[-\pi/2, \pi/2]$, assign values to y in the equation $\sin y = x$ to make a table of values.

y	$-\frac{\pi}{2}$	$-\frac{\pi}{4}$	$-\frac{\pi}{6}$	0	$\frac{\pi}{6}$	$\frac{\pi}{4}$	$\frac{\pi}{2}$
$x = \sin y$	-1	$-\frac{\sqrt{2}}{2}$	$-\frac{1}{2}$	0	$\frac{1}{2}$	$\frac{\sqrt{2}}{2}$	1

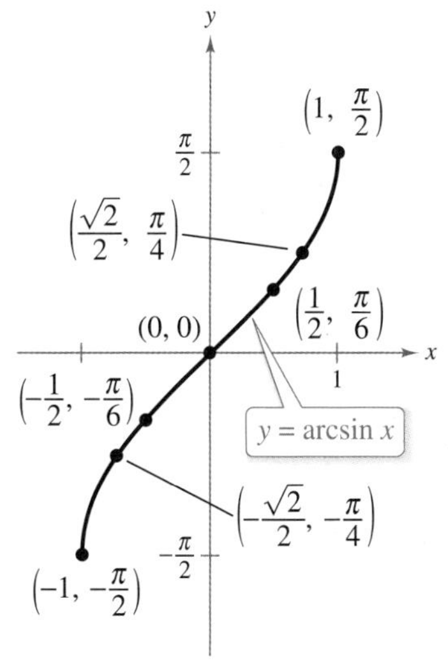

Figure 6.50

Then plot the points and connect them with a smooth curve. Figure 6.50 shows the graph of $y = \arcsin x$. Note that it is the reflection (in the line $y = x$) of the portion of the graph of $y = \sin x$ on the interval $[-\pi/2, \pi/2]$ in Figure 6.49. Be sure you see that Figure 6.50 shows the *entire* graph of the inverse sine function. Remember that the domain of $y = \arcsin x$ is the closed interval $[-1, 1]$ and the range is the closed interval $[-\pi/2, \pi/2]$.

✓ ***Checkpoint*** ▶ ***Audio-video solution in English & Spanish at LarsonPrecalculus.com***

Use a graphing utility to graph $f(x) = \sin x$, $g(x) = \arcsin x$, and $y = x$ in the same viewing window to verify geometrically that g is the inverse function of f. (Be sure to restrict the domain of f properly.) ■

GO DIGITAL

Other Inverse Trigonometric Functions

The cosine function is decreasing and one-to-one on the interval $0 \le x \le \pi$, as shown in the figure below.

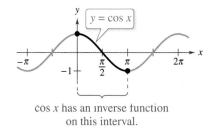

cos x has an inverse function
on this interval.

Consequently, on this interval the cosine function has an inverse function—the **inverse cosine function**—denoted by

$$y = \arccos x \quad \text{or} \quad y = \cos^{-1} x.$$

Similarly, to define an **inverse tangent function,** restrict the domain of $y = \tan x$ to the interval $(-\pi/2, \pi/2)$. The inverse tangent function is denoted by

$$y = \arctan x \quad \text{or} \quad y = \tan^{-1} x.$$

The list below summarizes the definitions of the three most common inverse trigonometric functions. Definitions of the other three inverse trigonometric functions are explored in Exercises 105–107.

Definitions of the Inverse Trigonometric Functions

Function	Domain	Range
$y = \arcsin x$ if and only if $\sin y = x$	$-1 \le x \le 1$	$-\dfrac{\pi}{2} \le y \le \dfrac{\pi}{2}$
$y = \arccos x$ if and only if $\cos y = x$	$-1 \le x \le 1$	$0 \le y \le \pi$
$y = \arctan x$ if and only if $\tan y = x$	$-\infty < x < \infty$	$-\dfrac{\pi}{2} < y < \dfrac{\pi}{2}$

The graphs of these three inverse trigonometric functions are shown below.

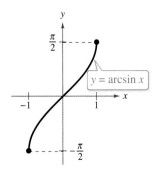

Domain: $[-1, 1]$

Range: $\left[-\dfrac{\pi}{2}, \dfrac{\pi}{2} \right]$

Intercept: $(0, 0)$

Symmetry: origin

Odd function

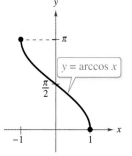

Domain: $[-1, 1]$

Range: $[0, \pi]$

y-intercept: $\left(0, \dfrac{\pi}{2} \right)$

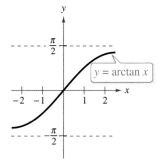

Domain: $(-\infty, \infty)$

Range: $\left(-\dfrac{\pi}{2}, \dfrac{\pi}{2} \right)$

Horizontal asymptotes: $y = \pm\dfrac{\pi}{2}$

Intercept: $(0, 0)$

Symmetry: origin

Odd function

GO DIGITAL

EXAMPLE 3 **Evaluating Inverse Trigonometric Functions**

Find the exact value of each expression.

a. $\arccos \dfrac{\sqrt{2}}{2}$

b. $\cos^{-1}(-1)$

c. $\arctan 0$

d. $\tan^{-1}(-1)$

Solution

a. You know that $\cos(\pi/4) = \sqrt{2}/2$ and $\pi/4$ lies in $[0, \pi]$, so

$$\arccos \frac{\sqrt{2}}{2} = \frac{\pi}{4}. \qquad \text{Angle whose cosine is } \sqrt{2}/2$$

b. You know that $\cos \pi = -1$ and π lies in $[0, \pi]$, so

$$\cos^{-1}(-1) = \pi. \qquad \text{Angle whose cosine is } -1$$

c. You know that $\tan 0 = 0$ and 0 lies in $(-\pi/2, \pi/2)$, so

$$\arctan 0 = 0. \qquad \text{Angle whose tangent is } 0$$

d. You know that $\tan(-\pi/4) = -1$ and $-\pi/4$ lies in $(-\pi/2, \pi/2)$, so

$$\tan^{-1}(-1) = -\frac{\pi}{4}. \qquad \text{Angle whose tangent is } -1$$

✓ *Checkpoint* *Audio-video solution in English & Spanish at LarsonPrecalculus.com*

Find the exact value of each expression.

a. $\cos^{-1}\left(-\dfrac{\sqrt{2}}{2}\right)$ **b.** $\arctan \sqrt{3}$

EXAMPLE 4 **Calculators and Inverse Trigonometric Functions**

Use a calculator to approximate the value of each expression, if possible.

a. $\arctan(-8.45)$ **b.** $\sin^{-1} 0.2447$ **c.** $\arccos 2$

Solution

Function	Mode	Calculator Keystrokes
a. $\arctan(-8.45)$	Radian	[TAN⁻¹] [(] [(−)] 8.45 [)] [ENTER]

From the display, it follows that $\arctan(-8.45) \approx -1.4530010$.

| **b.** $\sin^{-1} 0.2447$ | Radian | [SIN⁻¹] [(] 0.2447 [)] [ENTER] |

From the display, it follows that $\sin^{-1} 0.2447 \approx 0.2472103$.

| **c.** $\arccos 2$ | Radian | [COS⁻¹] [(] 2 [)] [ENTER] |

The calculator should display an *error message* because the domain of the inverse cosine function is $[-1, 1]$.

✓ *Checkpoint* *Audio-video solution in English & Spanish at LarsonPrecalculus.com*

Use a calculator to approximate the value of each expression, if possible.

a. $\arctan 4.84$ **b.** $\arcsin(-1.1)$ **c.** $\arccos(-0.349)$

GO DIGITAL

Compositions with Inverse Trigonometric Functions

Recall from Section 2.7 that for all x in the domains of f and f^{-1}, inverse functions have the properties

$$f(f^{-1}(x)) = x \quad \text{and} \quad f^{-1}(f(x)) = x.$$

Inverse Properties of Trigonometric Functions

If $-1 \le x \le 1$ and $-\pi/2 \le y \le \pi/2$, then

$$\sin(\arcsin x) = x \quad \text{and} \quad \arcsin(\sin y) = y.$$

If $-1 \le x \le 1$ and $0 \le y \le \pi$, then

$$\cos(\arccos x) = x \quad \text{and} \quad \arccos(\cos y) = y.$$

If x is a real number and $-\pi/2 < y < \pi/2$, then

$$\tan(\arctan x) = x \quad \text{and} \quad \arctan(\tan y) = y.$$

Keep in mind that these inverse properties do not apply for arbitrary values of x and y. For example,

$$\arcsin\left(\sin\frac{3\pi}{2}\right) = \arcsin(-1) = -\frac{\pi}{2} \ne \frac{3\pi}{2}.$$

In other words, the property $\arcsin(\sin y) = y$ is not valid for values of y outside the interval $[-\pi/2, \pi/2]$.

EXAMPLE 5   **Using Inverse Properties**

If possible, find the exact value of each expression.

a. $\tan[\arctan(-5)]$ **b.** $\arcsin\left(\sin\dfrac{5\pi}{3}\right)$ **c.** $\cos(\cos^{-1}\pi)$

Solution

a. You know that -5 lies in the domain of the arctangent function, so the inverse property applies, and you have

$$\tan[\arctan(-5)] = -5.$$

b. In this case, $5\pi/3$ does not lie in the range of the arcsine function, $-\pi/2 \le y \le \pi/2$. However, $5\pi/3$ is coterminal with

$$\frac{5\pi}{3} - 2\pi = -\frac{\pi}{3}$$

which does lie in the range of the arcsine function, and you have

$$\arcsin\left(\sin\frac{5\pi}{3}\right) = \arcsin\left[\sin\left(-\frac{\pi}{3}\right)\right] = -\frac{\pi}{3}.$$

c. The expression $\cos(\cos^{-1}\pi)$ is not defined because $\cos^{-1}\pi$ is not defined. Remember that the domain of the inverse cosine function is $[-1, 1]$.

✓ *Checkpoint* ▶ *Audio-video solution in English & Spanish at LarsonPrecalculus.com*

If possible, find the exact value of each expression.

a. $\tan[\tan^{-1}(-14)]$ **b.** $\sin^{-1}\left(\sin\dfrac{7\pi}{4}\right)$ **c.** $\cos(\arccos 0.54)$ ■

GO DIGITAL

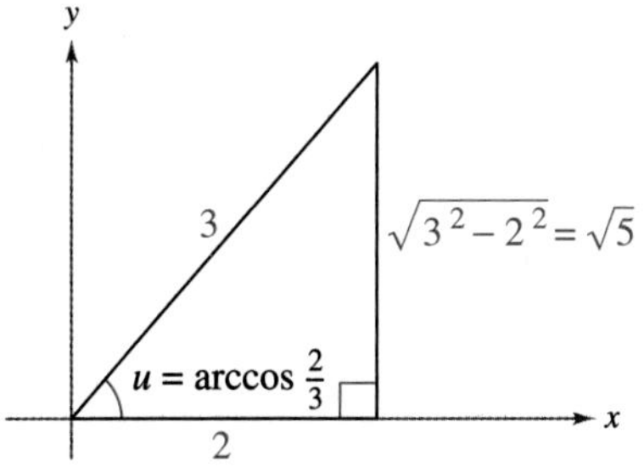

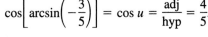

EXAMPLE 6 Evaluating Composite Functions

Find the exact value of each expression.

a. $\tan\left(\arccos \frac{2}{3}\right)$ **b.** $\cos\left[\arcsin\left(-\frac{3}{5}\right)\right]$

Solution

a. If you let $u = \arccos \frac{2}{3}$, then $\cos u = \frac{2}{3}$. The range of the inverse cosine function is $[0, \pi]$ and $\cos u$ is positive, so u is a *first*-quadrant angle. Sketch and label a right triangle with acute angle u, as shown in Figure 6.51. Consequently,

$$\tan\left(\arccos \frac{2}{3}\right) = \tan u = \frac{\text{opp}}{\text{adj}} = \frac{\sqrt{5}}{2}.$$

b. If you let $u = \arcsin\left(-\frac{3}{5}\right)$, then $\sin u = -\frac{3}{5}$. The range of the inverse sine function is $[-\pi/2, \pi/2]$ and $\sin u$ is negative, so u is a *fourth*-quadrant angle. Sketch and label a right triangle with acute angle u, as shown in Figure 6.52. Consequently,

$$\cos\left[\arcsin\left(-\frac{3}{5}\right)\right] = \cos u = \frac{\text{adj}}{\text{hyp}} = \frac{4}{5}.$$

✓ *Checkpoint* ▶ Audio-video solution in English & Spanish at *LarsonPrecalculus.com*

Find the exact value of $\cos\left[\arctan\left(-\frac{3}{4}\right)\right]$.

EXAMPLE 7 Some Problems from Calculus

Write an algebraic expression that is equivalent to each expression.

a. $\sin(\arccos 3x), \quad 0 \le x \le \frac{1}{3}$ **b.** $\cot(\arccos 3x), \quad 0 \le x < \frac{1}{3}$

Solution

If you let $u = \arccos 3x$, then $\cos u = 3x$, where $-1 \le 3x \le 1$. Write

$$\cos u = \frac{\text{adj}}{\text{hyp}} = \frac{3x}{1}$$

and sketch a right triangle with acute angle u, as shown in Figure 6.53. From this triangle, convert each expression to algebraic form.

a. $\sin(\arccos 3x) = \sin u = \dfrac{\text{opp}}{\text{hyp}} = \sqrt{1 - 9x^2}, \quad 0 \le x \le \dfrac{1}{3}$

b. $\cot(\arccos 3x) = \cot u = \dfrac{\text{adj}}{\text{opp}} = \dfrac{3x}{\sqrt{1 - 9x^2}}, \quad 0 \le x < \dfrac{1}{3}$

✓ *Checkpoint* ▶ Audio-video solution in English & Spanish at *LarsonPrecalculus.com*

Write an algebraic expression that is equivalent to $\sec(\arctan x)$. ■

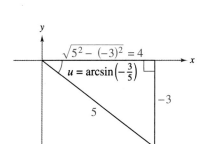

Angle whose cosine is $\frac{2}{3}$
Figure 6.51

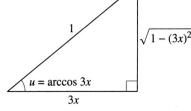

Angle whose sine is $-\frac{3}{5}$
Figure 6.52

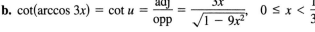

Angle whose cosine is $3x$
Figure 6.53

Summarize (Section 6.6)

1. State the definition of the inverse sine function *(page 478)*. For examples of evaluating and graphing the inverse sine function, see Examples 1 and 2.

2. State the definitions of the inverse cosine and inverse tangent functions *(page 480)*. For examples of evaluating inverse trigonometric functions, see Examples 3 and 4.

3. State the inverse properties of trigonometric functions *(page 482)*. For examples of finding composite functions involving inverse trigonometric functions, see Examples 5–7.

GO DIGITAL

6.6 Exercises

See CalcChat.com for tutorial help and worked-out solutions to odd-numbered exercises.

Vocabulary and Concept Check

In Exercises 1–4, fill in the blanks.

	Function	Alternative Notation	Domain	Range
1.	$y = \arcsin x$	_____	_____	$-\dfrac{\pi}{2} \le y \le \dfrac{\pi}{2}$
2.	_____	$y = \cos^{-1} x$	$-1 \le x \le 1$	_____
3.	$y = \arctan x$	_____	_____	_____

4. A trigonometric function has an _____ function only when its domain is restricted.

5. What notation can you use to represent the inverse cosecant function?

6. Does $\arccos x = 1/\cos x$?

Skills and Applications

Evaluating an Inverse Trigonometric Function   In Exercises 7–20, find the exact value of the expression, if possible.

7. $\arcsin \dfrac{1}{2}$

8. $\arcsin 0$

9. $\arccos 0$

10. $\arccos \dfrac{1}{2}$

11. $\arctan \dfrac{\sqrt{3}}{3}$

12. $\arctan 1$

13. $\arcsin 3$

14. $\arcsin \sqrt{3}$

15. $\tan^{-1}\left(-\sqrt{3}\right)$

16. $\cos^{-1}(-2)$

17. $\arccos\left(-\dfrac{1}{2}\right)$

18. $\arcsin \dfrac{\sqrt{2}}{2}$

19. $\sin^{-1}\left(-\dfrac{\sqrt{3}}{2}\right)$

20. $\tan^{-1}\left(-\dfrac{\sqrt{3}}{3}\right)$

 Graphing an Inverse Trigonometric Function   In Exercises 21 and 22, use a graphing utility to graph f, g, and $y = x$ in the same viewing window to verify geometrically that g is the inverse function of f. (Be sure to restrict the domain of f properly.)

21. $f(x) = \cos x$, $g(x) = \arccos x$

22. $f(x) = \tan x$, $g(x) = \arctan x$

Calculators and Inverse Trigonometric Functions   In Exercises 23–34, use a calculator to approximate the value of the expression, if possible. Round your result to two decimal places.

23. $\arccos 0.37$

24. $\arcsin 0.65$

25. $\arcsin(-0.75)$

26. $\arccos(-0.7)$

27. $\arctan(-3)$

28. $\arctan 25$

29. $\sin^{-1} 1.36$

30. $\cos^{-1} 0.26$

31. $\arccos\left(-\dfrac{4}{3}\right)$

32. $\arcsin \dfrac{7}{8}$

33. $\tan^{-1}\left(-\dfrac{95}{7}\right)$

34. $\tan^{-1}\left(-\sqrt{372}\right)$

Finding Missing Coordinates   In Exercises 35 and 36, determine the missing coordinates of the points on the graph of the function.

35.

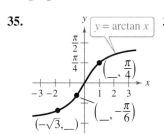

36.

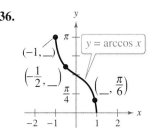

Using an Inverse Trigonometric Function   In Exercises 37–40, use an inverse trigonometric function to write θ as a function of x.

37.

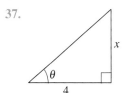

38.

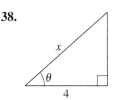

39.

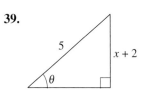

40.

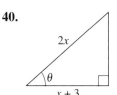

Using Inverse Properties   In Exercises 41–44, find the exact value of the expression, if possible.

41. $\sin(\arcsin 0.3)$

42. $\cos\left[\arccos\left(-\sqrt{3}\right)\right]$

43. $\arcsin[\sin(9\pi/4)]$

44. $\arccos[\cos(-3\pi/2)]$

Evaluating a Composite Function In Exercises 45–56, find the exact value of the expression, if possible.

45. $\sin\left(\arctan\frac{3}{4}\right)$

46. $\cos\left(\arcsin\frac{4}{5}\right)$

47. $\cos(\tan^{-1} 2)$

48. $\sin\left(\cos^{-1}\sqrt{5}\right)$

49. $\sec\left(\arcsin\frac{5}{13}\right)$

50. $\csc\left[\arctan\left(-\frac{5}{12}\right)\right]$

51. $\cot\left[\arctan\left(-\frac{3}{5}\right)\right]$

52. $\sec\left[\arccos\left(-\frac{3}{4}\right)\right]$

53. $\tan\left[\arccos\left(-\frac{2}{3}\right)\right]$

54. $\cot\left(\arctan\frac{5}{8}\right)$

55. $\csc\left(\cos^{-1}\frac{\sqrt{3}}{2}\right)$

56. $\tan\left[\sin^{-1}\left(-\frac{\sqrt{2}}{2}\right)\right]$

Writing an Expression In Exercises 57–66, write an algebraic expression that is equivalent to the given expression.

57. $\cos(\arcsin 2x)$

58. $\sin(\arctan x)$

59. $\cot(\arctan x)$

60. $\sec(\arctan 3x)$

61. $\sin(\arccos x)$

62. $\csc[\arccos(x - 1)]$

63. $\tan\left(\arccos\frac{x}{3}\right)$

64. $\cot\left(\arctan\frac{1}{x}\right)$

65. $\csc\left(\arctan\frac{x}{a}\right)$

66. $\cos\left(\arcsin\frac{x - h}{r}\right)$

Verifying Two Functions Are Equal In Exercises 67 and 68, use a graphing utility to graph f and g in the same viewing window to verify that the two functions are equal. Explain why they are equal. Identify any asymptotes of the graphs.

67. $f(x) = \sin(\arctan 2x), \quad g(x) = \dfrac{2x}{\sqrt{1 + 4x^2}}$

68. $f(x) = \tan\left(\arccos\dfrac{x}{2}\right), \quad g(x) = \dfrac{\sqrt{4 - x^2}}{x}$

Completing an Equation In Exercises 69–72, complete the equation.

69. $\arctan\dfrac{9}{x} = \arcsin(\ \rule{0.5cm}{0.3cm}\ ), \quad x > 0$

70. $\arcsin\dfrac{\sqrt{36 - x^2}}{6} = \arccos(\ \rule{0.5cm}{0.3cm}\ ), \quad 0 \le x \le 6$

71. $\arccos\dfrac{3}{\sqrt{x^2 - 2x + 10}} = \arcsin(\ \rule{0.5cm}{0.3cm}\ )$

72. $\arccos\dfrac{x - 2}{2} = \arctan(\ \rule{0.5cm}{0.3cm}\ ), \quad 2 < x < 4$

Sketching the Graph of a Function In Exercises 73–78, sketch the graph of the function and compare the graph to the graph of the parent inverse trigonometric function.

73. $y = 2 \arcsin x$

74. $f(x) = \arctan 2x$

75. $f(x) = \dfrac{\pi}{2} + \arctan x$

76. $g(t) = \arccos(t + 2)$

77. $h(v) = \arccos\dfrac{v}{2}$

78. $f(x) = \arcsin\dfrac{x}{4}$

Graphing an Inverse Trigonometric Function In Exercises 79–84, use a graphing utility to graph the function.

79. $f(x) = 2 \arccos 2x$

80. $f(x) = \pi \arcsin 4x$

81. $f(x) = \arctan(2x - 3)$

82. $f(x) = -3 + \arctan \pi x$

83. $f(x) = \pi - \sin^{-1}\dfrac{2x}{3} + \cos^{-1}\dfrac{2x}{3}$

84. $f(x) = \dfrac{\pi}{2} + \cos^{-1}\dfrac{x}{\pi} + \sin^{-1}\dfrac{x}{\pi}$

Using a Trigonometric Identity In Exercises 85 and 86, write the function in terms of the sine function by using the identity

$$A \cos \omega t + B \sin \omega t = \sqrt{A^2 + B^2}\, \sin\left(\omega t + \arctan\frac{A}{B}\right).$$

Use a graphing utility to graph both forms of the function. What does the graph imply?

85. $f(t) = 3 \cos 2t + 3 \sin 2t$

86. $f(t) = 4 \cos \pi t + 3 \sin \pi t$

Behavior of an Inverse Trigonometric Function In Exercises 87–92, fill in the blank. (*Note:* The notation $x \to c^+$ indicates that x approaches c from the right and $x \to c^-$ indicates that x approaches c from the left.)

87. As $x \to 1^-$, the value of $\arcsin x \to \ \rule{0.5cm}{0.3cm}\ $.

88. As $x \to 1^-$, the value of $\arccos x \to \ \rule{0.5cm}{0.3cm}\ $.

89. As $x \to \infty$, the value of $\arctan x \to \ \rule{0.5cm}{0.3cm}\ $.

90. As $x \to -1^+$, the value of $\arcsin x \to \ \rule{0.5cm}{0.3cm}\ $.

91. As $x \to -1^+$, the value of $\arccos x \to \ \rule{0.5cm}{0.3cm}\ $.

92. As $x \to -\infty$, the value of $\arctan x \to \ \rule{0.5cm}{0.3cm}\ $.

93. **Docking a Boat** A boat is pulled in by means of a winch located on a dock 5 feet above the deck of the boat (see figure). Let θ be the angle of elevation from the boat to the winch and let s be the length of the rope from the winch to the boat.

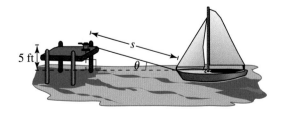

(a) Write θ as a function of s.

(b) Find θ when $s = 40$ feet and $s = 20$ feet.

94. Space Shuttle

A television camera at ground level films the liftoff of a space shuttle at a point 750 meters from the launch pad. Let θ be the angle of elevation to the shuttle and let s be the height of the shuttle.

(a) Write θ as a function of s.

(b) Find θ when $s = 300$ meters and $s = 1200$ meters.

95. Granular Angle of Repose Different types of granular substances naturally settle at different angles when stored in cone-shaped piles. This angle θ is called the *angle of repose* (see figure). When rock salt is stored in a cone-shaped pile 5.5 meters high, the diameter of the pile's base is about 17 meters.

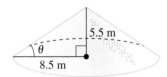

(a) Find the angle of repose for rock salt.

(b) How tall is a pile of rock salt that has a base diameter of 20 meters?

96. Granular Angle of Repose When shelled corn is stored in a cone-shaped pile 20 feet high, the diameter of the pile's base is about 94 feet.

(a) Draw a diagram that gives a visual representation of the problem. Label the known quantities.

(b) Find the angle of repose (see Exercise 95) for shelled corn.

(c) How tall is a pile of shelled corn that has a base diameter of 60 feet?

97. Angle of Elevation An airplane flies at an altitude of 6 miles toward a point directly over an observer. Consider θ and x as shown in the figure.

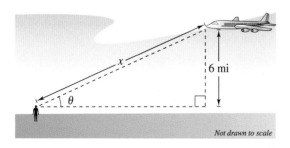

(a) Write θ as a function of x.

(b) Find θ when $x = 12$ miles and $x = 7$ miles.

98. Police Patrol A police car with its spotlight on is parked 20 meters from a warehouse. Consider θ and x as shown in the figure.

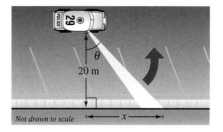

(a) Write θ as a function of x.

(b) Find θ when $x = 5$ meters and $x = 12$ meters.

Exploring the Concepts

True or False? In Exercises 99–102, determine whether the statement is true or false. Justify your answer.

99. $\arctan x = \dfrac{\arcsin x}{\arccos x}$

100. $\sin^{-1} x = \dfrac{1}{\sin x}$

101. $\sin \dfrac{5\pi}{6} = \dfrac{1}{2} \implies \arcsin \dfrac{1}{2} = \dfrac{5\pi}{6}$

102. $\tan\left(-\dfrac{\pi}{4}\right) = -1 \implies \arctan(-1) = -\dfrac{\pi}{4}$

103. Error Analysis Describe the error.

Because $\sin \dfrac{5\pi}{4} = -\dfrac{\sqrt{2}}{2}$, $\sin^{-1}\left(-\dfrac{\sqrt{2}}{2}\right) = \dfrac{5\pi}{4}$.

104. 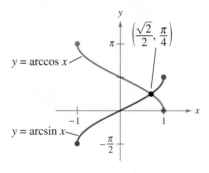 **HOW DO YOU SEE IT?** Use the figure below to determine the value(s) of x for which each statement is true.

(a) $\arcsin x < \arccos x$

(b) $\arcsin x = \arccos x$

(c) $\arcsin x > \arccos x$

105. Inverse Cotangent Function Define the inverse cotangent function by restricting the domain of the cotangent function to the interval $(0, \pi)$, and sketch the graph of the inverse trigonometric function.

106. Inverse Secant Function Define the inverse secant function by restricting the domain of the secant function to the intervals $[0, \pi/2)$ and $(\pi/2, \pi]$, and sketch the graph of the inverse trigonometric function.

107. Inverse Cosecant Function Define the inverse cosecant function by restricting the domain of the cosecant function to the intervals $[-\pi/2, 0)$ and $(0, \pi/2]$, and sketch the graph of the inverse trigonometric function.

108. Writing Use the results of Exercises 105–107 to explain how to graph (a) the inverse cotangent function, (b) the inverse secant function, and (c) the inverse cosecant function on a graphing utility.

Evaluating an Inverse Trigonometric Function In Exercises 109–118, use the results of Exercises 105–107 to find the exact value of the expression.

109. $\operatorname{arcsec} \sqrt{2}$

110. $\operatorname{arcsec} 1$

111. $\operatorname{arccot}(-1)$

112. $\operatorname{arccot}\left(-\sqrt{3}\right)$

113. $\operatorname{arccsc} 2$

114. $\operatorname{arccsc}(-1)$

115. $\operatorname{arccsc} \dfrac{2\sqrt{3}}{3}$

116. $\sec^{-1} \dfrac{2\sqrt{3}}{3}$

117. $\cot^{-1}\left(-\dfrac{\sqrt{3}}{3}\right)$

118. $\operatorname{arcsec}\left(-\dfrac{2\sqrt{3}}{3}\right)$

Calculators and Inverse Trigonometric Functions In Exercises 119–126, use the results of Exercises 105–107 and a calculator to approximate the value of the expression. Round your result to two decimal places.

119. $\operatorname{arcsec} 2.54$

120. $\operatorname{arcsec}(-1.52)$

121. $\operatorname{arccsc}\left(-\frac{25}{3}\right)$

122. $\operatorname{arccsc}(-12)$

123. $\operatorname{arccot} 5.25$

124. $\operatorname{arccot}(-10)$

125. $\operatorname{arccot} \frac{5}{3}$

126. $\operatorname{arccot}\left(-\frac{16}{7}\right)$

127. Think About It Consider the functions

$$f(x) = \sin x \quad \text{and} \quad f^{-1}(x) = \arcsin x.$$

(a) Use a graphing utility to graph the composite functions $f \circ f^{-1}$ and $f^{-1} \circ f$.

(b) Explain why the graphs in part (a) are not the graph of the line $y = x$. Why do the graphs of $f \circ f^{-1}$ and $f^{-1} \circ f$ differ?

128. Think About It Use a graphing utility to graph the functions

$$f(x) = \sqrt{x} \quad \text{and} \quad g(x) = 6 \arctan x$$

in a standard viewing window. For $x > 0$, it appears that $g > f$. Explain how you know that there exists a positive real number a such that $g < f$ for $x > a$. Adjust the viewing window until the intersection of the two curves is visible. Then use the *intersect* feature of the graphing utility to approximate a.

129. Area In calculus, it is shown that the area of the region bounded by the graphs of $y = 0$, $y = 1/(x^2 + 1)$, $x = a$, and $x = b$ (see figure) is given by

$$\text{Area} = \arctan b - \arctan a.$$

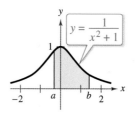

Find the area for each value of a and b.

(a) $a = 0, b = 1$ (b) $a = -1, b = 1$

(c) $a = 0, b = 3$ (d) $a = -1, b = 3$

130. Proof Prove each identity.

(a) $\arcsin(-x) = -\arcsin x$

(b) $\arctan(-x) = -\arctan x$

(c) $\arctan x + \arctan \dfrac{1}{x} = \dfrac{\pi}{2}, \quad x > 0$

(d) $\arcsin x + \arccos x = \dfrac{\pi}{2}$

(e) $\arcsin x = \arctan \dfrac{x}{\sqrt{1 - x^2}}$

Review & Refresh ▶ *Video solutions at LarsonPrecalculus.com*

Using the Pythagorean Theorem In Exercises 131–134, solve for x in the right triangle.

131.

132.

133.

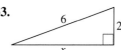

134.

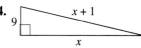

Writing an Equation In Exercises 135–138, write an equation that you can use to find c.

135.

136.

137.

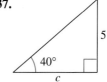

138.

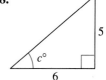

6.7 Applications and Models

Right triangles often occur in real-life situations. For example, in Exercise 25 on page 495, you will use right triangles to analyze the design of a new slide at a water park.

- ❯ Solve real-life problems involving right triangles.
- ❯ Solve real-life problems involving directional bearings.
- ❯ Solve real-life problems involving harmonic motion.

Applications Involving Right Triangles

In this section, the three angles of a right triangle are denoted by A, B, and C (where C is the right angle), and the lengths of the sides opposite these angles are denoted by a, b, and c, respectively (where c is the hypotenuse).

EXAMPLE 1   **Solving a Right Triangle**

▶▶▶ *See LarsonPrecalculus.com for an interactive version of this type of example.*

Solve the right triangle shown at the right for all unknown sides and angles. Round your results to two decimal places, if necessary.

Solution   Because $C = 90°$, it follows that

$$A + B = 90° \quad \text{and} \quad B = 90° - 34.2° = 55.8°.$$

To solve for a, use the fact that

$$\tan A = \frac{\text{opp}}{\text{adj}} = \frac{a}{b} \quad \Longrightarrow \quad a = b \tan A.$$

So, $a = 19.4 \tan 34.2° \approx 13.18$. In a similar manner, solve for c.

$$\cos A = \frac{\text{adj}}{\text{hyp}} = \frac{b}{c} \quad \Longrightarrow \quad c = \frac{b}{\cos A} = \frac{19.4}{\cos 34.2°} \approx 23.46.$$

✓ *Checkpoint* ▶ Audio-video solution in English & Spanish at LarsonPrecalculus.com

Solve the right triangle shown at the right for all unknown sides and angles. Round your results to two decimal places, if necessary.

EXAMPLE 2   **Finding a Side of a Right Triangle**

A safety regulation states that the maximum angle of elevation for a rescue ladder is 72°. A fire department's longest ladder is 110 feet. To the nearest foot, what is the maximum safe rescue height?

Solution   Represent this situation by drawing a sketch, as shown in Figure 6.54. From the equation $\sin A = a/c$, it follows that

$$a = c \sin A = 110 \sin 72° \approx 105.$$

So, the maximum safe rescue height is about 105 feet above the height of the fire truck.

✓ *Checkpoint* ▶ Audio-video solution in English & Spanish at LarsonPrecalculus.com

A ladder that is 16 feet long leans against the side of a house. The angle of elevation of the ladder is 80°. Find the height (to the nearest foot) from the top of the ladder to the ground. ■

Figure 6.54

GO DIGITAL

EXAMPLE 3 **Finding a Side of a Right Triangle**

At a point 200 feet from the base of a building, the angle of elevation to the *bottom* of a smokestack is 35°, whereas the angle of elevation to the *top* is 53°, as shown in Figure 6.55. Find the height *s* of the smokestack alone. Round your answer to one decimal place.

Solution

This problem involves two right triangles. For the smaller right triangle, use the fact that

$$\tan 35° = \frac{a}{200}$$

to find that the height of the building is

$$a = 200 \tan 35°.$$

For the larger right triangle, use the equation

$$\tan 53° = \frac{a + s}{200}$$

to find that $a + s = 200 \tan 53°$. So, the height of the smokestack is

$$s = 200 \tan 53° - a$$
$$= 200 \tan 53° - 200 \tan 35°$$
$$\approx 125.4 \text{ feet}.$$

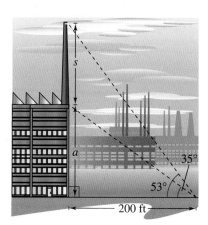

Figure 6.55

✓ *Checkpoint* 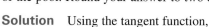 *Audio-video solution in English & Spanish at LarsonPrecalculus.com*

At a point 65 feet from the base of a building, the angle of elevation to the *bottom* of an antenna is 35°, whereas the angle of elevation to the *top* is 43°. Find the height of the antenna. Round your answer to one decimal place.

EXAMPLE 4 **Finding an Angle of Depression**

A swimming pool is 20 meters long and 12 meters wide. The bottom of the pool is slanted so that the water depth is 1.3 meters at the shallow end and 4 meters at the deep end, as shown in Figure 6.56. Find the angle of depression (in degrees) of the bottom of the pool. Round your answer to two decimal places.

Solution Using the tangent function,

$$\tan A = \frac{\text{opp}}{\text{adj}}$$
$$= \frac{2.7}{20}$$
$$= 0.135.$$

So, the angle of depression is

$$A = \arctan 0.135 \approx 7.69°.$$

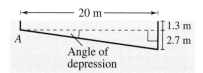

Figure 6.56

✓ *Checkpoint* *Audio-video solution in English & Spanish at LarsonPrecalculus.com*

From the time a small airplane is 100 feet high and 1600 ground feet from its landing runway, the plane descends in a straight line to the runway. Determine the angle of descent (in degrees) of the plane. Round your answer to two decimal places.

GO DIGITAL

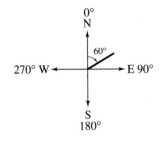
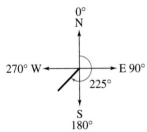
Trigonometry and Bearings

In surveying and navigation, directions can be given in terms of **bearings**. A bearing measures the acute angle that a path or line of sight makes with a fixed north-south line. For example, in the figures below, the bearing S 35° E means 35 degrees east of south, N 80° W means 80 degrees west of north, and N 45° E means 45 degrees east of north.

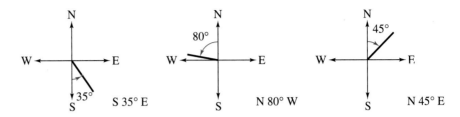

EXAMPLE 5   **Finding Directions in Terms of Bearings**

A ship leaves port at noon and heads due west at 20 knots, or 20 nautical miles (nm) per hour. At 2 P.M. the ship changes course to N 54° W, as shown in the figure below. Find the ship's bearing and distance from port at 3 P.M.

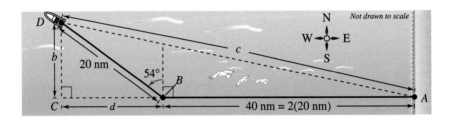

Solution

For triangle BCD, you have $B = 90° - 54° = 36°$. The two sides of this triangle are

$$b = 20 \sin 36° \quad \text{and} \quad d = 20 \cos 36°.$$

To find angle A in triangle ACD, note that

$$\tan A = \frac{b}{d + 40} = \frac{20 \sin 36°}{20 \cos 36° + 40}$$

which yields

$$A = \arctan \frac{20 \sin 36°}{20 \cos 36° + 40} \approx 11.82°. \qquad \text{Use a calculator in degree mode.}$$

The angle with the north-south line is $90° - 11.82° = 78.18°$. So, the bearing of the ship is N 78.18° W. Finally, from triangle ACD, you have

$$\sin A = \frac{b}{c}$$

which yields

$$c = \frac{b}{\sin A} = \frac{20 \sin 36°}{\sin 11.82°} \approx 57.4 \text{ nautical miles.} \qquad \text{Distance from port}$$

✓ **Checkpoint** ▶ *Audio-video solution in English & Spanish at LarsonPrecalculus.com*

A sailboat leaves a pier heading due west at 8 knots. After 15 minutes, the sailboat changes course to N 16° W at 10 knots. Find the sailboat's bearing and distance from the pier after 12 minutes on this course.

Harmonic Motion

The periodic nature of the trigonometric functions is useful for describing the motion of a point on an object that vibrates, oscillates, rotates, or is moved by wave motion.

For example, consider a ball that is bobbing up and down on the end of a spring. Assume that the maximum distance the ball moves vertically upward or downward from its equilibrium (at rest) position is 10 centimeters (see figure). Assume further that the time it takes for the ball to move from its maximum displacement above zero to its maximum displacement below zero and back again is $t = 4$ seconds. With the ideal conditions of perfect elasticity and no friction or air resistance, the ball would continue to move up and down in a uniform and regular manner.

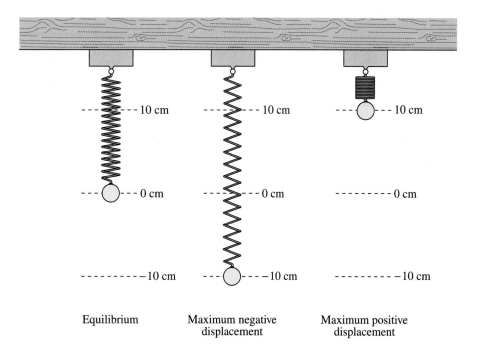

| Equilibrium | Maximum negative displacement | Maximum positive displacement |

The period (time for one complete cycle) of the motion is

Period = 4 seconds

the amplitude (maximum displacement from equilibrium) is

Amplitude = 10 centimeters

and the **frequency** (number of cycles per second) is

Frequency = $\frac{1}{4}$ cycle per second.

Motion of this nature can be described by a sine or cosine function and is called **simple harmonic motion.**

Definition of Simple Harmonic Motion

A point that moves on a coordinate line is in **simple harmonic motion** when its distance d from the origin at time t is given by either

$$d = a \sin \omega t \quad \text{or} \quad d = a \cos \omega t$$

where a and ω are real numbers such that $\omega > 0$. The motion has

amplitude $|a|$,   period $\dfrac{2\pi}{\omega}$,   and   frequency $\dfrac{\omega}{2\pi}$.

EXAMPLE 6   **Simple Harmonic Motion**

Write an equation for the simple harmonic motion of the ball described on the preceding page, where the period is 4 seconds. Verify that the frequency is $\frac{1}{4}$ cycle per second.

Solution   The spring is at equilibrium ($d = 0$) when $t = 0$, so use the equation $d = a \sin \omega t$. Moreover, because the maximum displacement from zero is 10, you know that

Amplitude $= |a| = 10$.

Also, the period is 4 seconds. So, use the expression for period to find ω.

$$\text{Period} = \frac{2\pi}{\omega} \qquad\qquad \text{Definition of period}$$

$$4 = \frac{2\pi}{\omega} \qquad\qquad \text{Substitute.}$$

$$\omega = \frac{\pi}{2}. \qquad\qquad \text{Solve for } \omega.$$

Consequently, an equation for the motion of the ball is

$$d = 10 \sin \frac{\pi}{2}t.$$

Note that the choice of $a = 10$ or $a = -10$ depends on whether the ball initially moves up or down. To verify the frequency, note that

$$\text{Frequency} = \frac{\omega}{2\pi} \qquad\qquad \text{Definition of frequency}$$

$$= \frac{\pi/2}{2\pi} \qquad\qquad \text{Substitute.}$$

$$= \frac{1}{4} \text{ cycle per second.} \qquad\qquad \text{Simplify.}$$

✓ *Checkpoint* ▶ *Audio-video solution in English & Spanish at LarsonPrecalculus.com*

Write an equation for the simple harmonic motion given $d = 0$ when $t = 0$, the amplitude is 6 centimeters, and the period is 3 seconds. Then find the frequency. ◼

One illustration of the relationship between sine waves and harmonic motion is in the wave motion that results when you drop a stone into a calm pool of water. The waves move outward in roughly the shape of sine (or cosine) waves, as shown in Figure 6.57. Now suppose you are fishing in the same pool of water and your fishing bobber does not move horizontally. As the waves move outward from the dropped stone, the fishing bobber moves up and down in simple harmonic motion, as shown Figure 6.58.

Figure 6.57

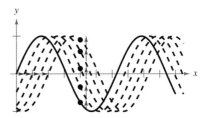

Figure 6.58

EXAMPLE 7   **Simple Harmonic Motion**

Consider the equation for simple harmonic motion $d = 6 \cos \dfrac{3\pi}{4} t$. Find (a) the maximum displacement, (b) the frequency, (c) the value of d when $t = 4$, and (d) the least positive value of t for which $d = 0$.

Algebraic Solution

The equation has the form $d = a \cos \omega t$, with $a = 6$ and $\omega = 3\pi/4$.

a. The maximum displacement (from the point of equilibrium) is the amplitude. So, the maximum displacement is 6.

b. Frequency $= \dfrac{\omega}{2\pi}$

$= \dfrac{3\pi/4}{2\pi}$

$= \dfrac{3}{8}$ cycle per unit of time

c. $d = 6 \cos \left[\dfrac{3\pi}{4}(4) \right] = 6 \cos 3\pi = 6(-1) = -6$

d. To find the least positive value of t for which $d = 0$, solve

$6 \cos \dfrac{3\pi}{4} t = 0.$

First divide each side by 6 to obtain

$\cos \dfrac{3\pi}{4} t = 0.$

This equation is satisfied when

$\dfrac{3\pi}{4} t = \dfrac{\pi}{2}, \dfrac{3\pi}{2}, \dfrac{5\pi}{2}, \ldots .$

Multiply these values by $4/(3\pi)$ to obtain

$t = \dfrac{2}{3}, 2, \dfrac{10}{3}, \ldots .$

So, the least positive value of t is $t = \dfrac{2}{3}$.

Graphical Solution

Use a graphing utility set in *radian* mode.

a.

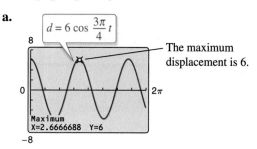

The maximum displacement is 6.

b.

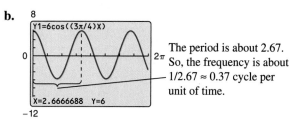

The period is about 2.67. So, the frequency is about $1/2.67 \approx 0.37$ cycle per unit of time.

c.

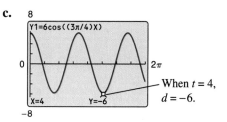

When $t = 4$, $d = -6$.

d. The least positive value of t for which $d = 0$ is $t \approx 0.67$.

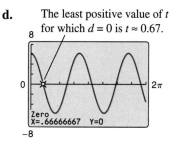

✔ *Checkpoint* ▶ *Audio-video solution in English & Spanish at LarsonPrecalculus.com*

Rework Example 7 for the equation $d = 4 \cos 6\pi t$. ■

Summarize   (Section 6.7)

1. Describe real-life applications of right triangles *(pages 488 and 489, Examples 1–4).*

2. Describe a real-life application of a directional bearing *(page 490, Example 5).*

3. Describe real-life applications of simple harmonic motion *(pages 492 and 493, Examples 6 and 7).*

GO DIGITAL

6.7 Exercises

See CalcChat.com for tutorial help and worked-out solutions to odd-numbered exercises.

GO DIGITAL

Vocabulary and Concept Check

In Exercises 1 and 2, fill in the blanks.

1. A _____ measures the acute angle that a path or line of sight makes with a fixed north-south line.

2. A point that moves on a coordinate line is in simple _____ _____ when its distance d from the origin at time t is given by either $d = a \sin \omega t$ or $d = a \cos \omega t$.

3. Does the bearing of N 20° E mean 20 degrees north of east?

4. Given the period of a point in harmonic motion, how do you find the frequency of the point?

Skills and Applications

Solving a Right Triangle In Exercises 5–12, solve the right triangle shown in the figure for all unknown sides and angles. Round your answers to two decimal places.

5. $A = 60°, \quad c = 12$
6. $B = 25°, \quad b = 4$
7. $B = 72.8°, \quad a = 4.4$
8. $A = 8.4°, \quad a = 40.5$
9. $a = 3, \quad b = 4$
10. $a = 25, \quad c = 35$
11. $b = 15.70, \quad c = 55.16$
12. $b = 1.32, \quad c = 9.45$

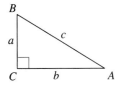

Figure for 5–12

Figure for 13–16

Finding an Altitude In Exercises 13–16, find the altitude of the isosceles triangle shown in the figure. Round your answers to two decimal places.

13. $\theta = 45°, \quad b = 6$
14. $\theta = 22°, \quad b = 14$
15. $\theta = 32°, \quad b = 8$
16. $\theta = 27°, \quad b = 11$

17. **Length** The sun is 25° above the horizon. Find the length of a shadow cast by a building that is 100 feet tall (see figure).

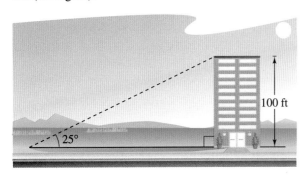

18. **Length** The sun is 20° above the horizon. Find the length of a shadow cast by a park statue that is 12 feet tall.

19. **Height** A ladder that is 20 feet long leans against the side of a house. The angle of elevation of the ladder is 80°. Find the height from the top of the ladder to the ground.

20. **Height** The length of a shadow of a tree is 125 feet when the angle of elevation of the sun is 33°. Approximate the height of the tree.

21. **Height** At a point 50 feet from the base of a castle, the angles of elevation to the bottom of a tower and the top of the tower are 35° and 48°, respectively. Find the height of the tower.

22. **Distance** An observer in a lighthouse 350 feet above sea level observes two ships directly offshore. The angles of depression to the ships are 4° and 6.5° (see figure). How far apart are the ships?

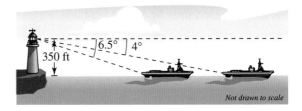

Not drawn to scale

23. **Distance** A passenger in an airplane at an altitude of 10 kilometers sees two towns directly to the east of the plane. The angles of depression to the towns are 28° and 55° (see figure). How far apart are the towns?

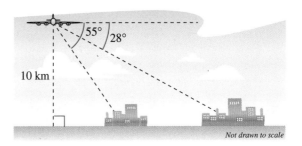

Not drawn to scale

24. Altitude You observe a plane approaching overhead and assume that its speed is 550 miles per hour. The angle of elevation of the plane is 16° at one time and 57° one minute later. Approximate the altitude of the plane.

25. Waterslide Design

The designers of a water park have sketched a preliminary drawing of a new slide (see figure).

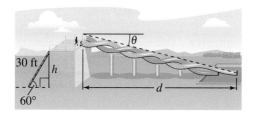

(a) Find the height h of the slide.

(b) Find the angle of depression θ from the top of the slide to the end of the slide at the ground in terms of the horizontal distance d a rider travels.

(c) Safety restrictions require the angle of depression to be no less than 25° and no more than 30°. Find an interval for how far a rider travels horizontally.

26. Speed Enforcement A police department has set up a speed enforcement zone on a straight length of highway. A patrol car is parked parallel to the zone, 200 feet from one end and 150 feet from the other end (see figure).

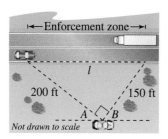

(a) Find the length l of the zone and the measures of angles A and B (in degrees).

(b) Find the minimum amount of time (in seconds) it takes for a vehicle to pass through the zone without exceeding the posted speed limit of 35 miles per hour.

27. Angle of Elevation An engineer designs a 75-foot cellular telephone tower. Find the angle of elevation to the top of the tower at a point on level ground 50 feet from its base.

28. Angle of Depression A cellular telephone tower that is 120 feet tall is placed on top of a mountain that is 1200 feet above sea level. What is the angle of depression from the top of the tower to a cell phone user who is 5 horizontal miles away and 400 feet above sea level?

29. Angle of Depression A Global Positioning System satellite orbits 12,500 miles above Earth's surface (see figure). Find the angle of depression from the satellite to the horizon. Assume the radius of Earth is 4000 miles.

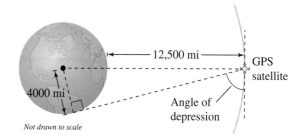

Not drawn to scale

30. Height You are holding one of the tethers attached to the top of a giant character balloon that is floating approximately 20 feet above ground level. You are standing approximately 100 feet ahead of the balloon (see figure).

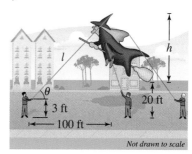

Not drawn to scale

(a) Find an equation for the length l of the tether you are holding in terms of h, the height of the balloon from top to bottom.

(b) Find an equation for the angle of elevation θ from you to the top of the balloon.

(c) The angle of elevation to the top of the balloon is 35°. Find the height h of the balloon.

31. Navigation A ship is 45 miles east and 30 miles south of port. The captain wants to sail directly to port. What bearing should the captain take?

32. Air Navigation An airplane is 160 miles north and 85 miles east of an airport. The pilot wants to fly directly to the airport. What bearing should the pilot take?

33. Air Navigation A jet flies 2472 miles from Reno, Nevada, to Miami, Florida, at a bearing of 100°.

(a) How far north and how far west is Reno relative to Miami?

(b) The jet is to return directly to Reno from Miami. At what bearing should it travel?

34. Air Navigation An airplane flying at 550 miles per hour has a bearing of 52°. After flying for 1.5 hours, how far north and how far east will the plane have traveled from its point of departure?

35. Airplane Ascent During takeoff, an airplane's angle of ascent is 18° and its speed is 260 feet per second.

(a) Find the plane's altitude after 1 minute.

(b) How long will it take for the plane to climb to an altitude of 10,000 feet?

36. Navigation A privately owned yacht leaves a dock in Myrtle Beach, South Carolina, and heads toward Freeport in the Bahamas at a bearing of S 1.4° E. The yacht averages a speed of 20 knots over the 428-nautical-mile trip.

(a) How long will it take the yacht to make the trip?

(b) How far east and south is the yacht after 12 hours?

(c) A plane leaves Myrtle Beach to fly to Freeport. At what bearing should it travel?

37. Surveying A surveyor wants to find the distance across a pond (see figure). The bearing from A to B is N 32° W. The surveyor walks 50 meters from A to C, and at the point C the bearing to B is N 68° W.

(a) Find the bearing from A to C.

(b) Find the distance from A to B.

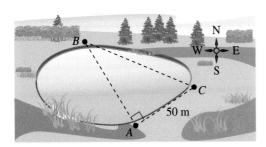

38. Location of a Fire Fire tower A is 30 kilometers due west of fire tower B. A fire is spotted from the towers, and the bearings from A and B are N 76° E and N 56° W, respectively (see figure). Find the distance d of the fire from the line segment AB.

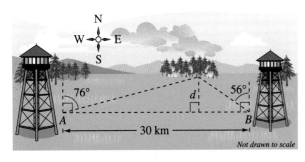

39. Geometry Find the length of the sides of a regular pentagon inscribed in a circle of radius 25 inches.

40. Geometry Find the length of the sides of a regular hexagon inscribed in a circle of radius 25 inches.

41. Geometry Determine the angle between the diagonal of a cube and the diagonal of its base (see figure).

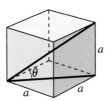

Figure for 41   Figure for 42

42. Geometry Determine the angle between the diagonal of a cube and its edge (see figure).

Simple Harmonic Motion In Exercises 43–46, find a model for simple harmonic motion satisfying the specified conditions.

Displacement ($t = 0$)	Amplitude	Period
43. 0 centimeters	4 centimeters	2 seconds
44. 0 meters	3 meters	6 seconds
45. 3 inches	3 inches	1.5 seconds
46. 2 feet	2 feet	10 seconds

47. Tuning Fork A point on the end of a tuning fork moves in simple harmonic motion described by $d = a \sin \omega t$. Find ω given that the tuning fork for middle C has a frequency of 262 vibrations per second.

48. Wave Motion A buoy oscillates in simple harmonic motion as waves go past. The buoy moves a total of 3.5 feet from its low point to its high point (see figure), and it returns to its high point every 10 seconds. Write an equation that describes the motion of the buoy where the high point corresponds to the time $t = 0$.

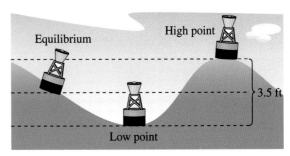

Simple Harmonic Motion In Exercises 49–52, for the simple harmonic motion described by the trigonometric function, find (a) the maximum displacement, (b) the frequency, (c) the value of d when $t = 5$, and (d) the least positive value of t for which $d = 0$. Use a graphing utility to verify your results.

49. $d = 9 \cos \dfrac{6\pi}{5} t$ **50.** $d = \dfrac{1}{2} \cos 20\pi t$

51. $d = \dfrac{1}{4} \sin 6\pi t$ **52.** $d = \dfrac{1}{64} \sin 792\pi t$

53. Oscillation of a Spring A ball that is bobbing up and down on the end of a spring has a maximum displacement of 3 inches. Its motion (in ideal conditions) is modeled by $y = \frac{1}{4}\cos 16t$, $t > 0$, where y is measured in feet and t is the time in seconds.

(a) Graph the function.

(b) What is the period of the oscillations?

(c) Determine the first time the weight passes the point of equilibrium ($y = 0$).

54. Hours of Daylight The numbers of hours H of daylight in Denver, Colorado, on the 15th of each month starting with January are: 9.68, 10.72, 11.92, 13.25, 14.35, 14.97, 14.72, 13.73, 12.47, 11.18, 10.00, and 9.37. A model for the data is

$$H(t) = 12.13 + 2.77 \sin(\pi t/6 - 1.60)$$

where t represents the month, with $t = 1$ corresponding to January. *(Source: United States Navy)*

(a) Use a graphing utility to graph the data and the model in the same viewing window.

(b) What is the period of the model? Is it what you expected? Explain.

(c) What is the amplitude of the model? What does it represent in the context of the problem?

55. Sales The table shows the average sales S (in millions of dollars) of an outerwear manufacturer for each month t, where $t = 1$ corresponds to January.

Time, t	1	2	3	4
Sales, S	13.46	11.15	8.00	4.85

Time, t	5	6	7	8
Sales, S	2.54	1.70	2.54	4.85

Time, t	9	10	11	12
Sales, S	8.00	11.15	13.46	14.30

(a) Create a scatter plot of the data.

(b) Find a trigonometric model that fits the data. Graph the model with your scatter plot. How well does the model fit the data?

(c) What is the period of the model? Do you think it is reasonable given the context? Explain.

(d) Interpret the meaning of the model's amplitude in the context of the problem.

Exploring the Concepts

True or False? In Exercises 56 and 57, determine whether the statement is true or false. Justify your answer.

56. The bearing N 24° E means 24 degrees north of east.

57. The Leaning Tower of Pisa is not vertical, but when you know the angle of elevation θ to the top of the tower as you stand d feet away from it, its height h can be found using the formula $h = d \tan \theta$.

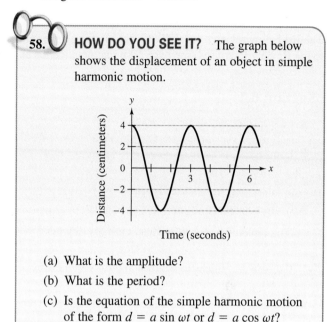

58. HOW DO YOU SEE IT? The graph below shows the displacement of an object in simple harmonic motion.

(a) What is the amplitude?

(b) What is the period?

(c) Is the equation of the simple harmonic motion of the form $d = a \sin \omega t$ or $d = a \cos \omega t$?

Review & Refresh ▶ *Video solutions at LarsonPrecalculus.com*

Factoring Polynomials In Exercises 59–64, factor the expression.

59. $6t^3 - 2t + 10t^2$

60. $d^2 + 7d - 18$

61. $10y^2 - 13y - 3$

62. $-3x^2 + 22x - 24$

63. $25z^4 - x^2$

64. $w^3 + 125v^3$

Applying Trigonometric Identities In Exercises 65 and 66, use the given function value and the trigonometric identities to find the exact value of each indicated trigonometric function. Assume that $0 \le \theta \le \pi/2$.

65. $\sin \theta = \dfrac{1}{4}$

(a) $\cos \theta$ (b) $\cot \theta$

(c) $\csc \theta$ (d) $\tan(90° - \theta)$

66. $\csc \theta = 6$

(a) $\sin \theta$ (b) $\sec \theta$

(c) $\tan \theta$ (d) $\cos(90° - \theta)$

Condensing a Logarithmic Expression In Exercises 67–70, condense the expression to the logarithm of a single quantity.

67. $\log_2 x + \log_2 5$

68. $3 \ln(1 + y) - \ln y$

69. $3 \log x - \log(x + 2) - \log(x - 2)$

70. $\frac{1}{2}\left[\log_5(z + 1) + \log_5(z - 1)\right] + 4 \log_5(z + 1)$

Summary and Study Strategies

GO DIGITAL

What Did You Learn?

The list below reviews the skills covered in the chapter and correlates each one
to the Review Exercises (see page 500) that practice the skill.

(see page 500)

Section 6.1	Review Exercises
■ Describe angles *(p. 422)*.	*1, 2*
■ Use degree measure *(p. 423)*.	*3–6*
■ Use radian measure *(p. 425)*.	*7–10*
■ Convert between degrees and radians *(p. 426)*.	*11–26*

To convert degrees to radians, multiply degrees by $\dfrac{\pi \text{ rad}}{180°}$.

To convert radians to degrees, multiply radians by $\dfrac{180°}{\pi \text{ rad}}$.

■ Use angles and their measure to model and solve real-life problems *(p. 427)*	*27–30*

Section 6.2

■ Evaluate trigonometric functions of acute angles *(p. 434)*.	*31–34*
■ Use fundamental trigonometric identities *(p. 437)*.	*35–38*

$\sin \theta = 1/\csc \theta \qquad \cos \theta = 1/\sec \theta \qquad \tan \theta = 1/\cot \theta$

$\tan \theta = (\sin \theta)/(\cos \theta) \qquad \cot \theta = (\cos \theta)/(\sin \theta)$

$\sin^2 \theta + \cos^2 \theta = 1 \qquad 1 + \tan^2 \theta = \sec^2 \theta \qquad 1 + \cot^2 \theta = \csc^2 \theta$

■ Use trigonometric functions to model and solve real-life problems *(p. 439)*	*39, 40*

Section 6.3

■ Evaluate trigonometric functions of any angle *(p. 445)*.	*41–50*
■ Use reference angles to evaluate trigonometric functions *(p. 447)*.	*51–62*
■ Evaluate trigonometric functions of real numbers *(p. 451)*.	*63–70*

The cosine and secant functions are *even*, so $\cos(-t) = \cos t$ and $\sec(-t) = \sec t$.

The sine, cosecant, tangent, and cotangent functions are *odd*, so $\sin(-t) = -\sin t$, $\csc(-t) = -\csc t$, $\tan(-t) = -\tan t$, and $\cot(-t) = -\cot t$.

Section 6.4

■ Sketch the graphs of sine and cosine functions *(p. 457)*.	*71, 72*
■ Use amplitude and period to help sketch the graphs of sine and cosine functions *(p. 459)*.	*73, 74*
■ Sketch translations of the graphs of sine and cosine functions *(p. 461)*.	*75–78*
■ Use sine and cosine functions to model real-life data *(p. 463)*.	*79, 80*

Section 6.5	Review Exercises
■ Sketch the graphs of tangent functions *(p. 468)*.	*81, 82*
■ Sketch the graphs of cotangent functions *(p. 470)*.	*83, 84*
■ Sketch the graphs of secant and cosecant functions *(p. 471)*.	*85–88*
■ Sketch the graphs of damped trigonometric functions *(p. 473)*.	*89, 90*
Section 6.6	
■ Evaluate and graph the inverse sine function *(p. 478)*.	*91, 92*
■ Evaluate and graph other inverse trigonometric functions *(p. 480)*	*93–108*
■ Evaluate composite functions involving inverse trigonometric functions *(p. 482)*.	*109–114*
Section 6.7	
■ Solve real-life problems involving right triangles *(p. 488)*.	*115, 116*
■ Solve real-life problems involving directional bearings *(p. 490)*.	*117*
■ Solve real-life problems involving harmonic motion *(p. 491)*.	*118*

Study Strategies

Using a Test-Taking Strategy   What do runners do before a race? They design a strategy for running their best. They get enough rest, eat sensibly, and get to the track early to warm up. In the same way, it is important for students to get a good night's sleep, eat a healthy meal, and get to class early to allow time to focus before a test.

The biggest difference between a runner's race and a math test is that a math student does not have to reach the finish line first! In fact, many students would increase their scores if they used all the test time instead of worrying about being the last student left in the class. This is why it is important to have a strategy for taking the test.

1. **Do a memory data dump.** When you get the test, turn it over and write down anything that you have trouble remembering, such as formulas, calculations, and rules.

2. **Preview the test.** Look over the test and mark the questions you know how to do easily. These are the problems you should do first.

3. **Do a second memory data dump.** As you previewed the test, you may have remembered other information. Write this information on the back of the test.

4. **Develop a test progress schedule.** Based on how many points each question is worth, decide on a progress schedule. You should always have more than half the test done before half the time has elapsed.

5. **Answer the easiest problems first.** Solve the problems you marked while previewing the test.

6. **Skip difficult problems.** Skip the problems that you suspect will give you trouble.

7. **Review the skipped problems.** After solving all the problems that you know how to do, go back and reread the problems you skipped.

8. **Try your best at the remaining problems.** Even if you cannot completely solve a problem, you may be able to get partial credit for a few correct steps.

9. **Review the test.** Look for any careless errors you may have made.

10. **Use all the allowed test time.** The test is not a race against the other students.

Review Exercises

See CalcChat.com for tutorial help and worked-out solutions to odd-numbered exercises.

6.1 **Estimating an Angle** In Exercises 1 and 2, estimate the number of degrees in the angle.

1.   **2.**

Using Degree or Radian Measure In Exercises 3–10, (a) sketch the angle in standard position, (b) determine the quadrant in which the angle lies, and (c) determine two coterminal angles (one positive and one negative).

3. $85°$

4. $310°$

5. $-110°$

6. $-405°$

7. $\dfrac{15\pi}{4}$

8. $\dfrac{2\pi}{9}$

9. $-\dfrac{4\pi}{3}$

10. $-\dfrac{23\pi}{3}$

Converting from Degrees to Radians In Exercises 11–18, convert the degree measure to radian measure. Round to three decimal places.

11. $450°$

12. $210°$

13. $-16°$

14. $-112°$

15. $20.36°$

16. $45.14°$

17. $-8.56°$

18. $-300.12°$

Converting from Radians to Degrees In Exercises 19–26, convert the radian measure to degree measure. Round to three decimal places, if necessary.

19. $\dfrac{3\pi}{10}$

20. $\dfrac{7\pi}{5}$

21. $-\dfrac{3\pi}{5}$

22. $-\dfrac{11\pi}{6}$

23. 5.2

24. 7

25. -2.15

26. -4.63

27. Arc Length Find the length of the arc on a circle with a radius of 20 inches intercepted by a central angle of $138°$.

28. Bicycle At what speed is a bicycle traveling when its 29-inch-diameter tires are rotating at an angular speed of 5π radians per second?

Area of a Sector of a Circle In Exercises 29 and 30, find the area of the sector of a circle of radius r and central angle θ.

Radius r	Central Angle θ
29. 20 inches	$150°$
30. 7.5 millimeters	$2\pi/3$ radians

6.2 **Evaluating Trigonometric Functions** In Exercises 31 and 32, find the exact values of the six trigonometric functions of the angle θ.

31. 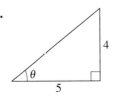 **32.**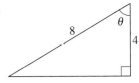

Using a Calculator In Exercises 33 and 34, use a calculator to evaluate each function. Round your answers to four decimal places. (Be sure the calculator is in the correct mode.)

33. (a) $\cos 38.9°$ (b) $\sec 79.3°$

34. (a) $\sin(\pi/18)$ (b) $\cot(2\pi/5)$

Applying Trigonometric Identities In Exercises 35–38, use the given function value and the trigonometric identities to find the exact value of each indicated trigonometric function.

35. $\sin\theta = \frac{1}{3}$

(a) $\csc\theta$ (b) $\cos\theta$

(c) $\sec\theta$ (d) $\tan\theta$

36. $\tan\theta = 4$

(a) $\cot\theta$ (b) $\sec\theta$

(c) $\cos\theta$ (d) $\csc\theta$

37. $\csc\theta = 4$

(a) $\sin\theta$ (b) $\cos\theta$

(c) $\sec\theta$ (d) $\tan\theta$

38. $\csc\theta = 5$

(a) $\sin\theta$ (b) $\cot\theta$

(c) $\tan\theta$ (d) $\sec(90° - \theta)$

39. Railroad Grade A train travels 3.5 kilometers on a straight track with a grade of $1.2°$ (see figure). What is the vertical rise of the train in that distance?

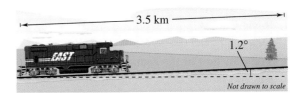

40. Guy Wire A guy wire runs from the ground to the top of a 25-foot telephone pole. The angle formed between the wire and the ground is $52°$. How far from the base of the pole is the guy wire anchored to the ground? Assume the pole is perpendicular to the ground.

6.3 **Evaluating Trigonometric Functions** In Exercises 41–44, the point is on the terminal side of an angle in standard position. Find the exact values of the six trigonometric functions of the angle.

41. $(12, 16)$ **42.** $(7, -24)$

43. $(-0.5, 4.5)$ **44.** $(0.2, 0.8)$

Evaluating Trigonometric Functions In Exercises 45–50, find the exact values of the remaining five trigonometric functions of θ satisfying the given conditions.

45. $\sec \theta = \frac{6}{5}$, $\tan \theta < 0$ **46.** $\csc \theta = \frac{3}{2}$, $\cos \theta < 0$

47. $\tan \theta = \frac{7}{3}$, $\cos \theta < 0$ **48.** $\sin \theta = \frac{3}{8}$, $\cos \theta < 0$

49. $\tan \theta = -\frac{40}{9}$, $\sin \theta > 0$

50. $\cos \theta = -\frac{2}{5}$, $\sin \theta > 0$

Finding a Reference Angle In Exercises 51–54, find the reference angle θ'. Then sketch θ in standard position and label θ'.

51. $\theta = 264°$ **52.** $\theta = 635°$

53. $\theta = -6\pi/5$ **54.** $\theta = 17\pi/3$

Using a Reference Angle In Exercises 55–62, evaluate the sine, cosine, and tangent of the angle without using a calculator.

55. $495°$ **56.** $120°$

57. $-150°$ **58.** $-420°$

59. $\pi/3$ **60.** $5\pi/6$

61. $-7\pi/3$ **62.** $-5\pi/4$

Using a Calculator In Exercises 63–66, use a calculator to evaluate the trigonometric function. Round your answer to four decimal places. (Be sure the calculator is in the correct mode.)

63. $\sin 106°$ **64.** $\tan 37°$

65. $\tan(-17\pi/15)$ **66.** $\cos(-25\pi/7)$

Evaluating Trigonometric Functions In Exercises 67–70, find the point (x, y) on the unit circle that corresponds to the real number t. Use the result to evaluate $\sin t$, $\cos t$, and $\tan t$.

67. $t = 2\pi/3$ **68.** $t = 7\pi/4$

69. $t = 7\pi/6$ **70.** $t = 3\pi/2$

6.4 **Sketching the Graph of a Sine or Cosine Function** In Exercises 71–78, sketch the graph of the function. (Include two full periods.)

71. $y = \sin 6x$ **72.** $y = -\cos 3x$

73. $f(x) = 5 \sin(2x/5)$ **74.** $f(x) = 8 \cos(-x/4)$

75. $y = 5 + \sin \pi x$ **76.** $y = -4 - \cos \pi x$

77. $g(t) = \frac{5}{2} \sin(t - \pi)$ **78.** $g(t) = 3 \cos(t + \pi)$

79. Sound Waves Sound waves can be modeled using sine functions of the form $y = a \sin bx$, where x is measured in seconds.

  (a) Write an equation of a sound wave whose amplitude is 2 and whose period is $\frac{1}{264}$ second.

  (b) What is the frequency of the sound wave described in part (a)?

80. Meteorology The local times S of sunset in London, United Kingdom on the 15th of each month starting with January are 16:20, 17:15, 18:06, 18:58, 19:47, 20:20, 20:11, 19:22, 18:14, 17:06, 16:11, and 15:52. A model (in which minutes have been converted to the decimal parts of an hour) for the data is

$$S(t) = 18.15 - 2.15 \sin\left(\frac{\pi t}{6} + 1.60\right)$$

where t represents the months, with $t = 1$ corresponding to January. *(Source: NOAA)*

  (a) Use a graphing utility to graph the data and the model in the same viewing window.

  (b) What is the period of the model? Is it what you expected? Explain.

  (c) What is the amplitude of the model? What does it represent in the context of the problem?

6.5 **Sketching the Graph of a Trigonometric Function** In Exercises 81–88, sketch the graph of the function. (Include two full periods.)

81. $f(x) = 3 \tan 2x$ **82.** $f(t) = \tan\left(t + \frac{\pi}{2}\right)$

83. $f(x) = \frac{1}{2} \cot x$ **84.** $g(t) = 2 \cot 2t$

85. $f(x) = 3 \sec x$ **86.** $h(t) = \sec\left(t - \frac{\pi}{4}\right)$

87. $f(x) = \frac{1}{2} \csc \frac{x}{2}$ **88.** $f(t) = 3 \csc\left(2t + \frac{\pi}{4}\right)$

Analyzing a Damped Trigonometric Graph In Exercises 89 and 90, use a graphing utility to graph the function and the damping factor of the function in the same viewing window. Describe the behavior of the function as x increases without bound.

89. $f(x) = x \cos x$ **90.** $g(x) = e^x \cos x$

6.6 **Evaluating an Inverse Trigonometric Function** In Exercises 91–96, find the exact value of the expression.

91. $\arcsin\left(-\frac{1}{2}\right)$ **92.** $\arcsin(-1)$

93. $\arccos\left(-\frac{\sqrt{2}}{2}\right)$ **94.** $\arccos \frac{\sqrt{2}}{2}$

95. $\cos^{-1} 1$ **96.** $\cos^{-1} \frac{\sqrt{3}}{2}$

Calculators and Inverse Trigonometric Functions
In Exercises 97–104, use a calculator to approximate the value of the expression, if possible. Round your result to two decimal places.

97. arcsin 3.26

98. arcsin(-0.363)

99. $\sin^{-1}(-0.17)$

100. $\sin^{-1} 0.15$

101. arccos 0.372

102. $\cos^{-1}(-1.754)$

103. $\tan^{-1}(-1.3)$

104. arctan 2.7

Graphing an Inverse Trigonometric Function In Exercises 105–108, use a graphing utility to graph the function.

105. $f(x) = 2\arcsin(x/2)$

106. $f(x) = 3\arccos x$

107. $f(x) = \arctan(x/2)$

108. $f(x) = -\arcsin 2x$

Evaluating a Composite Function In Exercises 109–112, find the exact value of the expression.

109. $\cos\left(\arctan \frac{3}{4}\right)$

110. $\tan\left(\arccos \frac{3}{5}\right)$

111. $\sec\left(\arctan \frac{12}{5}\right)$

112. $\cot\left[\arcsin\left(-\frac{12}{13}\right)\right]$

Writing an Expression In Exercises 113 and 114, write an algebraic expression that is equivalent to the given expression.

113. $\tan[\arccos(x/2)]$

114. $\sec[\arcsin(x-1)]$

6.7

115. Angle of Elevation The height of a radio transmission tower is 70 meters, and it casts a shadow of length 30 meters. Draw a right triangle that represents this situation. Label the known and unknown quantities. Then find the angle of elevation.

116. Ski Slope A ski slope on a mountain has an angle of elevation of 25.2°. The vertical height of the slope is 1808 feet. How long is the slope?

117. Navigation A ship leaves port at noon and has a bearing of N 45° E. The ship sails at 15 knots. How many nautical miles north and how many nautical miles east will the ship have traveled by 4:00 P.M.?

118. Wave Motion A fishing bobber oscillates in simple harmonic motion because of the waves in a lake. The bobber moves a total of 1.5 inches from its low point to its high point and returns to its high point every 3 seconds. Write an equation that describes the motion of the bobber, where the high point corresponds to the time $t = 0$.

Exploring the Concepts

True or False? In Exercises 119 and 120, determine whether the statement is true or false. Justify your answer.

119. $y = \sin \theta$ is not a function because $\sin 30° = \sin 150°$.

120. Because $\tan(3\pi/4) = -1$, $\arctan(-1) = 3\pi/4$.

121. Writing Describe the behavior of $f(\theta) = \sec \theta$ at the zeros of $g(\theta) = \cos \theta$. Explain.

122. Conjecture

(a) Use a graphing utility to complete the table.

θ	0.1	0.4	0.7	1.0	1.3
$\tan\left(\theta - \dfrac{\pi}{2}\right)$					
$-\cot \theta$					

(b) Make a conjecture about the relationship between $\tan\left(\theta - \dfrac{\pi}{2}\right)$ and $-\cot \theta$.

123. Writing When graphing a sine or cosine function, it is important to determine the amplitude of the function. Explain why this is not true for the other four trigonometric functions.

124. Oscillation of a Spring A weight is suspended from a ceiling by a steel spring. The weight is lifted (positive direction) from the equilibrium position and released. The resulting motion of the weight is modeled by

$$y = Ae^{-kt}\cos bt = \frac{1}{5}e^{-t/10}\cos 6t$$

where y is the distance (in feet) from equilibrium and t is the time (in seconds). The figure shows the graph of the function. For each of the following, describe the change in the graph without graphing the resulting function.

(a) A is changed from $\frac{1}{5}$ to $\frac{1}{3}$.

(b) k is changed from $\frac{1}{10}$ to $\frac{1}{3}$.

(c) b is changed from 6 to 9.

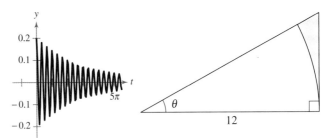

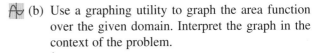
Figure for 124   Figure for 125

125. The base of the triangle shown in the figure is also the radius of a circular arc.

(a) Find the area A of the shaded region as a function of θ for $0 < \theta < \dfrac{\pi}{2}$.

(b) Use a graphing utility to graph the area function over the given domain. Interpret the graph in the context of the problem.

Chapter Test

See CalcChat.com for tutorial help and worked-out solutions to odd-numbered exercises.

Take this test as you would take a test in class. When you are finished, check your work against the answers given in the back of the book.

1. Consider an angle that measures $\dfrac{5\pi}{4}$ radians. *(Section 6.1)*

   (a) Sketch the angle in standard position.

   (b) Determine two coterminal angles (one positive and one negative).

   (c) Convert the radian measure to degree measure.

2. A truck is moving at a rate of 105 kilometers per hour, and the diameter of each of its wheels is 1 meter. Find the angular speed of the wheels in radians per minute. *(Section 6.1)*

3. A water sprinkler sprays water on a lawn over a distance of 25 feet and rotates through an angle of 130°. Find the area of the lawn watered by the sprinkler. *(Section 6.1)*

4. Given that θ is an acute angle and $\tan \theta = \frac{3}{2}$, find the exact values of the other five trigonometric functions of θ. *(Section 6.2)*

5. Find the exact values of the six trigonometric functions of the angle θ shown in the figure. *(Section 6.3)*

6. Evaluate the sine, cosine, and tangent of 210° without using a calculator. *(Section 6.3)*

7. Determine the quadrant in which θ lies when $\sec \theta < 0$ and $\tan \theta > 0$. *(Section 6.3)*

8. Find two exact values of θ in degrees ($0° \le \theta < 360°$) for which $\cos \theta = -\sqrt{3}/2$. Do not use a calculator. *(Section 6.3)*

In Exercises 9 and 10, find the exact values of the remaining five trigonometric functions of θ satisfying the given conditions. *(Section 6.3)*

9. $\cos \theta = \frac{3}{5}$, $\tan \theta < 0$
10. $\sec \theta = -\frac{29}{20}$, $\sin \theta > 0$

In Exercises 11–13, sketch the graph of the function. (Include two full periods.) *(Sections 6.4 and 6.5)*

11. $g(x) = -2 \sin\left(x - \dfrac{\pi}{4}\right)$
12. $f(t) = \cos\left(t + \dfrac{\pi}{2}\right) - 1$
13. $f(x) = \dfrac{1}{2} \tan 2x$

14. Find a, b, and c for the function $f(x) = a \sin(bx + c)$ such that the graph of f matches the figure. *(Section 6.4)*

In Exercises 15 and 16, use a graphing utility to graph the function. If the function is periodic, find its period. If not, describe the behavior of the function as x increases without bound. *(Section 6.5)*

15. $y = \sin 2\pi x + 2 \cos \pi x$
16. $y = 6e^{-0.12x} \cos(0.25x)$

17. Find the exact value of $\cot\left(\arcsin \frac{3}{8}\right)$. *(Section 6.6)*

18. Use a graphing utility to graph the function $f(x) = 5 \arcsin\left(\frac{1}{2}x\right)$. *(Section 6.6)*

19. An airplane is 90 miles south and 110 miles east of an airport. What bearing should the pilot take to fly directly to the airport? *(Section 6.7)*

20. A ball on a spring starts at its lowest point of 6 inches below equilibrium, bounces to its maximum height of 6 inches above equilibrium, and returns to its lowest point in a total of 2 seconds. Write an equation for the simple harmonic motion of the ball. *(Section 6.7)*

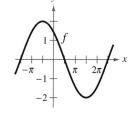

Figure for 5

(−2, 6)

θ

Figure for 14

f

Proofs in Mathematics

The Pythagorean Theorem

The Pythagorean Theorem is one of the most famous theorems in mathematics, with hundreds of different proofs. James A. Garfield, the twentieth president of the United States, developed a proof of the Pythagorean Theorem in 1876. His proof, shown below, involves the fact that a trapezoid can be formed from two congruent right triangles and an isosceles right triangle.

The Pythagorean Theorem

In a right triangle, the sum of the squares of the lengths of the legs is equal to the square of the length of the hypotenuse, where a and b are the legs and c is the hypotenuse.

$$a^2 + b^2 = c^2$$

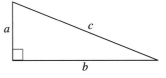

Proof

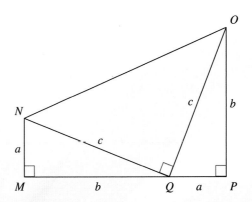

$$\text{Area of trapezoid } MNOP = \text{Area of } \triangle MNQ + \text{Area of } \triangle PQO + \text{Area of } \triangle NOQ$$

$$\frac{1}{2}(a + b)(a + b) = \frac{1}{2}ab + \frac{1}{2}ab + \frac{1}{2}c^2$$

$$\frac{1}{2}(a + b)(a + b) = ab + \frac{1}{2}c^2$$

$$(a + b)(a + b) = 2ab + c^2$$

$$a^2 + 2ab + b^2 = 2ab + c^2$$

$$a^2 + b^2 = c^2$$

P.S. Problem Solving

See CalcChat.com for tutorial help and worked-out solutions to odd-numbered exercises.

1. Angle of Rotation The restaurant at the top of the Space Needle in Seattle, Washington, is circular and has a radius of 47.25 feet. The dining part of the restaurant revolves, making about one complete revolution every 48 minutes. A dinner party, seated at the edge of the revolving restaurant at 6:45 P.M., finishes at 8:57 P.M.

(a) Find the angle through which the dinner party rotated.

(b) Find the distance the party traveled during dinner.

2. Bicycle Gears A bicycle's gear ratio is the number of times the freewheel turns for every one turn of the chainwheel (see figure). The table shows the numbers of teeth in the freewheel and the chainwheel for the first five gears of an 18-speed touring bicycle. The chainwheel completes one rotation for each gear. Find the angle through which the freewheel turns for each gear. Give your answers in both degrees and radians.

DATA	Gear Number	Number of Teeth in Freewheel	Number of Teeth in Chainwheel
	1	32	24
	2	26	24
	3	22	24
	4	32	40
	5	19	24

Spreadsheet at LarsonPrecalculus.com

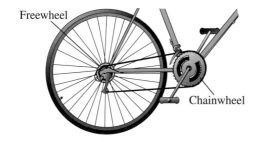

Freewheel

Chainwheel

3. Height of a Ferris Wheel Car A model for the height h (in feet) of a Ferris wheel car is

$$h = 50 + 50 \sin 8\pi t$$

where t is the time (in minutes). (The Ferris wheel has a radius of 50 feet.) This model yields a height of 50 feet when $t = 0$. Alter the model so that the car follows the same path and has a height of 1 foot when $t = 0$.

4. Periodic Function The function f is periodic, with period c. So, $f(t + c) = f(t)$. Determine whether each statement is true or false. Explain.

(a) $f(t - 2c) = f(t)$ (b) $f\left(t + \frac{1}{2}c\right) = f\left(\frac{1}{2}t\right)$

(c) $f\left(\frac{1}{2}[t + c]\right) = f\left(\frac{1}{2}t\right)$ (d) $f\left(\frac{1}{2}[t + 4c]\right) = f\left(\frac{1}{2}t\right)$

5. Surveying A surveyor in a helicopter is determining the width of an island, as shown in the figure.

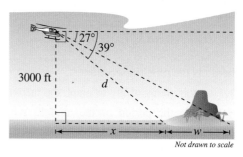

Not drawn to scale

(a) What is the shortest distance d the helicopter must travel to land on the island?

(b) What is the horizontal distance x the helicopter must travel before it is directly over the nearer end of the island?

(c) Find the width w of the island. Explain how you found your answer.

6. Similar Triangles and Trigonometric Functions Use the figure below.

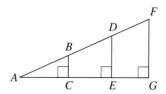

(a) Explain why $\triangle ABC$, $\triangle ADE$, and $\triangle AFG$ are similar triangles.

(b) What does similarity imply about the ratios

$$\frac{BC}{AB}, \quad \frac{DE}{AD}, \quad \text{and} \quad \frac{FG}{AF}?$$

(c) Does the value of $\sin A$ depend on which triangle from part (a) is used to calculate it? Does the value of $\sin A$ change when you use a different right triangle similar to the three given triangles?

(d) Do your conclusions from part (c) apply to the other five trigonometric functions? Explain.

7. Graphical Reasoning Use a graphing utility to graph h, and use the graph to determine whether h is even, odd, or neither.

(a) $h(x) = \cos^2 x$ (b) $h(x) = \sin^2 x$

8. Squares of Even and Odd Functions Given that f is an even function and g is an odd function, use the results of Exercise 7 to make a conjecture about each function h.

(a) $h(x) = [f(x)]^2$ (b) $h(x) = [g(x)]^2$

9. Blood Pressure The pressure P (in millimeters of mercury) against the walls of the blood vessels of a patient is modeled by

$$P = 100 - 20 \cos \frac{8\pi t}{3}$$

where t is the time (in seconds).

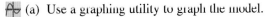 (a) Use a graphing utility to graph the model.

(b) What is the period of the model? What does it represent in the context of the problem?

(c) What is the amplitude of the model? What does it represent in the context of the problem?

(d) If one cycle of this model is equivalent to one heartbeat, what is the pulse of the patient?

(e) A physician wants the patient's pulse rate to be 64 beats per minute or less. What should the period be? What should the coefficient of t be?

10. Biorhythms A popular theory that attempts to explain the ups and downs of everyday life states that each person has three cycles, called biorhythms, which begin at birth. These three cycles can be modeled by the sine functions below, where t is the number of days since birth.

Physical (23 days): $P = \sin \dfrac{2\pi t}{23}, \quad t \geq 0$

Emotional (28 days): $E = \sin \dfrac{2\pi t}{28}, \quad t \geq 0$

Intellectual (33 days): $I = \sin \dfrac{2\pi t}{33}, \quad t \geq 0$

Consider a person who was born on July 20, 2000.

(a) Use a graphing utility to graph the three models in the same viewing window for $7300 \leq t \leq 7380$.

(b) Describe the person's biorhythms during the month of September 2020.

(c) Calculate the person's three energy levels on September 22, 2020.

11. Graphical Reasoning

(a) Use a graphing utility to graph the functions
$$f(x) = 2 \cos 2x + 3 \sin 3x$$
and
$$g(x) = 2 \cos 2x + 3 \sin 4x.$$

(b) Use the graphs from part (a) to find the period of each function.

(c) Is the function $h(x) = A \cos \alpha x + B \sin \beta x$, where α and β are positive integers, periodic? Explain.

12. Analyzing Trigonometric Functions Two trigonometric functions f and g have periods of 2, and their graphs intersect at $x = 5.35$.

(a) Give one positive value of x less than 5.35 and one value of x greater than 5.35 at which the functions have the same value.

(b) Determine one negative value of x at which the graphs intersect.

(c) Is it true that $f(13.35) = g(-4.65)$? Explain.

13. Refraction When you stand in shallow water and look at an object below the surface of the water, the object will look farther away from you than it really is. This is because when light rays pass between air and water, the water refracts, or bends, the light rays. The index of refraction for water is 1.333. This is the ratio of the sine of θ_1 and the sine of θ_2 (see figure).

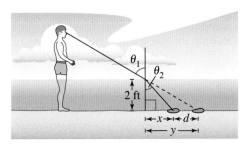

(a) While standing in water that is 2 feet deep, you look at a rock at angle $\theta_1 = 60°$ (measured from a line perpendicular to the surface of the water). Find θ_2.

(b) Find the distances x and y.

(c) Find the distance d between where the rock is and where it appears to be.

(d) What happens to d as you move closer to the rock? Explain.

14. Polynomial Approximation Using calculus, it can be shown that the arctangent function can be approximated by the polynomial

$$\arctan x \approx x - \frac{x^3}{3} + \frac{x^5}{5} - \frac{x^7}{7}$$

where x is in radians.

(a) Use a graphing utility to graph the arctangent function and its polynomial approximation in the same viewing window. How do the graphs compare?

(b) Predict the next term in the polynomial approximation of the arctangent function. Then repeat part (a). How does the accuracy of the approximation change when an additional term is added?

7 Analytic Trigonometry

GO DIGITAL

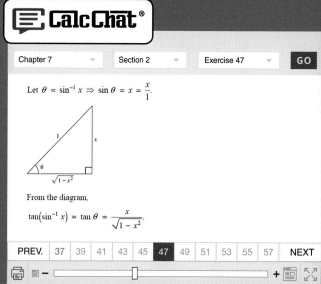

7.3 Ferris Wheel *(Exercise 94, p. 532)*

7.4 Standing Waves *(Exercise 82, p. 539)*

507

7.1 Using Fundamental Identities

Fundamental trigonometric identities are useful in simplifying trigonometric expressions. For example, in Exercise 59 on page 514, you will use trigonometric identities to simplify an expression for the coefficient of friction.

❯ **Recognize and write the fundamental trigonometric identities.**
❯ **Use the fundamental trigonometric identities to evaluate trigonometric functions, simplify trigonometric expressions, and rewrite trigonometric expressions.**

Introduction

In the preceding chapter, you studied the basic definitions, properties, graphs, and applications of the individual trigonometric functions. In this chapter, you will learn how to use the fundamental identities to perform the four tasks listed below.

1. Evaluate trigonometric functions.
2. Simplify trigonometric expressions.
3. Develop additional trigonometric identities.
4. Solve trigonometric equations.

Fundamental Trigonometric Identities

Reciprocal Identities

$$\sin u = \frac{1}{\csc u} \qquad \cos u = \frac{1}{\sec u} \qquad \tan u = \frac{1}{\cot u}$$

$$\csc u = \frac{1}{\sin u} \qquad \sec u = \frac{1}{\cos u} \qquad \cot u = \frac{1}{\tan u}$$

Quotient Identities

$$\tan u = \frac{\sin u}{\cos u} \qquad \cot u = \frac{\cos u}{\sin u}$$

Pythagorean Identities

$$\sin^2 u + \cos^2 u = 1 \qquad 1 + \tan^2 u = \sec^2 u \qquad 1 + \cot^2 u = \csc^2 u$$

Cofunction Identities

$$\sin\left(\frac{\pi}{2} - u\right) = \cos u \qquad \cos\left(\frac{\pi}{2} - u\right) = \sin u$$

$$\tan\left(\frac{\pi}{2} - u\right) = \cot u \qquad \cot\left(\frac{\pi}{2} - u\right) = \tan u$$

$$\sec\left(\frac{\pi}{2} - u\right) = \csc u \qquad \csc\left(\frac{\pi}{2} - u\right) = \sec u$$

Even/Odd Identities

$$\sin(-u) = -\sin u \qquad \cos(-u) = \cos u \qquad \tan(-u) = -\tan u$$

$$\csc(-u) = -\csc u \qquad \sec(-u) = \sec u \qquad \cot(-u) = -\cot u$$

ALGEBRA HELP

You should learn the fundamental trigonometric identities well, because you will use them frequently in trigonometry and they will also appear in calculus. Note that u can be an angle, a real number, or a variable.

Pythagorean identities are sometimes used in radical form such as

$$\sin u = \pm\sqrt{1 - \cos^2 u}$$

or

$$\tan u = \pm\sqrt{\sec^2 u - 1}$$

where the sign depends on the choice of u.

GO DIGITAL

Using the Fundamental Identities

One common application of trigonometric identities is to use given information about trigonometric functions to evaluate other trigonometric functions.

EXAMPLE 1 Using Identities to Evaluate a Function

Given $\sec u = -\frac{3}{2}$ and $\tan u > 0$, find the values of all six trigonometric functions.

Solution Using a reciprocal identity, you have

$$\cos u = \frac{1}{\sec u} = \frac{1}{-3/2} = -\frac{2}{3}.$$

Using the Pythagorean identity $\sin^2 u + \cos^2 u = 1$ in the equivalent form $\sin^2 u = 1 - \cos^2 u$, you have

$$\sin^2 u = 1 - \cos^2 u \qquad \text{Pythagorean identity}$$
$$= 1 - \left(-\frac{2}{3}\right)^2 \qquad \text{Substitute } -\frac{2}{3} \text{ for } \cos u.$$
$$= 1 - \frac{4}{9} \qquad \text{Evaluate power.}$$
$$= \frac{5}{9}. \qquad \text{Simplify.}$$

Because $\sec u < 0$ and $\tan u > 0$, it follows that u lies in Quadrant III. Moreover, $\sin u$ is negative when u is in Quadrant III, so choose the negative root and obtain $\sin u = -\sqrt{5}/3$. Knowing the values of the sine and cosine enables you to find the values of the remaining trigonometric functions.

$$\sin u = -\frac{\sqrt{5}}{3} \qquad\qquad \csc u = \frac{1}{\sin u} = -\frac{3}{\sqrt{5}} = -\frac{3\sqrt{5}}{5}$$

$$\cos u = -\frac{2}{3} \qquad\qquad \sec u = -\frac{3}{2}$$

$$\tan u = \frac{\sin u}{\cos u} = \frac{-\sqrt{5}/3}{-2/3} = \frac{\sqrt{5}}{2} \qquad \cot u = \frac{1}{\tan u} = \frac{2}{\sqrt{5}} = \frac{2\sqrt{5}}{5}$$

✓ *Checkpoint* Audio-video solution in English & Spanish at LarsonPrecalculus.com

Given $\tan x = \frac{1}{3}$ and $\cos x < 0$, find the values of all six trigonometric functions.

EXAMPLE 2 Simplifying a Trigonometric Expression

Simplify the expression $\sin x \cos^2 x - \sin x$.

Solution First factor out the common factor $\sin x$ and then use a Pythagorean identity.

$$\sin x \cos^2 x - \sin x = (\sin x)(\cos^2 x - 1) \qquad \text{Factor out common factor.}$$
$$= -(\sin x)(1 - \cos^2 x) \qquad \text{Factor out } -1.$$
$$= -(\sin x)(\sin^2 x) \qquad \text{Pythagorean identity}$$
$$= -\sin^3 x \qquad \text{Multiply.}$$

✓ *Checkpoint* Audio-video solution in English & Spanish at LarsonPrecalculus.com

Simplify the expression $\cos^2 x \csc x - \csc x$. ■

>>> **TECHNOLOGY**

Use a graphing utility to check the result of Example 2. To do this, let

$$y_1 = \sin x \cos^2 x - \sin x$$

and

$$y_2 = -\sin^3 x.$$

Select the *line* style for y_1 and the *path* style for y_2, then graph both equations in the same viewing window. The two graphs *appear* to coincide, so it is reasonable to assume that their expressions are equivalent. Note that the actual equivalence of the expressions can only be verified algebraically, as in Example 2.

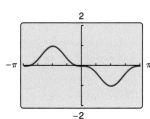

When factoring trigonometric expressions, it is helpful to find a polynomial form that fits the expression, as shown in Example 3.

EXAMPLE 3 **Factoring Trigonometric Expressions**

Factor each expression.

a. $\sec^2 \theta - 1$ **b.** $4 \tan^2 \theta + \tan \theta - 3$

Solution

a. This expression has the polynomial form $u^2 - v^2$, which is the difference of two squares. It factors as

$$\sec^2 \theta - 1 = (\sec \theta + 1)(\sec \theta - 1).$$

b. This expression has the polynomial form $ax^2 + bx + c$, and it factors as

$$4 \tan^2 \theta + \tan \theta - 3 = (4 \tan \theta - 3)(\tan \theta + 1).$$

✓ *Checkpoint* *Audio-video solution in English & Spanish at LarsonPrecalculus.com*

Factor each expression.

a. $1 - \cos^2 \theta$ **b.** $2 \csc^2 \theta - 7 \csc \theta + 6$

In some cases, when factoring or simplifying a trigonometric expression, it is helpful to first rewrite the expression in terms of just *one* trigonometric function or in terms of *sine and cosine only*. These strategies are demonstrated in Examples 4 and 5.

EXAMPLE 4 **Factoring a Trigonometric Expression**

Factor $\csc^2 x - \cot x - 3$.

Solution Use the identity $\csc^2 x = 1 + \cot^2 x$ to rewrite the expression.

$$\csc^2 x - \cot x - 3 = (1 + \cot^2 x) - \cot x - 3 \qquad \text{Pythagorean identity}$$
$$= \cot^2 x - \cot x - 2 \qquad \text{Combine like terms.}$$
$$= (\cot x - 2)(\cot x + 1) \qquad \text{Factor.}$$

✓ *Checkpoint* *Audio-video solution in English & Spanish at LarsonPrecalculus.com*

Factor $\sec^2 x + 3 \tan x + 1$.

EXAMPLE 5 **Simplifying a Trigonometric Expression**

 See LarsonPrecalculus.com for an interactive version of this type of example.

$$\sin t + \cot t \cos t = \sin t + \left(\frac{\cos t}{\sin t}\right) \cos t \qquad \text{Quotient identity}$$
$$= \frac{\sin^2 t + \cos^2 t}{\sin t} \qquad \text{Add fractions.}$$
$$= \frac{1}{\sin t} \qquad \text{Pythagorean identity}$$
$$= \csc t \qquad \text{Reciprocal identity}$$

✓ *Checkpoint* *Audio-video solution in English & Spanish at LarsonPrecalculus.com*

Simplify $\csc x - \cos x \cot x$.

 SKILLS REFRESHER

For a refresher on factoring polynomials, watch the video at *LarsonPrecalculus.com*.

ALGEBRA HELP

Remember that when adding fractions, you must first find the least common denominator (LCD). In Example 5, the LCD is $\sin t$.

EXAMPLE 6   **Adding Trigonometric Expressions**

Perform the addition and simplify: $\dfrac{\sin\theta}{1+\cos\theta}+\dfrac{\cos\theta}{\sin\theta}$.

Solution

$$\frac{\sin\theta}{1+\cos\theta}+\frac{\cos\theta}{\sin\theta}=\frac{(\sin\theta)(\sin\theta)+(\cos\theta)(1+\cos\theta)}{(1+\cos\theta)(\sin\theta)}$$

$$=\frac{\sin^2\theta+\cos^2\theta+\cos\theta}{(1+\cos\theta)(\sin\theta)}\qquad\text{Multiply.}$$

$$=\frac{1+\cos\theta}{(1+\cos\theta)(\sin\theta)}\qquad\text{Pythagorean identity}$$

$$=\frac{1}{\sin\theta}\qquad\text{Divide out common factor.}$$

$$=\csc\theta\qquad\text{Reciprocal identity}$$

✓ **Checkpoint** ▶ *Audio-video solution in English & Spanish at LarsonPrecalculus.com*

Perform the addition and simplify: $\dfrac{1}{1+\sin\theta}+\dfrac{1}{1-\sin\theta}$. ■

The next two examples involve techniques for rewriting expressions in forms that are used in calculus.

EXAMPLE 7   **Rewriting a Trigonometric Expression**

Rewrite $\dfrac{1}{1+\sin x}$ so that it is *not* in fractional form.

Solution   From the Pythagorean identity

$$\cos^2 x = 1 - \sin^2 x = (1-\sin x)(1+\sin x)$$

you can see that multiplying both the numerator and the denominator by $(1-\sin x)$ will produce a denominator with a single term.

$$\frac{1}{1+\sin x}=\frac{1}{1+\sin x}\left(\frac{1-\sin x}{1-\sin x}\right)\qquad\begin{array}{l}\text{Multiply numerator and}\\\text{denominator by }(1-\sin x).\end{array}$$

$$=\frac{1-\sin x}{1-\sin^2 x}\qquad\text{Multiply.}$$

$$=\frac{1-\sin x}{\cos^2 x}\qquad\text{Pythagorean identity}$$

$$=\frac{1}{\cos^2 x}-\frac{\sin x}{\cos^2 x}\qquad\text{Write as separate fractions.}$$

$$=\frac{1}{\cos^2 x}-\frac{\sin x}{\cos x}\cdot\frac{1}{\cos x}\qquad\text{Product of fractions}$$

$$=\sec^2 x-\tan x\sec x\qquad\text{Reciprocal and quotient identities}$$

✓ **Checkpoint** ▶ *Audio-video solution in English & Spanish at LarsonPrecalculus.com*

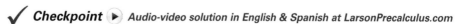

Rewrite $\dfrac{\cos^2\theta}{1-\sin\theta}$ so that it is *not* in fractional form. ■

EXAMPLE 8　**Trigonometric Substitution**　

Use the substitution $x = 2 \tan \theta$, $0 < \theta < \pi/2$, to write $\sqrt{4 + x^2}$ as a trigonometric function of θ.

Solution　Begin by letting $x = 2 \tan \theta$. Then, you obtain

$$\sqrt{4 + x^2} = \sqrt{4 + (2 \tan \theta)^2}$$　Substitute $2 \tan \theta$ for x.

$$= \sqrt{4 + 4 \tan^2 \theta}$$　Property of exponents

$$= \sqrt{4(1 + \tan^2 \theta)}$$　Factor.

$$= \sqrt{4 \sec^2 \theta}$$　Pythagorean identity

$$= 2 \sec \theta.$$　$\sec \theta > 0$ for $0 < \theta < \dfrac{\pi}{2}$

✔ *Checkpoint*　▶ Audio-video solution in English & Spanish at LarsonPrecalculus.com

Use the substitution $x = 3 \sin \theta$, $0 < \theta < \pi/2$, to write $\sqrt{9 - x^2}$ as a trigonometric function of θ.　■

Figure 7.1 shows the right triangle illustration of the trigonometric substitution $x = 2 \tan \theta$ in Example 8. You can use this triangle to check the solution to Example 8. For $0 < \theta < \pi/2$, you have

$$\text{opp} = x, \quad \text{adj} = 2, \quad \text{and} \quad \text{hyp} = \sqrt{4 + x^2}.$$

Using these expressions,

$$\sec \theta = \frac{\text{hyp}}{\text{adj}} = \frac{\sqrt{4 + x^2}}{2}.$$

So, $2 \sec \theta = \sqrt{4 + x^2}$, and the solution checks.

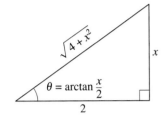

$$2 \tan \theta = x \quad \Longrightarrow \quad \tan \theta = \frac{x}{2}$$

Figure 7.1

EXAMPLE 9　**Rewriting a Logarithmic Expression**

Rewrite $\ln|\csc \theta| + \ln|\tan \theta|$ as a single logarithm and simplify the result.

Solution

$$\ln|\csc \theta| + \ln|\tan \theta| = \ln|\csc \theta \tan \theta|$$　Product Property of Logarithms

$$= \ln\left|\frac{1}{\sin \theta} \cdot \frac{\sin \theta}{\cos \theta}\right|$$　Reciprocal and quotient identities

$$= \ln\left|\frac{1}{\cos \theta}\right|$$　Simplify.

$$= \ln|\sec \theta|$$　Reciprocal identity

✔ *Checkpoint*　▶ Audio-video solution in English & Spanish at LarsonPrecalculus.com

Rewrite $\ln|\sec x| + \ln|\sin x|$ as a single logarithm and simplify the result.　■

ALGEBRA HELP
Recall from Section 5.3 that for positive real numbers u and v,

$$\ln u + \ln v = \ln(uv).$$

Summarize　*(Section 7.1)*

1. State the fundamental trigonometric identities *(page 508)*.
2. Explain how to use the fundamental trigonometric identities to evaluate trigonometric functions, simplify trigonometric expressions, and rewrite trigonometric expressions *(pages 509–512)*. For examples of these concepts, see Examples 1–9.

GO DIGITAL

7.1 Exercises

See CalcChat.com for tutorial help and worked-out solutions to odd-numbered exercises.

Vocabulary and Concept Check

In Exercises 1–3, fill in the blank to complete the trigonometric identity.

1. $\dfrac{1}{\tan u} = $ _____

2. $\sec\left(\dfrac{\pi}{2} - u\right) = $ _____

3. $1 + \cot^2 u = $ _____

4. Write three different expressions that are equivalent to $\cos u$.

Skills and Applications

Using Identities to Evaluate a Function In Exercises 5–10, use the given conditions to find the values of all six trigonometric functions.

5. $\sec x = -\dfrac{5}{2}$, $\tan x < 0$

6. $\csc x = -\dfrac{7}{6}$, $\tan x > 0$

7. $\sin \theta = -\dfrac{3}{4}$, $\cos \theta > 0$

8. $\cos \theta = \dfrac{2}{3}$, $\sin \theta < 0$

9. $\tan x = \dfrac{2}{3}$, $\cos x > 0$

10. $\cot x = \dfrac{7}{4}$, $\sin x < 0$

Matching Trigonometric Expressions In Exercises 11–16, match the trigonometric expression with its simplified form.

(a) $\csc x$ (b) -1 (c) 1

(d) $\sin x \tan x$ (e) $\sec^2 x$ (f) $\sec x$

11. $\sec x \cos x$

12. $\cot^2 x - \csc^2 x$

13. $\cos x(1 + \tan^2 x)$

14. $\cot x \sec x$

15. $\dfrac{\sec^2 x - 1}{\sin^2 x}$

16. $\dfrac{\cos^2[(\pi/2) - x]}{\cos x}$

Simplifying a Trigonometric Expression In Exercises 17–20, use the fundamental identities to simplify the expression. (There is more than one correct form of each answer.)

17. $\dfrac{\tan \theta \cot \theta}{\sec \theta}$

18. $\cos\left(\dfrac{\pi}{2} - x\right) \sec x$

19. $\tan^2 x - \tan^2 x \sin^2 x$

20. $\sin^2 x \sec^2 x - \sin^2 x$

Factoring a Trigonometric Expression In Exercises 21–30, factor the expression. Use the fundamental identities to simplify, if necessary. (There is more than one correct form of each answer.)

21. $\dfrac{\sec^2 x - 1}{\sec x - 1}$

22. $\dfrac{\cos x - 2}{\cos^2 x - 4}$

23. $1 - 2\cos^2 x + \cos^4 x$

24. $\sec^4 x - \tan^4 x$

25. $\cot^3 x + \cot^2 x + \cot x + 1$

26. $\sec^3 x - \sec^2 x - \sec x + 1$

27. $3\sin^2 x - 5\sin x - 2$

28. $6\cos^2 x + 5\cos x - 6$

29. $\cot^2 x + \csc x - 1$

30. $\sin^2 x + 3\cos x + 3$

Simplifying a Trigonometric Expression In Exercises 31–38, use the fundamental identities to simplify the expression. (There is more than one correct form of each answer.)

31. $\tan \theta \csc \theta$

32. $\tan(-x) \cos x$

33. $\sin \phi(\csc \phi - \sin \phi)$

34. $\cos x(\sec x - \cos x)$

35. $\sin \beta \tan \beta + \cos \beta$

36. $\cot u \sin u + \tan u \cos u$

37. $\dfrac{1 - \sin^2 x}{\csc^2 x - 1}$

38. $\dfrac{\sec^2 y - 1}{1 - \cos^2 y}$

Adding or Subtracting Trigonometric Expressions In Exercises 39–44, perform the addition or subtraction and use the fundamental identities to simplify. (There is more than one correct form of each answer.)

39. $\dfrac{1}{1 + \cos x} + \dfrac{1}{1 - \cos x}$

40. $\dfrac{1}{\sec x + 1} - \dfrac{1}{\sec x - 1}$

41. $\dfrac{\cos x}{1 + \sin x} - \dfrac{\cos x}{1 - \sin x}$

42. $\dfrac{\sin x}{1 + \cos x} + \dfrac{\sin x}{1 - \cos x}$

43. $\tan x - \dfrac{\sec^2 x}{\tan x}$

44. $\dfrac{\cos x}{1 + \sin x} + \dfrac{1 + \sin x}{\cos x}$

Rewriting a Trigonometric Expression In Exercises 45 and 46, rewrite the expression so that it is *not* in fractional form. (There is more than one correct form of each answer.)

45. $\dfrac{\sin^2 y}{1 - \cos y}$

46. $\dfrac{5}{\tan x + \sec x}$

Trigonometric Functions and Expressions In Exercises 47 and 48, use a graphing utility to determine which of the six trigonometric functions is equal to the expression. Verify your answer algebraically.

47. $\dfrac{\tan x + 1}{\sec x + \csc x}$

48. $\dfrac{1}{\sin x}\left(\dfrac{1}{\cos x} - \cos x\right)$

Solving a Trigonometric Equation In Exercises 49 and 50, use a graphing utility to solve the equation for θ, where $0 \le \theta < 2\pi$.

49. $\sin \theta = \sqrt{1 - \cos^2 \theta}$

50. $\sec \theta = \sqrt{1 + \tan^2 \theta}$

Trigonometric Substitution In Exercises 51–54, use the trigonometric substitution to write the algebraic expression as a trigonometric function of θ, where $0 < \theta < \pi/2.$

51. $\sqrt{9 - x^2}, \quad x = 3\cos\theta$

52. $\sqrt{49 - x^2}, \quad x = 7\sin\theta$

53. $\sqrt{x^2 - 4}, \quad x = 2\sec\theta$

54. $\sqrt{9x^2 + 25}, \quad 3x = 5\tan\theta$

Rewriting a Logarithmic Expression In Exercises 55–58, rewrite the expression as a single logarithm and simplify the result.

55. $\ln|\sin x| + \ln|\cot x|$

56. $\ln|\cos x| - \ln|\sin x|$

57. $\ln|\tan t| - \ln(1 - \cos^2 t)$

58. $\ln(\cos^2 t) + \ln(1 + \tan^2 t)$

59. **Friction**

The forces acting on an object weighing W units on an inclined plane positioned at an angle of θ with the horizontal (see figure) are modeled by $\mu W \cos\theta = W\sin\theta,$ where μ is the coefficient of friction. Solve the equation for μ and simplify the result.

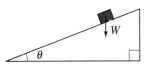

60. **Rate of Change** The rate of change of the function $f(x) = \sec x + \cos x$ is given by the expression $\sec x \tan x - \sin x.$ Show that this expression can also be written as $\sin x \tan^2 x.$

Exploring the Concepts

True or False? In Exercises 61–63, determine whether the statement is true or false. Justify your answer.

61. $\sin\theta \csc\theta = 1$

62. $\cos\theta \sec\varphi = 1$

63. The quotient identities and reciprocal identities can be used to write any trigonometric function in terms of sine and cosine.

64. **Error Analysis** Describe the error.

$$\frac{\sin\theta}{\cos(-\theta)} = \frac{\sin\theta}{-\cos\theta}$$
$$= -\tan\theta \quad \textbf{✗}$$

65. **Writing Trigonometric Functions in Terms of Sine** Write each of the other trigonometric functions of θ in terms of $\sin\theta.$

66. **HOW DO YOU SEE IT?**
Explain how to use the figure to derive the Pythagorean identities

$\sin^2\theta + \cos^2\theta = 1,$

$1 + \tan^2\theta = \sec^2\theta,$

and $1 + \cot^2\theta = \csc^2\theta.$

Discuss how to remember these identities and other fundamental trigonometric identities.

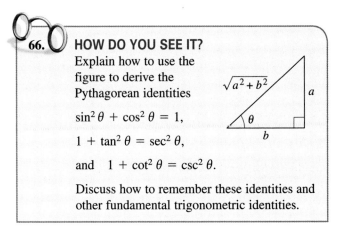

67. **Trigonometric Substitution** Using the trigonometric substitution $u = a\tan\theta,$ where $-\pi/2 < \theta < \pi/2$ and $a > 0,$ simplify the expression $\sqrt{a^2 + u^2}.$

68. **Rewriting a Trigonometric Expression** Rewrite the expression below in terms of $\sin\theta$ and $\cos\theta.$

$$\frac{\sec\theta(1 + \tan\theta)}{\sec\theta + \csc\theta}$$

Review & Refresh ▶ *Video solutions at LarsonPrecalculus.com*

Multiplying or Dividing Rational Expressions In Exercises 69–72, perform the multiplication or division and simplify.

69. $\dfrac{x}{x + 4} \cdot \dfrac{x^2 - 16}{2x}$

70. $\dfrac{x - 1}{x^2 + 3x} \cdot \dfrac{x^2 + x}{x - 1}$

71. $\dfrac{12}{x^3 - x} \div \dfrac{3x - 3}{x + x^2}$

72. $\dfrac{x^2 - 4x - 5}{x^2 + 2x + 1} \div \dfrac{x + 1}{30 - 6x}$

Multiplying Conjugates In Exercises 73–76, multiply the complex number by its complex conjugate.

73. $7 + 3i$

74. $2 - \sqrt{5}i$

75. $\sqrt{-32}$

76. $1 + \sqrt{-21}$

Sketching the Graph of a Sine or Cosine Function In Exercises 77–80, sketch the graph of the function. (Include two full periods.)

77. $y = -5\cos x$

78. $y = \sin(\pi x/2)$

79. $y = 3\sin(2x)$

80. $y = 2\cos(3x/2) + 2$

True or False? In Exercises 81 and 82, determine whether the statement is true or false given that $f(x) = \ln x$ and $x, y > 0.$ Justify your answer.

81. $f(xy) = f(x) + f(y)$

82. $[f(x)]^n = nf(x)$

7.2 Verifying Trigonometric Identities

Trigonometric identities enable you to rewrite trigonometric equations that model real-life situations. For example, in Exercise 51 on page 521, trigonometric identities can help you simplify an equation that models the length of a shadow cast by a gnomon (a device used to tell time).

❯ **Verify trigonometric identities.**

Verifying Trigonometric Identities

In the preceding section, you used identities to rewrite trigonometric expressions in equivalent forms. In this section, you will study techniques for verifying trigonometric identities. In the next section, you will study techniques for solving trigonometric equations. The key to both verifying identities *and* solving equations is your ability to use the fundamental identities and the rules of algebra to rewrite trigonometric expressions.

Remember that a *conditional equation* is an equation that is true for only some of the values in the domain of the variable. For example, the conditional equation

$$\sin x = 0 \qquad \text{Conditional equation}$$

is true only for

$$x = n\pi$$

where n is an integer. When you are finding the values of the variable for which the equation is true, you are *solving* the equation.

On the other hand, an equation that is true for all real values in the domain of the variable is an *identity*. For example, the familiar equation

$$\sin^2 x = 1 - \cos^2 x \qquad \text{Identity}$$

is true for all real numbers x. So, it is an identity.

Although there are similarities, verifying that a trigonometric equation is an identity is quite different from solving an equation. There is no well-defined set of rules to follow in verifying trigonometric identities. The process is best learned through practice.

Guidelines for Verifying Trigonometric Identities

1. Work with one side of the equation at a time. It is often better to work with the more complicated side first.
2. Look for opportunities to factor an expression, add fractions, or create a denominator with a single term.
3. Look for opportunities to use the fundamental identities. Note which functions are in the final expression you want. Sines and cosines pair up well, as do secants and tangents, and cosecants and cotangents.
4. When the preceding guidelines do not help, try converting all terms to sines and cosines.
5. Always try *something*. Even making an attempt that leads to a dead end can provide insight.

Verifying trigonometric identities is a useful process when you need to convert a trigonometric expression into a form that is more useful algebraically. When you verify an identity, you cannot *assume* that the two sides of the equation are equal because you are trying to verify that they *are* equal. As a result, when verifying identities, you cannot use operations such as adding the same quantity to each side of the equation or cross multiplication.

GO DIGITAL

© Leonard Zhukovsky/Shutterstock.com

EXAMPLE 1 **Verifying a Trigonometric Identity**

Verify the identity $\dfrac{\sec^2 \theta - 1}{\sec^2 \theta} = \sin^2 \theta$.

Solution The left side is more complicated, so start with it.

$$\frac{\sec^2 \theta - 1}{\sec^2 \theta} = \frac{\tan^2 \theta}{\sec^2 \theta} \qquad\qquad \text{Pythagorean identity}$$

$$= (\tan^2 \theta)(\cos^2 \theta) \qquad\quad \text{Reciprocal identity}$$

$$= \frac{\sin^2 \theta}{\cos^2 \theta}(\cos^2 \theta) \qquad\quad \text{Quotient identity}$$

$$= \sin^2 \theta \qquad\qquad\qquad \text{Simplify.}$$

Notice that you verify the identity by starting with the left side of the equation (the more complicated side) and using the fundamental trigonometric identities to simplify it until you obtain the right side.

✓ **Checkpoint**  *Audio-video solution in English & Spanish at LarsonPrecalculus.com*

Verify the identity $\dfrac{\sin^2 \theta + \cos^2 \theta}{\cos^2 \theta \sec^2 \theta} = 1$. ■

There can be more than one way to verify an identity. Here is another way to verify the identity in Example 1.

$$\frac{\sec^2 \theta - 1}{\sec^2 \theta} = \frac{\sec^2 \theta}{\sec^2 \theta} - \frac{1}{\sec^2 \theta} \qquad \text{Write as separate fractions.}$$

$$= 1 - \cos^2 \theta \qquad\qquad \text{Reciprocal identity}$$

$$= \sin^2 \theta \qquad\qquad\qquad \text{Pythagorean identity}$$

EXAMPLE 2 **Verifying a Trigonometric Identity**

Verify the identity $2 \sec^2 \alpha = \dfrac{1}{1 - \sin \alpha} + \dfrac{1}{1 + \sin \alpha}$.

Algebraic Solution

Because the right side is more complicated, start with it.

$$\frac{1}{1 - \sin \alpha} + \frac{1}{1 + \sin \alpha} = \frac{1 + \sin \alpha + 1 - \sin \alpha}{(1 - \sin \alpha)(1 + \sin \alpha)} \qquad \text{Add fractions.}$$

$$= \frac{2}{1 - \sin^2 \alpha} \qquad \text{Simplify.}$$

$$= \frac{2}{\cos^2 \alpha} \qquad \text{Pythagorean identity}$$

$$= 2 \sec^2 \alpha \qquad \text{Reciprocal identity}$$

Numerical Solution

Use a graphing utility to create a table that shows the values of $y_1 = 2/\cos^2 x$ and $y_2 = [1/(1 - \sin x)] + [1/(1 + \sin x)]$ for different values of x.

X	Y1	Y2
-.5	2.5969	2.5969
-.25	2.1304	2.1304
0	2	2
.25	2.1304	2.1304
.5	2.5969	2.5969
.75	3.7357	3.7357
1	6.851	6.851

X=-.5

The values in the table for y_1 and y_2 appear to be identical, so the equation appears to be an identity.

✓ **Checkpoint** *Audio-video solution in English & Spanish at LarsonPrecalculus.com*

Verify the identity $2 \csc^2 \beta = \dfrac{1}{1 - \cos \beta} + \dfrac{1}{1 + \cos \beta}$. ■

GO DIGITAL

EXAMPLE 3 **Verifying a Trigonometric Identity**

Verify the identity $(\tan^2 x + 1)(\cos^2 x - 1) = -\tan^2 x$.

Algebraic Solution

The left side is more complicated, so start with it by applying Pythagorean identities before multiplying.

$$(\tan^2 x + 1)(\cos^2 x - 1) = (\sec^2 x)(-\sin^2 x) \qquad \text{Pythagorean identities}$$

$$= -\frac{\sin^2 x}{\cos^2 x} \qquad \text{Reciprocal identity}$$

$$= -\left(\frac{\sin x}{\cos x}\right)^2 \qquad \text{Property of exponents}$$

$$= -\tan^2 x \qquad \text{Quotient identity}$$

Graphical Solution

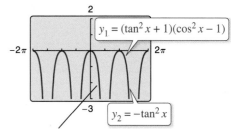

$y_1 = (\tan^2 x + 1)(\cos^2 x - 1)$

$y_2 = -\tan^2 x$

The graphs appear to coincide, so the given equation appears to be an identity.

✓ **Checkpoint** ▶ *Audio-video solution in English & Spanish at LarsonPrecalculus.com*

Verify the identity $(\sec^2 x - 1)(\sin^2 x - 1) = -\sin^2 x$.

EXAMPLE 4 **Converting to Sines and Cosines**

Verify each identity.
a. $\tan x \csc x = \sec x$

b. $\tan x + \cot x = \sec x \csc x$

Solution

a. Convert the left side into sines and cosines.

$$\tan x \csc x = \frac{\sin x}{\cos x} \cdot \frac{1}{\sin x} \qquad \text{Quotient and reciprocal identities}$$

$$= \frac{1}{\cos x} \qquad \text{Simplify.}$$

$$= \sec x \qquad \text{Reciprocal identity}$$

b. Convert the left side into sines and cosines.

$$\tan x + \cot x = \frac{\sin x}{\cos x} + \frac{\cos x}{\sin x} \qquad \text{Quotient identities}$$

$$= \frac{\sin^2 x + \cos^2 x}{\cos x \sin x} \qquad \text{Add fractions.}$$

$$= \frac{1}{\cos x \sin x} \qquad \text{Pythagorean identity}$$

$$= \frac{1}{\cos x} \cdot \frac{1}{\sin x} \qquad \text{Product of fractions}$$

$$= \sec x \csc x \qquad \text{Reciprocal identities}$$

✓ **Checkpoint** ▶ *Audio-video solution in English & Spanish at LarsonPrecalculus.com*

Verify each identity.

a. $\cot x \sec x = \csc x$

b. $\csc x - \sin x = \cos x \cot x$

GO DIGITAL

▶▶ **SKILLS REFRESHER**

For a refresher on rationalizing a denominator, watch the video at *LarsonPrecalculus.com.*

Recall from algebra that *rationalizing the denominator* using conjugates is, on occasion, a powerful simplification technique. A related form of this technique works for simplifying trigonometric expressions as well. For example, to simplify

$$\frac{1}{1 - \cos x}$$

multiply the numerator and the denominator by $1 + \cos x$.

$$\frac{1}{1 - \cos x} = \frac{1}{1 - \cos x}\left(\frac{1 + \cos x}{1 + \cos x}\right)$$

$$-\frac{1 + \cos x}{1 - \cos^2 x}$$

$$= \frac{1 + \cos x}{\sin^2 x}$$

$$= \csc^2 x(1 + \cos x)$$

The expression $\csc^2 x(1 + \cos x)$ is considered a simplified form of

$$\frac{1}{1 - \cos x}$$

because $\csc^2 x(1 + \cos x)$ does not contain fractions.

EXAMPLE 5   **Verifying a Trigonometric Identity**

▶▶ *See LarsonPrecalculus.com for an interactive version of this type of example.*

Verify the identity $\sec x + \tan x = \dfrac{\cos x}{1 - \sin x}$.

Algebraic Solution

Begin with the *right* side and create a denominator with a single term by multiplying the numerator and the denominator by $1 + \sin x$.

$$\frac{\cos x}{1 - \sin x} = \frac{\cos x}{1 - \sin x}\left(\frac{1 + \sin x}{1 + \sin x}\right) \qquad \text{Multiply numerator and denominator by } (1 + \sin x).$$

$$= \frac{\cos x + \cos x \sin x}{1 - \sin^2 x} \qquad \text{Multiply.}$$

$$= \frac{\cos x + \cos x \sin x}{\cos^2 x} \qquad \text{Pythagorean identity}$$

$$= \frac{\cos x}{\cos^2 x} + \frac{\cos x \sin x}{\cos^2 x} \qquad \text{Write as separate fractions.}$$

$$= \frac{1}{\cos x} + \frac{\sin x}{\cos x} \qquad \text{Simplify.}$$

$$= \sec x + \tan x \qquad \text{Identities}$$

Graphical Solution

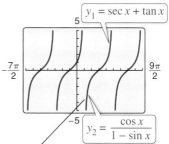

The graphs appear to coincide, so the given equation appears to be an identity.

✓ *Checkpoint* ▶ *Audio-video solution in English & Spanish at LarsonPrecalculus.com*

Verify the identity $\csc x + \cot x = \dfrac{\sin x}{1 - \cos x}$.

In Examples 1 through 5, you have been verifying trigonometric identities by working with one side of the equation and converting it to the form given on the other side. On occasion, it is practical to work with each side *separately* to obtain one common form that is equivalent to both sides. This is illustrated in Example 6.

EXAMPLE 6 **Working with Each Side Separately**

Verify the identity $\dfrac{\cot^2 \theta}{1 + \csc \theta} = \dfrac{1 - \sin \theta}{\sin \theta}$.

Algebraic Solution

Working with the left side, you have

$$\frac{\cot^2 \theta}{1 + \csc \theta} = \frac{\csc^2 \theta - 1}{1 + \csc \theta} \qquad \text{Pythagorean identity}$$

$$= \frac{(\csc \theta - 1)(\csc \theta + 1)}{1 + \csc \theta} \qquad \text{Factor.}$$

$$= \csc \theta - 1. \qquad \text{Simplify.}$$

Now, simplifying the right side, you have

$$\frac{1 - \sin \theta}{\sin \theta} = \frac{1}{\sin \theta} - \frac{\sin \theta}{\sin \theta} = \csc \theta - 1.$$

This verifies the identity because both sides are equal to $\csc \theta - 1$.

Numerical Solution

Use a graphing utility to create a table that shows the values of

$$y_1 = \frac{\cot^2 x}{1 + \csc x} \quad \text{and} \quad y_2 = \frac{1 - \sin x}{\sin x}$$

for different values of x.

X	Y1	Y2
-.5	-3.086	-3.086
-.25	-5.042	-5.042
0	ERROR	ERROR
.25	3.042	3.042
.5	1.0858	1.0858
.75	.46705	.46705
1	.1884	.1884
X=1		

The values for y_1 and y_2 appear to be identical, so the equation appears to be an identity.

✓ **Checkpoint** ▶ *Audio-video solution in English & Spanish at LarsonPrecalculus.com*

Verify the identity $\dfrac{\tan^2 \theta}{1 + \sec \theta} = \dfrac{1 - \cos \theta}{\cos \theta}$.

Example 7 shows powers of trigonometric functions rewritten as more complicated sums of products of trigonometric functions. This is a common procedure used in calculus.

EXAMPLE 7 **Two Examples from Calculus** ∫

Verify each identity.

a. $\tan^4 x = \tan^2 x \sec^2 x - \tan^2 x$ **b.** $\csc^4 x \cot x = \csc^2 x(\cot x + \cot^3 x)$

Solution

a. $\tan^4 x = (\tan^2 x)(\tan^2 x)$ Write as separate factors.

$ = \tan^2 x(\sec^2 x - 1)$ Pythagorean identity

$ = \tan^2 x \sec^2 x - \tan^2 x$ Multiply.

b. $\csc^4 x \cot x = \csc^2 x \csc^2 x \cot x$ Write as separate factors.

$ = \csc^2 x(1 + \cot^2 x) \cot x$ Pythagorean identity

$ = \csc^2 x(\cot x + \cot^3 x)$ Multiply.

✓ **Checkpoint** ▶ *Audio-video solution in English & Spanish at LarsonPrecalculus.com*

Verify each identity.

a. $\tan^3 x = \tan x \sec^2 x - \tan x$ **b.** $\sin^3 x \cos^4 x = (\cos^4 x - \cos^6 x) \sin x$ ■

Summarize (Section 7.2)

1. State the guidelines for verifying trigonometric identities *(page 515)*. For examples of verifying trigonometric identities, see Examples 1–7.

GO DIGITAL

7.2 Exercises

See CalcChat.com for tutorial help and worked-out solutions to odd-numbered exercises.

Vocabulary and Concept Check

In Exercises 1 and 2, fill in the blanks.

1. An equation that is true for all real values in the domain of the variable is an _____.
2. An equation that is true for only some values in the domain of the variable is a _____ _____.

3. Explain why a graphical or numerical solution is not sufficient to verify a trigonometric identity.
4. Explain how an attempt to verify a trigonometric identity that leads to a dead end can provide insight.

Skills and Applications

Verifying a Trigonometric Identity In Exercises 5–14, verify the identity.

5. $\tan t \cot t = 1$

6. $\dfrac{\tan x \cot x}{\cos x} = \sec x$

7. $1 + \sin^2 \alpha = 2 - \cos^2 \alpha$

8. $\cos^2 \beta - \sin^2 \beta = 2\cos^2 \beta - 1$

9. $\cos^2 \beta - \sin^2 \beta = 1 - 2\sin^2 \beta$

10. $\sin^2 \alpha - \sin^4 \alpha = \cos^2 \alpha - \cos^4 \alpha$

11. $\tan\left(\dfrac{\pi}{2} - \theta\right) \sin \theta = \cos \theta$

12. $\dfrac{\cos(\pi/2 - x)}{\sin(\pi/2 - x)}[\cot(\pi/2 - x)] = \tan^2 x$

13. $\sin t \csc\left(\dfrac{\pi}{2} - t\right) = \tan t$

14. $\sec^2 y - \cot^2\left(\dfrac{\pi}{2} - y\right) = 1$

Verifying a Trigonometric Identity In Exercises 15–20, verify the identity algebraically. Use the *table* feature of a graphing utility to check your result numerically.

15. $\dfrac{1}{\tan x} + \dfrac{1}{\cot x} = \tan x + \cot x$

16. $\dfrac{1}{\sin x} - \dfrac{1}{\csc x} = \csc x - \sin x$

17. $\dfrac{1 + \sin \theta}{\cos \theta} + \dfrac{\cos \theta}{1 + \sin \theta} = 2\sec \theta$

18. $\dfrac{\cos \theta \cot \theta}{1 - \sin \theta} - 1 = \csc \theta$

19. $\dfrac{1}{\cos x + 1} + \dfrac{1}{\cos x - 1} = -2\csc x \cot x$

20. $\cos x - \dfrac{\cos x}{1 - \tan x} = \dfrac{\sin x \cos x}{\sin x - \cos x}$

Verifying a Trigonometric Identity In Exercises 21–26, verify the identity algebraically. Use a graphing utility to check your result graphically.

21. $\sec^3 y \cos y = \sec^2 y$

22. $\cot^2 y(\sec^2 y - 1) = 1$

23. $\dfrac{\tan^2 \theta}{\sec \theta} = \sin \theta \tan \theta$

24. $\dfrac{\cot^3 t}{\csc t} = \cos t(\csc^2 t - 1)$

25. $\dfrac{1}{\tan \beta} + \tan \beta = \dfrac{\sec^2 \beta}{\tan \beta}$

26. $\dfrac{\sec \theta - 1}{1 - \cos \theta} = \sec \theta$

Converting to Sines and Cosines In Exercises 27–32, verify the identity by converting the left side into sines and cosines.

27. $\dfrac{\cot^2 t}{\csc t} = \dfrac{1 - \sin^2 t}{\sin t}$

28. $\cos x + \sin x \tan x = \sec x$

29. $\sec x - \cos x = \sin x \tan x$

30. $\cot x - \tan x = \sec x(\csc x - 2\sin x)$

31. $\dfrac{\cot x}{\sec x} + \dfrac{\tan x}{\csc x} = \sec x + \csc x - \sin x - \cos x$

32. $\dfrac{\csc(-x)}{\sec(-x)} + \dfrac{1}{\tan(-x)} = -2\cot x$

Verifying a Trigonometric Identity In Exercises 33–38, verify the identity.

33. $\sin^{1/2} x \cos x - \sin^{5/2} x \cos x = \cos^3 x \sqrt{\sin x}$

34. $\sec^6 x(\sec x \tan x) - \sec^4 x(\sec x \tan x) = \sec^5 x \tan^3 x$

35. $(1 + \sin y)[1 + \sin(-y)] = \cos^2 y$

36. $\dfrac{\tan x + \tan y}{1 - \tan x \tan y} = \dfrac{\cot x + \cot y}{\cot x \cot y - 1}$

37. $\sqrt{\dfrac{1 + \sin \theta}{1 - \sin \theta}} = \dfrac{1 + \sin \theta}{|\cos \theta|}$

38. $\dfrac{\cos x - \cos y}{\sin x + \sin y} + \dfrac{\sin x - \sin y}{\cos x + \cos y} = 0$

𝄑 Verifying a Trigonometric Identity In Exercises 39–42, verify the identity.

39. $\tan^5 x = \tan^3 x \sec^2 x - \tan^3 x$

40. $\sec^4 x \tan^2 x = (\tan^2 x + \tan^4 x)\sec^2 x$

41. $\cos^3 x \sin^2 x = (\sin^2 x - \sin^4 x)\cos x$

42. $\sin^4 x + \cos^4 x = 1 - 2\cos^2 x + 2\cos^4 x$

∿ Determining Identities In Exercises 43–46, (a) use a graphing utility to graph each side of the equation to determine whether the equation appears to be an identity, (b) use the *table* feature of the graphing utility to determine whether the equation appears to be an identity, and (c) determine whether the equation *is* an identity algebraically.

43. $(1 + \cot^2 x)(\cos^2 x) = \cot^2 x$

44. $\csc x(\csc x - \sin x) + \dfrac{\sin x - \cos x}{\sin x} + \cot x = \csc^2 x$

45. $2 + \cos^2 x - 3\cos^4 x = \sin^2 x(3 + 2\cos^2 x)$

46. $\tan^4 x + \tan^2 x - 3 = \sec^2 x(4\tan^2 x - 3)$

Verifying a Trigonometric Identity In Exercises 47–50, verify the identity.

47. $\tan(\sin^{-1} x) = \dfrac{x}{\sqrt{1 - x^2}}$ 48. $\cos(\sin^{-1} x) = \sqrt{1 - x^2}$

49. $\tan\left(\sin^{-1}\dfrac{x - 1}{4}\right) = \dfrac{x - 1}{\sqrt{16 - (x - 1)^2}}$

50. $\tan\left(\cos^{-1}\dfrac{x + 1}{2}\right) = \dfrac{\sqrt{4 - (x + 1)^2}}{x + 1}$

51. Shadow Length

The length s of a shadow cast by a gnomon (a device used to tell time) of height h when the angle of the sun above the horizon is θ can be modeled by the equation

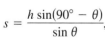

$$s = \dfrac{h\sin(90° - \theta)}{\sin\theta},$$

$0° < \theta \le 90°$.

(a) Verify that $s = h\cot\theta$.

(b) Use a graphing utility to create a table of the lengths s for different values of θ. Let $h = 5$ feet.

(c) Use your table from part (b) to determine the angle of the sun that results in the minimum length of the shadow.

(d) What time of day do you think it is when the angle of the sun above the horizon is 90°?

𝄑 52. Rate of Change The rate of change of the function $f(x) = \sin x + \csc x$ is given by the expression $\cos x - \csc x \cot x$. Show that the expression for the rate of change can also be written as $-\cos x \cot^2 x$.

Exploring the Concepts

True or False? In Exercises 53 and 54, determine whether the statement is true or false. Justify your answer.

53. $(1 + \csc x)(1 - \csc x) = -\cot x^2$

54. There can be more than one way to verify a trigonometric identity.

55. Error Analysis Describe the error.

$$\dfrac{1}{\tan x} + \cot(-x) = \cot x + \cot x = 2\cot x \quad \text{✗}$$

56.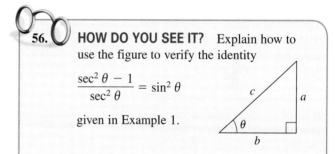

HOW DO YOU SEE IT? Explain how to use the figure to verify the identity

$$\dfrac{\sec^2\theta - 1}{\sec^2\theta} = \sin^2\theta$$

given in Example 1.

57. Think About It Explain why $\tan\theta = \sqrt{\sec^2\theta - 1}$ is not an identity, and find one value of the variable for which the equation is not true.

Review & Refresh ▶ *Video solutions at LarsonPrecalculus.com*

58. Angle of Elevation The height of an outdoor basketball backboard is $12\frac{1}{2}$ feet, and the backboard casts a shadow 17 feet long.

(a) Draw a right triangle that gives a visual representation of the problem. Label the known and unknown quantities.

(b) Use a trigonometric function to write an equation involving the unknown angle of elevation.

(c) Find the angle of elevation.

Solving an Equation In Exercises 59–62, solve the equation.

59. $x^2 - 15x - 34 = 0$ 60. $4x^2 - 5x - 9 = 0$

61. $(x + 1)^2 = 64$ 62. $(x - 11)^2 = 32$

Solving for θ In Exercises 63 and 64, find two solutions of each equation. Give your answers in degrees $(0° \le \theta < 360°)$ and in radians $(0 \le \theta < 2\pi)$. Do not use a calculator.

63. (a) $\sin\theta = 0$ 64. (a) $\cos\theta = -1/2$

   (b) $\sin\theta = \sqrt{2}/2$ (b) $\sec\theta = 2\sqrt{3}/3$

7.3 Solving Trigonometric Equations

Trigonometric equations have many applications in circular motion. For example, in Exercise 94 on page 532, you will solve a trigonometric equation to determine when a person riding a Ferris wheel will be at certain heights above the ground.

❯ Use standard algebraic techniques to solve trigonometric equations.
❯ Solve trigonometric equations of quadratic type.
❯ Solve trigonometric equations involving multiple angles.
❯ Use inverse trigonometric functions to solve trigonometric equations.

Introduction

To solve a trigonometric equation, use standard algebraic techniques (when possible) such as collecting like terms, extracting square roots, and factoring. Your preliminary goal in solving a trigonometric equation is to *isolate* the trigonometric function on one side of the equation. For example, to solve the equation $2 \sin x = 1$, divide each side by 2 to obtain $\sin x = \frac{1}{2}$. To solve for x, note in Figure 7.2 that the equation $\sin x = \frac{1}{2}$ has solutions $x = \pi/6$ and $x = 5\pi/6$ in the interval $[0, 2\pi)$. Moreover, because $\sin x$ has a period of 2π, there are infinitely many other solutions, which can be written as

$$x = \frac{\pi}{6} + 2n\pi \quad \text{and} \quad x = \frac{5\pi}{6} + 2n\pi \qquad \text{General solution}$$

where n is an integer. The solutions when $n = \pm 1$ and $n = 0$ are shown in Figure 7.2.

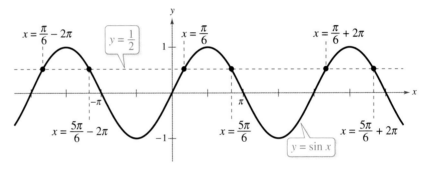

Figure 7.2

Another way to show that the equation $\sin x = \frac{1}{2}$ has infinitely many solutions is shown in the figure below. Any angles that are coterminal with $\pi/6$ or $5\pi/6$ are also solutions of the equation.

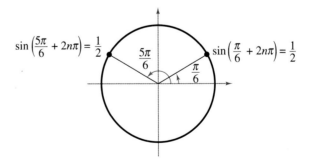

When solving some trigonometric equations, you will be able to find exact solutions because the angles involved are special angles. When solving more general trigonometric equations, you may need to write the solution using inverse trigonometric functions or a calculator.

EXAMPLE 1 **Collecting Like Terms**

Solve $\sin x + \sqrt{2} = -\sin x$.

Solution

Begin by isolating $\sin x$ on one side of the equation.

$$\sin x + \sqrt{2} = -\sin x \qquad \text{Write original equation.}$$

$$\sin x + \sin x + \sqrt{2} = 0 \qquad \text{Add } \sin x \text{ to each side.}$$

$$\sin x + \sin x = -\sqrt{2} \qquad \text{Subtract } \sqrt{2} \text{ from each side.}$$

$$2 \sin x = -\sqrt{2} \qquad \text{Combine like terms.}$$

$$\sin x = -\frac{\sqrt{2}}{2} \qquad \text{Divide each side by 2.}$$

The period of $\sin x$ is 2π, so first find all solutions in the interval $[0, 2\pi)$. These solutions are $x = 5\pi/4$ and $x = 7\pi/4$. Finally, add multiples of 2π to each of these solutions to obtain the general solution

$$x = \frac{5\pi}{4} + 2n\pi \quad \text{and} \quad x = \frac{7\pi}{4} + 2n\pi \qquad \text{General solution}$$

where n is an integer.

✓ *Checkpoint*  *Audio-video solution in English & Spanish at LarsonPrecalculus.com*

Solve $\sin x - \sqrt{2} = -\sin x$.

EXAMPLE 2 **Extracting Square Roots**

Solve $3 \tan^2 x - 1 = 0$.

Solution

Begin by isolating $\tan x$ on one side of the equation.

$$3 \tan^2 x - 1 = 0 \qquad \text{Write original equation.}$$

$$3 \tan^2 x = 1 \qquad \text{Add 1 to each side.}$$

$$\tan^2 x = \frac{1}{3} \qquad \text{Divide each side by 3.}$$

$$\tan x = \pm\frac{1}{\sqrt{3}} \qquad \text{Extract square roots.}$$

$$\tan x = \pm\frac{\sqrt{3}}{3} \qquad \text{Rationalize the denominator.}$$

ALGEBRA HELP

When you extract square roots, make sure you account for both the positive and negative solutions.

The period of $\tan x$ is π, so first find all solutions in the interval $[0, \pi)$. These solutions are $x = \pi/6$ and $x = 5\pi/6$. Finally, add multiples of π to each of these solutions to obtain the general solution

$$x = \frac{\pi}{6} + n\pi \quad \text{and} \quad x = \frac{5\pi}{6} + n\pi \qquad \text{General solution}$$

where n is an integer.

✓ *Checkpoint* *Audio-video solution in English & Spanish at LarsonPrecalculus.com*

Solve $4 \sin^2 x - 3 = 0$.

The equations in Examples 1 and 2 involved only one trigonometric function. When two or more functions occur in the same equation, collect all terms on one side and try to separate the functions by factoring or by using appropriate identities. This may produce factors that yield no solutions, as illustrated in Example 3.

EXAMPLE 3   Factoring

Solve $\cot x \cos^2 x = 2 \cot x$.

Solution   Begin by collecting all terms on one side of the equation and factoring.

$$\cot x \cos^2 x = 2 \cot x \qquad \text{Write original equation.}$$
$$\cot x \cos^2 x - 2 \cot x = 0 \qquad \text{Subtract 2 cot } x \text{ from each side.}$$
$$\cot x(\cos^2 x - 2) = 0 \qquad \text{Factor.}$$

Setting the first factor equal to zero, you obtain the equation

$$\cot x = 0. \qquad \text{Set first factor equal to 0.}$$

In the interval $(0, \pi)$, the equation $\cot x = 0$ has the solution $x = \pi/2$. Setting the second factor equal to zero, you obtain

$$\cos^2 x - 2 = 0 \qquad \text{Set second factor equal to 0.}$$
$$\cos^2 x = 2 \qquad \text{Add 2 to each side.}$$
$$\cos x = \pm\sqrt{2}. \qquad \text{Extract square roots.}$$

No solution exists for $\cos x = \pm\sqrt{2}$ because $\pm\sqrt{2}$ are outside the range of the cosine function. The period of $\cot x$ is π, so add multiples of π to $x = \pi/2$ to get the general solution

$$x = \frac{\pi}{2} + n\pi \qquad \text{General solution}$$

where n is an integer. The solutions when $n = -2, -1, 0,$ and 1 are

$$x = -\frac{3\pi}{2}, \quad x = -\frac{\pi}{2}, \quad x = \frac{\pi}{2}, \quad \text{and} \quad x = \frac{3\pi}{2}$$

respectively. Notice that these solutions correspond to the x-intercepts of the graph of $y = \cot x \cos^2 x - 2 \cot x$, as shown below.

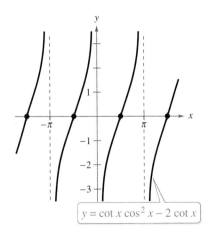

$y = \cot x \cos^2 x - 2 \cot x$

✓ *Checkpoint* ▶ *Audio-video solution in English & Spanish at LarsonPrecalculus.com*

Solve $\sin^2 x = 2 \sin x$.

Equations of Quadratic Type

Some trigonometric equations are of quadratic type $ax^2 + bx + c = 0$. Two examples, one in sine and one in secant, are shown below. To solve equations of this type, factor the quadratic or, when factoring is not possible, use the Quadratic Formula.

Quadratic in sin x	**Quadratic in sec x**
$2\sin^2 x - \sin x - 1 = 0$	$\sec^2 x - 3\sec x - 2 = 0$
$2(\sin x)^2 - (\sin x) - 1 = 0$	$(\sec x)^2 - 3(\sec x) - 2 = 0$

EXAMPLE 4 **Solving an Equation of Quadratic Type**

Find all solutions of $2\sin^2 x - \sin x - 1 = 0$ in the interval $[0, 2\pi)$.

Algebraic Solution

Treat the equation as quadratic in $\sin x$ and factor.

$2\sin^2 x - \sin x - 1 = 0$ Write original equation.

$(2\sin x + 1)(\sin x - 1) = 0$ Factor.

Setting each factor equal to zero, you obtain the following solutions in the interval $[0, 2\pi)$.

$2\sin x + 1 = 0$ or $\sin x - 1 = 0$

$\sin x = -\dfrac{1}{2}$ $\sin x = 1$

$x = \dfrac{7\pi}{6}, \dfrac{11\pi}{6}$ $x = \dfrac{\pi}{2}$

Graphical Solution

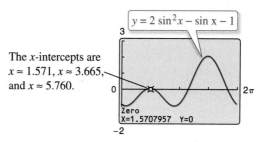

The x-intercepts are $x \approx 1.571$, $x \approx 3.665$, and $x \approx 5.760$.

Use the x-intercepts to conclude that the approximate solutions of $2\sin^2 x - \sin x - 1 = 0$ in the interval $[0, 2\pi)$ are

$x \approx 1.571 \approx \dfrac{\pi}{2}$, $x \approx 3.665 \approx \dfrac{7\pi}{6}$, and $x \approx 5.760 \approx \dfrac{11\pi}{6}$.

 ✓ *Checkpoint* ▶ Audio-video solution in English & Spanish at LarsonPrecalculus.com

Find all solutions of $2\sin^2 x - 3\sin x + 1 = 0$ in the interval $[0, 2\pi)$.

EXAMPLE 5 **Rewriting with a Single Trigonometric Function**

Solve $2\sin^2 x + 3\cos x - 3 = 0$.

Solution This equation contains both sine and cosine functions. Rewrite the equation so that it has only cosine functions by using the identity $\sin^2 x = 1 - \cos^2 x$.

$2\sin^2 x + 3\cos x - 3 = 0$	Write original equation.
$2(1 - \cos^2 x) + 3\cos x - 3 = 0$	Pythagorean identity
$-2\cos^2 x + 3\cos x - 1 = 0$	Combine like terms.
$2\cos^2 x - 3\cos x + 1 = 0$	Multiply each side by -1.
$(2\cos x - 1)(\cos x - 1) = 0$	Factor.

Setting each factor equal to zero, you obtain the solutions $x = 0$, $x = \pi/3$, and $x = 5\pi/3$ in the interval $[0, 2\pi)$. Because $\cos x$ has a period of 2π, the general solution is

$$x = 2n\pi, \quad x = \dfrac{\pi}{3} + 2n\pi, \quad \text{and} \quad x = \dfrac{5\pi}{3} + 2n\pi \qquad \text{General solution}$$

where n is an integer.

✓ *Checkpoint* ▶ Audio-video solution in English & Spanish at LarsonPrecalculus.com

Solve $3\sec^2 x - 2\tan^2 x - 4 = 0$.

In some cases, you can square each side of an equation to obtain an equation of quadratic type, as demonstrated in the next example. This procedure can introduce extraneous solutions, so check any solutions in the original equation to determine whether they are valid or extraneous.

 EXAMPLE 6 **Squaring and Converting to Quadratic Type**

▶〉〉 *See LarsonPrecalculus.com for an interactive version of this type of example.*

Find all solutions of $\cos x + 1 = \sin x$ in the interval $[0, 2\pi)$.

Solution It is not clear how to rewrite this equation in terms of a single trigonometric function. Because the squares of the sine and cosine functions are related by a Pythagorean identity, try squaring each side of the equation and then rewriting the equation so that it has only cosine functions.

$\cos x + 1 = \sin x$	Write original equation.
$\cos^2 x + 2\cos x + 1 = \sin^2 x$	Square each side.
$\cos^2 x + 2\cos x + 1 = 1 - \cos^2 x$	Pythagorean identity
$\cos^2 x + \cos^2 x + 2\cos x + 1 - 1 = 0$	Rewrite equation.
$2\cos^2 x + 2\cos x = 0$	Combine like terms.
$2\cos x(\cos x + 1) = 0$	Factor.

Set each factor equal to zero and solve for x.

$2\cos x = 0$ or $\cos x + 1 = 0$

$\cos x = 0$ $\qquad\qquad$ $\cos x = -1$

$x = \dfrac{\pi}{2}, \dfrac{3\pi}{2}$ $\qquad\qquad$ $x = \pi$

Because you squared the original equation, check for extraneous solutions.

Check $x = \dfrac{\pi}{2}$

$\cos \dfrac{\pi}{2} + 1 \overset{?}{=} \sin \dfrac{\pi}{2}$ $\qquad$ Substitute $\dfrac{\pi}{2}$ for x.

$0 + 1 = 1$ $\qquad$ Solution checks. ✓

Check $x = \dfrac{3\pi}{2}$

$\cos \dfrac{3\pi}{2} + 1 \overset{?}{=} \sin \dfrac{3\pi}{2}$ $\qquad$ Substitute $\dfrac{3\pi}{2}$ for x.

$0 + 1 \neq -1$ $\qquad$ Solution does not check.

Check $x = \pi$

$\cos \pi + 1 \overset{?}{=} \sin \pi$ $\qquad$ Substitute π for x.

$-1 + 1 = 0$ $\qquad$ Solution checks. ✓

Of the three possible solutions, $x = 3\pi/2$ is extraneous. So, in the interval $[0, 2\pi)$, the only two solutions are $x = \pi/2$ and $x = \pi$.

✓ *Checkpoint* ▶ *Audio-video solution in English & Spanish at LarsonPrecalculus.com*

Find all solutions of $\sin x + 1 = \cos x$ in the interval $[0, 2\pi)$. ■

ALGEBRA HELP

In some cases, after squaring each side, you can rewrite a trigonometric equation using the Pythagorean identities

$\sin^2 x + \cos^2 x = 1$

$\tan^2 x + 1 = \sec^2 x$

or

$\cot^2 x + 1 = \csc^2 x.$

GO DIGITAL

Equations Involving Multiple Angles

The next two examples involve trigonometric functions of multiple angles of the forms $\cos ku$ and $\tan ku$. To solve equations involving these forms, first solve the equation for ku, and then divide your result by k.

EXAMPLE 7 **Solving an Equation Involving a Multiple Angle**

Solve $2 \cos 3t - 1 = 0$.

Solution

$$2 \cos 3t - 1 = 0 \qquad \text{Write original equation.}$$

$$2 \cos 3t = 1 \qquad \text{Add 1 to each side.}$$

$$\cos 3t = \frac{1}{2} \qquad \text{Divide each side by 2.}$$

In the interval $[0, 2\pi)$, you know that $3t = \pi/3$ and $3t = 5\pi/3$ are the only solutions, so, in general, you have

$$3t = \frac{\pi}{3} + 2n\pi \quad \text{and} \quad 3t = \frac{5\pi}{3} + 2n\pi.$$

Dividing these results by 3, you obtain the general solution

$$t = \frac{\pi}{9} + \frac{2n\pi}{3} \quad \text{and} \quad t = \frac{5\pi}{9} + \frac{2n\pi}{3} \qquad \text{General solution}$$

where n is an integer.

✓ *Checkpoint* ▶ *Audio-video solution in English & Spanish at LarsonPrecalculus.com*

Solve $2 \sin 2t - \sqrt{3} = 0$.

EXAMPLE 8 **Solving an Equation Involving a Multiple Angle**

$$3 \tan \frac{x}{2} + 3 = 0 \qquad \text{Original equation}$$

$$3 \tan \frac{x}{2} = -3 \qquad \text{Subtract 3 from each side.}$$

$$\tan \frac{x}{2} = -1 \qquad \text{Divide each side by 3.}$$

In the interval $[0, \pi)$, you know that $x/2 = 3\pi/4$ is the only solution, so, in general, you have

$$\frac{x}{2} = \frac{3\pi}{4} + n\pi.$$

Multiplying each side by 2, you obtain the general solution

$$x = \frac{3\pi}{2} + 2n\pi \qquad \text{General solution}$$

where n is an integer.

✓ *Checkpoint* 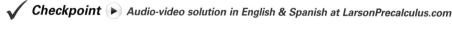 *Audio-video solution in English & Spanish at LarsonPrecalculus.com*

Solve $2 \tan \frac{x}{2} - 2 = 0$.

Using Inverse Functions

EXAMPLE 9 **Using Inverse Functions**

$$\sec^2 x - 2 \tan x = 4 \qquad \text{Original equation}$$
$$1 + \tan^2 x - 2 \tan x - 4 = 0 \qquad \text{Pythagorean identity}$$
$$\tan^2 x - 2 \tan x - 3 = 0 \qquad \text{Combine like terms.}$$
$$(\tan x - 3)(\tan x + 1) = 0 \qquad \text{Factor.}$$

Setting each factor equal to zero, you obtain two solutions in the interval $(-\pi/2, \pi/2)$. [Recall that the range of the inverse tangent function is $(-\pi/2, \pi/2)$.]

$$x = \arctan 3 \quad \text{and} \quad x = \arctan(-1) = -\pi/4$$

Finally, $\tan x$ has a period of π, so add multiples of π to obtain

$$x = \arctan 3 + n\pi \quad \text{and} \quad x = (-\pi/4) + n\pi \qquad \text{General solution}$$

where n is an integer. You can use a calculator to approximate the value of arctan 3.

✓ *Checkpoint* ▶ *Audio-video solution in English & Spanish at LarsonPrecalculus.com*

Solve $4 \tan^2 x + 5 \tan x - 6 = 0$.

EXAMPLE 10 **Using the Quadratic Formula**

Find all solutions of $\sin^2 x + 6 \sin x - 2 = 0$ in the interval $[0, 2\pi)$.

Solution The equation is quadratic in $\sin x$, but it is not factorable. Use the Quadratic Formula with $a = 1$, $b = 6$, and $c = -2$.

$$\sin x = \frac{-6 \pm \sqrt{6^2 - 4(1)(-2)}}{2(1)} \qquad \text{Quadratic Formula}$$

$$\sin x = \frac{-6 \pm \sqrt{44}}{2} \qquad \text{Simplify.}$$

$$\sin x = \frac{-6 \pm 2\sqrt{11}}{2} \qquad \sqrt{44} = \sqrt{4 \cdot 11} = \sqrt{2^2 \cdot 11} = 2\sqrt{11}$$

$$\sin x = -3 \pm \sqrt{11} \qquad \text{Divide out common factor 2.}$$

The range of the sine function is $[-1, 1]$. So, $\sin x = -3 + \sqrt{11} \approx 0.3$ has a solution, whereas $\sin x = -3 - \sqrt{11} \approx -6.3$ does not. To solve $\sin x = -3 + \sqrt{11}$, use a calculator and the inverse sine function to obtain

$$x = \arcsin(-3 + \sqrt{11}) \approx 0.3222. \qquad \text{Use a calculator.}$$

Because $\sin x = -3 + \sqrt{11}$ is positive, the angle x must lie in Quadrant I or Quadrant II. So, the solutions of $\sin^2 x + 6 \sin x - 2 = 0$ in the interval $[0, 2\pi)$ are

$$x \approx 0.3222 \quad \text{and} \quad x = \pi - \arcsin(-3 + \sqrt{11}) \approx 2.8194.$$

Check this with a graphing utility by graphing $y = \sin^2 x + 6 \sin x - 2$ and then identifying the x-intercepts in the interval $[0, 2\pi)$, as shown in Figure 7.3.

✓ *Checkpoint* ▶ *Audio-video solution in English & Spanish at LarsonPrecalculus.com*

Find all solutions of $\sin^2 x + 2 \sin x - 1 = 0$ in the interval $[0, 2\pi)$.

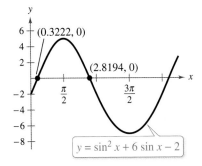

Figure 7.3

GO DIGITAL

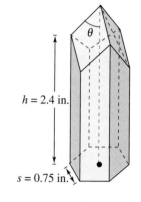

EXAMPLE 11   **Surface Area of a Honeycomb Cell**

The surface area S (in square inches) of a honeycomb cell is given by

$$S = 6hs + 1.5s^2\left(\frac{\sqrt{3} - \cos\theta}{\sin\theta}\right), \quad 0° < \theta \le 90°$$

where $h = 2.4$ inches, $s = 0.75$ inch, and θ is the angle shown in the figure at the right. What value of θ gives the minimum surface area?

$h = 2.4$ in.

$s = 0.75$ in.

Solution

Letting $h = 2.4$ and $s = 0.75$, you obtain

$$S = 10.8 + 0.84375\left(\frac{\sqrt{3} - \cos\theta}{\sin\theta}\right).$$

Graph this function using a graphing utility set in *degree* mode. Use the *minimum* feature to approximate the minimum point on the graph, as shown in the figure below.

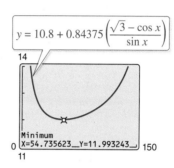

$y = 10.8 + 0.84375\left(\dfrac{\sqrt{3} - \cos x}{\sin x}\right)$

14

Minimum
X=54.735623 Y=11.993243

0   150
11

So, the minimum surface area occurs when $\theta \approx 54.7356°$. By using calculus, it can be shown that the *exact* minimum surface area occurs when $\theta = \arccos\left(1/\sqrt{3}\right)$.

✓ **Checkpoint** ▶ Audio-video solution in English & Spanish at *LarsonPrecalculus.com*

Use the equation for the surface area of a honeycomb cell given in Example 11 with $h = 3.2$ inches and $s = 0.75$ inch. What value of θ gives the minimum surface area?

Summarize (Section 7.3)

1. Explain how to use standard algebraic techniques to solve trigonometric equations *(page 522)*. For examples of using standard algebraic techniques to solve trigonometric equations, see Examples 1–3.

2. Explain how to solve a trigonometric equation of quadratic type *(page 525)*. For examples of solving trigonometric equations of quadratic type, see Examples 4–6.

3. Explain how to solve a trigonometric equation involving multiple angles *(page 527)*. For examples of solving trigonometric equations involving multiple angles, see Examples 7 and 8.

4. Explain how to use inverse trigonometric functions to solve trigonometric equations *(page 528)*. For examples of using inverse trigonometric functions to solve trigonometric equations, see Examples 9–11.

7.3 Exercises

See CalcChat.com for tutorial help and worked-out solutions to odd-numbered exercises.

GO DIGITAL

Vocabulary and Concept Check

In Exercises 1–4, fill in the blanks.

1. When solving a trigonometric equation, the preliminary goal is to _____ the trigonometric function on one side of the equation.

2. The _____ solution of the equation $2 \sin \theta + 1 = 0$ is $\theta = \dfrac{7\pi}{6} + 2n\pi$ and $\theta = \dfrac{11\pi}{6} + 2n\pi$, where n is an integer.

3. The equation $2 \tan^2 x - 3 \tan x + 1 = 0$ is a trigonometric equation of _____ type.

4. A solution of an equation that does not satisfy the original equation is an _____ solution.

5. Explain why $x = 0$ is not a solution of the equation $\cos x = 0$.

6. To solve $\sec x \sin^2 x = \sec x$, do you divide each side by $\sec x$?

Skills and Applications

Verifying Solutions In Exercises 7–12, verify that each x-value is a solution of the equation.

7. $\tan x - \sqrt{3} = 0$
   (a) $x = \dfrac{\pi}{3}$
   (b) $x = \dfrac{4\pi}{3}$

8. $\sec x - 2 = 0$
   (a) $x = \dfrac{\pi}{3}$
   (b) $x = \dfrac{5\pi}{3}$

9. $3 \tan^2 2x - 1 = 0$
   (a) $x = \dfrac{\pi}{12}$
   (b) $x = \dfrac{5\pi}{12}$

10. $2 \cos^2 4x - 1 = 0$
    (a) $x = \dfrac{\pi}{16}$
    (b) $x = \dfrac{3\pi}{16}$

11. $2 \sin^2 x - \sin x - 1 = 0$
    (a) $x = \dfrac{\pi}{2}$
    (b) $x = \dfrac{7\pi}{6}$

12. $\csc^4 x - 4 \csc^2 x = 0$
    (a) $x = \dfrac{\pi}{6}$
    (b) $x = \dfrac{5\pi}{6}$

Solving a Trigonometric Equation In Exercises 13–30, solve the equation.

13. $\sqrt{3} \csc x - 2 = 0$
14. $\tan x + \sqrt{3} = 0$
15. $\cos x + 1 = -\cos x$
16. $3 \sin x + 1 = \sin x$
17. $3 \sec^2 x - 4 = 0$
18. $3 \cot^2 x - 1 = 0$
19. $4 \cos^2 x - 1 = 0$
20. $2 - 4 \sin^2 x = 0$
21. $\sin x(\sin x + 1) = 0$
22. $(2 \sin^2 x - 1)(\tan^2 x - 3) = 0$
23. $\cos^3 x - \cos x = 0$
24. $\sec^2 x - 1 = 0$
25. $3 \tan^3 x = \tan x$
26. $\sec x \csc x = 2 \csc x$
27. $\sec^2 x - \sec x = 2$
28. $\csc^2 x + \csc x = 2$
29. $2 \cos^2 x + \cos x - 1 = 0$
30. $2 \sin^2 x + 3 \sin x + 1 = 0$

Solving a Trigonometric Equation In Exercises 31–40, find all solutions of the equation in the interval $[0, 2\pi)$.

31. $\sin x - 2 = \cos x - 2$
32. $\cos x + \sin x \tan x = 2$
33. $2 \sin^2 x = 2 + \cos x$
34. $\tan^2 x = \sec x - 1$
35. $\sin^2 x = 3 \cos^2 x$
36. $2 \sec^2 x + \tan^2 x - 3 = 0$
37. $2 \sin x + \csc x = 0$
38. $3 \sec x - 4 \cos x = 0$
39. $\csc x + \cot x = 1$
40. $\sec x + \tan x = 1$

Solving an Equation Involving a Multiple Angle In Exercises 41–48, solve the multiple-angle equation.

41. $2 \cos 2x - 1 = 0$
42. $2 \sin 2x + \sqrt{3} = 0$
43. $\tan 3x - 1 = 0$
44. $\sec 4x - 2 = 0$
45. $2 \cos \dfrac{x}{2} - \sqrt{2} = 0$
46. $2 \sin \dfrac{x}{2} + \sqrt{3} = 0$
47. $3 \tan \dfrac{x}{2} - \sqrt{3} = 0$
48. $\tan \dfrac{x}{2} + \sqrt{3} = 0$

Finding x-Intercepts In Exercises 49 and 50, find the x-intercepts of the graph.

49. $y = \sin \dfrac{\pi x}{2} + 1$

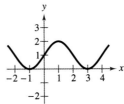

50. $y = \sin \pi x + \cos \pi x$

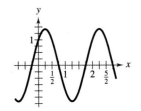

Approximating Solutions   In Exercises 51–60, use a graphing utility to approximate (to three decimal places) the solutions of the equation in the interval $[0, 2\pi)$.

51. $5 \sin x + 2 = 0$

52. $2 \tan x + 7 = 0$

53. $\sin x - 3 \cos x = 0$

54. $\sin x + 4 \cos x = 0$

55. $\cos x = x$

56. $\tan x = \csc x$

57. $\sec^2 x - 3 = 0$

58. $\csc^2 x - 5 = 0$

59. $2 \tan^2 x = 15$

60. $6 \sin^2 x = 5$

Using Inverse Functions   In Exercises 61–72, solve the equation.

61. $\tan^2 x + \tan x - 12 = 0$

62. $\tan^2 x - \tan x - 2 = 0$

63. $\sec^2 x - 6 \tan x = -4$

64. $\sec^2 x + \tan x = 3$

65. $2 \sin^2 x + 5 \cos x = 4$

66. $2 \cos^2 x + 7 \sin x = 5$

67. $\cot^2 x - 9 = 0$

68. $\cot^2 x - 6 \cot x + 5 = 0$

69. $\sec^2 x - 4 \sec x = 0$

70. $\sec^2 x + 2 \sec x - 8 = 0$

71. $\csc^2 x + 3 \csc x - 4 = 0$

72. $\csc^2 x - 5 \csc x = 0$

Using the Quadratic Formula   In Exercises 73–76, use the Quadratic Formula to find all solutions of the equation in the interval $[0, 2\pi)$. Round your result to four decimal places.

73. $12 \sin^2 x - 13 \sin x + 3 = 0$

74. $3 \tan^2 x + 4 \tan x - 4 = 0$

75. $\tan^2 x + 3 \tan x + 1 = 0$

76. $4 \cos^2 x - 4 \cos x - 1 = 0$

Approximating Solutions   In Exercises 77–80, use a graphing utility to approximate (to three decimal places) the solutions of the equation in the given interval.

77. $3 \tan^2 x + 5 \tan x - 4 = 0$, $\left[-\dfrac{\pi}{2}, \dfrac{\pi}{2} \right]$

78. $\cos^2 x - 2 \cos x - 1 = 0$, $[0, \pi]$

79. $4 \cos^2 x - 2 \sin x + 1 = 0$, $\left[-\dfrac{\pi}{2}, \dfrac{\pi}{2} \right]$

80. $2 \sec^2 x + \tan x - 6 = 0$, $\left[-\dfrac{\pi}{2}, \dfrac{\pi}{2} \right]$

Approximating Maximum and Minimum Points   In Exercises 81–86, (a) use a graphing utility to graph the function and approximate the maximum and minimum points on the graph in the interval $[0, 2\pi)$, and (b) solve the trigonometric equation and verify that its solutions are the x-coordinates of the maximum and minimum points of f. (Calculus is required to find the trigonometric equation.)

Function	Trigonometric Equation
81. $f(x) = \sin^2 x + \cos x$	$2 \sin x \cos x - \sin x = 0$
82. $f(x) = \cos^2 x - \sin x$	$-2 \sin x \cos x - \cos x = 0$
83. $f(x) = \sin x + \cos x$	$\cos x - \sin x = 0$
84. $f(x) = 2 \sin x + \cos 2x$	$2 \cos x - 4 \sin x \cos x = 0$
85. $f(x) = \sin x \cos x$	$-\sin^2 x + \cos^2 x = 0$
86. $f(x) = \sec x + \tan x - x$	$\sec x \tan x + \sec^2 x = 1$

Number of Points of Intersection   In Exercises 87 and 88, use the graph to approximate the number of points of intersection of the graphs of y_1 and y_2.

87. $y_1 = 2 \sin x$
$y_2 = 3x + 1$

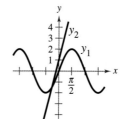

88. $y_1 = 2 \sin x$
$y_2 = \frac{1}{2}x + 1$

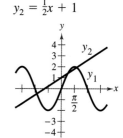

89. Harmonic Motion   A weight is oscillating on the end of a spring (see figure). The displacement from equilibrium of the weight relative to the point of equilibrium is given by $y = \frac{1}{12}(\cos 8t - 3 \sin 8t)$ where y is the displacement (in meters) and t is the time (in seconds). Find the times when the weight is at the point of equilibrium $(y = 0)$ for $0 \le t \le 1$.

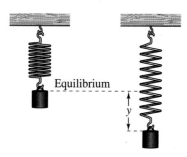

90. Damped Harmonic Motion   The displacement from equilibrium of a weight oscillating on the end of a spring is given by $y = 1.56e^{-0.22t} \cos 4.9t$ where y is the displacement (in feet) and t is the time (in seconds). Use a graphing utility to graph the displacement function for $0 \le t \le 10$. Find the time beyond which the distance between the weight and equilibrium does not exceed 1 foot.

91. Equipment Sales The monthly sales S (in hundreds of units) of skiing equipment at a sports store are approximated by

$$S = 58.3 + 32.5 \cos \frac{\pi t}{6}$$

where t is the time (in months), with $t = 1$ corresponding to January. Determine the months in which sales exceed 7500 units.

92. Projectile Motion A baseball is hit at an angle of θ with the horizontal and with an initial velocity of $v_0 = 100$ feet per second. An outfielder catches the ball 300 feet from home plate (see figure). Find θ when the range r of a projectile is given by

$$r = \frac{1}{32} v_0^2 \sin 2\theta.$$

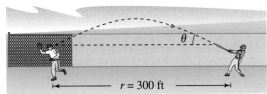

$\vdash$————— $r = 300$ ft ————— $\dashv$

Not drawn to scale

93. Meteorology The table shows the normal daily high temperatures C in Chicago (in degrees Fahrenheit) for month t, with $t = 1$ corresponding to January. *(Source: NOAA)*

DATA	Month, t	Chicago, C
	1	31.0
	2	35.3
	3	46.6
	4	59.0
	5	70.0
	6	79.7
	7	84.1
	8	81.9
	9	74.8
	10	62.3
	11	48.2
	12	34.8

Spreadsheet at LarsonPrecalculus.com

(a) Use a graphing utility to create a scatter plot of the data.

(b) Find a cosine model for the temperatures.

(c) Graph the model and the scatter plot in the same viewing window. How well does the model fit the data?

(d) Use the graphing utility to determine the months during which the normal daily high temperature is above 72°F and below 72°F.

94. Ferris Wheel

The height h (in feet) above ground of a seat on a Ferris wheel at time t (in minutes) can be modeled by

$$h(t) = 53 + 50 \sin\left(\frac{\pi}{16}t - \frac{\pi}{2}\right).$$

The wheel makes one revolution every 32 seconds. The ride begins when $t = 0$.

(a) During the first 32 seconds of the ride, when will a person's seat on the Ferris wheel be 53 feet above ground?

(b) When will a person's seat be at the top of the Ferris wheel for the first time during the ride? For a ride that lasts 160 seconds, how many times will a person's seat be at the top of the ride, and at what times?

95. Geometry The area of a rectangle inscribed in one arc of the graph of $y = \cos x$ (see figure) is given by

$$A = 2x \cos x, \ 0 < x < \pi/2.$$

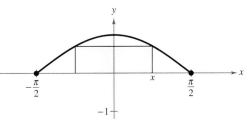

(a) Use a graphing utility to graph the area function, and approximate the area of the largest inscribed rectangle.

(b) Determine the values of x for which $A \geq 1$.

96. Graphical Reasoning Consider the function

$$f(x) = \cos \frac{1}{x}$$

and its graph, shown in the figure below.

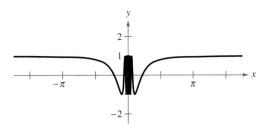

(a) What is the domain of the function?

(b) Identify any symmetry and any asymptotes of the graph.

(c) Describe the behavior of the function as $x \to 0$.

(d) How many solutions does the equation

$$\cos \frac{1}{x} = 0$$

have in the interval $[-1, 1]$? Find the solutions.

(e) Does the equation $\cos(1/x) = 0$ have a greatest solution? If so, then approximate the solution. If not, then explain why.

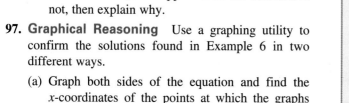 **97. Graphical Reasoning** Use a graphing utility to confirm the solutions found in Example 6 in two different ways.

(a) Graph both sides of the equation and find the x-coordinates of the points at which the graphs intersect.

Left side: $y = \cos x + 1$

Right side: $y = \sin x$

(b) Graph the equation $y = \cos x + 1 - \sin x$ and find the x-intercepts of the graph.

(c) Do both methods produce the same x-values? Which method do you prefer? Explain.

98. Quadratic Approximation Consider the function

$$f(x) = 3 \sin(0.6x - 2).$$

(a) Approximate the zero of the function in the interval $[0, 6]$.

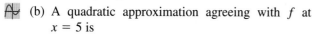 (b) A quadratic approximation agreeing with f at $x = 5$ is

$$g(x) = -0.45x^2 + 5.52x - 13.70.$$

Use a graphing utility to graph f and g in the same viewing window. Describe the result.

(c) Use the Quadratic Formula to find the zeros of g. Compare the zero of g in the interval $[0, 6]$ with the result of part (a).

Fixed Point In Exercises 99 and 100, find the least positive fixed point of the function f. [A *fixed point* of a function f is a real number c such that $f(c) = c$.]

99. $f(x) = \tan(\pi x/4)$

100. $f(x) = \cos x$

Exploring the Concepts

True or False? In Exercises 101 and 102, determine whether the statement is true or false. Justify your answer.

101. The equation $2 \sin 4t - 1 = 0$ has four times the number of solutions in the interval $[0, 2\pi)$ as the equation $2 \sin t - 1 = 0$.

102. The trigonometric equation $\sin x = 3.4$ can be solved using an inverse trigonometric function.

103. Think About It Explain what happens when you divide each side of the equation $\cot x \cos^2 x = 2 \cot x$ by $\cot x$. Is this a correct method to use when solving equations?

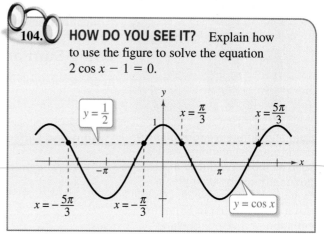 **104. HOW DO YOU SEE IT?** Explain how to use the figure to solve the equation $2 \cos x - 1 = 0$.

Review & Refresh ▶ *Video solutions at LarsonPrecalculus.com*

Completing the Square In Exercises 105–108, solve the equation by completing the square.

105. $x^2 - 4x = 8$

106. $23 = 3(3x^2 - 10x)$

107. $\dfrac{1}{x^2 + 2x - 20} = 1$

108. $\dfrac{13}{4x^2 - 12x + 11} = 4$

Solving an Equation In Exercises 109–112, solve the equation.

109. $(x - 4)^2 = 4x$

110. $\dfrac{-2x^2 + 2x - 4}{5} = -x^2$

111. $x^2 + 2x = \dfrac{2x + 2 - x^{-1}}{x^{-1}}$

112. $\sqrt{7 - 3x} = 2x$

Writing an Expression In Exercises 113–118, write an algebraic expression that is equivalent to the given expression.

113. $\tan\left(\arccos \dfrac{x}{2}\right)$

114. $\sec\left(\arctan \dfrac{4}{x}\right)$

115. $\csc\left(\arccos \dfrac{3}{x}\right)$

116. $\cos\left(\arcsin \dfrac{5}{x}\right)$

117. $\sqrt{4 - x^2} \tan\left(\arcsin \dfrac{x}{2}\right)$

118. $1 - [\sin(\arccos 3x)]^2$

Project: Meteorology To work an extended application analyzing the normal daily high temperatures in Phoenix, Arizona, and in Seattle, Washington, visit this text's website at *LarsonPrecalculus.com*. (*Source: NOAA*)

7.4 Sum and Difference Formulas

Sum and difference formulas are used to model standing waves, such as those produced in a guitar string. For example, in Exercise 82 on page 539, you will use a sum formula to write the equation of a standing wave.

> Use sum and difference formulas to evaluate trigonometric functions, verify identities, and solve trigonometric equations.

Using Sum and Difference Formulas

In this section and the next, you will study the uses of several trigonometric identities and formulas.

Sum and Difference Formulas

$$\sin(u + v) = \sin u \cos v + \cos u \sin v$$

$$\sin(u - v) = \sin u \cos v - \cos u \sin v$$

$$\cos(u + v) = \cos u \cos v - \sin u \sin v$$

$$\cos(u - v) = \cos u \cos v + \sin u \sin v$$

$$\tan(u + v) = \frac{\tan u + \tan v}{1 - \tan u \tan v} \qquad \tan(u - v) = \frac{\tan u - \tan v}{1 + \tan u \tan v}$$

For a proof of the sum and difference formulas for $\cos(u \pm v)$, see Proofs in Mathematics on page 555. The other formulas are left for you to verify (see Exercises 99 and 100).

Examples 1 and 2 show how **sum and difference formulas** enable you to find exact values of trigonometric functions involving sums or differences of special angles.

EXAMPLE 1   **Evaluating a Trigonometric Function**

Find the exact value of $\sin \dfrac{\pi}{12}$.

Solution   Using the fact that $\pi/12 = \pi/3 - \pi/4$ with the formula for $\sin(u - v)$ yields

$$\sin \frac{\pi}{12} = \sin\left(\frac{\pi}{3} - \frac{\pi}{4}\right)$$

$$= \sin \frac{\pi}{3} \cos \frac{\pi}{4} - \cos \frac{\pi}{3} \sin \frac{\pi}{4}$$

$$= \frac{\sqrt{3}}{2}\left(\frac{\sqrt{2}}{2}\right) - \frac{1}{2}\left(\frac{\sqrt{2}}{2}\right)$$

$$= \frac{\sqrt{6}}{4} - \frac{\sqrt{2}}{4}$$

$$= \frac{\sqrt{6} - \sqrt{2}}{4}.$$

Check this result on a calculator by comparing its value to $\sin(\pi/12) \approx 0.2588$.

✓ *Checkpoint* ▶ *Audio-video solution in English & Spanish at LarsonPrecalculus.com*

Find the exact value of $\cos \dfrac{\pi}{12}$.

GO DIGITAL

EXAMPLE 2 **Evaluating a Trigonometric Function**

Find the exact value of cos 75°.

Solution Use the fact that $75° = 30° + 45°$ with the formula for $\cos(u + v)$.

$$\cos 75° = \cos(30° + 45°)$$
$$= \cos 30° \cos 45° - \sin 30° \sin 45°$$
$$= \frac{\sqrt{3}}{2}\left(\frac{\sqrt{2}}{2}\right) - \frac{1}{2}\left(\frac{\sqrt{2}}{2}\right)$$
$$= \frac{\sqrt{6} - \sqrt{2}}{4}$$

✓ **Checkpoint** ▶ Audio-video solution in English & Spanish at LarsonPrecalculus.com

Find the exact value of sin 75°.

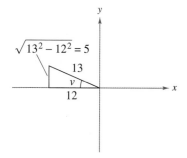

Figure 7.4

EXAMPLE 3 **Evaluating a Trigonometric Expression**

Find the exact value of $\sin(u + v)$ given $\sin u = 4/5$, where $0 < u < \pi/2$, and $\cos v = -12/13$, where $\pi/2 < v < \pi$.

Solution Because $\sin u = 4/5$ and u is in Quadrant I, $\cos u = 3/5$, as shown in Figure 7.4. Because $\cos v = -12/13$ and v is in Quadrant II, $\sin v = 5/13$, as shown in Figure 7.5. Use these values in the formula for $\sin(u + v)$.

$$\sin(u + v) = \sin u \cos v + \cos u \sin v$$
$$= \frac{4}{5}\left(-\frac{12}{13}\right) + \frac{3}{5}\left(\frac{5}{13}\right)$$
$$= -\frac{48}{65} + \frac{15}{65}$$
$$= -\frac{33}{65}$$

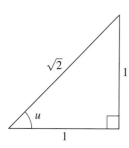

Figure 7.5

✓ **Checkpoint** ▶ Audio-video solution in English & Spanish at LarsonPrecalculus.com

Find the exact value of $\cos(u + v)$ given $\sin u = 12/13$, where $0 < u < \pi/2$, and $\cos v = -3/5$, where $\pi/2 < v < \pi$.

EXAMPLE 4 **An Application of a Sum Formula**

Write $\cos(\arctan 1 + \arccos x)$ as an algebraic expression.

Solution This expression fits the formula for $\cos(u + v)$. Figure 7.6 shows angles $u = \arctan 1$ and $v = \arccos x$.

$$\cos(u + v) = \cos(\arctan 1) \cos(\arccos x) - \sin(\arctan 1) \sin(\arccos x)$$
$$= \frac{1}{\sqrt{2}} \cdot x - \frac{1}{\sqrt{2}} \cdot \sqrt{1 - x^2}$$
$$= \frac{x - \sqrt{1 - x^2}}{\sqrt{2}}$$

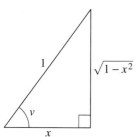

Figure 7.6

GO DIGITAL

✓ **Checkpoint** ▶ Audio-video solution in English & Spanish at LarsonPrecalculus.com

Write $\sin(\arctan 1 + \arccos x)$ as an algebraic expression. ■

HISTORICAL NOTE

Hipparchus, considered the most important of the Greek astronomers, was born about 190 B.C. in Nicaea. He is credited with the invention of trigonometry, and his work contributed to the derivation of the sum and difference formulas for $\sin(A \pm B)$ and $\cos(A \pm B)$.

EXAMPLE 5 **Verifying a Cofunction Identity**

▶▶▶ *See LarsonPrecalculus.com for an interactive version of this type of example.*

Verify the cofunction identity $\cos\left(\dfrac{\pi}{2} - x\right) = \sin x$.

Solution Use the formula for $\cos(u - v)$.

$$\cos\left(\frac{\pi}{2} - x\right) = \cos\frac{\pi}{2}\cos x + \sin\frac{\pi}{2}\sin x$$

$$= (0)(\cos x) + (1)(\sin x)$$

$$= \sin x$$

✓ *Checkpoint* ▶ Audio-video solution in English & Spanish at LarsonPrecalculus.com

Verify the cofunction identity $\sin\left(x - \dfrac{\pi}{2}\right) = -\cos x$. ■

When one of the angles in a sum or difference formula is a multiple of $\pi/2$ or $90°$, such as $\sin(\theta + 270°)$, the formula will reduce to a single-termed expression. The resulting formula is called a **reduction formula** because it reduces the complexity of the original expression, as shown in Example 6.

EXAMPLE 6 **Deriving Reduction Formulas**

Simplify each expression.

a. $\cos\left(\theta - \dfrac{3\pi}{2}\right)$

b. $\tan(\theta + 3\pi)$

Solution

a. Use the formula for $\cos(u - v)$.

$$\cos\left(\theta - \frac{3\pi}{2}\right) = \cos\theta\cos\frac{3\pi}{2} + \sin\theta\sin\frac{3\pi}{2}$$

$$= (\cos\theta)(0) + (\sin\theta)(-1)$$

$$= -\sin\theta$$

b. Use the formula for $\tan(u + v)$.

$$\tan(\theta + 3\pi) = \frac{\tan\theta + \tan 3\pi}{1 - \tan\theta\tan 3\pi}$$

$$= \frac{\tan\theta + 0}{1 - (\tan\theta)(0)}$$

$$= \tan\theta$$

✓ *Checkpoint* ▶ Audio-video solution in English & Spanish at LarsonPrecalculus.com

Simplify each expression.

a. $\sin\left(\dfrac{3\pi}{2} - \theta\right)$

b. $\tan(2\pi - \theta)$ ■

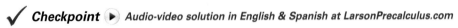

GO DIGITAL

EXAMPLE 7 **Solving a Trigonometric Equation**

Find all solutions of $\sin[x + (\pi/4)] + \sin[x - (\pi/4)] = -1$ in the interval $[0, 2\pi)$.

Algebraic Solution

Use sum and difference formulas to rewrite the equation.

$$\sin x \cos \frac{\pi}{4} + \cos x \sin \frac{\pi}{4} + \sin x \cos \frac{\pi}{4} - \cos x \sin \frac{\pi}{4} = -1$$

$$2 \sin x \cos \frac{\pi}{4} = -1$$

$$2(\sin x)\left(\frac{\sqrt{2}}{2}\right) = -1$$

$$\sin x = -\frac{1}{\sqrt{2}}$$

$$\sin x = -\frac{\sqrt{2}}{2}$$

So, the solutions in the interval $[0, 2\pi)$ are $x = \dfrac{5\pi}{4}$ and $x = \dfrac{7\pi}{4}$.

Graphical Solution

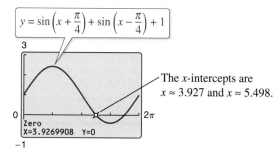

$y = \sin\left(x + \dfrac{\pi}{4}\right) + \sin\left(x - \dfrac{\pi}{4}\right) + 1$

The x-intercepts are $x \approx 3.927$ and $x \approx 5.498$.

Zero
X=3.9269908   Y=0

Use the x-intercepts of

$$y = \sin[x + (\pi/4)] + \sin[x - (\pi/4)] + 1$$

to conclude that the approximate solutions in the interval $[0, 2\pi)$ are

$$x \approx 3.927 \approx \frac{5\pi}{4} \quad \text{and} \quad x \approx 5.498 \approx \frac{7\pi}{4}.$$

✓ **Checkpoint** ▶ Audio-video solution in English & Spanish at LarsonPrecalculus.com

Find all solutions of $\sin[x + (\pi/2)] + \sin[x - (3\pi/2)] = 1$ in the interval $[0, 2\pi)$. ∎

The next example is an application from calculus.

EXAMPLE 8 **An Application from Calculus** ∫

Verify that $\dfrac{\sin(x + h) - \sin x}{h} = (\cos x)\left(\dfrac{\sin h}{h}\right) - (\sin x)\left(\dfrac{1 - \cos h}{h}\right)$, where $h \neq 0$.

Solution Use the formula for $\sin(u + v)$.

$$\frac{\sin(x + h) - \sin x}{h} = \frac{\sin x \cos h + \cos x \sin h - \sin x}{h}$$

$$= \frac{\cos x \sin h - \sin x(1 - \cos h)}{h}$$

$$= (\cos x)\left(\frac{\sin h}{h}\right) - (\sin x)\left(\frac{1 - \cos h}{h}\right)$$

✓ **Checkpoint** ▶ Audio-video solution in English & Spanish at LarsonPrecalculus.com

Verify that $\dfrac{\cos(x + h) - \cos x}{h} = (\cos x)\left(\dfrac{\cos h - 1}{h}\right) - (\sin x)\left(\dfrac{\sin h}{h}\right)$, where $h \neq 0$. ∎

Summarize (Section 7.4)

1. State the sum and difference formulas for sine, cosine, and tangent (*page 534*). For examples of using the sum and difference formulas to evaluate trigonometric functions, verify identities, and solve trigonometric equations, see Examples 1–8.

GO DIGITAL

7.4 Exercises

See CalcChat.com for tutorial help and worked-out solutions to odd-numbered exercises.

GO DIGITAL

Vocabulary and Concept Check

In Exercises 1–6, fill in the blank.

1. $\sin(u - v) =$ _____
2. $\cos(u + v) =$ _____
3. $\tan(u + v) =$ _____
4. $\sin(u + v) =$ _____
5. $\cos(u - v) =$ _____
6. $\tan(u - v) =$ _____

7. Rewrite $\sin(5\pi/12)$ so that you can use a sum formula.

8. Rewrite $\cos 15°$ so that you can use a difference formula.

Skills and Applications

Evaluating Trigonometric Expressions In Exercises 9–12, find the exact value of each expression.

9. (a) $\cos\left(\dfrac{\pi}{4} + \dfrac{\pi}{3}\right)$ (b) $\cos\dfrac{\pi}{4} + \cos\dfrac{\pi}{3}$

10. (a) $\sin\left(\dfrac{7\pi}{6} - \dfrac{\pi}{3}\right)$ (b) $\sin\dfrac{7\pi}{6} - \sin\dfrac{\pi}{3}$

11. (a) $\sin(135° - 30°)$ (b) $\sin 135° - \cos 30°$

12. (a) $\cos(120° + 45°)$ (b) $\cos 120° + \cos 45°$

Evaluating Trigonometric Functions In Exercises 13–28, find the exact values of the sine, cosine, and tangent of the angle.

13. $\dfrac{11\pi}{12} = \dfrac{3\pi}{4} + \dfrac{\pi}{6}$ 14. $\dfrac{7\pi}{12} = \dfrac{\pi}{3} + \dfrac{\pi}{4}$

15. $\dfrac{17\pi}{12} = \dfrac{9\pi}{4} - \dfrac{5\pi}{6}$ 16. $-\dfrac{\pi}{12} = \dfrac{\pi}{6} - \dfrac{\pi}{4}$

17. $105° = 60° + 45°$ 18. $165° = 135° + 30°$

19. $-195° = 30° - 225°$ 20. $255° = 300° - 45°$

21. $13\pi/12$ 22. $19\pi/12$

23. $-5\pi/12$ 24. $-7\pi/12$

25. $285°$ 26. $15°$

27. $-165°$ 28. $-105°$

Rewriting a Trigonometric Expression In Exercises 29–36, write the expression as the sine, cosine, or tangent of an angle.

29. $\sin 3 \cos 1 - \cos 3 \sin 1$

30. $\cos\dfrac{\pi}{7} \cos\dfrac{\pi}{5} - \sin\dfrac{\pi}{7} \sin\dfrac{\pi}{5}$

31. $\cos 130° \cos 40° - \sin 130° \sin 40°$

32. $\sin 60° \cos 15° + \cos 60° \sin 15°$

33. $\dfrac{\tan(\pi/15) + \tan(2\pi/5)}{1 - \tan(\pi/15)\tan(2\pi/5)}$ 34. $\dfrac{\tan 1.1 - \tan 4.6}{1 + \tan 1.1 \tan 4.6}$

35. $\cos 3x \cos 2y + \sin 3x \sin 2y$

36. $\sin x \cos 2x + \cos x \sin 2x$

Evaluating a Trigonometric Expression In Exercises 37–42, find the exact value of the expression.

37. $\sin\dfrac{\pi}{12} \cos\dfrac{\pi}{4} + \cos\dfrac{\pi}{12} \sin\dfrac{\pi}{4}$

38. $\cos\dfrac{\pi}{16} \cos\dfrac{3\pi}{16} - \sin\dfrac{\pi}{16} \sin\dfrac{3\pi}{16}$

39. $\cos 130° \cos 10° + \sin 130° \sin 10°$

40. $\sin 100° \cos 40° - \cos 100° \sin 40°$

41. $\dfrac{\tan(9\pi/8) - \tan(\pi/8)}{1 + \tan(9\pi/8)\tan(\pi/8)}$

42. $\dfrac{\tan 25° + \tan 110°}{1 - \tan 25° \tan 110°}$

Evaluating a Trigonometric Expression In Exercises 43–48, find the exact value of the trigonometric expression given that $\sin u = -\dfrac{3}{5}$, where $3\pi/2 < u < 2\pi$, and $\cos v = \dfrac{15}{17}$, where $0 < v < \pi/2$.

43. $\sin(u + v)$ 44. $\cos(u - v)$

45. $\tan(u + v)$ 46. $\csc(u - v)$

47. $\sec(v - u)$ 48. $\cot(u + v)$

Evaluating a Trigonometric Expression In Exercises 49–54, find the exact value of the trigonometric expression given that $\sin u = -\dfrac{7}{25}$ and $\cos v = -\dfrac{4}{5}$. (Both u and v are in Quadrant III.)

49. $\cos(u + v)$ 50. $\sin(u + v)$

51. $\tan(u - v)$ 52. $\cot(v - u)$

53. $\csc(u - v)$ 54. $\sec(v - u)$

An Application of a Sum or Difference Formula In Exercises 55–58, write the trigonometric expression as an algebraic expression.

55. $\sin(\arcsin x + \arccos x)$

56. $\sin(\arctan 2x - \arccos x)$

57. $\cos(\arccos x + \arcsin x)$

58. $\cos(\arccos x - \arctan x)$

Verifying a Trigonometric Identity In Exercises 59–66, verify the identity.

59. $\sin\left(\dfrac{\pi}{2} - x\right) = \cos x$ **60.** $\sin\left(\dfrac{\pi}{2} + x\right) = \cos x$

61. $\sin\left(\dfrac{\pi}{6} + x\right) = \dfrac{1}{2}\left(\cos x + \sqrt{3}\sin x\right)$

62. $\cos\left(\dfrac{5\pi}{4} - x\right) = -\dfrac{\sqrt{2}}{2}(\cos x + \sin x)$

63. $\tan\left(\theta - \dfrac{\pi}{4}\right) = \dfrac{\tan\theta - 1}{\tan\theta + 1}$

64. $\tan\left(\dfrac{\pi}{4} - \theta\right) = \dfrac{1 - \tan\theta}{1 + \tan\theta}$

65. $\cos(\pi - \theta) + \sin\left(\dfrac{\pi}{2} + \theta\right) = 0$

66. $\cos(x + y)\cos(x - y) = \cos^2 x - \sin^2 y$

Deriving a Reduction Formula In Exercises 67–70, simplify the expression. Use a graphing utility to confirm your answer graphically.

67. $\cos\left(\dfrac{3\pi}{2} - \theta\right)$ **68.** $\sin(\pi + \theta)$

69. $\csc\left(\dfrac{3\pi}{2} + \theta\right)$ **70.** $\cot(\theta - \pi)$

Verifying a Trigonometric Identity In Exercises 71–74, verify the identity.

71. $\cos(n\pi + \theta) = (-1)^n\cos\theta$, n is an integer

72. $\sin(n\pi + \theta) = (-1)^n\sin\theta$, n is an integer

73. $a\sin B\theta + b\cos B\theta = \sqrt{a^2 + b^2}\sin(B\theta + C)$, where $C = \arctan(b/a)$ and $a > 0$

74. $a\sin B\theta + b\cos B\theta = \sqrt{a^2 + b^2}\cos(B\theta - C)$, where $C = \arctan(a/b)$ and $b > 0$

Rewriting a Trigonometric Expression In Exercises 75–78, use the formulas given in Exercises 73 and 74 to write the trigonometric expression in the following forms.

(a) $\sqrt{a^2 + b^2}\sin(B\theta + C)$

(b) $\sqrt{a^2 + b^2}\cos(B\theta - C)$

75. $\sin\theta + \cos\theta$ **76.** $3\sin 2\theta + 4\cos 2\theta$

77. $12\sin 3\theta + 5\cos 3\theta$

78. $\sin 2\theta + \cos 2\theta$

Rewriting a Trigonometric Expression In Exercises 79 and 80, use the formulas given in Exercises 73 and 74 to write the trigonometric expression in the form $a\sin B\theta + b\cos B\theta$.

79. $2\sin[\theta + (\pi/4)]$

80. $5\cos[\theta - (\pi/4)]$

81. Harmonic Motion A weight is attached to a spring suspended vertically from a ceiling. When a driving force is applied to the system, the weight moves vertically from its equilibrium position, and this motion is modeled by

$$y = \dfrac{1}{3}\sin 2t + \dfrac{1}{4}\cos 2t$$

where y is the displacement (in feet) from equilibrium of the weight and t is the time (in seconds).

(a) Use the identity given in Exercise 73 to write the model in the form

$$y = \sqrt{a^2 + b^2}\sin(Bt + C).$$

(b) Find the amplitude of the oscillations of the weight.

(c) Find the frequency of the oscillations of the weight.

82. Standing Waves

The equation of a standing wave is obtained by adding the displacements of two waves traveling in opposite directions (see graphs shown below). Assume that each of the waves has amplitude A, period T, and wavelength λ. The models for two such waves are

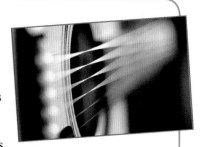

$$y_1 = A\cos 2\pi\left(\dfrac{t}{T} - \dfrac{x}{\lambda}\right) \text{ and } y_2 = A\cos 2\pi\left(\dfrac{t}{T} + \dfrac{x}{\lambda}\right).$$

(a) Show that

$$y_1 + y_2 = 2A\cos\dfrac{2\pi t}{T}\cos\dfrac{2\pi x}{\lambda}.$$

(b) Let $A = 1$ and $\lambda = 1$. For each value of t shown below, write and simplify the models y_1 and y_2 and the sum $y_1 + y_2$. Then use a graphing utility to verify the shapes of the graphs shown.

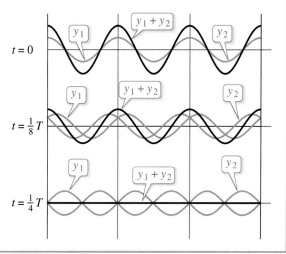

Solving a Trigonometric Equation In Exercises 83–88, find all solutions of the equation in the interval $[0, 2\pi)$.

83. $\sin(x + \pi) - \sin x + 1 = 0$

84. $\cos(x + \pi) - \cos x - 1 = 0$

85. $\cos\left(x + \dfrac{\pi}{4}\right) - \cos\left(x - \dfrac{\pi}{4}\right) = 1$

86. $\sin\left(x + \dfrac{\pi}{6}\right) - \sin\left(x - \dfrac{7\pi}{6}\right) = \dfrac{\sqrt{3}}{2}$

87. $\tan(x + \pi) + 2\sin(x + \pi) = 0$

88. $\sin\left(x + \dfrac{\pi}{2}\right) - \cos^2 x = 0$

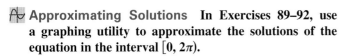 **Approximating Solutions** In Exercises 89–92, use a graphing utility to approximate the solutions of the equation in the interval $[0, 2\pi)$.

89. $\cos[x + (\pi/4)] + \cos[x - (\pi/4)] = 1$

90. $\tan(x + \pi) - \cos[x + (\pi/2)] = 0$

91. $\sin[x + (\pi/2)] + \cos^2 x = 0$

92. $\cos[x - (\pi/2)] - \sin^2 x = 0$

Exploring the Concepts

True or False? In Exercises 93 and 94, determine whether the statement is true or false. Justify your answer.

93. $\sin(u \pm v) = \sin u \cos v \pm \cos u \sin v$

94. $\cos(u \pm v) = \cos u \cos v \pm \sin u \sin v$

95. Error Analysis Describe the error.

$$\tan\left(x + \dfrac{\pi}{4}\right) = \dfrac{\tan x + \tan(\pi/4)}{1 + \tan x \tan(\pi/4)}$$
$$= \dfrac{\tan x + 1}{1 + \tan x}$$
$$= 1 \qquad \textbf{✗}$$

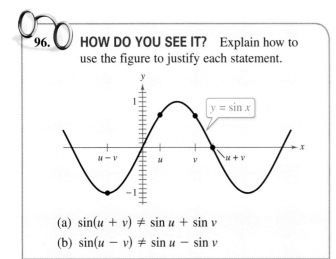

96. **HOW DO YOU SEE IT?** Explain how to use the figure to justify each statement.

(a) $\sin(u + v) \neq \sin u + \sin v$

(b) $\sin(u - v) \neq \sin u - \sin v$

Graphical Reasoning In Exercises 97 and 98, use a graphing utility to graph y_1 and y_2 in the same viewing window. Use the graphs to determine whether $y_1 = y_2$. Explain your reasoning.

97. $y_1 = \cos(x + 2)$, $y_2 = \cos x + \cos 2$

98. $y_1 = \sin(x + 4)$, $y_2 = \sin x + \sin 4$

99. Proof Write a proof of the formula for (a) $\sin(u + v)$ and (b) $\sin(u - v)$.

100. Proof Write a proof of the formula for (a) $\tan(u + v)$ and (b) $\tan(u - v)$.

101. An Application from Calculus Let $x = \pi/3$ in the identity in Example 8 and define the functions f and g as follows.

$$f(h) = \dfrac{\sin[(\pi/3) + h] - \sin(\pi/3)}{h}$$

$$g(h) = \cos\dfrac{\pi}{3}\left(\dfrac{\sin h}{h}\right) - \sin\dfrac{\pi}{3}\left(\dfrac{1 - \cos h}{h}\right)$$

(a) What are the domains of f and g?

(b) Use a graphing utility to complete a table that shows $f(h)$ and $g(h)$ for $h = 0.5, 0.2, 0.1, 0.05, 0.02,$ and 0.01.

(c) Use the graphing utility to graph f and g.

(d) Use the table and the graphs to make a conjecture about the values of f and g as $h \to 0^+$.

Review & Refresh ▶ *Video solutions at LarsonPrecalculus.com*

102. Using Cofunction Identities Use cofunction identities to evaluate each expression without using a calculator.

(a) $\sin^2 25° + \sin^2 65°$

(b) $\tan^2 63° + \cot^2 16° - \sec^2 74° - \csc^2 27°$

Evaluating Trigonometric Functions In Exercises 103–106, find the exact values of the remaining trigonometric functions of θ satisfying the given conditions.

103. $\cos \theta = \dfrac{5}{13}$, θ lies in Quadrant IV.

104. $\sin \theta = \dfrac{3}{5}$, θ lies in Quadrant I.

105. $\tan \theta = -\dfrac{1}{2}$, θ lies in Quadrant II.

106. $\sec \theta = -\dfrac{9}{4}$, θ lies in Quadrant III.

Solving a Trigonometric Equation In Exercises 107–112, solve the equation.

107. $2\cos x = 0$

108. $2\cos x = 1$

109. $1 + \sin x = 0$

110. $\cos x + 1 = 0$

111. $2\cos 3x = 0$

112. $\sin 4x = 0$

7.5 Multiple-Angle and Product-to-Sum Formulas

A variety of trigonometric formulas enable you to rewrite trigonometric equations in more convenient forms. For example, in Exercise 67 on page 549, you will use a half-angle formula to rewrite an equation relating the Mach number of a supersonic airplane to the apex angle of the cone formed by the sound waves behind the airplane.

- ❯ Use multiple-angle formulas to rewrite and evaluate trigonometric functions.
- ❯ Use power-reducing formulas to rewrite trigonometric expressions.
- ❯ Use half-angle formulas to rewrite and evaluate trigonometric functions.
- ❯ Use product-to-sum and sum-to-product formulas to rewrite and evaluate trigonometric expressions.
- ❯ Use trigonometric formulas to rewrite real-life models.

Multiple-Angle Formulas

In this section, you will study four other categories of trigonometric identities.

1. The first category involves *functions of multiple angles* such as $\sin ku$ and $\cos ku$.

2. The second category involves *squares of trigonometric functions* such as $\sin^2 u$.

3. The third category involves *functions of half-angles* such as $\sin(u/2)$.

4. The fourth category involves *products of trigonometric functions* such as $\sin u \cos v$.

You should learn the **double-angle formulas** listed below because they are used often in trigonometry and calculus. For proofs of these formulas, see Proofs in Mathematics on page 555.

Double-Angle Formulas

$$\sin 2u = 2 \sin u \cos u \qquad \cos 2u = \cos^2 u - \sin^2 u$$

$$\tan 2u = \frac{2 \tan u}{1 - \tan^2 u} \qquad\qquad = 2 \cos^2 u - 1$$

$$\qquad\qquad\qquad\qquad\qquad\qquad = 1 - 2 \sin^2 u$$

EXAMPLE 1   **Solving an Equation Involving a Multiple Angle**

Solve $2 \cos x + \sin 2x = 0$.

Solution   Begin by rewriting the equation so that it involves trigonometric functions of a single angle, x, rather than a multiple angle, $2x$. Then factor and solve.

$$2 \cos x + \sin 2x = 0 \qquad \text{Write original equation.}$$

$$2 \cos x + 2 \sin x \cos x = 0 \qquad \text{Double-angle formula}$$

$$2 \cos x(1 + \sin x) = 0 \qquad \text{Factor.}$$

$$2 \cos x = 0 \quad \text{and} \quad 1 + \sin x = 0 \qquad \text{Set factors equal to zero.}$$

$$x = \frac{\pi}{2}, \frac{3\pi}{2} \qquad\qquad x = \frac{3\pi}{2} \qquad \text{Solutions in } [0, 2\pi)$$

So, the general solution is

$$x = \frac{\pi}{2} + 2n\pi \quad \text{and} \quad x = \frac{3\pi}{2} + 2n\pi$$

where n is an integer.

✓ *Checkpoint* ▶ *Audio-video solution in English & Spanish at LarsonPrecalculus.com*

Solve $\cos 2x + \cos x = 0$.

GO DIGITAL

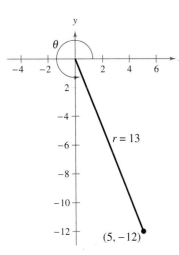

Figure 7.7

EXAMPLE 2   **Evaluating Functions Involving Double Angles**

Use the conditions below to find $\sin 2\theta$, $\cos 2\theta$, and $\tan 2\theta$.

$$\cos \theta = \frac{5}{13}, \quad \frac{3\pi}{2} < \theta < 2\pi$$

Solution   From Figure 7.7,

$$\sin \theta = \frac{y}{r} = -\frac{12}{13} \quad \text{and} \quad \tan \theta = \frac{y}{x} = -\frac{12}{5}.$$

Use these values with each of the double-angle formulas.

$$\sin 2\theta = 2 \sin \theta \cos \theta = 2\left(-\frac{12}{13}\right)\left(\frac{5}{13}\right) = -\frac{120}{169}$$

$$\cos 2\theta = 2 \cos^2 \theta - 1 = 2\left(\frac{25}{169}\right) - 1 = -\frac{119}{169}$$

$$\tan 2\theta = \frac{2 \tan \theta}{1 - \tan^2 \theta} = \frac{2\left(-\dfrac{12}{5}\right)}{1 - \left(-\dfrac{12}{5}\right)^2} = \frac{120}{119}$$

✓ *Checkpoint* ▶ *Audio-video solution in English & Spanish at LarsonPrecalculus.com*

Use the conditions below to find $\sin 2\theta$, $\cos 2\theta$, and $\tan 2\theta$.

$$\sin \theta = \frac{3}{5}, \quad 0 < \theta < \frac{\pi}{2} \qquad\qquad ■$$

The double-angle formulas are not restricted to the angles 2θ and θ. Other *double* combinations, such as 4θ and 2θ or 6θ and 3θ, are also valid. Here are two examples.

$$\sin 4\theta = 2 \sin 2\theta \cos 2\theta \quad \text{and} \quad \cos 6\theta = \cos^2 3\theta - \sin^2 3\theta$$

By using double-angle formulas together with the sum formulas given in the preceding section, you can derive other multiple-angle formulas.

EXAMPLE 3   **Deriving a Triple-Angle Formula**

Rewrite $\sin 3x$ in terms of $\sin x$.

Solution

$$\begin{aligned}
\sin 3x &= \sin(2x + x) &&\text{Rewrite the angle as a sum.}\\
&= \sin 2x \cos x + \cos 2x \sin x &&\text{Sum formula}\\
&= 2 \sin x \cos x \cos x + (1 - 2 \sin^2 x) \sin x &&\text{Double-angle formulas}\\
&= 2 \sin x \cos^2 x + \sin x - 2 \sin^3 x &&\text{Distributive Property}\\
&= 2 \sin x(1 - \sin^2 x) + \sin x - 2 \sin^3 x &&\text{Pythagorean identity}\\
&= 2 \sin x - 2 \sin^3 x + \sin x - 2 \sin^3 x &&\text{Distributive Property}\\
&= 3 \sin x - 4 \sin^3 x &&\text{Simplify.}
\end{aligned}$$

✓ *Checkpoint* ▶ *Audio-video solution in English & Spanish at LarsonPrecalculus.com*

Rewrite $\cos 3x$ in terms of $\cos x$. $\qquad\qquad\qquad\qquad ■$

Power-Reducing Formulas

The double-angle formulas can be used to obtain the **power-reducing formulas** listed below. For proofs of these formulas, see Proofs in Mathematics on page 556.

> **Power-Reducing Formulas**
>
> $$\sin^2 u = \frac{1 - \cos 2u}{2}$$
>
> $$\cos^2 u = \frac{1 + \cos 2u}{2}$$
>
> $$\tan^2 u = \frac{1 - \cos 2u}{1 + \cos 2u}$$

Example 4 shows a typical power reduction used in calculus.

EXAMPLE 4 **Reducing a Power**

Rewrite $\sin^4 x$ in terms of first powers of the cosines of multiple angles.

Solution Note the repeated use of power-reducing formulas.

$$\sin^4 x = (\sin^2 x)^2 \qquad \text{Property of exponents}$$

$$= \left(\frac{1 - \cos 2x}{2}\right)^2 \qquad \text{Power-reducing formula}$$

$$= \frac{1}{4}(1 - 2\cos 2x + \cos^2 2x) \qquad \text{Expand.}$$

$$= \frac{1}{4}\left(1 - 2\cos 2x + \frac{1 + \cos 4x}{2}\right) \qquad \text{Power-reducing formula}$$

$$= \frac{1}{4} - \frac{1}{2}\cos 2x + \frac{1}{8} + \frac{1}{8}\cos 4x \qquad \text{Distributive Property}$$

$$= \frac{3}{8} - \frac{1}{2}\cos 2x + \frac{1}{8}\cos 4x \qquad \text{Add fractions: } \tfrac{1}{4} + \tfrac{1}{8} = \tfrac{2}{8} + \tfrac{1}{8} = \tfrac{3}{8}.$$

$$= \frac{1}{8}(3 - 4\cos 2x + \cos 4x) \qquad \text{Factor out common factor.}$$

You can use a graphing utility to check this result, as shown below. Notice that the graphs coincide.

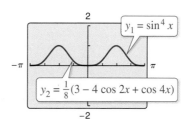

 Checkpoint ▶ *Audio-video solution in English & Spanish at LarsonPrecalculus.com*

Rewrite $\tan^4 x$ in terms of first powers of the cosines of multiple angles. ∎

GO DIGITAL

Half-Angle Formulas

You can derive some useful alternative forms of the power-reducing formulas by replacing u with $u/2$. The results are called **half-angle formulas.**

ALGEBRA HELP

To find the exact value of a trigonometric function with an angle measure in $D° M' S''$ form using a half-angle formula, first convert the angle measure to decimal degree form. Then multiply the resulting angle measure by 2.

Half-Angle Formulas

$$\sin \frac{u}{2} = \pm \sqrt{\frac{1 - \cos u}{2}} \qquad \cos \frac{u}{2} = \pm \sqrt{\frac{1 + \cos u}{2}}$$

$$\tan \frac{u}{2} = \frac{1 - \cos u}{\sin u} = \frac{\sin u}{1 + \cos u}$$

The signs of $\sin \dfrac{u}{2}$ and $\cos \dfrac{u}{2}$ depend on the quadrant in which $\dfrac{u}{2}$ lies.

TECHNOLOGY

Use your calculator to verify the result obtained in Example 5. That is, evaluate $\sin 105°$ and $\left(\sqrt{2 + \sqrt{3}}\right)/2$. Note that both values are approximately 0.9659258.

EXAMPLE 5   Using a Half-Angle Formula

Find the exact value of $\sin 105°$.

Solution   Begin by noting that $105°$ is half of $210°$. Then, use the half-angle formula for $\sin(u/2)$ and the fact that $105°$ lies in Quadrant II.

$$\sin 105° = \sqrt{\frac{1 - \cos 210°}{2}} = \sqrt{\frac{1 + \left(\sqrt{3}/2\right)}{2}} = \frac{\sqrt{2 + \sqrt{3}}}{2}$$

The positive square root is chosen because $\sin \theta$ is positive in Quadrant II.

✓ **Checkpoint** ▶ Audio-video solution in English & Spanish at LarsonPrecalculus.com

Find the exact value of $\cos 105°$.

EXAMPLE 6   Solving a Trigonometric Equation

Find all solutions of $1 + \cos^2 x = 2 \cos^2 \dfrac{x}{2}$ in the interval $[0, 2\pi)$.

Algebraic Solution

$$1 + \cos^2 x = 2 \cos^2 \frac{x}{2} \qquad \text{Write original equation.}$$

$$1 + \cos^2 x = 2 \left(\pm \sqrt{\frac{1 + \cos x}{2}} \right)^2 \qquad \text{Half-angle formula}$$

$$1 + \cos^2 x = 1 + \cos x \qquad \text{Simplify.}$$

$$\cos^2 x - \cos x = 0 \qquad \text{Simplify.}$$

$$\cos x(\cos x - 1) = 0 \qquad \text{Factor.}$$

By setting the factors $\cos x$ and $\cos x - 1$ equal to zero, you find that the solutions in the interval $[0, 2\pi)$ are

$$x = \frac{\pi}{2}, \quad x = \frac{3\pi}{2}, \quad \text{and} \quad x = 0.$$

Graphical Solution

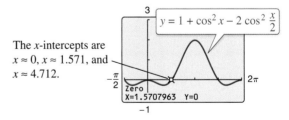

The x-intercepts are $x \approx 0$, $x \approx 1.571$, and $x \approx 4.712$.

Use the x-intercepts of $y = 1 + \cos^2 x - 2 \cos^2(x/2)$ to conclude that the approximate solutions of $1 + \cos^2 x = 2 \cos^2(x/2)$ in the interval $[0, 2\pi)$ are

$$x = 0, \quad x \approx 1.571 \approx \frac{\pi}{2}, \quad \text{and} \quad x \approx 4.712 \approx \frac{3\pi}{2}.$$

✓ **Checkpoint** ▶ Audio-video solution in English & Spanish at LarsonPrecalculus.com

Find all solutions of $\cos^2 x = \sin^2(x/2)$ in the interval $[0, 2\pi)$.

GO DIGITAL

Product-to-Sum and Sum-to-Product Formulas

Each of the **product-to-sum formulas** listed below can be proved using the sum and difference formulas discussed in the preceding section.

Product-to-Sum Formulas

$$\sin u \sin v = \frac{1}{2}[\cos(u - v) - \cos(u + v)]$$

$$\cos u \cos v = \frac{1}{2}[\cos(u - v) + \cos(u + v)]$$

$$\sin u \cos v = \frac{1}{2}[\sin(u + v) + \sin(u - v)]$$

$$\cos u \sin v = \frac{1}{2}[\sin(u + v) - \sin(u - v)]$$

Product-to-sum formulas are used in calculus to solve problems involving the products of sines and cosines of two different angles.

EXAMPLE 7 **Writing Products as Sums**

Rewrite the product $\cos 5x \sin 4x$ as a sum or difference.

Solution Use the product-to-sum formula for $\cos u \sin v$ with $u = 5x$ and $v = 4x$.

$$\cos 5x \sin 4x = \frac{1}{2}[\sin(5x + 4x) - \sin(5x - 4x)]$$

$$= \frac{1}{2} \sin 9x - \frac{1}{2} \sin x$$

✓ *Checkpoint* ▶ *Audio-video solution in English & Spanish at LarsonPrecalculus.com*

Rewrite the product $\sin 5x \cos 3x$ as a sum or difference. ■

Occasionally, it is useful to reverse the procedure and write a sum of trigonometric functions as a product. This can be accomplished with the **sum-to-product formulas** listed below.

Sum-to-Product Formulas

$$\sin u + \sin v = 2 \sin\left(\frac{u + v}{2}\right) \cos\left(\frac{u - v}{2}\right)$$

$$\sin u - \sin v = 2 \cos\left(\frac{u + v}{2}\right) \sin\left(\frac{u - v}{2}\right)$$

$$\cos u + \cos v = 2 \cos\left(\frac{u + v}{2}\right) \cos\left(\frac{u - v}{2}\right)$$

$$\cos u - \cos v = -2 \sin\left(\frac{u + v}{2}\right) \sin\left(\frac{u - v}{2}\right)$$

For a proof of the sum-to-product formulas, see Proofs in Mathematics on page 556.

GO DIGITAL

EXAMPLE 8 **Using a Sum-to-Product Formula**

Find the exact value of $\cos 195° + \cos 105°$.

Solution Use the sum-to-product formula for $\cos u + \cos v$ with $u = 195°$ and $v = 105°$.

$$\cos 195° + \cos 105° = 2 \cos\left(\frac{195° + 105°}{2}\right) \cos\left(\frac{195° - 105°}{2}\right)$$

$$= 2 \cos 150° \cos 45°$$

$$= 2\left(-\frac{\sqrt{3}}{2}\right)\left(\frac{\sqrt{2}}{2}\right)$$

$$= -\frac{\sqrt{6}}{2}$$

✓ *Checkpoint* ▶ Audio-video solution in English & Spanish at LarsonPrecalculus.com

Find the exact value of $\sin 195° + \sin 105°$.

EXAMPLE 9 **Solving a Trigonometric Equation**

▶▶▶ *See LarsonPrecalculus.com for an interactive version of this type of example.*

Solve $\sin 5x + \sin 3x = 0$.

Solution

$$\sin 5x + \sin 3x = 0 \qquad \text{Write original equation.}$$

$$2 \sin\left(\frac{5x + 3x}{2}\right) \cos\left(\frac{5x - 3x}{2}\right) = 0 \qquad \text{Sum-to-product formula}$$

$$2 \sin 4x \cos x = 0 \qquad \text{Simplify.}$$

Set the factor $2 \sin 4x$ equal to zero. The solutions in the interval $[0, 2\pi)$ are

$$x = 0, \frac{\pi}{4}, \frac{\pi}{2}, \frac{3\pi}{4}, \pi, \frac{5\pi}{4}, \frac{3\pi}{2}, \frac{7\pi}{4}.$$

The equation $\cos x = 0$ yields no additional solutions, so the solutions are of the form $x = n\pi/4$, where n is an integer. To check this graphically, sketch the graph of $y = \sin 5x + \sin 3x$, as shown below.

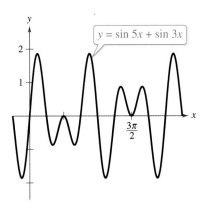

Notice from the graph that the x-intercepts occur at multiples of $\pi/4$.

✓ *Checkpoint* ▶ Audio-video solution in English & Spanish at LarsonPrecalculus.com

Solve $\sin 4x - \sin 2x = 0$.

Application

EXAMPLE 10 Projectile Motion

Ignoring air resistance, the range of a projectile fired at an angle θ with the horizontal and with an initial velocity of v_0 feet per second is given by

$$r = \frac{1}{16} v_0^2 \sin \theta \cos \theta$$

where r is the horizontal distance (in feet) that the projectile travels. A football player kicks a football from ground level with an initial velocity of 80 feet per second.

a. Rewrite the projectile motion model in terms of the first power of the sine of a multiple angle.

b. At what angle must the player kick the football so that the football travels 200 feet?

Solution

a. To rewrite the model in terms of the first power of the sine of a multiple angle, first rewrite it to obtain $2 \sin \theta \cos \theta$. Then use the double-angle formula for $\sin 2\theta$.

$$r = \frac{1}{32} v_0^2 (2 \sin \theta \cos \theta) \qquad \text{Rewrite original model: } \frac{1}{16} = \left(\frac{1}{32}\right)(2).$$

$$= \frac{1}{32} v_0^2 \sin 2\theta. \qquad \text{Double-angle formula}$$

b. $\quad r = \frac{1}{32} v_0^2 \sin 2\theta \qquad\qquad \text{Write projectile motion model.}$

$\quad 200 = \frac{1}{32}(80)^2 \sin 2\theta \qquad \text{Substitute 200 for } r \text{ and 80 for } v_0.$

$\quad 200 = 200 \sin 2\theta \qquad\qquad \text{Simplify.}$

$\quad\quad 1 = \sin 2\theta \qquad\qquad\quad \text{Divide each side by 200.}$

You know that $2\theta = \pi/2$. Dividing this result by 2 produces $\theta = \pi/4$, or 45°. So, the player must kick the football at an angle of 45° so that the football travels 200 feet.

✔ *Checkpoint* ▶ Audio-video solution in English & Spanish at LarsonPrecalculus.com

In Example 10, for what angle is the horizontal distance the football travels a maximum?

Summarize (Section 7.5)

1. State the double-angle formulas *(page 541)*. For examples of using multiple-angle formulas to rewrite and evaluate trigonometric functions, see Examples 1–3.

2. State the power-reducing formulas *(page 543)*. For an example of using power-reducing formulas to rewrite a trigonometric expression, see Example 4.

3. State the half-angle formulas *(page 544)*. For examples of using half-angle formulas to rewrite and evaluate trigonometric functions, see Examples 5 and 6.

4. State the product-to-sum and sum-to-product formulas *(page 545)*. For an example of using a product-to-sum formula to rewrite a trigonometric expression, see Example 7. For examples of using sum-to-product formulas to rewrite and evaluate trigonometric functions, see Examples 8 and 9.

5. Describe an example of how to use a trigonometric formula to rewrite a real-life model *(page 547, Example 10)*.

GO DIGITAL

7.5 Exercises

See CalcChat.com for tutorial help and worked-out solutions to odd-numbered exercises.

Vocabulary and Concept Check

In Exercises 1 and 2, fill in the blank to complete the trigonometric formula.

1. $1 - 2\sin^2 u = $ _____

2. $(1 + \cos 2u)/2 = $ _____

In Exercises 3 and 4, describe a situation in which the given type of formula is useful. Explain.

3. Product-to-sum formula

4. Half-angle formula

Skills and Applications

Solving an Equation Involving a Multiple Angle In Exercises 5–12, solve the equation.

5. $\sin 2x - \sin x = 0$

6. $\sin 2x \sin x = \cos x$

7. $\cos 2x - \cos x = 0$

8. $\cos 2x + \sin x = 0$

9. $\sin 4x = -2\sin 2x$

10. $(\sin 2x + \cos 2x)^2 = 1$

11. $\tan 2x - \cot x = 0$

12. $\tan 2x - 2\cos x = 0$

Using a Double-Angle Formula In Exercises 13–18, use a double-angle formula to rewrite the expression.

13. $6\sin x \cos x$

14. $\sin x \cos x$

15. $6\cos^2 x - 3$

16. $\cos^2 x - \frac{1}{2}$

17. $4 - 8\sin^2 x$

18. $10\sin^2 x - 5$

Evaluating Functions Involving Double Angles In Exercises 19–22, use the given conditions to find the exact values of $\sin 2u$, $\cos 2u$, and $\tan 2u$ using the double-angle formulas.

19. $\sin u = -3/5, \quad 3\pi/2 < u < 2\pi$

20. $\cos u = -4/5, \quad \pi/2 < u < \pi$

21. $\tan u = 3/5, \quad 0 < u < \pi/2$

22. $\sec u = -2, \quad \pi < u < 3\pi/2$

23. Deriving a Multiple-Angle Formula Rewrite $\cos 4x$ in terms of $\cos x$.

24. Deriving a Multiple-Angle Formula Rewrite $\tan 3x$ in terms of $\tan x$.

Reducing Powers In Exercises 25–32, use the power-reducing formulas to rewrite the expression in terms of first powers of the cosines of multiple angles.

25. $\cos^4 x$

26. $\sin^8 x$

27. $\sin^4 2x$

28. $\cos^4 2x$

29. $\tan^4 2x$

30. $\tan^2 2x \cos^4 2x$

31. $\sin^2 2x \cos^2 2x$

32. $\sin^4 x \cos^2 x$

Using Half-Angle Formulas In Exercises 33–38, use the half-angle formulas to determine the exact values of the sine, cosine, and tangent of the angle.

33. $75°$

34. $165°$

35. $112° \, 30'$

36. $67° \, 30'$

37. $\pi/8$

38. $7\pi/12$

Using Half-Angle Formulas In Exercises 39–42, use the given conditions to (a) determine the quadrant in which $u/2$ lies, and (b) find the exact values of $\sin(u/2)$, $\cos(u/2)$, and $\tan(u/2)$ using the half-angle formulas.

39. $\cos u = 7/25, \quad 0 < u < \pi/2$

40. $\sin u = 5/13, \quad \pi/2 < u < \pi$

41. $\tan u = -5/12, \quad 3\pi/2 < u < 2\pi$

42. $\cot u = 3, \quad \pi < u < 3\pi/2$

Solving a Trigonometric Equation In Exercises 43–46, find all solutions of the equation in the interval $[0, 2\pi)$. Use a graphing utility to verify the solutions.

43. $\sin \dfrac{x}{2} + \cos x = 0$

44. $\sin \dfrac{x}{2} + \cos x - 1 = 0$

45. $\cos \dfrac{x}{2} - \sin x = 0$

46. $\tan \dfrac{x}{2} - \sin x = 0$

Using Product-to-Sum Formulas In Exercises 47–50, use the product-to-sum formulas to rewrite the product as a sum or difference.

47. $\sin 5\theta \sin 3\theta$

48. $7\cos(-5\beta)\sin 3\beta$

49. $\cos 2\theta \cos 4\theta$

50. $\sin(x + y)\cos(x - y)$

Using Sum-to-Product Formulas In Exercises 51–54, use the sum-to-product formulas to rewrite the sum or difference as a product.

51. $\sin 5\theta - \sin 3\theta$

52. $\sin 3\theta + \sin \theta$

53. $\cos 6x + \cos 2x$

54. $\cos x + \cos 4x$

Using Sum-to-Product Formulas In Exercises 55–58, use the sum-to-product formulas to find the exact value of the expression.

55. $\sin 75° + \sin 15°$ **56.** $\cos 120° + \cos 60°$

57. $\cos(3\pi/4) - \cos(\pi/4)$ **58.** $\sin(5\pi/4) - \sin(3\pi/4)$

Solving a Trigonometric Equation In Exercises 59–62, find all solutions of the equation in the interval $[0, 2\pi)$. Use a graphing utility to verify the solutions.

59. $\sin 6x + \sin 2x = 0$ **60.** $\cos 2x - \cos 6x = 0$

61. $\dfrac{\cos 2x}{\sin 3x - \sin x} - 1 = 0$ **62.** $\sin^2 3x - \sin^2 x = 0$

Verifying a Trigonometric Identity In Exercises 63–66, verify the identity.

63. $\csc 2\theta = \dfrac{\csc \theta}{2 \cos \theta}$

64. $\cos^4 x - \sin^4 x = \cos 2x$

65. $\dfrac{\sin x \pm \sin y}{\cos x + \cos y} = \tan \dfrac{x \pm y}{2}$

66. $\cos\left(\dfrac{\pi}{3} + x\right) + \cos\left(\dfrac{\pi}{3} - x\right) = \cos x$

67. Mach Number

The Mach number M of a supersonic airplane is the ratio of its speed to the speed of sound. When an airplane travels faster than the speed of sound, the sound waves form a cone behind the airplane. The Mach number is related to the apex angle θ of the cone by $\sin(\theta/2) = 1/M$.

(a) Use a half-angle formula to rewrite the equation in terms of $\cos \theta$.

(b) Find the angle θ that corresponds to a Mach number of 2.

(c) The speed of sound is about 760 miles per hour. Determine the speed of an object with the Mach number from part (b).

68. Projectile Motion The range of a projectile fired at an angle θ with the horizontal and with an initial velocity of v_0 feet per second is $r = \frac{1}{32}v_0^2 \sin 2\theta$ where r is the horizontal distance (in feet) the projectile travels. An athlete throws a javelin at 75 feet per second. At what angle must the athlete throw the javelin so that the javelin travels 130 feet?

69. Complementary Angles Verify each identity for complementary angles ϕ and θ.

(a) $\sin(\phi - \theta) = \cos 2\theta$

(b) $\cos(\phi - \theta) = \sin 2\theta$

70. HOW DO YOU SEE IT? Explain how to use the figure to verify the double-angle formulas (a) $\sin 2u = 2 \sin u \cos u$ and (b) $\cos 2u = \cos^2 u - \sin^2 u$.

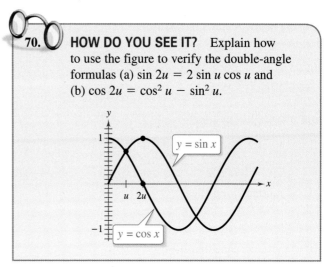

Exploring the Concepts

True or False? In Exercises 71 and 72, determine whether the statement is true or false. Justify your answer.

71. The sine function is an odd function, so

$$\sin(-2x) = -2 \sin x \cos x.$$

72. $\sin\dfrac{u}{2} = -\sqrt{\dfrac{1 - \cos u}{2}}$ when u is in the second quadrant.

Review & Refresh ▶ Video solutions at LarsonPrecalculus.com

Finding Angles of a Triangle In Exercises 73–76, use the given information to find the missing measures of the angles of $\triangle ABC$.

73. $A = 25°, C = 92°$ **74.** $B = C = 30°$

75. $A = 90°, B = C$ **76.** $A = \frac{1}{2}B, C = 3B$

Using a Calculator In Exercises 77–80, use a calculator to evaluate the trigonometric function. Round your answer to four decimal places.

77. $\cos 28°$ **78.** $\csc 140°$

79. $\sec(11\pi/15)$

80. $\tan(-6\pi/5)$

81. Navigation A ship leaves port at noon and has a bearing of S 29° W. The ship sails at 20 knots. (a) How many nautical miles south and how many nautical miles west will the ship have traveled by 6:00 P.M.? (b) At 6:00 P.M., the ship changes course to due west. Find the ship's bearing and distance from port at 7:00 P.M.

Summary and Study Strategies

GO DIGITAL

What Did You Learn?

The list below reviews the skills covered in the chapter and correlates each one to the
Review Exercises (see page 552) that practice the skill.

Section 7.1	Review Exercises

■ Recognize and write the fundamental trigonometric identities *(p. 508)*.   *1–4*

Reciprocal Identities

$\sin u = 1/\csc u \qquad \cos u = 1/\sec u \qquad \tan u = 1/\cot u$

$\csc u = 1/\sin u \qquad \sec u = 1/\cos u \qquad \cot u = 1/\tan u$

Quotient Identities

$\tan u = \dfrac{\sin u}{\cos u} \qquad \cot u = \dfrac{\cos u}{\sin u}$

Pythagorean Identities

$\sin^2 u + \cos^2 u = 1 \qquad 1 + \tan^2 u = \sec^2 u \qquad 1 + \cot^2 u = \csc^2 u$

Cofunction Identities

$\sin[(\pi/2) - u] = \cos u \qquad \cos[(\pi/2) - u] = \sin u$

$\tan[(\pi/2) - u] = \cot u \qquad \cot[(\pi/2) - u] = \tan u$

$\sec[(\pi/2) - u] = \csc u \qquad \csc[(\pi/2) - u] = \sec u$

Even/Odd Identities

$\sin(-u) = -\sin u \qquad \cos(-u) = \cos u \qquad \tan(-u) = -\tan u$

$\csc(-u) = -\csc u \qquad \sec(-u) = \sec u \qquad \cot(-u) = -\cot u$

■ Use the fundamental trigonometric identities to evaluate trigonometric   *5–18*
functions, simplify trigonometric expressions, and rewrite trigonometric
expressions *(p. 509)*.

Section 7.2

■ Verify trigonometric identities *(p. 515)*.   *19–26*

Section 7.3

■ Use standard algebraic techniques to solve trigonometric equations *(p. 522)*.   *27–32*

■ Solve trigonometric equations of quadratic type *(p. 525)*.   *33–36*

■ Solve trigonometric equations involving multiple angles *(p. 527)*.   *37–42*

■ Use inverse trigonometric functions to solve trigonometric equations *(p. 528)*.   *43–46*

Section 7.4

■ Use sum and difference formulas to evaluate trigonometric functions, verify   *47–64*
identities, and solve trigonometric equations *(p. 534)*.

$\sin(u \pm v) = \sin u \cos v \pm \cos u \sin v \qquad \cos(u \pm v) = \cos u \cos v \mp \sin u \sin v$

$\tan(u \pm v) = \dfrac{\tan u \pm \tan v}{1 \mp \tan u \tan v}$

Section 7.5	**Review Exercises**

■ Use multiple-angle formulas to rewrite and evaluate trigonometric functions *(p. 541)*.   *65–68*

$$\sin 2u = 2 \sin u \cos u \qquad \tan 2u = \frac{2 \tan u}{1 - \tan^2 u}$$

$$\cos 2u = \cos^2 u - \sin^2 u = 2 \cos^2 u - 1 = 1 - 2 \sin^2 u$$

■ Use power-reducing formulas to rewrite trigonometric expressions *(p. 543)*.   *69, 70*

$$\sin^2 u = \frac{1 - \cos 2u}{2} \qquad \cos^2 u = \frac{1 + \cos 2u}{2} \qquad \tan^2 u = \frac{1 - \cos 2u}{1 + \cos 2u}$$

■ Use half-angle formulas to rewrite and evaluate trigonometric functions *(p. 544)*.   *71–76*

$$\sin \frac{u}{2} = \pm \sqrt{\frac{1 - \cos u}{2}} \qquad \cos \frac{u}{2} = \pm \sqrt{\frac{1 + \cos u}{2}}$$

$$\tan \frac{u}{2} = \frac{1 - \cos u}{\sin u} = \frac{\sin u}{1 + \cos u}$$

■ Use product-to-sum and sum-to-product formulas to rewrite and evaluate trigonometric expressions *(p. 545)*.   *77–80*

Product-to-Sum Formulas

$$\sin u \sin v = \frac{1}{2}[\cos(u - v) - \cos(u + v)]$$

$$\cos u \cos v = \frac{1}{2}[\cos(u - v) + \cos(u + v)]$$

$$\sin u \cos v = \frac{1}{2}[\sin(u + v) + \sin(u - v)]$$

$$\cos u \sin v = \frac{1}{2}[\sin(u + v) - \sin(u - v)]$$

Sum-to-Product Formulas

$$\sin u + \sin v = 2 \sin\left(\frac{u + v}{2}\right) \cos\left(\frac{u - v}{2}\right)$$

$$\sin u - \sin v = 2 \cos\left(\frac{u + v}{2}\right) \sin\left(\frac{u - v}{2}\right)$$

$$\cos u + \cos v = 2 \cos\left(\frac{u + v}{2}\right) \cos\left(\frac{u - v}{2}\right)$$

$$\cos u - \cos v = -2 \sin\left(\frac{u + v}{2}\right) \sin\left(\frac{u - v}{2}\right)$$

■ Use trigonometric formulas to rewrite real-life models *(p. 547)*.   *81, 82*

Study Strategies

Viewing Math as a Foreign Language   Learning math requires more than just completing homework problems. For instance, learning the material in a chapter may require using approaches similar to those used for learning a foreign language in that you must

- understand and memorize vocabulary words,
- understand and memorize mathematical rules (as you would memorize grammatical rules), and
- apply rules to mathematical expressions or equations (like creating sentences using correct grammar rules).

You should understand the vocabulary words and rules in a chapter as well as memorize and say them out loud. Strive to speak the mathematical language with fluency, just as a student learning a foreign language must.

Review Exercises

See CalcChat.com for tutorial help and worked-out solutions to odd-numbered exercises.

GO DIGITAL

7.1 **Recognizing a Fundamental Identity** In Exercises 1–4, name the trigonometric function that is equivalent to the expression.

1. $\dfrac{\cos x}{\sin x}$

2. $\dfrac{1}{\cos x}$

3. $\sin\left(\dfrac{\pi}{2} - x\right)$

4. $\sqrt{\cot^2 x + 1}$

Using Identities to Evaluate a Function In Exercises 5 and 6, use the given conditions and fundamental trigonometric identities to find the values of all six trigonometric functions.

5. $\cos\theta = -\frac{2}{5}$, $\tan\theta > 0$ 6. $\cot x = -\frac{2}{3}$, $\cos x < 0$

Simplifying a Trigonometric Expression In Exercises 7–16, use the fundamental trigonometric identities to simplify the expression. (There is more than one correct form of each answer.)

7. $\dfrac{1}{\cot^2 x + 1}$

8. $\dfrac{\tan\theta}{1 - \cos^2\theta}$

9. $\tan^2 x(\csc^2 x - 1)$

10. $\cot^2 x(\sin^2 x)$

11. $\dfrac{\cot\left(\dfrac{\pi}{2} - u\right)}{\cos u}$

12. $\dfrac{\sec^2(-\theta)}{\csc^2\theta}$

13. $\cos^2 x + \cos^2 x \cot^2 x$

14. $(\tan x + 1)^2 \cos x$

15. $\dfrac{1}{\csc\theta + 1} - \dfrac{1}{\csc\theta - 1}$

16. $\dfrac{\tan^2 x}{1 + \sec x}$

Trigonometric Substitution In Exercises 17 and 18, use the trigonometric substitution to write the algebraic expression as a trigonometric function of θ, where $0 < \theta < \pi/2$.

17. $\sqrt{25 - x^2}$, $x = 5\sin\theta$ 18. $\sqrt{x^2 - 16}$, $x = 4\sec\theta$

7.2 **Verifying a Trigonometric Identity** In Exercises 19–26, verify the identity.

19. $\cos x(\tan^2 x + 1) = \sec x$

20. $\sec^2 x \cot x - \cot x = \tan x$

21. $\sin\left(\dfrac{\pi}{2} - \theta\right)\tan\theta = \sin\theta$

22. $\cot\left(\dfrac{\pi}{2} - x\right)\csc x = \sec x$

23. $\dfrac{1}{\tan\theta \csc\theta} = \cos\theta$ 24. $\dfrac{1}{\tan x \csc x \sin x} = \cot x$

25. $\sin^5 x \cos^2 x = (\cos^2 x - 2\cos^4 x + \cos^6 x)\sin x$

26. $\cos^3 x \sin^2 x = (\sin^2 x - \sin^4 x)\cos x$

7.3 **Solving a Trigonometric Equation** In Exercises 27–32, solve the equation.

27. $\sin x = \sqrt{3} - \sin x$ 28. $4\cos\theta = 1 + 2\cos\theta$

29. $3\sqrt{3}\tan u = 3$ 30. $\frac{1}{2}\sec x - 1 = 0$

31. $3\csc^2 x = 4$ 32. $4\tan^2 u - 1 = \tan^2 u$

Solving a Trigonometric Equation In Exercises 33–42, find all solutions of the equation in the interval $[0, 2\pi)$.

33. $\sin^3 x = \sin x$ 34. $2\cos^2 x + 3\cos x = 0$

35. $\cos^2 x + \sin x = 1$ 36. $\sin^2 x + 2\cos x = 2$

37. $2\sin 2x - \sqrt{2} = 0$ 38. $2\cos\dfrac{x}{2} + 1 = 0$

39. $3\tan^2\left(\dfrac{x}{3}\right) - 1 = 0$ 40. $\sqrt{3}\tan 3x = 0$

41. $\cos 4x(\cos x - 1) = 0$ 42. $3\csc^2 5x = -4$

Using Inverse Functions In Exercises 43–46, solve the equation.

43. $\tan^2 x - 2\tan x = 0$

44. $2\tan^2 x - 3\tan x = -1$

45. $\tan^2\theta + \tan\theta - 6 = 0$

46. $\sec^2 x + 6\tan x + 4 = 0$

7.4 **Evaluating Trigonometric Functions** In Exercises 47–52, find the exact values of the sine, cosine, and tangent of the angle.

47. $75° = 120° - 45°$ 48. $375° = 135° + 240°$

49. $\dfrac{25\pi}{12} = \dfrac{11\pi}{6} + \dfrac{\pi}{4}$ 50. $\dfrac{19\pi}{12} = \dfrac{11\pi}{6} - \dfrac{\pi}{4}$

51. $-\dfrac{19\pi}{12}$ 52. $195°$

Rewriting a Trigonometric Expression In Exercises 53 and 54, write the expression as the sine, cosine, or tangent of an angle.

53. $\sin 60° \cos 45° - \cos 60° \sin 45°$

54. $\dfrac{\tan 68° - \tan 115°}{1 + \tan 68° \tan 115°}$

Evaluating a Trigonometric Expression In Exercises 55–58, find the exact value of the trigonometric expression given that $\tan u = \frac{3}{4}$ and $\cos v = -\frac{4}{5}$. (u is in Quadrant I and v is in Quadrant III.)

55. $\sin(u + v)$ 56. $\tan(u + v)$

57. $\cos(u - v)$ 58. $\sin(u - v)$

Verifying a Trigonometric Identity In Exercises 59–62, verify the identity.

59. $\cos\left(x + \dfrac{\pi}{2}\right) = -\sin x$ **60.** $\tan\left(x - \dfrac{\pi}{2}\right) = -\cot x$

61. $\tan(\pi - x) = -\tan x$ **62.** $\sin(x - \pi) = -\sin x$

Solving a Trigonometric Equation In Exercises 63 and 64, find all solutions of the equation in the interval $[0, 2\pi)$.

63. $\sin\left(x + \dfrac{\pi}{4}\right) - \sin\left(x - \dfrac{\pi}{4}\right) = 1$

64. $\cos\left(x + \dfrac{\pi}{6}\right) - \cos\left(x - \dfrac{\pi}{6}\right) = 1$

7.5 Evaluating Functions Involving Double Angles In Exercises 65 and 66, use the given conditions to find the exact values of $\sin 2u$, $\cos 2u$, and $\tan 2u$ using the double-angle formulas.

65. $\sin u = \dfrac{4}{5}$, $0 < u < \pi/2$

66. $\cos u = \dfrac{-2}{\sqrt{5}}$, $\pi/2 < u < \pi$

67. Deriving a Multiple-Angle Formula Rewrite $\sin 4x$ in terms of $\sin x$ and $\cos x$.

68. Deriving a Multiple-Angle Formula Rewrite $\tan 4x$ in terms of $\tan x$.

f **Reducing Powers** In Exercises 69 and 70, use the power-reducing formulas to rewrite the expression in terms of first powers of the cosines of multiple angles.

69. $\tan^2 3x$ **70.** $\sin^2 x \cos^2 x$

Using Half-Angle Formulas In Exercises 71 and 72, use the half-angle formulas to determine the exact values of the sine, cosine, and tangent of the angle.

71. $-75°$ **72.** $\dfrac{5\pi}{12}$

Using Half-Angle Formulas In Exercises 73–76, use the given conditions to (a) determine the quadrant in which $u/2$ lies, and (b) find the exact values of $\sin(u/2)$, $\cos(u/2)$, and $\tan(u/2)$ using the half-angle formulas.

73. $\tan u = \dfrac{4}{3}$, $\pi < u < \dfrac{3\pi}{2}$

74. $\sin u = \dfrac{3}{5}$, $0 < u < \dfrac{\pi}{2}$

75. $\cos u = -\dfrac{2}{7}$, $\dfrac{\pi}{2} < u < \pi$

76. $\tan u = -\dfrac{\sqrt{21}}{2}$, $\dfrac{3\pi}{2} < u < 2\pi$

Using Product-to-Sum Formulas In Exercises 77 and 78, use the product-to-sum formulas to rewrite the product as a sum or difference.

77. $\cos 4\theta \sin 6\theta$

78. $2 \sin 7\theta \cos 3\theta$

Using Sum-to-Product Formulas In Exercises 79 and 80, use the sum-to-product formulas to rewrite the sum or difference as a product.

79. $\cos 6\theta + \cos 5\theta$

80. $\sin 3x - \sin x$

81. Projectile Motion A baseball leaves the hand of a player at first base at an angle of θ with the horizontal and at an initial velocity of $v_0 = 80$ feet per second. A player at second base 100 feet away catches the ball. Find θ when the range r of a projectile is

$$r = \dfrac{1}{32}v_0^2 \sin 2\theta.$$

82. Geometry A trough for feeding cattle is 4 meters long and its cross sections are isosceles triangles with the two equal sides being $\frac{1}{2}$ meter (see figure). The angle between the two sides is θ.

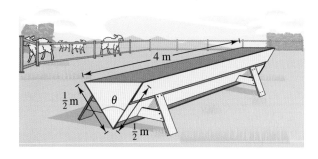

(a) Write the volume of the trough as a function of $\theta/2$.

(b) Write the volume of the trough as a function of θ and determine the value of θ such that the volume is maximized.

Exploring the Concepts

True or False? In Exercises 83–86, determine whether the statement is true or false. Justify your answer.

83. If $\dfrac{\pi}{2} < \theta < \pi$, then $\cos \dfrac{\theta}{2} < 0$.

84. $\cot x \sin^2 x = \cos x \sin x$

85. $4 \sin(-x) \cos(-x) = 2 \sin 2x$

86. $4 \sin 45° \cos 15° = 1 + \sqrt{3}$

87. Think About It Is it possible for a trigonometric equation that is not an identity to have an infinite number of solutions? Explain.

Chapter Test See CalcChat.com for tutorial help and worked-out solutions to odd-numbered exercises.

GO DIGITAL

Take this test as you would take a test in class. When you are finished, check your work against the answers given in the back of the book.

1. Use the conditions $\csc \theta = \frac{5}{2}$ and $\tan \theta < 0$ to find the values of all six trigonometric functions. *(Section 7.1)*

2. Use the fundamental identities to simplify $\csc^2 \beta (1 - \cos^2 \beta)$. *(Section 7.1)*

3. Factor and simplify $\dfrac{\sec^4 x - \tan^4 x}{\sec^2 x + \tan^2 x}$. *(Section 7.1)*

4. Add and simplify $\dfrac{\cos \theta}{\sin \theta} + \dfrac{\sin \theta}{\cos \theta}$. *(Section 7.1)*

In Exercises 5–10, verify the identity. *(Sections 7.2 and 7.5)*

5. $\sin \theta \sec \theta = \tan \theta$

6. $\sec^2 x \tan^2 x + \sec^2 x = \sec^4 x$

7. $\dfrac{\csc \alpha + \sec \alpha}{\sin \alpha + \cos \alpha} = \cot \alpha + \tan \alpha$

8. $\tan\left(x + \dfrac{\pi}{2}\right) = -\cot x$

9. $1 + \cos 10y = 2 \cos^2 5y$

10. $\sin \dfrac{\alpha}{3} \cos \dfrac{\alpha}{3} = \dfrac{1}{2} \sin \dfrac{2\alpha}{3}$

11. Rewrite $4 \sin 3\theta \cos 2\theta$ as a sum or difference. *(Section 7.5)*

12. Rewrite $\cos\left(\theta + \dfrac{\pi}{2}\right) - \cos\left(\theta - \dfrac{\pi}{2}\right)$ as a product. *(Section 7.5)*

In Exercises 13–16, find all solutions of the equation in the interval $[0, 2\pi)$. *(Section 7.3)*

13. $\tan^2 x + \tan x = 0$

14. $\sin 2\alpha - \cos \alpha = 0$

15. $4 \cos^2 x - 3 = 0$

16. $\csc^2 x - \csc x - 2 = 0$

17. Use a graphing utility to approximate (to three decimal places) the solutions of $5 \sin x - x = 0$ in the interval $[0, 2\pi)$. *(Section 7.3)*

18. Find the exact value of $\cos 105°$ using the fact that $105° = 135° - 30°$. *(Section 7.4)*

19. Use the figure to find the exact values of $\sin 2u$, $\cos 2u$, and $\tan 2u$. *(Section 7.5)*

20. Cheyenne, Wyoming, has a latitude of $41°$ N. At this latitude, the number of hours of daylight D can be modeled by

$$D = 2.914 \sin(0.017t - 1.321) + 12.134$$

where t represents the day, with $t = 1$ corresponding to January 1. Use a graphing utility to determine the days on which there are more than 10 hours of daylight. *(Source: U.S. Naval Observatory)* *(Section 7.3)*

21. The heights h_1 and h_2 (in feet) above ground of two people in different seats on a Ferris wheel can be modeled by

$$h_1 = 28 \cos 10t + 38$$

and

$$h_2 = 28 \cos\left[10\left(t - \dfrac{\pi}{6}\right)\right] + 38, \quad 0 \le t \le 2$$

where t represents the time (in minutes). When are the two people at the same height? *(Section 7.5)*

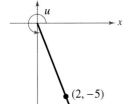

$(2, -5)$

Figure for 19

Proofs in Mathematics

> ### Sum and Difference Formulas (p. 534)
>
> $$\sin(u + v) = \sin u \cos v + \cos u \sin v \qquad \tan(u + v) = \frac{\tan u + \tan v}{1 - \tan u \tan v}$$
> $$\sin(u - v) = \sin u \cos v - \cos u \sin v$$
> $$\cos(u + v) = \cos u \cos v - \sin u \sin v \qquad \tan(u - v) = \frac{\tan u - \tan v}{1 + \tan u \tan v}$$
> $$\cos(u - v) = \cos u \cos v + \sin u \sin v$$

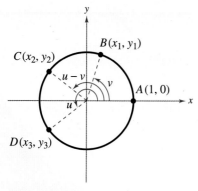

Figure 1

Proof

In the proofs of the formulas for $\cos(u \pm v)$, assume that $0 < v < u < 2\pi$. Figure 1 uses u and v to locate the points $B(x_1, y_1)$, $C(x_2, y_2)$, and $D(x_3, y_3)$ on the unit circle. So, $x_i^2 + y_i^2 = 1$ for $i = 1, 2,$ and 3. In Figure 2, note that arcs AC and BD have the same length. So, line segments AC and BD are also equal in length. This leads to the following.

$$\sqrt{(x_2 - 1)^2 + (y_2 - 0)^2} = \sqrt{(x_3 - x_1)^2 + (y_3 - y_1)^2}$$
$$x_2^2 - 2x_2 + 1 + y_2^2 = x_3^2 - 2x_1x_3 + x_1^2 + y_3^2 - 2y_1y_3 + y_1^2$$
$$(x_2^2 + y_2^2) + 1 - 2x_2 = (x_3^2 + y_3^2) + (x_1^2 + y_1^2) - 2x_1x_3 - 2y_1y_3$$
$$1 + 1 - 2x_2 = 1 + 1 - 2x_1x_3 - 2y_1y_3$$
$$x_2 = x_3x_1 + y_3y_1$$

Substitute the values $x_2 = \cos(u - v)$, $x_3 = \cos u$, $x_1 = \cos v$, $y_3 = \sin u$, and $y_1 = \sin v$ to obtain $\cos(u - v) = \cos u \cos v + \sin u \sin v$. To establish the formula for $\cos(u + v)$, consider $u + v = u - (-v)$ and use the formula just derived to obtain

$$\cos(u + v) = \cos[u - (-v)]$$
$$= \cos u \cos(-v) + \sin u \sin(-v)$$
$$= \cos u \cos v - \sin u \sin v. \qquad \blacksquare$$

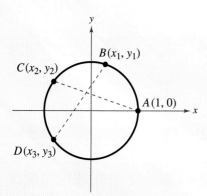

Figure 2

> ### Double-Angle Formulas (p. 541)
>
> $$\sin 2u = 2 \sin u \cos u \qquad\qquad \cos 2u = \cos^2 u - \sin^2 u$$
> $$\tan 2u = \frac{2 \tan u}{1 - \tan^2 u} \qquad\qquad\quad = 2 \cos^2 u - 1$$
> $$= 1 - 2 \sin^2 u$$

Proof
Prove each double-angle formula by letting $v = u$ in the corresponding sum formula.

$$\sin 2u = \sin(u + u) = \sin u \cos u + \cos u \sin u = 2 \sin u \cos u$$
$$\cos 2u = \cos(u + u) = \cos u \cos u - \sin u \sin u = \cos^2 u - \sin^2 u$$
$$\tan 2u = \tan(u + u) = \frac{\tan u + \tan u}{1 - \tan u \tan u} = \frac{2 \tan u}{1 - \tan^2 u} \qquad \blacksquare$$

HISTORICAL NOTE

Early astronomers used trigonometry to calculate measurements in the universe. For instance, they used trigonometry to calculate the circumference of Earth and the distance from Earth to the moon. Another major accomplishment in astronomy using trigonometry was computing distances to stars.

Power-Reducing Formulas   (p. 543)

$$\sin^2 u = \frac{1 - \cos 2u}{2} \qquad \cos^2 u = \frac{1 + \cos 2u}{2} \qquad \tan^2 u = \frac{1 - \cos 2u}{1 + \cos 2u}$$

Proof

Prove the first formula by solving for $\sin^2 u$ in $\cos 2u = 1 - 2 \sin^2 u$.

$$\cos 2u = 1 - 2 \sin^2 u \qquad \text{Write double-angle formula.}$$

$$2 \sin^2 u = 1 - \cos 2u \qquad \text{Subtract } \cos 2u \text{ from, and add } 2 \sin^2 u \text{ to, each side.}$$

$$\sin^2 u = \frac{1 - \cos 2u}{2} \qquad \text{Divide each side by 2.}$$

Similarly, to prove the second formula, solve for $\cos^2 u$ in $\cos 2u = 2 \cos^2 u - 1$. To prove the third formula, use a quotient identity.

$$\tan^2 u = \frac{\sin^2 u}{\cos^2 u} \qquad \text{Quotient identity}$$

$$= \frac{\dfrac{1 - \cos 2u}{2}}{\dfrac{1 + \cos 2u}{2}} \qquad \text{Power-reducing formulas for } \sin^2 u \text{ and } \cos^2 u$$

$$= \frac{1 - \cos 2u}{1 + \cos 2u} \qquad \text{Simplify.} \qquad ■$$

Sum-to-Product Formulas   (p. 545)

$$\sin u + \sin v = 2 \sin\left(\frac{u + v}{2}\right) \cos\left(\frac{u - v}{2}\right)$$

$$\sin u - \sin v = 2 \cos\left(\frac{u + v}{2}\right) \sin\left(\frac{u - v}{2}\right)$$

$$\cos u + \cos v = 2 \cos\left(\frac{u + v}{2}\right) \cos\left(\frac{u - v}{2}\right)$$

$$\cos u - \cos v = -2 \sin\left(\frac{u + v}{2}\right) \sin\left(\frac{u - v}{2}\right)$$

Proof

To prove the first formula, let $x = u + v$ and $y = u - v$. Then substitute $u = (x + y)/2$ and $v = (x - y)/2$ in the product-to-sum formula.

$$\sin u \cos v = \frac{1}{2}[\sin(u + v) + \sin(u - v)]$$

$$\sin\left(\frac{x + y}{2}\right) \cos\left(\frac{x - y}{2}\right) = \frac{1}{2}(\sin x + \sin y)$$

$$2 \sin\left(\frac{x + y}{2}\right) \cos\left(\frac{x - y}{2}\right) = \sin x + \sin y$$

The other sum-to-product formulas can be proved in a similar manner.   ■

P.S. Problem Solving

See CalcChat.com for tutorial help and worked-out solutions to odd-numbered exercises.

GO DIGITAL

1. Writing Trigonometric Functions in Terms of Cosine Write each of the other trigonometric functions of θ in terms of $\cos \theta$.

2. Verifying a Trigonometric Identity Verify that for all integers n,

$$\cos\left[\frac{(2n+1)\pi}{2}\right] = 0.$$

3. Verifying a Trigonometric Identity Verify that for all integers n,

$$\sin\left[\frac{(12n+1)\pi}{6}\right] = \frac{1}{2}.$$

4. Sound Wave A sound wave is modeled by

$$p(t) = \frac{1}{4\pi}[p_1(t) + 30p_2(t) + p_3(t) + p_5(t) + 30p_6(t)]$$

where $p_n(t) = \frac{1}{n}\sin(524n\pi t)$, and t represents the time (in seconds).

(a) Find the sine components $p_n(t)$ and use a graphing utility to graph the components. Then verify the graph of p shown below.

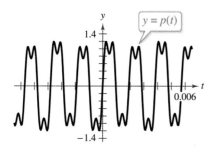

(b) Find the period of each sine component of p. Is p periodic? If so, then what is its period?

(c) Use the graphing utility to find the t-intercepts of the graph of p over one cycle.

(d) Use the graphing utility to approximate the absolute maximum and absolute minimum values of p over one cycle.

5. Geometry Three squares of side length s are placed side by side (see figure). Make a conjecture about the relationship between the sum $u + v$ and w. Prove your conjecture by using the identity for the tangent of the sum of two angles.

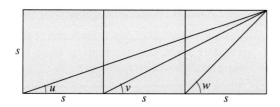

6. Projectile Motion The path traveled by an object (neglecting air resistance) that is projected at an initial height of h_0 feet, an initial velocity of v_0 feet per second, and an initial angle θ is given by

$$y = -\frac{16}{v_0^2 \cos^2 \theta}x^2 + (\tan \theta)x + h_0$$

where the horizontal distance x and the vertical distance y are measured in feet. Find a formula for the maximum height of an object projected from ground level at velocity v_0 and angle θ. To do this, find half of the horizontal distance

$$\frac{1}{32}v_0^2 \sin 2\theta$$

and then substitute it for x in the model for the path of a projectile (where $h_0 = 0$).

7. Geometry The length of each of the two equal sides of an isosceles triangle is 10 meters (see figure). The angle between the two sides is θ.

(a) Write the area of the triangle as a function of $\theta/2$.

(b) Write the area of the triangle as a function of θ. Determine the value of θ such that the area is a maximum.

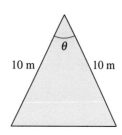

Figure for 7   Figure for 8

8. Geometry Use the figure to derive the formulas for

$$\sin \frac{\theta}{2}, \cos \frac{\theta}{2}, \text{ and } \tan \frac{\theta}{2}$$

where θ is an acute angle.

9. Force The force F (in pounds) on a person's back when he or she bends over at an angle θ from an upright position is modeled by

$$F = \frac{0.6W \sin(\theta + 90°)}{\sin 12°}$$

where W represents the person's weight (in pounds).

(a) Simplify the model.

(b) Use a graphing utility to graph the model, where $W = 185$ and $0° \leq \theta \leq 90°$.

(c) At what angle is the force maximized? At what angle is the force minimized?

10. Hours of Daylight The number of hours of daylight that occur at any location on Earth depends on the time of year and the latitude of the location. The equations below model the numbers of hours of daylight in Seward, Alaska (60° latitude), and New Orleans, Louisiana (30° latitude).

$$D = 12.2 - 6.4 \cos\left[\frac{\pi(t + 0.2)}{182.6}\right] \qquad \text{Seward}$$

$$D = 12.2 - 1.9 \cos\left[\frac{\pi(t + 0.2)}{182.6}\right] \qquad \text{New Orleans}$$

In these models, D represents the number of hours of daylight and t represents the day, with $t = 0$ corresponding to January 1.

(a) Use a graphing utility to graph both models in the same viewing window. Use a viewing window of $0 \le t \le 365$.

(b) Find the days of the year on which both cities receive the same amount of daylight.

(c) Which city has the greater variation in the number of hours of daylight? Which constant in each model would you use to determine the difference between the greatest and least numbers of hours of daylight?

(d) Determine the period of each model.

11. Ocean Tide The tide, or depth of the ocean near the shore, changes throughout the day. The water depth d (in feet) of a bay can be modeled by

$$d = 35 - 28 \cos \frac{\pi}{6.2} t$$

where t represents the time in hours, with $t = 0$ corresponding to 12:00 A.M.

(a) Algebraically find the times at which the high and low tides occur.

(b) If possible, algebraically find the time(s) at which the water depth is 3.5 feet.

(c) Use a graphing utility to verify your results from parts (a) and (b).

12. Piston Heights The heights h (in inches) of pistons 1 and 2 in an automobile engine can be modeled by

$$h_1 = 3.75 \sin 733t + 7.5$$

and

$$h_2 = 3.75 \sin 733\left(t + \frac{4\pi}{3}\right) + 7.5$$

respectively, where t is measured in seconds.

(a) Use a graphing utility to graph the heights of these pistons in the same viewing window for $0 \le t \le 1$.

(b) How often are the pistons at the same height?

13. Index of Refraction The index of refraction n of a transparent material is the ratio of the speed of light in a vacuum to the speed of light in the material. Some common materials and their indices of refraction are air (1.00), water (1.33), and glass (1.50). Triangular prisms are often used to measure the index of refraction based on the formula

$$n = \frac{\sin\left(\dfrac{\theta}{2} + \dfrac{\alpha}{2}\right)}{\sin\dfrac{\theta}{2}}.$$

For the prism shown in the figure, $\alpha = 60°$.

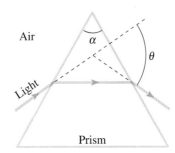

(a) Write the index of refraction as a function of $\cot(\theta/2)$.

(b) Find θ for a prism made of glass.

14. Sum Formulas

(a) Write a sum formula for $\sin(u + v + w)$.

(b) Write a sum formula for $\tan(u + v + w)$.

15. Solving Trigonometric Inequalities Find the solution of each inequality in the interval $[0, 2\pi)$.

(a) $\sin x \ge 0.5$ (b) $\cos x \le -0.5$

(c) $\tan x < \sin x$ (d) $\cos x \ge \sin x$

16. Sum of Fourth Powers Consider the function $f(x) = \sin^4 x + \cos^4 x$.

(a) Use the power-reducing formulas to write the function in terms of cosine to the first power.

(b) Determine another way of rewriting the original function. Use a graphing utility to rule out incorrectly rewritten functions.

(c) Add a trigonometric term to the original function so that it becomes a perfect square trinomial. Rewrite the function as a perfect square trinomial minus the term that you added. Use the graphing utility to rule out incorrectly rewritten functions.

(d) Rewrite the result of part (c) in terms of the sine of a double angle. Use the graphing utility to rule out incorrectly rewritten functions.

(e) When you rewrite a trigonometric expression, the result may not be the same as a friend's. Does this mean that one of you is wrong? Explain.

8 Additional Topics in Trigonometry

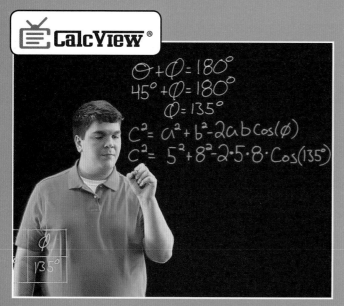

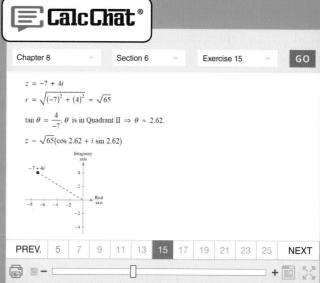

8.1 Distance *(Exercise 46, p. 567)*

8.5 Shipping *(Exercise 43, p. 604)*

559

left, © Drozdin Vladimir/Shutterstock.com; right, © d3sign/Getty Images

8.1 Law of Sines

The Law of Sines is a useful tool for solving real-life problems involving oblique triangles. For example, in Exercise 46 on page 567, you will use the Law of Sines to determine the distance from a boat to a shoreline.

❯ **Use the Law of Sines to solve oblique triangles (AAS or ASA).**
❯ **Use the Law of Sines to solve oblique triangles (SSA).**
❯ **Find the areas of oblique triangles.**
❯ **Use the Law of Sines to model and solve real-life problems.**

Introduction

In Chapter 6, you studied techniques for solving right triangles. In this section and the next, you will solve **oblique triangles**—triangles that have no right angles. As standard notation, the angles of a triangle are labeled A, B, and C, and their opposite sides are labeled a, b, and c, as shown in the triangles below.

Oblique Triangles

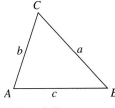

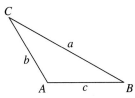

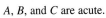

A, B, and C are acute.

A is obtuse; B and C are acute.

To solve an oblique triangle, you need to know one side length and any two of the other measures. So, there are four cases involving sides (S) and angles (A).

1. Two angles and any side (AAS or ASA)
2. Two sides and an angle opposite one of them (SSA)
3. Three sides (SSS)
4. Two sides and their included angle (SAS)

The first two cases can be solved using the **Law of Sines,** whereas the last two cases require the Law of Cosines (presented in the next section).

Law of Sines

If ABC is a triangle with sides a, b, and c, then $\dfrac{a}{\sin A} = \dfrac{b}{\sin B} = \dfrac{c}{\sin C}$.

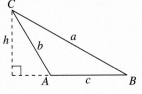

A is acute.

A is obtuse.

ALGEBRA HELP

The height h of each oblique triangle at the right can be found using the formula

$$\frac{h}{b} = \sin A$$

or $h = b \sin A$.

The Law of Sines can also be written in the reciprocal form

$$\frac{\sin A}{a} = \frac{\sin B}{b} = \frac{\sin C}{c}.$$ Reciprocal form

For a proof of the Law of Sines, see Proofs in Mathematics on page 622.

GO DIGITAL

Figure 8.1

EXAMPLE 1 **Given Two Angles and One Side—AAS**

For the triangle in Figure 8.1, $C = 102°$, $B = 29°$, and $b = 28$ feet. Find the remaining angle and sides.

Solution The third angle of the triangle is

$$A = 180° - B - C = 180° - 29° - 102° = 49°.$$

By the Law of Sines, you have

$$\frac{a}{\sin A} = \frac{b}{\sin B} = \frac{c}{\sin C}.$$

Using $b = 28$ produces

$$a = \frac{b}{\sin B}(\sin A) = \frac{28}{\sin 29°}(\sin 49°) \approx 43.59 \text{ feet}$$

and

$$c = \frac{b}{\sin B}(\sin C) = \frac{28}{\sin 29°}(\sin 102°) \approx 56.49 \text{ feet}.$$

✓ **Checkpoint** ▶ Audio-video solution in English & Spanish at LarsonPrecalculus.com

For the triangle shown, $A = 30°$, $B = 45°$, and $a = 32$ centimeters. Find the remaining angle and sides.

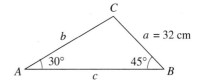

EXAMPLE 2 **Given Two Angles and One Side—ASA**

A pole tilts toward the sun at an 8° angle from the vertical, and it casts a 22-foot shadow. (See Figure 8.2.) The angle of elevation from the tip of the shadow to the top of the pole is 43°. How tall is the pole?

Solution In Figure 8.2, $A = 43°$ and

$$B = 90° + 8° = 98°.$$

So, the third angle is

$$C = 180° - A - B = 180° - 43° - 98° = 39°.$$

By the Law of Sines, you have

$$\frac{a}{\sin A} = \frac{c}{\sin C}.$$

Because $c = 22$ feet, the height of the pole is

$$a = \frac{c}{\sin C}(\sin A) = \frac{22}{\sin 39°}(\sin 43°) \approx 23.84 \text{ feet}.$$

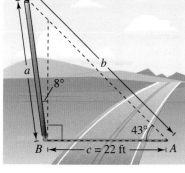

Figure 8.2

✓ **Checkpoint** ▶ Audio-video solution in English & Spanish at LarsonPrecalculus.com

Find the height of the tree shown in the figure.

GO DIGITAL

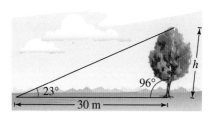

The Ambiguous Case (SSA)

In Examples 1 and 2, you saw that two angles and one side determine a unique triangle. However, if two sides and one opposite angle are given, then three possible situations can occur: (1) no such triangle exists, (2) one such triangle exists, or (3) two distinct triangles exist.

The Ambiguous Case (SSA)

Consider a triangle in which a, b, and A are given, and $h = b \sin A$.

	A is acute.	A is acute.	A is acute.	A is acute.	A is obtuse.	A is obtuse.
Sketch						
Necessary condition	$a < h$	$a = h$	$a \geq b$	$h < a < b$	$a \leq b$	$a > b$
Triangles possible	None	One	One	Two	None	One

EXAMPLE 3	**Single-Solution Case—SSA**

 See LarsonPrecalculus.com for an interactive version of this type of example.

For the triangle in Figure 8.3, $a = 22$ inches, $b = 12$ inches, and $A = 42°$. Find the remaining side and angles.

Solution By the Law of Sines, you have

$$\frac{\sin B}{b} = \frac{\sin A}{a} \qquad \text{Reciprocal form}$$

$$\sin B = b\left(\frac{\sin A}{a}\right) \qquad \text{Multiply each side by } b.$$

$$\sin B = 12\left(\frac{\sin 42°}{22}\right) \qquad \text{Substitute for } A, a, \text{ and } b.$$

$$B \approx 21.41°. \qquad \text{Solve for acute angle } B.$$

So, the third angle of the triangle is $C \approx 180° - 42° - 21.41° = 116.59°$. Finally, determine the remaining side.

$$\frac{c}{\sin C} = \frac{a}{\sin A} \qquad \text{Law of Sines}$$

$$c = \frac{a}{\sin A}(\sin C) \qquad \text{Multiply each side by } \sin C.$$

$$c \approx \frac{22}{\sin 42°}(\sin 116.59°) \qquad \text{Substitute for } a, A, \text{ and } C.$$

$$c \approx 29.40 \text{ inches} \qquad \text{Simplify.}$$

✓ **Checkpoint** ▶ *Audio-video solution in English & Spanish at LarsonPrecalculus.com*

Given $A = 31°$, $a = 12$ inches, and $b = 5$ inches, find the remaining side and angles of the triangle. ■

Left margin:

$b = 12$ in. $a = 22$ in.

C

A $42°$ c B

One solution: $a \geq b$

Figure 8.3

▶▶▶▶

ALGEBRA HELP

When using the Law of Sines, choose the form so that the unknown variable is in the numerator.

GO DIGITAL

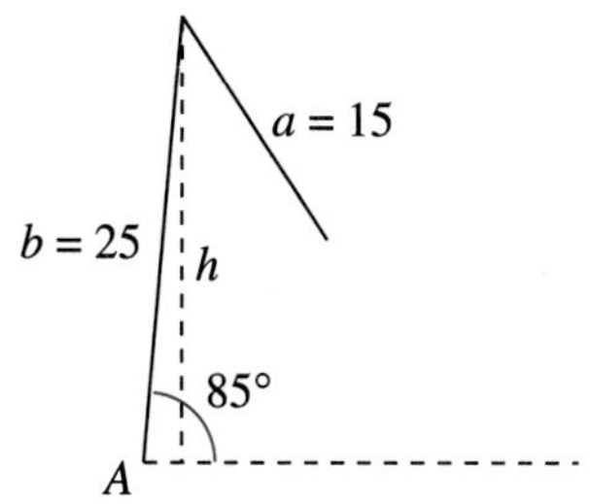

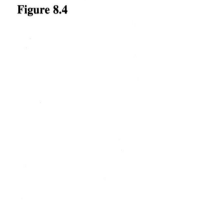

No solution: $a < h$
Figure 8.4

EXAMPLE 4 No-Solution Case—SSA

Show that there is no triangle for which $a = 15$, $b = 25$, and $A = 85°$.

Solution Begin by making the sketch shown in Figure 8.4. From this figure, it appears that no triangle is possible. Verify this using the Law of Sines.

$$\frac{\sin B}{b} = \frac{\sin A}{a} \qquad \text{Reciprocal form}$$

$$\sin B = b\left(\frac{\sin A}{a}\right) \qquad \text{Multiply each side by } b.$$

$$\sin B = 25\left(\frac{\sin 85°}{15}\right) \approx 1.6603 > 1$$

This contradicts the fact that $|\sin B| \le 1$. So, no triangle can be formed with sides $a = 15$ and $b = 25$ and angle $A = 85°$.

✓ *Checkpoint* ▶ *Audio-video solution in English & Spanish at LarsonPrecalculus.com*

Show that there is no triangle for which $a = 4$, $b = 14$, and $A = 60°$. ■

EXAMPLE 5 Two-Solution Case—SSA

Find two triangles for which $a = 12$ meters, $b = 31$ meters, and $A = 20.5°$.

Solution Because $h = b \sin A = 31(\sin 20.5°) \approx 10.86$ meters and $h < a < b$, there are two possible triangles. Use the Law of Sines to find B.

$$\frac{\sin B}{b} = \frac{\sin A}{a} \qquad \text{Reciprocal form}$$

$$\sin B = b\left(\frac{\sin A}{a}\right) = 31\left(\frac{\sin 20.5°}{12}\right)$$

There are two angles between $0°$ and $180°$ that satisfy this equation. One angle is

$$B_1 = \arcsin\left(\frac{31 \sin 20.5°}{12}\right) \implies B_1 \approx 64.8°.$$

Supplementary angles have the same sine value, so the second angle is

$$B_2 \approx 180° - 64.8° = 115.2°.$$

Now, you can solve each triangle. For $B_1 \approx 64.8°$, you obtain

$$C \approx 180° - 20.5° - 64.8° = 94.7°$$

and

$$c = \frac{a}{\sin A}(\sin C) \approx \frac{12}{\sin 20.5°}(\sin 94.7°) \approx 34.15 \text{ meters.}$$

For $B_2 \approx 115.2°$, you obtain

$$C \approx 180° - 20.5° - 115.2° = 44.3°$$

and

$$c = \frac{a}{\sin A}(\sin C) \approx \frac{12}{\sin 20.5°}(\sin 44.3°) \approx 23.93 \text{ meters.}$$

The resulting triangles are shown in Figure 8.5.

✓ *Checkpoint* ▶ *Audio-video solution in English & Spanish at LarsonPrecalculus.com*

Find two triangles for which $a = 4.5$ feet, $b = 5$ feet, and $A = 58°$. ■

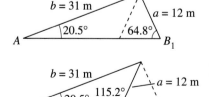

Two solutions: $h < a < b$
Figure 8.5

GO DIGITAL

Area of an Oblique Triangle

The procedure used to prove the Law of Sines leads to a formula for the area of an oblique triangle. Consider the two triangles below.

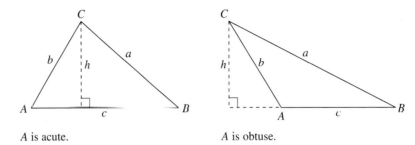

A is acute.   A is obtuse.

Note that each triangle has a height of $h = b \sin A$. Consequently, the area of each triangle is

$$\text{Area} = \frac{1}{2}(\text{base})(\text{height}) = \frac{1}{2}(c)(b \sin A) = \frac{1}{2}bc \sin A.$$

By similar arguments, you can develop the other two forms shown below.

Area of an Oblique Triangle

The area of any triangle is one-half the product of the lengths of two sides times the sine of their included angle. That is,

$$\text{Area} = \frac{1}{2}bc \sin A = \frac{1}{2}ab \sin C = \frac{1}{2}ac \sin B.$$

Note that when angle A is $90°$, the formula gives the area of a right triangle:

$$\text{Area} = \frac{1}{2}bc(\sin 90°) = \frac{1}{2}bc = \frac{1}{2}(\text{base})(\text{height}). \qquad \text{\small sin } 90° = 1$$

You obtain similar results for angles C and B equal to $90°$.

EXAMPLE 6   **Finding the Area of a Triangular Lot**

Find the area of a triangular lot with two sides of lengths 90 meters and 52 meters and an included angle of $102°$.

Solution   Draw a triangle to represent the problem, as shown at the right. The area is

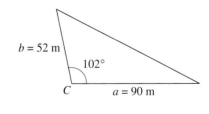

$$\text{Area} = \frac{1}{2}ab \sin C$$

$$= \frac{1}{2}(90)(52)(\sin 102°)$$

$$\approx 2289 \text{ square meters.}$$

✓ **Checkpoint** ▶ *Audio-video solution in English & Spanish at LarsonPrecalculus.com*

Find the area of a triangular lot with two sides of lengths 24 yards and 18 yards and an included angle of $80°$.

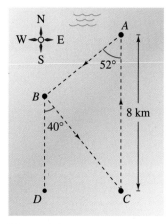

Figure 8.6

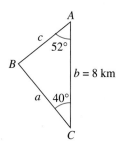

Figure 8.7

Application

EXAMPLE 7   **An Application of the Law of Sines**

The course for a boat race starts at point A and proceeds in the direction S 52° W to point B, then in the direction S 40° E to point C, and finally back to point A, as shown in Figure 8.6. Point C lies 8 kilometers directly south of point A. Approximate the total distance of the race course.

Solution   The lines BD and AC are parallel, so $\angle BCA \cong \angle CBD$. Consequently, triangle ABC has the measures shown in Figure 8.7. The measure of angle B is $180° - 52° - 40° = 88°$. Using the Law of Sines,

$$\frac{a}{\sin 52°} = \frac{8}{\sin 88°} = \frac{c}{\sin 40°}.$$

Solving for a and c, you have

$$a = \frac{8}{\sin 88°}(\sin 52°) \approx 6.31 \quad \text{and} \quad c = \frac{8}{\sin 88°}(\sin 40°) \approx 5.15.$$

So, the total distance of the course is approximately

$$8 + 6.31 + 5.15 = 19.46 \text{ kilometers.}$$

✓ *Checkpoint* ▶ Audio-video solution in English & Spanish at *LarsonPrecalculus.com*

On a small lake, a person swims from point A to point B at a bearing of N 28° E, then to point C at a bearing of N 58° W, and finally back to point A, as shown in the figure below. Point C lies 800 meters directly north of point A. Approximate the total distance the person swims.

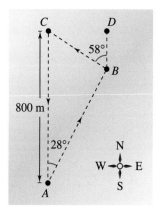

Summarize (Section 8.1)

1. State the Law of Sines *(page 560)*. For examples of using the Law of Sines to solve oblique triangles (AAS or ASA), see Examples 1 and 2.

2. List the necessary conditions and the corresponding numbers of possible triangles for the ambiguous case (SSA) *(page 562)*. For examples of using the Law of Sines to solve oblique triangles (SSA), see Examples 3–5.

3. State the formulas for the area of an oblique triangle *(page 564)*. For an example of finding the area of an oblique triangle, see Example 6.

4. Describe a real-life application of the Law of Sines *(page 565, Example 7)*.

GO DIGITAL

8.1 Exercises

See CalcChat.com for tutorial help and worked-out solutions to odd-numbered exercises.

GO DIGITAL

Vocabulary and Concept Check

In Exercises 1 and 2, fill in the blanks.

1. An _____ triangle is a triangle that has no right angle.
2. Two _____ and one _____ determine a unique triangle.
3. Which two cases can be solved using the Law of Sines?
4. List measures of sides a and b and angle A so that two distinct triangles ABC exist.

Skills and Applications

Using the Law of Sines In Exercises 5–22, use the Law of Sines to solve the triangle. Round your answers to two decimal places.

5.

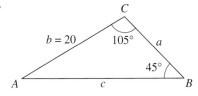

6.

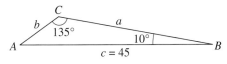

7.

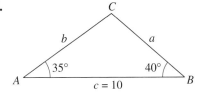

8.

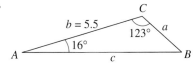

9. $A = 102.4°$, $C = 16.7°$, $a = 21.6$
10. $A = 24.3°$, $C = 54.6°$, $c = 2.68$
11. $A = 83° 20'$, $C = 54.6°$, $c = 18.1$
12. $A = 5° 40'$, $B = 8° 15'$, $b = 4.8$
13. $A = 35°$, $B = 65°$, $c = 10$
14. $A = 120°$, $B = 45°$, $c = 16$
15. $C = 55°$, $A = 42°$, $b = \frac{3}{4}$
16. $B = 28°$, $C = 104°$, $a = 3\frac{5}{8}$
17. $A = 36°$, $a = 8$, $b = 5$
18. $A = 30°$, $a = 7$, $b = 14$
19. $A = 145°$, $a = 14$, $b = 4$
20. $A = 100°$, $a = 125$, $c = 10$
21. $B = 15° 30'$, $a = 4.5$, $b = 6.8$
22. $B = 2° 45'$, $b = 6.2$, $c = 5.8$

Using the Law of Sines In Exercises 23–32, use the Law of Sines to solve (if possible) the triangle. If two solutions exist, find both. Round your answers to two decimal places.

23. $A = 110°$, $a = 125$, $b = 100$
24. $A = 110°$, $a = 125$, $b = 200$
25. $A = 76°$, $a = 18$, $b = 20$
26. $A = 76°$, $a = 34$, $b = 21$
27. $A = 58°$, $a = 11.4$, $b = 12.8$
28. $A = 58°$, $a = 4.5$, $b = 12.8$
29. $A = 120°$, $a = b = 25$
30. $A = 120°$, $a = 25$, $b = 24$
31. $A = 45°$, $a = b = 1$
32. $A = 25° 4'$, $a = 9.5$, $b = 22$

Using the Law of Sines In Exercises 33–36, find values for b such that the triangle has (a) one solution, (b) two solutions (if possible), and (c) no solution.

33. $A = 36°$, $a = 5$
34. $A = 60°$, $a = 10$
35. $A = 105°$, $a = 80$
36. $A = 132°$, $a = 215$

Finding the Area of a Triangle In Exercises 37–44, find the area of the triangle. Round your answers to one decimal place.

37. $A = 125°$, $b = 9$, $c = 6$
38. $C = 150°$, $a = 17$, $b = 10$
39. $B = 39°$, $a = 25$, $c = 12$
40. $A = 72°$, $b = 31$, $c = 44$
41. $C = 103° 15'$, $a = 16$, $b = 28$
42. $B = 54° 30'$, $a = 62$, $c = 35$
43. $A = 67°$, $B = 43°$, $a = 8$
44. $B = 118°$, $C = 29°$, $a = 52$

45. Environmental Science The bearing from the Pine Knob fire tower to the Colt Station fire tower is N 65° E, and the two towers are 30 kilometers apart. A fire spotted by rangers in each tower has a bearing of N 80° E from Pine Knob and S 70° E from Colt Station (see figure). Find the distance of the fire from each tower.

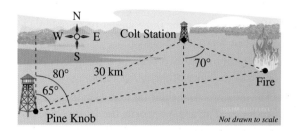

46. Distance

A boat is traveling due east parallel to the shoreline at a speed of 10 miles per hour. At a given time, the bearing to a lighthouse is S 70° E, and 15 minutes later the bearing is S 63° E (see figure). The lighthouse is located at the shoreline. What is the distance from the boat to the shoreline?

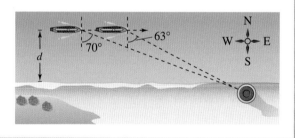

47. Bridge Design A bridge is built across a small lake from a gazebo to a dock (see figure). The bearing from the gazebo to the dock is S 41° W. From a tree 100 meters from the gazebo, the bearings to the gazebo and the dock are S 74° E and S 28° E, respectively. Find the distance from the gazebo to the dock.

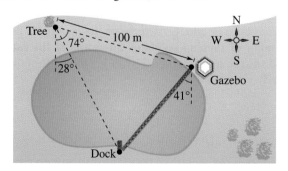

48. Flight Path A plane flies 500 kilometers with a bearing of 316° from Naples to Elgin (see figure). The plane then flies 720 kilometers from Elgin to Canton (Canton is due west of Naples). Find the bearing of the flight from Elgin to Canton.

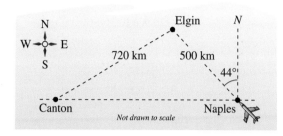

49. Height A tree grows at an angle of 4° from the vertical due to prevailing winds. At a point 40 meters from the base of the tree, the angle of elevation to the top of the tree is 30° (see figure).
(a) Write an equation that you can use to find the height h of the tree.
(b) Find the height of the tree.

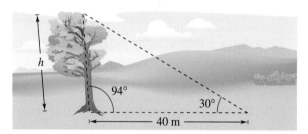

50. Angle of Elevation A 10-meter utility pole casts a 17-meter shadow directly down a slope when the angle of elevation of the sun is 42° (see figure). Find θ, the angle of elevation of the ground.

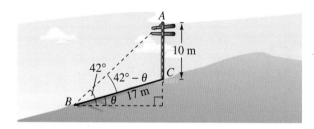

51. Height A flagpole at a right angle to the horizontal is located on a slope that makes an angle of 12° with the horizontal. The flagpole's shadow is 16 meters long and points directly up the slope. The angle of elevation from the tip of the shadow to the sun is 20°.
(a) Draw a diagram that represents the problem. Show the known quantities on the diagram and use a variable to indicate the height of the flagpole.
(b) Write an equation that you can use to find the height of the flagpole.
(c) Find the height of the flagpole.

52. Distance Air traffic controllers continuously monitor the angles of elevation θ and ϕ to an airplane from an airport control tower and from an observation post 2 miles away (see figure). Write an equation giving the distance d between the plane and the observation post in terms of θ and ϕ.

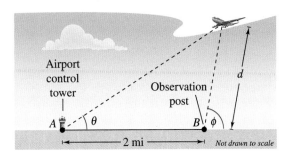

53. Numerical Analysis In the figure, α and β are positive angles.

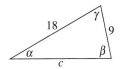

(a) Write α as a function of β.
(b) Use a graphing utility to graph the function in part (a). Determine its domain and range.
(c) Use the result of part (a) to write c as a function of β.
(d) Use the graphing utility to graph the function in part (c). Determine its domain and range.
(e) Complete the table. What can you infer?

β	0.4	0.8	1.2	1.6	2.0	2.4	2.8
α							
c							

54. HOW DO YOU SEE IT? In the figure, a triangle is to be formed by drawing a line segment of length a from $(4, 3)$ to the positive x-axis. For what value(s) of a can you form (a) one triangle, (b) two triangles, and (c) no triangles? Explain.

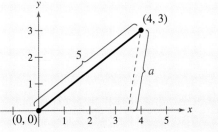

Exploring the Concepts

True or False? In Exercises 55–59, determine whether the statement is true or false. Justify your answer.

55. If a triangle contains an obtuse angle, then it must be oblique.

56. Two angles and one side of a triangle do not necessarily determine a unique triangle.

57. When you know the three angles of an oblique triangle, you can solve the triangle.

58. The ratio of any two sides of a triangle is equal to the ratio of the sines of the opposite angles of the two sides.

59. The area of a triangular lot with side lengths 11 feet and 16 feet and an angle opposite the 16-foot side of 58° can be found using

$$\text{Area} = \tfrac{1}{2}(11)(16)(\sin 58°).$$

60. Think About It

(a) Write the area A of the shaded region in the figure as a function of θ.
(b) Use a graphing utility to graph the function.
(c) Determine the domain of the function. Explain how decreasing the length of the eight-centimeter line segment affects the area of the region and the domain of the function.

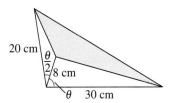

Review & Refresh ▶ Video solutions at LarsonPrecalculus.com

Finding the Slope of a Line Through Two Points In Exercises 61–64, find the slope of the line passing through the pair of points.

61. $(1, -6), (5, 0)$ **62.** $(-2, 3), (-2, -12)$
63. $(-5, 4), (-13, 4)$ **64.** $(-2, 11), (8, 9)$

Finding Real Zeros of a Polynomial Function In Exercises 65–68, (a) find all real zeros of the polynomial function, (b) determine whether the multiplicity of each zero is even or odd, (c) determine the maximum possible number of turning points of the graph of the function, and (d) use a graphing utility to graph the function and verify your answers.

65. $h(x) = x^2 - 10x + 25$
66. $f(x) = x^3 + 3x^2 - 4x$
67. $f(t) = -4t^4 - 10t^3 + 6t^2$
68. $g(t) = 2t^5 - 14t^3 - 36t$

8.2 Law of Cosines

The Law of Cosines is a useful tool for solving real-life problems involving oblique triangles. For example, in Exercise 52 on page 574, you will use the Law of Cosines to determine the total distance a piston moves in an engine.

❯ **Use the Law of Cosines to solve oblique triangles (SSS or SAS).**
❯ **Use the Law of Cosines to model and solve real-life problems.**
❯ **Use Heron's Area Formula to find areas of triangles.**

Introduction

Two cases remain in the list of conditions needed to solve an oblique triangle—SSS and SAS. When you are given three sides (SSS), or two sides and their included angle (SAS), you cannot solve the triangle using the Law of Sines alone. In such cases, use the **Law of Cosines.**

Law of Cosines	
Standard Form	**Alternative Form**
$a^2 = b^2 + c^2 - 2bc \cos A$	$\cos A = \dfrac{b^2 + c^2 - a^2}{2bc}$
$b^2 = a^2 + c^2 - 2ac \cos B$	$\cos B = \dfrac{a^2 + c^2 - b^2}{2ac}$
$c^2 = a^2 + b^2 - 2ab \cos C$	$\cos C = \dfrac{a^2 + b^2 - c^2}{2ab}$

For a proof of the Law of Cosines, see Proofs in Mathematics on page 622.

EXAMPLE 1 Given Three Sides—SSS

Find the three angles of the triangle shown below.

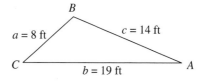

Solution   It is a good idea to find the angle opposite the longest side first—side b in this case. Using the alternative form of the Law of Cosines,

$$\cos B = \frac{a^2 + c^2 - b^2}{2ac} = \frac{8^2 + 14^2 - 19^2}{2(8)(14)} = -\frac{101}{224}.$$

Because $\cos B$ is negative, B is an *obtuse* angle given by $B \approx 116.80°$. At this point, use the Law of Sines to determine A.

$$\sin A = a\left(\frac{\sin B}{b}\right) \approx 8\left(\frac{\sin 116.80°}{19}\right) \implies A \approx \arcsin\left(\frac{8 \sin 116.80°}{19}\right)$$

Because B is obtuse and a triangle can have at most one obtuse angle, you know that A must be acute. So, $A \approx 22.08°$ and $C \approx 180° - 22.08° - 116.80° = 41.12°$.

✓ *Checkpoint* ▶ *Audio-video solution in English & Spanish at LarsonPrecalculus.com*

Find the three angles of the triangle whose sides have lengths $a = 6$ centimeters, $b = 8$ centimeters, and $c = 12$ centimeters. ■

ALGEBRA HELP ⫸⫸⫸⫸

When solving an oblique triangle given three sides, use the alternative form of the Law of Cosines to solve for an angle.

GO DIGITAL

Do you see why it was wise to find the largest angle *first* in Example 1? Knowing the cosine of an angle, you can determine whether the angle is acute or obtuse. That is,

$$\cos \theta > 0 \quad \text{for} \quad 0° < \theta < 90° \qquad \text{Acute}$$

and

$$\cos \theta < 0 \quad \text{for} \quad 90° < \theta < 180°. \qquad \text{Obtuse}$$

So, in Example 1, after you find that angle B is obtuse, you know that angles A and C must both be acute. Furthermore, if the largest angle is acute, then the remaining two angles must also be acute.

EXAMPLE 2   **Given Two Sides and Their Included Angle—SAS**

▶▶▶ *See LarsonPrecalculus.com for an interactive version of this type of example.*

Find the remaining angles and side of the triangle shown below.

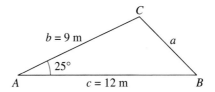

ALGEBRA HELP

When solving an oblique triangle given two sides and their included angle, use the standard form of the Law of Cosines to solve for the remaining side.

Solution   Use the standard form of the Law of Cosines to find side a.

$$a^2 = b^2 + c^2 - 2bc \cos A$$
$$a^2 = 9^2 + 12^2 - 2(9)(12) \cos 25°$$
$$a^2 = 225 - 216 \cos 25°$$
$$a \approx 5.4072 \text{ meters}$$

Of angles B and C, angle B must be the smaller angle because it is opposite the shorter side ($b < c$). So, B cannot be obtuse and must be acute. Because you know a, A, and b, use the reciprocal form of the Law of Sines to solve for acute angle B.

$$\frac{\sin B}{b} = \frac{\sin A}{a} \qquad \text{Reciprocal form}$$

$$\sin B = b\left(\frac{\sin A}{a}\right) \qquad \text{Multiply each side by } b.$$

$$\sin B \approx 9\left(\frac{\sin 25°}{5.4072}\right) \qquad \text{Substitute for } A, a, \text{ and } b.$$

$$B \approx 44.7° \qquad \text{Use a calculator.}$$

Now, find C.

$$C \approx 180° - 25° - 44.7° = 110.3°$$

✓ *Checkpoint* ▶ *Audio-video solution in English & Spanish at LarsonPrecalculus.com*

Find the remaining angles and side of the triangle shown below.

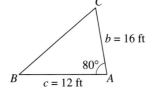

GO DIGITAL

Applications

EXAMPLE 3   **An Application of the Law of Cosines**

The pitcher's mound on a softball field is 43 feet from home plate and the distance between the bases is 60 feet, as shown in Figure 8.8. (The pitcher's mound is *not* halfway between home plate and second base.) How far is the pitcher's mound from first base?

Solution   In triangle *HPF*, $H = 45°$ (line segment *HP* bisects the right angle at *H*), $f = 43$, and $p = 60$. Using the standard form of the Law of Cosines for this SAS case,

$$h^2 = f^2 + p^2 - 2fp \cos H \qquad \text{Law of Cosines}$$
$$= 43^2 + 60^2 - 2(43)(60) \cos 45° \qquad \text{Substitute for } H, f, \text{ and } p.$$
$$= 5449 - 5160 \cos 45°.$$

So, the approximate distance from the pitcher's mound to first base is

$$h = \sqrt{5449 - 5160 \cos 45°} \approx 42.43 \text{ feet.}$$

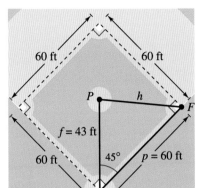

Figure 8.8

✓ *Checkpoint* ▶ *Audio-video solution in English & Spanish at LarsonPrecalculus.com*

In a softball game, a batter hits a ball directly over second base to center field, a distance of 240 feet from home plate. The center fielder then throws the ball to third base and gets a runner out. The distance between the bases is 60 feet. How far is the center fielder from third base?

EXAMPLE 4   **An Application of the Law of Cosines**

A ship travels 60 miles due east and then adjusts its course, as shown in the figure. After traveling 80 miles in this new direction, the ship is 139 miles from its point of departure. Describe the bearing from point *B* to point *C*.

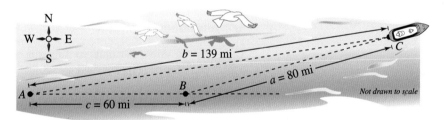

Solution   You have $a = 80$, $b = 139$, and $c = 60$. So, using the alternative form of the Law of Cosines,

$$\cos B = \frac{a^2 + c^2 - b^2}{2ac} \qquad \text{Alternative form}$$
$$= \frac{80^2 + 60^2 - 139^2}{2(80)(60)} \qquad \text{Substitute for } a, b, \text{ and } c.$$
$$= -\frac{9321}{9600}.$$

So, $B \approx 166.15°$, and the bearing measured from due north from point *B* to point *C* is approximately $166.15° - 90° = 76.15°$, or N 76.15° E.

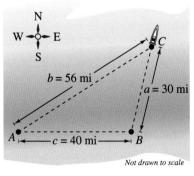

Figure 8.9

✓ *Checkpoint* ▶ *Audio-video solution in English & Spanish at LarsonPrecalculus.com*

A ship travels 40 miles due east and then changes direction, as shown in Figure 8.9. After traveling 30 miles in this new direction, the ship is 56 miles from its point of departure. Describe the bearing from point *B* to point *C*. ■

HISTORICAL NOTE

Heron of Alexandria (circa A.D. 10–75) was a Greek geometer and inventor. His works describe how to find areas of triangles, quadrilaterals, regular polygons with 3 to 12 sides, and circles, as well as surface areas and volumes of three-dimensional objects.

Heron's Area Formula

The Law of Cosines can be used to establish a formula for the area of a triangle. This formula is called **Heron's Area Formula** after the Greek mathematician Heron (circa A.D. 10–75).

Heron's Area Formula

Given any triangle with sides of lengths a, b, and c, the area of the triangle is

$$\text{Area} = \sqrt{s(s-a)(s-b)(s-c)}$$

where

$$s = \frac{a+b+c}{2}.$$

For a proof of Heron's Area Formula, see Proofs in Mathematics on page 623.

EXAMPLE 5 **Using Heron's Area Formula**

Use Heron's Area Formula to find the area of a triangle with sides of lengths $a = 43$ meters, $b = 53$ meters, and $c = 72$ meters.

Solution First, determine that $s = (a + b + c)/2 = 168/2 = 84$. Then Heron's Area Formula yields

$$\text{Area} = \sqrt{s(s-a)(s-b)(s-c)} \qquad \text{Heron's Area Formula}$$

$$= \sqrt{84(84-43)(84-53)(84-72)} \qquad \text{Substitute.}$$

$$= \sqrt{84(41)(31)(12)} \qquad \text{Subtract.}$$

$$\approx 1131.89 \text{ square meters.} \qquad \text{Use a calculator.}$$

✓ **Checkpoint** ▶ Audio-video solution in English & Spanish at LarsonPrecalculus.com

Use Heron's Area Formula to find the area of a triangle with sides of lengths $a = 5$ inches, $b = 9$ inches, and $c = 8$ inches. ∎

You have now studied three different formulas for the area of a triangle.

Standard Formula: $\text{Area} = \dfrac{1}{2}bh$

Oblique Triangle: $\text{Area} = \dfrac{1}{2}bc \sin A = \dfrac{1}{2}ab \sin C = \dfrac{1}{2}ac \sin B$

Heron's Area Formula: $\text{Area} = \sqrt{s(s-a)(s-b)(s-c)}$

Summarize (Section 8.2)

1. State the Law of Cosines *(page 569)*. For examples of using the Law of Cosines to solve oblique triangles (SSS or SAS), see Examples 1 and 2.

2. Describe real-life applications of the Law of Cosines *(page 571, Examples 3 and 4)*.

3. State Heron's Area Formula *(page 572)*. For an example of using Heron's Area Formula to find the area of a triangle, see Example 5.

GO DIGITAL

8.2 Exercises

See CalcChat.com for tutorial help and worked-out solutions to odd-numbered exercises.

Vocabulary and Concept Check

In Exercises 1 and 2, fill in the blanks.

1. When solving an oblique triangle given three sides, use the _____ form of the Law of Cosines to solve for an angle.

2. When solving an oblique triangle given two sides and their included angle, use the _____ form of the Law of Cosines to solve for the remaining side.

3. State the alternative form of the Law of Cosines for $b^2 = a^2 + c^2 - 2ac \cos B$.

4. State two other formulas, besides Heron's Area Formula, that you can use to find the area of a triangle.

Skills and Applications

Using the Law of Cosines In Exercises 5–24, use the Law of Cosines to solve the triangle. Round your answers to two decimal places.

5.
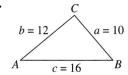

6.
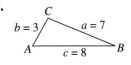

7.
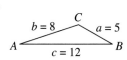

8.
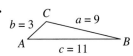

9.
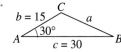

10.
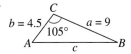

11.
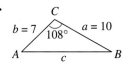

12.
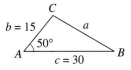

13. $a = 11$, $b = 15$, $c = 21$

14. $a = 55$, $b = 25$, $c = 72$

15. $a = 2.5$, $b = 1.8$, $c = 0.9$

16. $a = 75.4$, $b = 52.5$, $c = 52.5$

17. $A = 120°$, $b = 6$, $c = 7$

18. $A = 48°$, $b = 3$, $c = 14$

19. $B = 10° 35'$, $a = 40$, $c = 30$

20. $B = 75° 20'$, $a = 9$, $c = 6$

21. $B = 125° 40'$, $a = 37$, $c = 37$

22. $C = 15° 15'$, $a = 7.45$, $b = 2.15$

23. $C = 43°$, $a = \frac{4}{9}$, $b = \frac{7}{9}$

24. $C = 101°$, $a = \frac{3}{8}$, $b = \frac{3}{4}$

Finding Measures in a Parallelogram In Exercises 25–30, find the missing values by solving the parallelogram shown in the figure. (The lengths of the diagonals are given by c and d.)

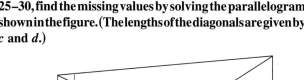

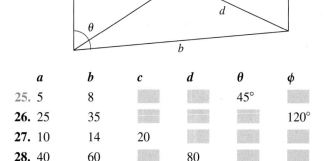

	a	b	c	d	θ	ϕ
25.	5	8			45°	
26.	25	35				120°
27.	10	14	20			
28.	40	60		80		
29.	15		25	20		
30.		25	50	35		

Solving a Triangle In Exercises 31–36, determine whether the Law of Sines or the Law of Cosines is needed to solve the triangle. Then solve (if possible) the triangle. If two solutions exist, find both. Round your answers to two decimal places.

31. $a = 8$, $c = 5$, $B = 40°$

32. $a = 10$, $b = 12$, $C = 70°$

33. $A = 24°$, $a = 4$, $b = 18$

34. $a = 11$, $b = 13$, $c = 7$

35. $A = 42°$, $B = 35°$, $c = 1.2$

36. $B = 12°$, $a = 160$, $b = 63$

Using Heron's Area Formula **In Exercises 37–44, use Heron's Area Formula to find the area of the triangle.**

37. $a = 6$, $b = 12$, $c = 17$

38. $a = 33$, $b = 36$, $c = 21$

39. $a = 2.5$, $b = 10.2$, $c = 8$

40. $a = 12.32$, $b = 8.46$, $c = 15.9$

41. $a = 1$, $b = \frac{1}{2}$, $c = \frac{5}{4}$

42. $a = \frac{3}{5}$, $b = \frac{4}{3}$, $c = \frac{7}{8}$

43. $A = 80°$, $b = 75$, $c = 41$

44. $C = 109°$, $a = 16$, $b = 3.5$

45. **Surveying** To approximate the length of a marsh, a surveyor walks 250 meters from point A to point B, then turns 75° and walks 220 meters to point C (see figure). Find the length AC of the marsh.

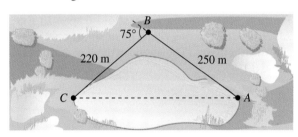

46. **Streetlight Design** Determine the angle θ in the design of the streetlight shown in the figure.

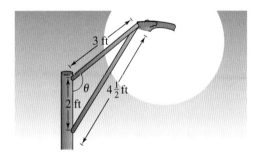

47. **Baseball** A baseball player in center field is approximately 330 feet from a television camera that is behind home plate. A batter hits a fly ball that goes to the wall 420 feet from the camera (see figure). The camera turns 8° to follow the play. How far does the center fielder have to run to make the catch?

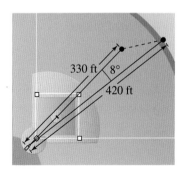

48. **Baseball** On a baseball diamond with 90-foot sides, the pitcher's mound is 60.5 feet from home plate. How far is the pitcher's mound from third base?

49. **Map** On a map, Minneapolis is 165 millimeters due west of Albany, Phoenix is 216 millimeters from Minneapolis, and Phoenix is 368 millimeters from Albany (see figure).

(a) Find the bearing of Minneapolis from Phoenix.

(b) Find the bearing of Albany from Phoenix.

50. **Air Navigation** A plane flies 810 miles from Franklin to Centerville with a bearing of 75°. Then it flies 648 miles from Centerville to Rosemount with a bearing of 32°. Draw a diagram that gives a visual representation of the problem. Then find the straight-line distance and bearing from Franklin to Rosemount.

51. **Surveying** A triangular parcel of land has 115 meters of frontage, and the other boundaries have lengths of 76 meters and 92 meters. What angles does the frontage make with the two other boundaries?

52. **Mechanical Engineering**

A piston in an engine has a seven-inch connecting rod fastened to a crank (see figure).

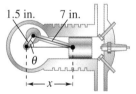

(a) Use the Law of Cosines to write an equation giving the relationship between x and θ.

(b) Write x as a function of θ. (Select the sign that yields positive values of x.)

(c) Use a graphing utility to graph the function in part (b).

(d) Use the graph in part (c) to determine the total distance the piston moves in one cycle.

53. Surveying   A triangular parcel of ground has sides of lengths 725 feet, 650 feet, and 575 feet. Find the measure of the largest angle.

54. Distance   Two ships leave a port at 9 A.M. One travels at a bearing of N 53° W at 12 miles per hour, and the other travels at a bearing of S 67° W at s miles per hour.

(a) Use the Law of Cosines to write an equation that relates s and the distance d between the two ships at noon.

(b) Find the speed s that the second ship must travel so that the ships are 43 miles apart at noon.

55. Geometry   A triangular parcel of land has sides of lengths 200 feet, 500 feet, and 600 feet. Use Heron's Area Formula to find the area of the parcel.

56. Geometry   A parking lot has the shape of a parallelogram (see figure). The lengths of two adjacent sides are 70 meters and 100 meters. The angle between the two sides is 70°. Use the Law of Cosines and Heron's Area Formula to find the area of the parking lot.

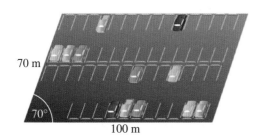

57. Geometry   You want to buy a triangular lot measuring 510 yards by 840 yards by 1120 yards. The price of the land is $2000 per acre. How much does the land cost? (*Hint:* 1 acre = 4840 square yards)

58. Geometry   You want to buy a triangular lot measuring 1350 feet by 1860 feet by 2490 feet. The cost of the land is $62,000. What is the price of the land per acre? (*Hint:* 1 acre = 43,560 square feet)

Exploring the Concepts

True or False?   In Exercises 59 and 60, determine whether the statement is true or false. Justify your answer.

59. In Heron's Area Formula, s is the average of the lengths of the three sides of the triangle.

60. In addition to SSS and SAS, the Law of Cosines can be used to solve triangles with AAS conditions.

61. Think About It   What familiar formula do you obtain when you use the standard form of the Law of Cosines, $c^2 = a^2 + b^2 - 2ab \cos C$, and you let $C = 90°$? What is the relationship between the Law of Cosines and this formula?

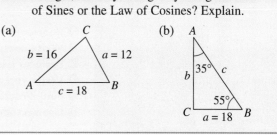

62.   **HOW DO YOU SEE IT?**   To solve the triangle, would you begin by using the Law of Sines or the Law of Cosines? Explain.

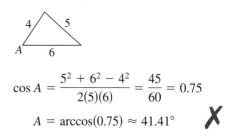

63. Proof   Use the Law of Cosines to prove each identity.

(a) $\dfrac{1}{2}bc(1 + \cos A) = \dfrac{a + b + c}{2} \cdot \dfrac{-a + b + c}{2}$

(b) $\dfrac{1}{2}bc(1 - \cos A) = \dfrac{a - b + c}{2} \cdot \dfrac{a + b - c}{2}$

64. Error Analysis   Describe the error.

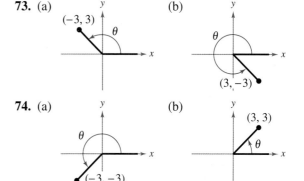

$$\cos A = \frac{5^2 + 6^2 - 4^2}{2(5)(6)} = \frac{45}{60} = 0.75$$

$$A = \arccos(0.75) \approx 41.41° \quad \textbf{X}$$

Review & Refresh ▶ Video solutions at LarsonPrecalculus.com

Finding Slope and Distance   In Exercises 65–72, find (a) the slope of the line passing through the points (if possible) and (b) the distance between the points.

65. $(1, 1), (5, 8)$ **66.** $(-2, 3), (-6, 9)$

67. $(-7, -4), (-3, -5)$ **68.** $(1, -2), (4, -1)$

69. $(1, 12), (-5, 12)$ **70.** $(-9, 5), (-9, -5)$

71. $(12, 10), (-12, -10)$ **72.** $(-5, 2), (5, -6)$

Evaluating Trigonometric Functions   In Exercises 73 and 74, find the exact values of the six trigonometric functions of each angle θ.

73. (a)

(b)

74. (a)

(b)

8.3 Vectors in the Plane

Vectors are useful tools for modeling and solving real-life problems involving magnitude and direction. For example, in Exercise 90 on page 587, you will use vectors to determine the speed and true direction of a commercial jet.

- ◉ **Represent vectors as directed line segments.**
- ◉ **Write component forms of vectors.**
- ◉ **Perform basic vector operations and represent vector operations graphically.**
- ◉ **Write vectors as linear combinations of unit vectors.**
- ◉ **Find direction angles of vectors.**
- ◉ **Use vectors to model and solve real-life problems.**

Introduction

Quantities such as force and velocity involve both *magnitude* and *direction* and cannot be completely characterized by a single real number. To represent such a quantity, you can use a **directed line segment,** as shown in Figure 8.10. The directed line segment $\overrightarrow{PQ}$ has **initial point** P and **terminal point** Q. Its **magnitude** (or **length**) is denoted by $\|\overrightarrow{PQ}\|$ and can be found using the Distance Formula.

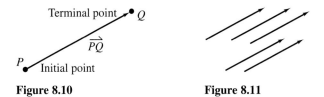

Figure 8.10   **Figure 8.11**

Two directed line segments that have the same magnitude and direction are *equivalent*. For example, the directed line segments in Figure 8.11 are all equivalent. The set of all directed line segments that are equivalent to the directed line segment $\overrightarrow{PQ}$ is a **vector v in the plane,** written $\mathbf{v} = \overrightarrow{PQ}$. Vectors are denoted by lowercase, boldface letters such as $\mathbf{u}$, $\mathbf{v}$, and $\mathbf{w}$.

| **EXAMPLE 1** | **Showing That Two Vectors Are Equivalent** |

Show that $\mathbf{u}$ and $\mathbf{v}$ in Figure 8.12 are equivalent.

Solution   From the Distance Formula, $\overrightarrow{PQ}$ and $\overrightarrow{RS}$ have the *same magnitude*.

$$\|\overrightarrow{PQ}\| = \sqrt{(3 - 0)^2 + (2 - 0)^2} = \sqrt{13}$$

$$\|\overrightarrow{RS}\| = \sqrt{(4 - 1)^2 + (4 - 2)^2} = \sqrt{13}$$

Moreover, both line segments have the *same direction* because they are both directed toward the upper right on lines that have the same slope.

$$\text{Slope of } \overrightarrow{PQ} = \frac{2 - 0}{3 - 0} = \frac{2}{3} \quad \text{and} \quad \text{Slope of } \overrightarrow{RS} = \frac{4 - 2}{4 - 1} = \frac{2}{3}$$

Because $\overrightarrow{PQ}$ and $\overrightarrow{RS}$ have the same magnitude and direction, $\mathbf{u}$ and $\mathbf{v}$ are equivalent.

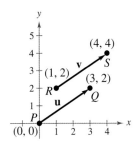

Figure 8.12

✓ *Checkpoint* ▶ Audio-video solution in English & Spanish at LarsonPrecalculus.com

Show that $\mathbf{u}$ and $\mathbf{v}$ in the figure at the right are equivalent.

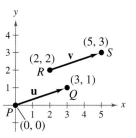

Component Form of a Vector

The directed line segment whose initial point is the origin is often the most convenient representative of a set of equivalent directed line segments. This representative of the vector **v** is in **standard position.**

A vector whose initial point is the origin $(0, 0)$ can be uniquely represented by the coordinates of its terminal point (v_1, v_2). This is the **component form of a vector v,** written as $\mathbf{v} = \langle v_1, v_2 \rangle$. The coordinates v_1 and v_2 are the *components* of **v.** If both the initial point and the terminal point lie at the origin, then **v** is the **zero vector** and is denoted by $\mathbf{0} = \langle 0, 0 \rangle$.

Component Form of a Vector

The component form of the vector with initial point $P(p_1, p_2)$ and terminal point $Q(q_1, q_2)$ is given by

$$\overrightarrow{PQ} = \langle q_1 - p_1, q_2 - p_2 \rangle = \langle v_1, v_2 \rangle = \mathbf{v}.$$

The **magnitude** (or **length**) of **v** is given by

$$\|\mathbf{v}\| = \sqrt{(q_1 - p_1)^2 + (q_2 - p_2)^2} = \sqrt{v_1^2 + v_2^2}.$$

If $\|\mathbf{v}\| = 1$, then **v** is a **unit vector.** Moreover, $\|\mathbf{v}\| = 0$ if and only if **v** is the zero vector **0.**

Two vectors $\mathbf{u} = \langle u_1, u_2 \rangle$ and $\mathbf{v} = \langle v_1, v_2 \rangle$ are *equal* if and only if $u_1 = v_1$ and $u_2 = v_2$. For instance, in Example 1, the vector **u** from $P(0, 0)$ to $Q(3, 2)$ is $\mathbf{u} = \overrightarrow{PQ} = \langle 3 - 0, 2 - 0 \rangle = \langle 3, 2 \rangle$, and the vector **v** from $R(1, 2)$ to $S(4, 4)$ is $\mathbf{v} = \overrightarrow{RS} = \langle 4 - 1, 4 - 2 \rangle = \langle 3, 2 \rangle$. So, the vectors **u** and **v** in Example 1 are equal.

EXAMPLE 2 **Finding the Component Form of a Vector**

Find the component form and magnitude of the vector **v** that has initial point $(4, -7)$ and terminal point $(-1, 5)$.

Algebraic Solution

Let

$$P(4, -7) = (p_1, p_2)$$

and

$$Q(-1, 5) = (q_1, q_2).$$

Then, the components of $\mathbf{v} = \langle v_1, v_2 \rangle$ are

$$v_1 = q_1 - p_1 = -1 - 4 = -5$$

and

$$v_2 = q_2 - p_2 = 5 - (-7) = 12.$$

So, $\mathbf{v} = \langle -5, 12 \rangle$ and the magnitude of **v** is

$$\|\mathbf{v}\| = \sqrt{(-5)^2 + 12^2}$$
$$= \sqrt{169}$$
$$= 13.$$

Graphical Solution

Use centimeter graph paper to plot the points $P(4, -7)$ and $Q(-1, 5)$. Carefully sketch the vector **v.** Use the sketch to find the components of $\mathbf{v} = \langle v_1, v_2 \rangle$. Then use a centimeter ruler to find the magnitude of **v.** The figure at the right shows that the components of **v** are $v_1 = -5$ and $v_2 = 12$, so $\mathbf{v} = \langle -5, 12 \rangle$. The figure also shows that the magnitude of **v** is $\|\mathbf{v}\| = 13$.

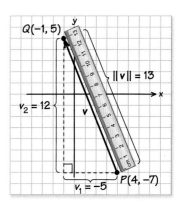

 Checkpoint ▶ *Audio-video solution in English & Spanish at LarsonPrecalculus.com*

Find the component form and magnitude of the vector **v** that has initial point $(-2, 3)$ and terminal point $(-7, 9)$. ■

GO DIGITAL

Vector Operations

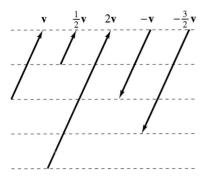

The scalar multiplication of **v**
Figure 8.13

The two basic vector operations are **scalar multiplication** and **vector addition.** In operations with vectors, numbers are usually referred to as **scalars.** In this text, scalars will always be real numbers. Geometrically, the product of a vector **v** and a scalar k is the vector that is $|k|$ times as long as **v**. When k is positive, k**v** has the same direction as **v**, and when k is negative, k**v** has the direction opposite that of **v**, as shown in Figure 8.13.

To add two vectors **u** and **v** geometrically, first position them (without changing their lengths or directions) so that the initial point of the second vector **v** coincides with the terminal point of the first vector **u**. The sum **u** + **v** is the vector formed by joining the initial point of the first vector **u** with the terminal point of the second vector **v**, as shown in the next two figures. This technique is called the **parallelogram law** for vector addition because the vector **u** + **v**, often called the **resultant** of vector addition, is the diagonal of a parallelogram with adjacent sides **u** and **v**.

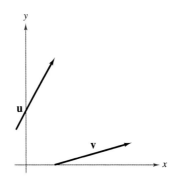
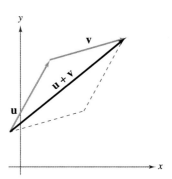

To find **u** + **v**, move the initial point of **v** to the terminal point of **u**, as shown in the figure on the right.

Definitions of Vector Addition and Scalar Multiplication

Let **u** = $\langle u_1, u_2 \rangle$ and **v** = $\langle v_1, v_2 \rangle$ be vectors and let k be a scalar (a real number). Then the **sum** of **u** and **v** is the vector

$$\mathbf{u} + \mathbf{v} = \langle u_1 + v_1, u_2 + v_2 \rangle \qquad \text{Sum}$$

and the **scalar multiple** of k times **u** is the vector

$$k\mathbf{u} = k\langle u_1, u_2 \rangle = \langle ku_1, ku_2 \rangle. \qquad \text{Scalar multiple}$$

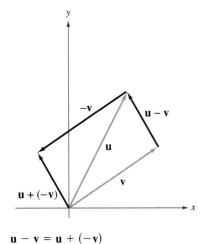

u − **v** = **u** + (−**v**)
Figure 8.14

The **negative** of **v** = $\langle v_1, v_2 \rangle$ is

$$-\mathbf{v} = (-1)\mathbf{v}$$
$$= \langle -v_1, -v_2 \rangle \qquad \text{Negative}$$

and the **difference** of **u** and **v** is

$$\mathbf{u} - \mathbf{v} = \mathbf{u} + (-\mathbf{v}) \qquad \text{Add } (-\mathbf{v}). \text{ See Figure 8.14.}$$
$$= \langle u_1 - v_1, u_2 - v_2 \rangle. \qquad \text{Difference}$$

To represent **u** − **v** geometrically, use directed line segments with the *same* initial point. The difference **u** − **v** is the vector from the terminal point of **v** to the terminal point of **u**, which is equal to **u** + (−**v**), as shown in Figure 8.14.

Example 3 on the next page illustrates the component definitions of vector addition and scalar multiplication. In this example, note the geometrical interpretations of each of the vector operations.

GO DIGITAL

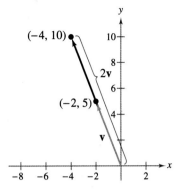

Figure 8.15

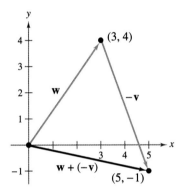

Figure 8.16

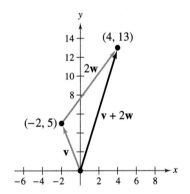

Figure 8.17

EXAMPLE 3   **Vector Operations**

▷▷▷ *See LarsonPrecalculus.com for an interactive version of this type of example.*

Let $\mathbf{v} = \langle -2, 5 \rangle$ and $\mathbf{w} = \langle 3, 4 \rangle$. Find each vector.

a. $2\mathbf{v}$ **b.** $\mathbf{w} - \mathbf{v}$ **c.** $\mathbf{v} + 2\mathbf{w}$

Solution

a. $2\mathbf{v} = 2\langle -2, 5 \rangle = \langle 2(-2), 2(5) \rangle = \langle -4, 10 \rangle$

A sketch of $2\mathbf{v}$ is shown in Figure 8.15.

b. $\mathbf{w} - \mathbf{v} = \langle 3, 4 \rangle - \langle -2, 5 \rangle = \langle 3 - (-2), 4 - 5 \rangle = \langle 5, -1 \rangle$

Note that Figure 8.16 shows the vector difference $\mathbf{w} - \mathbf{v}$ as the sum $\mathbf{w} + (-\mathbf{v})$.

c. $\mathbf{v} + 2\mathbf{w} = \langle -2, 5 \rangle + 2\langle 3, 4 \rangle = \langle -2, 5 \rangle + \langle 6, 8 \rangle = \langle 4, 13 \rangle$

A sketch of $\mathbf{v} + 2\mathbf{w}$ is shown in Figure 8.17.

✓ **Checkpoint** ▶ *Audio-video solution in English & Spanish at LarsonPrecalculus.com*

Let $\mathbf{u} = \langle 1, 4 \rangle$ and $\mathbf{v} = \langle 3, 2 \rangle$. Find each vector.

a. $\mathbf{u} + \mathbf{v}$ **b.** $\mathbf{u} - \mathbf{v}$ **c.** $2\mathbf{u} - 3\mathbf{v}$

Vector addition and scalar multiplication share many of the properties of ordinary arithmetic.

Properties of Vector Addition and Scalar Multiplication

Let $\mathbf{u}$, $\mathbf{v}$, and $\mathbf{w}$ be vectors and let c and d be scalars. Then the properties listed below are true.

1. $\mathbf{u} + \mathbf{v} = \mathbf{v} + \mathbf{u}$ **2.** $(\mathbf{u} + \mathbf{v}) + \mathbf{w} = \mathbf{u} + (\mathbf{v} + \mathbf{w})$

3. $\mathbf{u} + \mathbf{0} = \mathbf{u}$ **4.** $\mathbf{u} + (-\mathbf{u}) = \mathbf{0}$

5. $c(d\mathbf{u}) = (cd)\mathbf{u}$ **6.** $(c + d)\mathbf{u} = c\mathbf{u} + d\mathbf{u}$

7. $c(\mathbf{u} + \mathbf{v}) = c\mathbf{u} + c\mathbf{v}$ **8.** $1(\mathbf{u}) = \mathbf{u}, \quad 0(\mathbf{u}) = \mathbf{0}$

9. $\|c\mathbf{v}\| = |c| \|\mathbf{v}\|$

EXAMPLE 4   **Finding the Magnitude of a Scalar Multiple**

Let $\mathbf{u} = \langle 1, 3 \rangle$ and $\mathbf{v} = \langle -2, 5 \rangle$. Find the magnitude of each scalar multiple.

a. $\|2\mathbf{u}\|$ **b.** $\|-5\mathbf{u}\|$ **c.** $\|3\mathbf{v}\|$

Solution

a. $\|2\mathbf{u}\| = |2| \|\mathbf{u}\| = |2| \|\langle 1, 3 \rangle\| = |2| \sqrt{1^2 + 3^2} = 2\sqrt{10}$

b. $\|-5\mathbf{u}\| = |-5| \|\mathbf{u}\| = |-5| \|\langle 1, 3 \rangle\| = |-5| \sqrt{1^2 + 3^2} = 5\sqrt{10}$

c. $\|3\mathbf{v}\| = |3| \|\mathbf{v}\| = |3| \|\langle -2, 5 \rangle\| = |3| \sqrt{(-2)^2 + 5^2} = 3\sqrt{29}$

✓ **Checkpoint** ▶ *Audio-video solution in English & Spanish at LarsonPrecalculus.com*

Let $\mathbf{u} = \langle 4, -1 \rangle$ and $\mathbf{v} = \langle 3, 2 \rangle$. Find the magnitude of each scalar multiple.

a. $\|3\mathbf{u}\|$ **b.** $\|-2\mathbf{v}\|$ **c.** $\|5\mathbf{v}\|$

Unit Vectors

In many applications of vectors, it is useful to find a unit vector that has the same direction as a given nonzero vector **v**. To do this, divide **v** by its magnitude to obtain

$$\mathbf{u} = \text{unit vector} = \frac{\mathbf{v}}{\|\mathbf{v}\|} = \left(\frac{1}{\|\mathbf{v}\|}\right)\mathbf{v}. \qquad \text{Unit vector in direction of } \mathbf{v}$$

Note that **u** is a scalar multiple of **v**. The vector **u** has a magnitude of 1 and the same direction as **v**. The vector **u** is called a **unit vector in the direction of v.**

EXAMPLE 5 **Finding a Unit Vector**

Find a unit vector **u** in the direction of $\mathbf{v} = \langle -2, 5 \rangle$. Verify that $\|\mathbf{u}\| = 1$.

Solution The unit vector **u** in the direction of **v** is

$$\mathbf{u} = \frac{\mathbf{v}}{\|\mathbf{v}\|} = \frac{\langle -2, 5 \rangle}{\sqrt{(-2)^2 + 5^2}} = \frac{1}{\sqrt{29}}\langle -2, 5 \rangle = \left\langle \frac{-2}{\sqrt{29}}, \frac{5}{\sqrt{29}} \right\rangle.$$

The vector **u** has a magnitude of 1 because

$$\|\mathbf{u}\| = \sqrt{\left(\frac{-2}{\sqrt{29}}\right)^2 + \left(\frac{5}{\sqrt{29}}\right)^2} = \sqrt{\frac{4}{29} + \frac{25}{29}} = \sqrt{\frac{29}{29}} = 1.$$

✓ **Checkpoint** ▶ Audio-video solution in English & Spanish at LarsonPrecalculus.com

Find a unit vector **u** in the direction of $\mathbf{v} = \langle 6, -1 \rangle$. Verify that $\|\mathbf{u}\| = 1$. ∎

The unit vectors $\langle 1, 0 \rangle$ and $\langle 0, 1 \rangle$ are called the **standard unit vectors** and are denoted by

$$\mathbf{i} = \langle 1, 0 \rangle \quad \text{and} \quad \mathbf{j} = \langle 0, 1 \rangle$$

as shown in Figure 8.18. $\Big($Note that the lowercase letter **i** is in boldface and not italicized to distinguish it from the imaginary unit $i = \sqrt{-1}.\Big)$ These vectors can be used to represent any vector $\mathbf{v} = \langle v_1, v_2 \rangle$ because

$$\mathbf{v} = \langle v_1, v_2 \rangle = \langle v_1, 0 \rangle + \langle 0, v_2 \rangle = v_1\langle 1, 0 \rangle + v_2\langle 0, 1 \rangle = v_1\mathbf{i} + v_2\mathbf{j}.$$

The scalars v_1 and v_2 are the **horizontal** and **vertical components of v,** respectively. The vector sum $v_1\mathbf{i} + v_2\mathbf{j}$ is a **linear combination** of the vectors **i** and **j**. Any vector in the plane can be written as a linear combination of the standard unit vectors **i** and **j**.

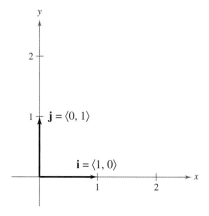

Standard unit vectors **i** and **j**
Figure 8.18

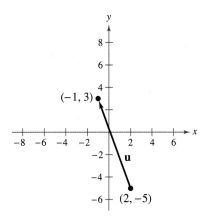 EXAMPLE 6   **Writing a Linear Combination of Unit Vectors**

Let **u** be the vector with initial point $(2, -5)$ and terminal point $(-1, 3)$. Write **u** as a linear combination of the standard unit vectors **i** and **j**.

Solution   Begin by writing the component form of the vector **u**. Then write the component form in terms of **i** and **j**.

$$\mathbf{u} = \langle -1 - 2, 3 - (-5) \rangle = \langle -3, 8 \rangle = -3\mathbf{i} + 8\mathbf{j}$$

This result is shown graphically below.

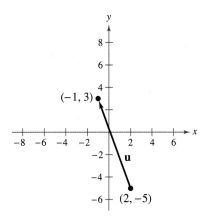

✓ *Checkpoint* ▶ *Audio-video solution in English & Spanish at LarsonPrecalculus.com*

Let **u** be the vector with initial point $(-2, 6)$ and terminal point $(-8, 3)$. Write **u** as a linear combination of the standard unit vectors **i** and **j**.

EXAMPLE 7   **Vector Operations**

Let $\mathbf{u} = -3\mathbf{i} + 8\mathbf{j}$ and $\mathbf{v} = 2\mathbf{i} - \mathbf{j}$. Find $2\mathbf{u} - 3\mathbf{v}$.

Solution   It is not necessary to convert **u** and **v** to component form to solve this problem. Just perform the operations with the vectors in unit vector form.

$$2\mathbf{u} - 3\mathbf{v} = 2(-3\mathbf{i} + 8\mathbf{j}) - 3(2\mathbf{i} - \mathbf{j})$$
$$= -6\mathbf{i} + 16\mathbf{j} - 6\mathbf{i} + 3\mathbf{j}$$
$$= -12\mathbf{i} + 19\mathbf{j}$$

✓ *Checkpoint* ▶ *Audio-video solution in English & Spanish at LarsonPrecalculus.com*

Let $\mathbf{u} = \mathbf{i} - 2\mathbf{j}$ and $\mathbf{v} = -3\mathbf{i} + 2\mathbf{j}$. Find $5\mathbf{u} - 2\mathbf{v}$.   ■

In Example 7, you could perform the operations in component form by writing

$$\mathbf{u} = -3\mathbf{i} + 8\mathbf{j} = \langle -3, 8 \rangle \quad \text{and} \quad \mathbf{v} = 2\mathbf{i} - \mathbf{j} = \langle 2, -1 \rangle.$$

The difference of $2\mathbf{u}$ and $3\mathbf{v}$ is

$$2\mathbf{u} - 3\mathbf{v} = 2\langle -3, 8 \rangle - 3\langle 2, -1 \rangle$$
$$= \langle -6, 16 \rangle - \langle 6, -3 \rangle$$
$$= \langle -6 - 6, 16 - (-3) \rangle$$
$$= \langle -12, 19 \rangle.$$

Compare this result with the solution to Example 7.

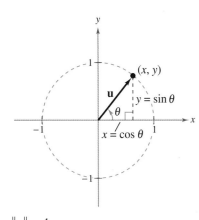

$\|\mathbf{u}\| = 1$
Figure 8.19

Direction Angles

If **u** is a unit vector such that θ is the angle (measured counterclockwise) from the positive x-axis to **u**, then the terminal point of **u** lies on the unit circle and you have

$$\mathbf{u} = \langle x, y \rangle$$
$$= \langle \cos \theta, \sin \theta \rangle$$
$$= (\cos \theta)\mathbf{i} + (\sin \theta)\mathbf{j}$$

as shown in Figure 8.19. The angle θ is the **direction angle** of the vector **u**.

Consider a unit vector **u** with direction angle θ. If $\mathbf{v} = a\mathbf{i} + b\mathbf{j}$ is any nonzero vector making an angle θ with the positive x-axis, then it has the same direction as **u** and you can write

$$\mathbf{v} = \|\mathbf{v}\|\langle \cos \theta, \sin \theta \rangle$$
$$= \|\mathbf{v}\|(\cos \theta)\mathbf{i} + \|\mathbf{v}\|(\sin \theta)\mathbf{j}.$$

Because $\mathbf{v} = a\mathbf{i} + b\mathbf{j} = \|\mathbf{v}\|(\cos \theta)\mathbf{i} + \|\mathbf{v}\|(\sin \theta)\mathbf{j}$, it follows that the direction angle θ for **v** is determined from

$$\tan \theta = \frac{\sin \theta}{\cos \theta} \qquad \text{Quotient identity}$$

$$= \frac{\|\mathbf{v}\| \sin \theta}{\|\mathbf{v}\| \cos \theta} \qquad \text{Multiply numerator and denominator by } \|\mathbf{v}\|.$$

$$= \frac{b}{a}. \qquad \text{Simplify.}$$

EXAMPLE 8   **Finding Direction Angles of Vectors**

Find the direction angle of each vector.

a. $\mathbf{u} = 3\mathbf{i} + 3\mathbf{j}$ **b.** $\mathbf{v} = 3\mathbf{i} - 4\mathbf{j}$

Solution

a. The direction angle is determined from

$$\tan \theta = \frac{b}{a} = \frac{3}{3} = 1.$$

So, $\theta = 45°$, as shown in Figure 8.20(a).

b. The direction angle is determined from

$$\tan \theta = \frac{b}{a} = \frac{-4}{3}.$$

Moreover, $\mathbf{v} = 3\mathbf{i} - 4\mathbf{j}$ lies in Quadrant IV, so θ lies in Quadrant IV, and its reference angle is

$$\theta' = \left| \arctan\left(-\frac{4}{3}\right) \right| \quad \Longrightarrow \quad \theta' \approx |-53.13°| = 53.13°.$$

So, the direction angle is $\theta = 360° - \theta' \approx 360° - 53.13° = 306.87°$, as shown in Figure 8.20(b).

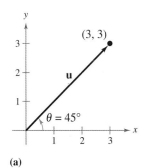

(a)

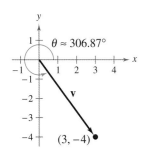

(b)
Figure 8.20

✓ **Checkpoint** ▶ Audio-video solution in English & Spanish at LarsonPrecalculus.com

Find the direction angle of each vector.

a. $\mathbf{v} = -6\mathbf{i} + 6\mathbf{j}$ **b.** $\mathbf{v} = -7\mathbf{i} - 4\mathbf{j}$

Applications

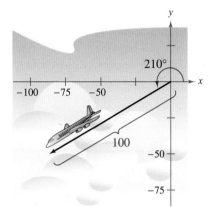

Figure 8.21

| EXAMPLE 9 | Finding a Velocity Vector |

Find the component form of the vector that represents the velocity of an airplane descending at a speed of 100 miles per hour at an angle 30° below the horizontal, as shown in Figure 8.21.

Solution The velocity vector **v** has a magnitude of 100 and a direction angle of $\theta = 210°$.

$$\mathbf{v} = \|\mathbf{v}\|(\cos \theta)\mathbf{i} + \|\mathbf{v}\|(\sin \theta)\mathbf{j}$$
$$= 100(\cos 210°)\mathbf{i} + 100(\sin 210°)\mathbf{j}$$
$$= 100\left(-\frac{\sqrt{3}}{2}\right)\mathbf{i} + 100\left(-\frac{1}{2}\right)\mathbf{j}$$
$$= -50\sqrt{3}\mathbf{i} - 50\mathbf{j}$$
$$= \langle -50\sqrt{3}, -50 \rangle$$

Note that **v** has a magnitude of 100 because

$$\|\mathbf{v}\| = \sqrt{\left(-50\sqrt{3}\right)^2 + (-50)^2} = \sqrt{7500 + 2500} = \sqrt{10,000} = 100.$$

✓ *Checkpoint* ▶ *Audio-video solution in English & Spanish at LarsonPrecalculus.com*

Find the component form of the vector that represents the velocity of an airplane descending at a speed of 100 miles per hour at an angle 15° below the horizontal ($\theta = 195°$).

| EXAMPLE 10 | Using Vectors to Determine Weight |

A force of 600 pounds is required to pull a boat and trailer up a ramp inclined at 15° from the horizontal. Find the combined weight of the boat and trailer.

Solution Use Figure 8.22 to make the observations below.

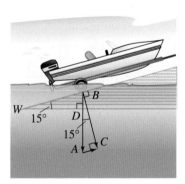

Figure 8.22

$\|\overrightarrow{BA}\| = $ force of gravity $ = $ combined weight of boat and trailer

$\|\overrightarrow{BC}\| = $ force against ramp

$\|\overrightarrow{AC}\| = $ force required to move boat up ramp $ = 600$ pounds

Notice that $\overrightarrow{AC}$ and the ramp are both perpendicular to $\overrightarrow{BC}$. So, $\overrightarrow{AC}$ is parallel to the ramp, and angle *DBW* is equal to angle *A* by alternate interior angles. Also, angles *WDB* and *C* are right angles. So, triangles *BWD* and *ABC* are similar, and angle *ABC* equals 15°. To find the combined weight, use the right triangle definition of the sine function to write an expression relating angle *ABC*, $\|\overrightarrow{AC}\|$, and $\|\overrightarrow{BA}\|$.

$$\sin 15° = \frac{\|\overrightarrow{AC}\|}{\|\overrightarrow{BA}\|} \quad \Longrightarrow \quad \sin 15° = \frac{600}{\|\overrightarrow{BA}\|}$$

Next solve for $\|\overrightarrow{BA}\|$.

$$\|\overrightarrow{BA}\| = \frac{600}{\sin 15°} \quad \Longrightarrow \quad \|\overrightarrow{BA}\| \approx 2318$$

So, the combined weight is approximately 2318 pounds.

✓ *Checkpoint* ▶ *Audio-video solution in English & Spanish at LarsonPrecalculus.com*

A force of 500 pounds is required to pull a boat and trailer up a ramp inclined at 12° from the horizontal. Find the combined weight of the boat and trailer. ■

GO DIGITAL

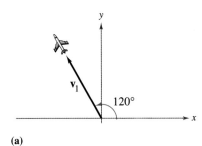

(a)

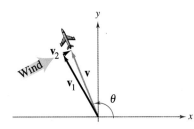

(b)
Figure 8.23

True direction of the airplane
Figure 8.24

EXAMPLE 11   **Using Vectors to Find Speed and Direction**

An airplane travels at a speed of 500 miles per hour with a bearing of 330° at a fixed altitude with a negligible wind velocity, as shown in Figure 8.23(a). (Note that a bearing of 330° corresponds to a direction angle of 120°.) The airplane encounters a wind with a velocity of 70 miles per hour in the direction N 45° E, as shown in Figure 8.23(b). What are the resultant speed and true direction of the airplane? (Recall from Section 6.7 that in air navigation, bearings are measured in degrees clockwise from north. In Figure 8.23, north is in the positive y-direction.)

Solution   Using Figure 8.23, the velocity of the airplane (alone) is

$$\mathbf{v}_1 = 500\langle \cos 120°, \sin 120° \rangle = \langle -250, 250\sqrt{3} \rangle$$

and the velocity of the wind is

$$\mathbf{v}_2 = 70\langle \cos 45°, \sin 45° \rangle = \langle 35\sqrt{2}, 35\sqrt{2} \rangle.$$

So, the velocity of the airplane (in the wind) is

$$\mathbf{v} = \mathbf{v}_1 + \mathbf{v}_2 = \langle -250 + 35\sqrt{2}, 250\sqrt{3} + 35\sqrt{2} \rangle$$

and the resultant speed of the airplane is

$$\|\mathbf{v}\| = \sqrt{\left(-250 + 35\sqrt{2}\right)^2 + \left(250\sqrt{3} + 35\sqrt{2}\right)^2} \approx 522.5 \text{ miles per hour.}$$

To find the direction angle θ of the flight path, you have

$$\tan \theta = \frac{250\sqrt{3} + 35\sqrt{2}}{-250 + 35\sqrt{2}}.$$

The flight path lies in Quadrant II, so θ lies in Quadrant II, and its reference angle is

$$\theta' = \left| \arctan\left(\frac{250\sqrt{3} + 35\sqrt{2}}{-250 + 35\sqrt{2}}\right) \right| \quad \Longrightarrow \quad \theta' \approx 67.4°.$$

So, the direction angle is $\theta = 180° - \theta' \approx 180° - 67.4° = 112.6°$. The true direction of the airplane is approximately $270° + (180° - 112.6°) = 337.4°$. (See Figure 8.24.)

✓ *Checkpoint* ▶ *Audio-video solution in English & Spanish at LarsonPrecalculus.com*

Repeat Example 11 for an airplane traveling at a speed of 450 miles per hour with a bearing of 300° that encounters a wind with a velocity of 40 miles per hour in the direction N 30° E. ∎

Summarize (Section 8.3)

1. Explain how to represent a vector as a directed line segment *(page 576)*. For an example involving vectors represented as directed line segments, see Example 1.

2. Explain how to find the component form of a vector *(page 577)*. For an example of finding the component form of a vector, see Example 2.

3. Explain how to perform basic vector operations *(page 578)*. For an example of performing basic vector operations, see Example 3.

4. Explain how to write a vector as a linear combination of unit vectors *(page 580)*. For examples involving unit vectors, see Examples 5–7.

5. Explain how to find the direction angle of a vector *(page 582)*. For an example of finding direction angles of vectors, see Example 8.

6. Describe real-life applications of vectors *(pages 583 and 584, Examples 9–11)*.

GO DIGITAL

8.3 Exercises

See CalcChat.com for tutorial help and worked-out solutions to odd-numbered exercises.

GO DIGITAL

Vocabulary and Concept Check

In Exercises 1–4, fill in the blanks.

1. You can use a _____ _____ _____ to represent a quantity that involves both magnitude and direction.

2. The set of all directed line segments that are equivalent to a given directed line segment $\vec{PQ}$ is a _____ **v** in the plane.

3. The two basic vector operations are scalar _____ and vector _____.

4. The vector sum $v_1\mathbf{i} + v_2\mathbf{j}$ is a _____ _____ of the vectors **i** and **j**, and the scalars v_1 and v_2 are the _____ and _____ components of **v**, respectively.

5. Explain what determines whether two vectors are equivalent.

6. What characteristic determines whether a vector is a unit vector?

Skills and Applications

Determining Whether Two Vectors Are Equivalent   In Exercises 7–12, determine whether u and v are equivalent. Explain.

7.
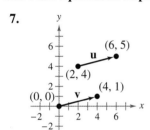

8.
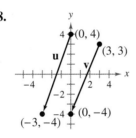

Vector	Initial Point	Terminal Point
9. **u**	$(2, 2)$	$(-1, 4)$
v	$(-3, -1)$	$(-5, 2)$
10. **u**	$(2, 0)$	$(7, 4)$
v	$(-8, 1)$	$(2, 9)$
11. **u**	$(2, -1)$	$(5, -10)$
v	$(6, 1)$	$(9, -8)$
12. **u**	$(8, 1)$	$(13, -1)$
v	$(-2, 4)$	$(-7, 6)$

Finding the Component Form of a Vector   In Exercises 13–22, find the component form and magnitude of the vector v.

13.
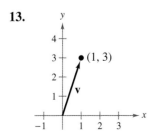

14.
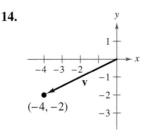

15.
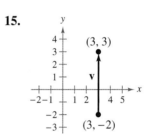

16.
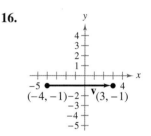

	Initial Point	Terminal Point
17.	$(-3, -5)$	$(-11, 1)$
18.	$(-2, 7)$	$(5, -17)$
19.	$(1, 3)$	$(-8, -9)$
20.	$(17, -5)$	$(9, 3)$
21.	$(-1, 5)$	$(15, -21)$
22.	$(-3, 11)$	$(9, 40)$

Sketching the Graph of a Vector   In Exercises 23–28, use the figure to sketch a graph of the specified vector. To print an enlarged copy of the graph, go to MathGraphs.com.

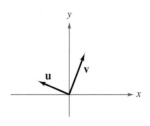

23. $-\mathbf{v}$

24. $5\mathbf{v}$

25. $\mathbf{u} + \mathbf{v}$

26. $\mathbf{u} + 2\mathbf{v}$

27. $\mathbf{u} - \mathbf{v}$

28. $\mathbf{v} - \dfrac{1}{2}\mathbf{u}$

Vector Operations In Exercises 29–34, find (a) **u** + **v**, (b) **u** − **v**, and (c) 2**u** − 3**v**. Then sketch each resultant vector.

29. **u** = $\langle 2, 1 \rangle$,   **v** = $\langle 1, 3 \rangle$
30. **u** = $\langle 2, 3 \rangle$,   **v** = $\langle 4, 0 \rangle$
31. **u** = $\langle -5, 3 \rangle$,   **v** = $\langle 0, 0 \rangle$
32. **u** = $\langle 0, 0 \rangle$,   **v** = $\langle 2, 1 \rangle$
33. **u** = $\langle 0, -7 \rangle$,   **v** = $\langle 1, -2 \rangle$
34. **u** = $\langle -3, 1 \rangle$,   **v** = $\langle 2, -5 \rangle$

Finding the Magnitude of a Scalar Multiple In Exercises 35–38, find the magnitude of the scalar multiple, where **u** = $\langle 2, 0 \rangle$ and **v** = $\langle -3, 6 \rangle$.

35. $\|5\mathbf{u}\|$
36. $\|4\mathbf{v}\|$
37. $\|-3\mathbf{v}\|$
38. $\left\|-\frac{3}{4}\mathbf{u}\right\|$

Finding a Unit Vector In Exercises 39–44, find a unit vector **u** in the direction of **v**. Verify that $\|\mathbf{u}\| = 1$.

39. **v** = $\langle 3, 0 \rangle$
40. **v** = $\langle 0, -2 \rangle$
41. **v** = $\langle -2, 2 \rangle$
42. **v** = $\langle -5, 12 \rangle$
43. **v** = $\langle 1, -6 \rangle$
44. **v** = $\langle -8, -4 \rangle$

Finding a Vector In Exercises 45–48, find the vector **v** with the given magnitude and the same direction as **u**.

45. $\|\mathbf{v}\| = 10$,   **u** = $\langle -3, 4 \rangle$
46. $\|\mathbf{v}\| = 3$,   **u** = $\langle -12, -5 \rangle$
47. $\|\mathbf{v}\| = 9$,   **u** = $\langle 2, 5 \rangle$
48. $\|\mathbf{v}\| = 8$,   **u** = $\langle 3, 3 \rangle$

Writing a Linear Combination of Unit Vectors In Exercises 49–52, the initial and terminal points of a vector are given. Write the vector as a linear combination of the standard unit vectors **i** and **j**.

Initial Point	Terminal Point
49. $(-2, 1)$	$(3, -2)$
50. $(0, -2)$	$(3, 6)$
51. $(0, 1)$	$(-6, 4)$
52. $(2, 3)$	$(-1, -5)$

Vector Operations In Exercises 53–58, find the component form of **v** and sketch the specified vector operations geometrically, where **u** = 2**i** − **j** and **w** = **i** + 2**j**.

53. $\mathbf{v} = \frac{3}{2}\mathbf{u}$
54. $\mathbf{v} = \frac{3}{4}\mathbf{w}$
55. $\mathbf{v} = \mathbf{u} + 2\mathbf{w}$
56. $\mathbf{v} = -\mathbf{u} + \mathbf{w}$
57. $\mathbf{v} = \mathbf{u} - 2\mathbf{w}$
58. $\mathbf{v} = \frac{1}{2}(3\mathbf{u} + \mathbf{w})$

Finding the Direction Angle of a Vector In Exercises 59–62, find the magnitude and direction angle of the vector **v**.

59. **v** = 6**i** − 6**j**
60. **v** = −5**i** + 4**j**
61. **v** = $3(\cos 60°\mathbf{i} + \sin 60°\mathbf{j})$
62. **v** = $8(\cos 135°\mathbf{i} + \sin 135°\mathbf{j})$

Finding the Component Form of a Vector In Exercises 63–68, find the component form of **v** given its magnitude and the angle it makes with the positive x-axis. Then sketch **v**.

Magnitude	Angle
63. $\|\mathbf{v}\| = 3$	$\theta = 0°$
64. $\|\mathbf{v}\| = 4\sqrt{3}$	$\theta = 90°$
65. $\|\mathbf{v}\| = \frac{7}{2}$	$\theta = 150°$
66. $\|\mathbf{v}\| = 2\sqrt{3}$	$\theta = 45°$
67. $\|\mathbf{v}\| = 3$	**v** in the direction $3\mathbf{i} + 4\mathbf{j}$
68. $\|\mathbf{v}\| = 2$	**v** in the direction $\mathbf{i} + 3\mathbf{j}$

Finding the Component Form of a Vector In Exercises 69 and 70, find the component form of the sum of **u** and **v** with direction angles $\theta_{\mathbf{u}}$ and $\theta_{\mathbf{v}}$.

69. $\|\mathbf{u}\| = 4$, $\theta_{\mathbf{u}} = 60°$
    $\|\mathbf{v}\| = 4$, $\theta_{\mathbf{v}} = 90°$
70. $\|\mathbf{u}\| = 20$, $\theta_{\mathbf{u}} = 45°$
    $\|\mathbf{v}\| = 50$, $\theta_{\mathbf{v}} = 180°$

Using the Law of Cosines In Exercises 71 and 72, use the Law of Cosines to find the angle α between the vectors. (Assume $0° \le \alpha \le 180°$.)

71. **v** = **i** + **j**,   **w** = 2**i** − 2**j**
72. **v** = **i** + 2**j**,   **w** = 2**i** − **j**

Resultant Force In Exercises 73 and 74, find the angle between the forces given the magnitude of their resultant. (*Hint:* Write force 1 as a vector in the direction of the positive x-axis and force 2 as a vector at an angle θ with the positive x-axis.)

Force 1	Force 2	Resultant Force
73. 45 pounds	60 pounds	90 pounds
74. 3000 pounds	1000 pounds	3750 pounds

75. **Velocity** A bee is flying with a velocity of 22 feet per second at an angle of 6° above the horizontal. Find the vertical and horizontal components of the velocity.

76. **Velocity** Pitcher Aroldis Chapman threw a pitch with a recorded velocity of 105 miles per hour. Assuming he threw the pitch at an angle of 3.5° below the horizontal, find the vertical and horizontal components of the velocity. (*Source: Guinness World Records*)

77. Resultant Force Forces with magnitudes of 125 newtons and 300 newtons act on a hook (see figure). The angle between the two forces is 45°. Find the direction and magnitude of the resultant of these forces. (*Hint:* Write the vector representing each force in component form, then add the vectors.)

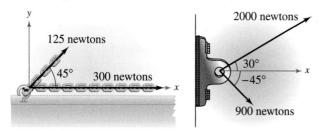

Figure for 77   Figure for 78

78. Resultant Force Forces with magnitudes of 2000 newtons and 900 newtons act on a machine part at angles of 30° and −45°, respectively, with the positive x-axis (see figure). Find the direction and magnitude of the resultant of these forces.

79. Resultant Force Three forces with magnitudes of 75 pounds, 100 pounds, and 125 pounds act on an object at angles of 30°, 45°, and 120°, respectively, with the positive x-axis. Find the direction and magnitude of the resultant of these forces.

80. Resultant Force Three forces with magnitudes of 70 pounds, 40 pounds, and 60 pounds act on an object at angles of −30°, 45°, and 135°, respectively, with the positive x-axis. Find the direction and magnitude of the resultant of these forces.

81. Cable Tension The cranes shown in the figure are lifting an object that weighs 20,240 pounds. Find the tension (in pounds) in the cable of each crane.

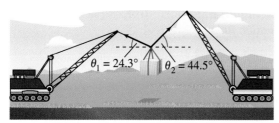

82. Cable Tension Repeat Exercise 81 for $\theta_1 = 35.6°$ and $\theta_2 = 40.4°$.

83. Rope Tension A tetherball weighing 1 pound is pulled outward from the pole by a horizontal force **u** until the rope makes a 45° angle with the pole (see figure). Determine the resulting tension (in pounds) in the rope and the magnitude of **u**.

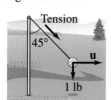

84. Physics Use the figure to determine the tension (in pounds) in each cable supporting the load.

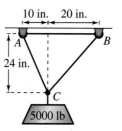

Inclined Ramp In Exercises 85–88, a force of F pounds is required to pull an object weighing W pounds up a ramp inclined at θ degrees from the horizontal.

85. Find F when $W = 100$ pounds and $\theta = 12°$.

86. Find W when $F = 600$ pounds and $\theta = 14°$.

87. Find θ when $F = 5000$ pounds and $W = 15,000$ pounds.

88. Find F when $W = 5000$ pounds and $\theta = 26°$.

89. Air Navigation An airplane travels in the direction of 148° with an airspeed of 875 kilometers per hour. Due to the wind, its ground speed and direction are 800 kilometers per hour and 140°, respectively (see figure). Find the direction and speed of the wind.

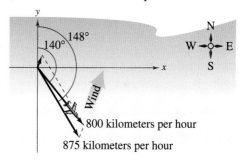

90. Air Navigation

A commercial jet travels from Miami to Seattle. The jet's velocity with respect to the air is 580 miles per hour, and its bearing is 332°. The jet encounters a wind with a velocity of 60 miles per hour from the southwest.

(a) Draw a diagram that gives a visual representation of the problem.

(b) Write the velocity of the wind as a vector in component form.

(c) Write the velocity of the jet relative to the air in component form.

(d) What is the speed of the jet with respect to the ground?

(e) What is the true direction of the jet?

Exploring the Concepts

True or False? In Exercises 91–94, determine whether the statement is true or false. Justify your answer.

91. If **u** and **v** have the same magnitude and direction, then **u** and **v** are equivalent.

92. If **u** is a unit vector in the direction of **v**, then $\mathbf{v} = \|\mathbf{v}\|\mathbf{u}$.

93. If $\mathbf{v} = a\mathbf{i} + b\mathbf{j} = \mathbf{0}$, then $a = -b$.

94. If $\mathbf{u} = a\mathbf{i} + b\mathbf{j}$ is a unit vector, then $a^2 + b^2 = 1$.

95. **Error Analysis** Describe the error in finding the component form of the vector **u** that has initial point $(-3, 4)$ and terminal point $(6, -1)$.

The components are $u_1 = -3 - 6 = -9$ and $u_2 = 4 - (-1) = 5$. So, $\mathbf{u} = \langle -9, 5 \rangle$.

96. **Error Analysis** Describe the error in finding the direction angle θ of the vector $\mathbf{v} = -5\mathbf{i} + 8\mathbf{j}$.

Because $\tan \theta = \dfrac{b}{a} = \dfrac{8}{-5}$, the reference angle

is $\theta' = \left| \arctan\left(-\dfrac{8}{5}\right) \right| \approx |-57.99°| = 57.99°$

and $\theta \approx 360° - 57.99° = 302.01°$.

97. **Proof** Prove that

$$(\cos \theta)\mathbf{i} + (\sin \theta)\mathbf{j}$$

is a unit vector for any value of θ.

98. **Technology** Write a program for your graphing utility that graphs two vectors and their difference given the vectors in component form.

 Finding the Difference of Two Vectors In Exercises 99 and 100, use the program in Exercise 98 to find the difference of the vectors shown in the figure.

99. 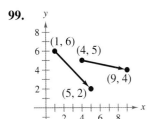   100.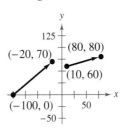

101. **Graphical Reasoning** Consider two forces

$$\mathbf{F}_1 = \langle 10, 0 \rangle \quad \text{and} \quad \mathbf{F}_2 = 5\langle \cos \theta, \sin \theta \rangle.$$

(a) Find $\|\mathbf{F}_1 + \mathbf{F}_2\|$ as a function of θ.

(b) Use a graphing utility to graph the function in part (a) for $0 \le \theta < 2\pi$.

(c) Use the graph in part (b) to determine the range of the function. What is its maximum, and for what value of θ does it occur? What is its minimum, and for what value of θ does it occur?

(d) Explain why the magnitude of the resultant is never 0.

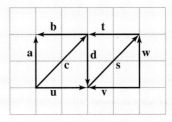

102. **HOW DO YOU SEE IT?** Use the figure to determine whether each statement is true or false. Justify your answer.

(a) $\mathbf{a} = -\mathbf{d}$ (b) $\mathbf{c} = \mathbf{s}$

(c) $\mathbf{a} + \mathbf{u} = \mathbf{c}$ (d) $\mathbf{v} + \mathbf{w} = -\mathbf{s}$

(e) $\mathbf{a} + \mathbf{w} = -2\mathbf{d}$ (f) $\mathbf{a} + \mathbf{d} = \mathbf{0}$

(g) $\mathbf{u} - \mathbf{v} = -2(\mathbf{b} + \mathbf{t})$ (h) $\mathbf{t} - \mathbf{w} = \mathbf{b} - \mathbf{a}$

Review & Refresh ▶ Video solutions at *LarsonPrecalculus.com*

Finding a Distance In Exercises 103–110, find the distance between the points. Round your answer to two decimal places.

103. $(2, 8), (8, 2)$

104. $(1, 4), (5, 18)$

105. $(-15, 16), (-14, 19)$

106. $(-14, 3), (-12, 9)$

107. $(5, -20), (11, -19)$

108. $(-16, -3), (-15, -10)$

109. $(-3, 0), (13, 6)$

110. $(0, -7), (4, 12)$

Solving an Equation In Exercises 111–114, solve the equation for x.

111. $\cos x = \dfrac{-3(3) + 1(9)}{\sqrt{10(90)}}$

112. $\cos x = \dfrac{40}{\sqrt{1600}}$

113. $\cos \dfrac{\pi}{4} = \dfrac{4x + 21}{\sqrt{1250}}$

114. $\cos \dfrac{\pi}{6} = \dfrac{15}{\sqrt{3(25 + x^2)}}$

Solving a Triangle In Exercises 115–118, solve the triangle. If two solutions exist, find both. Round your answers to two decimal places.

115. $a = 9, b = 8, c = 16$

116. $A = 27°, a = 28, c = 38$

117. $B = 59°, C = 84°, c = 20$

118. $B = 18°, a = 18, c = 37$

8.4 Vectors and Dot Products

The dot product of two vectors has many real-life applications. For example, in Exercise 74 on page 596, you will use the dot product to find the force necessary to keep a sport utility vehicle from rolling down a hill.

❯ Find the dot product of two vectors and use the properties of the dot product.
❯ Find the angle between two vectors and determine whether two vectors are orthogonal.
❯ Write a vector as the sum of two vector components.
❯ Use vectors to determine the work done by a force.

The Dot Product of Two Vectors

So far, you have studied two vector operations—vector addition and multiplication by a scalar—each of which yields another vector. In this section, you will study a third vector operation, the **dot product.** This operation yields a scalar, rather than a vector.

> **Definition of the Dot Product**
>
> The **dot product** of $\mathbf{u} = \langle u_1, u_2 \rangle$ and $\mathbf{v} = \langle v_1, v_2 \rangle$ is $\mathbf{u} \cdot \mathbf{v} = u_1 v_1 + u_2 v_2$.

EXAMPLE 1 Finding Dot Products

Find each dot product.

a. $\langle 4, 5 \rangle \cdot \langle 2, 3 \rangle$ **b.** $\langle 2, -1 \rangle \cdot \langle 1, 2 \rangle$ **c.** $\langle 0, 3 \rangle \cdot \langle 4, -2 \rangle$

Solution

a. $\langle 4, 5 \rangle \cdot \langle 2, 3 \rangle = 4(2) + 5(3) = 8 + 15 = 23$

b. $\langle 2, -1 \rangle \cdot \langle 1, 2 \rangle = 2(1) + (-1)(2) = 2 - 2 = 0$

c. $\langle 0, 3 \rangle \cdot \langle 4, -2 \rangle = 0(4) + 3(-2) = 0 - 6 = -6$

✓ **Checkpoint** ▶ *Audio-video solution in English & Spanish at LarsonPrecalculus.com*

Find each dot product.

a. $\langle 3, 4 \rangle \cdot \langle 2, -3 \rangle$ **b.** $\langle -3, -5 \rangle \cdot \langle 1, -8 \rangle$ **c.** $\langle -6, 5 \rangle \cdot \langle 5, 6 \rangle$ ∎

In Example 1, be sure you see that the dot product of two vectors is a scalar (a real number), *not* a vector. Moreover, notice that the dot product can be positive, zero, or negative.

> **Properties of the Dot Product**
>
> Let **u**, **v**, and **w** be vectors and let c be a scalar.
>
> **1.** $\mathbf{u} \cdot \mathbf{v} = \mathbf{v} \cdot \mathbf{u}$ Commutative Property
>
> **2.** $\mathbf{u} \cdot (\mathbf{v} + \mathbf{w}) = \mathbf{u} \cdot \mathbf{v} + \mathbf{u} \cdot \mathbf{w}$ Distributive Property
>
> **3.** $c(\mathbf{u} \cdot \mathbf{v}) = c\mathbf{u} \cdot \mathbf{v} = \mathbf{u} \cdot c\mathbf{v}$ Associative Property
>
> **4.** $\mathbf{0} \cdot \mathbf{v} = 0$
>
> **5.** $\mathbf{v} \cdot \mathbf{v} = \|\mathbf{v}\|^2$

For proofs of the properties of the dot product, see Proofs in Mathematics on page 624.

GO DIGITAL

EXAMPLE 2 **Using Properties of the Dot Product**

Let $\mathbf{u} = \langle -1, 3 \rangle$, $\mathbf{v} = \langle 2, -4 \rangle$, and $\mathbf{w} = \langle 1, -2 \rangle$. Use properties of the dot product to find (a) $(\mathbf{u} \cdot \mathbf{v})\mathbf{w}$, (b) $\mathbf{u} \cdot 2\mathbf{v}$, and (c) $\|\mathbf{u}\|$.

Solution Begin by finding the dot product of $\mathbf{u}$ and $\mathbf{v}$ and the dot product of $\mathbf{u}$ and $\mathbf{u}$.

$$\mathbf{u} \cdot \mathbf{v} = \langle -1, 3 \rangle \cdot \langle 2, -4 \rangle = -1(2) + 3(-4) = -14$$

$$\mathbf{u} \cdot \mathbf{u} = \langle -1, 3 \rangle \cdot \langle -1, 3 \rangle = -1(-1) + 3(3) = 10$$

a. $(\mathbf{u} \cdot \mathbf{v})\mathbf{w} = -14\langle 1, -2 \rangle = \langle -14, 28 \rangle$

b. $\mathbf{u} \cdot 2\mathbf{v} = 2(\mathbf{u} \cdot \mathbf{v}) = 2(-14) = -28$

c. Because $\|\mathbf{u}\|^2 = \mathbf{u} \cdot \mathbf{u} = 10$, it follows that $\|\mathbf{u}\| = \sqrt{\mathbf{u} \cdot \mathbf{u}} = \sqrt{10}$.

Notice that the product in part (a) is a vector, whereas the product in part (b) is a scalar. Can you see why?

✓ *Checkpoint* ▶ *Audio-video solution in English & Spanish at LarsonPrecalculus.com*

Let $\mathbf{u} = \langle 3, 4 \rangle$ and $\mathbf{v} = \langle -2, 6 \rangle$. Use properties of the dot product to find (a) $(\mathbf{u} \cdot \mathbf{v})\mathbf{v}$, (b) $\mathbf{u} \cdot (\mathbf{u} + \mathbf{v})$, and (c) $\|\mathbf{v}\|$.

The Angle Between Two Vectors

The **angle between two nonzero vectors** is the angle θ, $0 \le \theta \le \pi$, between their respective standard position vectors, as shown in Figure 8.25. This angle can be found using the dot product.

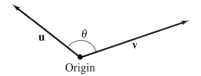

u θ v

Origin

Figure 8.25

> **Angle Between Two Vectors**
>
> If θ is the angle between two nonzero vectors $\mathbf{u}$ and $\mathbf{v}$, then
>
> $$\cos \theta = \frac{\mathbf{u} \cdot \mathbf{v}}{\|\mathbf{u}\| \|\mathbf{v}\|}.$$

For a proof of the angle between two vectors, see Proofs in Mathematics on page 624.

EXAMPLE 3 **Finding the Angle Between Two Vectors**

▶▶▶ *See LarsonPrecalculus.com for an interactive version of this type of example.*

Find the angle θ between $\mathbf{u} = \langle 4, 3 \rangle$ and $\mathbf{v} = \langle 3, 5 \rangle$.

Solution Begin by sketching a graph of the vectors, as shown in Figure 8.26.

$$\cos \theta = \frac{\mathbf{u} \cdot \mathbf{v}}{\|\mathbf{u}\| \|\mathbf{v}\|} = \frac{\langle 4, 3 \rangle \cdot \langle 3, 5 \rangle}{\|\langle 4, 3 \rangle\| \|\langle 3, 5 \rangle\|} = \frac{4(3) + 3(5)}{\sqrt{4^2 + 3^2}\sqrt{3^2 + 5^2}} = \frac{27}{5\sqrt{34}}$$

This implies that the angle between the two vectors is

$$\theta = \cos^{-1}\frac{27}{5\sqrt{34}} \approx 0.3869 \text{ radian.}$$ Use a calculator.

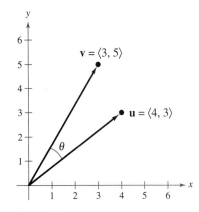

Figure 8.26

✓ *Checkpoint* ▶ *Audio-video solution in English & Spanish at LarsonPrecalculus.com*

Find the angle θ between $\mathbf{u} = \langle 2, 1 \rangle$ and $\mathbf{v} = \langle 1, 3 \rangle$.

GO DIGITAL

Rewriting the expression for the angle between two vectors in the form

$$\mathbf{u} \cdot \mathbf{v} = \|\mathbf{u}\| \, \|\mathbf{v}\| \cos \theta \qquad \text{Alternative form of dot product}$$

produces an alternative way to calculate the dot product. This form shows that $\mathbf{u} \cdot \mathbf{v}$ and $\cos \theta$ always have the same sign, because $\|\mathbf{u}\|$ and $\|\mathbf{v}\|$ are always positive. The figures below show the five possible orientations of two vectors.

Opposite direction	Obtuse angle $\mathbf{u} \cdot \mathbf{v} < 0$	Right angle $\mathbf{u} \cdot \mathbf{v} = 0$	Acute angle $\mathbf{u} \cdot \mathbf{v} > 0$	Same direction
$\theta = \pi$	$\dfrac{\pi}{2} < \theta < \pi$	$\theta = \dfrac{\pi}{2}$	$0 < \theta < \dfrac{\pi}{2}$	$\theta = 0$
$\cos \theta = -1$	$-1 < \cos \theta < 0$	$\cos \theta = 0$	$0 < \cos \theta < 1$	$\cos \theta = 1$

Definition of Orthogonal Vectors

The vectors $\mathbf{u}$ and $\mathbf{v}$ are **orthogonal** if and only if $\mathbf{u} \cdot \mathbf{v} = 0$.

The terms *orthogonal* and *perpendicular* have essentially the same meaning—meeting at right angles. Even though the angle between the zero vector and another vector is not defined, it is convenient to extend the definition of orthogonality to include the zero vector. In other words, the zero vector is orthogonal to every vector $\mathbf{u}$, because $\mathbf{0} \cdot \mathbf{u} = 0$.

>>> **TECHNOLOGY**

A graphing utility program that graphs two vectors and finds the angle between them is available at *LarsonPrecalculus.com*. Use this program, called "Finding the Angle Between Two Vectors," to verify the solutions to Examples 3 and 4.

EXAMPLE 4 Determining Orthogonal Vectors

Determine whether the vectors $\mathbf{u} = \langle 2, -3 \rangle$ and $\mathbf{v} = \langle 6, 4 \rangle$ are orthogonal.

Solution

Find the dot product of the two vectors.

$$\mathbf{u} \cdot \mathbf{v} = \langle 2, -3 \rangle \cdot \langle 6, 4 \rangle = 2(6) + (-3)(4) = 0$$

The dot product is 0, so the two vectors are orthogonal (see figure below).

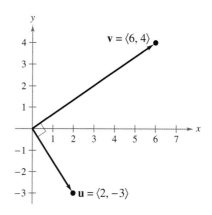

✓ Checkpoint ▶ Audio-video solution in English & Spanish at LarsonPrecalculus.com

Determine whether the vectors $\mathbf{u} = \langle 6, 10 \rangle$ and $\mathbf{v} = \left\langle -\frac{1}{3}, \frac{1}{5} \right\rangle$ are orthogonal.

GO DIGITAL

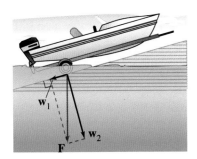

Figure 8.27

Finding Vector Components

You have seen applications in which you add two vectors to produce a resultant vector. Many applications in physics and engineering pose the reverse problem—decomposing a given vector into the sum of two **vector components.**

For example, consider a boat on an inclined ramp, as shown in Figure 8.27. The force **F** due to gravity pulls the boat *down* the ramp and *against* the ramp. These two orthogonal forces, $\mathbf{w}_1$ and $\mathbf{w}_2$, are vector components of **F**. That is,

$$\mathbf{F} = \mathbf{w}_1 + \mathbf{w}_2. \qquad \text{Vector components of } \mathbf{F}$$

The negative of component $\mathbf{w}_1$ represents the force needed to keep the boat from rolling down the ramp, whereas $\mathbf{w}_2$ represents the force that the tires must withstand against the ramp. A procedure for finding $\mathbf{w}_1$ and $\mathbf{w}_2$ is developed below.

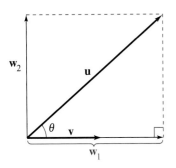

θ is acute.

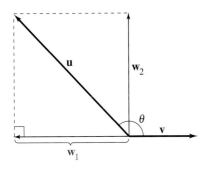

θ is obtuse.
Figure 8.28

Definition of Vector Components

Let **u** and **v** be nonzero vectors such that

$$\mathbf{u} = \mathbf{w}_1 + \mathbf{w}_2$$

where $\mathbf{w}_1$ and $\mathbf{w}_2$ are orthogonal and $\mathbf{w}_1$ is parallel to (or a scalar multiple of) **v**, as shown in Figure 8.28. The vectors $\mathbf{w}_1$ and $\mathbf{w}_2$ are **vector components** of **u**. The vector $\mathbf{w}_1$ is the **projection** of **u** onto **v** and is denoted by

$$\mathbf{w}_1 = \text{proj}_{\mathbf{v}}\mathbf{u}.$$

The vector $\mathbf{w}_2$ is given by

$$\mathbf{w}_2 = \mathbf{u} - \mathbf{w}_1.$$

From the definition of vector components, notice that you can find the component $\mathbf{w}_2$ once you have found the projection of **u** onto **v**. To find the projection, use the dot product, as shown below.

$$\mathbf{u} = \mathbf{w}_1 + \mathbf{w}_2$$

$$\mathbf{u} = c\mathbf{v} + \mathbf{w}_2 \qquad \text{$\mathbf{w}_1$ is a scalar multiple of } \mathbf{v}.$$

$$\mathbf{u} \cdot \mathbf{v} = (c\mathbf{v} + \mathbf{w}_2) \cdot \mathbf{v} \qquad \text{Take the dot product of each side with } \mathbf{v}.$$

$$\mathbf{u} \cdot \mathbf{v} = c\mathbf{v} \cdot \mathbf{v} + \mathbf{w}_2 \cdot \mathbf{v} \qquad \text{Distributive Property}$$

$$\mathbf{u} \cdot \mathbf{v} = c\|\mathbf{v}\|^2 + 0 \qquad \text{$\mathbf{w}_2$ and $\mathbf{v}$ are orthogonal.}$$

So, solving for c yields

$$c = \frac{\mathbf{u} \cdot \mathbf{v}}{\|\mathbf{v}\|^2} \qquad \text{Solve for } c.$$

and $\mathbf{w}_1$, the projection of **u** onto **v**, is

$$\mathbf{w}_1 = \text{proj}_{\mathbf{v}}\mathbf{u} = c\mathbf{v} = \left(\frac{\mathbf{u} \cdot \mathbf{v}}{\|\mathbf{v}\|^2}\right)\mathbf{v}.$$

Projection of u onto v

Let **u** and **v** be nonzero vectors. The projection of **u** onto **v** is given by

$$\text{proj}_{\mathbf{v}}\mathbf{u} = \left(\frac{\mathbf{u} \cdot \mathbf{v}}{\|\mathbf{v}\|^2}\right)\mathbf{v}.$$

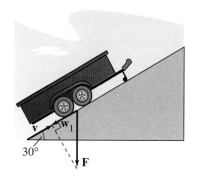

Figure 8.29

EXAMPLE 5 **Decomposing a Vector into Components**

Find the projection of $\mathbf{u} = \langle 3, -5 \rangle$ onto $\mathbf{v} = \langle 6, 2 \rangle$. Then write $\mathbf{u}$ as the sum of two orthogonal vectors, one of which is $\text{proj}_\mathbf{v}\mathbf{u}$.

Solution The projection of $\mathbf{u}$ onto $\mathbf{v}$ is

$$\mathbf{w}_1 = \text{proj}_\mathbf{v}\mathbf{u} = \left(\frac{\mathbf{u} \cdot \mathbf{v}}{\|\mathbf{v}\|^2}\right)\mathbf{v} = \left(\frac{8}{40}\right)\langle 6, 2 \rangle = \left\langle \frac{6}{5}, \frac{2}{5} \right\rangle$$

as shown in Figure 8.29. The component $\mathbf{w}_2$ is

$$\mathbf{w}_2 = \mathbf{u} - \mathbf{w}_1 = \langle 3, -5 \rangle - \left\langle \frac{6}{5}, \frac{2}{5} \right\rangle = \left\langle \frac{9}{5}, -\frac{27}{5} \right\rangle.$$

So,

$$\mathbf{u} = \mathbf{w}_1 + \mathbf{w}_2 = \left\langle \frac{6}{5}, \frac{2}{5} \right\rangle + \left\langle \frac{9}{5}, -\frac{27}{5} \right\rangle = \langle 3, -5 \rangle.$$

✓ *Checkpoint* ▶ *Audio-video solution in English & Spanish at LarsonPrecalculus.com*

Find the projection of $\mathbf{u} = \langle 3, 4 \rangle$ onto $\mathbf{v} = \langle 8, 2 \rangle$. Then write $\mathbf{u}$ as the sum of two orthogonal vectors, one of which is $\text{proj}_\mathbf{v}\mathbf{u}$. ∎

EXAMPLE 6 **Finding a Force**

A 200-pound cart is on a ramp inclined at $30°$, as shown in Figure 8.30. What force is required to keep the cart from rolling down the ramp?

Solution The force due to gravity is vertical and downward, so use the vector

$$\mathbf{F} = -200\mathbf{j} \qquad \text{Force due to gravity}$$

to represent the gravitational force. To find the force required to keep the cart from rolling down the ramp, project $\mathbf{F}$ onto a unit vector $\mathbf{v}$ in the direction of the ramp, where

$$\mathbf{v} = (\cos 30°)\mathbf{i} + (\sin 30°)\mathbf{j}$$

$$= \frac{\sqrt{3}}{2}\mathbf{i} + \frac{1}{2}\mathbf{j}. \qquad \text{Unit vector along ramp}$$

So, the projection of $\mathbf{F}$ onto $\mathbf{v}$ is

$$\mathbf{w}_1 = \text{proj}_\mathbf{v}\mathbf{F}$$

$$= \left(\frac{\mathbf{F} \cdot \mathbf{v}}{\|\mathbf{v}\|^2}\right)\mathbf{v}$$

$$= (\mathbf{F} \cdot \mathbf{v})\mathbf{v} \qquad \|\mathbf{v}\|^2 = 1$$

$$= (-200)\left(\frac{1}{2}\right)\mathbf{v}$$

$$= -100\left(\frac{\sqrt{3}}{2}\mathbf{i} + \frac{1}{2}\mathbf{j}\right).$$

The magnitude of this force is 100. So, a force of 100 pounds is required to keep the cart from rolling down the ramp.

✓ *Checkpoint* ▶ *Audio-video solution in English & Spanish at LarsonPrecalculus.com*

Rework Example 6 for a 150-pound cart that is on a ramp inclined at $15°$. ∎

Figure 8.30

Work

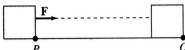

Force acts along the line of motion.
Figure 8.31

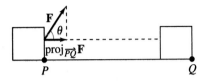

Force acts at angle θ with the line of motion.
Figure 8.32

The work W done by a constant force $\mathbf{F}$ acting along the line of motion of an object is given by

$$W = (\text{magnitude of force})(\text{distance}) = \|\mathbf{F}\|\|\overrightarrow{PQ}\|$$

as shown in Figure 8.31. When the constant force $\mathbf{F}$ is *not* directed along the line of motion, as shown in Figure 8.32, the work W done by the force is given by

$$W = \|\text{proj}_{\overrightarrow{PQ}}\mathbf{F}\|\ \|\overrightarrow{PQ}\| \qquad \text{Projection form for work}$$
$$= (\cos\theta)\|\mathbf{F}\|\ \|\overrightarrow{PQ}\| \qquad \|\text{proj}_{\overrightarrow{PQ}}\mathbf{F}\| = (\cos\theta)\|\mathbf{F}\|$$
$$= \mathbf{F}\cdot\overrightarrow{PQ}. \qquad \text{Dot product form for work}$$

The definition below summarizes the concept of work.

Definition of Work

The **work** W done by a constant force $\mathbf{F}$ as its point of application moves along the vector $\overrightarrow{PQ}$ is given by either formula below.

1. $W = \|\text{proj}_{\overrightarrow{PQ}}\mathbf{F}\|\ \|\overrightarrow{PQ}\|$ Projection form

2. $W = \mathbf{F}\cdot\overrightarrow{PQ}$ Dot product form

Figure 8.33

EXAMPLE 7   **Determining the Work Done**

To close a sliding barn door, a person pulls on a rope with a constant force of 50 pounds at a constant angle of 60°, as shown in Figure 8.33. Determine the work done in moving the barn door 12 feet to its closed position.

Solution   To determine the work done, project $\mathbf{F}$ onto $\overrightarrow{PQ}$.

$$W = \|\text{proj}_{\overrightarrow{PQ}}\mathbf{F}\|\ \|\overrightarrow{PQ}\| = (\cos 60°)\|\mathbf{F}\|\ \|\overrightarrow{PQ}\| = \frac{1}{2}(50)(12) = 300 \text{ foot-pounds}$$

So, the work done is 300 foot-pounds. Verify this result by finding the vectors $\mathbf{F}$ and $\overrightarrow{PQ}$ and calculating their dot product.

✓ *Checkpoint* ▶ *Audio-video solution in English & Spanish at LarsonPrecalculus.com*

A person pulls a wagon by exerting a constant force of 35 pounds on a handle that makes a 30° angle with the horizontal. Determine the work done in pulling the wagon 40 feet. ■

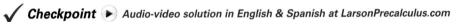

Summarize (Section 8.4)

1. State the definition of the dot product and list the properties of the dot product *(page 589)*. For examples of finding dot products and using the properties of the dot product, see Examples 1 and 2.

2. Explain how to find the angle between two vectors and how to determine whether two vectors are orthogonal *(page 590)*. For examples involving the angle between two vectors, see Examples 3 and 4.

3. Explain how to write a vector as the sum of two vector components *(page 592)*. For examples involving vector components, see Examples 5 and 6.

4. State the definition of work *(page 594)*. For an example of determining work, see Example 7.

8.4 Exercises

See CalcChat.com for tutorial help and worked-out solutions to odd-numbered exercises.

Vocabulary and Concept Check

1. Give an example of the Distributive Property of the dot product.

2. Is the dot product of two vectors an angle, a vector, or a scalar?

In Exercises 3 and 4, fill in the blanks.

3. The projection of **u** onto **v** is given by proj$_v$**u** = _____.

4. The work W done by a constant force **F** as its point of application moves along the vector $\overrightarrow{PQ}$ is given by $W =$ _____ or $W =$ _____.

Skills and Applications

Finding a Dot Product   In Exercises 5–10, find u · v.

5. $\mathbf{u} = \langle 7, 1 \rangle$
   $\mathbf{v} = \langle -3, 2 \rangle$

6. $\mathbf{u} = \langle 6, 10 \rangle$
   $\mathbf{v} = \langle -2, 3 \rangle$

7. $\mathbf{u} = \langle -6, 2 \rangle$
   $\mathbf{v} = \langle 1, 3 \rangle$

8. $\mathbf{u} = \langle -2, 5 \rangle$
   $\mathbf{v} = \langle -1, -8 \rangle$

9. $\mathbf{u} = 4\mathbf{i} - 2\mathbf{j}$
   $\mathbf{v} = \mathbf{i} - \mathbf{j}$

10. $\mathbf{u} = \mathbf{i} - 2\mathbf{j}$
    $\mathbf{v} = -2\mathbf{i} - \mathbf{j}$

Using Properties of the Dot Product   In Exercises 11–22, use the vectors u = ⟨3, 3⟩, v = ⟨−4, 2⟩, and w = ⟨3, −1⟩ to find the quantity. State whether the result is a vector or a scalar.

11. $\mathbf{u} \cdot \mathbf{u}$

12. $3\mathbf{u} \cdot \mathbf{v}$

13. $(\mathbf{u} \cdot \mathbf{v})\mathbf{v}$

14. $(\mathbf{u} \cdot 2\mathbf{v})\mathbf{w}$

15. $(\mathbf{v} \cdot \mathbf{0})\mathbf{w}$

16. $(\mathbf{u} + \mathbf{v}) \cdot \mathbf{0}$

17. $3\|\mathbf{v}\|^2$

18. $-2\|\mathbf{u}\|^2$

19. $\|\mathbf{w}\| - 1$

20. $2 - \|\mathbf{u}\|$

21. $(\mathbf{u} \cdot \mathbf{v}) - (\mathbf{u} \cdot \mathbf{w})$

22. $(\mathbf{v} \cdot \mathbf{u}) - (\mathbf{w} \cdot \mathbf{v})$

Finding the Magnitude of a Vector   In Exercises 23–28, use the dot product to find the magnitude of u.

23. $\mathbf{u} = \langle -8, 15 \rangle$

24. $\mathbf{u} = \langle 4, -6 \rangle$

25. $\mathbf{u} = 20\mathbf{i} + 25\mathbf{j}$

26. $\mathbf{u} = 12\mathbf{i} - 16\mathbf{j}$

27. $\mathbf{u} = 6\mathbf{j}$

28. $\mathbf{u} = -21\mathbf{i}$

Finding the Angle Between Two Vectors   In Exercises 29–38, find the angle θ (in radians) between the vectors.

29. $\mathbf{u} = \langle 1, 0 \rangle$
    $\mathbf{v} = \langle 0, -2 \rangle$

30. $\mathbf{u} = \langle 3, 2 \rangle$
    $\mathbf{v} = \langle 4, 0 \rangle$

31. $\mathbf{u} = 3\mathbf{i} + 4\mathbf{j}$
    $\mathbf{v} = -2\mathbf{j}$

32. $\mathbf{u} = 2\mathbf{i} - 3\mathbf{j}$
    $\mathbf{v} = \mathbf{i} - 2\mathbf{j}$

33. $\mathbf{u} = 2\mathbf{i} - \mathbf{j}$
    $\mathbf{v} = 6\mathbf{i} - 3\mathbf{j}$

34. $\mathbf{u} = 5\mathbf{i} + 5\mathbf{j}$
    $\mathbf{v} = -6\mathbf{i} + 6\mathbf{j}$

35. $\mathbf{u} = -6\mathbf{i} - 3\mathbf{j}$
    $\mathbf{v} = -8\mathbf{i} + 4\mathbf{j}$

36. $\mathbf{u} = 2\mathbf{i} - 3\mathbf{j}$
    $\mathbf{v} = 4\mathbf{i} + 3\mathbf{j}$

37. $\mathbf{u} = \cos\left(\dfrac{\pi}{3}\right)\mathbf{i} + \sin\left(\dfrac{\pi}{3}\right)\mathbf{j}$
    $\mathbf{v} = \cos\left(\dfrac{3\pi}{4}\right)\mathbf{i} + \sin\left(\dfrac{3\pi}{4}\right)\mathbf{j}$

38. $\mathbf{u} = \cos\left(\dfrac{\pi}{4}\right)\mathbf{i} + \sin\left(\dfrac{\pi}{4}\right)\mathbf{j}$
    $\mathbf{v} = \cos\left(\dfrac{5\pi}{4}\right)\mathbf{i} + \sin\left(\dfrac{5\pi}{4}\right)\mathbf{j}$

Finding the Angle Between Two Vectors   In Exercises 39–42, find the angle θ (in degrees) between the vectors.

39. $\mathbf{u} = 3\mathbf{i} + 4\mathbf{j}$
    $\mathbf{v} = -7\mathbf{i} + 5\mathbf{j}$

40. $\mathbf{u} = 6\mathbf{i} - 3\mathbf{j}$
    $\mathbf{v} = -4\mathbf{i} - 4\mathbf{j}$

41. $\mathbf{u} = -5\mathbf{i} - 5\mathbf{j}$
    $\mathbf{v} = -8\mathbf{i} + 8\mathbf{j}$

42. $\mathbf{u} = 2\mathbf{i} - 3\mathbf{j}$
    $\mathbf{v} = 8\mathbf{i} + 3\mathbf{j}$

Finding the Angles in a Triangle   In Exercises 43–46, use vectors to find the interior angles of the triangle with the given vertices.

43. $(1, 2), (3, 4), (2, 5)$

44. $(-3, -4), (1, 7), (8, 2)$

45. $(-3, 0), (2, 2), (0, 6)$

46. $(-3, 5), (-1, 9), (7, 9)$

Using the Angle Between Two Vectors   In Exercises 47–50, find u · v, where θ is the angle between u and v.

47. $\|\mathbf{u}\| = 4, \quad \|\mathbf{v}\| = 10, \quad \theta = 2\pi/3$

48. $\|\mathbf{u}\| = 4, \quad \|\mathbf{v}\| = 12, \quad \theta = \pi/3$

49. $\|\mathbf{u}\| = 100, \quad \|\mathbf{v}\| = 250, \quad \theta = \pi/6$

50. $\|\mathbf{u}\| = 9, \quad \|\mathbf{v}\| = 36, \quad \theta = 3\pi/4$

Determining Orthogonal Vectors In Exercises 51–56, determine whether **u** and **v** are orthogonal.

51. $\mathbf{u} = \langle 3, 15 \rangle$
    $\mathbf{v} = \langle -1, 5 \rangle$

52. $\mathbf{u} = \langle 30, 12 \rangle$
    $\mathbf{v} = \langle \frac{1}{2}, -\frac{5}{4} \rangle$

53. $\mathbf{u} = 2\mathbf{i} - 2\mathbf{j}$
    $\mathbf{v} = -\mathbf{i} - \mathbf{j}$

54. $\mathbf{u} = \frac{1}{4}(3\mathbf{i} - \mathbf{j})$
    $\mathbf{v} = 5\mathbf{i} + 6\mathbf{j}$

55. $\mathbf{u} = \mathbf{i}$
    $\mathbf{v} = -2\mathbf{i} + 2\mathbf{j}$

56. $\mathbf{u} = \langle \cos \theta, \sin \theta \rangle$
    $\mathbf{v} = \langle \sin \theta, -\cos \theta \rangle$

Decomposing a Vector into Components In Exercises 57–60, find the projection of **u** onto **v**. Then write **u** as the sum of two orthogonal vectors, one of which is proj$_\mathbf{v}$**u**.

57. $\mathbf{u} = \langle 2, 2 \rangle$
    $\mathbf{v} = \langle 6, 1 \rangle$

58. $\mathbf{u} = \langle 0, 3 \rangle$
    $\mathbf{v} = \langle 2, 15 \rangle$

59. $\mathbf{u} = \langle 4, 2 \rangle$
    $\mathbf{v} = \langle 1, -2 \rangle$

60. $\mathbf{u} = \langle -3, -2 \rangle$
    $\mathbf{v} = \langle -4, -1 \rangle$

Finding the Projection of u onto v In Exercises 61–64, use the graph to find the projection of **u** onto **v**. (The terminal points of the vectors in standard position are given.) Use the formula for the projection of **u** onto **v** to verify your result.

61.
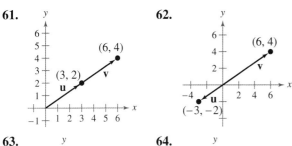

62.

63.
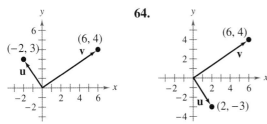

64.

Finding Orthogonal Vectors In Exercises 65–68, find two vectors in opposite directions that are orthogonal to the vector **u**. (There are many correct answers.)

65. $\mathbf{u} = \langle 3, 5 \rangle$

66. $\mathbf{u} = \langle -8, 3 \rangle$

67. $\mathbf{u} = \frac{1}{2}\mathbf{i} - \frac{2}{3}\mathbf{j}$

68. $\mathbf{u} = -\frac{5}{2}\mathbf{i} - 3\mathbf{j}$

Determining the Work Done In Exercises 69 and 70, determine the work done in moving a particle from P to Q when the magnitude and direction of the force are given by **v**.

69. $P(0, 0)$, $Q(4, 7)$, $\mathbf{v} = \langle 1, 4 \rangle$

70. $P(1, 3)$, $Q(-3, 5)$, $\mathbf{v} = -2\mathbf{i} + 3\mathbf{j}$

71. **Business** The vector $\mathbf{u} = \langle 1225, 2445 \rangle$ gives the numbers of hours worked by employees of a temporary work agency at two pay levels. The vector $\mathbf{v} = \langle 12.20, 8.50 \rangle$ gives the hourly wage (in dollars) paid at each level, respectively.

    (a) Find the dot product $\mathbf{u} \cdot \mathbf{v}$ and interpret the result in the context of the problem.

    (b) Identify the vector operation used to increase wages by 2%.

72. **Revenue** The vector $\mathbf{u} = \langle 3140, 2750 \rangle$ gives the numbers of hamburgers and hot dogs, respectively, sold at a fast-food stand in one month. The vector $\mathbf{v} = \langle 2.25, 1.75 \rangle$ gives the prices (in dollars) of the food items, respectively.

    (a) Find the dot product $\mathbf{u} \cdot \mathbf{v}$ and interpret the result in the context of the problem.

    (b) Identify the vector operation used to increase the prices by 2.5%.

73. **Physics** A truck with a gross weight of 30,000 pounds is parked on a slope of $d°$ (see figure). Assume that the only force to overcome is the force of gravity.

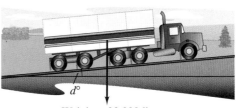

Weight = 30,000 lb

    (a) Find the force required to keep the truck from rolling down the hill in terms of d.

    (b) Use a graphing utility to complete the table.

d	0°	1°	2°	3°	4°	5°
Force						

d	6°	7°	8°	9°	10°
Force					

    (c) Find the force perpendicular to the hill when $d = 5°$.

74. **Braking Load**

    A sport utility vehicle with a gross weight of 5400 pounds is parked on a slope of 10°. Assume that the only force to overcome is the force of gravity. Find the force required to keep the vehicle from rolling down the hill. Find the force perpendicular to the hill.

75. Determining the Work Done Determine the work done by a person lifting a 245-newton bag of sugar 3 meters.

76. Determining the Work Done Determine the work done by a crane lifting a 2400-pound car 5 feet.

77. Determining the Work Done A constant force of 45 pounds, exerted at an angle of 30° with the horizontal, is required to slide a table across a floor. Determine the work done in sliding the table 20 feet.

78. Determining the Work Done A constant force of 50 pounds, exerted at an angle of 25° with the horizontal, is required to slide a desk across a floor. Determine the work done in sliding the desk 15 feet.

79. Determining the Work Done A child pulls a toy wagon by exerting a constant force of 25 pounds on a handle that makes a 20° angle with the horizontal (see figure). Determine the work done in pulling the wagon 50 feet.

80. Determining the Work Done One of the events in a strength competition is to pull a cement block 100 feet. One competitor pulls the block by exerting a constant force of 250 pounds on a rope attached to the block at an angle of 30° with the horizontal. Determine the work done in pulling the block.

Exploring the Concepts

True or False? In Exercises 81 and 82, determine whether the statement is true or false. Justify your answer.

81. The work W done by a constant force $\mathbf{F}$ acting along the line of motion of an object is represented by a vector.

82. A sliding door moves along the line of vector $\overrightarrow{PQ}$. If a force is applied to the door along a vector that is orthogonal to $\overrightarrow{PQ}$, then no work is done.

Finding an Unknown Vector Component In Exercises 83 and 84, find the value of k such that vectors u and v are orthogonal.

83. $\mathbf{u} = 8\mathbf{i} + 4\mathbf{j}$
$\mathbf{v} = 2\mathbf{i} - k\mathbf{j}$

84. $\mathbf{u} = -3k\mathbf{i} + 5\mathbf{j}$
$\mathbf{v} = 2\mathbf{i} - 4\mathbf{j}$

85. Error Analysis Describe the error.
$$\langle 5, 8 \rangle \cdot \langle -2, 7 \rangle = \langle -10, 56 \rangle \quad \text{✗}$$

86. **HOW DO YOU SEE IT?** What is known about θ, the angle between two nonzero vectors u and v, under each condition (see figure)?

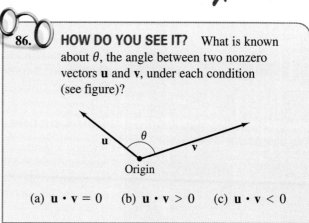

(a) $\mathbf{u} \cdot \mathbf{v} = 0$ (b) $\mathbf{u} \cdot \mathbf{v} > 0$ (c) $\mathbf{u} \cdot \mathbf{v} < 0$

87. Think About It Let u be a unit vector. What is the value of $\mathbf{u} \cdot \mathbf{u}$? Explain.

88. Think About It What can be said about the vectors u and v under each condition?

(a) The projection of u onto v equals u.

(b) The projection of u onto v equals **0**.

89. Proof Use vectors to prove that the diagonals of a rhombus are perpendicular.

90. Proof Prove that $\|\mathbf{u} - \mathbf{v}\|^2 = \|\mathbf{u}\|^2 + \|\mathbf{v}\|^2 - 2\mathbf{u} \cdot \mathbf{v}$.

Review & Refresh ▶ *Video solutions at LarsonPrecalculus.com*

Finding a Distance and Midpoint In Exercises 91–94, find (a) the distance between the points and (b) the midpoint of the line segment joining the points.

91. $(0, 9)$, $(3, 13)$ **92.** $(-5, -12)$, $(5, 16)$

93. $(-2.5, 3.5)$, $(2.5, -3.5)$ **94.** $\left(\frac{1}{2}, -\frac{3}{2}\right)$, $\left(-\frac{5}{2}, \frac{5}{2}\right)$

Adding or Subtracting Complex Numbers In Exercises 95–98, perform the operation and write the result in standard form.

95. $(6 + 11i) + (-18 + 3i)$

96. $(-5 - 2i) + (8 + 7i)$

97. $(4 - 9i) - (10 - 5i)$

98. $(-3 + 4i) - (-11 - 11i)$

Finding a Complex Conjugate In Exercises 99–102, find the complex conjugate of the complex number.

99. $1 + 3i$ **100.** $9 - 2i$

101. 4 **102.** $-3i$

Vector Operations In Exercises 103 and 104, find (a) $\mathbf{u} + \mathbf{v}$ and (b) $\mathbf{u} - \mathbf{v}$. Then sketch each resultant vector.

103. $\mathbf{u} = \langle -3, 2 \rangle$, $\mathbf{v} = \langle 4, -3 \rangle$

104. $\mathbf{u} = 6\mathbf{i} + 2\mathbf{j}$, $\mathbf{v} = -10\mathbf{i} + 10\mathbf{j}$

8.5 The Complex Plane

The complex plane has many practical applications. For example, in Exercise 43 on page 604, you will use the complex plane to write complex numbers that represent the positions of two ships.

❯ Plot complex numbers in the complex plane and find absolute values of complex numbers.
❯ Perform operations with complex numbers in the complex plane.
❯ Use the Distance and Midpoint Formulas in the complex plane.

The Complex Plane

Just as a real number can be represented by a point on the real number line, a complex number $z = a + bi$ can be represented by the point (a, b) in a coordinate plane called the **complex plane.** In the complex plane, the horizontal axis is the **real axis** and the vertical axis is the **imaginary axis,** as shown in the figure below.

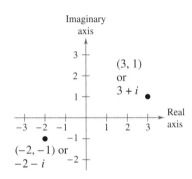

The **absolute value,** or **modulus,** of the complex number $z = a + bi$ is the distance between the origin $(0, 0)$ and the point (a, b). (The plural of modulus is *moduli.*)

Definition of the Absolute Value of a Complex Number

The **absolute value** of the complex number $z = a + bi$ is

$$|a + bi| = \sqrt{a^2 + b^2}.$$

When the complex number $z = a + bi$ is a real number (that is, when $b = 0$), this definition agrees with that given for the absolute value of a real number

$$|a + 0i| = \sqrt{a^2 + 0^2} = |a|.$$

EXAMPLE 1   **Finding the Absolute Value of a Complex Number**

❯❯❯ *See LarsonPrecalculus.com for an interactive version of this type of example.*

Plot $z = -2 + 5i$ in the complex plane and find its absolute value.

Solution   In the complex plane, $z = -2 + 5i$ is represented by the point $(-2, 5)$, as shown in Figure 8.34. The absolute value of z is

$$|z| = \sqrt{(-2)^2 + 5^2}$$
$$= \sqrt{29}.$$

Figure 8.34

✓ *Checkpoint* ▶ *Audio-video solution in English & Spanish at LarsonPrecalculus.com*

Plot $z = 3 - 4i$ in the complex plane and find its absolute value.

GO DIGITAL

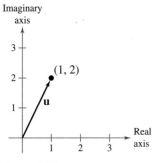

Figure 8.35

Operations with Complex Numbers in the Complex Plane

In Section 8.3, you learned how to add and subtract vectors geometrically in the coordinate plane. In a similar way, you can add and subtract complex numbers geometrically in the complex plane.

The complex number $z = a + bi$ can be represented by the vector $\mathbf{u} = \langle a, b \rangle$. For example, in Figure 8.35, the complex number $z = 1 + 2i$ is represented by the vector $\mathbf{u} = \langle 1, 2 \rangle$.

To add two complex numbers $z_1 = a + bi$ and $z_2 = c + di$ geometrically, first represent the numbers as vectors $\mathbf{u} = \langle a, b \rangle$ and $\mathbf{v} = \langle c, d \rangle$, respectively (see Figure 8.36). Then use the parallelogram law for vector addition to find $\mathbf{u} + \mathbf{v}$, as shown in Figure 8.37. Because $\mathbf{u} + \mathbf{v}$ represents $z_1 + z_2$, it follows that

$$z_1 + z_2 = (a + c) + (b + d)i.$$

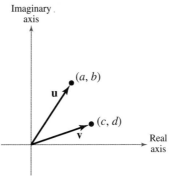

Represent $z_1 = a + bi$ and $z_2 = c + di$ as vectors $\mathbf{u} = \langle a, b \rangle$ and $\mathbf{v} = \langle c, d \rangle$

Figure 8.36

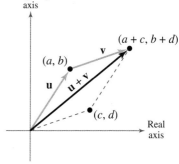

Use parallelogram law for vector addition

Figure 8.37

EXAMPLE 2   **Adding in the Complex Plane**

Find $(1 + 3i) + (2 + i)$ in the complex plane.

Solution

Begin by writing a vector to represent each complex number.

Complex number	Vector
$1 + 3i$	$\mathbf{u} = \langle 1, 3 \rangle$
$2 + i$	$\mathbf{v} = \langle 2, 1 \rangle$

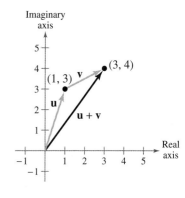

Graph $\mathbf{u}$ and $\mathbf{v}$, where the initial point of $\mathbf{v}$ coincides with the terminal point of $\mathbf{u}$, as shown in the figure. Next, graph $\mathbf{u} + \mathbf{v}$. From the figure, $\mathbf{u} + \mathbf{v} = \langle 3, 4 \rangle$, which implies that

$$(1 + 3i) + (2 + i) = 3 + 4i.$$

✓ **Checkpoint** ▶ Audio-video solution in English & Spanish at LarsonPrecalculus.com

Find $(3 + i) + (1 + 2i)$ in the complex plane. ∎

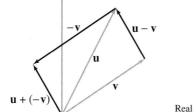

Figure 8.38

To subtract two complex numbers geometrically, first represent the numbers as vectors $\mathbf{u}$ and $\mathbf{v}$. Then subtract the vectors, as shown in Figure 8.38. The difference of the vectors represents the difference of the complex numbers.

GO DIGITAL

EXAMPLE 3 **Subtracting in the Complex Plane**

Find $(4 + 2i) - (3 - i)$ in the complex plane.

Solution

Begin by writing a vector to represent each complex number.

Complex number	Vector
$4 + 2i$	$\mathbf{u} = \langle 4, 2 \rangle$
$3 - i$	$\mathbf{v} = \langle 3, -1 \rangle$

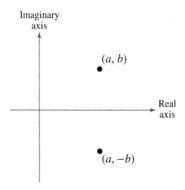

Because $\mathbf{u} - \mathbf{v} = \mathbf{u} + (-\mathbf{v})$, graph $\mathbf{u}$ and $-\mathbf{v} = \langle -3, 1 \rangle$, where the initial point of $-\mathbf{v}$ coincides with the terminal point of $\mathbf{u}$, as shown in the figure. Next, graph $\mathbf{u} + (-\mathbf{v})$. From the figure, $\mathbf{u} + (-\mathbf{v}) = \langle 1, 3 \rangle$, which implies that

$$(4 + 2i) - (3 - i) = 1 + 3i.$$

✓ *Checkpoint* ▶ *Audio-video solution in English & Spanish at LarsonPrecalculus.com*

Find $(2 - 4i) - (1 + i)$ in the complex plane.

Recall that the complex numbers $a + bi$ and $a - bi$ are *complex conjugates*. The points (a, b) and $(a, -b)$ are reflections of each other in the real axis, as shown in the figure below. This information enables you to find a complex conjugate geometrically.

EXAMPLE 4 **Complex Conjugates in the Complex Plane**

Plot $z = -3 + i$ and its complex conjugate in the complex plane. Write the conjugate as a complex number.

Solution

Begin by plotting $-3 + i$. As shown in Figure 8.39, the reflection $(-3, 1)$ in the real axis is $(-3, -1)$. So, the complex numbers

$$-3 + i \qquad \text{and} \qquad -3 - i$$

are complex conjugates.

Figure 8.39

✓ *Checkpoint* ▶ *Audio-video solution in English & Spanish at LarsonPrecalculus.com*

Plot $z = 2 - 3i$ and its complex conjugate in the complex plane. Write the conjugate as a complex number.

GO DIGITAL

Distance and Midpoint Formulas in the Complex Plane

For two points in the complex plane, the distance between the points is the modulus (or absolute value) of the difference of the two corresponding complex numbers. Let (a, b) and (s, t) be points in the complex plane. One way to write the difference of the corresponding complex numbers is $(s + ti) - (a + bi) = (s - a) + (t - b)i$. The modulus of the difference is

$$|(s - a) + (t - b)i| = \sqrt{(s - a)^2 + (t - b)^2}.$$

So, $d = \sqrt{(s - a)^2 + (t - b)^2}$ is the distance between the points in the complex plane.

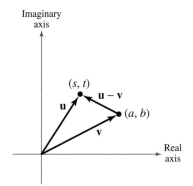

Imaginary axis

(s, t)

$\mathbf{u}$ $\mathbf{u} - \mathbf{v}$

(a, b)

$\mathbf{v}$

Real axis

Figure 8.40

Distance Formula in the Complex Plane

The distance d between the points (a, b) and (s, t) in the complex plane is

$$d = \sqrt{(s - a)^2 + (t - b)^2}.$$

Figure 8.40 shows the points represented as vectors. The magnitude of the vector $\mathbf{u} - \mathbf{v}$ is the distance between (a, b) and (s, t).

$$\mathbf{u} - \mathbf{v} = \langle s - a, t - b \rangle$$

$$\|\mathbf{u} - \mathbf{v}\| = \sqrt{(s - a)^2 + (t - b)^2}$$

EXAMPLE 5   **Finding Distance in the Complex Plane**

Find the distance between $2 + 3i$ and $5 - 2i$ in the complex plane.

Solution

Let $a + bi = 2 + 3i$ and $s + ti = 5 - 2i$. The distance is

$$\begin{aligned}
d &= \sqrt{(s - a)^2 + (t - b)^2} \\
&= \sqrt{(5 - 2)^2 + (-2 - 3)^2} \\
&= \sqrt{3^2 + (-5)^2} \\
&= \sqrt{34} \\
&\approx 5.83 \text{ units}
\end{aligned}$$

as shown in the figure below.

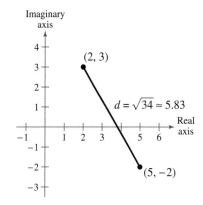

Imaginary axis

$(2, 3)$

$d = \sqrt{34} \approx 5.83$

Real axis

$(5, -2)$

✓ **Checkpoint** ▶ *Audio-video solution in English & Spanish at LarsonPrecalculus.com*

Find the distance between $5 - 4i$ and $6 + 5i$ in the complex plane.

GO DIGITAL

To find the midpoint of the line segment joining two points in the complex plane, find the average values of the respective coordinates of the two endpoints.

Midpoint Formula in the Complex Plane

The midpoint of the line segment joining the points (a, b) and (s, t) in the complex plane is

$$\text{Midpoint} = \left(\frac{a + s}{2}, \frac{b + t}{2}\right).$$

EXAMPLE 6   **Finding a Midpoint in the Complex Plane**

Find the midpoint of the line segment joining the points corresponding to $4 - 3i$ and $2 + 2i$ in the complex plane.

Solution

Let the points $(4, -3)$ and $(2, 2)$ represent the complex numbers $4 - 3i$ and $2 + 2i$, respectively. Apply the Midpoint Formula.

$$\text{Midpoint} = \left(\frac{a + s}{2}, \frac{b + t}{2}\right) = \left(\frac{4 + 2}{2}, \frac{-3 + 2}{2}\right) = \left(3, -\frac{1}{2}\right)$$

The midpoint is $\left(3, -\frac{1}{2}\right)$, as shown in the figure below.

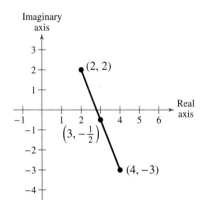

✓ *Checkpoint* ▶ *Audio-video solution in English & Spanish at LarsonPrecalculus.com*

Find the midpoint of the line segment joining the points corresponding to $2 + i$ and $5 - 5i$ in the complex plane.   ■

Summarize (Section 8.5)

1. State the definition of the absolute value, or modulus, of a complex number *(page 598)*. For an example of finding the absolute value of a complex number, see Example 1.

2. Explain how to add, subtract, and find complex conjugates of complex numbers in the complex plane *(page 599)*. For examples of performing operations with complex numbers in the complex plane, see Examples 2–4.

3. Explain how to use the Distance and Midpoint Formulas in the complex plane *(page 601)*. For examples of using the Distance and Midpoint Formulas in the complex plane, see Examples 5 and 6.

8.5 Exercises

See CalcChat.com for tutorial help and worked-out solutions to odd-numbered exercises.

GO DIGITAL

Vocabulary and Concept Check

In Exercises 1 and 2, fill in the blanks.

1. In the complex plane, the horizontal axis is the _____ axis and the vertical axis is the _____ axis.

2. The distance between two points in the complex plane is the _____ of the difference of the two corresponding complex numbers.

3. How do you find the complex conjugate of a complex number?

4. Compare the absolute value of a complex number with the absolute value of a real number.

Skills and Applications

Matching In Exercises 5–10, match the complex number with its representation in the complex plane. [The representations are labeled (a)–(f).]

(a)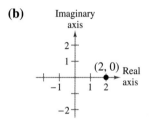

(b)

(c)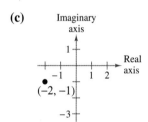

(d)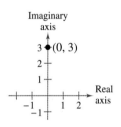

(e)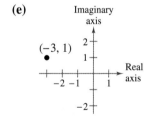

(f)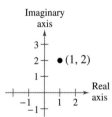

5. 2

6. $3i$

7. $1 + 2i$

8. $3 - i$

9. $-3 + i$

10. $-2 - i$

Finding the Absolute Value of a Complex Number In Exercises 11–16, plot the complex number and find its absolute value.

11. $-7i$

12. -7

13. $-6 + 8i$

14. $5 - 12i$

15. $4 - 6i$

16. $-8 + 3i$

Adding in the Complex Plane In Exercises 17–22, find the sum of the complex numbers in the complex plane.

17. $(3 + i) + (2 + 5i)$

18. $(5 + 2i) + (3 + 4i)$

19. $(8 - 2i) + (2 + 6i)$

20. $(3 - i) + (-7 + i)$

21.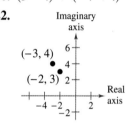

22.

Subtracting in the Complex Plane In Exercises 23–28, find the difference of the complex numbers in the complex plane.

23. $(4 + 2i) - (6 + 4i)$

24. $(-3 + i) - (3 + i)$

25. $(5 - i) - (-5 + 2i)$

26. $(2 - 3i) - (3 + 2i)$

27. $2 - (2 + 6i)$

28. $-2i - (3 - 5i)$

Complex Conjugates in the Complex Plane In Exercises 29–34, plot the complex number and its complex conjugate. Write the conjugate as a complex number.

29. $2 + 3i$

30. $5 - 4i$

31. $-1 - 2i$

32. $-7 + 3i$

33. $8i$

34. -3

Finding Distance in the Complex Plane In Exercises 35–38, find the distance between the complex numbers in the complex plane.

35. $1 + 2i, -1 + 4i$

36. $-5 + i, -2 + 5i$

37. $6i, 3 - 4i$

38. $-7 - 3i, 3 + 5i$

Finding a Midpoint in the Complex Plane In Exercises 39–42, find the midpoint of the line segment joining the points corresponding to the complex numbers in the complex plane.

39. $2 + i, 6 + 5i$

40. $-3 + 4i, 1 - 2i$

41. $7i, 9 - 10i$

42. $-1 - \frac{3}{4}i, \frac{1}{2} + \frac{1}{4}i$

43. Shipping

Ship A is 3 miles east and 4 miles north of port. Ship B is 5 miles west and 2 miles north of port (see figure).

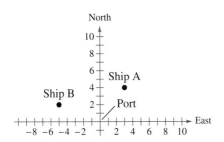

(a) Using the positive imaginary axis as north and the positive real axis as east, write complex numbers that represent the positions of Ship A and Ship B relative to port.

(b) How can you use the complex numbers in part (a) to find the distance between Ship A and Ship B?

44. Force Two forces are acting on a point. The first force has a horizontal component of 5 newtons and a vertical component of 3 newtons. The second force has a horizontal component of 4 newtons and a vertical component of 2 newtons.

(a) Plot the vectors that represent the two forces in the complex plane.

(b) Find the horizontal and vertical components of the resultant force acting on the point using the complex plane.

Exploring the Concepts

True or False? In Exercises 45–47, determine whether the statement is true or false. Justify your answer.

45. The modulus of a complex number can be real or imaginary.

46. The distance between two points in the complex plane is always real.

47. The modulus of the sum of two complex numbers is equal to the sum of their moduli.

48. **HOW DO YOU SEE IT?** Determine which graph represents each expression.

(a) $(a + bi) + (a - bi)$

(b) $(a + bi) - (a - bi)$

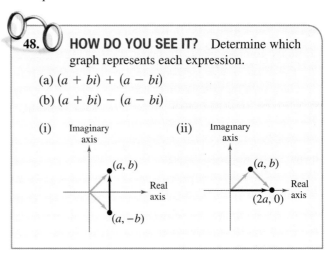

49. **Think About It** What does the set of all points with the same modulus represent in the complex plane? Explain.

50. **Think About It** The points corresponding to a complex number and its complex conjugate are plotted in the complex plane. What type of triangle do these points form with the origin?

Review & Refresh ▶ *Video solutions at LarsonPrecalculus.com*

Finding the Zeros of a Polynomial Function In Exercises 51–54, use the given zero to find all the zeros of the function.

Function	Zero
51. $f(x) = x^3 + 2x^2 + x + 2$	i
52. $f(x) = 2x^3 - x^2 + 18x - 9$	$-3i$
53. $f(x) = x^3 - 9x^2 + 28x - 40$	$2 - 2i$
54. $f(x) = x^4 - 6x^3 + 6x^2 + 24x - 40$	$3 + i$

Evaluating Trigonometric Functions In Exercises 55–58, find the exact values of the remaining trigonometric values of θ satisfying the given conditions.

55. $\cos \theta = -\frac{3}{5}, \tan \theta < 0$

56. $\cot \theta = \frac{5}{4}, \sin \theta > 0$

57. $\tan \theta = 5, \sec \theta < 0$

58. $\csc \theta$ is undefined, $\frac{\pi}{2} \le \theta \le \frac{3\pi}{2}$

Solving a Trigonometric Equation In Exercises 59–62, solve the equation.

59. $1 + 2 \sin \theta = 0$

60. $\sec \theta - \sqrt{2} = 0$

61. $2 \cot^2 \theta - 1 = 5$

62. $\tan^3 \theta = 3 \tan \theta$

8.6 Trigonometric Form of a Complex Number

Trigonometric forms of complex numbers have applications in circuit analysis. For example, in Exercises 79 and 80 on page 613, you will use trigonometric forms of complex numbers to find the voltage and current of an alternating current circuit.

> ❯ Write trigonometric forms of complex numbers.
> ❯ Multiply and divide complex numbers written in trigonometric form.
> ❯ Use DeMoivre's Theorem to find powers of complex numbers.
> ❯ Find *n*th roots of complex numbers.

Trigonometric Form of a Complex Number

In Section 1.5, you learned how to add, subtract, multiply, and divide complex numbers. To work effectively with *powers* and *roots* of complex numbers, it is helpful to write complex numbers in trigonometric form. Consider the nonzero complex number $a + bi$, plotted at the right. By letting θ be the angle from the positive real axis (measured counterclockwise) to the line segment connecting the origin and the point (a, b), you can write $a = r \cos \theta$ and $b = r \sin \theta$, where $r = \sqrt{a^2 + b^2}$. Consequently, you have $a + bi = (r \cos \theta) + (r \sin \theta)i$, from which you can obtain the **trigonometric form of a complex number.**

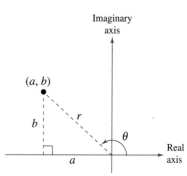

ALGEBRA HELP

For $0 \le \theta < 2\pi$, use the guidelines below. When z lies in Quadrant I, $\theta = \arctan(b/a)$. When z lies in Quadrant II or Quadrant III, $\theta = \pi + \arctan(b/a)$. When z lies in Quadrant IV, $\theta = 2\pi + \arctan(b/a)$.

❱❱❱❱

Trigonometric Form of a Complex Number

The **trigonometric form** of the complex number $z = a + bi$ is

$$z = r(\cos \theta + i \sin \theta)$$

where $a = r \cos \theta$, $b = r \sin \theta$, $r = \sqrt{a^2 + b^2}$, and $\tan \theta = b/a$. The number r is the **modulus** of z, and θ is an **argument** of z.

The trigonometric form of a complex number is also called the *polar form*. There are infinitely many choices for θ, so the trigonometric form of a complex number is not unique. Normally, θ is restricted to the interval $0 \le \theta < 2\pi$, although on occasion it is convenient to use $\theta < 0$.

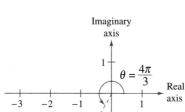

Figure 8.41

GO DIGITAL

EXAMPLE 1 Trigonometric Form of a Complex Number

Write the complex number $z = -2 - 2\sqrt{3}i$ in trigonometric form.

Solution The modulus of z is $r = \sqrt{(-2)^2 + \left(-2\sqrt{3}\right)^2} = \sqrt{16} = 4$, and the argument θ is determined from

$$\tan \theta = \frac{b}{a} = \frac{-2\sqrt{3}}{-2} = \sqrt{3}.$$

Because $z = -2 - 2\sqrt{3}i$ lies in Quadrant III, as shown in Figure 8.41, you have $\theta = \pi + \arctan \sqrt{3} = \pi + (\pi/3) = 4\pi/3$. So, the trigonometric form of z is

$$z = r(\cos \theta + i \sin \theta) = 4\left(\cos \frac{4\pi}{3} + i \sin \frac{4\pi}{3}\right).$$

✓ *Checkpoint* ▶ Audio-video solution in English & Spanish at LarsonPrecalculus.com

Write the complex number $z = 6 - 6i$ in trigonometric form. ∎

EXAMPLE 2   **Writing a Complex Number in Standard Form**

Write $z = \sqrt{8}[\cos(-\pi/3) + i\sin(-\pi/3)]$ in standard form $a + bi$.

Solution   Because $\cos(-\pi/3) = 1/2$ and $\sin(-\pi/3) = -\sqrt{3}/2$, you can write

$$z = \sqrt{8}\left[\cos\left(-\frac{\pi}{3}\right) + i\sin\left(-\frac{\pi}{3}\right)\right] = 2\sqrt{2}\left(\frac{1}{2} - \frac{\sqrt{3}}{2}i\right) = \sqrt{2} - \sqrt{6}i.$$

✓ **Checkpoint** Audio-video solution in English & Spanish at LarsonPrecalculus.com

Write $z = 8[\cos(2\pi/3) + i\sin(2\pi/3)]$ in standard form $a + bi$. ∎

Multiplication and Division of Complex Numbers

The trigonometric form adapts nicely to multiplication and division of complex numbers. Consider two complex numbers $z_1 = r_1(\cos\theta_1 + i\sin\theta_1)$ and $z_2 = r_2(\cos\theta_2 + i\sin\theta_2)$. The product of z_1 and z_2 is

$$z_1 z_2 = r_1 r_2(\cos\theta_1 + i\sin\theta_1)(\cos\theta_2 + i\sin\theta_2)$$
$$= r_1 r_2[(\cos\theta_1\cos\theta_2 - \sin\theta_1\sin\theta_2) + i(\sin\theta_1\cos\theta_2 + \cos\theta_1\sin\theta_2)].$$

Using the sum and difference formulas for cosine and sine, this equation is equivalent to

$$z_1 z_2 = r_1 r_2[\cos(\theta_1 + \theta_2) + i\sin(\theta_1 + \theta_2)].$$

This establishes the first part of the rule below. The second part is left for you to verify (see Exercise 83).

Product and Quotient of Two Complex Numbers

Let $z_1 = r_1(\cos\theta_1 + i\sin\theta_1)$ and $z_2 = r_2(\cos\theta_2 + i\sin\theta_2)$ be complex numbers.

$$z_1 z_2 = r_1 r_2[\cos(\theta_1 + \theta_2) + i\sin(\theta_1 + \theta_2)] \qquad \text{Product}$$

$$\frac{z_1}{z_2} = \frac{r_1}{r_2}[\cos(\theta_1 - \theta_2) + i\sin(\theta_1 - \theta_2)], \quad z_2 \neq 0 \qquad \text{Quotient}$$

EXAMPLE 3   **Multiplying Complex Numbers**

Find the product $z_1 z_2$ of $z_1 = 3\left(\cos\dfrac{\pi}{4} + i\sin\dfrac{\pi}{4}\right)$ and $z_2 = 2\left(\cos\dfrac{3\pi}{4} + i\sin\dfrac{3\pi}{4}\right)$.

Solution

$$z_1 z_2 = 3\left(\cos\frac{\pi}{4} + i\sin\frac{\pi}{4}\right) \cdot 2\left(\cos\frac{3\pi}{4} + i\sin\frac{3\pi}{4}\right)$$

$$= 6\left[\cos\left(\frac{\pi}{4} + \frac{3\pi}{4}\right) + i\sin\left(\frac{\pi}{4} + \frac{3\pi}{4}\right)\right] \qquad \text{Multiply moduli and add arguments.}$$

$$= 6(\cos\pi + i\sin\pi)$$

$$= 6[-1 + i(0)]$$

$$= -6$$

✓ **Checkpoint**  Audio-video solution in English & Spanish at LarsonPrecalculus.com

Find the product $z_1 z_2$ of $z_1 = 2\left(\cos\dfrac{5\pi}{6} + i\sin\dfrac{5\pi}{6}\right)$ and $z_2 = 5\left(\cos\dfrac{7\pi}{6} + i\sin\dfrac{7\pi}{6}\right)$. ∎

EXAMPLE 4   Multiplying Complex Numbers

$$2\left(\cos\frac{2\pi}{3} + i\sin\frac{2\pi}{3}\right) \cdot 8\left(\cos\frac{11\pi}{6} + i\sin\frac{11\pi}{6}\right)$$

$$= 16\left[\cos\left(\frac{2\pi}{3} + \frac{11\pi}{6}\right) + i\sin\left(\frac{2\pi}{3} + \frac{11\pi}{6}\right)\right]$$ Multiply moduli and add arguments.

$$= 16\left(\cos\frac{5\pi}{2} + i\sin\frac{5\pi}{2}\right)$$

$$= 16\left(\cos\frac{\pi}{2} + i\sin\frac{\pi}{2}\right)$$ $\frac{5\pi}{2}$ and $\frac{\pi}{2}$ are coterminal.

$$= 16[0 + i]$$

$$= 16i$$

✓ **Checkpoint** Audio-video solution in English & Spanish at LarsonPrecalculus.com

Find the product z_1z_2 of $z_1 = 3\left(\cos\frac{\pi}{3} + i\sin\frac{\pi}{3}\right)$ and $z_2 = 4\left(\cos\frac{\pi}{6} + i\sin\frac{\pi}{6}\right)$.

EXAMPLE 5   Dividing Complex Numbers

$$\frac{24(\cos 300° + i\sin 300°)}{8(\cos 75° + i\sin 75°)}$$

$$= 3[\cos(300° - 75°) + i\sin(300° - 75°)]$$ Divide moduli and subtract arguments.

$$= 3(\cos 225° + i\sin 225°)$$

$$= -\frac{3\sqrt{2}}{2} - \frac{3\sqrt{2}}{2}i$$

✓ **Checkpoint** Audio-video solution in English & Spanish at LarsonPrecalculus.com

Find the quotient z_1/z_2 of $z_1 = \cos 40° + i\sin 40°$ and $z_2 = \cos 10° + i\sin 10°$. ■

In the preceding section, you added, subtracted, and found complex conjugates of complex numbers geometrically in the complex plane. In a similar way, you can multiply complex numbers geometrically in the complex plane.

EXAMPLE 6   Multiplying in the Complex Plane

Find the product z_1z_2 of $z_1 = 2(\cos 30° + i\sin 30°)$ and $z_2 = 2(\cos 60° + i\sin 60°)$ in the complex plane.

Solution

Let $\mathbf{u} = 2\langle\cos 30°, \sin 30°\rangle = \langle\sqrt{3}, 1\rangle$ and $\mathbf{v} = 2\langle\cos 60°, \sin 60°\rangle = \langle 1, \sqrt{3}\rangle$. Then

$$\|\mathbf{u}\| = \sqrt{\left(\sqrt{3}\right)^2 + 1^2} = \sqrt{4} = 2 \quad\text{and}\quad \|\mathbf{v}\| = \sqrt{1^2 + \left(\sqrt{3}\right)^2} = \sqrt{4} = 2.$$

So, the magnitude of the product vector is $2(2) = 4$. The sum of the direction angles is $30° + 60° = 90°$. So, the product vector lies on the imaginary axis and is represented in vector form as $\langle 0, 4\rangle$, as shown in Figure 8.42. This implies that $z_1z_2 = 4i$.

✓ **Checkpoint** Audio-video solution in English & Spanish at LarsonPrecalculus.com

Find the product z_1z_2 of $z_1 = 2(\cos 45° + i\sin 45°)$ and $z_2 = 4(\cos 135° + i\sin 135°)$ in the complex plane. ■

ALGEBRA HELP

Check the solution to Example 4 by first converting the complex numbers to the standard forms $-1 + \sqrt{3}i$ and $4\sqrt{3} - 4i$ and then multiplying algebraically, as in Section 1.5.

▶▶▶▶

▶▶▶ **TECHNOLOGY**

Some graphing utilities can multiply and divide complex numbers in trigonometric form. If you have access to such a graphing utility, use it to check the solutions to Examples 3–5.

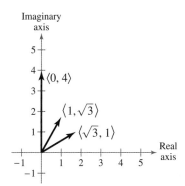

Figure 8.42

GO DIGITAL

Powers of Complex Numbers

The trigonometric form of a complex number is used to raise a complex number to a power. To accomplish this, consider repeated use of the multiplication rule.

$$z = r(\cos \theta + i \sin \theta)$$

$$z^2 = r(\cos \theta + i \sin \theta)r(\cos \theta + i \sin \theta) = r^2(\cos 2\theta + i \sin 2\theta)$$

$$z^3 = r^2(\cos 2\theta + i \sin 2\theta)r(\cos \theta + i \sin \theta) = r^3(\cos 3\theta + i \sin 3\theta)$$

$$z^4 = r^4(\cos 4\theta + i \sin 4\theta)$$

$$z^5 = r^5(\cos 5\theta + i \sin 5\theta)$$

$$\vdots$$

This pattern leads to **DeMoivre's Theorem,** which is named after the French mathematician Abraham DeMoivre (1667–1754).

> ### DeMoivre's Theorem
>
> If $z = r(\cos \theta + i \sin \theta)$ is a complex number and n is a positive integer, then
>
> $$z^n = [r(\cos \theta + i \sin \theta)]^n$$
> $$= r^n(\cos n\theta + i \sin n\theta).$$

EXAMPLE 7 **Finding a Power of a Complex Number**

Use DeMoivre's Theorem to find $\left(-1 + \sqrt{3}i\right)^{12}$.

Solution The modulus of $z = -1 + \sqrt{3}i$ is

$$r = \sqrt{(-1)^2 + \left(\sqrt{3}\right)^2} = 2$$

and the argument θ is determined from $\tan \theta = \sqrt{3}/(-1)$. Because $z = -1 + \sqrt{3}i$ lies in Quadrant II,

$$\theta = \pi + \arctan \frac{\sqrt{3}}{-1} = \pi + \left(-\frac{\pi}{3}\right) = \frac{2\pi}{3}.$$

So, the trigonometric form of z is

$$z = -1 + \sqrt{3}i = 2\left(\cos \frac{2\pi}{3} + i \sin \frac{2\pi}{3}\right).$$

Then, by DeMoivre's Theorem, you have

$$\left(-1 + \sqrt{3}i\right)^{12} = \left[2\left(\cos \frac{2\pi}{3} + i \sin \frac{2\pi}{3}\right)\right]^{12}$$

$$= 2^{12}\left[\cos \frac{12(2\pi)}{3} + i \sin \frac{12(2\pi)}{3}\right]$$

$$= 4096(\cos 8\pi + i \sin 8\pi)$$

$$= 4096(1 + 0)$$

$$= 4096.$$

✓ *Checkpoint* ▶ Audio-video solution in English & Spanish at LarsonPrecalculus.com

Use DeMoivre's Theorem to find $(-1 - i)^4$.

>>> **SKILLS REFRESHER**

For a refresher on finding
nth roots of real numbers and
simplifying radicals, watch the
video at *LarsonPrecalculus.com*.

Roots of Complex Numbers

Recall that a consequence of the Fundamental Theorem of Algebra is that a polynomial equation of degree n has n solutions in the complex number system. For example, the equation $x^6 = 1$ has six solutions. To find these solutions, use factoring and the Quadratic Formula.

$$x^6 - 1 = 0$$
$$(x^3 - 1)(x^3 + 1) = 0$$
$$(x - 1)(x^2 + x + 1)(x + 1)(x^2 - x + 1) = 0$$

Consequently, the solutions are

$$x = \pm 1, \quad x = \frac{-1 \pm \sqrt{3}i}{2}, \quad \text{and} \quad x = \frac{1 \pm \sqrt{3}i}{2}.$$

Each of these numbers is a sixth root of 1. In general, an **nth root of a complex number** is defined as follows.

Definition of an nth Root of a Complex Number

The complex number $u = a + bi$ is an **nth root** of the complex number z when

$$z = u^n$$
$$= (a + bi)^n.$$

To find a formula for an nth root of a complex number, let u be an nth root of z, where

$$u = s(\cos \beta + i \sin \beta)$$

and

$$z = r(\cos \theta + i \sin \theta).$$

By DeMoivre's Theorem and the fact that $u^n = z$, you have

$$s^n(\cos n\beta + i \sin n\beta) = r(\cos \theta + i \sin \theta).$$

Taking the absolute value of each side of this equation, it follows that $s^n = r$. Substituting back into the previous equation and dividing by r gives

$$\cos n\beta + i \sin n\beta = \cos \theta + i \sin \theta.$$

So, it follows that

$$\cos n\beta = \cos \theta$$

and

$$\sin n\beta = \sin \theta.$$

Both sine and cosine have a period of 2π, so these last two equations have solutions if and only if the angles differ by a multiple of 2π. Consequently, there must exist an integer k such that

$$n\beta = \theta + 2\pi k$$
$$\beta = \frac{\theta + 2\pi k}{n}.$$

Substituting this value of β and $s = \sqrt[n]{r}$ into the trigonometric form of u gives the result stated on the next page.

GO DIGITAL

> **Finding *n*th Roots of a Complex Number**
>
> For a positive integer n, the complex number $z = r(\cos\theta + i\sin\theta)$ has exactly n distinct nth roots given by
>
> $$z_k = \sqrt[n]{r}\left(\cos\frac{\theta + 2\pi k}{n} + i\sin\frac{\theta + 2\pi k}{n}\right)$$
>
> where $k = 0, 1, 2, \ldots, n-1$.

ALGEBRA HELP

When θ is in degrees, replace 2π with $360°$ in z_k.

When $k > n - 1$, the roots begin to repeat. For example, when $k = n$, the angle

$$\frac{\theta + 2\pi n}{n} = \frac{\theta}{n} + 2\pi$$

is coterminal with θ/n, which is also obtained when $k = 0$.

The formula for the nth roots of a complex number z has a geometrical interpretation, as shown in Figure 8.43. Note that the nth roots of z all have the same magnitude $\sqrt[n]{r}$, so they all lie on a circle of radius $\sqrt[n]{r}$ with center at the origin. Furthermore, successive nth roots have arguments that differ by $2\pi/n$, so the n roots are equally spaced around the circle.

On the preceding page, you found the sixth roots of 1 by factoring and using the Quadratic Formula. Example 8 shows how to solve the same problem with the formula for nth roots.

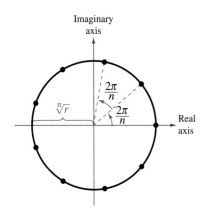

Figure 8.43

EXAMPLE 8   **Finding the *n*th Roots of a Complex Number**

Find all sixth roots of 1.

Solution   First, write 1 in the trigonometric form $z = 1(\cos 0 + i\sin 0)$. Then, by the nth root formula with $n = 6$, $r = 1$, and $\theta = 0$, the roots have the form

$$z_k = \sqrt[6]{1}\left(\cos\frac{0 + 2\pi k}{6} + i\sin\frac{0 + 2\pi k}{6}\right) = \cos\frac{\pi k}{3} + i\sin\frac{\pi k}{3}.$$

So, for $k = 0$, 1, 2, 3, 4, and 5, the sixth roots of 1 are shown below. (See Figure 8.44.)

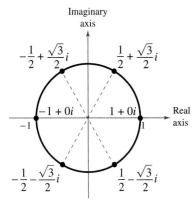

Figure 8.44

$z_0 = \cos 0 + i\sin 0 = 1$

$z_1 = \cos\dfrac{\pi}{3} + i\sin\dfrac{\pi}{3} = \dfrac{1}{2} + \dfrac{\sqrt{3}}{2}i$ Increment by $\dfrac{2\pi}{n} = \dfrac{2\pi}{6} = \dfrac{\pi}{3}$

$z_2 = \cos\dfrac{2\pi}{3} + i\sin\dfrac{2\pi}{3} = -\dfrac{1}{2} + \dfrac{\sqrt{3}}{2}i$

$z_3 = \cos\pi + i\sin\pi = -1$

$z_4 = \cos\dfrac{4\pi}{3} + i\sin\dfrac{4\pi}{3} = -\dfrac{1}{2} - \dfrac{\sqrt{3}}{2}i$

$z_5 = \cos\dfrac{5\pi}{3} + i\sin\dfrac{5\pi}{3} = \dfrac{1}{2} - \dfrac{\sqrt{3}}{2}i$

✓ *Checkpoint* ▶ Audio-video solution in English & Spanish at LarsonPrecalculus.com

Find all fourth roots of 1.   ◼

In Figure 8.44, notice that the roots obtained in Example 8 all have a magnitude of 1 and are equally spaced around the unit circle. Also notice that the complex roots occur in conjugate pairs, as discussed in Section 3.4. The n distinct nth roots of 1 are called the ***n*th roots of unity.**

GO DIGITAL

| EXAMPLE 9 | **Finding the *n*th Roots of a Complex Number** |

▶▶▶ *See LarsonPrecalculus.com for an interactive version of this type of example.*

Find the three cube roots of $z = -2 + 2i$. Write the roots in trigonometric form using degree measure.

Solution The modulus of z is $r = \sqrt{(-2)^2 + 2^2} = \sqrt{8}$. Note that z lies in Quadrant II and

$$\tan \theta = \frac{b}{a} = \frac{2}{-2} = -1.$$

In Quadrant II the reference angle is $45°$, so $\theta = 135°$. The trigonometric form of z is

$$z = -2 + 2i = \sqrt{8}(\cos 135° + i \sin 135°).$$

By the *n*th root formula, the roots have the form

$$z_k = \sqrt[6]{8}\left(\cos \frac{135° + 360°k}{3} + i \sin \frac{135° + 360°k}{3}\right).$$

So, for $k = 0, 1,$ and 2, the cube roots of z are shown below. (See Figure 8.45.)

$$z_0 = \sqrt[6]{8}\left(\cos \frac{135° + 360°(0)}{3} + i \sin \frac{135° + 360°(0)}{3}\right)$$
$$= \sqrt{2}(\cos 45° + i \sin 45°)$$
$$= 1 + i$$

$$z_1 = \sqrt[6]{8}\left(\cos \frac{135° + 360°(1)}{3} + i \sin \frac{135° + 360°(1)}{3}\right)$$
$$= \sqrt{2}(\cos 165° + i \sin 165°)$$
$$\approx -1.3660 + 0.3660i$$

$$z_2 = \sqrt[6]{8}\left(\cos \frac{135° + 360°(2)}{3} + i \sin \frac{135° + 360°(2)}{3}\right)$$
$$= \sqrt{2}(\cos 285° + i \sin 285°)$$
$$\approx 0.3660 - 1.3660i$$

✓ **Checkpoint** ▶ Audio-video solution in English & Spanish at LarsonPrecalculus.com

Find the three cube roots of $z = -6 + 6i$. Write the roots in trigonometric form using degree measure. ◼

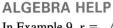

ALGEBRA HELP

In Example 9, $r = \sqrt{8}$, so it follows that

$$\sqrt[n]{r} = \sqrt[3]{\sqrt{8}}$$
$$= \sqrt[3 \cdot 2]{8}$$
$$= \sqrt[6]{8}.$$

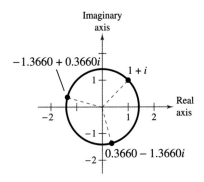

Figure 8.45

| Summarize | (Section 8.6) |

1. State the trigonometric form of a complex number *(page 605)*. For examples of writing complex numbers in trigonometric form and standard form, see Examples 1 and 2.

2. Explain how to multiply and divide complex numbers written in trigonometric form *(page 606)*. For examples of multiplying and dividing complex numbers written in trigonometric form, see Examples 3–6.

3. Explain how to use DeMoivre's Theorem to find a power of a complex number *(page 608)*. For an example of using DeMoivre's Theorem, see Example 7.

4. Explain how to find the *n*th roots of a complex number *(page 609)*. For examples of finding *n*th roots of complex numbers, see Examples 8 and 9.

GO DIGITAL

8.6 Exercises

See CalcChat.com for tutorial help and worked-out solutions to odd-numbered exercises.

Vocabulary and Concept Check

In Exercises 1 and 2, fill in the blanks.

1. _____ Theorem states that if $z = r(\cos \theta + i \sin \theta)$ is a complex number and n is a positive integer, then $z^n = r^n(\cos n\theta + i \sin n\theta)$.

2. The complex number $u = a + bi$ is an _____ _____ of the complex number z when $z = u^n = (a + bi)^n$.

3. In the trigonometric form of a complex number $z = r(\cos \theta + i \sin \theta)$, what do the variables r and θ represent?

4. Successive nth roots of a complex number have arguments that differ by what amount?

Skills and Applications

Trigonometric Form of a Complex Number In Exercises 5–20, plot the complex number. Then write the trigonometric form of the complex number.

5. $1 + i$
6. $5 - 5i$
7. $1 - \sqrt{3}i$
8. $4 - 4\sqrt{3}i$
9. $-2(1 + \sqrt{3}i)$
10. $\frac{5}{2}(\sqrt{3} - i)$
11. $-5i$
12. $12i$
13. 2
14. 4
15. $-7 + 4i$
16. $3 - i$
17. $3 + \sqrt{3}i$
18. $3\sqrt{2} - 7i$
19. $-8 - 5\sqrt{3}i$
20. $-9 - 2\sqrt{10}i$

Writing a Complex Number in Standard Form In Exercises 21–26, write the standard form of the complex number. Then plot the complex number.

21. $2(\cos 60° + i \sin 60°)$
22. $5(\cos 135° + i \sin 135°)$
23. $\frac{9}{4}\left(\cos \frac{3\pi}{4} + i \sin \frac{3\pi}{4}\right)$
24. $6\left(\cos \frac{5\pi}{12} + i \sin \frac{5\pi}{12}\right)$
25. $\sqrt{48}[\cos(-30°) + i \sin(-30°)]$
26. $\sqrt{8}(\cos 225° + i \sin 225°)$

Writing a Complex Number in Standard Form In Exercises 27–30, use a graphing utility to write the complex number in standard form.

27. $5\left(\cos \frac{\pi}{9} + i \sin \frac{\pi}{9}\right)$
28. $10\left(\cos \frac{2\pi}{5} + i \sin \frac{2\pi}{5}\right)$
29. $2(\cos 155° + i \sin 155°)$
30. $9(\cos 58° + i \sin 58°)$

Multiplying Complex Numbers In Exercises 31–34, find the product. Leave the result in trigonometric form.

31. $\left[2\left(\cos \frac{\pi}{4} + i \sin \frac{\pi}{4}\right)\right]\left[6\left(\cos \frac{\pi}{12} + i \sin \frac{\pi}{12}\right)\right]$

32. $\left[\frac{3}{4}\left(\cos \frac{\pi}{3} + i \sin \frac{\pi}{3}\right)\right]\left[4\left(\cos \frac{3\pi}{4} + i \sin \frac{3\pi}{4}\right)\right]$

33. $\left[\frac{5}{3}(\cos 120° + i \sin 120°)\right]\left[\frac{2}{3}(\cos 30° + i \sin 30°)\right]$

34. $\left[\frac{1}{2}(\cos 100° + i \sin 100°)\right]\left[\frac{4}{5}(\cos 300° + i \sin 300°)\right]$

Dividing Complex Numbers In Exercises 35–38, find the quotient. Leave the result in trigonometric form.

35. $\dfrac{3(\cos 50° + i \sin 50°)}{9(\cos 20° + i \sin 20°)}$
36. $\dfrac{\cos 120° + i \sin 120°}{2(\cos 40° + i \sin 40°)}$

37. $\dfrac{\cos \pi + i \sin \pi}{\cos(\pi/3) + i \sin(\pi/3)}$
38. $\dfrac{5(\cos 4.3 + i \sin 4.3)}{4(\cos 2.1 + i \sin 2.1)}$

Multiplying in the Complex Plane In Exercises 39 and 40, find the product in the complex plane.

39. $\left[2\left(\cos \frac{2\pi}{3} + i \sin \frac{2\pi}{3}\right)\right]\left[\frac{1}{2}\left(\cos \frac{\pi}{3} + i \sin \frac{\pi}{3}\right)\right]$

40. $\left[2\left(\cos \frac{\pi}{4} + i \sin \frac{\pi}{4}\right)\right]\left[3\left(\cos \frac{\pi}{4} + i \sin \frac{\pi}{4}\right)\right]$

Finding a Power of a Complex Number In Exercises 41–54, use DeMoivre's Theorem to find the power of the complex number. Write the result in standard form.

41. $[5(\cos 20° + i \sin 20°)]^3$
42. $[3(\cos 60° + i \sin 60°)]^4$

43. $\left(\cos \frac{\pi}{4} + i \sin \frac{\pi}{4}\right)^{12}$
44. $\left[2\left(\cos \frac{\pi}{2} + i \sin \frac{\pi}{2}\right)\right]^8$

45. $[5(\cos 3.2 + i \sin 3.2)]^4$
46. $(\cos 0 + i \sin 0)^{20}$

47. $[3(\cos 15° + i \sin 15°)]^4$
48. $\left[2\left(\cos \frac{\pi}{8} + i \sin \frac{\pi}{8}\right)\right]^6$

49. $(1 + i)^5$
50. $(2 + 2i)^6$

51. $(-1 + i)^6$

52. $(3 - 2i)^8$

53. $2(\sqrt{3} + i)^{10}$

54. $4(1 - \sqrt{3}i)^3$

Finding the *n*th Roots of a Complex Number In Exercises 55–58, (a) find the roots of the complex number given in trigonometric form, (b) write each of the roots in standard form, and (c) represent each of the roots graphically.

55. Square roots of $5(\cos 120° + i \sin 120°)$

56. Square roots of $16(\cos 60° + i \sin 60°)$

57. Cube roots of $8\left(\cos \dfrac{2\pi}{3} + i \sin \dfrac{2\pi}{3}\right)$

58. Fifth roots of $32\left(\cos \dfrac{5\pi}{6} + i \sin \dfrac{5\pi}{6}\right)$

Finding the *n*th Roots of a Complex Number In Exercises 59–70, (a) find the roots of the complex number, (b) write each of the roots in standard form, and (c) represent each of the roots graphically.

59. Cube roots of 1000

60. Cube roots of -125

61. Fourth roots of -4

62. Fourth roots of 16

63. Fourth roots of i

64. Fifth roots of 1

65. Square roots of $-25i$

66. Fourth roots of $625i$

67. Fifth roots of $4(1 - i)$

68. Sixth roots of $64i$

69. Cube roots of $-\dfrac{125}{2}(1 + \sqrt{3}i)$

70. Cube roots of $-4\sqrt{2}(-1 + i)$

Solving an Equation In Exercises 71–78, find all solutions of the equation. Represent the solutions graphically.

71. $x^4 + i = 0$

72. $x^3 + 1 = 0$

73. $x^5 + 243 = 0$

74. $x^3 - 27 = 0$

75. $x^4 + 16i = 0$

76. $x^6 + 64i = 0$

77. $x^3 - (1 - i) = 0$

78. $x^4 + (1 + i) = 0$

Ohm's Law

In Exercises 79 and 80, use the following information. Ohm's Law for alternating current circuits is given by the formula $E = IZ$, where E is the voltage, I is the current (in amperes), and Z is the impedance (in ohms). Each variable is a complex number.

79. Write E in trigonometric form when $I = 6(\cos 41° + i \sin 41°)$ amperes and $Z = 4[\cos(-11°) + i \sin(-11°)]$ ohms. Then write E in standard form.

80. Write I in standard form when $E = 12 + 24i$ volts and $Z = 12 + 20i$ ohms. Then write I in trigonometric form.

Exploring the Concepts

True or False? In Exercises 81 and 82, determine whether the statement is true or false. Justify your answer.

81. Geometrically, the *n*th roots of any complex number z are all equally spaced around the unit circle.

82. The product of two complex numbers is zero only when the modulus of one (or both) of the complex numbers is zero.

83. Quotient of Two Complex Numbers Given two complex numbers $z_1 = r_1(\cos \theta_1 + i \sin \theta_1)$ and $z_2 = r_2(\cos \theta_2 + i \sin \theta_2)$, $z_2 \neq 0$, show that

$$\dfrac{z_1}{z_2} = \dfrac{r_1}{r_2}[\cos(\theta_1 - \theta_2) + i \sin(\theta_1 - \theta_2)].$$

84. HOW DO YOU SEE IT? The figure shows one of the fourth roots of a complex number z.

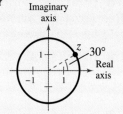

(a) How many roots are not shown?

(b) Describe the other roots.

85. Error Analysis Describe the error.

$$\dfrac{20(\cos 90° + i \sin 90°)}{5(\cos 60° + i \sin 60°)} = 2\sqrt{3} - 2i \quad \textbf{✗}$$

Review & Refresh ▶ *Video solutions at LarsonPrecalculus.com*

86. Siblings Three siblings are three different ages. The oldest is twice the age of the middle sibling, and the middle sibling is six years older than one-half the age of the youngest. If the oldest sibling is 16 years old, find the ages of the other two siblings.

Satisfying an Equation In Exercises 87–90, determine whether the ordered pair satisfies the linear equation in two variables.

87. $(2, 1)$; $2x + y = 5$

88. $(2, 1)$; $3x - 2y = 4$

89. $(-2, 1)$; $3y - 2x = -8$

90. $(2, -1)$; $2x - y = 3$

Solving for *y* In Exercises 91–94, solve the equation for *y*.

91. $x + y = 4$

92. $x - y = 2$

93. $5x - 3y = 6$

94. $3x + 4y = 5$

Solving a Quadratic Equation In Exercises 95–98, solve the quadratic equation.

95. $3x^2 + 4x - (2x + 1) = 7$

96. $x^2 - (5 + 2x) + 3x = 1$

97. $x^2 + x + 4 = 3$

98. $2x^2 + 4x - (2x + 3)^2 = 0$

Summary and Study Strategies

GO DIGITAL

What Did You Learn?

The list below reviews the skills covered in the chapter and correlates each one to the Review Exercises (see page 616) that practice the skill.

Section 8.1	Review Exercises
■ Use the Law of Sines to solve oblique triangles (AAS or ASA) *(p. 560)*.	*1–8*
If *ABC* is a triangle with sides *a*, *b*, and *c*, then $\dfrac{a}{\sin A} = \dfrac{b}{\sin B} = \dfrac{c}{\sin C}$.	
■ Use the Law of Sines to solve oblique triangles (SSA) *(p. 562)*.	*9–12*
■ Find the areas of oblique triangles *(p. 564)*.	*13–16*
The area of any triangle is one-half the product of the lengths of two sides times the sine of their included angle.	
■ Use the Law of Sines to model and solve real-life problems *(p. 565)*.	*17, 18*

Section 8.2

	Review Exercises
■ Use the Law of Cosines to solve oblique triangles (SSS or SAS) *(p. 569)*.	*19–30*

Standard Form

$$a^2 = b^2 + c^2 - 2bc \cos A$$

$$b^2 = a^2 + c^2 - 2ac \cos B$$

$$c^2 = a^2 + b^2 - 2ab \cos C$$

Alternative Form

$$\cos A = \frac{b^2 + c^2 - a^2}{2bc}$$

$$\cos B = \frac{a^2 + c^2 - b^2}{2ac}$$

$$\cos C = \frac{a^2 + b^2 - c^2}{2ab}$$

	Review Exercises
■ Use the Law of Cosines to model and solve real-life problems *(p. 571)*.	*31, 32*
■ Use Heron's Area Formula to find areas of triangles *(p. 572)*.	*33–36*
Area $= \sqrt{s(s - a)(s - b)(s - c)}$, where $s = (a + b + c)/2$	

Section 8.3

	Review Exercises
■ Represent vectors as directed line segments *(p. 576)*.	*37, 38*
■ Write component forms of vectors *(p. 577)*.	*39, 40, 65, 66*
■ Perform basic vector operations and represent vector operations graphically *(p. 578)*.	*41–48, 53–58*
Let $\mathbf{u} = \langle u_1, u_2 \rangle$ and $\mathbf{v} = \langle v_1, v_2 \rangle$ be vectors and let k be a scalar.	
$\mathbf{u} + \mathbf{v} = \langle u_1 + v_1, u_2 + v_2 \rangle \qquad k\mathbf{u} = k\langle u_1, u_2 \rangle = \langle ku_1, ku_2 \rangle$	
■ Write vectors as linear combinations of unit vectors *(p. 580)*.	*49–52*
■ Find direction angles of vectors *(p. 582)*.	*59–64*
■ Use vectors to model and solve real-life problems *(p. 583)*.	*67, 68*

Section 8.4	Review Exercises

■ Find the dot product of two vectors and use the properties of the dot product *(p. 589)*.

69–80

The dot product of $\mathbf{u} = \langle u_1, u_2 \rangle$ and $\mathbf{v} = \langle v_1, v_2 \rangle$ is $\mathbf{u} \cdot \mathbf{v} = u_1 v_1 + u_2 v_2$.

1. $\mathbf{u} \cdot \mathbf{v} = \mathbf{v} \cdot \mathbf{u}$
2. $\mathbf{u} \cdot (\mathbf{v} + \mathbf{w}) = \mathbf{u} \cdot \mathbf{v} + \mathbf{u} \cdot \mathbf{w}$
3. $c(\mathbf{u} \cdot \mathbf{v}) = c\mathbf{u} \cdot \mathbf{v} = \mathbf{u} \cdot c\mathbf{v}$
4. $\mathbf{0} \cdot \mathbf{v} = 0$
5. $\mathbf{v} \cdot \mathbf{v} = \|\mathbf{v}\|^2$

■ Find the angle between two vectors and determine whether two vectors are orthogonal *(p. 590)*.

81–88

If θ is the angle between two nonzero vectors $\mathbf{u}$ and $\mathbf{v}$, then $\cos \theta = \dfrac{\mathbf{u} \cdot \mathbf{v}}{\|\mathbf{u}\|\|\mathbf{v}\|}$.

■ Write a vector as the sum of two vector components *(p. 592)*.

89–92

■ Use vectors to determine the work done by a force *(p. 594)*.

93–96

Section 8.5	

■ Plot complex numbers in the complex plane and find absolute values of complex numbers *(p. 598)*.

97–100

■ Perform operations with complex numbers in the complex plane *(p. 599)*.

101–106

■ Use the Distance and Midpoint Formulas in the complex plane *(p. 601)*.

107–110

Section 8.6	

■ Write trigonometric forms of complex numbers *(p. 605)*.

111–116

The trigonometric form of the complex number $z = a + bi$ is

$z = r(\cos \theta + i \sin \theta)$, where $a = r \cos \theta$, $b = r \sin \theta$, $r = \sqrt{a^2 + b^2}$, and $\tan \theta = b/a$.

■ Multiply and divide complex numbers written in trigonometric form *(p. 606)*.

117–120

$z_1 z_2 = r_1 r_2 [\cos(\theta_1 + \theta_2) + i \sin(\theta_1 + \theta_2)]$

$\dfrac{z_1}{z_2} = \dfrac{r_1}{r_2}[\cos(\theta_1 - \theta_2) + i \sin(\theta_1 - \theta_2)]$, $z_2 \neq 0$

■ Use DeMoivre's Theorem to find powers of complex numbers *(p. 608)*.

121–124

If $z = r(\cos \theta + i \sin \theta)$ is a complex number and n is a positive integer, then

$z^n = [r(\cos \theta + i \sin \theta)]^n = r^n(\cos n\theta + i \sin n\theta)$.

■ Find nth roots of complex numbers *(p. 609)*.

125–132

Study Strategies

Studying in a Group   Many students endure unnecessary frustration because they study by themselves. Studying in a group or with a partner has many benefits. First, the combined memory and comprehension of the members minimizes the likelihood of any member getting "stuck" on a particular problem. Second, discussing math often helps clarify unclear areas. Third, regular study groups keep many students from procrastinating. Finally, study groups often build a camaraderie that helps students stick with the course when it gets tough.

Review Exercises

See CalcChat.com for tutorial help and worked-out solutions to odd-numbered exercises.

8.1 Using the Law of Sines In Exercises 1–12, use the Law of Sines to solve (if possible) the triangle. If two solutions exist, find both. Round your answers to two decimal places.

1. $A = 38°$, $B = 70°$, $a = 8$
2. $A = 22°$, $B = 121°$, $a = 19$
3. $B = 72°$, $C = 82°$, $b = 54$
4. $B = 10°$, $C = 20°$, $c = 33$
5. $A = 16°$, $B = 98°$, $c = 8.4$
6. $A = 95°$, $B = 45°$, $c = 104.8$
7. $A = 24°$, $C = 48°$, $b = 27.5$
8. $B = 64°$, $C = 36°$, $a = 367$
9. $B = 150°$, $b = 30$, $c = 10$
10. $B = 150°$, $a = 10$, $b = 3$
11. $A = 75°$, $a = 51.2$, $b = 33.7$
12. $B = 25°$, $a = 6.2$, $b = 4$

Finding the Area of a Triangle In Exercises 13–16, find the area of the triangle. Round your answers to one decimal place.

13. $A = 33°$, $b = 7$, $c = 10$
14. $B = 80°$, $a = 4$, $c = 8$
15. $C = 119°$, $a = 18$, $b = 6$
16. $A = 11°$, $B = 89°$, $c = 21$

17. **Height** From a certain distance, the angle of elevation to the top of a building is 17°. At a point 50 meters closer to the building, the angle of elevation is 31°. Find the height of the building.

18. **River Width** A surveyor finds that a tree on the opposite bank of a river flowing due east has a bearing of N 22° 30′ E from a certain point and a bearing of N 15° W from a point 400 feet downstream. Find the width of the river.

8.2 Using the Law of Cosines In Exercises 19–26, use the Law of Cosines to solve the triangle. Round your answers to two decimal places.

19. $a = 6$, $b = 9$, $c = 14$
20. $a = 75$, $b = 50$, $c = 110$
21. $a = 2.5$, $b = 5.0$, $c = 4.5$
22. $a = 16.4$, $b = 8.8$, $c = 12.2$
23. $B = 108°$, $a = 11$, $c = 11$
24. $B = 150°$, $a = 10$, $c = 20$
25. $C = 43°$, $a = 22.5$, $b = 31.4$
26. $A = 62°$, $b = 11.34$, $c = 19.52$

Solving a Triangle In Exercises 27–30, determine whether the Law of Sines or the Law of Cosines is needed to solve the triangle. Then solve (if possible) the triangle. If two solutions exist, find both. Round your answers to two decimal places.

27. $B = 38°$, $a = 15$, $b = 6$
28. $B = 52°$, $a = 4$, $c = 5$
29. $a = 13$, $b = 15$, $c = 24$
30. $A = 44°$, $B = 31°$, $c = 2.8$

31. **Geometry** The lengths of the diagonals of a parallelogram are 10 feet and 16 feet. Find the lengths of the sides of the parallelogram when the diagonals intersect at an angle of 28°.

32. **Air Navigation** Two planes leave an airport at approximately the same time. One flies 425 miles per hour at a bearing of 355°, and the other flies 530 miles per hour at a bearing of 67°. Draw a diagram that gives a visual representation of the problem and determine the distance between the planes after they fly for 2 hours.

Using Heron's Area Formula In Exercises 33–36, use Heron's Area Formula to find the area of the triangle.

33. $a = 3$, $b = 6$, $c = 8$
34. $a = 15$, $b = 8$, $c = 10$
35. $a = 12.3$, $b = 15.8$, $c = 3.7$
36. $a = \dfrac{4}{5}$, $b = \dfrac{3}{4}$, $c = \dfrac{5}{8}$

8.3 Determining Whether Two Vectors Are Equivalent In Exercises 37 and 38, determine whether **u** and **v** are equivalent. Explain.

37. 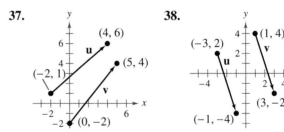   38.

Finding the Component Form of a Vector In Exercises 39 and 40, find the component form and magnitude of the vector **v**.

39. Initial point: $(0, 10)$

   Terminal point: $(7, 3)$

40. Initial point: $(1, 5)$

   Terminal point: $(15, 9)$

Vector Operations In Exercises 41–48, find (a) $\mathbf{u} + \mathbf{v}$, (b) $\mathbf{u} - \mathbf{v}$, (c) $4\mathbf{u}$, and (d) $3\mathbf{v} + 5\mathbf{u}$. Then sketch each resultant vector.

41. $\mathbf{u} = \langle -1, -3 \rangle$, $\mathbf{v} = \langle -3, 6 \rangle$

42. $\mathbf{u} = \langle 4, 5 \rangle$, $\mathbf{v} = \langle 0, -1 \rangle$

43. $\mathbf{u} = \langle -5, 2 \rangle$, $\mathbf{v} = \langle 4, 4 \rangle$

44. $\mathbf{u} = \langle 1, -8 \rangle$, $\mathbf{v} = \langle 3, -2 \rangle$

45. $\mathbf{u} = 2\mathbf{i} - \mathbf{j}$, $\mathbf{v} = 5\mathbf{i} + 3\mathbf{j}$

46. $\mathbf{u} = -7\mathbf{i} - 3\mathbf{j}$, $\mathbf{v} = 4\mathbf{i} - \mathbf{j}$

47. $\mathbf{u} = 4\mathbf{i}$, $\mathbf{v} = -\mathbf{i} + 6\mathbf{j}$

48. $\mathbf{u} = -6\mathbf{j}$, $\mathbf{v} = \mathbf{i} + \mathbf{j}$

Writing a Linear Combination of Unit Vectors In Exercises 49–52, the initial and terminal points of a vector are given. Write the vector as a linear combination of the standard unit vectors $\mathbf{i}$ and $\mathbf{j}$.

	Initial Point	Terminal Point
49.	$(2, 3)$	$(1, 8)$
50.	$(4, -2)$	$(-2, -10)$
51.	$(3, 4)$	$(9, 8)$
52.	$(-2, 7)$	$(5, -9)$

Vector Operations In Exercises 53–58, find the component form of $\mathbf{w}$ and sketch the specified vector operations geometrically, where $\mathbf{u} = 6\mathbf{i} - 5\mathbf{j}$ and $\mathbf{v} = 10\mathbf{i} + 3\mathbf{j}$.

53. $\mathbf{w} = 3\mathbf{v}$

54. $\mathbf{w} = \frac{1}{2}\mathbf{v}$

55. $\mathbf{w} = 2\mathbf{u} + \mathbf{v}$

56. $\mathbf{w} = 4\mathbf{u} - 5\mathbf{v}$

57. $\mathbf{w} = 5\mathbf{u} - 4\mathbf{v}$

58. $\mathbf{w} = -3\mathbf{u} + 2\mathbf{v}$

Finding the Direction Angle of a Vector In Exercises 59–64, find the magnitude and direction angle of the vector $\mathbf{v}$.

59. $\mathbf{v} = 5\mathbf{i} + 4\mathbf{j}$

60. $\mathbf{v} = -4\mathbf{i} + 7\mathbf{j}$

61. $\mathbf{v} = -3\mathbf{i} - 3\mathbf{j}$

62. $\mathbf{v} = 8\mathbf{i} - \mathbf{j}$

63. $\mathbf{v} = 7(\cos 60°\mathbf{i} + \sin 60°\mathbf{j})$

64. $\mathbf{v} = 3(\cos 150°\mathbf{i} + \sin 150°\mathbf{j})$

Finding the Component Form of a Vector In Exercises 65 and 66, find the component form of $\mathbf{v}$ given its magnitude and the angle it makes with the positive x-axis. Then sketch $\mathbf{v}$.

	Magnitude	Angle
65.	$\|\mathbf{v}\| = 8$	$\theta = 120°$
66.	$\|\mathbf{v}\| = \frac{1}{2}$	$\theta = 225°$

67. Resultant Force Forces with magnitudes of 85 pounds and 50 pounds act on a single point at angles of 45° and 60°, respectively, with the positive x-axis. Find the direction and magnitude of the resultant of these forces.

68. Rope Tension Two ropes support a 180-pound weight, as shown in the figure. Find the tension in each rope.

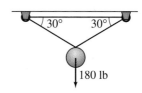

180 lb

8.4 **Finding a Dot Product** In Exercises 69–72, find $\mathbf{u} \cdot \mathbf{v}$.

69. $\mathbf{u} = \langle 6, 7 \rangle$
$\mathbf{v} = \langle -3, 9 \rangle$

70. $\mathbf{u} = \langle -7, 12 \rangle$
$\mathbf{v} = \langle -4, -14 \rangle$

71. $\mathbf{u} = 3\mathbf{i} + 7\mathbf{j}$
$\mathbf{v} = 11\mathbf{i} - 5\mathbf{j}$

72. $\mathbf{u} = -7\mathbf{i} + 2\mathbf{j}$
$\mathbf{v} = 16\mathbf{i} - 12\mathbf{j}$

Using Properties of the Dot Product In Exercises 73–80, use the vectors $\mathbf{u} = \langle -4, 2 \rangle$ and $\mathbf{v} = \langle 5, 1 \rangle$ to find the quantity. State whether the result is a vector or a scalar.

73. $2\mathbf{u} \cdot \mathbf{u}$

74. $3\mathbf{u} \cdot \mathbf{v}$

75. $4 - \|\mathbf{u}\|$

76. $\|\mathbf{v}\|^2$

77. $\mathbf{u}(\mathbf{u} \cdot \mathbf{v})$

78. $(\mathbf{u} \cdot \mathbf{v})\mathbf{v}$

79. $(\mathbf{u} \cdot \mathbf{u}) - (\mathbf{u} \cdot \mathbf{v})$

80. $(\mathbf{v} \cdot \mathbf{v}) - (\mathbf{v} \cdot \mathbf{u})$

Finding the Angle Between Two Vectors In Exercises 81–84, find the angle θ (in degrees) between the vectors.

81. $\mathbf{u} = \langle 2\sqrt{2}, -4 \rangle$, $\mathbf{v} = \langle -\sqrt{2}, 1 \rangle$

82. $\mathbf{u} = \langle 3, \sqrt{3} \rangle$, $\mathbf{v} = \langle 4, 3\sqrt{3} \rangle$

83. $\mathbf{u} = \cos\frac{7\pi}{4}\mathbf{i} + \sin\frac{7\pi}{4}\mathbf{j}$, $\mathbf{v} = \cos\frac{5\pi}{6}\mathbf{i} + \sin\frac{5\pi}{6}\mathbf{j}$

84. $\mathbf{u} = \cos 45°\mathbf{i} + \sin 45°\mathbf{j}$, $\mathbf{v} = \cos 300°\mathbf{i} + \sin 300°\mathbf{j}$

Determining Orthogonal Vectors In Exercises 85–88, determine whether $\mathbf{u}$ and $\mathbf{v}$ are orthogonal.

85. $\mathbf{u} = \langle -3, 8 \rangle$
$\mathbf{v} = \langle 8, 3 \rangle$

86. $\mathbf{u} = \langle \frac{1}{4}, -\frac{1}{2} \rangle$
$\mathbf{v} = \langle -2, 4 \rangle$

87. $\mathbf{u} = -\mathbf{i}$
$\mathbf{v} = \mathbf{i} + 2\mathbf{j}$

88. $\mathbf{u} = -2\mathbf{i} + \mathbf{j}$
$\mathbf{v} = 3\mathbf{i} + 6\mathbf{j}$

Decomposing a Vector into Components In Exercises 89–92, find the projection of $\mathbf{u}$ onto $\mathbf{v}$. Then write $\mathbf{u}$ as the sum of two orthogonal vectors, one of which is $\text{proj}_\mathbf{v}\,\mathbf{u}$.

89. $\mathbf{u} = \langle -4, 3 \rangle$, $\mathbf{v} = \langle -8, -2 \rangle$

90. $\mathbf{u} = \langle 5, 6 \rangle$, $\mathbf{v} = \langle 10, 0 \rangle$

91. $\mathbf{u} = \langle 2, 7 \rangle$, $\mathbf{v} = \langle 1, -1 \rangle$

92. $\mathbf{u} = \langle -3, 5 \rangle$, $\mathbf{v} = \langle -5, 2 \rangle$

Determining the Work Done In Exercises 93 and 94, determine the work done in moving a particle from P to Q when the magnitude and direction of the force are given by **v**.

93. $P(5, 3)$, $Q(8, 9)$, $\mathbf{v} = \langle 2, 7 \rangle$

94. $P(-2, -9)$, $Q(-12, 8)$, $\mathbf{v} = 3\mathbf{i} - 6\mathbf{j}$

95. Determining the Work Done Determine the work done by a crane lifting an 18,000-pound truck 4 feet.

96. Determining the Work Done A constant force of 25 pounds, exerted at an angle of 20° with the horizontal, is required to slide a crate across a floor. Determine the work done in sliding the crate 12 feet.

8.5 **Finding the Absolute Value of a Complex Number** In Exercises 97–100, plot the complex number and find its absolute value.

97. 4 **98.** $-6i$

99. $5 + 3i$ **100.** $-10 - 4i$

Adding in the Complex Plane In Exercises 101 and 102, find the sum of the complex numbers in the complex plane.

101. $(2 + 3i) + (1 - 2i)$ **102.** $(-4 + 2i) + (2 + i)$

Subtracting in the Complex Plane In Exercises 103 and 104, find the difference of the complex numbers in the complex plane.

103. $(1 + 2i) - (3 + i)$ **104.** $(-2 + i) - (1 + 4i)$

Complex Conjugates in the Complex Plane In Exercises 105 and 106, plot the complex number and its complex conjugate. Write the conjugate as a complex number.

105. $3 + i$ **106.** $2 - 5i$

Finding Distance in the Complex Plane In Exercises 107 and 108, find the distance between the complex numbers in the complex plane.

107. $3 + 2i, 2 - i$ **108.** $1 + 5i, -1 + 3i$

Finding a Midpoint in the Complex Plane In Exercises 109 and 110, find the midpoint of the line segment joining the points corresponding to the complex numbers in the complex plane.

109. $1 + i, 4 + 3i$ **110.** $2 - i, 1 + 4i$

8.6 **Trigonometric Form of a Complex Number** In Exercises 111–116, plot the complex number. Then write the trigonometric form of the complex number.

111. $4i$ **112.** -7

113. $7 - 7i$ **114.** $5 + 12i$

115. $-5 - 12i$ **116.** $-3\sqrt{3} + 3i$

Multiplying Complex Numbers In Exercises 117 and 118, find the product. Leave the result in trigonometric form.

117. $\left[2\left(\cos \dfrac{\pi}{4} + i \sin \dfrac{\pi}{4} \right) \right]\left[2\left(\cos \dfrac{\pi}{3} + i \sin \dfrac{\pi}{3} \right) \right]$

118. $\left[4\left(\cos \dfrac{\pi}{3} + i \sin \dfrac{\pi}{3} \right) \right]\left[3\left(\cos \dfrac{5\pi}{6} + i \sin \dfrac{5\pi}{6} \right) \right]$

Dividing Complex Numbers In Exercises 119 and 120, find the quotient. Leave the result in trigonometric form.

119. $\dfrac{2(\cos 60° + i \sin 60°)}{3(\cos 15° + i \sin 15°)}$ **120.** $\dfrac{\cos 150° + i \sin 150°}{2(\cos 50° + i \sin 50°)}$

Finding a Power of a Complex Number In Exercises 121–124, use DeMoivre's Theorem to find the power of the complex number. Write the result in standard form.

121. $\left[5\left(\cos \dfrac{\pi}{12} + i \sin \dfrac{\pi}{12} \right) \right]^4$

122. $\left[2\left(\cos \dfrac{4\pi}{15} + i \sin \dfrac{4\pi}{15} \right) \right]^5$

123. $(2 + 3i)^6$

124. $(1 - i)^8$

Finding the nth Roots of a Complex Number In Exercises 125–128, (a) find the roots of the complex number, (b) write each of the roots in standard form, and (c) represent each of the roots graphically.

125. Sixth roots of $-729i$ **126.** Fourth roots of $256i$

127. Cube roots of 8

128. Fifth roots of -1024

Solving an Equation In Exercises 129–132, find all solutions of the equation. Represent the solutions graphically.

129. $x^4 + 81 = 0$

130. $x^5 - 32 = 0$

131. $x^3 + 8i = 0$

132. $x^4 - 64i = 0$

Exploring the Concepts

True or False? In Exercises 133 and 134, determine whether the statement is true or false. Justify your answer.

133. The Law of Sines is true when one of the angles in the triangle is a right angle.

134. When the Law of Sines is used, the solution is always unique.

135. Writing Describe the characteristics of a vector in the plane.

GO DIGITAL

Chapter Test

See CalcChat.com for tutorial help and worked-out solutions to odd-numbered exercises.

Take this test as you would take a test in class. When you are finished, check your work against the answers given in the back of the book.

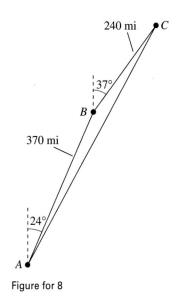

240 mi • C

37°

B •

370 mi

24°

A •

Figure for 8

In Exercises 1–6, determine whether the Law of Sines or the Law of Cosines is needed to solve the triangle. Then solve (if possible) the triangle. If two solutions exist, find both. Round your answers to two decimal places. *(Sections 8.1 and 8.2)*

1. $A = 24°$, $B = 68°$, $a = 12.2$
2. $B = 110°$, $C = 28°$, $a = 15.6$
3. $A = 24°$, $a = 11.2$, $b = 13.4$
4. $a = 6.0$, $b = 7.3$, $c = 12.4$
5. $B = 100°$, $a = 23$, $b = 15$
6. $C = 121°$, $a = 34$, $b = 55$

7. A triangular parcel of land has sides of lengths 60 meters, 70 meters, and 82 meters. Find the area of the parcel of land. *(Section 8.2)*

8. An airplane flies 370 miles from point A to point B with a bearing of 24°. Then it flies 240 miles from point B to point C with a bearing of 37° (see figure). Find the straight-line distance and bearing from point A to point C. *(Sections 8.1 and 8.2)*

In Exercises 9 and 10, find the component form of the vector v. *(Section 8.3)*

9. Initial point of **v**: $(-3, 7)$; terminal point of **v**: $(11, -16)$
10. Magnitude of **v**: $\|\mathbf{v}\| = 12$; direction of **v**: $\mathbf{u} = \langle 3, -5 \rangle$

In Exercises 11–14, u = $\langle 2, 7 \rangle$ and v = $\langle -6, 5 \rangle$. Find the resultant vector and sketch its graph. *(Section 8.3)*

11. $\mathbf{u} + \mathbf{v}$
12. $\mathbf{u} - \mathbf{v}$
13. $5\mathbf{u} - 3\mathbf{v}$
14. $4\mathbf{u} + 2\mathbf{v}$

15. Find the distance between $4 + 3i$ and $1 - i$ in the complex plane. *(Section 8.5)*

16. Forces with magnitudes of 250 pounds and 130 pounds act on an object at angles of 45° and $-60°$, respectively, with the positive x-axis. Find the direction and magnitude of the resultant of these forces. *(Section 8.3)*

17. Find the angle θ (in degrees) between the vectors $\mathbf{u} = \langle -1, 5 \rangle$ and $\mathbf{v} = \langle 3, -2 \rangle$. *(Section 8.4)*

18. Determine whether the vectors $\mathbf{u} = \langle 6, -10 \rangle$ and $\mathbf{v} = \langle 5, 3 \rangle$ are orthogonal. *(Section 8.4)*

19. Find the projection of $\mathbf{u} = \langle 6, 7 \rangle$ onto $\mathbf{v} = \langle -5, -1 \rangle$. Then write $\mathbf{u}$ as the sum of two orthogonal vectors, one of which is $\text{proj}_\mathbf{v}\mathbf{u}$. *(Section 8.4)*

20. A 500-pound motorcycle is stopped at a red light on a hill inclined at 12°. Find the force required to keep the motorcycle from rolling down the hill. *(Section 8.4)*

21. Write the complex number $z = 4 - 4i$ in trigonometric form. *(Section 8.6)*

22. Write the complex number $z = 6(\cos 120° + i \sin 120°)$ in standard form. *(Section 8.6)*

In Exercises 23 and 24, use DeMoivre's Theorem to find the power of the complex number. Write the result in standard form. *(Section 8.6)*

23. $\left[3\left(\cos \dfrac{7\pi}{6} + i \sin \dfrac{7\pi}{6} \right) \right]^8$
24. $(3 - 3i)^6$

25. Find the fourth roots of 256. *(Section 8.6)*

26. Find all solutions of the equation $x^3 - 27i = 0$ and represent the solutions graphically. *(Section 8.6)*

Take this test as you would take a test in class. When you are finished, check your work against the answers given in the back of the book.

1. Consider the angle $\theta = -120°$. *(Sections 6.1 and 6.2)*
   (a) Sketch the angle in standard position.
   (b) Determine a coterminal angle in the interval $[0°, 360°)$.
   (c) Rewrite the angle in radian measure as a multiple of π. Do not use a calculator.
   (d) Find the reference angle θ'.
   (e) Find the exact values of the six trigonometric functions of θ.

2. Convert -1.45 radians to degrees. Round to three decimal places. *(Section 6.1)*

3. Find $\cos\theta$ when $\tan\theta = -\frac{21}{20}$ and $\sin\theta < 0$. *(Section 6.3)*

In Exercises 4–6, sketch the graph of the function. (Include two full periods.) *(Sections 6.4 and 6.5)*

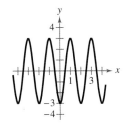

Figure for 7

4. $f(x) = 3 - 2\sin\pi x$ 5. $g(x) = \dfrac{1}{2}\tan\left(x - \dfrac{\pi}{2}\right)$ 6. $h(x) = -\sec(x + \pi)$

7. Find a, b, and c for the function $h(x) = a\cos(bx + c)$ such that the graph of h matches the figure. *(Section 6.4)*

8. Sketch the graph of the function $f(x) = \frac{1}{2}x\sin x$ on the interval $[-3\pi, 3\pi]$. *(Section 6.5)*

In Exercises 9 and 10, find the exact value of the expression. *(Section 6.6)*

9. $\tan(\arctan 4.9)$ 10. $\tan\left(\arcsin\frac{3}{5}\right)$

11. Write an algebraic expression that is equivalent to $\sin(\arccos 2x)$. *(Section 6.6)*

12. Use the fundamental identities to simplify: $\cos\left(\dfrac{\pi}{2} - x\right)\csc x$. *(Section 7.1)*

13. Subtract and simplify: $\dfrac{\sin\theta - 1}{\cos\theta} - \dfrac{\cos\theta}{\sin\theta - 1}$. *(Section 7.1)*

In Exercises 14–16, verify the identity. *(Section 7.2)*

14. $\cot^2\alpha(\sec^2\alpha - 1) = 1$

15. $\sin(x + y)\sin(x - y) = \sin^2 x - \sin^2 y$

16. $\sin^2 x \cos^2 x = \frac{1}{8}(1 - \cos 4x)$

In Exercises 17 and 18, find all solutions of the equation in the interval $[0, 2\pi)$. *(Section 7.3)*

17. $2\cos^2\beta - \cos\beta = 0$ 18. $3\tan\theta - \cot\theta = 0$

19. Use the Quadratic Formula to find all solutions of the equation in the interval $[0, 2\pi)$: $\sin^2 x + 2\sin x + 1 = 0$. *(Section 7.3)*

20. Given that $\sin u = \frac{12}{13}$, $\cos v = \frac{3}{5}$, and angles u and v are both in Quadrant I, find $\tan(u - v)$. *(Section 7.4)*

21. Given that $\tan u = \dfrac{1}{2}$ and $0 < u < \dfrac{\pi}{2}$, find the exact value of $\tan(2u)$. *(Section 7.5)*

22. Given that $\tan u = \dfrac{4}{3}$ and $0 < u < \dfrac{\pi}{2}$, find the exact value of $\sin\dfrac{u}{2}$. *(Section 7.5)*

23. Rewrite $5 \sin \dfrac{3\pi}{4} \cdot \cos \dfrac{7\pi}{4}$ as a sum or difference.   *(Section 7.5)*

24. Rewrite $\cos 9x - \cos 7x$ as a product.   *(Section 7.5)*

In Exercises 25–30, determine whether the Law of Sines or the Law of Cosines is needed to solve the triangle at the left. Then solve the triangle. Round your answers to two decimal places.   *(Sections 8.1 and 8.2)*

25. $A = 30°$, $a = 9$, $b = 8$ **26.** $A = 30°$, $b = 8$, $c = 10$

27. $A = 30°$, $C = 90°$, $b = 10$ **28.** $a = 4.7$, $b = 8.1$, $c = 10.3$

29. $A = 45°$, $B = 26°$, $c = 20$ **30.** $C = 80°$, $a = 1.2$, $b = 10$

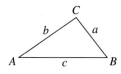

Figure for 25–30

31. Find the area of a triangle with two sides of lengths 7 inches and 12 inches and an included angle of 99°.   *(Section 8.1)*

32. Use Heron's Area Formula to find the area of a triangle with sides of lengths 30 meters, 41 meters, and 45 meters.   *(Section 8.2)*

33. Write the vector with initial point $(-1, 2)$ and terminal point $(6, 10)$ as a linear combination of the standard unit vectors $\mathbf{i}$ and $\mathbf{j}$.   *(Section 8.3)*

34. Find a unit vector $\mathbf{u}$ in the direction of $\mathbf{v} = \mathbf{i} + \mathbf{j}$.   *(Section 8.3)*

35. Find $\mathbf{u} \cdot \mathbf{v}$ for $\mathbf{u} = 3\mathbf{i} + 4\mathbf{j}$ and $\mathbf{v} = \mathbf{i} - 2\mathbf{j}$.   *(Section 8.4)*

36. Find the projection of $\mathbf{u} = \langle 8, -2 \rangle$ onto $\mathbf{v} = \langle 1, 5 \rangle$. Then write $\mathbf{u}$ as the sum of two orthogonal vectors, one of which is $\text{proj}_{\mathbf{v}}\,\mathbf{u}$.   *(Section 8.4)*

37. Plot $3 - 2i$ and its complex conjugate. Write the conjugate as a complex number.   *(Section 8.5)*

38. Write the complex number $-2 + 2i$ in trigonometric form.   *(Section 8.6)*

39. Find the product of $[4(\cos 30° + i \sin 30°)]$ and $[6(\cos 120° + i \sin 120°)]$. Leave the result in trigonometric form.   *(Section 8.6)*

40. Find the three cube roots of 1.   *(Section 8.6)*

41. Find all solutions of the equation $x^4 + 625 = 0$ and represent the solutions graphically.   *(Section 8.6)*

42. A ceiling fan with 21-inch blades makes 63 revolutions per minute. Find the angular speed of the fan in radians per minute. Find the linear speed (in inches per minute) of the tips of the blades.   *(Section 6.1)*

43. Find the area of the sector of a circle with a radius of 12 yards and a central angle of 105°.   *(Section 6.1)*

44. From a point 200 feet from a flagpole, the angles of elevation to the bottom and top of the flag are 16° 45′ and 18°, respectively. Approximate the height of the flag to the nearest foot.   *(Section 6.2)*

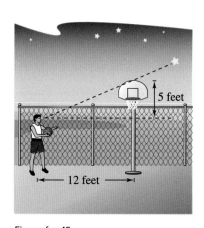

Figure for 45

45. To determine the angle of elevation of a star in the sky, a student aligns the star and the top of the backboard of a basketball hoop that is 5 feet higher than his eyes in his line of vision (see figure). His horizontal distance from the backboard is 12 feet. What is the angle of elevation of the star?   *(Section 6.3)*

46. Find a model for a particle in simple harmonic motion with a displacement (at $t = 0$) of 4 inches, an amplitude of 4 inches, and a period of 8 seconds.   *(Section 6.4)*

47. An airplane has a speed of 500 kilometers per hour at a bearing of 30°. The wind velocity is 50 kilometers per hour in the direction N 60° E. Find the resultant speed and true direction of the airplane.   *(Section 8.3)*

48. A constant force of 85 pounds, exerted at an angle of 60° with the horizontal, is required to slide an object across a floor. Determine the work done in sliding the object 10 feet.   *(Section 8.4)*

Proofs in Mathematics

HISTORICAL NOTE

Besides the Law of Sines and the Law of Cosines, there is also a Law of Tangents, developed by French mathematician François Viète (1540–1603). The Law of Tangents follows from the Law of Sines and the sum-to-product formulas for sine and is defined as

$$\frac{a + b}{a - b} = \frac{\tan[(A + B)/2]}{\tan[(A - B)/2]}.$$

The Law of Tangents can be used to solve a triangle when two sides and the included angle (SAS) are given. Before the invention of calculators, it was easier to use the Law of Tangents to solve the SAS case instead of the Law of Cosines because the computations by hand were not as tedious.

Law of Sines (p. 560)

If ABC is a triangle with sides a, b, and c, then

$$\frac{a}{\sin A} = \frac{b}{\sin B} = \frac{c}{\sin C}.$$

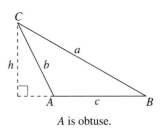

A is acute. A is obtuse.

Proof

Let h be an altitude of either triangle shown above. Next, write two expressions for h involving $\sin A$ and $\sin B$. From right triangle trigonometry, $\sin B = h/a$ in both triangles. In the acute triangle, $\sin A = h/b$, and in the obtuse triangle, $\sin(180° - A) = h/b$. But $\sin(180° - A) = \sin A$. So, in either triangle, $\sin A = h/b$. Now, solve both equations for h.

$$\sin A = \frac{h}{b} \implies h = b \sin A \qquad \text{Solve for } h.$$

$$\sin B = \frac{h}{a} \implies h = a \sin B \qquad \text{Solve for } h.$$

Equating these two values of h, you have

$$a \sin B = b \sin A \implies \frac{a}{\sin A} = \frac{b}{\sin B}. \qquad \text{Divide each side by } \sin A \sin B.$$

Note that $\sin A \neq 0$ and $\sin B \neq 0$ because no angle of a triangle can have a measure of 0° or 180°. In a similar manner, construct an altitude h from vertex B to side AC (extended in the obtuse triangle), as shown at the left. Then you have

$$\sin A = \frac{h}{c} \implies h = c \sin A \quad \text{and} \quad \sin C = \frac{h}{a} \implies h = a \sin C.$$

Equating these two values of h, you have

$$a \sin C = c \sin A \implies \frac{a}{\sin A} = \frac{c}{\sin C}. \qquad \text{Divide each side by } \sin A \sin C.$$

By the Transitive Property of Equality, you can conclude that

$$\frac{a}{\sin A} = \frac{b}{\sin B} = \frac{c}{\sin C}.$$

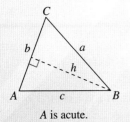

A is acute.

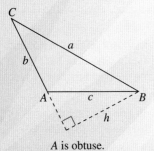

A is obtuse.

Law of Cosines (p. 569)

Standard Form	Alternative Form
$a^2 = b^2 + c^2 - 2bc \cos A$	$\cos A = \dfrac{b^2 + c^2 - a^2}{2bc}$
$b^2 = a^2 + c^2 - 2ac \cos B$	$\cos B = \dfrac{a^2 + c^2 - b^2}{2ac}$
$c^2 = a^2 + b^2 - 2ab \cos C$	$\cos C = \dfrac{a^2 + b^2 - c^2}{2ab}$

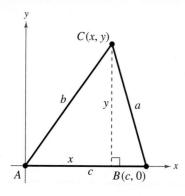

Proof To prove the first formula, consider the triangle at the left, which has three acute angles. Note that vertex B has coordinates $(c, 0)$. Furthermore, C has coordinates (x, y), where $x = b \cos A$ and $y = b \sin A$. Because a is the distance from C to B, it follows that

$$a = \sqrt{(x - c)^2 + (y - 0)^2}$$ Distance Formula

$$a^2 = (x - c)^2 + (y - 0)^2$$ Square each side.

$$a^2 = (b \cos A - c)^2 + (b \sin A)^2$$ Substitute for x and y.

$$a^2 = b^2 \cos^2 A - 2bc \cos A + c^2 + b^2 \sin^2 A$$ Expand.

$$a^2 = b^2(\sin^2 A + \cos^2 A) + c^2 - 2bc \cos A$$ Factor out b^2.

$$a^2 = b^2 + c^2 - 2bc \cos A.$$ $\sin^2 A + \cos^2 A = 1$

Similar arguments are used to establish the second and third formulas. ■

Heron's Area Formula (p. 572)

Given any triangle with sides of lengths a, b, and c, the area of the triangle is

$$\text{Area} = \sqrt{s(s - a)(s - b)(s - c)}, \text{ where } s = \frac{a + b + c}{2}.$$

ALGEBRA HELP

$$\frac{1}{2}bc(1 + \cos A)$$

$$= \frac{1}{2}bc\left(1 + \frac{b^2 + c^2 - a^2}{2bc}\right)$$

$$= \frac{1}{2}bc\left(\frac{2bc + b^2 + c^2 - a^2}{2bc}\right)$$

$$= \frac{1}{4}(2bc + b^2 + c^2 - a^2)$$

$$= \frac{1}{4}(b^2 + 2bc + c^2 - a^2)$$

$$= \frac{1}{4}[(b + c)^2 - a^2]$$

$$= \frac{1}{4}(b + c + a)(b + c - a)$$

$$= \frac{a + b + c}{2} \cdot \frac{-a + b + c}{2}$$

▷▷▷▷▷

Proof From Section 8.1, you know that

$$\text{Area} = \frac{1}{2}bc \sin A$$ Formula for the area of an oblique triangle

$$(\text{Area})^2 = \frac{1}{4}b^2c^2 \sin^2 A$$ Square each side.

$$\text{Area} = \sqrt{\frac{1}{4}b^2c^2 \sin^2 A}$$ Take the square root of each side.

$$= \sqrt{\frac{1}{4}b^2c^2(1 - \cos^2 A)}$$ Pythagorean identity

$$= \sqrt{\left[\frac{1}{2}bc(1 + \cos A)\right]\left[\frac{1}{2}bc(1 - \cos A)\right]}.$$ Factor.

Using the alternative form of the Law of Cosines,

$$\frac{1}{2}bc(1 + \cos A) = \frac{a + b + c}{2} \cdot \frac{-a + b + c}{2}$$

and

$$\frac{1}{2}bc(1 - \cos A) = \frac{a - b + c}{2} \cdot \frac{a + b - c}{2}.$$

Letting $s = (a + b + c)/2$, rewrite these two equations as

$$\frac{1}{2}bc(1 + \cos A) = s(s - a) \quad \text{and} \quad \frac{1}{2}bc(1 - \cos A) = (s - b)(s - c).$$

Substitute into the last formula for area to conclude that

$$\text{Area} = \sqrt{s(s - a)(s - b)(s - c)}.$$ ■

> **Properties of the Dot Product** *(p. 589)*
>
> Let $\mathbf{u}$, $\mathbf{v}$, and $\mathbf{w}$ be vectors and let c be a scalar.
>
> **1.** $\mathbf{u} \cdot \mathbf{v} = \mathbf{v} \cdot \mathbf{u}$ **2.** $\mathbf{u} \cdot (\mathbf{v} + \mathbf{w}) = \mathbf{u} \cdot \mathbf{v} + \mathbf{u} \cdot \mathbf{w}$
>
> **3.** $c(\mathbf{u} \cdot \mathbf{v}) = c\mathbf{u} \cdot \mathbf{v} = \mathbf{u} \cdot c\mathbf{v}$ **4.** $\mathbf{0} \cdot \mathbf{v} = 0$
>
> **5.** $\mathbf{v} \cdot \mathbf{v} = \|\mathbf{v}\|^2$

Proof

Let $\mathbf{u} = \langle u_1, u_2 \rangle$, $\mathbf{v} = \langle v_1, v_2 \rangle$, $\mathbf{w} = \langle w_1, w_2 \rangle$, $\mathbf{0} = \langle 0, 0 \rangle$, and let c be a scalar.

1. $\mathbf{u} \cdot \mathbf{v} = u_1 v_1 + u_2 v_2 = v_1 u_1 + v_2 u_2 = \mathbf{v} \cdot \mathbf{u}$

2. $\mathbf{u} \cdot (\mathbf{v} + \mathbf{w}) = \mathbf{u} \cdot \langle v_1 + w_1, v_2 + w_2 \rangle$

$$= u_1(v_1 + w_1) + u_2(v_2 + w_2)$$

$$= u_1 v_1 + u_1 w_1 + u_2 v_2 + u_2 w_2$$

$$= (u_1 v_1 + u_2 v_2) + (u_1 w_1 + u_2 w_2)$$

$$= \mathbf{u} \cdot \mathbf{v} + \mathbf{u} \cdot \mathbf{w}$$

3. $c(\mathbf{u} \cdot \mathbf{v}) = c(\langle u_1, u_2 \rangle \cdot \langle v_1, v_2 \rangle)$

$$= c(u_1 v_1 + u_2 v_2)$$

$$= (cu_1)v_1 + (cu_2)v_2$$

$$= \langle cu_1, cu_2 \rangle \cdot \langle v_1, v_2 \rangle$$

$$= c\mathbf{u} \cdot \mathbf{v}$$

4. $\mathbf{0} \cdot \mathbf{v} = 0(v_1) + 0(v_2) = 0$

5. $\mathbf{v} \cdot \mathbf{v} = v_1^2 + v_2^2 = \left(\sqrt{v_1^2 + v_2^2} \right)^2 = \|\mathbf{v}\|^2$ ■

> **Angle Between Two Vectors** *(p. 590)*
>
> If θ is the angle between two nonzero vectors $\mathbf{u}$ and $\mathbf{v}$, then $\cos \theta = \dfrac{\mathbf{u} \cdot \mathbf{v}}{\|\mathbf{u}\|\|\mathbf{v}\|}$.

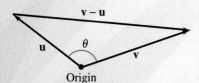

Proof

Consider the triangle determined by vectors $\mathbf{u}$, $\mathbf{v}$, and $\mathbf{v} - \mathbf{u}$, as shown at the left. By the Law of Cosines,

$$\|\mathbf{v} - \mathbf{u}\|^2 = \|\mathbf{u}\|^2 + \|\mathbf{v}\|^2 - 2\|\mathbf{u}\|\|\mathbf{v}\| \cos \theta$$

$$(\mathbf{v} - \mathbf{u}) \cdot (\mathbf{v} - \mathbf{u}) = \|\mathbf{u}\|^2 + \|\mathbf{v}\|^2 - 2\|\mathbf{u}\|\|\mathbf{v}\| \cos \theta$$

$$(\mathbf{v} - \mathbf{u}) \cdot \mathbf{v} - (\mathbf{v} - \mathbf{u}) \cdot \mathbf{u} = \|\mathbf{u}\|^2 + \|\mathbf{v}\|^2 - 2\|\mathbf{u}\|\|\mathbf{v}\| \cos \theta$$

$$\mathbf{v} \cdot \mathbf{v} - \mathbf{u} \cdot \mathbf{v} - \mathbf{v} \cdot \mathbf{u} + \mathbf{u} \cdot \mathbf{u} = \|\mathbf{u}\|^2 + \|\mathbf{v}\|^2 - 2\|\mathbf{u}\|\|\mathbf{v}\| \cos \theta$$

$$\|\mathbf{v}\|^2 - 2\mathbf{u} \cdot \mathbf{v} + \|\mathbf{u}\|^2 = \|\mathbf{u}\|^2 + \|\mathbf{v}\|^2 - 2\|\mathbf{u}\|\|\mathbf{v}\| \cos \theta$$

$$\cos \theta = \frac{\mathbf{u} \cdot \mathbf{v}}{\|\mathbf{u}\|\|\mathbf{v}\|}.$$ ■

P.S. Problem Solving

See CalcChat.com for tutorial help and worked-out solutions to odd-numbered exercises.

1. **Distance** In the figure, a beam of light is directed at the blue mirror, reflected to the red mirror, and then reflected back to the blue mirror. Find *PT*, the distance that the light travels from the red mirror back to the blue mirror.

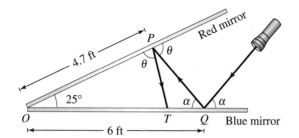

2. **Correcting a Course** A triathlete sets a course to swim S 25° E from a point on shore to a buoy $\frac{3}{4}$ mile away. After swimming 300 yards through a strong current, the triathlete is off course at a bearing of S 35° E. Find the bearing and distance the triathlete needs to swim to correct her course.

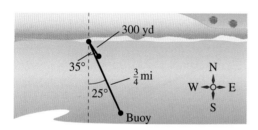

3. **Locating Lost Hikers** A group of hikers is lost in a national park. Two ranger stations receive an emergency SOS signal from the hikers. Station B is 75 miles due east of station A. The bearing from station A to the signal is S 60° E and the bearing from station B to the signal is S 75° W.

   (a) Draw a diagram that gives a visual representation of the problem.

   (b) Find the distance from each station to the SOS signal.

   (c) A rescue party is in the park 20 miles from station A at a bearing of S 80° E. Find the distance and the bearing the rescue party must travel to reach the lost hikers.

4. **Seeding a Courtyard** You are seeding a triangular courtyard. One side of the courtyard is 52 feet long and another side is 46 feet long. The angle opposite the 52-foot side is 65°.

   (a) Draw a diagram that gives a visual representation of the problem.

   (b) How long is the third side of the courtyard?

   (c) One bag of grass seed covers an area of 50 square feet. How many bags of grass seed do you need to cover the courtyard?

5. **Finding Magnitudes** For each pair of vectors, find the value of each expression.

   (i) $\|\mathbf{u}\|$ (ii) $\|\mathbf{v}\|$ (iii) $\|\mathbf{u} + \mathbf{v}\|$

   (iv) $\left\|\dfrac{\mathbf{u}}{\|\mathbf{u}\|}\right\|$ (v) $\left\|\dfrac{\mathbf{v}}{\|\mathbf{v}\|}\right\|$ (vi) $\left\|\dfrac{\mathbf{u} + \mathbf{v}}{\|\mathbf{u} + \mathbf{v}\|}\right\|$

   (a) $\mathbf{u} = \langle 1, -1 \rangle$
       $\mathbf{v} = \langle -1, 2 \rangle$

   (b) $\mathbf{u} = \langle 0, 1 \rangle$
       $\mathbf{v} = \langle 3, -3 \rangle$

   (c) $\mathbf{u} = \langle 1, \frac{1}{2} \rangle$
       $\mathbf{v} = \langle 2, 3 \rangle$

   (d) $\mathbf{u} = \langle 2, -4 \rangle$
       $\mathbf{v} = \langle 5, 5 \rangle$

6. **Writing a Vector in Terms of Other Vectors** Write the vector **w** in terms of **u** and **v**, given that the terminal point of **w** bisects the line segment (see figure).

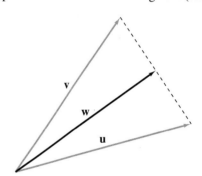

7. **Proof** Prove that if **u** is orthogonal to **v** and **w**, then **u** is orthogonal to

   $c\mathbf{v} + d\mathbf{w}$

   for any scalars *c* and *d*.

8. **Comparing Work Done** Two forces of the same magnitude $\mathbf{F}_1$ and $\mathbf{F}_2$ act at angles θ_1 and θ_2, respectively. Use a diagram to compare the work done by $\mathbf{F}_1$ with the work done by $\mathbf{F}_2$ in moving along the vector $\overrightarrow{PQ}$ when

   (a) $\theta_1 = -\theta_2$

   (b) $\theta_1 = 60°$ and $\theta_2 = 30°$.

9. **Writing the Roots of a Complex Number** For each graph of the roots of a complex number, write each of the roots in trigonometric form.

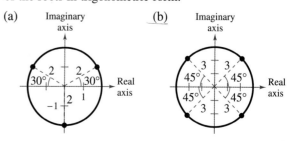

10. Skydiving A skydiver falls at a constant downward velocity of 120 miles per hour. In the figure, vector **u** represents the skydiver's velocity. A steady breeze pushes the skydiver to the east at 40 miles per hour. Vector **v** represents the wind velocity.

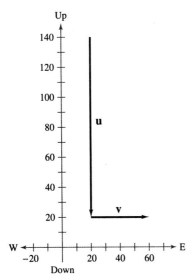

(a) Write the vectors **u** and **v** in component form.

(b) Let

$$s = u + v.$$

Use the figure to sketch **s**.

(c) Find the magnitude of **s**. What information does the magnitude give you about the skydiver's fall?

(d) Without wind, the skydiver would fall in a path perpendicular to the ground. At what angle to the ground is the path of the skydiver when affected by the 40-mile-per-hour wind from due west?

(e) The next day, the skydiver falls at a constant downward velocity of 120 miles per hour and a steady breeze pushes the skydiver to the west at 30 miles per hour. Draw a new figure that gives a visual representation of the problem and find the skydiver's new velocity.

11. Comparing Vector Magnitudes The vectors **u** and **v** have the same magnitudes in the two figures. In which figure is the magnitude of the sum greater? Explain.

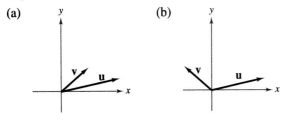

12. Speed and Velocity of an Airplane Four basic forces are in action during flight: weight, lift, thrust, and drag. To fly through the air, an object must overcome its own *weight*. To do this, it must create an upward force called *lift*. To generate lift, a forward motion called *thrust* is needed. The thrust must be great enough to overcome air resistance, which is called *drag*.

For a commercial jet aircraft, a quick climb is important to maximize efficiency because the performance of an aircraft is enhanced at high altitudes. In addition, it is necessary to clear obstacles such as buildings and mountains and to reduce noise in residential areas. In the diagram, the angle θ is called the climb angle. The velocity of the plane can be represented by a vector **v** with a vertical component $\|v\| \sin \theta$ (called climb speed) and a horizontal component $\|v\| \cos \theta$, where $\|v\|$ is the speed of the plane.

When taking off, a pilot must decide how much of the thrust to apply to each component. The more the thrust is applied to the horizontal component, the faster the airplane gains speed. The more the thrust is applied to the vertical component, the quicker the airplane climbs.

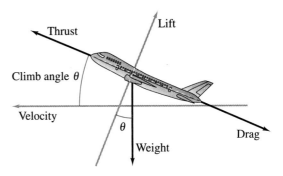

(a) Complete the table for an airplane that has a speed of $\|v\| = 100$ miles per hour.

θ	0.5°	1.0°	1.5°	2.0°	2.5°	3.0°
$\|v\| \sin \theta$						
$\|v\| \cos \theta$						

(b) Does an airplane's speed equal the sum of the vertical and horizontal components of its velocity? If not, how could you find the speed of an airplane whose velocity components were known?

(c) Use the result of part (b) to find the speed of an airplane with the given velocity components.

  (i) $\|v\| \sin \theta = 5.235$ miles per hour

  $\|v\| \cos \theta = 149.909$ miles per hour

  (ii) $\|v\| \sin \theta = 10.463$ miles per hour

  $\|v\| \cos \theta = 149.634$ miles per hour

9 Systems of Equations and Inequalities

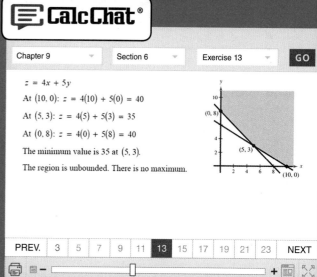

9.4 Thermodynamics *(Exercise 52, p. 669)*

9.6 Agriculture *(Exercise 36, p. 687)*

627

9.1 Linear and Nonlinear Systems of Equations

Graphs of systems of equations can help you solve real-life problems. For example, in Exercise 67 on page 636, you will use the graph of a system of equations to compare the consumption of biofuels and the consumption of wind energy.

❯ Use the method of substitution to solve systems of linear equations in two variables.
❯ Use the method of substitution to solve systems of nonlinear equations in two variables.
❯ Use a graphical method to solve systems of equations in two variables.
❯ Use systems of equations to model and solve real-life problems.

The Method of Substitution

Up to this point in the text, most problems have involved either a function of one variable or a single equation in two variables. However, many problems in science, business, and engineering involve two or more equations in two or more variables. To solve such a problem, you need to find the solutions of a **system of equations.** Here is an example of a system of two equations in two variables, x and y.

$$\begin{cases} 2x + y = 5 & \text{Equation 1} \\ 3x - 2y = 4 & \text{Equation 2} \end{cases}$$

A **solution** of this system is an ordered pair that satisfies each equation in the system. Finding the set of all solutions is called **solving the system of equations.** For example, the ordered pair $(2, 1)$ is a solution of this system. To check this, substitute 2 for x and 1 for y in *each* equation.

Check (2, 1) in Equation 1 and Equation 2:

$$2x + y = 5 \qquad \text{Write Equation 1.}$$
$$2(2) + 1 \overset{?}{=} 5 \qquad \text{Substitute 2 for } x \text{ and 1 for } y.$$
$$4 + 1 = 5 \qquad \text{Solution checks in Equation 1. ✓}$$
$$3x - 2y = 4 \qquad \text{Write Equation 2.}$$
$$3(2) - 2(1) \overset{?}{=} 4 \qquad \text{Substitute 2 for } x \text{ and 1 for } y.$$
$$6 - 2 = 4 \qquad \text{Solution checks in Equation 2. ✓}$$

In this chapter, you will study four ways to solve systems of equations, beginning with the **method of substitution.**

Method	Section	Type of System
1. Substitution	9.1	Linear or nonlinear, two variables
2. Graphical	9.1	Linear or nonlinear, two variables
3. Elimination	9.2	Linear, two variables
4. Gaussian elimination	9.3	Linear, three or more variables

Method of Substitution

1. *Solve* one of the equations for one variable in terms of the other.

2. *Substitute* the expression found in Step 1 into the other equation to obtain an equation in one variable.

3. *Solve* the equation obtained in Step 2.

4. *Back-substitute* the value obtained in Step 3 into the expression obtained in Step 1 to find the value of the other variable.

5. *Check* that the solution satisfies *each* of the original equations.

GO DIGITAL

 **EXAMPLE 1**   **Solving a System of Equations by Substitution**

Solve the system of equations.

$$\begin{cases} x + y = 4 & \text{Equation 1} \\ x - y = 2 & \text{Equation 2} \end{cases}$$

Solution   Begin by solving for y in Equation 1.

$$x + y = 4 \qquad \text{Write Equation 1.}$$

$$y = 4 - x \qquad \text{Subtract } x \text{ from each side.}$$

Next, substitute this expression for y into Equation 2 and solve the resulting single-variable equation for x.

$$x - y = 2 \qquad \text{Write Equation 2.}$$

$$x - (4 - x) = 2 \qquad \text{Substitute } 4 - x \text{ for } y.$$

$$x - 4 + x = 2 \qquad \text{Distributive Property}$$

$$2x - 4 = 2 \qquad \text{Combine like terms.}$$

$$2x = 6 \qquad \text{Add 4 to each side.}$$

$$x = 3 \qquad \text{Divide each side by 2.}$$

Finally, solve for y by *back-substituting* $x = 3$ into the equation $y = 4 - x$.

$$y = 4 - x \qquad \text{Write revised Equation 1.}$$

$$y = 4 - 3 \qquad \text{Substitute 3 for } x.$$

$$y = 1 \qquad \text{Solve for } y.$$

So, the solution of the system is the ordered pair $(3, 1)$. Check this solution algebraically, as shown below.

Check

Substitute $(3, 1)$ into Equation 1:

$$x + y = 4 \qquad \text{Write Equation 1.}$$

$$3 + 1 \stackrel{?}{=} 4 \qquad \text{Substitute for } x \text{ and } y.$$

$$4 = 4 \qquad \text{Solution checks in Equation 1. } \checkmark$$

Substitute $(3, 1)$ into Equation 2:

$$x - y = 2 \qquad \text{Write Equation 2.}$$

$$3 - 1 \stackrel{?}{=} 2 \qquad \text{Substitute for } x \text{ and } y.$$

$$2 = 2 \qquad \text{Solution checks in Equation 2. } \checkmark$$

The point $(3, 1)$ satisfies both equations in the system. This confirms that $(3, 1)$ is a solution of the system of equations.

✓ *Checkpoint* 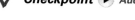 *Audio-video solution in English & Spanish at LarsonPrecalculus.com*

Solve the system of equations.

$$\begin{cases} x - y = 0 \\ 5x - 3y = 6 \end{cases}$$

The term *back-substitution* implies that you work *backwards*. First you solve for one of the variables, and then you substitute that value *back* into one of the equations in the system to find the value of the other variable.

> **ALGEBRA HELP**
>
> When using the method of substitution, it does not matter which variable you choose to solve for first. Whether you solve for y first or x first, you will obtain the same solution. When making your choice, you should choose the variable and equation that are easier to work with.

> **ALGEBRA HELP**
>
> Many steps are required to solve a system of equations, so there are many possible ways to make errors in arithmetic. You should always check your solution by substituting it into *each* equation in the original system.

GO DIGITAL

EXAMPLE 2 **Solving a System by Substitution**

A total of \$12,000 is invested in two funds paying 5% and 3% simple interest. The total annual interest is \$500. How much is invested at each rate?

Solution

Recall that the formula for simple interest is $I = Prt$, where P is the principal, r is the annual interest rate (in decimal form), and t is the time.

Verbal model:

Amount in 5% fund	+	Amount in 3% fund	=	Total investment

Interest for 5% fund	+	Interest for 3% fund	=	Total interest

Labels:

Amount in 5% fund $= x$	(dollars)
Interest for 5% fund $= 0.05x$	(dollars)
Amount in 3% fund $= y$	(dollars)
Interest for 3% fund $= 0.03y$	(dollars)
Total investment $= 12{,}000$	(dollars)
Total interest $= 500$	(dollars)

System:
$$\begin{cases} x + y = 12{,}000 & \text{Equation 1} \\ 0.05x + 0.03y = 500 & \text{Equation 2} \end{cases}$$

To begin, it is convenient to multiply each side of Equation 2 by 100. This eliminates the need to work with decimals.

$$100(0.05x + 0.03y) = 100(500) \qquad \text{Multiply each side of Equation 2 by 100.}$$
$$5x + 3y = 50{,}000 \qquad \text{Revised Equation 2}$$

To solve this system, you can solve for x in Equation 1.

$$x = 12{,}000 - y \qquad \text{Revised Equation 1}$$

Then, substitute this expression for x into revised Equation 2 and solve for y.

$$5(12{,}000 - y) + 3y = 50{,}000 \qquad \text{Substitute } 12{,}000 - y \text{ for } x \text{ in revised Equation 2.}$$
$$60{,}000 - 5y + 3y = 50{,}000 \qquad \text{Distributive Property}$$
$$-2y = -10{,}000 \qquad \text{Combine like terms.}$$
$$y = 5000 \qquad \text{Divide each side by } -2.$$

Next, back-substitute $y = 5000$ to solve for x.

$$x = 12{,}000 - y \qquad \text{Write revised Equation 1.}$$
$$x = 12{,}000 - 5000 \qquad \text{Substitute 5000 for } y.$$
$$x = 7000 \qquad \text{Subtract.}$$

The solution is (7000, 5000). So, \$7000 is invested at 5% and \$5000 is invested at 3%. Check this in the original system.

 Checkpoint ▶ *Audio-video solution in English & Spanish at LarsonPrecalculus.com*

A total of \$25,000 is invested in two funds paying 6.5% and 8.5% simple interest. The total annual interest is \$2000. How much is invested at each rate?

GO DIGITAL

Nonlinear Systems of Equations

The equations in Examples 1 and 2 are linear. The method of substitution can also be used to solve systems in which one or both of the equations are nonlinear.

> **EXAMPLE 3**   **Substitution: Two-Solution Case**

Solve the system of equations.

$$\begin{cases} 3x^2 + 4x - y = 7 & \text{Equation 1} \\ 2x - y = -1 & \text{Equation 2} \end{cases}$$

Solution   Begin by solving for y in Equation 2 to obtain $y = 2x + 1$. Next, substitute this expression for y into Equation 1 and solve for x.

$$3x^2 + 4x - (2x + 1) = 7 \qquad \text{Substitute } 2x + 1 \text{ for } y \text{ in Equation 1.}$$

$$3x^2 + 2x - 1 = 7 \qquad \text{Simplify.}$$

$$3x^2 + 2x - 8 = 0 \qquad \text{Write in general form.}$$

$$(3x - 4)(x + 2) = 0 \qquad \text{Factor.}$$

$$x = \frac{4}{3}, -2 \qquad \text{Solve for } x.$$

▶▶▶ SKILLS REFRESHER

For a refresher on solving polynomial equations, watch the video at *LarsonPrecalculus.com*.

Back-substituting these values of x to solve for the corresponding values of y produces the solutions $\left(\frac{4}{3}, \frac{11}{3}\right)$ and $(-2, -3)$. Check these in the original system.

✓ *Checkpoint* Audio-video solution in English & Spanish at *LarsonPrecalculus.com*

Solve the system of equations.

$$\begin{cases} -2x + y = 5 \\ x^2 - y + 3x = 1 \end{cases}$$

> **EXAMPLE 4**   **Substitution: No-Real-Solution Case**

Solve the system of equations.

$$\begin{cases} -x + y = 4 & \text{Equation 1} \\ x^2 + y = 3 & \text{Equation 2} \end{cases}$$

Solution   Begin by solving for y in Equation 1 to obtain $y = x + 4$. Next, substitute this expression for y into Equation 2 and solve for x.

$$x^2 + (x + 4) = 3 \qquad \text{Substitute } x + 4 \text{ for } y \text{ in Equation 2.}$$

$$x^2 + x + 1 = 0 \qquad \text{Write in general form.}$$

$$x = \frac{-1 \pm \sqrt{-3}}{2} \qquad \text{Use the Quadratic Formula.}$$

Because the discriminant is negative, the equation

$$x^2 + x + 1 = 0$$

has no real solution. So, the original system has no real solution.

✓ *Checkpoint* Audio-video solution in English & Spanish at *LarsonPrecalculus.com*

Solve the system of equations.

$$\begin{cases} 2x - y = -3 \\ 2x^2 + 4x - y^2 = 0 \end{cases}$$

GO DIGITAL

Graphical Method for Finding Solutions

Notice from Examples 2, 3, and 4 that a system of two equations in two variables can have exactly one solution, more than one solution, or no solution. A **graphical method** helps you to gain insight about the number of solutions of a system of equations. When you graph each equation in the same coordinate plane, the solutions of the system correspond to the **points of intersection** of the graphs. For example, the two equations in Figure 9.1 graph as two lines with a *single point* of intersection; the two equations in Figure 9.2 graph as a parabola and a line with *two points* of intersection; and the two equations in Figure 9.3 graph as a parabola and a line with *no points* of intersection.

TECHNOLOGY

Most graphing utilities have built-in features for approximating any points of intersection of two graphs. Use a graphing utility to find the points of intersection of the graphs in Figures 9.1 through 9.3. Be sure to adjust the viewing window so that you see all the points of intersection.

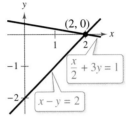

One intersection point
Figure 9.1

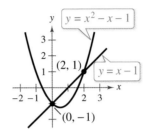

Two intersection points
Figure 9.2

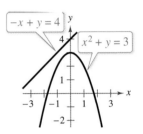

No intersection points
Figure 9.3

EXAMPLE 5   **Solving a System of Equations Graphically**

See LarsonPrecalculus.com for an interactive version of this type of example.

Solve the system of equations.

$$\begin{cases} y = \ln x & \text{Equation 1} \\ x + y = 1 & \text{Equation 2} \end{cases}$$

Solution   There is only one point of intersection of the graphs of the two equations, and $(1, 0)$ is the solution point (see Figure 9.4). Check this solution as follows.

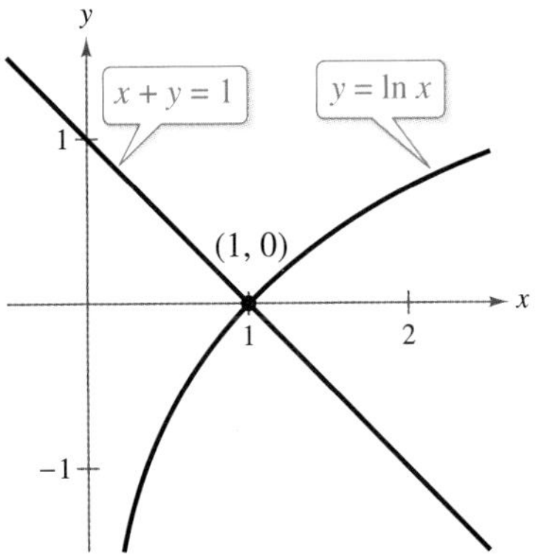

Check (1, 0) in Equation 1:

$y = \ln x$	Write Equation 1.
$0 = \ln 1$	Substitute for x and y.
$0 = 0$	Solution checks in Equation 1. ✓

Check (1, 0) in Equation 2:

$x + y = 1$	Write Equation 2.
$1 + 0 = 1$	Substitute for x and y.
$1 = 1$	Solution checks in Equation 2. ✓

✓ **Checkpoint** ▶ Audio-video solution in English & Spanish at LarsonPrecalculus.com

Solve the system of equations.

$$\begin{cases} y = 3 - \log x \\ -2x + y = 1 \end{cases}$$

■

Example 5 shows the benefit of solving systems of equations in two variables graphically. Note that using the substitution method in Example 5 produces $x + \ln x = 1$. It is difficult to solve this equation for x using standard algebraic techniques.

Figure 9.4

GO DIGITAL

Applications

The total cost C of producing x units of a product typically has two components—the initial cost and the cost per unit. When enough units have been sold so that the total revenue R equals the total cost C, the sales are said to have reached the **break-even point.** The break-even point corresponds to the point of intersection of the cost and revenue curves.

EXAMPLE 6 **Break-Even Analysis**

A shoe company invests \$300,000 in equipment to produce a new line of athletic footwear. Each pair of shoes costs \$15 to produce and sells for \$70. How many pairs of shoes must the company sell to break even?

Solution

The total cost of producing x units is

$$\boxed{\text{Total cost}} = \boxed{\text{Cost per unit}} \cdot \boxed{\text{Number of units}} + \boxed{\text{Initial cost}}$$

$$C = 15x + 300,000. \qquad \text{Equation 1}$$

The revenue obtained by selling x units is

$$\boxed{\text{Total revenue}} = \boxed{\text{Price per unit}} \cdot \boxed{\text{Number of units}}$$

$$R = 70x. \qquad \text{Equation 2}$$

The break-even point occurs when $R = C$, so you have $C = 70x$. This gives you the system of equations below to solve.

$$\begin{cases} C = 15x + 300,000 \\ C = 70x \end{cases}$$

Solve by substitution.

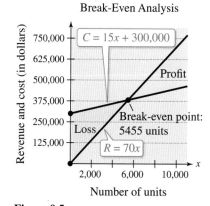

Break-Even Analysis

$C = 15x + 300,000$

$R = 70x$

Profit

Loss

Break-even point: 5455 units

Revenue and cost (in dollars)

Number of units

Figure 9.5

$70x = 15x + 300,000 \qquad$ Substitute $70x$ for C in Equation 1.

$55x = 300,000 \qquad$ Subtract $15x$ from each side.

$x \approx 5455 \qquad$ Divide each side by 55.

The company must sell about 5455 pairs of shoes to break even. Note in Figure 9.5 that revenue less than the break-even point corresponds to an overall loss, whereas revenue greater than the break-even point corresponds to a profit. Verify the break-even point using the *intersect* feature or the *zoom* and *trace* features of a graphing utility.

✓ *Checkpoint*  *Audio-video solution in English & Spanish at LarsonPrecalculus.com*

In Example 6, each pair of shoes costs \$12 to produce. How many pairs of shoes must the company sell to break even? ■

Another way to view the solution to Example 6 is to consider the profit function

$$P = R - C.$$

The break-even point occurs when the profit P is 0, which is the same as saying that

$$R = C.$$

GO DIGITAL

EXAMPLE 7 **Movie Ticket Sales**

Two new movies, a comedy and a drama, are released in the same week. In the first six weeks, the weekly ticket sales S (in millions of dollars) decrease for the comedy and increase for the drama according to the models

$$\begin{cases} S = 60 - 8x & \text{Comedy (Equation 1)} \\ S = 10 + 4.5x & \text{Drama (Equation 2)} \end{cases}$$

where x represents the time (in weeks), with $x = 1$ corresponding to the first week of release. According to the models, in what week are the ticket sales of the two movies equal?

Algebraic Solution

Both equations are already solved for S in terms of x, so substitute the expression for S from Equation 2 into Equation 1 and solve for x.

$10 + 4.5x = 60 - 8x$	Substitute for S in Equation 1.
$4.5x + 8x = 60 - 10$	Add $8x$ and -10 to each side.
$12.5x = 50$	Combine like terms.
$x = 4$	Divide each side by 12.5.

According to the models, the weekly ticket sales for the two movies are equal in the fourth week.

Numerical Solution

Create a table of values for each model.

Number of Weeks, x	1	2	3	4	5	6
Sales, S (comedy)	52	44	36	28	20	12
Sales, S (drama)	14.5	19	23.5	28	32.5	37

According to the table, the weekly ticket sales for the two movies are equal in the fourth week.

✓ *Checkpoint* ▶ *Audio-video solution in English & Spanish at LarsonPrecalculus.com*

Two new movies, an animated movie and a horror movie, are released in the same week. In the first eight weeks, the weekly ticket sales S (in millions of dollars) decrease for the animated movie and increase for the horror movie according to the models

$$\begin{cases} S = 108 - 9.4x & \text{Animated} \\ S = 16 + 9x & \text{Horror} \end{cases}$$

where x represents the time (in weeks), with $x = 1$ corresponding to the first week of release. According to the models, in what week are the ticket sales of the two movies equal? ▪

Summarize (Section 9.1)

1. Explain how to use the method of substitution to solve a system of linear equations in two variables *(page 628)*. For examples of using the method of substitution to solve systems of linear equations in two variables, see Examples 1 and 2.

2. Explain how to use the method of substitution to solve a system of nonlinear equations in two variables *(page 631)*. For examples of using the method of substitution to solve systems of nonlinear equations in two variables, see Examples 3 and 4.

3. Explain how to use a graphical method to solve a system of equations in two variables *(page 632)*. For an example of using a graphical approach to solve a system of equations in two variables, see Example 5.

4. Describe examples of how to use systems of equations to model and solve real-life problems *(pages 633 and 634, Examples 6 and 7)*.

GO DIGITAL

9.1 Exercises

See CalcChat.com for tutorial help and worked-out solutions to odd-numbered exercises.

GO DIGITAL

Vocabulary and Concept Check

In Exercises 1–4, fill in the blanks.

1. A set of two or more equations in two or more variables is called a _____ of _____.
2. A _____ of a system of equations is an ordered pair that satisfies each equation in the system.
3. The first step in solving a system of equations by the method of _____ is to solve one of the equations for one variable in terms of the other.
4. The ordered pair described in Exercise 2 corresponds to a point of _____ of the graphs of the equations.

5. What is the point of intersection of the graphs of the cost and revenue functions called?
6. The graphs of the equations of a system do not intersect. What can you conclude about the system?

Skills and Applications

Checking Solutions In Exercises 7 and 8, determine whether each ordered pair is a solution of the system.

7. $\begin{cases} 2x - y = 4 \\ 8x + y = -9 \end{cases}$ (a) $(0, -4)$ (b) $(3, -1)$ (c) $\left(\frac{3}{2}, -1\right)$ (d) $\left(-\frac{1}{2}, -5\right)$

8. $\begin{cases} 4x^2 + y = 3 \\ -x - y = 11 \end{cases}$ (a) $(2, -13)$ (b) $(1, -2)$ (c) $\left(-\frac{3}{2}, -\frac{31}{3}\right)$ (d) $\left(-\frac{7}{4}, -\frac{37}{4}\right)$

Solving a System by Substitution In Exercises 9–16, solve the system by the method of substitution. Check your solution(s) graphically.

9. $\begin{cases} 2x + y = 6 \\ -x + y = 0 \end{cases}$

10. $\begin{cases} x - 4y = -11 \\ x + 3y = 3 \end{cases}$

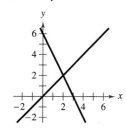
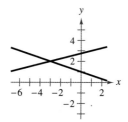

11. $\begin{cases} x - y = -4 \\ x^2 - y = -2 \end{cases}$

12. $\begin{cases} 3x + y = 2 \\ x^3 - 2 + y = 0 \end{cases}$

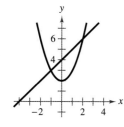
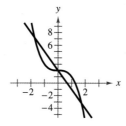

13. $\begin{cases} x^2 + y = 0 \\ x^2 - 4x - y = 0 \end{cases}$

14. $\begin{cases} x + y = 0 \\ x^3 - 5x - y = 0 \end{cases}$

15. $\begin{cases} y = x^3 - 3x^2 + 1 \\ y = x^2 - 3x + 1 \end{cases}$

16. $\begin{cases} -\frac{1}{2}x + y = -\frac{5}{2} \\ x^2 + y^2 = 25 \end{cases}$

Solving a System by Substitution In Exercises 17–32, solve the system by the method of substitution.

17. $\begin{cases} x - y = 2 \\ 6x - 5y = 16 \end{cases}$

18. $\begin{cases} 2x + y = 9 \\ 3x - 5y = 20 \end{cases}$

19. $\begin{cases} x + 4y = 3 \\ 2x - 7y = -24 \end{cases}$

20. $\begin{cases} x + y = 2 \\ 5x - 2y = 0 \end{cases}$

21. $\begin{cases} 2x - y + 2 = 0 \\ 4x + y - 5 = 0 \end{cases}$

22. $\begin{cases} 6x - 3y - 4 = 0 \\ x + 2y - 4 = 0 \end{cases}$

23. $\begin{cases} 1.5x + 0.8y = 2.3 \\ 0.3x - 0.2y = 0.1 \end{cases}$

24. $\begin{cases} 0.5x + y = -3.5 \\ x - 3.2y = 3.4 \end{cases}$

25. $\begin{cases} 0.5x + 3.2y = 9.0 \\ 0.2x - 1.6y = -3.6 \end{cases}$

26. $\begin{cases} 0.5x + 3.2y = 9.0 \\ 0.2x - 1.6y = 3.6 \end{cases}$

27. $\begin{cases} \frac{1}{5}x + \frac{1}{2}y = 8 \\ x + y = 20 \end{cases}$

28. $\begin{cases} \frac{1}{2}x + \frac{3}{4}y = 10 \\ \frac{3}{4}x - y = 4 \end{cases}$

29. $\begin{cases} 6x + 5y = -3 \\ -x - \frac{5}{6}y = -7 \end{cases}$

30. $\begin{cases} -\frac{2}{3}x + y = 2 \\ 2x - 3y = 6 \end{cases}$

31. $\begin{cases} \frac{1}{2}x + y = 5 \\ x - \frac{1}{2}y = 3 \end{cases}$

32. $\begin{cases} \frac{1}{2}x + 1 + y = 2 \\ x - 1 + \frac{1}{2}y = -2 \end{cases}$

Solving a System by Substitution In Exercises 33–36, the given amount of annual interest is earned from a total of $12,000 invested in two funds paying simple interest. Write and solve a system of equations to find the amount invested at each given rate.

	Annual Interest	Rate 1	Rate 2
33.	$500	2%	6%
34.	$630	4%	7%
35.	$396	2.8%	3.8%
36.	$254	1.75%	2.25%

Solving a System with a Nonlinear Equation In Exercises 37–40, solve the system by the method of substitution.

37. $\begin{cases} x^2 - y = 0 \\ 2x + y = 0 \end{cases}$ 38. $\begin{cases} x - 2y = 0 \\ 3x - y^2 = 0 \end{cases}$

39. $\begin{cases} x - y = -1 \\ x^2 - y = -4 \end{cases}$ 40. $\begin{cases} y = -x \\ y = x^3 + 3x^2 + 2x \end{cases}$

Solving a System of Equations Graphically In Exercises 41–50, solve the system graphically.

41. $\begin{cases} -x + 2y = -2 \\ 3x + y = 20 \end{cases}$ 42. $\begin{cases} x + y = 0 \\ 2x - 7y = -18 \end{cases}$

43. $\begin{cases} x - 3y = -3 \\ 5x + 3y = -6 \end{cases}$ 44. $\begin{cases} -x + 2y = -7 \\ x - y = 2 \end{cases}$

45. $\begin{cases} x + y = 4 \\ x^2 + y^2 - 4x = 0 \end{cases}$ 46. $\begin{cases} -x + y = 3 \\ x^2 - 6x - 27 + y^2 = 0 \end{cases}$

47. $\begin{cases} 3x - 2y = 0 \\ x^2 - y^2 = 4 \end{cases}$ 48. $\begin{cases} 2x - y + 3 = 0 \\ x^2 + y^2 - 4x = 0 \end{cases}$

49. $\begin{cases} x^2 + y^2 = 25 \\ 3x^2 - 16y = 0 \end{cases}$ 50. $\begin{cases} x^2 + y^2 = 25 \\ (x - 8)^2 + y^2 = 25 \end{cases}$

Solving a System of Equations Graphically In Exercises 51–54, use a graphing utility to solve the system of equations graphically. Round your solution(s) to two decimal places, if necessary.

51. $\begin{cases} y = e^x \\ x - y + 1 = 0 \end{cases}$ 52. $\begin{cases} y = -4e^{-x} \\ y + 3x + 8 = 0 \end{cases}$

53. $\begin{cases} y + 2 = \ln(x - 1) \\ 3y + 2x = 9 \end{cases}$ 54. $\begin{cases} x^2 + y^2 = 4 \\ 2x^2 - y = 2 \end{cases}$

Choosing a Solution Method In Exercises 55–62, solve the system graphically or algebraically. Explain your choice of method.

55. $\begin{cases} y = 2x \\ y = x^2 + 1 \end{cases}$ 56. $\begin{cases} x^2 + y^2 = 9 \\ x - y = -3 \end{cases}$

57. $\begin{cases} x - 2y = 4 \\ x^2 - y = 0 \end{cases}$ 58. $\begin{cases} y = x^3 - 2x^2 + x - 1 \\ y = -x^2 + 3x - 1 \end{cases}$

59. $\begin{cases} y - e^{-x} = 1 \\ y - \ln x = 3 \end{cases}$ 60. $\begin{cases} x^2 + y = 4 \\ e^x - y = 0 \end{cases}$

61. $\begin{cases} xy - 1 = 0 \\ 2x - 4y + 7 = 0 \end{cases}$ 62. $\begin{cases} x - 2y = 1 \\ y = \sqrt{x - 1} \end{cases}$

Break-Even Analysis In Exercises 63 and 64, use the equations for the total cost C and total revenue R to find the number x of units a company must sell to break even. (Round to the nearest whole unit.)

63. $C = 8650x + 250{,}000, \quad R = 9502x$

64. $C = 5.5\sqrt{x} + 10{,}000, \quad R = 4.22x$

65. **Break-Even Analysis** A company invests $16,000 to produce a product that will sell for $55.95. Each unit costs $9.45 to produce.

(a) How many units must the company sell to break even?

(b) How many units must the company sell to make a profit of $100,000?

66. **Supply and Demand** The supply and demand curves for a business dealing with oats are

Supply: $p = 1.45 + 0.00014x^2$

Demand: $p = (2.388 - 0.007x)^2$

where p is the price (in dollars) per bushel and x is the quantity in bushels per day. Use a graphing utility to graph the supply and demand equations and find the market equilibrium. (The market equilibrium is the point of intersection of the graphs for $x > 0$.)

67. **Environmental Science**

The table shows the consumption C_1 (in trillions of Btus) of biofuels and the consumption C_2 (in trillions of Btus) of wind energy in the United States from 2013 through 2019. *(Source: U.S. Energy Information Administration)*

Year	Biofuels, C_1	Wind, C_2
2013	2014	1601
2014	2077	1728
2015	2153	1777
2016	2287	2096
2017	2304	2343
2018	2283	2486
2019	2254	2736

Spreadsheet at LarsonPrecalculus.com

(a) Use a graphing utility to find a cubic model for the biofuels consumption data and a quadratic model for the wind energy consumption data. Let t represent the year, with $t = 13$ corresponding to 2013.

(b) Use the graphing utility to graph the data and the two models in the same viewing window.

(c) Use the graph from part (b) to approximate the point of intersection of the graphs of the models. Interpret your answer in the context of the problem.

(d) Describe the behavior of each model.

68. Choice of Two Jobs You are offered two jobs selling college textbooks. One company offers an annual salary of $33,000 plus a year-end bonus of 1% of your total sales. The other company offers an annual salary of $30,000 plus a year-end bonus of 2.5% of your total sales. How much would you have to sell in a year to make the second offer the better offer?

Geometry **In Exercises 69 and 70, use a system of equations to find the dimensions of the rectangle meeting the specified conditions.**

69. The perimeter is 56 meters and the length is 4 meters greater than the width.

70. The perimeter is 42 inches and the width is three-fourths the length.

Exploring the Concepts

True or False? **In Exercises 71 and 72, determine whether the statement is true or false. Justify your answer.**

71. To solve a system of equations by substitution, you must always solve for y in one of the two equations and then back-substitute.

72. If the graph of a system consists of a parabola and a circle, then the system can have at most two solutions.

73. Think About It When solving a system of equations by substitution, how do you recognize that the system has no real solution?

74. **HOW DO YOU SEE IT?** The cost C of producing x units and the revenue R obtained by selling x units are shown in the figure.

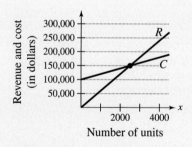

Number of units

(a) Estimate the point of intersection. What does this point represent?

(b) Use the figure to identify the x-values that correspond to (i) an overall loss and (ii) a profit. Explain.

75. Think About It Find equations of lines whose graphs intersect the graph of the parabola $y = x^2$ at (a) two points, (b) one point, and (c) no point. (There are many correct answers.)

76. Think About It Create a system of linear equations in two variables that has the solution $(2, -1)$ as its only solution. (There are many correct answers.)

77. Think About It Consider the system of equations

$$\begin{cases} ax + by = c \\ dx + ey = f \end{cases}.$$

(a) Find values for a, b, c, d, e, and f so that the system has one distinct solution. (There is more than one correct answer.)

(b) Explain how to solve the system in part (a) by the method of substitution and graphically.

78. Error Analysis Describe the error in solving the system of equations.

$$\begin{cases} x^2 + 2x - y = 3 \\ 2x - y = 2 \end{cases}$$

$$x^2 + 2x - (-2x + 2) = 3$$
$$x^2 + 4x - 2 = 3$$
$$x^2 + 4x - 5 = 0$$
$$(x + 5)(x - 1) = 0$$
$$x = -5, 1$$

When $x = -5$, $y = -2(-5) + 2 = 12$, and when $x = 1$, $y = -2(1) + 2 = 0$. So, the solutions are $(-5, 12)$ and $(1, 0)$.

Review & Refresh ▶ *Video solutions at LarsonPrecalculus.com*

Rewriting an Equation **In Exercises 79–84, use the Distributive Property and multiplication to rewrite each side of the equation without parentheses.**

79. $4(5x + y) = 4(-1)$ **80.** $3(5x - y) = 3(17)$

81. $3(2x - 4y) = 3(14)$ **82.** $4(5x + 3y) = 4(9)$

83. $100(0.02x - 0.05y) = 100(-0.38)$

84. $100(0.03x + 0.04y) = 100(1.04)$

Solving for x **In Exercises 85–92, solve the equation for x.**

85. $3x + 5x + 2y - 2y = 4 + 12$

86. $2x + 2x + y - y = 4 - 1$

87. $2x - 4y + 20x + 4y = -7 + (-4)$

88. $2x + 3y + 15x - 3y = 17 + 3(17)$

89. $20x + 12y + 6x - 12y = 4(9) + 3(14)$

90. $9x + 12y - (4x + 12y) = 3(75) - 2(90)$

91. $2(3y + 2x) - 3(2y + 5x) = 7(2) - 3$

92. $3(2y - 5x) - 2(3y + 4x) = -2(104) + 3(-38)$

Sketching the Graph of an Equation **In Exercises 93–96, sketch the graph of the equation.**

93. $2x - 3y = 3$ **94.** $-2x + 3y = 6$

95. $x + 2y = 5$ **96.** $-4x + 6y = -6$

9.2 Two-Variable Linear Systems

❯ Use the method of elimination to solve systems of linear equations in two variables.
❯ Interpret graphically the numbers of solutions of systems of linear equations in two variables.
❯ Use systems of linear equations in two variables to model and solve real-life problems.

The Method of Elimination

In the preceding section, you studied two methods for solving a system of equations: substitution and graphing. Now, you will study the **method of elimination.** The key step in this method is to obtain, for one of the variables, coefficients that differ only in sign so that *adding* the equations eliminates the variable.

$$3x + 5y = 7 \qquad \text{Equation 1}$$
$$\underline{-3x - 2y = -1} \qquad \text{Equation 2}$$
$$3y = 6 \qquad \text{Add equations.}$$

Note that by adding the two equations, you eliminate the x-terms and obtain a single equation in y. Solving this equation for y produces $y = 2$, which you can back-substitute into one of the original equations to solve for x.

Systems of equations in two variables can help you model and solve mixture problems. For example, in Exercise 52 on page 648, you will write, graph, and solve a system of equations to find the numbers of gallons of 87- and 92-octane gasoline that must be mixed to obtain 500 gallons of 89-octane gasoline.

EXAMPLE 1   Solving a System of Equations by Elimination

Solve the system of linear equations.

$$\begin{cases} 3x + 2y = 4 & \text{Equation 1} \\ 5x - 2y = 12 & \text{Equation 2} \end{cases}$$

Solution   The coefficients of y differ only in sign, so eliminate the y-terms by adding the two equations.

$$3x + 2y = 4 \qquad \text{Write Equation 1.}$$
$$\underline{5x - 2y = 12} \qquad \text{Write Equation 2.}$$
$$8x = 16 \qquad \text{Add equations.}$$
$$x = 2 \qquad \text{Solve for } x.$$

Solve for y by back-substituting $x = 2$ into Equation 1.

$$3x + 2y = 4 \qquad \text{Write Equation 1.}$$
$$3(2) + 2y = 4 \qquad \text{Substitute 2 for } x.$$
$$y = -1 \qquad \text{Solve for } y.$$

The solution is $(2, -1)$. Check this in the original system, as follows.

Check

$$3(2) + 2(-1) = 4 \qquad \text{Solution checks in Equation 1. } \checkmark$$
$$5(2) - 2(-1) = 12 \qquad \text{Solution checks in Equation 2. } \checkmark$$

✓ *Checkpoint* ▶ *Audio-video solution in English & Spanish at LarsonPrecalculus.com*

Solve the system of linear equations.

$$\begin{cases} 2x + y = 4 \\ 2x - y = -1 \end{cases}$$

> **Method of Elimination**
>
> To use the **method of elimination** to solve a system of two linear equations in x and y, perform the following steps.
>
> 1. *Obtain coefficients* for x (or y) that differ only in sign by multiplying all terms of one or both equations by suitably chosen constants.
>
> 2. *Add* the equations to eliminate one variable.
>
> 3. *Solve* the equation obtained in Step 2.
>
> 4. *Back-substitute* the value obtained in Step 3 into either of the original equations and solve for the other variable.
>
> 5. *Check* that the solution satisfies *each* of the original equations.

EXAMPLE 2   **Solving a System of Equations by Elimination**

Solve the system of linear equations.

$$\begin{cases} 2x - 4y = -7 & \text{Equation 1} \\ 5x + \ \ y = -1 & \text{Equation 2} \end{cases}$$

Solution   To obtain coefficients for y that differ only in sign, multiply Equation 2 by 4.

$2x - 4y = -7$	$\Longrightarrow$	$2x - 4y = -7$	Write Equation 1.
$5x + \ \ y = -1$	$\Longrightarrow$	$\underline{20x + 4y = -4}$	Multiply Equation 2 by 4.
		$22x \qquad\ \ = -11$	Add equations.
		$x \qquad\quad = -\frac{1}{2}$	Solve for x.

Solve for y by back-substituting $x = -\frac{1}{2}$ into Equation 1.

$2x - 4y = -7$	Write Equation 1.
$2\left(-\frac{1}{2}\right) - 4y = -7$	Substitute $-\frac{1}{2}$ for x.
$-4y = -6$	Simplify.
$y = \frac{3}{2}$	Solve for y.

The solution is $\left(-\frac{1}{2}, \frac{3}{2}\right)$. Check this in the original system, as follows.

Check

$2x - 4y = -7$	Write Equation 1.
$2\left(-\frac{1}{2}\right) - 4\left(\frac{3}{2}\right) \overset{?}{=} -7$	Substitute for x and y.
$-1 - 6 = -7$	Solution checks in Equation 1. ✓
$5x + y = -1$	Write Equation 2.
$5\left(-\frac{1}{2}\right) + \frac{3}{2} \overset{?}{=} -1$	Substitute for x and y.
$-\frac{5}{2} + \frac{3}{2} = -1$	Solution checks in Equation 2. ✓

✓ *Checkpoint* 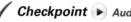 *Audio-video solution in English & Spanish at LarsonPrecalculus.com*

Solve the system of linear equations.

$$\begin{cases} 2x + 3y = 17 \\ 5x - \ \ y = 17 \end{cases}$$

GO DIGITAL

In Example 2, the two systems of linear equations (the original system and the system obtained by multiplying Equation 2 by a constant)

$$\begin{cases} 2x - 4y = -7 \\ 5x + y = -1 \end{cases} \quad \text{and} \quad \begin{cases} 2x - 4y = -7 \\ 20x + 4y = -4 \end{cases}$$

are **equivalent systems** because they have the same solution set. The operations that can be performed on a system of linear equations to produce an equivalent system are (1) interchange two equations, (2) multiply one of the equations by a nonzero constant, and (3) add a multiple of one equation to another equation to replace the latter equation. You will study these operations in more depth in the next section.

EXAMPLE 3 Solving a System of Linear Equations

Solve the system of linear equations.

$$\begin{cases} 5x + 3y = 9 & \text{Equation 1} \\ 2x - 4y = 14 & \text{Equation 2} \end{cases}$$

Algebraic Solution

To obtain coefficients of y that differ only in sign, multiply Equation 1 by 4 and multiply Equation 2 by 3.

$$\begin{array}{rll} 5x + 3y = 9 & \Rightarrow \quad 20x + 12y = 36 & \text{Multiply Equation 1 by 4.} \\ 2x - 4y = 14 & \Rightarrow \quad \underline{6x - 12y = 42} & \text{Multiply Equation 2 by 3.} \\ & \qquad\quad 26x \quad = 78 & \text{Add equations.} \\ & \qquad\qquad x \quad = 3 & \text{Solve for } x. \end{array}$$

Solve for y by back-substituting $x = 3$ into Equation 2.

$$\begin{array}{rl} 2x - 4y = 14 & \text{Write Equation 2.} \\ 2(3) - 4y = 14 & \text{Substitute 3 for } x. \\ -4y = 8 & \text{Simplify.} \\ y = -2 & \text{Solve for } y. \end{array}$$

The solution is $(3, -2)$. Check this in the original system.

Graphical Solution

Solve each equation for y and use a graphing utility to graph the equations in the same viewing window.

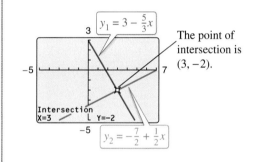

$y_1 = 3 - \frac{5}{3}x$

The point of intersection is $(3, -2)$.

Intersection X=3 Y=-2

$y_2 = -\frac{7}{2} + \frac{1}{2}x$

From the graph, the solution is $(3, -2)$. Check this in the original system.

✓ *Checkpoint* ▶ Audio-video solution in English & Spanish at LarsonPrecalculus.com

Solve the system of linear equations.

$$\begin{cases} 3x + 2y = 7 \\ 2x + 5y = 1 \end{cases}$$

In Example 3, note that the graphical solution can be used to check the algebraic solution. Of course, you can check either solution algebraically, as shown below.

$$\begin{array}{rl} 5x + 3y = 9 & \text{Write Equation 1.} \\ 5(3) + 3(-2) \overset{?}{=} 9 & \text{Substitute 3 for } x \text{ and } -2 \text{ for } y. \\ 15 - 6 = 9 & \text{Solution checks in Equation 1. } ✓ \\ 2x - 4y = 14 & \text{Write Equation 2.} \\ 2(3) - 4(-2) \overset{?}{=} 14 & \text{Substitute 3 for } x \text{ and } -2 \text{ for } y. \\ 6 + 8 = 14 & \text{Solution checks in Equation 2. } ✓ \end{array}$$

GO DIGITAL

The solution of the linear system

$$\begin{cases} ax + by = c \\ dx + ey = f \end{cases}$$

is given by

$$x = \frac{ce - bf}{ae - bd}$$

and

$$y = \frac{af - cd}{ae - bd}.$$

If $ae - bd = 0$, then the system does not have a unique solution. A graphing utility program for solving such a system is available at *LarsonPrecalculus.com.* Use this program, called "Systems of Linear Equations," to solve the system in Example 4.

Example 4 illustrates a strategy for solving a system of linear equations that has decimal coefficients.

EXAMPLE 4   A Linear System Having Decimal Coefficients

Solve the system of linear equations.

$$\begin{cases} 0.02x - 0.05y = -0.38 & \text{Equation 1} \\ 0.03x + 0.04y = 1.04 & \text{Equation 2} \end{cases}$$

Solution   The coefficients in this system have two decimal places, so multiply each equation by 100 to produce a system in which the coefficients are all integers.

$$\begin{cases} 2x - 5y = -38 & \text{Revised Equation 1} \\ 3x + 4y = 104 & \text{Revised Equation 2} \end{cases}$$

Now, to obtain coefficients that differ only in sign, multiply revised Equation 1 by 3 and multiply revised Equation 2 by -2.

$2x - 5y = -38$ ⟹	$6x - 15y = -114$	Multiply revised Equation 1 by 3.
$3x + 4y = 104$ ⟹	$-6x - 8y = -208$	Multiply revised Equation 2 by -2.
	$-23y = -322$	Add equations.
	$y = \dfrac{-322}{-23}$	Divide each side by -23.
	$y = 14$	Simplify.

Solve for x by back-substituting $y = 14$ into revised Equation 2.

$3x + 4y = 104$	Write revised Equation 2.
$3x + 4(14) = 104$	Substitute 14 for y.
$3x + 56 = 104$	Multiply.
$3x = 48$	Subtract 56 from each side.
$x = 16$	Solve for x.

The solution is

$(16, 14).$

Check this in the original system, as follows.

Check

$0.02x - 0.05y = -0.38$	Write Equation 1.
$0.02(16) - 0.05(14) \overset{?}{=} -0.38$	Substitute for x and y.
$0.32 - 0.70 = -0.38$	Solution checks in Equation 1. ✓
$0.03x + 0.04y = 1.04$	Write Equation 2.
$0.03(16) + 0.04(14) \overset{?}{=} 1.04$	Substitute for x and y.
$0.48 + 0.56 = 1.04$	Solution checks in Equation 2. ✓

 ✓ *Checkpoint* ▶ Audio-video solution in English & Spanish at LarsonPrecalculus.com

Solve the system of linear equations.

$$\begin{cases} 0.03x + 0.04y = 0.75 \\ 0.02x + 0.06y = 0.90 \end{cases}$$

GO DIGITAL

Graphical Interpretation of Solutions

It is possible for a system of equations to have exactly one solution, two or more solutions, or no solution. In a system of *linear* equations, however, if the system has two different solutions, then it must have an *infinite* number of solutions. To see why this is true, consider the following graphical interpretation of a system of two linear equations in two variables.

Graphical Interpretations of Solutions

For a system of two linear equations in two variables, the number of solutions is one of the following.

Number of Solutions	Graphical Interpretation	Slopes of Lines
1. Exactly one solution	The two lines intersect at one point.	The slopes of the two lines are not equal.
2. Infinitely many solutions	The two lines coincide (are identical).	The slopes of the two lines are equal.
3. No solution	The two lines are parallel.	The slopes of the two lines are equal.

A system of linear equations is **consistent** when it has at least one solution. A system is **inconsistent** when it has no solution.

EXAMPLE 5   **Recognizing Graphs of Linear Systems**

▶▷▷ *See LarsonPrecalculus.com for an interactive version of this type of example.*

Match each system of linear equations with its graph. Describe the number of solutions and state whether the system is consistent or inconsistent.

a. $\begin{cases} 2x - 3y = 3 \\ -4x + 6y = 6 \end{cases}$ **b.** $\begin{cases} 2x - 3y = 3 \\ x + 2y = 5 \end{cases}$ **c.** $\begin{cases} 2x - 3y = 3 \\ -4x + 6y = -6 \end{cases}$

i.    **ii.** 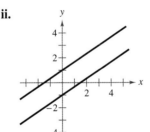   **iii.**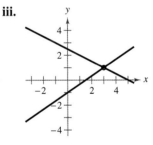

ALGEBRA HELP

When solving a system of linear equations graphically, it helps to begin by writing the equations in slope-intercept form, so you can compare the slopes and *y*-intercepts of their graphs. Do this for the systems in Example 5.

ALGEBRA HELP

A consistent system that has exactly one solution is **independent**. A consistent system that has infinitely many solutions is **dependent**. So, the system in Example 5(b) is independent and the system in Example 5(c) is dependent.

Solution

a. The graph of system (a) is a pair of parallel lines (ii). The lines have no point of intersection, so the system has no solution. The system is inconsistent.

b. The graph of system (b) is a pair of intersecting lines (iii). The lines have one point of intersection, so the system has exactly one solution. The system is consistent.

c. The graph of system (c) is a pair of lines that coincide (i). The lines have infinitely many points of intersection, so the system has infinitely many solutions. The system is consistent.

✓ **Checkpoint** ▶ *Audio-video solution in English & Spanish at LarsonPrecalculus.com*

Sketch the graph of the system of linear equations. Then describe the number of solutions and state whether the system is consistent or inconsistent.

$$\begin{cases} -2x + 3y = 6 \\ 4x - 6y = -9 \end{cases}$$

In Examples 6 and 7, note how the method of elimination is used to determine that a system of linear equations has no solution or infinitely many solutions.

EXAMPLE 6 Method of Elimination: No Solution

Solve the system of linear equations.

$$\begin{cases} x - 2y = 3 & \text{Equation 1} \\ -2x + 4y = 1 & \text{Equation 2} \end{cases}$$

Solution To obtain coefficients that differ only in sign, multiply Equation 1 by 2.

$$
\begin{array}{lll}
x - 2y = 3 & \Longrightarrow & 2x - 4y = 6 & \text{Multiply Equation 1 by 2.} \\
-2x + 4y = 1 & \Longrightarrow & \underline{-2x + 4y = 1} & \text{Write Equation 2.} \\
& & \quad\quad\;\; 0 = 7 & \text{False statement}
\end{array}
$$

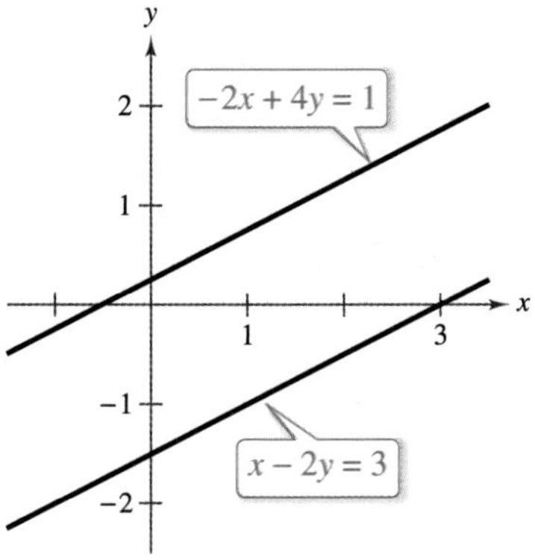

Adding the equations yields the false statement $0 = 7$. There are no values of x and y for which $0 = 7$, so the system is inconsistent and has no solution. Figure 9.6 shows that the graph of the system is two parallel lines with no point of intersection.

✓ **Checkpoint** Audio-video solution in English & Spanish at LarsonPrecalculus.com

Solve the system of linear equations.

$$\begin{cases} 6x - 5y = 3 \\ -12x + 10y = 5 \end{cases}$$

■

In Example 6, note that the occurrence of a false statement, such as $0 = 7$, indicates that the system has no solution. In the next example, note that the occurrence of a statement that is true for all values of the variables, such as $0 = 0$, indicates that the system has infinitely many solutions.

EXAMPLE 7 Method of Elimination: Infinitely Many Solutions

Solve the system of linear equations.

$$\begin{cases} 2x - y = 1 & \text{Equation 1} \\ 4x - 2y = 2 & \text{Equation 2} \end{cases}$$

Solution To obtain coefficients that differ only in sign, multiply Equation 1 by -2.

$$
\begin{array}{lll}
2x - y = 1 & \Longrightarrow & -4x + 2y = -2 & \text{Multiply Equation 1 by } -2. \\
4x - 2y = 2 & \Longrightarrow & \underline{4x - 2y = \;\; 2} & \text{Write Equation 2.} \\
& & \quad\quad\; 0 = \;\; 0 & \text{Add equations.}
\end{array}
$$

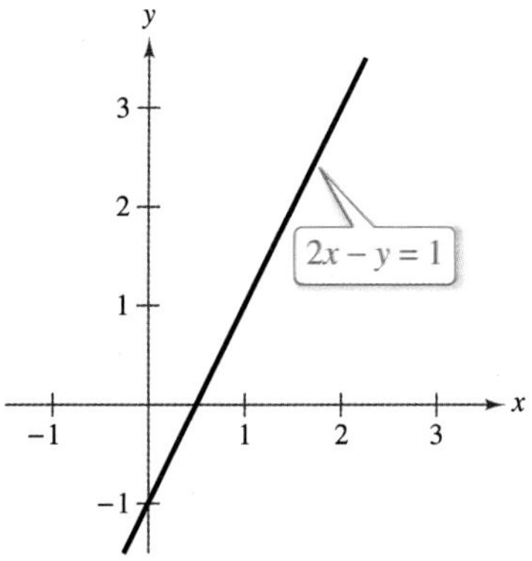

Because $0 = 0$ for all values of x and y, the two equations are equivalent (have the same solution set). You can conclude that the system is consistent and has infinitely many solutions. The solution set consists of all points (x, y) lying on the line $2x - y = 1$, as shown in Figure 9.7. Letting $x = a$, where a is any real number, then $y = 2a - 1$. So, the solutions of the system are all ordered pairs of the form $(a, 2a - 1)$.

✓ **Checkpoint** Audio-video solution in English & Spanish at LarsonPrecalculus.com

Solve the system of linear equations.

$$\begin{cases} \dfrac{1}{2}x - \dfrac{1}{8}y = -\dfrac{3}{8} \\ -4x + y = \;\; 3 \end{cases}$$

■

Figure 9.6

Figure 9.7

GO DIGITAL

Applications

At this point, you may be asking the question "How can I tell whether I can solve an application problem using a system of linear equations?" To answer this question, start with the following considerations.

1. Does the problem involve more than one unknown quantity?

2. Are there two (or more) equations or conditions to be satisfied?

When the answer to one or both of these questions is "yes," the appropriate model for the problem may be a system of linear equations.

Original flight

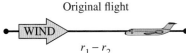

$r_1 - r_2$

Return flight

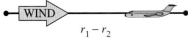

$r_1 + r_2$

Figure 9.8

EXAMPLE 8   Aviation

An airplane flying into a headwind travels the 2000-mile flying distance between Tallahassee, Florida, and Los Angeles, California, in 4 hours and 24 minutes. On the return flight, the airplane travels this distance in 4 hours. Find the airspeed of the plane and the speed of the wind, assuming that both remain constant.

Solution   The two unknown quantities are the airspeed of the plane and the speed of the wind. Let r_1 be the airspeed of the plane and r_2 be the speed of the wind (see Figure 9.8).

$$r_1 - r_2 = \text{speed of the plane against the wind}$$
$$r_1 + r_2 = \text{speed of the plane with the wind}$$

Use the formula

$$\text{distance} = (\text{rate})(\text{time})$$

to write equations involving these two speeds.

$$2000 = (r_1 - r_2)\left(4 + \frac{24}{60}\right)$$
$$2000 = (r_1 + r_2)(4)$$

These two equations simplify as follows.

$$\begin{cases} 5000 = 11r_1 - 11r_2 & \text{Equation 1} \\ 500 = r_1 + r_2 & \text{Equation 2} \end{cases}$$

To solve this system by elimination, multiply Equation 2 by 11.

$$
\begin{array}{rcll}
5000 = 11r_1 - 11r_2 & \Longrightarrow & 5000 = 11r_1 - 11r_2 & \text{Write Equation 1.} \\
500 = r_1 + r_2 & \Longrightarrow & 5500 = 11r_1 + 11r_2 & \text{Multiply Equation 2 by 11.} \\
\hline
& & 10{,}500 = 22r_1 & \text{Add equations.}
\end{array}
$$

So,

$$r_1 = \frac{10{,}500}{22} = \frac{5250}{11} \approx 477.27 \text{ miles per hour} \qquad \text{Airspeed of plane}$$

and

$$r_2 = 500 - \frac{5250}{11} = \frac{250}{11} \approx 22.73 \text{ miles per hour.} \qquad \text{Speed of wind}$$

Check this solution in the original statement of the problem.

✓ Checkpoint ▶ *Audio-video solution in English & Spanish at LarsonPrecalculus.com*

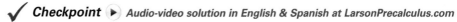

In Example 8, the return flight takes 4 hours and 6 minutes. Find the airspeed of the plane and the speed of the wind, assuming that both remain constant.   ■

In a free market, the demands for many products are related to the prices of the products. As the prices decrease, the quantities demanded by consumers increase and the quantities that producers are able or willing to supply decrease. The **equilibrium point** is the price p and number of units x that satisfy both the demand and supply equations.

EXAMPLE 9 Finding the Equilibrium Point

The demand and supply equations for a video game console are

$$\begin{cases} p = 180 - 0.00001x & \text{Demand equation} \\ p = 90 + 0.00002x & \text{Supply equation} \end{cases}$$

where p is the price per unit (in dollars) and x is the number of units. Find the equilibrium point for this market.

Solution Because p is written in terms of x, begin by substituting the expression for p from the supply equation into the demand equation.

$$p = 180 - 0.00001x \qquad \text{Write demand equation.}$$
$$90 + 0.00002x = 180 - 0.00001x \qquad \text{Substitute } 90 + 0.00002x \text{ for } p.$$
$$0.00003x = 90 \qquad \text{Combine like terms.}$$
$$x = 3{,}000{,}000 \qquad \text{Solve for } x.$$

So, the equilibrium point occurs when the demand and supply are each 3 million units. (See Figure 9.9.) To find the price that corresponds to this x-value, back-substitute $x = 3{,}000{,}000$ into either of the original equations. Using the demand equation produces the following.

$$p = 180 - 0.00001(3{,}000{,}000)$$
$$= 180 - 30$$
$$= \$150$$

The equilibrium point is $(3{,}000{,}000, 150)$. Check this by substituting $(3{,}000{,}000, 150)$ into the demand and supply equations.

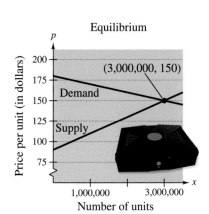

Equilibrium

(3,000,000, 150)

Demand

Supply

Price per unit (in dollars)

Number of units

Figure 9.9

✓ ***Checkpoint*** *Audio-video solution in English & Spanish at LarsonPrecalculus.com*

The demand and supply equations for a flat-screen television are

$$\begin{cases} p = 567 - 0.00002x & \text{Demand equation} \\ p = 492 + 0.00003x & \text{Supply equation} \end{cases}$$

where p is the price per unit (in dollars) and x is the number of units. Find the equilibrium point for this market. ■

Summarize (Section 9.2)

1. Explain how to use the method of elimination to solve a system of linear equations in two variables *(page 638)*. For examples of using the method of elimination to solve systems of linear equations in two variables, see Examples 1–4.

2. Explain how to interpret graphically the numbers of solutions of systems of linear equations in two variables *(page 642)*. For examples of interpreting graphically the numbers of solutions of systems of linear equations in two variables, see Examples 5–7.

3. Describe real-life applications of systems of linear equations in two variables *(pages 644 and 645, Examples 8 and 9)*.

GO DIGITAL

9.2 Exercises

See CalcChat.com for tutorial help and worked-out solutions to odd-numbered exercises.

GO DIGITAL

Vocabulary and Concept Check

In Exercises 1–3, fill in the blanks.

1. The first step in solving a system of equations by the method of _____ is to obtain coefficients for x (or y) that differ only in sign.

2. Two systems of equations that have the same solution set are _____ systems.

3. In business applications, the _____ _____ (x, p) is the price p and the number of units x that satisfy both the demand and supply equations.

4. Is a system of linear equations with at least one solution consistent or inconsistent?

5. Is a system of linear equations with no solution consistent or inconsistent?

6. When a system of linear equations has no solution, do the lines intersect?

Skills and Applications

Solving a System by Elimination In Exercises 7–14, solve the system by the method of elimination. Label each line with its equation.

7. $\begin{cases} 2x + y = 7 \\ x - y = -4 \end{cases}$

8. $\begin{cases} x + 3y = 1 \\ -x + 2y = 4 \end{cases}$

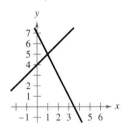

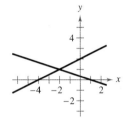

9. $\begin{cases} x + y = 0 \\ 3x + 2y = 1 \end{cases}$

10. $\begin{cases} \frac{1}{2}x - y = -2 \\ x + \frac{1}{3}y = 3 \end{cases}$

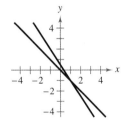

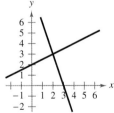

11. $\begin{cases} x - y = 2 \\ -2x + 2y = 5 \end{cases}$

12. $\begin{cases} 3x + 2y = 3 \\ 6x + 4y = 14 \end{cases}$

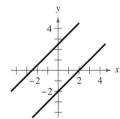

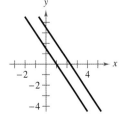

13. $\begin{cases} 3x - 2y = 5 \\ -6x + 4y = -10 \end{cases}$

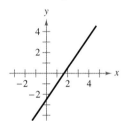

14. $\begin{cases} 9x - 3y = -15 \\ -3x + y = 5 \end{cases}$

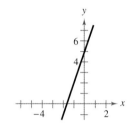

Solving a System by Elimination In Exercises 15–32, solve the system by the method of elimination and check any solutions algebraically.

15. $\begin{cases} x + 2y = 6 \\ x - 2y = 2 \end{cases}$

16. $\begin{cases} 3x - 5y = 8 \\ 2x + 5y = 22 \end{cases}$

17. $\begin{cases} 5x + 3y = 6 \\ 3x - y = 5 \end{cases}$

18. $\begin{cases} x + 5y = 10 \\ 3x - 10y = -5 \end{cases}$

19. $\begin{cases} 2u + 3v = -1 \\ 7u + 15v = 4 \end{cases}$

20. $\begin{cases} 2r + 4s = 5 \\ 16r + 50s = 55 \end{cases}$

21. $\begin{cases} 3x + 2y = 10 \\ 2x + 5y = 3 \end{cases}$

22. $\begin{cases} 3x + 11y = 4 \\ -2x - 5y = 9 \end{cases}$

23. $\begin{cases} 4b + 3m = 3 \\ 3b + 11m = 13 \end{cases}$

24. $\begin{cases} 2x + 5y = 8 \\ 5x + 8y = 10 \end{cases}$

25. $\begin{cases} 0.2x - 0.5y = -27.8 \\ 0.3x + 0.4y = 68.7 \end{cases}$

26. $\begin{cases} 0.5x - 0.3y = 6.5 \\ 0.7x + 0.2y = 6.0 \end{cases}$

27. $\begin{cases} 3x + 2y = 4 \\ 9x + 6y = 3 \end{cases}$

28. $\begin{cases} -6x + 4y = 7 \\ 15x - 10y = -16 \end{cases}$

29. $\begin{cases} -5x + 6y = -3 \\ 20x - 24y = 12 \end{cases}$

30. $\begin{cases} 7x + 8y = 6 \\ -14x - 16y = -12 \end{cases}$

31. $\begin{cases} \dfrac{x + 3}{4} + \dfrac{y - 1}{3} = 1 \\ 2x - y = 12 \end{cases}$

32. $\begin{cases} \dfrac{x - 1}{2} + \dfrac{y + 2}{3} = 4 \\ x - 2y = 5 \end{cases}$

Matching a System with Its Graph In Exercises 33–36, match the system of linear equations with its graph. Describe the number of solutions and state whether the system is consistent or inconsistent. [The graphs are labeled (a), (b), (c) and (d).]

(a)
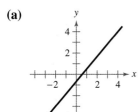

(b)
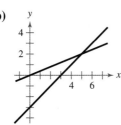

(c)
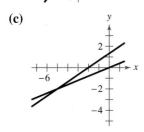

(d)
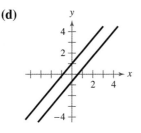

33. $\begin{cases} -7x + 6y = -4 \\ 14x - 12y = 8 \end{cases}$ 34. $\begin{cases} 2x - 5y = 0 \\ 2x - 3y = -4 \end{cases}$

35. $\begin{cases} 7x - 6y = -6 \\ -7x + 6y = -4 \end{cases}$ 36. $\begin{cases} 2x - 5y = 0 \\ x - y = 3 \end{cases}$

Choosing a Solution Method In Exercises 37–42, use any method to solve the system. Explain your choice of method.

37. $\begin{cases} 3x - 5y = 7 \\ 2x + y = 9 \end{cases}$ 38. $\begin{cases} -x + 3y = 17 \\ 4x + 3y = 7 \end{cases}$

39. $\begin{cases} -2x + 8y = 20 \\ y = x - 5 \end{cases}$ 40. $\begin{cases} -5x + 9y = 13 \\ y = x - 4 \end{cases}$

41. $\begin{cases} y = -2x - 17 \\ y = 2 - 3x \end{cases}$ 42. $\begin{cases} y = -3x - 8 \\ y = 15 - 2x \end{cases}$

43. **Airplane Speed** An airplane flying into a headwind travels the 1500-mile flying distance between Indianapolis, Indiana, and Phoenix, Arizona, in 3 hours. On the return flight, the airplane travels this distance in 2 hours and 30 minutes. Find the airspeed of the plane and the speed of the wind, assuming that both remain constant.

44. **Airplane Speed** Two planes start from Los Angeles International Airport and fly in opposite directions. The second plane starts $\frac{1}{2}$ hour after the first plane, but its speed is 80 kilometers per hour faster. Find the speed of each plane when 2 hours after the first plane departs the planes are 3200 kilometers apart.

45. **Nutrition** Two cheeseburgers and one small order of fries contain a total of 1460 calories. Three cheeseburgers and two small orders of fries contain a total of 2350 calories. Find the caloric content of each item.

46. **Nutrition** One eight-ounce glass of apple juice and one eight-ounce glass of orange juice contain a total of 179.2 milligrams of vitamin C. Two eight-ounce glasses of apple juice and three eight-ounce glasses of orange juice contain a total of 442.1 milligrams of vitamin C. How much vitamin C is in an eight-ounce glass of each type of juice?

47. **Finding the Equilibrium Point** The demand and supply equations for a laptop computer are

$$\begin{cases} p = 500 - 0.4x & \text{Demand equation} \\ p = 380 + 0.1x & \text{Supply equation} \end{cases}$$

where p is the price per unit (in dollars) and x is the number of units. Find the equilibrium point for this market.

48. **Finding the Equilibrium Point** The demand and supply equations for a set of large candles are

$$\begin{cases} p = 100 - 0.05x & \text{Demand equation} \\ p = 25 + 0.1x & \text{Supply equation} \end{cases}$$

where p is the price per unit (in dollars) and x is the number of units. Find the equilibrium point for this market.

49. **Investment Portfolio** A total of $24,000 is invested in two corporate bonds that pay 3.5% and 5% simple interest. The investor wants an annual interest income of $930 from the investments. What amount should be invested in the 3.5% bond?

50. **Investment Portfolio** A total of $32,000 is invested in two municipal bonds that pay 5.75% and 6.25% simple interest. The investor wants an annual interest income of $1900 from the investments. What amount should be invested in the 5.75% bond?

51. **Daily Sales** A store manager wants to know the demand for a product as a function of the price. The table shows the daily sales y for different prices x of the product.

Price, x	Demand, y
$1.00	45
$1.20	37
$1.50	23

(a) Find the least squares regression line $y = ax + b$ for the data by solving the system

$$\begin{cases} 3.00b + 3.70a = 105.00 \\ 3.70b + 4.69a = 123.90 \end{cases}$$

for a and b. Use a graphing utility to confirm the result.

(b) Use the linear model from part (a) to predict the demand when the price is $1.75.

52. Fuel Mixture

Five hundred gallons of 89-octane gasoline is obtained by mixing 87-octane gasoline with 92-octane gasoline.

(a) Write a system of equations in which one equation represents the total amount of final mixture required and the other represents the amounts of 87- and 92-octane gasoline in the final mixture. Let x and y represent the numbers of gallons of 87- and 92-octane gasoline, respectively.

(b) Use a graphing utility to graph the two equations in part (a) in the same viewing window. As the amount of 87-octane gasoline increases, how does the amount of 92-octane gasoline change?

(c) How much of each type of gasoline is required to obtain the 500 gallons of 89-octane gasoline?

Fitting a Line to Data One way to find the least squares regression line $y = ax + b$ for a set of points

$$(x_1, y_1), (x_2, y_2), \ldots, (x_n, y_n)$$

is by solving the system below for a and b.

$$\begin{cases} nb + \left(\displaystyle\sum_{i=1}^{n} x_i\right) a = \left(\displaystyle\sum_{i=1}^{n} y_i\right) \\ \left(\displaystyle\sum_{i=1}^{n} x_i\right) b + \left(\displaystyle\sum_{i=1}^{n} x_i^2\right) a = \left(\displaystyle\sum_{i=1}^{n} x_i y_i\right) \end{cases}$$

In Exercises 53 and 54, the sums have been evaluated. Solve the simplified system for a and b to find the least squares regression line for the points. Use a graphing utility to confirm the result. (*Note:* The symbol Σ is used to denote a sum of the terms of a sequence. You will learn how to use this notation in Section 11.1.)

53. $\begin{cases} 5b + 10a = 20.2 \\ 10b + 30a = 50.1 \end{cases}$ **54.** $\begin{cases} 6b + 15a = 23.6 \\ 15b + 55a = 48.8 \end{cases}$

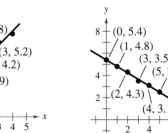

55. Agriculture

An agricultural scientist used four test plots to determine the relationship between wheat yield y (in bushels per acre) and the amount of fertilizer x (in hundreds of pounds per acre). The table shows the results.

Fertilizer, x	1.0	1.5	2.0	2.5
Yield, y	32	41	48	53

(a) Find the least squares regression line $y = ax + b$ for the data by solving the system for a and b.

$$\begin{cases} 4b + 7a = 174 \\ 7b + 13.5a = 322 \end{cases}$$

(b) Use the linear model from part (a) to estimate the yield for a fertilizer application of 160 pounds per acre.

56. Gross Domestic Product

The table shows the total gross domestic products y (in billions of dollars) of the United States for the years 2013 through 2019. (*Source: U.S. Office of Management and Budget*)

DATA	Year	GDP, y
	2013	16,603.8
	2014	17,335.6
	2015	18,099.6
	2016	18,554.8
	2017	19,287.7
	2018	20,335.5
	2019	21,215.7

Spreadsheet at LarsonPrecalculus.com

(a) Find the least squares regression line $y = at + b$ for the data, where t represents the year with $t = 13$ corresponding to 2013, by solving the system

$$\begin{cases} 7b + 112a = 131,432.7 \\ 112b + 1820a = 2,123,946.8 \end{cases}$$

for a and b. Use the *regression* feature of a graphing utility to confirm the result.

(b) Use the linear model to create a table of estimated values of y. Compare the estimated values with the actual data.

(c) Use the linear model to estimate the gross domestic product for 2020.

(d) Use the Internet, your school's library, or some other reference source to find the total national outlay for 2020. How does this value compare with your answer in part (c)?

(e) Is the linear model valid for long-term predictions of gross domestic products? Explain.

Exploring the Concepts

True or False? **In Exercises 57 and 58, determine whether the statement is true or false. Justify your answer.**

57. If two lines do not have exactly one point of intersection, then they must be parallel.

58. Solving a system of equations graphically will always give an exact solution.

Finding the Value of a Constant **In Exercises 59 and 60, find the value of k such that the system of linear equations is inconsistent.**

59. $\begin{cases} 4x - 8y = -3 \\ 2x + ky = 16 \end{cases}$
60. $\begin{cases} 15x + 3y = 6 \\ -10x + ky = 9 \end{cases}$

61. Comparing Methods Use the method of substitution to solve the system in Example 1. Do you prefer the method of substitution or the method of elimination? Explain.

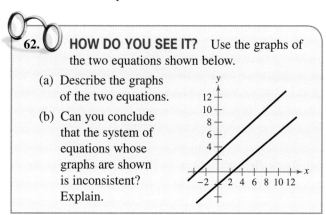

62. **HOW DO YOU SEE IT?** Use the graphs of the two equations shown below.

(a) Describe the graphs of the two equations.

(b) Can you conclude that the system of equations whose graphs are shown is inconsistent? Explain.

Comparing Methods **In Exercises 63 and 64, the graphs of the two equations appear to be parallel. Yet, when you solve the system algebraically, you find that the system does have a solution. Find the solution and explain why it does not appear on the portion of the graph shown.**

63. $\begin{cases} 100y - x = 200 \\ 99y - x = -198 \end{cases}$
64. $\begin{cases} 21x - 20y = 0 \\ 13x - 12y = 120 \end{cases}$

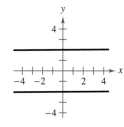
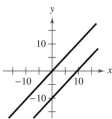

65. Writing Briefly explain whether it is possible for a consistent system of linear equations to have exactly two solutions.

66. Writing a System of Equations Give examples of systems of linear equations that have (a) no solution and (b) infinitely many solutions.

Advanced Applications **In Exercises 67 and 68, solve the system of equations for u and v. While solving for these variables, consider the trigonometric functions as constants. (Systems of this type appear in a course in differential equations.)**

67. $\begin{cases} u \sin x + v \cos x = 0 \\ u \cos x - v \sin x = \sec x \end{cases}$

68. $\begin{cases} u \cos 2x + v \sin 2x = 0 \\ u(-2 \sin 2x) + v(2 \cos 2x) = \csc 2x \end{cases}$

Review & Refresh ▶ *Video solutions at LarsonPrecalculus.com*

Solving for x **In Exercises 69–74, solve for x.**

69. $2x^2 - 3x = \dfrac{8x^2 + 12}{4}$

70. $1 = \dfrac{3x^2 - 4}{3x^2 + 2x}$

71. $6x + 5 = x(2x + 3)$

72. $7x^2 + 10 = 5x^2 + 12x - 6$

73. $5x^2 - 6 - 3x^4 = 3 - 5x^2 - 2x^4$

74. $\dfrac{7x^4 - 6}{x^2(3 - 4x^2)} = -1$

Rewriting an Equation **In Exercises 75–78, rewrite the equation in slope-intercept form.**

75. $11x - 4y = 19$

76. $3x - 10y = -26$

77. $\frac{1}{5}x + 1 = \frac{1}{2}x + \frac{1}{5}y$

78. $y + 12x = -1 - \frac{5}{3}y$

Finding Values for Which $f(x) = 0$ **In Exercises 79–82, find all real values of x for which $f(x) = 0$.**

79. $f(x) = x^2 - 12x + 24$

80. $f(x) = -\dfrac{8x^3 + 1}{13}$

81. $f(x) = 6x^3 - 7x^2 - 6x + 7$

82. $f(x) = x^4 - 8x^2 - 9$

Project: College Expenses To work an extended application analyzing the average undergraduate tuition, room, and board charges at private degree-granting institutions in the United States from 1998 through 2018, visit this text's website at *LarsonPrecalculus.com*. (*Source: U.S. Department of Education*)

9.3 Multivariable Linear Systems

Systems of linear equations in three or more variables can help you model and solve real-life problems. For example, in Exercise 68 on page 661, you will use a system of linear equations in three variables to analyze the reproductive rates of deer in a wildlife preserve.

❯ Use back-substitution to solve linear systems in row-echelon form.
❯ Use Gaussian elimination to solve systems of linear equations.
❯ Solve nonsquare systems of linear equations.
❯ Use systems of linear equations in three or more variables to model and solve real-life problems.

Row-Echelon Form and Back-Substitution

The method of elimination can be applied to a system of linear equations in more than two variables. In fact, this method adapts to computer use for solving linear systems with dozens of variables.

When using the method of elimination to solve a system of linear equations, the goal is to rewrite the system in a form to which back-substitution can be applied. To see how this works, consider the following two systems of linear equations.

$$\begin{cases} x - 2y + 3z = 9 \\ -x + 3y = -4 \\ 2x - 5y + 5z = 17 \end{cases}$$ System of three linear equations in three variables (See Example 3.)

$$\begin{cases} x - 2y + 3z = 9 \\ y + 3z = 5 \\ z = 2 \end{cases}$$ Equivalent system in row-echelon form (See Example 1.)

The second system is in **row-echelon form,** which means that it has a "stair-step" pattern with leading coefficients of 1. In comparing the two systems, notice that the row-echelon form can readily be solved using back-substitution.

EXAMPLE 1 Using Back-Substitution in Row-Echelon Form

Solve the system of linear equations.

$$\begin{cases} x - 2y + 3z = 9 & \text{Equation 1} \\ y + 3z = 5 & \text{Equation 2} \\ z = 2 & \text{Equation 3} \end{cases}$$

Solution From Equation 3, you know the value of z. To solve for y, back-substitute $z = 2$ into Equation 2.

$$y + 3(2) = 5 \qquad \text{Substitute 2 for } z.$$
$$y = -1 \qquad \text{Solve for } y.$$

To solve for x, back-substitute $y = -1$ and $z = 2$ into Equation 1.

$$x - 2(-1) + 3(2) = 9 \qquad \text{Substitute } -1 \text{ for } y \text{ and } 2 \text{ for } z.$$
$$x = 1 \qquad \text{Solve for } x.$$

The solution is $x = 1$, $y = -1$, and $z = 2$, which can be written as the **ordered triple** $(1, -1, 2)$. Check this in the original system of equations.

✓ *Checkpoint* ▶ Audio-video solution in English & Spanish at LarsonPrecalculus.com

Solve the system of linear equations.

$$\begin{cases} x - y + 5z = 22 \\ y + 3z = 6 \\ z = 3 \end{cases}$$

GO DIGITAL

© sitayi/Shutterstock.com

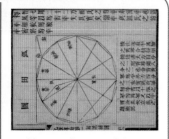

Historically, one of the most influential Chinese mathematics books was the *Chui-chang suan-shu,* or *Nine Chapters on the Mathematical Art,* a compilation of ancient Chinese problems published in A.D. 263. Chapter Eight of the *Nine Chapters* contained solutions of systems of linear equations such as the system below.

$$\begin{cases} 3x + 2y + z = 39 \\ 2x + 3y + z = 34 \\ x + 2y + 3z = 26 \end{cases}$$

This system was solved by performing column operations on a matrix, using the same strategies as Gaussian elimination. Matrices (plural for matrix) are discussed in the next chapter.

Gaussian Elimination

Recall from the preceding section that two systems of equations are *equivalent* when they have the same solution set. To solve a system that is not in row-echelon form, first convert it to an *equivalent* system that is in row-echelon form by using one or more of the row operations given below.

Operations That Produce Equivalent Systems

Each of the following **row operations** on a system of linear equations produces an *equivalent* system of linear equations.

1. Interchange two equations.

2. Multiply one of the equations by a nonzero constant.

3. Add a multiple of one equation to another equation to replace the latter equation.

To see how to use row operations, take another look at the method of elimination, as applied to a system of two linear equations.

EXAMPLE 2   **Using Row Operations to Solve a System**

Solve the system of linear equations.

$$\begin{cases} 3x - 2y = -1 & \text{Equation 1} \\ x - y = 0 & \text{Equation 2} \end{cases}$$

Solution   Two strategies seem reasonable: eliminate the variable x or eliminate the variable y. The following steps show one way to eliminate x in the second equation. Start by interchanging the equations to obtain a leading coefficient of 1 in the first equation.

$$\begin{cases} x - y = 0 \\ 3x - 2y = -1 \end{cases} \qquad \text{Interchange the two equations in the system.}$$

$$-3x + 3y = 0 \qquad \text{Multiply the first equation by } -3.$$

$$\begin{array}{r} -3x + 3y = 0 \\ \underline{3x - 2y = -1} \\ y = -1 \end{array} \qquad \begin{array}{l} \text{Add the multiple of the first equation to the} \\ \text{second equation to obtain a new second equation.} \end{array}$$

$$\begin{cases} x - y = 0 \\ y = -1 \end{cases} \qquad \text{New system in row-echelon form}$$

Now back-substitute $y = -1$ into the first equation of the new system in row-echelon form and solve for x.

$$x - (-1) = 0 \qquad \text{Substitute } -1 \text{ for } y.$$

$$x = -1 \qquad \text{Solve for } x.$$

The solution is $(-1, -1)$. Check this in the original system.

✓ *Checkpoint* ▶ Audio-video solution in English & Spanish at LarsonPrecalculus.com

Solve the system of linear equations.

$$\begin{cases} 2x + y = 3 \\ x + 2y = 3 \end{cases}$$

Rewriting a system of linear equations in row-echelon form usually involves a chain of equivalent systems, each of which is obtained by using one of the three basic row operations listed on the preceding page. This process is called **Gaussian elimination,** after the German mathematician Carl Friedrich Gauss (1777–1855).

EXAMPLE 3 **Using Gaussian Elimination to Solve a System**

Solve the system of linear equations.

$$\begin{cases} x - 2y + 3z = 9 & \text{Equation 1} \\ -x + 3y = -4 & \text{Equation 2} \\ 2x - 5y + 5z = 17 & \text{Equation 3} \end{cases}$$

Solution The leading coefficient of the first equation is 1, so begin by keeping the x in the upper left position and eliminating the other x-terms from the first column.

> **ALGEBRA HELP**
>
> Arithmetic errors are common when performing row operations. You should note the operation performed in each step to make checking your work easier.

$$\begin{array}{ll} x - 2y + 3z = 9 & \text{Write Equation 1.} \\ \underline{-x + 3y = -4} & \text{Write Equation 2.} \\ y + 3z = 5 & \text{Add.} \end{array}$$

$$\begin{cases} x - 2y + 3z = 9 \\ y + 3z = 5 \\ 2x - 5y + 5z = 17 \end{cases}$$ Adding the first equation to the second equation produces a new second equation.

$$\begin{array}{ll} -2x + 4y - 6z = -18 & \text{Multiply Equation 1 by } -2. \\ \underline{2x - 5y + 5z = 17} & \text{Write Equation 3.} \\ -y - z = -1 & \text{Add.} \end{array}$$

$$\begin{cases} x - 2y + 3z = 9 \\ y + 3z = 5 \\ -y - z = -1 \end{cases}$$ Adding -2 times the first equation to the third equation produces a new third equation.

Now that all but the first x have been eliminated from the first column, begin to work on the second column. (You need to eliminate y from the third equation.)

$$\begin{array}{ll} y + 3z = 5 & \text{Write second equation.} \\ \underline{-y - z = -1} & \text{Write third equation.} \\ 2z = 4 & \text{Add.} \end{array}$$

$$\begin{cases} x - 2y + 3z = 9 \\ y + 3z = 5 \\ 2z = 4 \end{cases}$$ Adding the second equation to the third equation produces a new third equation.

Finally, you need a coefficient of 1 for z in the third equation.

$$\begin{cases} x - 2y + 3z = 9 \\ y + 3z = 5 \\ z = 2 \end{cases}$$ Multiplying the third equation by $\frac{1}{2}$ produces a new third equation.

This is the same system that was solved in Example 1. So, as in that example, the solution is $(1, -1, 2)$.

✓ **Checkpoint** ▶ *Audio-video solution in English & Spanish at LarsonPrecalculus.com*

Solve the system of linear equations.

$$\begin{cases} x + y + z = 6 \\ 2x - y + z = 3 \\ 3x + y - z = 2 \end{cases}$$

The next example involves an inconsistent system—one that has no solution. The key to recognizing an inconsistent system is that at some stage in the elimination process, you obtain a false statement such as $0 = -2$.

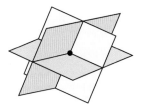

Solution: one point
Figure 9.10

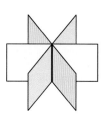

Solution: one line
Figure 9.11

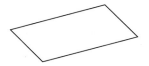

Solution: one plane
Figure 9.12

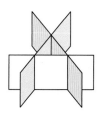

Solution: none
Figure 9.13

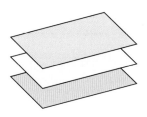

Solution: none
Figure 9.14

GO DIGITAL

| EXAMPLE 4 | **An Inconsistent System** |

Solve the system of linear equations.

$$\begin{cases} x - 3y + z = 1 \\ 2x - y - 2z = 2 \\ x + 2y - 3z = -1 \end{cases}$$
    Equation 1
    Equation 2
    Equation 3

Solution

$$\begin{cases} x - 3y + z = 1 \\ 5y - 4z = 0 \\ x + 2y - 3z = -1 \end{cases}$$

> Adding -2 times the first equation to the second equation produces a new second equation.

$$\begin{cases} x - 3y + z = 1 \\ 5y - 4z = 0 \\ 5y - 4z = -2 \end{cases}$$

> Adding -1 times the first equation to the third equation produces a new third equation.

$$\begin{cases} x - 3y + z = 1 \\ 5y - 4z = 0 \\ 0 = -2 \end{cases}$$

> Adding -1 times the second equation to the third equation produces a new third equation.

You obtain the false statement $0 = -2$, so this system is inconsistent and has no solution. Moreover, this system is equivalent to the original system, so the original system also has no solution.

✓ **Checkpoint** ▶ *Audio-video solution in English & Spanish at LarsonPrecalculus.com*

Solve the system of linear equations.

$$\begin{cases} x + y - 2z = 3 \\ 3x - 2y + 4z = 1 \\ 2x - 3y + 6z = 8 \end{cases}$$

As with a system of linear equations in two variables, the number of solutions of a system of linear equations in more than two variables must fall into one of three categories.

The Number of Solutions of a Linear System

For a system of linear equations, exactly one of the following is true.

1. There is exactly one solution.
2. There are infinitely many solutions.
3. There is no solution.

In the preceding section, you learned that the graph of a system of two linear equations in two variables is a pair of lines that intersect, coincide, or are parallel. Similarly, the graph of a system of three linear equations in three variables consists of three planes in space. These planes must intersect in one point (exactly one solution, see Figure 9.10), intersect in a line or a plane (infinitely many solutions, see Figures 9.11 and 9.12), or have no points common to all three planes (no solution, see Figures 9.13 and 9.14).

EXAMPLE 5 **A System with Infinitely Many Solutions**

Solve the system of linear equations.

$$\begin{cases} x + y - 3z = -1 & \text{Equation 1} \\ \quad\quad y - z = 0 & \text{Equation 2} \\ -x + 2y \quad\quad = 1 & \text{Equation 3} \end{cases}$$

Solution

$$\begin{cases} x + y - 3z = -1 \\ \quad\quad y - z = 0 \\ \quad\quad 3y - 3z = 0 \end{cases}$$

> Adding the first equation to the third equation produces a new third equation.

$$\begin{cases} x + y - 3z = -1 \\ \quad\quad y - z = 0 \\ \quad\quad\quad 0 = 0 \end{cases}$$

> Adding -3 times the second equation to the third equation produces a new third equation.

You have $0 = 0$, so Equation 3 depends on Equations 1 and 2 in the sense that it gives no additional information about the variables. So, the original system is equivalent to

$$\begin{cases} x + y - 3z = -1 \\ \quad\quad y - z = 0 \end{cases}.$$

In the second equation, solve for y in terms of z to obtain $y = z$. Back-substituting for y in the first equation yields $x = 2z - 1$. So, the original system has infinitely many solutions because for every value of z, you can use $y = z$ and $x = 2z - 1$ to find corresponding values of x and y. Letting $z = a$, where a is any real number, the solutions of the original system are all ordered triples of the form

$$(2a - 1, a, a).$$

✓ **Checkpoint** ▶ *Audio-video solution in English & Spanish at LarsonPrecalculus.com*

Solve the system of linear equations.

$$\begin{cases} x + 2y - 7z = -4 \\ 2x + 3y + z = 5 \\ 3x + 7y - 36z = -25 \end{cases}$$

In Example 5, there are other ways to write the same infinite set of solutions. For instance, letting $x = b$, the solutions could have been written as

$$\left(b, \tfrac{1}{2}(b + 1), \tfrac{1}{2}(b + 1)\right). \quad\quad b \text{ is a real number.}$$

To convince yourself that this form produces the same set of solutions, consider the following.

Substitution	Solution	
$a = 0$	$(2(0) - 1, 0, 0) = (-1, 0, 0)$	Same solution
$b = -1$	$\left(-1, \tfrac{1}{2}(-1 + 1), \tfrac{1}{2}(-1 + 1)\right) = (-1, 0, 0)$	
$a = 1$	$(2(1) - 1, 1, 1) = (1, 1, 1)$	Same solution
$b = 1$	$\left(1, \tfrac{1}{2}(1 + 1), \tfrac{1}{2}(1 + 1)\right) = (1, 1, 1)$	
$a = 2$	$(2(2) - 1, 2, 2) = (3, 2, 2)$	Same solution
$b = 3$	$\left(3, \tfrac{1}{2}(3 + 1), \tfrac{1}{2}(3 + 1)\right) = (3, 2, 2)$	

ALGEBRA HELP

There are an infinite number of solutions to Example 5, but they are all of a specific form. For instance, by selecting a-values of 0, 1, and 3, you can verify that $(-1, 0, 0)$, $(1, 1, 1)$, and $(5, 3, 3)$ are specific solutions. It is incorrect to say simply that the solution to Example 5 is "infinite." You must also specify the form of the solutions.

GO DIGITAL

Nonsquare Systems

So far, each system of linear equations you have looked at has been *square*, which means that the number of equations is equal to the number of variables. In a **nonsquare** system, the number of equations differs from the number of variables. A system of linear equations cannot have a unique solution unless there are at least as many equations as there are variables in the system.

EXAMPLE 6   A System with Fewer Equations than Variables

▶▶▶ *See LarsonPrecalculus.com for an interactive version of this type of example.*

Solve the system of linear equations.

$$\begin{cases} x - 2y + z = 2 & \text{Equation 1} \\ 2x - y - z = 1 & \text{Equation 2} \end{cases}$$

Solution   The system has three variables and only two equations, so the system does not have a unique solution. Begin by rewriting the system in row-echelon form.

$$\begin{cases} x - 2y + z = 2 \\ 3y - 3z = -3 \end{cases}$$

Adding -2 times the first equation to the second equation produces a new second equation.

$$\begin{cases} x - 2y + z = 2 \\ y - z = -1 \end{cases}$$

Multiplying the second equation by $\frac{1}{3}$ produces a new second equation.

Solve the new second equation for y in terms of z to obtain

$$y = z - 1.$$

Solve for x by back-substituting $y = z - 1$ into Equation 1.

$$\begin{aligned} x - 2y + z &= 2 & \text{Write Equation 1.} \\ x - 2(z - 1) + z &= 2 & \text{Substitute } z - 1 \text{ for } y. \\ x - 2z + 2 + z &= 2 & \text{Distributive Property} \\ x &= z & \text{Solve for } x. \end{aligned}$$

Finally, by letting $z = a$, where a is any real number, you have the solution

$$x = a, \quad y = a - 1, \quad \text{and} \quad z = a.$$

So, the solution set of the system consists of all ordered triples of the form $(a, a - 1, a)$, where a is a real number.

✓ *Checkpoint*  *Audio-video solution in English & Spanish at LarsonPrecalculus.com*

Solve the system of linear equations.

$$\begin{cases} x - y + 4z = 3 \\ 4x - z = 0 \end{cases}$$

In Example 6, you can choose several values of a to obtain different solutions of the system, such as

$$(1, 0, 1), \quad (2, 1, 2), \quad \text{and} \quad (3, 2, 3).$$

Then check each ordered triple in the original system to verify that it is a solution of the system.

Applications

EXAMPLE 7 Modeling Vertical Motion

The height at time t of an object that is moving in a (vertical) line with constant acceleration a is given by the **position equation**

$$s = \tfrac{1}{2}at^2 + v_0 t + s_0$$

where s is the height in feet, a is the acceleration in feet per second squared, t is the time in seconds, v_0 is the initial velocity (at $t = 0$) in feet per second, and s_0 is the initial height in feet. Find the values of a, v_0, and s_0 when $s = 52$ at $t = 1$, $s = 52$ at $t = 2$, and $s = 20$ at $t = 3$, and interpret the result. (See Figure 9.15.)

Solution Substitute the three sets of values for t and s into the position equation to obtain three linear equations in a, v_0, and s_0.

When $t = 1$: $\tfrac{1}{2}a(1)^2 + v_0(1) + s_0 = 52$ ⟹ $a + 2v_0 + 2s_0 = 104$
When $t = 2$: $\tfrac{1}{2}a(2)^2 + v_0(2) + s_0 = 52$ ⟹ $2a + 2v_0 + s_0 = 52$
When $t = 3$: $\tfrac{1}{2}a(3)^2 + v_0(3) + s_0 = 20$ ⟹ $9a + 6v_0 + 2s_0 = 40$

Solve the resulting system of linear equations using Gaussian elimination.

$$\begin{cases} a + 2v_0 + 2s_0 = 104 \\ 2a + 2v_0 + s_0 = 52 \\ 9a + 6v_0 + 2s_0 = 40 \end{cases}$$

$$\begin{cases} a + 2v_0 + 2s_0 = 104 \\ -2v_0 - 3s_0 = -156 \\ 9a + 6v_0 + 2s_0 = 40 \end{cases}$$

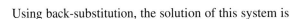 Adding -2 times the first equation to the second equation produces a new second equation.

$$\begin{cases} a + 2v_0 + 2s_0 = 104 \\ -2v_0 - 3s_0 = -156 \\ -12v_0 - 16s_0 = -896 \end{cases}$$

Adding -9 times the first equation to the third equation produces a new third equation.

$$\begin{cases} a + 2v_0 + 2s_0 = 104 \\ -2v_0 - 3s_0 = -156 \\ 2s_0 = 40 \end{cases}$$

Adding -6 times the second equation to the third equation produces a new third equation.

$$\begin{cases} a + 2v_0 + 2s_0 = 104 \\ v_0 + \tfrac{3}{2}s_0 = 78 \\ s_0 = 20 \end{cases}$$

Multiplying the second equation by $-\tfrac{1}{2}$ produces a new second equation and multiplying the third equation by $\tfrac{1}{2}$ produces a new third equation.

Using back-substitution, the solution of this system is

$$a = -32, \quad v_0 = 48, \quad \text{and} \quad s_0 = 20.$$

So, the position equation for the object is $s = -16t^2 + 48t + 20$, which implies that the object was thrown upward at a velocity of 48 feet per second from a height of 20 feet.

✓ *Checkpoint* ▶ *Audio-video solution in English & Spanish at LarsonPrecalculus.com*

Use the position equation

$$s = \tfrac{1}{2}at^2 + v_0 t + s_0$$

from Example 7 to find the values of a, v_0, and s_0 when $s = 104$ at $t = 1$, $s = 76$ at $t = 2$, and $s = 16$ at $t = 3$, and interpret the result. ∎

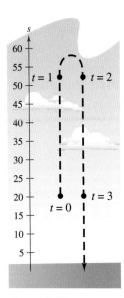

Figure 9.15

GO DIGITAL

EXAMPLE 8   **Data Analysis: Curve-Fitting**

Find a quadratic equation $y = ax^2 + bx + c$ whose graph passes through the points $(-1, 3)$, $(1, 1)$, and $(2, 6)$.

Solution   Because the graph of $y = ax^2 + bx + c$ passes through the points $(-1, 3)$, $(1, 1)$, and $(2, 6)$, you can write the following.

When $x = -1$, $y = 3$: $a(-1)^2 + b(-1) + c = 3$
When $x = 1$, $y = 1$: $a(1)^2 + b(1) + c = 1$
When $x = 2$, $y = 6$: $a(2)^2 + b(2) + c = 6$

Solve the resulting system of linear equations using Gaussian elimination.

$$\begin{cases} a - b + c = 3 \\ a + b + c = 1 \\ 4a + 2b + c = 6 \end{cases}$$

$$\begin{cases} a - b + c = 3 \\ \quad\ 2b \quad\ = -2 \\ 4a + 2b + c = 6 \end{cases}$$

Adding -1 times the first equation to the second equation produces a new second equation.

$$\begin{cases} a - b + c = 3 \\ \quad\ 2b \quad\ = -2 \\ \quad\ 6b - 3c = -6 \end{cases}$$

Adding -4 times the first equation to the third equation produces a new third equation.

$$\begin{cases} a - b + c = 3 \\ \quad\ 2b \quad\ = -2 \\ \quad\ -3c = 0 \end{cases}$$

Adding -3 times the second equation to the third equation produces a new third equation.

$$\begin{cases} a - b + c = 3 \\ \quad\ b \quad\ = -1 \\ \quad\ c = 0 \end{cases}$$

Multiplying the second equation by $\frac{1}{2}$ produces a new second equation and multiplying the third equation by $-\frac{1}{3}$ produces a new third equation.

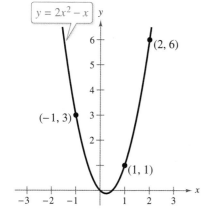

$y = 2x^2 - x$

(2, 6)

(−1, 3)

(1, 1)

Figure 9.16

The solution of this system is $a = 2$, $b = -1$, and $c = 0$. So, the equation of the parabola is $y = 2x^2 - x$. (See Figure 9.16.)

✓ *Checkpoint* ▶ *Audio-video solution in English & Spanish at LarsonPrecalculus.com*

Find a quadratic equation $y = ax^2 + bx + c$ whose graph passes through the points $(0, 0)$, $(3, -3)$, and $(6, 0)$. ■

Summarize (Section 9.3)

1. Explain what row-echelon form is *(page 650)*. For an example of solving a linear system in row-echelon form, see Example 1.

2. Describe the process of Gaussian elimination *(pages 651 and 652)*. For examples of using Gaussian elimination to solve systems of linear equations, see Examples 2–5.

3. Explain the difference between a square system of linear equations and a nonsquare system of linear equations *(page 655)*. For an example of solving a nonsquare system of linear equations, see Example 6.

4. Describe examples of how to use systems of linear equations in three or more variables to model and solve real-life problems *(pages 656 and 657, Examples 7 and 8)*.

GO DIGITAL

9.3 Exercises

See CalcChat.com for tutorial help and worked-out solutions to odd-numbered exercises.

GO DIGITAL

Vocabulary and Concept Check

In Exercises 1–4, fill in the blanks.

1. A system of equations in _____ form has a "stair-step" pattern with leading coefficients of 1.
2. A solution of a system of three linear equations in three variables can be written as an _____ _____, which has the form (x, y, z).
3. The process used to write a system of linear equations in row echelon form is called _____ elimination.
4. In a _____ system, the number of equations differs from the number of variables in the system.

5. What three row operations can you use on a system of linear equations to produce an equivalent system?

6. When a system of three linear equations in three variables has infinitely many solutions, is any ordered triple a solution of the system? Explain.

Skills and Applications

Checking Solutions In Exercises 7–10, determine whether each ordered triple is a solution of the system of equations.

7. $\begin{cases} 6x - y + z = -1 \\ 4x - 3z = -19 \\ 2y + 5z = 25 \end{cases}$

   (a) $(0, 3, 1)$ (b) $(-3, 0, 5)$
   (c) $(0, -1, 4)$ (d) $(-1, 0, 5)$

8. $\begin{cases} 3x + 4y - z = 17 \\ 5x - y + 2z = -2 \\ 2x - 3y + 7z = -21 \end{cases}$

   (a) $(3, -1, 2)$ (b) $(1, 3, -2)$
   (c) $(1, 5, 6)$ (d) $(1, -2, 2)$

9. $\begin{cases} 4x + y - z = 0 \\ -8x - 6y + z = -\frac{7}{4} \\ 3x - y = -\frac{9}{4} \end{cases}$

   (a) $\left(\frac{1}{2}, -\frac{3}{4}, -\frac{7}{4}\right)$ (b) $\left(\frac{3}{2}, -\frac{2}{5}, \frac{3}{5}\right)$
   (c) $\left(-\frac{1}{2}, \frac{3}{4}, -\frac{5}{4}\right)$ (d) $\left(-\frac{1}{2}, \frac{1}{6}, -\frac{3}{4}\right)$

10. $\begin{cases} 4x - 5y + 3z = 6 \\ y + z = 0 \\ 2x - 4y = 3 \end{cases}$

    (a) $\left(-\frac{7}{2}, -2, 2\right)$ (b) $\left(3, \frac{3}{2}, \frac{1}{2}\right)$
    (c) $\left(\frac{9}{4}, \frac{3}{4}, -\frac{3}{4}\right)$ (d) $\left(-\frac{1}{4}, -\frac{7}{8}, \frac{7}{8}\right)$

Using Back-Substitution in Row-Echelon Form In Exercises 11–16, use back-substitution to solve the system of linear equations.

11. $\begin{cases} x - y + 5z = 37 \\ y + 2z = 6 \\ z = 8 \end{cases}$

12. $\begin{cases} x - 2y + 2z = 20 \\ y - z = 8 \\ z = -1 \end{cases}$

13. $\begin{cases} x + y - 3z = 7 \\ y + z = 12 \\ z = 2 \end{cases}$

14. $\begin{cases} x - y + 2z = 22 \\ y - 8z = 13 \\ z = -3 \end{cases}$

15. $\begin{cases} x - 2y + z = -\frac{1}{4} \\ y - z = -4 \\ z = 11 \end{cases}$

16. $\begin{cases} x - 8z = \frac{1}{2} \\ y - 5z = 22 \\ z = -4 \end{cases}$

Performing Row Operations In Exercises 17 and 18, perform the row operation and write the equivalent system.

17. Add Equation 1 to Equation 2.

$\begin{cases} x - 2y + 3z = 5 & \text{Equation 1} \\ -x + 3y - 5z = 4 & \text{Equation 2} \\ 2x - 3z = 0 & \text{Equation 3} \end{cases}$

What did this operation accomplish?

18. Add -2 times Equation 1 to Equation 3.

$\begin{cases} x - 2y + 3z = 5 & \text{Equation 1} \\ -x + 3y - 5z = 4 & \text{Equation 2} \\ 2x - 3z = 0 & \text{Equation 3} \end{cases}$

What did this operation accomplish?

Solving a System of Linear Equations In Exercises 19–22, solve the system of linear equations and check any solutions algebraically.

19. $\begin{cases} -2x + 3y = 10 \\ x + y = 0 \end{cases}$

20. $\begin{cases} 2x - y = 0 \\ x - y = 7 \end{cases}$

21. $\begin{cases} 3x - y = 9 \\ x - 2y = -2 \end{cases}$

22. $\begin{cases} x + 2y = 1 \\ 5x - 4y = -23 \end{cases}$

Solving a System of Linear Equations In Exercises 23–40, solve the system of linear equations and check any solutions algebraically.

23. $\begin{cases} x + y + z = 7 \\ 2x - y + z = 9 \\ 3x \quad\;\; - z = 10 \end{cases}$
24. $\begin{cases} x + y + z = 5 \\ x - 2y + 4z = 13 \\ 3y + 4z = 13 \end{cases}$

25. $\begin{cases} 2x + 4y - z = 7 \\ 2x - 4y + 2z = -6 \\ x + 4y + z = 0 \end{cases}$
26. $\begin{cases} 2x + 4y + z = 1 \\ x - 2y - 3z = 2 \\ x + y - z = -1 \end{cases}$

27. $\begin{cases} 2x + y - z = 7 \\ x - 2y + 2z = -9 \\ 3x - y + z = 5 \end{cases}$
28. $\begin{cases} 5x - 3y + 2z = 3 \\ 2x + 4y - z = 7 \\ x - 11y + 4z = 3 \end{cases}$

29. $\begin{cases} 3x - 5y + 5z = 1 \\ 2x - 2y + 3z = 0 \\ 7x - y + 3z = 0 \end{cases}$
30. $\begin{cases} 2x + y + 3z = 1 \\ 2x + 6y + 8z = 3 \\ 6x + 8y + 18z = 5 \end{cases}$

31. $\begin{cases} 2x + 3y \quad\;\; = 0 \\ 4x + 3y - z = 0 \\ 8x + 3y + 3z = 0 \end{cases}$
32. $\begin{cases} 4x + 3y + 17z = 0 \\ 5x + 4y + 22z = 0 \\ 4x + 2y + 19z = 0 \end{cases}$

33. $\begin{cases} x \quad\;\; + 4z = 1 \\ x + y + 10z = 10 \\ 2x - y + 2z = -5 \end{cases}$
34. $\begin{cases} 2x - 2y - 6z = -4 \\ -3x + 2y + 6z = 1 \\ x - y - 5z = -3 \end{cases}$

35. $\begin{cases} 3x - 3y + 6z = 6 \\ x + 2y - z = 5 \\ 5x - 8y + 13z = 7 \end{cases}$
36. $\begin{cases} x \quad\;\; + 2z = 5 \\ 3x - y - z = 1 \\ 6x - y + 5z = 16 \end{cases}$

37. $\begin{cases} x + 2y - 7z = -4 \\ 2x + y + z = 13 \\ 3x + 9y - 36z = -33 \end{cases}$

38. $\begin{cases} 2x + y - 3z = 4 \\ 4x \quad\;\; + 2z = 10 \\ -2x + 3y - 13z = -8 \end{cases}$

39. $\begin{cases} x \quad\quad\;\; + 3w = 4 \\ 2y - z - w = 0 \\ 3y \quad\;\; - 2w = 1 \\ 2x - y + 4z \quad\;\; = 5 \end{cases}$

40. $\begin{cases} x + y + z + w = 6 \\ 2x + 3y \quad\;\; - w = 0 \\ -3x + 4y + z + 2w = 4 \\ x + 2y - z + w = 0 \end{cases}$

Solving a Nonsquare System In Exercises 41–44, solve the system of linear equations and check any solutions algebraically.

41. $\begin{cases} x - 2y + 5z = 2 \\ 4x \quad\;\; - z = 0 \end{cases}$
42. $\begin{cases} x - 3y + 2z = 18 \\ 5x - 13y + 12z = 80 \end{cases}$

43. $\begin{cases} 2x - 3y + z = -2 \\ -4x + 9y = 7 \end{cases}$

44. $\begin{cases} 2x + 3y + 3z = 7 \\ 4x + 18y + 15z = 44 \end{cases}$

Modeling Vertical Motion In Exercises 45–48, an object moving vertically is at the given heights at the specified times. Find the position equation $s = \frac{1}{2}at^2 + v_0t + s_0$ for the object.

45. At $t = 1$ second, $s = 128$ feet
At $t = 2$ seconds, $s = 80$ feet
At $t = 3$ seconds, $s = 0$ feet

46. At $t = 1$ second, $s = 32$ feet
At $t = 2$ seconds, $s = 32$ feet
At $t = 3$ seconds, $s = 0$ feet

47. At $t = 1$ second, $s = 352$ feet
At $t = 2$ seconds, $s = 272$ feet
At $t = 3$ seconds, $s = 160$ feet

48. At $t = 1$ second, $s = 132$ feet
At $t = 2$ seconds, $s = 100$ feet
At $t = 3$ seconds, $s = 36$ feet

Finding the Equation of a Parabola In Exercises 49–54, find the equation

$$y = ax^2 + bx + c$$

of the parabola that passes through the points. To verify your result, use a graphing utility to plot the points and graph the parabola.

49. $(0, 0), (2, -2), (4, 0)$
50. $(0, 3), (1, 4), (2, 3)$
51. $(2, 0), (3, -1), (4, 0)$
52. $(1, 3), (2, 2), (3, -3)$
53. $\left(\frac{1}{2}, 1\right), (1, 3), (2, 13)$
54. $(-2, -3), (-1, 0), \left(\frac{1}{2}, -3\right)$

Finding the Equation of a Circle In Exercises 55–58, find the equation

$$x^2 + y^2 + Dx + Ey + F = 0$$

of the circle that passes through the points. To verify your result, use a graphing utility to plot the points and graph the circle.

55. $(0, 0), (5, 5), (10, 0)$
56. $(0, 0), (0, 6), (3, 3)$
57. $(-3, -1), (2, 4), (-6, 8)$
58. $(0, 0), (0, -2), (3, 0)$

59. **Agriculture** A mixture of 5 pounds of fertilizer A, 13 pounds of fertilizer B, and 4 pounds of fertilizer C provides the optimal nutrients for a plant. Commercial brand X contains equal parts of fertilizer B and fertilizer C. Commercial brand Y contains one part of fertilizer A and two parts of fertilizer B. Commercial brand Z contains two parts of fertilizer A, five parts of fertilizer B, and two parts of fertilizer C. How much of each fertilizer brand is needed to obtain the desired mixture?

60. Finance A college student borrowed $30,000 to pay for tuition, room, and board. Some of the money was borrowed at 4%, some at 6%, and some at 8%. How much was borrowed at each rate, given that the annual interest was $1550 and the amount borrowed at 8% was three times the amount borrowed at 6%?

Geometry In Exercises 61 and 62, find the values of x, y, and z in the figure.

61.

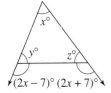

62.

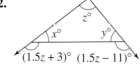

63. Electrical Network Applying Kirchhoff's Laws to the electrical network in the figure, the currents I_1, I_2, and I_3, are the solution of the system

$$\begin{cases} I_1 - I_2 + I_3 = 0 \\ 3I_1 + 2I_2 \quad\;\; = 7. \\ \qquad 2I_2 + 4I_3 = 8 \end{cases}$$

Find the currents.

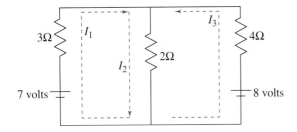

64. Pulley System A system of pulleys is loaded with 128-pound and 32-pound weights (see figure). The tensions t_1 and t_2 in the ropes and the acceleration a of the 32-pound weight are found by solving the system

$$\begin{cases} t_1 - 2t_2 \qquad = 0 \\ t_1 \qquad - 2a = 128 \\ \quad\; t_2 + a = 32 \end{cases}$$

where t_1 and t_2 are in pounds and a is in feet per second squared. Solve this system.

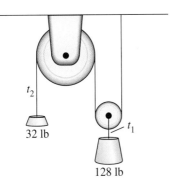

Fitting a Parabola One way to find the least squares regression parabola $y = ax^2 + bx + c$ for a set of points

$(x_1, y_1), (x_2, y_2), \ldots, (x_n, y_n)$

is by solving the system below for a, b, and c.

$$\begin{cases} nc + \left(\sum_{i=1}^{n} x_i\right)b + \left(\sum_{i=1}^{n} x_i^2\right)a = \sum_{i=1}^{n} y_i \\ \left(\sum_{i=1}^{n} x_i\right)c + \left(\sum_{i=1}^{n} x_i^2\right)b + \left(\sum_{i=1}^{n} x_i^3\right)a = \sum_{i=1}^{n} x_i y_i \\ \left(\sum_{i=1}^{n} x_i^2\right)c + \left(\sum_{i=1}^{n} x_i^3\right)b + \left(\sum_{i=1}^{n} x_i^4\right)a = \sum_{i=1}^{n} x_i^2 y_i \end{cases}$$

In Exercises 65 and 66, the sums have been evaluated. Solve the simplified system for a, b, and c to find the least squares regression parabola for the points. Use a graphing utility to confirm the result. *(Note:* The symbol Σ is used to denote a sum of the terms of a sequence. You will learn how to use this notation in Section 11.1.)

65.
$$\begin{cases} 4c + 9b + 29a = 20 \\ 9c + 29b + 99a = 70 \\ 29c + 99b + 353a = 254 \end{cases}$$

66.
$$\begin{cases} 4c \qquad + 40a = 19 \\ \quad\; 40b \qquad = -12 \\ 40c \qquad + 544a = 160 \end{cases}$$

67. Stopping Distance In testing a new automobile braking system, engineers recorded the speed x (in miles per hour) and the stopping distance y (in feet). The table shows the results.

Speed, x	30	40	50	60	70
Stopping Distance, y	75	118	175	240	315

(a) Find the least squares regression parabola $y = ax^2 + bc + c$ for the data by solving the system.

$$\begin{cases} 5c + 250b + 13{,}500a = 923 \\ 250c + 13{,}500b + 775{,}000a = 52{,}170 \\ 13{,}500c + 775{,}000b + 46{,}590{,}000a = 3{,}101{,}300 \end{cases}$$

(b) Use a graphing utility to graph the model you found in part (a) and the data in the same viewing window. How well does the model fit the data? Explain.

(c) Use the model to estimate the stopping distance when the speed is 75 miles per hour.

68. Wildlife Preservation

A wildlife management team studied the reproductive rates of deer in four tracts of a wildlife preserve. In each tract, the number of females x and the percent of females y that had offspring the following year were recorded. The table shows the results.

Number, x	100	120	140	160
Percent, y	75	68	55	30

(a) Find the least squares regression parabola $y = ax^2 + bx + c$ for the data by solving the system.

$$\begin{cases} 4c + 520b + 69{,}600a = 228 \\ 520c + 69{,}600b + 9{,}568{,}000a = 28{,}160 \\ 69{,}600c + 9{,}568{,}000b + 1{,}346{,}880{,}000a = 3{,}575{,}200 \end{cases}$$

(b) Use a graphing utility to graph the model you found in part (a) and the data in the same viewing window. How well does the model fit the data? Explain.

(c) Use the model to estimate the percent of females that had offspring when there were 170 females.

(d) Use the model to estimate the number of females when 40% of the females had offspring.

♪ Advanced Applications   In Exercises 69 and 70, find values of x, y, and λ that satisfy the system. These systems arise in certain optimization problems in calculus, and λ is called a Lagrange multiplier.

69. $\begin{cases} 2x - 2x\lambda = 0 \\ -2y + \lambda = 0 \\ y - x^2 = 0 \end{cases}$ 70. $\begin{cases} 2 + 2y + 2\lambda = 0 \\ 2x + 1 + \lambda = 0 \\ 2x + y - 100 = 0 \end{cases}$

Exploring the Concepts

True or False?   In Exercises 71 and 72, determine whether the statement is true or false. Justify your answer.

71. Every nonsquare system of linear equations has a unique solution.

72. If a system of three linear equations is inconsistent, then there are no points common to the graphs of all three equations of the system.

73. **Writing a System of Equations**   Write a system of equations with infinitely many solutions of the form $(-3a + 1, 4a - 2, a)$, where a is any real number.

74.   **HOW DO YOU SEE IT?**   The number of sides x and the combined number of sides and diagonals y for each of three regular polygons are shown below. Write a system of linear equations to find an equation of the form $y = ax^2 + bx + c$ that represents the relationship between x and y for the three polygons.

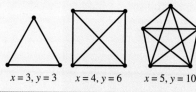

$x = 3, y = 3$ $x = 4, y = 6$ $x = 5, y = 10$

Finding Systems of Linear Equations   In Exercises 75–78, find two systems of linear equations that have the ordered triple as a solution. (There are many correct answers.)

75. $(2, 0, -1)$ 76. $(-5, 3, -2)$
77. $\left(\frac{1}{2}, -3, 0\right)$ 78. $\left(4, \frac{2}{5}, \frac{1}{2}\right)$

Review & Refresh ▶ *Video solutions at LarsonPrecalculus.com*

Simplifying a Rational Expression   In Exercises 79–82, write the rational expression in simplest form.

79. $\dfrac{x^2 + 2x - 8}{-20 - 5x}$ 80. $\dfrac{2x^2 - 15x - 27}{x^2 - 81}$

81. $\dfrac{3x^2 + 9x + 6}{x^3 - 2x^2 - x - 2}$ 82. $\dfrac{-2x^2 + 16x - 32}{x^3 - 64}$

Dividing Polynomials   In Exercises 83–86, use long division or synthetic division to divide.

83. $(-3x^2 + 19x - 6) \div (x - 6)$
84. $(2x^2 - 4x - 5) \div (x + 1)$
85. $(x^3 - 2x^2 - x + 7) \div (x - 3)$
86. $(3x^3 + 5x^2 + 6x + 1) \div (x - 2)$

Sketching the Graph of a Trigonometric Function   In Exercises 87 and 88, sketch the graph of the function. (Include two full periods.)

87. $y = 2 \tan \pi x$

88. $y = 1 + \sec \dfrac{x}{3}$

Project: Earnings per Share   To work an extended application analyzing the earnings per share for The Home Depot, Inc., from 2004 through 2019, visit this text's website at *LarsonPrecalculus.com*. (*Source: The Home Depot, Inc.*)

9.4 Partial Fractions

❯ Recognize partial fraction decompositions of rational expressions.
❯ Find partial fraction decompositions of rational expressions.

Introduction

In this section, you will learn to write a rational expression as the sum of two or more simpler rational expressions. For example, the rational expression

$$\frac{x + 7}{x^2 - x - 6}$$

can be written as the sum of two fractions with first-degree denominators. That is,

Partial fraction decomposition
of $\dfrac{x + 7}{x^2 - x - 6}$

$$\frac{x + 7}{x^2 - x - 6} = \overbrace{\underbrace{\frac{2}{x - 3}}_{\substack{\text{Partial} \\ \text{fraction}}} + \underbrace{\frac{-1}{x + 2}}_{\substack{\text{Partial} \\ \text{fraction}}}}.$$

Partial fractions can help you analyze the behavior of a rational function. For example, in Exercise 52 on page 669, you will use partial fractions to analyze the exhaust temperatures of a diesel engine.

Each fraction on the right side of the equation is a **partial fraction,** and together they make up the **partial fraction decomposition** of the left side.

Decomposition of $N(x)/D(x)$ into Partial Fractions

1. *Divide when improper:* When $N(x)/D(x)$ is an improper fraction [degree of $N(x) \geq$ degree of $D(x)$], divide the denominator into the numerator to obtain

   $$\frac{N(x)}{D(x)} = (\text{polynomial}) + \frac{N_1(x)}{D(x)}$$

   and apply Steps 2, 3, and 4 to the proper rational expression

   $$\frac{N_1(x)}{D(x)}.$$

   Note that $N_1(x)$ is the remainder from the division of $N(x)$ by $D(x)$.

2. *Factor the denominator:* Completely factor the denominator into factors of the form

   $$(px + q)^m \quad \text{and} \quad (ax^2 + bx + c)^n$$

   where $(ax^2 + bx + c)$ is irreducible.

3. *Linear factors:* For *each* factor of the form $(px + q)^m$, the partial fraction decomposition must include the following sum of m fractions.

   $$\frac{A_1}{(px + q)} + \frac{A_2}{(px + q)^2} + \cdot \cdot \cdot + \frac{A_m}{(px + q)^m}$$

4. *Quadratic factors:* For *each* factor of the form $(ax^2 + bx + c)^n$, the partial fraction decomposition must include the following sum of n fractions.

   $$\frac{B_1 x + C_1}{ax^2 + bx + c} + \frac{B_2 x + C_2}{(ax^2 + bx + c)^2} + \cdot \cdot \cdot + \frac{B_n x + C_n}{(ax^2 + bx + c)^n}$$

ALGEBRA HELP

Section P.5 shows you how to combine expressions such as

$$\frac{1}{x - 2} + \frac{-1}{x + 3} = \frac{5}{(x - 2)(x + 3)}.$$

The method of partial fraction decomposition shows you how to reverse this process and write

$$\frac{5}{(x - 2)(x + 3)} = \frac{1}{x - 2} + \frac{-1}{x + 3}.$$

⟫⟫⟫

GO DIGITAL

© VFXartist/Shutterstock.com

Partial Fraction Decomposition

The examples in this section demonstrate algebraic techniques for determining the constants in the numerators of partial fractions. Note that the techniques vary slightly, depending on the type of factors of the denominator: linear or quadratic, distinct or repeated.

EXAMPLE 1   **Distinct Linear Factors**

Write the partial fraction decomposition of

$$\frac{x+7}{x^2-x-6}.$$

Solution   The expression is proper, so begin by factoring the denominator.

$$x^2 - x - 6 = (x-3)(x+2)$$

Include one partial fraction with a constant numerator for each linear factor of the denominator. Write the form of the decomposition with A and B as the unknown constants.

$$\frac{x+7}{x^2-x-6} = \frac{A}{x-3} + \frac{B}{x+2} \qquad \text{Write form of decomposition.}$$

Multiply each side of this equation by the least common denominator, $(x-3)(x+2)$, to obtain the **basic equation.**

$$x + 7 = A(x+2) + B(x-3) \qquad \text{Basic equation}$$

This equation is true for all x, so substitute any *convenient* values of x that will help determine the constants A and B. Values of x that are especially convenient are those that make the factors $(x+2)$ and $(x-3)$ equal to zero. For example, to solve for B, let $x = -2$.

$$-2 + 7 = A(-2+2) + B(-2-3) \qquad \text{Substitute } -2 \text{ for } x.$$
$$5 = A(0) + B(-5)$$
$$5 = -5B$$
$$-1 = B$$

To solve for A, let $x = 3$.

$$3 + 7 = A(3+2) + B(3-3) \qquad \text{Substitute } 3 \text{ for } x.$$
$$10 = A(5) + B(0)$$
$$10 = 5A$$
$$2 = A$$

So, the partial fraction decomposition is

$$\frac{x+7}{x^2-x-6} = \frac{2}{x-3} + \frac{-1}{x+2}.$$

Check this result by combining the two partial fractions on the right side of the equation, or by using a graphing utility.

✓ *Checkpoint* ▶ Audio-video solution in English & Spanish at LarsonPrecalculus.com

Write the partial fraction decomposition of

$$\frac{x+5}{2x^2-x-1}.$$

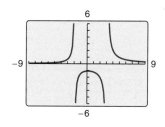

GO DIGITAL

EXAMPLE 2 **Repeated Linear Factors**

Write the partial fraction decomposition of $\dfrac{x^4 + 2x^3 + 6x^2 + 20x + 6}{x^3 + 2x^2 + x}$.

Solution This rational expression is improper, so begin by dividing the numerator by the denominator.

$$
\begin{array}{r}
x \\
x^3 + 2x^2 + x \overline{)\,x^4 + 2x^3 + 6x^2 + 20x + 6\,} \\
\underline{x^4 + 2x^3 + x^2} \qquad\qquad\qquad \\
5x^2 + 20x + 6
\end{array}
$$

       Multiply: $x(x^3 + 2x^2 + x)$.

       Subtract.

Use the result of the long division to rewrite the original expression as

$$\frac{x^4 + 2x^3 + 6x^2 + 20x + 6}{x^3 + 2x^2 + x} = x + \frac{5x^2 + 20x + 6}{x^3 + 2x^2 + x}.$$

The denominator of the remainder factors as

$$x^3 + 2x^2 + x = x(x^2 + 2x + 1) = x(x + 1)^2$$

so include a partial fraction with a constant numerator for each power of x and $(x + 1)$.

$$\frac{5x^2 + 20x + 6}{x(x + 1)^2} = \frac{A}{x} + \frac{B}{x + 1} + \frac{C}{(x + 1)^2}$$

       Write form of decomposition.

Multiply each side by the LCD, $x(x + 1)^2$, to obtain the basic equation.

$$5x^2 + 20x + 6 = A(x + 1)^2 + Bx(x + 1) + Cx$$

       Basic equation

ALGEBRA HELP

To obtain the basic equation, be sure to multiply *each* fraction by the LCD.

Let $x = -1$ to eliminate the A- and B-terms.

$$5(-1)^2 + 20(-1) + 6 = A(-1 + 1)^2 + B(-1)(-1 + 1) + C(-1)$$
$$5 - 20 + 6 = 0 + 0 - C$$
$$9 = C$$

Let $x = 0$ to eliminate the B- and C-terms.

$$5(0)^2 + 20(0) + 6 = A(0 + 1)^2 + B(0)(0 + 1) + C(0)$$
$$6 = A(1) + 0 + 0$$
$$6 = A$$

SKILLS REFRESHER

For a refresher on finding the LCD, watch the video at *LarsonPrecalculus.com*.

At this point, you have exhausted the most convenient values of x, so to find the value of B, use *any other value* of x along with the known values of A and C. So, let $x = 1$, $A = 6$, and $C = 9$.

$$5(1)^2 + 20(1) + 6 = 6(1 + 1)^2 + B(1)(1 + 1) + 9(1)$$
$$31 = 6(4) + 2B + 9$$
$$-2 = 2B$$
$$-1 = B$$

So, the partial fraction decomposition is

$$\frac{x^4 + 2x^3 + 6x^2 + 20x + 6}{x^3 + 2x^2 + x} = x + \frac{6}{x} + \frac{-1}{x + 1} + \frac{9}{(x + 1)^2}.$$

✓ *Checkpoint* ▶ *Audio-video solution in English & Spanish at LarsonPrecalculus.com*

Write the partial fraction decomposition of $\dfrac{x^4 + x^3 + x + 4}{x^3 + x^2}$.

The procedure used to solve for the constants in Examples 1 and 2 works well when the factors of the denominator are linear. When the denominator contains irreducible quadratic factors, a better process is to write the right side of the basic equation in polynomial form, *equate the coefficients* of like terms to form a system of equations, and solve the resulting system for the constants.

EXAMPLE 3 Distinct Linear and Quadratic Factors

Write the partial fraction decomposition of

$$\frac{3x^2 + 4x + 4}{x^3 + 4x}.$$

Solution This expression is proper, so begin by factoring the denominator. The denominator factors as

$$x^3 + 4x = x(x^2 + 4)$$

so when writing the form of the decomposition, include one partial fraction with a constant numerator and one partial fraction with a linear numerator.

$$\frac{3x^2 + 4x + 4}{x^3 + 4x} = \frac{A}{x} + \frac{Bx + C}{x^2 + 4} \qquad \text{Write form of decomposition.}$$

Multiply each side by the LCD, $x(x^2 + 4)$, to obtain the basic equation.

$$3x^2 + 4x + 4 = A(x^2 + 4) + (Bx + C)x \qquad \text{Basic equation}$$

Expand this basic equation and collect like terms.

$$3x^2 + 4x + 4 = Ax^2 + 4A + Bx^2 + Cx$$
$$= (A + B)x^2 + Cx + 4A \qquad \text{Polynomial form}$$

Use the fact that two polynomials are equal if and only if the coefficients of like terms are equal to write a system of linear equations.

$$3x^2 + 4x + 4 = (A + B)x^2 + Cx + 4A \qquad \text{Equate coefficients of like terms.}$$

$$\begin{cases} A + B & = 3 & \text{Equation 1} \\ \qquad C = 4 & \text{Equation 2} \\ 4A \qquad = 4 & \text{Equation 3} \end{cases}$$

From Equation 3 and Equation 2, you have

$$A = 1 \quad \text{and} \quad C = 4.$$

Back-substituting $A = 1$ into Equation 1 yields

$$1 + B = 3 \implies B = 2.$$

So, the partial fraction decomposition is

$$\frac{3x^2 + 4x + 4}{x^3 + 4x} = \frac{1}{x} + \frac{2x + 4}{x^2 + 4}.$$

✓ *Checkpoint* ▶ *Audio-video solution in English & Spanish at LarsonPrecalculus.com*

Write the partial fraction decomposition of

$$\frac{2x^2 - 5}{x^3 + x}.$$

HISTORICAL NOTE

Johann Bernoulli (1667–1748), a Swiss mathematician, introduced the method of partial fractions and was instrumental in the early development of calculus. Bernoulli was a professor at the University of Basel and taught many outstanding students, including the renowned Leonhard Euler.

GO DIGITAL

The next example shows how to find the partial fraction decomposition of a rational expression whose denominator has a *repeated* quadratic factor.

EXAMPLE 4 **Repeated Quadratic Factors**

▶▶▶ *See LarsonPrecalculus.com for an interactive version of this type of example.*

Write the partial fraction decomposition of

$$\frac{8x^3 + 13x}{(x^2 + 2)^2}.$$

Solution Include one partial fraction with a linear numerator for each power of $(x^2 + 2)$.

$$\frac{8x^3 + 13x}{(x^2 + 2)^2} = \frac{Ax + B}{x^2 + 2} + \frac{Cx + D}{(x^2 + 2)^2} \qquad \text{Write form of decomposition.}$$

Multiply each side by the LCD, $(x^2 + 2)^2$, to obtain the basic equation.

$$8x^3 + 13x = (Ax + B)(x^2 + 2) + Cx + D \qquad \text{Basic equation}$$

$$= Ax^3 + 2Ax + Bx^2 + 2B + Cx + D$$

$$= Ax^3 + Bx^2 + (2A + C)x + (2B + D) \qquad \text{Polynomial form}$$

Equate coefficients of like terms on opposite sides of the equation to write a system of linear equations.

$$8x^3 + 0x^2 + 13x + 0 = Ax^3 + Bx^2 + (2A + C)x + (2B + D)$$

$$\begin{cases} A & = 8 & \text{Equation 1} \\ B & = 0 & \text{Equation 2} \\ 2A + \quad C & = 13 & \text{Equation 3} \\ 2B + D & = 0 & \text{Equation 4} \end{cases}$$

Use the values $A = 8$ and $B = 0$ to obtain the values of C and D.

$$2(8) + C = 13 \qquad \text{Substitute 8 for } A \text{ in Equation 3.}$$

$$C = -3$$

$$2(0) + D = 0 \qquad \text{Substitute 0 for } B \text{ in Equation 4.}$$

$$D = 0$$

So, using

$$A = 8, \quad B = 0, \quad C = -3, \quad \text{and} \quad D = 0$$

the partial fraction decomposition is

$$\frac{8x^3 + 13x}{(x^2 + 2)^2} = \frac{8x}{x^2 + 2} + \frac{-3x}{(x^2 + 2)^2}.$$

Check this result by combining the two partial fractions on the right side of the equation, or by using a graphing utility.

✓ **Checkpoint** ▶ *Audio-video solution in English & Spanish at LarsonPrecalculus.com*

Write the partial fraction decomposition of $\dfrac{x^3 + 3x^2 - 2x + 7}{(x^2 + 4)^2}$.

GO DIGITAL

EXAMPLE 5 **Repeated Linear and Quadratic Factors**

Write the partial fraction decomposition of $\dfrac{x+5}{x^2(x^2+1)^2}$.

Solution Include one partial fraction with a constant numerator for each power of x and one partial fraction with a linear numerator for each power of (x^2+1).

$$\frac{x+5}{x^2(x^2+1)^2} = \frac{A}{x} + \frac{B}{x^2} + \frac{Cx+D}{x^2+1} + \frac{Ex+F}{(x^2+1)^2} \qquad \text{Write form of decomposition.}$$

Multiply each side by the LCD, $x^2(x^2+1)^2$, to obtain the basic equation.

$$x+5 = Ax(x^2+1)^2 + B(x^2+1)^2 + (Cx+D)x^2(x^2+1) + (Ex+F)x^2 \qquad \text{Basic equation}$$
$$= (A+C)x^5 + (B+D)x^4 + (2A+C+E)x^3 + (2B+D+F)x^2 + Ax + B$$

Write and solve the system of equations formed by equating coefficients on opposite sides of the equation to show that $A=1$, $B=5$, $C=-1$, $D=-5$, $E=-1$, and $F=-5$, and that the partial fraction decomposition is

$$\frac{x+5}{x^2(x^2+1)^2} = \frac{1}{x} + \frac{5}{x^2} - \frac{x+5}{x^2+1} - \frac{x+5}{(x^2+1)^2}.$$

✓ *Checkpoint* *Audio-video solution in English & Spanish at LarsonPrecalculus.com*

Write the partial fraction decomposition of $\dfrac{4x-8}{x^2(x^2+2)^2}$. ■

Guidelines for Solving the Basic Equation

Linear Factors

1. Substitute the *zeros* of the distinct linear factors into the basic equation.
2. For repeated linear factors, use the coefficients determined in Step 1 to rewrite the basic equation. Then substitute *other* convenient values of x and solve for the remaining coefficients.

Quadratic Factors

1. Expand the basic equation.
2. Collect terms according to powers of x.
3. Equate the coefficients of like terms to obtain a system of equations involving the constants, A, B, C, ...
4. Use the system of linear equations to solve for A, B, C, ...

Keep in mind that for *improper* rational expressions, you must first divide before applying partial fraction decomposition.

Summarize (Section 9.4)

1. Explain what is meant by the partial fraction decomposition of a rational expression *(page 662)*.
2. Explain how to find the partial fraction decomposition of a rational expression *(pages 662–667)*. For examples of finding partial fraction decompositions of rational expressions, see Examples 1–5.

GO DIGITAL

9.4 Exercises

See CalcChat.com for tutorial help and worked-out solutions to odd-numbered exercises.

GO DIGITAL

Vocabulary and Concept Check

In Exercises 1 and 2, fill in the blanks.

1. If the degree of the numerator of a rational expression is greater than or equal to the degree of the denominator, then the fraction is _____.

2. You obtain the _____ _____ by multiplying each side of the partial fraction decomposition form by the least common denominator.

3. The partial fraction decomposition of $2x/[(x - 1)(x + 6)]$ involves the sum of how many linear factors?

4. Explain how to determine the constants A and B in the basic equation $5x - 10 = A(2x - 1) + B(x + 2)$.

Skills and Applications

Writing the Form of the Decomposition In Exercises 5–10, write the form of the partial fraction decomposition of the rational expression. Do not solve for the constants.

5. $\dfrac{3}{x^2 - 2x}$

6. $\dfrac{x - 2}{x^2 + 4x + 3}$

7. $\dfrac{6x + 5}{(x + 2)^4}$

8. $\dfrac{5x^2 + 3}{x^2(x - 4)^2}$

9. $\dfrac{8x}{x^2(x^2 + 3)^2}$

10. $\dfrac{2x - 3}{x^3 + 10x}$

Writing the Partial Fraction Decomposition In Exercises 11–34, write the partial fraction decomposition of the rational expression. Check your result algebraically.

11. $\dfrac{1}{x^2 + x}$

12. $\dfrac{3}{x^2 - 3x}$

13. $\dfrac{3}{x^2 + x - 2}$

14. $\dfrac{x + 1}{x^2 - x - 6}$

15. $\dfrac{1}{x^2 - 1}$

16. $\dfrac{1}{4x^2 - 9}$

17. $\dfrac{x^2 + 12x + 12}{x^3 - 4x}$

18. $\dfrac{x + 2}{x(x^2 - 9)}$

19. $\dfrac{3x}{(x - 3)^2}$

20. $\dfrac{2x - 3}{(x - 1)^2}$

21. $\dfrac{4x^2 + 2x - 1}{x^2(x + 1)}$

22. $\dfrac{6x^2 + 1}{x^2(x - 1)^2}$

23. $\dfrac{2x}{x^3 - 1}$

24. $\dfrac{x^2 + 2x + 3}{x^3 + x}$

25. $\dfrac{x}{x^3 - x^2 - 2x + 2}$

26. $\dfrac{x + 6}{x^3 - 3x^2 - 4x + 12}$

27. $\dfrac{x}{16x^4 - 1}$

28. $\dfrac{3}{x^4 + x}$

29. $\dfrac{x^2 + 5}{(x + 1)(x^2 - 2x + 3)}$

30. $\dfrac{x^2 - 4x + 7}{(x + 1)(x^2 - 2x + 3)}$

31. $\dfrac{2x^2 + x + 8}{(x^2 + 4)^2}$

32. $\dfrac{5x^2 - 2}{(x^2 + 3)^3}$

33. $\dfrac{8x - 12}{x^2(x^2 + 2)^2}$

34. $\dfrac{x + 1}{x^3(x^2 + 1)^2}$

Improper Rational Expression Decomposition In Exercises 35–42, write the partial fraction decomposition of the improper rational expression.

35. $\dfrac{x^2 - x}{x^2 + x + 1}$

36. $\dfrac{x^2 - 4x}{x^2 + x + 6}$

37. $\dfrac{2x^3 - x^2 + x + 5}{x^2 + 3x + 2}$

38. $\dfrac{x^3 + 2x^2 - x + 1}{x^2 + 3x - 4}$

39. $\dfrac{x^4}{(x - 1)^3}$

40. $\dfrac{16x^4}{(2x - 1)^3}$

41. $\dfrac{x^4 + 2x^3 + 4x^2 + 8x + 2}{x^3 + 2x^2 + x}$

42. $\dfrac{2x^4 + 8x^3 + 7x^2 - 7x - 12}{x^3 + 4x^2 + 4x}$

Writing the Partial Fraction Decomposition In Exercises 43–50, write the partial fraction decomposition of the rational expression. Use a graphing utility to check your result.

43. $\dfrac{5 - x}{2x^2 + x - 1}$

44. $\dfrac{4x^2 - 1}{2x(x + 1)^2}$

45. $\dfrac{3x^2 - 7x - 2}{x^3 - x}$

46. $\dfrac{3x + 6}{x^3 + 2x}$

47. $\dfrac{x^2 + x + 2}{(x^2 + 2)^2}$

48. $\dfrac{x^3}{(x + 2)^2(x - 2)^2}$

49. $\dfrac{2x^3 - 4x^2 - 15x + 5}{x^2 - 2x - 8}$

50. $\dfrac{x^3 - x + 3}{x^2 + x - 2}$

51. Environmental Science The predicted cost C (in thousands of dollars) for a company to remove $p\%$ of a chemical from its waste water is given by the model

$$C = \frac{120p}{10{,}000 - p^2}, \quad 0 \le p < 100.$$

Write the partial fraction decomposition for the rational function. Use a graphing utility to verify your result.

52. Thermodynamics

The magnitude of the range R of exhaust temperatures (in degrees Fahrenheit) in an experimental diesel engine is approximated by the model

$$R = \frac{5000(4 - 3x)}{(11 - 7x)(7 - 4x)}, \quad 0 < x \le 1$$

where x is the relative load.

(a) Write the partial fraction decomposition of the equation.

(b) The decomposition in part (a) is the difference of two fractions.
The absolute values of the terms give the expected maximum and minimum temperatures of the exhaust gases for different loads.

$$\text{Ymax} = |1\text{st term}| \qquad \text{Ymin} = |2\text{nd term}|$$

Write the equations for Ymax and Ymin.

(c) Use a graphing utility to graph each equation from part (b) in the same viewing window.

(d) Determine the expected maximum and minimum temperatures for a relative load of 0.5.

Exploring the Concepts

True or False? In Exercises 53 and 54, determine whether the statement is true or false. Justify your answer.

53. When writing the partial fraction decomposition of

$$\frac{x^3 + x - 2}{x^2 - 5x - 14}$$

the first step is to divide the numerator by the denominator.

54. In the partial fraction decomposition of a rational expression, the denominators of each partial fraction always have a lower degree than the denominator of the original expression.

55. Error Analysis Describe the error in writing the basic equation for the partial fraction decomposition of the rational expression.

$$\frac{x^2 + 1}{x(x - 1)} = \frac{A}{x} + \frac{B}{x - 1}$$

$$x^2 + 1 = A(x - 1) + Bx \qquad ✗$$

56. **HOW DO YOU SEE IT?** Identify the graph of the rational function and the graph representing each partial fraction of its partial fraction decomposition. Explain.

(a) $y = \dfrac{x - 12}{x(x - 4)}$ (b) $y = \dfrac{2(4x - 3)}{x^2 - 9}$

$\quad\;\; = \dfrac{3}{x} - \dfrac{2}{x - 4}$ $\quad\;\; = \dfrac{3}{x - 3} + \dfrac{5}{x + 3}$

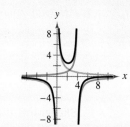 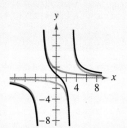

57. Writing Describe two ways of solving for the constants in a partial fraction decomposition.

Review & Refresh ▶ *Video solutions at LarsonPrecalculus.com*

58. Think About It Find values of a, b, and c such that $x \ge 10$ is the solution of the inequality $ax - b \le c$. (There are many correct answers.)

Sketching the Graph of an Equation In Exercises 59–68, sketch the graph of the equation.

59. $x - y = 2$　　　**60.** $x + 2y = 3$

61. $|2x| - y = 1$　　**62.** $|x| + y = 1$

63. $x = -2$　　　　 **64.** $y = 3$

65. $2x - y^2 = 0$　　 **66.** $y = x^2 - 1$

67. $(x + 2)^2 + (y - 2)^2 = 16$

68. $x^2 + (y + 5)^2 = 8$

Checking Solutions In Exercises 69–72, determine whether each value of x is a solution of the inequality.

Inequality	Value of x	
69. $1 \ge x^2 - 1$	(a) -1	(b) $\sqrt{2}$
70. $(x + 2)^2 < 16$	(a) 2	(b) -6
71. $|x - 1| < 2$	(a) 0	(b) 4
72. $|x + 1| > 3$	(a) -2	(b) -3

9.5 Systems of Inequalities

> ❯ Sketch the graphs of inequalities in two variables.
> ❯ Solve systems of inequalities.
> ❯ Use systems of inequalities in two variables to model and solve real-life problems.

Systems of inequalities in two variables can help you model and solve real-life problems. For example, in Exercise 68 on page 678, you will use a system of inequalities to analyze a person's recommended target heart rate during exercise.

The Graph of an Inequality

The statements $3x - 2y < 6$ and $2x^2 + 3y^2 > 6$ are inequalities in two variables. An ordered pair (a, b) is a **solution of an inequality** in x and y when the inequality is true after a and b are substituted for x and y, respectively. The **graph of an inequality** is the collection of all solutions of the inequality. To sketch the graph of an inequality, begin by sketching the graph of the *corresponding equation*. The graph of the equation will usually separate the plane into two or more regions. In each such region, one of the following must be true.

1. *All* points in the region are solutions of the inequality.

2. *No* point in the region is a solution of the inequality.

So, you can determine whether the points in an entire region satisfy the inequality by testing *one* point in the region.

Sketching the Graph of an Inequality in Two Variables

1. Replace the inequality sign by an equal sign and sketch the graph of the equation. Use a dashed curve for $<$ or $>$ and a solid curve for $\leq$ or $\geq$. (A dashed curve means that all points on the curve *are not* solutions of the inequality. A solid curve means that all points on the curve *are* solutions of the inequality.)

2. Test one point in each of the regions formed by the graph in Step 1. If the point satisfies the inequality, then shade the entire region to denote that every point in the region satisfies the inequality.

EXAMPLE 1   Sketching the Graph of an Inequality

❯❯❯ *See LarsonPrecalculus.com for an interactive version of this type of example.*

Sketch the graph of $y \geq x^2 - 1$.

Solution   Begin by graphing the corresponding *equation* $y = x^2 - 1$, which is a parabola. The inequality sign is $\geq$, so use a solid curve to draw the parabola, as shown in Figure 9.17. Next, test a point *above* the parabola, such as $(0, 0)$, and a point *below* the parabola, such as $(0, -2)$.

Test $(0, 0)$: $0 \overset{?}{\geq} 0^2 - 1$ $\quad\Longrightarrow\quad$ $0 \geq -1$ $\qquad$ (0, 0) is a solution.

Test $(0, -2)$: $-2 \overset{?}{\geq} 0^2 - 1$ $\quad\Longrightarrow\quad$ $-2 \ngeq -1$ $\qquad$ (0, -2) is *not* a solution.

So, the points that satisfy the inequality are those lying above the parabola *and* those lying on the parabola. This solution is indicated graphically by the shaded region *and* the solid parabola in Figure 9.17.

Figure 9.17

$y \geq x^2 - 1$ $y = x^2 - 1$

$(0, 0)$

Test point above parabola

Test point below parabola

$(0, -2)$

✓ *Checkpoint* ❯ **Audio-video solution in English & Spanish at LarsonPrecalculus.com**

Sketch the graph of $(x + 2)^2 + (y - 2)^2 < 16$. ■

GO DIGITAL

The inequality in Example 1 is a nonlinear inequality in two variables. Many of the examples in this section involve **linear inequalities** such as $ax + by < c$, where a and b are not both zero. The graph of a linear inequality is a half-plane lying on one side of the line $ax + by = c$.

Sketching Graphs of Linear Inequalities

Sketch the graph of each linear inequality.

a. $x > -2$ **b.** $y \le 3$

Solution

a. The graph of the corresponding equation $x = -2$ is a vertical line. The points that satisfy the inequality $x > -2$ are those lying to the right of (but not on) this line, as shown in Figure 9.18(a).

b. The graph of the corresponding equation $y = 3$ is a horizontal line. The points that satisfy the inequality $y \le 3$ are those lying below this line *and* those lying on the line, as shown in Figure 9.18(b).

> **TECHNOLOGY**
>
> A graphing utility can be used to graph an inequality or a system of inequalities. For example, to graph $y \ge x - 2$, enter $y = x - 2$ and use the *shade* feature of the graphing utility to shade the solution region as shown below. Consult the user's guide for your graphing utility for specific keystrokes.

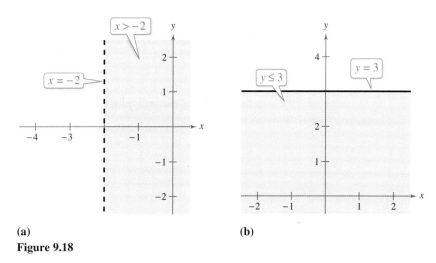

(a)   **(b)**
Figure 9.18

✓ *Checkpoint* ▶ Audio-video solution in English & Spanish at LarsonPrecalculus.com

Sketch the graph of $x \ge 3$.

Sketching the Graph of a Linear Inequality

Sketch the graph of $x - y < 2$.

Solution The graph of the corresponding equation $x - y = 2$ is a line, as shown in Figure 9.19. The origin $(0, 0)$ satisfies the inequality, so the graph consists of the half-plane lying above (but not on) the line. (Check a point below the line. Regardless of which point you choose, you will find that it does not satisfy the inequality.)

✓ *Checkpoint* ▶ Audio-video solution in English & Spanish at LarsonPrecalculus.com

Sketch the graph of $x + y > -2$.

Figure 9.19

GO DIGITAL

To graph a linear inequality, it sometimes helps to write the inequality in slope-intercept form. For example, writing $x - y < 2$ as

$$y > x - 2$$

helps you to see that the solution points lie *above* the line $x - y = 2$, or $y = x - 2$, as shown in Figure 9.19.

Systems of Inequalities

Many practical problems in business, science, and engineering involve systems of linear inequalities. A **solution** of a system of inequalities in x and y is a point (x, y) that satisfies each inequality in the system.

To sketch the graph of a system of inequalities in two variables, first sketch the graph of each individual inequality (on the same coordinate system) and then find the region that is *common* to every graph in the system. This region represents the **solution set** of the system. For a system of *linear inequalities,* it is helpful to find the vertices of the solution region.

EXAMPLE 4 Solving a System of Inequalities

Sketch the graph of the solution set of the system of inequalities. Label the vertices of the region.

$$\begin{cases} x - y < 2 \\ x > -2 \\ y \le 3 \end{cases}$$

Solution The graphs of these inequalities are shown in Figures 9.19, 9.18(a), and 9.18(b), respectively, on the preceding page. The triangular region common to all three graphs can be found by superimposing the graphs on the same coordinate system, as shown in Figure 9.20(a). To find the vertices of the region, solve the three systems of corresponding equations obtained by taking *pairs* of equations representing the boundaries of the individual regions.

Vertex A: $(-2, -4)$	Vertex B: $(5, 3)$	Vertex C: $(-2, 3)$
$\begin{cases} x - y = 2 \\ x = -2 \end{cases}$	$\begin{cases} x - y = 2 \\ y = 3 \end{cases}$	$\begin{cases} x = -2 \\ y = 3 \end{cases}$

ALGEBRA HELP

Using a different colored pencil to shade the solution of each inequality in a system will make identifying the solution of the system of inequalities easier.

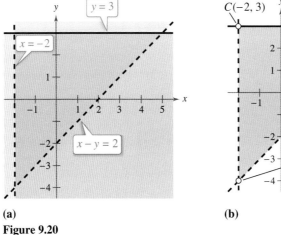

(a)   (b)

Figure 9.20

Note in Figure 9.20(b) that the vertices of the region are represented by open dots. This means that the vertices *are not* solutions of the system of inequalities.

✓ *Checkpoint* ▶ *Audio-video solution in English & Spanish at LarsonPrecalculus.com*

Sketch the graph of the solution set of the system of inequalities. Label the vertices of the region.

$$\begin{cases} x + y \ge 1 \\ -x + y \ge 1 \\ y \le 2 \end{cases}$$

GO DIGITAL

For the triangular region shown in Example 4, each pair of boundary lines intersects at a vertex of the region. With more complicated regions, two boundary lines can sometimes intersect at a point that is not a vertex of the region, as shown below. As you sketch the graph of a solution set, use your sketch along with the inequalities of the system to determine which points of intersection are actually vertices of the region.

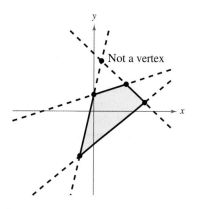

EXAMPLE 5 Solving a System of Inequalities

Sketch the graph of the solution set of the system of inequalities.

$$\begin{cases} x^2 - y \le 1 & \text{Inequality 1} \\ -x + y \le 1 & \text{Inequality 2} \end{cases}$$

Solution The points that satisfy the inequality

$$x^2 - y \le 1 \qquad \text{Inequality 1}$$

are those lying above *and* those lying on the parabola

$$y = x^2 - 1. \qquad \text{Parabola}$$

The points satisfying the inequality

$$-x + y \le 1 \qquad \text{Inequality 2}$$

are those lying below *and* those lying on the line

$$y = x + 1. \qquad \text{Line}$$

To find the points of intersection of the parabola and the line, solve the system of corresponding equations.

$$\begin{cases} x^2 - y = 1 \\ -x + y = 1 \end{cases}$$

Using the method of substitution, you find that the solutions are $(-1, 0)$ and $(2, 3)$. These points are both solutions of the original system, so they are represented by closed dots in the graph of the solution region shown at the right.

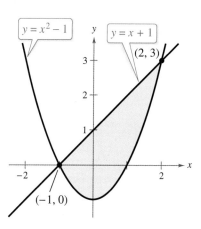

✓ *Checkpoint* ▶ Audio-video solution in English & Spanish at LarsonPrecalculus.com

Sketch the graph of the solution set of the system of inequalities.

$$\begin{cases} x - y^2 > 0 \\ x + y < 2 \end{cases}$$

When solving a system of inequalities, be aware that the system might have no solution *or* its graph might be an unbounded region in the plane. Examples 6 and 7 show these two possibilities.

EXAMPLE 6   A System with No Solution

Sketch the solution set of the system of inequalities.

$$\begin{cases} x + y > 3 \\ x + y < -1 \end{cases}$$

Solution   It should be clear from the way it is written that the system has no solution, because the quantity $(x + y)$ cannot be both less than -1 and greater than 3. The graph of the inequality $x + y > 3$ is the half-plane lying above the line $x + y = 3$, and the graph of the inequality $x + y < -1$ is the half-plane lying below the line $x + y = -1$, as shown below. These two half-planes have no points in common. So, the system of inequalities has no solution.

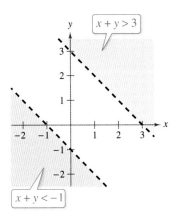

✓ *Checkpoint* ▶ *Audio-video solution in English & Spanish at LarsonPrecalculus.com*

Sketch the solution set of the system of inequalities.

$$\begin{cases} 2x - y < -3 \\ 2x - y > 1 \end{cases}$$

EXAMPLE 7   An Unbounded Solution Set

Sketch the solution set of the system of inequalities.

$$\begin{cases} x + y < 3 \\ x + 2y > 3 \end{cases}$$

Solution   The graph of the inequality $x + y < 3$ is the half-plane that lies below the line $x + y = 3$. The graph of the inequality $x + 2y > 3$ is the half-plane that lies above the line $x + 2y = 3$. The intersection of these two half-planes is an *infinite wedge* that has a vertex at $(3, 0)$, as shown in Figure 9.21. So, the solution set of the system of inequalities is unbounded.

✓ *Checkpoint* ▶ *Audio-video solution in English & Spanish at LarsonPrecalculus.com*

Sketch the solution set of the system of inequalities.

$$\begin{cases} x^2 - y < 0 \\ x - y < -2 \end{cases}$$

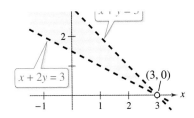

Figure 9.21

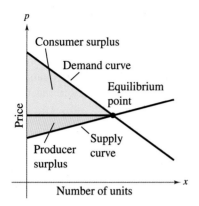

Figure 9.22

Applications

Example 9 in Section 9.2 discussed the equilibrium point for a system of demand and supply equations. The next example discusses two related concepts that economists call *consumer surplus* and *producer surplus*. As shown in Figure 9.22, the **consumer surplus** is the area of the region formed by the demand curve, the horizontal line passing through the equilibrium point, and the *p*-axis. Similarly, the **producer surplus** is the area of the region formed by the supply curve, the horizontal line passing through the equilibrium point, and the *p*-axis. The consumer surplus is a measure of the amount that consumers would have been willing to pay *above* what they actually paid, whereas the producer surplus is a measure of the amount that producers would have been willing to receive *below* what they actually received.

EXAMPLE 8   **Consumer Surplus and Producer Surplus**

The demand and supply equations for a new type of video game console are

$$\begin{cases} p = 180 - 0.00001x & \text{Demand equation} \\ p = 90 + 0.00002x & \text{Supply equation} \end{cases}$$

where p is the price per unit (in dollars) and x is the number of units. Find the consumer surplus and producer surplus for these two equations.

Solution   Begin by finding the equilibrium point (when supply and demand are equal) by solving the equation

$$90 + 0.00002x = 180 - 0.00001x.$$

In Example 9 in Section 9.2, you saw that the solution is $x = 3{,}000{,}000$ units, which corresponds to a price of $p = \$150$. So, the consumer surplus and producer surplus are the areas of the solution sets of the following systems of inequalities.

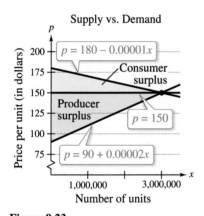

Figure 9.23

Consumer Surplus

$$\begin{cases} p \le 180 - 0.00001x \\ p \ge 150 \\ x \ge 0 \end{cases}$$

Producer Surplus

$$\begin{cases} p \ge 90 + 0.00002x \\ p \le 150 \\ x \ge 0 \end{cases}$$

In other words, the consumer and producer surpluses are the areas of the shaded triangles shown in Figure 9.23.

$$\begin{aligned} \text{Consumer surplus} &= \tfrac{1}{2}(\text{base})(\text{height}) \\ &= \tfrac{1}{2}(3{,}000{,}000)(30) \\ &= \$45{,}000{,}000 \end{aligned}$$

$$\begin{aligned} \text{Producer surplus} &= \tfrac{1}{2}(\text{base})(\text{height}) \\ &= \tfrac{1}{2}(3{,}000{,}000)(60) \\ &= \$90{,}000{,}000 \end{aligned}$$

✓ **Checkpoint** ▶ *Audio-video solution in English & Spanish at LarsonPrecalculus.com*

The demand and supply equations for a flat-screen television are

$$\begin{cases} p = 567 - 0.00002x & \text{Demand equation} \\ p = 492 + 0.00003x & \text{Supply equation} \end{cases}$$

where p is the price per unit (in dollars) and x is the number of units. Find the consumer surplus and producer surplus for these two equations.

GO DIGITAL

EXAMPLE 9   **Nutrition**

The liquid portion of a diet is to provide at least 300 calories, 36 units of vitamin A, and 90 units of vitamin C. A cup of dietary drink X provides 60 calories, 12 units of vitamin A, and 10 units of vitamin C. A cup of dietary drink Y provides 60 calories, 6 units of vitamin A, and 30 units of vitamin C. Write a system of linear inequalities that describes how many cups of each drink must be consumed each day to meet or exceed the minimum daily requirements for calories and vitamins.

Solution   Begin by letting x represent the number of cups of dietary drink X and y represent the number of cups of dietary drink Y. To meet or exceed the minimum daily requirements, the following inequalities must be satisfied.

$$\begin{cases} 60x + 60y \geq 300 & \text{Calories} \\ 12x + 6y \geq 36 & \text{Vitamin A} \\ 10x + 30y \geq 90 & \text{Vitamin C} \\ \quad\quad x \geq 0 \\ \quad\quad y \geq 0 \end{cases}$$

The last two inequalities are included because x and y cannot be negative. The graph of this system of inequalities is shown below. (More is said about this application in Example 6 in the next section.)

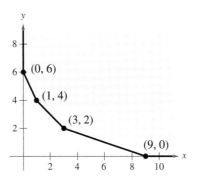

✓ *Checkpoint* ▶ *Audio-video solution in English & Spanish at LarsonPrecalculus.com*

A public aquarium is adding coral nutrients to a large reef tank. A bottle of brand X nutrients contains 8 units of nutrient A, 1 unit of nutrient B, and 2 units of nutrient C. A bottle of brand Y nutrients contains 2 units of nutrient A, 1 unit of nutrient B, and 7 units of nutrient C. The minimum amounts of nutrients A, B, and C that need to be added to the tank are 16 units, 5 units, and 20 units, respectively. Set up a system of linear inequalities that describes how many bottles of each brand must be added to meet or exceed the needs.   ■

Summarize   (Section 9.5)

1. Explain how to sketch the graph of an inequality in two variables *(page 670)*. For examples of sketching the graphs of inequalities in two variables, see Examples 1–3.

2. Explain how to solve a system of inequalities *(page 672)*. For examples of solving systems of inequalities, see Examples 4–7.

3. Describe examples of how to use systems of inequalities in two variables to model and solve real-life problems *(pages 675 and 676, Examples 8 and 9)*.

GO DIGITAL

9.5 Exercises

See CalcChat.com for tutorial help and worked-out solutions to odd-numbered exercises.

GO DIGITAL

Vocabulary and Concept Check

In Exercises 1 and 2, fill in the blanks.

1. The _____ of an inequality is the collection of all solutions of the inequality.

2. A _____ of a system of inequalities in x and y is a point (x, y) that satisfies each inequality in the system.

3. Compare the graphs of an inequality with a $<$ sign and an inequality with a $\leq$ sign.

4. Describe the appearances of a few possible solutions sets of a system of two linear inequalities.

Skills and Applications

Graphing an Inequality In Exercises 5–18, sketch the graph of the inequality.

5. $y < 5 - x^2$

6. $y^2 - x < 0$

7. $x \geq 6$

8. $x < -4$

9. $y > -7$

10. $10 \geq y$

11. $y < 2 - x$

12. $y > 4x - 3$

13. $2y - x \geq 4$

14. $5x + 3y \geq -15$

15. $x^2 + (y - 3)^2 < 4$

16. $(x + 2)^2 + y^2 > 9$

17. $y > -\dfrac{2}{x^2 + 1}$

18. $y \leq \dfrac{3}{x^2 + x + 1}$

 Graphing an Inequality In Exercises 19–26, use a graphing utility to graph the inequality.

19. $y \leq 2 - \frac{1}{5}x$

20. $y > -2.4x + 3.3$

21. $\frac{2}{3}y + 2x^2 - 5 \geq 0$

22. $-\frac{1}{6}x^2 - \frac{2}{7}y < -\frac{1}{3}$

23. $y \geq -\ln(x - 1)$

24. $y < \ln(x + 3) - 1$

25. $y < 2^x$

26. $y \geq 3^{-x} - 2$

Writing an Inequality In Exercises 27–30, write an inequality for the shaded region shown in the figure.

27.

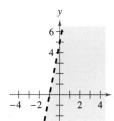

28.

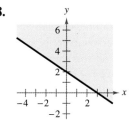

29.

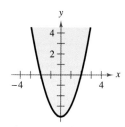

30.

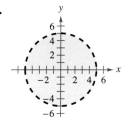

Solving a System of Inequalities In Exercises 31–38, sketch the graph of the solution set of the system of inequalities. Label the vertices of the region.

31. $\begin{cases} x + y \leq 1 \\ -x + y \leq 1 \\ y \geq 0 \end{cases}$

32. $\begin{cases} 3x + 4y < 12 \\ x > 0 \\ y > 0 \end{cases}$

33. $\begin{cases} -3x + 2y < 6 \\ x - 4y > -2 \\ 2x + y < 3 \end{cases}$

34. $\begin{cases} x - 7y > -36 \\ 5x + 2y > 5 \\ 6x - 5y > 6 \end{cases}$

35. $\begin{cases} 2x + y > 2 \\ 6x + 3y < 2 \end{cases}$

36. $\begin{cases} x - 2y \leq -6 \\ 5x - 3y \geq -9 \end{cases}$

37. $\begin{cases} 2x - 3y > 7 \\ 5x + y \leq 9 \end{cases}$

38. $\begin{cases} 4x - 6y > 2 \\ -2x + 3y \geq 5 \end{cases}$

Solving a System of Inequalities In Exercises 39–44, sketch the graph of the solution set of the system of inequalities.

39. $\begin{cases} x^2 + y \leq 7 \\ x \geq -2 \\ y \geq 0 \end{cases}$

40. $\begin{cases} 4x^2 + y \geq 2 \\ x \leq 1 \\ y \leq 1 \end{cases}$

41. $\begin{cases} x - y^2 > 0 \\ x - y > 2 \end{cases}$

42. $\begin{cases} 3x + 4 \geq y^2 \\ x - y \leq 0 \end{cases}$

43. $\begin{cases} x^2 + y^2 \leq 25 \\ 4x - 3y < 0 \end{cases}$

44. $\begin{cases} x^2 - y^2 \geq 7 \\ 3x - 4y < 0 \end{cases}$

 Solving a System of Inequalities In Exercises 45–50, use a graphing utility to graph the solution set of the system of inequalities.

45. $\begin{cases} y \leq \sqrt{3x} + 1 \\ y \geq x^2 + 1 \end{cases}$

46. $\begin{cases} y < 2\sqrt{x} - 1 \\ y \geq x^2 - 1 \end{cases}$

47. $\begin{cases} y < -x^2 + 2x + 3 \\ y > x^2 - 4x + 3 \end{cases}$

48. $\begin{cases} y \geq x^4 - 2x^2 + 1 \\ y + x^2 \leq 1 \end{cases}$

49. $\begin{cases} x^2 y \geq 1 \\ 0 < x \leq 4 \\ y \leq 4 \end{cases}$

50. $\begin{cases} y \leq e^{-x^2/2} \\ y \geq 0 \\ -2 \leq x \leq 2 \end{cases}$

Writing a System of Inequalities In Exercises 51–58, write a system of inequalities that describes the region.

51.
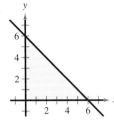

52.
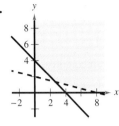

53.
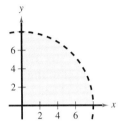

54.
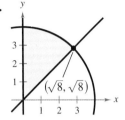

55. Rectangle: vertices at $(4, 3)$, $(9, 3)$, $(9, 9)$, $(4, 9)$

56. Parallelogram: vertices at $(0, 0)$, $(4, 0)$, $(1, 4)$, $(5, 4)$

57. Triangle: vertices at $(0, 0)$, $(6, 0)$, $(1, 5)$

58. Trapezoid: vertices at $(-1, 0)$, $(3, 0)$, $(0, 2)$, $(2, 2)$

Consumer Surplus and Producer Surplus In Exercises 59–62, (a) graph the systems of inequalities representing the consumer surplus and producer surplus for the supply and demand equations and (b) find the consumer surplus and producer surplus.

Demand	Supply
59. $p = 50 - 0.5x$	$p = 0.125x$
60. $p = 100 - 0.05x$	$p = 25 + 0.1x$
61. $p = 140 - 0.00002x$	$p = 80 + 0.00001x$
62. $p = 400 - 0.0002x$	$p = 225 + 0.0005x$

63. Investment Analysis A person plans to invest up to $20,000 in two different interest-bearing accounts. Each account must contain at least $5000. The amount in one account is to be at least twice the amount in the other account. Write and graph a system of inequalities that describes the various amounts that can be deposited in each account.

64. Ticket Sales For a concert event, there are $30 reserved seat tickets and $20 general admission tickets. There are 2000 reserved seats available, and fire regulations limit the number of paid ticket holders to 3000. The promoter must take in at least $75,000 in ticket sales. Write and graph a system of inequalities that describes the different numbers of tickets that can be sold.

65. Production A furniture company produces tables and chairs. Each table requires 1 hour in the assembly center and $1\frac{1}{3}$ hours in the finishing center. Each chair requires $1\frac{1}{2}$ hours in the assembly center and $1\frac{1}{2}$ hours in the finishing center. The assembly center is available 12 hours per day, and the finishing center is available 15 hours per day. Write and graph a system of inequalities that describes all possible production levels.

66. Inventory A store sells two models of laptop computers. The store stocks at least twice as many units of model A as of model B. The costs to the store for the two models are $800 and $1200, respectively. The management does not want more than $20,000 in computer inventory at any one time, and it wants at least four model A laptop computers and two model B laptop computers in inventory at all times. Write and graph a system of inequalities that describes all possible inventory levels.

67. Nutrition A dietician prescribes a special dietary plan using two different foods. Each ounce of food X contains 180 milligrams of calcium, 6 milligrams of iron, and 220 milligrams of magnesium. Each ounce of food Y contains 100 milligrams of calcium, 1 milligram of iron, and 40 milligrams of magnesium. The minimum daily requirements of the diet are 1000 milligrams of calcium, 18 milligrams of iron, and 400 milligrams of magnesium.

(a) Write and graph a system of inequalities that describes the different amounts of food X and food Y that can be prescribed.

(b) Find two solutions of the system and interpret their meanings in the context of the problem.

68. Target Heart Rate

One formula for a person's maximum heart rate is $y = 220 - x$, where x is the person's age in years for $20 \le x \le 70$. The American Heart Association recommends that when a person exercises, the person should strive for a heart rate that is at least 50% of the maximum and at most 85% of the maximum. (*Source: American Heart Association*)

(a) Write and graph a system of inequalities that describes the exercise target heart rate region.

(b) Find two solutions of the system and interpret their meanings in the context of the problem.

69. Shipping A warehouse supervisor has instructions to ship at least 50 bags of gravel that weigh 55 pounds each and at least 40 bags of stone that weigh 70 pounds each. The maximum weight capacity of the truck being used is 7500 pounds.

(a) Write and graph a system that describes the numbers of bags of stone and gravel that can be shipped.

(b) Find two solutions of the system and interpret their meanings in the context of the problem.

70. Athletic Center A college athletic center is constructing an indoor running track with space for exercise equipment inside the track (see figure). The track must be at least 125 meters long, and the exercise space must have an area of at least 500 square meters.

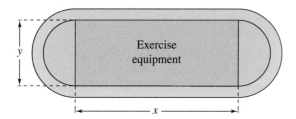

(a) Write and graph a system of inequalities that describes the requirements of the athletic center.

(b) Find two solutions of the system and interpret their meanings in the context of the problem.

Exploring the Concepts

True or False? In Exercises 71–73, determine whether the statement is true or false. Justify your answer.

71. The area of the figure described by the system

$$\begin{cases} x \ge -3 \\ x \le 6 \\ y \le 5 \\ y \ge -6 \end{cases}$$

is 99 square units.

72. The graph shows the solution of the system

$$\begin{cases} y \le 6 \\ -4x - 9y > 6. \\ 3x + y^2 \ge 2 \end{cases}$$

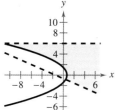

73. The system of inequalities has no solution.

$$\begin{cases} x - 3y < 8 \\ 2x - 6y > -10 \end{cases}$$

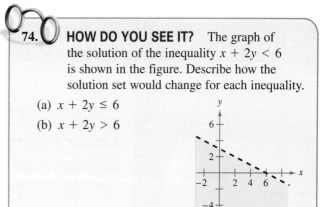

74. **HOW DO YOU SEE IT?** The graph of the solution of the inequality $x + 2y < 6$ is shown in the figure. Describe how the solution set would change for each inequality.

(a) $x + 2y \le 6$

(b) $x + 2y > 6$

75. Writing After graphing the boundary line of the inequality $x + y < 3$, explain how to determine the region that you need to shade.

76. Graphical Reasoning Two concentric circles have radii x and y, where $y > x$. The area between the circles is at least 10 square units.

(a) Write a system of inequalities that describes the constraints on the circles.

(b) Use a graphing utility to graph the system of inequalities in part (a). Graph the line $y = x$ in the same viewing window.

(c) Identify the graph of the line in relation to the boundary of the inequality. Explain its meaning in the context of the problem.

Review & Refresh ▶ *Video solutions at LarsonPrecalculus.com*

Graphing a Linear Equation In Exercises 77–80, find the slope and y-intercept (if possible) of the line. Sketch the line.

77. $y = -3x + 6$

78. $y = \frac{2}{3}x - 5$

79. $8x + 2y = -3$

80. $7x + 4 = 0$

Approximating Relative Minima or Maxima In Exercises 81–84, use a graphing utility to approximate (to two decimal places) any relative minima or maxima of the function.

81. $f(x) = 2x^2 + 5x$

82. $h(x) = -x^3 + 4x^2 - x$

83. $v(x) = x\sqrt{x + 1}$

84. $g(x) = (4 - x)\sqrt{x}$

Using Half-Angle Formulas In Exercises 85–90, use the half-angle formulas to determine the exact values of the sine, cosine, and tangent of the angle.

85. $15°$

86. $-67.5°$

87. $-\dfrac{5\pi}{12}$

88. $\dfrac{5\pi}{8}$

89. $-22° 30'$

90. $247° 30'$

9.6 Linear Programming

Linear programming is often used to make real-life decisions. For example, in Exercise 36 on page 687, you will use linear programming to determine the optimal acreage, yield, and profit for two fruit crops.

> Solve linear programming problems.
> Use linear programming to model and solve real-life problems.

Linear Programming: A Graphical Approach

Many applications in business and economics involve a process called **optimization,** in which you find the minimum or maximum value of a quantity. In this section, you will study an optimization strategy called **linear programming.**

A two-dimensional linear programming problem consists of a linear **objective function** and a system of linear inequalities called **constraints.** The objective function gives the quantity to be maximized (or minimized), and the constraints determine the set of **feasible solutions.** For example, one such problem is to maximize the value of

$$z = ax + by \qquad \text{Objective function}$$

subject to a set of constraints that determines the shaded region shown below. Every point in the shaded region satisfies each constraint, so it is not clear how you should find the point that yields a maximum value of z. Fortunately, it can be shown that when there is an optimal solution, it must occur at one of the vertices. So, *to find the maximum value of z, evaluate z at each of the vertices* and compare the resulting z-values.

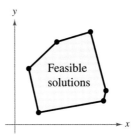

Optimal Solution of a Linear Programming Problem

If a linear programming problem has an optimal solution, then it must occur at a vertex of the set of feasible solutions.

A linear programming problem can include hundreds, and sometimes even thousands, of variables. However, in this section, you will solve linear programming problems that involve only two variables. The guidelines for solving a linear programming problem in two variables are listed below.

Solving a Linear Programming Problem

1. Sketch the region corresponding to the system of constraints. (The points inside or on the boundary of the region are *feasible solutions*.)
2. Find the vertices of the region.
3. Evaluate the objective function at each of the vertices and select the values of the variables that optimize the objective function. For a bounded region, both a minimum and a maximum value will exist. (For an unbounded region, *if* an optimal solution exists, then it will occur at a vertex.)

ALGEBRA HELP

Remember that a vertex of a region can be found using a system of linear equations. The system will consist of the equations of the lines passing through the vertex.

GO DIGITAL

© Marton Roux / EyeEm/Getty Images

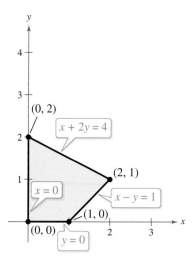

Figure 9.24

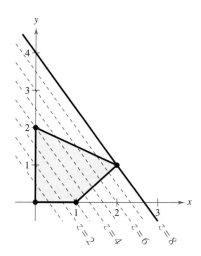

Figure 9.25

EXAMPLE 1 **Solving a Linear Programming Problem**

Find the maximum value of

$$z = 3x + 2y \qquad \text{Objective function}$$

subject to the following constraints.

$$\left.\begin{array}{r} x \geq 0 \\ y \geq 0 \\ x + 2y \leq 4 \\ x - y \leq 1 \end{array}\right\} \quad \text{Constraints}$$

Solution The constraints form the region shown in Figure 9.24. At the four vertices of this region, the objective function has the following values.

At $(0, 0)$: $z = 3(0) + 2(0) = 0$

At $(0, 2)$: $z = 3(0) + 2(2) = 4$

At $(2, 1)$: $z = 3(2) + 2(1) = 8$ $\qquad$ Maximum value of z

At $(1, 0)$: $z = 3(1) + 2(0) = 3$

So, the maximum value of z is 8, and this occurs when $x = 2$ and $y = 1$.

✓ **Checkpoint** ▶ *Audio-video solution in English & Spanish at LarsonPrecalculus.com*

Find the maximum value of

$$z = 4x + 5y \qquad \text{Objective function}$$

subject to the following constraints.

$$\left.\begin{array}{r} x \geq 0 \\ y \geq 0 \\ x + y \leq 6 \end{array}\right\} \quad \text{Constraints}$$

In Example 1, consider some of the *interior* points in the region. You will see that the corresponding values of z are less than 8. Here are some examples.

At $(1, 1)$: $z = 3(1) + 2(1) = 5$

At $\left(\dfrac{1}{2}, \dfrac{3}{2}\right)$: $z = 3\left(\dfrac{1}{2}\right) + 2\left(\dfrac{3}{2}\right) = \dfrac{9}{2}$

At $\left(\dfrac{3}{2}, 1\right)$: $z = 3\left(\dfrac{3}{2}\right) + 2(1) = \dfrac{13}{2}$

To see why the maximum value of the objective function in Example 1 must occur at a vertex, consider writing the objective function in slope-intercept form.

$$y = -\frac{3}{2}x + \frac{z}{2} \qquad \text{Family of lines}$$

Notice that the y-intercept $b = z/2$ varies according to the value of z. This equation represents a family of lines, each of slope $-\frac{3}{2}$. Of these infinitely many lines, you want the one that has the largest z-value while still intersecting the region determined by the constraints. In other words, of all the lines whose slope is $-\frac{3}{2}$, you want the one that has the largest y-intercept *and* intersects the region, as shown in Figure 9.25. Notice from the graph that this line will pass through one point of the region, the vertex $(2, 1)$.

The next example shows that the same basic procedure can be used to solve a problem in which the objective function is to be *minimized*.

EXAMPLE 2 Minimizing an Objective Function

▶▷▷ *See LarsonPrecalculus.com for an interactive version of this type of example.*

Find the minimum value of

$$z = 5x + 7y \qquad \text{Objective function}$$

where $x \geq 0$ and $y \geq 0$, subject to the following constraints.

$$\left. \begin{array}{r} 2x + 3y \geq 6 \\ 3x - y \leq 15 \\ -x + y \leq 4 \\ 2x + 5y \leq 27 \end{array} \right\} \quad \text{Constraints}$$

Solution Figure 9.26 shows the region bounded by the constraints. Evaluate the objective function at each vertex.

At $(0, 2)$: $z = 5(0) + 7(2) = 14$ Minimum value of z

At $(0, 4)$: $z = 5(0) + 7(4) = 28$

At $(1, 5)$: $z = 5(1) + 7(5) = 40$

At $(6, 3)$: $z = 5(6) + 7(3) = 51$

At $(5, 0)$: $z = 5(5) + 7(0) = 25$

At $(3, 0)$: $z = 5(3) + 7(0) = 15$

The minimum value of z is 14, and this occurs when $x = 0$ and $y = 2$.

✓ **Checkpoint** ▶ Audio-video solution in English & Spanish at LarsonPrecalculus.com

Find the minimum value of

$$z = 12x + 8y \qquad \text{Objective function}$$

where $x \geq 0$ and $y \geq 0$, subject to the following constraints.

$$\left. \begin{array}{r} 5x + 6y \leq 420 \\ -x + 6y \leq 240 \\ -2x + y \geq -100 \end{array} \right\} \quad \text{Constraints}$$

EXAMPLE 3 Maximizing an Objective Function

Find the maximum value of

$$z = 5x + 7y \qquad \text{Objective function}$$

where $x \geq 0$ and $y \geq 0$, subject to the constraints given in Example 2.

Solution Using the values of z found at the vertices in Example 2, you can conclude that the maximum value of z is 51, which occurs when $x = 6$ and $y = 3$.

✓ **Checkpoint** ▶ Audio-video solution in English & Spanish at LarsonPrecalculus.com

Find the maximum value of

$$z = 12x + 8y \qquad \text{Objective function}$$

where $x \geq 0$ and $y \geq 0$, subject to the constraints given in the Checkpoint with Example 2. ■

Figure 9.26

HISTORICAL NOTE

George Dantzig (1914–2005) was the first to propose the simplex method for linear programming in 1947. This technique defined the steps needed to find the optimal solution of a complex multivariable problem.

GO DIGITAL

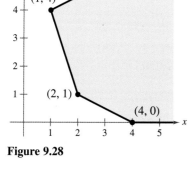

Figure 9.27

It is possible for the maximum (or minimum) value in a linear programming problem to occur at *two* different vertices. For example, at the vertices of the region shown in Figure 9.27, the objective function

$$z = 2x + 2y \qquad \text{Objective function}$$

has the following values.

At $(0, 0)$: $z = 2(0) + 2(0) = 0$

At $(0, 4)$: $z = 2(0) + 2(4) = 8$

At $(2, 4)$: $z = 2(2) + 2(4) = 12$ Maximum value of z

At $(5, 1)$: $z = 2(5) + 2(1) = 12$ Maximum value of z

At $(5, 0)$: $z = 2(5) + 2(0) = 10$

In this case, the objective function has a maximum value not only at the vertices $(2, 4)$ and $(5, 1)$; it also has a maximum value (of 12) at *any point on the line segment connecting these two vertices*. Note that the objective function in slope-intercept form $y = -x + \frac{1}{2}z$ has the same slope as the line through the vertices $(2, 4)$ and $(5, 1)$.

Some linear programming problems have no optimal solution. This can occur when the region determined by the constraints is *unbounded*. Example 4 illustrates such a problem.

EXAMPLE 4   An Unbounded Region

Find the maximum value of

$$z = 4x + 2y \qquad \text{Objective function}$$

where $x \geq 0$ and $y \geq 0$, subject to the following constraints.

$$\left. \begin{array}{r} x + 2y \geq 4 \\ 3x + y \geq 7 \\ -x + 2y \leq 7 \end{array} \right\} \text{Constraints}$$

Solution Figure 9.28 shows the region determined by the constraints. For this unbounded region, there is no maximum value of z. To see this, note that the point $(x, 0)$ lies in the region for all values of $x \geq 4$. Substituting this point into the objective function, you get

$$z = 4(x) + 2(0) = 4x.$$

You can choose values of x to obtain values of z that are as large as you want. So, there is no maximum value of z. However, there *is* a minimum value of z.

At $(1, 4)$: $z = 4(1) + 2(4) = 12$

At $(2, 1)$: $z = 4(2) + 2(1) = 10$ Minimum value of z

At $(4, 0)$: $z = 4(4) + 2(0) = 16$

So, the minimum value of z is 10, and this occurs when $x = 2$ and $y = 1$.

Figure 9.28

 Checkpoint 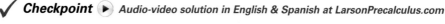 *Audio-video solution in English & Spanish at LarsonPrecalculus.com*

Find the minimum value of

$$z = 3x + 7y \qquad \text{Objective function}$$

where $x \geq 0$ and $y \geq 0$, subject to the following constraints.

$$\left. \begin{array}{r} x + y \geq 8 \\ 3x + 5y \geq 30 \end{array} \right\} \text{Constraints}$$

GO DIGITAL

Applications

Example 5 shows how linear programming can help you find the maximum profit in a business application.

EXAMPLE 5 **Optimal Profit**

A candy manufacturer wants to maximize the combined profit for two types of boxed chocolates. A box of chocolate-covered creams yields a profit of $1.50 per box, and a box of chocolate-covered nuts yields a profit of $2.00 per box. Market tests and available resources indicate the following constraints.

1. The combined production level must not exceed 1200 boxes per month.
2. The demand for chocolate-covered nuts is no more than half the demand for chocolate-covered creams.
3. The production level for chocolate-covered creams must be less than or equal to 600 boxes plus three times the production level for chocolate-covered nuts.

What is the maximum monthly profit? How many boxes of each type are produced per month to yield the maximum profit?

Solution Let x be the number of boxes of chocolate-covered creams and let y be the number of boxes of chocolate-covered nuts. Then, the objective function (for the combined profit) is

$$P = 1.5x + 2y. \qquad \text{Objective function}$$

The three constraints yield the following linear inequalities.

1. $x + y \le 1200 \qquad \Longrightarrow \qquad x + y \le 1200$
2. $y \le \frac{1}{2}x \qquad \Longrightarrow \qquad -x + 2y \le 0$
3. $x \le 600 + 3y \qquad \Longrightarrow \qquad x - 3y \le 600$

Neither x nor y can be negative, so there are two additional constraints of

$$x \ge 0 \quad \text{and} \quad y \ge 0.$$

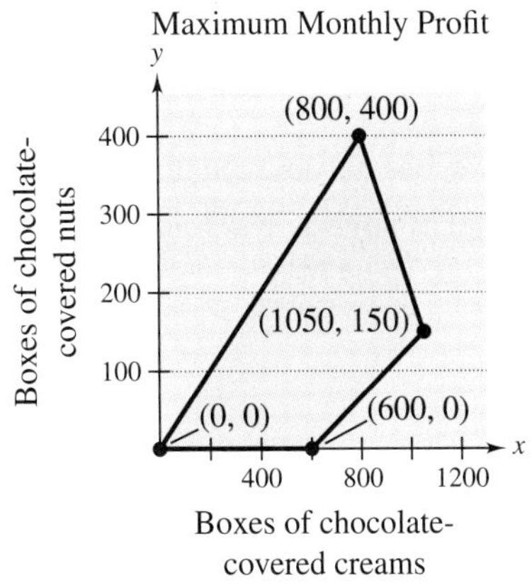

Figure 9.29 shows the region determined by the constraints. To find the maximum monthly profit, evaluate P at the vertices of the region.

At $(0, 0)$: $\qquad P = 1.5(0) \quad + 2(0) \quad = \quad 0$

At $(800, 400)$: $\quad P = 1.5(800) \quad + 2(400) = 2000 \qquad \text{Maximum profit}$

At $(1050, 150)$: $\quad P = 1.5(1050) + 2(150) = 1875$

At $(600, 0)$: $\qquad P = 1.5(600) \quad + 2(0) \quad = \quad 900$

So, the maximum monthly profit is $2000, and it occurs when the monthly production consists of 800 boxes of chocolate-covered creams and 400 boxes of chocolate-covered nuts.

Maximum Monthly Profit

Boxes of chocolate-covered nuts

(800, 400)

(1050, 150)

(0, 0)   (600, 0)

Boxes of chocolate-covered creams

Figure 9.29

✓ *Checkpoint* 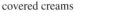 *Audio-video solution in English & Spanish at LarsonPrecalculus.com*

In Example 5, the candy manufacturer improves the production of chocolate-covered creams so that the profit is $2.50 per box. The constraints do not change. What is the maximum monthly profit? How many boxes of each type are produced per month to yield the maximum profit?

Example 6 shows how linear programming can help you find the optimal cost in a real-life application.

> [!NOTE] EXAMPLE 6 **Optimal Cost**

The liquid portion of a daily diet is to provide at least 300 calories, 36 units of vitamin A, and 90 units of vitamin C. Two dietary drinks (drink X and drink Y) will be used to meet these requirements. Information about one cup of each drink is shown below. How many cups of each dietary drink must be consumed each day to satisfy the daily requirements at the minimum possible cost?

	Cost	Calories	Vitamin A	Vitamin C
Drink X	$0.72	60	12 units	10 units
Drink Y	$0.90	60	6 units	30 units

Solution   As in Example 9 in the preceding section, let x be the number of cups of dietary drink X and let y be the number of cups of dietary drink Y.

$$
\left. \begin{array}{rl}
\text{For calories:} & 60x + 60y \geq 300 \\
\text{For vitamin A:} & 12x + 6y \geq 36 \\
\text{For vitamin C:} & 10x + 30y \geq 90 \\
& x \geq 0 \\
& y \geq 0
\end{array} \right\} \text{Constraints}
$$

The cost C is given by

$$C = 0.72x + 0.90y. \qquad \text{Objective function}$$

Figure 9.30 shows the graph of the region corresponding to the constraints. You want to incur as little cost as possible, so you want to determine the *minimum* cost. Evaluate C at each vertex of the region.

At $(0, 6)$: $C = 0.72(0) + 0.90(6) = 5.40$

At $(1, 4)$: $C = 0.72(1) + 0.90(4) = 4.32$

At $(3, 2)$: $C = 0.72(3) + 0.90(2) = 3.96$ Minimum value of C

At $(9, 0)$: $C = 0.72(9) + 0.90(0) = 6.48$

You find that the minimum cost is $3.96 per day, and this occurs when 3 cups of drink X and 2 cups of drink Y are consumed each day.

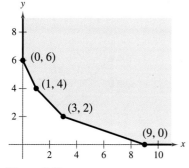

y

$(0, 6)$

$(1, 4)$

$(3, 2)$

$(9, 0)$

x

Figure 9.30

✓ *Checkpoint* *Audio-video solution in English & Spanish at LarsonPrecalculus.com*

A public aquarium is adding coral nutrients to a large reef tank. The minimum amounts of nutrients A, B, and C that need to be added to the tank are 16 units, 5 units, and 20 units, respectively. Information about each bottle of brand X and brand Y additives is shown below. How many bottles of each brand must be added to satisfy the needs of the reef tank at the minimum possible cost?

	Cost	Nutrient A	Nutrient B	Nutrient C
Brand X	$15	8 units	1 unit	2 units
Brand Y	$30	2 units	1 unit	7 units

> [!NOTE] **Summarize**   (Section 9.6)
>
> 1. State the guidelines for solving a linear programming problem *(page 680)*. For examples of solving linear programming problems, see Examples 1–4.
> 2. Describe examples of real-life applications of linear programming *(pages 684 and 685, Examples 5 and 6)*.

GO DIGITAL

9.6 Exercises

See CalcChat.com for tutorial help and worked-out solutions to odd-numbered exercises.

Vocabulary and Concept Check

In Exercises 1 and 2, fill in the blanks.

1. The _____ function of a linear programming problem gives the quantity to be maximized (or minimized).

2. The _____ of a linear programming problem determine the set of _____ _____.

3. Explain how to find the vertices of the region corresponding to the system of constraints in a linear programming problem.

4. When can a linear programming problem have no optimal solution?

Skills and Applications

Solving a Linear Programming Problem In Exercises 5–10, use the graph of the region corresponding to the system of constraints to find the minimum and maximum values of the objective function subject to the constraints. Identify the points where the optimal values occur.

5. Objective function:

   $z = 4x + 3y$

   Constraints:

   $x \geq 0$

   $y \geq 0$

   $x + y \leq 5$

   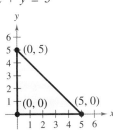

6. Objective function:

   $z = 2x + 8y$

   Constraints:

   $x \geq 0$

   $y \geq 0$

   $2x + y \leq 4$

   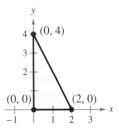

7. Objective function:

   $z = 2x + 5y$

   Constraints:

   $y \geq 0$

   $5x + y \geq 5$

   $x + 3y \leq 15$

   $4x + y \leq 16$

   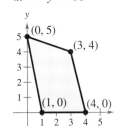

8. Objective function:

   $z = 4x + 5y$

   Constraints:

   $x \geq 0$

   $2x + 3y \geq 6$

   $3x - y \leq 9$

   $x + 4y \leq 16$

   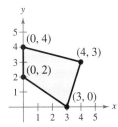

9. Objective function:

   $z = 10x + 7y$

   Constraints:

   $0 \leq x \leq 60$

   $y \leq 45$

   $5x + 6y \leq 420$

   $x + 3y \geq 60$

   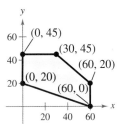

10. Objective function:

    $z = 40x + 45y$

    Constraints:

    $x \geq 0$

    $y \geq 0$

    $8x + 9y \leq 7200$

    $8x + 9y \geq 3600$

Solving a Linear Programming Problem In Exercises 11–14, sketch the region corresponding to the system of constraints. Then find the minimum and maximum values of the objective function (if possible) and the points where they occur, subject to the constraints.

11. Objective function:

    $z = 3x + 2y$

    Constraints:

    $x \geq 0$

    $y \geq 0$

    $3x + 2y \leq 24$

    $4x + y \geq 12$

12. Objective function:

    $z = 5x + \frac{1}{2}y$

    Constraints:

    $x \geq 0$

    $y \geq 0$

    $\frac{1}{2}x + y \leq 8$

    $x + \frac{1}{2}y \geq 4$

13. Objective function:

    $z = 4x + 5y$

    Constraints:

    $x \geq 0$

    $y \geq 0$

    $x + y \geq 8$

    $3x + 5y \geq 30$

14. Objective function:

    $z = 5x + 4y$

    Constraints:

    $x \geq 0$

    $y \geq 0$

    $2x + 2y \geq 10$

    $x + 2y \geq 6$

Solving a Linear Programming Problem In Exercises 15–18, use a graphing utility to graph the region corresponding to the system of constraints. Then find the minimum and maximum values of the objective function and the points where they occur, subject to the constraints.

15. Objective function:

$z = 3x + y$

Constraints:

$x \geq 0$

$y \geq 0$

$x + 4y \leq 60$

$3x + 2y \geq 48$

16. Objective function:

$z = 6x + 3y$

Constraints:

$x \geq 0$

$y \geq 0$

$2x + 3y \leq 60$

$2x + y \leq 28$

$4x + y \leq 48$

17. Objective function:

$z = x$

Constraints:

(See Exercise 15.)

18. Objective function:

$z = y$

Constraints:

(See Exercise 16.)

Finding Minimum and Maximum Values In Exercises 19–22, find the minimum and maximum values of the objective function and the points where they occur, subject to the constraints $x \geq 0$, $y \geq 0$, $x + 4y \leq 20$, $x + y \leq 18$, and $2x + 2y \leq 21$.

19. $z = x + 5y$

20. $z = 2x + 4y$

21. $z = 4x + 5y$

22. $z = 4x + y$

Finding Minimum and Maximum Values In Exercises 23–26, find the minimum and maximum values (if possible) of the objective function and the points where they occur, subject to the constraints $x \geq 0$, $3x + y \geq 15$, $-x + 4y \geq 8$, and $-2x + y \geq -19$.

23. $z = x + 2y$

24. $z = 5x + 3y$

25. $z = x - y$

26. $z = y - x$

Describing an Unusual Characteristic In Exercises 27–34, the linear programming problem has an unusual characteristic. Sketch a graph of the solution region for the problem and describe the unusual characteristic. Find the minimum and maximum values of the objective function (if possible) and the points where they occur.

27. Objective function:

$z = 2.5x + y$

Constraints:

$x \geq 0$

$y \geq 0$

$3x + 5y \leq 15$

$5x + 2y \leq 10$

28. Objective function:

$z = x + y$

Constraints:

$x \geq 0$

$y \geq 0$

$-x + y \leq 1$

$-x + 2y \leq 4$

29. Objective function:

$z = -x + 2y$

Constraints:

$x \geq 0$

$y \geq 0$

$x \leq 10$

$x + y \leq 7$

30. Objective function:

$z = x + y$

Constraints:

$x \geq 0$

$y \geq 0$

$-x + y \leq 0$

$-3x + y \geq 3$

31. Objective function:

$z = 3x + 4y$

Constraints:

$x \geq 0$

$y \geq 0$

$x + y \leq 1$

$2x + y \geq 4$

32. Objective function:

$z = x + 2y$

Constraints:

$x \geq 0$

$y \geq 0$

$x + 2y \leq 4$

$2x + y \leq 4$

33. Objective function:

$z = x + y$

Constraints:

$x \geq 9$

$0 \leq y \leq 7$

$-x + 3y \leq -6$

34. Objective function:

$z = 2x - y$

Constraints:

$0 \leq x \leq 9$

$0 \leq y \leq 11$

$5x + 2y \leq 67$

35. Optimal Revenue An accounting firm has 780 hours of staff time and 272 hours of reviewing time available each week. The firm charges $1600 for an audit and $250 for a tax return. Each audit requires 60 hours of staff time and 16 hours of review time. Each tax return requires 10 hours of staff time and 4 hours of review time. What numbers of audits and tax returns will yield an optimal revenue? What is the optimal revenue?

36. Agriculture

A fruit grower raises crops A and B. The yield is 300 bushels per acre for crop A and 500 bushels per acre for crop B. Research and available resources indicate the following constraints.

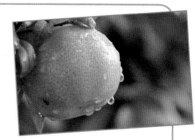

- The fruit grower has 150 acres of land available.
- It takes 1 day to trim the trees on an acre of crop A and 2 days to trim an acre of crop B, and there are 240 days per year available for trimming.
- It takes 0.3 day to pick an acre of crop A and 0.1 day to pick an acre of crop B, and there are 30 days per year available for picking.

(a) What is the optimal acreage for each fruit? What is the optimal yield?

(b) The profit is $185 per acre for crop A and $245 per acre for crop B. What is the optimal profit?

37. Optimal Cost A public aquarium is adding coral nutrients to a large reef tank. The minimum amounts of nutrients A, B, and C that need to be added to the tank are 30 units, 16 units, and 24 units, respectively. Information about each bottle of brand X and brand Y additives is shown below. How many bottles of each brand must be added to satisfy the needs of the reef tank at the minimum possible cost?

	Cost	Nutrient A	Nutrient B	Nutrient C
Brand X	$25	3 units	3 units	7 units
Brand Y	$15	9 units	2 units	2 units

38. Optimal Labor A manufacturer has two factories that produce structural steel, rail steel, and pipe steel. They must produce 32 tons of structural, 26 tons of rail, and 30 tons of pipe steel to fill an order. The table shows the number of employees at each factory and the amounts of steel they produce hourly. How many employee hours should each factory operate to fill the order at the minimum labor?

	Factory X	Factory Y
Employees	120	80
Structural steel	2	5
Rail steel	8	2
Pipe steel	3	3

39. Production A manufacturer produces two models of exercise machines. The times for assembling, finishing, and packaging model X are 3 hours, 3 hours, and 0.8 hour, respectively. The times for model Y are 4 hours, 2.5 hours, and 0.4 hour. The total times available for assembling, finishing, and packaging are 6000 hours, 4200 hours, and 950 hours, respectively. The profits per unit are $300 for model X and $375 for model Y. What is the optimal production level for each model? What is the optimal profit?

40. Investment Portfolio A college has up to $450,000 to invest in two types of investments. Type A pays 6% annually and type B pays 10% annually. At least one-half of the total portfolio is to be allocated to type A investments and at least one-fourth of the portfolio is to be allocated to type B investments. What is the optimal amount that should be invested in each type of investment? What is the optimal return?

Exploring the Concepts

True or False? **In Exercises 41 and 42, determine whether the statement is true or false. Justify your answer.**

41. If an objective function has a maximum value at the vertices $(4, 7)$ and $(8, 3)$, then it also has a maximum value at the points $(4.5, 6.5)$ and $(7.8, 3.2)$.

42. If an objective function has a minimum value at the vertex $(20, 0)$, then it also has a minimum value at $(0, 0)$.

43. Error Analysis Describe the error.

For the objective function $z = x - y$, subject to the constraints $x \geq 0$, $y \geq 0$, and $x + y \geq 1$, the vertices occur at $(0, 0)$, $(1, 0)$, and $(0, 1)$. The maximum value of z is 1, and this occurs when $x = 1$ and $y = 0$.

44. 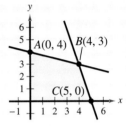 **HOW DO YOU SEE IT?** Using the constraint region shown below, determine which of the following objective functions has (a) a maximum at vertex A, (b) a maximum at vertex B, (c) a maximum at vertex C, and (d) a minimum at vertex C.

(i) $z = 2x + y$

(ii) $z = 2x - y$

(iii) $z = -x + 2y$

45. Think About It A linear programming problem has an objective function $z = 3x + 5y$ and an infinite number of optimal solutions that lie on the line segment connecting two points. What is the slope between the points?

Review & Refresh ▶ *Video solutions at LarsonPrecalculus.com*

46. Think About It Are the following two systems of equations equivalent? Give reasons for your answer.

$$\begin{cases} x + 3y - z = 6 \\ 2x - y + 2z = 1 \\ 3x + 2y - z = 2 \end{cases} \quad \begin{cases} x + 3y - z = 6 \\ -7y + 4z = 1 \\ -7y - 4z = -16 \end{cases}$$

Solving a System of Linear Equations **In Exercises 47–51, solve the system of linear equations and check any solutions algebraically.**

47. $\begin{cases} -x + y + z = -3 \\ 2x - 3y + 2z = -4 \\ x - y - 2z = 7 \end{cases}$

48. $\begin{cases} 3x + y + 3z = 420 \\ x + 3y + 3z = 420 \\ 3x + 3y + z = 420 \end{cases}$

49. $\begin{cases} x - y - z = 0 \\ x + 2y - z = 8 \\ 2x - z = 5 \end{cases}$

50. $\begin{cases} 3x - 2y + 3z = 22 \\ 3y - z = 24 \\ 6x - 7y = -22 \end{cases}$

51. $\begin{cases} -x + 2w = 1 \\ 4y - z - w = 2 \\ y - w = 0 \\ 3x - 2y + 3z = 4 \end{cases}$

Summary and Study Strategies

What Did You Learn?

The list below reviews the skills covered in the chapter and correlates each one to the Review Exercises (see page 691) that practice the skill.

Section 9.1	Review Exercises
■ Use the method of substitution to solve systems of linear equations in two variables *(p. 628)*.	*1–6*
1. *Solve* one of the equations for one variable in terms of the other.	
2. *Substitute* the expression found in Step 1 into the other equation to obtain an equation in one variable.	
3. *Solve* the equation obtained in Step 2.	
4. *Back-substitute* the value obtained in Step 3 into the expression obtained in Step 1 to find the value of the other variable.	
5. *Check* that the solution satisfies *each* of the original equations.	
■ Use the method of substitution to solve systems of nonlinear equations in two variables *(p. 631)*.	*7–10*
■ Use a graphical method to solve systems of equations in two variables *(p. 632)*.	*11–18*
■ Use systems of equations to model and solve real-life problems *(p. 633)*.	*19–22*

Section 9.2	
■ Use the method of elimination to solve systems of linear equations in two variables *(p. 638)*.	*23–28*
1. *Obtain coefficients* for x (or y) that differ only in sign.	
2. *Add* the equations to eliminate one variable.	
3. *Solve* the equation obtained in Step 2.	
4. *Back-substitute* the value obtained in Step 3 into either of the original equations and solve for the other variable.	
5. *Check* that the solution satisfies *each* of the original equations.	
■ Interpret graphically the numbers of solutions of systems of linear equations in two variables *(p. 642)*.	*29–32*
■ Use systems of linear equations in two variables to model and solve real-life problems *(p. 644)*.	*33, 34*

Section 9.3	
■ Use back-substitution to solve linear systems in row-echelon form *(p. 650)*.	*35, 36*
■ Use Gaussian elimination to solve systems of linear equations *(p. 651)*.	*37–42, 45–48*
■ Solve nonsquare systems of linear equations *(p. 655)*.	*43, 44*
■ Use systems of linear equations in three or more variables to model and solve real-life problems *(p. 656)*.	*49–54*

Section 9.4	**Review Exercises**
■ Recognize partial fraction decompositions of rational expressions *(p. 662)*.	*55–58*
■ Find partial fraction decompositions of rational expressions *(p. 663)*.	*59–66*

Section 9.5	
■ Sketch the graphs of inequalities in two variables *(p. 670)*.	*67–72*

   1. Replace the inequality sign by an equal sign and sketch the graph of the equation. Use a dashed curve for < or > and a solid curve for ≤ or ≥ .

   2. Test one point in each of the regions formed by the graph in Step 1. If the point satisfies the inequality, then shade the entire region to denote that every point in the region satisfies the inequality.

■ Solve systems of inequalities *(p. 672)*.	*73–80*
■ Use systems of inequalities in two variables to model and solve real-life problems *(p. 675)*.	*81–86*

Section 9.6	
■ Solve linear programming problems *(p. 680)*.	*87–90*

   1. Sketch the region corresponding to the system of constraints.

   2. Find the vertices of the region.

   3. Evaluate the objective function at each of the vertices and select the values of the variables that optimize the objective function. For a bounded region, both a minimum and a maximum value will exist. (For an unbounded region, *if* an optimal solution exists, then it will occur at a vertex.)

■ Use linear programming to model and solve real-life problems *(p. 684)*.	*91, 92*

Study Strategies

Improving Your Memory   Improve your memory by learning how to focus during class and while studying on your own. Here are some suggestions.

During class

- When you sit down at your desk, get all other issues out of your mind by reviewing your notes from the last class and focusing just on math.

- Repeat in your mind what you are writing in your notes.

- When the math is particularly difficult, ask your instructor for another example.

While completing homework

- Before doing homework, review the concept boxes and examples. Talk through the examples out loud.

- Complete homework as though you were also preparing for a quiz. Review the different types of problems, formulas, rules, and so on.

Between classes

- Review the concept boxes and check your memory using the checkpoint exercises, Summarize feature, Vocabulary and Concept Check exercises, and What Did You Learn? feature.

Preparing for a test

- Review all your notes that pertain to the upcoming test. Review examples of each type of problem that could appear on the test.

Review Exercises

See CalcChat.com for tutorial help and worked-out solutions to odd-numbered exercises.

GO DIGITAL

9.1 **Solving a System by Substitution** **In Exercises 1–10, solve the system by the method of substitution.**

1. $\begin{cases} x + y = 2 \\ x - y = 0 \end{cases}$

2. $\begin{cases} 2x - 3y = 3 \\ x - y = 0 \end{cases}$

3. $\begin{cases} 4x - y - 1 = 0 \\ 8x + y - 17 = 0 \end{cases}$

4. $\begin{cases} 10x + 6y + 14 = 0 \\ x + 9y + 7 = 0 \end{cases}$

5. $\begin{cases} 0.5x + y = 0.75 \\ 1.25x - 4.5y = -2.5 \end{cases}$

6. $\begin{cases} -x + \frac{2}{5}y = \frac{3}{5} \\ -x + \frac{1}{5}y = -\frac{4}{5} \end{cases}$

7. $\begin{cases} x^2 - y^2 = 9 \\ x - y = 1 \end{cases}$

8. $\begin{cases} x^2 + y^2 = 169 \\ 3x + 2y = 39 \end{cases}$

9. $\begin{cases} y = 2x^2 \\ y = x^4 - 2x^2 \end{cases}$

10. $\begin{cases} x = y + 3 \\ x = y^2 + 1 \end{cases}$

Solving a System of Equations Graphically **In Exercises 11–14, solve the system graphically.**

11. $\begin{cases} 2x - y = 10 \\ x + 5y = -6 \end{cases}$

12. $\begin{cases} 8x - 3y = -3 \\ 2x + 5y = 28 \end{cases}$

13. $\begin{cases} y = 2x^2 - 4x + 1 \\ y = x^2 - 4x + 3 \end{cases}$

14. $\begin{cases} y^2 - 2y + x = 0 \\ x + y = 0 \end{cases}$

Solving a System of Equations Graphically **In Exercises 15–18, use a graphing utility to solve the systems of equations graphically. Round your solution(s) to two decimal places, if necessary.**

15. $\begin{cases} y = -2e^{-x} \\ 2e^x + y = 0 \end{cases}$

16. $\begin{cases} x^2 + y^2 = 100 \\ 2x - 3y = -12 \end{cases}$

17. $\begin{cases} y = 2 + \log x \\ y = \frac{3}{4}x + 5 \end{cases}$

18. $\begin{cases} y = \ln(x + 2) - 3 \\ y = 4 - \frac{1}{2}x \end{cases}$

19. **Body Mass Index** Body Mass Index (BMI) is a measure of body fat based on height and weight. The 85th percentile BMI for females, ages 9 to 20, increases more slowly than that for males of the same age range. Models that represent the 85th percentile BMI for males and females, ages 9 to 20, are

$\begin{cases} B = 0.78a + 11.7 & \text{Males} \\ B = 0.68a + 13.5 & \text{Females} \end{cases}$

where B is the BMI (kg/m²) and a represents the age, with $a = 9$ corresponding to 9 years old. Use a graphing utility to determine when the BMI for males exceeds the BMI for females. *(Source: Centers for Disease Control and Prevention)*

20. **Choice of Two Jobs** You receive two sales job offers. One company offers an annual salary of $55,000 plus a year-end bonus of 1.5% of your total sales. The other company offers an annual salary of $52,000 plus a year-end bonus of 2% of your total sales. How much would you have to sell to make the second job offer better?

21. **Geometry** The perimeter of a rectangle is 68 feet and its width is $\frac{8}{9}$ times its length. Use a system of equations to find the dimensions of the rectangle.

22. **Geometry** The perimeter of a rectangle is 40 inches. The area of the rectangle is 96 square inches. Use a system of equations to find the dimensions of the rectangle.

9.2 **Solving a System by Elimination** **In Exercises 23–28, solve the system by the method of elimination and check any solutions algebraically.**

23. $\begin{cases} 2x - y = 2 \\ 6x + 8y = 39 \end{cases}$

24. $\begin{cases} 12x + 42y = -17 \\ 30x - 18y = 19 \end{cases}$

25. $\begin{cases} 3x - 2y = 0 \\ 3x + 2y = 0 \end{cases}$

26. $\begin{cases} 7x + 12y = 63 \\ 2x + 3y = 15 \end{cases}$

27. $\begin{cases} 1.25x - 2y = 3.5 \\ 5x - 8y = 14 \end{cases}$

28. $\begin{cases} 1.5x + 2.5y = 8.5 \\ 6x + 10y = 24 \end{cases}$

Matching a System with Its Graph **In Exercises 29–32, match the system of linear equations with its graph. Describe the number of solutions and state whether the system is consistent or inconsistent. [The graphs are labeled (a), (b), (c), and (d).]**

(a)

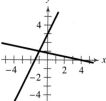

(b)
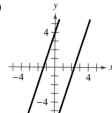

(c)

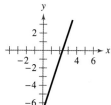

(d)
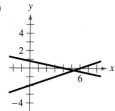

29. $\begin{cases} x + 5y = 4 \\ x - 3y = 6 \end{cases}$

30. $\begin{cases} -3x + y = -7 \\ 9x - 3y = 21 \end{cases}$

31. $\begin{cases} 3x - y = 7 \\ -6x + 2y = 8 \end{cases}$

32. $\begin{cases} 2x - y = -3 \\ x + 5y = 4 \end{cases}$

Finding the Equilibrium Point   In Exercises 33 and 34, find the equilibrium point of the demand and supply equations.

	Demand	Supply
33.	$p = 43 - 0.0002x$	$p = 22 + 0.00001x$
34.	$p = 120 - 0.0001x$	$p = 45 + 0.0002x$

9.3 **Using Back-Substitution in Row-Echelon Form**   In Exercises 35 and 36, use back-substitution to solve the system of linear equations.

35. $\begin{cases} x - 4y + 3z = 3 \\ \quad\ y - z = 1 \\ \quad\quad\quad z = -5 \end{cases}$

36. $\begin{cases} x - 7y + 8z = 85 \\ \quad\ y - 9z = -35 \\ \quad\quad\quad z = 3 \end{cases}$

Solving a System of Linear Equations   In Exercises 37–42, solve the system of linear equations and check any solutions algebraically.

37. $\begin{cases} 4x - 3y - 2z = -65 \\ \quad\quad\ 8y - 7z = -14 \\ 4x \quad\quad\ - 2z = -44 \end{cases}$

38. $\begin{cases} 5x \quad\quad - 7z = 9 \\ \quad\ 3y - 8z = -4 \\ 5x - 3y \quad\quad = 20 \end{cases}$

39. $\begin{cases} \quad\ x + 2y + 6z = 4 \\ -3x + 2y - z = -4 \\ \ 4x \quad\quad + 2z = 16 \end{cases}$

40. $\begin{cases} \quad\ x - 2y + z = -6 \\ \ 2x - 3y \quad\quad = -7 \\ -x + 3y - 3z = 11 \end{cases}$

41. $\begin{cases} 2x \quad\quad + 6z = -9 \\ 3x - 2y + 11z = -16 \\ 3x - y + 7z = -11 \end{cases}$

42. $\begin{cases} x \quad\quad\quad\ + 4w = 1 \\ \quad 3y + z - w = 4 \\ \quad 2y \quad\ - 3w = 2 \\ 4x - y + 2z \quad\ = 5 \end{cases}$

Solving a Nonsquare System   In Exercises 43 and 44, solve the system of linear equations and check any solutions algebraically.

43. $\begin{cases} 5x - 12y + 7z = 16 \\ 3x - 7y + 4z = 9 \end{cases}$

44. $\begin{cases} 2x + 5y - 19z = 34 \\ 3x + 8y - 31z = 54 \end{cases}$

Finding the Equation of a Parabola   In Exercises 45 and 46, find the equation

$$y = ax^2 + bx + c$$

of the parabola that passes through the points. To verify your result, use a graphing utility to plot the points and graph the parabola.

45.   **46.**

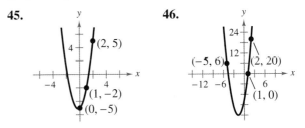

Finding the Equation of a Circle   In Exercises 47 and 48, find the equation

$$x^2 + y^2 + Dx + Ey + F = 0$$

of the circle that passes through the points. To verify your result, use a graphing utility to plot the points and graph the circle.

47.   **48.**

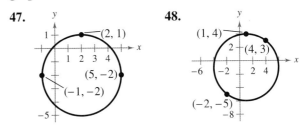

49. Agriculture   A mixture of 6 gallons of chemical A, 8 gallons of chemical B, and 13 gallons of chemical C is required to kill a destructive crop insect. Commercial spray X contains one, two, and two parts, respectively, of these chemicals. Commercial spray Y contains only chemical C. Commercial spray Z contains chemicals A, B, and C in equal amounts. How much of each type of commercial spray gives the desired mixture?

50. Sports   The Old Course at St Andrews Links in St Andrews, Scotland, is one of the oldest golf courses in the world. It is an 18-hole course that consists of par-3 holes, par-4 holes, and par-5 holes. There are seven times as many par-4 holes as par-5 holes, and the sum of the numbers of par-3 and par-5 holes is four. Find the numbers of par-3, par-4, and par-5 holes on the course.   *(Source: St Andrews Links Trust)*

51. Investment   An inheritance of $40,000 is divided among three investments yielding $3500 in interest per year. The interest rates for the three investments are 7%, 9%, and 11% simple interest. Find the amount placed in each investment when the second and third amounts are $3000 and $5000 less than the first, respectively.

52. Investment An amount of $46,000 is divided among three investments yielding $3020 in interest per year. The interest rates for the three investments are 5%, 7%, and 8% simple interest. Find the amount placed in each investment when the second and third amounts are $2000 and $3000 less than the first, respectively.

Modeling Vertical Motion In Exercises 53 and 54, an object moving vertically is at the given heights at the specified times. Find the position equation

$$s = \tfrac{1}{2}at^2 + v_0 t + s_0$$

for the object.

53. At $t = 1$ second, $s = 134$ feet

At $t = 2$ seconds, $s = 86$ feet

At $t = 3$ seconds, $s = 6$ feet

54. At $t = 1$ second, $s = 184$ feet

At $t = 2$ seconds, $s = 116$ feet

At $t = 3$ seconds, $s = 16$ feet

9.4 **Writing the Form of the Decomposition** In Exercises 55–58, write the form of the partial fraction decomposition of the rational expression. Do not solve for the constants.

55. $\dfrac{3}{x^2 + 20x}$

56. $\dfrac{x - 8}{x^2 - 3x - 28}$

57. $\dfrac{3x - 4}{x^3 - 5x^2}$

58. $\dfrac{x - 2}{x(x^2 + 2)^2}$

Writing the Partial Fraction Decomposition In Exercises 59–66, write the partial fraction decomposition of the rational expression. Check your result algebraically.

59. $\dfrac{4 - x}{x^2 + 6x + 8}$

60. $\dfrac{-x}{x^2 + 3x + 2}$

61. $\dfrac{x^2}{x^2 + 2x - 15}$

62. $\dfrac{9}{x^2 - 9}$

63. $\dfrac{x^2 + 2x}{x^3 - x^2 + x - 1}$

64. $\dfrac{4x}{3(x - 1)^2}$

65. $\dfrac{3x^2 + 4x}{(x^2 + 1)^2}$

66. $\dfrac{4x^2}{(x - 1)(x^2 + 1)}$

9.5 **Graphing an Inequality** In Exercises 67–72, sketch the graph of the inequality.

67. $y \geq 5$

68. $x < -3$

69. $y \leq 5 - 2x$

70. $3y - x \geq 7$

71. $(x - 1)^2 + (y - 3)^2 < 16$

72. $x^2 + (y + 5)^2 > 1$

Solving a System of Inequalities In Exercises 73–76, sketch the graph of the solution set of the system of inequalities. Label the vertices of the region.

73. $\begin{cases} x + 2y \leq 2 \\ -x + 2y \leq 2 \\ \qquad y \geq 0 \end{cases}$

74. $\begin{cases} 2x + 3y < 6 \\ x \qquad > 0 \\ \qquad y > 0 \end{cases}$

75. $\begin{cases} 2x - y < -1 \\ -3x + 2y > 4 \\ \qquad y > 0 \end{cases}$

76. $\begin{cases} 3x - 2y > -4 \\ 6x - y < 5 \\ \qquad y < 1 \end{cases}$

Solving a System of Inequalities In Exercises 77–80, sketch the graph of the solution set of the system of inequalities.

77. $\begin{cases} y < x + 1 \\ y > x^2 - 1 \end{cases}$

78. $\begin{cases} y \leq 6 - 2x - x^2 \\ y \geq x + 6 \end{cases}$

79. $\begin{cases} x^2 + y^2 > 4 \\ x^2 + y^2 \leq 9 \end{cases}$

80. $\begin{cases} x^2 + y^2 \leq 169 \\ x + y \leq 7 \end{cases}$

81. Geometry Write a system of inequalities to describe the region of a rectangle with vertices at $(3, 1)$, $(7, 1)$, $(7, 10)$, and $(3, 10)$.

82. Geometry Write a system of inequalities that describes the triangular region with vertices $(0, 5)$, $(5, 0)$, and $(0, 0)$.

Consumer Surplus and Producer Surplus In Exercises 83 and 84, (a) graph the systems of inequalities representing the consumer surplus and producer surplus for the supply and demand equations and (b) find the consumer surplus and producer surplus.

Demand	Supply
83. $p = 160 - 0.0001x$	$p = 70 + 0.0002x$
84. $p = 130 - 0.0002x$	$p = 30 + 0.0003x$

85. Inventory Costs A warehouse operator has 24,000 square feet of floor space in which to store two products. Each unit of product I requires 20 square feet of floor space and costs $12 per day to store. Each unit of product II requires 30 square feet of floor space and costs $8 per day to store. The total storage cost per day cannot exceed $12,400. Write and graph a system that describes all possible inventory levels.

86. Nutrition A dietician prescribes a special dietary plan using two different foods. Each ounce of food X contains 200 milligrams of calcium, 3 milligrams of iron, and 100 milligrams of magnesium. Each ounce of food Y contains 150 milligrams of calcium, 2 milligrams of iron, and 80 milligrams of magnesium. The minimum daily requirements of the diet are 700 milligrams of calcium, 7 milligrams of iron, and 80 milligrams of magnesium.

(a) Write and graph a system of inequalities that describes the different amounts of food X and food Y that can be prescribed.

(b) Find two solutions to the system and interpret their meanings in the context of the problem.

9.6 Solving a Linear Programming Problem In Exercises 87–90, sketch the region corresponding to the system of constraints. Then find the minimum and maximum values of the objective function (if possible) and the points where they occur, subject to the constraints.

87. Objective function:
$z = 3x + 4y$
Constraints:
$x \geq 0$
$y \geq 0$
$2x + 5y \leq 50$
$4x + y \leq 28$

88. Objective function:
$z = 10x + 7y$
Constraints:
$x \geq 0$
$y \geq 0$
$2x + y \geq 100$
$x + y \geq 75$

89. Objective function:
$z = 1.75x + 2.25y$
Constraints:
$x \geq 0$
$y \geq 0$
$2x + y \geq 25$
$3x + 2y \geq 45$

90. Objective function:
$z = 50x + 70y$
Constraints:
$x \geq 0$
$y \geq 0$
$x + 2y \leq 1500$
$5x + 2y \leq 3500$

91. Optimal Revenue A student is working part time as a hairdresser to pay college expenses. The student may work no more than 24 hours per week. Haircuts cost $25 and require an average of 20 minutes, and permanents cost $70 and require an average of 1 hour and 10 minutes. How many haircuts and/or permanents will yield an optimal revenue? What is the optimal revenue?

92. Production A manufacturer produces two models of bicycles. The table shows the times (in hours) required for assembling, painting, and packaging each model.

Process	Hours, Model A	Hours, Model B
Assembling	2	2.5
Painting	4	1
Packaging	1	0.75

The total times available for assembling, painting, and packaging are 4000 hours, 4800 hours, and 1500 hours, respectively. The profits per unit are $45 for model A and $50 for model B. What is the optimal production level for each model? What is the optimal profit?

Exploring the Concepts

True or False? In Exercises 93 and 94, determine whether the statement is true or false. Justify your answer.

93. The system
$$\begin{cases} y \leq 2 \\ y \leq -2 \\ y \leq 4x - 10 \\ y \leq -4x + 26 \end{cases}$$
represents a region in the shape of an isosceles trapezoid.

94. For the rational expression $\dfrac{2x + 3}{x^2(x + 2)^2}$, the partial fraction decomposition is of the form $\dfrac{Ax + B}{x^2} + \dfrac{Cx + D}{(x + 2)^2}$.

Writing a System of Linear Equations In Exercises 95–98, write a system of linear equations that has the ordered pair as a solution. (There are many correct answers.)

95. $(-8, 10)$　　**96.** $(5, -4)$

97. $\left(\frac{4}{3}, 3\right)$　　**98.** $\left(-2, \frac{11}{5}\right)$

Writing a System of Linear Equations In Exercises 99–102, write a system of linear equations that has the ordered triple as a solution. (There are many correct answers.)

99. $(4, -1, 3)$　　**100.** $(-3, 5, 6)$

101. $\left(5, \frac{3}{2}, 2\right)$　　**102.** $\left(-\frac{1}{2}, -2, -\frac{3}{4}\right)$

103. Writing Explain what is meant by an inconsistent system of linear equations.

104. Graphical Reasoning How can you tell graphically that a system of linear equations in two variables has no solution? Give an example.

Chapter Test

See CalcChat.com for tutorial help and worked-out solutions to odd-numbered exercises.

Take this test as you would take a test in class. When you are finished, check your work against the answers given in the back of the book.

In Exercises 1–3, solve the system of equations by the method of substitution. *(Section 9.1)*

1. $\begin{cases} x + y = -9 \\ 5x - 8y = 20 \end{cases}$

2. $\begin{cases} y = x - 1 \\ y = (x - 1)^3 \end{cases}$

3. $\begin{cases} 2x - y^2 = 0 \\ x - y = 4 \end{cases}$

In Exercises 4–6, solve the system of equations graphically. *(Section 9.1)*

4. $\begin{cases} 3x - 6y = 0 \\ 2x + 5y = 18 \end{cases}$

5. $\begin{cases} y = 9 - x^2 \\ y = x + 3 \end{cases}$

6. $\begin{cases} y - \ln x = 4 \\ 7x - 2y - 5 = -6 \end{cases}$

In Exercises 7 and 8, solve the system of equations by the method of elimination. *(Section 9.2)*

7. $\begin{cases} 3x + 4y = -26 \\ 7x - 5y = 11 \end{cases}$

8. $\begin{cases} 1.4x - y = 17 \\ 0.8x + 6y = -10 \end{cases}$

In Exercises 9 and 10, solve the system of linear equations and check any solutions algebraically. *(Section 9.3)*

9. $\begin{cases} x - 2y + 3z = 11 \\ 2x \quad - z = 3 \\ \quad 3y + z = -8 \end{cases}$

10. $\begin{cases} 3x + 2y + z = 17 \\ -x + y + z = 4 \\ x - y - z = 3 \end{cases}$

In Exercises 11–14, write the partial fraction decomposition of the rational expression. Check your result algebraically. *(Section 9.4)*

11. $\dfrac{2x + 5}{x^2 - x - 2}$

12. $\dfrac{3x^2 - 2x + 4}{x^2(2 - x)}$

13. $\dfrac{x^4 + 5}{x^3 - x}$

14. $\dfrac{x^2 - 4}{x^3 + 2x}$

In Exercises 15–17, sketch the graph of the solution set of the system of inequalities. *(Section 9.5)*

15. $\begin{cases} 2x + y \le 4 \\ 2x - y \ge 0 \\ x \ge 0 \end{cases}$

16. $\begin{cases} y < -x^2 + x + 4 \\ y > 4x \end{cases}$

17. $\begin{cases} x^2 + y^2 \le 36 \\ x \ge 2 \\ y \ge -4 \end{cases}$

18. Find the minimum and maximum values of the objective function $z = 20x + 12y$ and the points where they occur, subject to the following constraints. *(Section 9.6)*

$$\left. \begin{array}{r} x \ge 0 \\ y \ge 0 \\ x + 4y \le 32 \\ 3x + 2y \le 36 \end{array} \right\} \text{Constraints}$$

19. A total of $50,000 is invested in two funds that pay 4% and 5.5% simple interest. The yearly interest is $2390. How much is invested at each rate? *(Section 9.2)*

20. Find the equation of the parabola $y = ax^2 + bx + c$ that passes through the points $(0, 6)$, $(-2, 2)$, and $\left(3, \frac{9}{2}\right)$. *(Section 9.3)*

21. A manufacturer produces two models of television stands. The table at the left shows the times (in hours) required for assembling, staining, and packaging the two models. The total times available for assembling, staining, and packaging are 3750 hours, 8950 hours, and 2650 hours, respectively. The profits per unit are $30 for model I and $40 for model II. What is the optimal inventory level for each model? What is the optimal profit? *(Section 9.6)*

	Model I	Model II
Assembling	0.5	0.75
Staining	2.0	1.5
Packaging	0.5	0.5

Table for 21

Proofs in Mathematics

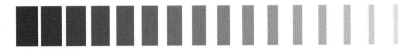

An **indirect proof** can be useful in proving statements of the form "p implies q." Recall that the conditional statement $p \rightarrow q$ is false only when p is true and q is false. To prove a conditional statement indirectly, assume that p is true and q is false. If this assumption leads to an impossibility, then you have proved that the conditional statement is true. An indirect proof is also called a **proof by contradiction.**

For instance, you can use an indirect proof to prove the conditional statement "if a is a positive integer and a^2 is divisible by 2, then a is divisible by 2." Let p represent "a is a positive integer and a^2 is divisible by 2" and let q represent "a is divisible by 2." Then assume that p is true and q is false. This means that a is not divisible by 2. If so, then a is odd and can be written as $a = 2n + 1$, where n is an integer. Next, find a^2.

$a = 2n + 1$	Definition of an odd integer
$a^2 = 4n^2 + 4n + 1$	Square each side.
$a^2 = 2(2n^2 + 2n) + 1$	Distributive Property

So, by the definition of an odd integer, a^2 is odd. This contradicts the assumption, and you can conclude that a is divisible by 2.

EXAMPLE Using an Indirect Proof

Use an indirect proof to prove that $\sqrt{2}$ is an irrational number.

Solution Begin by assuming that $\sqrt{2}$ is *not* an irrational number. Then $\sqrt{2}$ can be written as the quotient of two integers a and b ($b \neq 0$) that have no common factors.

$\sqrt{2} = \dfrac{a}{b}$	Assume that $\sqrt{2}$ is a rational number.
$2 = \dfrac{a^2}{b^2}$	Square each side.
$2b^2 = a^2$	Multiply each side by b^2.

This implies that 2 is a factor of a^2. So, 2 is also a factor of a, and a can be written as $2c$, where c is an integer.

$2b^2 = (2c)^2$	Substitute $2c$ for a.
$2b^2 = 4c^2$	Simplify.
$b^2 = 2c^2$	Divide each side by 2.

This implies that 2 is a factor of b^2 and also a factor of b. So, 2 is a factor of both a and b. This contradicts the assumption that a and b have no common factors. So, you can conclude that $\sqrt{2}$ is an irrational number.

P.S. Problem Solving

See CalcChat.com for tutorial help and worked-out solutions to odd-numbered exercises.

GO DIGITAL

1. Geometry A theorem from geometry states that if a triangle is inscribed in a circle such that one side of the triangle is a diameter of the circle, then the triangle is a right triangle. Show that this theorem is true for the circle

$$x^2 + y^2 = 100$$

and the triangle formed by the lines

$$y = 0$$

$$y = \tfrac{1}{2}x + 5$$

and

$$y = -2x + 20.$$

2. Finding Values of Constants Find values of k_1 and k_2 such that the system of equations has an infinite number of solutions.

$$\begin{cases} 3x - 5y = 8 \\ 2x + k_1 y = k_2 \end{cases}$$

3. Finding Conditions on Constants Under what condition(s) will the system of equations in x and y have exactly one solution?

$$\begin{cases} ax + by = e \\ cx + dy = f \end{cases}$$

4. Finding Values of Constants Find values of a, b, and c (if possible) such that the system of linear equations has (a) a unique solution, (b) no solution, and (c) an infinite number of solutions.

$$\begin{cases} x + y \quad\;\; = 2 \\ \quad\;\; y + z = 2 \\ x \quad\;\; + z = 2 \\ ax + by + cz = 0 \end{cases}$$

5. Graphical Analysis Graph the lines determined by each system of linear equations. Then use Gaussian elimination to solve each system. At each step of the elimination process, graph the corresponding lines. How do the graphs at the different steps compare?

(a) $\begin{cases} x - 4y = -3 \\ 5x - 6y = 13 \end{cases}$

(b) $\begin{cases} 2x - 3y = \quad 7 \\ -4x + 6y = -14 \end{cases}$

6. Maximum Numbers of Solutions A system of two equations in two variables has a finite number of solutions. Determine the maximum number of solutions of the system satisfying each condition.

(a) Both equations are linear.

(b) One equation is linear and the other is quadratic.

(c) Both equations are quadratic.

7. Vietnam Veterans Memorial The Vietnam Veterans Memorial (or "The Wall") in Washington, D.C., was designed by Maya Ying Lin when she was a student at Yale University. This monument has two vertical, triangular sections of black granite with a common side (see figure). The bottom of each section is level with the ground. The tops of the two sections can be approximately modeled by the equations

$$-2x + 50y = 505$$

and

$$2x + 50y = 505$$

when the x-axis is superimposed at the base of the wall. Each unit in the coordinate system represents 1 foot. How high is the memorial at the point where the two sections meet? How long is each section?

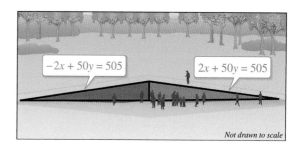

$-2x + 50y = 505$ $\quad$ $2x + 50y = 505$

Not drawn to scale

8. Finding Atomic Weights Weights of atoms and molecules are measured in atomic mass units (u). A molecule of C_2H_6 (ethane) is made up of two carbon atoms and six hydrogen atoms and weighs 30.069 u. A molecule of C_3H_8 (propane) is made up of three carbon atoms and eight hydrogen atoms and weighs 44.096 u. Find the weights of a carbon atom and a hydrogen atom.

9. Media Player Connector Cables Connecting a digital media player to a television set requires a cable with special connectors at both ends. You buy a six-foot cable for $15.50 and a three-foot cable for $10.25. Assuming that the cost of a cable is the sum of the cost of the two connectors and the cost of the cable itself, what is the cost of a four-foot cable?

10. Distance A hotel 35 miles from an airport runs a shuttle bus service to and from the airport. The 9:00 A.M. bus leaves for the airport traveling at 30 miles per hour. The 9:15 A.M. bus leaves for the airport traveling at 40 miles per hour.

(a) Write a system of linear equations that represents distance as a function of time for the buses.

(b) Graph and solve the system.

(c) How far from the airport will the 9:15 A.M. bus catch up to the 9:00 A.M. bus?

11. Systems with Rational Expressions Solve each system of equations by letting $X = 1/x$, $Y = 1/y$, and $Z = 1/z$.

(a) $\begin{cases} \dfrac{12}{x} - \dfrac{12}{y} = 7 \\[2mm] \dfrac{3}{x} + \dfrac{4}{y} = 0 \end{cases}$

(b) $\begin{cases} \dfrac{2}{x} + \dfrac{1}{y} - \dfrac{3}{z} = 4 \\[2mm] \dfrac{4}{x} \phantom{+ \dfrac{1}{y}} + \dfrac{2}{z} = 10 \\[2mm] -\dfrac{2}{x} + \dfrac{3}{y} - \dfrac{13}{z} = -8 \end{cases}$

12. Finding Values of Constants For what values of a, b, and c does the linear system have $(-1, 2, -3)$ as its only solution?

$\begin{cases} x + 2y - 3z = a & \text{Equation 1} \\ -x - y + z = b & \text{Equation 2} \\ 2x + 3y - 2z = c & \text{Equation 3} \end{cases}$

13. System of Linear Equations The following system has one solution: $x = 1$, $y = -1$, and $z = 2$.

$\begin{cases} 4x - 2y + 5z = 16 & \text{Equation 1} \\ x + y = 0 & \text{Equation 2} \\ -x - 3y + 2z = 6 & \text{Equation 3} \end{cases}$

Solve each system of two equations that consists of (a) Equation 1 and Equation 2, (b) Equation 1 and Equation 3, and (c) Equation 2 and Equation 3. (d) How many solutions does each of these systems have?

14. System of Linear Equations Solve the system of linear equations algebraically.

$\begin{cases} x_1 - x_2 + 2x_3 + 2x_4 + 6x_5 = 6 \\ 3x_1 - 2x_2 + 4x_3 + 4x_4 + 12x_5 = 14 \\ -x_2 - x_3 - x_4 - 3x_5 = -3 \\ 2x_1 - 2x_2 + 4x_3 + 5x_4 + 15x_5 = 10 \\ 2x_1 - 2x_2 + 4x_3 + 4x_4 + 13x_5 = 13 \end{cases}$

15. Biology Each day, an average adult moose can process about 32 kilograms of terrestrial vegetation (twigs and leaves) and aquatic vegetation. From this food, it needs to obtain about 1.9 grams of sodium and 11,000 calories of energy. Aquatic vegetation has about 0.15 gram of sodium per kilogram and about 193 calories of energy per kilogram, whereas terrestrial vegetation has minimal sodium and about four times as much energy as aquatic vegetation. Write and graph a system of inequalities that describes the amounts t and a of terrestrial and aquatic vegetation, respectively, for the daily diet of an average adult moose. *(Source: Biology by Numbers)*

16. Height and Weight For a healthy person who is 4 feet 10 inches tall, the recommended minimum weight is about 91 pounds and increases by about 3.6 pounds for each additional inch of height. The recommended maximum weight is about 115 pounds and increases by about 4.5 pounds for each additional inch of height. *(Source: National Institutes of Health)*

(a) Let x be the number of inches by which a person's height exceeds 4 feet 10 inches ($x \geq 0$) and let y be the person's weight (in pounds). Write a system of inequalities that describes the recommended values of x and y for a healthy person.

(b) Use a graphing utility to graph the system of inequalities from part (a).

(c) What is the recommended weight range for a healthy person who is 6 feet tall?

17. Cholesterol Cholesterol in human blood is necessary, but too much can lead to health problems. There are three main types of cholesterol: HDL (high-density lipoproteins), LDL (low-density lipoproteins), and VLDL (very low-density lipoproteins). HDL is considered "good" cholesterol; LDL and VLDL are considered "bad" cholesterol.

A standard fasting cholesterol blood test measures total cholesterol, HDL cholesterol, and triglycerides. These numbers are used to estimate LDL and VLDL, which are difficult to measure directly. Your doctor recommends that your combined LDL/VLDL cholesterol level be less than 130 milligrams per deciliter, your HDL cholesterol level be at least 60 milligrams per deciliter, and your total cholesterol level be no more than 200 milligrams per deciliter.

(a) Write a system of linear inequalities for the recommended cholesterol levels. Let x represent the HDL cholesterol level, and let y represent the combined LDL/VLDL cholesterol level.

(b) Graph the system of inequalities from part (a). Label any vertices of the solution region.

(c) Is the following set of cholesterol levels within the recommendations? Explain.

LDL/VLDL: 120 milligrams per deciliter

HDL: 90 milligrams per deciliter

Total: 210 milligrams per deciliter

(d) Give an example of cholesterol levels in which the LDL/VLDL cholesterol level is too high but the HDL cholesterol level is acceptable.

(e) Another recommendation is that the ratio of total cholesterol to HDL cholesterol be less than 4 (that is, less than 4 to 1). Identify a point in the solution region from part (b) that meets this recommendation, and explain why it meets the recommendation.

10 Matrices and Determinants

GO DIGITAL

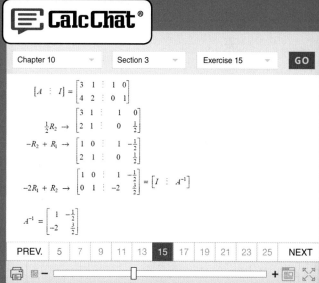

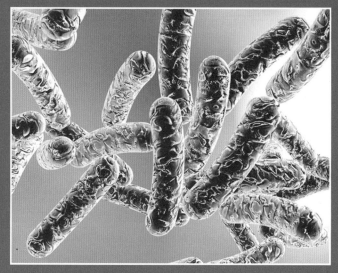

10.1 Waterborne Disease *(Exercise 89, p. 712)*

10.3 Circuit Analysis *(Exercises 55–58, p. 735)*

10.1 Matrices and Systems of Equations

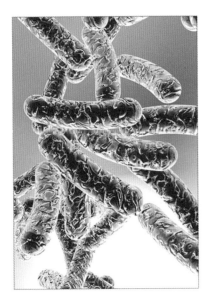

Matrices can help you solve real-life problems that are represented by systems of equations. For example, in Exercise 89 on page 712, you will use a matrix to find a model for the numbers of new cases of a waterborne disease in a small city.

- ❯ Write matrices and determine their dimensions.
- ❯ Perform elementary row operations on matrices.
- ❯ Use matrices and Gaussian elimination to solve systems of linear equations.
- ❯ Use matrices and Gauss-Jordan elimination to solve systems of linear equations.

Matrices

In this section, you will study a streamlined technique for solving systems of linear equations. This technique involves the use of a rectangular array of numbers called a **matrix.** The plural of matrix is *matrices.*

Definition of Matrix

If m and n are positive integers, then an $m \times n$ (read "m by n") matrix is a rectangular array

$$
\begin{array}{c}
\quad\quad \text{Column 1} \quad \text{Column 2} \quad \text{Column 3} \quad \ldots \quad \text{Column } n \\
\begin{array}{c}
\text{Row 1} \\
\text{Row 2} \\
\text{Row 3} \\
\vdots \\
\text{Row } m
\end{array}
\begin{bmatrix}
a_{11} & a_{12} & a_{13} & \cdots & a_{1n} \\
a_{21} & a_{22} & a_{23} & \cdots & a_{2n} \\
a_{31} & a_{32} & a_{33} & \cdots & a_{3n} \\
\vdots & \vdots & \vdots & & \vdots \\
a_{m1} & a_{m2} & a_{m3} & \cdots & a_{mn}
\end{bmatrix}
\end{array}
$$

in which each **entry** a_{ij} of the matrix is a number. An $m \times n$ matrix has m rows and n columns.

The entry in the ith row and jth column of a matrix is denoted by the *double subscript* notation a_{ij}. For example, a_{23} refers to the entry in the second row, third column. A matrix having m rows and n columns is said to be of **dimension** $m \times n$. If $m = n$, then the matrix is **square** of dimension $m \times m$ (or $n \times n$). For a square matrix, the entries $a_{11}, a_{22}, a_{33}, \ldots$ are the **main diagonal** entries. A matrix with only one row is called a **row matrix,** and a matrix with only one column is called a **column matrix.**

ALGEBRA HELP

A matrix having m rows and n columns is also said to be of **size** $m \times n$.

EXAMPLE 1 Dimensions of Matrices

Determine the dimension of each matrix.

a. $\begin{bmatrix} 2 \end{bmatrix}$ **b.** $\begin{bmatrix} 1 & -3 & 0 & \frac{1}{2} \end{bmatrix}$ **c.** $\begin{bmatrix} 0 & 0 \\ 0 & 0 \end{bmatrix}$ **d.** $\begin{bmatrix} 5 & 0 \\ 2 & -2 \\ -7 & 4 \end{bmatrix}$

Solution

a. This matrix has *one* row and *one* column. The dimension of the matrix is 1×1.

b. This matrix has *one* row and *four* columns. The dimension of the matrix is 1×4.

c. This matrix has *two* rows and *two* columns. The dimension of the matrix is 2×2.

d. This matrix has *three* rows and *two* columns. The dimension of the matrix is 3×2.

✓ **Checkpoint** ▶ Audio-video solution in English & Spanish at LarsonPrecalculus.com

Determine the dimension of the matrix $\begin{bmatrix} 14 & 7 & 10 \\ -2 & -3 & -8 \end{bmatrix}$. ■

GO DIGITAL

A matrix derived from a system of linear equations (each written in standard form with the constant term on the right) is the **augmented matrix** of the system. Moreover, the matrix derived from the coefficients of the system (but not including the constant terms) is the **coefficient matrix** of the system.

$$\text{System: } \begin{cases} x - 4y + 3z = 5 \\ -x + 3y - z = -3 \\ 2x \qquad - 4z = 6 \end{cases}$$

$$\text{Augmented matrix: } \begin{bmatrix} 1 & -4 & 3 & \vdots & 5 \\ -1 & 3 & -1 & \vdots & -3 \\ 2 & 0 & -4 & \vdots & 6 \end{bmatrix} \qquad \text{Coefficient matrix: } \begin{bmatrix} 1 & -4 & 3 \\ -1 & 3 & -1 \\ 2 & 0 & -4 \end{bmatrix}$$

Note the use of 0 for the coefficient of the missing y-variable in the third equation, and also note the fourth column of constant terms in the augmented matrix.

When forming either the coefficient matrix or the augmented matrix of a system, you should begin by vertically aligning the variables in the equations and using zeros for the coefficients of the missing variables.

> **ALGEBRA HELP**
>
> The vertical dots in an augmented matrix separate the coefficients of the linear system from the constant terms.

EXAMPLE 2 Writing an Augmented Matrix

Write the augmented matrix for the system of linear equations.

$$\begin{cases} x + 3y - w = 9 \\ -y + 4z + 2w = -2 \\ x - 5z - 6w = 0 \\ 2x + 4y - 3z = 4 \end{cases}$$

What is the dimension of the augmented matrix?

Solution

Begin by rewriting the linear system and aligning the variables.

$$\begin{cases} x + 3y \qquad - w = 9 \\ -y + 4z + 2w = -2 \\ x \qquad - 5z - 6w = 0 \\ 2x + 4y - 3z \qquad = 4 \end{cases}$$

Next, use the coefficients and constant terms as the matrix entries. Include zeros for the coefficients of the missing variables.

$$\begin{matrix} R_1 \\ R_2 \\ R_3 \\ R_4 \end{matrix} \begin{bmatrix} 1 & 3 & 0 & -1 & \vdots & 9 \\ 0 & -1 & 4 & 2 & \vdots & -2 \\ 1 & 0 & -5 & -6 & \vdots & 0 \\ 2 & 4 & -3 & 0 & \vdots & 4 \end{bmatrix}$$

The augmented matrix has four rows and five columns, so it is a 4×5 matrix. The notation R_n is used to designate each row in the matrix. For example, Row 1 is represented by R_1.

✓ **Checkpoint**  *Audio-video solution in English & Spanish at LarsonPrecalculus.com*

Write the augmented matrix for the system of linear equations. What is the dimension of the augmented matrix?

$$\begin{cases} x + y + z = 2 \\ 2x - y + 3z = -1 \\ -x + 2y - z = 4 \end{cases}$$

Elementary Row Operations

In Section 9.3, you studied three operations that can be used on a system of linear equations to produce an equivalent system.

1. Interchange two equations.

2. Multiply an equation by a nonzero constant.

3. Add a multiple of an equation to another equation.

In matrix terminology, these three operations correspond to **elementary row operations.** An elementary row operation on an augmented matrix of a given system of linear equations produces a new augmented matrix corresponding to a new (but equivalent) system of linear equations. Two matrices are **row-equivalent** when one can be obtained from the other by a sequence of elementary row operations.

Elementary Row Operations

Operation	Notation
1. Interchange two rows.	$R_a \leftrightarrow R_b$
2. Multiply a row by a nonzero constant.	$cR_a \quad (c \neq 0)$
3. Add a multiple of a row to another row.	$cR_a + R_b$

EXAMPLE 3 Elementary Row Operations

a. Interchange the first and second rows of the original matrix.

Original Matrix
$$\begin{bmatrix} 0 & 1 & 3 & 4 \\ -1 & 2 & 0 & 3 \\ 2 & -3 & 4 & 1 \end{bmatrix}$$

New Row-Equivalent Matrix
$$\begin{matrix} R_2 \\ R_1 \end{matrix} \begin{bmatrix} -1 & 2 & 0 & 3 \\ 0 & 1 & 3 & 4 \\ 2 & -3 & 4 & 1 \end{bmatrix}$$

b. Multiply the first row of the original matrix by $\frac{1}{2}$.

Original Matrix
$$\begin{bmatrix} 2 & -4 & 6 & -2 \\ 1 & 3 & -3 & 0 \\ 5 & -2 & 1 & 2 \end{bmatrix}$$

New Row-Equivalent Matrix
$$\frac{1}{2}R_1 \rightarrow \begin{bmatrix} 1 & -2 & 3 & -1 \\ 1 & 3 & -3 & 0 \\ 5 & -2 & 1 & 2 \end{bmatrix}$$

c. Add -2 times the first row of the original matrix to the third row.

Original Matrix
$$\begin{bmatrix} 1 & 2 & -4 & 3 \\ 0 & 3 & -2 & -1 \\ 2 & 1 & 5 & -2 \end{bmatrix}$$

New Row-Equivalent Matrix
$$-2R_1 + R_3 \rightarrow \begin{bmatrix} 1 & 2 & -4 & 3 \\ 0 & 3 & -2 & -1 \\ 0 & -3 & 13 & -8 \end{bmatrix}$$

Note that the elementary row operation is written beside the row that is *changed.*

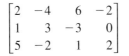 **Checkpoint** ▶ *Audio-video solution in English & Spanish at LarsonPrecalculus.com*

Identify the elementary row operation performed to obtain the new row-equivalent matrix.

Original Matrix
$$\begin{bmatrix} 1 & 0 & 2 \\ 3 & 1 & 7 \\ 2 & -6 & 14 \end{bmatrix}$$

New Row-Equivalent Matrix
$$\begin{bmatrix} 1 & 0 & 2 \\ 0 & 1 & 1 \\ 2 & -6 & 14 \end{bmatrix}$$

Gaussian Elimination with Back-Substitution

In Example 3 in Section 9.3, you used Gaussian elimination with back-substitution to solve a system of linear equations. The next example demonstrates the matrix version of Gaussian elimination. The two methods are essentially the same. The basic difference is that with matrices you do not need to keep writing the variables.

EXAMPLE 4   Comparing Linear Systems and Matrix Operations

Linear System

$$\begin{cases} x - 2y + 3z = 9 \\ -x + 3y \quad\quad = -4 \\ 2x - 5y + 5z = 17 \end{cases}$$

Associated Augmented Matrix

$$\begin{bmatrix} 1 & -2 & 3 & \vdots & 9 \\ -1 & 3 & 0 & \vdots & -4 \\ 2 & -5 & 5 & \vdots & 17 \end{bmatrix}$$

Add the first equation to the second equation.

$$\begin{cases} x - 2y + 3z = 9 \\ y + 3z = 5 \\ 2x - 5y + 5z = 17 \end{cases}$$

Add the first row to the second row: $R_1 + R_2$.

$$R_1 + R_2 \rightarrow \begin{bmatrix} 1 & -2 & 3 & \vdots & 9 \\ 0 & 1 & 3 & \vdots & 5 \\ 2 & -5 & 5 & \vdots & 17 \end{bmatrix}$$

Add -2 times the first equation to the third equation.

$$\begin{cases} x - 2y + 3z = 9 \\ y + 3z = 5 \\ -y - z = -1 \end{cases}$$

Add -2 times the first row to the third row: $-2R_1 + R_3$.

$$-2R_1 + R_3 \rightarrow \begin{bmatrix} 1 & -2 & 3 & \vdots & 9 \\ 0 & 1 & 3 & \vdots & 5 \\ 0 & -1 & -1 & \vdots & -1 \end{bmatrix}$$

Add the second equation to the third equation.

$$\begin{cases} x - 2y + 3z = 9 \\ y + 3z = 5 \\ 2z = 4 \end{cases}$$

Add the second row to the third row: $R_2 + R_3$.

$$R_2 + R_3 \rightarrow \begin{bmatrix} 1 & -2 & 3 & \vdots & 9 \\ 0 & 1 & 3 & \vdots & 5 \\ 0 & 0 & 2 & \vdots & 4 \end{bmatrix}$$

Multiply the third equation by $\frac{1}{2}$.

$$\begin{cases} x - 2y + 3z = 9 \\ y + 3z = 5 \\ z = 2 \end{cases}$$

Multiply the third row by $\frac{1}{2}$: $\frac{1}{2}R_3$.

$$\frac{1}{2}R_3 \rightarrow \begin{bmatrix} 1 & -2 & 3 & \vdots & 9 \\ 0 & 1 & 3 & \vdots & 5 \\ 0 & 0 & 1 & \vdots & 2 \end{bmatrix}$$

ALGEBRA HELP

Remember that you should check a solution by substituting the values of x, y, and z into each equation of the original system. For example, check the solution to Example 4 as shown below.

Equation 1:
$1 - 2(-1) + 3(2) = 9$ ✓

Equation 2:
$-1 + 3(-1) = -4$ ✓

Equation 3:
$2(1) - 5(-1) + 5(2) = 17$ ✓

At this point, use back-substitution to find x and y.

$y + 3(2) = 5$	Substitute 2 for z.
$y = -1$	Solve for y.
$x - 2(-1) + 3(2) = 9$	Substitute -1 for y and 2 for z.
$x = 1$	Solve for x.

The solution is $(1, -1, 2)$.

 **Checkpoint** ▶ *Audio-video solution in English & Spanish at LarsonPrecalculus.com*

Compare solving the linear system below to solving it using its associated augmented matrix.

$$\begin{cases} 2x + y - z = -3 \\ 4x - 2y + 2z = -2 \\ -6x + 5y + 4z = 10 \end{cases}$$

GO DIGITAL

The last matrix in Example 4 is in *row-echelon form.* The term *echelon* refers to the stair-step pattern formed by the nonzero entries of the matrix. The row-echelon form and *reduced row-echelon form* of matrices are described below.

Row-Echelon Form and Reduced Row-Echelon Form

A matrix in **row-echelon form** has the following properties.

1. Any rows consisting entirely of zeros occur at the bottom of the matrix.
2. For each row that does not consist entirely of zeros, the first nonzero entry is 1, called a **leading 1**.
3. For two successive (nonzero) rows, the leading 1 in the higher row is farther to the left than the leading 1 in the lower row.

A matrix in *row-echelon form* is in **reduced row-echelon form** when every column that has a leading 1 has zeros in every position above and below its leading 1.

It is worth noting that the row-echelon form of a matrix is not unique. That is, two different sequences of elementary row operations may yield different row-echelon forms. The *reduced* row-echelon form of a matrix, however, is unique.

EXAMPLE 5 **Row-Echelon Form**

Determine whether each matrix is in row-echelon form. If it is, determine whether it is in reduced row-echelon form.

a. $\begin{bmatrix} 1 & 2 & -1 & 4 \\ 0 & 1 & 0 & 3 \\ 0 & 0 & 1 & -2 \end{bmatrix}$
b. $\begin{bmatrix} 1 & 2 & -1 & 2 \\ 0 & 0 & 0 & 0 \\ 0 & 1 & 2 & -4 \end{bmatrix}$

c. $\begin{bmatrix} 1 & -5 & 2 & -1 & 3 \\ 0 & 0 & 1 & 3 & -2 \\ 0 & 0 & 0 & 1 & 4 \\ 0 & 0 & 0 & 0 & 1 \end{bmatrix}$
d. $\begin{bmatrix} 1 & 0 & 0 & -1 \\ 0 & 1 & 0 & 2 \\ 0 & 0 & 1 & 3 \\ 0 & 0 & 0 & 0 \end{bmatrix}$

e. $\begin{bmatrix} 1 & 2 & -3 & 4 \\ 0 & 2 & 1 & -1 \\ 0 & 0 & 1 & -3 \end{bmatrix}$
f. $\begin{bmatrix} 0 & 1 & 0 & 5 \\ 0 & 0 & 1 & 3 \\ 0 & 0 & 0 & 0 \end{bmatrix}$

Solution The matrices in (a), (c), (d), and (f) are in row-echelon form. The matrices in (d) and (f) are in *reduced* row-echelon form because every column that has a leading 1 has zeros in every position above and below its leading 1. The matrix in (b) is not in row-echelon form because a row of all zeros occurs above a row that is not all zeros. The matrix in (e) is not in row-echelon form because the first nonzero entry in Row 2 is not a leading 1.

✓ *Checkpoint* 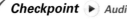 Audio-video solution in English & Spanish at LarsonPrecalculus.com

Determine whether the matrix is in row-echelon form. If it is, determine whether it is in reduced row-echelon form.

$\begin{bmatrix} 1 & 0 & -2 & 4 \\ 0 & 1 & 11 & 3 \\ 0 & 0 & 0 & 0 \end{bmatrix}$

Every matrix is row-equivalent to a matrix in row-echelon form. For instance, in Example 5, you can change the matrix in part (e) to row-echelon form by multiplying its second row by $\frac{1}{2}$.

Gaussian elimination with back-substitution works well for solving systems of linear equations by hand or with a computer. For this algorithm, the order in which the elementary row operations are performed is important. You should operate from left to right by columns, using elementary row operations to obtain zeros in all entries directly below the leading 1's.

EXAMPLE 6　Gaussian Elimination with Back-Substitution

Solve the system

$$\begin{cases} \quad\quad y + z - 2w = -3 \\ x + 2y - z \quad\quad\quad = 2 \\ 2x + 4y + z - 3w = -2 \\ x - 4y - 7z - w = -19 \end{cases}$$

Solution

$$\begin{bmatrix} 0 & 1 & 1 & -2 & \vdots & -3 \\ 1 & 2 & -1 & 0 & \vdots & 2 \\ 2 & 4 & 1 & -3 & \vdots & -2 \\ 1 & -4 & -7 & -1 & \vdots & -19 \end{bmatrix}$$
Write augmented matrix.

$$\begin{matrix} R_2 \\ R_1 \end{matrix} \begin{bmatrix} 1 & 2 & -1 & 0 & \vdots & 2 \\ 0 & 1 & 1 & -2 & \vdots & -3 \\ 2 & 4 & 1 & -3 & \vdots & -2 \\ 1 & -4 & -7 & -1 & \vdots & -19 \end{bmatrix}$$
Interchange R_1 and R_2 so first column has leading 1 in upper left corner.

$$\begin{matrix} \\ \\ -2R_1 + R_3 \rightarrow \\ -R_1 + R_4 \rightarrow \end{matrix} \begin{bmatrix} 1 & 2 & -1 & 0 & \vdots & 2 \\ 0 & 1 & 1 & -2 & \vdots & -3 \\ 0 & 0 & 3 & -3 & \vdots & -6 \\ 0 & -6 & -6 & -1 & \vdots & -21 \end{bmatrix}$$
Perform operations on R_3 and R_4 so first column has zeros below its leading 1.

$$\begin{matrix} \\ \\ \\ 6R_2 + R_4 \rightarrow \end{matrix} \begin{bmatrix} 1 & 2 & -1 & 0 & \vdots & 2 \\ 0 & 1 & 1 & -2 & \vdots & -3 \\ 0 & 0 & 3 & -3 & \vdots & -6 \\ 0 & 0 & 0 & -13 & \vdots & -39 \end{bmatrix}$$
Perform operations on R_4 so second column has zeros below its leading 1.

$$\begin{matrix} \\ \\ \frac{1}{3}R_3 \rightarrow \\ -\frac{1}{13}R_4 \rightarrow \end{matrix} \begin{bmatrix} 1 & 2 & -1 & 0 & \vdots & 2 \\ 0 & 1 & 1 & -2 & \vdots & -3 \\ 0 & 0 & 1 & -1 & \vdots & -2 \\ 0 & 0 & 0 & 1 & \vdots & 3 \end{bmatrix}$$
Perform operations on R_3 and R_4 so third and fourth columns have leading 1's.

The matrix is now in row-echelon form, and the corresponding system is

$$\begin{cases} x + 2y - z \quad\quad = 2 \\ \quad y + z - 2w = -3 \\ \quad\quad z - w = -2 \\ \quad\quad\quad w = 3 \end{cases}$$

Using back-substitution, the solution is $(-1, 2, 1, 3)$.

✓ Checkpoint ▶ *Audio-video solution in English & Spanish at LarsonPrecalculus.com*

Solve the system

$$\begin{cases} -3x + 5y + 3z = -19 \\ 3x + 4y + 4z = 8. \\ 4x - 8y - 6z = 26 \end{cases}$$

GO DIGITAL

▶▶▶▶

ALGEBRA HELP

Note that the order of the variables in the system of equations is x, y, z, and w. The coordinates of the solution are given in this order.

The steps below summarize the procedure used in Example 6.

> **Gaussian Elimination with Back-Substitution**
>
> 1. Write the augmented matrix of the system of linear equations.
> 2. Use elementary row operations to rewrite the augmented matrix in row-echelon form.
> 3. Write the system of linear equations corresponding to the matrix in row-echelon form and use back-substitution to find the solution.

When solving a system of linear equations, remember that it is possible for the system to have no solution. If, in the elimination process, you obtain a row of all zeros except for the last entry, then the system has no solution, or is *inconsistent*.

EXAMPLE 7   **A System with No Solution**

Solve the system $\begin{cases} x - y + 2z = 4 \\ x \qquad + z = 6 \\ 2x - 3y + 5z = 4 \\ 3x + 2y - z = 1 \end{cases}$.

Solution

$$\begin{bmatrix} 1 & -1 & 2 & \vdots & 4 \\ 1 & 0 & 1 & \vdots & 6 \\ 2 & -3 & 5 & \vdots & 4 \\ 3 & 2 & -1 & \vdots & 1 \end{bmatrix}$$ Write augmented matrix.

$$\begin{matrix} \\ -R_1 + R_2 \rightarrow \\ -2R_1 + R_3 \rightarrow \\ -3R_1 + R_4 \rightarrow \end{matrix}\begin{bmatrix} 1 & -1 & 2 & \vdots & 4 \\ 0 & 1 & -1 & \vdots & 2 \\ 0 & -1 & 1 & \vdots & -4 \\ 0 & 5 & -7 & \vdots & -11 \end{bmatrix}$$ Perform row operations.

$$\begin{matrix} \\ \\ R_2 + R_3 \rightarrow \\ \\ \end{matrix}\begin{bmatrix} 1 & -1 & 2 & \vdots & 4 \\ 0 & 1 & -1 & \vdots & 2 \\ 0 & 0 & 0 & \vdots & -2 \\ 0 & 5 & -7 & \vdots & -11 \end{bmatrix}$$ Perform row operations.

Note that the third row of this matrix consists entirely of zeros except for the last entry. This means that the original system of linear equations is inconsistent. You can see why this is true by converting back to a system of linear equations.

$$\begin{cases} x - y + 2z = 4 \\ \quad y - z = 2 \\ \qquad\quad 0 = -2 \leftarrow \text{Not possible} \\ \quad 5y - 7z = -11 \end{cases}$$

The third equation is not possible, so the system has no solution.

✓ **Checkpoint** ▶ *Audio-video solution in English & Spanish at LarsonPrecalculus.com*

Solve the system $\begin{cases} x + y + z = 1 \\ x + 2y + 2z = 2 \\ x - y - z = 1 \end{cases}$.

GO DIGITAL

Gauss-Jordan Elimination

With Gaussian elimination, elementary row operations are applied to a matrix to obtain a (row-equivalent) row-echelon form of the matrix. A second method of elimination, called **Gauss-Jordan elimination,** after Carl Friedrich Gauss and Wilhelm Jordan (1842–1899), continues the reduction process until the *reduced* row-echelon form is obtained. This procedure is demonstrated in Example 8.

EXAMPLE 8   Gauss-Jordan Elimination

 See LarsonPrecalculus.com for an interactive version of this type of example.

Use Gauss-Jordan elimination to solve the system $\begin{cases} x - 2y + 3z = \ \ 9 \\ -x + 3y \qquad = -4. \\ 2x - 5y + 5z = \ 17 \end{cases}$

Solution   In Example 4, Gaussian elimination was used to obtain the row-echelon form of the linear system above.

$$\begin{bmatrix} 1 & -2 & 3 & \vdots & 9 \\ 0 & 1 & 3 & \vdots & 5 \\ 0 & 0 & 1 & \vdots & 2 \end{bmatrix}$$

Now, rather than using back-substitution, apply elementary row operations until you obtain zeros above each of the leading 1's.

$$\begin{matrix} 2R_2 + R_1 \rightarrow \end{matrix} \begin{bmatrix} 1 & 0 & 9 & \vdots & 19 \\ 0 & 1 & 3 & \vdots & 5 \\ 0 & 0 & 1 & \vdots & 2 \end{bmatrix}$$

Perform operations on R_1 so second column has a zero above its leading 1.

$$\begin{matrix} -9R_3 + R_1 \rightarrow \\ -3R_3 + R_2 \rightarrow \end{matrix} \begin{bmatrix} 1 & 0 & 0 & \vdots & 1 \\ 0 & 1 & 0 & \vdots & -1 \\ 0 & 0 & 1 & \vdots & 2 \end{bmatrix}$$

Perform operations on R_1 and R_2 so third column has zeros above its leading 1.

The matrix is now in reduced row-echelon form. Converting back to a system of linear equations, you have

$$\begin{cases} x = \ \ 1 \\ y = -1. \\ z = \ \ 2 \end{cases}$$

So, the solution is $(1, -1, 2)$.

✓ *Checkpoint* ▶ Audio-video solution in English & Spanish at LarsonPrecalculus.com

Use Gauss-Jordan elimination to solve the system $\begin{cases} -3x + 7y + 2z = \ \ 1 \\ -5x + 3y - 5z = -8. \\ \ \ 2x - 2y - 3z = \ 15 \end{cases}$ ■

The elimination procedures described in this section sometimes result in fractional coefficients. For example, consider the system

$$\begin{cases} 2x - 5y + 5z = \ 17 \\ 3x - 2y + 3z = \ 11. \\ -3x + 3y \qquad = -6 \end{cases}$$

Multiplying the first row by $\frac{1}{2}$ to produce a leading 1 results in fractional coefficients. You can sometimes avoid fractions by judiciously choosing the order in which you apply elementary row operations.

GO DIGITAL

EXAMPLE 9   **A System with an Infinite Number of Solutions**

Solve the system $\begin{cases} 2x + 4y - 2z = 0 \\ 3x + 5y \quad\quad = 1 \end{cases}$.

Solution

$$\begin{bmatrix} 2 & 4 & -2 & \vdots & 0 \\ 3 & 5 & 0 & \vdots & 1 \end{bmatrix}$$

$$\frac{1}{2}R_1 \rightarrow \begin{bmatrix} 1 & 2 & -1 & \vdots & 0 \\ 3 & 5 & 0 & \vdots & 1 \end{bmatrix}$$

$$-3R_1 + R_2 \rightarrow \begin{bmatrix} 1 & 2 & -1 & \vdots & 0 \\ 0 & -1 & 3 & \vdots & 1 \end{bmatrix}$$

$$-R_2 \rightarrow \begin{bmatrix} 1 & 2 & -1 & \vdots & 0 \\ 0 & 1 & -3 & \vdots & -1 \end{bmatrix}$$

$$-2R_2 + R_1 \rightarrow \begin{bmatrix} 1 & 0 & 5 & \vdots & 2 \\ 0 & 1 & -3 & \vdots & -1 \end{bmatrix}$$

The corresponding system of equations is

$$\begin{cases} x + 5z = \quad 2 \\ y - 3z = -1 \end{cases}.$$

Solving for x and y in terms of z, you have

$$x = -5z + 2 \quad \text{and} \quad y = 3z - 1.$$

To write a solution of the system that does not use any of the three variables of the system, let a represent any real number and let $z = a$. Substitute a for z in the equations for x and y.

$$x = -5z + 2 = -5a + 2 \quad \text{and} \quad y = 3z - 1 = 3a - 1$$

So, the solution set can be written as an ordered triple of the form

$$(-5a + 2, 3a - 1, a)$$

where a is any real number. Remember that a solution set of this form represents an infinite number of solutions. Substitute values for a to obtain a few solutions. Then check each solution in the original system of equations.

✓ *Checkpoint*  *Audio-video solution in English & Spanish at LarsonPrecalculus.com*

Solve the system $\begin{cases} 2x - 6y + 6z = 46 \\ 2x - 3y \quad\quad = 31 \end{cases}$. ∎

Summarize (Section 10.1)

1. State the definition of a matrix *(page 700)*. For examples of writing matrices and determining their dimensions, see Examples 1 and 2.

2. List the elementary row operations *(page 702)*. For an example of performing elementary row operations, see Example 3.

3. Explain how to use matrices and Gaussian elimination to solve systems of linear equations *(page 703)*. For examples of using Gaussian elimination, see Examples 4, 6, and 7.

4. Explain how to use matrices and Gauss-Jordan elimination to solve systems of linear equations *(page 707)*. For examples of using Gauss-Jordan elimination, see Examples 8 and 9.

GO DIGITAL

10.1 Exercises

See CalcChat.com for tutorial help and worked-out solutions to odd-numbered exercises.

GO DIGITAL

Vocabulary and Concept Check

In Exercises 1 and 2, fill in the blanks.

1. Two matrices are _____ when one can be obtained from the other by a sequence of elementary row operations.

2. A matrix in row-echelon form is in _____ _____ _____ when every column that has a leading 1 has zeros in every position above and below its leading 1.

In Exercises 3 and 4, refer to the system of linear equations

$$\begin{cases} -2x + 3y = 5 \\ 6x + 7y = 4 \end{cases}.$$

3. Describe the coefficient matrix for the system.

4. Explain why the augmented matrix is row equivalent to its reduced row-echelon form.

Skills and Applications

Dimension of a Matrix   In Exercises 5–12, determine the dimension of the matrix.

5. $\begin{bmatrix} 7 & 0 \end{bmatrix}$

6. $\begin{bmatrix} 5 & -3 & 8 & 7 \end{bmatrix}$

7. $\begin{bmatrix} 2 \\ 36 \\ 3 \end{bmatrix}$

8. $\begin{bmatrix} -3 & 7 & 15 & 0 \\ 0 & 0 & 3 & 3 \\ 1 & 1 & 6 & 7 \end{bmatrix}$

9. $\begin{bmatrix} 33 & 45 \\ -9 & 20 \end{bmatrix}$

10. $\begin{bmatrix} -7 & 6 & 4 \\ 0 & -5 & 1 \end{bmatrix}$

11. $\begin{bmatrix} 1 & 6 & -1 \\ 8 & 0 & 3 \\ 3 & -9 & 9 \end{bmatrix}$

12. $\begin{bmatrix} 3 & -1 \\ 4 & 1 \\ -5 & 9 \end{bmatrix}$

Writing an Augmented Matrix   In Exercises 13–18, write the augmented matrix for the system of linear equations.

13. $\begin{cases} 2x - y = 7 \\ x + y = 2 \end{cases}$

14. $\begin{cases} 5x + 2y = 13 \\ -3x + 4y = -24 \end{cases}$

15. $\begin{cases} x - y + 2z = 2 \\ 4x - 3y + z = -1 \\ 2x + y = 0 \end{cases}$

16. $\begin{cases} -2x - 4y + z = 13 \\ 6x - 7z = 22 \\ 3x - y + z = 9 \end{cases}$

17. $\begin{cases} 3x - 5y + 2z = 12 \\ 12x - 7z = 10 \end{cases}$

18. $\begin{cases} 9x + y - 3z = 21 \\ -15y + 13z = -8 \end{cases}$

Writing a System of Equations   In Exercises 19–24, write a system of linear equations represented by the augmented matrix.

19. $\begin{bmatrix} 1 & 1 & \vdots & 3 \\ 5 & -3 & \vdots & -1 \end{bmatrix}$

20. $\begin{bmatrix} 5 & 2 & \vdots & 9 \\ 3 & -8 & \vdots & 0 \end{bmatrix}$

21. $\begin{bmatrix} 2 & 0 & 5 & \vdots & -12 \\ 0 & 1 & -2 & \vdots & 7 \\ 6 & 3 & 0 & \vdots & 2 \end{bmatrix}$

22. $\begin{bmatrix} 4 & -5 & -1 & \vdots & 18 \\ -11 & 0 & 6 & \vdots & 25 \\ 3 & 8 & 0 & \vdots & -29 \end{bmatrix}$

23. $\begin{bmatrix} 9 & 12 & 3 & 0 & \vdots & 0 \\ -2 & 18 & 5 & 2 & \vdots & 10 \\ 1 & 7 & -8 & 0 & \vdots & -4 \\ 3 & 0 & 2 & 0 & \vdots & -10 \end{bmatrix}$

24. $\begin{bmatrix} 6 & 2 & -1 & -5 & \vdots & -25 \\ -1 & 0 & 7 & 3 & \vdots & 7 \\ 4 & -1 & -10 & 6 & \vdots & 23 \\ 0 & 8 & 1 & -11 & \vdots & -21 \end{bmatrix}$

Identifying Elementary Row Operations   In Exercises 25–28, identify the elementary row operation(s) performed to obtain the row-equivalent matrix.

Original Matrix	Row-Equivalent Matrix

25. $\begin{bmatrix} -2 & 5 & 1 \\ 3 & -1 & -8 \end{bmatrix}$ $\begin{bmatrix} 13 & 0 & -39 \\ 3 & -1 & -8 \end{bmatrix}$

26. $\begin{bmatrix} 3 & -1 & -4 \\ -4 & 3 & 7 \end{bmatrix}$ $\begin{bmatrix} 3 & -1 & -4 \\ 5 & 0 & -5 \end{bmatrix}$

27. $\begin{bmatrix} 0 & -1 & -5 & 5 \\ -1 & 3 & -7 & 6 \\ 4 & -5 & 1 & 3 \end{bmatrix}$ $\begin{bmatrix} -1 & 3 & -7 & 6 \\ 0 & -1 & -5 & 5 \\ 0 & 7 & -27 & 27 \end{bmatrix}$

28. $\begin{bmatrix} -1 & -2 & 3 & -2 \\ 2 & -5 & 1 & -7 \\ 5 & 4 & -7 & 6 \end{bmatrix}$ $\begin{bmatrix} -1 & -2 & 3 & -2 \\ 0 & -9 & 7 & -11 \\ 0 & -6 & 8 & -4 \end{bmatrix}$

Elementary Row Operations In Exercises 29–36, fill in the blank(s) using elementary row operations to form a row-equivalent matrix.

29. $\begin{bmatrix} 3 & 6 & 8 \\ 4 & -3 & 6 \end{bmatrix}$

$\begin{bmatrix} 1 & \rule{1cm}{0.4pt} & \frac{8}{3} \\ 4 & -3 & 6 \end{bmatrix}$

30. $\begin{bmatrix} 1 & 4 & 3 \\ 2 & 10 & 5 \end{bmatrix}$

$\begin{bmatrix} 1 & 4 & 3 \\ 0 & \rule{1cm}{0.4pt} & -1 \end{bmatrix}$

31. $\begin{bmatrix} 1 & 1 & 1 \\ 5 & -2 & 4 \end{bmatrix}$

$\begin{bmatrix} 1 & 1 & 1 \\ 0 & \rule{1cm}{0.4pt} & -1 \end{bmatrix}$

32. $\begin{bmatrix} -3 & 3 & 12 \\ 18 & -8 & 4 \end{bmatrix}$

$\begin{bmatrix} 1 & -1 & \rule{1cm}{0.4pt} \\ 18 & -8 & 4 \end{bmatrix}$

33. $\begin{bmatrix} 1 & 5 & 4 & -1 \\ 0 & 1 & -2 & 2 \\ 0 & 0 & 1 & -7 \end{bmatrix}$

$\begin{bmatrix} 1 & 0 & \rule{1cm}{0.4pt} & \rule{1cm}{0.4pt} \\ 0 & 1 & -2 & 2 \\ 0 & 0 & 1 & -7 \end{bmatrix}$

34. $\begin{bmatrix} 1 & 0 & 6 & 1 \\ 0 & -1 & 0 & 7 \\ 0 & 0 & -1 & 3 \end{bmatrix}$

$\begin{bmatrix} 1 & 0 & 6 & 1 \\ 0 & 1 & 0 & \rule{1cm}{0.4pt} \\ 0 & 0 & 1 & \rule{1cm}{0.4pt} \end{bmatrix}$

35. $\begin{bmatrix} 1 & 1 & 4 & -1 \\ 3 & 8 & 10 & 3 \\ -2 & 1 & 12 & 6 \end{bmatrix}$

$\begin{bmatrix} 1 & 1 & 4 & -1 \\ 0 & 5 & \rule{1cm}{0.4pt} & \rule{1cm}{0.4pt} \\ 0 & 3 & \rule{1cm}{0.4pt} & \rule{1cm}{0.4pt} \end{bmatrix}$

$\begin{bmatrix} 1 & 1 & 4 & -1 \\ 0 & 1 & -\frac{2}{5} & \frac{6}{5} \\ 0 & 3 & \rule{1cm}{0.4pt} & \rule{1cm}{0.4pt} \end{bmatrix}$

36. $\begin{bmatrix} 2 & 4 & 8 & 3 \\ 1 & -1 & -3 & 2 \\ 2 & 6 & 4 & 9 \end{bmatrix}$

$\begin{bmatrix} 1 & \rule{1cm}{0.4pt} & \rule{1cm}{0.4pt} & \rule{1cm}{0.4pt} \\ 1 & -1 & -3 & 2 \\ 2 & 6 & 4 & 9 \end{bmatrix}$

$\begin{bmatrix} 1 & 2 & 4 & \frac{3}{2} \\ 0 & \rule{1cm}{0.4pt} & -7 & \frac{1}{2} \\ 0 & 2 & \rule{1cm}{0.4pt} & \rule{1cm}{0.4pt} \end{bmatrix}$

Comparing Linear Systems and Matrix Operations In Exercises 37 and 38, (a) perform the row operations on the augmented matrix, (b) write and solve a corresponding system of linear equations represented by the augmented matrix, and (c) compare the results of parts (a) and (b). Which method do you prefer?

37. $\begin{bmatrix} -3 & 4 & \vdots & 22 \\ 6 & -4 & \vdots & -28 \end{bmatrix}$

   (i) Add R_2 to R_1.

   (ii) Add -2 times R_1 to R_2.

   (iii) Multiply R_2 by $-\frac{1}{4}$.

   (iv) Multiply R_1 by $\frac{1}{3}$.

38. $\begin{bmatrix} 7 & 13 & 1 & \vdots & -4 \\ -3 & -5 & -1 & \vdots & -4 \\ 3 & 6 & 1 & \vdots & -2 \end{bmatrix}$

   (i) Add R_2 to R_1.

   (ii) Multiply R_1 by $\frac{1}{4}$.

   (iii) Add R_3 to R_2.

   (iv) Add -3 times R_1 to R_3.

   (v) Add -2 times R_2 to R_1.

Row-Echelon Form In Exercises 39–42, determine whether the matrix is in row-echelon form. If it is, determine whether it is in reduced row-echelon form.

39. $\begin{bmatrix} 1 & 0 & 0 & 0 \\ 0 & 1 & 1 & 5 \\ 0 & 0 & 0 & 0 \end{bmatrix}$

40. $\begin{bmatrix} 1 & 0 & 0 & 5 \\ 0 & 1 & 0 & 3 \\ 0 & 1 & 0 & 0 \end{bmatrix}$

41. $\begin{bmatrix} 1 & 0 & 0 & 1 \\ 0 & 1 & 0 & -1 \\ 0 & 0 & 0 & 2 \end{bmatrix}$

42. $\begin{bmatrix} 1 & 0 & 0 & 5 \\ 0 & 1 & -2 & 3 \\ 0 & 0 & 1 & 0 \end{bmatrix}$

Writing a Matrix in Row-Echelon Form In Exercises 43–48, write the matrix in row-echelon form. (Remember that the row-echelon form of a matrix is not unique.)

43. $\begin{bmatrix} 1 & 1 & 0 & 5 \\ -2 & -1 & 2 & -10 \\ 3 & 6 & 7 & 14 \end{bmatrix}$

44. $\begin{bmatrix} 1 & 2 & -1 & 3 \\ 3 & 7 & -5 & 14 \\ -2 & -1 & -3 & 8 \end{bmatrix}$

45. $\begin{bmatrix} 1 & -1 & -1 & 1 \\ 5 & -4 & 1 & 8 \\ -6 & 8 & 18 & 0 \end{bmatrix}$

46. $\begin{bmatrix} 1 & -3 & 0 & -7 \\ -3 & 10 & 1 & 23 \\ 4 & -10 & 2 & -24 \end{bmatrix}$

47. $\begin{bmatrix} 1 & 2 & 3 & -5 \\ 1 & 2 & 4 & -9 \\ -2 & -4 & -4 & 3 \\ 4 & 8 & 11 & -14 \end{bmatrix}$

48. $\begin{bmatrix} -2 & 3 & -1 & -2 \\ 4 & -2 & 5 & 8 \\ 1 & 5 & -2 & 0 \\ 3 & 8 & -10 & -30 \end{bmatrix}$

Writing a Matrix in Reduced Row-Echelon Form In Exercises 49–52, use the matrix capabilities of a graphing utility to write the matrix in reduced row-echelon form.

49. $\begin{bmatrix} -1 & 2 & 1 \\ 3 & 4 & 9 \\ 2 & 1 & -2 \end{bmatrix}$

50. $\begin{bmatrix} 1 & 3 & 2 \\ 5 & 15 & 9 \\ 2 & 6 & 10 \end{bmatrix}$

51. $\begin{bmatrix} -3 & 5 & 1 & 12 \\ 1 & -1 & 1 & 4 \end{bmatrix}$

52. $\begin{bmatrix} 5 & 1 & 2 & 4 \\ -1 & 5 & 10 & -32 \end{bmatrix}$

Using Back-Substitution In Exercises 53–56, write a system of linear equations represented by the augmented matrix. Then use back-substitution to solve the system.

53. $\begin{bmatrix} 1 & -2 & \vdots & 4 \\ 0 & 1 & \vdots & -1 \end{bmatrix}$

54. $\begin{bmatrix} 1 & 5 & \vdots & 0 \\ 0 & 1 & \vdots & 6 \end{bmatrix}$

55. $\begin{bmatrix} 1 & -1 & 2 & \vdots & 4 \\ 0 & 1 & -1 & \vdots & 2 \\ 0 & 0 & 1 & \vdots & -2 \end{bmatrix}$

56. $\begin{bmatrix} 1 & 2 & -2 & \vdots & -1 \\ 0 & 1 & 1 & \vdots & 9 \\ 0 & 0 & 1 & \vdots & -3 \end{bmatrix}$

Gaussian Elimination with Back-Substitution In Exercises 57–66, use matrices to solve the system of linear equations, if possible. Use Gaussian elimination with back-substitution.

57. $\begin{cases} x + 2y = 7 \\ -x + y = 8 \end{cases}$ 58. $\begin{cases} 2x + 6y = 16 \\ 2x + 3y = 7 \end{cases}$

59. $\begin{cases} 3x - 2y = -27 \\ x + 3y = 13 \end{cases}$ 60. $\begin{cases} -x + y = 4 \\ 2x - 4y = -34 \end{cases}$

61. $\begin{cases} x + 2y - 3z = -28 \\ 4y + 2z = 0 \\ -x + y - z = -5 \end{cases}$

62. $\begin{cases} 3x - 2y + z = 15 \\ -x + y + 2z = -10 \\ x - y - 4z = 14 \end{cases}$

63. $\begin{cases} -3x + 2y = -22 \\ 3x + 4y = 4 \\ 4x - 8y = 32 \end{cases}$

64. $\begin{cases} x + 2y = 0 \\ x + y = 6 \\ 3x - 2y = 8 \end{cases}$

65. $\begin{cases} 3x + 2y - z + w = 0 \\ x - y + 4z + 2w = 25 \\ -2x + y + 2z - w = 2 \\ x + y + z + w = 6 \end{cases}$

66. $\begin{cases} x - 4y + 3z - 2w = 9 \\ 3x - 2y + z - 4w = -13 \\ -4x + 3y - 2z + w = -4 \\ -2x + y - 4z + 3w = -10 \end{cases}$

Interpreting Reduced Row-Echelon Form   In Exercises 67 and 68, an augmented matrix that represents a system of linear equations has been reduced using Gauss-Jordan elimination. Write a solution represented by the augmented matrix.

67. $\begin{bmatrix} 1 & 0 & \vdots & 3 \\ 0 & 1 & \vdots & -4 \end{bmatrix}$ 68. $\begin{bmatrix} 1 & 0 & 0 & \vdots & 5 \\ 0 & 1 & 0 & \vdots & -3 \\ 0 & 0 & 1 & \vdots & 0 \end{bmatrix}$

Gauss-Jordan Elimination In Exercises 69–78, use matrices to solve the system of linear equations, if possible. Use Gauss-Jordan elimination.

69. $\begin{cases} -2x + 6y = -22 \\ x + 2y = -9 \end{cases}$ 70. $\begin{cases} 5x - 5y = -5 \\ -2x - 3y = 7 \end{cases}$

71. $\begin{cases} x - 3z = -2 \\ 3x + y - 2z = 5 \\ 2x + 2y + z = 4 \end{cases}$ 72. $\begin{cases} 2x - y + 3z = 24 \\ 2y - z = 14 \\ 7x - 5y = 6 \end{cases}$

73. $\begin{cases} x + 2y + z = 8 \\ 3x + 7y + 6z = 26 \end{cases}$ 74. $\begin{cases} x + y + 4z = 5 \\ 2x + y - z = 9 \end{cases}$

75. $\begin{cases} -x + y - z = -14 \\ 2x - y + z = 21 \\ 3x + 2y + z = 19 \end{cases}$ 76. $\begin{cases} 2x + 2y - z = 2 \\ x - 3y + z = -28 \\ -x + y = 14 \end{cases}$

77. $\begin{cases} 3x + 3y + 12z = 6 \\ x + y + 4z = 2 \\ 2x + 5y + 20z = 10 \\ -x + 2y + 8z = 4 \end{cases}$ 78. $\begin{cases} 2x + 10y + 2z = 6 \\ x + 5y + 2z = 6 \\ x + 5y + z = 3 \\ -3x - 15y - 3z = -9 \end{cases}$

Solving a System of Linear Equations In Exercises 79–82, use the matrix capabilities of a graphing utility to solve the system of linear equations.

79. $\begin{cases} 2x + y - z + 2w = -6 \\ 3x + 4y + w = 1 \\ x + 5y + 2z + 6w = -3 \\ 5x + 2y - z - w = 3 \end{cases}$

80. $\begin{cases} x + 2y + 2z + 4w = 11 \\ 3x + 6y + 5z + 12w = 30 \\ x + 3y - 3z + 2w = -5 \\ 6x - y - z + w = -9 \end{cases}$

81. $\begin{cases} x + y + z + w = 0 \\ 2x + 3y + z - 2w = 0 \\ 3x + 5y + z = 0 \end{cases}$

82. $\begin{cases} x + 2y + z + 3w = 0 \\ x - y + w = 0 \\ y - z + 2w = 0 \end{cases}$

Curve Fitting In Exercises 83–86, use a system of linear equations to find the quadratic function $f(x) = ax^2 + bx + c$ that satisfies the given conditions. Solve the system using matrices.

83. $f(1) = 1, f(2) = -1, f(3) = -5$

84. $f(1) = 2, f(2) = 9, f(3) = 20$

85. $f(-2) = -15, f(-1) = 7, f(1) = -3$

86. $f(-2) = -3, f(1) = -3, f(2) = -11$

87. **Borrowing Money** A city government borrows $2,000,000 at simple annual interest. Some of the money is borrowed at 8%, some at 9%, and some at 12%. Use a system of linear equations to determine how much is borrowed at each rate given that the total annual interest is $186,000 and the amount borrowed at 8% is twice the amount borrowed at 12%. Solve the system of linear equations using matrices.

88. **Borrowing Money** A state government borrows $2,000,000 at simple annual interest. Some of the money is borrowed at 7%, some at 8.5%, and some at 9.5%. Use a system of linear equations to determine how much is borrowed at each rate given that the total annual interest is $169,750 and the amount borrowed at 8.5% is four times the amount borrowed at 9.5%. Solve the system of linear equations using matrices.

89. Waterborne Disease

From 2009 through 2020, the numbers of new cases of a waterborne disease in a small city increased in a pattern that was approximately linear (see figure). Find the least squares regression line

$$y = at + b$$

for the data shown in the figure by solving the system below using matrices. Let t represent the year, with $t = 0$ corresponding to 2009.

$$\begin{cases} 12b + 66a = 831 \\ 66b + 506a = 5643 \end{cases}$$

Use the result to predict the number of new cases of the waterborne disease in 2024. Is the estimate reasonable? Explain.

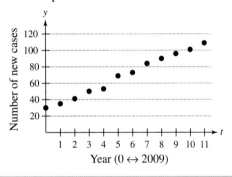

Year ($0 \leftrightarrow 2009$)

 90. Mathematical Modeling Software was used to analyze a video of the path of a thrown ball. The position of the ball was measured three times. The coordinates obtained are shown in the table. (x and y are measured in feet.)

Horizontal Distance, x	0	15	30
Height, y	5.0	9.6	12.4

(a) Write a system of equations to find the equation of the parabola $y = ax^2 + bx + c$ that passes through the three points.

(b) Solve the system using matrices.

(c) Use a graphing utility to graph the parabola.

(d) Graphically approximate the maximum height of the ball and the point at which the ball struck the ground.

(e) Analytically find the maximum height of the ball and the point at which the ball struck the ground.

(f) Compare your results from parts (d) and (e).

Exploring the Concepts

True or False? In Exercises 91 and 92, determine whether the statement is true or false. Justify your answer.

91. $\begin{bmatrix} 5 & 0 & -2 & 7 \\ -1 & 3 & -6 & 0 \end{bmatrix}$ is a 4×2 matrix.

92. The method of Gaussian elimination reduces a matrix until a reduced row-echelon form is obtained.

93. Error Analysis Describe the error.

The matrix

$$\begin{bmatrix} 1 & 2 & 7 & 2 \\ 0 & 1 & 4 & 0 \\ 0 & 0 & 1 & \frac{1}{3} \end{bmatrix}$$

is in reduced row-echelon form.

94. **HOW DO YOU SEE IT?** Determine whether the matrix below is in row-echelon form, reduced row-echelon form, or neither when it satisfies the given conditions.

$$\begin{bmatrix} 1 & b \\ c & 1 \end{bmatrix}$$

(a) $b = 0, c = 0$ (b) $b \neq 0, c = 0$

(c) $b = 0, c \neq 0$ (d) $b \neq 0, c \neq 0$

Review & Refresh ▶ Video solutions at LarsonPrecalculus.com

Order of Operations In Exercises 95 and 96, simplify the expression.

95. $1(-2) + 0(1) + 3(-1)$

96. $2(-2) + (-1)(1) + (-2)(-1)$

Equality of Complex Numbers In Exercises 97–100, find real numbers a and b such that the equation is true.

97. $a + bi = -3 + 8i$ **98.** $a + bi = 10 - 5i$

99. $(a - 2) + (b + 1)i = 6 + 5i$

100. $(a + 2) + (b - 3)i = 4 + 7i$

Solving for x In Exercises 101–104, solve for x.

101. $3x + a = b$ **102.** $2x - a = b$

103. $6b - x = a$ **104.** $5a + 2x = b$

Vector Operations In Exercises 105–108, find the component form of u and sketch the specified vector operations geometrically.

105. $\mathbf{u} = \mathbf{v} + \mathbf{w}, \mathbf{v} = \langle 2, 4 \rangle, \mathbf{w} = \langle 6, 2 \rangle$

106. $\mathbf{u} = 3\mathbf{v} + \mathbf{w}, \mathbf{v} = \langle 3, 6 \rangle, \mathbf{w} = \langle 8, 5 \rangle$

107. $\mathbf{u} = \mathbf{v} - 2\mathbf{w}, \mathbf{v} = \langle 2, 4 \rangle, \mathbf{w} = \langle 6, 2 \rangle$

108. $\mathbf{u} = \mathbf{v} - \mathbf{w}, \mathbf{v} = \langle 3, 6 \rangle, \mathbf{w} = \langle 8, 5 \rangle$

10.2 Operations with Matrices

Matrix operations have many practical applications. For example, in Exercise 73 on page 726, you will use matrix multiplication to analyze the calories burned by individuals of different body weights while performing different types of exercises.

- Determine whether two matrices are equal.
- Add and subtract matrices, and multiply matrices by scalars.
- Multiply two matrices.
- Use matrices to transform vectors.
- Use matrix operations to model and solve real-life problems.

Equality of Matrices

In the preceding section, you used matrices to solve systems of linear equations. There is a rich mathematical theory of matrices, and its applications are numerous. This section and the next two sections introduce some fundamental concepts of matrix theory. It is standard mathematical convention to represent matrices in any of the three ways listed below.

Representation of Matrices

1. A matrix can be denoted by an uppercase letter such as A, B, or C.
2. A matrix can be denoted by a representative element enclosed in brackets, such as $[a_{ij}]$, $[b_{ij}]$, or $[c_{ij}]$.
3. A matrix can be denoted by a rectangular array of numbers such as

$$A = [a_{ij}] = \begin{bmatrix} a_{11} & a_{12} & a_{13} & \cdots & a_{1n} \\ a_{21} & a_{22} & a_{23} & \cdots & a_{2n} \\ a_{31} & a_{32} & a_{33} & \cdots & a_{3n} \\ \vdots & \vdots & \vdots & & \vdots \\ a_{m1} & a_{m2} & a_{m3} & \cdots & a_{mn} \end{bmatrix}.$$

Two matrices $A = [a_{ij}]$ and $B = [b_{ij}]$ are **equal** when they have the same dimension $(m \times n)$ and $a_{ij} = b_{ij}$ for $1 \le i \le m$ and $1 \le j \le n$. In other words, two matrices are equal when their corresponding entries are equal.

EXAMPLE 1 **Equality of Matrices**

Solve for a_{11}, a_{12}, a_{21}, and a_{22} in the matrix equation $\begin{bmatrix} a_{11} & a_{12} \\ a_{21} & a_{22} \end{bmatrix} = \begin{bmatrix} 2 & -1 \\ -3 & 0 \end{bmatrix}$.

Solution Two matrices are equal when their corresponding entries are equal, so $a_{11} = 2$, $a_{12} = -1$, $a_{21} = -3$, and $a_{22} = 0$.

✓ *Checkpoint* ▶ *Audio-video solution in English & Spanish at LarsonPrecalculus.com*

Solve for a_{11}, a_{12}, a_{21}, and a_{22} in the matrix equation $\begin{bmatrix} a_{11} & a_{12} \\ a_{21} & a_{22} \end{bmatrix} = \begin{bmatrix} 6 & 3 \\ -2 & 4 \end{bmatrix}$. ■

Be sure you see that for two matrices to be equal, they must have the same dimension *and* their corresponding entries must be equal. For example,

$$\begin{bmatrix} 2 & -1 \\ \sqrt{4} & \frac{1}{2} \end{bmatrix} = \begin{bmatrix} 2 & -1 \\ 2 & 0.5 \end{bmatrix} \quad \text{but} \quad \begin{bmatrix} 2 & -1 & 0 \\ 3 & 4 & 0 \end{bmatrix} \ne \begin{bmatrix} 2 & -1 \\ 3 & 4 \end{bmatrix}.$$

GO DIGITAL

HISTORICAL NOTE

Arthur Cayley (1821–1895), a British mathematician, is credited with introducing matrix theory in 1858. Cayley was a Cambridge University graduate and a lawyer by profession. He began his groundbreaking work on matrices as he studied the theory of transformations. Cayley also was instrumental in the development of determinants, which are discussed later in this chapter. Cayley and two American mathematicians, Benjamin Peirce (1809–1880) and his son Charles S. Peirce (1839–1914), are credited with developing "matrix algebra."

Matrix Addition and Scalar Multiplication

Two basic matrix operations are matrix addition and scalar multiplication. With matrix addition, you add two matrices (of the same dimension) by adding their corresponding entries.

Definition of Matrix Addition

If $A = [a_{ij}]$ and $B = [b_{ij}]$ are matrices of dimension $m \times n$, then their sum is the $m \times n$ matrix

$$A + B = [a_{ij} + b_{ij}].$$

The sum of two matrices of different dimensions is undefined.

EXAMPLE 2 **Addition of Matrices**

a. $\begin{bmatrix} -1 & 2 \\ 0 & 1 \end{bmatrix} + \begin{bmatrix} 1 & 3 \\ -1 & 2 \end{bmatrix} = \begin{bmatrix} -1+1 & 2+3 \\ 0+(-1) & 1+2 \end{bmatrix} = \begin{bmatrix} 0 & 5 \\ -1 & 3 \end{bmatrix}$

b. $\begin{bmatrix} 0 & 1 & -2 \\ 1 & 2 & 3 \end{bmatrix} + \begin{bmatrix} 0 & 0 & 0 \\ 0 & 0 & 0 \end{bmatrix} = \begin{bmatrix} 0 & 1 & -2 \\ 1 & 2 & 3 \end{bmatrix}$

c. $\begin{bmatrix} 1 \\ -3 \\ -2 \end{bmatrix} + \begin{bmatrix} -1 \\ 3 \\ 2 \end{bmatrix} = \begin{bmatrix} 0 \\ 0 \\ 0 \end{bmatrix}$

d. The sum of

$$A = \begin{bmatrix} 2 & 1 & 0 \\ 4 & 0 & -1 \\ 3 & -2 & 2 \end{bmatrix} \quad \text{and} \quad B = \begin{bmatrix} 0 & 1 \\ -1 & 3 \\ 2 & 4 \end{bmatrix}$$

is undefined because A is of dimension 3×3 and B is of dimension 3×2.

✓ *Checkpoint* ▶ *Audio-video solution in English & Spanish at LarsonPrecalculus.com*

Find each sum, if possible.

a. $\begin{bmatrix} 4 & -1 \\ 2 & -3 \end{bmatrix} + \begin{bmatrix} 2 & -1 \\ 0 & 6 \end{bmatrix}$ **b.** $\begin{bmatrix} 2 & -1 \\ 3 & 4 \\ 0 & -2 \end{bmatrix} + \begin{bmatrix} -2 & 1 \\ -3 & -4 \\ 0 & 2 \end{bmatrix}$

c. $\begin{bmatrix} 3 & 9 & 6 \\ 0 & 4 & -2 \\ 1 & -1 & 0 \end{bmatrix} + \begin{bmatrix} 3 & 9 & 6 \\ 0 & 2 & -4 \end{bmatrix}$ **d.** $\begin{bmatrix} 1 \\ -1 \\ 1 \end{bmatrix} + \begin{bmatrix} -1 \\ 1 \\ 1 \end{bmatrix}$

In operations with matrices, numbers are usually referred to as **scalars**. In this text, scalars will always be real numbers. To multiply a matrix A by a scalar c, multiply each entry in A by c.

Definition of Scalar Multiplication

If $A = [a_{ij}]$ is an $m \times n$ matrix and c is a scalar, then the **scalar multiple** of A by c is the $m \times n$ matrix

$$cA = [ca_{ij}].$$

The symbol $-A$ represents the **negation** of A, which is the scalar product $(-1)A$. Moreover, if A and B are of the same dimension, then $A - B$ represents the sum of A and $(-1)B$. That is,

$$A - B = A + (-1)B.$$ Subtraction of matrices

EXAMPLE 3 **Operations with Matrices**

For the matrices below, find (a) $3A$, (b) $-B$, and (c) $3A - B$.

$$A = \begin{bmatrix} 2 & 2 & 4 \\ -3 & 0 & -1 \\ 2 & 1 & 2 \end{bmatrix} \quad \text{and} \quad B = \begin{bmatrix} 2 & 0 & 0 \\ 1 & -4 & 3 \\ -1 & 3 & 2 \end{bmatrix}$$

Solution

a. $3A = 3\begin{bmatrix} 2 & 2 & 4 \\ -3 & 0 & -1 \\ 2 & 1 & 2 \end{bmatrix}$ Scalar multiplication

$$= \begin{bmatrix} 3(2) & 3(2) & 3(4) \\ 3(-3) & 3(0) & 3(-1) \\ 3(2) & 3(1) & 3(2) \end{bmatrix}$$ Multiply each entry by 3.

$$= \begin{bmatrix} 6 & 6 & 12 \\ -9 & 0 & -3 \\ 6 & 3 & 6 \end{bmatrix}$$ Simplify.

b. $-B = (-1)\begin{bmatrix} 2 & 0 & 0 \\ 1 & -4 & 3 \\ -1 & 3 & 2 \end{bmatrix}$ Definition of negation

$$= \begin{bmatrix} -2 & 0 & 0 \\ -1 & 4 & -3 \\ 1 & -3 & -2 \end{bmatrix}$$ Multiply each entry by -1.

c. $3A - B = \begin{bmatrix} 6 & 6 & 12 \\ -9 & 0 & -3 \\ 6 & 3 & 6 \end{bmatrix} + \begin{bmatrix} -2 & 0 & 0 \\ -1 & 4 & -3 \\ 1 & -3 & -2 \end{bmatrix}$ $3A - B = 3A + (-1)B$

$$= \begin{bmatrix} 4 & 6 & 12 \\ -10 & 4 & -6 \\ 7 & 0 & 4 \end{bmatrix}$$ Add corresponding entries.

> **ALGEBRA HELP**
>
> The order of operations for matrix expressions is similar to that for real numbers. As shown in Example 3(c), you perform scalar multiplication before matrix addition and subtraction.

✓ *Checkpoint* *Audio-video solution in English & Spanish at LarsonPrecalculus.com*

For the matrices below, find (a) $A - B$, (b) $3A$, and (c) $3A - 2B$.

$$A = \begin{bmatrix} 4 & -1 \\ 0 & 4 \\ -3 & 8 \end{bmatrix} \quad \text{and} \quad B = \begin{bmatrix} 0 & 4 \\ -1 & 3 \\ 1 & 7 \end{bmatrix}$$

It is often convenient to rewrite the scalar multiple cA by factoring c out of every entry in the matrix. The example below shows factoring the scalar $\frac{1}{2}$ out of a matrix.

$$\begin{bmatrix} \frac{1}{2} & -\frac{3}{2} \\ \frac{5}{2} & \frac{1}{2} \end{bmatrix} = \begin{bmatrix} \frac{1}{2}(1) & \frac{1}{2}(-3) \\ \frac{1}{2}(5) & \frac{1}{2}(1) \end{bmatrix} = \frac{1}{2}\begin{bmatrix} 1 & -3 \\ 5 & 1 \end{bmatrix}$$

GO DIGITAL

▶▶ **SKILLS REFRESHER**

For a refresher on the properties of addition and multiplication of real numbers (and other properties of real numbers), watch the video at *LarsonPrecalculus.com*.

The properties of matrix addition and scalar multiplication are similar to those of addition and multiplication of real numbers.

Properties of Matrix Addition and Scalar Multiplication

Let A, B, and C be $m \times n$ matrices and let c and d be scalars.

1. $A + B = B + A$ Commutative Property of Matrix Addition

2. $A + (B + C) = (A + B) + C$ Associative Property of Matrix Addition

3. $(cd)A = c(dA)$ Associative Property of Scalar Multiplication

4. $1A = A$ Scalar Identity Property

5. $c(A + B) = cA + cB$ Distributive Property

6. $(c + d)A = cA + dA$ Distributive Property

Note that the Associative Property of Matrix Addition allows you to write expressions such as $A + B + C$ without ambiguity because the same sum occurs no matter how the matrices are grouped. This same reasoning applies to sums of four or more matrices.

▶▶ TECHNOLOGY

Most graphing utilities can perform matrix operations. Consult the user's guide for your graphing utility for specific keystrokes. Use a graphing utility to find the sum of the matrices

$$A = \begin{bmatrix} 2 & -3 \\ -1 & 0 \end{bmatrix}$$

and

$$B = \begin{bmatrix} -1 & 4 \\ 2 & -5 \end{bmatrix}.$$

EXAMPLE 4 **Addition of More than Two Matrices**

$$\begin{bmatrix} 1 \\ 2 \\ -3 \end{bmatrix} + \begin{bmatrix} -1 \\ -1 \\ 2 \end{bmatrix} + \begin{bmatrix} 0 \\ 1 \\ 4 \end{bmatrix} + \begin{bmatrix} 2 \\ -3 \\ -2 \end{bmatrix} = \begin{bmatrix} 2 \\ -1 \\ 1 \end{bmatrix}$$ Add corresponding entries.

✓ **Checkpoint** ▶ Audio-video solution in English & Spanish at *LarsonPrecalculus.com*

Evaluate the expression.

$$\begin{bmatrix} 3 & -8 \\ 0 & 2 \end{bmatrix} + \begin{bmatrix} -2 & 3 \\ 6 & -5 \end{bmatrix} + \begin{bmatrix} 0 & 7 \\ 4 & -1 \end{bmatrix}$$

EXAMPLE 5 **Using the Distributive Property**

Note how the Distributive Property is used to evaluate the expression below.

$$3\left(\begin{bmatrix} -2 & 0 \\ 4 & 1 \end{bmatrix} + \begin{bmatrix} 4 & -2 \\ 3 & 7 \end{bmatrix}\right) = 3\begin{bmatrix} -2 & 0 \\ 4 & 1 \end{bmatrix} + 3\begin{bmatrix} 4 & -2 \\ 3 & 7 \end{bmatrix}$$

$$= \begin{bmatrix} -6 & 0 \\ 12 & 3 \end{bmatrix} + \begin{bmatrix} 12 & -6 \\ 9 & 21 \end{bmatrix}$$

$$= \begin{bmatrix} 6 & -6 \\ 21 & 24 \end{bmatrix}$$

✓ **Checkpoint** ▶ Audio-video solution in English & Spanish at *LarsonPrecalculus.com*

Evaluate the expression using the Distributive Property.

$$2\left(\begin{bmatrix} 1 & 3 \\ -2 & 2 \end{bmatrix} + \begin{bmatrix} -4 & 0 \\ -3 & 1 \end{bmatrix}\right)$$

In Example 5, you could add the two matrices first and then multiply the resulting matrix by 3. The result would be the same.

GO DIGITAL

One important property of addition of real numbers is that the number 0 is the additive identity. That is, $c + 0 = c$ for any real number c. For matrices, a similar property holds. That is, if A is an $m \times n$ matrix and O is the $m \times n$ **zero matrix** consisting entirely of zeros, then

$$A + O = A.$$

In other words, O is the **additive identity** for the set of all $m \times n$ matrices. For example, the matrices below are the additive identities for the sets of all 2×3 and 2×2 matrices.

$$O = \begin{bmatrix} 0 & 0 & 0 \\ 0 & 0 & 0 \end{bmatrix} \quad \text{and} \quad O = \begin{bmatrix} 0 & 0 \\ 0 & 0 \end{bmatrix}$$

$\underbrace{}_{2 \times 3 \text{ zero matrix}}$ $\underbrace{}_{2 \times 2 \text{ zero matrix}}$

The algebra of real numbers and the algebra of matrices have many similarities. For example, compare the solutions below.

Real Numbers (Solve for x.)	$m \times n$ **Matrices** (Solve for X.)
$x + a = b$	$X + A = B$
$x + a + (-a) = b + (-a)$	$X + A + (-A) = B + (-A)$
$x + 0 = b - a$	$X + O = B - A$
$x = b - a$	$X = B - A$

ALGEBRA HELP

When you solve for X in a matrix equation, you are solving for a *matrix X* that makes the equation true.

The algebra of real numbers and the algebra of matrices also have important differences (see Example 9 and Exercises 77–82).

EXAMPLE 6   **Solving a Matrix Equation**

Solve for X in the equation $3X + A = B$, where

$$A = \begin{bmatrix} 1 & -2 \\ 0 & 3 \end{bmatrix} \quad \text{and} \quad B = \begin{bmatrix} -3 & 4 \\ 2 & 1 \end{bmatrix}.$$

Solution   Begin by solving the matrix equation for X.

$$3X + A = B$$
$$3X = B - A$$
$$X = \tfrac{1}{3}(B - A)$$

Now, substituting the matrices A and B, you have

$$X = \tfrac{1}{3}\left(\begin{bmatrix} -3 & 4 \\ 2 & 1 \end{bmatrix} - \begin{bmatrix} 1 & -2 \\ 0 & 3 \end{bmatrix} \right) \qquad \text{Substitute the matrices.}$$

$$= \tfrac{1}{3}\begin{bmatrix} -4 & 6 \\ 2 & -2 \end{bmatrix} \qquad \text{Subtract matrix } A \text{ from matrix } B.$$

$$= \begin{bmatrix} -\tfrac{4}{3} & 2 \\ \tfrac{2}{3} & -\tfrac{2}{3} \end{bmatrix}. \qquad \text{Multiply the resulting matrix by } \tfrac{1}{3}.$$

✓ **Checkpoint** ▶ *Audio-video solution in English & Spanish at LarsonPrecalculus.com*

Solve for X in the equation $2X - A = B$, where

$$A = \begin{bmatrix} 6 & 1 \\ 0 & 3 \end{bmatrix} \quad \text{and} \quad B = \begin{bmatrix} 4 & -1 \\ -2 & 5 \end{bmatrix}.$$

GO DIGITAL

Matrix Multiplication

Another basic matrix operation is **matrix multiplication.** At first glance, the definition may seem unusual. You will see later, however, that this definition of the product of two matrices has many practical applications.

> ### Definition of Matrix Multiplication
>
> If $A = [a_{ij}]$ is an $m \times n$ matrix and $B = [b_{ij}]$ is an $n \times p$ matrix, then the product AB is an $m \times p$ matrix given by $AB = [c_{ij}]$, where
> $$c_{ij} = a_{i1}b_{1j} + a_{i2}b_{2j} + a_{i3}b_{3j} + \cdots + a_{in}b_{nj}.$$

$$\underset{m \times n}{A} \quad \times \quad \underset{n \times p}{B} \quad = \quad \underset{m \times p}{AB}$$

Equal

Dimension of AB

The definition of matrix multiplication uses a *row-by-column* multiplication, where the entry in the ith row and jth column of the product AB is obtained by multiplying the entries in the ith row of A by the corresponding entries in the jth column of B and then adding the results. So, for the product of two matrices to be defined, the number of columns of the first matrix must equal the number of rows of the second matrix. That is, the middle two indices must be the same. The outside two indices give the dimension of the product, as shown at the left. The general pattern for matrix multiplication is shown below.

$$\begin{bmatrix} a_{11} & a_{12} & a_{13} & \cdots & a_{1n} \\ a_{21} & a_{22} & a_{23} & \cdots & a_{2n} \\ a_{31} & a_{32} & a_{33} & \cdots & a_{3n} \\ \vdots & \vdots & \vdots & & \vdots \\ a_{i1} & a_{i2} & a_{i3} & \cdots & a_{in} \\ \vdots & \vdots & \vdots & & \vdots \\ a_{m1} & a_{m2} & a_{m3} & \cdots & a_{mn} \end{bmatrix} \begin{bmatrix} b_{11} & b_{12} & \cdots & b_{1j} & \cdots & b_{1p} \\ b_{21} & b_{22} & \cdots & b_{2j} & \cdots & b_{2p} \\ b_{31} & b_{32} & \cdots & b_{3j} & \cdots & b_{3p} \\ \vdots & \vdots & & \vdots & & \vdots \\ b_{n1} & b_{n2} & \cdots & b_{nj} & \cdots & b_{np} \end{bmatrix} = \begin{bmatrix} c_{11} & c_{12} & \cdots & c_{1j} & \cdots & c_{1p} \\ c_{21} & c_{22} & \cdots & c_{2j} & \cdots & c_{2p} \\ \vdots & \vdots & & \vdots & & \vdots \\ c_{i1} & c_{i2} & \cdots & c_{ij} & \cdots & c_{ip} \\ \vdots & \vdots & & \vdots & & \vdots \\ c_{m1} & c_{m2} & \cdots & c_{mj} & \cdots & c_{mp} \end{bmatrix}$$

$$a_{i1}b_{1j} + a_{i2}b_{2j} + a_{i3}b_{3j} + \cdots + a_{in}b_{nj} = c_{ij}$$

EXAMPLE 7   Finding the Product of Two Matrices

Find the product AB, where $A = \begin{bmatrix} -1 & 3 \\ 4 & -2 \\ 5 & 0 \end{bmatrix}$ and $B = \begin{bmatrix} -3 & 2 \\ -4 & 1 \end{bmatrix}$.

Solution   To find the entries of the product, multiply each row of A by each column of B.

$$AB = \begin{bmatrix} -1 & 3 \\ 4 & -2 \\ 5 & 0 \end{bmatrix} \begin{bmatrix} -3 & 2 \\ -4 & 1 \end{bmatrix}$$

$$= \begin{bmatrix} (-1)(-3) + 3(-4) & (-1)(2) + 3(1) \\ 4(-3) + (-2)(-4) & 4(2) + (-2)(1) \\ 5(-3) + 0(-4) & 5(2) + 0(1) \end{bmatrix}$$

$$= \begin{bmatrix} -9 & 1 \\ -4 & 6 \\ -15 & 10 \end{bmatrix}$$

ALGEBRA HELP

In Example 7, the product AB is defined because the number of columns of A is equal to the number of rows of B. Also, note that the product AB has dimension 3×2.

 Checkpoint ▶ Audio-video solution in English & Spanish at LarsonPrecalculus.com

Find the product AB, where $A = \begin{bmatrix} -1 & 4 \\ 2 & 0 \\ 1 & 2 \end{bmatrix}$ and $B = \begin{bmatrix} 1 & -2 \\ 0 & 7 \end{bmatrix}$.   ■

EXAMPLE 8   **Finding the Product of Two Matrices**

Find the product AB, where $A = \begin{bmatrix} 1 & 0 & 3 \\ 2 & -1 & -2 \end{bmatrix}$ and $B = \begin{bmatrix} -2 & 4 \\ 1 & 0 \\ -1 & 1 \end{bmatrix}$.

Solution   Note that the dimension of A is 2×3 and the dimension of B is 3×2. So, the product AB has dimension 2×2.

$$AB = \begin{bmatrix} 1 & 0 & 3 \\ 2 & -1 & -2 \end{bmatrix} \begin{bmatrix} -2 & 4 \\ 1 & 0 \\ -1 & 1 \end{bmatrix}$$

$$= \begin{bmatrix} 1(-2) + 0(1) + 3(-1) & 1(4) + 0(0) + 3(1) \\ 2(-2) + (-1)(1) + (-2)(-1) & 2(4) + (-1)(0) + (-2)(1) \end{bmatrix}$$

$$= \begin{bmatrix} -5 & 7 \\ -3 & 6 \end{bmatrix}$$

✓ *Checkpoint* ▶ Audio-video solution in English & Spanish at LarsonPrecalculus.com

Find the product AB, where $A = \begin{bmatrix} 0 & 4 & -3 \\ 2 & 1 & 7 \\ 3 & -2 & 1 \end{bmatrix}$ and $B = \begin{bmatrix} -2 & 0 \\ 0 & -4 \\ 1 & 2 \end{bmatrix}$.

EXAMPLE 9   **Matrix Multiplication**

 See LarsonPrecalculus.com for an interactive version of this type of example.

a. $\underset{2 \times 2}{\begin{bmatrix} 3 & 4 \\ -2 & 5 \end{bmatrix}} \underset{2 \times 2}{\begin{bmatrix} 1 & 0 \\ 0 & 1 \end{bmatrix}} = \underset{2 \times 2}{\begin{bmatrix} 3 & 4 \\ -2 & 5 \end{bmatrix}}$

b. $\underset{3 \times 3}{\begin{bmatrix} 6 & 2 & 0 \\ 3 & -1 & 2 \\ 1 & 4 & 6 \end{bmatrix}} \underset{3 \times 1}{\begin{bmatrix} 1 \\ 2 \\ -3 \end{bmatrix}} = \underset{3 \times 1}{\begin{bmatrix} 10 \\ -5 \\ -9 \end{bmatrix}}$

c. $\underset{1 \times 3}{\begin{bmatrix} 1 & -2 & -3 \end{bmatrix}} \underset{3 \times 1}{\begin{bmatrix} 2 \\ -1 \\ 1 \end{bmatrix}} = \underset{1 \times 1}{\begin{bmatrix} 1 \end{bmatrix}}$

d. $\underset{3 \times 1}{\begin{bmatrix} 2 \\ -1 \\ 1 \end{bmatrix}} \underset{1 \times 3}{\begin{bmatrix} 1 & -2 & -3 \end{bmatrix}} = \underset{3 \times 3}{\begin{bmatrix} 2 & -4 & -6 \\ -1 & 2 & 3 \\ 1 & -2 & -3 \end{bmatrix}}$

e. The product $\underset{3 \times 2}{\begin{bmatrix} -2 & 1 \\ 1 & -3 \\ 1 & 4 \end{bmatrix}} \underset{3 \times 4}{\begin{bmatrix} -2 & 3 & 1 & 4 \\ 0 & 1 & -1 & 2 \\ 2 & -1 & 0 & 1 \end{bmatrix}}$ is not defined.

> **ALGEBRA HELP**
>
> In Examples 9(c) and 9(d), note that the two products are different. Even when both AB and BA are defined, matrix multiplication is not, in general, commutative. That is, for most matrices, $AB \neq BA$. This is one way in which the algebra of real numbers and the algebra of matrices differ.

✓ *Checkpoint* ▶ Audio-video solution in English & Spanish at LarsonPrecalculus.com

Find each product, if possible.

a. $\begin{bmatrix} 1 \\ -3 \end{bmatrix} \begin{bmatrix} 3 & -1 \end{bmatrix}$ **b.** $\begin{bmatrix} 3 & -1 \end{bmatrix} \begin{bmatrix} 1 \\ -3 \end{bmatrix}$ **c.** $\begin{bmatrix} 3 & 1 & 2 \\ 7 & 0 & -2 \end{bmatrix} \begin{bmatrix} 6 & 4 \\ 2 & -1 \end{bmatrix}$ ■

GO DIGITAL

EXAMPLE 10 **Squaring a Matrix**

Find A^2, where $A = \begin{bmatrix} 3 & 1 \\ -1 & 2 \end{bmatrix}$. (*Note:* $A^2 = AA$.)

Solution

$$A^2 = \begin{bmatrix} 3 & 1 \\ -1 & 2 \end{bmatrix}\begin{bmatrix} 3 & 1 \\ -1 & 2 \end{bmatrix}$$

$$= \begin{bmatrix} 8 & 5 \\ -5 & 3 \end{bmatrix}$$

✓ *Checkpoint* ▶ *Audio-video solution in English & Spanish at LarsonPrecalculus.com*

Find A^2, where $A = \begin{bmatrix} 2 & 1 \\ 3 & -2 \end{bmatrix}$.

Properties of Matrix Multiplication

Let A, B, and C be matrices and let c be a scalar.

1. $A(BC) = (AB)C$ Associative Property of Matrix Multiplication

2. $A(B + C) = AB + AC$ Left Distributive Property

3. $(A + B)C = AC + BC$ Right Distributive Property

4. $c(AB) = (cA)B = A(cB)$ Associative Property of Scalar Multiplication

---

Definition of the Identity Matrix

The $n \times n$ matrix that consists of 1's on its main diagonal and 0's elsewhere is called the **identity matrix of dimension $n \times n$** and is denoted by

$$I_n = \begin{bmatrix} 1 & 0 & 0 & \cdots & 0 \\ 0 & 1 & 0 & \cdots & 0 \\ 0 & 0 & 1 & \cdots & 0 \\ \vdots & \vdots & \vdots & & \vdots \\ 0 & 0 & 0 & \cdots & 1 \end{bmatrix}. \qquad \text{Identity matrix}$$

Note that an identity matrix must be *square*. When the dimension is understood to be $n \times n$, you can denote I_n simply by I.

If A is an $n \times n$ matrix, then the identity matrix has the property that $AI_n = A$ and $I_nA = A$. For example,

$$\begin{bmatrix} 3 & -2 & 5 \\ 1 & 0 & 4 \\ -1 & 2 & -3 \end{bmatrix}\begin{bmatrix} 1 & 0 & 0 \\ 0 & 1 & 0 \\ 0 & 0 & 1 \end{bmatrix} = \begin{bmatrix} 3 & -2 & 5 \\ 1 & 0 & 4 \\ -1 & 2 & -3 \end{bmatrix} \qquad AI = A$$

and

$$\begin{bmatrix} 1 & 0 & 0 \\ 0 & 1 & 0 \\ 0 & 0 & 1 \end{bmatrix}\begin{bmatrix} 3 & -2 & 5 \\ 1 & 0 & 4 \\ -1 & 2 & -3 \end{bmatrix} = \begin{bmatrix} 3 & -2 & 5 \\ 1 & 0 & 4 \\ -1 & 2 & -3 \end{bmatrix}. \qquad IA = A$$

GO DIGITAL

Using Matrices to Transform Vectors

In Section 8.3, you performed vector operations with vectors written in component form and with vectors written as linear combinations of the standard unit vectors **i** and **j**. Another way to perform vector operations is with the vectors written as column matrices.

EXAMPLE 11   **Vector Operations**

Let $\mathbf{v} = \langle 2, 4 \rangle$ and $\mathbf{w} = \langle 6, 2 \rangle$. Use matrices to find each vector.

a. $\mathbf{v} + \mathbf{w}$ **b.** $\mathbf{w} - 2\mathbf{v}$

Solution   Begin by writing **v** and **w** as column matrices.

$$\mathbf{v} = \begin{bmatrix} 2 \\ 4 \end{bmatrix}, \qquad \mathbf{w} = \begin{bmatrix} 6 \\ 2 \end{bmatrix}$$

a. $\mathbf{v} + \mathbf{w} = \begin{bmatrix} 2 \\ 4 \end{bmatrix} + \begin{bmatrix} 6 \\ 2 \end{bmatrix} = \begin{bmatrix} 8 \\ 6 \end{bmatrix} = \langle 8, 6 \rangle$

Figure 10.1 shows a sketch of $\mathbf{v} + \mathbf{w}$.

b. $\mathbf{w} - 2\mathbf{v} = \begin{bmatrix} 6 \\ 2 \end{bmatrix} - 2\begin{bmatrix} 2 \\ 4 \end{bmatrix} = \begin{bmatrix} 6 \\ 2 \end{bmatrix} - \begin{bmatrix} 4 \\ 8 \end{bmatrix} = \begin{bmatrix} 2 \\ -6 \end{bmatrix} = \langle 2, -6 \rangle$

Figure 10.2 shows a sketch of $\mathbf{w} - 2\mathbf{v} = \mathbf{w} + (-2\mathbf{v})$.

✓ *Checkpoint* 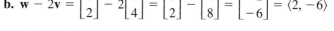 Audio-video solution in English & Spanish at LarsonPrecalculus.com

Let $\mathbf{v} = \langle 3, 6 \rangle$ and $\mathbf{w} = \langle 8, 5 \rangle$. Use matrices to find each vector.

a. $\mathbf{v} - \mathbf{w}$ **b.** $3\mathbf{v} + \mathbf{w}$

One way to transform a vector **v** is to multiply **v** by a square **transformation matrix** A to produce another vector $A\mathbf{v}$. A column matrix with two rows can represent a vector **v**, so the transformation matrix must have two columns (and also two rows) for $A\mathbf{v}$ to be defined.

EXAMPLE 12   **Describing a Vector Transformation**

Let $A = \begin{bmatrix} 1 & 0 \\ 0 & -1 \end{bmatrix}$ and $\mathbf{v} = \langle 1, 3 \rangle$. Find $A\mathbf{v}$ and describe the transformation.

Solution   First note that A has two columns and **v**, written as the column matrix

$$\mathbf{v} = \begin{bmatrix} 1 \\ 3 \end{bmatrix},$$

has two rows, so $A\mathbf{v}$ is defined.

$$A\mathbf{v} = \begin{bmatrix} 1 & 0 \\ 0 & -1 \end{bmatrix}\begin{bmatrix} 1 \\ 3 \end{bmatrix} = \begin{bmatrix} 1 \\ -3 \end{bmatrix} = \langle 1, -3 \rangle$$

Figure 10.3 shows a sketch of the vectors **v** and $A\mathbf{v}$. The matrix A transforms **v** by reflecting **v** in the x-axis.

✓ *Checkpoint*  Audio-video solution in English & Spanish at LarsonPrecalculus.com

Let $A = \begin{bmatrix} -1 & 0 \\ 0 & 1 \end{bmatrix}$ and $\mathbf{v} = \langle 3, 1 \rangle$. Find $A\mathbf{v}$ and describe the transformation.

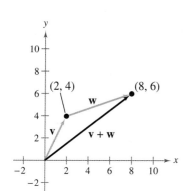

Figure 10.1

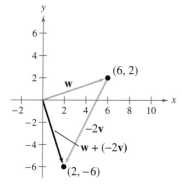

Figure 10.2

Figure 10.3

GO DIGITAL

Applications

Matrix multiplication can be used to represent a system of linear equations. Note how the system below can be written as the matrix equation $AX = B$, where A is the *coefficient matrix* of the system and X and B are column matrices. The column matrix B is also called a *constant matrix*. Its entries are the constant terms in the system of equations.

System

$$\begin{cases} a_{11}x_1 + a_{12}x_2 + a_{13}x_3 = b_1 \\ a_{21}x_1 + a_{22}x_2 + a_{23}x_3 = b_2 \\ a_{31}x_1 + a_{32}x_2 + a_{33}x_3 = b_3 \end{cases}$$

Matrix Equation $AX = B$

$$\underbrace{\begin{bmatrix} a_{11} & a_{12} & a_{13} \\ a_{21} & a_{22} & a_{23} \\ a_{31} & a_{32} & a_{33} \end{bmatrix}}_{A} \underbrace{\begin{bmatrix} x_1 \\ x_2 \\ x_3 \end{bmatrix}}_{X} = \underbrace{\begin{bmatrix} b_1 \\ b_2 \\ b_3 \end{bmatrix}}_{B}$$

In Example 13, $[A \vdots B]$ represents the augmented matrix formed when you *adjoin* matrix B to matrix A. Also, $[I \vdots X]$ represents the reduced row-echelon form of the augmented matrix that yields the solution of the system.

EXAMPLE 13　Solving a System of Linear Equations

For the system of linear equations, (a) write the system as a matrix equation, $AX = B$, and (b) use Gauss-Jordan elimination on $[A \vdots B]$ to solve for the matrix X.

$$\begin{cases} x_1 - 2x_2 + x_3 = -4 \\ x_2 + 2x_3 = 4 \\ 2x_1 + 3x_2 - 2x_3 = 2 \end{cases}$$

Solution

a. In matrix form, $AX = B$, the system is

$$\begin{bmatrix} 1 & -2 & 1 \\ 0 & 1 & 2 \\ 2 & 3 & -2 \end{bmatrix} \begin{bmatrix} x_1 \\ x_2 \\ x_3 \end{bmatrix} = \begin{bmatrix} -4 \\ 4 \\ 2 \end{bmatrix}.$$

b. Form the augmented matrix by adjoining matrix B to matrix A.

$$[A \vdots B] = \begin{bmatrix} 1 & -2 & 1 & \vdots & -4 \\ 0 & 1 & 2 & \vdots & 4 \\ 2 & 3 & -2 & \vdots & 2 \end{bmatrix}$$

Using Gauss-Jordan elimination, rewrite this matrix as

$$[I \vdots X] = \begin{bmatrix} 1 & 0 & 0 & \vdots & -1 \\ 0 & 1 & 0 & \vdots & 2 \\ 0 & 0 & 1 & \vdots & 1 \end{bmatrix}.$$

So, the solution of the matrix equation is

$$X = \begin{bmatrix} x_1 \\ x_2 \\ x_3 \end{bmatrix} = \begin{bmatrix} -1 \\ 2 \\ 1 \end{bmatrix}.$$

✓ *Checkpoint* ▶ *Audio-video solution in English & Spanish at LarsonPrecalculus.com*

For the system of linear equations, (a) write the system as a matrix equation, $AX = B$, and (b) use Gauss-Jordan elimination on $[A \vdots B]$ to solve for the matrix X.

$$\begin{cases} -2x_1 - 3x_2 = -4 \\ 6x_1 + x_2 = -36 \end{cases}$$

GO DIGITAL

EXAMPLE 14 **Softball Team Expenses**

Two softball teams submit equipment lists to their sponsors.

Equipment	Women's Team	Men's Team
Bats	12	15
Balls	45	38
Gloves	15	17

Each bat costs $80, each ball costs $4, and each glove costs $90. Use matrices to find the total cost of equipment for each team.

Solution Write the equipment lists E and the costs per item C in matrix form as

$$E = \begin{bmatrix} 12 & 15 \\ 45 & 38 \\ 15 & 17 \end{bmatrix}$$

and

$$C = \begin{bmatrix} 80 & 4 & 90 \end{bmatrix}.$$

The total cost of equipment for each team is the product

$$CE = \begin{bmatrix} 80 & 4 & 90 \end{bmatrix} \begin{bmatrix} 12 & 15 \\ 45 & 38 \\ 15 & 17 \end{bmatrix}$$

$$= \begin{bmatrix} 80(12) + 4(45) + 90(15) & 80(15) + 4(38) + 90(17) \end{bmatrix}$$

$$= \begin{bmatrix} 2490 & 2882 \end{bmatrix}.$$

So, the total cost of equipment for the women's team is $2490, and the total cost of equipment for the men's team is $2882.

✓ *Checkpoint*  *Audio-video solution in English & Spanish at LarsonPrecalculus.com*

Repeat Example 14 when each bat costs $100, each ball costs $3, and each glove costs $65. ■

ALGEBRA HELP

Notice in Example 14 that it is not possible to find the total cost using the product EC because EC is not defined. That is, the number of columns of E (2 columns) does not equal the number of rows of C (1 row).

Summarize (Section 10.2)

1. State the conditions under which two matrices are equal *(page 713)*. For an example involving matrix equality, see Example 1.

2. Explain how to add matrices *(page 714)*. For an example of matrix addition, see Example 2.

3. Explain how to multiply a matrix by a scalar *(page 714)*. For an example of scalar multiplication, see Example 3.

4. List the properties of matrix addition and scalar multiplication *(page 716)*. For examples of using these properties, see Examples 4–6.

5. Explain how to multiply two matrices *(page 718)*. For examples of matrix multiplication, see Examples 7–10.

6. Explain how to use matrices to transform vectors *(page 721)*. For examples involving matrices and vectors, see Examples 11 and 12.

7. Describe real-life applications of matrix operations *(pages 722 and 723, Examples 13 and 14)*.

GO DIGITAL

10.2 Exercises

See CalcChat.com for tutorial help and worked-out solutions to odd-numbered exercises.

Vocabulary and Concept Check

In Exercises 1 and 2, fill in the blanks.

1. When performing matrix operations, real numbers are usually referred to as _____.

2. A matrix consisting entirely of zeros is called a _____ matrix and is denoted by _____.

3. Matrix A is 2×4 and Matrix B is 3×2. Which product is defined, AB or BA?

4. Write the 4×4 identity matrix I_4.

Skills and Applications

Equality of Matrices In Exercises 5–8, solve for x and y.

5. $\begin{bmatrix} x & -2 \\ 7 & 23 \end{bmatrix} = \begin{bmatrix} -4 & -2 \\ 7 & y \end{bmatrix}$ 6. $\begin{bmatrix} -5 & x \\ 3y & 8 \end{bmatrix} = \begin{bmatrix} -5 & 13 \\ 12 & 8 \end{bmatrix}$

7. $\begin{bmatrix} 16 & 4 & x & 4 \\ 0 & 2 & 4 & 0 \end{bmatrix} = \begin{bmatrix} 16 & 4 & 2x+1 & 4 \\ 0 & 2 & 3y-5 & 0 \end{bmatrix}$

8. $\begin{bmatrix} x+2 & 8 & -3 \\ 1 & 18 & -8 \\ 7 & -2 & y+2 \end{bmatrix} = \begin{bmatrix} 2x+6 & 8 & -3 \\ 1 & 18 & -8 \\ 7 & -2 & x \end{bmatrix}$

Operations with Matrices In Exercises 9–16, if possible, find (a) $A + B$, (b) $A - B$, (c) $3A$, and (d) $3A - 2B$.

9. $A = \begin{bmatrix} 1 & -1 \\ 2 & -1 \end{bmatrix}$, $B = \begin{bmatrix} 2 & -1 \\ -1 & 8 \end{bmatrix}$

10. $A = \begin{bmatrix} 1 & 2 \\ 2 & 1 \end{bmatrix}$, $B = \begin{bmatrix} -3 & -2 \\ 4 & 2 \end{bmatrix}$

11. $A = \begin{bmatrix} 6 & 0 & 3 \\ -1 & -4 & 0 \end{bmatrix}$, $B = \begin{bmatrix} 8 & -1 \\ 4 & -3 \end{bmatrix}$

12. $A = \begin{bmatrix} 3 \\ 2 \\ -1 \end{bmatrix}$, $B = \begin{bmatrix} -4 & 6 & 2 \end{bmatrix}$

13. $A = \begin{bmatrix} 8 & -1 \\ 2 & 3 \\ -4 & 5 \end{bmatrix}$, $B = \begin{bmatrix} 1 & 6 \\ -1 & -5 \\ 1 & 10 \end{bmatrix}$

14. $A = \begin{bmatrix} 1 & -1 & 3 \\ 0 & 6 & 9 \end{bmatrix}$, $B = \begin{bmatrix} -2 & 0 & -5 \\ -3 & 4 & -7 \end{bmatrix}$

15. $A = \begin{bmatrix} 4 & 5 & -1 & 3 & 4 \\ 1 & 2 & -2 & -1 & 0 \end{bmatrix}$,

$B = \begin{bmatrix} 1 & 0 & -1 & 1 & 0 \\ -6 & 8 & 2 & -3 & -7 \end{bmatrix}$

16. $A = \begin{bmatrix} -1 & 4 & 0 \\ 3 & -2 & 2 \\ 5 & 4 & -1 \\ 0 & 8 & -6 \\ -4 & -1 & 0 \end{bmatrix}$, $B = \begin{bmatrix} -3 & 5 & 1 \\ 2 & -4 & -7 \\ 10 & -9 & -1 \\ 3 & 2 & -4 \\ 0 & 1 & -2 \end{bmatrix}$

Evaluating an Expression In Exercises 17–22, evaluate the expression.

17. $\begin{bmatrix} -5 & 0 \\ 3 & -6 \end{bmatrix} + \begin{bmatrix} 7 & 1 \\ -2 & -1 \end{bmatrix} + \begin{bmatrix} -10 & -8 \\ 14 & 6 \end{bmatrix}$

18. $\begin{bmatrix} 6 & 8 \\ -1 & 0 \end{bmatrix} + \begin{bmatrix} 0 & 5 \\ -3 & -1 \end{bmatrix} + \begin{bmatrix} -11 & -7 \\ 2 & -1 \end{bmatrix}$

19. $4\left(\begin{bmatrix} -4 & 0 & 1 \\ 0 & 2 & 3 \end{bmatrix} - \begin{bmatrix} 2 & 1 & -2 \\ 3 & -6 & 0 \end{bmatrix} \right)$

20. $\frac{1}{2}\left(\begin{bmatrix} 5 & -2 & 4 & 0 \end{bmatrix} + \begin{bmatrix} 14 & 6 & -18 & 9 \end{bmatrix} \right)$

21. $-3\left(\begin{bmatrix} 0 & -3 \\ 7 & 2 \end{bmatrix} + \begin{bmatrix} -6 & 3 \\ 8 & 1 \end{bmatrix} \right) - 2\begin{bmatrix} 4 & -4 \\ 7 & -9 \end{bmatrix}$

22. $-1\begin{bmatrix} 4 & 11 \\ -2 & -1 \\ 9 & 3 \end{bmatrix} + \frac{1}{6}\left(\begin{bmatrix} -5 & -1 \\ 3 & 4 \\ 0 & 13 \end{bmatrix} + \begin{bmatrix} 7 & 5 \\ -9 & -1 \\ 6 & -1 \end{bmatrix} \right)$

Solving a Matrix Equation In Exercises 23–30, solve for X in the equation, where

$$A = \begin{bmatrix} -2 & 1 & 3 \\ -1 & 0 & 4 \end{bmatrix} \text{ and } B = \begin{bmatrix} 0 & 2 & -4 \\ 3 & 0 & 1 \end{bmatrix}.$$

23. $X = 2A + 2B$ 24. $X = 3A - 2B$

25. $2X = 2A - B$ 26. $2X = A + B$

27. $2X + 3A = B$ 28. $3X - 4A = 2B$

29. $4B = -2X - 2A$ 30. $5A = 6B - 3X$

Finding the Product of Two Matrices In Exercises 31–36, if possible, find AB and state the dimension of the result.

31. $A = \begin{bmatrix} -1 & 6 \\ -4 & 5 \\ 0 & 3 \end{bmatrix}$, $B = \begin{bmatrix} 2 & 3 \\ 0 & 9 \end{bmatrix}$

32. $A = \begin{bmatrix} 0 & -1 & 2 \\ 6 & 0 & 3 \\ 7 & -1 & 8 \end{bmatrix}$, $B = \begin{bmatrix} 2 & -1 \\ 4 & -5 \\ 1 & 6 \end{bmatrix}$

33. $A = \begin{bmatrix} 2 & 1 \\ -3 & 4 \\ 1 & 6 \end{bmatrix}$, $B = \begin{bmatrix} 0 & -1 & 0 \\ 4 & 0 & 2 \\ 8 & -1 & 7 \end{bmatrix}$

34. $A = \begin{bmatrix} 1 & 0 & 3 & -2 \\ 6 & 13 & 8 & -17 \end{bmatrix}$, $B = \begin{bmatrix} 1 & 6 \\ 4 & 2 \end{bmatrix}$

35. $A = \begin{bmatrix} 5 & 0 & 0 \\ 0 & -8 & 0 \\ 0 & 0 & 7 \end{bmatrix}$, $B = \begin{bmatrix} \frac{1}{5} & 0 & 0 \\ 0 & -\frac{1}{8} & 0 \\ 0 & 0 & \frac{1}{2} \end{bmatrix}$

36. $A = \begin{bmatrix} 0 & 0 & 5 \\ 0 & 0 & -3 \\ 0 & 0 & 4 \end{bmatrix}$, $B = \begin{bmatrix} 6 & -11 & 4 & 2 \\ 8 & 16 & 4 & -4 \\ 0 & 0 & 0 & 0 \end{bmatrix}$

Operations with Matrices In Exercises 37–44, if possible, find (a) AB, (b) BA, and (c) A^2.

37. $A = \begin{bmatrix} 1 & 2 \\ 4 & 2 \end{bmatrix}$, $B = \begin{bmatrix} 2 & -1 \\ -1 & 8 \end{bmatrix}$

38. $A = \begin{bmatrix} 6 & 3 \\ -2 & -4 \end{bmatrix}$, $B = \begin{bmatrix} -2 & 0 \\ 2 & 4 \end{bmatrix}$

39. $A = \begin{bmatrix} 2 & -2 \\ -3 & 0 \\ 7 & 6 \end{bmatrix}$, $B = \begin{bmatrix} 1 & 0 \\ 0 & 1 \end{bmatrix}$

40. $A = \begin{bmatrix} 5 & -9 & 0 \\ 3 & 0 & -8 \\ -1 & 4 & 11 \end{bmatrix}$, $B = \begin{bmatrix} 1 & 0 & 0 \\ 0 & 1 & 0 \\ 0 & 0 & 1 \end{bmatrix}$

41. $A = \begin{bmatrix} -4 & -1 \\ 2 & 12 \end{bmatrix}$, $B = \begin{bmatrix} -6 \\ 5 \end{bmatrix}$

42. $A = \begin{bmatrix} 1 & 3 & -2 \\ -5 & 10 & 1 \end{bmatrix}$, $B = \begin{bmatrix} 3 \\ 3 \\ 3 \end{bmatrix}$

43. $A = \begin{bmatrix} 7 \\ 8 \\ -1 \end{bmatrix}$, $B = \begin{bmatrix} 1 & 1 & 2 \end{bmatrix}$

44. $A = \begin{bmatrix} 3 & 2 & 1 & 4 \end{bmatrix}$, $B = \begin{bmatrix} 2 \\ 3 \\ 0 \\ 1 \end{bmatrix}$

Operations with Matrices In Exercises 45–50, evaluate the expression. Use the matrix capabilities of a graphing utility to verify your answer.

45. $\begin{bmatrix} 7 & 5 & -4 \\ -2 & 5 & 1 \\ 10 & -4 & -7 \end{bmatrix}\begin{bmatrix} 2 & -2 & 3 \\ 8 & 1 & 4 \\ -4 & 2 & -8 \end{bmatrix}$

46. $-3\left(\begin{bmatrix} 6 & 5 & -1 \\ 1 & -2 & 0 \end{bmatrix}\begin{bmatrix} 0 & 3 \\ -1 & -3 \\ 4 & 1 \end{bmatrix}\right)$

47. $\begin{bmatrix} 3 & 1 \\ 0 & -2 \end{bmatrix}\begin{bmatrix} 1 & 0 \\ -2 & 2 \end{bmatrix}\begin{bmatrix} 1 & 0 \\ 2 & 4 \end{bmatrix}$

48. $\begin{bmatrix} 3 \\ -1 \\ 5 \\ 7 \end{bmatrix}(\begin{bmatrix} 5 & -6 \end{bmatrix} + \begin{bmatrix} 7 & -1 \end{bmatrix} + \begin{bmatrix} -8 & 9 \end{bmatrix})$

49. $\begin{bmatrix} 0 & 2 & -2 \\ 4 & 1 & 2 \end{bmatrix}\left(\begin{bmatrix} 4 & 0 \\ 0 & -1 \\ -1 & 2 \end{bmatrix} + \begin{bmatrix} -2 & 3 \\ -3 & 5 \\ 0 & -3 \end{bmatrix}\right)$

50. $0.5\left(\begin{bmatrix} -2 & -5 & 0 \\ 14 & 10 & 1 \\ 0 & 6 & 2 \end{bmatrix} + \begin{bmatrix} 3 & -4 & 3 \\ -1 & 0 & 0 \\ 7 & 4 & 1 \end{bmatrix}\right)^2$

Vector Operations In Exercises 51–54, use matrices to find (a) $\mathbf{u} + \mathbf{v}$, (b) $\mathbf{u} - \mathbf{v}$, and (c) $3\mathbf{v} - \mathbf{u}$.

51. $\mathbf{u} = \langle 1, 5 \rangle$, $\mathbf{v} = \langle 3, 2 \rangle$

52. $\mathbf{u} = \langle 4, 2 \rangle$, $\mathbf{v} = \langle 6, -3 \rangle$

53. $\mathbf{u} = \langle -2, 2 \rangle$, $\mathbf{v} = \langle 5, 4 \rangle$

54. $\mathbf{u} = \langle 7, -4 \rangle$, $\mathbf{v} = \langle 2, 1 \rangle$

Describing a Vector Transformation In Exercises 55–60, find $A\mathbf{v}$, where $\mathbf{v} = \langle 4, 2 \rangle$, and describe the transformation.

55. $A = \begin{bmatrix} 1 & 0 \\ 0 & -1 \end{bmatrix}$ 　　 56. $A = \begin{bmatrix} -1 & 0 \\ 0 & 1 \end{bmatrix}$

57. $A = \begin{bmatrix} 0 & 1 \\ 1 & 0 \end{bmatrix}$ 　　 58. $A = \begin{bmatrix} 0 & -1 \\ -1 & 0 \end{bmatrix}$

59. $A = \begin{bmatrix} 2 & 0 \\ 0 & 1 \end{bmatrix}$ 　　 60. $A = \begin{bmatrix} 1 & 0 \\ 0 & 3 \end{bmatrix}$

Solving a System of Linear Equations In Exercises 61–66, (a) write the system of linear equations as a matrix equation, $AX = B$, and (b) use Gauss-Jordan elimination on $[A \vdots B]$ to solve for the matrix X.

61. $\begin{cases} 2x_1 + 3x_2 = 5 \\ x_1 + 4x_2 = 10 \end{cases}$

62. $\begin{cases} -2x_1 - 3x_2 = -4 \\ 6x_1 + x_2 = -36 \end{cases}$

63. $\begin{cases} x_1 - 2x_2 + 3x_3 = 9 \\ -x_1 + 3x_2 - x_3 = -6 \\ 2x_1 - 5x_2 + 5x_3 = 17 \end{cases}$

64. $\begin{cases} x_1 + x_2 - 3x_3 = -1 \\ -x_1 + 2x_2 = 1 \\ x_1 - x_2 + x_3 = 2 \end{cases}$

65. $\begin{cases} x_1 - 5x_2 + 2x_3 = -20 \\ -3x_1 + x_2 - x_3 = 8 \\ -2x_2 + 5x_3 = -16 \end{cases}$

66. $\begin{cases} x_1 - x_2 + 4x_3 = 17 \\ x_1 + 3x_2 = -11 \\ -6x_2 + 5x_3 = 40 \end{cases}$

67. Manufacturing A car manufacturer has four factories that produce sport utility vehicles and pickup trucks. The production levels are represented by A.

	Factory				
	1	2	3	4	
$A =$	100	90	70	30	SUV ⎫ Vehicle
	40	20	60	60	Pickup ⎭ Type

Find the production levels after they increase by 10%.

68. Vacation Packages A travel agent identifies four resorts with special all-inclusive packages. The current rates for two types of rooms (double and quadruple occupancy) at the four resorts are represented by A.

	Resort w	Resort x	Resort y	Resort z	
$A =$	615	670	740	990	Double ⎫ Occupancy
	995	1030	1180	1105	Quadruple ⎭

The rates are expected to increase by no more than 12% by next season. Find the maximum rate per package per resort.

69. Agriculture A farmer grows apples and peaches. Each crop is shipped to three different outlets. The shipment levels are represented by A.

	Outlet			
	1	2	3	
$A =$	125	100	75	Apples ⎫ Crop
	100	175	125	Peaches ⎭

The profits per unit are represented by the matrix $B = [\$3.50 \quad \$6.00]$. Compute and interpret BA.

70. Revenue An electronics manufacturer produces three models of high-definition televisions, which are shipped to two warehouses. The shipment levels are represented by A.

	Warehouse		
	1	2	
$A =$	5,000	4,000	A ⎫
	6,000	10,000	B ⎬ Model
	8,000	5,000	C ⎭

The prices per unit are represented by the matrix $B = [\$699.95 \quad \$899.95 \quad \$1099.95]$. Compute and interpret BA.

71. Labor and Wages A company has two factories that manufacture three sizes of boats. The numbers of hours of labor required to manufacture each size are represented by S.

	Department			
	Cutting	Assembly	Packaging	
$S =$	1.0	0.5	0.2	Small ⎫
	1.6	1.0	0.2	Medium ⎬ Boat size
	2.5	2.0	1.4	Large ⎭

The wages of the workers are represented by T.

	Factory		
	A	B	
$T =$	\$15	\$13	Cutting ⎫
	\$12	\$11	Assembly ⎬ Department
	\$11	\$10	Packaging ⎭

Compute and interpret ST.

72. Profit At a grocery store, the numbers of gallons of skim milk, 2% milk, and whole milk sold on Friday, Saturday, and Sunday are represented by A.

	Skim milk	2% milk	Whole milk	
$A =$	40	64	52	Friday
	60	82	76	Saturday
	76	96	84	Sunday

The selling prices per gallon and the profits per gallon for the three types of milk are represented by B.

	Selling price	Profit	
$B =$	\$3.45	\$1.20	Skim milk
	\$3.65	\$1.30	2% milk
	\$3.85	\$1.45	Whole milk

(a) Compute and interpret AB.

(b) Find the store's total profit from milk sales for the three days.

73. Exercise

The numbers of calories burned by individuals of different body weights while performing different types of exercises for a one-hour time period are represented by A.

	Calories burned		
	130-lb person	155-lb person	
$A =$	472	563	Basketball
	590	704	Jumping rope
	177	211	Weight lifting

(a) On Saturday, a 130-pound person and a 155-pound person play basketball for 2 hours, jump rope for 15 minutes, and lift weights for 30 minutes. Organize the times spent exercising in a matrix B.

(b) Compute BA and interpret the result.

(c) On Sunday, the same two people play basketball for 1 hour, jump rope for 18 minutes, and lift weights for 45 minutes. How many fewer calories does each person burn exercising than on Saturday?

74. Voting Preferences The matrix

$$P = \begin{bmatrix} 0.6 & 0.1 & 0.1 \\ 0.2 & 0.7 & 0.1 \\ 0.2 & 0.2 & 0.8 \end{bmatrix} \begin{matrix} R \\ D \\ I \end{matrix} \bigg\} \text{To}$$

From
R D I

is called a *stochastic matrix*. Each entry p_{ij} ($i \neq j$) represents the proportion of the voting population that changes from party i to party j, and p_{ii} represents the proportion that remains loyal to the party from one election to the next. Compute and interpret P^2.

Exploring the Concepts

True or False? **In Exercises 75 and 76, determine whether the statement is true or false. Justify your answer.**

75. Matrix multiplication is commutative.

76. If $A + B$ is defined for two matrices A and B, then AB is defined.

Think About It **In Exercises 77–80, use the matrices**

$$A = \begin{bmatrix} 2 & -1 \\ 1 & 3 \end{bmatrix} \quad \text{and} \quad B = \begin{bmatrix} -1 & 1 \\ 0 & -2 \end{bmatrix}.$$

77. Show that $(A + B)^2 \neq A^2 + 2AB + B^2$.

78. Show that $(A - B)^2 \neq A^2 - 2AB + B^2$.

79. Show that $(A + B)(A - B) \neq A^2 - B^2$.

80. Show that $(A + B)^2 = A^2 + AB + BA + B^2$.

81. Think About It If a, b, and c are real numbers such that $c \neq 0$ and $ac = bc$, then $a = b$. However, if A, B, and C are nonzero matrices such that $AC = BC$, then A is *not necessarily* equal to B. Illustrate this using the following matrices.

$$A = \begin{bmatrix} 0 & 1 \\ 0 & 1 \end{bmatrix}, \quad B = \begin{bmatrix} 1 & 0 \\ 1 & 0 \end{bmatrix}, \quad C = \begin{bmatrix} 2 & 3 \\ 2 & 3 \end{bmatrix}$$

82. Think About It If a and b are real numbers such that $ab = 0$, then $a = 0$ or $b = 0$. However, if A and B are matrices such that $AB = O$, it is *not necessarily* true that $A = O$ or $B = O$. Illustrate this using the following matrices.

$$A = \begin{bmatrix} 3 & 3 \\ 4 & 4 \end{bmatrix}, \quad B = \begin{bmatrix} 1 & -1 \\ -1 & 1 \end{bmatrix}$$

83. Conjecture Let A and B be unequal diagonal matrices of the same dimension. (A **diagonal matrix** is a square matrix in which each entry not on the main diagonal is zero.) Determine the products AB for several pairs of such matrices. Make a conjecture about a rule that can be used to calculate AB without using row-by-column multiplication.

84. HOW DO YOU SEE IT? A corporation has three factories that manufacture acoustic guitars and electric guitars. The production levels are represented by A.

Factory
A B C

$$A = \begin{bmatrix} 70 & 50 & 25 \\ 35 & 100 & 70 \end{bmatrix} \begin{matrix} \text{Acoustic} \\ \text{Electric} \end{matrix} \bigg\} \text{Guitar type}$$

(a) Interpret the value of a_{22}.

(b) How could you find the production levels when production increases by 20%?

(c) Each acoustic guitar sells for \$80 and each electric guitar sells for \$120. How could you use matrices to find the total sales value of the guitars produced at each factory?

85. Finding Matrices Find two matrices A and B such that $AB = BA$.

86. Transforming a Vector Find a matrix such that a vector is reflected about the origin when multiplied by the matrix.

87. Matrices with Complex Entries Let $i = \sqrt{-1}$ and let

$$A = \begin{bmatrix} i & 0 \\ 0 & i \end{bmatrix} \quad \text{and} \quad B = \begin{bmatrix} 0 & -i \\ i & 0 \end{bmatrix}.$$

(a) Find A^2, A^3, and A^4. Identify any similarities with i^2, i^3, and i^4.

(b) Find and identify B^2.

Review & Refresh ▶ Video solutions at LarsonPrecalculus.com

88. Navigation A cargo ship is 120 miles west and 80 miles north of a wharf. The captain wants to sail directly to the wharf. What bearing should the captain take?

Expanding a Logarithmic Expression **In Exercises 89–92, use the properties of logarithms to expand the expression as a sum, difference, and/or constant multiple of logarithms. (Assume all variables are positive.)**

89. $\log_3 15x$

90. $\ln \sqrt{4pq}$

91. $\ln \dfrac{a}{b^3}$

92. $\log \dfrac{x^2 + x}{y^2 z}$

Solving a System of Linear Equations **In Exercises 93–96, solve the system of linear equations and check any solutions algebraically.**

93. $\begin{cases} 5x - 2y = 2 \\ 5x + 4y = 8 \end{cases}$

94. $\begin{cases} x + 3y = 14 \\ -4x - y = 10 \end{cases}$

95. $\begin{cases} 4x + y - z = -5 \\ -x - 6y + 2z = -16 \\ x + 3y + z = 7 \end{cases}$

96. $\begin{cases} 9x + 3y + z = 2 \\ 2x - 2y = 12 \\ 3x + 4z = 35 \end{cases}$

10.3 The Inverse of a Square Matrix

Inverse matrices are used to model and solve real-life problems. For example, in Exercises 55–58 on page 735, you will use an inverse matrix to find the currents in a circuit.

❯ Verify that two matrices are inverses of each other.
❯ Use Gauss-Jordan elimination to find the inverses of matrices.
❯ Use a formula to find the inverses of 2 × 2 matrices.
❯ Use inverse matrices to solve systems of linear equations.

The Inverse of a Matrix

This section further develops the algebra of matrices. To begin, consider the real number equation $ax = b$. To solve this equation for x, multiply each side of the equation by a^{-1} (provided that $a \neq 0$).

$$ax = b$$
$$(a^{-1}a)x = a^{-1}b$$
$$(1)x = a^{-1}b$$
$$x = a^{-1}b$$

The number a^{-1} is called the *multiplicative inverse* of a because $a^{-1}a = 1$. The multiplicative **inverse of a matrix** is defined in a similar way.

Definition of the Inverse of a Square Matrix

Let A be an $n \times n$ matrix and let I_n be the $n \times n$ identity matrix. If there exists a matrix A^{-1} such that

$$AA^{-1} = I_n = A^{-1}A$$

then A^{-1} is the **inverse** of A. The symbol A^{-1} is read as "A inverse."

EXAMPLE 1 The Inverse of a Matrix

Show that $B = \begin{bmatrix} 1 & -2 \\ 1 & -1 \end{bmatrix}$ is the inverse of $A = \begin{bmatrix} -1 & 2 \\ -1 & 1 \end{bmatrix}$.

Solution To show that B is the inverse of A, show that $AB = I = BA$.

$$AB = \begin{bmatrix} -1 & 2 \\ -1 & 1 \end{bmatrix}\begin{bmatrix} 1 & -2 \\ 1 & -1 \end{bmatrix} = \begin{bmatrix} -1+2 & 2-2 \\ -1+1 & 2-1 \end{bmatrix} = \begin{bmatrix} 1 & 0 \\ 0 & 1 \end{bmatrix}$$

$$BA = \begin{bmatrix} 1 & -2 \\ 1 & -1 \end{bmatrix}\begin{bmatrix} -1 & 2 \\ -1 & 1 \end{bmatrix} = \begin{bmatrix} -1+2 & 2-2 \\ -1+1 & 2-1 \end{bmatrix} = \begin{bmatrix} 1 & 0 \\ 0 & 1 \end{bmatrix}$$

So, B is the inverse of A because $AB = I = BA$. This is an example of a square matrix that has an inverse. Note that not all square matrices have inverses.

✓ **Checkpoint** ▶ Audio-video solution in English & Spanish at LarsonPrecalculus.com

Show that $B = \begin{bmatrix} -1 & -1 \\ -3 & -2 \end{bmatrix}$ is the inverse of $A = \begin{bmatrix} 2 & -1 \\ -3 & 1 \end{bmatrix}$. ■

Recall that it is not always true that $AB = BA$, even when both products are defined. However, if A and B are both square matrices and $AB = I_n$, then it can be shown that $BA = I_n$. So, in Example 1, you need only to check that $AB = I_2$.

GO DIGITAL

Finding Inverse Matrices

If a matrix A has an inverse, then A is **invertible,** or **nonsingular;** otherwise, A is **singular.** A nonsquare matrix cannot have an inverse. To see this, note that when A is of dimension $m \times n$ and B is of dimension $n \times m$ (where $m \neq n$), the products AB and BA are of different dimensions and so cannot be equal to each other. Not all square matrices have inverses (see the matrix at the bottom of page 731). When a matrix does have an inverse, however, that inverse is unique. Example 2 shows how to use a system of equations to find the inverse of a matrix.

EXAMPLE 2   Finding the Inverse of a Matrix

Find the inverse of $A = \begin{bmatrix} 1 & 4 \\ -1 & -3 \end{bmatrix}$.

Solution   To find the inverse of A, solve the matrix equation $AX = I$ for X.

$$\overset{A}{\begin{bmatrix} 1 & 4 \\ -1 & -3 \end{bmatrix}} \overset{X}{\begin{bmatrix} x_{11} & x_{12} \\ x_{21} & x_{22} \end{bmatrix}} \overset{=}{=} \overset{I}{\begin{bmatrix} 1 & 0 \\ 0 & 1 \end{bmatrix}} \qquad \text{Write matrix equation.}$$

$$\begin{bmatrix} x_{11} + 4x_{21} & x_{12} + 4x_{22} \\ -x_{11} - 3x_{21} & -x_{12} - 3x_{22} \end{bmatrix} = \begin{bmatrix} 1 & 0 \\ 0 & 1 \end{bmatrix} \qquad \text{Multiply.}$$

Equating corresponding entries, you obtain two systems of linear equations.

$$\begin{cases} x_{11} + 4x_{21} = 1 \\ -x_{11} - 3x_{21} = 0 \end{cases} \qquad \begin{cases} x_{12} + 4x_{22} = 0 \\ -x_{12} - 3x_{22} = 1 \end{cases}$$

Solve the first system using elementary row operations to determine that

$$x_{11} = -3 \quad \text{and} \quad x_{21} = 1.$$

Solve the second system to determine that

$$x_{12} = -4 \quad \text{and} \quad x_{22} = 1.$$

So, the inverse of A is

$$X = A^{-1}$$

$$= \begin{bmatrix} -3 & -4 \\ 1 & 1 \end{bmatrix}.$$

Use matrix multiplication to check this result in two ways.

Check

$$AA^{-1} = \begin{bmatrix} 1 & 4 \\ -1 & -3 \end{bmatrix} \begin{bmatrix} -3 & -4 \\ 1 & 1 \end{bmatrix}$$

$$= \begin{bmatrix} 1 & 0 \\ 0 & 1 \end{bmatrix} \checkmark$$

$$A^{-1}A = \begin{bmatrix} -3 & -4 \\ 1 & 1 \end{bmatrix} \begin{bmatrix} 1 & 4 \\ -1 & -3 \end{bmatrix}$$

$$= \begin{bmatrix} 1 & 0 \\ 0 & 1 \end{bmatrix} \checkmark$$

✓ **Checkpoint**  *Audio-video solution in English & Spanish at LarsonPrecalculus.com*

Find the inverse of $A = \begin{bmatrix} 1 & -2 \\ -1 & 3 \end{bmatrix}$. ■

GO DIGITAL

In Example 2, note that the two systems of linear equations have the *same coefficient matrix A*. Rather than solve the two systems represented by

$$\begin{bmatrix} 1 & 4 & \vdots & 1 \\ -1 & -3 & \vdots & 0 \end{bmatrix}$$

and

$$\begin{bmatrix} 1 & 4 & \vdots & 0 \\ -1 & -3 & \vdots & 1 \end{bmatrix}$$

separately, you can solve them *simultaneously* by *adjoining* the identity matrix to the coefficient matrix to obtain

$$\begin{matrix} A & & & I \\ \begin{bmatrix} 1 & 4 & \vdots & 1 & 0 \\ -1 & -3 & \vdots & 0 & 1 \end{bmatrix} \end{matrix}.$$

This "doubly augmented" matrix can be represented as

$$[A \;\vdots\; I].$$

By applying Gauss-Jordan elimination to this matrix, you can solve *both* systems with a single elimination process.

$$\begin{bmatrix} 1 & 4 & \vdots & 1 & 0 \\ -1 & -3 & \vdots & 0 & 1 \end{bmatrix}$$

$$\begin{matrix} & \begin{bmatrix} 1 & 4 & \vdots & 1 & 0 \\ R_1 + R_2 \rightarrow & 0 & 1 & \vdots & 1 & 1 \end{bmatrix} \end{matrix}$$

$$\begin{matrix} -4R_2 + R_1 \rightarrow & \begin{bmatrix} 1 & 0 & \vdots & -3 & -4 \\ 0 & 1 & \vdots & 1 & 1 \end{bmatrix} \end{matrix}$$

So, from the "doubly augmented" matrix $[A \;\vdots\; I]$, you obtain the matrix $[I \;\vdots\; A^{-1}]$.

$$\begin{matrix} A & & & I \\ \begin{bmatrix} 1 & 4 & \vdots & 1 & 0 \\ -1 & -3 & \vdots & 0 & 1 \end{bmatrix} \end{matrix} \implies \begin{matrix} I & & & A^{-1} \\ \begin{bmatrix} 1 & 0 & \vdots & -3 & -4 \\ 0 & 1 & \vdots & 1 & 1 \end{bmatrix} \end{matrix}$$

This procedure (or algorithm) works for any square matrix that has an inverse.

TECHNOLOGY

Most graphing utilities can find the inverse of a square matrix. To do so, you may have to use the inverse key $\boxed{x^{-1}}$. Consult the user's guide for your graphing utility for specific keystrokes.

Finding an Inverse Matrix

Let A be a square matrix of dimension $n \times n$.

1. Write the $n \times 2n$ matrix that consists of the given matrix A on the left and the $n \times n$ identity matrix I on the right to obtain

   $$[A \;\vdots\; I].$$

2. If possible, row reduce A to I using elementary row operations on the *entire* matrix

   $$[A \;\vdots\; I].$$

   The result will be the matrix

   $$[I \;\vdots\; A^{-1}].$$

   If this is not possible, then A is not invertible.

3. Check your work by multiplying to see that

   $$AA^{-1} = I = A^{-1}A.$$

EXAMPLE 3   **Finding the Inverse of a Matrix**

Find the inverse of

$$A = \begin{bmatrix} 1 & -1 & 0 \\ 1 & 0 & -1 \\ 6 & -2 & -3 \end{bmatrix}.$$

Solution   Begin by adjoining the identity matrix to A to form the matrix

$$[A \;\vdots\; I] = \begin{bmatrix} 1 & -1 & 0 & \vdots & 1 & 0 & 0 \\ 1 & 0 & -1 & \vdots & 0 & 1 & 0 \\ 6 & -2 & -3 & \vdots & 0 & 0 & 1 \end{bmatrix}.$$

Use elementary row operations to obtain the form $[I \;\vdots\; A^{-1}]$.

$$\begin{matrix} \\ -R_1 + R_2 \rightarrow \\ -6R_1 + R_3 \rightarrow \end{matrix} \begin{bmatrix} 1 & -1 & 0 & \vdots & 1 & 0 & 0 \\ 0 & 1 & -1 & \vdots & -1 & 1 & 0 \\ 0 & 4 & -3 & \vdots & -6 & 0 & 1 \end{bmatrix}$$

$$\begin{matrix} R_2 + R_1 \rightarrow \\ \\ -4R_2 + R_3 \rightarrow \end{matrix} \begin{bmatrix} 1 & 0 & -1 & \vdots & 0 & 1 & 0 \\ 0 & 1 & -1 & \vdots & -1 & 1 & 0 \\ 0 & 0 & 1 & \vdots & -2 & -4 & 1 \end{bmatrix}$$

$$\begin{matrix} R_3 + R_1 \rightarrow \\ R_3 + R_2 \rightarrow \\ \\ \end{matrix} \begin{bmatrix} 1 & 0 & 0 & \vdots & -2 & -3 & 1 \\ 0 & 1 & 0 & \vdots & -3 & -3 & 1 \\ 0 & 0 & 1 & \vdots & -2 & -4 & 1 \end{bmatrix} = [I \;\vdots\; A^{-1}]$$

So, the matrix A is invertible and its inverse is

$$A^{-1} = \begin{bmatrix} -2 & -3 & 1 \\ -3 & -3 & 1 \\ -2 & -4 & 1 \end{bmatrix}.$$

Check

$$AA^{-1} = \begin{bmatrix} 1 & -1 & 0 \\ 1 & 0 & -1 \\ 6 & -2 & -3 \end{bmatrix}\begin{bmatrix} -2 & -3 & 1 \\ -3 & -3 & 1 \\ -2 & -4 & 1 \end{bmatrix} = \begin{bmatrix} 1 & 0 & 0 \\ 0 & 1 & 0 \\ 0 & 0 & 1 \end{bmatrix} = I$$

✓ *Checkpoint* *Audio-video solution in English & Spanish at LarsonPrecalculus.com*

Find the inverse of

$$A = \begin{bmatrix} 1 & -2 & -1 \\ 0 & -1 & 2 \\ 1 & -2 & 0 \end{bmatrix}.$$

The process shown in Example 3 applies to any $n \times n$ matrix A. When using this algorithm, if the matrix A does not reduce to the identity matrix, then A does not have an inverse. For example, the matrix below has no inverse.

$$A = \begin{bmatrix} 1 & 2 & 0 \\ 3 & -1 & 2 \\ -2 & 3 & -2 \end{bmatrix}$$

To confirm that this matrix has no inverse, adjoin the identity matrix to A to form $[A \;\vdots\; I]$ and try to apply Gauss-Jordan elimination to the matrix. You will find that it is impossible to obtain the identity matrix I on the left. So, A is not invertible.

ALGEBRA HELP

Be sure to check your solution, because it is not uncommon to make arithmetic errors when using elementary row operations.

GO DIGITAL

The Inverse of a 2 × 2 Matrix

Using Gauss-Jordan elimination to find the inverse of a matrix works well (even as a computer technique) for matrices of dimension 3 × 3 or greater. For 2 × 2 matrices, however, many people prefer to use a formula for the inverse rather than Gauss-Jordan elimination. This simple formula, which works *only* for 2 × 2 matrices, is explained as follows. A 2 × 2 matrix A given by

$$A = \begin{bmatrix} a & b \\ c & d \end{bmatrix}$$

is invertible if and only if

$$ad - bc \neq 0.$$

Moreover, if $ad - bc \neq 0$, then the inverse is given by

$$A^{-1} = \frac{1}{ad - bc} \begin{bmatrix} d & -b \\ -c & a \end{bmatrix}. \qquad \text{Formula for the inverse of a 2 × 2 matrix}$$

The denominator

$$ad - bc$$

is the **determinant** of the 2 × 2 matrix A. You will study determinants in the next section.

EXAMPLE 4   Finding the Inverse of a 2 × 2 Matrix

▶▶▶ *See LarsonPrecalculus.com for an interactive version of this type of example.*

If possible, find the inverse of each matrix.

a. $A = \begin{bmatrix} 3 & -1 \\ -2 & 2 \end{bmatrix}$ **b.** $B = \begin{bmatrix} 3 & -1 \\ -6 & 2 \end{bmatrix}$

Solution

a. The determinant of matrix A is

$$ad - bc = 3(2) - (-1)(-2) = 4.$$

This quantity is not zero, so the matrix is invertible. The inverse is formed by interchanging the entries on the main diagonal, changing the signs of the other two entries, and multiplying by the scalar $\frac{1}{4}$.

$$A^{-1} = \frac{1}{ad - bc} \begin{bmatrix} d & -b \\ -c & a \end{bmatrix} \qquad \text{Formula for the inverse of a 2 × 2 matrix}$$

$$= \frac{1}{4} \begin{bmatrix} 2 & 1 \\ 2 & 3 \end{bmatrix} \qquad \text{Substitute for } a, b, c, d, \text{ and the determinant.}$$

$$= \begin{bmatrix} \frac{1}{2} & \frac{1}{4} \\ \frac{1}{2} & \frac{3}{4} \end{bmatrix} \qquad \text{Multiply by the scalar } \frac{1}{4}.$$

b. The determinant of matrix B is

$$ad - bc = 3(2) - (-1)(-6) = 0.$$

Because $ad - bc = 0$, B is not invertible.

✓ **Checkpoint** ▶ *Audio-video solution in English & Spanish at LarsonPrecalculus.com*

If possible, find the inverse of $A = \begin{bmatrix} 5 & -1 \\ 3 & 4 \end{bmatrix}$.   ■

Systems of Linear Equations

You know that a system of linear equations can have exactly one solution, infinitely many solutions, or no solution. If the coefficient matrix A of a *square* system (a system that has the same number of equations as variables) is invertible, then the system has a unique solution, which can be found using an inverse matrix as follows.

> ### A System of Equations with a Unique Solution
>
> If A is an invertible matrix, then the system of linear equations represented by $AX = B$ has a unique solution given by $X = A^{-1}B$.

EXAMPLE 5   Solving a System Using an Inverse Matrix

 TECHNOLOGY

On most graphing utilities, to solve a linear system that has an invertible coefficient matrix, you can use the formula $X = A^{-1}B$. That is, enter the $n \times n$ coefficient matrix $[A]$ and the $n \times 1$ column matrix $[B]$. The solution matrix X is given by

$$[A]^{-1}[B].$$

Use an inverse matrix to solve the system

$$\begin{cases} x + & y + & z = 10{,}000 \\ 0.06x + & 0.075y + & 0.095z = 730. \\ x & & - \; 2z = 0 \end{cases}$$

Solution   Begin by writing the system in the matrix form $AX = B$.

$$\begin{bmatrix} 1 & 1 & 1 \\ 0.06 & 0.075 & 0.095 \\ 1 & 0 & -2 \end{bmatrix} \begin{bmatrix} x \\ y \\ z \end{bmatrix} = \begin{bmatrix} 10{,}000 \\ 730 \\ 0 \end{bmatrix}$$

Then, use Gauss-Jordan elimination to find A^{-1}.

$$A^{-1} = \begin{bmatrix} 15 & -200 & -2 \\ -21.5 & 300 & 3.5 \\ 7.5 & -100 & -1.5 \end{bmatrix}$$

Finally, multiply B by A^{-1} on the left to obtain the solution.

$$X = A^{-1}B = \begin{bmatrix} 15 & -200 & -2 \\ -21.5 & 300 & 3.5 \\ 7.5 & -100 & -1.5 \end{bmatrix} \begin{bmatrix} 10{,}000 \\ 730 \\ 0 \end{bmatrix} = \begin{bmatrix} 4000 \\ 4000 \\ 2000 \end{bmatrix}$$

The solution of the system is $x = 4000$, $y = 4000$, and $z = 2000$, or $(4000, 4000, 2000)$.

✓ *Checkpoint* 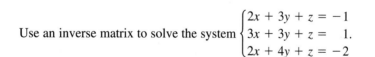 Audio-video solution in English & Spanish at *LarsonPrecalculus.com*

Use an inverse matrix to solve the system $\begin{cases} 2x + 3y + z = -1 \\ 3x + 3y + z = 1. \\ 2x + 4y + z = -2 \end{cases}$ ∎

Summarize   (Section 10.3)

1. State the definition of the inverse of a square matrix *(page 728)*. For an example of how to show that a matrix is the inverse of another matrix, see Example 1.

2. Explain how to find an inverse matrix *(pages 729 and 730)*. For examples of finding inverse matrices, see Examples 2 and 3.

3. State the formula for the inverse of a 2×2 matrix *(page 732)*. For an example of using this formula to find an inverse matrix, see Example 4.

4. Explain how to use an inverse matrix to solve a system of linear equations *(page 733)*. For an example of using an inverse matrix to solve a system of linear equations, see Example 5.

10.3 Exercises

See CalcChat.com for tutorial help and worked-out solutions to odd-numbered exercises.

GO DIGITAL

Vocabulary and Concept Check

In Exercises 1 and 2, fill in the blanks.

1. If there exists an $n \times n$ matrix A^{-1} such that $AA^{-1} = I_n = A^{-1}A$, then A^{-1} is the _____ of A.

2. A matrix that has an inverse is invertible or _____. A matrix that does not have an inverse is _____.

3. Why might a square matrix not be invertible?

4. If A is an invertible matrix, then what is the unique solution of the system of linear equations represented by $AX = B$?

Skills and Applications

The Inverse of a Matrix   In Exercises 5–12, show that B is the inverse of A.

5. $A = \begin{bmatrix} 2 & 1 \\ 5 & 3 \end{bmatrix}$, $B = \begin{bmatrix} 3 & -1 \\ -5 & 2 \end{bmatrix}$

6. $A = \begin{bmatrix} 1 & -1 \\ -1 & 2 \end{bmatrix}$, $B = \begin{bmatrix} 2 & 1 \\ 1 & 1 \end{bmatrix}$

7. $A = \begin{bmatrix} 3 & 2 \\ 1 & 4 \end{bmatrix}$, $B = \dfrac{1}{10}\begin{bmatrix} 4 & -2 \\ -1 & 3 \end{bmatrix}$

8. $A = \begin{bmatrix} 1 & -1 \\ 2 & 3 \end{bmatrix}$, $B = \dfrac{1}{5}\begin{bmatrix} 3 & 1 \\ -2 & 1 \end{bmatrix}$

9. $A = \begin{bmatrix} 2 & -17 & 11 \\ -1 & 11 & -7 \\ 0 & 3 & -2 \end{bmatrix}$, $B = \begin{bmatrix} 1 & 1 & 2 \\ 2 & 4 & -3 \\ 3 & 6 & -5 \end{bmatrix}$

10. $A = \begin{bmatrix} -4 & 1 & 5 \\ -1 & 2 & 4 \\ 0 & -1 & -1 \end{bmatrix}$, $B = \dfrac{1}{4}\begin{bmatrix} -2 & 4 & 6 \\ 1 & -4 & -11 \\ -1 & 4 & 7 \end{bmatrix}$

11. $A = \begin{bmatrix} 2 & 0 & 2 & 1 \\ 3 & 0 & 0 & 1 \\ -1 & 1 & -2 & 1 \\ 3 & -1 & 1 & 0 \end{bmatrix}$,

$B = \dfrac{1}{3}\begin{bmatrix} -1 & 3 & -2 & -2 \\ -2 & 9 & -7 & -10 \\ 1 & 0 & -1 & -1 \\ 3 & -6 & 6 & 6 \end{bmatrix}$

12. $A = \begin{bmatrix} -1 & 1 & 0 & -1 \\ 1 & -1 & 1 & 0 \\ -1 & 1 & 2 & 0 \\ 0 & -1 & 1 & 1 \end{bmatrix}$,

$B = \dfrac{1}{3}\begin{bmatrix} -3 & 1 & 1 & -3 \\ -3 & -1 & 2 & -3 \\ 0 & 1 & 1 & 0 \\ -3 & -2 & 1 & 0 \end{bmatrix}$

Finding the Inverse of a Matrix   In Exercises 13–22, find the inverse of the matrix, if possible.

13. $\begin{bmatrix} 1 & -2 \\ 2 & -3 \end{bmatrix}$

14. $\begin{bmatrix} -7 & 33 \\ 4 & -19 \end{bmatrix}$

15. $\begin{bmatrix} 3 & 1 \\ 4 & 2 \end{bmatrix}$

16. $\begin{bmatrix} 4 & -1 \\ -3 & 1 \end{bmatrix}$

17. $\begin{bmatrix} 1 & 1 & 1 \\ 3 & 5 & 4 \\ 3 & 6 & 5 \end{bmatrix}$

18. $\begin{bmatrix} 1 & 2 & 2 \\ 3 & 7 & 9 \\ -1 & -4 & -7 \end{bmatrix}$

19. $\begin{bmatrix} -5 & 0 & 0 \\ 2 & 0 & 0 \\ -1 & 5 & 7 \end{bmatrix}$

20. $\begin{bmatrix} 1 & 0 & 0 \\ 3 & 0 & 0 \\ 2 & 5 & 5 \end{bmatrix}$

21. $\begin{bmatrix} -8 & 0 & 0 & 0 \\ 0 & 1 & 0 & 0 \\ 0 & 0 & 4 & 0 \\ 0 & 0 & 0 & -5 \end{bmatrix}$

22. $\begin{bmatrix} 1 & 3 & -2 & 0 \\ 0 & 2 & 4 & 6 \\ 0 & 0 & -2 & 1 \\ 0 & 0 & 0 & 5 \end{bmatrix}$

Finding the Inverse of a Matrix   In Exercises 23–30, use the matrix capabilities of a graphing utility to find the inverse of the matrix, if possible.

23. $\begin{bmatrix} 1 & 2 & -1 \\ 3 & 7 & -10 \\ -5 & -7 & -15 \end{bmatrix}$

24. $\begin{bmatrix} 10 & 5 & -7 \\ -5 & 1 & 4 \\ 3 & 2 & -2 \end{bmatrix}$

25. $\begin{bmatrix} -\frac{1}{2} & \frac{3}{4} & \frac{1}{4} \\ 1 & 0 & -\frac{3}{2} \\ 0 & -1 & \frac{1}{2} \end{bmatrix}$

26. $\begin{bmatrix} -\frac{5}{6} & \frac{1}{3} & \frac{11}{6} \\ 0 & \frac{2}{3} & 2 \\ 1 & -\frac{1}{2} & -\frac{5}{2} \end{bmatrix}$

27. $\begin{bmatrix} 0.1 & 0.2 & 0.3 \\ -0.3 & 0.2 & 0.2 \\ 0.5 & 0.4 & 0.4 \end{bmatrix}$

28. $\begin{bmatrix} 0.6 & 0 & -0.3 \\ 0.7 & -1 & 0.2 \\ 1 & 0 & -0.9 \end{bmatrix}$

29. $\begin{bmatrix} -1 & 0 & 1 & 0 \\ 0 & 2 & 0 & -1 \\ 2 & 0 & -1 & 0 \\ 0 & -1 & 0 & 1 \end{bmatrix}$

30. $\begin{bmatrix} 1 & -2 & -1 & -2 \\ 3 & -5 & -2 & -3 \\ 2 & -5 & -2 & -5 \\ -1 & 4 & 4 & 11 \end{bmatrix}$

Finding the Inverse of a 2 × 2 Matrix In Exercises 31–36, use the formula on page 732 to find the inverse of the 2 × 2 matrix, if possible.

31. $\begin{bmatrix} 2 & 3 \\ -1 & 5 \end{bmatrix}$ 32. $\begin{bmatrix} 1 & -2 \\ -3 & 2 \end{bmatrix}$

33. $\begin{bmatrix} -4 & -6 \\ 2 & 3 \end{bmatrix}$ 34. $\begin{bmatrix} -12 & 3 \\ 5 & -2 \end{bmatrix}$

35. $\begin{bmatrix} 0.5 & 0.3 \\ 1.5 & 0.6 \end{bmatrix}$ 36. $\begin{bmatrix} -1.25 & 0.625 \\ 0.16 & 0.32 \end{bmatrix}$

Solving a System Using an Inverse Matrix In Exercises 37–40, use the inverse matrix found in Exercise 13 to solve the system of linear equations.

37. $\begin{cases} x - 2y = 5 \\ 2x - 3y = 10 \end{cases}$ 38. $\begin{cases} x - 2y = 0 \\ 2x - 3y = 3 \end{cases}$

39. $\begin{cases} x - 2y = 4 \\ 2x - 3y = 2 \end{cases}$ 40. $\begin{cases} x - 2y = 1 \\ 2x - 3y = -2 \end{cases}$

Solving a System Using an Inverse Matrix In Exercises 41 and 42, use the inverse matrix found in Exercise 17 to solve the system of linear equations.

41. $\begin{cases} x + y + z = 0 \\ 3x + 5y + 4z = 5 \\ 3x + 6y + 5z = 2 \end{cases}$ 42. $\begin{cases} x + y + z = -1 \\ 3x + 5y + 4z = 2 \\ 3x + 6y + 5z = 0 \end{cases}$

Solving a System Using an Inverse Matrix In Exercises 43 and 44, use the inverse matrix found in Exercise 30 to solve the system of linear equations.

43. $\begin{cases} x_1 - 2x_2 - x_3 - 2x_4 = 0 \\ 3x_1 - 5x_2 - 2x_3 - 3x_4 = 1 \\ 2x_1 - 5x_2 - 2x_3 - 5x_4 = -1 \\ -x_1 + 4x_2 + 4x_3 + 11x_4 = 2 \end{cases}$

44. $\begin{cases} x_1 - 2x_2 - x_3 - 2x_4 = 1 \\ 3x_1 - 5x_2 - 2x_3 - 3x_4 = -2 \\ 2x_1 - 5x_2 - 2x_3 - 5x_4 = 0 \\ -x_1 + 4x_2 + 4x_3 + 11x_4 = -3 \end{cases}$

Solving a System Using an Inverse Matrix In Exercises 45–52, use an inverse matrix to solve the system of linear equations, if possible.

45. $\begin{cases} 5x + 4y = -1 \\ 2x + 5y = 3 \end{cases}$ 46. $\begin{cases} 18x + 12y = 13 \\ 30x + 24y = 23 \end{cases}$

47. $\begin{cases} -0.4x + 0.8y = 1.6 \\ 2x - 4y = 5 \end{cases}$ 48. $\begin{cases} 0.2x - 0.6y = 2.4 \\ -x + 1.4y = -8.8 \end{cases}$

49. $\begin{cases} 2.3x - 1.9y = 6 \\ 1.5x + 0.75y = -12 \end{cases}$ 50. $\begin{cases} 5.1x - 3.4y = -20 \\ 0.9x - 0.6y = -51 \end{cases}$

51. $\begin{cases} 4x - y + z = -5 \\ 2x + 2y + 3z = 10 \\ 5x - 2y + 6z = 1 \end{cases}$ 52. $\begin{cases} 4x - 2y + 3z = -2 \\ 2x + 2y + 5z = 16 \\ 8x - 5y - 2z = 4 \end{cases}$

Using a Graphing Utility In Exercises 53 and 54, use the matrix capabilities of a graphing utility to solve the system of linear equations, if possible.

53. $\begin{cases} 5x - 3y + 2z = 2 \\ 2x + 2y - 3z = 3 \\ x - 7y + 7z = -4 \end{cases}$ 54. $\begin{cases} 2x + 3y + 5z = 4 \\ 3x + 5y + 9z = 7 \\ 5x + 9y + 16z = 13 \end{cases}$

Circuit Analysis

In Exercises 55–58, consider the circuit shown in the figure. The currents I_1, I_2, and I_3 (in amperes) are the solution of the system

$$\begin{cases} 2I_1 + 4I_3 = E_1 \\ I_2 + 4I_3 = E_2 \\ I_1 + I_2 - I_3 = 0 \end{cases}$$

where E_1 and E_2 are voltages. Use the inverse of the coefficient matrix of this system to find the unknown currents for the given voltages.

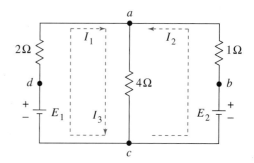

55. $E_1 = 15$ volts
    $E_2 = 17$ volts

56. $E_1 = 10$ volts
    $E_2 = 10$ volts

57. $E_1 = 28$ volts
    $E_2 = 21$ volts

58. $E_1 = 24$ volts
    $E_2 = 23$ volts

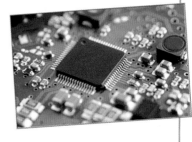

Raw Materials In Exercises 59 and 60, find the numbers of bags of potting soil that a company can produce for seedlings, general potting, and hardwood plants with the given amounts of raw materials. The raw materials used in one bag of each type of potting soil are shown below.

	Sand	Loam	Peat Moss
Seedlings	2 units	1 unit	1 unit
General	1 unit	2 units	1 unit
Hardwoods	2 units	2 units	2 units

59. 500 units of sand
    500 units of loam
    400 units of peat moss

60. 500 units of sand
    750 units of loam
    450 units of peat moss

61. International Travel The table shows the numbers of international visitors y (in thousands) to France from 2016 through 2018. *(Source: World Tourism Organization (UNWTO))*

Year	Visitors, y (in thousands)
2016	82,700
2017	86,918
2018	89,400

(a) The data can be modeled by the quadratic function $y = at^2 + bt + c$. Write a system of linear equations for the data. Let t represent the year, with $t = 16$ corresponding to 2016.

(b) Use the matrix capabilities of a graphing utility to find the inverse of the coefficient matrix of the system from part (a).

(c) Use the result of part (b) to solve the system and write the model $y = at^2 + bt + c$.

(d) Use the graphing utility to graph the model with the data.

Exploring the Concepts

62. **HOW DO YOU SEE IT?** Consider the matrix

$$A = \begin{bmatrix} x & y \\ 0 & z \end{bmatrix}.$$

Use the determinant of A to state the conditions for which (a) A^{-1} exists and (b) $A^{-1} = A$.

True or False? In Exercises 63 and 64, determine whether the statement is true or false. Justify your answer.

63. Multiplication of an invertible matrix and its inverse is commutative.

64. When the product of two square matrices is the identity matrix, the matrices are inverses of one another.

65. Writing Explain how to determine whether the inverse of a 2×2 matrix exists, as well as how to find the inverse when it exists.

66. Writing Explain how to write a system of three linear equations in three variables as a matrix equation $AX = B$, as well as how to solve the system using an inverse matrix.

67. Verifying a Formula Verify that the inverse of an invertible 2×2 matrix

$$A = \begin{bmatrix} a & b \\ c & d \end{bmatrix}$$

is given by $A^{-1} = \dfrac{1}{ad - bc}\begin{bmatrix} d & -b \\ -c & a \end{bmatrix}$.

68. Conjecture Consider matrices of the form

$$A = \begin{bmatrix} a_{11} & 0 & 0 & 0 & \cdots & 0 \\ 0 & a_{22} & 0 & 0 & \cdots & 0 \\ 0 & 0 & a_{33} & 0 & \cdots & 0 \\ \vdots & \vdots & \vdots & \vdots & & \vdots \\ 0 & 0 & 0 & 0 & \cdots & a_{nn} \end{bmatrix}.$$

(a) Write a 2×2 matrix and a 3×3 matrix in the form of A. Find the inverse of each.

(b) Use the result of part (a) to make a conjecture about the inverses of matrices in the form of A.

Think About It In Exercises 69 and 70, find the value of k that makes the matrix singular.

69. $\begin{bmatrix} 4 & 3 \\ -2 & k \end{bmatrix}$

70. $\begin{bmatrix} 2k + 1 & 3 \\ -7 & 1 \end{bmatrix}$

Review & Refresh ▶ *Video solutions at LarsonPrecalculus.com*

Finding a Dot Product In Exercises 71–76, find $\mathbf{u} \cdot \mathbf{v}$.

71. $\mathbf{u} = \langle -5, -7 \rangle$
    $\mathbf{v} = \langle -6, 1 \rangle$

72. $\mathbf{u} = \langle 1, 5 \rangle$
    $\mathbf{v} = \langle 8, 3 \rangle$

73. $\mathbf{u} = \langle 7, -12 \rangle$
    $\mathbf{v} = \langle -8, 8 \rangle$

74. $\mathbf{u} = \langle 2, -6 \rangle$
    $\mathbf{v} = \langle -12, 11 \rangle$

75. $\mathbf{u} = 2\mathbf{i} - \mathbf{j}$
    $\mathbf{v} = 7\mathbf{i} + 6\mathbf{j}$

76. $\mathbf{u} = -6\mathbf{i} + 4\mathbf{j}$
    $\mathbf{v} = -3\mathbf{i} - 11\mathbf{j}$

Evaluating an Expression In Exercises 77 and 78, evaluate the expression.

77. $-\dfrac{1}{3}\begin{bmatrix} 3 & 6 \\ 0 & -3 \end{bmatrix} + 2\begin{bmatrix} 6 & 8 \\ 5 & -5 \end{bmatrix}$

78. $-\dfrac{3}{2}\begin{bmatrix} 8 & -2 \\ 6 & 4 \end{bmatrix} - \begin{bmatrix} 3 & -3 \\ 7 & 9 \end{bmatrix}$

Solving a Matrix Equation In Exercises 79 and 80, solve for X in the equation.

79. $\dfrac{2}{3}\begin{bmatrix} 3 & -21 \\ -9 & 12 \end{bmatrix} + \dfrac{1}{3}\begin{bmatrix} 15 & 24 \\ 18 & 0 \end{bmatrix} = -2X$

80. $14X = -1\begin{bmatrix} -3 & -4 \\ -\frac{1}{2} & \frac{9}{10} \end{bmatrix} + \dfrac{4}{3}\begin{bmatrix} 4 & \frac{6}{7} \\ \frac{1}{2} & \frac{1}{3} \end{bmatrix}$

Project: Consumer Credit To work an extended application analyzing the outstanding consumer credit in the United States, visit this text's website at *LarsonPrecalculus.com*. *(Source: Board of Governors of the Federal Reserve System)*

10.4 The Determinant of a Square Matrix

Determinants are often used in other branches of mathematics. For example, the types of determinants in Exercises 81–86 on page 744 occur when changes of variables are made in calculus.

- ◉ Find the determinants of 2 × 2 matrices.
- ◉ Find minors and cofactors of square matrices.
- ◉ Find the determinants of square matrices.

The Determinant of a 2 × 2 Matrix

Every *square* matrix can be associated with a real number called its **determinant.** Determinants have many uses, and several will be discussed in this section and the next section. Historically, the use of determinants arose from special number patterns that occur when systems of linear equations are solved. For example, the system

$$\begin{cases} a_1x + b_1y = c_1 \\ a_2x + b_2y = c_2 \end{cases}$$

has a solution

$$x = \frac{c_1b_2 - c_2b_1}{a_1b_2 - a_2b_1} \quad \text{and} \quad y = \frac{a_1c_2 - a_2c_1}{a_1b_2 - a_2b_1}$$

provided that $a_1b_2 - a_2b_1 \neq 0$. Note that the denominators of the two fractions are the same. This denominator is called the *determinant* of the coefficient matrix of the system.

Coefficient Matrix   **Determinant**

$$A = \begin{bmatrix} a_1 & b_1 \\ a_2 & b_2 \end{bmatrix} \qquad \det(A) = a_1b_2 - a_2b_1$$

The determinant of matrix A can also be denoted by vertical bars on both sides of the matrix, as shown in the definition below.

Definition of the Determinant of a 2 × 2 Matrix

The **determinant** of the matrix

$$A = \begin{bmatrix} a_1 & b_1 \\ a_2 & b_2 \end{bmatrix}$$

is given by

$$\det(A) = |A| = \begin{vmatrix} a_1 & b_1 \\ a_2 & b_2 \end{vmatrix} = a_1b_2 - a_2b_1.$$

In this text, $\det(A)$ and $|A|$ are used interchangeably to represent the determinant of A. Although vertical bars are also used to denote the absolute value of a real number, the context will show which use is intended.

A convenient method for remembering the formula for the determinant of a 2 × 2 matrix is shown below.

$$\det(A) = \begin{vmatrix} a_1 & b_1 \\ a_2 & b_2 \end{vmatrix} = a_1b_2 - a_2b_1$$

Note that the determinant is the difference of the products of the two diagonals of the matrix.

GO DIGITAL

In Example 1, you will see that the determinant of a matrix can be positive, zero, or negative.

EXAMPLE 1 **The Determinant of a 2 × 2 Matrix**

Find the determinant of each matrix.

a. $A = \begin{bmatrix} 2 & -3 \\ 1 & 2 \end{bmatrix}$

b. $B = \begin{bmatrix} 2 & 1 \\ 4 & 2 \end{bmatrix}$

c. $C = \begin{bmatrix} 0 & \frac{3}{2} \\ 2 & 4 \end{bmatrix}$

Solution

a. $\det(A) = \begin{vmatrix} 2 & -3 \\ 1 & 2 \end{vmatrix} = 2(2) - 1(-3) = 4 + 3 = 7$

b. $\det(B) = \begin{vmatrix} 2 & 1 \\ 4 & 2 \end{vmatrix} = 2(2) - 4(1) = 4 - 4 = 0$

c. $\det(C) = \begin{vmatrix} 0 & \frac{3}{2} \\ 2 & 4 \end{vmatrix} = 0(4) - 2\left(\frac{3}{2}\right) = 0 - 3 = -3$

✓ *Checkpoint* 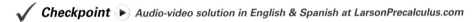 Audio-video solution in English & Spanish at LarsonPrecalculus.com

Find the determinant of each matrix.

a. $A = \begin{bmatrix} 1 & 2 \\ 3 & -1 \end{bmatrix}$

b. $B = \begin{bmatrix} 5 & 0 \\ -4 & 2 \end{bmatrix}$

c. $C = \begin{bmatrix} 3 & 6 \\ 2 & 4 \end{bmatrix}$

The determinant of a matrix of dimension 1×1 is defined simply as the entry of the matrix. For example, if $A = \begin{bmatrix} -2 \end{bmatrix}$, then $\det(A) = -2$.

▶▶▶ **TECHNOLOGY**

Most graphing utilities can find the determinant of a matrix. For example, to find the determinant of

$$A = \begin{bmatrix} 2.4 & 0.8 \\ -0.6 & -3.2 \end{bmatrix}$$

use the *matrix editor* to enter the matrix as $\begin{bmatrix} A \end{bmatrix}$ and then choose the *determinant* feature. The result is -7.2, as shown below.

```
[A]
        [2.4   .8]
        [-.6  -3.2]
det([A])
                -7.2
```

Consult the user's guide for your graphing utility for specific keystrokes.

Minors and Cofactors

To define the determinant of a square matrix of dimension 3×3 or greater, it is helpful to introduce the concepts of **minors** and **cofactors.**

Sign Pattern for Cofactors

$$\begin{bmatrix} + & - & + \\ - & + & - \\ + & - & + \end{bmatrix}$$

3×3 matrix

$$\begin{bmatrix} + & - & + & - \\ - & + & - & + \\ + & - & + & - \\ - & + & - & + \end{bmatrix}$$

4×4 matrix

$$\begin{bmatrix} + & - & + & - & + & \cdots \\ - & + & - & + & - & \cdots \\ + & - & + & - & + & \cdots \\ - & + & - & + & - & \cdots \\ + & - & + & - & + & \cdots \\ \vdots & \vdots & \vdots & \vdots & \vdots & \end{bmatrix}$$

$n \times n$ matrix

> ### Minors and Cofactors of a Square Matrix
>
> If A is a square matrix, then the **minor** M_{ij} of the entry a_{ij} is the determinant of the matrix obtained by deleting the ith row and jth column of A. The **cofactor** C_{ij} of the entry a_{ij} is
>
> $$C_{ij} = (-1)^{i+j}M_{ij}.$$

In the sign pattern for cofactors at the left, notice that *odd* positions (where $i + j$ is odd) have negative signs and *even* positions (where $i + j$ is even) have positive signs.

EXAMPLE 2 **Finding the Minors and Cofactors of a Matrix**

Find all the minors and cofactors of

$$A = \begin{bmatrix} 0 & 2 & 1 \\ 3 & -1 & 2 \\ 4 & 0 & 1 \end{bmatrix}.$$

Solution To find the minor M_{11}, delete the first row and first column of A and find the determinant of the resulting matrix.

$$\begin{bmatrix} 0 & 2 & 1 \\ 3 & -1 & 2 \\ 4 & 0 & 1 \end{bmatrix}, \quad M_{11} = \begin{vmatrix} -1 & 2 \\ 0 & 1 \end{vmatrix} = -1(1) - 0(2) = -1$$

Similarly, to find M_{12}, delete the first row and second column.

$$\begin{bmatrix} 0 & 2 & 1 \\ 3 & -1 & 2 \\ 4 & 0 & 1 \end{bmatrix}, \quad M_{12} = \begin{vmatrix} 3 & 2 \\ 4 & 1 \end{vmatrix} = 3(1) - 4(2) = -5$$

Continuing this pattern, you obtain the minors.

$$\begin{array}{lll} M_{11} = -1 & M_{12} = -5 & M_{13} = 4 \\ M_{21} = 2 & M_{22} = -4 & M_{23} = -8 \\ M_{31} = 5 & M_{32} = -3 & M_{33} = -6 \end{array}$$

Now, to find the cofactors, combine these minors with the sign pattern for a 3×3 matrix shown at the upper left.

$$\begin{array}{lll} C_{11} = -1 & C_{12} = 5 & C_{13} = 4 \\ C_{21} = -2 & C_{22} = -4 & C_{23} = 8 \\ C_{31} = 5 & C_{32} = 3 & C_{33} = -6 \end{array}$$

✓ *Checkpoint* 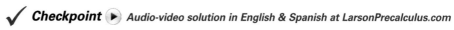 *Audio-video solution in English & Spanish at LarsonPrecalculus.com*

Find all the minors and cofactors of

$$A = \begin{bmatrix} 1 & 2 & 3 \\ 0 & -1 & 5 \\ 2 & 1 & 4 \end{bmatrix}.$$

GO DIGITAL

The Determinant of a Square Matrix

The definition below is *inductive* because it uses determinants of matrices of dimension $(n-1) \times (n-1)$ to define determinants of matrices of dimension $n \times n$.

Determinant of a Square Matrix

If A is a square matrix (of dimension 2×2 or greater), then the determinant of A is the sum of the entries in any row (or column) of A multiplied by their respective cofactors. For example, expanding along the first row yields

$$|A| = a_{11}C_{11} + a_{12}C_{12} + \cdots + a_{1n}C_{1n}.$$

Applying this definition to find a determinant is called **expanding by cofactors.**

Verify that for a 2×2 matrix

$$A = \begin{bmatrix} a_1 & b_1 \\ a_2 & b_2 \end{bmatrix}$$

this definition of the determinant yields

$$|A| = a_1 b_2 - a_2 b_1$$

as previously defined.

EXAMPLE 3   The Determinant of a 3×3 Matrix

▶▶▶ *See LarsonPrecalculus.com for an interactive version of this type of example.*

Find the determinant of $A = \begin{bmatrix} 0 & 2 & 1 \\ 3 & -1 & 2 \\ 4 & 0 & 1 \end{bmatrix}$.

Solution Note that this is the same matrix used in Example 2. There you found that the cofactors of the entries in the first row are

$$C_{11} = -1, \quad C_{12} = 5, \quad \text{and} \quad C_{13} = 4.$$

Use the definition of the determinant of a square matrix to expand along the first row.

$$
\begin{aligned}
|A| &= a_{11}C_{11} + a_{12}C_{12} + a_{13}C_{13} &&\text{First-row expansion} \\
&= 0(-1) + 2(5) + 1(4) &&\text{Substitute.} \\
&= 14 &&\text{Simplify.}
\end{aligned}
$$

✓ **Checkpoint** ▶ Audio-video solution in English & Spanish at LarsonPrecalculus.com

Find the determinant of $A = \begin{bmatrix} 3 & 4 & -2 \\ 3 & 5 & 0 \\ -1 & 4 & 1 \end{bmatrix}$.   ∎

In Example 3, it was efficient to expand by cofactors along the first row, but any row or column can be used. For example, expanding along the second row gives the same result.

$$
\begin{aligned}
|A| &= a_{21}C_{21} + a_{22}C_{22} + a_{23}C_{23} &&\text{Second-row expansion} \\
&= 3(-2) + (-1)(-4) + 2(8) &&\text{Substitute.} \\
&= 14 &&\text{Simplify.}
\end{aligned}
$$

GO DIGITAL

When expanding by cofactors, you do not need to find cofactors of zero entries, because zero times its cofactor is zero. So, the row (or column) containing the most zeros is usually the best choice for expansion by cofactors. This is demonstrated in the next example.

EXAMPLE 4   **The Determinant of a 4 × 4 Matrix**

Find the determinant of $A = \begin{bmatrix} 1 & -2 & 3 & 0 \\ -1 & 1 & 0 & 2 \\ 0 & 2 & 0 & 3 \\ 3 & 4 & 0 & 2 \end{bmatrix}$.

Solution Notice that three of the entries in the third column are zeros. So, to eliminate some of the work in the expansion, expand along the third column.

$$|A| = 3(C_{13}) + 0(C_{23}) + 0(C_{33}) + 0(C_{43})$$

The cofactors C_{23}, C_{33}, and C_{43} have zero coefficients, so the only cofactor you need to find is C_{13}. Start by deleting the first row and third column of A to form the determinant that gives the minor M_{13}.

$$C_{13} = (-1)^{1+3} \begin{vmatrix} -1 & 1 & 2 \\ 0 & 2 & 3 \\ 3 & 4 & 2 \end{vmatrix} \qquad \text{Delete 1st row and 3rd column.}$$

$$= \begin{vmatrix} -1 & 1 & 2 \\ 0 & 2 & 3 \\ 3 & 4 & 2 \end{vmatrix} \qquad \text{Simplify.}$$

Now, expand by cofactors along the second row.

$$C_{13} = 0(-1)^3 \begin{vmatrix} 1 & 2 \\ 4 & 2 \end{vmatrix} + 2(-1)^4 \begin{vmatrix} -1 & 2 \\ 3 & 2 \end{vmatrix} + 3(-1)^5 \begin{vmatrix} -1 & 1 \\ 3 & 4 \end{vmatrix}$$

$$= 0 + 2(1)(-8) + 3(-1)(-7)$$

$$= 5$$

So, $|A| = 3C_{13} = 3(5) = 15$.

✓ **Checkpoint** ▶ Audio-video solution in English & Spanish at LarsonPrecalculus.com

Find the determinant of $A = \begin{bmatrix} 2 & 6 & -4 & 2 \\ 2 & -2 & 3 & 6 \\ 1 & 5 & 0 & 1 \\ 3 & 1 & 0 & -5 \end{bmatrix}$.

Summarize (Section 10.4)

1. State the definition of the determinant of a 2 × 2 matrix *(page 737)*. For an example of finding the determinants of 2 × 2 matrices, see Example 1.

2. State the definitions of minors and cofactors of a square matrix *(page 739)*. For an example of finding the minors and cofactors of a square matrix, see Example 2.

3. State the definition of the determinant of a square matrix using expanding by cofactors *(page 740)*. For examples of finding determinants using expanding by cofactors, see Examples 3 and 4.

10.4 Exercises

See CalcChat.com for tutorial help and worked-out solutions to odd-numbered exercises.

Vocabulary and Concept Check

In Exercises 1 and 2, fill in the blank.

1. Both $\det(A)$ and $|A|$ represent the _____ of the matrix A.

2. The _____ M_{ij} of the entry a_{ij} is the determinant of the matrix obtained by deleting the ith row and jth column of the square matrix A.

3. For a square matrix B, the minor $M_{23} - 5$. What is the cofactor C_{23} of matrix B?

4. When finding the determinant of a matrix using expanding by cofactors, explain why you do not need to find all the cofactors.

Skills and Applications

Finding the Determinant of a Matrix In Exercises 5–18, find the determinant of the matrix.

5. $\begin{bmatrix} 4 \end{bmatrix}$

6. $\begin{bmatrix} -10 \end{bmatrix}$

7. $\begin{bmatrix} 8 & 4 \\ 2 & 3 \end{bmatrix}$

8. $\begin{bmatrix} -9 & 0 \\ 6 & -2 \end{bmatrix}$

9. $\begin{bmatrix} 6 & -3 \\ -5 & 2 \end{bmatrix}$

10. $\begin{bmatrix} 3 & -3 \\ 4 & -8 \end{bmatrix}$

11. $\begin{bmatrix} -7 & 0 \\ 3 & 0 \end{bmatrix}$

12. $\begin{bmatrix} 4 & -3 \\ 0 & 0 \end{bmatrix}$

13. $\begin{bmatrix} 3 & 4 \\ -2 & 1 \end{bmatrix}$

14. $\begin{bmatrix} 5 & -9 \\ 7 & 16 \end{bmatrix}$

15. $\begin{bmatrix} -3 & -2 \\ -6 & -4 \end{bmatrix}$

16. $\begin{bmatrix} -7 & 6 \\ 0.5 & 3 \end{bmatrix}$

17. $\begin{bmatrix} -\frac{1}{2} & \frac{1}{3} \\ -6 & \frac{1}{3} \end{bmatrix}$

18. $\begin{bmatrix} \frac{2}{3} & -\frac{4}{3} \\ -1 & \frac{1}{3} \end{bmatrix}$

Finding the Determinant of a Matrix In Exercises 19–22, use the matrix capabilities of a graphing utility to find the determinant of the matrix.

19. $\begin{bmatrix} 19 & 20 \\ 43 & -56 \end{bmatrix}$

20. $\begin{bmatrix} 101 & 197 \\ -253 & 172 \end{bmatrix}$

21. $\begin{bmatrix} \frac{1}{10} & \frac{1}{5} \\ -\frac{3}{10} & \frac{1}{5} \end{bmatrix}$

22. $\begin{bmatrix} 0.1 & 0.1 \\ 7.5 & 6.2 \end{bmatrix}$

Finding the Minors and Cofactors of a Matrix In Exercises 23–28, find all the (a) minors and (b) cofactors of the matrix.

23. $\begin{bmatrix} 4 & 5 \\ 3 & -6 \end{bmatrix}$

24. $\begin{bmatrix} 0 & 10 \\ 3 & -4 \end{bmatrix}$

25. $\begin{bmatrix} 4 & 0 & 2 \\ -3 & 2 & 1 \\ 1 & -1 & 1 \end{bmatrix}$

26. $\begin{bmatrix} 1 & -1 & 0 \\ 3 & 2 & 5 \\ 4 & -6 & 4 \end{bmatrix}$

27. $\begin{bmatrix} -4 & 6 & 3 \\ 7 & -2 & 8 \\ 1 & 0 & -5 \end{bmatrix}$

28. $\begin{bmatrix} -2 & 9 & 4 \\ 7 & -6 & 0 \\ 6 & 7 & -6 \end{bmatrix}$

Finding the Determinant of a Matrix In Exercises 29–38, find the determinant of the matrix. Expand by cofactors using the indicated row or column.

29. $\begin{bmatrix} 2 & 5 \\ 6 & -3 \end{bmatrix}$
   (a) Row 1
   (b) Column 1

30. $\begin{bmatrix} 7 & -1 \\ -4 & 10 \end{bmatrix}$
   (a) Row 2
   (b) Column 2

31. $\begin{bmatrix} 5 & 0 & -3 \\ 0 & 12 & 4 \\ 1 & 6 & 3 \end{bmatrix}$
   (a) Row 2
   (b) Column 2

32. $\begin{bmatrix} 3 & -2 & 5 \\ 1 & 0 & 3 \\ 0 & 4 & -1 \end{bmatrix}$
   (a) Row 3
   (b) Column 1

33. $\begin{bmatrix} -3 & 2 & 1 \\ 4 & 5 & 6 \\ 2 & -3 & 1 \end{bmatrix}$
   (a) Row 1
   (b) Column 2

34. $\begin{bmatrix} -3 & 4 & 2 \\ 6 & 3 & 1 \\ 4 & -7 & -8 \end{bmatrix}$
   (a) Row 2
   (b) Column 3

35. $\begin{bmatrix} 6 & 0 & -3 & 5 \\ 4 & 0 & 6 & -8 \\ -1 & 0 & 7 & 4 \\ 8 & 0 & 0 & 2 \end{bmatrix}$
   (a) Row 4
   (b) Column 2

36. $\begin{bmatrix} 10 & 8 & 3 & -7 \\ 4 & 0 & 5 & -6 \\ 0 & 3 & 2 & 7 \\ 0 & 0 & 0 & 0 \end{bmatrix}$
   (a) Row 4
   (b) Column 1

37. $\begin{bmatrix} -2 & 4 & 7 & 1 \\ 3 & 0 & 0 & 0 \\ 8 & 5 & 10 & 5 \\ 6 & 0 & 5 & 0 \end{bmatrix}$
   (a) Row 2
   (b) Column 4

38. $\begin{bmatrix} 7 & 0 & 0 & -6 \\ 6 & 0 & 1 & -2 \\ 1 & -2 & 3 & 2 \\ -3 & 0 & -1 & 4 \end{bmatrix}$
   (a) Row 1
   (b) Column 2

Finding the Determinant of a Matrix In Exercises 39–52, find the determinant of the matrix. Expand by cofactors using the row or column that appears to make the computations easiest.

39. $\begin{bmatrix} -1 & 8 & -3 \\ 0 & 3 & -6 \\ 0 & 0 & 3 \end{bmatrix}$ 40. $\begin{bmatrix} 1 & 0 & 0 \\ -1 & -1 & 0 \\ 4 & 11 & 5 \end{bmatrix}$

41. $\begin{bmatrix} 6 & 3 & -7 \\ 0 & 0 & 0 \\ 4 & -6 & 3 \end{bmatrix}$ 42. $\begin{bmatrix} 0 & 1 & 2 \\ 3 & 1 & 0 \\ -2 & 0 & 3 \end{bmatrix}$

43. $\begin{bmatrix} 2 & -1 & 0 \\ 4 & 2 & 1 \\ 4 & 2 & 1 \end{bmatrix}$ 44. $\begin{bmatrix} -2 & 2 & 3 \\ 1 & -1 & 0 \\ 0 & 1 & 4 \end{bmatrix}$

45. $\begin{bmatrix} 1 & 4 & -2 \\ 3 & 2 & 0 \\ -1 & 4 & 3 \end{bmatrix}$ 46. $\begin{bmatrix} 2 & -1 & 3 \\ -4 & 2 & -6 \\ 1 & 0 & 2 \end{bmatrix}$

47. $\begin{bmatrix} 2 & 6 & 0 & 2 \\ 2 & 7 & 3 & 6 \\ 1 & 0 & 0 & 1 \\ 3 & 7 & 0 & 7 \end{bmatrix}$ 48. $\begin{bmatrix} 1 & 4 & 3 & 2 \\ -5 & 6 & 2 & 1 \\ 0 & 0 & 0 & 0 \\ 3 & -2 & 1 & 5 \end{bmatrix}$

49. $\begin{bmatrix} 5 & 3 & 0 & 6 \\ 4 & 6 & 4 & 12 \\ 0 & 2 & -3 & 4 \\ 0 & 1 & -2 & 2 \end{bmatrix}$ 50. $\begin{bmatrix} 3 & 6 & -5 & 4 \\ -2 & 0 & 6 & 0 \\ 1 & 1 & 2 & 2 \\ 0 & 3 & -1 & -1 \end{bmatrix}$

51. $\begin{bmatrix} 3 & 2 & 4 & -1 & 5 \\ -2 & 0 & 1 & 3 & 2 \\ 1 & 0 & 0 & 4 & 0 \\ 6 & 0 & 2 & -1 & 0 \\ 3 & 0 & 5 & 1 & 0 \end{bmatrix}$

52. $\begin{bmatrix} 5 & 2 & 0 & 0 & -2 \\ 0 & 1 & 4 & 3 & \frac{1}{2} \\ 0 & 0 & 2 & 6 & 3 \\ 0 & 0 & 3 & \frac{3}{2} & 1 \\ 0 & 0 & 0 & 0 & 2 \end{bmatrix}$

Finding the Determinant of a Matrix In Exercises 53–56, use the matrix capabilities of a graphing utility to find the determinant.

53. $\begin{vmatrix} 3 & 8 & -7 \\ 0 & -5 & 4 \\ 8 & 1 & 6 \end{vmatrix}$ 54. $\begin{vmatrix} 5 & -8 & 0 \\ 9 & 7 & 4 \\ -8 & 7 & 1 \end{vmatrix}$

55. $\begin{vmatrix} 1 & -1 & 8 & 4 \\ 2 & 6 & 0 & -4 \\ 2 & 0 & 2 & 6 \\ 0 & 2 & 8 & 0 \end{vmatrix}$ 56. $\begin{vmatrix} 0 & -3 & 8 & 2 \\ 8 & 1 & -1 & 6 \\ -4 & 6 & 0 & 9 \\ -7 & 0 & 0 & 14 \end{vmatrix}$

The Determinant of a Matrix Product In Exercises 57–62, find (a) $|A|$, (b) $|B|$, (c) AB, and (d) $|AB|$.

57. $A = \begin{bmatrix} -1 & 0 \\ 0 & 3 \end{bmatrix}$, $B = \begin{bmatrix} 2 & 0 \\ 0 & -1 \end{bmatrix}$

58. $A = \begin{bmatrix} -2 & 1 \\ 4 & -2 \end{bmatrix}$, $B = \begin{bmatrix} 1 & 2 \\ 0 & -1 \end{bmatrix}$

59. $A = \begin{bmatrix} 4 & 0 \\ 3 & -2 \end{bmatrix}$, $B = \begin{bmatrix} -1 & 1 \\ -2 & 2 \end{bmatrix}$

60. $A = \begin{bmatrix} 5 & 4 \\ 3 & -1 \end{bmatrix}$, $B = \begin{bmatrix} 0 & 6 \\ 1 & -2 \end{bmatrix}$

61. $A = \begin{bmatrix} -1 & 2 & 1 \\ 1 & 0 & 1 \\ 0 & 1 & 0 \end{bmatrix}$, $B = \begin{bmatrix} -1 & 0 & 0 \\ 0 & 2 & 0 \\ 0 & 0 & 3 \end{bmatrix}$

62. $A = \begin{bmatrix} 2 & 0 & 1 \\ 1 & -1 & 2 \\ 3 & 1 & 0 \end{bmatrix}$, $B = \begin{bmatrix} 2 & -1 & 4 \\ 0 & 1 & 3 \\ 3 & -2 & 1 \end{bmatrix}$

Creating a Matrix In Exercises 63–68, create a matrix A with the given characteristics. (There are many correct answers.)

63. Dimension: 2×2, $|A| = 3$

64. Dimension: 2×2, $|A| = -5$

65. Dimension: 3×3, $|A| = -1$

66. Dimension: 3×3, $|A| = 4$

67. Dimension: 2×2, $|A| = 0$, $A \neq O$

68. Dimension: 3×3, $|A| = 0$, $A \neq O$

Verifying an Equation In Exercises 69–74, find the determinant(s) to verify the equation.

69. $\begin{vmatrix} w & x \\ y & z \end{vmatrix} = -\begin{vmatrix} y & z \\ w & x \end{vmatrix}$ 70. $\begin{vmatrix} w & cx \\ y & cz \end{vmatrix} = c\begin{vmatrix} w & x \\ y & z \end{vmatrix}$

71. $\begin{vmatrix} w & x \\ y & z \end{vmatrix} = \begin{vmatrix} w & x + cw \\ y & z + cy \end{vmatrix}$ 72. $\begin{vmatrix} w & x \\ cw & cx \end{vmatrix} = 0$

73. $\begin{vmatrix} 1 & x & x^2 \\ 1 & y & y^2 \\ 1 & z & z^2 \end{vmatrix} = (y - x)(z - x)(z - y)$

74. $\begin{vmatrix} a + b & a & a \\ a & a + b & a \\ a & a & a + b \end{vmatrix} = b^2(3a + b)$

Solving an Equation In Exercises 75–80, solve for x.

75. $\begin{vmatrix} x & 2 \\ 1 & x \end{vmatrix} = 2$ 76. $\begin{vmatrix} x & 4 \\ -1 & x \end{vmatrix} = 20$

77. $\begin{vmatrix} x + 1 & 2 \\ -1 & x \end{vmatrix} = 4$ 78. $\begin{vmatrix} x - 2 & -1 \\ -3 & x \end{vmatrix} = 0$

79. $\begin{vmatrix} x + 3 & 2 \\ 1 & x + 2 \end{vmatrix} = 0$ 80. $\begin{vmatrix} x + 4 & -2 \\ 7 & x - 5 \end{vmatrix} = 0$

Entries Involving Expressions

In Exercises 81–86, find the determinant in which the entries are functions. Determinants of this type occur when changes of variables are made in calculus.

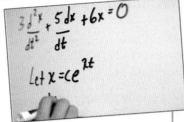

81. $\begin{vmatrix} 4u & -1 \\ -1 & 2v \end{vmatrix}$

82. $\begin{vmatrix} 3x^2 & -3y^2 \\ 1 & 1 \end{vmatrix}$

83. $\begin{vmatrix} e^{2x} & e^{3x} \\ 2e^{2x} & 3e^{3x} \end{vmatrix}$

84. $\begin{vmatrix} e^{-x} & xe^{-x} \\ -e^{-x} & (1-x)e^{-x} \end{vmatrix}$

85. $\begin{vmatrix} x & \ln x \\ 1 & 1/x \end{vmatrix}$

86. $\begin{vmatrix} x & x \ln x \\ 1 & 1 + \ln x \end{vmatrix}$

Exploring the Concepts

True or False? In Exercises 87 and 88, determine whether the statement is true or false. Justify your answer.

87. If a square matrix has an entire row of zeros, then the determinant of the matrix is zero.

88. If the rows of a 2×2 matrix are the same, then the determinant of the matrix is zero.

89. **Think About It** Find square matrices A and B such that $|A + B| \neq |A| + |B|$.

90. **Think About It** Let A be a 3×3 matrix such that $|A| = 5$. Is it possible to find $|2A|$? Explain.

91. **Properties of Determinants** Explain why each equation is an example of the given property of determinants (A and B are square matrices). Use a graphing utility to verify the results.

(a) If B is obtained from A by interchanging two rows of A or interchanging two columns of A, then $|B| = -|A|$.

$$\begin{vmatrix} 1 & 3 & 4 \\ -7 & 2 & -5 \\ 6 & 1 & 2 \end{vmatrix} = -\begin{vmatrix} 1 & 4 & 3 \\ -7 & -5 & 2 \\ 6 & 2 & 1 \end{vmatrix}$$

(b) If B is obtained from A by adding a multiple of a row of A to another row of A or by adding a multiple of a column of A to another column of A, then $|B| = |A|$.

$$\begin{vmatrix} 1 & -3 \\ 5 & 2 \end{vmatrix} = \begin{vmatrix} 1 & -3 \\ 0 & 17 \end{vmatrix}$$

(c) If B is obtained from A by multiplying a row by a nonzero constant c or by multiplying a column by a nonzero constant c, then $|B| = c|A|$.

$$\begin{vmatrix} 5 & 10 \\ 2 & -3 \end{vmatrix} = 5\begin{vmatrix} 1 & 2 \\ 2 & -3 \end{vmatrix}$$

92. **HOW DO YOU SEE IT?** Explain why the determinant of each matrix is equal to zero.

(a) $\begin{bmatrix} 2 & -4 & 5 \\ 1 & -2 & 3 \\ 0 & 0 & 0 \end{bmatrix}$

(b) $\begin{bmatrix} 4 & -4 & 5 & 7 \\ 2 & -2 & 3 & 1 \\ 4 & -4 & 5 & 7 \\ 6 & 1 & -3 & -3 \end{bmatrix}$

93. **Conjecture** A **diagonal matrix** is a square matrix in which each entry not on the main diagonal is zero. Find the determinant of each diagonal matrix. Make a conjecture based on your results.

(a) $\begin{bmatrix} 7 & 0 \\ 0 & 4 \end{bmatrix}$

(b) $\begin{bmatrix} -1 & 0 & 0 \\ 0 & 5 & 0 \\ 0 & 0 & 2 \end{bmatrix}$

(c) $\begin{bmatrix} 2 & 0 & 0 & 0 \\ 0 & -2 & 0 & 0 \\ 0 & 0 & 1 & 0 \\ 0 & 0 & 0 & 3 \end{bmatrix}$

94. **Error Analysis** Describe the error.

$$\begin{vmatrix} 1 & 1 & 4 \\ 3 & 2 & 0 \\ 2 & 1 & 3 \end{vmatrix} = 3(1)\begin{vmatrix} 1 & 4 \\ 1 & 3 \end{vmatrix} + 2(-1)\begin{vmatrix} 1 & 4 \\ 2 & 3 \end{vmatrix}$$

$$+ 0(1)\begin{vmatrix} 1 & 1 \\ 2 & 1 \end{vmatrix}$$

$$= 3(-1) - 2(-5) + 0$$

$$= 7 \qquad \textbf{✗}$$

Review & Refresh ▶ *Video solutions at LarsonPrecalculus.com*

Order of Operations In Exercises 95–100, simplify the expression.

95. $\dfrac{1(3) - 9(4)}{3(3) - 5(4)}$

96. $\dfrac{10(-5) - 11(-2)}{4(-5) - 3(-2)}$

97. $4(2)(6) + (-1)(3)(5) + 1(2)(-2)$

98. $-1(2)(5) - 4(3)(-2) - (-1)(2)(6)$

99. $4(-1)^{1+1}[2(6) - (-2)(3)]$

100. $(-1)(-1)^{1+2}[2(6) - 5(3)] + 1(-1)^{1+3}[2(-2) - 5(2)]$

Choosing a Solution Method In Exercises 101–104, use any method to solve the system. Explain your choice of method.

101. $\begin{cases} 4x - 2y = 10 \\ 3x - 5y = 11 \end{cases}$

102. $\begin{cases} 3x + 4y = 1 \\ 5x + 3y = 9 \end{cases}$

103. $\begin{cases} 4x - y + z = 12 \\ 2x + 2y + 3z = 1 \\ 5x - 2y + 6z = 22 \end{cases}$

104. $\begin{cases} -x + 2y - 3z = 1 \\ 2x + z = 0 \\ 3x - 4y + 4z = 2 \end{cases}$

Finding the Product of Two Matrices In Exercises 105 and 106, find the product of the matrices.

105. $\begin{bmatrix} -1 & 0 \\ 0 & 1 \end{bmatrix}\begin{bmatrix} 2 \\ 0 \end{bmatrix}$

106. $\begin{bmatrix} -1 & 0 \\ 0 & 1 \end{bmatrix}\begin{bmatrix} 2 \\ 2 \end{bmatrix}$

10.5 Applications of Matrices and Determinants

- Use Cramer's Rule to solve systems of linear equations.
- Use determinants to find areas of triangles.
- Use determinants to test for collinear points and find equations of lines passing through two points.
- Use 2 × 2 matrices to perform transformations in the plane and find areas of parallelograms.
- Use matrices to encode and decode messages.

Cramer's Rule

So far, you have studied four methods for solving a system of linear equations: substitution, graphing, elimination with equations, and elimination with matrices. In this section, you will study one more method, **Cramer's Rule,** named after the Swiss mathematician Gabriel Cramer (1704–1752). This rule uses determinants to write the solution of a system of linear equations. To see how Cramer's Rule works, consider the system described at the beginning of the preceding section, which is shown below.

$$\begin{cases} a_1x + b_1y = c_1 \\ a_2x + b_2y = c_2 \end{cases}$$

This system has a solution

$$x = \frac{c_1b_2 - c_2b_1}{a_1b_2 - a_2b_1} \quad \text{and} \quad y = \frac{a_1c_2 - a_2c_1}{a_1b_2 - a_2b_1}$$

provided that

$$a_1b_2 - a_2b_1 \neq 0.$$

Each numerator and denominator in this solution can be expressed as a determinant.

$$x = \frac{c_1b_2 - c_2b_1}{a_1b_2 - a_2b_1} = \frac{\begin{vmatrix} c_1 & b_1 \\ c_2 & b_2 \end{vmatrix}}{\begin{vmatrix} a_1 & b_1 \\ a_2 & b_2 \end{vmatrix}} \qquad y = \frac{a_1c_2 - a_2c_1}{a_1b_2 - a_2b_1} = \frac{\begin{vmatrix} a_1 & c_1 \\ a_2 & c_2 \end{vmatrix}}{\begin{vmatrix} a_1 & b_1 \\ a_2 & b_2 \end{vmatrix}}$$

Relative to the original system, the denominators for x and y are the determinant of the *coefficient* matrix of the system. This determinant is denoted by D. The numerators for x and y are denoted by D_x and D_y, respectively, and are formed by using the column of constants as replacements for the coefficients of x and y.

Coefficient Matrix	D	D_x	D_y
$\begin{bmatrix} a_1 & b_1 \\ a_2 & b_2 \end{bmatrix}$	$\begin{vmatrix} a_1 & b_1 \\ a_2 & b_2 \end{vmatrix}$	$\begin{vmatrix} c_1 & b_1 \\ c_2 & b_2 \end{vmatrix}$	$\begin{vmatrix} a_1 & c_1 \\ a_2 & c_2 \end{vmatrix}$

For example, given the system

$$\begin{cases} 2x - 5y = 3 \\ -4x + 3y = 8 \end{cases}$$

the coefficient matrix, D, D_x, and D_y are as follows.

Coefficient Matrix	D	D_x	D_y
$\begin{bmatrix} 2 & -5 \\ -4 & 3 \end{bmatrix}$	$\begin{vmatrix} 2 & -5 \\ -4 & 3 \end{vmatrix}$	$\begin{vmatrix} 3 & -5 \\ 8 & 3 \end{vmatrix}$	$\begin{vmatrix} 2 & 3 \\ -4 & 8 \end{vmatrix}$

Determinants have many applications in real life. For example, in Exercise 19 on page 755, you will use a determinant to find the area of a region of forest infested with gypsy moths.

GO DIGITAL

Cramer's Rule generalizes to systems of n equations in n variables. The value of each variable is given as the quotient of two determinants. The denominator is the determinant of the coefficient matrix, and the numerator is the determinant of the matrix formed by replacing the column in the coefficient matrix corresponding to the variable being solved for with the column representing the constants. For example, the solution for x_3 in the system below is shown.

$$\begin{cases} a_{11}x_1 + a_{12}x_2 + a_{13}x_3 = b_1 \\ a_{21}x_1 + a_{22}x_2 + a_{23}x_3 = b_2 \\ a_{31}x_1 + a_{32}x_2 + a_{33}x_3 = b_3 \end{cases} \qquad x_3 = \frac{|A_3|}{|A|} = \frac{\begin{vmatrix} a_{11} & a_{12} & b_1 \\ a_{21} & a_{22} & b_2 \\ a_{31} & a_{32} & b_3 \end{vmatrix}}{\begin{vmatrix} a_{11} & a_{12} & a_{13} \\ a_{21} & a_{22} & a_{23} \\ a_{31} & a_{32} & a_{33} \end{vmatrix}}$$

Cramer's Rule

If a system of n linear equations in n variables has a coefficient matrix A with a nonzero determinant $|A|$, then the solution of the system is

$$x_1 = \frac{|A_1|}{|A|}, \quad x_2 = \frac{|A_2|}{|A|}, \quad \dots, \quad x_n = \frac{|A_n|}{|A|}$$

where the ith column of A_i is the column of constants in the system of equations. If the determinant of the coefficient matrix is zero, then the system has either no solution or infinitely many solutions.

EXAMPLE 1   Using Cramer's Rule for a 2 × 2 System

Use Cramer's Rule (if possible) to solve the system

$$\begin{cases} 4x - 2y = 10 \\ 3x - 5y = 11 \end{cases}.$$

Solution   To begin, find the determinant of the coefficient matrix.

$$D = \begin{vmatrix} 4 & -2 \\ 3 & -5 \end{vmatrix} = -20 - (-6) = -14$$

This determinant is not zero, so you can apply Cramer's Rule.

$$x = \frac{D_x}{D} = \frac{\begin{vmatrix} 10 & -2 \\ 11 & -5 \end{vmatrix}}{-14} = \frac{-50 - (-22)}{-14} = \frac{-28}{-14} = 2$$

$$y = \frac{D_y}{D} = \frac{\begin{vmatrix} 4 & 10 \\ 3 & 11 \end{vmatrix}}{-14} = \frac{44 - 30}{-14} = \frac{14}{-14} = -1$$

The solution is $(2, -1)$. Check this in the original system.

✓ *Checkpoint* ▶ Audio-video solution in English & Spanish at LarsonPrecalculus.com

Use Cramer's Rule (if possible) to solve the system

$$\begin{cases} 3x + 4y = 1 \\ 5x + 3y = 9 \end{cases}.$$

EXAMPLE 2 **Using Cramer's Rule for a 3 × 3 System**

Use Cramer's Rule (if possible) to solve the system $\begin{cases} -x + 2y - 3z = 1 \\ 2x \qquad + z = 0. \\ 3x - 4y + 4z = 2 \end{cases}$

Solution   To find the determinant of the coefficient matrix

$$\begin{bmatrix} -1 & 2 & -3 \\ 2 & 0 & 1 \\ 3 & -4 & 4 \end{bmatrix}$$

expand along the second row.

$$D = 2(-1)^3 \begin{vmatrix} 2 & -3 \\ -4 & 4 \end{vmatrix} + 0(-1)^4 \begin{vmatrix} -1 & -3 \\ 3 & 4 \end{vmatrix} + 1(-1)^5 \begin{vmatrix} -1 & 2 \\ 3 & -4 \end{vmatrix}$$

$$= -2(-4) + 0 - 1(-2)$$

$$= 10$$

This determinant is not zero, so you can apply Cramer's Rule.

$$x = \frac{D_x}{D} = \frac{\begin{vmatrix} 1 & 2 & -3 \\ 0 & 0 & 1 \\ 2 & -4 & 4 \end{vmatrix}}{10} = \frac{8}{10} = \frac{4}{5}$$

$$y = \frac{D_y}{D} = \frac{\begin{vmatrix} -1 & 1 & -3 \\ 2 & 0 & 1 \\ 3 & 2 & 4 \end{vmatrix}}{10} = \frac{-15}{10} = -\frac{3}{2}$$

$$z = \frac{D_z}{D} = \frac{\begin{vmatrix} -1 & 2 & 1 \\ 2 & 0 & 0 \\ 3 & -4 & 2 \end{vmatrix}}{10} = \frac{-16}{10} = -\frac{8}{5}$$

The solution is

$$\left(\frac{4}{5}, -\frac{3}{2}, -\frac{8}{5} \right).$$

Check this in the original system.

✓ *Checkpoint* ▶ *Audio-video solution in English & Spanish at LarsonPrecalculus.com*

Use Cramer's Rule (if possible) to solve the system $\begin{cases} 4x - y + z = 12 \\ 2x + 2y + 3z = 1. \\ 5x - 2y + 6z = 22 \end{cases}$ ■

Remember that Cramer's Rule does not apply when the determinant of the coefficient matrix is zero. This would create division by zero, which is undefined. For example, consider the system of linear equations below.

$$\begin{cases} -x \qquad + z = 4 \\ 2x - y + z = -3 \\ \qquad y - 3z = 1 \end{cases}$$

The determinant of the coefficient matrix is zero, so you cannot apply Cramer's Rule.

GO DIGITAL

Area of a Triangle

Another application of matrices and determinants is finding the area of a triangle whose vertices are given as three points in a coordinate plane.

> **Area of a Triangle**
>
> The area of a triangle with vertices (x_1, y_1), (x_2, y_2), and (x_3, y_3) is
>
> $$\text{Area} = +\frac{1}{2} \begin{vmatrix} x_1 & y_1 & 1 \\ x_2 & y_2 & 1 \\ x_3 & y_3 & 1 \end{vmatrix}$$
>
> where you choose the sign $(\pm)$ so that the area is positive.

For a proof of this formula for the area of a triangle, see Proofs in Mathematics on page 765.

EXAMPLE 3 **Finding the Area of a Triangle**

▶▶▶ *See LarsonPrecalculus.com for an interactive version of this type of example.*

Find the area of the triangle whose vertices are $(1, 0)$, $(2, 2)$, and $(4, 3)$, as shown at the right.

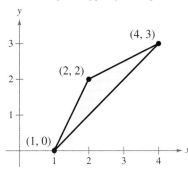

Solution Letting $(x_1, y_1) = (1, 0)$, $(x_2, y_2) = (2, 2)$, and $(x_3, y_3) = (4, 3)$, you have

$$\begin{vmatrix} x_1 & y_1 & 1 \\ x_2 & y_2 & 1 \\ x_3 & y_3 & 1 \end{vmatrix} = \begin{vmatrix} 1 & 0 & 1 \\ 2 & 2 & 1 \\ 4 & 3 & 1 \end{vmatrix}$$

$$= 1(-1)^2 \begin{vmatrix} 2 & 1 \\ 3 & 1 \end{vmatrix} + 0(-1)^3 \begin{vmatrix} 2 & 1 \\ 4 & 1 \end{vmatrix} + 1(-1)^4 \begin{vmatrix} 2 & 2 \\ 4 & 3 \end{vmatrix}$$

$$= -3.$$

Using this value, the area of the triangle is

$$\text{Area} = -\frac{1}{2} \begin{vmatrix} 1 & 0 & 1 \\ 2 & 2 & 1 \\ 4 & 3 & 1 \end{vmatrix} \qquad \text{Choose } (-) \text{ so that the area is positive.}$$

$$= -\frac{1}{2}(-3)$$

$$= \frac{3}{2} \text{ square units.}$$

ALGEBRA HELP

Recall from Section 8.2 that another way to find the area of a triangle is to use Heron's Area Formula. Verify the result of Example 3 using Heron's Area Formula. Which method do you prefer?

▶▶▶▶

✓ *Checkpoint* ▶ *Audio-video solution in English & Spanish at LarsonPrecalculus.com*

Find the area of the triangle whose vertices are $(0, 0)$, $(4, 1)$, and $(2, 5)$. ■

GO DIGITAL

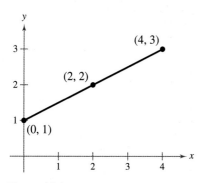

Figure 10.4

Lines in a Plane

In Example 3, what would have happened if the three points were collinear (lying on the same line)? The answer is that the determinant would have been zero. Consider, for example, the three collinear points $(0, 1)$, $(2, 2)$, and $(4, 3)$, as shown in Figure 10.4. The area of the "triangle" that has these three points as vertices is

$$\frac{1}{2}\begin{vmatrix} 0 & 1 & 1 \\ 2 & 2 & 1 \\ 4 & 3 & 1 \end{vmatrix} = \frac{1}{2}\left[0(-1)^2\begin{vmatrix} 2 & 1 \\ 3 & 1 \end{vmatrix} + 1(-1)^3\begin{vmatrix} 2 & 1 \\ 4 & 1 \end{vmatrix} + 1(-1)^4\begin{vmatrix} 2 & 2 \\ 4 & 3 \end{vmatrix} \right]$$

$$= \frac{1}{2}[0 - 1(-2) + 1(-2)]$$

$$= 0.$$

A generalization of this result is below.

Test for Collinear Points

Three points

$$(x_1, y_1), \quad (x_2, y_2), \quad \text{and} \quad (x_3, y_3)$$

are **collinear** (lie on the same line) if and only if

$$\begin{vmatrix} x_1 & y_1 & 1 \\ x_2 & y_2 & 1 \\ x_3 & y_3 & 1 \end{vmatrix} = 0.$$

EXAMPLE 4 **Testing for Collinear Points**

Determine whether the points

$$(-2, -2), \quad (1, 1), \quad \text{and} \quad (7, 5)$$

are collinear. (See Figure 10.5.)

Solution Let $(x_1, y_1) = (-2, -2)$, $(x_2, y_2) = (1, 1)$, and $(x_3, y_3) = (7, 5)$. Then expand along row 1 to obtain

$$\begin{vmatrix} x_1 & y_1 & 1 \\ x_2 & y_2 & 1 \\ x_3 & y_3 & 1 \end{vmatrix} = \begin{vmatrix} -2 & -2 & 1 \\ 1 & 1 & 1 \\ 7 & 5 & 1 \end{vmatrix}$$

$$= -2(-1)^2\begin{vmatrix} 1 & 1 \\ 5 & 1 \end{vmatrix} + (-2)(-1)^3\begin{vmatrix} 1 & 1 \\ 7 & 1 \end{vmatrix} + 1(-1)^4\begin{vmatrix} 1 & 1 \\ 7 & 5 \end{vmatrix}$$

$$= -2(-4) + 2(-6) + 1(-2)$$

$$= -6.$$

The value of this determinant is *not* zero, so the three points are not collinear. Note that the area of the triangle with vertices at these points is $\left(-\frac{1}{2}\right)(-6) = 3$ square units.

Figure 10.5

✓ **Checkpoint** ▶ *Audio-video solution in English & Spanish at LarsonPrecalculus.com*

Determine whether the points

$$(-2, 4), \quad (3, -1), \quad \text{and} \quad (6, -4)$$

are collinear.

GO DIGITAL

The test for collinear points can be adapted for another use. Given two points on a rectangular coordinate system, you can find an equation of the line passing through the two points.

> **Two-Point Form of the Equation of a Line**
>
> An equation of the line passing through the distinct points (x_1, y_1) and (x_2, y_2) is given by
> $$\begin{vmatrix} x & y & 1 \\ x_1 & y_1 & 1 \\ x_2 & y_2 & 1 \end{vmatrix} = 0.$$

EXAMPLE 5   **Finding an Equation of a Line**

Find an equation of the line passing through the points $(2, 4)$ and $(-1, 3)$, as shown in the figure.

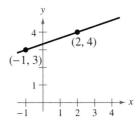

Solution   Let $(x_1, y_1) = (2, 4)$ and $(x_2, y_2) = (-1, 3)$. Applying the determinant formula for the equation of a line produces

$$\begin{vmatrix} x & y & 1 \\ 2 & 4 & 1 \\ -1 & 3 & 1 \end{vmatrix} = 0.$$

Evaluate this determinant to find an equation of the line.

$$x(-1)^2 \begin{vmatrix} 4 & 1 \\ 3 & 1 \end{vmatrix} + y(-1)^3 \begin{vmatrix} 2 & 1 \\ -1 & 1 \end{vmatrix} + 1(-1)^4 \begin{vmatrix} 2 & 4 \\ -1 & 3 \end{vmatrix} = 0$$

$$x(1) - y(3) + (1)(10) = 0$$

$$x - 3y + 10 = 0$$

✓ **Checkpoint** ▶ *Audio-video solution in English & Spanish at LarsonPrecalculus.com*

Find an equation of the line passing through the points $(-3, -1)$ and $(3, 5)$. ■

Note that this method of finding an equation of a line works for all lines, including horizontal and vertical lines. For example, an equation of the vertical line passing through $(2, 0)$ and $(2, 2)$ is

$$\begin{vmatrix} x & y & 1 \\ 2 & 0 & 1 \\ 2 & 2 & 1 \end{vmatrix} = 0$$

$$-2x + 4 = 0$$

$$x = 2.$$

Further Applications of 2 × 2 Matrices

In addition to transforming vectors (see Section 10.2), you can use transformation matrices to transform figures in the coordinate plane. Several transformations and their corresponding transformation matrices are listed below.

Transformation Matrices

Reflection in the y-axis	Reflection in the x-axis	Horizontal stretch ($k > 1$) or shrink ($0 < k < 1$)	Vertical stretch ($k > 1$) or shrink ($0 < k < 1$)
$\begin{bmatrix} -1 & 0 \\ 0 & 1 \end{bmatrix}$	$\begin{bmatrix} 1 & 0 \\ 0 & -1 \end{bmatrix}$	$\begin{bmatrix} k & 0 \\ 0 & 1 \end{bmatrix}$	$\begin{bmatrix} 1 & 0 \\ 0 & k \end{bmatrix}$

EXAMPLE 6　Transforming a Square

A square has vertices at $(0, 0)$, $(2, 0)$, $(0, 2)$, and $(2, 2)$. To find the image of the square after a reflection in the y-axis, first write the vertices as column matrices. Then multiply each column matrix by the appropriate transformation matrix on the left.

$$\begin{bmatrix} -1 & 0 \\ 0 & 1 \end{bmatrix}\begin{bmatrix} 0 \\ 0 \end{bmatrix} = \begin{bmatrix} 0 \\ 0 \end{bmatrix} \qquad \begin{bmatrix} -1 & 0 \\ 0 & 1 \end{bmatrix}\begin{bmatrix} 2 \\ 0 \end{bmatrix} = \begin{bmatrix} -2 \\ 0 \end{bmatrix}$$

$$\begin{bmatrix} -1 & 0 \\ 0 & 1 \end{bmatrix}\begin{bmatrix} 0 \\ 2 \end{bmatrix} = \begin{bmatrix} 0 \\ 2 \end{bmatrix} \qquad \begin{bmatrix} -1 & 0 \\ 0 & 1 \end{bmatrix}\begin{bmatrix} 2 \\ 2 \end{bmatrix} = \begin{bmatrix} -2 \\ 2 \end{bmatrix}$$

So, the vertices of the image are $(0, 0)$, $(-2, 0)$, $(0, 2)$, and $(-2, 2)$. Figure 10.6 shows a sketch of the square and its image.

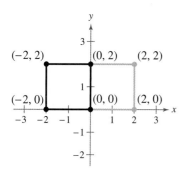

Figure 10.6

✓ **Checkpoint** ▶ *Audio-video solution in English & Spanish at LarsonPrecalculus.com*

Find the image of the square in Example 6 after a vertical stretch by a factor of $k = 2$. ■

You can find the area of a parallelogram using the determinant of a 2 × 2 matrix.

Area of a Parallelogram

The area of a parallelogram with vertices $(0, 0)$, (a, b), (c, d), and $(a + c, b + d)$ is

$$\text{Area} = |\det(A)| \qquad \text{\small $|\det(A)|$ is the absolute value of the determinant.}$$

where $A = \begin{bmatrix} a & b \\ c & d \end{bmatrix}$.

EXAMPLE 7　Finding the Area of a Parallelogram

To find the area of the parallelogram shown in Figure 10.7 using the formula above, let $(a, b) = (2, 0)$ and $(c, d) = (1, 3)$. Then

$$A = \begin{bmatrix} 2 & 0 \\ 1 & 3 \end{bmatrix}$$

and the area of the parallelogram is

$$\text{Area} = |\det(A)| = |6| = 6 \text{ square units.}$$

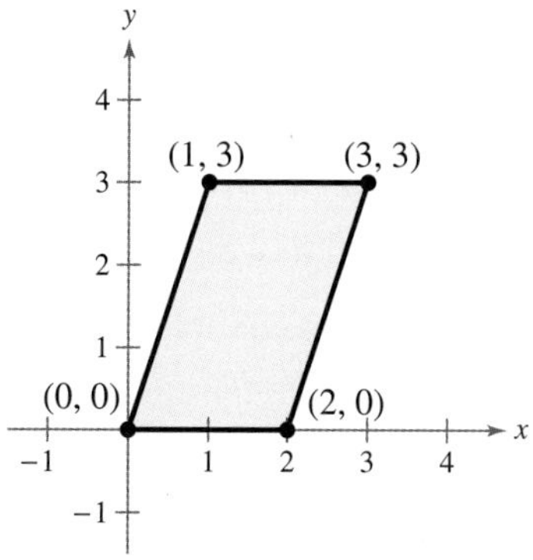

Figure 10.7

✓ **Checkpoint** ▶ *Audio-video solution in English & Spanish at LarsonPrecalculus.com*

Find the area of the parallelogram with vertices $(0, 0)$, $(5, 5)$, $(2, 4)$, and $(7, 9)$. ■

Information security is of the utmost importance when conducting business online, and can include the use of data *encryption.* This is the process of encoding information so that the only way to decode it, apart from an "exhaustion attack," is to use a *key.* Data encryption technology uses algorithms based on the material presented here, but on a much more sophisticated level.

Cryptography

A **cryptogram** is a message written according to a secret code. (The Greek word *kryptos* means "hidden.") Matrix multiplication can be used to encode and decode messages. To begin, assign a number to each letter in the alphabet (with 0 assigned to a blank space), as listed below.

0 = _	9 = I	18 = R
1 = A	10 = J	19 = S
2 = B	11 = K	20 = T
3 = C	12 = L	21 = U
4 = D	13 = M	22 = V
5 = E	14 = N	23 = W
6 = F	15 = O	24 = X
7 = G	16 = P	25 = Y
8 = H	17 = Q	26 = Z

Then convert the message to numbers and partition the numbers into **uncoded row matrices,** each having *n* entries, as demonstrated in Example 8.

EXAMPLE 8 **Forming Uncoded Row Matrices**

Write the uncoded 1×3 row matrices for the message

    MEET ME MONDAY.

Solution Partitioning the message (including blank spaces, but ignoring punctuation) into groups of three produces the uncoded row matrices below.

$$[13 \quad 5 \quad 5] \quad [20 \quad 0 \quad 13] \quad [5 \quad 0 \quad 13] \quad [15 \quad 14 \quad 4] \quad [1 \quad 25 \quad 0]$$
$$\text{M} \quad \text{E} \quad \text{E} \quad \text{T} \qquad \text{M} \quad \text{E} \qquad \text{M} \quad \text{O} \quad \text{N} \quad \text{D} \quad \text{A} \quad \text{Y}$$

Note the use of a blank space to fill out the last uncoded row matrix.

✓ *Checkpoint* ▶ *Audio-video solution in English & Spanish at LarsonPrecalculus.com*

Write the uncoded 1×3 row matrices for the message

    OWLS ARE NOCTURNAL.

To encode a message, create an $n \times n$ invertible matrix A, called an **encoding matrix,** such as

$$A = \begin{bmatrix} 1 & -2 & 2 \\ -1 & 1 & 3 \\ 1 & -1 & -4 \end{bmatrix}.$$

Multiply the uncoded row matrices by A (on the right) to obtain the **coded row matrices.** Here is an example.

Uncoded Matrix	Encoding Matrix A	Coded Matrix
$[13 \quad 5 \quad 5]$	$\begin{bmatrix} 1 & -2 & 2 \\ -1 & 1 & 3 \\ 1 & -1 & -4 \end{bmatrix}$	$= [13 \quad -26 \quad 21]$

© Andrea Danti/Shutterstock.com

EXAMPLE 9   **Encoding a Message**

Use the invertible matrix below to encode the message MEET ME MONDAY.

$$A = \begin{bmatrix} 1 & -2 & 2 \\ -1 & 1 & 3 \\ 1 & -1 & -4 \end{bmatrix}$$

Solution   Obtain the coded row matrices by multiplying each of the uncoded row matrices found in Example 8 by the matrix A.

Uncoded Matrix	Encoding Matrix A	Coded Matrix

$$\begin{bmatrix} 13 & 5 & 5 \end{bmatrix} \begin{bmatrix} 1 & -2 & 2 \\ -1 & 1 & 3 \\ 1 & -1 & -4 \end{bmatrix} = \begin{bmatrix} 13 & -26 & 21 \end{bmatrix}$$

$$\begin{bmatrix} 20 & 0 & 13 \end{bmatrix} \begin{bmatrix} 1 & -2 & 2 \\ -1 & 1 & 3 \\ 1 & -1 & -4 \end{bmatrix} = \begin{bmatrix} 33 & -53 & -12 \end{bmatrix}$$

$$\begin{bmatrix} 5 & 0 & 13 \end{bmatrix} \begin{bmatrix} 1 & -2 & 2 \\ -1 & 1 & 3 \\ 1 & -1 & -4 \end{bmatrix} = \begin{bmatrix} 18 & -23 & -42 \end{bmatrix}$$

$$\begin{bmatrix} 15 & 14 & 4 \end{bmatrix} \begin{bmatrix} 1 & -2 & 2 \\ -1 & 1 & 3 \\ 1 & -1 & -4 \end{bmatrix} = \begin{bmatrix} 5 & -20 & 56 \end{bmatrix}$$

$$\begin{bmatrix} 1 & 25 & 0 \end{bmatrix} \begin{bmatrix} 1 & -2 & 2 \\ -1 & 1 & 3 \\ 1 & -1 & -4 \end{bmatrix} = \begin{bmatrix} -24 & 23 & 77 \end{bmatrix}$$

So, the sequence of coded row matrices is

$$\begin{bmatrix} 13 & -26 & 21 \end{bmatrix} \begin{bmatrix} 33 & -53 & -12 \end{bmatrix} \begin{bmatrix} 18 & -23 & -42 \end{bmatrix} \begin{bmatrix} 5 & -20 & 56 \end{bmatrix} \begin{bmatrix} -24 & 23 & 77 \end{bmatrix}.$$

Finally, removing the matrix notation produces the cryptogram

13 −26 21 33 −53 −12 18 −23 −42 5 −20 56 −24 23 77.

 Checkpoint ▶ *Audio-video solution in English & Spanish at LarsonPrecalculus.com*

Use the invertible matrix below to encode the message OWLS ARE NOCTURNAL.

$$A = \begin{bmatrix} 1 & -1 & 0 \\ 1 & 0 & -1 \\ 6 & -2 & -3 \end{bmatrix}$$

If you do not know the encoding matrix A, decoding a cryptogram such as the one found in Example 9 can be difficult. But if you know the encoding matrix A, decoding is straightforward. You just multiply the coded row matrices by A^{-1} (on the right) to obtain the uncoded row matrices. Here is an example.

$$\underbrace{\begin{bmatrix} 13 & -26 & 21 \end{bmatrix}}_{\text{Coded}} \underbrace{\begin{bmatrix} -1 & -10 & -8 \\ -1 & -6 & -5 \\ 0 & -1 & -1 \end{bmatrix}}_{A^{-1}} = \underbrace{\begin{bmatrix} 13 & 5 & 5 \end{bmatrix}}_{\text{Uncoded}}$$

GO DIGITAL

EXAMPLE 10 **Decoding a Message**

Use the inverse of A in Example 9 to decode the cryptogram

$$13 \quad -26 \quad 21 \quad 33 \quad -53 \quad -12 \quad 18 \quad -23 \quad -42 \quad 5 \quad -20 \quad 56 \quad -24 \quad 23 \quad 77.$$

Solution Find the decoding matrix A^{-1}, partition the message into groups of three to form the coded row matrices and multiply each coded row matrix by A^{-1} (on the right).

Coded Matrix	Decoding Matrix A^{-1}	Decoded Matrix

$$\begin{bmatrix} 13 & -26 & 21 \end{bmatrix} \begin{bmatrix} -1 & -10 & 8 \\ -1 & -6 & -5 \\ 0 & -1 & -1 \end{bmatrix} = \begin{bmatrix} 13 & 5 & 5 \end{bmatrix}$$

$$\begin{bmatrix} 33 & -53 & -12 \end{bmatrix} \begin{bmatrix} -1 & -10 & -8 \\ -1 & -6 & -5 \\ 0 & -1 & -1 \end{bmatrix} = \begin{bmatrix} 20 & 0 & 13 \end{bmatrix}$$

$$\begin{bmatrix} 18 & -23 & -42 \end{bmatrix} \begin{bmatrix} -1 & -10 & -8 \\ -1 & -6 & -5 \\ 0 & -1 & -1 \end{bmatrix} = \begin{bmatrix} 5 & 0 & 13 \end{bmatrix}$$

$$\begin{bmatrix} 5 & -20 & 56 \end{bmatrix} \begin{bmatrix} -1 & -10 & -8 \\ -1 & -6 & -5 \\ 0 & -1 & -1 \end{bmatrix} = \begin{bmatrix} 15 & 14 & 4 \end{bmatrix}$$

$$\begin{bmatrix} -24 & 23 & 77 \end{bmatrix} \begin{bmatrix} -1 & -10 & -8 \\ -1 & -6 & -5 \\ 0 & -1 & -1 \end{bmatrix} = \begin{bmatrix} 1 & 25 & 0 \end{bmatrix}$$

So, the message is

$$\begin{bmatrix} 13 & 5 & 5 \end{bmatrix} \begin{bmatrix} 20 & 0 & 13 \end{bmatrix} \begin{bmatrix} 5 & 0 & 13 \end{bmatrix} \begin{bmatrix} 15 & 14 & 4 \end{bmatrix} \begin{bmatrix} 1 & 25 & 0 \end{bmatrix}.$$
M E E T   M E   M O N   D A Y

✓ ***Checkpoint*** ▶ *Audio-video solution in English & Spanish at LarsonPrecalculus.com*

Use the inverse of A in the Checkpoint with Example 9 to decode the cryptogram

$$110 \quad -39 \quad -59 \quad 25 \quad -21 \quad -3 \quad 23 \quad -18 \quad -5 \quad 47 \quad -20 \quad -24$$
$$149 \quad -56 \quad -75 \quad 87 \quad -38 \quad -37.$$ ∎

Summarize (Section 10.5)

1. Explain how to use Cramer's Rule to solve systems of linear equations *(page 746)*. For examples of using Cramer's Rule, see Examples 1 and 2.

2. State the formula for finding the area of a triangle using a determinant *(page 748)*. For an example of using this formula to find the area of a triangle, see Example 3.

3. Explain how to use determinants to test for collinear points *(page 749)* and find equations of lines passing through two points *(page 750)*. For examples of these applications, see Examples 4 and 5.

4. Explain how to use 2×2 matrices to perform transformations in the plane and find areas of parallelograms *(page 751)*. For examples of these applications, see Examples 6 and 7.

5. Explain how to use matrices to encode and decode messages *(pages 752–754)*. For examples involving encoding and decoding messages, see Examples 8–10.

10.5 Exercises

See CalcChat.com for tutorial help and worked-out solutions to odd-numbered exercises.

Vocabulary and Concept Check

In Exercises 1 and 2, fill in the blanks.

1. The method of using determinants to solve a system of linear equations is called _____ _____.

2. A message written according to a secret code is a _____.

In Exercises 3 and 4, consider the system
$$\begin{cases} 2x + 3y = 8 \\ 3x + 2y = 7 \end{cases}.$$

3. Find the determinant of the coefficient Matrix D.

4. Find the determinants D_x and D_y.

Skills and Applications

Using Cramer's Rule **In Exercises 5–12, use Cramer's Rule (if possible) to solve the system of equations.**

5. $\begin{cases} -5x + 9y = -14 \\ 3x - 7y = 10 \end{cases}$ 6. $\begin{cases} 4x - 3y = -10 \\ 6x + 9y = 12 \end{cases}$

7. $\begin{cases} 3x + 2y = -2 \\ 6x + 4y = 4 \end{cases}$ 8. $\begin{cases} 12x - 7y = -4 \\ -11x + 8y = 10 \end{cases}$

9. $\begin{cases} 4x - y + z = -5 \\ 2x + 2y + 3z = 10 \\ 5x - 2y + 6z = 1 \end{cases}$ 10. $\begin{cases} 4x - 2y + 3z = -2 \\ 2x + 2y + 5z = 16 \\ 8x - 5y - 2z = 4 \end{cases}$

11. $\begin{cases} x + 2y + 3z = -3 \\ -2x + y - z = 6 \\ 3x - 3y + 2z = -11 \end{cases}$ 12. $\begin{cases} 5x - 4y + z = -14 \\ -x + 2y - 2z = 10 \\ 3x + y + z = 1 \end{cases}$

Finding the Area of a Triangle **In Exercises 13–16, use a determinant to find the area of the triangle with the given vertices.**

13.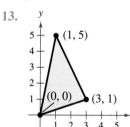

14.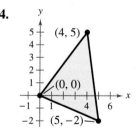

15. $(0, 4), (-2 -3), (2, -3)$
16. $(-2, 1), (1, 6), (3, -1)$

Finding a Coordinate **In Exercises 17 and 18, find a value of y such that the triangle with the given vertices has an area of 4 square units.**

17. $(-5, 1), (0, 2), (-2, y)$
18. $(-4, 2), (-3, 5), (-1, y)$

19. **Area of Infestation**

A large region of forest is infested with gypsy moths. The region is triangular, as shown in the figure. From vertex A, the distances to the other vertices are 25 miles south

and 10 miles east (for vertex B), and 20 miles south and 28 miles east (for vertex C). Use a graphing utility to find the area (in square miles) of the region.

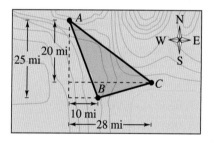

20. **Botany** A botanist is studying the plants growing in the triangular region shown in the figure. Starting at vertex A, the botanist walks 65 feet east and 50 feet north to vertex B, and then walks 85 feet west and 30 feet north to vertex C. Use a graphing utility to find the area (in square feet) of the region.

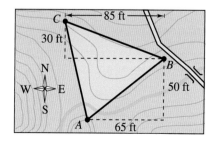

Testing for Collinear Points In Exercises 21–26, use a determinant to determine whether the points are collinear.

21. $(2, -6), (0, -2), (3, -8)$
22. $(3, -5), (6, 1), (4, 2)$
23. $\left(2, -\frac{1}{2}\right), (-4, 4), (6, -3)$
24. $(0, 1), \left(-2, \frac{7}{2}\right), \left(1, -\frac{1}{4}\right)$
25. $(0, 2), (1, 2.4), (-1, 1.6)$
26. $(3, 7), (4, 9.5), (-1, -5)$

Finding a Coordinate In Exercises 27 and 28, find the value of y such that the points are collinear.

27. $(2, -5), (4, y), (5, -2)$ 28. $(-6, 2), (-5, y), (-3, 5)$

Finding an Equation of a Line In Exercises 29–34, use a determinant to find an equation of the line passing through the points.

29. $(0, 0), (5, 3)$
30. $(0, 0), (-2, 2)$
31. $(-4, 3), (2, 1)$
32. $(10, 7), (-2, -7)$
33. $\left(-\frac{1}{2}, 3\right), \left(\frac{5}{2}, 1\right)$
34. $\left(\frac{2}{3}, 4\right), (6, 12)$

Transforming a Square In Exercises 35–38, use matrices to find the vertices of the image of the square with the given vertices after the given transformation. Then sketch the square and its image.

35. $(0, 0), (0, 3), (3, 0), (3, 3)$; horizontal stretch, $k = 2$
36. $(1, 2), (3, 2), (1, 4), (3, 4)$; reflection in the x-axis
37. $(4, 3), (5, 3), (4, 4), (5, 4)$; reflection in the y-axis
38. $(1, 1), (3, 2), (0, 3), (2, 4)$; vertical shrink, $k = \frac{1}{2}$

Finding the Area of a Parallelogram In Exercises 39–42, use a determinant to find the area of the parallelogram with the given vertices.

39. $(0, 0), (1, 0), (2, 2), (3, 2)$
40. $(0, 0), (3, 0), (4, 1), (7, 1)$
41. $(0, 0), (-2, 0), (3, 5), (1, 5)$
42. $(0, 0), (0, 8), (8, -6), (8, 2)$

Encoding a Message In Exercises 43 and 44, (a) write the uncoded 1×2 row matrices for the message, and then (b) encode the message using the encoding matrix.

Message	Encoding Matrix
43. COME HOME SOON	$\begin{bmatrix} 1 & 2 \\ 3 & 5 \end{bmatrix}$
44. HELP IS ON THE WAY	$\begin{bmatrix} -2 & 3 \\ -1 & 1 \end{bmatrix}$

Encoding a Message In Exercises 45 and 46, (a) write the uncoded 1×3 row matrices for the message, and then (b) encode the message using the encoding matrix.

Message	Encoding Matrix
45. TEXT ME TOMORROW	$\begin{bmatrix} 1 & -1 & 0 \\ 1 & 0 & -1 \\ -6 & 2 & 3 \end{bmatrix}$
46. I USED THE APP TO SEND MONEY	$\begin{bmatrix} 4 & 2 & 1 \\ -3 & -3 & -1 \\ 3 & 2 & 1 \end{bmatrix}$

Encoding a Message In Exercises 47–50, write a cryptogram for the message using the matrix

$$A = \begin{bmatrix} 1 & 2 & 2 \\ 3 & 7 & 9 \\ -1 & -4 & -7 \end{bmatrix}.$$

47. DOWNLOAD SUCCESSFUL
48. SEND ME A PIC
49. HAPPY BIRTHDAY
50. OPERATION TERMINATED

Decoding a Message In Exercises 51–54, use A^{-1} to decode the cryptogram.

51. $A = \begin{bmatrix} 1 & 2 \\ 3 & 5 \end{bmatrix}$

11 21 64 112 25 50 29 53 23 46 40
75 55 92

52. $A = \begin{bmatrix} 2 & 3 \\ 3 & 4 \end{bmatrix}$

85 120 6 8 10 15 84 117 42 56 90
125 60 80 30 45 19 26

53. $A = \begin{bmatrix} 1 & -1 & 0 \\ 1 & 0 & -1 \\ -6 & 2 & 3 \end{bmatrix}$

9 -1 -9 38 -19 -19 28 -9 -19
-80 25 41 -64 21 31 9 -5 -4

54. $A = \begin{bmatrix} 3 & -4 & 2 \\ 0 & 2 & 1 \\ 4 & -5 & 3 \end{bmatrix}$

112 -140 83 19 -25 13 72 -76 61
95 -118 71 20 21 38 35 -23 36 42
-48 32

55. **Decoding a Message** The cryptogram below was encoded with a 2×2 matrix.

8 21 -15 -10 -13 -13 5 10 5 25
5 19 -1 6 20 40 -18 -18 1 16

The last word of the message is _RON. What is the message?

56. Decoding a Message   The cryptogram below was encoded with a 2×2 matrix.

$$5 \quad 2 \quad 25 \quad 11 \quad -2 \quad -7 \quad -15 \quad -15 \quad 32 \quad 14$$
$$-8 \quad -13 \quad 38 \quad 19 \quad -19 \quad -19 \quad 37 \quad 16$$

The last word of the message is _SUE. What is the message?

57. Circuit Analysis   Consider the circuit shown in the figure. The currents I_1, I_2, and I_3 (in amperes) are the solution of

$$\begin{cases} 4I_1 \quad\quad + 8I_3 = 2 \\ \quad\quad 2I_2 + 8I_3 = 6. \\ I_1 + I_2 - I_3 = 0 \end{cases}$$

Use Cramer's Rule to find the three currents.

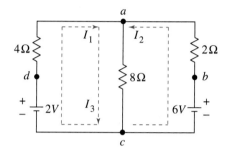

58. Pulley System   A system of pulleys is loaded with 192-pound and 64-pound weights (see figure). The tensions t_1 and t_2 (in pounds) in the ropes and the acceleration a (in feet per second squared) of the 64-pound weight are found by solving

$$\begin{cases} t_1 - 2t_2 \quad\quad = \quad 0 \\ t_1 \quad\quad - 3a = 192. \\ \quad\quad t_2 + 2a = \quad 64 \end{cases}$$

Use Cramer's Rule to find t_1, t_2, and a.

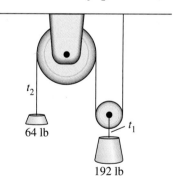

Exploring the Concepts

True or False?   In Exercises 59 and 60, determine whether the statement is true or false. Justify your answer.

59. If three points are not collinear, then the test for collinear points gives a nonzero determinant.

60. Cramer's Rule cannot be used to solve a system of linear equations when the determinant of the coefficient matrix is zero.

61. Error Analysis   Describe the error. Consider the system

$$\begin{cases} 2x - 3y = 0 \\ 4x - 6y = 0 \end{cases}.$$

The system has no solution because the determinant of the coefficient matrix is

$$D = \begin{vmatrix} 2 & -3 \\ 4 & -6 \end{vmatrix} = -12 - (-12) = 0. \quad$$

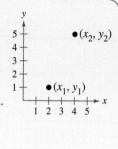

62.   HOW DO YOU SEE IT?   Briefly describe two methods that can be used to find an equation of the line that passes through the points shown. Discuss the advantages and disadvantages of each method.

63. Finding the Area of a Triangle   Use a determinant to find the area of the triangle whose vertices are $(3, -1)$, $(7, -1)$, and $(7, 5)$. Confirm your answer by plotting the points in a coordinate plane and using the formula Area $= \frac{1}{2}$(base)(height).

Review & Refresh ▶ *Video solutions at LarsonPrecalculus.com*

64. Electronics   The resistance y (in ohms) of 1000 feet of solid copper wire at 68 degrees Fahrenheit is given by

$$y = \frac{10,370}{x^2}$$

where x is the diameter of the wire in mils (0.001 inch).

(a) Complete the table.

x	20	30	40	50	60	70	80	90
y								

(b) Use the table of values in part (a) to sketch a graph of the model. Then use your graph to estimate the resistance when $x = 85.5$.

(c) Use the model to confirm algebraically the estimate you found in part (b).

(d) What can you conclude about the relationship between the diameter of the copper wire and the resistance?

Evaluating an Expression   In Exercises 65–68, evaluate the expression when $n = 1$, $n = 2$, $n = 3$, and $n = 4$.

65. $3 + (-1)^n$

66. $2 + (-1)^{n+1}$

67. $(-1)^{n+1}(n^2 + 1)$

68. $3^n + 1$

GO DIGITAL

What Did You Learn?

The list below reviews the skills covered in the chapter and correlates each
one to the Review Exercises (see page 760) that practice the skill.

Section 10.1	Review Exercises
■ Write matrices and determine their dimensions *(p. 700)*.	*1–8*
■ Perform elementary row operations on matrices *(p. 702)*.	*9, 10*
1. Interchange two rows.	
2. Multiply a row by a nonzero constant.	
3. Add a multiple of a row to another row.	
■ Use matrices and Gaussian elimination to solve systems of linear equations *(p. 703)*.	*11–24*
1. Write the augmented matrix of the system of linear equations.	
2. Use elementary row operations to rewrite the augmented matrix in row-echelon form.	
3. Write the system of linear equations corresponding to the matrix in row-echelon form and use back-substitution to find the solution.	
■ Use matrices and Gauss-Jordan elimination to solve systems of linear equations *(p. 707)*.	*25–30*

Section 10.2

	Review Exercises
■ Determine whether two matrices are equal *(p. 713)*.	*31–34*
■ Add and subtract matrices, and multiply matrices by scalars *(p. 714)*.	*35–46*

If $A = [a_{ij}]$ and $B = [b_{ij}]$ are matrices of dimension $m \times n$, then their sum is the $m \times n$ matrix $A + B = [a_{ij} + b_{ij}]$. If $A = [a_{ij}]$ is an $m \times n$ matrix and c is a scalar, then the scalar multiple of A by c is the $m \times n$ matrix $cA = [ca_{ij}]$.

■ Multiply two matrices *(p. 718)*.	*47–56*

If $A = [a_{ij}]$ is an $m \times n$ matrix and $B = [b_{ij}]$ is an $n \times p$ matrix, then the product AB is an $m \times p$ matrix given by $AB = [c_{ij}]$, where $c_{ij} = a_{i1}b_{1j} + a_{i2}b_{2j} + a_{i3}b_{3j} + \ldots + a_{in}b_{nj}$.

■ Use matrices to transform vectors *(p. 721)*.	*57–60*
■ Use matrix operations to model and solve real-life problems *(p. 722)*.	*61, 62*

Section 10.3

	Review Exercises
■ Verify that two matrices are inverses of each other *(p. 728)*.	*63–66*

Let A be an $n \times n$ matrix and let I_n be the $n \times n$ identity matrix. If there exists a matrix A^{-1} such that $AA^{-1} = I_n = A^{-1}A$, then A^{-1} is the inverse of A.

■ Use Gauss-Jordan elimination to find the inverses of matrices *(p. 729)*.	*67–72*
■ Use a formula to find the inverses of 2×2 matrices *(p. 732)*.	*73–76*
■ Use inverse matrices to solve systems of linear equations *(p. 733)*.	*77–88*

Section 10.4 | Review Exercises

- Find the determinants of 2×2 matrices *(p. 737)*.   *89–92*

$$\det(A) = |A| = \begin{vmatrix} a_1 & b_1 \\ a_2 & b_2 \end{vmatrix} = a_1b_2 - a_2b_1$$

- Find minors and cofactors of square matrices *(p. 739)*.   *93–96*

- Find the determinants of square matrices *(p. 740)*.   *97–102*

Section 10.5

- Use Cramer's Rule to solve systems of linear equations *(p. 745)*.   *103–106*

- Use determinants to find areas of triangles *(p. 748)*.   *107, 108*

$$\text{Area} = \pm\frac{1}{2}\begin{vmatrix} x_1 & y_1 & 1 \\ x_2 & y_2 & 1 \\ x_3 & y_3 & 1 \end{vmatrix}$$

- Use determinants to test for collinear points and find equations of lines passing through two points *(p. 749)*.   *109–114*

- Use 2×2 matrices to perform transformations in the plane and find areas of parallelograms *(p. 751)*.   *115, 116*

- Use matrices to encode and decode messages *(p. 752)*.   *117, 118*

Study Strategies

Make the Most of Class Time   Learning math in college is a team effort between instructor and student. The more you understand in class, the more you will be able to learn while studying outside of class.

1. **Sit where you can easily see and hear the instructor, and the instructor can see you.** The instructor may be able to tell when you are confused just by the look on your face and may adjust the lesson accordingly. In addition, sitting in this strategic place will keep your mind from wandering.

2. **Pay attention to what the instructor says about the math, not just what is written on the board.** Write problems on the left side of your notes and what the instructor says about the problems on the right side.

3. **When the instructor is moving through the material too fast, ask a question.** Questions help to slow the pace for a few minutes and also to clarify what is confusing to you.

4. **Try to memorize new information while learning it.** Repeat in your head what you are writing in your notes. That way you are reviewing the information twice.

5. **Ask for clarification.** When you do not understand something at all and do not even know how to phrase a question, just ask for clarification. You might say something like, "Could you please explain the steps in this problem one more time?"

6. **Think as intensely as if you were going to take a quiz on the material at the end of class.** This kind of mindset will help you to process new information.

7. **If the instructor asks for someone to go up to the board, volunteer.** The student at the board often receives additional attention and instruction to complete the problem.

8. **At the end of class, identify concepts or problems on which you still need clarification.** Make sure you see the instructor or a tutor as soon as possible.

Review Exercises
See CalcChat.com for tutorial help and worked-out solutions to odd-numbered exercises.

GO DIGITAL

10.1 **Dimension of a Matrix** In Exercises 1–4, determine the dimension of the matrix.

1. $\begin{bmatrix} -1 & 3 \end{bmatrix}$

2. $\begin{bmatrix} 3 & 1 \\ 5 & -2 \end{bmatrix}$

3. $\begin{bmatrix} 2 & 1 & 0 & 4 & -1 \\ 6 & 2 & 1 & 8 & 0 \end{bmatrix}$

4. $\begin{bmatrix} 5 \end{bmatrix}$

Writing an Augmented Matrix In Exercises 5 and 6, write the augmented matrix for the system of linear equations.

5. $\begin{cases} 3x - 10y = 15 \\ 5x + 4y = 22 \end{cases}$

6. $\begin{cases} 8x - 7y + 4z = 12 \\ 3x - 5y = 20 \end{cases}$

Writing a System of Equations In Exercises 7 and 8, write a system of linear equations represented by the augmented matrix.

7. $\begin{bmatrix} 1 & 0 & 2 & \vdots & -8 \\ 2 & -2 & 3 & \vdots & 12 \\ 4 & 7 & 1 & \vdots & 3 \end{bmatrix}$

8. $\begin{bmatrix} 2 & 10 & 8 & 5 & \vdots & -1 \\ -3 & 4 & 0 & 9 & \vdots & 2 \end{bmatrix}$

Writing a Matrix in Row-Echelon Form In Exercises 9 and 10, write the matrix in row-echelon form. (Remember that the row-echelon form of a matrix is not unique.)

9. $\begin{bmatrix} 0 & 1 & 1 \\ 1 & 2 & 3 \\ 2 & 2 & 2 \end{bmatrix}$

10. $\begin{bmatrix} 4 & 8 & 16 \\ 3 & -1 & 2 \\ -2 & 10 & 12 \end{bmatrix}$

Using Back-Substitution In Exercises 11–14, write a system of linear equations represented by the augmented matrix. Then use back-substitution to solve the system.

11. $\begin{bmatrix} 1 & -3 & \vdots & 9 \\ 0 & 1 & \vdots & -1 \end{bmatrix}$

12. $\begin{bmatrix} 1 & 3 & -9 & \vdots & 4 \\ 0 & 1 & -1 & \vdots & 10 \\ 0 & 0 & 1 & \vdots & -2 \end{bmatrix}$

13. $\begin{bmatrix} 1 & 3 & 4 & \vdots & 1 \\ 0 & 1 & 2 & \vdots & 3 \\ 0 & 0 & 1 & \vdots & 4 \end{bmatrix}$

14. $\begin{bmatrix} 1 & -8 & 0 & \vdots & -2 \\ 0 & 1 & -1 & \vdots & -7 \\ 0 & 0 & 1 & \vdots & 1 \end{bmatrix}$

Gaussian Elimination with Back-Substitution In Exercises 15–24, use matrices to solve the system of linear equations, if possible. Use Gaussian elimination with back-substitution.

15. $\begin{cases} 5x + 4y = 2 \\ -x + y - -22 \end{cases}$

16. $\begin{cases} 2x - 5y = 2 \\ 3x - 7y = 1 \end{cases}$

17. $\begin{cases} -x + 2y = 3 \\ 2x - 4y = 6 \end{cases}$

18. $\begin{cases} -x + 2y = 3 \\ 2x - 4y = -6 \end{cases}$

19. $\begin{cases} 2x + y + 2z = 4 \\ 2x + 2y = 5 \\ 2x - y + 6z = 2 \end{cases}$

20. $\begin{cases} x + 2y + 6z = 1 \\ 2x + 5y + 15z = 4 \\ 3x + y + 3z = -6 \end{cases}$

21. $\begin{cases} 2x + 3y + z = 10 \\ 2x - 3y - 3z = 22 \\ 4x - 2y + 3z = -2 \end{cases}$

22. $\begin{cases} 2x + 3y + 3z = 3 \\ 6x + 6y + 12z = 13 \\ 12x + 9y - z = 2 \end{cases}$

23. $\begin{cases} x - 2y + z - w = 11 \\ 2x + y - 2z + 3w = -16 \\ -x + 3y + 2z - w = 1 \\ -y + 5z + w = 8 \end{cases}$

24. $\begin{cases} 3x - y - z - 6w = 0 \\ x + 2y + z - w = 9 \\ -x + 2w = 1 \\ 5y + 4z - 2w = -8 \end{cases}$

Gauss-Jordan Elimination In Exercises 25–28, use matrices to solve the system of linear equations, if possible. Use Gauss-Jordan elimination.

25. $\begin{cases} x + 2y - z = 3 \\ x - y - z = -3 \\ 2x + y + 3z = 10 \end{cases}$

26. $\begin{cases} x - 3y + z = 2 \\ 3x - y - z = -6 \\ -x + y - 3z = -2 \end{cases}$

27. $\begin{cases} -x + y + 2z = 1 \\ 2x + 3y + z = -2 \\ 5x + 4y + 2z = 4 \end{cases}$

28. $\begin{cases} 4x + 4y + 4z = 5 \\ 4x - 2y - 8z = 1 \\ 5x + 3y + 8z = 6 \end{cases}$

Solving a System of Linear Equations In Exercises 29 and 30, use the matrix capabilities of a graphing utility to write the augmented matrix corresponding to the system of linear equations in reduced row-echelon form. Then solve the system, if possible.

29. $\begin{cases} 3x - y + 5z - 2w = -44 \\ x + 6y + 4z - w = 1 \\ 5x - y + z + 3w = -15 \\ 4y - z - 8w = 58 \end{cases}$

30. $\begin{cases} 4x + 12y + 2z = 20 \\ x + 6y + 4z = 12 \\ x + 6y + z = 8 \\ -2x - 10y - 2z = -10 \end{cases}$

10.2 **Equality of Matrices** In Exercises 31–34, solve for *x* and *y*.

31. $\begin{bmatrix} -1 & x \\ y & 9 \end{bmatrix} = \begin{bmatrix} -1 & 12 \\ 11 & 9 \end{bmatrix}$

32. $\begin{bmatrix} -1 & 0 \\ x & 5 \\ -4 & -3 \end{bmatrix} = \begin{bmatrix} -1 & 0 \\ 8 & 5 \\ -4 & y \end{bmatrix}$

33. $\begin{bmatrix} x+3 & -4 & 44 \\ 0 & -3 & 2 \\ -2 & y+5 & 6 \end{bmatrix} = \begin{bmatrix} 5x-1 & -4 & 44 \\ 0 & -3 & 2 \\ -2 & 16 & 6 \end{bmatrix}$

34. $\begin{bmatrix} -9 & 4 & 2 & -5 \\ 0 & -3 & 7 & 2y \\ 6 & -1 & 1 & 0 \end{bmatrix} = \begin{bmatrix} -9 & 4 & x-10 & -5 \\ 0 & -3 & 7 & -6 \\ 6 & -1 & 1 & 0 \end{bmatrix}$

Operations with Matrices In Exercises 35–38, if possible, find (a) $A + B$, (b) $A - B$, (c) $4A$, and (d) $2A + 2B$.

35. $A = \begin{bmatrix} 2 & -2 \\ 3 & 5 \end{bmatrix}$, $B = \begin{bmatrix} -3 & 10 \\ 12 & 8 \end{bmatrix}$

36. $A = \begin{bmatrix} 4 & 3 \\ -6 & 1 \\ 10 & 1 \end{bmatrix}$, $B = \begin{bmatrix} 3 & 11 \\ 15 & 25 \\ 20 & 29 \end{bmatrix}$

37. $A = \begin{bmatrix} 5 & 4 \\ -7 & 2 \\ 11 & 2 \end{bmatrix}$, $B = \begin{bmatrix} 0 & 3 \\ 4 & 12 \\ 20 & 40 \end{bmatrix}$

38. $A = \begin{bmatrix} 6 & -5 & 7 \end{bmatrix}$, $B = \begin{bmatrix} -1 \\ 4 \\ 8 \end{bmatrix}$

Evaluating an Expression In Exercises 39–42, evaluate the expression.

39. $\begin{bmatrix} 7 & 3 \\ -1 & 5 \end{bmatrix} + \begin{bmatrix} 10 & -20 \\ 14 & -3 \end{bmatrix} + \begin{bmatrix} 5 & 0 \\ 1 & 9 \end{bmatrix}$

40. $\begin{bmatrix} -11 & -7 \\ 16 & -2 \\ 19 & 1 \end{bmatrix} - \begin{bmatrix} 6 & 0 \\ 8 & -4 \\ -2 & 10 \end{bmatrix} + \begin{bmatrix} -3 & 1 \\ 2 & 28 \\ 12 & -2 \end{bmatrix}$

41. $-2\left(\begin{bmatrix} 1 & 2 \\ 5 & -4 \\ 6 & 0 \end{bmatrix} + \begin{bmatrix} 7 & 1 \\ 1 & 2 \\ 1 & 4 \end{bmatrix} \right)$

42. $5\left(\begin{bmatrix} 8 & -1 & 8 \\ -2 & 4 & 12 \\ 0 & -6 & 0 \end{bmatrix} - \begin{bmatrix} -2 & 0 & -4 \\ 3 & -1 & 1 \\ 6 & 12 & -8 \end{bmatrix} \right)$

Solving a Matrix Equation In Exercises 43–46, solve for *X* in the equation, where

$A = \begin{bmatrix} -4 & 0 \\ 1 & -5 \\ -3 & 2 \end{bmatrix}$ and $B = \begin{bmatrix} 1 & 2 \\ -2 & 1 \\ 4 & 4 \end{bmatrix}$.

43. $X = 2A - 3B$ **44.** $6X = 4A + 3B$

45. $3X + 2A = B$

46. $2A = 3X + 5B$

Finding the Product of Two Matrices In Exercises 47–50, if possible, find *AB* and state the dimension of the result.

47. $A = \begin{bmatrix} 2 & -2 \\ 3 & 5 \end{bmatrix}$, $B = \begin{bmatrix} -3 & 10 \\ 12 & 8 \end{bmatrix}$

48. $A = \begin{bmatrix} 5 & 4 \\ -7 & 2 \\ 11 & 2 \end{bmatrix}$, $B = \begin{bmatrix} 4 & 12 \\ 20 & 40 \\ 15 & 30 \end{bmatrix}$

49. $A = \begin{bmatrix} 5 & 4 \\ -7 & 2 \\ 11 & 2 \end{bmatrix}$, $B = \begin{bmatrix} 4 & 12 \\ 20 & 40 \end{bmatrix}$

50. $A = \begin{bmatrix} 6 & -5 & 7 \end{bmatrix}$, $B = \begin{bmatrix} -1 \\ 4 \\ 8 \end{bmatrix}$

Operations with Matrices In Exercises 51 and 52, if possible, find (a) *AB*, (b) *BA*, and (c) A^2.

51. $A = \begin{bmatrix} 1 & 3 \\ 4 & 1 \end{bmatrix}$, $B = \begin{bmatrix} 5 & -1 \\ -2 & 0 \end{bmatrix}$

52. $A = \begin{bmatrix} 2 & 3 \\ 8 & -1 \\ 0 & 2 \end{bmatrix}$, $B = \begin{bmatrix} 4 \\ 1 \end{bmatrix}$

Operations with Matrices In Exercises 53–56, evaluate the expression, if possible. Use the matrix capabilities of a graphing utility to verify your answer.

53. $\begin{bmatrix} 4 & 1 \\ 11 & -7 \\ 12 & 3 \end{bmatrix} \begin{bmatrix} 3 & -5 & 6 \\ 2 & -2 & -2 \end{bmatrix}$

54. $\begin{bmatrix} 1 & 2 & -1 \\ 0 & 4 & -2 \\ 1 & 1 & 3 \end{bmatrix} \begin{bmatrix} 1 & -1 & 2 \end{bmatrix}$

55. $\begin{bmatrix} 3 & 5 \\ 0 & -1 \end{bmatrix} \begin{bmatrix} 2 & 0 \\ 1 & 1 \end{bmatrix} \begin{bmatrix} -2 & 4 & 1 \\ 4 & -5 & 2 \end{bmatrix}$

56. $\left(\begin{bmatrix} 8 & -6 & 0 \\ 2 & 3 & -1 \\ 4 & 12 & -2 \end{bmatrix} + \begin{bmatrix} 5 & 1 & 4 \\ 0 & -10 & -3 \\ 6 & -6 & 2 \end{bmatrix} \right) \begin{bmatrix} 0.5 \\ 1.5 \\ -0.5 \end{bmatrix}$

Describing a Vector Transformation In Exercises 57–60, find *A*v, where v = $\langle 2, 5 \rangle$, and describe the transformation.

57. $A = \begin{bmatrix} 1 & 0 \\ 0 & -1 \end{bmatrix}$ **58.** $A = \begin{bmatrix} 0 & -1 \\ -1 & 0 \end{bmatrix}$

59. $A = \begin{bmatrix} \frac{1}{2} & 0 \\ 0 & 1 \end{bmatrix}$ **60.** $A = \begin{bmatrix} 1 & 0 \\ 0 & 6 \end{bmatrix}$

61. Manufacturing A tire corporation has three factories that manufacture two models of tires. The production levels are represented by A.

$$A = \begin{matrix} & \text{Factory} \\ & \begin{matrix} 1 & \;\;2 & \;\;\;3 \end{matrix} \\ \begin{bmatrix} 80 & 120 & 140 \\ 40 & 100 & 80 \end{bmatrix} \begin{matrix} \text{A} \\ \text{B} \end{matrix} \end{matrix} \Big\} \text{ Model}$$

Find the production levels when production decreases by 5%.

62. Cell Phone Charges The pay-as-you-go charges (per minute) of two cell phone companies for calls inside the coverage area, regional roaming calls, and calls outside the coverage area are represented by C.

$$C = \begin{matrix} & \text{Company} \\ & \begin{matrix} \text{A} & \;\;\;\text{B} \end{matrix} \\ \begin{bmatrix} \$0.07 & \$0.095 \\ \$0.10 & \$0.08 \\ \$0.28 & \$0.25 \end{bmatrix} \begin{matrix} \text{Inside} \\ \text{Regional Roaming} \\ \text{Outside} \end{matrix} \end{matrix} \Big\} \text{ Coverage area}$$

The numbers of minutes you plan to use in the coverage areas per month are represented by the matrix

$$T = \begin{bmatrix} 120 & 80 & 20 \end{bmatrix}.$$

Compute and interpret TC.

10.3 **The Inverse of a Matrix** In Exercises 63–66, show that B is the inverse of A.

63. $A = \begin{bmatrix} -4 & -1 \\ 7 & 2 \end{bmatrix}$, $B = \begin{bmatrix} -2 & -1 \\ 7 & 4 \end{bmatrix}$

64. $A = \begin{bmatrix} 5 & -1 \\ 11 & -2 \end{bmatrix}$, $B = \begin{bmatrix} -2 & 1 \\ -11 & 5 \end{bmatrix}$

65. $A = \begin{bmatrix} 1 & 1 & 0 \\ 1 & 0 & 1 \\ 6 & 2 & 3 \end{bmatrix}$, $B = \begin{bmatrix} -2 & -3 & 1 \\ 3 & 3 & -1 \\ 2 & 4 & -1 \end{bmatrix}$

66. $A = \begin{bmatrix} 1 & -1 & 0 \\ -1 & 0 & -1 \\ 8 & -4 & 2 \end{bmatrix}$,

$B = \begin{bmatrix} -2 & 1 & \frac{1}{2} \\ -3 & 1 & \frac{1}{2} \\ 2 & -2 & -\frac{1}{2} \end{bmatrix}$

Finding the Inverse of a Matrix In Exercises 67–70, find the inverse of the matrix, if possible.

67. $\begin{bmatrix} -6 & 5 \\ -5 & 4 \end{bmatrix}$

68. $\begin{bmatrix} 3 & 4 \\ 6 & 8 \end{bmatrix}$

69. $\begin{bmatrix} 2 & 0 & 3 \\ -1 & 1 & 1 \\ 2 & -2 & 1 \end{bmatrix}$

70. $\begin{bmatrix} 0 & -2 & 1 \\ -5 & -2 & -3 \\ 7 & 3 & 4 \end{bmatrix}$

Finding the Inverse of a Matrix In Exercises 71 and 72, use the matrix capabilities of a graphing utility to find the inverse of the matrix, if possible.

71. $\begin{bmatrix} -1 & -2 & -2 \\ 3 & 7 & 9 \\ 1 & 4 & 7 \end{bmatrix}$

72. $\begin{bmatrix} 8 & 0 & 2 & 8 \\ 4 & -2 & 0 & -2 \\ 1 & 2 & 1 & 4 \\ -1 & 4 & 1 & 1 \end{bmatrix}$

Finding the Inverse of a 2 × 2 Matrix In Exercises 73–76, use the formula on page 732 to find the inverse of the 2 × 2 matrix, if possible.

73. $\begin{bmatrix} -7 & 2 \\ -8 & 2 \end{bmatrix}$

74. $\begin{bmatrix} 10 & 4 \\ 7 & 3 \end{bmatrix}$

75. $\begin{bmatrix} -12 & 6 \\ 10 & -5 \end{bmatrix}$

76. $\begin{bmatrix} -18 & -15 \\ -6 & -5 \end{bmatrix}$

Solving a System Using an Inverse Matrix In Exercises 77–86, use an inverse matrix to solve the system of linear equations, if possible.

77. $\begin{cases} -x + 4y = 8 \\ 2x - 7y = -5 \end{cases}$

78. $\begin{cases} 5x - y = 13 \\ -9x + 2y = -24 \end{cases}$

79. $\begin{cases} -3x + 10y = 8 \\ 5x - 17y = -13 \end{cases}$

80. $\begin{cases} 4x - 2y = -10 \\ -19x + 9y = 47 \end{cases}$

81. $\begin{cases} \frac{1}{2}x + \frac{1}{3}y = 2 \\ -6x - 4y = 0 \end{cases}$

82. $\begin{cases} -\frac{5}{6}x + \frac{3}{8}y = -2 \\ 4x - 3y = 0 \end{cases}$

83. $\begin{cases} 0.3x + 0.7y = 10.2 \\ 0.4x + 0.6y = 7.6 \end{cases}$

84. $\begin{cases} 3.5x - 4.5y = 8 \\ 2.5x - 7.5y = 25 \end{cases}$

85. $\begin{cases} 3x + 2y - z = 6 \\ x - y + 2z = -1 \\ 5x + y + z = 7 \end{cases}$

86. $\begin{cases} 4x + 5y - 6z = -6 \\ 3x + 2y + 2z = 8 \\ 2x + y + z = 3 \end{cases}$

Using a Graphing Utility In Exercises 87 and 88, use the matrix capabilities of a graphing utility to solve the system of linear equations, if possible.

87. $\begin{cases} 5x + 10y = 7 \\ 2x + y = -98 \end{cases}$

88. $\begin{cases} x + 3y + 3z = -1 \\ 6x + 2y - 4z = 1 \\ 3x - y - z = 7 \end{cases}$

10.4 **Finding the Determinant of a Matrix** In Exercises 89–92, find the determinant of the matrix.

89. $\begin{bmatrix} 2 & 5 \\ -4 & 3 \end{bmatrix}$

90. $\begin{bmatrix} -3 & 1 \\ 5 & -2 \end{bmatrix}$

91. $\begin{bmatrix} 10 & -2 \\ 18 & 8 \end{bmatrix}$

92. $\begin{bmatrix} -30 & 10 \\ 5 & 2 \end{bmatrix}$

Finding the Minors and Cofactors of a Matrix
In Exercises 93–96, find all the (a) minors and (b) cofactors of the matrix.

93. $\begin{bmatrix} 2 & -1 \\ 7 & 4 \end{bmatrix}$ 94. $\begin{bmatrix} 3 & 6 \\ 5 & -4 \end{bmatrix}$

95. $\begin{bmatrix} 3 & 2 & -1 \\ -2 & 5 & 0 \\ 1 & 8 & 6 \end{bmatrix}$ 96. $\begin{bmatrix} 8 & 3 & 4 \\ 6 & 5 & -9 \\ -4 & 1 & 2 \end{bmatrix}$

Finding the Determinant of a Matrix In Exercises 97–102, find the determinant of the matrix. Expand by cofactors using the row or column that appears to make the computations easiest.

97. $\begin{bmatrix} -2 & 0 & 0 \\ 2 & -1 & 0 \\ -1 & 1 & -3 \end{bmatrix}$ 98. $\begin{bmatrix} 0 & 1 & -2 \\ 0 & 1 & 2 \\ -1 & -1 & 3 \end{bmatrix}$

99. $\begin{bmatrix} 4 & 1 & -1 \\ 2 & 3 & 2 \\ 1 & -1 & 0 \end{bmatrix}$ 100. $\begin{bmatrix} -1 & 5 & 1 \\ 2 & 3 & 0 \\ -5 & -1 & 1 \end{bmatrix}$

101. $\begin{bmatrix} -2 & 4 & 1 \\ -6 & 0 & 2 \\ 5 & 3 & 4 \end{bmatrix}$ 102. $\begin{bmatrix} 1 & 1 & 4 \\ -4 & 1 & 2 \\ 0 & 1 & -1 \end{bmatrix}$

10.5 **Using Cramer's Rule** In Exercises 103–106, use Cramer's Rule (if possible) to solve the system of equations.

103. $\begin{cases} 5x - 2y = 6 \\ -11x + 3y = -23 \end{cases}$ 104. $\begin{cases} 3x + 8y = -7 \\ 9x - 5y = 37 \end{cases}$

105. $\begin{cases} -2x + 3y - 5z = -11 \\ 4x - y + z = -3 \\ -x - 4y + 6z = 15 \end{cases}$

106. $\begin{cases} 5x - 2y + z = 15 \\ 3x - 3y - z = -7 \\ 2x - y - 7z = -3 \end{cases}$

Finding the Area of a Triangle In Exercises 107 and 108, use a determinant to find the area of the triangle with the given vertices.

107.   108.

Testing for Collinear Points In Exercises 109 and 110, use a determinant to determine whether the points are collinear.

109. $(-1, 7), (3, -9), (-3, 15)$

110. $(0, -5), (-2, -6), (8, -1)$

Finding an Equation of a Line In Exercises 111–114, use a determinant to find an equation of the line passing through the points.

111. $(-4, 0), (4, 4)$ 112. $(2, 5), (6, -1)$

113. $\left(-\frac{5}{2}, 3\right), \left(\frac{7}{2}, 1\right)$ 114. $(-0.8, 0.2), (0.7, 3.2)$

Finding the Area of a Parallelogram In Exercises 115 and 116, use a determinant to find the area of the parallelogram with the given vertices.

115. $(0, 0), (2, 0), (1, 4), (3, 4)$

116. $(0, 0), (-3, 0), (1, 3), (-2, 3)$

Decoding a Message In Exercises 117 and 118, use A^{-1} to decode the cryptogram.

$$A = \begin{bmatrix} -5 & 4 & -3 \\ 10 & -7 & 6 \\ 8 & -6 & 5 \end{bmatrix}$$

117. -5 11 -2 370 -265 225 -57 48 -33 32 -15 20 245 -171 147

118. 145 -105 92 264 -188 160 23 -16 15 129 -84 78 -9 8 -5 159 -118 100 219 -152 133 370 -265 225 -105 84 -63

Exploring the Concepts

True or False? In Exercises 119 and 120, determine whether the statement is true or false. Justify your answer.

119. It is possible to find the determinant of a 4×5 matrix.

120. $\begin{vmatrix} a_{11} & a_{12} & a_{13} \\ a_{21} & a_{22} & a_{23} \\ a_{31} + c_1 & a_{32} + c_2 & a_{33} + c_3 \end{vmatrix}$

$$= \begin{vmatrix} a_{11} & a_{12} & a_{13} \\ a_{21} & a_{22} & a_{23} \\ a_{31} & a_{32} & a_{33} \end{vmatrix} + \begin{vmatrix} a_{11} & a_{12} & a_{13} \\ a_{21} & a_{22} & a_{23} \\ c_1 & c_2 & c_3 \end{vmatrix}$$

121. **Writing** What is the cofactor of an entry of a matrix? How are cofactors used to find the determinant of the matrix?

122. **Think About It** Three people are solving a system of equations using an augmented matrix. Each person writes the matrix in row-echelon form. Their reduced matrices are shown below.

$$\begin{bmatrix} 1 & 2 & \vdots & 3 \\ 0 & 1 & \vdots & 1 \end{bmatrix}$$

$$\begin{bmatrix} 1 & 0 & \vdots & 1 \\ 0 & 1 & \vdots & 1 \end{bmatrix}$$

$$\begin{bmatrix} 1 & 2 & \vdots & 3 \\ 0 & 0 & \vdots & 0 \end{bmatrix}$$

Can all three be correct? Explain.

See CalcChat.com for tutorial help and worked-out solutions to odd-numbered exercises.

Chapter Test

GO DIGITAL

Take this test as you would take a test in class. When you are finished, check your work against the answers given in the back of the book.

In Exercises 1 and 2, write the matrix in reduced row-echelon form. *(Section 10.1)*

1. $\begin{bmatrix} 1 & -1 & 5 \\ 6 & 2 & 3 \\ 5 & 3 & -3 \end{bmatrix}$

2. $\begin{bmatrix} 1 & 0 & -1 & 2 \\ -1 & 1 & 1 & -3 \\ 1 & 1 & -1 & 1 \\ 3 & 2 & -3 & 4 \end{bmatrix}$

3. Write the augmented matrix for the system of equations and solve the system. *(Section 10.1)*

$$\begin{cases} 4x + 3y - 2z = 14 \\ -x - y + 2z = -5 \\ 3x + y - 4z = 8 \end{cases}$$

4. If possible, find (a) $A - B$, (b) $3C$, (c) $3A - 2B$, (d) BC, and (e) C^2. *(Section 10.2)*

$$A = \begin{bmatrix} 6 & 5 \\ -5 & -5 \end{bmatrix}, \quad B = \begin{bmatrix} 5 & 0 \\ -5 & -1 \end{bmatrix}, \quad C = \begin{bmatrix} 2 & -1 & 4 \\ 0 & 6 & -3 \end{bmatrix}$$

5. Find the product $A\mathbf{v}$, where $A = \begin{bmatrix} 0 & -1 \\ -1 & 0 \end{bmatrix}$ and $\mathbf{v} = \langle 2, 3 \rangle$, and describe the transformation. *(Section 10.2)*

In Exercises 6 and 7, find the inverse of the matrix, if possible. *(Section 10.3)*

6. $\begin{bmatrix} -4 & 3 \\ 5 & -2 \end{bmatrix}$

7. $\begin{bmatrix} -2 & 4 & -6 \\ 2 & 1 & 0 \\ 4 & -2 & 5 \end{bmatrix}$

8. Use the result of Exercise 6 to solve the system. *(Section 10.3)*

$$\begin{cases} -4x + 3y = 6 \\ 5x - 2y = 24 \end{cases}$$

In Exercises 9–11, find the determinant of the matrix. *(Section 10.4)*

9. $\begin{bmatrix} -6 & 4 \\ 10 & 12 \end{bmatrix}$

10. $\begin{bmatrix} \frac{5}{2} & -\frac{3}{8} \\ -8 & \frac{6}{5} \end{bmatrix}$

11. $\begin{bmatrix} 6 & -7 & 2 \\ 3 & -2 & 0 \\ 1 & 5 & 1 \end{bmatrix}$

In Exercises 12 and 13, use Cramer's Rule to solve the system. *(Section 10.5)*

12. $\begin{cases} 7x + 6y = 9 \\ -2x - 11y = -49 \end{cases}$

13. $\begin{cases} 6x - y + 2z = -4 \\ -2x + 3y - z = 10 \\ 4x - 4y + z = -18 \end{cases}$

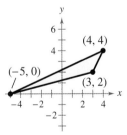

Figure for 14

14. Use a determinant to find the area of the triangle at the left. *(Section 10.5)*

15. Write the uncoded 1×3 row matrices for the message KNOCK ON WOOD. Then encode the message using the encoding matrix A at the right. *(Section 10.5)*

$$A = \begin{bmatrix} 1 & -1 & 0 \\ 1 & 0 & -1 \\ 6 & -2 & -3 \end{bmatrix}$$

16. One hundred liters of a 50% solution is obtained by mixing a 60% solution with a 20% solution. Use a system of linear equations to determine how many liters of each solution are required to obtain the desired mixture. Solve the system using matrices. *(Sections 10.1, 10.3, and 10.5)*

Proofs in Mathematics

Area of a Triangle (p. 748)

The area of a triangle with vertices (x_1, y_1), (x_2, y_2), and (x_3, y_3) is

$$\text{Area} = \pm\frac{1}{2}\begin{vmatrix} x_1 & y_1 & 1 \\ x_2 & y_2 & 1 \\ x_3 & y_3 & 1 \end{vmatrix}$$

where you choose the sign $(\pm)$ so that the area is positive.

Proof

Prove the case for $y_i > 0$. Assume that

$$x_1 \leq x_3 \leq x_2$$

and that (x_3, y_3) lies above the line segment connecting (x_1, y_1) and (x_2, y_2), as shown in the figure below.

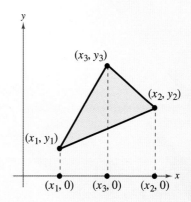

Consider the three trapezoids whose vertices are

Trapezoid 1: $(x_1, 0)$, (x_1, y_1), (x_3, y_3), $(x_3, 0)$

Trapezoid 2: $(x_3, 0)$, (x_3, y_3), (x_2, y_2), $(x_2, 0)$

Trapezoid 3: $(x_1, 0)$, (x_1, y_1), (x_2, y_2), $(x_2, 0)$.

The area of the triangle is the sum of the areas of the first two trapezoids minus the area of the third trapezoid. So,

$$\text{Area} = \frac{1}{2}(y_1 + y_3)(x_3 - x_1) + \frac{1}{2}(y_3 + y_2)(x_2 - x_3) - \frac{1}{2}(y_1 + y_2)(x_2 - x_1)$$

$$= \frac{1}{2}(x_1 y_2 + x_2 y_3 + x_3 y_1 - x_1 y_3 - x_2 y_1 - x_3 y_2)$$

$$= \frac{1}{2}\begin{vmatrix} x_1 & y_1 & 1 \\ x_2 & y_2 & 1 \\ x_3 & y_3 & 1 \end{vmatrix}.$$

If the vertices do not occur in the order

$$x_1 \leq x_3 \leq x_2$$

or if the vertex (x_3, y_3) does not lie above the line segment connecting the other two vertices, then the formula above may yield the negative of the area. So, use $\pm$ and choose the correct sign so that the area is positive. ∎

A proof without words is a picture or diagram that gives a visual understanding of why a theorem or statement is true. It can also provide a starting point for writing a formal proof.

In Section 10.5 (page 751), you learned that the area of a parallelogram with vertices $(0, 0)$, (a, b), (c, d), and $(a + c, b + d)$ is the absolute value of the determinant of the matrix A, where

$$A = \begin{bmatrix} a & b \\ c & d \end{bmatrix}.$$

The color-coded visual proof below shows this for a case in which the determinant is positive. Also shown is a brief explanation of why this proof works.

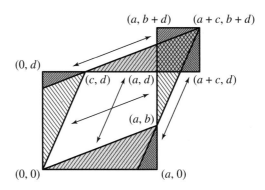

$$\begin{vmatrix} a & b \\ c & d \end{vmatrix} = ad - bc = \|\square\| - \|\square\| = \|\square\|$$

Area of $\square$ = Area of orange $\triangle$ + Area of yellow $\triangle$ + Area of blue $\triangle$
  + Area of pink $\triangle$ + Area of white quadrilateral

Area of $\square$ = Area of orange $\triangle$ + Area of pink $\triangle$ + Area of green quadrilateral

Area of $\square$ = Area of white quadrilateral + Area of blue $\triangle$ + Area of yellow $\triangle$
  − Area of green quadrilateral

  = Area of $\square$ − Area of $\square$

The formula in Section 10.5 is a generalization, taking into consideration the possibility that the coordinates could yield a negative determinant. Area is always positive, which is the reason the formula uses absolute value. Verify the formula using values of a, b, c, and d that produce a negative determinant.

From "Proof Without Words: A 2 × 2 Determinant Is the Area of a Parallelogram" by Solomon W. Golomb, *Mathematics Magazine,* Vol. 58, No. 2, pg. 107.

P.S. Problem Solving

See CalcChat.com for tutorial help and worked-out solutions to odd-numbered exercises.

GO DIGITAL

1. Multiplying by a Transformation Matrix The columns of matrix T show the coordinates of the vertices of a triangle. Matrix A is a transformation matrix.

$$A = \begin{bmatrix} 0 & -1 \\ 1 & 0 \end{bmatrix} \qquad T = \begin{bmatrix} 1 & 2 & 3 \\ 1 & 4 & 2 \end{bmatrix}$$

(a) Find AT and AAT. Then sketch the original triangle and the two images of the triangle. What transformation does A represent?

(b) Given the triangle determined by AAT, describe the transformation that produces the triangle determined by AT and then the triangle determined by T.

2. Population The matrices show the male and female populations in the United States in 2015 and 2018. The male and female populations are separated into three age groups. *(Source: U.S. Census Bureau)*

2015

	0–19	20–64	65+
Male	41,936,220	94,930,166	21,044,389
Female	40,146,685	96,061,906	26,623,307

2018

	0–19	20–64	65+
Male	41,881,940	95,939,921	23,306,818
Female	40,100,725	96,813,655	29,124,375

(a) The total population in 2015 was 320,742,673, and the total population in 2018 was 327,167,434. Rewrite the matrices to give the information as percents of the total population.

(b) Write a matrix that gives the change in the percent of the population for each gender and age group from 2015 to 2018.

(c) Based on the result of part (b), which gender(s) and age group(s) had percents that decreased from 2015 to 2018?

3. Determining Whether Matrices are Idempotent A square matrix is **idempotent** when $A^2 = A$. Determine whether each matrix is idempotent.

(a) $\begin{bmatrix} 1 & 0 \\ 0 & 0 \end{bmatrix}$ (b) $\begin{bmatrix} 0 & 1 \\ 1 & 0 \end{bmatrix}$

(c) $\begin{bmatrix} 2 & 3 \\ -1 & -2 \end{bmatrix}$ (d) $\begin{bmatrix} 2 & 3 \\ 1 & 2 \end{bmatrix}$

(e) $\begin{bmatrix} 0 & 0 & 1 \\ 0 & 1 & 0 \\ 1 & 0 & 0 \end{bmatrix}$ (f) $\begin{bmatrix} 0 & 1 & 0 \\ 1 & 0 & 0 \\ 0 & 0 & 1 \end{bmatrix}$

4. Finding a Matrix Find a singular 2×2 matrix satisfying $A^2 = A$.

5. Quadratic Matrix Equation Let

$$A = \begin{bmatrix} 1 & 2 \\ -2 & 1 \end{bmatrix}.$$

(a) Show that $A^2 - 2A + 5I = O$, where I is the identity matrix of dimension 2×2.

(b) Show that $A^{-1} = \frac{1}{5}(2I - A)$.

(c) Show that for any square matrix satisfying

$$A^2 - 2A + 5I = O$$

the inverse of A is given by

$$A^{-1} = \frac{1}{5}(2I - A).$$

6. Satellite Television Two competing companies offer satellite television to a city with 100,000 households. Gold Satellite System has 25,000 subscribers and Galaxy Satellite Network has 30,000 subscribers. (The other 45,000 households do not subscribe.) The matrix shows the percent changes in satellite subscriptions each year.

Percent Changes

Percent Changes		From Gold	From Galaxy	From Non-subscriber
	To Gold	0.70	0.15	0.15
	To Galaxy	0.20	0.80	0.15
	To Nonsubscriber	0.10	0.05	0.70

(a) Find the number of subscribers each company will have in 1 year using matrix multiplication. Explain how you obtained your answer.

(b) Find the number of subscribers each company will have in 2 years using matrix multiplication. Explain how you obtained your answer.

(c) Find the number of subscribers each company will have in 3 years using matrix multiplication. Explain how you obtained your answer.

(d) What is happening to the number of subscribers to each company? What is happening to the number of nonsubscribers?

7. The Transpose of a Matrix The **transpose** of a matrix, denoted A^T, is formed by writing its rows as columns. Find the transpose of each matrix and verify that $(AB)^T = B^T A^T$.

$$A = \begin{bmatrix} -1 & 1 & -2 \\ 2 & 0 & 1 \end{bmatrix}, \quad B = \begin{bmatrix} -3 & 0 \\ 1 & 2 \\ 1 & -1 \end{bmatrix}$$

8. Finding a Value Find x such that the matrix is equal to its own inverse.

$$A = \begin{bmatrix} 3 & x \\ -2 & -3 \end{bmatrix}$$

9. Finding a Value Find x such that the matrix is singular.

$$A = \begin{bmatrix} 4 & x \\ -2 & -3 \end{bmatrix}$$

10. Verifying an Equation Verify the following equation.

$$\begin{vmatrix} 1 & 1 & 1 \\ a & b & c \\ a^2 & b^2 & c^2 \end{vmatrix} = (a - b)(b - c)(c - a)$$

11. Verifying an Equation Verify the following equation.

$$\begin{vmatrix} 1 & 1 & 1 \\ a & b & c \\ a^3 & b^3 & c^3 \end{vmatrix} = (a - b)(b - c)(c - a)(a + b + c)$$

12. Verifying an Equation Verify the following equation.

$$\begin{vmatrix} x & 0 & c \\ -1 & x & b \\ 0 & -1 & a \end{vmatrix} = ax^2 + bx + c$$

13. Finding a Matrix Find a 4×4 matrix whose determinant is equal to $ax^3 + bx^2 + cx + d$. (*Hint:* Use the equation in Exercise 12 as a model.)

14. Finding the Determinant of a Matrix Let A be an $n \times n$ matrix each of whose rows sum to zero. Find $|A|$.

15. Finding Atomic Masses The table shows the masses (in atomic mass units) of three compounds. Use a linear system and Cramer's Rule to find the atomic masses of sulfur (S), nitrogen (N), and fluorine (F).

Compound	Formula	Mass
Tetrasulfur tetranitride	S_4N_4	184
Sulfur hexafluoride	SF_6	146
Dinitrogen tetrafluoride	N_2F_4	104

16. Finding the Costs of Items A walkway lighting package includes a transformer, a certain length of wire, and a certain number of lights on the wire. The price of each lighting package depends on the length of wire and the number of lights on the wire. Use the information below to find the cost of a transformer, the cost per foot of wire, and the cost of a light. Assume that the cost of each item is the same in each lighting package.

- A package that contains a transformer, 25 feet of wire, and 5 lights costs $20.
- A package that contains a transformer, 50 feet of wire, and 15 lights costs $35.
- A package that contains a transformer, 100 feet of wire, and 20 lights costs $50.

17. Decoding a Message Use the inverse of A to decode the cryptogram.

$$A = \begin{bmatrix} 1 & -2 & 2 \\ 1 & 1 & -3 \\ 1 & -1 & 4 \end{bmatrix}$$

23 13 −34 31 −34 63 25 −17 61
24 14 −37 41 −17 −8 20 −29 40 38
−56 116 13 −11 1 22 −3 −6 41
−53 85 28 −32 16

18. Decoding a Message A code breaker intercepts the encoded message below.

45 −35 38 −30 18 −18 35 −30 81 −60
42 −28 75 −55 2 −2 22 −21 15 −10

Let $A^{-1} = \begin{bmatrix} w & x \\ y & z \end{bmatrix}$.

(a) You know that

$$[45 \ \ -35]A^{-1} = [10 \ \ \ 15]$$
$$[38 \ \ -30]A^{-1} = [8 \ \ \ 14]$$

where A^{-1} is the inverse of the encoding matrix A. Write and solve two systems of equations to find w, x, y, and z.

(b) Decode the message.

19. Conjecture Let

$$A = \begin{bmatrix} 6 & 4 & 1 \\ 0 & 2 & 3 \\ 1 & 1 & 2 \end{bmatrix}.$$

Use a graphing utility to find A^{-1}. Compare $|A^{-1}|$ with $|A|$. Make a conjecture about the determinant of the inverse of a matrix.

20. Conjecture Consider matrices of the form

$$A = \begin{bmatrix} 0 & a_{12} & a_{13} & a_{14} & \cdots & a_{1n} \\ 0 & 0 & a_{23} & a_{24} & \cdots & a_{2n} \\ 0 & 0 & 0 & a_{34} & \cdots & a_{3n} \\ \vdots & \vdots & \vdots & \vdots & & \vdots \\ 0 & 0 & 0 & 0 & \cdots & a_{(n-1)n} \\ 0 & 0 & 0 & 0 & \cdots & 0 \end{bmatrix}.$$

(a) Write a 2×2 matrix and a 3×3 matrix in the form of A.

(b) Use a graphing utility to raise each of the matrices to higher powers. Describe the result.

(c) Use the result of part (b) to make a conjecture about powers of A when A is a 4×4 matrix. Use the graphing utility to test your conjecture.

(d) Use the results of parts (b) and (c) to make a conjecture about powers of A when A is an $n \times n$ matrix.

11 Sequences, Series, and Probability

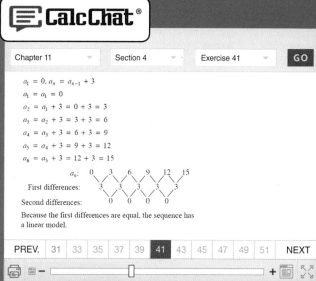

11.2 Falling Object *(Exercise 75, p. 787)*

11.7 Backup System *(Exercise 59, p. 836)*

11.1 Sequences and Series

Sequences and series model many real-life situations over time. For example, in Exercise 94 on page 778, a sequence models the percent of U.S. adults who met federal physical activity guidelines from 2007 through 2018.

● Use sequence notation to write the terms of sequences.
● Use factorial notation.
● Use summation notation to write sums.
● Find the sums of series.
● Use sequences and series to model and solve real-life problems.

Sequences

In mathematics, the word *sequence* is used in much the same way as in ordinary English. Saying that a collection is listed in *sequence* means that it is ordered so that it has a first member, a second member, a third member, and so on. Two examples are 1, 2, 3, 4, . . . and 1, 3, 5, 7, ...

Mathematically, you can think of a sequence as a *function* whose domain is the set of positive integers. Rather than using function notation, however, sequences are usually written using subscript notation, as shown in the following definition.

> ### Definition of Sequence
>
> An **infinite sequence** is a function whose domain is the set of positive integers. The function values
>
> $$a_1, a_2, a_3, a_4, \ldots, a_n, \ldots$$
>
> are the **terms** of the sequence. When the domain of the function consists of the first n positive integers only, the sequence is a **finite sequence.**

On occasion, it is convenient to begin subscripting a sequence with 0 instead of 1 so that the terms of the sequence become

$$a_0, a_1, a_2, a_3, \ldots .$$

When this is the case, the domain includes 0.

ALGEBRA HELP

The subscripts of a sequence make up the domain of the sequence and serve to identify the positions of terms within the sequence. For example, a_4 is the fourth term of the sequence, and a_n is the nth term of the sequence.

EXAMPLE 1 **Writing the Terms of a Sequence**

a. The first four terms of the sequence given by $a_n = 3n - 2$ are listed below.

$a_1 = 3(1) - 2 = 1$ 1st term

$a_2 = 3(2) - 2 = 4$ 2nd term

$a_3 = 3(3) - 2 = 7$ 3rd term

$a_4 = 3(4) - 2 = 10$ 4th term

b. The first four terms of the sequence given by $a_n = 3 + (-1)^n$ are listed below.

$a_1 = 3 + (-1)^1 = 3 - 1 = 2$ 1st term

$a_2 = 3 + (-1)^2 = 3 + 1 = 4$ 2nd term

$a_3 = 3 + (-1)^3 = 3 - 1 = 2$ 3rd term

$a_4 = 3 + (-1)^4 = 3 + 1 = 4$ 4th term

✓ *Checkpoint* ▶ Audio-video solution in English & Spanish at LarsonPrecalculus.com

Write the first four terms of the sequence given by $a_n = 2n + 1$. ■

© Mezzotint/Shutterstock.com

EXAMPLE 2   A Sequence Whose Terms Alternate in Sign

Write the first four terms of the sequence given by $a_n = \dfrac{(-1)^n}{2n + 1}$.

Solution   The first four terms of the sequence are listed below.

$$a_1 = \frac{(-1)^1}{2(1) + 1} = \frac{-1}{2 + 1} = -\frac{1}{3} \qquad \text{1st term}$$

$$a_2 = \frac{(-1)^2}{2(2) + 1} = \frac{1}{4 + 1} = \frac{1}{5} \qquad \text{2nd term}$$

$$a_3 = \frac{(-1)^3}{2(3) + 1} = \frac{-1}{6 + 1} = -\frac{1}{7} \qquad \text{3rd term}$$

$$a_4 = \frac{(-1)^4}{2(4) + 1} = \frac{1}{8 + 1} = \frac{1}{9} \qquad \text{4th term}$$

✓ **Checkpoint** 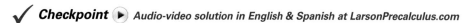 Audio-video solution in English & Spanish at LarsonPrecalculus.com

Write the first four terms of the sequence given by $a_n = \dfrac{2 + (-1)^n}{n}$. ■

Simply listing the first few terms is not sufficient to define a unique sequence—the *n*th term *must be given*. To see this, consider the two sequences below, both of which have the same first three terms.

$$\frac{1}{2}, \frac{1}{4}, \frac{1}{8}, \frac{1}{16}, \cdots \cdots \frac{1}{2^n}, \cdots$$

$$\frac{1}{2}, \frac{1}{4}, \frac{1}{8}, \frac{1}{15}, \cdots \cdots \frac{6}{(n + 1)(n^2 - n + 6)}, \cdots$$

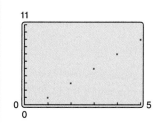 ▶▶▶ **TECHNOLOGY**

To graph a sequence using a graphing utility, set the mode to *sequence* and *dot* and enter the *n*th term of the sequence. The graph of the sequence in Example 3(a) is shown below. To identify the terms, use the *trace* feature.

EXAMPLE 3   Finding the *n*th Term of a Sequence

Write an expression for the apparent *n*th term a_n of each sequence.

a. $1, 3, 5, 7, \ldots$ **b.** $2, -5, 10, -17, \ldots$

Solution

a. n: 1   2   3   4 . . . n

Terms: 1   3   5   7 . . . a_n

Apparent pattern: Each term is 1 less than twice *n*. So, the apparent *n*th term is

$$a_n = 2n - 1.$$

b. n: 1   2   3   4 . . . n

Terms: 2   −5   10   −17 . . . a_n

Apparent pattern: The absolute value of each term is 1 more than the square of *n*, and the terms have alternating signs, with those in the even positions being negative. So, the apparent *n*th term is

$$a_n = (-1)^{n+1}(n^2 + 1).$$

✓ **Checkpoint**  Audio-video solution in English & Spanish at LarsonPrecalculus.com

Write an expression for the apparent *n*th term a_n of each sequence.

a. $1, 5, 9, 13, \ldots$ **b.** $2, -4, 6, -8, \ldots$ ■

Some sequences are defined **recursively.** To define a sequence recursively, you need to be given one or more of the first few terms. All other terms of the sequence are then defined using previous terms.

EXAMPLE 4 A Recursive Sequence

Write the first five terms of the sequence defined recursively as

$$a_1 = 3$$
$$a_k = 2a_{k-1} + 1, \quad \text{where } k \geq 2.$$

Solution

$a_1 = 3$	1st term is given.
$a_2 = 2a_{2-1} + 1 = 2a_1 + 1 = 2(3) + 1 = 7$	Use recursion formula.
$a_3 = 2a_{3-1} + 1 = 2a_2 + 1 = 2(7) + 1 = 15$	Use recursion formula.
$a_4 = 2a_{4-1} + 1 = 2a_3 + 1 = 2(15) + 1 = 31$	Use recursion formula.
$a_5 = 2a_{5-1} + 1 = 2a_4 + 1 = 2(31) + 1 = 63$	Use recursion formula.

 Checkpoint ▶ Audio-video solution in English & Spanish at *LarsonPrecalculus.com*

Write the first five terms of the sequence defined recursively as

$$a_1 = 6$$
$$a_{k+1} = a_k + 1, \quad \text{where} \quad k \geq 1.$$

In the next example, you will study a well-known recursive sequence, the Fibonacci sequence.

EXAMPLE 5 The Fibonacci Sequence: A Recursive Sequence

The Fibonacci sequence is defined recursively, as follows.

$$a_0 = 1$$
$$a_1 = 1$$
$$a_k = a_{k-2} + a_{k-1}, \quad \text{where} \quad k \geq 2$$

Write the first six terms of this sequence.

Solution

$a_0 = 1$	0th term is given.
$a_1 = 1$	1st term is given.
$a_2 = a_{2-2} + a_{2-1} = a_0 + a_1 = 1 + 1 = 2$	Use recursion formula.
$a_3 = a_{3-2} + a_{3-1} = a_1 + a_2 = 1 + 2 = 3$	Use recursion formula.
$a_4 = a_{4-2} + a_{4-1} = a_2 + a_3 = 2 + 3 = 5$	Use recursion formula.
$a_5 = a_{5-2} + a_{5-1} = a_3 + a_4 = 3 + 5 = 8$	Use recursion formula.

✓ **Checkpoint** ▶ Audio-video solution in English & Spanish at *LarsonPrecalculus.com*

Write the first five terms of the sequence defined recursively as

$$a_0 = 1, \quad a_1 = 3, \quad a_k = a_{k-2} + a_{k-1}, \quad \text{where} \quad k \geq 2.$$

Factorial Notation

Some sequences involve terms defined using special products called **factorials.**

ALGEBRA HELP

The value of n does not have to be very large before the value of $n!$ becomes extremely large. For example, $10! = 3,628,800$.

> **Definition of Factorial**
>
> If n is a positive integer, then ***n*** **factorial** is defined as
>
> $$n! = 1 \cdot 2 \cdot 3 \cdot 4 \cdots (n - 1) \cdot n.$$
>
> As a special case, zero factorial is defined as $0! = 1$.

Notice from the definition of factorial that both $0! = 1$ and $1! = 1$. Some other values of $n!$ are $2! = 1 \cdot 2 = 2$, $3! = 1 \cdot 2 \cdot 3 = 6$, and $4! = 1 \cdot 2 \cdot 3 \cdot 4 = 24$.

Factorials follow the same conventions for order of operations as exponents. For example,

$$2n! = 2(n!) = 2(1 \cdot 2 \cdot 3 \cdot 4 \cdots n), \quad \text{whereas} \quad (2n)! = 1 \cdot 2 \cdot 3 \cdot 4 \cdots 2n.$$

EXAMPLE 6 **Writing the Terms of a Sequence Involving Factorials**

Write the first five terms of the sequence given by $a_n = \dfrac{2^n}{n!}$. Begin with $n = 0$.

Algebraic Solution

$$a_0 = \frac{2^0}{0!} = \frac{1}{1} = 1 \qquad \text{0th term}$$

$$a_1 = \frac{2^1}{1!} = \frac{2}{1} = 2 \qquad \text{1st term}$$

$$a_2 = \frac{2^2}{2!} = \frac{4}{2} = 2 \qquad \text{2nd term}$$

$$a_3 = \frac{2^3}{3!} = \frac{8}{6} = \frac{4}{3} \qquad \text{3rd term}$$

$$a_4 = \frac{2^4}{4!} = \frac{16}{24} = \frac{2}{3} \qquad \text{4th term}$$

Graphical Solution

Using a graphing utility set to *dot* and *sequence* modes, enter the nth term of the sequence. Next, graph the sequence. Use the graph to estimate the first five terms.

1	0th term
2	1st term
2	2nd term
$1.333 \approx \frac{4}{3}$	3rd term
$0.667 \approx \frac{2}{3}$	4th term

Use the *trace* feature to approximate the first five terms.

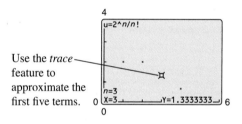

✓ **Checkpoint** ▶ Audio-video solution in English & Spanish at *LarsonPrecalculus.com*

Write the first five terms of the sequence given by $a_n = \dfrac{3^n + 1}{n!}$. Begin with $n = 0$.

EXAMPLE 7 **Simplifying Factorial Expressions**

ALGEBRA HELP

Here is another way to simplify the expression in Example 7(a).

$$\frac{8!}{2! \cdot 6!} = \frac{8 \cdot 7 \cdot 6!}{2 \cdot 1 \cdot 6!} = 28$$

a. $\dfrac{8!}{2! \cdot 6!} = \dfrac{1 \cdot 2 \cdot 3 \cdot 4 \cdot 5 \cdot 6 \cdot 7 \cdot 8}{1 \cdot 2 \cdot 1 \cdot 2 \cdot 3 \cdot 4 \cdot 5 \cdot 6} = \dfrac{7 \cdot 8}{2} = 28$

b. $\dfrac{n!}{(n - 1)!} = \dfrac{1 \cdot 2 \cdot 3 \cdots (n - 1) \cdot n}{1 \cdot 2 \cdot 3 \cdots (n - 1)} = n$

✓ **Checkpoint** ▶ Audio-video solution in English & Spanish at *LarsonPrecalculus.com*

Simplify the factorial expression $\dfrac{4!(n + 1)!}{3!n!}$.

GO DIGITAL

Summation Notation

A convenient notation for the sum of the terms of a finite sequence is called **summation notation** or **sigma notation.** It involves the use of the uppercase Greek letter sigma, written as Σ.

> ### Definition of Summation Notation
>
> The sum of the first n terms of a sequence is represented by
>
> $$\sum_{i=1}^{n} a_i = a_1 + a_2 + a_3 + a_4 + \cdots + a_n$$
>
> where i is the **index of summation,** n is the **upper limit of summation,** and 1 is the **lower limit of summation.**

ALGEBRA HELP

Summation notation is an instruction to add the terms of a sequence. Note that the upper limit of summation tells you the last term of the sum. Summation notation helps you generate the terms of the sequence prior to finding the sum.

EXAMPLE 8   Summation Notation for a Sum

a. $\displaystyle\sum_{i=1}^{5} 3i = 3(1) + 3(2) + 3(3) + 3(4) + 3(5)$

$\qquad\qquad = 3(1 + 2 + 3 + 4 + 5)$

$\qquad\qquad = 3(15)$

$\qquad\qquad = 45$

b. $\displaystyle\sum_{k=3}^{6} (1 + k^2) = (1 + 3^2) + (1 + 4^2) + (1 + 5^2) + (1 + 6^2)$

$\qquad\qquad\qquad = 10 + 17 + 26 + 37$

$\qquad\qquad\qquad = 90$

c. $\displaystyle\sum_{i=0}^{8} \frac{1}{i!} = \frac{1}{0!} + \frac{1}{1!} + \frac{1}{2!} + \frac{1}{3!} + \frac{1}{4!} + \frac{1}{5!} + \frac{1}{6!} + \frac{1}{7!} + \frac{1}{8!}$

$\qquad\quad = 1 + 1 + \frac{1}{2} + \frac{1}{6} + \frac{1}{24} + \frac{1}{120} + \frac{1}{720} + \frac{1}{5040} + \frac{1}{40,320}$

$\qquad\quad \approx 2.71828$

Note that this sum is very close to the irrational number $e \approx 2.718281828$. It can be shown that as more terms of the sequence whose nth term is $1/n!$ are added, the sum becomes closer and closer to e.

ALGEBRA HELP

In Example 8, note that the lower limit of a summation does not have to be 1, and the index of summation does not have to be the letter i. For instance, in part (b), the lower limit of summation is 3, and the index of summation is k.

✓ *Checkpoint* *Audio-video solution in English & Spanish at LarsonPrecalculus.com*

Find the sum $\displaystyle\sum_{i=1}^{4} (4i + 1)$.

> ### Properties of Sums
>
> **1.** $\displaystyle\sum_{i=1}^{n} c = cn,\quad c$ is a constant.   **2.** $\displaystyle\sum_{i=1}^{n} ca_i = c\sum_{i=1}^{n} a_i,\quad c$ is a constant.
>
> **3.** $\displaystyle\sum_{i=1}^{n} (a_i + b_i) = \sum_{i=1}^{n} a_i + \sum_{i=1}^{n} b_i$ **4.** $\displaystyle\sum_{i=1}^{n} (a_i - b_i) = \sum_{i=1}^{n} a_i - \sum_{i=1}^{n} b_i$

For proofs of these properties, see Proofs in Mathematics on page 846.

Series

Many applications involve the sum of the terms of a finite or infinite sequence. Such a sum is called a **series.**

Definition of Series

Consider the infinite sequence $a_1, a_2, a_3, \ldots, a_i, \ldots$

1. The sum of the first n terms of the sequence is called a **finite series** or the **nth partial sum** of the sequence and is denoted by

$$a_1 + a_2 + a_3 + \cdots + a_n = \sum_{i=1}^{n} a_i.$$

2. The sum of all the terms of the infinite sequence is called an **infinite series** and is denoted by

$$a_1 + a_2 + a_3 + \cdots + a_i + \cdots = \sum_{i=1}^{\infty} a_i.$$

EXAMPLE 9   Finding the Sum of a Series

▶▶▶ *See LarsonPrecalculus.com for an interactive version of this type of example.*

For the series

$$\sum_{i=1}^{\infty} \frac{3}{10^i}$$

find (a) the third partial sum and (b) the sum.

Solution

a. The third partial sum is

$$\sum_{i=1}^{3} \frac{3}{10^i} = \frac{3}{10^1} + \frac{3}{10^2} + \frac{3}{10^3}$$

$$= 0.3 + 0.03 + 0.003$$

$$= 0.333.$$

b. The sum of the series is

$$\sum_{i=1}^{\infty} \frac{3}{10^i} = \frac{3}{10^1} + \frac{3}{10^2} + \frac{3}{10^3} + \frac{3}{10^4} + \frac{3}{10^5} + \cdots$$

$$= 0.3 + 0.03 + 0.003 + 0.0003 + 0.00003 + \cdots$$

$$= 0.33333\ldots$$

$$= \frac{1}{3}.$$

✔ *Checkpoint* ▶ **Audio-video solution in English & Spanish at LarsonPrecalculus.com**

For the series

$$\sum_{i=1}^{\infty} \frac{5}{10^i}$$

find (a) the fourth partial sum and (b) the sum. ■

Notice in Example 9(b) that the sum of an infinite series can be a finite number.

Application

Sequences have many applications in business and science. Example 10 illustrates one such application.

EXAMPLE 10 Compound Interest

An investor deposits $5000 in an account that earns 3% interest compounded quarterly. The balance in the account after n quarters is given by

$$A_n = 5000\left(1 + \frac{0.03}{4}\right)^n, \quad n - 0, 1, 2, \dots.$$

a. Write the first three terms of the sequence.

b. Find the balance in the account after 10 years by computing the 40th term of the sequence.

Solution

a. The first three terms of the sequence are as follows.

$$A_0 = 5000\left(1 + \frac{0.03}{4}\right)^0 = \$5000.00 \qquad \text{Original deposit}$$

$$A_1 = 5000\left(1 + \frac{0.03}{4}\right)^1 = \$5037.50 \qquad \text{First-quarter balance}$$

$$A_2 = 5000\left(1 + \frac{0.03}{4}\right)^2 \approx \$5075.28 \qquad \text{Second-quarter balance}$$

b. The 40th term of the sequence is

$$A_{40} = 5000\left(1 + \frac{0.03}{4}\right)^{40} \approx \$6741.74. \qquad \text{Ten-year balance}$$

 ✓ **Checkpoint** ▶ *Audio-video solution in English & Spanish at LarsonPrecalculus.com*

An investor deposits $1000 in an account that earns 3% interest compounded monthly. The balance in the account after n months is given by

$$A_n = 1000\left(1 + \frac{0.03}{12}\right)^n, \quad n = 0, 1, 2, \dots.$$

a. Write the first three terms of the sequence.

b. Find the balance in the account after four years by computing the 48th term of the sequence. ■

Summarize (Section 11.1)

1. State the definition of a sequence *(page 770)*. For examples of writing the terms of sequences, see Examples 1–5.

2. State the definition of a factorial *(page 773)*. For examples of using factorial notation, see Examples 6 and 7.

3. State the definition of summation notation *(page 774)*. For an example of using summation notation, see Example 8.

4. State the definition of a series *(page 775)*. For an example of finding the sum of a series, see Example 9.

5. Describe an example of how to use a sequence to model and solve a real-life problem *(page 776, Example 10)*.

GO DIGITAL

11.1 Exercises

See CalcChat.com for tutorial help and worked-out solutions to odd-numbered exercises.

GO DIGITAL

Vocabulary and Concept Check

In Exercises 1–3, fill in the blanks.

1. When you are given one or more of the first few terms of a sequence, and all other terms of the sequence are defined using previous terms, the sequence is defined _____.

2. For the sum $\sum_{i=1}^{n} a_i$, i is the _____ of summation, n is the _____ limit of summation, and 1 is the _____ limit of summation.

3. The sum of the terms of a finite or infinite sequence is called a _____.

4. Which is the domain of an infinite sequence? a finite sequence?
   (a) the first n positive integers (b) the set of positive integers

5. Write $1 \cdot 2 \cdot 3 \cdot 4 \cdot 5 \cdot 6$ in factorial notation.

6. What is another name for a finite series with 10 terms?

Skills and Applications

Writing the Terms of a Sequence In Exercises 7–22, write the first five terms of the sequence. (Assume that n begins with 1.)

7. $a_n = 4n - 7$
8. $a_n = -2n + 8$
9. $a_n = (-1)^{n+1} + 4$
10. $a_n = 1 - (-1)^n$
11. $a_n = \frac{2}{3}$
12. $a_n = \left(\frac{1}{2}\right)^n$
13. $a_n = (-2)^n$
14. $a_n = 6(-1)^{n+1}$
15. $a_n = \frac{1}{3}n^3$
16. $a_n = \frac{1}{n^2}$
17. $a_n = \frac{n}{n+2}$
18. $a_n = \frac{6n}{3n^2 - 1}$
19. $a_n = n(n-1)(n-2)$
20. $a_n = n(n^2 - 6)$
21. $a_n = (-1)^n\left(\frac{n}{n+1}\right)$
22. $a_n = \frac{(-1)^{n+1}}{n^2 + 1}$

Finding a Term of a Sequence In Exercises 23–26, find the missing term of the sequence.

23. $a_n = (-1)^n(3n - 2)$

    $a_{25} = $ ▢

24. $a_n = (-1)^{n-1}[n(n-1)]$

    $a_{16} = $ ▢

25. $a_n = \frac{4n}{2n^2 - 3}$

    $a_{11} = $ ▢

26. $a_n = \frac{4n^2 - n + 3}{n(n-1)(n+2)}$

    $a_{13} = $ ▢

Graphing the Terms of a Sequence In Exercises 27–32, use a graphing utility to graph the first 10 terms of the sequence. (Assume that n begins with 1.)

27. $a_n = \frac{2}{3}n$
28. $a_n = 3n + 3(-1)^n$
29. $a_n = 16(-0.5)^{n-1}$
30. $a_n = 8(0.75)^{n-1}$
31. $a_n = \frac{2n}{n+1}$
32. $a_n = \frac{3n^2}{n^2 + 1}$

Finding the nth Term of a Sequence In Exercises 33–46, write an expression for the apparent nth term a_n of the sequence. (Assume that n begins with 1.)

33. $3, 7, 11, 15, 19, \ldots$
34. $91, 82, 73, 64, 55, \ldots$
35. $3, 10, 29, 66, 127, \ldots$
36. $0, 3, 8, 15, 24, \ldots$
37. $1, -1, 1, -1, 1, \ldots$
38. $1, 3, 1, 3, 1, \ldots$
39. $-\frac{2}{3}, \frac{3}{4}, -\frac{4}{5}, \frac{5}{6}, -\frac{6}{7}, \ldots$
40. $\frac{1}{2}, -\frac{1}{4}, \frac{1}{8}, -\frac{1}{16}, \ldots$
41. $\frac{2}{1}, \frac{3}{3}, \frac{4}{5}, \frac{5}{7}, \frac{6}{9}, \ldots$
42. $\frac{1}{3}, \frac{2}{9}, \frac{4}{27}, \frac{8}{81}, \ldots$
43. $1, \frac{1}{2}, \frac{1}{6}, \frac{1}{24}, \frac{1}{120}, \ldots$
44. $2, 3, 7, 25, 121, \ldots$
45. $\frac{1}{1}, \frac{3}{1}, \frac{9}{2}, \frac{27}{6}, \frac{81}{24}, \ldots$
46. $\frac{2}{1}, \frac{6}{3}, \frac{24}{7}, \frac{120}{15}, \frac{720}{31}, \ldots$

Writing the Terms of a Recursive Sequence In Exercises 47–52, write the first five terms of the sequence defined recursively.

47. $a_1 = 28, \quad a_{k+1} = a_k - 4$
48. $a_1 = 3, \quad a_{k+1} = 2(a_k - 1)$
49. $a_1 = 81, \quad a_{k+1} = \frac{1}{3}a_k$
50. $a_1 = 14, \quad a_{k+1} = (-2)a_k$
51. $a_0 = 1, \quad a_1 = 2, \quad a_k = a_{k-2} + \frac{1}{2}a_{k-1}$
52. $a_0 = -1, \quad a_1 = 1, \quad a_k = a_{k-2} + a_{k-1}$

Fibonacci Sequence In Exercises 53 and 54, use the Fibonacci sequence. (See Example 5.)

53. Write the first 12 terms of the Fibonacci sequence whose nth term is a_n and the first 10 terms of the sequence given by $b_n = a_{n+1}/a_n$, $n \geq 1$.

54. Using the definition for b_n in Exercise 53, show that b_n can be defined recursively by $b_n = 1 + 1/b_{n-1}$.

Writing the Terms of a Sequence Involving Factorials In Exercises 55–58, write the first five terms of the sequence. (Assume that n begins with 0.)

55. $a_n = \dfrac{5}{n!}$

56. $a_n = \dfrac{1}{(n+1)!}$

57. $a_n = \dfrac{(-1)^n(n+3)!}{n!}$

58. $a_n = \dfrac{(-1)^{2n+1}}{(2n+1)!}$

Simplifying a Factorial Expression In Exercises 59–62, simplify the factorial expression.

59. $\dfrac{4!}{6!}$

60. $\dfrac{12!}{4! \cdot 8!}$

61. $\dfrac{(n+1)!}{n!}$

62. $\dfrac{(2n-1)!}{(2n+1)!}$

Finding a Sum In Exercises 63–70, find the sum.

63. $\displaystyle\sum_{i=1}^{5}(2i-1)$

64. $\displaystyle\sum_{k=1}^{4}10$

65. $\displaystyle\sum_{j=3}^{5}\dfrac{1}{j^2-3}$

66. $\displaystyle\sum_{i=0}^{4}(3i^2+5)$

67. $\displaystyle\sum_{k=2}^{5}(k+1)^2(k-3)$

68. $\displaystyle\sum_{i=1}^{4}[(i-1)^2+(i+1)^3]$

69. $\displaystyle\sum_{i=1}^{4}\dfrac{i!}{2^i}$

70. $\displaystyle\sum_{j=0}^{5}\dfrac{(-1)^j}{j!}$

Finding a Sum In Exercises 71–74, use a graphing utility to find the sum.

71. $\displaystyle\sum_{k=0}^{4}\dfrac{(-1)^k}{k!}$

72. $\displaystyle\sum_{k=0}^{4}\dfrac{(-1)^k}{k+1}$

73. $\displaystyle\sum_{n=0}^{25}\dfrac{1}{4^n}$

74. $\displaystyle\sum_{n=0}^{10}\dfrac{n!}{2^n}$

Using Sigma Notation to Write a Sum In Exercises 75–84, use sigma notation to write the sum.

75. $\dfrac{1}{3(1)} + \dfrac{1}{3(2)} + \dfrac{1}{3(3)} + \cdots + \dfrac{1}{3(9)}$

76. $\dfrac{5}{1+1} + \dfrac{5}{1+2} + \dfrac{5}{1+3} + \cdots + \dfrac{5}{1+15}$

77. $\left[2\left(\tfrac{1}{8}\right)+3\right] + \left[2\left(\tfrac{2}{8}\right)+3\right] + \cdots + \left[2\left(\tfrac{8}{8}\right)+3\right]$

78. $\left[1-\left(\tfrac{1}{6}\right)^2\right] + \left[1-\left(\tfrac{2}{6}\right)^2\right] + \cdots + \left[1-\left(\tfrac{6}{6}\right)^2\right]$

79. $3 - 9 + 27 - 81 + 243 - 729$

80. $1 - \dfrac{1}{2} + \dfrac{1}{4} - \dfrac{1}{8} + \cdots - \dfrac{1}{128}$

81. $\dfrac{1^2}{2} + \dfrac{2^2}{6} + \dfrac{3^2}{24} + \dfrac{4^2}{120} + \cdots + \dfrac{7^2}{40,320}$

82. $\dfrac{1}{1\cdot3} + \dfrac{1}{2\cdot4} + \dfrac{1}{3\cdot5} + \cdots + \dfrac{1}{10\cdot12}$

83. $\dfrac{1}{4} + \dfrac{3}{8} + \dfrac{7}{16} + \dfrac{15}{32} + \dfrac{31}{64}$

84. $\dfrac{1}{2} + \dfrac{2}{4} + \dfrac{6}{8} + \dfrac{24}{16} + \dfrac{120}{32} + \dfrac{720}{64}$

Finding a Partial Sum of a Series In Exercises 85–88, find the (a) third, (b) fourth, and (c) fifth partial sums of the series.

85. $\displaystyle\sum_{i=1}^{\infty}\left(\tfrac{1}{2}\right)^i$

86. $\displaystyle\sum_{i=1}^{\infty}2\left(\tfrac{1}{3}\right)^i$

87. $\displaystyle\sum_{n=1}^{\infty}4\left(-\tfrac{1}{2}\right)^n$

88. $\displaystyle\sum_{n=1}^{\infty}5\left(-\tfrac{1}{4}\right)^n$

Finding the Sum of an Infinite Series In Exercises 89–92, find the sum of the infinite series.

89. $\displaystyle\sum_{i=1}^{\infty}\dfrac{6}{10^i}$

90. $\displaystyle\sum_{k=1}^{\infty}\left(\dfrac{1}{10}\right)^k$

91. $\displaystyle\sum_{k=1}^{\infty}7\left(\dfrac{1}{10}\right)^k$

92. $\displaystyle\sum_{i=1}^{\infty}\dfrac{2}{10^i}$

93. Compound Interest An investor deposits $10,000 in an account that earns 3.5% interest compounded quarterly. The balance in the account after n quarters is given by

$$A_n = 10{,}000\left(1 + \dfrac{0.035}{4}\right)^n, \quad n = 1, 2, 3, \ldots .$$

(a) Write the first eight terms of the sequence.

(b) Find the balance in the account after 10 years by computing the 40th term of the sequence.

(c) Is the balance after 20 years twice the balance after 10 years? Explain.

94. Physical Activity

The percent p_n of U.S. adults who met federal physical activity guidelines from 2007 through 2018 can be approximated by

$$p_n = 0.0251n^3 - 0.989n^2 + 13.34n - 12.2,$$

$n = 7, 8, \ldots, 18$

where n is the year, with $n = 7$ corresponding to 2007. (*Source: National Center for Health Statistics*)

(a) Write the terms of this finite sequence. Use a graphing utility to construct a bar graph that represents the sequence.

(b) What can you conclude from the bar graph in part (a)?

Exploring the Concepts

True or False? In Exercises 95 and 96, determine whether the statement is true or false. Justify your answer.

95. $\displaystyle\sum_{i=1}^{4}(i^2 + 2i) = \sum_{i=1}^{4}i^2 + 2\sum_{i=1}^{4}i$

96. $\displaystyle\sum_{j=1}^{4}2^j = \sum_{j=3}^{6}2^{j-2}$

Arithmetic Mean In Exercises 97–99, use the following definition of the arithmetic mean $\bar{x}$ of a set of n measurements $x_1, x_2, x_3, \ldots, x_n$.

$$\bar{x} = \frac{1}{n}\sum_{i=1}^{n}x_i$$

97. Find the arithmetic mean of the six checking account balances $327.15, $785.69, $433.04, $265.38, $604.12, and $590.30. Use the statistical capabilities of a graphing utility to verify your result.

98. Proof Prove that $\displaystyle\sum_{i=1}^{n}(x_i - \bar{x}) = 0$.

99. Proof Prove that

$$\sum_{i=1}^{n}(x_i - \bar{x})^2 = \sum_{i=1}^{n}x_i^2 - \frac{1}{n}\left(\sum_{i=1}^{n}x_i\right)^2.$$

100. HOW DO YOU SEE IT? The graph represents the first six terms of a sequence.

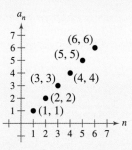

(a) Write the first six terms of the sequence.

(b) Write an expression for the apparent nth term a_n of the sequence.

(c) Use sigma notation to represent the partial sum of the first 50 terms of the sequence.

Error Analysis In Exercises 101 and 102, describe the error in finding the sum.

101. $\displaystyle\sum_{k=1}^{4}(3 + 2k^2) = \sum_{k=1}^{4}3 + \sum_{k=1}^{4}2k^2$

$$= 3 + (2 + 8 + 18 + 32)$$

$$= 63 \qquad ✗$$

102. $\displaystyle\sum_{n=0}^{3}(-1)^n n! = (-1)(1) + (1)(2) + (-1)(6)$

$$= -5 \qquad ✗$$

103. Write the first four terms of the sequence given by

$$b_n = \frac{(-1)^{n+1}}{2n + 1}.$$

Are the terms the same as the first four terms of a_n given in Example 2? Explain.

104. Cube A $3 \times 3 \times 3$ cube is made up of 27 unit cubes (a unit cube has a length, width, and height of 1 unit), and only the faces of each cube that are visible are painted blue, as shown in the figure.

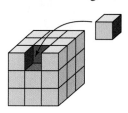

(a) Determine how many unit cubes of the $3 \times 3 \times 3$ cube have 0 blue faces, 1 blue face, 2 blue faces, and 3 blue faces.

(b) Repeat part (a) for a $4 \times 4 \times 4$ cube, a $5 \times 5 \times 5$ cube, and a $6 \times 6 \times 6$ cube.

(c) Write formulas you could use to repeat part (a) for an $n \times n \times n$ cube.

Review & Refresh ▶ Video solutions at LarsonPrecalculus.com

Using the Point-Slope Form In Exercises 105–108, find the slope-intercept form of the equation of the line that has the given slope and passes through the given point. Sketch the line.

105. $m = -\frac{1}{2}, (0, 3)$ **106.** $m = -2, (-4, 0)$

107. $m = 5, (1, 1)$ **108.** $m = \frac{4}{3}, (-3, 5)$

Sketching a Conic In Exercises 109–112, find the vertices of the ellipse or hyperbola. Then sketch the conic.

109. $\dfrac{y^2}{9} - x^2 = 1$ **110.** $\dfrac{x^2}{25} + \dfrac{y^2}{49} = 1$

111. $x^2 + 27y^2 = 9$ **112.** $4x^2 - 12y^2 = 16$

Writing a Linear Combination of Unit Vectors In Exercises 113–116, the initial and terminal points of a vector are given. Write the vector as a linear combination of the standard unit vectors **i** and **j**.

Initial Point	Terminal Point
113. $(4, 1)$	$(6, -3)$
114. $(-2, 0)$	$(-8, -1)$
115. $(5, -5)$	$(-4, 0)$
116. $(-6, -9)$	$(-2, -7)$

11.2 Arithmetic Sequences and Partial Sums

Arithmetic sequences have many real-life applications. For example, in Exercise 75 on page 787, you will use an arithmetic sequence to determine how far an object falls in 7 seconds when dropped from the top of the Willis Tower in Chicago.

❯ Recognize, write, and find the nth terms of arithmetic sequences.
❯ Find nth partial sums of arithmetic sequences.
❯ Use arithmetic sequences to model and solve real-life problems.

Arithmetic Sequences

A sequence whose consecutive terms have a common difference is an **arithmetic sequence.**

> **Definition of Arithmetic Sequence**
>
> A sequence is **arithmetic** when the differences between consecutive terms are the same. So, the sequence
>
> $$a_1, a_2, a_3, a_4, \ldots, a_n, \ldots$$
>
> is arithmetic when there is a number d such that
>
> $$a_2 - a_1 = a_3 - a_2 = a_4 - a_3 = \cdots = d.$$
>
> The number d is the **common difference** of the arithmetic sequence.

EXAMPLE 1   **Examples of Arithmetic Sequences**

a. The sequence whose nth term is $4n + 3$ is arithmetic. The common difference between consecutive terms is 4.

$$7, 11, 15, 19, \ldots, 4n + 3, \ldots \qquad \text{Begin with } n = 1.$$
$$\underbrace{}_{11 - 7 = 4}$$

b. The sequence whose nth term is $7 - 5n$ is arithmetic. The common difference between consecutive terms is -5.

$$2, -3, -8, -13, \ldots, 7 - 5n, \ldots \qquad \text{Begin with } n = 1.$$
$$\underbrace{}_{-3 - 2 = -5}$$

c. The sequence whose nth term is $\frac{1}{4}(n + 3)$ is arithmetic. The common difference between consecutive terms is $\frac{1}{4}$.

$$1, \frac{5}{4}, \frac{3}{2}, \frac{7}{4}, \ldots, \frac{n + 3}{4}, \ldots \qquad \text{Begin with } n = 1.$$
$$\underbrace{}_{\frac{5}{4} - 1 = \frac{1}{4}}$$

✓ *Checkpoint* ▶ *Audio-video solution in English & Spanish at LarsonPrecalculus.com*

Write the first four terms of the arithmetic sequence whose nth term is $3n - 1$. Then find the common difference between consecutive terms. ■

The sequence 1, 4, 9, 16, $\ldots$, whose nth term is n^2, is *not* arithmetic. The difference between the first two terms is

$$a_2 - a_1 = 4 - 1 = 3$$

but the difference between the second and third terms is

$$a_3 - a_2 = 9 - 4 = 5.$$

GO DIGITAL

The nth term of an arithmetic sequence can be derived from the pattern below.

$$a_1 = a_1 \qquad \text{1st term}$$
$$a_2 = a_1 + d \qquad \text{2nd term}$$
$$a_3 = a_1 + 2d \qquad \text{3rd term}$$
$$a_4 = a_1 + 3d \qquad \text{4th term}$$
$$a_5 = a_1 + 4d \qquad \text{5th term}$$

1 less

$$\vdots$$

$$a_n = a_1 + (n - 1)d \qquad n\text{th term}$$

1 less

The next definition summarizes this result.

The nth Term of an Arithmetic Sequence

The nth term of an arithmetic sequence has the form

$$a_n = a_1 + (n - 1)d$$

where d is the common difference between consecutive terms of the sequence and a_1 is the first term.

EXAMPLE 2 **Finding the nth Term of an Arithmetic Sequence**

Find a formula for the nth term of the arithmetic sequence whose common difference is 3 and whose first term is 2.

Solution You know that the formula for the nth term is of the form $a_n = a_1 + (n - 1)d$. Moreover, the common difference is $d = 3$ and the first term is $a_1 = 2$, so the formula must have the form

$$a_n = 2 + 3(n - 1). \qquad \text{Substitute 2 for } a_1 \text{ and 3 for } d.$$

So, the formula for the nth term is $a_n = 3n - 1$.

✓ *Checkpoint* ▶ *Audio-video solution in English & Spanish at LarsonPrecalculus.com*

Find a formula for the nth term of the arithmetic sequence whose common difference is 5 and whose first term is -1. ■

The sequence in Example 2 is as follows.

$$2, 5, 8, 11, 14, \ldots, 3n - 1, \ldots$$

The figure below shows a graph of the first 15 terms of this sequence. Notice that the points lie on a line. This makes sense because a_n is a linear function of n. In other words, the terms "arithmetic" and "linear" are closely connected.

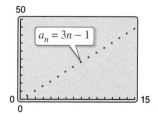

EXAMPLE 3 **Writing the Terms of an Arithmetic Sequence**

The 4th term of an arithmetic sequence is 20, and the 13th term is 65. Write the first 11 terms of this sequence.

Solution You know that $a_4 = 20$ and $a_{13} = 65$. So, you must add the common difference d nine times to the 4th term to obtain the 13th term. Therefore, the 4th and 13th terms of the sequence are related by

$$a_{13} = a_4 + 9d. \qquad \text{\small a_4 and a_{13} are nine terms apart.}$$

Using $a_4 = 20$ and $a_{13} = 65$, you have $65 = 20 + 9d$. Solve for d to find that the common difference is $d = 5$. Use the common difference with the known term a_4 to write the other terms of the sequence.

a_1	a_2	a_3	a_4	a_5	a_6	a_7	a_8	a_9	a_{10}	a_{11} ...
5	10	15	20	25	30	35	40	45	50	55 ...

✓ *Checkpoint* Audio-video solution in English & Spanish at LarsonPrecalculus.com

The 8th term of an arithmetic sequence is 25, and the 12th term is 41. Write the first 11 terms of this sequence.

When you know the nth term of an arithmetic sequence *and* you know the common difference of the sequence, you can find the $(n + 1)$th term by using the *recursion formula*

$$a_{n+1} = a_n + d. \qquad \text{\small Recursion formula}$$

With this formula, you can find any term of an arithmetic sequence, *provided* that you know the preceding term. For example, when you know the first term, you can find the second term. Then, knowing the second term, you can find the third term, and so on.

EXAMPLE 4 **Using a Recursion Formula**

Find the ninth term of the arithmetic sequence whose first two terms are 2 and 9.

Solution You know that the sequence is arithmetic. Also, $a_1 = 2$ and $a_2 = 9$. So, the common difference for this sequence is

$$d = 9 - 2 = 7.$$

There are two ways to find the ninth term. One way is to write the first nine terms (by repeatedly adding 7).

$$2, 9, 16, 23, 30, 37, 44, 51, 58 \qquad \text{\small First nine terms of the sequence}$$

Another way to find the ninth term is to first find a formula for the nth term. The common difference is $d = 7$ and the first term is $a_1 = 2$, so the formula must have the form

$$a_n = 2 + 7(n - 1). \qquad \text{\small Substitute 2 for a_1 and 7 for d.}$$

Therefore, a formula for the nth term is

$$a_n = 7n - 5 \qquad \text{\small nth term}$$

which implies that the ninth term is

$$a_9 = 7(9) - 5 = 58. \qquad \text{\small 9th term}$$

✓ *Checkpoint* Audio-video solution in English & Spanish at LarsonPrecalculus.com

Find the 10th term of the arithmetic sequence that begins with 7 and 15.

The Sum of a Finite Arithmetic Sequence

ALGEBRA HELP

Note that this formula works only for *arithmetic* sequences.

>>>>>

> ### The Sum of a Finite Arithmetic Sequence
>
> The sum of a finite arithmetic sequence with n terms is given by $S_n = \dfrac{n}{2}(a_1 + a_n)$.

For a proof of this formula, see Proofs in Mathematics on page 847.

EXAMPLE 5   Sum of a Finite Arithmetic Sequence

Find the sum: $1 + 3 + 5 + 7 + 9 + 11 + 13 + 15 + 17 + 19$.

Solution   To begin, notice that the sequence is arithmetic (with a common difference of 2). Moreover, the sequence has 10 terms. So, the sum of the sequence is

$$S_n = \frac{n}{2}(a_1 + a_n) \qquad \text{Sum of a finite arithmetic sequence}$$

$$= \frac{10}{2}(1 + 19) \qquad \text{Substitute 10 for } n, \text{ 1 for } a_1, \text{ and 19 for } a_n.$$

$$= 5(20) \qquad \text{Simplify.}$$

$$= 100. \qquad \text{Sum of the sequence}$$

✓ **Checkpoint** ▶ Audio-video solution in English & Spanish at LarsonPrecalculus.com

Find the sum: $40 + 37 + 34 + 31 + 28 + 25 + 22$.

EXAMPLE 6   Sum of a Finite Arithmetic Sequence

Find the sum of the integers (a) from 1 to 100 and (b) from 1 to N.

Solution

a. The integers from 1 to 100 form an arithmetic sequence that has 100 terms. So, use the formula for the sum of a finite arithmetic sequence.

$$S_n = 1 + 2 + 3 + 4 + 5 + 6 + \cdots + 99 + 100$$

$$= \frac{n}{2}(a_1 + a_n) \qquad \text{Sum of a finite arithmetic sequence}$$

$$= \frac{100}{2}(1 + 100) \qquad \text{Substitute 100 for } n, \text{ 1 for } a_1, \text{ and 100 for } a_n.$$

$$= 50(101) \qquad \text{Simplify.}$$

$$= 5050 \qquad \text{Sum of the sequence}$$

b. $S_n = 1 + 2 + 3 + 4 + \cdots + N$

$$= \frac{n}{2}(a_1 + a_n) \qquad \text{Sum of a finite arithmetic sequence}$$

$$= \frac{N}{2}(1 + N) \qquad \text{Substitute } N \text{ for } n, \text{ 1 for } a_1, \text{ and } N \text{ for } a_n.$$

✓ **Checkpoint** ▶ Audio-video solution in English & Spanish at LarsonPrecalculus.com

Find the sum of the integers (a) from 1 to 35 and (b) from 1 to $2N$.

HISTORICAL NOTE

A teacher of Carl Friedrich Gauss (1777–1855) asked him to add all the integers from 1 to 100. When Gauss returned with the correct answer after only a few moments, the teacher could only look at him in astounded silence. How Gauss solved the problem is shown below:

$$\begin{array}{r} S_n = 1 + 2 + 3 + \cdots + 100 \\ S_n = 100 + 99 + 98 + \cdots + 1 \\ \hline 2S_n = 101 + 101 + 101 + \cdots + 101 \end{array}$$

$$S_n = \frac{100 \times 101}{2} = 5050$$

GO DIGITAL

© Bettmann/Getty Images

Recall from the preceding section that the sum of the first n terms of an infinite sequence is the *nth partial sum*. The **nth partial sum of an arithmetic sequence** can be found by using the formula for the sum of a finite arithmetic sequence.

EXAMPLE 7 **Partial Sum of an Arithmetic Sequence**

Find the 150th partial sum of the arithmetic sequence

5, 16, 27, 38, 49, ...

Solution For this arithmetic sequence, $a_1 = 5$ and $d = 16 - 5 = 11$. So,

$$a_n = 5 + 11(n - 1)$$

and the nth term is

$$a_n = 11n - 6.$$

Therefore, $a_{150} = 11(150) - 6 = 1644$, and the sum of the first 150 terms is

$$S_{150} = \frac{n}{2}(a_1 + a_{150}) \qquad \text{\textit{n}th partial sum formula}$$

$$= \frac{150}{2}(5 + 1644) \qquad \text{Substitute 150 for \textit{n}, 5 for } a_1 \text{, and 1644 for } a_{150}.$$

$$= 75(1649) \qquad \text{Simplify.}$$

$$= 123{,}675. \qquad \text{\textit{n}th partial sum}$$

✓ *Checkpoint* ▶ Audio-video solution in English & Spanish at LarsonPrecalculus.com

Find the 120th partial sum of the arithmetic sequence

6, 12, 18, 24, 30, ...

EXAMPLE 8 **Partial Sum of an Arithmetic Sequence**

Find the 16th partial sum of the arithmetic sequence

100, 95, 90, 85, 80, ...

Solution For this arithmetic sequence, $a_1 = 100$ and $d = 95 - 100 = -5$. So,

$$a_n = 100 + (-5)(n - 1)$$

and the nth term is

$$a_n = -5n + 105.$$

Therefore, $a_{16} = -5(16) + 105 = 25$, and the sum of the first 16 terms is

$$S_{16} = \frac{n}{2}(a_1 + a_{16}) \qquad \text{\textit{n}th partial sum formula}$$

$$= \frac{16}{2}(100 + 25) \qquad \text{Substitute 16 for \textit{n}, 100 for } a_1 \text{, and 25 for } a_{16}.$$

$$= 8(125) \qquad \text{Simplify.}$$

$$= 1000. \qquad \text{\textit{n}th partial sum}$$

✓ *Checkpoint* ▶ Audio-video solution in English & Spanish at LarsonPrecalculus.com

Find the 30th partial sum of the arithmetic sequence

78, 76, 74, 72, 70, ...

GO DIGITAL

Application

| EXAMPLE 9 | **Total Sales** |

▶▷▷▷ *See LarsonPrecalculus.com for an interactive version of this type of example.*

A small business sells $10,000 worth of skin care products during its first year. The owner of the business has set a goal of increasing annual sales by $7500 each year for 9 years. Assuming that this goal is met, find the total sales during the first 10 years this business is in operation.

Solution   When the goal is met, the annual sales form an arithmetic sequence with

$$a_1 = 10,000 \quad \text{and} \quad d = 7500.$$

So,

$$a_n = 10,000 + 7500(n - 1)$$

and the nth term of the sequence is

$$a_n = 7500n + 2500.$$

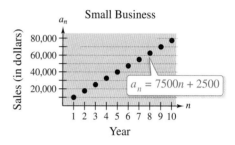

Therefore, the 10th term of the sequence is

$$a_{10} = 7500(10) + 2500$$

$$= 77,500. \qquad \text{See figure.}$$

The sum of the first 10 terms of the sequence is

$$S_{10} = \frac{n}{2}(a_1 + a_{10}) \qquad \text{\textit{n}th partial sum formula}$$

$$= \frac{10}{2}(10,000 + 77,500) \qquad \text{Substitute 10 for } n, \text{ 10,000 for } a_1, \text{ and 77,500 for } a_{10}.$$

$$= 5(87,500) \qquad \text{Simplify.}$$

$$= 437,500. \qquad \text{Multiply.}$$

So, the total sales for the first 10 years will be $437,500.

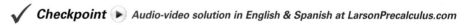

✓ *Checkpoint* ▶ *Audio-video solution in English & Spanish at LarsonPrecalculus.com*

A company sells $160,000 worth of printing paper during its first year. The sales manager has set a goal of increasing annual sales of printing paper by $20,000 each year for 9 years. Assuming that this goal is met, find the total sales of printing paper during the first 10 years this company is in operation. ■

Summarize   (Section 11.2)

1. State the definition of an arithmetic sequence *(page 780)*, and state the formula for the nth term of an arithmetic sequence *(page 781)*. For examples of recognizing, writing, and finding the nth terms of arithmetic sequences, see Examples 1–4.

2. State the formula for the sum of a finite arithmetic sequence and explain how to use it to find the nth partial sum of an arithmetic sequence *(pages 783 and 784)*. For examples of finding sums of arithmetic sequences, see Examples 5–8.

3. Describe an example of how to use an arithmetic sequence to model and solve a real-life problem *(page 785, Example 9)*.

GO DIGITAL

11.2 Exercises

See CalcChat.com for tutorial help and worked-out solutions to odd-numbered exercises.

GO DIGITAL

Vocabulary and Concept Check

In Exercises 1 and 2, fill in the blanks.

1. The nth term of an arithmetic sequence has the form $a_n =$ _____.

2. The formula $S_n = \dfrac{n}{2}(a_1 + a_n)$ gives the sum of a _____ _____ _____ with n terms.

3. How do you know when a sequence is arithmetic?

4. Explain how you can use the first two terms of an arithmetic sequence to write a formula for the nth term of the sequence.

Skills and Applications

Determining Whether a Sequence Is Arithmetic In Exercises 5–12, determine whether the sequence is arithmetic. If so, find the common difference.

5. $1, 2, 4, 8, 16, \ldots$
6. $4, 9, 14, 19, 24, \ldots$
7. $10, 8, 6, 4, 2, \ldots$
8. $80, 40, 20, 10, 5, \ldots$
9. $\frac{5}{4}, \frac{3}{2}, \frac{7}{4}, 2, \frac{9}{4}, \ldots$
10. $6.6, 5.9, 5.2, 4.5, 3.8, \ldots$
11. $1^2, 2^2, 3^2, 4^2, 5^2, \ldots$
12. $\ln 1, \ln 2, \ln 4, \ln 8, \ln 16, \ldots$

Writing the Terms of a Sequence In Exercises 13–20, write the first five terms of the sequence. Determine whether the sequence is arithmetic. If so, find the common difference. (Assume that n begins with 1.)

13. $a_n = 5 + 3n$
14. $a_n = 100 - 3n$
15. $a_n = 3 - 4(n - 2)$
16. $a_n = 1 + (n - 1)n$
17. $a_n = (-1)^n$
18. $a_n = n - (-1)^n$
19. $a_n = (2^n)n$
20. $a_n = \dfrac{3(-1)^n}{n}$

Finding the nth Term In Exercises 21–30, find a formula for the nth term of the arithmetic sequence.

21. $a_1 = 1, d = 3$
22. $a_1 = 15, d = 4$
23. $a_1 = 100, d = -8$
24. $a_1 = 0, d = -\frac{2}{3}$
25. $4, \frac{3}{2}, -1, -\frac{7}{2}, \ldots$
26. $10, 5, 0, -5, -10, \ldots$
27. $a_1 = 5, a_4 = 15$
28. $a_1 = -4, a_5 = 16$
29. $a_3 = 94, a_6 = 103$
30. $a_5 = 190, a_{10} = 115$

Writing the Terms of an Arithmetic Sequence In Exercises 31–36, write the first five terms of the arithmetic sequence.

31. $a_1 = 5, d = 6$
32. $a_1 = 5, d = -\frac{3}{4}$
33. $a_1 = 2, a_{12} = -64$
34. $a_4 = 16, a_{10} = 46$
35. $a_8 = 26, a_{12} = 42$
36. $a_3 = 19, a_{15} = -1.7$

Writing the Terms of an Arithmetic Sequence In Exercises 37–40, write the first five terms of the arithmetic sequence defined recursively.

37. $a_1 = 15, a_{n+1} = a_n + 4$
38. $a_1 = 200, a_{n+1} = a_n - 10$
39. $a_5 = 7, a_{n+1} = a_n - 2$
40. $a_3 = 0.5, a_{n+1} = a_n + 0.75$

Using a Recursion Formula In Exercises 41–44, the first two terms of the arithmetic sequence are given. Find the missing term.

41. $a_1 = 5, a_2 = -1, a_{10} = $ ▨
42. $a_1 = 3, a_2 = 13, a_9 = $ ▨
43. $a_1 = \dfrac{1}{8}, a_2 = \dfrac{3}{4}, a_7 = $ ▨
44. $a_1 = -0.7, a_2 = -13.8, a_8 = $ ▨

Sum of a Finite Arithmetic Sequence In Exercises 45–50, find the sum of the finite arithmetic sequence.

45. $2 + 4 + 6 + 8 + 10 + 12 + 14 + 16 + 18 + 20$
46. $1 + 4 + 7 + 10 + 13 + 16 + 19$
47. $-1 + (-3) + (-5) + (-7) + (-9)$
48. $-5 + (-3) + (-1) + 1 + 3 + 5$
49. Sum of the first 100 positive odd integers
50. Sum of the integers from -100 to 30

Partial Sum of an Arithmetic Sequence In Exercises 51–54, find the nth partial sum of the arithmetic sequence for the given value of n.

51. $8, 20, 32, 44, \ldots, \quad n = 50$
52. $-6, -2, 2, 6, \ldots, \quad n = 100$
53. $0, -9, -18, -27, \ldots, \quad n = 40$
54. $75, 70, 65, 60, \ldots, \quad n = 25$

Finding a Sum In Exercises 55–60, find the partial sum.

55. $\displaystyle\sum_{n=1}^{50} n$

56. $\displaystyle\sum_{n=51}^{100} 7n$

57. $\displaystyle\sum_{n=1}^{500} (n + 8)$

58. $\displaystyle\sum_{n=1}^{250} (1000 - n)$

59. $\displaystyle\sum_{n=1}^{100} (-6n + 20)$

60. $\displaystyle\sum_{n=1}^{75} (12n - 9)$

Matching an Arithmetic Sequence with Its Graph In Exercises 61–64, match the arithmetic sequence with its graph. [The graphs are labeled (a)–(d).]

(a)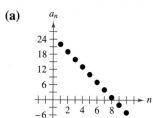

(b)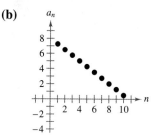

(c)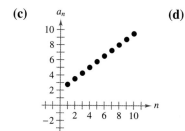

(d)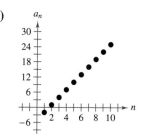

61. $a_n = -\frac{3}{4}n + 8$

62. $a_n = 3n - 5$

63. $a_n = 2 + \frac{3}{4}n$

64. $a_n = 25 - 3n$

Graphing the Terms of a Sequence In Exercises 65–68, use a graphing utility to graph the first 10 terms of the sequence. (Assume that n begins with 1.)

65. $a_n = 15 - \frac{3}{2}n$

66. $a_n = -5 + 2n$

67. $a_n = 0.2n + 3$

68. $a_n = -0.3n + 8$

Job Offer In Exercises 69 and 70, consider a job offer with the given starting salary and annual raise. (a) Determine the salary during the sixth year of employment. (b) Determine the total compensation from the company through six full years of employment.

	Starting Salary	Annual Raise
69.	$32,500	$1500
70.	$36,800	$1750

71. **Seating Capacity** Determine the seating capacity of an auditorium with 36 rows of seats when there are 15 seats in the first row, 18 seats in the second row, 21 seats in the third row, and so on.

72. **Brick Pattern** A triangular brick wall is made by cutting some bricks in half to use in the first column of every other row. The wall has 28 rows. The top row is one-half brick wide and the bottom row is 14 bricks wide. How many bricks are in the finished wall?

73. **Business** The table shows the net numbers of new stores opened by H&M from 2016 through 2019. (*Source: H&M Hennes & Mauritz AB*)

DATA	Year	New Stores
	2016	427
	2017	388
	2018	229
	2019	108

Spreadsheet at LarsonPrecalculus.com

(a) Construct a bar graph showing the annual net numbers of new stores opened by H&M from 2016 through 2019.

(b) Find the nth term (a_n) of an arithmetic sequence that approximates the data. Let n represent the year, with $n = 1$ corresponding to 2016. (*Hint:* Use the average change per year for d.)

(c) Use a graphing utility to graph the terms of the finite sequence you found in part (b).

(d) Use summation notation to represent the *total* number of new stores opened from 2016 through 2019. Use this sum to approximate the total number of new stores opened during these years.

74. **Business** In Exercise 73, there are a total number of 3924 stores at the end of 2015. Write the terms of a sequence that represents the total number of stores at the end of each year from 2016 through 2019. Is the sequence approximately arithmetic? Explain.

75. **Falling Object**

An object with negligible air resistance is dropped from the top of the Willis Tower in Chicago at a height of 1451 feet. During the first second of fall, the object falls 16 feet; during the second second, it falls 48 feet; during the third second, it falls 80 feet; during the fourth second, it falls 112 feet. Assuming this pattern continues, how many feet does the object fall in the first 7 seconds after it is dropped?

76. Pattern Recognition

(a) Compute the following sums of consecutive positive odd integers.

$1 + 3 =$

$1 + 3 + 5 =$

$1 + 3 + 5 + 7 =$

$1 + 3 + 5 + 7 + 9 =$

$1 + 3 + 5 + 7 + 9 + 11 =$

(b) Use the sums in part (a) to make a conjecture about the sums of consecutive positive odd integers. Check your conjecture for the sum

$1 + 3 + 5 + 7 + 9 + 11 + 13 =$.

(c) Verify your conjecture algebraically.

Exploring the Concepts

True or False? In Exercises 77 and 78, determine whether the statement is true or false. Justify your answer.

77. Given an arithmetic sequence for which only the first two terms are known, it is possible to find the nth term.

78. When the first term, the nth term, and n are known for an arithmetic sequence, you have enough information to find the nth partial sum of the sequence.

79. Comparing Graphs of a Sequence and a Line

(a) Graph the first 10 terms of the arithmetic sequence $a_n = 2 + 3n$.

(b) Graph the equation of the line $y = 3x + 2$.

(c) Discuss any differences between the graph of

$$a_n = 2 + 3n$$

and the graph of

$$y = 3x + 2.$$

(d) Compare the slope of the line in part (b) with the common difference of the sequence in part (a). What can you conclude about the slope of a line and the common difference of an arithmetic sequence?

80. Writing Describe two ways to use the first two terms of an arithmetic sequence to find the 13th term.

Finding the Terms of a Sequence In Exercises 81 and 82, find the first 10 terms of the sequence.

81. $a_1 = x, d = 2x$

82. $a_1 = -y, d = 5y$

83. Error Analysis Describe the error in finding the sum of the first 50 odd integers.

$$S_n = \frac{n}{2}(a_1 + a_n) = \frac{50}{2}(1 + 101) = 2550 \quad ✗$$

84. HOW DO YOU SEE IT?

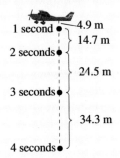

A steel ball with negligible air resistance is dropped from an airplane. The figure shows the distance that the ball falls during each of the first four seconds after it is dropped.

1 second •⎤ 4.9 m

14.7 m

2 seconds •

21.5 m

3 seconds •

34.3 m

4 seconds •

(a) Describe a pattern in the distances shown. Explain why the distances form a finite arithmetic sequence.

(b) Assume the pattern described in part (a) continues. Describe the steps and formulas involved in using the sum of a finite sequence to find the total distance the ball falls in n seconds, where n is a whole number.

Review & Refresh ▶ *Video solutions at LarsonPrecalculus.com*

Solving an Exponential Equation In Exercises 85–88, solve the equation algebraically. Approximate the result to three decimal places, if necessary.

85. $(5/7)^x = 15/14$

86. $(8/7)^x = 31/28$

87. $(3/5)(1 - 4^x) = -3$

88. $1 = 4(1 - 2^{-x})$

Solving a Logarithmic Equation In Exercises 89–92, solve the logarithmic equation algebraically. Approximate the result to three decimal places, if necessary.

89. $3 \log_5 x = 6$

90. $11 \log_8 x = 22$

91. $\log_2 5 = x - 7$

92. $x \ln 9 = \ln 4 + 2$

Writing the Partial Fraction Decomposition In Exercises 93–96, write the partial fraction decomposition of the rational expression.

93. $\dfrac{14(1 - x^3)}{1 - x^2}$

94. $\dfrac{11(1 - x^3)}{(1 - x)^2}$

95. $\dfrac{1 - x}{4 - x^2}$

96. $\dfrac{1 - x - x^2}{(x^2 - 9)^2}$

Project: Net Sales To work an extended application analyzing the net sales for Dollar Tree from 2001 through 2019, visit the text's website at *LarsonPrecalculus.com.* *(Source: Dollar Tree, Inc.)*

11.3 Geometric Sequences and Series

- ⟩ Recognize, write, and find the *n*th terms of geometric sequences.
- ⟩ Find the sum of a finite geometric sequence.
- ⟩ Find the sum of an infinite geometric series.
- ⟩ Use geometric sequences to model and solve real-life problems.

Geometric sequences can help you model and solve real-life problems. For example, in Exercise 84 on page 796, you will use a geometric sequence to model the population of Argentina from 2011 through 2020.

Geometric Sequences

In the preceding section, you learned that a sequence whose consecutive terms have a common *difference* is an arithmetic sequence. In this section, you will study another important type of sequence called a **geometric sequence**. Consecutive terms of a geometric sequence have a common *ratio*.

> **Definition of Geometric Sequence**
>
> A sequence is **geometric** when the ratios of consecutive terms are the same. So, the sequence $a_1, a_2, a_3, a_4, \ldots, a_n, \ldots$ is geometric when there is a number r such that
>
> $$\frac{a_2}{a_1} = \frac{a_3}{a_2} = \frac{a_4}{a_3} = \cdots = r, \quad r \neq 0.$$
>
> The number r is the **common ratio** of the geometric sequence.

ALGEBRA HELP

Be sure you understand that a sequence such as $1, 4, 9, 16, \ldots$, whose *n*th term is n^2, is *not* geometric. The ratio of the second term to the first term is

$$\frac{a_2}{a_1} = \frac{4}{1} = 4$$

but the ratio of the third term to the second term is

$$\frac{a_3}{a_2} = \frac{9}{4}.$$

| EXAMPLE 1 | **Examples of Geometric Sequences** |

a. The sequence whose *n*th term is 2^n is geometric. The common ratio of consecutive terms is 2.

$$2, 4, 8, 16, \ldots, 2^n, \ldots \qquad \text{Begin with } n = 1.$$
$$\tfrac{4}{2} = 2$$

b. The sequence whose *n*th term is $4(3^n)$ is geometric. The common ratio of consecutive terms is 3.

$$12, 36, 108, 324, \ldots, 4(3^n), \ldots \qquad \text{Begin with } n = 1.$$
$$\tfrac{36}{12} = 3$$

c. The sequence whose *n*th term is $\left(-\frac{1}{3}\right)^n$ is geometric. The common ratio of consecutive terms is $-\frac{1}{3}$.

$$-\frac{1}{3}, \frac{1}{9}, -\frac{1}{27}, \frac{1}{81}, \ldots, \left(-\frac{1}{3}\right)^n, \ldots \qquad \text{Begin with } n = 1.$$
$$\tfrac{1/9}{-1/3} = -\tfrac{1}{3}$$

✓ *Checkpoint* *Audio-video solution in English & Spanish at LarsonPrecalculus.com*

Write the first four terms of the geometric sequence whose *n*th term is $6(-2)^n$. Then find the common ratio of the consecutive terms. ∎

In Example 1, notice that each of the geometric sequences has an *n*th term that is of the form ar^n, where the common ratio of the sequence is r. A geometric sequence may be thought of as an exponential function whose domain is the set of natural numbers.

© iStockphoto.com/holgs

> ### The *n*th Term of a Geometric Sequence
>
> The *n*th term of a geometric sequence has the form
>
> $$a_n = a_1 r^{n-1}$$
>
> where *r* is the common ratio of consecutive terms of the sequence. So, every geometric sequence can be written in the form below.
>
> $$a_1, \quad a_2, \quad a_3, \quad a_4, \quad a_5, \quad \ldots, \quad a_n, \ldots$$
>
> $$a_1, a_1 r, a_1 r^2, a_1 r^3, a_1 r^4, \ldots, a_1 r^{n-1}, \ldots$$

When you know the *n*th term of a geometric sequence, multiply by *r* to find the $(n + 1)$th term. That is, $a_{n+1} = a_n r$.

EXAMPLE 2 Writing the Terms of a Geometric Sequence

Write the first five terms of the geometric sequence whose first term is $a_1 = 3$ and whose common ratio is $r = 2$. Then graph the terms on a set of coordinate axes.

Solution Starting with 3, repeatedly multiply by 2 to obtain the terms below.

$a_1 = 3$	1st term	$a_4 = 3(2^3) = 24$	4th term
$a_2 = 3(2^1) = 6$	2nd term	$a_5 = 3(2^4) = 48$	5th term
$a_3 = 3(2^2) = 12$	3rd term		

Figure 11.1

Figure 11.1 shows the graph of the first five terms of this geometric sequence.

✓ *Checkpoint* ▶ Audio-video solution in English & Spanish at LarsonPrecalculus.com

Write the first five terms of the geometric sequence whose first term is $a_1 = 2$ and whose common ratio is $r = 4$. Then graph the terms on a set of coordinate axes.

EXAMPLE 3 Finding a Term of a Geometric Sequence

Find the 15th term of the geometric sequence whose first term is 20 and whose common ratio is 1.05.

Algebraic Solution

$$a_n = a_1 r^{n-1}$$ — Formula for *n*th term of a geometric sequence

$$a_{15} = 20(1.05)^{15-1}$$ — Substitute 20 for a_1, 1.05 for *r*, and 15 for *n*.

$$\approx 39.60$$ — Use a calculator.

So, the 15th term of the sequence is about 39.60.

Numerical Solution

For this sequence, $r = 1.05$ and $a_1 = 20$. So, $a_n = 20(1.05)^{n-1}$. Use a graphing utility to create a table that shows the terms of the sequence.

n	u(n)
9	29.549
10	31.027
11	32.578
12	34.207
13	35.917
14	37.713
15	39.599
u(n)=39.59863199	

The number in the 15th row is the 15th term of the sequence.

So, the 15th term of the sequence is about 39.60.

✓ *Checkpoint* ▶ Audio-video solution in English & Spanish at LarsonPrecalculus.com

Find the 12th term of the geometric sequence whose first term is 14 and whose common ratio is 1.2.

GO DIGITAL

EXAMPLE 4   **Writing the *n*th Term of a Geometric Sequence**

Find a formula for the *n*th term of the geometric sequence

$$5, 15, 45, \ldots .$$

What is the 12th term of the sequence?

Solution   The common ratio of this sequence is $r = 15/5 = 3$. The first term is $a_1 = 5$, so the formula for the *n*th term is

$$a_n = a_1 r^{n-1}$$

$$= 5(3)^{n-1}.$$

Use the formula for a_n to find the 12th term of the sequence.

$$a_{12} = 5(3)^{12-1} \qquad \text{Substitute 12 for } n.$$

$$= 5(177,147) \qquad \text{Use a calculator.}$$

$$= 885,735. \qquad \text{Multiply.}$$

✓ *Checkpoint*  *Audio-video solution in English & Spanish at LarsonPrecalculus.com*

Find a formula for the *n*th term of the geometric sequence

$$4, 20, 100, \ldots .$$

What is the 12th term of the sequence?   ■

When you know *any* two terms of a geometric sequence, you can use that information to find *any other* term of the sequence.

EXAMPLE 5   **Finding a Term of a Geometric Sequence**

The 4th term of a geometric sequence is 125, and the 10th term is 125/64. Find the 14th term. (Assume that the terms of the sequence are positive.)

Solution   The 10th term is related to the 4th term by the equation

$$a_{10} = a_4 r^6. \qquad \text{Multiply fourth term by } r^{10-4}.$$

Use $a_{10} = 125/64$ and $a_4 = 125$ to solve for r.

$$\frac{125}{64} = 125r^6 \qquad \text{Substitute } \tfrac{125}{64} \text{ for } a_{10} \text{ and 125 for } a_4.$$

$$\frac{1}{64} = r^6 \qquad \text{Divide each side by 125.}$$

$$\frac{1}{2} = r \qquad \text{Take the sixth root of each side.}$$

Multiply the 10th term by $r^{14-10} = r^4$ to obtain the 14th term.

$$a_{14} = a_{10} r^4 = \frac{125}{64}\left(\frac{1}{2}\right)^4 = \frac{125}{64}\left(\frac{1}{16}\right) = \frac{125}{1024}$$

✓ *Checkpoint* 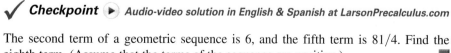 *Audio-video solution in English & Spanish at LarsonPrecalculus.com*

The second term of a geometric sequence is 6, and the fifth term is 81/4. Find the eighth term. (Assume that the terms of the sequence are positive.)   ■

ALGEBRA HELP

Remember that r is the common ratio of consecutive terms of a geometric sequence. So, in Example 5

$$a_{10} = a_1 r^9$$

$$= a_1 \cdot r \cdot r \cdot r \cdot r^6$$

$$= a_1 \cdot \frac{a_2}{a_1} \cdot \frac{a_3}{a_2} \cdot \frac{a_4}{a_3} \cdot r^6$$

$$= a_4 r^6.$$

GO DIGITAL

The Sum of a Finite Geometric Sequence

The formula for the sum of a *finite* geometric sequence is as follows.

> **The Sum of a Finite Geometric Sequence**
>
> The sum of the finite geometric sequence
>
> $$a_1, a_1r, a_1r^2, a_1r^3, a_1r^4, \ldots, a_1r^{n-1}$$
>
> with common ratio $r \neq 1$ is given by $S_n = \sum_{i=1}^{n} a_1 r^{i-1} = a_1\left(\dfrac{1 - r^n}{1 - r}\right)$.

For a proof of this formula for the sum of a finite geometric sequence, see Proofs in Mathematics on page 847.

▶▶▶ TECHNOLOGY

Using the *summation* feature or the *sum sequence* feature of a graphing utility, the sum of the sequence in Example 6 is about 5.714, as shown below.

```
12
Σ(4*0.3^(I-1))
I=1
            5.714282677
sum(seq(4*0.3^(I-1),I,1,12,1))
            5.714282677
```

EXAMPLE 6 **Sum of a Finite Geometric Sequence**

Find the sum $\displaystyle\sum_{i=1}^{12} 4(0.3)^{i-1}$.

Solution You have

$$\sum_{i=1}^{12} 4(0.3)^{i-1} = 4(0.3)^0 + 4(0.3)^1 + 4(0.3)^2 + \cdots + 4(0.3)^{11}.$$

Using $a_1 = 4$, $r = 0.3$, and $n = 12$, apply the formula for the sum of a finite geometric sequence.

$$S_n = a_1\left(\frac{1 - r^n}{1 - r}\right) \qquad \text{Sum of a finite geometric sequence}$$

$$\sum_{i=1}^{12} 4(0.3)^{i-1} = 4\left[\frac{1 - (0.3)^{12}}{1 - 0.3}\right] \qquad \text{Substitute 4 for } a_1, \text{ 0.3 for } r, \text{ and 12 for } n.$$

$$\approx 5.714 \qquad \text{Use a calculator.}$$

✓ *Checkpoint* ▶ Audio-video solution in English & Spanish at LarsonPrecalculus.com

Find the sum $\displaystyle\sum_{i=1}^{10} 2(0.25)^{i-1}$.

When using the formula for the sum of a finite geometric sequence, make sure that the sum is of the form

$$\sum_{i=1}^{n} a_1 r^{i-1}. \qquad \text{Exponent for } r \text{ is } i - 1.$$

For a sum that is not of this form, you must rewrite the sum before applying the formula. For example, the sum $\displaystyle\sum_{i=1}^{12} 4(0.3)^i$ is evaluated as follows.

$$\sum_{i=1}^{12} 4(0.3)^i = \sum_{i=1}^{12} 4[(0.3)(0.3)^{i-1}] \qquad \text{Property of exponents}$$

$$= \sum_{i=1}^{12} 4(0.3)(0.3)^{i-1} \qquad \text{Associative Property}$$

$$= 4(0.3)\left[\frac{1 - (0.3)^{12}}{1 - 0.3}\right] \qquad a_1 = 4(0.3), r = 0.3, n = 12$$

$$\approx 1.714$$

GO DIGITAL

Geometric Series

The sum of the terms of an infinite geometric *sequence* is called an **infinite geometric series** or simply a **geometric series.**

The formula for the sum of a *finite geometric sequence* can, depending on the value of r, be extended to produce a formula for the sum of an *infinite geometric series*. Specifically, if the common ratio r has the property that $|r| < 1$, then it can be shown that r^n approaches zero as n increases without bound. Consequently,

$$a_1\left(\frac{1 - r^n}{1 - r}\right) \rightarrow a_1\left(\frac{1 - 0}{1 - r}\right) \quad \text{as} \quad n \rightarrow \infty.$$

The following summarizes this result.

The Sum of an Infinite Geometric Sequence

If $|r| < 1$, then the infinite geometric series

$$a_1 + a_1 r + a_1 r^2 + a_1 r^3 + \cdots + a_1 r^{n-1} + \cdots$$

has the sum

$$S = \sum_{i=0}^{\infty} a_1 r^i = \frac{a_1}{1 - r}.$$

Note that when $|r| \geq 1$, the series does not have a sum.

EXAMPLE 7   **Finding the Sum of an Infinite Geometric Series**

Find each sum.

a. $\displaystyle\sum_{n=0}^{\infty} 4(0.6)^n$

b. $3 + 0.3 + 0.03 + 0.003 + \cdots$

Solution

a. $\displaystyle\sum_{n=0}^{\infty} 4(0.6)^n = 4 + 4(0.6) + 4(0.6)^2 + 4(0.6)^3 + \cdots + 4(0.6)^n + \cdots$

$$= \frac{4}{1 - 0.6} \qquad \frac{a_1}{1 - r}$$

$$= 10$$

b. $3 + 0.3 + 0.03 + 0.003 + \cdots = 3 + 3(0.1) + 3(0.1)^2 + 3(0.1)^3 + \cdots$

$$= \frac{3}{1 - 0.1} \qquad \frac{a_1}{1 - r}$$

$$= \frac{10}{3}$$

$$\approx 3.33$$

✓ **Checkpoint**  *Audio-video solution in English & Spanish at LarsonPrecalculus.com*

Find each sum.

a. $\displaystyle\sum_{n=0}^{\infty} 5(0.5)^n$

b. $5 + 1 + 0.2 + 0.04 + \cdots$

GO DIGITAL

Application

EXAMPLE 8 **Compound Interest**

 See LarsonPrecalculus.com for an interactive version of this type of example.

An investor deposits $50 on the first day of each month in an account that pays 3% interest, compounded monthly. What is the balance at the end of 2 years?

Solution To find the balance in the account after 24 months, consider each of the 24 deposits separately. The first deposit will gain interest for 24 months, and its balance will be

$$A_{24} = 50\left(1 + \frac{0.03}{12}\right)^{24}$$
$$= 50(1.0025)^{24}.$$

The second deposit will gain interest for 23 months, and its balance will be

$$A_{23} = 50\left(1 + \frac{0.03}{12}\right)^{23}$$
$$= 50(1.0025)^{23}.$$

The last deposit will gain interest for only 1 month, and its balance will be

$$A_1 = 50\left(1 + \frac{0.03}{12}\right)^{1}$$
$$= 50(1.0025).$$

The total balance in the annuity will be the sum of the balances of the 24 deposits. Using the formula for the sum of a finite geometric sequence, with $A_1 = 50(1.0025)$, $r = 1.0025$, and $n = 24$, you have

$$S_n = A_1\left(\frac{1 - r^n}{1 - r}\right) \qquad \text{Sum of a finite geometric sequence}$$

$$S_{24} = 50(1.0025)\left[\frac{1 - (1.0025)^{24}}{1 - 1.0025}\right] \qquad \text{Substitute } 50(1.0025) \text{ for } A_1, 1.0025 \text{ for } r, \text{ and } 24 \text{ for } n.$$

$$\approx \$1238.23. \qquad \text{Use a calculator.}$$

 Checkpoint ▶ *Audio-video solution in English & Spanish at LarsonPrecalculus.com*

An investor deposits $70 on the first day of each month in an account that pays 2% interest, compounded monthly. What is the balance at the end of 4 years? ∎

 Note: The Algebra Help sidebar:

ALGEBRA HELP

Recall from Section 5.1 that the formula for compound interest (for *n* compoundings per year) is

$$A = P\left(1 + \frac{r}{n}\right)^{nt}.$$

So, in Example 8, $50 is the principal *P*, 0.03 is the annual interest rate *r*, 12 is the number *n* of compoundings per year, and 2 is the time *t* in years. When you substitute these values into the formula, you obtain

$$A = 50\left(1 + \frac{0.03}{12}\right)^{12(2)}$$
$$= 50\left(1 + \frac{0.03}{12}\right)^{24}.$$

Summarize (Section 11.3)

1. State the definition of a geometric sequence *(page 789)* and state the formula for the *n*th term of a geometric sequence *(page 790)*. For examples of recognizing, writing, and finding the *n*th terms of geometric sequences, see Examples 1–5.

2. State the formula for the sum of a finite geometric sequence *(page 792)*. For an example of finding the sum of a finite geometric sequence, see Example 6.

3. State the formula for the sum of an infinite geometric series *(page 793)*. For an example of finding the sums of infinite geometric series, see Example 7.

4. Describe an example of how to use a geometric sequence to model and solve a real-life problem *(page 794, Example 8)*.

GO DIGITAL

11.3 Exercises

See CalcChat.com for tutorial help and worked-out solutions to odd-numbered exercises.

GO DIGITAL

Vocabulary and Concept Check

In Exercises 1 and 2, fill in the blanks.

1. Consecutive terms of a geometric sequence have a common _____.

2. The common _____ of a geometric sequence can be any real number except _____.

3. When you know the nth term and the common ratio of a geometric sequence, how can you find the $(n + 1)$th term?

4. For what values of the common ratio r is it possible to find the sum of an infinite geometric series?

Skills and Applications

Determining Whether a Sequence Is Geometric In Exercises 5–12, determine whether the sequence is geometric. If so, find the common ratio.

5. $3, 6, 12, 24, \ldots$

6. $5, 10, 15, 20, \ldots$

7. $\frac{1}{27}, \frac{1}{9}, \frac{1}{3}, 1, \ldots$

8. $27, -9, 3, -1, \ldots$

9. $1, \frac{1}{2}, \frac{1}{3}, \frac{1}{4}, \ldots$

10. $5, 1, 0.2, 0.04, \ldots$

11. $1, -\sqrt{7}, 7, -7\sqrt{7}, \ldots$

12. $2, \frac{4}{\sqrt{3}}, \frac{8}{3}, \frac{16}{3\sqrt{3}}, \ldots$

Writing the Terms of a Geometric Sequence In Exercises 13–22, write the first five terms of the geometric sequence.

13. $a_1 = 4, r = 3$

14. $a_1 = 7, r = 4$

15. $a_1 = 1, r = \frac{1}{2}$

16. $a_1 = 6, r = -\frac{1}{4}$

17. $a_1 = 1, r = e$

18. $a_1 = 2, r = \pi$

19. $a_1 = 3, r = \sqrt{5}$

20. $a_1 = 4, r = -1/\sqrt{2}$

21. $a_1 = 2, r = 3x$

22. $a_1 = 4, r = x/5$

Finding a Term of a Geometric Sequence In Exercises 23–30, write an expression for the nth term of the geometric sequence. Then find the missing term.

23. $a_1 = 4, r = \frac{1}{2}, a_{10} =$ ▢

24. $a_1 = 5, r = \frac{7}{2}, a_8 =$ ▢

25. $a_1 = 6, r = -\frac{1}{3}, a_{12} =$ ▢

26. $a_1 = 64, r = -\frac{1}{4}, a_{10} =$ ▢

27. $a_1 = 100, r = e^x, a_9 =$ ▢

28. $a_1 = 1, r = e^{-x}, a_4 =$ ▢

29. $a_1 = 1, r = \sqrt{2}, a_{12} =$ ▢

30. $a_1 = 1, r = \sqrt{3}, a_8 =$ ▢

Writing the nth Term of a Geometric Sequence In Exercises 31–36, find a formula for the nth term of the geometric sequence.

31. $64, 32, 16, \ldots$

32. $81, 27, 9, \ldots$

33. $9, 18, 36, \ldots$

34. $5, -10, 20, \ldots$

35. $6, -9, \frac{27}{2}, \ldots$

36. $80, -40, 20, \ldots$

Finding a Term of a Geometric Sequence In Exercises 37–44, find the specified term of the geometric sequence.

37. $a_8: a_1 = 6, a_2 = 18$

38. $a_7: a_1 = 5, a_2 = 20$

39. $a_9: a_1 = \frac{1}{3}, a_2 = -\frac{1}{6}$

40. $a_8: a_1 = \frac{3}{2}, a_2 = -1$

41. $a_3: a_1 = 16, a_4 = \frac{27}{4}$

42. $a_1: a_2 = 3, a_5 = \frac{3}{64}$

43. $a_6: a_4 = -18, a_7 = \frac{2}{3}$

44. $a_5: a_2 = 2, a_3 = -\sqrt{2}$

Matching a Geometric Sequence with Its Graph In Exercises 45–48, match the geometric sequence with its graph. [The graphs are labeled (a), (b), (c), and (d).]

(a)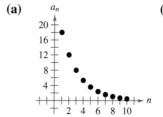

(b)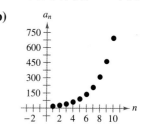

(c)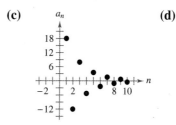

(d)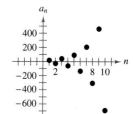

45. $a_n = 18\left(\frac{2}{3}\right)^{n-1}$

46. $a_n = 18\left(-\frac{2}{3}\right)^{n-1}$

47. $a_n = 18\left(\frac{3}{2}\right)^{n-1}$

48. $a_n = 18\left(-\frac{3}{2}\right)^{n-1}$

Graphing the Terms of a Sequence In Exercises 49–52, use a graphing utility to graph the first 10 terms of the sequence.

49. $a_n = 14(1.4)^{n-1}$

50. $a_n = 18(0.7)^{n-1}$

51. $a_n = 8(-0.3)^{n-1}$

52. $a_n = 11(-1.9)^{n-1}$

Sum of a Finite Geometric Sequence **In Exercises 53–62, find the sum of the finite geometric sequence.**

53. $\sum_{n=1}^{7} 4^{n-1}$

54. $\sum_{n=1}^{10} \left(\frac{3}{2}\right)^{n-1}$

55. $\sum_{n=1}^{6} (-7)^{n-1}$

56. $\sum_{n=1}^{8} 5\left(-\frac{5}{2}\right)^{n-1}$

57. $\sum_{n=0}^{20} 3\left(\frac{3}{2}\right)^{n}$

58. $\sum_{n=0}^{40} 5\left(\frac{3}{5}\right)^{n}$

59. $\sum_{n=0}^{5} 200(1.05)^{n}$

60. $\sum_{n=0}^{6} 500(1.04)^{n}$

61. $\sum_{n=0}^{40} 2\left(-\frac{1}{4}\right)^{n}$

62. $\sum_{n=0}^{50} 10\left(\frac{2}{3}\right)^{n-1}$

Using Summation Notation **In Exercises 63–66, use summation notation to write the sum.**

63. $10 + 30 + 90 + \cdots + 7290$

64. $15 - 3 + \frac{3}{5} - \cdots - \frac{3}{625}$

65. $0.1 + 0.4 + 1.6 + \cdots + 102.4$

66. $32 + 24 + 18 + 13.5 + 10.125$

Sum of an Infinite Geometric Series **In Exercises 67–76, find the sum of the infinite geometric series.**

67. $\sum_{n=0}^{\infty} \left(\frac{1}{2}\right)^{n}$

68. $\sum_{n=0}^{\infty} 2\left(\frac{3}{4}\right)^{n}$

69. $\sum_{n=0}^{\infty} 2\left(-\frac{2}{3}\right)^{n}$

70. $\sum_{n=0}^{\infty} \left(-\frac{1}{2}\right)^{n}$

71. $\sum_{n=0}^{\infty} (0.8)^{n}$

72. $\sum_{n=0}^{\infty} 4(0.2)^{n}$

73. $8 + 6 + \frac{9}{2} + \frac{27}{8} + \cdots$

74. $9 + 6 + 4 + \frac{8}{3} + \cdots$

75. $\frac{1}{9} - \frac{1}{3} + 1 - 3 + \cdots$

76. $-\frac{125}{36} + \frac{25}{6} - 5 + 6 - \cdots$

Writing a Repeating Decimal as a Rational Number **In Exercises 77–80, find the rational number representation of the repeating decimal.**

77. $0.\overline{36}$

78. $0.\overline{297}$

79. $0.3\overline{18}$

80. $1.3\overline{8}$

Graphical Reasoning **In Exercises 81 and 82, use a graphing utility to graph the function. Identify the horizontal asymptote of the graph and determine its relationship to the sum.**

81. $f(x) = 6\left[\dfrac{1 - (0.5)^{x}}{1 - (0.5)}\right]$, $\quad \sum_{n=0}^{\infty} 6\left(\frac{1}{2}\right)^{n}$

82. $f(x) = 2\left[\dfrac{1 - (0.8)^{x}}{1 - (0.8)}\right]$, $\quad \sum_{n=0}^{\infty} 2\left(\frac{4}{5}\right)^{n}$

83. Depreciation A manufacturing facility buys a machine for \$175,000 and it depreciates at a rate of 30% per year. (In other words, at the end of each year the depreciated value is 70% of what it was at the beginning of the year.) Find the depreciated value of the machine after 5 full years.

84. Population

The table shows the populations of Argentina (in millions) from 2011 through 2020. (*Source: Worldometer*)

DATA

Spreadsheet at LarsonPrecalculus.com

Year	Population
2011	41.3
2012	41.8
2013	42.2
2014	42.6
2015	43.1
2016	43.5
2017	43.9
2018	44.4
2019	44.8
2020	45.2

(a) Use the *exponential regression* feature of a graphing utility to find the nth term (a_n) of a geometric sequence that models the data. Let n represent the year, with $n = 11$ corresponding to 2011.

(b) Use the sequence from part (a) to describe the rate at which the population of Argentina is growing.

(c) Use the sequence from part (a) to predict the population of Argentina in 2025. Worldometer predicts the population of Argentina will be 47.2 million in 2025. How does this value compare with your prediction?

(d) Use the sequence from part (a) to predict when the population of Argentina will reach 50.0 million.

85. Compound Interest An investor deposits P dollars on the first day of each month in an account with an annual interest rate r, compounded monthly. The balance A after t years is

$$A = P\left(1 + \frac{r}{12}\right) + \cdots + P\left(1 + \frac{r}{12}\right)^{12t}.$$

Show that the balance is

$$A = P\left[\left(1 + \frac{r}{12}\right)^{12t} - 1\right]\left(1 + \frac{12}{r}\right).$$

86. Compound Interest A person saving to buy a car deposits $100 on the first day of each month in an account that pays 2% interest, compounded monthly. The balance A in the account at the end of 5 years is

$$A = 100\left(1 + \frac{0.02}{12}\right)^1 + \cdots + 100\left(1 + \frac{0.02}{12}\right)^{60}.$$

Use the result of Exercise 85 to find A.

87. Geometry The sides of a square are 27 inches in length. New squares are formed by dividing the original square into nine squares. The center square is then shaded (see figure). This process is repeated three more times. Determine the total area of the shaded region.

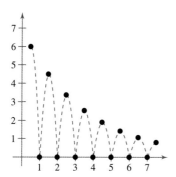

88. Distance A ball is dropped from a height of 6 feet and begins bouncing as shown in the figure. The height of each bounce is three-fourths the height of the previous bounce. Find the total vertical distance the ball travels before coming to rest.

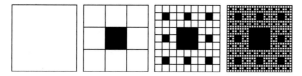

Exploring the Concepts

True or False? In Exercises 89 and 90, determine whether the statement is true or false. Justify your answer.

89. A sequence is geometric when the ratios of consecutive differences of consecutive terms are the same.

90. To find the nth term of a geometric sequence, multiply its common ratio by the first term of the sequence raised to the $(n - 1)$th power.

91. Graphical Reasoning Consider the graph of

$$y = \frac{1 - r^x}{1 - r}.$$

(a) Use a graphing utility to graph y for $r = \frac{1}{2}, \frac{2}{3},$ and $\frac{4}{5}$. What happens as $x \to \infty$?

(b) Use the graphing utility to graph y for $r = 1.5, 2,$ and 3. What happens as $x \to \infty$?

92. **HOW DO YOU SEE IT?** Use the figures shown below.

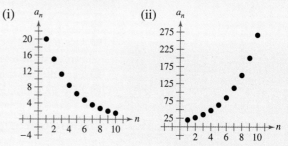

(a) Without performing any calculations, determine which figure shows terms of a sequence given by $a_n = 20\left(\frac{4}{3}\right)^{n-1}$ and which shows terms of a sequence given by $a_n = 20\left(\frac{3}{4}\right)^{n-1}$. Explain.

(b) Which infinite sequence has terms that can be summed? Explain.

93. Error Analysis Describe the error.

The sequence 21, 42, 84, 168, . . . is a geometric sequence with a common ratio of $\frac{1}{2}$.

Review & Refresh ▶ *Video solutions at LarsonPrecalculus.com*

94. Error Analysis Describe the error.

$$\frac{6 - x}{x(x + 2)} + \frac{x + 2}{x^2} + \frac{8}{x^2(x + 2)}$$

$$= \frac{6 - x + (x + 2)^2 + 8}{x^2(x + 2)}$$

$$= \frac{6 - x + x^2 + 4x + 4 + 8}{x^2(x + 2)}$$

$$= \frac{x^2 + 3x + 18}{x^2(x + 2)} \qquad \text{✗}$$

Determining Whether a Number Is Prime or Composite In Exercises 95–98, determine whether the number is prime or composite. Explain your reasoning.

95. 97

96. 257

97. 65,537

98. 33,291

Showing that a Statement Is True In Exercises 99–102, show that the statement is true.

99. $1 + 3 + 5 + 7 + 9 = 5^2$

100. $1 + 3 + 5 + 7 + 9 + 11 = 6^2$

101. $k^2 + (2k + 2 - 1) = (k + 1)^2$

102. $\{4[(k + 1) - 1] - 3\} + [4(k + 1) - 3]$
$$= (4k - 3) + (4k + 1)$$

Project: Population To work an extended application analyzing the population of Delaware, visit this text's website at *LarsonPrecalculus.com*. (*Source: U.S. Census Bureau*)

11.4 Mathematical Induction

Finite differences can help you determine what type of model to use to represent a sequence. For example, in Exercises 47 and 48 on page 807, you will use finite differences to find a model for the populations of New York state from 2014 through 2019.

❯ Use mathematical induction to prove statements involving a positive integer n.
❯ Use pattern recognition and mathematical induction to write a formula for the nth term of a sequence.
❯ Find the sums of powers of integers.
❯ Find finite differences of sequences.

Introduction

In this section, you will study a form of mathematical proof called **mathematical induction.** It is important that you see the logical need for it, so take a closer look at the problem discussed in Example 5 in Section 11.2.

$$S_1 = 1 = 1^2$$
$$S_2 = 1 + 3 = 2^2$$
$$S_3 = 1 + 3 + 5 = 3^2$$
$$S_4 = 1 + 3 + 5 + 7 = 4^2$$
$$S_5 = 1 + 3 + 5 + 7 + 9 = 5^2$$
$$S_6 = 1 + 3 + 5 + 7 + 9 + 11 = 6^2$$

Judging from the pattern formed by these first six sums, it appears that the sum of the first n odd integers is

$$S_n = 1 + 3 + 5 + 7 + 9 + 11 + \cdots + (2n - 1) = n^2.$$

Although this particular formula *is* valid, it is important for you to see that recognizing a pattern and then simply *jumping to the conclusion* that the pattern must be true for all values of n is *not* a logically valid method of proof. There are many examples in which a pattern appears to be developing for small values of n, but then at some point the pattern fails. One of the most famous cases of this was the conjecture by the French mathematician Pierre de Fermat (1601–1665), who speculated that all numbers of the form

$$F_n = 2^{2^n} + 1, \quad n = 0, 1, 2, \ldots$$

are prime. For $n = 0, 1, 2, 3,$ and 4, the conjecture is true.

$$F_0 = 3$$
$$F_1 = 5$$
$$F_2 = 17$$
$$F_3 = 257$$
$$F_4 = 65{,}537$$

The size of the next Fermat number ($F_5 = 4{,}294{,}967{,}297$) is so great that it was difficult for Fermat to determine whether it was prime or not. However, another well-known mathematician, Leonhard Euler (1707–1783), later found the factorization

$$F_5 = 4{,}294{,}967{,}297 = 641(6{,}700{,}417)$$

which proved that F_5 is not prime and therefore Fermat's conjecture was false.

Just because a rule, pattern, or formula seems to work for several values of n, you cannot simply decide that it is valid for all values of n without going through a *legitimate proof.* Mathematical induction is one method of proof.

GO DIGITAL

> **The Principle of Mathematical Induction**
>
> Let P_n be a statement involving the positive integer n. If
>
> **(1)** P_1 is true, and
>
> **(2)** for every positive integer k, the truth of P_k implies the truth of P_{k+1}
>
> then the statement P_n must be true for all positive integers n.

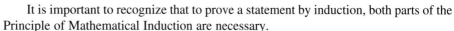

ALGEBRA HELP

In other words, the second condition says that P_k implies P_{k+1} for all $k \geq 1$.

It is important to recognize that to prove a statement by induction, both parts of the Principle of Mathematical Induction are necessary.

To apply the Principle of Mathematical Induction, you need to be able to determine the statement P_{k+1} for a given statement P_k. To determine P_{k+1}, substitute the quantity $k + 1$ for k in the statement P_k.

EXAMPLE 1 **A Preliminary Example**

Find the statement P_{k+1} for each given statement P_k.

a. P_k: $S_k = \dfrac{k^2(k + 1)^2}{4}$

b. P_k: $S_k = 1 + 5 + 9 + \cdots + [4(k - 1) - 3] + (4k - 3)$

c. P_k: $k + 3 < 5k^2$

d. P_k: $3^k \geq 2k + 1$

Solution

a. P_{k+1}: $S_{k+1} = \dfrac{(k + 1)^2(k + 1 + 1)^2}{4}$ Replace k with $k + 1$.

$\qquad\qquad = \dfrac{(k + 1)^2(k + 2)^2}{4}$ Simplify.

b. P_{k+1}: $S_{k+1} = 1 + 5 + 9 + \cdots + \{4[(k + 1) - 1] - 3\} + [4(k + 1) - 3]$

$\qquad\qquad = 1 + 5 + 9 + \cdots + (4k - 3) + (4k + 1)$

c. P_{k+1}: $(k + 1) + 3 < 5(k + 1)^2$

$\qquad\qquad k + 4 < 5(k^2 + 2k + 1)$

d. P_{k+1}: $3^{k+1} \geq 2(k + 1) + 1$

$\qquad\qquad 3^{k+1} \geq 2k + 3$

✓ **_Checkpoint_** ▶ Audio-video solution in English & Spanish at LarsonPrecalculus.com

Find the statement P_{k+1} for each given statement P_k.

a. P_k: $S_k = \dfrac{6}{k(k + 3)}$ **b.** P_k: $k + 2 \leq 3(k - 1)^2$ **c.** P_k: $2^{4k-2} + 1 > 5k$ ■

An unending line of dominoes can illustrate how the Principle of Mathematical Induction works.
Figure 11.2

A well-known illustration of how the Principle of Mathematical Induction works is an unending line of dominoes (see Figure 11.2). It is clear that you could not knock down an infinite number of dominoes *one domino* at a time. However, if it were true that each domino would knock down the next one as it fell, then you could knock them all down by pushing the first one and starting a chain reaction. Mathematical induction works in the same way. If the truth of P_k implies the truth of P_{k+1} and if P_1 is true, then the chain reaction proceeds as follows: P_1 implies P_2, P_2 implies P_3, P_3 implies P_4, and so on.

GO DIGITAL

When using mathematical induction to prove a *summation* formula (such as the one in Example 2), it is helpful to think of

$$S_{k+1} \quad \text{as} \quad S_{k+1} = S_k + a_{k+1},$$

where a_{k+1} is the $(k+1)$th term of the original sum.

EXAMPLE 2   Using Mathematical Induction

Use mathematical induction to prove the formula

$$S_n = 1 + 3 + 5 + 7 + \cdots + (2n - 1) = n^2$$

for all integers $n \geq 1$.

Solution   Mathematical induction consists of two distinct parts.

1. First, you must show that the formula is true when $n = 1$.

$$S_1 = 1 = 1^2. \hspace{3cm} \text{True statement}$$

   This is a true statement, so you have verified the first part of the Principle of Mathematical Induction.

2. For the second part of mathematical induction, *assume* that the formula is valid for some positive integer k. So, assuming

$$S_k = 1 + 3 + 5 + 7 + \cdots + (2k - 1) = k^2 \hspace{1cm} \text{Assume } S_k \text{ is true.}$$

   is true, then the kth term is $a_k = 2k - 1$. Next, you must show that S_{k+1} is true. Note that the $(k+1)$th term is $a_{k+1} = 2(k+1) - 1$.

$$
\begin{aligned}
S_{k+1} &= 1 + 3 + 5 + 7 + \cdots + (2k - 1) + [2(k + 1) - 1] \\
&= [1 + 3 + 5 + 7 + \cdots + (2k - 1)] + (2k + 2 - 1) \\
&= S_k + (2k + 1) \hspace{2cm} \text{Group terms to form } S_k. \\
&= k^2 + 2k + 1 \hspace{2.2cm} \text{By assumption} \\
&= (k + 1)^2 \hspace{2.6cm} S_k \text{ implies } S_{k+1}.
\end{aligned}
$$

   So, assuming S_k is true, this result implies that S_{k+1} is true. You have verified the second part of the Principle of Mathematical Induction.

Combining the results of parts (1) and (2), you can conclude by mathematical induction that the formula is valid for all integers $n \geq 1$.

✓ *Checkpoint*   ▶ *Audio-video solution in English & Spanish at LarsonPrecalculus.com*

Use mathematical induction to prove the formula

$$S_n = 5 + 7 + 9 + 11 + \cdots + (2n + 3) = n(n + 4)$$

for all integers $n \geq 1$. ■

It occasionally happens that a statement involving natural numbers is not true for the first $k - 1$ positive integers but is true for all values of $n \geq k$. In these instances, you use a slight variation of the Principle of Mathematical Induction in which you verify P_k rather than P_1. This variation is called the *Extended Principle of Mathematical Induction*. To see the validity of this principle, note in the unending line of dominoes in Figure 11.2 that all but the first $k - 1$ dominoes can be knocked down by knocking over the kth domino. This suggests that you can prove a statement P_n to be true for $n \geq k$ by showing that P_k is true and that P_k implies P_{k+1}. In Exercises 19–22 of this section, you will apply the Extended Principle of Mathematical Induction.

GO DIGITAL

EXAMPLE 3 **Using Mathematical Induction**

Use mathematical induction to prove the formula

$$S_n = 1^2 + 2^2 + 3^2 + 4^2 + \cdots + n^2 = \frac{n(n+1)(2n+1)}{6}$$

for all integers $n \geq 1$.

Solution

1. When $n = 1$, the formula is valid, because

$$S_1 = 1^2$$

$$= \frac{1(1+1)(2 \cdot 1 + 1)}{6}$$

$$= \frac{1(2)(3)}{6}. \qquad \text{True statement}$$

2. Assuming that the formula

$$S_k = 1^2 + 2^2 + 3^2 + 4^2 + \cdots + k^2 \qquad a_k = k^2$$

$$= \frac{k(k+1)(2k+1)}{6}$$

is true for some positive integer k, you must show that

$$S_{k+1} = \frac{(k+1)(k+1+1)[2(k+1)+1]}{6}$$

$$= \frac{(k+1)(k+2)(2k+3)}{6}$$

is true.

ALGEBRA HELP

Remember that when adding rational expressions, you must first find a common denominator. Example 3 uses the *least* common denominator of 6.

$$S_{k+1} = 1^2 + 2^2 + 3^2 + 4^2 + \cdots + k^2 + (k+1)^2 \qquad S_{k+1} = S_k + a_{k+1}$$

$$= s_k + (k+1)^2 \qquad \text{Group terms to form } S_k.$$

$$= \frac{k(k+1)(2k+1)}{6} + (k+1)^2 \qquad \text{By assumption}$$

$$= \frac{k(k+1)(2k+1) + 6(k+1)^2}{6} \qquad \text{Combine fractions.}$$

$$= \frac{(k+1)[k(2k+1) + 6(k+1)]}{6} \qquad \text{Factor.}$$

$$= \frac{(k+1)(2k^2 + 7k + 6)}{6} \qquad \text{Simplify.}$$

$$= \frac{(k+1)(k+2)(2k+3)}{6} \qquad S_k \text{ implies } S_{k+1}.$$

Combining the results of parts (1) and (2), you can conclude by mathematical induction that the formula is valid for all integers $n \geq 1$.

✓ **Checkpoint** ▶ *Audio-video solution in English & Spanish at LarsonPrecalculus.com*

Use mathematical induction to prove the formula

$$S_n = 1(1-1) + 2(2-1) + 3(3-1) + \cdots + n(n-1) = \frac{n(n-1)(n+1)}{3}$$

for all integers $n \geq 1$.

GO DIGITAL

When proving a formula using mathematical induction, the only statement that you *need* to verify is P_1. As a check, however, it is a good idea to try verifying some of the other statements. For instance, in Example 3, try verifying P_2 and P_3.

EXAMPLE 4 **Proving an Inequality**

Prove that $n < 2^n$ for all integers $n \geq 1$.

Solution

1. For $n = 1$, the statement is true because

   $$1 < 2^1. \qquad \text{True statement}$$

2. Assuming that $k < 2^k$ is true for some positive integer k, you need to show that $k + 1 < 2^{k+1}$. Note that $2(2^k) = 2^{k+1}$, so start by multiplying each side of $k < 2^k$ by 2.

   $$k < 2^k \qquad \text{Assume } k < 2^k \text{ is true.}$$

   $$2(k) < 2(2^k) \qquad \text{Multiply each side by 2.}$$

   $$2k < 2^{k+1} \qquad \text{Property of exponents}$$

   Because $k + 1 \leq k + k = 2k$ for all $k \geq 1$, it follows that

   $$k + 1 \leq 2k < 2^{k+1}.$$

   So, you can conclude that

   $$k + 1 < 2^{k+1} \qquad \text{True statement}$$

   is a true statement.

Combining the results of parts (1) and (2), you can conclude by mathematical induction that $n < 2^n$ for all integers $n \geq 1$.

✓ *Checkpoint* ▶ *Audio-video solution in English & Spanish at LarsonPrecalculus.com*

Prove that $n! \geq n$ for all integers $n \geq 1$.

EXAMPLE 5 **Proving a Property**

Prove that 3 is a factor of $4^n - 1$ for all integers $n \geq 1$.

Solution

1. For $n = 1$, the statement is true because $4^1 - 1 = 3$. So, 3 is a factor.

2. Assuming that 3 is a factor of $4^k - 1$, you must show that 3 is a factor of $4^{k+1} - 1$. To do this, write the following.

   $$4^{k+1} - 1 = 4^{k+1} - 4^k + 4^k - 1 \qquad \text{Subtract and add } 4^k.$$

   $$= 4^k(4 - 1) + (4^k - 1) \qquad \text{Regroup terms.}$$

   $$= 4^k \cdot 3 + (4^k - 1) \qquad \text{Simplify.}$$

   Because 3 is a factor of $4^k \cdot 3$ and 3 is also a factor of $4^k - 1$, it follows that 3 is a factor of $4^{k+1} - 1$.

Combining the results of parts (1) and (2), you can conclude by mathematical induction that 3 is a factor of $4^n - 1$ for all integers $n \geq 1$.

✓ *Checkpoint* ▶ *Audio-video solution in English & Spanish at LarsonPrecalculus.com*

Prove that 2 is a factor of $3^n + 1$ for all integers $n \geq 1$. ■

Pattern Recognition

Although choosing a formula on the basis of a few observations does *not* guarantee the validity of the formula, pattern recognition *is* important. Once you have a pattern or formula that you think works, try using mathematical induction to prove your formula.

> ### Finding a Formula for the *n*th Term of a Sequence
>
> To find a formula for the *n*th term of a sequence, consider these guidelines.
>
> 1. Calculate the first several terms of the sequence. It is often a good idea to write the terms in both simplified and factored forms.
> 2. Try to find a recognizable pattern for the terms and write a formula for the *n*th term of the sequence. This is your *hypothesis* or *conjecture*. You might compute one or two more terms in the sequence to test your hypothesis.
> 3. Use mathematical induction to prove your hypothesis.

EXAMPLE 6   Finding a Formula for a Finite Sum

Find a formula for the finite sum and prove its validity.

$$\frac{1}{1 \cdot 2} + \frac{1}{2 \cdot 3} + \frac{1}{3 \cdot 4} + \frac{1}{4 \cdot 5} + \cdots + \frac{1}{n(n + 1)}$$

Solution   Begin by writing the first few sums.

$$S_1 = \frac{1}{1 \cdot 2} = \frac{1}{2} = \frac{1}{1 + 1}$$

$$S_2 = \frac{1}{1 \cdot 2} + \frac{1}{2 \cdot 3} = \frac{4}{6} = \frac{2}{3} = \frac{2}{2 + 1}$$

$$S_3 = \frac{1}{1 \cdot 2} + \frac{1}{2 \cdot 3} + \frac{1}{3 \cdot 4} = \frac{9}{12} = \frac{3}{4} = \frac{3}{3 + 1}$$

From this sequence, it appears that the formula for the *k*th sum is

$$S_k = \frac{1}{1 \cdot 2} + \frac{1}{2 \cdot 3} + \frac{1}{3 \cdot 4} + \frac{1}{4 \cdot 5} + \cdots + \frac{1}{k(k + 1)} = \frac{k}{k + 1}.$$

To prove the validity of this hypothesis, use mathematical induction. Note that you have already verified the formula for $n = 1$, so begin by assuming that the formula is valid for $n = k$ and trying to show that it is valid for $n = k + 1$.

$$S_{k+1} = \left[\frac{1}{1 \cdot 2} + \frac{1}{2 \cdot 3} + \frac{1}{3 \cdot 4} + \frac{1}{4 \cdot 5} + \cdots + \frac{1}{k(k + 1)} \right] + \frac{1}{(k + 1)(k + 2)}$$

$$= \frac{k}{k + 1} + \frac{1}{(k + 1)(k + 2)} \qquad \text{By assumption}$$

$$= \frac{k(k + 2) + 1}{(k + 1)(k + 2)} = \frac{k^2 + 2k + 1}{(k + 1)(k + 2)} = \frac{(k + 1)^2}{(k + 1)(k + 2)} = \frac{k + 1}{k + 2}$$

So, by mathematical induction the hypothesis is valid.

✓ *Checkpoint* ▶ *Audio-video solution in English & Spanish at LarsonPrecalculus.com*

Find a formula for the finite sum and prove its validity.

$$3 + 7 + 11 + 15 + \cdots + 4n - 1$$

Sums of Powers of Integers

The formula in Example 3 is one of a collection of useful summation formulas. This and other formulas dealing with the sums of various powers of the first n positive integers are summarized below.

Sums of Powers of Integers

1. $1 + 2 + 3 + 4 + \cdots + n = \dfrac{n(n + 1)}{2}$

2. $1^2 + 2^2 + 3^2 + 4^2 + \cdots + n^2 = \dfrac{n(n + 1)(2n + 1)}{6}$

3. $1^3 + 2^3 + 3^3 + 4^3 + \cdots + n^3 = \dfrac{n^2(n + 1)^2}{4}$

4. $1^4 + 2^4 + 3^4 + 4^4 + \cdots + n^4 = \dfrac{n(n + 1)(2n + 1)(3n^2 + 3n - 1)}{30}$

5. $1^5 + 2^5 + 3^5 + 4^5 + \cdots + n^5 = \dfrac{n^2(n + 1)^2(2n^2 + 2n - 1)}{12}$

EXAMPLE 7 **Finding Sums**

Find each sum.

a. $\displaystyle\sum_{i=1}^{7} i^3 = 1^3 + 2^3 + 3^3 + 4^3 + 5^3 + 6^3 + 7^3$ **b.** $\displaystyle\sum_{i=1}^{4}(6i - 4i^2)$

Solution

a. Using the formula for the sum of the cubes of the first n positive integers, you obtain

$$\sum_{i=1}^{7} i^3 = 1^3 + 2^3 + 3^3 + 4^3 + 5^3 + 6^3 + 7^3$$

$$= \frac{7^2(7 + 1)^2}{4} \qquad \text{Formula 3}$$

$$= \frac{49(64)}{4}$$

$$= 784.$$

b. $\displaystyle\sum_{i=1}^{4}(6i - 4i^2) = \sum_{i=1}^{4} 6i - \sum_{i=1}^{4} 4i^2$ $\displaystyle\sum_{i=1}^{n}(a_i - b_i) = \sum_{i=1}^{n} a_i - \sum_{i=1}^{n} b_i$

$$= 6\sum_{i=1}^{4} i - 4\sum_{i=1}^{4} i^2 \qquad \sum_{i=1}^{n} ca_i = c\sum_{i=1}^{n} a_i$$

$$= 6\left[\frac{4(4 + 1)}{2}\right] - 4\left[\frac{4(4 + 1)(2 \cdot 4 + 1)}{6}\right] \quad \text{Formulas 1 and 2}$$

$$= 6(10) - 4(30)$$

$$= -60$$

✓ *Checkpoint* ▶ Audio-video solution in English & Spanish at LarsonPrecalculus.com

Find each sum.

a. $\displaystyle\sum_{i=1}^{20} i$ **b.** $\displaystyle\sum_{i=1}^{5}(2i^2 + 3i^3)$

Finite Differences

The **first differences** of a sequence are found by subtracting consecutive terms. The **second differences** are found by subtracting consecutive first differences. The first and second differences of the sequence 3, 5, 8, 12, 17, 23, . . . are shown below.

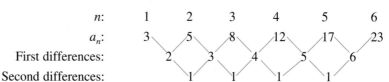

For this sequence, the second differences are all the same nonzero number. When this happens, the sequence has a perfect *quadratic* model. When the first differences are all the same nonzero number, the sequence has a perfect *linear* model. That is, the sequence is arithmetic.

EXAMPLE 8   Finding a Quadratic Model

▶▶▶ *See LarsonPrecalculus.com for an interactive version of this type of example.*

Find the quadratic model for the sequence 3, 5, 8, 12, 17, 23, ...

Solution   You know from the second differences shown above that the model is quadratic and has the form $a_n = an^2 + bn + c$. By substituting 1, 2, and 3 for n, you obtain a system of three linear equations in three variables.

$a_1 = a(1)^2 + b(1) + c = 3$ Substitute 1 for n.

$a_2 = a(2)^2 + b(2) + c = 5$ Substitute 2 for n.

$a_3 = a(3)^2 + b(3) + c = 8$ Substitute 3 for n.

You now have a system of three equations in a, b, and c.

$$\begin{cases} a + b + c = 3 & \text{Equation 1} \\ 4a + 2b + c = 5 & \text{Equation 2} \\ 9a + 3b + c = 8 & \text{Equation 3} \end{cases}$$

Using the techniques discussed in Chapter 9, you find that the solution of the system is $a = \frac{1}{2}$, $b = \frac{1}{2}$, and $c = 2$. So, the quadratic model is $a_n = \frac{1}{2}n^2 + \frac{1}{2}n + 2$. Check the values of a_1, a_2, and a_3 in this model.

✓ **Checkpoint** ▶ Audio-video solution in English & Spanish at LarsonPrecalculus.com

Find a quadratic model for the sequence -2, 0, 4, 10, 18, 28, ... ■

Summarize   (Section 11.4)

1. State the Principle of Mathematical Induction *(page 799)*. For examples of using mathematical induction to prove statements involving a positive integer n, see Examples 2–5.

2. Explain how to use pattern recognition and mathematical induction to write a formula for the nth term of a sequence *(page 803)*. For an example of using pattern recognition and mathematical induction to write a formula for the nth term of a sequence, see Example 6.

3. State the formulas for the sums of powers of integers *(page 804)*. For an example of finding sums of powers of integers, see Example 7.

4. Explain how to find finite differences of sequences *(page 805)*. For an example of using finite differences to find a quadratic model, see Example 8.

GO DIGITAL

11.4 Exercises

See CalcChat.com for tutorial help and worked-out solutions to odd-numbered exercises.

GO DIGITAL

Vocabulary and Concept Check

In Exercises 1 and 2, fill in the blanks.

1. The first step in proving a formula by _____ _____ is to show that the formula is true when $n = 1$.

2. A sequence is an _____ sequence when the first differences are all the same nonzero number.

3. How do you find the first differences of a sequence?

4. What can you conclude when the second differences of a sequence are all the same nonzero number?

Skills and Applications

Finding P_{k+1} Given P_k In Exercises 5–8, find the statement P_{k+1} for the given statement P_k.

5. $P_k = \dfrac{5}{k(k+1)}$

6. $P_k = \dfrac{1}{2(k+2)}$

7. $P_k = k^2(k+3)^2$

8. $P_k = \frac{1}{3}k(2k+1)$

Using Mathematical Induction In Exercises 9–18, use mathematical induction to prove the formula for all integers $n \geq 1$.

9. $2 + 4 + 6 + 8 + \cdots + 2n = n(n+1)$

10. $1 + 4 + 7 + 10 + \cdots + (3n-2) = \dfrac{n}{2}(3n-1)$

11. $1 + 2 + 2^2 + 2^3 + \cdots + 2^{n-1} = 2^n - 1$

12. $2(1 + 3 + 3^2 + 3^3 + \cdots + 3^{n-1}) = 3^n - 1$

13. $1 + 2 + 3 + 4 + \cdots + n = \dfrac{n(n+1)}{2}$

14. $1^3 + 2^3 + 3^3 + 4^3 + \cdots + n^3 = \dfrac{n^2(n+1)^2}{4}$

15. $1^2 + 3^2 + 5^2 + \cdots + (2n-1)^2 = \dfrac{n(2n-1)(2n+1)}{3}$

16. $\left(1 + \dfrac{1}{1}\right)\left(1 + \dfrac{1}{2}\right)\left(1 + \dfrac{1}{3}\right) \cdots \left(1 + \dfrac{1}{n}\right) = n + 1$

17. $\displaystyle\sum_{i=1}^{n} i^5 = \dfrac{n^2(n+1)^2(2n^2 + 2n - 1)}{12}$

18. $\displaystyle\sum_{i=1}^{n} i^4 = \dfrac{n(n+1)(2n+1)(3n^2 + 3n - 1)}{30}$

Proving an Inequality In Exercises 19–22, use mathematical induction to prove the inequality for the specified integer values.

19. $n! > 2^n, \quad n \geq 4$

20. $\left(\frac{4}{3}\right)^n > n, \quad n \geq 7$

21. $\dfrac{1}{\sqrt{1}} + \dfrac{1}{\sqrt{2}} + \dfrac{1}{\sqrt{3}} + \cdots + \dfrac{1}{\sqrt{n}} > \sqrt{n}, \quad n \geq 2$

22. $2n^2 > (n+1)^2, \quad n \geq 3$

Proving a Property In Exercises 23–26, use mathematical induction to prove the property for all integers $n \geq 1$.

23. A factor of $n^3 + 3n^2 + 2n$ is 3.

24. A factor of $n^4 - n + 4$ is 2.

25. A factor of $2^{2n+1} + 1$ is 3.

26. A factor of $2^{2n-1} + 3^{2n-1}$ is 5.

Finding a Formula for a Finite Sum In Exercises 27–30, find a formula for the sum of the first n terms of the sequence. Prove the validity of your formula.

27. $1, 5, 9, 13, \ldots$

28. $3, -\frac{9}{2}, \frac{27}{4}, -\frac{81}{8}, \ldots$

29. $\dfrac{1}{4}, \dfrac{1}{12}, \dfrac{1}{24}, \dfrac{1}{40}, \cdots, \dfrac{1}{2n(n+1)}, \cdots$

30. $\dfrac{1}{2 \cdot 3}, \dfrac{1}{3 \cdot 4}, \dfrac{1}{4 \cdot 5}, \dfrac{1}{5 \cdot 6}, \cdots, \dfrac{1}{(n+1)(n+2)}, \cdots$

Finding a Sum In Exercises 31–36, find the sum using the formulas for the sums of powers of integers.

31. $\displaystyle\sum_{n=1}^{15} n$

32. $\displaystyle\sum_{n=1}^{6} n^2$

33. $\displaystyle\sum_{n=1}^{5} n^4$

34. $\displaystyle\sum_{n=1}^{8} n^5$

35. $\displaystyle\sum_{n=1}^{6} (n^2 - n)$

36. $\displaystyle\sum_{i=1}^{6} (6i - 8i^3)$

Finding a Linear or Quadratic Model In Exercises 37–40, decide whether the sequence can be represented perfectly by a linear or a quadratic model. Then find the model.

37. $5, 14, 23, 32, 41, 50, \ldots$

38. $3, 9, 15, 21, 27, 33, \ldots$

39. $4, 10, 20, 34, 52, 74, \ldots$

40. $-2, 13, 38, 73, 118, 173, \ldots$

Linear Model, Quadratic Model, or Neither? In Exercises 41–44, write the first six terms of the sequence beginning with the term a_1. Then calculate the first and second differences of the sequence. State whether the sequence has a perfect linear model, a perfect quadratic model, or neither.

41. $a_1 = 0$
$a_n = a_{n-1} + 3$

42. $a_1 = 2$
$a_n = a_{n-1} + 2$

43. $a_1 = 4$
$a_n = a_{n-1} + 3n$

44. $a_1 = 3$
$a_n = 2a_{n-1}$

Finding a Quadratic Model In Exercises 45 and 46, find the quadratic model for the sequence with the given terms.

45. $a_0 = 3, a_1 = 3, a_4 = 15$ **46.** $a_1 = 0, a_2 = 7, a_4 = 27$

Population of New York ─────

In Exercises 47 and 48, the table shows the populations a_n (in millions) of New York state from 2014 through 2019, where $n = 1$ corresponds to 2014. *(Source: U.S. Census Bureau)*

Year	Number of Residents, a_n
2014	19.65
2015	19.65
2016	19.63
2017	19.59
2018	19.53
2019	19.45

Spreadsheet at LarsonPrecalculus.com

DATA

47. Find the first differences of the sequence of data a_n. Is a linear model appropriate to approximate the data? Explain. If a linear model is appropriate, find a model and compare it with the linear model found using a graphing utility.

WELCOME TO NEW YORK The Empire State
ROCKLAND COUNTY

48. Find the second differences of the sequence of data a_n. Is a quadratic model appropriate to approximate the data? Explain. If a quadratic model is appropriate, find a model and compare it with the quadratic model found using a graphing utility.

Exploring the Concepts

True or False? In Exercises 49 and 50, determine whether the statement is true or false. Justify your answer.

49. A sequence with n terms has $n - 1$ second differences.

50. If the truth of statement P_k implies the truth of statement P_{k+1}, then the statement P_1 is also true.

51. Error Analysis Describe the error.

$$1^2 + 2^2 + 3^2 + \ldots + 9^2 = \frac{9^2(9+1)^2}{4} = 2025 \quad \text{✗}$$

52. **HOW DO YOU SEE IT?** Find a formula for the sum of the angles (in degrees) of a regular polygon. Then use mathematical induction to prove this formula for a general n-sided polygon.

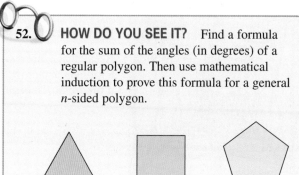

Equilateral triangle (180°)   Square (360°)   Regular pentagon (540°)

53. Think About It Explain why recognizing a pattern in a sequence and then concluding that the pattern must be true for all integer values of n is not a logically valid method of proof.

Review & Refresh ▶ *Video solutions at LarsonPrecalculus.com*

54. True or False? Determine whether each statement is true or false. Justify your answer.

(a) $(a + b)^2 = a^2 + b^2$ (b) $(a^n)^k = a^{n^k}$

Raising a Polynomial to a Power In Exercises 55–62, simplify the expression.

55. $(x + y)^2$

56. $(x + y)^3$

57. $(y - 2)^4$

58. $(2x - y)^5$

59. $(x^2 + 4)^3$

60. $(5 + y^2)^3$

61. $(x + y)^0$, where $x + y \ne 0$

62. $(x + y)^1$

Simplifying a Factorial Expression In Exercises 63–66, simplify the factorial expression.

63. $\dfrac{10!}{7!3!}$

64. $\dfrac{7!}{7!0!}$

65. $\dfrac{8!}{0!8!}$

66. $\dfrac{9!}{7!2!}$

Solving a Rational Equation In Exercises 67 and 68, solve the equation and check your solution.

67. $\dfrac{5x}{4} + \dfrac{1}{2} = x - \dfrac{1}{2}$

68. $\dfrac{6}{x} - \dfrac{2}{x+3} = \dfrac{3(x+5)}{x^2 + 3x}$

11.5 The Binomial Theorem

Binomial coefficients have many applications in real life. For example, in Exercise 84 on page 814, you will use binomial coefficients to write the expansion of a model that represents the average prices of residential electricity in the United States.

◉ Use the Binomial Theorem to find binomial coefficients.
◉ Use Pascal's Triangle to find binomial coefficients.
◉ Use binomial coefficients to write binomial expansions.

Binomial Coefficients

Recall that a *binomial* is a polynomial that has two terms. In this section, you will study a formula that provides a quick method of raising a binomial to a power, or **expanding a binomial.** To begin, look at the expansion of

$$(x + y)^n$$

for several values of n.

$$(x + y)^0 = 1$$
$$(x + y)^1 = x + y$$
$$(x + y)^2 = x^2 + 2xy + y^2$$
$$(x + y)^3 = x^3 + 3x^2y + 3xy^2 + y^3$$
$$(x + y)^4 = x^4 + 4x^3y + 6x^2y^2 + 4xy^3 + y^4$$
$$(x + y)^5 = x^5 + 5x^4y + 10x^3y^2 + 10x^2y^3 + 5xy^4 + y^5$$

There are several observations you can make about these expansions.

1. In each expansion, there are $n + 1$ terms.

2. In each expansion, x and y have symmetric roles. The powers of x decrease by 1 in successive terms, whereas the powers of y increase by 1.

3. The sum of the powers of each term is n. For example, in the expansion of $(x + y)^5$, the sum of the powers of each term is 5.

$$4 + 1 = 5 \quad 3 + 2 = 5$$
$$(x + y)^5 = x^5 + 5x^4y^1 + 10x^3y^2 + 10x^2y^3 + 5x^1y^4 + y^5$$

4. The coefficients increase and then decrease in a symmetric pattern.

The coefficients of a binomial expansion are called **binomial coefficients.** To find them, you can use the **Binomial Theorem.**

ALGEBRA HELP

Another way to represent the Binomial Theorem is by using summation notation, as shown below.

$$(x + y)^n = \sum_{r=0}^{n} \binom{n}{r} x^{n-r} y^r$$

> ### The Binomial Theorem
>
> Let n be a positive integer, and let $r = 0, 1, 2, 3, \ldots, n$. In the expansion of $(x + y)^n$
>
> $$(x + y)^n = x^n + nx^{n-1}y + \cdots + {}_nC_r x^{n-r}y^r + \cdots + nxy^{n-1} + y^n$$
>
> the coefficient of $x^{n-r}y^r$ is
>
> $${}_nC_r = \frac{n!}{(n-r)!r!}.$$
>
> The symbol $\binom{n}{r}$ is often used in place of ${}_nC_r$ to denote binomial coefficients.

For a proof of the Binomial Theorem, see Proofs in Mathematics on page 848.

GO DIGITAL

>>> **TECHNOLOGY**

Most graphing utilities can evaluate $_nC_r$. If yours can, use it to check Example 1.

EXAMPLE 1   **Finding Binomial Coefficients**

Find each binomial coefficient.

a. $_8C_2$ **b.** $\begin{pmatrix} 10 \\ 3 \end{pmatrix}$ **c.** $_7C_0$ **d.** $\begin{pmatrix} 8 \\ 8 \end{pmatrix}$

Solution

a. $_8C_2 = \dfrac{8!}{6! \cdot 2!} = \dfrac{(8 \cdot 7) \cdot 6!}{6! \cdot 2!} = \dfrac{8 \cdot 7}{2 \cdot 1} = 28$

b. $\begin{pmatrix} 10 \\ 3 \end{pmatrix} = \dfrac{10!}{7! \cdot 3!} = \dfrac{(10 \cdot 9 \cdot 8) \cdot 7!}{7! \cdot 3!} = \dfrac{10 \cdot 9 \cdot 8}{3 \cdot 2 \cdot 1} = 120$

c. $_7C_0 = \dfrac{7!}{7! \cdot 0!} = 1$ **d.** $\begin{pmatrix} 8 \\ 8 \end{pmatrix} = \dfrac{8!}{0! \cdot 8!} = 1$

✓ *Checkpoint* ▶ *Audio-video solution in English & Spanish at LarsonPrecalculus.com*

Find each binomial coefficient.

a. $\begin{pmatrix} 11 \\ 5 \end{pmatrix}$ **b.** $_9C_2$ **c.** $\begin{pmatrix} 5 \\ 0 \end{pmatrix}$ **d.** $_{15}C_{15}$ ■

When $r \neq 0$ and $r \neq n$, as in parts (a) and (b) above, there is a pattern for evaluating binomial coefficients that works because there will always be factorial terms that divide out of the expression.

$$\underset{\text{2 factors}}{\underbrace{}} \qquad \underset{\text{3 factors}}{\overbrace{}}$$
$$_8C_2 = \dfrac{\overbrace{8 \cdot 7}^{\text{2 factors}}}{\underbrace{2 \cdot 1}_{\text{2 factors}}} \quad \text{and} \quad \begin{pmatrix} 10 \\ 3 \end{pmatrix} = \dfrac{\overbrace{10 \cdot 9 \cdot 8}^{\text{3 factors}}}{\underbrace{3 \cdot 2 \cdot 1}_{\text{3 factors}}}$$

EXAMPLE 2   **Finding Binomial Coefficients**

a. $_7C_3 = \dfrac{7 \cdot 6 \cdot 5}{3 \cdot 2 \cdot 1} = 35$

b. $\begin{pmatrix} 7 \\ 4 \end{pmatrix} = \dfrac{7 \cdot 6 \cdot 5 \cdot 4}{4 \cdot 3 \cdot 2 \cdot 1} = 35$

c. $_{12}C_1 = \dfrac{12}{1} = 12$

d. $\begin{pmatrix} 12 \\ 11 \end{pmatrix} = \dfrac{12 \cdot 11 \cdot 10 \cdot 9 \cdot 8 \cdot 7 \cdot 6 \cdot 5 \cdot 4 \cdot 3 \cdot 2}{11 \cdot 10 \cdot 9 \cdot 8 \cdot 7 \cdot 6 \cdot 5 \cdot 4 \cdot 3 \cdot 2 \cdot 1} = \dfrac{12}{1} = 12$

✓ *Checkpoint* ▶ *Audio-video solution in English & Spanish at LarsonPrecalculus.com*

Find each binomial coefficient.

a. $_7C_5$ **b.** $\begin{pmatrix} 7 \\ 2 \end{pmatrix}$ **c.** $_{14}C_{13}$ **d.** $\begin{pmatrix} 14 \\ 1 \end{pmatrix}$ ■

It is not a coincidence that the results in parts (a) and (b) of Example 2 are the same and that the results in parts (c) and (d) are the same. In general, it is true that $_nC_r = {}_nC_{n-r}$.

GO DIGITAL

Pascal's Triangle

There is a convenient way to remember the pattern for binomial coefficients. By arranging the coefficients in a triangular pattern, you obtain the array called **Pascal's Triangle.** The first eight rows of Pascal's Triangle are shown below. This triangle is named after the French mathematician Blaise Pascal (1623–1662).

$$
\begin{array}{ccccccccccccccc}
& & & & & & & 1 & & & & & & & \\
& & & & & & 1 & & 1 & & & & & & \\
& & & & & 1 & & 2 & & 1 & & & & & \\
& & & & 1 & & 3 & & 3 & & 1 & & & & \\
& & & 1 & & 4 & & 6 & & 4 & & 1 & & & \quad 4 + 6 = 10 \\
& & 1 & & 5 & & 10 & & 10 & & 5 & & 1 & & \\
& 1 & & 6 & & 15 & & 20 & & 15 & & 6 & & 1 & \\
1 & & 7 & & 21 & & 35 & & 35 & & 21 & & 7 & & 1 \quad 15 + 6 = 21
\end{array}
$$

In each row of Pascal's Triangle, the first and last numbers are 1's. Also, each number between the 1's is the sum of the two numbers immediately above that number. Pascal noticed that numbers in this triangle are precisely the same numbers as the coefficients of binomial expansions, as shown below for $n = 0, 1, 2, \ldots, 7$.

$$
\begin{array}{ll}
(x + y)^0 = 1 & \text{0th row} \\
(x + y)^1 = 1x + 1y & \text{1st row} \\
(x + y)^2 = 1x^2 + 2xy + 1y^2 & \text{2nd row} \\
(x + y)^3 = 1x^3 + 3x^2y + 3xy^2 + 1y^3 & \text{3rd row} \\
(x + y)^4 = 1x^4 + 4x^3y + 6x^2y^2 + 4xy^3 + 1y^4 & \vdots \\
(x + y)^5 = 1x^5 + 5x^4y + 10x^3y^2 + 10x^2y^3 + 5xy^4 + 1y^5 \\
(x + y)^6 = 1x^6 + 6x^5y + 15x^4y^2 + 20x^3y^3 + 15x^2y^4 + 6xy^5 + 1y^6 \\
(x + y)^7 = 1x^7 + 7x^6y + 21x^5y^2 + 35x^4y^3 + 35x^3y^4 + 21x^2y^5 + 7xy^6 + 1y^7
\end{array}
$$

The top row in Pascal's Triangle is called the *zeroth row* because it corresponds to the binomial expansion $(x + y)^0 = 1$. Similarly, the next row is called the *first row* because it corresponds to the binomial expansion

$$(x + y)^1 = 1x + 1y.$$

In general, the *nth row* in Pascal's Triangle gives the coefficients of $(x + y)^n$.

EXAMPLE 3   **Using Pascal's Triangle**

Use the seventh row of Pascal's Triangle to find the binomial coefficients.

$$_8C_0, \; _8C_1, \; _8C_2, \; _8C_3, \; _8C_4, \; _8C_5, \; _8C_6, \; _8C_7, \; _8C_8$$

Solution

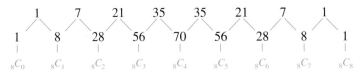

✓ *Checkpoint* ▶ *Audio-video solution in English & Spanish at LarsonPrecalculus.com*

Use the eighth row of Pascal's Triangle (see Example 3) to find the binomial coefficients.

$$_9C_0, \; _9C_1, \; _9C_2, \; _9C_3, \; _9C_4, \; _9C_5, \; _9C_6, \; _9C_7, \; _9C_8, \; _9C_9$$

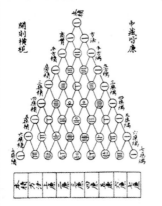
Binomial Expansions

The formula for binomial coefficients and Pascal's Triangle give you a systematic way to write the coefficients of a binomial expansion, as demonstrated in the next four examples.

EXAMPLE 4   Expanding a Binomial

Write the expansion of the expression

$(x + 1)^3$.

Solution   The binomial coefficients from the third row of Pascal's Triangle are

1, 3, 3, 1.

So, the expansion is

$$(x + 1)^3 = (1)x^3 + (3)x^2(1) + (3)x(1^2) + (1)(1^3)$$
$$= x^3 + 3x^2 + 3x + 1.$$

✓ *Checkpoint* ▶ *Audio-video solution in English & Spanish at LarsonPrecalculus.com*

Write the expansion of the expression

$(x + 2)^4$.   ■

To expand binomials representing *differences* rather than sums, you alternate signs. Here are two examples.

Alternate signs

$$(x - 1)^2 = x^2 - 2x + 1$$

Alternate signs

$$(x - 1)^3 = x^3 - 3x^2 + 3x - 1$$

EXAMPLE 5   Expanding a Binomial

▶▶▶   *See LarsonPrecalculus.com for an interactive version of this type of example.*

Write the expansion of each expression.

a. $(2x - 3)^4$ **b.** $(x - 2y)^4$

Solution   The binomial coefficients from the fourth row of Pascal's Triangle are

1, 4, 6, 4, 1.

The expansions are given below.

a. $(2x - 3)^4 = (1)(2x)^4 - (4)(2x)^3(3) + (6)(2x)^2(3^2) - (4)(2x)(3^3) + (1)(3^4)$
$$= 16x^4 - 96x^3 + 216x^2 - 216x + 81$$

b. $(x - 2y)^4 = (1)x^4 - (4)x^3(2y) + (6)x^2(2y)^2 - (4)x(2y)^3 + (1)(2y)^4$
$$= x^4 - 8x^3y + 24x^2y^2 - 32xy^3 + 16y^4$$

✓ *Checkpoint* ▶ *Audio-video solution in English & Spanish at LarsonPrecalculus.com*

Write the expansion of each expression.

a. $(y - 2)^4$

b. $(2x - y)^5$ ■

▶▶ TECHNOLOGY

Use a graphing utility to check the expansion in Example 6. Graph the original binomial expression and the expansion in the same viewing window. The graphs should coincide, as shown in the figure below.

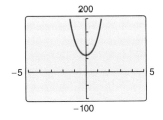

EXAMPLE 6   Expanding a Binomial

Write the expansion of $(x^2 + 4)^3$.

Solution   Use the third row of Pascal's Triangle.

$$(x^2 + 4)^3 = (1)(x^2)^3 + (3)(x^2)^2(4) + (3)x^2(4^2) + (1)(4^3)$$
$$= x^6 + 12x^4 + 48x^2 + 64$$

✓ **Checkpoint** ▶ *Audio-video solution in English & Spanish at LarsonPrecalculus.com*

Write the expansion of $(5 + y^2)^3$. ■

Sometimes you will need to find a specific term in a binomial expansion. Instead of writing the entire expansion, use the fact that, from the Binomial Theorem, the $(r + 1)$th term is $_nC_r x^{n-r} y^r$.

EXAMPLE 7   Finding a Term or Coefficient

a. Find the sixth term of $(a + 2b)^8$.

b. Find the coefficient of the term a^6b^5 in the expansion of $(3a - 2b)^{11}$.

Solution

a. Remember that the formula is for the $(r + 1)$th term, so r is one less than the number of the term you need. So, to find the sixth term in this binomial expansion, use $r = 5$, $n = 8$, $x = a$, and $y = 2b$.

$$_nC_r x^{n-r} y^r = {_8C_5} a^3(2b)^5$$
$$= 56a^3(32b^5)$$
$$= 1792a^3b^5$$

b. In this case, $n = 11$, $r = 5$, $x = 3a$, and $y = -2b$. Substitute these values to obtain

$$_nC_r x^{n-r} y^r = {_{11}C_5}(3a)^6(-2b)^5$$
$$= (462)(729a^6)(-32b^5)$$
$$= -10{,}777{,}536a^6b^5.$$

So, the coefficient is $-10{,}777{,}536$.

✓ **Checkpoint** ▶ *Audio-video solution in English & Spanish at LarsonPrecalculus.com*

a. Find the fifth term of $(a + 2b)^8$.

b. Find the coefficient of the term a^4b^7 in the expansion of $(3a - 2b)^{11}$. ■

Summarize   (Section 11.5)

1. State the Binomial Theorem *(page 808)*. For examples of using the Binomial Theorem to find binomial coefficients, see Examples 1 and 2.

2. Explain how to use Pascal's Triangle to find binomial coefficients *(page 810)*. For an example of using Pascal's Triangle to find binomial coefficients, see Example 3.

3. Explain how to use binomial coefficients to write a binomial expansion *(page 811)*. For examples of using binomial coefficients to write binomial expansions, see Examples 4–6.

GO DIGITAL

11.5 Exercises   See CalcChat.com for tutorial help and worked-out solutions
to odd-numbered exercises.

GO DIGITAL

Vocabulary and Concept Check

In Exercises 1 and 2, fill in the blanks.

1. When you find the terms that result from raising a binomial to a power, you are _____ the binomial.

2. The symbol used to denote a binomial coefficient is _____ or _____.

3. List two ways to find binomial coefficients.

4. In the expression of $(x + y)^3$, what is the sum of the powers of the third term?

Skills and Applications

Finding a Binomial Coefficient In Exercises 5–12, find the binomial coefficient.

5. $_5C_3$

6. $_7C_6$

7. $_{12}C_0$

8. $_{20}C_{20}$

9. $\binom{10}{4}$

10. $\binom{10}{6}$

11. $\binom{100}{98}$

12. $\binom{100}{2}$

Using Pascal's Triangle In Exercises 13–16, evaluate using Pascal's Triangle.

13. $_6C_3$

14. $_4C_2$

15. $\binom{5}{1}$

16. $\binom{7}{4}$

Expanding a Binomial In Exercises 17–24, use the Binomial Theorem to write the expansion of the expression.

17. $(x + 1)^6$

18. $(x + 1)^4$

19. $(y - 3)^3$

20. $(y - 2)^5$

21. $(r + 3s)^3$

22. $(x + 2y)^4$

23. $(3a - 4b)^5$

24. $(2x - 5y)^5$

Expanding an Expression In Exercises 25–38, expand the expression by using Pascal's Triangle to determine the coefficients.

25. $(a + 6)^4$

26. $(a + 5)^5$

27. $(y - 1)^6$

28. $(y - 4)^4$

29. $(3 - 2z)^4$

30. $(3v + 2)^6$

31. $(x + 2y)^5$

32. $(2t - s)^5$

33. $(x^2 + y^2)^4$

34. $(x^2 + y^2)^6$

35. $\left(\dfrac{1}{x} + y\right)^5$

36. $\left(\dfrac{1}{x} + 2y\right)^6$

37. $2(x - 3)^4 + 5(x - 3)^2$

38. $(4x - 1)^3 - 2(4x - 1)^4$

Finding a Term In Exercises 39–46, find the specified nth term in the expansion of the binomial.

39. $(x + y)^{10}$, $n = 4$

40. $(x - y)^6$, $n = 2$

41. $(x - 6y)^5$, $n = 3$

42. $(x + 2z)^7$, $n = 4$

43. $(4x + 3y)^9$, $n = 8$

44. $(5a + 6b)^5$, $n = 5$

45. $(10x - 3y)^{12}$, $n = 10$

46. $(7x + 2y)^{15}$, $n = 7$

Finding a Coefficient In Exercises 47–54, find the coefficient a of the term in the expansion of the binomial.

Binomial	Term
47. $(x + 2)^6$	ax^3
48. $(x - 2)^6$	ax^3
49. $(4x - y)^{10}$	ax^2y^8
50. $(x - 2y)^{10}$	ax^8y^2
51. $(2x - 5y)^9$	ax^4y^5
52. $(3x + 4y)^8$	ax^6y^2
53. $(x^2 + y)^{10}$	ax^8y^6
54. $(z^2 - t)^{10}$	az^4t^8

Expanding an Expression In Exercises 55–60, use the Binomial Theorem to write the expansion of the expression.

55. $\left(\sqrt{x} + 5\right)^3$

56. $\left(2\sqrt{t} - 1\right)^3$

57. $(x^{2/3} - y^{1/3})^3$

58. $(u^{3/5} + 2)^5$

59. $\left(3\sqrt{t} + \sqrt[4]{t}\right)^4$

60. $(x^{3/4} - 2x^{5/4})^4$

∫ Simplifying a Difference Quotient In Exercises 61–66, simplify the difference quotient, using the Binomial Theorem if necessary.

$$\dfrac{f(x + h) - f(x)}{h}$$ Difference quotient

61. $f(x) = x^3$

62. $f(x) = x^4$

63. $f(x) = x^6$

64. $f(x) = x^7$

65. $f(x) = \sqrt{x}$

66. $f(x) = \dfrac{1}{x}$

Expanding a Complex Number   In Exercises 67–72, use the Binomial Theorem to expand the complex number. Simplify your result.

67. $(1 + i)^4$

68. $(2 - i)^5$

69. $\left(2 - \sqrt{-4}\right)^6$

70. $\left(5 + \sqrt{-9}\right)^3$

71. $\left(-\dfrac{1}{2} + \dfrac{\sqrt{3}}{2}i\right)^3$

72. $\left(5 - \sqrt{3}i\right)^4$

Approximation   In Exercises 73–76, use the Binomial Theorem to approximate the quantity accurate to three decimal places. For example, in Exercise 73, use the first four terms of the expansion

$$(1.02)^8 = (1 + 0.02)^8$$

$$= 1 + 8(0.02) + 28(0.02)^2 + \cdots + (0.02)^8.$$

73. $(1.02)^8$

74. $(2.005)^{10}$

75. $(2.99)^{12}$

76. $(1.98)^9$

77. Finding a Pattern   Describe the pattern formed by the sums of the numbers along the diagonal line segments shown in Pascal's Triangle (see figure).

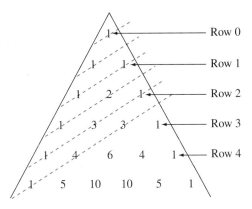

78. Error Analysis   Describe the error.

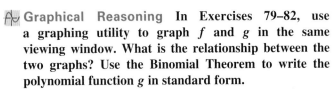

$$(x - 3)^3 = {}_3C_0x^3 + {}_3C_1x^2(3) + {}_3C_2x(3)^2$$
$$+ {}_3C_3(3)^3$$
$$= 1x^3 + 3x^2(3) + 3x(3)^2 + 1(3)^3$$
$$= x^3 + 9x^2 + 27x + 27 \quad ✗$$

Graphical Reasoning   In Exercises 79–82, use a graphing utility to graph f and g in the same viewing window. What is the relationship between the two graphs? Use the Binomial Theorem to write the polynomial function g in standard form.

79. $f(x) = x^3 - 4x$
  $g(x) = f(x + 4)$

80. $f(x) = x^4 - 5x^2$
  $g(x) = f(x - 2)$

81. $f(x) = -x^4 + 4x^2 - 1$
  $g(x) = f(x - 3)$

82. $f(x) = -x^3 + 3x^2 - 4$
  $g(x) = f(x + 5)$

83. Social Media   The numbers $f(t)$ (in millions) of monthly active Twitter users worldwide from 2012 through 2019 can be approximated by the model

$$f(t) = -5.893t^2 + 91.35t - 18.0, \; 2 \le t \le 9$$

where t represents the year, with $t = 2$ corresponding to 2012. *(Source: Twitter, Inc.)*

(a) You want to adjust the model so that $t = 2$ corresponds to 2014 rather than 2012. To do this, you shift the graph of f two units *to the left* to obtain $g(t) = f(t + 2)$. Use binomial coefficients to write $g(t)$ in standard form.

(b) Use a graphing utility to graph f and g in the same viewing window.

(c) Use the graphs to estimate when Twitter exceeded 300 million monthly active users.

84. Electricity

The ordered pairs show the average prices $f(t)$ (in cents per kilowatt-hour) of residential electricity in the United States from 2009 through 2018. *(Source: U.S. Energy Information Administration)*

(2009, 11.51)	(2014, 12.52)
(2010, 11.54)	(2015, 12.65)
(2011, 11.72)	(2016, 12.55)
(2012, 11.88)	(2017, 12.89)
(2013, 12.13)	(2018, 12.87)

(a) Use the regression feature of a graphing utility to find a cubic model for the data. Let t represent the year, with $t = 9$ corresponding to 2009.

(b) Use the graphing utility to plot the data and the model in the same viewing window.

(c) You want to adjust the model so that $t = 9$ corresponds to 2017 rather than 2009. To do this, you shift the graph of f eight units *to the left* to obtain $g(t) = f(t + 8)$. Use binomial coefficients to write $g(t)$ in standard form.

(d) Use the graphing utility to graph g in the same viewing window as f.

(e) Use both models to predict the average price in 2020. Do you obtain the same answer?

(f) Do your answers to part (e) seem reasonable? Explain.

(g) What factors do you think contributed to the change in the average price?

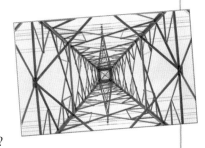

Exploring the Concepts

True or False? In Exercises 85–88, determine whether the statement is true or false. Justify your answer.

85. The Binomial Theorem can be used to produce each row of Pascal's Triangle.

86. A binomial that represents a difference cannot always be accurately expanded using the Binomial Theorem.

87. The n-term and the n^2-term of the expansion of $(n - 4)^9$ have identical coefficients.

88. The x^{10}-term and the x^{14}-term of the expansion of $(x^2 + 3)^{12}$ have identical coefficients.

89. **Writing** Explain how to form the rows of Pascal's Triangle.

90. **Forming Rows of Pascal's Triangle** Form rows 8–10 of Pascal's Triangle.

91. **Graphical Reasoning** Use a graphing utility to graph the functions in the same viewing window. Which two functions have identical graphs, and why?

$f(x) = (1 - x)^3$

$g(x) = 1 - x^3$

$h(x) = 1 + 3x + 3x^2 + x^3$

$k(x) = 1 - 3x + 3x^2 - x^3$

$p(x) = 1 + 3x - 3x^2 + x^3$

92. **HOW DO YOU SEE IT?** The expansions of $(x + y)^4$, $(x + y)^5$, and $(x + y)^6$ are shown below.

$(x + y)^4 = 1x^4 + 4x^3y + 6x^2y^2 + 4xy^3 + 1y^4$

$(x + y)^5 = 1x^5 + 5x^4y + 10x^3y^2 + 10x^2y^3$
$\qquad + 5xy^4 + 1y^5$

$(x + y)^6 = 1x^6 + 6x^5y + 15x^4y^2 + 20x^3y^3 + 15x^2y^4$
$\qquad + 6xy^5 + 1y^6$

(a) Explain how the exponent of a binomial is related to the number of terms in its expansion.

(b) How many terms are in the expansion of $(x + y)^n$?

93. **Finding a Probability** Consider n independent trials of an experiment in which each trial has two possible outcomes, success or failure. The probability of a success on each trial is p and the probability of a failure is $q = 1 - p$. In this context, the term $_nC_k p^k q^{n-k}$ in the expansion of $(p + q)^n$ gives the probability of k successes in the n trials of the experiment. Find the probability that seven coin tosses result in four heads by evaluating the term $_7C_4(0.5)^4(0.5)^3$ in the expansion of $(0.5 + 0.5)^7$.

94. **Binomial Coefficients and Pascal's Triangle** Complete the table. What characteristic of Pascal's Triangle does this table illustrate?

n	r	$_nC_r$	$_nC_{n-r}$
9	5		
7	1		
12	4		
6	0		
10	7		

Proof In Exercises 95–98, prove the property for all integers r and n, where $0 \le r \le n$.

95. $_nC_r = {_nC_{n-r}}$

96. $_nC_0 - {_nC_1} + {_nC_2} - \cdots \pm {_nC_n} = 0$

97. $_{n+1}C_r = {_nC_r} + {_nC_{r-1}}$

98. The sum of the numbers in the nth row of Pascal's Triangle is 2^n.

Review & Refresh ▶ Video solutions at LarsonPrecalculus.com

Average Rate of Change of a Function In Exercises 99–104, find the average rate of change of the function from x_1 to x_2.

Function	x-Values
99. $f(x) = \frac{1}{3}x - 4$	$x_1 = -2, x_2 = 2$
100. $f(x) = -5x + 1$	$x_1 = -3, x_2 = 0$
101. $f(x) = 6 - 3x^2$	$x_1 = 0, x_2 = 3$
102. $f(x) = -x^2 + 2x + 8$	$x_1 = -3, x_2 = 1$
103. $f(x) = 2x^3 - x^2 - x$	$x_1 = 1, x_2 = 4$
104. $f(x) = \dfrac{x + 1}{x - 4}$	$x_1 = 5, x_2 = 9$

Simplifying a Factorial Expression In Exercises 105–108, simplify the factorial expression.

105. $\dfrac{6!}{7!}$

106. $\dfrac{3!5!}{8!}$

107. $\dfrac{(n + 9)!}{(n + 7)!}$

108. $\dfrac{(n - 4)!}{(n - 2)!}$

Using a Reference Angle In Exercises 109–114, evaluate the sine, cosine, and tangent of the angle without using a calculator.

109. $330°$

110. $-225°$

111. $-\dfrac{3\pi}{4}$

112. $\dfrac{15\pi}{2}$

113. $\dfrac{11\pi}{3}$

114. $-\dfrac{35\pi}{6}$

11.6 Counting Principles

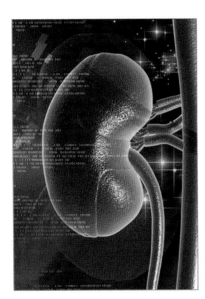

Counting principles are useful for helping you solve counting problems that occur in real life. For example, in Exercise 35 on page 824, you will use counting principles to determine the number of possible orders there are for best match, second-best match, and third-best match kidney donors.

> ○ Solve simple counting problems.
> ○ Use the Fundamental Counting Principle to solve counting problems.
> ○ Use permutations to solve counting problems.
> ○ Use combinations to solve counting problems.

Simple Counting Problems

This section and the next present a brief introduction to some of the basic counting principles and their applications to probability. In the next section, you will see that much of probability has to do with counting the number of ways an event can occur. The two examples below describe simple counting problems.

EXAMPLE 1   Selecting Pairs of Numbers at Random

You place eight pieces of paper, numbered from 1 to 8, in a box. You draw one piece of paper at random from the box, record its number, and *replace* the paper in the box. Then, you draw a second piece of paper at random from the box and record its number. Finally, you add the two numbers. How many different ways can you obtain a sum of 12?

Solution   To solve this problem, count the different ways to obtain a sum of 12 using two numbers from 1 to 8.

First number	4	5	6	7	8
Second number	8	7	6	5	4

So, a sum of 12 can occur in five different ways.

✓ *Checkpoint* ▶ *Audio-video solution in English & Spanish at LarsonPrecalculus.com*

In Example 1, how many different ways can you obtain a sum of 14?

EXAMPLE 2   Selecting Pairs of Numbers at Random

You place eight pieces of paper, numbered from 1 to 8, in a box. You draw one piece of paper at random from the box, record its number, and *do not* replace the paper in the box. Then, you draw a second piece of paper at random from the box and record its number. Finally, you add the two numbers. How many different ways can you obtain a sum of 12?

Solution   To solve this problem, count the different ways to obtain a sum of 12 using two *different* numbers from 1 to 8.

First number	4	5	7	8
Second number	8	7	5	4

So, a sum of 12 can occur in four different ways.

✓ *Checkpoint* ▶ *Audio-video solution in English & Spanish at LarsonPrecalculus.com*

In Example 2, how many different ways can you obtain a sum of 14?

Notice the difference between the counting problems in Examples 1 and 2. The random selection in Example 1 occurs **with replacement,** whereas the random selection in Example 2 occurs **without replacement,** which eliminates the possibility of choosing two 6's.

GO DIGITAL

The Fundamental Counting Principle

Examples 1 and 2 describe simple counting problems and *list* each possible way that an event can occur. When it is possible, this is always the best way to solve a counting problem. However, some events can occur in so many different ways that it is not feasible to write the entire list. In such cases, you must rely on formulas and counting principles. The most important of these is the **Fundamental Counting Principle.**

> **Fundamental Counting Principle**
>
> Let E_1 and E_2 be two events. The first event E_1 can occur in m_1 different ways. After E_1 has occurred, E_2 can occur in m_2 different ways. The number of ways the two events can occur is $m_1 \cdot m_2$.

The Fundamental Counting Principle can be extended to three or more events. For example, the number of ways that three events E_1, E_2, and E_3 can occur is

$$m_1 \cdot m_2 \cdot m_3.$$

EXAMPLE 3 **Using the Fundamental Counting Principle**

How many different pairs of letters from the English alphabet are possible?

Solution There are two events in this situation. The first event is the choice of the first letter, and the second event is the choice of the second letter. The English alphabet contains 26 letters, so it follows that the number of two-letter pairs is

$$26 \cdot 26 = 676.$$

✓ *Checkpoint* ▶ Audio-video solution in English & Spanish at LarsonPrecalculus.com

A combination lock will open when you select the right choice of three numbers (from 1 to 30, inclusive). How many different lock combinations are possible?

EXAMPLE 4 **Using the Fundamental Counting Principle**

Telephone numbers in the United States have 10 digits. The first three digits are the *area code* and the next seven digits are the *local telephone number*. How many different telephone numbers are possible within each area code? (Note that a local telephone number cannot begin with 0 or 1.)

Solution The first digit of a local telephone number cannot be 0 or 1, so there are only eight choices for the first digit. For each of the other six digits, there are 10 choices.

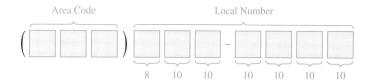

So, the number of telephone numbers that are possible within each area code is

$$8 \cdot 10 \cdot 10 \cdot 10 \cdot 10 \cdot 10 \cdot 10 = 8,000,000.$$

✓ *Checkpoint* ▶ Audio-video solution in English & Spanish at LarsonPrecalculus.com

A product's catalog number is made up of one letter from the English alphabet followed by a five-digit number. How many different catalog numbers are possible? ∎

GO DIGITAL

Permutations

One important application of the Fundamental Counting Principle is in determining the number of ways that n elements can be arranged (in order). An ordering of n elements is called a **permutation** of the elements.

Definition of a Permutation

A **permutation** of n different elements is an ordering of the elements such that one element is first, one is second, one is third, and so on.

EXAMPLE 5   **Finding the Number of Permutations**

How many permutations of the letters

   A,   B,   C,   D,   E,   and   F

are possible?

Solution   Consider the reasoning below.

  First position: Any of the *six* letters

  Second position: Any of the remaining *five* letters

  Third position: Any of the remaining *four* letters

  Fourth position: Any of the remaining *three* letters

  Fifth position: Either of the remaining *two* letters

  Sixth position: The *one* remaining letter

So, the numbers of choices for the six positions are as shown in the figure.

Permutations of six letters

6   5   4   3   2   1

The total number of permutations of the six letters is

   $6! = 6 \cdot 5 \cdot 4 \cdot 3 \cdot 2 \cdot 1 = 720.$

✓ *Checkpoint* ▶ *Audio-video solution in English & Spanish at LarsonPrecalculus.com*

How many permutations of the letters

   W,   X,   Y,   and   Z

are possible?

   Generalizing the result in Example 5, the number of permutations of n different elements is $n!$.

Number of Permutations of n Elements

The number of permutations of n elements is

   $n \cdot (n - 1) \cdot \cdot \cdot 4 \cdot 3 \cdot 2 \cdot 1 = n!.$

In other words, there are $n!$ different ways of ordering n elements.

GO DIGITAL

It is useful, on occasion, to order a *subset* of a collection of elements rather than the entire collection. For example, you may want to order *r* elements out of a collection of *n* elements. Such an ordering is called a **permutation of *n* elements taken *r* at a time.** The next example demonstrates this ordering.

EXAMPLE 6 **Counting Horse Race Finishes**

Eight horses are running in a race. In how many different ways can these horses come in first, second, and third? (Assume that there are no ties.)

Solution Here are the different possibilities.

Win (first position): *Eight* choices

Place (second position): *Seven* choices

Show (third position): *Six* choices

The numbers of choices for the three positions are as shown in the figure.

Different orders of horses

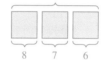

So, using the Fundamental Counting Principle, there are

$$8 \cdot 7 \cdot 6 = 336$$

different ways in which the eight horses can come in first, second, and third.

 ✓ *Checkpoint* ▶ **Audio-video solution in English & Spanish at LarsonPrecalculus.com**

A coin club has five members. In how many different ways can there be a president and a vice president? ∎

Generalizing the result in Example 6 gives the formula below.

Permutations of *n* Elements Taken *r* at a Time

The number of permutations of *n* elements taken *r* at a time is

$$_nP_r = \frac{n!}{(n-r)!} = n(n-1)(n-2) \cdots (n-r+1).$$

▶▶▶ **TECHNOLOGY**

Most graphing utilities can evaluate $_nP_r$. If yours can, use it to evaluate several permutations. Check your results algebraically by hand.

Using this formula, rework Example 6 to find that the number of permutations of eight horses taken three at a time is

$$_8P_3 = \frac{8!}{(8-3)!}$$

$$= \frac{8!}{5!}$$

$$= \frac{8 \cdot 7 \cdot 6 \cdot 5!}{5!}$$

$$= 336$$

which is the same answer obtained in the example.

Remember that for permutations, order is important. For example, to find the possible permutations of the letters A, B, C, and D taken three at a time, count A, B, D and B, A, D as different because the *order* of the elements is different.

Consider, however, the possible permutations of the letters A, A, B, and C. The total number of permutations of the four letters is $_4P_4 = 4!$. However, not all of these arrangements are *distinguishable* because there are two A's in the list. To find the number of distinguishable permutations, use the formula below.

Distinguishable Permutations

Consider a set of n objects that has n_1 of one kind of object, n_2 of a second kind, n_3 of a third kind, and so on, with

$$n = n_1 + n_2 + n_3 + \cdots + n_k.$$

The number of **distinguishable permutations** of the n objects is

$$\frac{n!}{n_1! \cdot n_2! \cdot n_3! \cdot \cdots \cdot n_k!}.$$

EXAMPLE 7   **Distinguishable Permutations**

▶⟩⟩⟩   *See LarsonPrecalculus.com for an interactive version of this type of example.*

In how many distinguishable ways can the letters in BANANA be written?

Solution   This word has six letters, of which three are A's, two are N's, and one is a B. So, the number of distinguishable ways the letters can be written is

$$\frac{n!}{n_1! \cdot n_2! \cdot n_3!} = \frac{6!}{3! \cdot 2! \cdot 1!}$$

$$= \frac{6 \cdot 5 \cdot 4 \cdot 3!}{3! \cdot 2!}$$

$$= 60.$$

The 60 different distinguishable permutations are listed below.

AAABNN	AAANBN	AAANNB	AABANN
AABNAN	AABNNA	AANABN	AANANB
AANBAN	AANBNA	AANNAB	AANNBA
ABAANN	ABANAN	ABANNA	ABNAAN
ABNANA	ABNNAA	ANAABN	ANAANB
ANABAN	ANABNA	ANANAB	ANANBA
ANBAAN	ANBANA	ANBNAA	ANNAAB
ANNABA	ANNBAA	BAAANN	BAANAN
BAANNA	BANAAN	BANANA	BANNAA
BNAAAN	BNAANA	BNANAA	BNNAAA
NAAABN	NAAANB	NAABAN	NAABNA
NAANAB	NAANBA	NABAAN	NABANA
NABNAA	NANAAB	NANABA	NANBAA
NBAAAN	NBAANA	NBANAA	NBNAAA
NNAAAB	NNAABA	NNABAA	NNBAAA

✓ **Checkpoint** ▶ *Audio-video solution in English & Spanish at LarsonPrecalculus.com*

In how many distinguishable ways can the letters in MITOSIS be written?   ■

Combinations

When you count the number of possible permutations of a set of elements, order is important. As a final topic in this section, you will look at a method of selecting subsets of a larger set in which order is *not* important. Such subsets are called **combinations of *n* elements taken *r* at a time.** For example, the combinations

$$\{A, B, C\} \quad \text{and} \quad \{B, A, C\}$$

are equivalent because both sets contain the same three elements, and the order in which the elements are listed is not important. So, you would count only one of the two sets. Another example of how a combination occurs is in a card game in which players are free to reorder the cards after they have been dealt.

EXAMPLE 8   Combinations of *n* Elements Taken *r* at a Time

In how many different ways can three letters be chosen from the letters

A,   B,   C,   D,   and   E?

(The order of the three letters is not important.)

Solution   The subsets listed below represent the different combinations of three letters that can be chosen from the five letters.

$\{A, B, C\}$	$\{A, B, D\}$
$\{A, B, E\}$	$\{A, C, D\}$
$\{A, C, E\}$	$\{A, D, E\}$
$\{B, C, D\}$	$\{B, C, E\}$
$\{B, D, E\}$	$\{C, D, E\}$

So, when order is not important, there are 10 different ways that three letters can be chosen from five letters.

✓ **Checkpoint**  Audio-video solution in English & Spanish at LarsonPrecalculus.com

In how many different ways can two letters be chosen from the letters A, B, C, D, E, F, and G? (The order of the two letters is not important.) ■

> **Combinations of *n* Elements Taken *r* at a Time**
>
> The number of combinations of *n* elements taken *r* at a time is
>
> $$_nC_r = \frac{n!}{(n - r)!r!}$$
>
> which is equivalent to $_nC_r = \dfrac{_nP_r}{r!}$.

ALGEBRA HELP

Note that the formula for $_nC_r$ is the same one given for binomial coefficients.

To see how to use this formula, rework the counting problem in Example 8. In that problem, you want to find the number of combinations of five elements taken three at a time. So, $n = 5$, $r = 3$, and the number of combinations is

$$_5C_3 = \frac{5!}{2!3!} = \frac{5 \cdot \overset{2}{\cancel{4}} \cdot \cancel{3!}}{\cancel{2} \cdot 1 \cdot \cancel{3!}} = 10$$

which is the same answer obtained in the example.

GO DIGITAL

Ranks and suits in a standard deck of playing cards
Figure 11.3

ALGEBRA HELP

When solving a problem involving counting principles, you need to distinguish among the various counting principles to determine which is necessary to solve the problem. To do this, consider the questions below.

1. Is the order of the elements important? *Permutation*

2. Is the order of the elements not important? *Combination*

3. Does the problem involve two or more separate events? *Fundamental Counting Principle*

EXAMPLE 9 **Counting Card Hands**

A standard poker hand consists of five cards dealt from a deck of 52 (see Figure 11.3). How many different poker hands are possible? (Order is not important.)

Solution To determine the number of different poker hands, find the number of combinations of 52 elements taken five at a time.

$$_{52}C_5 = \frac{52!}{(52-5)!5!}$$

$$= \frac{52!}{47!5!}$$

$$= \frac{52 \cdot 51 \cdot 50 \cdot 49 \cdot 48 \cdot \cancel{47!}}{\cancel{47!} \cdot 5 \cdot 4 \cdot 3 \cdot 2 \cdot 1}$$

$$= 2{,}598{,}960$$

✓ *Checkpoint* ▶ Audio-video solution in English & Spanish at LarsonPrecalculus.com

In three-card poker, a hand consists of three cards dealt from a deck of 52. How many different three-card poker hands are possible? (Order is not important.)

EXAMPLE 10 **Forming a Team**

You are forming a 12-member high school swim team from 10 juniors and 15 seniors. The team must consist of five juniors and seven seniors. How many different 12-member teams are possible?

Solution There are $_{10}C_5$ ways of choosing five juniors. There are $_{15}C_7$ ways of choosing seven seniors. By the Fundamental Counting Principle, there are $_{10}C_5 \cdot {}_{15}C_7$ ways of choosing five juniors and seven seniors.

$$_{10}C_5 \cdot {}_{15}C_7 = \frac{10!}{5! \cdot 5!} \cdot \frac{15!}{8! \cdot 7!} = 252 \cdot 6435 = 1{,}621{,}620$$

So, the possible number of 12-member swim teams is 1,621,620.

✓ *Checkpoint* ▶ Audio-video solution in English & Spanish at LarsonPrecalculus.com

In Example 10, the team must consist of six juniors and six seniors. How many different 12-member teams are possible? ■

Summarize (Section 11.6)

1. Explain how to solve a simple counting problem (*page 816*). For examples of solving simple counting problems, see Examples 1 and 2.

2. State the Fundamental Counting Principle (*page 817*). For examples of using the Fundamental Counting Principle to solve counting problems, see Examples 3 and 4.

3. Explain how to find the number of permutations of *n* elements (*page 818*), the number of permutations of *n* elements taken *r* at a time (*page 819*), and the number of distinguishable permutations (*page 820*). For examples of using permutations to solve counting problems, see Examples 5–7.

4. Explain how to find the number of combinations of *n* elements taken *r* at a time (*page 821*). For examples of using combinations to solve counting problems, see Examples 8–10.

11.6 Exercises

See CalcChat.com for tutorial help and worked-out solutions to odd-numbered exercises.

Vocabulary and Concept Check

In Exercises 1 and 2, fill in the blanks.

1. The number of _____ _____ of n objects is given by $\dfrac{n!}{n_1! \cdot n_2! \cdot n_3! \cdot \cdots \cdot n_k!}$.

2. The number of combinations of n elements taken r at a time is given by _____.

3. What is the difference between a permutation of n elements and a combination of n elements?

4. What do n and r represent in the formula $_nP_r = \dfrac{n!}{(n-r)!}$?

Skills and Applications

Random Selection   In Exercises 5–12, determine the number of ways a computer can randomly generate one or more such integers from 1 through 12.

5. An odd integer

6. An even integer

7. A prime integer

8. An integer that is greater than 9

9. An integer that is divisible by 4

10. An integer that is divisible by 3

11. Two *distinct* integers whose sum is 9

12. Two *distinct* integers whose sum is 8

13. **Entertainment Systems**   A customer can choose one of three amplifiers, one of two compact disc players, and one of five speaker models for an entertainment system. Determine the number of possible system configurations.

14. **Job Applicants**   A small college needs two additional faculty members: a chemist and a statistician. There are five applicants for the chemistry position and three applicants for the statistics position. In how many ways can the college fill these positions?

15. **Course Schedule**   A college student is preparing a course schedule for the next semester. The student may select one of two mathematics courses, one of three science courses, and one of five courses from the social sciences. How many schedules are possible?

16. **Physiology**   In a physiology class, a student must dissect three different specimens. The student can select one of nine earthworms, one of four frogs, and one of seven fetal pigs. In how many ways can the student select the specimens?

17. **True-False Exam**   In how many ways can you answer a six-question true-false exam? (Assume that you do not omit any questions.)

18. **Combination Lock**   A combination lock will open when you select the right choice of three numbers (from 1 to 50, inclusive). How many different lock combinations are possible?

19. **License Plate Numbers**   In the state of Pennsylvania, each standard automobile license plate number consists of three letters followed by a four-digit number. How many distinct license plate numbers are possible in Pennsylvania?

20. **License Plate Numbers**   In a certain state, each automobile license plate number consists of two letters followed by a four-digit number. To avoid confusion between "O" and "zero" and between "I" and "one," the letters "O" and "I" are not used. How many distinct license plate numbers are possible in this state?

21. **Three-Digit Numbers**   How many three-digit numbers are possible under each condition?

    (a) The leading digit cannot be zero and no repetition of digits is allowed.

    (b) The leading digit cannot be zero and the number must be a multiple of 5.

    (c) The number is at least 400.

22. **Four-Digit Numbers**   How many four-digit numbers are possible under each condition?

    (a) The leading digit cannot be zero and no repetition of digits is allowed.

    (b) The leading digit cannot be zero and the number must be less than 5000.

    (c) The leading digit cannot be zero and the number must be even.

23. **Concert Seats**   Four couples reserve seats in one row for a concert. In how many different ways can they sit when

    (a) there are no seating restrictions?

    (b) the two members of each couple wish to sit together?

24. Single File   In how many orders can four cats and four dogs walk through a doorway single file when

(a) there are no restrictions?

(b) the cats walk through before the dogs?

25. Posing for a Photograph   In how many ways can five children posing for a photograph line up in a row?

26. Riding in a Car   In how many ways can six people sit in a six-passenger car?

Evaluating $_nP_r$ In Exercises 27–30, evaluate $_nP_r$.

27. $_5P_2$ **28.** $_6P_6$ **29.** $_{12}P_2$ **30.** $_6P_5$

Evaluating $_nP_r$ In Exercises 31–34, use a graphing utility to evaluate $_nP_r$.

31. $_{15}P_3$ **32.** $_{100}P_4$ **33.** $_{50}P_4$ **34.** $_{10}P_5$

35. Kidney Donors

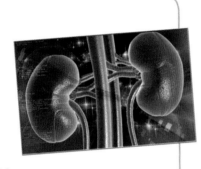

A patient with end-stage kidney disease has nine family members who are potential kidney donors. How many possible orders are there for a best match, a second-best match, and a third-best match?

36. Choosing Officers   From a pool of 12 candidates, the offices of president, vice president, secretary, and treasurer need to be filled. In how many different ways can the offices be filled?

37. Batting Order   A baseball coach is creating a nine-player batting order by selecting from a team of 15 players. How many different batting orders are possible?

38. Athletics   Eight sprinters qualify for the finals in the 100-meter dash at the NCAA national track meet. In how many ways can the sprinters come in first, second, and third? (Assume there are no ties.)

Distinguishable Permutations   In Exercises 39–42, find the number of distinguishable permutations of the group of letters.

39. A, A, G, E, E, E, M   **40.** B, B, B, T, T, T, T, T

41. A, L, G, E, B, R, A   **42.** M, I, S, S, I, S, S, I, P, P, I

43. Writing Permutations   Write all permutations of the letters A, B, C, and D.

44. Writing Permutations   Write all permutations of the letters A, B, C, and D when letters B and C must remain between A and D.

Evaluating $_nC_r$ In Exercises 45–48, evaluate $_nC_r$ using the formula from this section.

45. $_6C_4$ **46.** $_5C_4$ **47.** $_9C_9$ **48.** $_{12}C_0$

Evaluating $_nC_r$ In Exercises 49–52, use a graphing utility to evaluate $_nC_r$.

49. $_{16}C_2$ **50.** $_{17}C_5$ **51.** $_{20}C_6$ **52.** $_{50}C_8$

53. Writing Combinations   Write all combinations of two letters that can be formed from the letters A, B, C, D, E, and F. (Order is not important.)

54. Forming an Experimental Group   To conduct an experiment, researchers randomly select five students from a class of 20. How many different groups of five students are possible?

55. Jury Selection   In how many different ways can a jury of 12 people be randomly selected from a group of 40 people?

56. Committee Members   A U.S. Senate Committee has 14 members. Assuming party affiliation is not a factor in selection, how many different committees are possible from the 100 U.S. senators?

57. Defective Units   A shipment of 25 television sets contains three defective units. In how many ways can a vending company purchase four of these units and receive (a) all good units, (b) two good units, and (c) at least two good units?

58. Interpersonal Relationships   The complexity of interpersonal relationships increases dramatically as the size of a group increases. Determine the numbers of different two-person relationships in groups of people of sizes (a) 3, (b) 8, (c) 12, and (d) 20.

59. Poker Hand   You are dealt five cards from a standard deck of 52 playing cards. In how many ways can you get (a) a full house and (b) a five-card combination containing two jacks and three aces? (A full house consists of three of one kind and two of another. For example, A-A-A-5-5 and K-K-K-10-10 are full houses.)

60. Job Applicants   An employer interviews 12 people for four openings at a company. Five of the 12 people are women. All 12 applicants are qualified. In how many ways can the employer fill the four positions when (a) the selection is random and (b) exactly two selections are women?

61. Forming a Committee   A local college is forming a six-member research committee with one administrator, three faculty members, and two students. There are seven administrators, 12 faculty members, and 20 students in contention for the committee. How many six-member committees are possible?

62. Lottery Powerball is a lottery game that is operated by the Multi-State Lottery Association and is played in 45 states, Washington D.C., Puerto Rico, and the U.S. Virgin Islands. The game is played by drawing five white balls out of a drum of 69 white balls (numbered 1–69) and one red powerball out of a drum of 26 red balls (numbered 1–26). The jackpot is won by matching all five white balls in any order and the red powerball.

(a) Find the possible number of winning Powerball numbers.

(b) Find the possible number of winning Powerball numbers when you win the jackpot by matching all five white balls in order and the red powerball.

Geometry In Exercises 63–66, find the number of diagonals of the polygon. (A *diagonal* is a line segment connecting any two nonadjacent vertices of a polygon.)

63. Pentagon

64. Hexagon

65. Octagon

66. Decagon (10 sides)

Solving an Equation In Exercises 67–74, solve for n.

67. $4 \cdot {}_{n+1}P_2 = {}_{n+2}P_3$

68. $5 \cdot {}_{n-1}P_1 = {}_nP_2$

69. ${}_{n+1}P_3 = 4 \cdot {}_nP_2$

70. ${}_{n+2}P_3 = 6 \cdot {}_{n+2}P_1$

71. $14 \cdot {}_nP_3 = {}_{n+2}P_4$

72. ${}_nP_5 = 18 \cdot {}_{n-2}P_4$

73. ${}_nP_4 = 10 \cdot {}_{n-1}P_3$

74. ${}_nP_6 = 12 \cdot {}_{n-1}P_5$

75. Geometry Three points that are not collinear determine three lines. How many lines are determined by nine points, no three of which are collinear?

76. **HOW DO YOU SEE IT?** Without calculating, determine whether the value of ${}_nP_r$ is greater than the value of ${}_nC_r$ for the values of n and r given in the table. Complete the table using yes (Y) or no (N). Is the value of ${}_nP_r$ always greater than the value of ${}_nC_r$? Explain.

$\frac{r}{n}$	0	1	2	3	4	5	6	7
1								
2								
3								
4								
5								
6								
7								

Exploring the Concepts

True or False? In Exercises 77 and 78, determine whether the statement is true or false. Justify your answer.

77. The number of letter pairs that can be formed in any order from any two of the first 13 letters in the alphabet (A–M) is an example of a permutation.

78. The number of permutations of n elements can be determined by using the Fundamental Counting Principle.

79. Think About It Without calculating, determine which of the following is greater. Explain.

(a) The number of combinations of 10 elements taken six at a time

(b) The number of permutations of 10 elements taken six at a time

80. Think About It Can your graphing utility evaluate ${}_{100}P_{80}$? If not, explain why.

Proof In Exercises 81–84, prove the identity.

81. ${}_nP_{n-1} = {}_nP_n$

82. ${}_nC_n = {}_nC_0$

83. ${}_nC_{n-1} = {}_nC_1$

84. ${}_nC_r = \dfrac{{}_nP_r}{r!}$

Review & Refresh ▶ *Video solutions at LarsonPrecalculus.com*

Simplifying a Factorial Expression In Exercises 85–88, simplify the factorial expression.

85. $\dfrac{6!(16!)}{22!}$

86. $\dfrac{2!(11!)}{13!}$

87. $3\left(\dfrac{20!(9!)}{29!}\right)$

88. $\dfrac{3!(25!)}{26!}$

Finding a Term In Exercises 89–94, find the specified nth term in the expansion of the binomial.

89. $(2x + 3y)^5$, $n = 3$

90. $(x + 4y)^6$, $n = 2$

91. $(8x - 5y)^8$, $n = 7$

92. $(-4x + 12y)^9$, $n = 4$

93. $(-3x - 9y)^{13}$, $n = 6$

94. $(-10x - 2y)^{14}$, $n = 9$

Writing a Quadratic Equation In Exercises 95 and 96, using a system of equations, find the quadratic equation whose graph passes through the three points given in the table.

95.

x	-2	4	8
y	-8	10	12

96.

x	-6	3	9
y	36	3	24

11.7 Probability

Probability applies to many real-life applications. For example, in Exercise 59 on page 836, you will find probabilities that relate to a communication network and an independent backup system for a space vehicle.

⟩ **Find probabilities of events.**
⟩ **Find probabilities of mutually exclusive events.**
⟩ **Find probabilities of independent events.**
⟩ **Find the probability of the complement of an event**

The Probability of an Event

Any happening for which the result is uncertain is an **experiment.** The possible results of the experiment are **outcomes,** the set of all possible outcomes of the experiment is the **sample space** of the experiment, and any subcollection of a sample space is an **event.**

For example, when you toss a six-sided die, the numbers 1 through 6 can represent the sample space. For the experiment to be fair, each of the outcomes must be *equally likely.*

To describe sample spaces in such a way that each outcome is equally likely, you must sometimes distinguish between or among various outcomes in ways that appear artificial. Example 1 illustrates such a situation.

> **EXAMPLE 1** **Finding a Sample Space**

Find the sample space for each experiment.

a. You toss one coin.

b. You toss two coins.

c. You toss three coins.

Solution

a. The coin will land either heads up (denoted by H) or tails up (denoted by T), so the sample space is $S = \{H, T\}$.

b. Either coin can land heads up or tails up. Here is the list of possible outcomes.

HH = heads up on both coins

HT = heads up on the first coin and tails up on the second coin

TH = tails up on the first coin and heads up on the second coin

TT = tails up on both coins

So, the sample space is

$S = \{HH, HT, TH, TT\}$.

Note that this list distinguishes between the two cases HT and TH, even though these two outcomes appear to be similar.

c. Using notation similar to that used in part (b), the sample space is

$S = \{HHH, HHT, HTH, HTT, THH, THT, TTH, TTT\}$.

Note that this list distinguishes among the cases HHT, HTH, and THH, and among the cases HTT, THT, and TTH.

✓ **Checkpoint** ⟩ *Audio-video solution in English & Spanish at LarsonPrecalculus.com*

Find the sample space for the experiment.

You toss a coin twice and a six-sided die once.

GO DIGITAL

To find the probability of an event, count the number of outcomes in the event and in the sample space. The *number of equally likely outcomes* in event E is denoted by $n(E)$, and the number of equally likely outcomes in the sample space S is denoted by $n(S)$. The probability that event E will occur is given by $n(E)/n(S)$.

The Probability of an Event

If an event E has $n(E)$ equally likely outcomes and its sample space S has $n(S)$ equally likely outcomes, then the **probability** of event E is

$$P(E) = \frac{n(E)}{n(S)}.$$

The number of outcomes in an event must be less than or equal to the number of outcomes in the sample space, so the probability of an event must be a number from 0 to 1, inclusive. That is,

$$0 \le P(E) \le 1$$

as shown in the figure. If $P(E) = 0$, then event E *cannot occur,* and E is an **impossible event.** If $P(E) = 1$, then event E *must occur,* and E is a **certain event.**

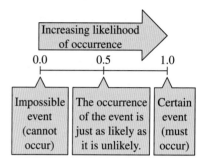

| Impossible event (cannot occur) | The occurrence of the event is just as likely as it is unlikely. | Certain event (must occur) |

EXAMPLE 2 **Finding the Probability of an Event**

▶▶▶ *See LarsonPrecalculus.com for an interactive version of this type of example.*

a. You toss two coins. What is the probability that both land heads up?

b. You draw one card at random from a standard deck of 52 playing cards. What is the probability that it is an ace?

Solution

a. Using the results of Example 1(b), let

$$E = \{HH\} \quad \text{and} \quad S = \{HH, HT, TH, TT\}.$$

The probability of getting two heads is

$$P(E) = \frac{n(E)}{n(S)} = \frac{1}{4}.$$

b. The deck has four aces (one in each suit), so the probability of drawing an ace is

$$P(E) = \frac{n(E)}{n(S)} = \frac{4}{52} = \frac{1}{13}.$$

✓ *Checkpoint* ▶ *Audio-video solution in English & Spanish at LarsonPrecalculus.com*

a. You toss three coins. What is the probability that all three land tails up?

b. You draw one card at random from a standard deck of 52 playing cards. What is the probability that it is a diamond? ■

In some cases, the number of outcomes in the sample space may not be given. In these cases, either write out the sample space or use the counting principles discussed in Section 11.6. Example 3 on the next page uses the Fundamental Counting Principle.

▶▶▶

ALGEBRA HELP

You can write a probability as a fraction, a decimal, or a percent. For instance, in Example 2(a), the probability of getting two heads can be written as $\frac{1}{4}$, 0.25, or 25%.

GO DIGITAL

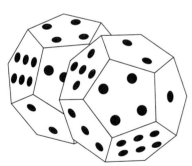

Figure 11.4

EXAMPLE 3 **Finding the Probability of an Event**

You toss two six-sided dice. What is the probability that the total of the two dice is 7?

Solution There are six possible outcomes on each die, so by the Fundamental Counting Principle, there are 6 · 6 or 36 different outcomes when you toss two dice. To find the probability of rolling a total of 7, you must first count the number of ways in which this can occur.

First Die	Second Die
1	6
2	5
3	4
4	3
5	2
6	1

So, a total of 7 can be rolled in six ways, which means that the probability of rolling a total of 7 is

$$P(E) = \frac{n(E)}{n(S)} = \frac{6}{36} = \frac{1}{6}.$$

✓ *Checkpoint* ▶ Audio-video solution in English & Spanish at LarsonPrecalculus.com

You toss two six-sided dice. What is the probability that the total of the two dice is 5?

EXAMPLE 4 **Finding the Probability of an Event**

Twelve-sided dice, as shown in Figure 11.4, can be constructed (in the shape of regular dodecahedrons) such that each of the numbers from 1 to 6 occurs twice on each die. Show that these dice can be used in any game requiring ordinary six-sided dice without changing the probabilities of the various events.

Solution For an ordinary six-sided die, each of the numbers

1, 2, 3, 4, 5, and 6

occurs once, so the probability of rolling any one of these numbers is

$$P(E) = \frac{n(E)}{n(S)} = \frac{1}{6}.$$

For one of the 12-sided dice, each number occurs twice, so the probability of rolling each number is

$$P(E) = \frac{n(E)}{n(S)} = \frac{2}{12} = \frac{1}{6}.$$

✓ *Checkpoint* ▶ Audio-video solution in English & Spanish at LarsonPrecalculus.com

Show that the probability of drawing a club at random from a standard deck of 52 playing cards is the same as the probability of drawing the ace of hearts at random from a set of four cards consisting of the aces of hearts, diamonds, clubs, and spades.

GO DIGITAL

EXAMPLE 5 **Random Selection**

The figure shows the numbers of degree-granting postsecondary institutions in various regions of the United States in 2018. What is the probability that an institution selected at random is in one of the three southern regions? *(Source: National Center for Education Statistics)*

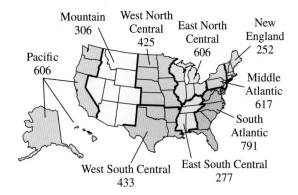

Solution From the figure, the total number of institutions is 4313. There are $791 + 277 + 433 = 1501$ institutions in the three southern regions, so the probability that the institution is in one of these regions is

$$P(E) = \frac{n(E)}{n(S)} = \frac{1501}{4313} \approx 0.348.$$

✓ *Checkpoint* Audio-video solution in English & Spanish at LarsonPrecalculus.com

In Example 5, what is the probability that an institution selected at random is in the Pacific region?

EXAMPLE 6 **Finding the Probability of Winning a Lottery**

In Arizona's The Pick game, a player chooses six different numbers from 1 to 44. When these six numbers match the six numbers drawn (in any order), the player wins (or shares) the top prize. What is the probability of winning the top prize when the player buys one ticket?

Solution To find the number of outcomes in the sample space, use the formula for the number of combinations of 44 numbers taken six at a time.

$$n(S) = {}_{44}C_6$$
$$= \frac{44 \cdot 43 \cdot 42 \cdot 41 \cdot 40 \cdot 39}{6 \cdot 5 \cdot 4 \cdot 3 \cdot 2 \cdot 1}$$
$$= 7{,}059{,}052$$

When a player buys one ticket, the probability of winning is

$$P(E) = \frac{1}{7{,}059{,}052}.$$

✓ *Checkpoint* Audio-video solution in English & Spanish at LarsonPrecalculus.com

In Pennsylvania's Cash 5 game, a player chooses five different numbers from 1 to 43. When these five numbers match the five numbers drawn (in any order), the player wins (or shares) the top prize. What is the probability of winning the top prize when the player buys one ticket?

GO DIGITAL

Mutually Exclusive Events

Two events A and B (from the same sample space) are **mutually exclusive** when A and B have no outcomes in common. In the terminology of sets, the intersection of A and B is the empty set, which implies that

$$P(A \cap B) = 0.$$

For example, when you toss two dice, the event A of rolling a total of 6 and the event B of rolling a total of 9 are mutually exclusive. To find the probability that one or the other of two mutually exclusive events will occur, *add* their individual probabilities.

Probability of the Union of Two Events

If A and B are events in the same sample space, then the probability of A or B occurring is given by

$$P(A \cup B) = P(A) + P(B) - P(A \cap B).$$

If A and B are mutually exclusive, then

$$P(A \cup B) = P(A) + P(B).$$

EXAMPLE 7 **Probability of a Union of Events**

You draw one card at random from a standard deck of 52 playing cards. What is the probability that the card is either a heart or a face card?

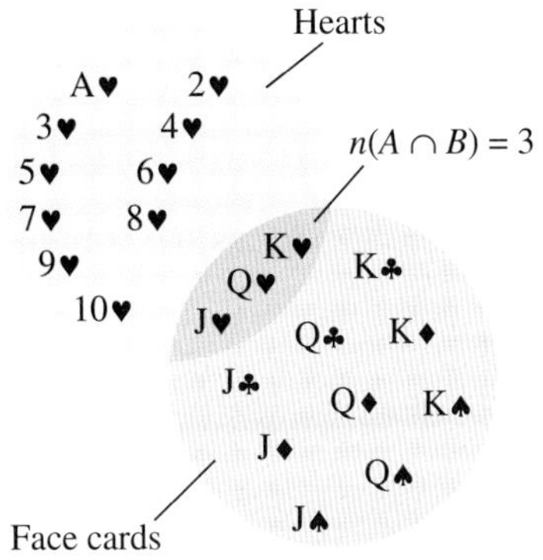

Solution The deck has 13 hearts, so the probability of drawing a heart (event A) is

$$P(A) = \frac{13}{52}. \qquad \text{Probability of drawing a heart}$$

Similarly, the deck has 12 face cards, so the probability of drawing a face card (event B) is

$$P(B) = \frac{12}{52}. \qquad \text{Probability of drawing a face card}$$

Three of the cards are hearts *and* face cards (see Figure 11.5), so it follows that

$$P(A \cap B) = \frac{3}{52}. \qquad \text{Probability of drawing a heart that is a face card}$$

Finally, applying the formula for the probability of the union of two events, the probability of drawing either a heart or a face card is

$$P(A \cup B) = P(A) + P(B) - P(A \cap B)$$
$$= \frac{13}{52} + \frac{12}{52} - \frac{3}{52}$$
$$= \frac{22}{52}$$
$$\approx 0.423.$$

✓ **Checkpoint** ▶ Audio-video solution in English & Spanish at LarsonPrecalculus.com

You draw one card at random from a standard deck of 52 playing cards. What is the probability that the card is either an ace or a spade? ∎

Hearts

A♥ 2♥
3♥ 4♥
5♥ 6♥ $n(A \cap B) = 3$
7♥ 8♥
9♥ K♥ K♣
10♥ Q♥ K♦
J♥ Q♣ K♦
J♣ Q♦ K♠
J♦ Q♠
J♠

Face cards

Figure 11.5

EXAMPLE 8   **Probability of Mutually Exclusive Events**

The human resources department of a company has compiled data showing the number of years of service for each employee. The table shows the results.

DATA Years of Service	Number of Employees
0–4	157
5–9	89
10–14	74
15–19	63
20–24	42
25–29	38
30–34	35
35–39	21
40–44	8
45 or more	2

Spreadsheet at LarsonPrecalculus.com

a. What is the probability that an employee chosen at random has 4 or fewer years of service?

b. What is the probability that an employee chosen at random has 9 or fewer years of service?

Solution

a. To begin, find the total number of employees.

$$157 + 89 + 74 + 63 + 42 + 38 + 35 + 21 + 8 + 2 = 529$$

Next, let event A represent choosing an employee with 0 to 4 years of service. Then the probability of choosing an employee who has 4 or fewer years of service is

$$P(A) = \frac{157}{529} \approx 0.297.$$

b. Let event B represent choosing an employee with 5 to 9 years of service. Then

$$P(B) = \frac{89}{529}.$$

Event A from part (a) and event B have no outcomes in common, so these two events are mutually exclusive and

$$P(A \cup B) = P(A) + P(B)$$
$$= \frac{157}{529} + \frac{89}{529}$$
$$= \frac{246}{529}$$
$$\approx 0.465.$$

So, the probability of choosing an employee who has 9 or fewer years of service is about 0.465.

✓ *Checkpoint* *Audio-video solution in English & Spanish at LarsonPrecalculus.com*

In Example 8, what is the probability that an employee chosen at random has 30 or more years of service?

GO DIGITAL

Independent Events

Two events are **independent** when the occurrence of one has no effect on the occurrence of the other. For example, rolling a total of 12 with two six-sided dice has no effect on the outcome of future rolls of the dice. To find the probability that two independent events will occur, *multiply* the probabilities of each.

> **Probability of Independent Events**
>
> If A and B are independent events, then the probability that both A and B will occur is
>
> $$P(A \text{ and } B) = P(A) \cdot P(B).$$
>
> This rule can be extended to any number of independent events.

EXAMPLE 9 **Probability of Independent Events**

A random number generator selects three integers from 1 to 20. What is the probability that all three numbers are less than or equal to 5?

Solution Let event A represent selecting a number from 1 to 5. Then the probability of selecting a number from 1 to 5 is

$$P(A) = \frac{5}{20} = \frac{1}{4}.$$

So, the probability that all three numbers are less than or equal to 5 is

$$P(A) \cdot P(A) \cdot P(A) = \left(\frac{1}{4}\right)\left(\frac{1}{4}\right)\left(\frac{1}{4}\right)$$

$$= \frac{1}{64}.$$

✓ *Checkpoint* ▶ Audio-video solution in English & Spanish at LarsonPrecalculus.com

A random number generator selects two integers from 1 to 30. What is the probability that both numbers are less than 12?

EXAMPLE 10 **Probability of Independent Events**

Approximately 65% of Americans expect much of the workforce to be automated within 50 years. In a survey, researchers selected 10 people at random from the population. What is the probability that all 10 people expected much of the workforce to be automated within 50 years? *(Source: Pew Research Center)*

Solution Let event A represent selecting a person who expected much of the workforce to be automated within 50 years. The probability of event A is 0.65. Each of the 10 occurrences of event A is an independent event, so the probability that all 10 people expected much of the workforce to be automated within 50 years is

$$[P(A)]^{10} = (0.65)^{10}$$

$$\approx 0.013.$$

✓ *Checkpoint* ▶ Audio-video solution in English & Spanish at LarsonPrecalculus.com

In Example 10, researchers selected five people at random from the population. What is the probability that all five people expected much of the workforce to be automated within 50 years?

GO DIGITAL

The Complement of an Event

The **complement of an event** A is the collection of all outcomes in the sample space that are *not* in A. The complement of event A is denoted by A'. Because $P(A \text{ or } A') = 1$ and A and A' are mutually exclusive, it follows that $P(A) + P(A') = 1$. So, the probability of A' is

$$P(A') = 1 - P(A).$$

> **Probability of a Complement**
>
> Let A be an event and let A' be its complement. If the probability of A is $P(A)$, then the probability of the complement is
>
> $$P(A') = 1 - P(A).$$

For example, if the probability of *winning* a game is $P(A) = \frac{1}{4}$, then the probability of *losing* the game is $P(A') = 1 - \frac{1}{4} = \frac{3}{4}$.

EXAMPLE 11 Probability of a Complement

A manufacturer has determined that a machine averages one faulty unit for every 1000 it produces. What is the probability that an order of 200 units will have one or more faulty units?

Solution To solve this problem as stated, you would need to find the probabilities of having exactly one faulty unit, exactly two faulty units, exactly three faulty units, and so on. However, using complements, it is much less tedious to find the probability that all units are perfect and then subtract this value from 1. The probability that any given unit is perfect is $999/1000$, so the probability that all 200 units are perfect is

$$P(A) = \left(\frac{999}{1000}\right)^{200} \approx 0.819$$

and the probability that at least one unit is faulty is

$$P(A') = 1 - P(A) \approx 1 - 0.819 = 0.181.$$

✓ **Checkpoint** ▶ Audio-video solution in English & Spanish at LarsonPrecalculus.com

A manufacturer has determined that a machine averages one faulty unit for every 500 it produces. What is the probability that an order of 300 units will have one or more faulty units? ■

> ### Summarize (Section 11.7)
>
> 1. State the definition of the probability of an event *(page 827)*. For examples of finding the probabilities of events, see Examples 2–6.
> 2. State the definition of mutually exclusive events and explain how to find the probability of the union of two events *(page 830)*. For examples of finding the probabilities of the unions of two events, see Examples 7 and 8.
> 3. State the definition of, and explain how to find the probability of, independent events *(page 832)*. For examples of finding the probabilities of independent events, see Examples 9 and 10.
> 4. State the definition of, and explain how to find the probability of, the complement of an event *(page 833)*. For an example of finding the probability of the complement of an event, see Example 11.

11.7 Exercises

See CalcChat.com for tutorial help and worked-out solutions to odd-numbered exercises.

Vocabulary and Concept Check

In Exercises 1–5, fill in the blanks.

1. An _____ is any happening for which the result is uncertain, and the possible results are called _____.
2. The set of all possible outcomes of an experiment is the _____ _____.
3. Any subcollection of a sample space is an _____.
4. For an experiment to be fair, each of the outcomes must be _____ _____.
5. The number of equally likely outcomes in an event E is denoted by _____, and the number of equally likely outcomes in the corresponding sample space S is denoted by _____.

6. What is the probability of an impossible event?
7. What is the probability of a certain event?
8. Match the probability formula with the correct probability name.
   - (a) Probability of the union of two events
   - (b) Probability of mutually exclusive events
   - (c) Probability of independent events
   - (d) Probability of a complement
   - (i) $P(A \cup B) = P(A) + P(B)$
   - (ii) $P(A') = 1 - P(A)$
   - (iii) $P(A \cup B) = P(A) + P(B) - P(A \cap B)$
   - (iv) $P(A \text{ and } B) = P(A) \cdot P(B)$

Skills and Applications

Finding a Sample Space **In Exercises 9–14, find the sample space for the experiment.**

9. You toss a coin and a six-sided die.
10. You toss a six-sided die twice and record the sum.
11. A taste tester ranks three varieties of yogurt, A, B, and C, according to preference.
12. You select two marbles (without replacement) from a bag containing two red marbles, two blue marbles, and one yellow marble. You record the color of each marble.
13. Two county supervisors are selected from five supervisors, A, B, C, D, and E, to study a recycling plan.
14. A sales representative visits three homes per day. In each home, there may be a sale (denote by S) or there may be no sale (denote by F).

Tossing a Coin **In Exercises 15–20, find the probability for the experiment of tossing a coin three times.**

15. The probability of getting exactly one tail
16. The probability of getting exactly two tails
17. The probability of getting a head on the first toss
18. The probability of getting a tail on the last toss
19. The probability of getting at least one head
20. The probability of getting at least two heads

Drawing a Card **In Exercises 21–24, find the probability for the experiment of drawing a card at random from a standard deck of 52 playing cards.**

21. The card is a face card.
22. The card is not a face card.
23. The card is a red face card.
24. The card is a 9 or lower. (Aces are low.)

Tossing a Die **In Exercises 25–30, find the probability for the experiment of tossing a six-sided die twice.**

25. The sum is 6.
26. The sum is at least 8.
27. The sum is less than 11.
28. The sum is 2, 3, or 12.
29. The sum is odd and no more than 7.
30. The sum is odd or prime.

Drawing Marbles **In Exercises 31–34, find the probability for the experiment of drawing two marbles at random (without replacement) from a bag containing one green, two yellow, and three red marbles.**

31. Both marbles are red.
32. Both marbles are yellow.
33. Neither marble is yellow.
34. The marbles are different colors.

35. Unemployment   In 2019, there were approximately 6 million unemployed in the U.S. civilian labor force. The circle graph shows the age profile of the unemployed. *(Source: U.S. Bureau of Labor Statistics)*

Ages of Unemployed

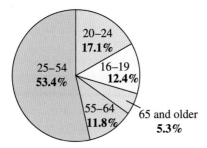

(a) Estimate the number of unemployed in the 16–19 age group.

(b) What is the probability that a person selected at random from the population of unemployed is in the 20–24 age group?

(c) What is the probability that a person selected at random from the population of unemployed is in the 25–54 age group?

(d) What is the probability that a person selected at random from the population of unemployed is 55 or older?

36. Poll   A campus-based polling group surveyed 100 juniors and seniors to determine whether they favor requiring a doctoral degree for professors in their field of study. The table lists the results of the survey. In the table, *J* represents juniors and *S* represents seniors.

	Yes	No	Unsure	Total
J	23	25	7	55
S	32	9	4	45
Total	55	34	11	100

Find the probability that a student selected at random from the sample is as described.

(a) A student who responded "no"

(b) A senior

(c) A junior who responded "yes"

37. Education   In a high school graduating class of 128 students, 52 are on the honor roll. Of these, 48 are going on to college. Of the 76 students not on the honor roll, 56 are going on to college. What is the probability that a student selected at random from the class is (a) going to college, (b) not going to college, and (c) not going to college and on the honor roll?

38. Alumni Association   A college surveys the 1254 alumni from 2019 and 2020. Of the 672 alumni from 2019, 124 went on to graduate school. Of the 582 alumni from 2020, 198 went on to graduate school. Find the probability that an alumnus selected at random from 2019 and 2020 is as described.

(a) a 2019 alumnus

(b) a 2020 alumnus

(c) a 2019 alumnus who did not attend graduate school

39. Winning an Election   Three people are running for president of a class. The results of a poll show that the first candidate has an estimated 37% chance of winning and the second candidate has an estimated 44% chance of winning. What is the probability that the third candidate will win?

40. Payroll Error   The employees of a company work in six departments: 31 are in sales, 54 are in research, 42 are in marketing, 20 are in engineering, 47 are in finance, and 58 are in production. The payroll clerk loses one employee's paycheck. What is the probability that the employee works in the research department?

41. Exam Questions   A class receives a list of 20 study problems, from which 10 will be part of an upcoming exam. A student knows how to solve 15 of the problems. Find the probability that the student will be able to answer (a) all 10 questions on the exam, (b) exactly eight questions on the exam, and (c) at least nine questions on the exam.

42. Payroll Error   A payroll clerk addresses five paychecks and envelopes to five different people and randomly inserts the paychecks into the envelopes. Find the probability of each event.

(a) Exactly one paycheck is inserted in the correct envelope.

(b) At least one paycheck is inserted in the correct envelope.

43. Game Show   On a game show, you are given five digits to arrange in the proper order to form the price of a car. If you are correct, you win the car. What is the probability of winning, given the following conditions?

(a) You guess the position of each digit.

(b) You know the first digit and guess the positions of the other digits.

44. Card Game   The deck for a card game contains 108 cards. Twenty-five each are red, yellow, blue, and green, and eight are wild cards. Each player is randomly dealt a seven-card hand.

(a) What is the probability that a hand will contain exactly two wild cards?

(b) What is the probability that a hand will contain two wild cards, two red cards, and three blue cards?

45. Drawing a Card You draw one card at random from a standard deck of 52 playing cards. Find the probability that (a) the card is an even-numbered card, (b) the card is a heart or a diamond, and (c) the card is a nine or a face card.

46. Drawing Cards You draw five cards at random from a standard deck of 52 playing cards. What is the probability that the hand drawn is a full house? (A full house consists of three of one kind and two of another.)

47. Shipment A shipment of 12 microwave ovens contains three defective units. A vending company purchases four units at random. What is the probability that (a) all four units are good, (b) exactly two units are good, and (c) at least two units are good?

48. PIN Code ATM personal identification number (PIN) codes typically consist of four-digit sequences of numbers. Find the probability that if you forget your PIN, you can guess the correct sequence (a) at random and (b) when you recall the first two digits.

49. Random Number Generator A random number generator selects two integers from 1 through 40. What is the probability that (a) both numbers are even, (b) one number is even and one number is odd, (c) both numbers are less than 30, and (d) the same number is selected twice?

50. Flexible Work Hours In a recent survey, people were asked whether they would prefer to work flexible hours—even when it meant slower career advancement—so they could spend more time with their families. The figure shows the results of the survey. What is the probability that three people chosen at random would prefer flexible work hours?

Flexible Work Hours

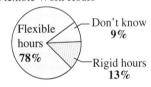

Probability of a Complement **In Exercises 51–54, you are given the probability that an event *will* happen. Find the probability that the event *will not* happen.**

51. $P(E) = 0.73$ **52.** $P(E) = 0.28$

53. $P(E) = \dfrac{1}{5}$ **54.** $P(E) = \dfrac{2}{7}$

Probability of a Complement **In Exercises 55–58, you are given the probability that an event *will not* happen. Find the probability that the event *will* happen.**

55. $P(E') = 0.29$ **56.** $P(E') = 0.89$

57. $P(E') = \dfrac{14}{25}$ **58.** $P(E') = \dfrac{79}{100}$

59. Backup System

A space vehicle has an independent backup system for one of its communication networks. The probability that either system will function satisfactorily during a flight is 0.985. What is the probability that during a given flight (a) both systems function satisfactorily, (b) both systems fail, and (c) at least one system functions satisfactorily?

60. Backup Vehicle A fire department keeps two rescue vehicles. Due to the demand on the vehicles and the chance of mechanical failure, the probability that a specific vehicle is available when needed is 90%. The availability of one vehicle is independent of the availability of the other. Find the probability that (a) both vehicles are available at a given time, (b) neither vehicle is available at a given time, and (c) at least one vehicle is available at a given time.

61. Roulette American roulette is a game in which a wheel turns on a spindle and is divided into 38 pockets. Thirty-six of the pockets are numbered 1–36, of which half are red and half are black. Two of the pockets are green and are numbered 0 and 00 (see figure). The dealer spins the wheel and a small ball in opposite directions. As the ball slows to a stop, it has an equal probability of landing in any of the numbered pockets.

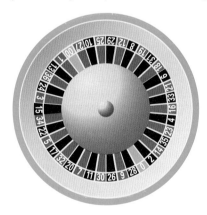

(a) Find the probability of landing in the number 00 pocket.

(b) Find the probability of landing in a red pocket.

(c) Find the probability of landing in a green pocket or a black pocket.

(d) Find the probability of landing in the number 14 pocket on two consecutive spins.

(e) Find the probability of landing in a red pocket on three consecutive spins.

62. Male or Female? Assume that the probability of the birth of a blue whale calf of a particular sex is 50%. A female blue whale gives birth to four calves. Find the probability of each event.

(a) All the calves are males.

(b) All the calves are the same sex.

(c) There is at least one male.

Exploring the Concepts

True or False? **In Exercises 63 and 64, determine whether the statement is true or false. Justify your answer.**

63. If A and B are independent events with nonzero probabilities, then A can occur when B occurs.

64. Rolling a number less than 3 on a six-sided die has a probability of $\frac{1}{3}$. The complement of this event is rolling a number greater than 3, which has a probability of $\frac{1}{2}$.

65. Error Analysis Describe the error.

A random number generator selects two integers from 1 to 15. The generator permits duplicates.

The probability that both numbers are greater than 12 is $\left(\frac{4}{15}\right)\left(\frac{4}{15}\right) = \frac{16}{225}$.

66. Pattern Recognition Consider a group of n people.

(a) Explain why the pattern below gives the probabilities that the n people have distinct birthdays.

$$n = 2: \quad \frac{365}{365} \cdot \frac{364}{365} = \frac{365 \cdot 364}{365^2}$$

$$n = 3: \quad \frac{365}{365} \cdot \frac{364}{365} \cdot \frac{363}{365} = \frac{365 \cdot 364 \cdot 363}{365^3}$$

(b) Write an expression for the probability that $n = 4$ people have distinct birthdays.

(c) Let P_n be the probability that the n people have distinct birthdays. Verify that this probability can be obtained recursively by

$$P_1 = 1 \quad \text{and} \quad P_n = \frac{365 - (n - 1)}{365} P_{n-1}.$$

(d) Explain why $Q_n = 1 - P_n$ gives the probability that at least two people in a group of n people have the same birthday.

(e) Complete the table.

n	10	15	20	23	30	40	50
P_n							
Q_n							

(f) How many people must be in a group so that the probability of at least two of them having the same birthday is greater than $\frac{1}{2}$? Explain.

67. Think About It Let A and B be two events from the same sample space such that $P(A) = 0.76$ and $P(B) = 0.58$. Is it possible that A and B are mutually exclusive? Explain.

68. HOW DO YOU SEE IT? The circle graphs show the percents of undergraduates by class level at two colleges. A student is chosen at random from the combined undergraduate population of the two colleges. The probability that the student is a freshman, sophomore, or junior is 81%. Which college has a greater number of undergraduates? Explain.

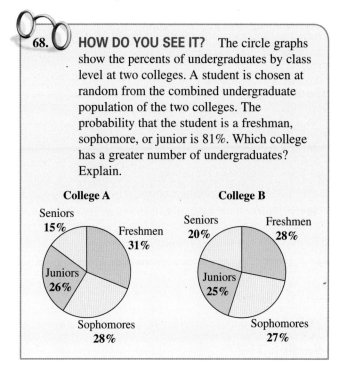

69. Think About It A weather forecast indicates that the probability of rain is 40%. What does this mean?

Review & Refresh ▶ *Video solutions at LarsonPrecalculus.com*

70. Error Analysis Describe the error.

In the equation for the surface area of a sphere, $S = 4\pi r^2$, the surface area S varies jointly with π and the square of the radius r. ✗

Trigonometric Substitution **In Exercises 71 and 72, use the trigonometric substitution to write the algebraic equation as a trigonometric equation of θ, where $-\pi/2 < \theta < \pi/2$. Then find $\sin \theta$ and $\cos \theta$.**

71. $\sqrt{2} = \sqrt{4 - x^2}, \quad x = 2 \sin \theta$

72. $5\sqrt{3} = \sqrt{100 - x^2}, \quad x = 10 \cos \theta$

Analyzing Trigonometric Functions **In Exercises 73 and 74, fill in the blanks.** (*Note:* The notation $x \to c^+$ indicates that x approaches c from the right and $x \to c^-$ indicates that x approaches c from the left.)

73. As $x \to \left(\frac{\pi}{2}\right)^-$, $\tan x \to$ ▨ and $\cot x \to$ ▨.

74. As $x \to \pi^+$, $\sin x \to$ ▨ and $\csc x \to$ ▨.

Simplifying a Complex Number **In Exercises 75 and 76, simplify the complex number and write it in standard form.**

75. $-6i^3 + i^2$

76. $4i^2 - 2i^3$

Summary and Study Strategies

GO DIGITAL

What Did You Learn?

The list below reviews the skills covered in the chapter and correlates each
one to the Review Exercises (see page 840) that practice the skill.

Section 11.1	Review Exercises
■ Use sequence notation to write the terms of sequences *(p. 770)*.	*1–8*
■ Use factorial notation *(p. 773)*. If n is a positive integer, then $n! = 1 \cdot 2 \cdot 3 \cdot 4 \cdots (n - 1) \cdot n$. As a special case, zero factorial is defined as $0! = 1$.	*9–12*
■ Use summation notation to write sums *(p. 774)*. The sum of the first n terms of a sequence is represented by $$\sum_{i=1}^{n} a_i = a_1 + a_2 + a_3 + a_4 + \cdots + a_n.$$	*13–16*
■ Find the sums of series *(p. 775)*.	*17, 18*
■ Use sequences and series to model and solve real-life problems *(p. 776)*.	*19, 20*

Section 11.2	
■ Recognize, write, and find the *n*th terms of arithmetic sequences *(p. 780)*. *n*th term of an arithmetic sequence: $a_n = a_1 + (n - 1)d$, where d is the common difference between consecutive terms of the sequence and a_1 is the first term	*21–30*
■ Find *n*th partial sums of arithmetic sequences *(p. 783)*.	*31–36*
■ Use arithmetic sequences to model and solve real-life problems *(p. 785)*.	*37, 38*

Section 11.3	
■ Recognize, write, and find the *n*th terms of geometric sequences *(p. 789)*. *n*th term of a geometric sequence: $a_n = a_1 r^{n-1}$, where r is the common ratio of consecutive terms of the sequence	*39–50*
■ Find the sum of a finite geometric sequence *(p. 792)*.	*51–58*
■ Find the sum of an infinite geometric series *(p. 793)*.	*59–62*
■ Use geometric sequences to model and solve real-life problems *(p. 794)*.	*63, 64*

Section 11.4	
■ Use mathematical induction to prove statements involving a positive integer *n* *(p. 798)*.	*65–68*
■ Use pattern recognition and mathematical induction to write a formula for the *n*th term of a sequence *(p. 803)*.	*69–72*
■ Find the sums of powers of integers *(p. 804)*.	*73, 74*
■ Find finite differences of sequences *(p. 805)*.	*75, 76*

Section 11.5

	Review Exercises
■ Use the Binomial Theorem to find binomial coefficients *(p. 808)*.	*77, 78*

In the expansion of $(x + y)^n$, the coefficient of $x^{n-r}y^r$ is $_nC_r = \dfrac{n!}{(n-r)!r!}$.

■ Use Pascal's Triangle to find binomial coefficients *(p. 810)*.	*79, 80*
■ Use binomial coefficients to write binomial expansions *(p. 811)*.	*81–84*

Section 11.6

■ Solve simple counting problems *(p. 816)*.	*85, 86*
■ Use the Fundamental Counting Principle to solve counting problems *(p. 817)*.	*87, 88*

Event E_1 can occur in m_1 different ways. After E_1 has occurred, event E_2 can occur in m_2 different ways. The number of ways the two events can occur is $m_1 \cdot m_2$.

■ Use permutations to solve counting problems *(p. 818)*.	*89, 90*

Number of permutations of n elements taken r at a time: $_nP_r = \dfrac{n!}{(n-r)!}$

■ Use combinations to solve counting problems *(p. 821)*.	*91, 92*

Number of combinations of n elements taken r at a time: $_nC_r = \dfrac{n!}{(n-r)!r!}$.

Section 11.7

■ Find probabilities of events *(p. 826)*.	*93, 94*
■ Find probabilities of mutually exclusive events *(p. 830)*.	*95, 96*
■ Find probabilities of independent events *(p. 832)*.	*97, 98*
■ Find the probability of the complement of an event *(p. 833)*.	*99, 100*

Study Strategies

Preparing for the Final Exam   As the semester ends, many students are inundated with projects, papers, and tests. This is why it is important to plan your review time for the final exam at least three weeks before the exam.

1. **Form a study group of three or four students several weeks before the final exam.** Review what you have learned while continuing to learn new material.
2. **Find out what material you must know for the final exam, even if the instructor has not yet covered it.** As a group, meet with the instructor outside of class. A group is likely to receive more attention and can ask more questions.
3. **Ask for or create a practice final exam and have the instructor look at it.** Look for sample problems in old tests and in cumulative tests in the text.
4. **Have each group member take the practice final exam.** Then have each member identify what he or she needs to study.
5. **Decide when the group is going to meet during the next couple of weeks and what you will cover during each session.** Set up several study times for each week.
6. **Keep the study session on track.** Prepare for each session by knowing what material you are going to cover and having the class notes for that material.

Review Exercises

See CalcChat.com for tutorial help and worked-out solutions to odd-numbered exercises.

GO DIGITAL

11.1 **Writing the Terms of a Sequence** **In Exercises 1–4, write the first five terms of the sequence. (Assume that n begins with 1.)**

1. $a_n = 3 + \dfrac{12}{n}$

2. $a_n = \dfrac{(-1)^n 5n}{2n - 1}$

3. $a_n = \dfrac{120}{n!}$

4. $a_n = (n + 1)(n + 2)$

Finding the nth Term of a Sequence **In Exercises 5–8, write an expression for the apparent nth term a_n of the sequence. (Assume that n begins with 1.)**

5. $-2, 2, -2, 2, -2, \ldots$

6. $-1, 2, 7, 14, 23, \ldots$

7. $3, \frac{9}{2}, 9, \frac{81}{4}, \frac{243}{5}, \ldots$

8. $1, -\frac{1}{2}, \frac{1}{3}, -\frac{1}{4}, \frac{1}{5}, \ldots$

Simplifying a Factorial Expression **In Exercises 9–12, simplify the factorial expression.**

9. $\dfrac{3!}{5!}$

10. $\dfrac{7!}{3! \cdot 4!}$

11. $\dfrac{(n - 1)!}{(n + 1)!}$

12. $\dfrac{n!}{(n + 2)!}$

Finding a Sum **In Exercises 13 and 14, find the sum.**

13. $\displaystyle\sum_{j=1}^{4} \dfrac{6}{j^2}$

14. $\displaystyle\sum_{k=1}^{5} 2k^3$

Using Sigma Notation to Write a Sum **In Exercises 15 and 16, use sigma notation to write the sum.**

15. $\dfrac{1}{2(1)} + \dfrac{1}{2(2)} + \dfrac{1}{2(3)} + \cdots + \dfrac{1}{2(20)}$

16. $\dfrac{1}{2} + \dfrac{2}{3} + \dfrac{3}{4} + \cdots + \dfrac{9}{10}$

Finding the Sum of an Infinite Series **In Exercises 17 and 18, find the sum of the infinite series.**

17. $\displaystyle\sum_{i=1}^{\infty} \dfrac{4}{10^i}$

18. $\displaystyle\sum_{k=1}^{\infty} 8\left(\dfrac{1}{10}\right)^k$

19. Compound Interest An investor deposits $10,000 in an account that earns 2.25% interest compounded monthly. The balance in the account after n months is given by

$$A_n = 10{,}000\left(1 + \dfrac{0.0225}{12}\right)^n, \quad n = 1, 2, 3, \ldots.$$

(a) Write the first 10 terms of the sequence.

(b) Find the balance in the account after 10 years by computing the 120th term of the sequence.

20. Population The population a_n (in thousands) of Miami, Florida, from 2014 through 2018 can be approximated by

$$a_n = -0.68n^2 + 18.7n + 365$$

where n is the year with $n = 4$ corresponding to 2014. Write the terms of this finite sequence. Use a graphing utility to construct a bar graph that represents the sequence. *(Source: U.S. Census Bureau)*

11.2 **Determining Whether a Sequence Is Arithmetic** **In Exercises 21–24, determine whether the sequence is arithmetic. If so, find the common difference.**

21. $5, -1, -7, -13, -19, \ldots$

22. $0, 1, 3, 6, 10, \ldots$

23. $\frac{1}{8}, \frac{1}{4}, \frac{1}{2}, 1, 2, \ldots$

24. $1, \frac{15}{16}, \frac{7}{8}, \frac{13}{16}, \frac{3}{4}, \ldots$

Finding the nth Term **In Exercises 25–28, find a formula for the nth term of the arithmetic sequence.**

25. $a_1 = 7, d = 12$

26. $a_1 = 34, d = -4$

27. $a_3 = 96, a_7 = 24$

28. $a_7 = 8, a_{13} = 6$

Writing the Terms of an Arithmetic Sequence **In Exercises 29 and 30, write the first five terms of the arithmetic sequence.**

29. $a_1 = 4, d = 17$

30. $a_1 = 25, a_{n+1} = a_n + 3$

31. Sum of a Finite Arithmetic Sequence Find the sum of the first 100 positive multiples of 9.

32. Sum of a Finite Arithmetic Sequence Find the sum of the integers from 30 to 80.

Finding a Sum **In Exercises 33–36, find the partial sum.**

33. $\displaystyle\sum_{j=1}^{40} 2j$

34. $\displaystyle\sum_{j=1}^{200} (20 - 3j)$

35. $\displaystyle\sum_{k=1}^{100} \left(\dfrac{2}{3}k + 4\right)$

36. $\displaystyle\sum_{k=1}^{25} \left(\dfrac{3k + 1}{4}\right)$

37. Job Offer The starting salary for a job is $43,800 with a guaranteed increase of $1950 per year. Determine (a) the salary during the fifth year and (b) the total compensation through five full years of employment.

38. Baling Hay In the first two trips baling hay around a large field, a farmer obtains 123 bales and 112 bales, respectively. Each round gets shorter, so the farmer estimates that the same pattern will continue. Estimate the total number of bales made after the farmer takes another six trips around the field.

11.3 Determining Whether a Sequence Is Geometric   In Exercises 39–42, determine whether the sequence is geometric. If so, find the common ratio.

39. 2, 6, 18, 54, 162, . . .   **40.** 48, -24, 12, -6, . . .

41. $\frac{1}{5}, -\frac{3}{5}, \frac{9}{5}, -\frac{27}{5}, . . .$ **42.** $\frac{1}{4}, \frac{2}{5}, \frac{3}{6}, \frac{4}{7}, . . .$

Writing the Terms of a Geometric Sequence   In Exercises 43–46, write the first five terms of the geometric sequence.

43. $a_1 = 2, \quad r = 15$

44. $a_1 = 6, \quad r = -\frac{1}{3}$

45. $a_1 = 9, \quad a_3 = 4$

46. $a_1 = 2, \quad a_3 = 12$

Finding a Term of a Geometric Sequence   In Exercises 47–50, write an expression for the nth term of the geometric sequence. Then find the 10th term of the sequence.

47. $a_1 = 100, \quad r = 1.05$

48. $a_1 = 5, \quad r = 0.2$

49. $a_1 = 18, \quad a_2 = -9$

50. $a_3 = 6, \quad a_4 = 1$

Sum of a Finite Geometric Sequence   In Exercises 51–58, find the sum of the finite geometric sequence.

51. $\sum_{i=1}^{7} 2^{i-1}$ **52.** $\sum_{i=1}^{5} 3^{i-1}$

53. $\sum_{i=1}^{4} \left(\frac{1}{2}\right)^{i}$ **54.** $\sum_{i=1}^{6} \left(\frac{1}{3}\right)^{i-1}$

55. $\sum_{i=1}^{5} (2)^{i-1}$ **56.** $\sum_{i=1}^{4} 6(3)^{i}$

57. $\sum_{i=1}^{5} 10(0.6)^{i-1}$ **58.** $\sum_{i=1}^{4} 20(0.2)^{i-1}$

Sum of an Infinite Geometric Series   In Exercises 59–62, find the sum of the infinite geometric series.

59. $\sum_{i=0}^{\infty} \left(\frac{7}{8}\right)^{i}$ **60.** $\sum_{k=1}^{\infty} 1.3\left(\frac{1}{10}\right)^{k-1}$

61. $1 - \frac{1}{2} + \frac{1}{4} - \frac{1}{8} + . . .$

62. $4 + \frac{8}{3} + \frac{16}{9} + \frac{32}{27} + . . .$

63. Depreciation   A paper manufacturer buys a machine for \$120,000. It depreciates at a rate of 30% per year. (In other words, at the end of each year the depreciated value is 70% of what it was at the beginning of the year.)

(a) Find the formula for the nth term of a geometric sequence that gives the value of the machine t full years after it is purchased.

(b) Find the depreciated value of the machine after 5 full years.

64. Compound Interest   An investor deposits \$800 in an account on the first day of each month for 10 years. The account pays 3%, compounded monthly. What is the balance at the end of 10 years?

11.4 Using Mathematical Induction   In Exercises 65–68, use mathematical induction to prove the formula for all integers $n \geq 1$.

65. $3 + 5 + 7 + \cdot\cdot\cdot + (2n + 1) = n(n + 2)$

66. $1 + \frac{3}{2} + 2 + \frac{5}{2} + \cdot\cdot\cdot + \frac{1}{2}(n + 1) = \frac{n}{4}(n + 3)$

67. $\sum_{i=0}^{n-1} ar^{i} = \frac{a(1 - r^{n})}{1 - r}$

68. $\sum_{k=0}^{n-1} (a + kd) = \frac{n}{2}[2a + (n - 1)d]$

Finding a Formula for a Finite Sum   In Exercises 69–72, find a formula for the sum of the first n terms of the sequence. Prove the validity of your formula.

69. 9, 13, 17, 21, . . .

70. 68, 60, 52, 44, . . .

71. $1, \frac{3}{5}, \frac{9}{25}, \frac{27}{125}, . . .$

72. $12, -1, \frac{1}{12}, -\frac{1}{144}, . . .$

Finding a Sum   In Exercises 73 and 74, find the sum using the formulas for the sums of powers of integers.

73. $\sum_{n=1}^{75} n$ **74.** $\sum_{n=1}^{6} (n^5 - n^2)$

Linear Model, Quadratic Model, or Neither?   In Exercises 75 and 76, write the first five terms of the sequence beginning with the term a_1. Then calculate the first and second differences of the sequence. State whether the sequence has a perfect linear model, a perfect quadratic model, or neither.

75. $a_1 = 5$ **76.** $a_1 = -3$
$a_n = a_{n-1} + 5$ $a_n = a_{n-1} - 2n$

11.5 Finding a Binomial Coefficient   In Exercises 77 and 78, find the binomial coefficient.

77. $_6C_4$ **78.** $_{12}C_3$

Using Pascal's Triangle   In Exercises 79 and 80, evaluate using Pascal's Triangle.

79. $\binom{7}{2}$ **80.** $\binom{10}{4}$

Expanding a Binomial   In Exercises 81–84, use the Binomial Theorem to write the expansion of the expression.

81. $(x + 4)^4$ **82.** $(5 + 2z)^4$

83. $(4 - 5x)^3$ **84.** $(a - 3b)^5$

11.6 Random Selection In Exercises 85 and 86, determine the number of ways a computer can randomly generate one or more such integers from 1 through 14.

85. A composite number

86. Two *distinct* integers whose sum is 12

87. Telephone Numbers All of the landline telephone numbers in a small town use the same three-digit prefix. How many different telephone numbers are possible by changing only the last four digits?

88. Course Schedule A college student is preparing a course schedule for the next semester. The student may select one of three mathematics courses, one of four science courses, and one of six history courses. How many schedules are possible?

89. Genetics A geneticist is using gel electrophoresis to analyze five DNA samples. The geneticist treats each sample with a different restriction enzyme and then injects it into one of five wells formed in a bed of gel. In how many orders can the geneticist inject the five samples into the wells?

90. Race There are 10 bicyclists entered in a race. In how many different ways can the top three places be decided?

91. Jury Selection In how many different ways can a jury of 12 people be randomly selected from a group of 32 people?

92. Menu Choices A local sandwich shop offers five different breads, four different meats, three different cheeses, and six different vegetables. A customer can choose one bread, one or no meat, one or no cheese, and up to three vegetables. Find the total number of combinations of sandwiches possible.

11.7

93. Apparel A drawer contains six white socks, two blue socks, and two gray socks.

    (a) What is the probability of randomly selecting one blue sock?

    (b) What is the probability of randomly selecting one white sock?

94. Bookshelf Order A child returns a five-volume set of books to a bookshelf. The child is not able to read, and so cannot distinguish one volume from another. What is the probability that the child shelves the books in the correct order?

95. Students by Class At a university, 31% of the students are freshmen, 26% are sophomores, 25% are juniors, and 18% are seniors. One student receives a cash scholarship randomly by lottery. Find the probability that the scholarship winner is as described.

    (a) A junior or senior

    (b) A freshman, sophomore, or junior

96. Opinion Poll In a survey, a sample of college students, faculty members, and administrators were asked whether they favor a proposed increase in the annual activity fee to enhance student life on campus. The table lists the results of the survey.

	Students	Faculty	Admin.	Total
Favor	237	37	18	292
Oppose	163	38	7	208
Total	400	75	25	500

Find the probability that a person selected at random from the sample is as described.

    (a) A person who opposes the proposal

    (b) A student

    (c) A faculty member who favors the proposal

97. Tossing a Die You toss a six-sided die four times. What is the probability of getting four 5's?

98. Tossing a Die You toss a six-sided die six times. What is the probability of getting each number exactly once?

99. Drawing a Card You draw one card at random from a standard deck of 52 playing cards. What is the probability that the card is not a club?

100. Tossing a Coin You toss a coin five times. What is the probability of getting at least one tail?

Exploring the Concepts

True or False? In Exercises 101–104, determine whether the statement is true or false. Justify your answer.

101. $\dfrac{(n+2)!}{n!} = \dfrac{n+2}{n}$

102. $\displaystyle\sum_{i=1}^{5}(i^3 + 2i) = \sum_{i=1}^{5}i^3 + \sum_{i=1}^{5}2i$

103. $\displaystyle\sum_{k=1}^{8}3k = 3\sum_{k=1}^{8}k$

104. $\displaystyle\sum_{j=1}^{6}2^j = \sum_{j=3}^{8}2^{j-2}$

105. Think About It An infinite sequence beginning with a_1 is a function. What is the domain of the function?

106. Think About It How do the two sequences differ?

    (a) $a_n = \dfrac{(-1)^n}{n}$ (b) $a_n = \dfrac{(-1)^{n+1}}{n}$

107. Writing Explain what is meant by a recursion formula.

108. Writing Write a brief paragraph explaining how to identify the graph of an arithmetic sequence and the graph of a geometric sequence.

Chapter Test

See CalcChat.com for tutorial help and worked-out solutions to odd-numbered exercises.

GO DIGITAL

Take this test as you would take a test in class. When you are finished, check your work against the answers given in the back of the book.

1. Write the first five terms of the sequence $a_n = \dfrac{(-1)^n}{3n + 2}$. (Assume that n begins with 1.) *(Section 11.1)*

2. Write an expression for the apparent nth term a_n of the sequence. (Assume that n begins with 1.) *(Section 11.1)*

   $$\dfrac{3}{1!}, \dfrac{4}{2!}, \dfrac{5}{3!}, \dfrac{6}{4!}, \dfrac{7}{5!}, \cdots$$

3. Write the next three terms of the series $8 + 21 + 34 + 47 + \cdots$. Then find the seventh partial sum of the series. *(Section 11.2)*

4. The 5th term of an arithmetic sequence is 45, and the 12th term is 24. Find the nth term. *(Section 11.2)*

5. The second term of a geometric sequence is 14, and the sixth term is 224. Find the nth term. (Assume that the terms of the sequence are positive.) *(Section 11.3)*

In Exercises 6–9, find the sum. *(Sections 11.2, 11.3, and 11.4)*

6. $\displaystyle\sum_{i=1}^{50} (2i^2 + 5)$

7. $\displaystyle\sum_{n=1}^{9} (12n - 7)$

8. $\displaystyle\sum_{i=1}^{\infty} 4\left(\dfrac{1}{2}\right)^i$

9. $\displaystyle\sum_{n=1}^{\infty} \left(-\dfrac{1}{3}\right)^n$

10. Use mathematical induction to prove the formula for all integers $n \geq 1$. *(Section 11.4)*

    $$5 + 10 + 15 + \cdots + 5n = 5n(n + 1)/2$$

11. Use the Binomial Theorem to write the expansion of $(x + 6y)^4$. *(Section 11.5)*

12. Expand $3(x - 2)^5 + 4(x - 2)^3$ by using Pascal's Triangle to determine the coefficients. *(Section 11.5)*

13. Find the coefficient of the term $a^4 b^3$ in the expansion of $(3a - 2b)^7$. *(Section 11.5)*

In Exercises 14 and 15, evaluate each expression. *(Section 11.6)*

14. (a) $_9P_2$ (b) $_{70}P_3$

15. (a) $_{11}C_4$ (b) $_{66}C_4$

16. How many distinct license plate numbers consisting of one letter followed by a three-digit number are possible? *(Section 11.6)*

17. Eight people are going for a ride in a boat that seats eight people. One person will drive, and only three of the remaining people are willing to ride in the two bow seats. How many seating arrangements are possible? *(Section 11.6)*

18. You attend a karaoke night and hope to hear your favorite song. The karaoke song book has 300 different songs (your favorite song is among them). Assuming that the singers are equally likely to pick any song and no song repeats, what is the probability that your favorite song is one of the 20 that you hear that night? *(Section 11.7)*

19. You and three of your friends are at a party. Names of all of the 30 guests are placed in a hat and drawn randomly to award four door prizes. Each guest can win only one prize. What is the probability that you and your friends win all four prizes? *(Section 11.7)*

20. The weather report calls for a 90% chance of snow. According to this report, what is the probability that it will *not* snow? *(Section 11.7)*

Cumulative Test for Chapters 9–11

See CalcChat.com for tutorial help and worked-out solutions to odd-numbered exercises.

GO DIGITAL

Take this test as you would take a test in class. When you are finished, check your work against the answers given in the back of the book.

In Exercises 1–4, solve the system by the specified method. *(Sections 9.1, 9.2, 9.3, and 10.1)*

1. Substitution
$$\begin{cases} y = 3 - x^2 \\ 2(y - 2) = x - 1 \end{cases}$$

2. Elimination
$$\begin{cases} x + 3y = -6 \\ 2x + 4y = -10 \end{cases}$$

3. Gaussian Elimination
$$\begin{cases} -2x + 4y - z = -16 \\ x - 2y + 2z = 5 \\ x - 3y - z = 13 \end{cases}$$

4. Gauss-Jordan Elimination
$$\begin{cases} x + 3y - 2z = -7 \\ -2x + y - z = -5 \\ 4x + y + z = 3 \end{cases}$$

5. A custom-blend bird seed is made by mixing two types of bird seeds costing $0.75 per pound and $1.25 per pound. How many pounds of each type of seed mixture are used to make 200 pounds of custom-blend bird seed costing $0.95 per pound? *(Section 9.2)*

6. Find a quadratic equation $y = ax^2 + bx + c$ whose graph passes through the points $(0, 6)$, $(2, 3)$, and $(4, 2)$. *(Section 9.3)*

7. Write the partial fraction decomposition of the rational expression $\dfrac{2x^2 - x - 6}{x^3 + 2x}$. *(Section 9.4)*

In Exercises 8 and 9, sketch the graph of the solution set of the system of inequalities. Label the vertices of the region. *(Section 9.5)*

8. $\begin{cases} 2x + y \geq -3 \\ x - 3y \leq 2 \end{cases}$

9. $\begin{cases} x - y > 6 \\ 5x + 2y < 10 \end{cases}$

10. Sketch the region corresponding to the system of constraints. Then find the minimum and maximum values of the objective function $z = 3x + 2y$ and the points where they occur, subject to the constraints. *(Section 9.6)*

$$\begin{aligned} x + 4y &\leq 20 \\ 2x + y &\leq 12 \\ x &\geq 0 \\ y &\geq 0 \end{aligned}$$

$$\begin{cases} -x + 2y - z = 9 \\ 2x - y + 2z = -9 \\ 3x + 3y - 4z = 7 \end{cases}$$

System for 11 and 12

In Exercises 11 and 12, use the system of linear equations shown at the left. *(Section 10.1)*

11. Write the augmented matrix for the system.

12. Solve the system using the matrix found in Exercise 11 and Gauss-Jordan elimination.

In Exercises 13–18, perform the operation(s) using the matrices below, if possible. *(Section 10.2)*

$$A = \begin{bmatrix} -1 & 3 \\ 6 & 2 \end{bmatrix}, \quad B = \begin{bmatrix} -2 & 5 \\ 0 & -1 \end{bmatrix}, \quad C = \begin{bmatrix} 4 & 0 & 1 \\ -3 & 2 & -1 \end{bmatrix}$$

13. $A + B$

14. $2A - 5B$

15. AC

16. CB

17. A^2

18. $BA - B^2$

19. Find the inverse of the matrix, if possible: $\begin{bmatrix} 1 & 2 & -1 \\ 3 & 7 & -10 \\ -5 & -7 & -15 \end{bmatrix}$. *(Section 10.3)*

$\begin{bmatrix} 7 & 1 & 0 \\ -2 & 4 & -1 \\ 3 & 8 & 5 \end{bmatrix}$

Matrix for 20

20. Find the determinant of the matrix shown at the left. *(Section 10.4)*

21. Use matrices to find the vertices of the image of the square with vertices $(0, 2)$, $(0, 5)$, $(3, 2)$, and $(3, 5)$ after a reflection in the *x*-axis. *(Section 10.5)*

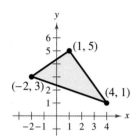

$$\text{Age} \atop \text{group} \begin{matrix} & \text{Gym} & \text{Jogging} & \text{Walking} \\ & \text{shoes} & \text{shoes} & \text{shoes} \\ 14{-}17 \\ 18{-}24 \\ 25{-}34 \end{matrix} \begin{bmatrix} 0.079 & 0.064 & 0.029 \\ 0.050 & 0.060 & 0.022 \\ 0.103 & 0.259 & 0.085 \end{bmatrix}$$

Matrix for 22

22. The matrix at the left shows the percents (in decimal form) of the total amounts spent on three types of footwear in a recent year. The total amounts (in millions of dollars) spent by the age groups on the three types of footwear were \$479.88 (14–17 age group), \$365.88 (18–24 age group), and \$1248.89 (25–34 age group). How many dollars worth of gym shoes, jogging shoes, and walking shoes were sold that year? *(Source: National Sporting Goods Association)* *(Section 10.2)*

In Exercises 23 and 24, use Cramer's Rule to solve the system of equations. (Section 10.5)

23. $\begin{cases} 8x - 3y = -52 \\ 3x + 5y = 5 \end{cases}$

24. $\begin{cases} 5x + 4y + 3z = 7 \\ -3x - 8y + 7z = -9 \\ 7x - 5y - 6z = -53 \end{cases}$

25. Use a determinant to find the area of the triangle shown at the left. *(Section 10.5)*

Figure for 25

26. Write the first five terms of the sequence $a_n = \dfrac{(-1)^{n+1}}{2n+3}$. *(Section 11.1)*

27. Write an expression for the apparent *n*th term a_n of the sequence. *(Section 11.1)*

$$\frac{2!}{4}, \frac{3!}{5}, \frac{4!}{6}, \frac{5!}{7}, \frac{6!}{8}, \ldots$$

28. Find the 16th partial sum of the arithmetic sequence $6, 18, 30, 42, \ldots$ *(Section 11.2)*

29. The sixth term of an arithmetic sequence is 20.6, and the ninth term is 30.2. *(Section 11.2)*

(a) Find the 20th term.   (b) Find the *n*th term.

30. Write the first five terms of the sequence $a_n = 3(2)^{n-1}$. *(Section 11.3)*

31. Find the sum: $\displaystyle\sum_{i=0}^{\infty} 1.9\left(\tfrac{1}{10}\right)^{i-1}$. *(Section 11.3)*

32. Use mathematical induction to prove the inequality

$(n + 1)! > 2^n, \quad n \ge 2.$ *(Section 11.4)*

33. Use the Binomial Theorem to write the expansion of $(w - 9)^4$. *(Section 11.5)*

In Exercises 34–37, evaluate the expression. (Section 11.6)

34. $_{14}P_3$ **35.** $_{25}P_2$ **36.** $_8C_4$ **37.** $_{11}C_6$

In Exercises 38 and 39, find the number of distinguishable permutations of the group of letters. (Section 11.6)

38. B, A, S, K, E, T, B, A, L, L   **39.** A, N, T, A, R, C, T, I, C, A

40. You and your friend are among 10 applicants for two sales positions at a department store. All of the applicants are qualified. What is the probability that the store hires both of you? *(Section 11.7)*

41. On a game show, a contestant is given the digits 3, 4, and 5 to arrange in the proper order to form the price of an appliance. If the contestant is correct, he or she wins the appliance. What is the probability of winning when the contestant knows that the price is at least \$400? *(Section 11.7)*

Proofs in Mathematics

Properties of Sums *(p. 774)*

1. $\displaystyle\sum_{i=1}^{n} c = cn,$ c is a constant.

2. $\displaystyle\sum_{i=1}^{n} ca_i = c\sum_{i=1}^{n} a_i,$ c is a constant.

3. $\displaystyle\sum_{i=1}^{n} (a_i + b_i) = \sum_{i=1}^{n} a_i + \sum_{i=1}^{n} b_i$

4. $\displaystyle\sum_{i=1}^{n} (a_i - b_i) = \sum_{i=1}^{n} a_i - \sum_{i=1}^{n} b_i$

Proof

Each of these properties follows directly from the properties of real numbers.

1. $\displaystyle\sum_{i=1}^{n} c = c + c + c + \cdots + c = cn$ *n* terms

The proof of Property 2 uses the Distributive Property.

2. $\displaystyle\sum_{i=1}^{n} ca_i = ca_1 + ca_2 + ca_3 + \cdots + ca_n$

$$= c(a_1 + a_2 + a_3 + \cdots + a_n)$$

$$= c\sum_{i=1}^{n} a_i$$

The proof of Property 3 uses the Commutative and Associative Properties of Addition.

3. $\displaystyle\sum_{i=1}^{n} (a_i + b_i) = (a_1 + b_1) + (a_2 + b_2) + (a_3 + b_3) + \cdots + (a_n + b_n)$

$$= (a_1 + a_2 + a_3 + \cdots + a_n) + (b_1 + b_2 + b_3 + \cdots + b_n)$$

$$= \sum_{i=1}^{n} a_i + \sum_{i=1}^{n} b_i$$

The proof of Property 4 uses the Commutative and Associative Properties of Addition and the Distributive Property.

4. $\displaystyle\sum_{i=1}^{n} (a_i - b_i) = (a_1 - b_1) + (a_2 - b_2) + (a_3 - b_3) + \cdots + (a_n - b_n)$

$$= (a_1 + a_2 + a_3 + \cdots + a_n) + (-b_1 - b_2 - b_3 - \cdots - b_n)$$

$$= (a_1 + a_2 + a_3 + \cdots + a_n) - (b_1 + b_2 + b_3 + \cdots + b_n)$$

$$= \sum_{i=1}^{n} a_i - \sum_{i=1}^{n} b_i$$

The Sum of a Finite Arithmetic Sequence *(p. 783)*

The sum of a finite arithmetic sequence with n terms is given by

$$S_n = \frac{n}{2}(a_1 + a_n).$$

Proof

Begin by generating the terms of the arithmetic sequence in two ways. In the first way, repeatedly add d to the first term.

$$S_n = a_1 + a_2 + a_3 + \cdots + a_{n-2} + a_{n-1} + a_n$$
$$= a_1 + [a_1 + d] + [a_1 + 2d] + \cdots + [a_1 + (n-1)d]$$

In the second way, repeatedly subtract d from the nth term.

$$S_n = a_n + a_{n-1} + a_{n-2} + \cdots + a_3 + a_2 + a_1$$
$$= a_n + [a_n - d] + [a_n - 2d] + \cdots + [a_n - (n-1)d]$$

Add these two versions of S_n. The multiples of d sum to zero and you obtain the formula.

$$2S_n = (a_1 + a_n) + (a_1 + a_n) + (a_1 + a_n) + \cdots + (a_1 + a_n) \qquad n \text{ terms}$$
$$2S_n = n(a_1 + a_n)$$
$$S_n = \frac{n}{2}(a_1 + a_n)$$

The Sum of a Finite Geometric Sequence *(p. 792)*

The sum of the finite geometric sequence

$$a_1, a_1r, a_1r^2, a_1r^3, a_1r^4, \ldots, a_1r^{n-1}$$

with common ratio $r \neq 1$ is given by

$$S_n = \sum_{i=1}^{n} a_1 r^{i-1} = a_1 \left(\frac{1 - r^n}{1 - r} \right).$$

Proof

$$S_n = a_1 + a_1r + a_1r^2 + \cdots + a_1r^{n-2} + a_1r^{n-1}$$
$$rS_n = a_1r + a_1r^2 + a_1r^3 + \cdots + a_1r^{n-1} + a_1r^n \qquad \text{Multiply by } r.$$

Subtracting the second equation from the first yields

$$S_n - rS_n = a_1 - a_1r^n.$$

So, $S_n(1 - r) = a_1(1 - r^n)$ and, because $r \neq 1$, you have

$$S_n = a_1 \left(\frac{1 - r^n}{1 - r} \right).$$

> ## The Binomial Theorem   (p. 808)
>
> Let n be a positive integer, and let $r = 0, 1, 2, 3, \ldots, n$. In the expansion of $(x + y)^n$
>
> $$(x + y)^n = x^n + nx^{n-1}y + \cdots + {}_nC_r\, x^{n-r}y^r + \cdots + nxy^{n-1} + y^n$$
>
> the coefficient of $x^{n-r}y^r$ is
>
> $${}_nC_r - \frac{n!}{(n-r)!r!}.$$

Proof

The Binomial Theorem can be proved quite nicely using mathematical induction. The steps are straightforward but look a little messy, so only an outline of the proof is given here.

1. For $n = 1$, you have $(x + y)^1 = x^1 + y^1 = {}_1C_0 x + {}_1C_1 y$, and the formula is valid.

2. Assuming that the formula is true for $n = k$, the coefficient of $x^{k-r}y^r$ is

$$ {}_kC_r = \frac{k!}{(k-r)!r!} = \frac{k(k-1)(k-2)\cdots(k-r+1)}{r!}.$$

To show that the formula is true for $n = k + 1$, look at the coefficient of $x^{k+1-r}y^r$ in the expansion of

$$(x + y)^{k+1} = (x + y)^k(x + y).$$

From the right-hand side of this equation, you can determine that the term involving $x^{k+1-r}y^r$ is the sum of two products, as shown below.

$$({}_kC_r x^{k-r}y^r)(x) + ({}_kC_{r-1}x^{k+1-r}y^{r-1})(y)$$

$$= \left[\frac{k!}{(k-r)!r!} + \frac{k!}{(k+1-r)!(r-1)!} \right]x^{k+1-r}y^r$$

$$= \left[\frac{(k+1-r)k!}{(k+1-r)!r!} + \frac{k!r}{(k+1-r)!r!} \right]x^{k+1-r}y^r$$

$$= \left[\frac{k!(k+1-r+r)}{(k+1-r)!r!} \right]x^{k+1-r}y^r$$

$$= \left[\frac{(k+1)!}{(k+1-r)!r!} \right]x^{k+1-r}y^r$$

$$= {}_{k+1}C_r x^{k+1-r}y^r$$

So, by mathematical induction, the Binomial Theorem is valid for all positive integers n.

P.S. Problem Solving

See CalcChat.com for tutorial help and worked-out solutions to odd-numbered exercises.

GO DIGITAL

1. Decreasing Sequence Consider the sequence

$$a_n = \frac{n+1}{n^2+1}.$$

(a) Use a graphing utility to graph the first 10 terms of the sequence.

(b) Use the graph from part (a) to estimate the value of a_n as n approaches infinity.

(c) Complete the table.

n	1	10	100	1000	10,000
a_n					

(d) Use the table from part (c) to determine (if possible) the value of a_n as n approaches infinity.

2. Alternating Sequence Consider the sequence

$$a_n = 3 + (-1)^n.$$

(a) Use a graphing utility to graph the first 10 terms of the sequence.

(b) Use the graph from part (a) to describe the behavior of the graph of the sequence.

(c) Complete the table.

n	1	10	101	1000	10,001
a_n					

(d) Use the table from part (c) to determine (if possible) the value of a_n as n approaches infinity.

3. Greek Mythology Can the Greek hero Achilles, running at 20 feet per second, ever catch a tortoise, starting 20 feet ahead of Achilles and running at 10 feet per second? The Greek mathematician Zeno said no. When Achilles runs 20 feet, the tortoise will be 10 feet ahead. Then, when Achilles runs 10 feet, the tortoise will be 5 feet ahead. Achilles will keep cutting the distance in half but will never catch the tortoise. The table shows Zeno's reasoning. In the table, both the distances and the times required to achieve them form infinite geometric series. Using the table, show that both series have finite sums. What do these sums represent?

DATA	Distance (in feet)	Time (in seconds)
	20	1
	10	0.5
	5	0.25
	2.5	0.125
	1.25	0.0625
	0.625	0.03125

Spreadsheet at LarsonPrecalculus.com

4. Conjecture Let $x_0 = 1$ and consider the sequence x_n given by

$$x_n = \frac{1}{2}x_{n-1} + \frac{1}{x_{n-1}}, \quad n = 1, 2, \ldots.$$

Use a graphing utility to compute the first 10 terms of the sequence and make a conjecture about the value of x_n as n approaches infinity.

5. Operations on an Arithmetic Sequence Determine whether performing each operation on an arithmetic sequence results in another arithmetic sequence. If so, state the common difference.

(a) A constant C is added to each term.

(b) Each term is multiplied by a nonzero constant C.

(c) Each term is squared.

6. Sequences of Powers The following sequence of perfect squares is not arithmetic.

$$1, 4, 9, 16, 25, 36, 49, 64, 81, \ldots$$

The related sequence formed from the first differences of this sequence, however, is arithmetic.

(a) Write the first eight terms of the related arithmetic sequence described above. What is the nth term of this sequence?

(b) Explain how to find an arithmetic sequence that is related to the following sequence of perfect cubes.

$$1, 8, 27, 64, 125, 216, 343, 512, 729, \ldots$$

(c) Write the first seven terms of the related arithmetic sequence in part (b) and find the nth term of the sequence.

(d) Explain how to find an arithmetic sequence that is related to the following sequence of perfect fourth powers.

$$1, 16, 81, 256, 625, 1296, 2401, 4096, 6561, \ldots$$

(e) Write the first six terms of the related arithmetic sequence in part (d) and find the nth term of the sequence.

7. Piecewise-Defined Sequence A sequence can be defined using a piecewise formula. An example of a piecewise-defined sequence is given below.

$$a_1 = 7, a_n = \begin{cases} \frac{1}{2}a_{n-1}, & \text{when } a_{n-1} \text{ is even.} \\ 3a_{n-1} + 1, & \text{when } a_{n-1} \text{ is odd.} \end{cases}$$

(a) Write the first 20 terms of the sequence.

(b) Write the first 10 terms of the sequences for which $a_1 = 4$, $a_1 = 5$, and $a_1 = 12$ (using a_n as defined above). What conclusion can you make about the behavior of each sequence?

849

8. Fibonacci Sequence Let $f_1, f_2, \ldots, f_n, \ldots$ be the Fibonacci sequence.

(a) Use mathematical induction to prove that
$$f_1 + f_2 + \cdots + f_n = f_{n+2} - 1.$$

(b) Find the sum of the first 20 terms of the Fibonacci sequence.

9. Pentagonal Numbers The numbers 1, 5, 12, 22, 35, 51, . . . are called pentagonal numbers because they represent the numbers of dots in the sequence of figures shown below. Use mathematical induction to prove that the nth pentagonal number P_n is given by

$$P_n = \frac{n(3n - 1)}{2}.$$

10. Think About It What conclusion can be drawn about the sequence of statements P_n for each situation?

(a) P_3 is true and P_k implies P_{k+1}.

(b) $P_1, P_2, P_3, \ldots, P_{50}$ are all true.

(c) $P_1, P_2,$ and P_3 are all true, but the truth of P_k does not imply that P_{k+1} is true.

(d) P_2 is true and P_{2k} implies P_{2k+2}.

11. Sierpinski Triangle Recall that a *fractal* is a geometric figure that consists of a pattern that is repeated infinitely on a smaller and smaller scale. One well-known fractal is the *Sierpinski Triangle*. In the first stage, the midpoints of the three sides are used to create the vertices of a new triangle, which is then removed, leaving three triangles. The figure below shows the first two stages. Note that each remaining triangle is similar to the original triangle. Assume that the length of each side of the original triangle is one unit. Write a formula that describes the side length of the triangles generated in the nth stage. Write a formula for the area of the triangles generated in the nth stage.

12. Job Offer You work for a company that pays $0.01 the first day, $0.02 the second day, $0.04 the third day, and so on. If the daily wage keeps doubling, what will your total income be for working 30 days?

13. Multiple Choice A multiple-choice question has five possible answers. You know that the answer is not B or D, but you are not sure about answers A, C, and E. What is the probability that you will get the right answer when you take a guess?

14. Throwing a Dart You throw a dart at the circular target shown below. The dart is equally likely to hit any point inside the target. What is the probability that it hits the region outside the triangle?

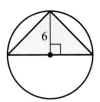

15. Odds The odds in favor of an event occurring is the ratio of the probability that the event will occur to the probability that the event will not occur. The reciprocal of this ratio represents the odds against the event occurring.

(a) A bag contains three blue marbles and seven yellow marbles. What are the odds in favor of choosing a blue marble? What are the odds against choosing a blue marble?

(b) Six of the marbles in a bag are red. The odds against choosing a red marble are 4 to 1. How many marbles are in the bag?

(c) Write a formula for converting the odds in favor of an event to the probability of the event.

(d) Write a formula for converting the probability of an event to the odds in favor of the event.

16. Expected Value An event A has n possible outcomes, which have the values $x_1, x_2, \ldots, x_n$. The probabilities of the n outcomes occurring are $p_1, p_2, \ldots, p_n$. The **expected value** V of an event A is the sum of the products of the outcomes' probabilities and their values,

$$V = p_1 x_1 + p_2 x_2 + \cdots + p_n x_n.$$

(a) To win California's SuperLotto Plus game, you must match five different numbers chosen from the numbers 1 to 47, plus one Mega number chosen from the numbers 1 to 27. You purchase a ticket for $1. If the jackpot for the next drawing is $12,000,000, what is the expected value of the ticket?

(b) You are playing a dice game in which you need to score 60 points to win. On each turn, you toss two six-sided dice. Your score for the turn is 0 when the dice do not show the same number. Your score for the turn is the product of the numbers on the dice when they do show the same number. What is the expected value of each turn? How many turns will it take on average to score 60 points?

Appendix A   Errors and the Algebra of Calculus

> ❯ Avoid common algebraic errors.
> ❯ Recognize and use algebraic techniques that are common in calculus.

Algebraic Errors to Avoid

This appendix contains five lists of common algebraic errors: errors involving parentheses, errors involving fractions, errors involving exponents, errors involving radicals, and errors involving dividing out common factors. Many of these errors occur because they seem to be the *easiest* things to do. For example, students often believe that the operations of subtraction and division are commutative and associative. The examples below illustrate the fact that subtraction and division are neither commutative nor associative.

Not commutative	Not associative
$4 - 3 \neq 3 - 4$	$8 - (6 - 2) \neq (8 - 6) - 2$
$15 \div 5 \neq 5 \div 15$	$20 \div (4 \div 2) \neq (20 \div 4) \div 2$

Errors Involving Parentheses

Potential Error	Correct Form	Comment
$a - (x - b) = a - x - b$ ✗	$a - (x - b) = a - x + b$	Distribute negative sign to each term in parentheses.
$(a + b)^2 = a^2 + b^2$ ✗	$(a + b)^2 = a^2 + 2ab + b^2$	Remember the middle term when squaring binomials.
$\left(\frac{1}{2}a\right)\left(\frac{1}{2}b\right) = \frac{1}{2}(ab)$ ✗	$\left(\frac{1}{2}a\right)\left(\frac{1}{2}b\right) = \frac{1}{4}(ab) = \frac{ab}{4}$	$\frac{1}{2}$ occurs twice as a factor.
$(3x + 6)^2 = 3(x + 2)^2$ ✗	$(3x + 6)^2 = [3(x + 2)]^2$ $= 3^2(x + 2)^2$	When factoring, raise all factors to the power.

Errors Involving Fractions

Potential Error	Correct Form	Comment
$\frac{2}{x + 4} = \frac{2}{x} + \frac{2}{4}$ ✗	Leave as $\frac{2}{x + 4}$.	The fraction is already in simplest form.
$\frac{\left(\frac{x}{a}\right)}{b} = \frac{bx}{a}$ ✗	$\frac{\left(\frac{x}{a}\right)}{b} = \left(\frac{x}{a}\right)\left(\frac{1}{b}\right) = \frac{x}{ab}$	Multiply by the reciprocal when dividing fractions.
$\frac{1}{a} + \frac{1}{b} = \frac{1}{a + b}$ ✗	$\frac{1}{a} + \frac{1}{b} = \frac{b + a}{ab}$	Use the property for adding fractions with unlike denominators.
$\frac{1}{3x} = \frac{1}{3}x$ ✗	$\frac{1}{3x} = \frac{1}{3} \cdot \frac{1}{x}$	Use the property for multiplying fractions.
$(1/3)x = \frac{1}{3x}$ ✗	$(1/3)x = \frac{1}{3} \cdot x = \frac{x}{3}$	Be careful when expressing fractions in the form $1/a$.
$(1/x) + 2 = \frac{1}{x + 2}$ ✗	$(1/x) + 2 = \frac{1}{x} + 2 = \frac{1 + 2x}{x}$	Be careful when expressing fractions in the form $1/a$. Be sure to find a common denominator before adding fractions.

Errors Involving Exponents

Potential Error	Correct Form	Comment
$(x^2)^3 = x^5$ ✗	$(x^2)^3 = x^{2 \cdot 3} = x^6$	Multiply exponents when raising a power to a power.
$x^2 \cdot x^3 = x^6$ ✗	$x^2 \cdot x^3 = x^{2+3} = x^5$	Add exponents when multiplying powers with like bases.
$(2x)^3 = 2x^3$ ✗	$(2x)^3 = 2^3x^3 = 8x^3$	Raise each factor to the power.
$\dfrac{1}{x^2 - x^3} = x^{-2} - x^{-3}$ ✗	Leave as $\dfrac{1}{x^2 - x^3}$.	Do not move term-by-term from denominator to numerator.

Errors Involving Radicals

Potential Error	Correct Form	Comment
$\sqrt{5x} = 5\sqrt{x}$ ✗	$\sqrt{5x} = \sqrt{5}\sqrt{x}$	Radicals apply to every factor inside the radical.
$\sqrt{x^2 + a^2} = x + a$ ✗	Leave as $\sqrt{x^2 + a^2}$.	Do not apply radicals term-by-term when adding or subtracting terms.
$\sqrt{-x + a} = -\sqrt{x - a}$ ✗	Leave as $\sqrt{-x + a}$.	Do not factor negative signs out of square roots.

Errors Involving Dividing Out

Potential Error	Correct Form	Comment
$\dfrac{a + bx}{a} = 1 + bx$ ✗	$\dfrac{a + bx}{a} = \dfrac{a}{a} + \dfrac{bx}{a} = 1 + \dfrac{b}{a}x$	Divide out common factors, not common terms.
$\dfrac{a + ax}{a} = a + x$ ✗	$\dfrac{a + ax}{a} = \dfrac{a(1 + x)}{a} = 1 + x$	Factor before dividing out common factors.
$1 + \dfrac{x}{2x} = 1 + \dfrac{1}{x}$ ✗	$1 + \dfrac{x}{2x} = 1 + \dfrac{1}{2} = \dfrac{3}{2}$	Divide out common factors.

A good way to avoid errors is to *work slowly, write neatly,* and *talk to yourself about each step.* Each time you write a step, ask yourself why the step is algebraically legitimate. For example, the step below is legitimate because *dividing the numerator and denominator by the same nonzero number produces an equivalent fraction.*

$$\frac{2x}{6} = \frac{\cancel{2} \cdot x}{\cancel{2} \cdot 3} = \frac{x}{3}$$

EXAMPLE 1   **Describing and Correcting an Error**

Describe and correct the error. $\dfrac{1}{2x} + \dfrac{1}{3x} = \dfrac{1}{5x}$ ✗

Solution Use the property for adding fractions with unlike denominators. (See Section P.1.)

$$\frac{1}{2x} + \frac{1}{3x} = \frac{3x + 2x}{(2x)(3x)} = \frac{5x}{6x^2} = \frac{5}{6x}$$

✓ *Checkpoint*  *Audio-video solution in English & Spanish at LarsonPrecalculus.com*

Describe and correct the error. $\sqrt{x^2 + 4} = x + 2$ ✗   ■

GO DIGITAL

Some Algebra of Calculus

In calculus it is often necessary to rewrite a simplified algebraic expression, or "unsimplify" it. See the expressions in the next four lists, which are from a standard calculus text. Note how the useful calculus form is an "unsimplified" form of the given expression.

Unusual Factoring

Expression	Useful Calculus Form	Comment
$\dfrac{5x^4}{8}$	$\dfrac{5}{8}x^4$	Write with fractional coefficient.
$\dfrac{x^2 + 3x}{-6}$	$-\dfrac{1}{6}(x^2 + 3x)$	Write with fractional coefficient.
$2x^2 - x - 3$	$2\left(x^2 - \dfrac{x}{2} - \dfrac{3}{2}\right)$	Factor out the leading coefficient.
$\dfrac{x}{2}(x + 1)^{-1/2} + (x + 1)^{1/2}$	$\dfrac{(x + 1)^{-1/2}}{2}[x + 2(x + 1)]$	Factor out the fractional coefficient and the variable expression with the lesser exponent.

Writing with Negative Exponents

Expression	Useful Calculus Form	Comment
$\dfrac{9}{5x^3}$	$\dfrac{9}{5}x^{-3}$	Move the factor to the numerator and change the sign of the exponent.
$\dfrac{7}{\sqrt{2x - 3}}$	$7(2x - 3)^{-1/2}$	Move the factor to the numerator and change the sign of the exponent.

Writing a Fraction as a Sum

Expression	Useful Calculus Form	Comment
$\dfrac{x + 2x^2 + 1}{\sqrt{x}}$	$x^{1/2} + 2x^{3/2} + x^{-1/2}$	Divide each term of the numerator by $x^{1/2}$.
$\dfrac{1 + x}{x^2 + 1}$	$\dfrac{1}{x^2 + 1} + \dfrac{x}{x^2 + 1}$	Rewrite the fraction as a sum of fractions.
$\dfrac{2x}{x^2 + 2x + 1}$	$\dfrac{2x + 2 - 2}{x^2 + 2x + 1}$	Add and subtract the same term.
	$= \dfrac{2x + 2}{x^2 + 2x + 1} - \dfrac{2}{(x + 1)^2}$	Rewrite the fraction as a difference of fractions.
$\dfrac{x^2 - 2}{x + 1}$	$x - 1 - \dfrac{1}{x + 1}$	Use polynomial long division. (See Section 3.3.)
$\dfrac{x + 7}{x^2 - x - 6}$	$\dfrac{2}{x - 3} - \dfrac{1}{x + 2}$	Use the method of partial fractions. (See Section 9.4.)

GO DIGITAL

Inserting Factors and Terms

Expression	Useful Calculus Form	Comment
$(2x - 1)^3$	$\dfrac{1}{2}(2x - 1)^3(2)$	Multiply and divide by 2.
$7x^2(4x^3 - 5)^{1/2}$	$\dfrac{7}{12}(4x^3 - 5)^{1/2}(12x^2)$	Multiply and divide by 12.
$\dfrac{4x^2}{9} - 4y^2 = 1$	$\dfrac{x^2}{9/4} - \dfrac{y^2}{1/4} = 1$	Write with fractional denominators.
$\dfrac{x}{x + 1}$	$\dfrac{x + 1 - 1}{x + 1} = 1 - \dfrac{1}{x + 1}$	Add and subtract the same term.

The next five examples demonstrate many of the steps in the preceding lists.

EXAMPLE 2   Factors Involving Negative Exponents

Factor $x(x + 1)^{-1/2} + (x + 1)^{1/2}$.

Solution When multiplying powers with like bases, you add exponents. When factoring, you are undoing multiplication, and so you *subtract* exponents.

$$\begin{aligned} x(x + 1)^{-1/2} + (x + 1)^{1/2} &= (x + 1)^{-1/2}[x(x + 1)^0 + (x + 1)^1] \\ &= (x + 1)^{-1/2}[x + (x + 1)] \\ &= (x + 1)^{-1/2}(2x + 1) \end{aligned}$$

✓ *Checkpoint* ▶ *Audio-video solution in English & Spanish at LarsonPrecalculus.com*

Factor $x(x - 2)^{-1/2} + 6(x - 2)^{1/2}$. ∎

Another way to simplify the expression in Example 2 is to multiply the expression by a fractional form of 1 and then use the Distributive Property.

$$\begin{aligned} [x(x + 1)^{-1/2} + (x + 1)^{1/2}] \cdot \frac{(x + 1)^{1/2}}{(x + 1)^{1/2}} &= \frac{x(x + 1)^0 + (x + 1)^1}{(x + 1)^{1/2}} \\ &= \frac{2x + 1}{\sqrt{x + 1}} \end{aligned}$$

EXAMPLE 3   Inserting Factors in an Expression

Insert the required factor: $\dfrac{x + 2}{(x^2 + 4x - 3)^2} = ()\dfrac{1}{(x^2 + 4x - 3)^2}(2x + 4)$.

Solution The expression on the right side of the equation is twice the expression on the left side. To make both sides equal, insert a factor of $\frac{1}{2}$.

$$\frac{x + 2}{(x^2 + 4x - 3)^2} = \left(\frac{1}{2}\right)\frac{1}{(x^2 + 4x - 3)^2}(2x + 4)$$

✓ *Checkpoint* ▶ *Audio-video solution in English & Spanish at LarsonPrecalculus.com*

Insert the required factor: $\dfrac{6x - 3}{(x^2 - x + 4)^2} = ()\dfrac{1}{(x^2 - x + 4)^2}(2x - 1)$. ∎

GO DIGITAL

EXAMPLE 4   **Rewriting Fractions**

Show that the two expressions are equivalent.

$$\frac{16x^2}{25} - 9y^2 = \frac{x^2}{25/16} - \frac{y^2}{1/9}$$

Solution   To write the expression on the left side of the equation in the form given on the right side, multiply the numerator and denominator of the first term by $1/16$ and multiply the numerator and denominator of the second term by $1/9$.

$$\frac{16x^2}{25} - 9y^2 = \frac{16x^2}{25}\left(\frac{1/16}{1/16}\right) - 9y^2\left(\frac{1/9}{1/9}\right) = \frac{x^2}{25/16} - \frac{y^2}{1/9}$$

✓ *Checkpoint* *Audio-video solution in English & Spanish at LarsonPrecalculus.com*

Show that the two expressions are equivalent.

$$\frac{9x^2}{16} + 25y^2 = \frac{x^2}{16/9} + \frac{y^2}{1/25}$$

EXAMPLE 5   **Rewriting with Negative Exponents**

Rewrite each expression using negative exponents.

a. $\dfrac{-4x}{(1 - 2x^2)^2}$ **b.** $\dfrac{2}{5x^3} - \dfrac{1}{\sqrt{x}} + \dfrac{3}{5(4x)^2}$

Solution

a. $\dfrac{-4x}{(1 - 2x^2)^2} = -4x(1 - 2x^2)^{-2}$

b. $\dfrac{2}{5x^3} - \dfrac{1}{\sqrt{x}} + \dfrac{3}{5(4x)^2} = \dfrac{2}{5x^3} - \dfrac{1}{x^{1/2}} + \dfrac{3}{5(4x)^2} = \dfrac{2}{5}x^{-3} - x^{-1/2} + \dfrac{3}{5}(4x)^{-2}$

✓ *Checkpoint* *Audio-video solution in English & Spanish at LarsonPrecalculus.com*

Rewrite $\dfrac{-6x}{(1 - 3x^2)^2} + \dfrac{1}{\sqrt[3]{x}}$ using negative exponents.

EXAMPLE 6   **Rewriting Fractions as Sums of Terms**

Rewrite each fraction as the sum of three terms.

a. $\dfrac{x^2 - 4x + 8}{2x}$ **b.** $\dfrac{x + 2x^2 + 1}{\sqrt{x}}$

Solution

a. $\dfrac{x^2 - 4x + 8}{2x} = \dfrac{x^2}{2x} - \dfrac{4x}{2x} + \dfrac{8}{2x} = \dfrac{x}{2} - 2 + \dfrac{4}{x}$

b. $\dfrac{x + 2x^2 + 1}{\sqrt{x}} = \dfrac{x}{x^{1/2}} + \dfrac{2x^2}{x^{1/2}} + \dfrac{1}{x^{1/2}} = x^{1/2} + 2x^{3/2} + x^{-1/2}$

✓ *Checkpoint* *Audio-video solution in English & Spanish at LarsonPrecalculus.com*

Rewrite each fraction as the sum of three terms.

a. $\dfrac{x^4 - 2x^3 + 5}{x^3}$ **b.** $\dfrac{x^2 - x + 5}{\sqrt{x}}$

| A | **Exercises** | See CalcChat.com for tutorial help and worked-out solutions to odd-numbered exercises. |

Vocabulary and Concept Check

In Exercises 1 and 2, fill in the blank.

1. To rewrite $3/x^5$ using negative exponents, move x^5 to the _____ and change the sign of the exponent.

2. When dividing fractions, multiply by the _____.

In Exercises 3 and 4, let $x = 5$ and $y = 4$ to verify the statement.

3. $(x - y)^2 \neq x^2 - y^2$

4. $\sqrt{x^2 - y^2} \neq x - y$

Skills and Applications

Describing and Correcting an Error In Exercises 5–12, describe and correct the error.

5. $(x + 3)^2 = x^2 + 9$ ✗

6. $3(x - 2) = 3x - 2$ ✗

7. $\sqrt{x + 9} = \sqrt{x} + 3$ ✗

8. $\sqrt{25 - x^2} = 5 - x$ ✗

9. $\dfrac{2x^2 + 1}{5x} = \dfrac{2x + 1}{5}$ ✗

10. $\dfrac{6x + y}{6x - y} = \dfrac{x + y}{x - y}$ ✗

11. $\dfrac{3}{x} + \dfrac{4}{y} = \dfrac{7}{x + y}$ ✗

12. $5 + (1/y) = \dfrac{1}{5 + y}$ ✗

Factors Involving Negative Exponents In Exercises 13–16, factor the expression.

13. $2x(x + 2)^{-1/2} + (x + 2)^{1/2}$

14. $x^2(x^2 + 1)^{-5} - (x^2 + 1)^{-4}$

15. $4x^3(2x - 1)^{3/2} - 2x(2x - 1)^{-1/2}$

16. $x(x + 1)^{-4/3} + (x + 1)^{2/3}$

Unusual Factoring In Exercises 17–24, complete the factored form of the expression.

17. $\dfrac{5x + 3}{4} = \dfrac{1}{4}()$

18. $\dfrac{7x^2}{10} = \dfrac{7}{10}()$

19. $\frac{2}{3}x^2 + \frac{1}{3}x + 5 = \frac{1}{3}()$

20. $\frac{3}{4}x + \frac{1}{2} = \frac{1}{4}()$

21. $x^{1/3} - 5x^{4/3} = x^{1/3}()$

22. $3(2x + 1)x^{1/2} + 4x^{3/2} = x^{1/2}()$

23. $\dfrac{1}{10}(2x + 1)^{5/2} - \dfrac{1}{6}(2x + 1)^{3/2} = \dfrac{(2x + 1)^{3/2}}{15}()$

24. $\dfrac{3}{7}(t + 1)^{7/3} - \dfrac{3}{4}(t + 1)^{4/3} = \dfrac{3(t + 1)^{4/3}}{28}()$

Inserting a Factor in an Expression In Exercises 25–28, insert the required factor in the parentheses.

25. $x^2(x^3 - 1)^4 = ()(x^3 - 1)^4(3x^2)$

26. $x(1 - 2x^2)^3 = ()(1 - 2x^2)^3(-4x)$

27. $\dfrac{4x + 6}{(x^2 + 3x + 7)^3} = ()\dfrac{1}{(x^2 + 3x + 7)^3}(2x + 3)$

28. $\dfrac{x + 1}{(x^2 + 2x - 3)^2} = ()\dfrac{1}{(x^2 + 2x - 3)^2}(2x + 2)$

Rewriting Fractions In Exercises 29–32, show that the two expressions are equivalent.

29. $4x^2 + \dfrac{6y^2}{10} = \dfrac{x^2}{1/4} + \dfrac{3y^2}{5}$

30. $\dfrac{4x^2}{14} - 2y^2 = \dfrac{2x^2}{7} - \dfrac{y^2}{1/2}$

31. $\dfrac{25x^2}{36} + \dfrac{4y^2}{9} = \dfrac{x^2}{36/25} + \dfrac{y^2}{9/4}$

32. $\dfrac{5x^2}{9} - \dfrac{16y^2}{49} = \dfrac{x^2}{9/5} - \dfrac{y^2}{49/16}$

Rewriting with Negative Exponents In Exercises 33–38, rewrite the expression using negative exponents.

33. $\dfrac{7}{(x + 3)^5}$

34. $\dfrac{2 - x}{(x + 1)^{3/2}}$

35. $\dfrac{2x^5}{(3x + 5)^4}$

36. $\dfrac{x + 1}{x(6 - x)^{1/2}}$

37. $\dfrac{4}{3x} + \dfrac{4}{x^4} - \dfrac{7x}{\sqrt[3]{2x}}$

38. $\dfrac{x}{x - 2} + \dfrac{1}{x^2} + \dfrac{8}{3(9x)^3}$

Rewriting a Fraction as a Sum of Terms In Exercises 39–44, rewrite the fraction as a sum of terms.

39. $\dfrac{x^2 + 6x + 12}{3x}$

40. $\dfrac{x^3 - 5x^2 + 4}{x^2}$

41. $\dfrac{4x^3 - 7x^2 + 1}{x^{1/3}}$

42. $\dfrac{2x^5 - 3x^3 + 5x - 1}{x^{3/2}}$

43. $\dfrac{3 - 5x^2 - x^4}{\sqrt{x}}$

44. $\dfrac{x^3 - 5x^4}{3x^2}$

Simplifying an Expression In Exercises 45–56, simplify the expression.

45. $\dfrac{-2(x^2 - 3)^{-3}(2x)(x + 1)^3 - 3(x + 1)^2(x^2 - 3)^{-2}}{[(x + 1)^3]^2}$

46. $\dfrac{x^5(-3)(x^2 + 1)^{-4}(2x) - (x^2 + 1)^{-3}(5)x^4}{(x^5)^2}$

47. $\dfrac{(6x + 1)^3(27x^2 + 2) - (9x^3 + 2x)(3)(6x + 1)^2(6)}{[(6x + 1)^3]^2}$

48. $\dfrac{(4x^2 + 9)^{1/2}(2) - (2x + 3)\left(\dfrac{1}{2}\right)(4x^2 + 9)^{-1/2}(8x)}{[(4x^2 + 9)^{1/2}]^2}$

49. $\dfrac{(x + 2)^{3/4}(x + 3)^{-2/3} - (x + 3)^{1/3}(x + 2)^{-1/4}}{[(x + 2)^{3/4}]^2}$

50. $(2x - 1)^{1/2} - (x + 2)(2x - 1)^{-1/2}$

51. $\dfrac{2(3x - 1)^{1/3} - (2x + 1)\left(\dfrac{1}{3}\right)(3x - 1)^{-2/3}(3)}{(3x - 1)^{2/3}}$

52. $\dfrac{(x + 1)\left(\dfrac{1}{2}\right)(2x - 3x^2)^{-1/2}(2 - 6x) - (2x - 3x^2)^{1/2}}{(x + 1)^2}$

53. $\dfrac{1}{(x^2 + 4)^{1/2}} \cdot \dfrac{1}{2}(x^2 + 4)^{-1/2}(2x)$

54. $\dfrac{1}{x^2 - 6}(2x) + \dfrac{1}{2x + 5}(2)$

55. $(x^2 + 5)^{1/2}\left(\dfrac{3}{2}\right)(3x - 2)^{1/2}(3)$
$\qquad + (3x - 2)^{3/2}\left(\dfrac{1}{2}\right)(x^2 + 5)^{-1/2}(2x)$

56. $(3x + 2)^{-1/2}(3)(x - 6)^{1/2}(1)$
$\qquad + (x - 6)^3\left(-\dfrac{1}{2}\right)(3x + 2)^{-3/2}(3)$

57. Verifying an Equation

(a) Verify that $y_1 = y_2$ analytically.
$$y_1 = \frac{1}{2}(x^2 + 4)^{1/3} + \frac{1}{6}(x^2 + 4)^{4/3}$$
$$y_2 = \frac{1}{6}(x^2 + 4)^{1/3}(x^2 + 7)$$

(b) Complete the table and demonstrate the equality in part (a) numerically.

x	-2	-1	$-\frac{1}{2}$	0	1	2	$\frac{5}{2}$
y_1							
y_2							

(c) Use a graphing utility to verify the equality in part (a) graphically.

58. Verifying an equation Repeat Exercise 57 using the following for y_1 and y_2.

$$y_1 = x^2\left(\frac{1}{3}\right)(x^2 + 1)^{-2/3}(2x) + (x^2 + 1)^{1/3}(2x)$$

$$y_2 = \frac{2x(4x^2 + 3)}{3(x^2 + 1)^{2/3}}$$

Exploring the Concepts

59. Writing Write a paragraph explaining why
$$\frac{1}{(x - 2)^{1/2} + x^4} \neq (x - 2)^{-1/2} + x^{-4}.$$

60. Think About It You are taking a course in calculus, and for one of the homework problems you obtain the following answer.

$$\frac{2}{3}x(2x - 3)^{3/2} - \frac{2}{15}(2x - 3)^{5/2}$$

The answer in the back of the book is
$$\frac{2}{5}(2x - 3)^{3/2}(x + 1).$$

Show how the second answer can be obtained from the first. Then use the same technique to simplify the expression
$$\frac{2}{3}x(4 + x)^{3/2} - \frac{2}{15}(4 + x)^{5/2}.$$

Review & Refresh ▶ Video solutions at LarsonPrecalculus.com

Error Analysis In Exercises 61–64, describe the error.

61.

x	1	2	8
y	0	1	3

From the table, you can conclude that y is an exponential function of x. ✗

62. $20° = (20 \text{ deg})\left(\dfrac{180 \text{ rad}}{\pi \text{ deg}}\right) = \dfrac{3600}{\pi} \text{ rad}$ ✗

63. A circle has a radius of 6 millimeters. The length of the arc intercepted by a central angle of 72° is
$s = r\theta$
$\quad = 6(72)$
$\quad = 432$ millimeters. ✗

64. Error Analysis Describe the error.
The system
$$\begin{cases} x - 2y + 3z = 12 \\ y + 3z = 5 \\ 2z = 4 \end{cases}$$ ✗
is in row-echelon form.

Answers to Odd-Numbered Exercises and Tests

Chapter P

Section P.1 *(page 12)*

1. irrational   3. terms   5. Yes; $|-7| = |7|$
6. (a) iii (b) iv (c) ii (d) v (e) i
7. (a) $5, 8, 2$ (b) $5, 0, 8, 2$ (c) $-9, 5, 0, 8, -4, 2, -11$
   (d) $-9, -\frac{7}{2}, 5, \frac{2}{3}, 0, 8, -4, 2, -11$ (e) $\sqrt{3}$
9. (a) 1 (b) 1 (c) $-13, 1, -6$
   (d) $2.01, 0.\overline{6}, -13, 1, -6$ (e) $0.010110111\ldots$
11.

$\begin{array}{c} \bullet\!\!-\!\!-\!\!-\!\!-\!\!\bullet \to x \\ {\scriptstyle -8\ -7\ -6\ -5\ -4} \end{array}$

$-4 > -8$

13.

$\begin{array}{c} {\scriptstyle \frac{2}{3}\ \frac{5}{6}} \\ \xrightarrow{\ \ \bullet\bullet\ \ } x \\ {\scriptstyle 0\qquad 1} \end{array}$

$\frac{5}{6} > \frac{2}{3}$

15.

$\begin{array}{c} {\scriptstyle -8.5\qquad -5.2} \\ \xleftarrow{\ \bullet\ \ \ \ \ \bullet\ } x \\ {\scriptstyle -9\ -8\ -7\ -6\ -5} \end{array}$

$-5.2 > -8.5$

17. $x \le 5$ denotes the set of all real numbers less than or equal to 5.
19. $-2 < x < 2$ denotes the set of all real numbers greater than -2 and less than 2.
21. $[4, \infty)$ denotes the set of all real numbers greater than or equal to 4; $x \ge 4$

$\xrightarrow[\ 1\ \ 2\ \ 3\ \ 4\ \ 5\ \ 6\ \ 7\ ]{\hspace{1.2cm}[\hspace{1.5cm}} x$

23. $[-5, 2)$ denotes the set of all real numbers greater than or equal to -5 and less than 2; $-5 \le x < 2$

$\xrightarrow[\ -5\ \ -3\ \ -1\ \ \ 1\ \ \ 3\ ]{[\hspace{2cm})} x$

25. $(-\infty, 0]; \ y \le 0$

$\xleftarrow[\ -2\ \ -1\ \ \ 0\ \ \ 1\ \ \ 2\ ]{\hspace{2cm}]} y$

27. $[10, 22]; \ 10 \le t \le 22$

$\xrightarrow[\ 10\ 12\ 14\ 16\ 18\ 20\ 22\ 24\ ]{[\hspace{2.5cm}]} t$

29. 10 31. 5 33. -1 35. 25 37. -1
39. $|-4| = |4|$ 41. $-|-6| < |-6|$ 43. 51 45. $\frac{5}{2}$
47. $\$2450.0$ billion; $\$1076.6$ billion
49. $\$3268.0$ billion; $\$584.6$ billion
51. $7x$ and 4 are the terms; 7 is the coefficient.
53. $4x^3, 0.5x,$ and -5 are the terms; 4 and 0.5 are the coefficients.
55. (a) 2 (b) 6 57. $\dfrac{5x}{12}$ 59. $\dfrac{x}{4}$
61. False. Zero is nonnegative but not positive.
63. The 5 was not distributed to the 3.
65. $5/n$ increases or decreases without bound; As n approaches 0, the quotients $5/n$ increase when n is positive and decrease when n is negative.
67. (a) 2 (b) 24 69. (a) 9 (b) 108 71. -9
73. -5 75. $2^4 \cdot 3$ 77. $2^3 \cdot 3^2 \cdot 11$
79. $37{,}850$ 81. 0.00609

Section P.2 *(page 24)*

1. exponent; base
3. When all possible factors are removed from the radical, all fractions have radical-free denominators, and the index of the radical is reduced.

5. 625 7. 729 9. $\frac{1}{25}$ 11. -9 13. $\frac{16}{3}$ 15. 17
17. -24 19. 0.06 21. $125z^3$ 23. $24y^2$ 25. $\dfrac{7}{x}$
27. $\dfrac{5184}{y^7}$ 29. $\dfrac{x^2}{y^2}$ 31. $\dfrac{1}{4x^4}$ 33. $\dfrac{125x^9}{y^{12}}$ 35. $\dfrac{1}{9}$
37. 1.02504×10^4 39. 0.000314 41. 6.8×10^5
43. 2.0×10^{11} 45. (a) 3 (b) $\frac{3}{2}$ 47. (a) 2 (b) $2x$
49. $2\sqrt{5}$ 51. $\dfrac{2\sqrt[3]{2}}{3}$ 53. $6x\sqrt{2x}$ 55. $\dfrac{18}{z^2}$
57. $\dfrac{5|x|\sqrt{3}}{y^2}$ 59. $29|x|\sqrt{5}$ 61. $11x\sqrt[3]{2}$ 63. $\dfrac{\sqrt{3}}{3}$
65. $\dfrac{\sqrt{14} + 2}{2}$ 67. $\dfrac{2}{3(\sqrt{5} - \sqrt{3})}$ 69. $64^{1/3}$ 71. $\dfrac{3}{\sqrt[3]{x^2}}$
73. $\frac{1}{8}$ 75. $\frac{2}{3}$ 77. $\sqrt{3}$ 79. $2\sqrt[4]{2}$ 81. $x - 1$
83. $(4x + 3)^{5/6}, x \ne -\dfrac{3}{4}$

85.

h	0	1	2	3	4	5	6
t	0	2.93	5.48	7.67	9.53	11.08	12.32

h	7	8	9	10	11	12
t	13.29	14.00	14.50	14.80	14.93	14.96

87. False. When $x = 0$, the expressions are not equal.
89. $(1)^{-3} = 1$, not -1
91. $1 = \dfrac{a^n}{a^n} = a^{n-n} = a^0$ 93. (a) 94 cm^2 (b) 60 cm^3
95. (a) 308.82 ft^2 (b) 365.366 ft^3
97. (a) -12 (b) -7 99. Terms: $2x, -3$; Coefficient: 2

Section P.3 *(page 31)*

1. $n; a_n; a_0$ 3. Yes. *Sample answer:* $x^3 + 1, x^3 + x + 5$
4. (a) iv (b) iii (c) v (d) ii (e) i
5. (a) $7x$ (b) Degree: 1; Leading coefficient: 7
   (c) Monomial
7. (a) $-\frac{1}{2}x^5 + 14x$ (b) Degree: 5; Leading coefficient: $-\frac{1}{2}$
   (c) Binomial
9. (a) $-4x^5 + 6x^4 + 1$
   (b) Degree: 5; Leading coefficient: -4 (c) Trinomial
11. Polynomial: $-3x^3 + 2x + 8$
13. Not a polynomial because it includes a term with a negative exponent
15. Polynomial: $-y^4 + y^3 + y^2$ 17. $-2x - 10$
19. $-3y^2 + 6$ 21. $-8.3x^3 + 0.3x^2 - 23$ 23. $12z + 8$
25. $3x^3 - 6x^2 + 3x$ 27. $-15z^2 + 5z$ 29. $-4.5t^3 - 15t$
31. $6x^2 - 7x - 5$ 33. $x^3 + 9x^2 + 19x + 35$
35. $x^4 + 2x^2 + x + 2$ 37. $x^2 - 100$ 39. $x^2 - 4y^2$
41. $4x^2 + 12x + 9$ 43. $16x^6 - 24x^3 + 9$
45. $x^3 + 9x^2 + 27x + 27$ 47. $8x^3 - 12x^2y + 6xy^2 - y^3$
49. $\frac{1}{25}x^2 - 9$ 51. $\frac{1}{16}x^2 - \frac{5}{2}x + 25$
53. $x^2 + 2xy + y^2 - 6x - 6y + 9$

CHAPTER P

55. $36x^2 - 9y^2$ **57.** $m^2 - n^2 - 6m + 9$ **59.** $u^4 - 16$
61. $-3x^3 - 3x^2 + 14$ **63.** $y^4 - 3y^3 - 19y^2 + 42y - 20$
65. $x - y, x \geq 0, y \geq 0$ **67.** $x^2 - 2x\sqrt{y} + y$
69. (a) 25%   (b) $0.25N^2 + 0.5Na + 0.25a^2$
(c) 25%; Answers will vary.
71. $30x^2$
73. (a) $V = 4x^3 - 88x^2 + 468x$
(b)

x (cm)	1	2	3
V (cm³)	384	616	720

75. (a) Approximations will vary.
(b) The difference of the safe loads decreases in magnitude as the span increases.
77. False. $(4x^2 + 1)(3x + 1) = 12x^3 + 4x^2 + 3x + 1$
79. False. $(4x + 3) + (-4x + 6) = 4x + 3 - 4x + 6$
$$= 3 + 6$$
$$= 9$$
81. $m + n$
83. The middle term was omitted when squaring the binomial.
$(x - 3)^2 = x^2 - 6x + 9 \neq x^2 + 9$
85. $5x^2 + 2x + 7$
87. 3   **89.** -15 **91.** x **93.** x^{10} **95.** $27x^6$
97. Multiplicative Inverse Property
99. Associative and Commutative Properties of Multiplication
101. Distributive Property

Section P.4   *(page 39)*

1. factoring   **3.** When each of its factors is prime
5. $2x(x^2 - 3)$ **7.** $(x - 5)(3x + 8)$ **9.** $(x + 9)(x - 9)$
11. $(5y + 2)(5y - 2)$ **13.** $(8 + 3z)(8 - 3z)$
15. $(x + 1)(x - 3)$ **17.** $(3u - 1)(3u + 1)(9u^2 + 1)$
19. $(x - 2)^2$ **21.** $(5z - 3)^2$ **23.** $(2y - 3)^2$
25. $(x - 2)(x^2 + 2x + 4)$ **27.** $(2t - 1)(4t^2 + 2t + 1)$
29. $(3x + 2)(9x^2 - 6x + 4)$ **31.** $(x + 2)(x - 1)$
33. $(s - 3)(s - 2)$ **35.** $(3x - 2)(x + 4)$
37. $(5x + 1)(x + 6)$ **39.** $-(5y - 2)(y + 2)$
41. $(x - 1)(x^2 + 2)$ **43.** $(2x - 1)(x^2 - 3)$
45. $(3 + x)(2 - x^3)$ **47.** $(3x^3 - 2)(x^2 + 2)$
49. $(x + 3)(2x + 3)$ **51.** $(2x + 3)(3x - 5)$
53. $6(x + 3)(x - 3)$ **55.** $x^2(x - 1)$ **57.** $(1 - 2x)^2$
59. $-2x(x + 1)(x - 2)$ **61.** $(x + 3)(x + 1)(x - 3)(x - 1)$
63. $(x - 2)(x + 2)(2x + 1)$ **65.** $(3x + 1)(5x + 1)$
67. $-(x - 2)(x + 1)(x - 8)$ **69.** $(x + 1)^2(2x + 1)(11x + 6)$
71. $\left(4x + \frac{1}{3}\right)\left(4x - \frac{1}{3}\right)$ **73.** $\left(z + \frac{1}{2}\right)^2$
75. $\left(y + \frac{2}{3}\right)\left(y^2 - \frac{2}{3}y + \frac{4}{9}\right)$
77.

79. $\pi h(R + r)(R - r); \ V = 2\pi\left[\left(\dfrac{R + r}{2}\right)(R - r)\right]h$
81. $-14, 14, -2, 2$ **83.** Two possible answers: $2, -12$
85. True. $a^2 - b^2 = (a + b)(a - b)$ **87.** No; $3(x - 2)(x + 1)$
89. 3 should be factored out of the second binomial to yield $9(x + 2)(x - 3)$.
91. $-\dfrac{x}{30}$ **93.** $\dfrac{3x}{20}$
95. (a) Division by 0 is undefined.   (b) 0   **97.** x^3
99. $x(x + 1)(x - 1)$

Section P.5   *(page 48)*

1. rational expression   **3.** equivalent
5. When its numerator and denominator have no factors in common aside from ± 1
7. All real numbers x
9. All real numbers x such that $x \neq 3$
11. All real numbers x such that $x \neq -\frac{2}{3}$
13. All real numbers x such that $x \neq -4, -2$
15. All real numbers x such that $x \geq 7$
17. All real numbers x such that $x > 3$
19. $\dfrac{3x}{2}, x \neq 0$ **21.** $-\dfrac{1}{2}, x \neq 5$ **23.** $y - 4, y \neq -4$
25. $\dfrac{3y}{4}, y \neq -\dfrac{2}{3}$ **27.** $\dfrac{x - 1}{x + 3}, x \neq -5$
29. $-\dfrac{x + 1}{x + 5}, x \neq 2$ **31.** $\dfrac{1}{x + 1}, x \neq \pm 4$
33. $\dfrac{1}{5(x - 2)}, x \neq 1$ **35.** $-\dfrac{(x + 2)^2}{6}, x \neq \pm 2$
37. $-\dfrac{8}{5}, y \neq -3, 4$ **39.** $\dfrac{x - y}{x(x + y)^2}, x \neq -2y$
41. $\dfrac{3}{x + 2}$ **43.** $\dfrac{3x^2 + 3x + 1}{(x + 1)(3x + 2)}$ **45.** $\dfrac{3 - 2x}{2(x + 2)}$
47. $\dfrac{-x^2 - 3}{(x + 1)(x - 2)(x - 3)}$ **49.** $\dfrac{2 - x}{x^2 + 1}, x \neq 0$
51. $\dfrac{1}{2}, x \neq 2$ **53.** $x(x + 1), x \neq -1, 0$
55. $\dfrac{2x - 1}{2x}, x > 0$ **57.** $\dfrac{x^2 + (x^2 + 3)^7}{(x^2 + 3)^4}$
59. $\dfrac{2x^3 - 2x^2 - 5}{(x - 1)^{1/2}}$ **61.** $\dfrac{3x - 1}{3}, x \neq 0$
63. $\dfrac{-1}{x(x + h)}, h \neq 0$ **65.** $\dfrac{-1}{(x - 4)(x + h - 4)}, h \neq 0$
67. $\dfrac{1}{\sqrt{x + 2} + \sqrt{x}}$ **69.** $\dfrac{1}{\sqrt{t + 3} + \sqrt{3}}, t \neq 0$
71. $\dfrac{1}{\sqrt{x + h + 1} + \sqrt{x + 1}}, h \neq 0$
73.

t	0	2	4	6	8	10	12
T	75	55.9	48.3	45	43.3	42.3	41.7

t	14	16	18	20	22
T	41.3	41.1	40.9	40.7	40.6

75. $\dfrac{x}{2(2x+1)}$, $x \neq 0$

77. (a) 4.57%   (b) $\dfrac{288(MN-P)}{N(MN+12P)}$, 4.57%

79. False. For the simplified expression to be equivalent to the original expression, the domain of the simplified expression needs to be restricted. If n is even, $x \neq -1, 1$. If n is odd, $x \neq 1$.

81.

x	0	1	2	3	4	5	6
$\dfrac{x-3}{x^2-x-6}$	$\dfrac{1}{2}$	$\dfrac{1}{3}$	$\dfrac{1}{4}$	Undef.	$\dfrac{1}{6}$	$\dfrac{1}{7}$	$\dfrac{1}{8}$
$\dfrac{1}{x+2}$	$\dfrac{1}{2}$	$\dfrac{1}{3}$	$\dfrac{1}{4}$	$\dfrac{1}{5}$	$\dfrac{1}{6}$	$\dfrac{1}{7}$	$\dfrac{1}{8}$

The expressions are not equivalent at $x = 3$.

83. When simplifying fractions, only common factors can be divided out, not terms.

85. (a) [number line: point at 3, marks −2 −1 0 1 2 3 4]   (b) [number line: point at $\frac{7}{2}$, marks −1 0 1 2 3 4 5]

(c) [number line: point at $-\frac{5}{2}$, marks −5 −4 −3 −2 −1 0 1]   (d) [number line: point at −5.2, marks −7 −6 −5 −4 −3 −2 −1]

87. $|x - 5| \le 3$ **89.** 605   **91.** $3\sqrt{5}$ **93.** Yes

Section P.6   *(page 57)*

1. Cartesian   **3.** c   **4.** f   **5.** a   **6.** d

7. e   **8.** b

9. $A: (2, 6)$, $B: (-6, -2)$, $C: (4, -4)$, $D: (-3, 2)$

11.

[graph with points plotted]

13. $(-3, 4)$ **15.** Quadrant IV   **17.** Quadrant II

19. Quadrant II or IV

21.

[scatter plot: Number of stores vs Year 2012–2018]

23. 13   **25.** $\sqrt{61}$ **27.** $\dfrac{\sqrt{277}}{6}$

29. (a) 5, 12, 13   (b) $5^2 + 12^2 = 13^2$

31. $\left(\sqrt{5}\right)^2 + \left(\sqrt{45}\right)^2 = \left(\sqrt{50}\right)^2$

33. Distances between the points: $\sqrt{29}$, $\sqrt{58}$, $\sqrt{29}$

35. (a)

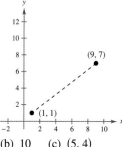

(b) 8   (c) (6, 1)

37. (a)

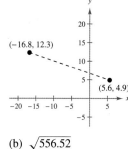

(b) 10   (c) (5, 4)

39. (a)

[graph with points $(-1, 2)$ and $(5, 4)$]

(b) $2\sqrt{10}$ (c) (2, 3)

41. (a)

[graph with points $(-16.8, 12.3)$ and $(5.6, 4.9)$]

(b) $\sqrt{556.52}$

(c) $(-5.6, 8.6)$

43. $2\sqrt{505} \approx 45$ yd   **45.** \$500.15 billion

47. $(0, 1), (4, 2), (1, 4)$ **49.** $(-3, 6), (2, 10), (2, 4), (-3, 4)$

51. True. Because $x < 0$ and $y > 0$, $2x < 0$ and $-3y < 0$, which is located in Quadrant III.

53. True. Two sides of the triangle have lengths of $\sqrt{149}$, and the third side has a length of $\sqrt{18}$.

55. Point on x-axis: $y = 0$; Point on y-axis: $x = 0$

57. $(2x_m - x_1, 2y_m - y_1)$

59. Use the Midpoint Formula to prove that the diagonals of the parallelogram bisect each other.

$$\left(\frac{b+a}{2}, \frac{c+0}{2}\right) = \left(\frac{a+b}{2}, \frac{c}{2}\right)$$

$$\left(\frac{a+b+0}{2}, \frac{c+0}{2}\right) = \left(\frac{a+b}{2}, \frac{c}{2}\right)$$

61. (a)

	First Set	Second Set
Distance A to B	3	$\sqrt{10}$
Distance B to C	5	$\sqrt{10}$
Distance A to C	4	$\sqrt{40}$
	Right triangle	Isosceles triangle

(b)

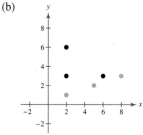

The first set of points is not collinear. The second set of points is collinear.

(c) A set of three points is collinear when the sum of two distances among the points is exactly equal to the third distance.

63. (a) -10 (b) -6 **65.** (a) -54 (b) 0
67. $x^2 - 2$ **69.** 0
71. (a) $P = 42x - 35{,}000$ (b) $\$175{,}000$

Review Exercises (page 62)

1. (a) 11 (b) 11 (c) $11, -14$
 (d) $11, -14, -\frac{8}{9}, \frac{5}{2}, 0.4$ (e) $\sqrt{6}$
3.
$$\frac{5}{4} > \frac{7}{8}$$

5. $x \geq 6$ denotes the set of all real numbers greater than or equal to 6.
7. $[-3, 4)$ denotes the set of all real numbers greater than or equal to -3 and less than 4; $-3 \leq x < 4$

9. 5 **11.** 15 **13.** -18 **15.** 122
17. (a) -7 (b) -19 **19.** (a) -1 (b) -3
21. Additive Identity Property
23. Associative Property of Addition
25. Commutative Property of Addition
27. 0 **29.** 32 **31.** $\dfrac{47x}{60}$
33. $\dfrac{x}{2}$ **35.** $192x^{11}$ **37.** $\dfrac{y^5}{2}$, $y \neq 0$ **39.** $-8z^3$
41. $\dfrac{1}{y^2}$ **43.** a^2b^2 **45.** $\dfrac{3u^5}{v^4}$ **47.** $\dfrac{1}{625a^4}$ **49.** $\dfrac{y}{xy + 1}$
51. 2.744×10^8 **53.** $484{,}000{,}000$ **55.** 9 **57.** $\frac{4}{5}$
59. 216 **61.** $\dfrac{x}{3}\sqrt[3]{2}$ **63.** $(2x + 1)\sqrt{3x}$
65. $(2x + 1)\sqrt{2x}$
67. Radicals cannot be combined by addition or subtraction unless the index and the radicand are the same.
69. $\dfrac{\sqrt{3}}{4}$ **71.** $2 + \sqrt{3}$ **73.** $\dfrac{3}{\sqrt{7} - 1}$ **75.** 64
77. $6x^2$
79. $-11x^2 + 3$; Degree: 2; Leading coefficient: -11
81. $-12x^2 - 4$; Degree: 2; Leading coefficient: -12
83. $-3x^2 - 7x + 1$ **85.** $2x^3 - 10x^2 + 12x$
87. $15x^2 - 27x - 6$ **89.** $36x^2 - 25$ **91.** $4x^2 - 12x + 9$
93. $x^4 - 6x^3 - 4x^2 - 37x - 10$
95. $2500r^2 + 5000r + 2500$ **97.** $x^2 + 28x + 192$
99. $x(x + 1)(x - 1)$ **101.** $(5x + 7)(5x - 7)$
103. $(x - 4)(x^2 + 4x + 16)$ **105.** $(x + 10)(2x + 1)$
107. $x(x - 4)(x + 3)$ **109.** $(x + 1)(x^2 - 2)$
111. All real numbers x such that $x \neq -1$
113. All real numbers x such that $x \geq -2$
115. $\dfrac{x - 8}{15}$, $x \neq -8$ **117.** $\dfrac{x - 4}{(x + 3)(x + 4)}$, $x \neq 3$
119. $\dfrac{2(x + 7)}{(x + 4)(x - 4)}$ **121.** $\dfrac{3ax^2}{(a^2 - x)(a - x)}$, $x \neq 0$
123. $\dfrac{-1}{2x(x + h)}$, $h \neq 0$

125.

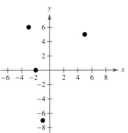

127. Quadrant IV
129. (a)

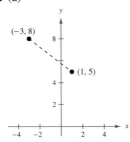

 (b) 5 (c) $\left(-1, \frac{13}{2}\right)$

131. (a)

 (b) $\sqrt{98.6}$ (c) $(2.8, 4.1)$

133. $(0, 0), (2, 0), (0, -5), (2, -5)$ **135.** 1823
137. False. There is also a cross-product term when a binomial sum is squared.
$(x + a)^2 = x^2 + 2ax + a^2$

Chapter Test (page 65)

1. $-\frac{10}{3} < -\frac{5}{3}$ **2.** 3 **3.** Additive Identity Property
4. (a) $-\frac{27}{125}$ (b) $\frac{8}{729}$ (c) $\frac{5}{49}$ (d) $\frac{1}{64}$
5. (a) 25 (b) $\dfrac{3\sqrt{6}}{2}$ (c) 1.8×10^5 (d) 9.6×10^5
6. (a) $12z^8$ (b) $(u - 2)^{-7}$ (c) $\dfrac{3x^2}{y^2}$
7. (a) $15z\sqrt{2z}$ (b) $4x^{14/15}$ (c) $\dfrac{2\sqrt[3]{2v}}{v^2}$
8. $-2x^5 - x^4 + 3x^3 + 3$; Degree: 5; Leading coefficient: -2
9. $2x^2 - 3x - 5$ **10.** $x^2 - 5$ **11.** $2x^3 - x^2 + 3$
12. $\dfrac{4}{y + 4}$, $y \neq 2$ **13.** $x^2(x - 1)$, $x \neq 0, 1$
14. $\dfrac{5(x + 2)^2}{(x - 4)(x + 4)}$
15. (a) $x^2(2x + 1)(x - 2)$ (b) $(x - 2)(x + 2)^2$
16. (a) $4\sqrt[3]{4}$ (b) $-4\left(1 + \sqrt{2}\right)$
17. All real numbers x such that $x \neq -2, 1$ **18.** $\$545$
19.

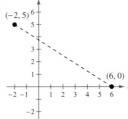

 Midpoint: $\left(2, \frac{5}{2}\right)$
 Distance: $\sqrt{89}$

20. $\frac{5}{2}x^2 + 3x + \frac{1}{2}$

Problem Solving   *(page 67)*

1. (a) Men's: 1,150,347 mm³; 696,910 mm³
   Women's: 696,910 mm³; 448,921 mm³
   (b) Men's: 1.04×10^{-5} kg/mm³; 6.31×10^{-6} kg/mm³
   Women's: 8.91×10^{-6} kg/mm³; 5.74×10^{-6} kg/mm³
   (c) No. Iron has a greater density than cork.
3. Answers will vary.
5. *Sample answers:* Man: 2,801,788,920 beats;
   Woman: 2,985,874,920 beats
7. $r \approx 0.28$
9. $SA = 10x^2 + 4x - 8$; 376 in.²
11. $y_1(0) = 0$, $y_2(0) = 2$
$$y_2 = \frac{x(2 - 3x^2)}{\sqrt{1 - x^2}}$$
13. (a) The second graph is misleading. The vertical axis does not have a break between 0 and 32.
    (b) It could show increases or decreases in a dramatic way.

Chapter 1

Section 1.1   *(page 78)*

1. solution or solution point   3. intercepts
5. origin   7. Algebraic, graphical
9. (a) Yes   (b) No
11. (a) No   (b) Yes
13.

x	-1	0	1	2	$\frac{5}{2}$
y	7	5	3	1	0
(x, y)	$(-1, 7)$	$(0, 5)$	$(1, 3)$	$(2, 1)$	$\left(\frac{5}{2}, 0\right)$

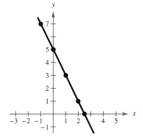

15.

x	-1	0	1	2	3
y	4	0	-2	-2	0
(x, y)	$(-1, 4)$	$(0, 0)$	$(1, -2)$	$(2, -2)$	$(3, 0)$

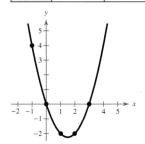

17. x-intercept: $(-2, 0)$
    y-intercept: $(0, 2)$
19. x-intercept: $(3, 0)$
    y-intercept: $(0, 9)$
21. x-intercept: $(1, 0)$
    y-intercept: $(0, 2)$
23. y-axis symmetry
25. Origin symmetry
27. Origin symmetry
29. x-axis symmetry
31.
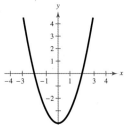
33.
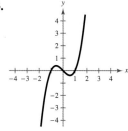

35. No symmetry
37. No symmetry

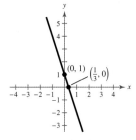
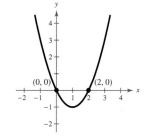

x-intercept: $\left(\frac{1}{3}, 0\right)$
y-intercept: $(0, 1)$

x-intercepts: $(0, 0)$, $(2, 0)$
y-intercept: $(0, 0)$

39. No symmetry
41. No symmetry

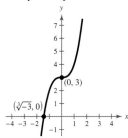
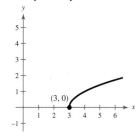

x-intercept: $\left(\sqrt[3]{-3}, 0\right)$
y-intercept: $(0, 3)$

x-intercept: $(3, 0)$
y-intercept: None

43. No symmetry
45. x-axis symmetry

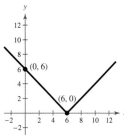
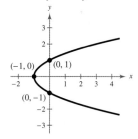

x-intercept: $(6, 0)$
y-intercept: $(0, 6)$

x-intercept: $(-1, 0)$
y-intercepts: $(0, \pm 1)$

47.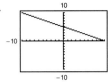

Intercepts: $(10, 0), (0, 5)$

49.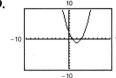

Intercepts:
$(3, 0), (1, 0), (0, 3)$

51.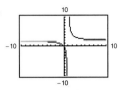

Intercept: $(0, 0)$

53.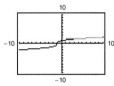

Intercepts: $(-1, 0), (0, 1)$

55.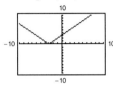

Intercepts: $(-3, 0), (0, 3)$

57. $x^2 + y^2 = 9$ **59.** $(x + 4)^2 + (y - 5)^2 = 4$
61. $(x - 3)^2 + (y - 8)^2 = 169$
63. $(x + 3)^2 + (y + 3)^2 = 61$
65. Center: $(0, 0)$; Radius: 5   **67.** Center: $(1, -3)$; Radius: 3

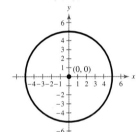 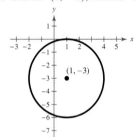

69. Center: $\left(\frac{1}{2}, \frac{1}{2}\right)$; Radius: $\frac{3}{2}$

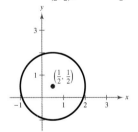

71.

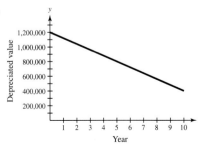

73. (a)    (b) Answers will vary.

(c)   (d) $x = 86\frac{2}{3}, y = 86\frac{2}{3}$

(e) A regulation NFL playing field is 120 yards long and $53\frac{1}{3}$ yards wide. The actual area is 6400 square yards.
75. False. $y = x$ is symmetric with respect to the origin.
77. When replacing y with $-y$ yields an equivalent equation, the graph is symmetric with respect to the x-axis.
79. (a) $a = 1, b = 0$ (b) $a = 0, b = 1$
81. $21x + 3$ **83.** $6x - 2$ **85.** 12   **87.** $x - 4$
89. $27\sqrt{2}$ **91.** 823,543   **93.** $2x^2 + 8x + 11$
95. $2x^2 - 5x - 63$

Section 1.2   *(page 87)*

1. equation   **3.** $ax + b = 0$ **5.** rational
7. Yes   **9.** Identity   **11.** Conditional equation
13. Contradiction   **15.** Identity   **17.** 2   **19.** -9
21. 12   **23.** 1   **25.** No solution   **27.** 9
29. $-\frac{96}{23}$ **31.** 4   **33.** 3   **35.** 0
37. No solution; The variable is divided out.
39. No solution; The solution is extraneous.   **41.** 5
43. x-intercept: $\left(\frac{12}{5}, 0\right)$ **45.** x-intercept: $\left(-\frac{1}{2}, 0\right)$
   y-intercept: $(0, 12)$ y-intercept: $(0, -3)$
47. x-intercept: $(5, 0)$ **49.** x-intercept: $(1.6, 0)$
   y-intercept: $\left(0, \frac{10}{3}\right)$ y-intercept: $(0, -0.3)$
51. x-intercept: $(-20, 0)$
   y-intercept: $\left(0, \frac{8}{3}\right)$
53. 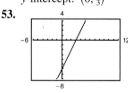   **55.**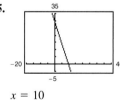

$x = 3$ $x = 10$

57.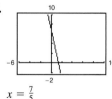

$x = \frac{7}{5}$
59. 19.993   **61.** 138.889   **63.** $h = 10$ ft   **65.** 63.7 in.
67. (a) About $(0, 407)$
   (b) $(0, 406.6)$; In 2010, the population of Raleigh was about 406,600.
   (c) 2015
69. 20,000 mi

71. False. $x(3 - x) = 10$
$3x - x^2 = 10$
The equation cannot be written in the form $ax + b = 0$.

73. False. The equation is a contradiction.

75. Yes. Both equations have the same solution of $x = 11$.

77. (a)

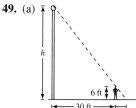

(b) $(2, 0)$

(c) The x-intercept is the solution of the equation $3x - 6 = 0$.

79. $3x^2 + 8x$ **81.** 220   **83.** 63   **85.** ± 5

87. 144   **89.** 38%   **91.** $2x\sqrt[3]{2x^2}$ **93.** $\sqrt{x}$

Section 1.3   (page 97)

1. mathematical modeling

3. A statement that two algebraic expressions are equal

5. A number increased by 2   **7.** A number divided by 6

9. A number decreased by 2 then divided by 3

11. The product of -2 and a number increased by 5

13. $(2n - 1)(2n + 1) = 4n^2 - 1$ **15.** $55t$

17. $0.20x$ **19.** $6x$ **21.** $2500 + 40x$ **23.** $d = 0.30L$

25. $N = \dfrac{p}{100} \cdot 672$ **27.** $4x + 8x = 12x$

29. 262, 263   **31.** 37, 185   **33.** $-5, -4$

35. First salesperson: \$516.89; Second salesperson: \$608.11

37. \$49,000

39. (a)

(b) $l = 1.5w$; $P = 5w$
(c) 7.5 m × 5 m

41. 97   **43.** 5 h   **45.** About 8.33 min   **47.** 945 ft

49. (a)

(b) 42 ft

51. \$4000 at $4\frac{1}{2}$%, \$8000 at 5%

53. Red maple: \$25,000; Dogwood: \$15,000

55. About 1.09 gal

57. $\dfrac{2A}{b}$ **59.** $\dfrac{S}{1 + R}$ **61.** $\dfrac{A - P}{Pt}$ **63.** 37°C

65. $\sqrt[3]{\dfrac{4.47}{\pi}} \approx 1.12$ in.   **67.** True

69. *Sample answer:* $\dfrac{5}{3n}, \left(\dfrac{5}{n}\right) \cdot 3$

The phrase "the quotient of 5 and a number" indicates that the variable is in the denominator.

71. The equation should be $t = d/r$.   **73.** $x^2 - 3$

75. $x^2 - 6x + 2$ **77.** $(2x + 1)^2$

79. $(u + 3v)(u^2 - 3uv + 9v^2)$ **81.** $(2x + 1)(x + 4)$

83. $\dfrac{-3 + 3\sqrt{5}}{2}$ **85.** 0   **87.** 9,460,000,000,000

89. -0.000375

Section 1.4   (page 110)

1. quadratic equation   **3.** discriminant

5. Factoring, extracting square roots, completing the square, using the Quadratic Formula

7. $0, -\frac{1}{2}$ **9.** $3, -\frac{1}{2}$ **11.** -5 **13.** $\pm\frac{3}{4}$

15. $-\frac{3}{2}, 11$ **17.** $-\frac{20}{3}, -4$ **19.** ± 7

21. $\pm\sqrt{19} \approx 4.36$ **23.** $\pm 3\sqrt{3} \approx \pm 5.20$ **25.** $-3, 11$

27. $-2 \pm \sqrt{14} \approx 1.74, -5.74$

29. $\dfrac{1 \pm 3\sqrt{2}}{2} \approx 2.62, -1.62$ **31.** 2   **33.** $4, -8$

35. $-2 \pm \sqrt{2}$ **37.** $1 \pm \dfrac{\sqrt{2}}{2}$

39. $1 \pm 2\sqrt{2}$ **41.** $\dfrac{-5 \pm \sqrt{89}}{4}$

43. (a)

45. (a)

(b) and (c) $-1, -5$ (b) and (c) 3, 1
(d) The answers are the same.   (d) The answers are the same.

47. (a)

49. (a)

(b) and (c) $-\frac{1}{2}, \frac{3}{2}$ (b) and (c) 1, -4
(d) The answers are the same.   (d) The answers are the same.

51. One repeated real solution   **53.** No real solution

55. Two real solutions   **57.** No real solution

59. Two real solutions

61. $\dfrac{1}{2}, -1$ **63.** $\dfrac{1}{4}, -\dfrac{3}{4}$ **65.** $-4 \pm 2\sqrt{5}$

67. $\dfrac{7}{4} \pm \dfrac{\sqrt{41}}{4}$ **69.** $1 \pm \sqrt{3}$ **71.** $-6 \pm 2\sqrt{5}$

73. $-\dfrac{3}{4} \pm \dfrac{\sqrt{41}}{4}$ **75.** $\dfrac{2}{7}$ **77.** $2 \pm \dfrac{\sqrt{6}}{2}$

79. $0.976, -0.643$ **81.** $-1.107, 1.853$

83. $-0.290, -2.200$ **85.** $1 \pm \sqrt{2}$ **87.** $-10, 6$

89. $\dfrac{1}{2} \pm \sqrt{3}$ **91.** $\dfrac{3}{4} \pm \dfrac{\sqrt{97}}{4}$ **93.** $\dfrac{1}{(x - 1)^2 + 4}$

95. $\dfrac{4}{(x + 5)^2 + 49}$ **97.** $\dfrac{1}{\sqrt{4 - (x - 1)^2}}$

99. $\dfrac{1}{\sqrt{16 - (x - 2)^2}}$

101. (a) $w(w + 14) = 1632$ (b) $w = 34$ ft, $l = 48$ ft

103. 6 in. × 6 in. × 3 in.

105. (a) About 16.51 ft × 15.51 ft   (b) 63,897.6 lb

107. (a) $s = -16t^2 + 984$ (b) 728 ft   (c) About 7.84 sec

CHAPTER 1

109. (a) $s = -16t^2 + 1815$

(b)

t	0	2	4	6	8	10	12
s	1815	1751	1559	1239	791	215	-489

(c) $10 < t < 12$; About 10.7 sec

(d) About 10.65 sec   (e) Answers will vary.

111. (a)

t	14	15	16	17
D	17.37	18.06	18.86	19.77

t	18	19	20
D	20.78	21.90	23.13

The public debt reached $20 trillion sometime in 2017.

(b) $t \approx 17.24$ (2017)

(c)

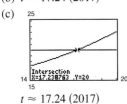

$t \approx 17.24$ (2017)

113. (a) Minimum: 89.2% at 2:00 P.M.

Maximum: 95.0% at 6:38 P.M.

(b) 4:00 P.M.

115. 500 mi/h   **117.** True. $b^2 - 4ac = 61 > 0$

119. Yes, the vertex of the parabola would be on the x-axis.

121. $a = 3, b = 1, c = -3$ **123.** $x^2 - 4x = 0$

125. $x^2 - 22x + 112 = 0$ **127.** $x^2 - 2x - 1 = 0$

129. $3x^2 - 8x + 27$ **131.** $11x^2 + 19x - 10$

133. $3x^2 + 13x - 30$ **135.** $4x^2 - 81$

137. $6x^4 + 5x^2y - 4y^2$ **139.** 1   **141.** $77 + 8\sqrt{5}$

143. $\dfrac{4\sqrt{3}}{5}$ **145.** $\dfrac{24 - 3\sqrt{11}}{53}$

Section 1.5 *(page 119)*

1. $\sqrt{-1}; -1$ **3.** (a) ii (b) iii (c) i

5. $-2 + 4i$ **7.** $a = 9, b = 8$ **9.** $7 + 4i$ **11.** 1

13. $3 - 3\sqrt{2}i$ **15.** $-12 + 9i$ **17.** $5 + i$

19. $108 + 12i$ **21.** 11   **23.** $-13 + 84i$ **25.** 85

27. 6   **29.** 20   **31.** 7   **33.** $\dfrac{8}{41} + \dfrac{10}{41}i$

35. $\dfrac{12}{13} + \dfrac{5}{13}i$ **37.** $-4 - 9i$ **39.** $-\dfrac{120}{1681} - \dfrac{27}{1681}i$

41. $-2\sqrt{3}$ **43.** -15 **45.** $7\sqrt{2}i$

47. $\left(21 + 5\sqrt{2}\right) + \left(7\sqrt{5} - 3\sqrt{10}\right)i$ **49.** $1 \pm i$

51. $-2 \pm \dfrac{1}{2}i$ **53.** $-2 \pm \dfrac{\sqrt{5}}{2}i$ **55.** $2 \pm \sqrt{2}i$

57. $\dfrac{5}{7} \pm \dfrac{5\sqrt{13}}{7}i$

59. $z_1 = 5 + 2i, z_2 = 3 - 4i, z = \dfrac{53}{17} - \dfrac{33}{34}i$

61. False. *Sample answer:* $(1 + i) + (3 + i) = 4 + 2i$

63. True.

$$x^4 - x^2 + 14 = 56$$
$$\left(-i\sqrt{6}\right)^4 - \left(-i\sqrt{6}\right)^2 + 14 \overset{?}{=} 56$$
$$36 + 6 + 14 \overset{?}{=} 56$$
$$56 = 56$$

65. $i, -1, -i, 1, i, -1, -i, 1$; The pattern repeats the first four results. Divide the exponent by 4.

When the remainder is 1, the result is i.

When the remainder is 2, the result is -1.

When the remainder is 3, the result is $-i$.

When the remainder is 0, the result is 1.

67. $\sqrt{-6}\sqrt{-6} = \sqrt{6}i\sqrt{6}i = 6i^2 = -6$

69. Proofs   **71.** $3x^2(x + 4)(x - 4)$

73. $(x - 3)(x^2 + 3)$ **75.** $3x(2x + 3)(x - 6)$

77. $(x + 1)(x - 1)(x^2 - 2)$

79. $(3x + 1)(3x - 1)(x + 2)(x - 2)$ **81.** (a) 4   (b) 2

83. (a) 0   (b) 0   **85.** (a) 25   (b) 25

87. (a) 0   (b) 0   **89.** (a) 0   (b) 4

Section 1.6 *(page 128)*

1. polynomial   **3.** Square both sides of the equation.

5. $0, \pm 3$ **7.** $-3, 0$ **9.** $\pm 3, \pm 3i$

11. $-8, 4 \pm 4\sqrt{3}i$ **13.** $-2, \pm\sqrt{3}i$ **15.** $\pm\sqrt{3}, \pm 1$

17. $\pm\dfrac{1}{2}, \pm 4$ **19.** $\dfrac{1}{4}$ **21.** $-\dfrac{64}{27}$ **23.** $-\dfrac{1}{5}, -\dfrac{1}{3}$

25. $-\dfrac{2}{3}, -4$ **27.** 20   **29.** $-\dfrac{55}{2}$ **31.** 1   **33.** 4

35. 3   **37.** $\dfrac{7}{4}$ **39.** 9   **41.** $\pm\sqrt{14}$ **43.** 1

45. $\dfrac{-3 \pm \sqrt{21}}{6}$ **47.** $-\dfrac{1}{3}, 5$ **49.** ± 1 **51.** $8, -3$

53. $2\sqrt{6}, -6$ **55.** $3, \dfrac{-1 - \sqrt{17}}{2}$

57. (a)

(b) and (c) $0, 3, -1$

(d) The x-intercepts and the solutions are the same.

59. (a)

(b) and (c) $5, 6$

(d) The x-intercepts and the solutions are the same.

61. (a)

(b) and (c) -1

(d) The x-intercept and the solution are the same.

63. (a)

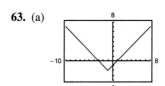

(b) and (c) 1, −3

(d) The *x*-intercepts and the solutions are the same.

65. 3, ±1.1   **67.** ±1.038   **69.** −1.143, 0.968

71. 4.217   **73.** 16.756   **75.** −2.280, −0.320

77. −1.603, 2.058   **79.** $x^2 - 3x - 28 = 0$

81. $21x^2 + 31x - 42 = 0$ **83.** $x^3 - 4x^2 - 3x + 12 = 0$

85. $x^2 + 1 = 0$ **87.** $x^4 - 1 = 0$ **89.** 34 students

91. 191.5 mi/h   **93.** About 1.5%   **95.** 26,250 passengers

97. (a) About 15 lb/in.^{2} (b) Answers will vary.

99. 500 units   **101.** 3 h

103. $U = \dfrac{kd^2}{2}$ **105.** False. See Example 7 on page 125.

107. The quadratic equation was not written in general form. As a result, the substitutions in the Quadratic Formula are incorrect.

109. $x = -4, -\dfrac{6}{5}$

111.
$$-12 \quad -11 \quad -10 \quad -9 \quad -8 \quad -7 \quad -6 \quad -5$$
$$-5 > -12$$

113.
$$0 \qquad \tfrac{3}{4} \, \tfrac{4}{5} \qquad 1$$
$$\tfrac{3}{4} < \tfrac{4}{5}$$

115.
$$-12 \, -11 \, -10 \, -9 \, -8 \, -7 \, -6 \, -5$$
$$-6.5 > -9.2$$

117. All real numbers less than or equal to 3

119. All real numbers greater than −5 and less than 5

121. All real numbers greater than 2; $x > 2$
$$0 \quad 1 \quad 2 \quad 3 \quad 4$$

123. All real numbers greater than or equal to −10 and less than 0; $-10 \le x < 0$
$$-10 \quad -8 \quad -6 \quad -4 \quad -2 \quad 0$$

125. 8   **127.** 2

Section 1.7 *(page 137)*

1. solution set   **3.** double   **5.** No

7. Bounded; $-2 \le x < 6$
$$-2 \quad 0 \quad 2 \quad 4 \quad 6$$

9. Bounded; $-1 \le x \le 5$
$$-2 \, -1 \, 0 \, 1 \, 2 \, 3 \, 4 \, 5 \, 6$$

11. Unbounded; $x > 11$
$$9 \quad 10 \quad 11 \quad 12 \quad 13$$

13. Unbounded; $x \le 7$
$$4 \quad 5 \quad 6 \quad 7 \quad 8 \quad 9 \quad 10$$

15. $x < 3$
$$1 \quad 2 \quad 3 \quad 4 \quad 5$$

17. $x < \tfrac{3}{2}$
$$-2 \quad -1 \quad 0 \quad 1 \quad 2 \quad 3$$

19. $x \ge 6$
$$4 \quad 5 \quad 6 \quad 7 \quad 8$$

21. $x > 2$
$$0 \quad 1 \quad 2 \quad 3 \quad 4$$

23. $x \ge 1$
$$-1 \quad 0 \quad 1 \quad 2 \quad 3 \quad 4$$

25. $x < 5$
$$3 \quad 4 \quad 5 \quad 6 \quad 7$$

27. $x \ge 4$
$$2 \quad 3 \quad 4 \quad 5 \quad 6$$

29. $x \ge 2$
$$0 \quad 1 \quad 2 \quad 3 \quad 4$$

31. $x \ge -4$
$$-6 \quad -5 \quad -4 \quad -3 \quad -2$$

33. $-1 < x < 3$
$$-1 \quad 0 \quad 1 \quad 2 \quad 3$$

35. $-3 < x \le 5$
$$-3 \, -2 \, -1 \, 0 \, 1 \, 2 \, 3 \, 4 \, 5$$

37. $-\tfrac{9}{2} < x < \tfrac{15}{2}$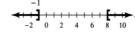
$$-6 \, -4 \, -2 \, 0 \, 2 \, 4 \, 6 \, 8$$

39. $-5 \le x < 1$
$$-6 \, -5 \, -4 \, -3 \, -2 \, -1 \, 0 \, 1 \, 2$$

41. $-5 < x < 5$
$$-6 \, -4 \, -2 \, 0 \, 2 \, 4 \, 6$$

43. $x < -2, x > 2$
$$-3 \, -2 \, -1 \, 0 \, 1 \, 2 \, 3$$

45. No solution

47. $x \le -1, x \ge 8$
$$-2 \, 0 \, 2 \, 4 \, 6 \, 8 \, 10$$

49. $x \le -5, x \ge 11$
$$-15 \, -10 \, -5 \, 0 \, 5 \, 10 \, 15$$

51.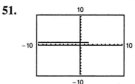
$x \le 2$

53.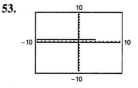
$x \le 4$

55.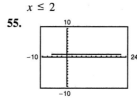
$-6 \le x \le 22$

57.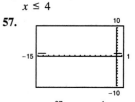
$x \le -\tfrac{27}{2}, x \ge -\tfrac{1}{2}$

59.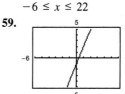
(a) $x \ge 1$ (b) $x \le \tfrac{1}{3}$

61.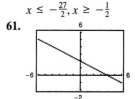
(a) $-2 \le x \le 4$ (b) $x \le 4$

63.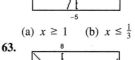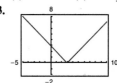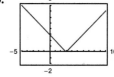

(a) $1 \le x \le 5$

(b) $x \le -1, x \ge 7$

65. $|x| \le 3$ **67.** $|x - 7| \ge 3$ **69.** $|x - 7| \ge 3$

71. $|x + 3| < 4$ **73.** $7.25 \le P \le 7.75$

75. $r < 0.08$ **77.** $100 \le r \le 170$

79. More than 6 units per hour

81. Greater than 10%   **83.** $x \ge 36$ **85.** $87 \le x \le 210$

CHAPTER 1

87. (a)
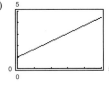

(b) $x \geq 2.9$

(c) $x \geq 2.908$

89. (a) $11.21 \leq t \leq 14.10$ (Between 2011 and 2014)

(b) $t > 16.98$ (2016)

91. $13.7 \leq t \leq 17.5$

93. $0.28

95. True by the Addition of a Constant Property of Inequalities.

97. False. If $-10 \leq x \leq 8$, then $10 \geq -x$ and $-x \geq -8$.

99. *Sample answer:* $x < x + 1$

101. Absolute value is always nonnegative, so the solution is all real numbers.

103. $-2, 3$ **105.** $-\frac{3}{4}, 2$ **107.** $\frac{3}{2}, \pm 4$ **109.** 8

111. No symmetry   **113.** *y*-axis symmetry

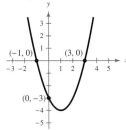
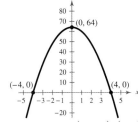

x-intercepts: $(-1, 0), (3, 0)$ *x*-intercepts: $(-4, 0), (4, 0)$

y-intercept: $(0, -3)$ *y*-intercept: $(0, 64)$

Section 1.8   *(page 147)*

1. positive; negative   **3.** $-2, 5$

5. (a) No   (b) Yes   (c) Yes   (d) No

7. (a) Yes   (b) No   (c) No   (d) Yes

9. $-3, 6$ **11.** $4, 5$

13. $(-2, 0)$ **15.** $(-3, 3)$

17. $[-7, 3]$ **19.** $(-\infty, -4] \cup [-2, \infty)$

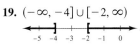

21. $(-3, 2)$ **23.** $(-3, 1)$

25. $\left(-\infty, -\frac{4}{3}\right) \cup (5, \infty)$

27. $(-1, 1) \cup (3, \infty)$ **29.** $(-\infty, -3) \cup (3, 7)$

31. $(-\infty, 0) \cup \left(0, \frac{3}{2}\right)$ **33.** $[-2, \infty)$

35. The solution set consists of the single real number $\frac{1}{2}$.

37. The solution set is empty.

39. $(-\infty, 0) \cup \left(\frac{1}{4}, \infty\right)$ **41.** $(-7, 1)$

43. $(-5, 3) \cup (11, \infty)$ **45.** $(-3, -2] \cup [0, 3)$

47. $(-\infty, -1) \cup (1, \infty)$

49.
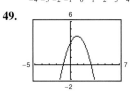
51.
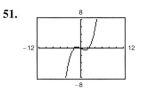

(a) $x \leq -1, x \geq 3$ (a) $-2 \leq x \leq 0,$

(b) $0 \leq x \leq 2$ $2 \leq x \leq \infty$

                         (b) $x \leq 4$

53.
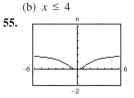
55.

(a) $0 \leq x < 2$ (a) $|x| \geq 2$

(b) $2 < x \leq 4$ (b) $-\infty < x < \infty$

57. $(-3.89, 3.89)$ **59.** $(-0.13, 25.13)$ **61.** $(2.26, 2.39)$

63. (a) $t = 10$ sec   (b) 4 sec $< t < 6$ sec

65. $40,000 \leq x \leq 50,000$; $50.00 \leq p \leq 55.00$

67. $[-2, 2]$ **69.** $(-\infty, 4] \cup [5, \infty)$

71. $(-5, 0] \cup (7, \infty)$

73. (a)
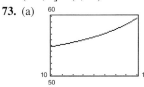

(b) $t < 14.9$ (2014)

(c) *Sample answer:* No. For $t > 22$, the model rapidly increases then decreases.

75. (a)

d	4	6	8	10	12
Load	2223.9	5593.9	10,312	16,378	23,792

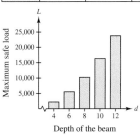

(b) 3.83 in.

77. 13.8 m $\leq L \leq 36.2$ m

79. False. There are four test intervals.

81.

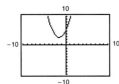

For part (b), the *y*-values that are less than or equal to 0 occur only at *x* = −1.

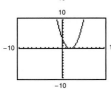

For part (c), there are no *y*-values that are less than 0.

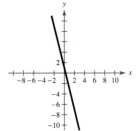

For part (d), the *y*-values that are greater than 0 occur for all values of *x* except 2.

83. $1/x$ is undefined when $x = 0$, so the solution set is $(0, \infty)$.

85. (a) $(-\infty, -6] \cup [6, \infty)$
(b) When $a > 0$ and $c > 0$, $b \le -2\sqrt{ac}$ or $b \ge 2\sqrt{ac}$.

87. (a) $\left(-\infty, -2\sqrt{30}\right] \cup \left[2\sqrt{30}, \infty\right)$
(b) When $a > 0$ and $c > 0$, $b \le -2\sqrt{ac}$ or $b \ge 2\sqrt{ac}$.

89. $\frac{1}{3}$ **91.** 0

93. *x*-intercept: $(-1, 0)$
y-intercept: $(0, 1)$

95. No symmetry   **97.** *y*-axis symmetry

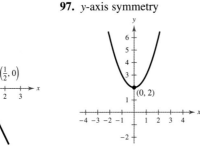

x-intercept: $\left(\frac{1}{2}, 0\right)$ No *x*-intercepts
y-intercept: $(0, 1)$ *y*-intercept: $(0, 2)$

Review Exercises   *(page 152)*

1.

x	−2	−1	0	1	2
y	9	5	1	−3	−7

3. *x*-intercepts: $(2, 0), (-4, 0)$
y-intercept: $(0, -2)$

5. No symmetry   **7.** *x*-axis symmetry

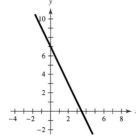   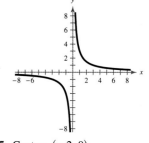

9. *y*-axis symmetry   **11.** Origin symmetry

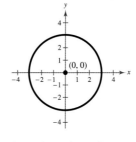   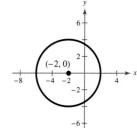

13. Center: $(0, 0)$;   **15.** Center: $(-2, 0)$;
Radius: 3   Radius: 4

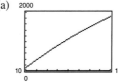

17. $(x - 2)^2 + (y + 3)^2 = 13$

19. (a)   (b) 2013

21. Identity   **23.** Conditional equation   **25.** 5   **27.** 6

29. -30 **31.** 13

33. *x*-intercept: $\left(\frac{1}{3}, 0\right)$ **35.** *x*-intercept: $(4, 0)$
y-intercept: $(0, -1)$ *y*-intercept: $(0, -8)$

37. *x*-intercept: $\left(\frac{4}{3}, 0\right)$
y-intercept: $\left(0, \frac{2}{3}\right)$

39. $h = 10$ in.   **41.** 2018: $5.30 billion; 2019: $6.80 billion

43. $90,000   **45.** $\frac{20}{7}$ L ≈ 2.857 L   **47.** $\frac{3V}{\pi r^2}$ **49.** $-\frac{5}{2}, 3$

51. $\pm\sqrt{2}$ **53.** $-8, -18$ **55.** $-6 \pm \sqrt{11}$

57. $-\frac{5}{4} \pm \frac{\sqrt{241}}{4}$

59. (a) $x = 0, 20$

(b)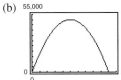

(c) $x = 10$

61. $-6\sqrt{3}$ **63.** $50 + 5\sqrt{2} + (10\sqrt{10} - 5\sqrt{5})i$

65. $-3 - 3i$ **67.** $15 + 6i$ **69.** 50 **71.** $\frac{4}{5} + \frac{8}{5}i$

73. $\frac{17}{26} + \frac{7}{26}i$ **75.** $\frac{21}{13} - \frac{1}{13}i$ **77.** $1 \pm 3i$

79. $-\frac{1}{2} \pm \frac{\sqrt{6}}{2}i$ **81.** $0, \dfrac{12}{5}$ **83.** $7, \pm 2i$

85. $-1, 2, \frac{1}{2} \pm \frac{\sqrt{3}}{2}i, -1 \pm \sqrt{3}i$ **87.** -1

89. $-124, 126$ **91.** $\pm\sqrt{10}$ **93.** $-5, 15$ **95.** $1, 3$

97. $143,203$ units

99. Bounded; $-7 < x \le 2$ **101.** Unbounded; $x \le -10$

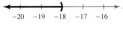   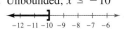

103. $x < -18$ **105.** $x \ge \frac{32}{15}$

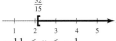

107. $10.5 \le x \le 13.5$ **109.** $-11 < x < -1$

111. $x > 37$ units

113. $(-3, 9)$ **115.** $(-\infty, -3) \cup (0, 3)$

117. $[-5, -1) \cup (1, \infty)$

119. 4.9%

121. False. $\sqrt{-18}\sqrt{-2} = (3\sqrt{2}i)(\sqrt{2}i) = 6i^2 = -6$
and $\sqrt{(-18)(-2)} = \sqrt{36} = 6$

123. Some solutions to certain types of equations may be extraneous solutions, which do not satisfy the original equations. So, checking is crucial.

Chapter Test *(page 155)*

1. No symmetry   **2.** y-axis symmetry

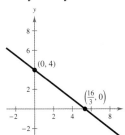   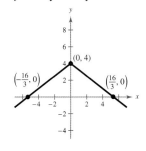

3. No symmetry   **4.** Origin symmetry

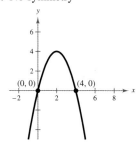   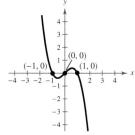

5. No symmetry   **6.** x-axis symmetry

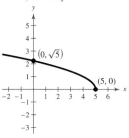   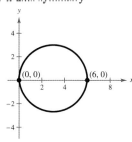

7. $\frac{128}{11}$ **8.** $-3, 5$

9. No solution. The variable is divided out.

10. $\pm\sqrt{2}, \pm\sqrt{3}i$ **11.** 4 **12.** $-2, \frac{8}{3}$

13. $-\frac{11}{2} \le x < 3$ **14.** $x < -6$ or $0 < x < 4$

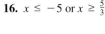   

15. $x < -4$ or $x > \frac{3}{2}$ **16.** $x \le -5$ or $x \ge \frac{5}{3}$

17. (a) -14 (b) $19 + 17i$ **18.** $\frac{8}{5} - \frac{16}{5}i$ **19.** $\frac{3}{2} \pm \frac{1}{2}i$

20. (a)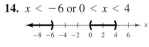

(b) and (c) About $433.9 million

21. 4.774 in.   **22.** $93\frac{3}{4}$ km/h   **23.** $a = 80, b = 20$

Problem Solving *(page 157)*

1.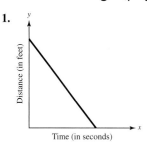

3. (a) Answers will vary.

(b)

a	4	7	10	13	16
A	64π	91π	100π	91π	64π

(c) $\dfrac{10\pi + 10\sqrt{\pi(\pi - 3)}}{\pi} \approx 12.12$ or

$\dfrac{10\pi - 10\sqrt{\pi(\pi - 3)}}{\pi} \approx 7.88$

(d)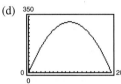

(e) 0, 20; They represent the minimum and maximum values of a.

(f) 100π; $a = 10$; $b = 10$

5. (a) 88.4 mi/h

(b) No, the maximum wind speed that the library can survive is 125 miles per hour.

(c) Answers will vary.

7. (a) $mn = 9$, $m + n = 0$

(b) Answers will vary. m and n are imaginary numbers.

9. (a) *Sample answer:* 5, 12, 13; 8, 15, 17

(b) Yes; yes; yes

(c) The product of the three numbers in a Pythagorean Triple is divisible by 60.

11. Proof

13. (a) $\frac{1}{2} - \frac{1}{2}i$ (b) $\frac{3}{10} + \frac{1}{10}i$ (c) $-\frac{1}{34} - \frac{2}{17}i$

15. (a) Yes   (b) No   (c) Yes

Chapter 2

Section 2.1   *(page 169)*

1. linear   **3.** point-slope   **5.** rate or rate of change

7. They are perpendicular.   **9.** (a) L_2 (b) L_3 (c) L_1

11.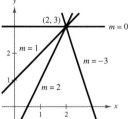

13. $\frac{3}{2}$

15. $m = 5$
y-intercept: $(0, 3)$

17. $m = -\frac{3}{4}$
y-intercept: $(0, -1)$

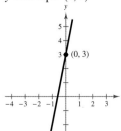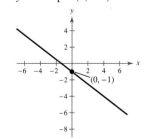

19. $m = 0$
y-intercept: $(0, 5)$

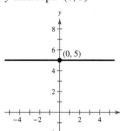

21. m is undefined.
y-intercept: none

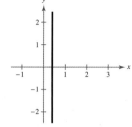

23. $m = \frac{7}{6}$
y-intercept: $(0, -5)$

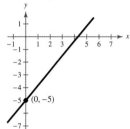

25. $m = -\frac{3}{2}$ **27.** $m = 2$ **29.** $m = 0$

31. m is undefined.   **33.** $m = 0.15$

35. $(-1, 7), (0, 7), (4, 7)$ **37.** $(-4, 6), (-3, 8), (-2, 10)$

39. $(-2, 7), \left(0, \frac{19}{3}\right), (1, 6)$ **41.** $(-4, -5), (-4, 0), (-4, 2)$

43. $y = 3x - 2$ **45.** $y = -2x$

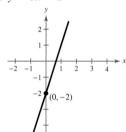

47. $y = -\frac{1}{3}x + \frac{4}{3}$ **49.** $y = -\frac{1}{2}x - 2$

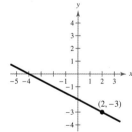

51. $y = \frac{5}{2}$ **53.** $y = 5x + 27.3$

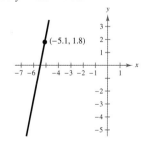

55. $y = -\frac{3}{5}x + 2$ **57.** $x = -7$

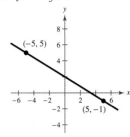

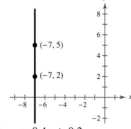

59. $y = -\frac{1}{2}x + \frac{3}{2}$ **61.** $y = 0.4x + 0.2$

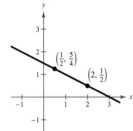

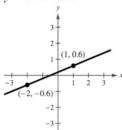

63. $y = -1$

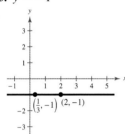

65. Parallel   **67.** Neither   **69.** Perpendicular
71. Parallel   **73.** (a) $y = 2x - 3$ (b) $y = -\frac{1}{2}x + 2$
75. (a) $y = -\frac{3}{4}x + \frac{3}{8}$ (b) $y = \frac{4}{3}x + \frac{127}{72}$
77. (a) $y = 4$ (b) $x = -2$
79. (a) $y = x + 4.3$ (b) $y = -x + 9.3$
81. $5x + 3y - 15 = 0$ **83.** $12x + 3y + 2 = 0$
85. $x + y - 3 = 0$
87. (a) Sales increasing 135 units/yr
       (b) No change in sales
       (c) Sales decreasing 40 units/yr
89. 12 ft   **91.** $F = 1.8C + 32$ or $C = \frac{5}{9}F - \frac{160}{9}$
93. $V = -166t + 830, 0 \le t \le 5$
95. (a) $C = 21t + 42,000$ (b) $R = 45t$
       (c) $P = 24t - 42,000$ (d) 1750 h
97. False. The slope with the greatest magnitude corresponds to the steepest line.
99. Find the slopes of the lines containing each two points and use the relationship $m_1 = -\frac{1}{m_2}$.
101. (a) The slope is $\frac{1}{2}$, not 2.
       (b) The y-intercept is -3, not 4.
103. The line $y = 4x$ rises most quickly, and the line $y = -4x$ falls most quickly. The greater the magnitude of the slope (the absolute value of the slope), the faster the line rises or falls.
105. 6; 0; 6; 30   **107.** $\frac{3}{2}; \frac{7}{10}; -2; -\frac{209}{10}$ **109.** $\pm 1, \frac{5}{2}$
111. $-\sqrt[3]{5}, 1$ **113.** 50   **115.** $-1, 2$

117. No real solution   **119.** $\dfrac{-2x + 1}{x}$ **121.** $x^2 + 3, x \ne 1$

Section 2.2   *(page 182)*

1. independent; dependent
3. A relation is a rule of correspondence between an input and an output. A function is a relation in which each input is matched with exactly one output.
5. Explicitly described   **7.** Function   **9.** Not a function
11. (a) Function
       (b) Function
       (c) Not a function, because not every element in A is matched with an element in B.
13. Not a function   **15.** Function   **17.** Function
19. Function   **21.** (a) 15   (b) 12   (c) $4t^2 + 13t + 15$
23. (a) 1   (b) 2.5   (c) $3 - 2|x|$
25. (a) $-\dfrac{1}{9}$ (b) Undefined   (c) $\dfrac{1}{y^2 + 6y}$
27. (a) 1   (b) -1 (c) $\dfrac{|x - 1|}{x - 1}$
29. (a) -1 (b) 2   (c) 6
31.

x	-2	-1	0	1	2
$f(x)$	1	4	5	4	1

33.

x	-2	-1	0	1	2
$f(x)$	5	$\frac{9}{2}$	4	1	0

35. 5   **37.** $\frac{4}{3}$ **39.** ± 9 **41.** $0, \pm 1$ **43.** $-1, 2$
45. $0, \pm 2$ **47.** All real numbers x
49. All real numbers y such that $y \ge -6$
51. All real numbers x except $x = 0, -2$
53. All real numbers s such that $s \ge 1$ except $s = 4$
55. (a) The maximum volume is 1024 cubic centimeters.
       (b)

       Yes, V is a function of x; $V = x(24 - 2x)^2$, $0 < x < 12$
57. No, the ball will be at a height of 18.5 feet.
59. $A = \dfrac{a^2}{2(a - 2)}$, $a > 2$
61. 2012: 79.42%
       2013: 81.18%
       2014: 82.94%
       2015: 84.70%
       2016: 85.90%
       2017: 86.80%
       2018: 87.70%

63. (a)

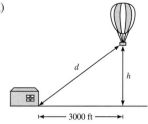

(b) $h = \sqrt{d^2 - 3000^2}, \quad d \ge 3000$

65. (a) $C = 12.30x + 98,000$ (b) $R = 17.98x$

(c) $P = 5.68x - 98,000$

67. (a) $R = \dfrac{240n - n^2}{20}, \quad n \ge 80$

(b)

n	90	100	110	120	130	140	150
$R(n)$	\$675	\$700	\$715	\$720	\$715	\$700	\$675

The revenue is maximum when 120 people take the trip.

69. $2 + h, \quad h \ne 0$ **71.** $3x^2 + 3xh + h^2 + 3, \quad h \ne 0$

73. $-\dfrac{x + 3}{9x^2}, \quad x \ne 3$ **75.** $\dfrac{\sqrt{5x - 5}}{x - 5}$

77. False. A function is a special type of relation.

79. False. The range is $[-1, \infty)$.

81. $g(x) = cx^2; c = -2$ **83.** $r(x) = \dfrac{c}{x}; c = 32$

85. The domain of $f(x)$ includes $x = 1$, and the domain of $g(x)$ does not because you cannot divide by 0. So, the functions do not have the same domain.

87. No; x is the independent variable, f is the name of the function.

89. $\pm 1, 3$ **91.** $-5, \frac{3}{2}$

93. x-intercepts: $(-2, 0), (1, 0)$ y-intercept: $(0, -2)$

95. $(0, 0), \left(\frac{3}{2}, 0\right), (6, 0)$ **97.** $(1.17, 0), (2.44, 0)$

99. 4   **101.** $-\frac{1}{12}$ **103.** 12

Section 2.3   *(page 194)*

1. zeros   **3.** average rate of change; secant   **5.** No

7. Domain: $(-2, 2]$; range: $[-1, 8]$

(a) -1 (b) 0   (c) -1 (d) 8

9. Domain: $(-\infty, \infty)$; range: $(-2, \infty)$

(a) 0   (b) 1   (c) 2   (d) 3

11. Function   **13.** Not a function   **15.** $-\frac{5}{2}, 6$

17. $0, \pm\sqrt{6}$ **19.** $\pm 3, 4$ **21.** $\frac{1}{2}$ **23.** ± 1 **25.** -3

27. (a)   0, 6

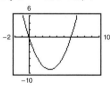

(b) 0, 6

29. (a) -5.5

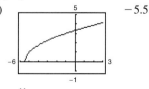

(b) $-\frac{11}{2}$

31. (a)   0.3333

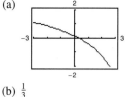

(b) $\frac{1}{3}$

33. Decreasing on $(-\infty, \infty)$

35. Increasing on $(1, \infty)$; Decreasing on $(-\infty, -1)$

37. Increasing on $(1, \infty)$; Decreasing on $(-\infty, -1)$
Constant on $(-1, 1)$

39. Increasing on $(-\infty, -1), (0, \infty)$; Decreasing on $(-1, 0)$

41.

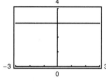

Constant on $(-\infty, \infty)$

43.

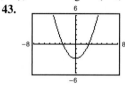

Decreasing on $(-\infty, 0)$
Increasing on $(0, \infty)$

45.

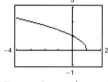

Decreasing on $(-\infty, 1)$

47.

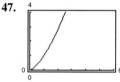

Increasing on $(0, \infty)$

49. Relative minimum: $(-1.5, -2.25)$

51. Relative maximum: $(0, 15)$
Relative minimum: $(4, -17)$

53. Relative minimum: $(0.33, -0.38)$

55.

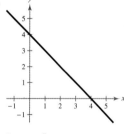

$(-\infty, 4]$

57.

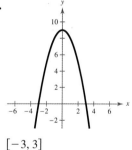

$[-3, 3]$

59.

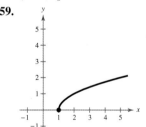

$[1, \infty)$

61. -2 **63.** -1

65. (a)

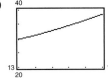

(b) About 1.73; The amount the U.S. federal government spent on applied research increased by about $1.73 billion each year from 2013 to 2018.

67. (a) $s = -16t^2 + 64t + 6$

(b)

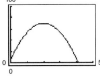

(c) 16 ft/sec

(d) The slope of the secant line is positive.

(e) $y = 16t + 6$

(f)

69. (a) $s = -16t^2 + 120t$

(b)

(c) -8 ft/sec

(d) The slope of the secant line is negative.

(e) $y = -8t + 240$

(f)

71. Even; y-axis symmetry   **73.** Neither; no symmetry

75. Neither; no symmetry

77.

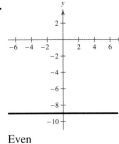

Even

79.

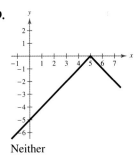

Neither

81.

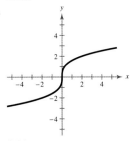

Odd

83. $h = 3 - 4x + x^2$ **85.** $L = 2 - \sqrt[3]{2y}$

87. (a)

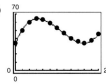

(b) The model fits the data very well.

(c) The temperature was increasing from 6 A.M. $(x = 0)$ to noon $(x = 6)$, and again from 2 A.M. $(x = 20)$ to 6 A.M. $(x = 24)$. The temperature was decreasing from noon to 2 A.M.

(d) The maximum temperature was 63.93°F and the minimum temperature was 33.98°F.

(e) Answers will vary.

89. False. The function $f(x) = \sqrt{x^2 + 1}$ has a domain of all real numbers.

91. The negative symbol should be divided out of each term, which yields $f(-x) = -(2x^3 + 5)$. So, the function is neither even nor odd.

93. (a) $\left(\frac{5}{3}, -7\right)$ (b) $\left(\frac{5}{3}, 7\right)$

95. (a)

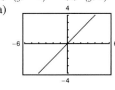

(b)

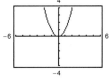

(c)

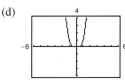

(d)

(e)

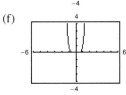

(f)

All the graphs pass through the origin. The graphs of the odd powers of x are symmetric with respect to the origin, and the graphs of the even powers are symmetric with respect to the y-axis. As the powers increase, the graphs become flatter in the interval $-1 < x < 1$.

97. $y = -x + 4$ **99.** $x = 5$

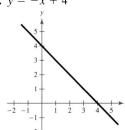

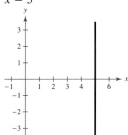

101. (a) -18 (b) 12   (c) $5x + 12$

103. (a) 3   (b) 5   (c) 2   **105.** $2n + 1$

Section 2.4 *(page 203)*

1. Greatest integer function   3. Reciprocal function
5. Square root function   7. Absolute value function
9. Linear function
11. (a) $f(x) = -2x + 6$ 13. (a) $f(x) = \frac{2}{3}x - 2$
    (b)   (b)

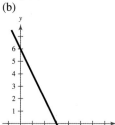   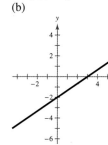

15.   17.

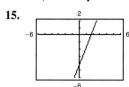   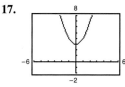

19.   21.

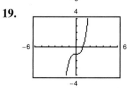   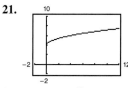

23.   25.

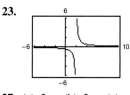   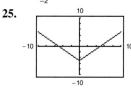

27. (a) 2   (b) 2   (c) -4 (d) 3
29. (a) 1   (b) -4 (c) 3   (d) 2
31.   33.

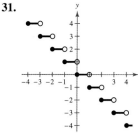   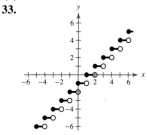

35.   37.

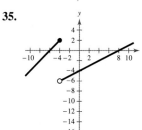   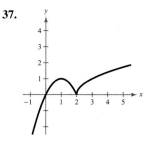

39.

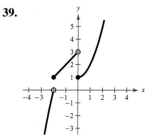

41. (a)

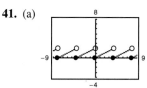

(b) Domain: $(-\infty, \infty)$
    Range: $[0, 2)$
43. (a) $W(30) = 420$; $W(40) = 560$;
        $W(45) = 665$; $W(50) = 770$

(b) $W(h) = \begin{cases} 14h, & 0 < h \le 36 \\ 21(h - 36) + 504, & h > 36 \end{cases}$

(c) $W(h) = \begin{cases} 16h, & 0 < h \le 40 \\ 24(h - 40) + 640, & h > 40 \end{cases}$

45. $f(t) = \begin{cases} t, & 0 \le t \le 2 \\ 2t - 2, & 2 < t \le 8 \\ \frac{1}{2}t + 10, & 8 < t \le 9 \end{cases}$

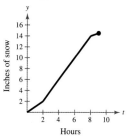

Total accumulation = 14.5 in.
47. False. A piecewise-defined function is a function that is defined by two or more equations over a specified domain. That domain may or may not include x- and y-intercepts.
49. The domains should be $x \le 3$ and $x > 3$.
51. (a) Yes. The amount that you pay in sales tax will increase as the price of the item purchased increases.
    (b) No. The length of time that you study the night before an exam does not necessarily determine your score on the exam.

53.   55.

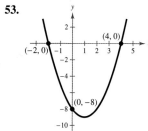   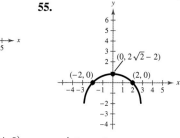

x-intercepts: $(-2, 0), (4, 0)$ x-intercepts:
y-intercept: $(0, -8)$ $(-2, 0), (2, 0)$
                                           y-intercept: $\left(0, 2\sqrt{2} - 2\right)$

CHAPTER 2

57. $-\dfrac{\sqrt[3]{x^2} - \sqrt[3]{x}}{x}$ **59.** $x^2 - 9x + 20$

Section 2.5 *(page 210)*

1. $-f(x); f(-x)$

3. Horizontal shifts, vertical shifts, reflections

5. (a) 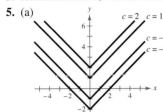   (b)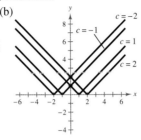

7. (a) 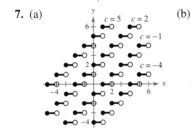   (b)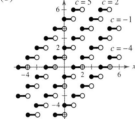

9. (a) 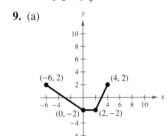   (b)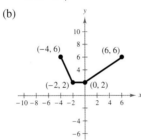

(c) 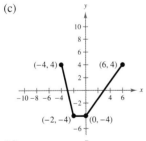   (d)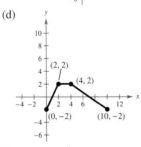

(e) 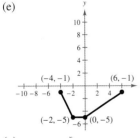   (f)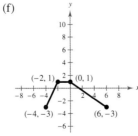

(g)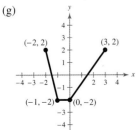

11. (a) $y = x^2 - 1$ (b) $y = -(x + 1)^2 + 1$

13. (a) $y = -|x + 3|$ (b) $y = |x - 2| - 4$

15. Right shift of $y = x^3$; $y = (x - 2)^3$

17. Reflection in the x-axis of $y = x^2$; $y = -x^2$

19. Reflection in the x-axis and upward shift of $y = \sqrt{x}$; $y = 1 - \sqrt{x}$

21. (a) $f(x) = x^2$
   (b) Upward shift of six units
   (c) 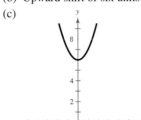   (d) $g(x) = f(x) + 6$

23. (a) $f(x) = x^3$
   (b) Reflection in the x-axis and a right shift of two units
   (c) 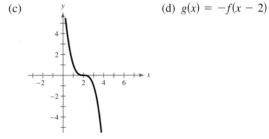   (d) $g(x) = -f(x - 2)$

25. (a) $f(x) = x^2$
   (b) Reflection in the x-axis, a left shift of one unit, and a downward shift of three units
   (c) 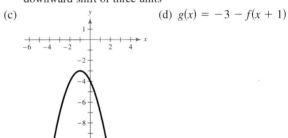   (d) $g(x) = -3 - f(x + 1)$

27. (a) $f(x) = x^2$
   (b) Reflection in the x-axis, a vertical stretch, and an upward shift of one unit
   (c) 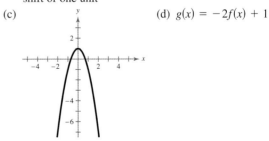   (d) $g(x) = -2f(x) + 1$

29. (a) $f(x) = |x|$
(b) Right shift of one unit and an upward shift of two units
(c) 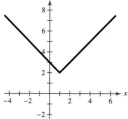   (d) $g(x) = f(x - 1) + 2$

31. (a) $f(x) = |x|$
(b) Horizontal shrink
(c) 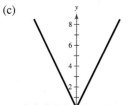   (d) $g(x) = f(2x)$

33. (a) $f(x) = |x|$
(b) Vertical stretch, a right shift of one unit, and an upward shift of two units
(c) 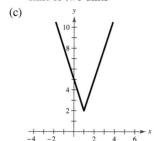   (d) $g(x) = 3f(x - 1) + 2$

35. (a) $f(x) = \sqrt{x}$ (b) Vertical stretch
(c) 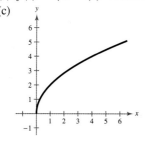   (d) $g(x) = 2f(x)$

37. (a) $f(x) = \sqrt{x}$ (b) Horizontal shift nine units to the right
(c) 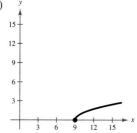   (d) $g(x) = f(x - 9)$

39. (a) $f(x) = \sqrt{x}$
(b) Reflection in the y-axis, a right shift of seven units, and a downward shift of two units
(c) 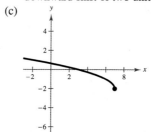   (d) $g(x) = f(7 - x) - 2$

41. (a) $f(x) = [\![x]\!]$
(b) Vertical stretch and a downward shift of one unit
(c) 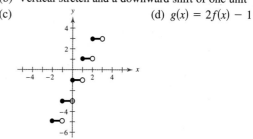   (d) $g(x) = 2f(x) - 1$

43. (a) $f(x) = [\![x]\!]$
(b) Reflection in the x-axis and an upward shift of three units
(c) 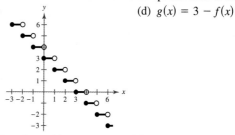   (d) $g(x) = 3 - f(x)$

45. $g(x) = (x - 3)^2 - 7$ **47.** $g(x) = (x - 13)^3$
49. $g(x) = -|x| - 12$ **51.** $g(x) = -\sqrt{-x + 6}$
53. (a) $y = -3x^2$ (b) $y = 4x^2 + 3$
55. (a) $y = -\frac{1}{2}|x|$ (b) $y = 3|x| - 3$
57. Vertical stretch of $y = x^3$; $y = 2x^3$
59. Reflection in the x-axis and vertical shrink of $y = x^2$; $y = -\frac{1}{2}x^2$
61. Reflection in the y-axis and vertical shrink of $y = \sqrt{x}$; $y = \frac{1}{2}\sqrt{-x}$
63. $y = -(x - 2)^3 + 2$ **65.** $y = -\sqrt{x} - 3$
67. (a)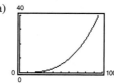

(b) $H\left(\dfrac{x}{1.6}\right) = 0.00001132x^3$; Horizontal stretch
69. False. The graph of $y = f(-x)$ is a reflection of the graph of $f(x)$ in the y-axis.
71. True. $|-x| = |x|$
73. $(-2, 0), (-1, 1), (0, 2)$
75. The equation should be $g(x) = (x - 1)^3$.

CHAPTER 2

77. No, $g(x) = -x^2 - 2$; Yes, $h(x) = -(x-3)^2$.
79. $x^2 + 4x$ **81.** $3x^2 + 2x - 2$ **83.** $x^3 - 3x^2$
85. $-0.2x^2 - 34x$ **87.** $\dfrac{1}{2x}$, $x \ne -\dfrac{5}{3}$

Section 2.6 *(page 219)*

1. addition; subtraction; multiplication; division
3. $g(x) = 2x$
5.

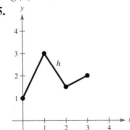

7. (a) $2x$ (b) 4 (c) $x^2 - 4$
  (d) $\dfrac{x+2}{x-2}$; all real numbers x except $x = 2$
9. (a) $x^2 + 4x - 5$ (b) $x^2 - 4x + 5$ (c) $4x^3 - 5x^2$
  (d) $\dfrac{x^2}{4x-5}$; all real numbers x except $x = \dfrac{5}{4}$
11. (a) $x^2 + 6 + \sqrt{1-x}$ (b) $x^2 + 6 - \sqrt{1-x}$
  (c) $(x^2+6)\sqrt{1-x}$
  (d) $\dfrac{(x^2+6)\sqrt{1-x}}{1-x}$; all real numbers x such that $x < 1$
13. (a) $\dfrac{x^4 + x^3 + x}{x+1}$ (b) $\dfrac{-x^4 - x^3 + x}{x+1}$ (c) $\dfrac{x^4}{x+1}$
  (d) $\dfrac{1}{x^2(x+1)}$; all real numbers x except $x = 0, -1$
15. 7 **17.** $-9t^2 + 3t + 5$ **19.** 306 **21.** $\dfrac{8}{23}$
23.   **25.**

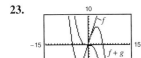   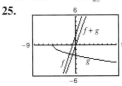

  $f(x), g(x)$ $f(x), f(x)$
27. (a) $x + 5$ (b) $x + 5$ (c) $x - 6$
29. (a) x (b) x (c) $x^9 + 3x^6 + 3x^3 + 2$
31. (a) $\sqrt{x^2 + 4}$ (b) $x + 4$
  Domains of f and $g \circ f$: all real numbers x such that $x \ge -4$
  Domains of g and $f \circ g$: all real numbers x
33. (a) $|x + 6|$ (b) $|x| + 6$
  Domains of $f, g, f \circ g$, and $g \circ f$: all real numbers x
35. (a) $\dfrac{1}{x+3}$ (b) $\dfrac{1}{x} + 3$
  Domains of f and $g \circ f$: all real numbers x except $x = 0$
  Domain of g: all real numbers x
  Domain of $f \circ g$: all real numbers x except $x = -3$

37. (a)   (b)

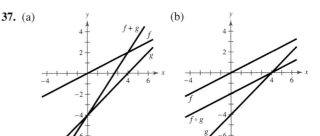

39. (a) 3 (b) 0 **41.** (a) 0 (b) 4
43. $f(x) = x^2$, $g(x) = 2x + 1$ **45.** $f(x) = \sqrt[3]{x}$, $g(x) = x^2 - 4$
47. $f(x) = \dfrac{1}{x}$, $g(x) = x + 2$ **49.** $f(x) = \dfrac{x+3}{4+x}$, $g(x) = -x^2$
51. $f(x) = \sqrt{x}$, $g(x) = \dfrac{1}{x^2 + 1}$
53. (a) $T = \dfrac{3}{4}x + \dfrac{1}{15}x^2$
  (b)

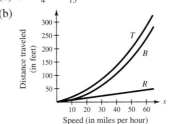

  (c) The braking function $B(x)$; As x increases, $B(x)$ increases at a faster rate than $R(x)$.
55. (a) $c(t) = \dfrac{b(t) - d(t)}{p(t)} \times 100$
  (b) $c(20)$ is the percent change in the population due to births and deaths in the year 2020.
57. (a) $r(x) = \dfrac{x}{2}$ (b) $A(r) = \pi r^2$
  (c) $(A \circ r)(x) = \pi\left(\dfrac{x}{2}\right)^2$;
  $(A \circ r)(x)$ represents the area of the circular base of the tank on the square foundation with side length x.
59. False. $(f \circ g)(x) = 6x + 1$ and $(g \circ f)(x) = 6x + 6$
61. (a) *Sample answer:* $f(x) = x + 1$, $g(x) = x + 3$
  (b) *Sample answer:* $f(x) = x^2$, $g(x) = x^3$
63. Proof
65. Proof
67. Symmetric with respect to x-axis
69. Symmetric with respect to y-axis
71.   **73.**

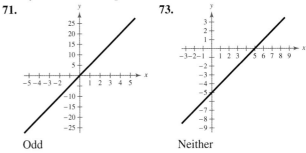

  Odd   Neither
75. $y = \dfrac{5 - 2x}{3}$ **77.** $y = x^2 - 1$, $x \ge 0$ **79.** $4\pi(r + 1)$

Section 2.7   *(page 228)*

1. inverse   **3.** range; domain   **5.** x **7.** One

9. $f^{-1}(x) = \frac{1}{6}x$ **11.** $f^{-1}(x) = \frac{x-1}{3}$

13. $f^{-1}(x) = \sqrt[3]{x-1}$ **15.** $f^{-1}(x) = \sqrt{x+4}$

17. $f(g(x)) = f(4x + 9) = \frac{(4x+9)-9}{4} = x$

$g(f(x)) = g\left(\frac{x-9}{4}\right) = 4\left(\frac{x-9}{4}\right) + 9 = x$

19. $f(g(x)) = f(\sqrt[3]{4x}) = \frac{(\sqrt[3]{4x})^3}{4} = x$

$g(f(x)) = g\left(\frac{x^3}{4}\right) = \sqrt[3]{4\left(\frac{x^3}{4}\right)} = x$

21.

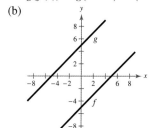

23. (a) $f(g(x)) = f(x+5) = (x+5) - 5 = x$
  $g(f(x)) = g(x-5) = (x-5) + 5 = x$
  (b)

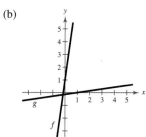

25. (a) $f(g(x)) = f\left(\frac{x-1}{7}\right) = 7\left(\frac{x-1}{7}\right) + 1 = x$

$g(f(x)) = g(7x+1) = \frac{(7x+1)-1}{7} = x$

  (b)

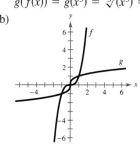

27. (a) $f(g(x)) = f(\sqrt[3]{x}) = (\sqrt[3]{x})^3 = x$
  $g(f(x)) = g(x^3) = \sqrt[3]{(x^3)} = x$
  (b)

29. (a) $f(g(x)) = f(x^2 - 5) = \sqrt{(x^2-5)+5} = x,\ x \geq 0$
  $g(f(x)) = g(\sqrt{x+5}) = (\sqrt{x+5})^2 - 5 = x$
  (b)

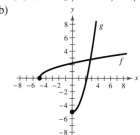

31. (a) $f(g(x)) = f\left(-\frac{5x+1}{x-1}\right) = \dfrac{-\left(\dfrac{5x+1}{x-1}\right) - 1}{-\left(\dfrac{5x+1}{x-1}\right) + 5}$

$= \dfrac{-5x - 1 - x + 1}{-5x - 1 + 5x - 5} = x$

$g(f(x)) = g\left(\dfrac{x-1}{x+5}\right) = \dfrac{-5\left(\dfrac{x-1}{x+5}\right) - 1}{\dfrac{x-1}{x+5} - 1}$

$= \dfrac{-5x + 5 - x - 5}{x - 1 - x - 5} = x$

  (b)

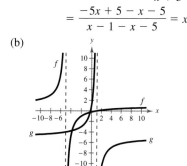

33. No

35.

x	3	5	7	9	11	13
$f^{-1}(x)$	-1	0	1	2	3	4

37. Yes   **39.** No

41.

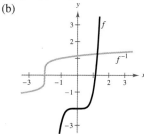

The function does not have an inverse function.

43.

The function does not have an inverse function.

45. (a) $f^{-1}(x) = \sqrt[5]{x+2}$
  (b)

  (c) The graph of f^{-1} is the reflection of the graph of f in the line $y = x$.
  (d) The domains and ranges of f and f^{-1} are all real numbers x.

CHAPTER 2

47. (a) $f^{-1}(x) = \sqrt{4 - x^2}$, $0 \le x \le 2$

(b)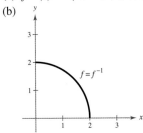

(c) The graph of f^{-1} is the same as the graph of f.

(d) The domains and ranges of f and f^{-1} are all real numbers x such that $0 \le x \le 2$.

49. (a) $f^{-1}(x) = \dfrac{4}{x}$

(b)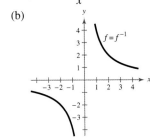

(c) The graph of f^{-1} is the same as the graph of f.

(d) The domains and ranges of f and f^{-1} are all real numbers x except $x = 0$.

51. (a) $f^{-1}(x) = \dfrac{2x + 1}{x - 1}$

(b)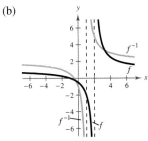

(c) The graph of f^{-1} is the reflection of the graph of f in the line $y = x$.

(d) The domain of f and the range of f^{-1} are all real numbers x except $x = 2$. The domain of f^{-1} and the range of f are all real numbers x except $x = 1$.

53. (a) $f^{-1}(x) = x^3 + 1$

(b)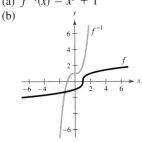

(c) The graph of f^{-1} is the reflection of the graph of f in the line $y = x$.

(d) The domains and ranges of f and f^{-1} are all real numbers x.

55. No inverse function

57. $g^{-1}(x) = 6x - 1$

59. No inverse function

61. $f^{-1}(x) = \dfrac{x^2 - 3}{2}$, $x \ge 0$

63. $f^{-1}(x) = \dfrac{5x - 4}{6 - 4x}$

65. $f^{-1}(x) = \sqrt{x} - 3$

67. No inverse function

69. (a) $y = \dfrac{x - 10}{0.75}$

x = hourly wage; y = number of units produced

(b) 19 units

71. $\dfrac{x + 1}{2}$ **73.** $\dfrac{x + 1}{2}$ **75.** 32   **77.** 472

79. $2\sqrt[3]{x + 3}$ **81.** False. $f(x) = x^2$ has no inverse function.

83. (a)

x	1	3	4	6
y	1	2	6	7

x	1	2	6	7
$f^{-1}(x)$	1	3	4	6

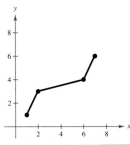

(b)

x	-4	-2	0	3
y	3	4	0	-1

The graph of f does not pass the Horizontal Line Test, so $f^{-1}(x)$ does not exist.

85. Proofs

87.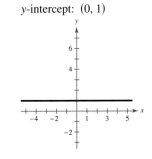

There is an inverse function $f^{-1}(x) = \sqrt{x - 1}$ because the domain of f is equal to the range of f^{-1} and the range of f is equal to the domain of f^{-1}.

89. $y = -(x - 6)(x - 4)$ **91.** $y = -(x - 9)(x - 4)$

93. Increasing on $(2, \infty)$
Decreasing on $(-\infty, 2)$

95. Increasing on $(-\infty, -1)$, $(1, \infty)$
Decreasing on $(-1, 1)$

97. Relative minimum: $(0, -3)$
Relative maximum: $(2, 0)$

Review Exercises *(page 233)*

1. $m = -\dfrac{1}{2}$
y-intercept: $(0, 1)$

3. $m = 0$
y-intercept: $(0, 1)$

5. $m = -1$

7. $y = \frac{1}{3}x - 7$

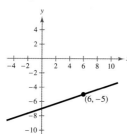

9. $y = \frac{1}{2}x + 7$

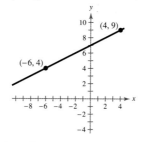

11. (a) $y = \frac{5}{4}x - \frac{23}{4}$ (b) $y = -\frac{4}{5}x + \frac{2}{5}$ **13.** $S = 0.80L$

15. Not a function **17.** Function

19. (a) 16 (b) $(t + 1)^{4/3}$ (c) 81 (d) $x^{4/3}$

21. All real numbers x such that $-5 \le x \le 5$

23. 16 ft/sec **25.** $4x + 2h + 3$, $h \ne 0$ **27.** Function

29. $-1, \frac{1}{5}$

31.

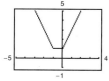

Increasing on $(0, \infty)$

Decreasing on $(-\infty, -1)$

Constant on $(-1, 0)$

33. Relative maximum: $(1, 2)$

35. 4 **37.** Neither even nor odd; no symmetry

39. Odd; origin symmetry

41. (a) $f(x) = -3x$

(b)

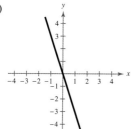

43.

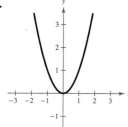

45.

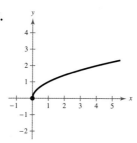

47.

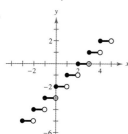

49.

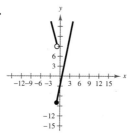

51. (a) $f(x) = x^2$ (b) Downward shift of nine units

(c)

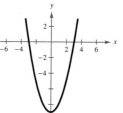

(d) $h(x) = f(x) - 9$

53. (a) $f(x) = |x|$

(b) Left shift of three units and a downward shift of five units

(c)

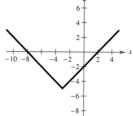

(d) $h(x) = f(x + 3) - 5$

55. (a) $f(x) = x^2$

(b) Reflection in the x-axis, a left shift of two units, and an upward shift of three units

(c)

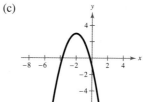

(d) $h(x) = -f(x + 2) + 3$

57. (a) $f(x) = [\![x]\!]$

(b) Reflection in the x-axis and an upward shift of six units

(c)

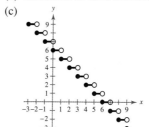

(d) $h(x) = -f(x) + 6$

59. (a) $f(x) = \sqrt{x}$

(b) Reflection in the x-axis, a left shift of one unit, and an upward shift of nine units

(c)

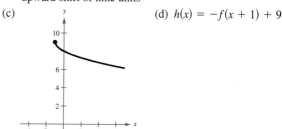

(d) $h(x) = -f(x + 1) + 9$

61. (a) $x^2 + 2x + 2$ (b) $x^2 - 2x + 4$

(c) $2x^3 - x^2 + 6x - 3$

(d) $\dfrac{x^2 + 3}{2x - 1}$; all real numbers x except $x = \dfrac{1}{2}$

CHAPTER 2

63. (a) $x - \frac{8}{3}$ (b) $x - 8$
Domains of $f, g, f \circ g,$ and $g \circ f$: all real numbers x

65. $(f \circ g)(x) = 0.95x - 100$; $(f \circ g)(x)$ represents the 5% discount before the $100 rebate.

67. $f^{-1}(x) = 5x + 4$

$$f(f^{-1}(x)) = \frac{5x + 4 - 4}{5} = x$$

$$f^{-1}(f(x)) = 5\left(\frac{x - 4}{5}\right) + 4 = x$$

69. 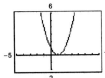   The function does not have an inverse function.

71. (a) $f^{-1}(x) = 2x + 6$ (b)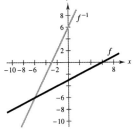
(c) The graphs are reflections of each other in the line $y = x$.
(d) Both f and f^{-1} have domains and ranges that are all real numbers x.

73. $x > 4$; $f^{-1}(x) = \sqrt{\dfrac{x}{2} + 4}, x \neq 0$

75. False. The graph is reflected in the x-axis, shifted 9 units to the left, and then shifted 13 units down.

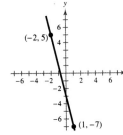

Chapter Test   *(page 235)*

1. $y = -4x - 3$ **2.** $y = \frac{5}{3}x - \frac{1}{3}$

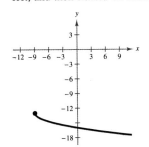 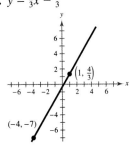

3. (a) $y = -\frac{5}{2}x + 4$ (b) $y = \frac{2}{5}x + 4$

4. (a) -9 (b) 1 (c) $|x - 4| - 15$

5. (a) $-\dfrac{1}{8}$ (b) $-\dfrac{1}{28}$ (c) $\dfrac{\sqrt{x}}{x^2 - 18x}$

6. (a) All real numbers x such that $x \neq 0, \frac{1}{2}$ (b) 5

7. (a) All real numbers x such that $x \leq 3$ (b) -97

8. (a)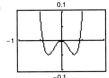
(b) Increasing on $(-0.31, 0), (0.31, \infty)$
Decreasing on $(-\infty, -0.31), (0, 0.31)$
(c) Even

9. (a)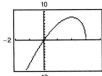
(b) Increasing on $(-\infty, 2)$
Decreasing on $(2, 3)$
(c) Neither

10. (a)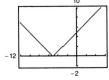
(b) Increasing on $(-5, \infty)$
Decreasing on $(-\infty, -5)$
(c) Neither

11. Relative maximum: $(0.82, 0.09)$ **12.** -3
Relative minimum: $(-0.82, -2.09)$

13.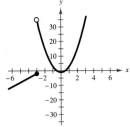

14. (a) $f(x) = [\![x]\!]$ (b) Vertical stretch
(c) 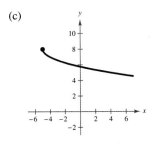   (d) $h(x) = 4f(x)$

15. (a) $f(x) = \sqrt{x}$
(b) Reflection in the x-axis, a left shift of five units, and an upward shift of eight units
(c)

16. (a) $f(x) = x^3$
(b) Vertical stretch, a reflection in the x-axis, a right shift of five units, and an upward shift of three units
(c)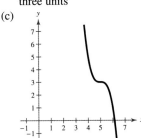

(d) $h(x) = -f(x + 5) + 8$ (d) $h(x) = -2f(x - 5) + 3$

17. (a) $2x^2 - 4x - 2$ (b) $4x^2 + 4x - 12$
(c) $-3x^4 - 12x^3 + 22x^2 + 28x - 35$ (d) $\dfrac{3x^2 - 7}{-x^2 - 4x + 5}$
(e) $3x^4 + 24x^3 + 18x^2 - 120x + 68$ (f) $-9x^4 + 30x^2 - 16$
The domain of f/g is all real numbers x except $x = -5$ and $x = 1$.

18. (a) $\dfrac{1 + 2x^{3/2}}{x}$ (b) $\dfrac{1 - 2x^{3/2}}{x}$ (c) $\dfrac{2\sqrt{x}}{x}$

(d) $\dfrac{1}{2x^{3/2}}$ (e) $\dfrac{\sqrt{x}}{2x}$ (f) $\dfrac{2\sqrt{x}}{x}$

The domain of f/g is all real numbers x such that $x > 0$.

19. $f^{-1}(x) = \sqrt[3]{x - 9}$ **20.** No inverse function

21. $f^{-1}(x) = \left(\tfrac{1}{3}x\right)^{2/3}$, $x \geq 0$ **22.** \$153

Cumulative Test for Chapters P–2   *(page 236)*

1. $\dfrac{4x^3}{15y^5}$, $x \neq 0$ **2.** $2|x|y\sqrt{5y}$ **3.** $5x - 6$

4. $x^3 - x^2 - 5x + 6$ **5.** $\dfrac{s - 19}{(s - 3)(s + 5)}$

6. $(x + 7)(5 - x)$ **7.** $x(x + 1)(1 - 6x)$

8. $2(3x + 2)(9x^2 - 6x + 4)$

9. $4x^2 + 5x + 1$ **10.** $\tfrac{5}{2}x^2 + 2x + \tfrac{1}{2}$

11.   **12.**

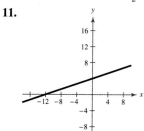   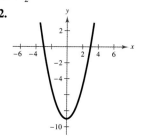

13.

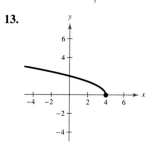

14. $-\dfrac{13}{3}$ **15.** $\dfrac{27}{5}$ **16.** $\dfrac{23}{6}$ **17.** $1, 3$ **18.** $-1, 3$

19. $-\dfrac{3}{2} \pm \dfrac{\sqrt{69}}{6}$ **20.** $\dfrac{-5 \pm \sqrt{97}}{6}$ **21.** ± 62 **22.** ± 8

23. $0, -12, \pm 2i$ **24.** $0, 3$ **25.** No solution

26. 6 **27.** $-3, 5$ **28.** No solution

29. $[-7, 5]$ **30.** $\left(-\infty, -\tfrac{4}{3}\right) \cup \left(-\tfrac{1}{3}, \infty\right)$

31. $\left(-\infty, -\tfrac{7}{5}\right] \cup [-1, \infty)$ **32.** $\left(-\tfrac{1}{2}, \tfrac{3}{4}\right)$

33. $y = 2x + 2$

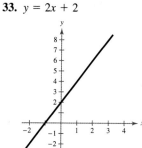

34. (a) $\dfrac{3}{2}$

(b) Division by 0 is undefined.

(c) $\dfrac{s + 2}{s}$

35. For some values of x there correspond two values of y.

36. Neither; no symmetry   **37.** Odd; origin symmetry

38. Even; y-axis symmetry

39. (a) Vertical shrink   (b) Upward shift of two units

(c) Left shift of two units

40. (a) $4x - 3$ (b) $-2x - 5$ (c) $3x^2 - 11x - 4$

(d) $\dfrac{x - 4}{3x + 1}$; Domain: all real numbers x except $x = -\dfrac{1}{3}$

41. (a) $\sqrt{x - 1} + x^2 + 1$ (b) $\sqrt{x - 1} - x^2 - 1$

(c) $x^2\sqrt{x - 1} + \sqrt{x - 1}$

(d) $\dfrac{\sqrt{x - 1}}{x^2 + 1}$; Domain: all real numbers x such that $x \geq 1$

42. (a) $2x + 12$ (b) $\sqrt{2x^2 + 6}$

Domain of f: all real numbers x

Domain of g: all real numbers x such that $x \geq -6$

Domain of $f \circ g$: all real numbers x such that $x \geq -6$

Domain of $g \circ f$: all real numbers x

43. (a) $|x| - 2$ (b) $|x - 2|$

Domains of $f, g, f \circ g,$ and $g \circ f$: all real numbers x

44. $h(x)^{-1} = \tfrac{1}{3}(x + 4)$ **45.** $n = 9$

46. (a) $R(n) = -0.05n^2 + 13n$, $n \geq 60$

(b)

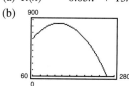

130 people

47. 4; Answers will vary.

Problem Solving   *(page 239)*

1. (a) $W_1 = 2000 + 0.07S$ (b) $W_2 = 2300 + 0.05S$

(c)

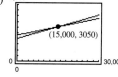

Both jobs pay the same monthly salary when sales equal \$15,000.

(d) No. Job 1 would pay \$3400 and job 2 would pay \$3300.

3. (a) The function will be even.   (b) The function will be odd.

(c) The function will be neither even nor odd.

5. $f(x) = a_{2n}x^{2n} + a_{2n-2}x^{2n-2} + \cdots + a_2x^2 + a_0$

$f(-x) = a_{2n}(-x)^{2n} + a_{2n-2}(-x)^{2n-2}$
$+ \cdots + a_2(-x)^2 + a_0$

$= f(x)$

CHAPTER 2

7. (a) $81\frac{2}{3}$ (b) $25\frac{5}{7}$ mi/h

(c) $y = \dfrac{-180}{7}x + 3400$ (d)

Domain: $0 \le x \le \dfrac{245}{3}$

Range:
$1300 \le y \le 3400$

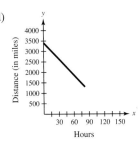

9. (a) $(f \circ g)(x) = 4x + 24$

(b) $(f \circ g)^{-1}(x) = \frac{1}{4}x - 6$

(c) $f^{-1}(x) = \frac{1}{4}x;\ g^{-1}(x) = x - 6$

(d) $(g^{-1} \circ f^{-1})(x) = \frac{1}{4}x - 6$; They are the same.

(e) $(f \circ g)(x) = 8x^3 + 1;\ (f \circ g)^{-1}(x) = \frac{1}{2}\sqrt[3]{x - 1}$;

$f^{-1}(x) = \sqrt[3]{x - 1};\ g^{-1}(x) = \frac{1}{2}x$;

$(g^{-1} \circ f^{-1})(x) = \frac{1}{2}\sqrt[3]{x - 1}$

(f) Answers will vary.

(g) $(f \circ g)^{-1}(x) = (g^{-1} \circ f^{-1})(x)$

11. (a)   (b)

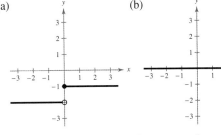

(c)   (d)

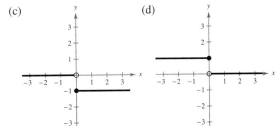

(e)   (f)

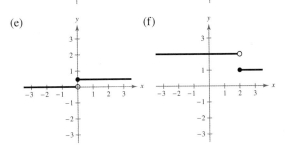

13. Proof

15. (a)

x	-4	-2	0	4
$f(f^{-1}(x))$	-4	-2	0	4

(b)

x	-3	-2	0	1
$(f + f^{-1})(x)$	5	1	-3	-5

(c)

x	-3	-2	0	1
$(f \cdot f^{-1})(x)$	4	0	2	6

(d)

x	-4	-3	0	4
$|f^{-1}(x)|$	2	1	1	3

Chapter 3

Section 3.1 *(page 248)*

1. nonnegative integer; real   **3.** Yes; $\left(\frac{1}{2}, 3\right)$ **5.** b

6. a   **7.** c   **8.** d

9. (a)   (b)

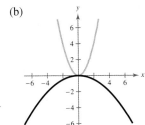

Vertical shrink   Vertical shrink and a
reflection in the x-axis

(c)   (d)

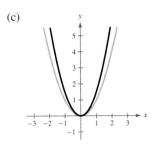
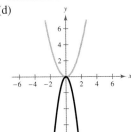

Vertical stretch   Vertical stretch and a
reflection in the x-axis

11. (a)   (b)

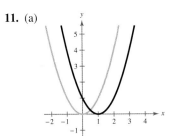
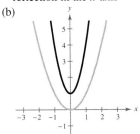

Right shift of one unit   Horizontal shrink and an
upward shift of one unit

(c)   (d)

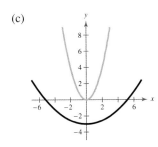
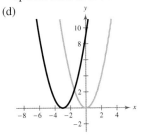

Horizontal stretch and a
downward shift of three
units   Left shift of three units

13. $f(x) = (x - 3)^2 - 9$

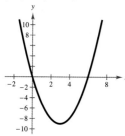

Vertex: $(3, -9)$
Axis of symmetry: $x = 3$
x-intercepts: $(0, 0), (6, 0)$

17. $f(x) = (x - 3)^2 - 7$

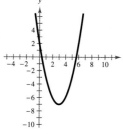

Vertex: $(3, -7)$
Axis of symmetry: $x = 3$
x-intercepts: $\left(3 \pm \sqrt{7}, 0\right)$

21. $f(x) = -\left(x - \frac{1}{2}\right)^2 - 1$

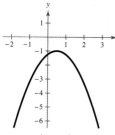

Vertex: $\left(\frac{1}{2}, -1\right)$
Axis of symmetry: $x = \frac{1}{2}$
No x-intercept

25. $h(x) = 4\left(x - \frac{1}{2}\right)^2 + 20$

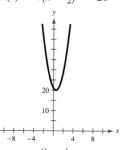

Vertex: $\left(\frac{1}{2}, 20\right)$
Axis of symmetry: $x = \frac{1}{2}$
No x-intercept

15. $h(x) = (x - 4)^2$

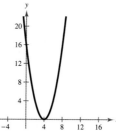

Vertex: $(4, 0)$
Axis of symmetry: $x = 4$
x-intercept: $(4, 0)$

19. $f(x) = (x + 6)^2 + 4$

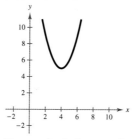

Vertex: $(-6, 4)$
Axis of symmetry: $x = -6$
No x-intercept

23. $f(x) = -(x - 1)^2 + 6$

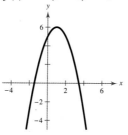

Vertex: $(1, 6)$
Axis of symmetry: $x = 1$
x-intercepts: $\left(1 \pm \sqrt{6}, 0\right)$

27.

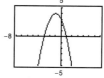

Vertex: $(-1, 4)$
Axis of symmetry: $x = -1$
x-intercepts: $(1, 0), (-3, 0)$

29.

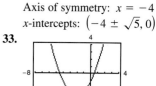

Vertex: $(-4, -5)$
Axis of symmetry: $x = -4$
x-intercepts: $\left(-4 \pm \sqrt{5}, 0\right)$

31.

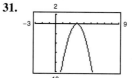

Vertex: $(3, 0)$
Axis of symmetry: $x = 3$
x-intercept: $(3, 0)$

33.

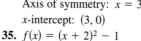

Vertex: $(-2, -3)$
Axis of symmetry: $x = -2$
x-intercepts: $\left(-2 \pm \sqrt{6}, 0\right)$

35. $f(x) = (x + 2)^2 - 1$ **37.** $f(x) = (x + 2)^2 + 5$

39. $f(x) = 4(x - 1)^2 - 2$ **41.** $f(x) = \frac{3}{4}(x - 5)^2 + 12$

43. $f(x) = -\frac{24}{49}\left(x + \frac{1}{4}\right)^2 + \frac{3}{2}$ **45.** $(-1, 0), (3, 0)$

47. $(-3, 0), \left(\frac{1}{2}, 0\right)$

49.

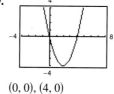

$(0, 0), (4, 0)$

51.

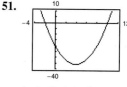

$(-2, 0), (10, 0)$

53. (a) $f(x) = x^2 - 9$
(b) $g(x) = -x^2 + 9$

55. (a) $f(x) = x^2 - 2x + 1$
(b) $g(x) = -x^2 + 1$

57. 55, 55   **59.** 12, 6   **61.** 16 ft   **63.** 20 fixtures

65. (a) $A = \dfrac{8x(50 - x)}{3}$

(b) $x = 25$ ft, $y = 33\frac{1}{3}$ ft

67. True. The equation has no real solutions, so the graph has no x-intercepts.

69. $b = \pm 10$

71. $f(x) = a\left(x + \dfrac{b}{2a}\right)^2 + \dfrac{4ac - b^2}{4a}$

73. (a) $3x^4$ (b) Degree: 4; Leading coefficient: 3
(c) Monomial

75. (a) $-4x^2 + x + 1$ (b) Degree: 2; Leading coefficient: -4
(c) Trinomial

77. (a) $-29x^3 + \frac{1}{4}x$ (b) Degree: 3; Leading coefficient: -29
(c) Binomial

79. $y^2 - 10y + 9$ **81.** $x^4 - 3x^3 - 9x^2 + 10x + 6$

83. x-intercept: $(-2, 0)$; y-intercept: $(0, 4)$

85. x-intercepts: $(2, 0), (6, 0)$; y-intercept: $(0, 2)$

Section 3.2 *(page 260)*

1. continuous   **3.** touches; crosses   **5.** No
7. $f(x_2) > 0$

CHAPTER 3

9. (a) 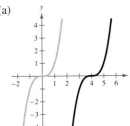   (b)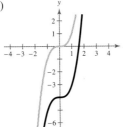

(c) 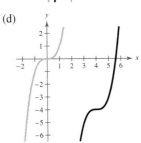   (d)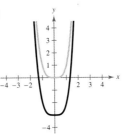

11. (a) 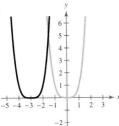   (b)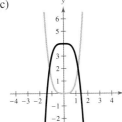

(c) 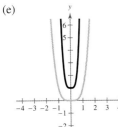   (d)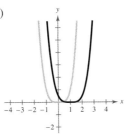

(e) 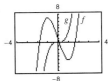   (f)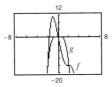

13. c   **14.** f   **15.** a   **16.** e   **17.** d   **18.** b

19. Falls to the left, rises to the right

21. Falls to the left and to the right

23. Rises to the left, falls to the right

25. Rises to the left and to the right

27. Rises to the left, falls to the right

29.    **31.**

33. (a) ± 6
(b) Odd multiplicity
(c) 1
(d)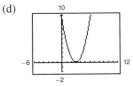

35. (a) 3
(b) Even multiplicity
(c) 1
(d)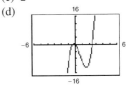

37. (a) $-2, 1$
(b) Odd multiplicity
(c) 1
(d)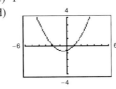

39. (a) $0, 1 \pm \sqrt{2}$
(b) Odd multiplicity
(c) 2
(d)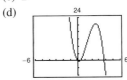

41. (a) $0, 2 \pm \sqrt{3}$
(b) Odd multiplicity
(c) 2
(d)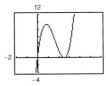

43. (a) $0, \pm\sqrt{3}$
(b) 0, odd multiplicity;
$\pm\sqrt{3}$, even multiplicity
(c) 4
(d)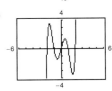

45. (a) No real zero
(b) No multiplicity
(c) 1
(d)

47. (a) $\pm 2, -3$
(b) Odd multiplicity
(c) 2
(d)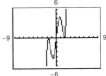

49. (a)

(b) and (c) $0, \frac{5}{2}$
(d) The answers are the same.

51. (a)

(b) and (c) $0, \pm 1, \pm 2$
(d) The answers are the same.

53. $f(x) = x^2 - 7x$ **55.** $f(x) = x^3 + 6x^2 + 8x$

57. $f(x) = x^4 - 4x^3 - 9x^2 + 36x$ **59.** $f(x) = x^2 - 2x - 1$

61. $f(x) = x^3 - 6x^2 + 7x + 2$ **63.** $f(x) = x^2 + 6x + 9$

65. $f(x) = x^3 + 4x^2 - 5x$

67. $f(x) = x^4 + x^3 - 15x^2 + 23x - 10$

69. (a) Rises to the left and to the right   (b) No zeros
(c) Answers will vary.   (d)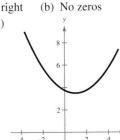

71. (a) Falls to the left, rises to the right
(b) $0, 5, -5$ (c) Answers will vary.
(d)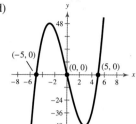

73. (a) Rises to the left and to the right
(b) $-2, 2$ (c) Answers will vary.
(d)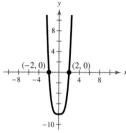

75. (a) Falls to the left, rises to the right
(b) $0, 2, 3$ (c) Answers will vary.
(d)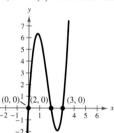

77. (a) Rises to the left, falls to the right
(b) $-5, 0$ (c) Answers will vary.
(d)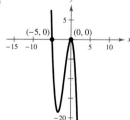

79. (a) Falls to the left, rises to the right
(b) $-2, 0$ (c) Answers will vary.
(d)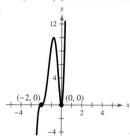

81. (a) Falls to the left and to the right
(b) ± 2 (c) Answers will vary.
(d)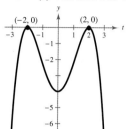

83.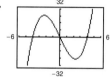

Zeros: $0, \pm 4$,
odd multiplicity

85.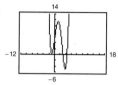

Zeros: -1,
even multiplicity;
$3, \frac{9}{2}$, odd multiplicity

87. (a) $[-1, 0], [1, 2], [2, 3]$ (b) $-0.879, 1.347, 2.532$
89. (a) $[-2, -1], [0, 1]$ (b) $-1.585, 0.779$
91. (a) $V(x) = x(36 - 2x)^2$ (b) Domain: $0 < x < 18$
(c)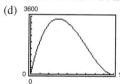

6 in. × 24 in. × 24 in.
(d)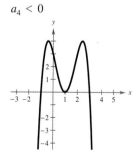

$x = 6$; The results are the same.
93. $x \approx 200$
95. False. The graph will always cross the x-axis.
97. True. A polynomial function falls to the right only when the leading coefficient is negative.
99. False. The graph falls to the left and to the right, or the graph rises to the left and to the right.
101. Answers will vary. *Sample answers:*

$a_4 < 0$ 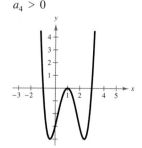 $a_4 > 0$

The x-intercepts of the graphs are the same. The first graph falls to the left and to the right. The second graph rises to the left and to the right. The second graph is a reflection of the first graph in the x-axis.

CHAPTER 3

103.

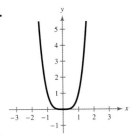

(a) Upward shift of two units; Even
(b) Left shift of two units; Neither
(c) Reflection in the y-axis; Even
(d) Reflection in the x-axis; Even
(e) Horizontal stretch; Even   (f) Vertical shrink; Even
(g) $g(x) = x^3, x \geq 0$; Neither   (h) $g(x) = x^{16}$; Even

105. 27 **107.** $56\frac{7}{10}$ **109.** $28\frac{108}{151}$ **111.** $3n^2 + 3n - 1$
113. $5x^2 + 7x + 4$ **115.** $-2 + 5i$ **117.** $21 - 52i$
119. $\frac{1}{10} - \frac{7}{10}i$ **121.** $-\frac{1}{2} - \frac{5}{2}i$

Section 3.3 *(page 270)*

1. $f(x)$: dividend; $d(x)$: divisor;
   $q(x)$: quotient; $r(x)$: remainder
3. proper   **5.** synthetic division   **7.** Answers will vary.
9. (a) and (b)

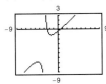

(c) Answers will vary.

11. $6 - \dfrac{1}{x+1}$ **13.** $2x + 4, \ x \neq -3$
15. $x^3 + 3x^2 - 1, \ x \neq -2$ **17.** $x - \dfrac{x+9}{x^2+1}$
19. $2x - 8 + \dfrac{x-1}{x^2+1}$ **21.** $2x^2 - 2x + 6, \ x \neq 4$
23. $6x^2 + 25x + 74 + \dfrac{248}{x-3}$ **25.** $4x^2 - 9, \ x \neq -2$
27. $-x^2 + 10x - 25, \ x \neq -10$ **29.** $x^2 + x + 4 + \dfrac{21}{x-4}$
31. $10x^3 + 10x^2 + 60x + 360 + \dfrac{1360}{x-6}$
33. $x^2 - 8x + 64, \ x \neq -8$ **35.** $4x^2 + 14x - 30, \ x \neq -\frac{1}{2}$
37. $f(x) = (x - 3)(x^2 + 2x - 4) - 5, \ f(3) = -5$
39. $f(x) = \left(x + \frac{2}{3}\right)(15x^3 - 6x + 4) + \frac{34}{3}, \ f\left(-\frac{2}{3}\right) = \frac{34}{3}$
41. (a) -2 (b) 1 (c) 36 (d) 5
43. (a) 15 (b) 240 (c) 695 (d) $56{,}720$
45. $(x + 3)(x + 2)(x + 1)$; Solutions: $-3, -2, -1$
47. $(2x - 1)(x - 5)(x - 2)$; Solutions: $\frac{1}{2}, 5, 2$
49. $\left(x + \sqrt{3}\right)\left(x - \sqrt{3}\right)(x + 2)$; Solutions: $-\sqrt{3}, \sqrt{3}, -2$

51. (a) Answers will vary.   (b) $2x - 1$
   (c) $f(x) = (2x - 1)(x + 2)(x - 1)$ (d) $\frac{1}{2}, -2, 1$
   (e)

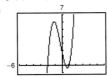

53. (a) Answers will vary.   (b) $x + 7$
   (c) $f(x) = (x + 7)(2x + 1)(3x - 2)$ (d) $-7, -\frac{1}{2}, \frac{2}{3}$
   (e)

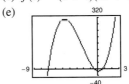

55. (a) $2, \pm 2.236$ (b) 2
   (c) $f(x) = (x - 2)\left(x - \sqrt{5}\right)\left(x + \sqrt{5}\right)$
57. (a) $0, 3, 4, \pm 1.414$ (b) 0
   (c) $h(x) = x(x - 4)(x - 3)\left(x + \sqrt{2}\right)\left(x - \sqrt{2}\right)$
59. $2x^2 - x - 1, \ x \neq \frac{3}{2}$
61. (a)

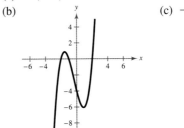

   (b) $\$250{,}366$
   (c) Answers will vary.

63. False. $-\frac{4}{7}$ is a zero of f.
65. True. The degree of the numerator is greater than the degree of the denominator.
67. $k = -1$, not 1.
69. $x^{2n} + 6x^n + 9, \ x^n \neq -3$
71. $c = -210$ **73.** $k = 7$
75. $-\frac{5}{2}$ **77.** 4 **79.** $-\frac{5}{3}, \frac{3}{2}$ **81.** $\pm 5, 1$
83. $\pm \sqrt{2}, 0, \frac{1}{2}$ **85.** $2 + 5i$ **87.** $-1 - 6i$ **89.** $\pm 2i$
91. $-1 \pm \sqrt{6}i$ **93.** $-1 \pm \dfrac{\sqrt{5}}{2}i$ **95.** $(x^n + y^n)(x^n - y^n)$
97. *Sample answer:* $x^2 - 3$

Section 3.4 *(page 283)*

1. Fundamental Theorem of Algebra   **3.** Rational Zero
5. five, three, or one   **7.** 3 **9.** 5 **11.** 2
13. $\pm 1, \pm 2$
15. $\pm 1, \pm 3, \pm 5, \pm 9, \pm 15, \pm 45, \pm \frac{1}{2}, \pm \frac{3}{2}, \pm \frac{5}{2}, \pm \frac{9}{2}, \pm \frac{15}{2}, \pm \frac{45}{2}$
17. $-2, -1, 3$ **19.** No rational zeros   **21.** $-6, -1$
23. $-1, \frac{1}{2}$ **25.** $-2, 3, \pm \frac{2}{3}$ **27.** $1, \dfrac{3}{5} \pm \dfrac{\sqrt{19}}{5}$
29. $-3, 1, -2 \pm \sqrt{6}$
31. (a) $\pm 1, \pm 2, \pm 4$
   (b)   (c) $-2, -1, 2$

33. (a) $\pm 1, \pm 3, \pm\frac{1}{2}, \pm\frac{3}{2}, \pm\frac{1}{4}, \pm\frac{3}{4}$

(b)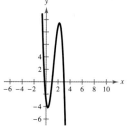

(c) $-\frac{1}{4}, 1, 3$

35. (a) $\pm 1, \pm 2, \pm 4, \pm 8, \pm\frac{1}{2}$

(b)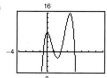

(c) $-\frac{1}{2}, 1, 2, 4$

37. (a) $\pm 1, \pm 3, \pm\frac{1}{2}, \pm\frac{3}{2}, \pm\frac{1}{4}, \pm\frac{3}{4}, \pm\frac{1}{8}, \pm\frac{3}{8}, \pm\frac{1}{16}, \pm\frac{3}{16}, \pm\frac{1}{32}, \pm\frac{3}{32}$

(b)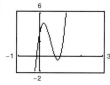

(c) $1, \frac{3}{4}, -\frac{1}{8}$

39. $f(x) = x^3 - x^2 + 25x - 25$

41. $f(x) = x^4 - 6x^3 + 14x^2 - 16x + 8$

43. $f(x) = 3x^4 - 17x^3 + 25x^2 + 23x - 22$

45. $f(x) = 2x^4 + 2x^3 - 2x^2 + 2x - 4$

47. $f(x) = x^3 + x^2 - 2x + 12$

49. (a) $(x^2 + 4)(x^2 - 2)$

(b) $(x^2 + 4)\left(x + \sqrt{2}\right)\left(x - \sqrt{2}\right)$

(c) $(x + 2i)(x - 2i)\left(x + \sqrt{2}\right)\left(x - \sqrt{2}\right)$

51. (a) $(x^2 - 6)(x^2 - 2x + 3)$

(b) $\left(x + \sqrt{6}\right)\left(x - \sqrt{6}\right)(x^2 - 2x + 3)$

(c) $\left(x + \sqrt{6}\right)\left(x - \sqrt{6}\right)\left(x - 1 - \sqrt{2}i\right)\left(x - 1 + \sqrt{2}i\right)$

53. $\pm 2i, 1$ **55.** $2, 3 \pm 2i$ **57.** $1, 3, 1 \pm \sqrt{2}i$

59. $(x + 6i)(x - 6i); \pm 6i$

61. $(x - 1 - 4i)(x - 1 + 4i); 1 \pm 4i$

63. $(x - 2)(x + 2)(x - 2i)(x + 2i); \pm 2, \pm 2i$

65. $(z - 1 + i)(z - 1 - i); 1 \pm i$

67. $(x + 1)(x - 2 + i)(x - 2 - i); -1, 2 \pm i$

69. $(x - 2)^2(x + 2i)(x - 2i); 2, \pm 2i$

71. $-10, -7 \pm 5i$ **73.** $-\frac{3}{4}, 1 \pm \frac{1}{2}i$ **75.** $-2, -\frac{1}{2}, \pm i$

77. One positive real zero, no negative real zeros

79. No positive real zeros, one negative real zero

81. Two or no positive real zeros, two or no negative real zeros

83. Two or no positive real zeros, one negative real zero

85. Answers will vary.   **87.** Answers will vary.

89. $\frac{3}{4}, \pm\frac{1}{2}$ **91.** $-\frac{3}{4}$ **93.** $\pm 2, \pm\frac{3}{2}$ **95.** $\pm 1, \frac{1}{4}$

97. (a)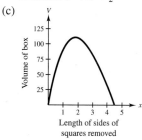

(b) $V(x) = x(9 - 2x)(15 - 2x)$

Domain: $0 < x < \frac{9}{2}$

(c)

$1.82 \text{ cm} \times 5.36 \text{ cm} \times 11.36 \text{ cm}$

(d) $\frac{1}{2}, \frac{7}{2}, 8$; 8 is not in the domain of V.

99. (a) $V(x) = x^3 + 9x^2 + 26x + 24 = 120$

(b) $4 \text{ ft} \times 5 \text{ ft} \times 6 \text{ ft}$

101. False. The most complex zeros it can have is two, and the Linear Factorization Theorem guarantees that there are three linear factors, so one zero must be real.

103. r_1, r_2, r_3 **105.** $5 + r_1, 5 + r_2, 5 + r_3$

107. The zeros cannot be determined.

109. Answers will vary. There are infinitely many possible functions for f. *Sample equation and graph:*

$f(x) = -2x^3 + 3x^2 + 11x - 6$

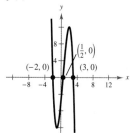

111. $f(x) = x^3 - 3x^2 + 4x - 2$

113. The function should be

$f(x) = (x + 2)(x - 3.5)(x + i)(x - i)$.

115. $-7 < x < 7$ **117.** $-\frac{5}{2} \le x \le 3$

119. $-1 < x < 3$

121. 4   **123.** $-\frac{1}{10}$ **125.** $\frac{8}{7}$ **127.** $-\frac{3}{2}$ **129.** $\frac{3}{2}$

131. $-\frac{4}{9}$ **133.** $\frac{3}{8}$

Section 3.5   *(page 294)*

1. sum; squared differences   **3.** correlation coefficient

5. z varies jointly as x and y; z is jointly proportional to x and y.

7. (a)

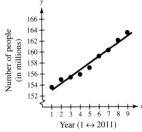

Year (1 ↔ 2011)

(b) The model is a good fit for the data.

9.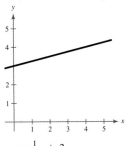

$y = \frac{1}{4}x + 3$

11.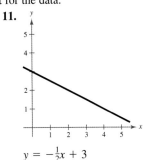

$y = -\frac{1}{2}x + 3$

13.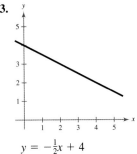

$y = -\frac{1}{2}x + 4$

15. (a) and (b)

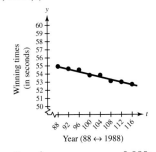

Year (88 ↔ 1988)

Sample answer: $y = -0.080t + 61.94$

(c) $y = -0.083t + 62.30$ (d) The models are similar.

17. $y = 7x$ **19.** $y = -\frac{1}{8}x$ **21.** $y = 2\pi x$

23.

x	2	4	6	8	10
$y = x^2$	4	16	36	64	100

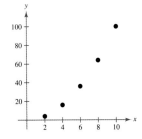

25.

x	2	4	6	8	10
$y = \frac{1}{2}x^3$	4	32	108	256	500

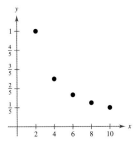

27.

x	2	4	6	8	10
$y = \dfrac{2}{x}$	1	$\frac{1}{2}$	$\frac{1}{3}$	$\frac{1}{4}$	$\frac{1}{5}$

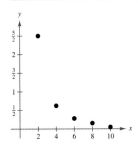

29.

x	2	4	6	8	10
$y = \dfrac{10}{x^2}$	$\frac{5}{2}$	$\frac{5}{8}$	$\frac{5}{18}$	$\frac{5}{32}$	$\frac{1}{10}$

31. $y = \dfrac{5}{x}$ **33.** $y = -\dfrac{7}{10}x$ **35.** $A = kr^2$ **37.** $y = \dfrac{k}{x^2}$

39. $F = \dfrac{kg}{r^2}$ **41.** $R = k(T - T_e)$ **43.** $P = kVI$

45. y is directly proportional to the square of x.

47. A is jointly proportional to b and h.

49. $y = 18x$ **51.** $y = \dfrac{75}{x}$ **53.** $z = 2xy$ **55.** $P = \dfrac{18x}{y^2}$

57. $I = 0.035P$ **59.** Model: $y = \frac{33}{13}x$; 25.4 cm, 50.8 cm

61. $293\frac{1}{3}$ N   **63.** About 39.47 lb

65. (a)

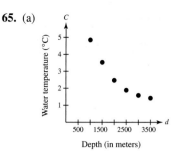

(b) Inverse variation

(c) About 4919.9

(d) 1640 m

67. About 0.61 mi/h

69. (a) 200 Hz   (b) 50 Hz   (c) 100 Hz

71. True. If $y = k_1 x$ and $x = k_2 z$, then $y = k_1(k_2 z) = (k_1 k_2)z$.

73. False. π is not a variable.

75. True. $f(2) = (2 - 2)p(2) = 0$.

77. $\dfrac{-x - 1}{5x - 4}$ **79.** $\dfrac{2x^2 + 3x + 15}{(x - 1)(x + 3)}$ **81.** $\dfrac{4}{3(2x + 1)}, x \ne -3$

83. Yes; All real numbers x **85.** No

87. Yes; All real numbers x such that $x \ge -5$

Review Exercises   *(page 300)*

1. (a)

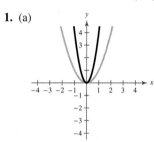

Vertical stretch

(b)

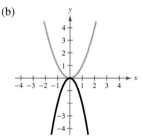

Vertical stretch and a reflection in the x-axis

(c)

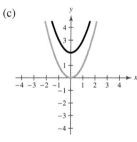

Upward shift of two units

(d)

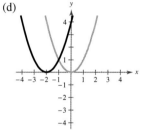

Left shift of two units

3. $g(x) = (x - 1)^2 - 1$

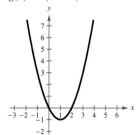
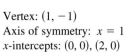

Vertex: $(1, -1)$
Axis of symmetry: $x = 1$
x-intercepts: $(0, 0), (2, 0)$

5. $f(x) = (x - 3)^2 - 8$

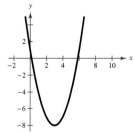

Vertex: $(3, -8)$
Axis of symmetry: $x = 3$
x-intercepts: $(3 \pm 2\sqrt{2}, 0)$

7. $f(x) = (x + 4)^2 - 6$

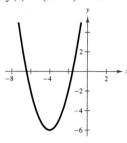

Vertex: $(-4, -6)$
Axis of symmetry: $x = -4$
x-intercepts: $(-4 \pm \sqrt{6}, 0)$

9. $h(x) = -(x - 2)^2 + 7$

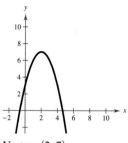

Vertex: $(2, 7)$
Axis of symmetry: $x = 2$
x-intercepts: $(2 \pm \sqrt{7}, 0)$

11. $h(x) = 4\left(x + \frac{1}{2}\right)^2 + 12$

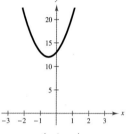

Vertex: $\left(-\frac{1}{2}, 12\right)$
Axis of symmetry: $x = -\frac{1}{2}$
No x-intercept

13. $f(x) = \frac{1}{3}\left(x + \frac{5}{2}\right)^2 - \frac{41}{12}$

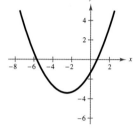

Vertex: $\left(-\frac{5}{2}, -\frac{41}{12}\right)$
Axis of symmetry: $x = -\frac{5}{2}$
x-intercepts:
$$\left(\frac{\pm\sqrt{41} - 5}{2}, 0\right)$$

15. $f(x) = -\frac{1}{2}(x - 4)^2 + 1$ **17.** $f(x) = -(x - 6)^2$

19. $f(x) = \frac{3}{4}(x - 2)^2 - \frac{5}{2}$

21. (a) $A = x\left(\dfrac{8 - x}{2}\right)$ (b) $0 < x < 8$

(c)

x	1	2	3	4	5	6
A	$\frac{7}{2}$	6	$\frac{15}{2}$	8	$\frac{15}{2}$	6

$x = 4, y = 2$

(d)

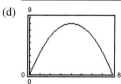

$x = 4, y = 2$

(e) $A = -\frac{1}{2}(x - 4)^2 + 8$; $x = 4, y = 2$

23. (a) $12,000; $13,750; $15,000

(b) Maximum revenue at $40; $16,000; Any price greater or less than $40 per unit will not yield as much revenue.

25. 1091 units

27.

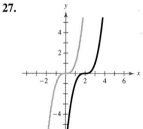

29.

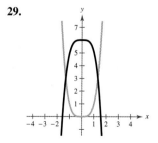

31.

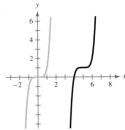

33. Falls to the left and to the right
35. Falls to the left, rises to the right
37. (a) $-8, \frac{4}{3}$ (b) Odd multiplicity   (c) 1
(d)

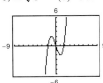

39. (a) $0, \pm\sqrt{3}$ (b) Odd multiplicity   (c) 2
(d)

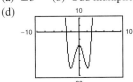

41. (a) ± 3 (b) Odd multiplicity   (c) 3
(d)

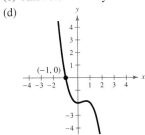

43. (a) Rises to the left, falls to the right   (b) -1
(c) Answers will vary.
(d)

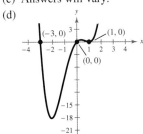

45. (a) Rises to the left and to the right   (b) $-3, 0, 1$
(c) Answers will vary.
(d)

47. (a) $[-1, 0]$ (b) -0.719
49. (a) $[-1, 0], [1, 2]$ (b) $-0.200, 1.772$
51. $6x + 3 + \dfrac{17}{5x - 3}$ **53.** $5x + 4,\ x \neq \dfrac{5}{2} \pm \dfrac{\sqrt{29}}{2}$

55. $x^2 - 3x + 2 - \dfrac{1}{x^2 + 2}$
57. $2x^2 - 9x - 6,\ x \neq 8$ **59.** $x^3 - 3x^2 + 7x - 12 + \dfrac{36}{x + 3}$
61. (a) -421 (b) -9
63. (a) Yes   (b) Yes   (c) Yes   (d) No
65. (a) Answers will vary.   (b) $(x + 7), (x + 1)$
(c) $f(x) = (x + 7)(x + 1)(x - 4)$ (d) $-7, -1, 4$
(e)

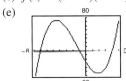

67. (a) Answers will vary.   (b) $(x + 1), (x - 4)$
(c) $f(x) = (x + 1)(x - 4)(x + 2)(x - 3)$
(d) $-2, -1, 3, 4$
(e)

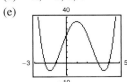

69. 1   **71.** 7   **73.** 3   **75.** $-6, -2, 5$ **77.** $-2, \frac{4}{3}$
79. $-4, 3$ **81.** $f(x) = 3x^4 - 14x^3 + 17x^2 - 42x + 24$
83. $2, \pm 4i$ **85.** $x(x - 1)(x + 5);\ 0, 1, -5$
87. $(x + 4)^2(x - 2 - 3i)(x - 2 + 3i);\ -4, 2 \pm 3i$
89. $16, \pm i$ **91.** $-3, 6, \pm 2i$
93. Two or no positive real zeros, one negative real zero
95. Answers will vary.   **97.** Radius: 1.82 in., height: 10.82 in.
99.

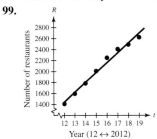

The model fits the data well.
101. $y = \frac{8}{5}x$; 8 km; 40 km   **103.** A factor of 4
105. 909 pastries
107. True. The leading coefficient is negative and the degree is odd.
109. True. If y is directly proportional to x, then $y = kx$, so $x = (1/k)y$. Therefore, x is directly proportional to y.
111. Answers will vary. *Sample answer:*

A polynomial of degree $n > 0$ with real coefficients can be written as the product of linear and quadratic factors with real coefficients, where the quadratic factors have no real zeros. Setting the factors equal to zero and solving for the variable can find the zeros of a polynomial function.

To solve an equation is to find all the values of the variable for which the equation is true.

Chapter Test *(page 304)*

1. (a)

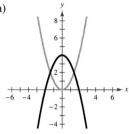

(b)

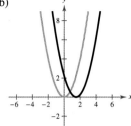

Reflection in the x-axis and an upward shift of four units

Right shift of $\frac{3}{2}$ units

2. Vertex: $(3, -1)$; Intercepts: $(2, 0)$, $(4, 0)$, $(0, 8)$

3. $y = (x - 3)^2 - 6$

4. (a) 50 ft

(b) 5. Yes, changing the constant term results in a vertical shift of the graph, so the maximum height changes.

5. Rises to the left, falls to the right

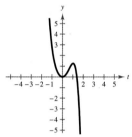

6. $3x + \dfrac{x - 1}{x^2 + 1}$ **7.** $2x^3 - 4x^2 + 5x - 6 + \dfrac{11}{x + 2}$

8. $(2x - 5)(x - \sqrt{3})(x + \sqrt{3})$; Zeros: $\pm\sqrt{3}, \frac{5}{2}$

9. $-2, \frac{3}{2}$ **10.** $\pm 1, -\frac{2}{3}$

11. $f(x) = x^4 - 2x^3 + 9x^2 - 18x$

12. $f(x) = x^4 - 6x^3 + 16x^2 - 18x + 7$

13. $1, -5, -\frac{2}{3}$ **14.** $-2, 4, -1 \pm \sqrt{2}i$ **15.** $v = 6\sqrt{s}$

16. $A = \dfrac{25}{6}xy$ **17.** $b = \dfrac{48}{a}$

18. $y = -227.7t + 8791$; The model fits that data well.

Problem Solving *(page 307)*

1. (a) (i) $6, -2$ (ii) $0, -5$ (iii) $-5, 2$

(iv) 2 (v) $1 \pm \sqrt{7}$ (vi) $\dfrac{-3 \pm \sqrt{7}i}{2}$

(b) (i)

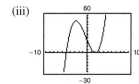

(ii)

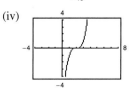

(iii)

(iv)

(v)

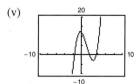

(vi)

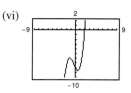

Graph (iii) touches the x-axis at $(2, 0)$, and all the other graphs pass through the x-axis at $(2, 0)$.

(c) (i) $(6, 0), (-2, 0)$ (ii) $(0, 0), (-5, 0)$

(iii) $(-5, 0)$ (iv) No other x-intercepts

(v) $(-1.6, 0), (3.6, 0)$ (vi) No other x-intercepts

(d) When the function has two real zeros, the results are the same. When the function has one real zero, the graph touches the x-axis at the zero. When there are no real zeros, there is no x-intercept.

3. (a) $l = \dfrac{600 - \pi r^2}{\pi r}$ (b) $V(r) = 300r - \dfrac{\pi}{2}r^3$

(c) About 1595.8 ft^3; $r \approx 8$ ft, $l \approx 16$ ft

5. (a)

y	1	2	3	4	5
$y^3 + y^2$	2	12	36	80	150

y	6	7	8	9	10
$y^3 + y^2$	252	392	576	810	1100

(b) (i) $x = 6$ (ii) $x = 6$ (iii) $x = 3$ (iv) $x = 10$

(v) $x = 6$ (vi) $x = 3$

(c) Answers will vary.

7. (a)

Function	Zeros	Sum of zeros	Product of zeros
$f_1(x)$	$2, 3$	5	6
$f_2(x)$	$-3, 1, 2$	0	-6
$f_3(x)$	$-3, 1, \pm 2i$	-2	-12
$f_4(x)$	$0, 2, -3, 2 \pm \sqrt{3}$	3	0

(b) The sum of the zeros is equal to the opposite of the coefficient of the $(n - 1)$th term.

(c) The product of the zeros is equal to the constant term when the function is of an even degree and to the opposite of the constant term when the function is of an odd degree.

9. (a) and (b) $y = -x^2 + 5x - 4$

11. (a) $A(x) = x\left(\dfrac{100 - x}{2}\right)$; Domain: $0 < x < 100$

(b)

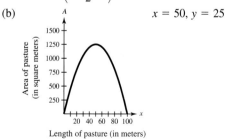

$x = 50, y = 25$

Area of pasture (in square meters)

Length of pasture (in meters)

(c) $A(x) = -\frac{1}{2}(x - 50)^2 + 1250$; $x = 50, y = 25$

13. 2 in. $\times$ 2 in. $\times$ 5 in.

CHAPTER 3

Chapter 4

Section 4.1 *(page 315)*

1. rational function   **3.** horizontal asymptote

5. vertical asymptote

7. Domain: all real numbers x except $x = 1$

$f(x) \to -\infty$ as $x \to 1^-$, $f(x) \to \infty$ as $x \to 1^+$

9. Domain: all real numbers x except $x = -4, -3$

$f(x) \to -\infty$ as $x \to -4^-$ and as $x \to -3^+$,

$f(x) \to \infty$ as $x \to -4^+$ and as $x \to -3^-$

11. Domain: all real numbers x except $x = \pm 1$

$f(x) \to \infty$ as $x \to -1^-$ and as $x \to 1^+$,

$f(x) \to -\infty$ as $x \to -1^+$ and as $x \to 1^-$

13. Domain: all real numbers x except $x = 1$

$f(x) \to \infty$ as $x \to 1^-$ and as $x \to 1^+$

15. Vertical asymptote: $x = 0$

Horizontal asymptote: $y = 0$

17. Vertical asymptote: $x = 5$

Horizontal asymptote: $y = -1$

19. Vertical asymptotes: $x = \pm 1$

21. Horizontal asymptote: $y = -4$

23. Vertical asymptote: $x = -4$

Horizontal asymptote: $y = 0$

Hole at $\left(4, \frac{1}{8}\right)$

25. Vertical asymptote: $x = 3$

Horizontal asymptote: $y = 1$

Hole at $\left(-1, \frac{1}{2}\right)$

27. Vertical asymptote: $x = \frac{1}{2}$

Horizontal asymptote: $y = \frac{1}{2}$

Hole at $\left(-1, \frac{5}{3}\right)$

29. d   **30.** e   **31.** b   **32.** a   **33.** c   **34.** f

35. (a) Domain of f: all real numbers x except $x = -2$

Domain of g: all real numbers x

(b) $x - 2$; Vertical asymptote: none

(c)

x	-4	-3	-2.5	-2	-1.5	-1	0
$f(x)$	-6	-5	-4.5	Undef.	-3.5	-3	-2
$g(x)$	-6	-5	-4.5	-4	-3.5	-3	-2

(d) The functions differ only at $x = -2$, where f is undefined.

37. (a) Domain of f: all real numbers x except $x = 0, \frac{1}{2}$

Domain of g: all real numbers x except $x = 0$

(b) $\dfrac{1}{x}$; Vertical asymptote: $x = 0$

(c)

x	-1	-0.5	0	0.5	2	3	4
$f(x)$	-1	-2	Undef.	Undef.	$\frac{1}{2}$	$\frac{1}{3}$	$\frac{1}{4}$
$g(x)$	-1	-2	Undef.	2	$\frac{1}{2}$	$\frac{1}{3}$	$\frac{1}{4}$

(d) The functions differ only at $x = 0.5$ where f is undefined.

39. (a)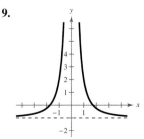

(b) \$4411.76; \$25,000; \$225,000

(c) No. The function is undefined at $p = 100$.

41. (a)

(b) 333 deer, 500 deer, 800 deer   (c) 1500 deer

43. (a)

n	1	2	3	4	5
P	0.50	0.74	0.82	0.86	0.89

n	6	7	8	9	10
P	0.91	0.92	0.93	0.94	0.95

P approaches 1 as n increases.

(b) 100%

45. True. Vertical asymptotes can only occur where the function is undefined.

47. The denominator could have a factor of $(x - c)$, so $x = c$ could be undefined for the rational function.

49. Answers will vary.

Sample answers: $f(x) = \dfrac{1}{x^2 + 15}$; $f(x) = \dfrac{1}{x - 15}$

51. $\dfrac{x + 3}{x + 1}, x \neq 3$ **53.** $\dfrac{x + 5}{x - 5}, x \neq -6$

55. x-intercepts: $(-10, 0), (-2, 0)$ **57.** x-intercept: $\left(-\frac{3}{2}, 0\right)$

y-intercept: $(0, 20)$ y-intercept: $(0, 3)$

59. Intercept: $(0, 0)$ **61.** x-intercept: $(-3, 0)$

y-intercept: $\left(0, \frac{3}{2}\right)$

Section 4.2 *(page 323)*

1. slant asymptote   **3.** $y = x - 1$

5.

7.

9.

11.

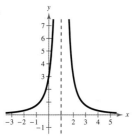

13. (a) Domain: all real numbers x except $x = -1$

(b) y-intercept: $(0, 1)$

(c) Vertical asymptote: $x = -1$

Horizontal asymptote: $y = 0$

(d)

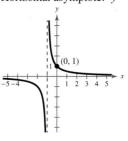

15. (a) Domain: all real numbers x except $x = -4$

(b) y-intercept: $\left(0, -\frac{1}{4}\right)$

(c) Vertical asymptote: $x = -4$

Horizontal asymptote: $y = 0$

(d)

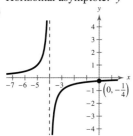

17. (a) Domain: all real numbers x except $x = -2$

(b) x-intercept: $\left(-\frac{3}{2}, 0\right)$

y-intercept: $\left(0, \frac{3}{2}\right)$

(c) Vertical asymptote: $x = -2$

Horizontal asymptote: $y = 2$

(d)

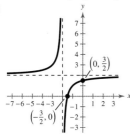

19. (a) Domain: all real numbers x except $x = -2$

(b) x-intercept: $\left(-\frac{5}{2}, 0\right)$

y-intercept: $\left(0, \frac{5}{2}\right)$

(c) Vertical asymptote: $x = -2$

Horizontal asymptote: $y = 2$

(d)

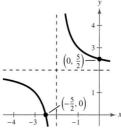

21. (a) Domain: all real numbers x

(b) Intercept: $(0, 0)$

(c) Horizontal asymptote: $y = 1$

(d)

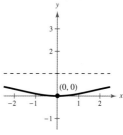

23. (a) Domain: all real numbers x except $x = \pm 3$

(b) Intercept: $(0, 0)$

(c) Vertical asymptotes: $x = \pm 3$

Horizontal asymptote: $y = 1$

(d)

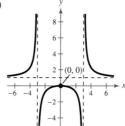

25. (a) Domain: all real numbers s

(b) Intercept: $(0, 0)$

(c) Horizontal asymptote: $y = 0$

(d)

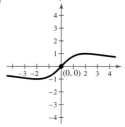

27. (a) Domain: all real numbers x except $x = 0, 4$

(b) x-intercept: $(-1, 0)$

(c) Vertical asymptotes: $x = 0$, $x = 4$

Horizontal asymptote: $y = 0$

(d)

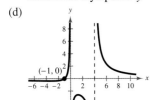

29. (a) Domain: all real numbers x except $x = 4, -1$

(b) Intercept: $(0, 0)$

(c) Vertical asymptotes: $x = -1$, $x = 4$

Horizontal asymptote: $y = 0$

(d)

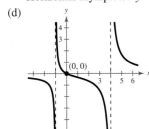

31. (a) Domain: all real numbers x except $x = 3, -4$

(b) y-intercept: $\left(0, -\frac{5}{3}\right)$

(c) Vertical asymptote: $x = 3$

Horizontal asymptote: $y = 0$

Hole: $\left(-4, -\frac{5}{7}\right)$

(d)

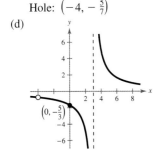

33. (a) Domain: all real numbers t except $t = 1$

(b) t-intercept: $(-1, 0)$

y-intercept: $(0, 1)$

(c) Hole: $(1, 2)$

(d)

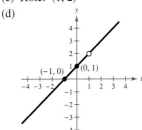

35. (a) Domain: all real numbers x except $x = \pm 2$

(b) x-intercepts: $(1, 0)$, $(4, 0)$

y-intercept: $(0, -1)$

(c) Vertical asymptotes: $x = \pm 2$

Horizontal asymptote: $y = 1$

(d)

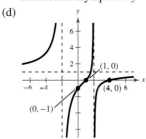

37. (a) Domain: all real numbers x except $x = -4, 1$

(b) y-intercept: $(0, -1)$

(c) Vertical asymptotes: $x = -4$, $x = 1$

Horizontal asymptote: $y = 1$

(d)

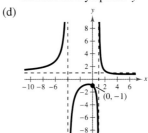

39. (a) Domain: all real numbers x except $x = -\frac{3}{2}, 2$

(b) x-intercept: $\left(\frac{1}{2}, 0\right)$

y-intercept: $\left(0, -\frac{1}{3}\right)$

(c) Vertical asymptote: $x = -\frac{3}{2}$

Horizontal asymptote: $y = 1$

Hole: $\left(2, \frac{3}{7}\right)$

(d)

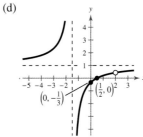

41. (a) Domain: all real numbers x except $x = \pm 1, 2$

(b) x-intercepts: $(3, 0)$, $\left(-\frac{1}{2}, 0\right)$

y-intercept: $\left(0, -\frac{3}{2}\right)$

(c) Vertical asymptotes: $x = 2$, $x = \pm 1$

Horizontal asymptote: $y = 0$

(d)

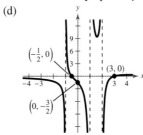

43. (a) Domain of f: all real numbers x except $x = -1$
    Domain of g: all real numbers x

(b)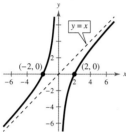

(c) Because there are only finitely many pixels, the graphing utility may not attempt to evaluate the function where it does not exist.

45. (a) Domain: all real numbers x except $x = 0$

(b) x-intercepts: $(\pm 2, 0)$

(c) Vertical asymptote: $x = 0$
    Slant asymptote: $y = x$

(d)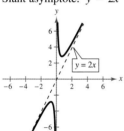

47. (a) Domain: all real numbers x except $x = 0$

(b) No intercepts

(c) Vertical asymptote: $x = 0$
    Slant asymptote: $y = 2x$

(d)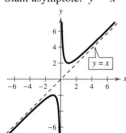

49. (a) Domain: all real numbers x except $x = 0$

(b) No intercepts

(c) Vertical asymptote: $x = 0$
    Slant asymptote: $y = x$

(d)

51. (a) Domain: all real numbers t except $t = -5$

(b) y-intercept: $\left(0, -\frac{1}{5}\right)$

(c) Vertical asymptote: $t = -5$
    Slant asymptote: $y = -t + 5$

(d)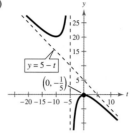

53. (a) Domain: all real numbers x except $x = \pm 2$

(b) Intercept: $(0, 0)$

(c) Vertical asymptotes: $x = \pm 2$
    Slant asymptote: $y = x$

(d)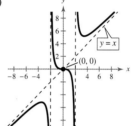

55. (a) Domain: all real numbers x except $x = 0, 1$

(b) No intercepts

(c) Vertical asymptote: $x = 0$
    Slant asymptote: $y = x + 1$
    Hole: $(1, 3)$

(d)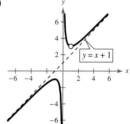

57. (a) Domain: all real numbers x except $x = 1$

(b) y-intercept: $(0, -1)$

(c) Vertical asymptote: $x = 1$
    Slant asymptote: $y = x$

(d)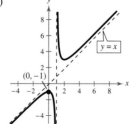

CHAPTER 4

59. (a) Domain: all real numbers x except $x = -1, -2$

(b) y-intercept: $\left(0, \frac{1}{2}\right)$

x-intercepts: $\left(\frac{1}{2}, 0\right), (1, 0)$

(c) Vertical asymptote: $x = -2$

Slant asymptote: $y = 2x - 7$

Hole: $(-1, 6)$

(d)

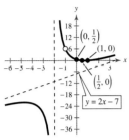

61. (a) $(-1, 0)$ (b) -1 **63.** (a) $(0, 0)$ (b) 0

65. (a) $(1, 0), (-1, 0)$ (b) ± 1

67. (a) $(-1, 0)$ (b) -1

69. (a) Answers will vary.   (b) $(4, \infty)$

(c)

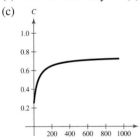

11.75 in. $\times$ 5.87 in.

71. (a) Answers will vary.   (b) $[0, 950]$

(c)

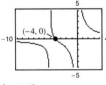

(d) Increases more slowly; 75%

73. (a)

$(-4, 0)$

(b) -4

75. (a)

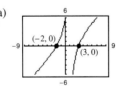

$(3, 0), (-2, 0)$

(b) $3, -2$

Relative minimum: $(-2, -1)$

Relative maximum: $(0, 3)$

77.

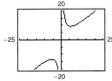

79.

Relative minimum: $(5.657, 9.314)$

Relative maximum:

$(-5.657, -13.314)$

81.

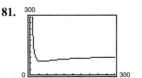

$x \approx 40.45$, or 4045 components

83. (a) Answers will vary.

(b) Vertical asymptote: $x = 25$

Horizontal asymptote: $y = 25$

(c)

x	30	35	40	45	50	55	60
y	150	87.5	66.7	56.3	50	45.8	42.9

(d)

(e) *Sample answer:* No. You might expect the average speed for the round trip to be the average of the average speeds for the two parts of the trip.

(f) No. At 20 miles per hour you would use more time in one direction than is required for the round trip at an average speed of 50 miles per hour.

85. False. The graph of $f(x) = \dfrac{x}{x^2 + 1}$ crosses $y = 0$, which is a horizontal asymptote.

87. True. Vertical asymptotes can only occur at the zeros of the denominator.

89. $h(x) = \dfrac{6 - 2x}{3 - x} = 2$, $x \neq 3$, so the graph has a hole at $x = 3$.

91. (a) If the degree of the numerator is exactly one more than the degree of the denominator, then the graph of the function has a slant asymptote.

(b) To find the equation of a slant asymptote, use long division to expand the function.

93. *Sample answer:* $f(x) = \dfrac{x^2 + 2x - 15}{x + 4}$

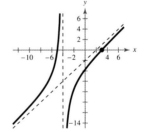

95. $f(x) = (x + 1)^2 - 1$

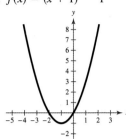

Vertex: $(-1, -1)$
Axis of Symmetry: $x = -1$
x-intercepts: $(-2, 0), (0, 0)$

97. $f(x) = -(x + 8)^2 + 51$

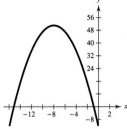

Vertex: $(-8, 51)$
Axis of Symmetry: $x = -8$
x-intercept:
$(-8 \pm \sqrt{51}, 0)$

99. $x^2 + (y - 3)^2 = 49$ **101.** $(x + 2)^2 + (y + 1)^2 = 25$

103. This situation could be represented by a one-to-one function if the runner does not stop to rest. The inverse function would represent the time in hours for a given number of miles completed.

Section 4.3   *(page 337)*

1. conic or conic section   **3.** axis   **5.** hyperbola; foci

7. Given the major and minor axis lengths, $2a$ and $2b$, respectively, use the equation $c^2 = a^2 - b^2$ to find c. The eccentricity is the ratio c/a.

9. b   **10.** c   **11.** f   **12.** d   **13.** a   **14.** e

15. Focus: $\left(0, \frac{1}{2}\right)$
Directrix: $y = -\frac{1}{2}$

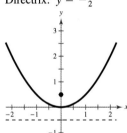

17. Focus: $\left(-\frac{3}{2}, 0\right)$
Directrix: $x = \frac{3}{2}$

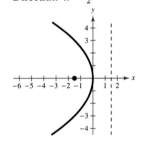

19. Focus: $(0, -3)$
Directrix: $y = 3$

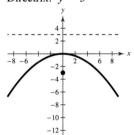

21. $y^2 = 12x$ **23.** $x^2 = -8y$ **25.** $y^2 = -9x$
27. $x^2 = \frac{3}{2}y$; Focus: $\left(0, \frac{3}{8}\right)$ **29.** $y^2 = 6x$

31. (a)

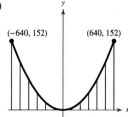

(b) $y = \dfrac{19x^2}{51,200}$

33. $\dfrac{x^2}{1} + \dfrac{y^2}{4} = 1$ **35.** $\dfrac{x^2}{25} + \dfrac{y^2}{21} = 1$ **37.** $\dfrac{x^2}{49} + \dfrac{y^2}{24} = 1$

39. $\dfrac{x^2}{81} + \dfrac{y^2}{9} = 1$ **41.** $\dfrac{x^2}{49} + \dfrac{y^2}{196} = 1$

43. Vertices: $(\pm 5, 0)$
$e = \dfrac{3}{5}$

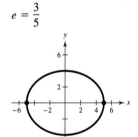

45. Vertices: $\left(\pm\dfrac{5}{3}, 0\right)$
$e = \dfrac{3}{5}$

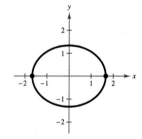

47. Vertices: $(\pm 6, 0)$
$e = \dfrac{\sqrt{29}}{6}$

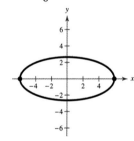

49. Vertices: $(0, \pm 3)$
$e = \dfrac{2}{3}$

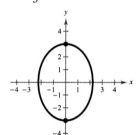

51. Vertices: $(0, \pm 1)$
$e = \dfrac{\sqrt{3}}{2}$

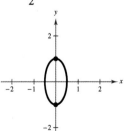

53. $\dfrac{x^2}{25} + \dfrac{y^2}{9} = 1$ **55.** $(\pm\sqrt{5}, 0)$; Length of string: 6 ft

57. (a)

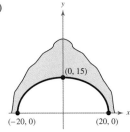

(b) $y = \frac{3}{4}\sqrt{400 - x^2}$

(c) Yes, with clearance of 0.52 foot.

59.

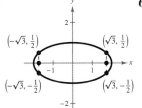

61.

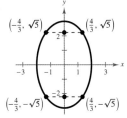

63. $\dfrac{y^2}{4} - \dfrac{x^2}{32} = 1$ **65.** $\dfrac{x^2}{1} - \dfrac{y^2}{9} = 1$ **67.** $\dfrac{x^2}{64} - \dfrac{y^2}{36} = 1$

69. $\dfrac{y^2}{9} - \dfrac{x^2}{9/4} = 1$

71. Vertices: $(\pm 5, 0)$

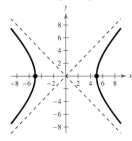

73. Vertices: $(0, \pm 6)$

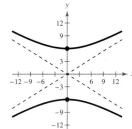

75. Vertices: $(0, \pm 1)$

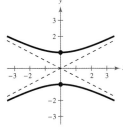

77. Vertices: $(0, \pm 3)$

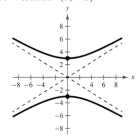

79. Vertices: $\left(\pm\frac{1}{3}, 0\right)$

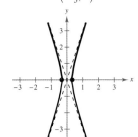

81. (a) $\dfrac{x^2}{1} - \dfrac{y^2}{169/3} = 1$ (b) 2.403 ft   **83.** 10 mi

85. False. The equation represents a hyperbola:
$$\frac{x^2}{144} - \frac{y^2}{144} = 1.$$

87. True. The directrix is perpendicular to the axis.

89. True. When the asymptotes are perpendicular, the slopes are 1 and -1, so $a = b$.

91. No. If it were a hyperbola, the equation would have to be second degree.

93. *Sample answer:* Solve the equation for y and graph the functions in the same viewing window; $y_1 = \sqrt{16 - 16x^2/25}$, $y_2 = -\sqrt{16 - 16x^2/25}$

95. (a) Left half of ellipse (portion to the left of the y-axis)

(b) Top half of ellipse (portion above the x-axis)

97. Answers will vary.   **99.** $-8, 2$ **101.** $\dfrac{6 \pm \sqrt{30}}{2}$

103. $2x^3, x \neq 0$ **105.** $\dfrac{16b^8}{a^{12}}, b \neq 0$ **107.** $f(x) = (x - 3)^2$

Section 4.4   *(page 347)*

1. b   **2.** d   **3.** e   **4.** c   **5.** a   **6.** f

7. Circle; shifted two units to the left and one unit up

9. Hyperbola; shifted one unit to the right and three units down

11. Parabola; shifted one unit to the left and two units up

13. Ellipse; shifted four units to the left and two units down

15. Center: $(0, 0)$
Radius: 7

17. Center: $(1, 0)$
Radius: $\sqrt{10}$

19. Center: $(4, 5)$
Radius: 6

21. $x^2 + (y - 4)^2 = 16$
Center: $(0, 4)$
Radius: 4

23. $(x - 1)^2 + (y + 3)^2 = 1$
Center: $(1, -3)$
Radius: 1

25. $\left(x + \frac{3}{2}\right)^2 + (y - 3)^2 = 1$
Center: $\left(-\frac{3}{2}, 3\right)$
Radius: 1

27. Vertex: $\left(5, -\frac{1}{2}\right)$
Focus: $\left(\frac{11}{2}, -\frac{1}{2}\right)$
Directrix: $x = \frac{9}{2}$

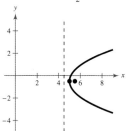

29. Vertex: $(1, -2)$
Focus: $(1, -4)$
Directrix: $y = 0$

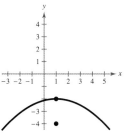

31. Vertex: $(1, 1)$
Focus: $(1, 2)$
Directrix: $y = 0$

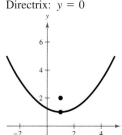

33. Vertex: $\left(-\frac{1}{2}, 1\right)$
Focus: $\left(-\frac{5}{2}, 1\right)$
Directrix: $x = \frac{3}{2}$

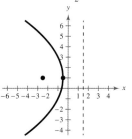

35. $(y - 2)^2 = -8(x - 3)$ **37.** $x^2 = 8(y - 4)$

39. $(y - 4)^2 = 16x$

41. (a) $17,500\sqrt{2}$ mi/h   (b) $x^2 = -16,400(y - 4100)$

43. (a) $x^2 = -49(y - 100)$ (b) 70 ft

45. Center: $(1, 5)$

Foci: $(1, 9), (1, 1)$

Vertices: $(1, 10), (1, 0)$

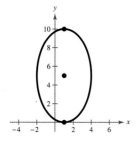

47. Center: $(-2, -4)$

Foci: $\left(-2 \pm \dfrac{\sqrt{3}}{2}, -4\right)$

Vertices:

$(-3, -4), (-1, -4)$

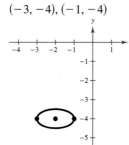

49. Center: $(2, 1)$

Foci: $\left(\frac{14}{5}, 1\right), \left(\frac{6}{5}, 1\right)$

Vertices: $(1, 1), (3, 1)$

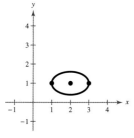

51. Center: $(-1, 0)$

Foci: $\left(-1, \pm\sqrt{21}\right)$

Vertices: $(-1, \pm 5)$

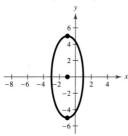

53. $\dfrac{(x - 3)^2}{1} + \dfrac{y^2}{9} = 1$ **55.** $\dfrac{(x - 2)^2}{16} + \dfrac{y^2}{12} = 1$

57. $\dfrac{(x - 1)^2}{12} + \dfrac{(y - 4)^2}{16} = 1$ **59.** $\dfrac{(x - 2)^2}{25} + \dfrac{y^2}{21} = 1$

61. $\dfrac{(x + 3)^2}{8} + \dfrac{(y + 2)^2}{9} = 1$ **63.** $(x - 2)^2 + \dfrac{(y - 1)^2}{4} = 1$

65. 2,756,170,000 mi; 4,583,830,000 mi

67. Center: $(2, -1)$

Foci: $(7, -1), (-3, -1)$

Vertices: $(6, -1), (-2, -1)$

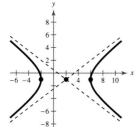

69. Center: $(2, -6)$

Foci: $\left(2, -6 \pm \sqrt{2}\right)$

Vertices: $(2, -5), (2, -7)$

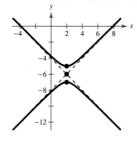

71. Center: $(-1, -3)$

Foci: $\left(-1 \pm \dfrac{5\sqrt{2}}{3}, -3\right)$

Vertices: $\left(-1 \pm \sqrt{5}, -3\right)$

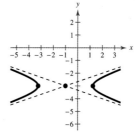

73. Center: $(-3, 2)$

Foci: $\left(-3, 2 \pm \sqrt{10}\right)$

Vertices: $(-3, 1), (-3, 3)$

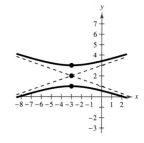

75. $\dfrac{(y - 1)^2}{1} - \dfrac{x^2}{3} = 1$ **77.** $\dfrac{(x + 4)^2}{4} - \dfrac{(y - 5)^2}{12} = 1$

79. $\dfrac{y^2}{9} - \dfrac{4(x - 2)^2}{9} = 1$ **81.** $\dfrac{(x - 3)^2}{9} - \dfrac{(y - 2)^2}{4} = 1$

83. $\dfrac{(y + 2)^2}{4} - \dfrac{x^2}{4} = 1$

Hyperbola

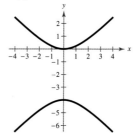

Shifted two units down

85. $\left(y + \dfrac{1}{4}\right)^2 = -8\left(x - \dfrac{1}{16}\right)$

Parabola

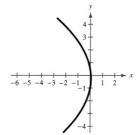

Shifted $\frac{1}{16}$ unit right and $\frac{1}{4}$ unit down

87. $\dfrac{(x + 2)^2}{16} + \dfrac{(y + 4)^2}{9} = 1$

Ellipse

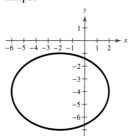

Shifted two units left and four units down

CHAPTER 4

89. False. The conic is an ellipse.

91. $\dfrac{(x-6)^2}{324} + \dfrac{(y-2)^2}{308} = 1$ **93.** The center is $(-4, 2)$.

95. All real numbers x

97. All real numbers x such that $x \geq -1$

99. All real numbers x such that $x \geq 0$

101.

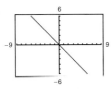

Decreasing on $(-\infty, \infty)$

103.

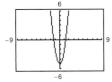

Increasing on $(0, \infty)$
Decreasing on $(-\infty, 0)$

105.

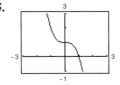

Decreasing on $(-\infty, \infty)$

107.

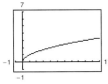

Increasing on $(0, \infty)$

109.

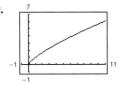

Increasing on $(0, \infty)$

Review Exercises *(page 352)*

1. Domain: all real numbers x except $x = -10$
$f(x) \to \infty$ as $x \to -10^-$, $f(x) \to -\infty$ as $x \to -10^+$

3. Domain: all real numbers x except $x = 6, 4$
$f(x) \to \infty$ as $x \to 4^-$ and as $x \to 6^+$,
$f(x) \to -\infty$ as $x \to 4^+$ and as $x \to 6^-$

5. Vertical asymptote: $x = -3$

7. Vertical asymptotes: $x = \pm 2$
Horizontal asymptote: $y = 1$

9. Vertical asymptote: $x = 6$
Horizontal asymptote: $y = 0$

11. $0.50 is the horizontal asymptote of the function.

13. (a) Domain: all real numbers x except $x = 0$
(b) No intercepts
(c) Vertical asymptote: $x = 0$
Horizontal asymptote: $y = 0$
(d)

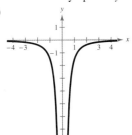

15. (a) Domain: all real numbers x except $x = 1$
(b) x-intercept: $(-2, 0)$
y-intercept: $(0, 2)$
(c) Vertical asymptote: $x = 1$
Horizontal asymptote: $y = -1$
(d)

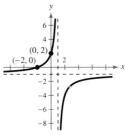

17. (a) Domain: all real numbers x
(b) Intercept: $(0, 0)$
(c) Horizontal asymptote: $y = \frac{5}{4}$
(d)

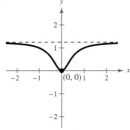

19. (a) Domain: all real numbers x except $x = \pm 4$
(b) Intercept: $(0, 0)$
(c) Vertical asymptotes: $x = \pm 4$
Horizontal asymptote: $y = 0$
(d)

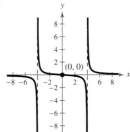

21. (a) Domain: all real numbers x
(b) Intercept: $(0, 0)$
(c) Horizontal asymptote: $y = -6$
(d)

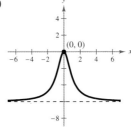

23. (a) Domain: all real numbers x except $x = 1, 3$
  (b) x-intercept: $\left(\frac{1}{3}, 0\right)$
    y-intercept: $(0, 1)$
  (c) Vertical asymptote: $x = 1$
    Horizontal asymptote: $y = 3$
    Hole: $(3, 4)$
  (d)

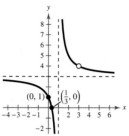

25. (a) Domain: all real numbers x
  (b) Intercept: $(0, 0)$
  (c) Slant asymptote: $y = 2x$
  (d)

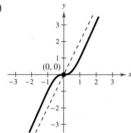

27. (a) Domain: all real numbers x except $x = -2$
  (b) x-intercepts: $(2, 0), (-5, 0)$
    y-intercept: $(0, -5)$
  (c) Vertical asymptote: $x = -2$
    Slant asymptote: $y = x + 1$
  (d)

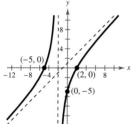

29. (a) Domain: all real numbers x except $x = \frac{4}{3}, -1$
  (b) x-intercepts: $\left(\frac{2}{3}, 0\right), (1, 0)$
    y-intercept: $\left(0, -\frac{1}{2}\right)$
  (c) Vertical asymptote: $x = \frac{4}{3}$
    Slant asymptote: $y = x - \frac{1}{3}$
    Hole: $\left(-1, -\frac{5}{4}\right)$
  (d)

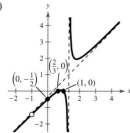

31. (a)

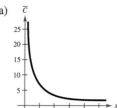

  (b) $100.90, $10.90, $1.90
  (c) $0.90 is the horizontal asymptote of the function.

33. (a)

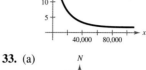

  (b) $N(5) = 304{,}000$ fish;
    $N(10) \approx 453{,}333$ fish;
    $N(25) \approx 702{,}222$ fish
  (c) $1{,}200{,}000$ fish; The graph has $N = 1200$ as a horizontal asymptote.

35. Parabola   **37.** Hyperbola   **39.** Ellipse   **41.** Circle
43. $y^2 = -24x$ **45.** $x^2 = 12y$ **47.** $y^2 = 12x$
49. $(0, 50)$ **51.** $\dfrac{x^2}{25} + \dfrac{y^2}{36} = 1$ **53.** $\dfrac{x^2}{221} + \dfrac{y^2}{25} = 1$
55. $\dfrac{x^2}{13} + \dfrac{y^2}{49} = 1$
57. The foci are 3 feet from the center and have the same height as the pillars.
59. $\dfrac{y^2}{1} - \dfrac{x^2}{24} = 1$ **61.** $\dfrac{x^2}{1} - \dfrac{y^2}{4} = 1$
63. $(x - 3)^2 = -2y$ **65.** $(x - 1)^2 + (y - 2)^2 = 4$
  Parabola   Circle

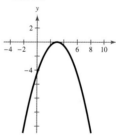

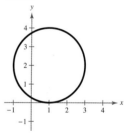

  Shifted three units right   Shifted one unit right and two units up

67. $\dfrac{(x + 5)^2}{9} + \dfrac{(y - 1)^2}{1} = 1$ **69.** $\dfrac{(x - 4)^2}{1} - \dfrac{(y - 4)^2}{9} = 1$
  Ellipse   Hyperbola

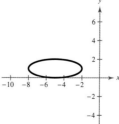

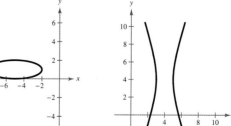

  Shifted five units left and one unit up   Shifted four units right and four units up

CHAPTER 4

71. $(x - 4)^2 = -8(y - 2)$ **73.** $(y + 8)^2 = 28(x - 8)$

75. $\dfrac{(x - 2)^2}{4} + \dfrac{(y - 2)^2}{1} = 1$ **77.** $\dfrac{(x - 2)^2}{16} + \dfrac{(y - 3)^2}{25} = 1$

79. $\dfrac{x^2}{5} + \dfrac{(y + 2)^2}{9} = 1$ **81.** $\dfrac{(x + 2)^2}{64} - \dfrac{(y - 3)^2}{36} = 1$

83. $\dfrac{y^2}{16} - \dfrac{(x - 3)^2}{4/5} = 1$ **85.** $8\sqrt{6}$ m

87. True. It could be a degenerate conic.

89. (a) Answers will vary.

(b) 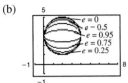   As e approaches 0, the ellipse becomes a circle.

Chapter Test *(page 355)*

1. Domain: all real numbers x except $x = -1$
Vertical asymptote: $x = -1$
Horizontal asymptote: $y = 3$

2. Domain: all real numbers x
Vertical asymptote: none
Horizontal asymptote: $y = -1$

3. Domain: all real numbers x except $x = 4, 5$
Vertical asymptote: $x = 5$
Horizontal asymptote: $y = 0$

4. x-intercepts: $\left(\pm\sqrt{3}, 0\right)$
Vertical asymptote: $x = 0$
Horizontal asymptote: $y = -1$

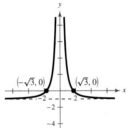

5. y-intercept: $(0, -2)$
Vertical asymptote: $x = 1$
Slant asymptote: $y = x + 1$

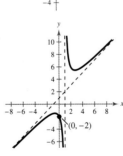

6. x-intercept: $(-1, 0)$
y-intercept: $\left(0, -\dfrac{1}{12}\right)$
Vertical asymptotes:
$x = 3, x = -4$
Horizontal asymptote: $y = 0$

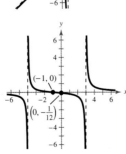

7. x-intercept: $\left(-\dfrac{3}{2}, 0\right)$
y-intercept: $\left(0, \dfrac{3}{4}\right)$
Vertical asymptote: $x = -4$
Horizontal asymptote: $y = 2$

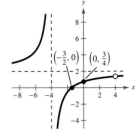

8. y-intercept: $(0, 1)$
Horizontal asymptote: $y = \dfrac{2}{5}$

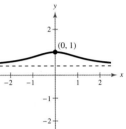

9. x-intercepts: $(-2, 0), \left(-\dfrac{3}{2}, 0\right)$
y-intercept: $(0, 6)$
Vertical asymptote: $x = -1$
Slant asymptote: $y = 2x + 5$

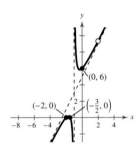

10. 6.24 in. × 12.49 in.

11. (a) $A = \dfrac{x^2}{2(x - 2)}, x > 2$

(b)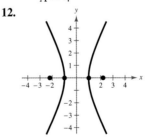

$A = 4$

12.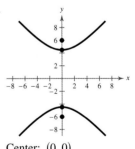

Center: $(0, 0)$
Vertices: $(\pm 1, 0)$
Foci: $\left(\pm\sqrt{5}, 0\right)$

13.

Center: $(0, 0)$
Vertices: $\left(0, \pm 2\sqrt{5}\right)$
Foci: $(0, \pm 6)$

14.

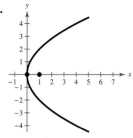

Vertex: $(0, 0)$
Focus: $(1, 0)$

15.

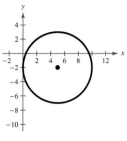

Center: $(5, -2)$

16.

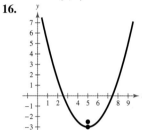

Vertex: $(5, -3)$
Focus: $\left(5, -\frac{5}{2}\right)$

17.

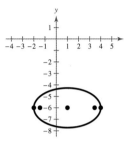

Center: $(1, -6)$
Vertices: $(4, -6), (-2, -6)$
Foci: $\left(1 \pm \sqrt{6}, -6\right)$

18. $\dfrac{(x-4)^2}{16} + \dfrac{(y-2)^2}{4} = 1$

19. $\dfrac{y^2}{9} - \dfrac{x^2}{4} = 1$ **20.** About 34 m

21. Least distance: About 363,301 km
Greatest distance: About 405,499 km

Problem Solving *(page 357)*

1. (a) iii   (b) ii   (c) iv   (d) i
3. (a) $y_1 \approx 0.031x^2 - 1.59x + 21.0$

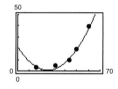

(b) $y_2 \approx \dfrac{1}{-0.007x + 0.44}$

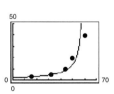

(c) The models are a good fit for the original data.
(d) $y_1(25) = 0.625$; $y_2(25) = 3.774$
The rational model is the better fit for the original data.
(e) The reciprocal model should not be used to predict the near point for a person who is 70 years old because a negative value is obtained. The quadratic model is a better fit.
5. Answers will vary.
7. (a) Answers will vary.
(b) Island 1: $(-6, 0)$; Island 2: $(6, 0)$
(c) 20 mi; Vertex: $(10, 0)$ (d) $\dfrac{x^2}{100} + \dfrac{y^2}{64} = 1$

9. (a)

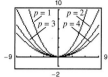

As p increases, the graph becomes wider.
(b) $p = 1$: focus $(0, 1)$ (c) $p = 1$: 4 units
   $p = 2$: focus $(0, 2)$ $p = 2$: 8 units
   $p = 3$: focus $(0, 3)$ $p = 3$: 12 units
   $p = 4$: focus $(0, 4)$ $p = 4$: 16 units
                                 Length of chord $= 4|p|$ units
(d) Answers will vary.
11. Proof

Chapter 5

Section 5.1   *(page 368)*

1. transcendental
3. Horizontal translation one unit to the left
5. 0.863   **7.** 1767.767
9. d   **10.** c   **11.** a   **12.** b
13.

x	-2	-1	0	1	2
$f(x)$	0.020	0.143	1	7	49

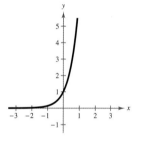

15.

x	-2	-1	0	1	2
$f(x)$	0.063	0.25	1	4	16

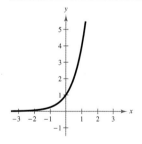

17.

x	−2	−1	0	1	2
f(x)	0.016	0.063	0.25	1	4

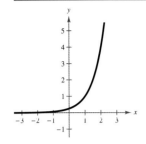

19.

x	−3	−2	−1	0	1
f(x)	3.25	3.5	4	5	7

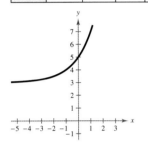

21. 2   **23.** −5   **25.** Shift the graph of f one unit up.
27. Reflect the graph of f in the y-axis and shift three units to the right.
29. 6.686   **31.** 7166.647

33.

x	−8	−7	−6	−5	−4
f(x)	0.055	0.149	0.406	1.104	3

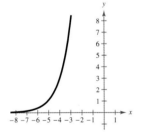

35.

x	−2	−1	0	1	2
f(x)	4.037	4.100	4.271	4.736	6

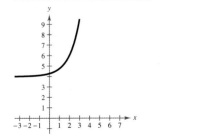

37. 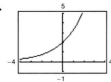   **39.**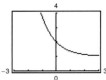

41. $\frac{1}{3}$ **43.** 3, −1
45.

n	1	2	4	12
A	$1828.49	$1830.29	$1831.19	$1831.80

n	365	Continuous
A	$1832.09	$1832.10

47.

n	1	2	4	12
A	$5477.81	$5520.10	$5541.79	$5556.46

n	365	Continuous
A	$5563.61	$5563.85

49.

t	10	20	30
A	$17,901.90	$26,706.49	$39,841.40

t	40	50
A	$59,436.39	$88,668.67

51.

t	10	20	30
A	$22,986.49	$44,031.56	$84,344.25

t	40	50
A	$161,564.86	$309,484.08

53. $104,710.29   **55.** $44.23
57. (a) 16 g   (b) 1.85 g
   (c)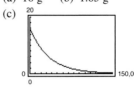

59. (a) $V(t) = 52,490\left(\frac{7}{8}\right)^t$ (b) About 30,769 infections
61. True. As $x \to -\infty$, $f(x) \to -2$ but never reaches -2.
63. $f(x) = h(x)$ **65.** $f(x) = g(x) = h(x)$
67.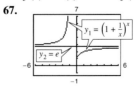

As the x-value increases, y_1 approaches the value of e.

69. (a) 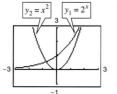   (b)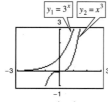

In both viewing windows, the constant raised to a variable power increases more rapidly than the variable raised to a constant power.

71. (a)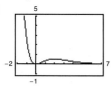

Decreasing: $(-\infty, 0), (2, \infty)$
Increasing: $(0, 2)$
Relative maximum: $(2, 4e^{-2})$
Relative minimum: $(0, 0)$

(b)

Decreasing: $(1.44, \infty)$
Increasing: $(-\infty, 1.44)$
Relative maximum: $(1.44, 4.25)$

73. c, d

75. (a) $f(g(x)) = f(x) = x$
$g(f(x)) = g(x) = x$

(b)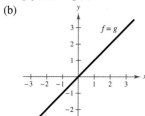

77. (a) $f(g(x)) = f\left(\dfrac{1}{x}\right) = \dfrac{1}{(1/x)} = x$

$g(f(x)) = g\left(\dfrac{1}{x}\right) = \dfrac{1}{(1/x)} = x$

(b)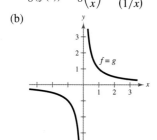

79. (a) $f(g(x)) = f\left(\dfrac{1-x}{x-2}\right) = \dfrac{2\left(\dfrac{1-x}{x-2}\right)+1}{\dfrac{1-x}{x-2}+1} = x$

$g(f(x)) = g\left(\dfrac{2x+1}{x+1}\right) = \dfrac{1-\dfrac{2x+1}{x+1}}{\dfrac{2x+1}{x+1}-2} = x$

(b)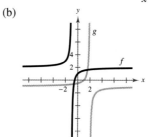

81. $f^{-1}(x) = x - 2$
The domain of f and the range of f^{-1} are all real numbers x such that $x \geq -2$. The domain of f^{-1} and the range of f are all real numbers x such that $x \geq 0$.

83. $f^{-1}(x) = \sqrt{x} - 6$
The domain of f and the range of f^{-1} are all real numbers x such that $x \geq -6$. The domain of f^{-1} and the range of f are all real numbers x such that $x \geq 0$.

85. $f^{-1}(x) = \dfrac{\sqrt{-2(x-5)}}{2}$
The domain of f and the range of f^{-1} are all real numbers x such that $x \geq 0$. The domain of f^{-1} and the range of f are all real numbers x such that $x \leq 5$.

87. $f^{-1}(x) = x + 3$
The domain of f and the range of f^{-1} are all real numbers x such that $x \geq 4$. The domain of f^{-1} and the range of f are all real numbers x such that $x \geq 1$.

Section 5.2   (page 378)

1. logarithmic   **3.** natural; e **5.** (a) x (b) y^2
7. $4^2 = 16$ **9.** $9^{-2} = \frac{1}{81}$ **11.** $\log_5 125 = 3$
13. $\log_4 \frac{1}{64} = -3$ **15.** 6   **17.** 1   **19.** $\frac{1}{2}$
21. 1.097   **23.** -0.058 **25.** 1   **27.** 2
29. 5   **31.** ± 2
33.   **35.**

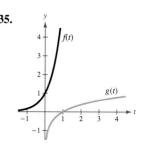

37. a; Upward shift of two units
39. b; Reflection in the y-axis and a right shift of one unit

41.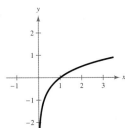

Domain: $(0, \infty)$
x-intercept: $(1, 0)$
Vertical asymptote: $x = 0$

43.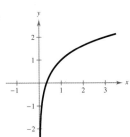

Domain: $(0, \infty)$
x-intercept: $\left(\frac{1}{3}, 0\right)$
Vertical asymptote: $x = 0$

45.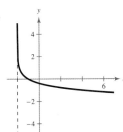

Domain: $(-2, \infty)$
x-intercept: $(-1, 0)$
Vertical asymptote: $x = -2$

47.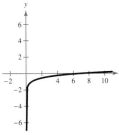

Domain: $(0, \infty)$
x-intercept: $(7, 0)$
Vertical asymptote: $x = 0$

49. $e^{-0693 \cdots} = \frac{1}{2}$ **51.** $e^{5.521 \cdots} = 250$
53. $\ln 7.3890 \ldots = 2$ **55.** $\ln \frac{1}{2} = -4x$ **57.** 23.123
59. 6.438   **61.** 4   **63.** 0   **65.** 1
67.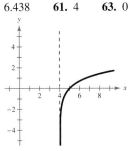

Domain: $(4, \infty)$
x-intercept: $(5, 0)$
Vertical asymptote: $x = 4$

69.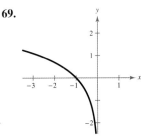

Domain: $(-\infty, 0)$
x-intercept: $(-1, 0)$
Vertical asymptote: $x = 0$

71.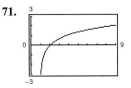

73. 8   **75.** $-2, 3$
77. (a) 30 yr; 10 yr
(b) \$323,179; \$199,109; \$173,179; \$49,109
(c) $x = 750$; The monthly payment must be greater than \$750.
79. (a)

r	0.005	0.010	0.015	0.020	0.025	0.030
t	138.6	69.3	46.2	34.7	27.7	23.1

As the rate of increase r increases, the time t in years for the population to double decreases.

(b)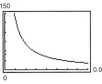

81. (a) 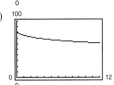   (b) 80   (c) 68.1   (d) 62.3

83. True.

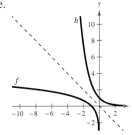

85. $\frac{1}{9}$ **87.** $y = \log_2 x$, so y is a logarithmic function of x.
89. (a) 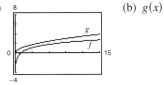   (b) $g(x)$

(c)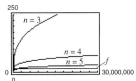

As n increases, $n = 2, 3, 4, \ldots, y = x^{1/n}$ increases at a slower rate.

91. $x^3 y^5, x \neq 0$ **93.** $\dfrac{m^3}{n}, m \neq 0$ **95.** $5x + 3, x \neq 4$

97. $-x^3 - 6x^2 - 36x - 36 - \dfrac{216}{x - 6}$

99. (a) $4x - 2$ (b) $4x - 11$ (c) $16x + 5$

101. (a) $\sqrt{\dfrac{1}{x^2} - 9}$ (b) $-\dfrac{1}{\sqrt{x^2 - 9}}$ (c) $x, x \neq 0$

Section 5.3 *(page 385)*

1. change-of-base **3.** $\log_3 24 = \dfrac{\ln 24}{\ln 3}$

5. (a) $\dfrac{\log 16}{\log 5}$ (b) $\dfrac{\ln 16}{\ln 5}$ **7.** (a) $\dfrac{\log \frac{3}{10}}{\log x}$ (b) $\dfrac{\ln \frac{3}{10}}{\ln x}$

9. 2.579 **11.** -0.606 **13.** $\log_3 5 + \log_3 7$

15. $\log_3 7 - 2\log_3 5$ **17.** $1 + \log_3 7 - \log_3 5$

19. 2 **21.** $-\frac{1}{3}$ **23.** -2 is not in the domain of $\log_2 x$.

25. $\frac{3}{4}$ **27.** 7 **29.** 2 **31.** $\frac{3}{2}$ **33.** 1.1833

35. -1.6542 **37.** 1.9563 **39.** -2.7124

41. $\ln 7 + \ln x$ **43.** $4\log_8 x$ **45.** $1 - \log_5 x$

47. $\frac{1}{2}\ln z$ **49.** $\ln x + \ln y + 2\ln z$ **51.** $\ln z + 2\ln(z - 1)$

53. $\frac{1}{2}\log_2(a + 2) + \frac{1}{2}\log_2(a - 2) - \log_2 7$

55. $2\log_5 x - 2\log_5 y - 3\log_5 z$ **57.** $\ln x - \frac{3}{2}\ln y$

59. $2\ln x + \frac{1}{2}\ln y - \frac{1}{2}\ln z$ **61.** $\frac{3}{4}\ln x + \frac{1}{4}\ln(x^2 + 3)$

63. $\ln 3x$ **65.** $\log_7(z - 2)^{2/3}$ **67.** $\log_3 \dfrac{5}{x^3}$

69. $\log x(x + 1)^2$ **71.** $\log \dfrac{xz^3}{y^2}$ **73.** $\ln \dfrac{x}{(x + 1)(x - 1)}$

75. $\ln \sqrt{\dfrac{x(x + 3)^2}{x^2 - 1}}$ **77.** $\ln y = -\dfrac{1}{4}\ln x + \ln \dfrac{5}{2}$

79. $\ln y = -0.14 \ln x + 5.7$

81. (a) and (b)

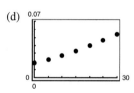

(c)

$T = 21 + e^{-0.037t + 3.997}$
The results are similar.

(d)

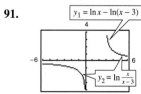

$T = 21 + \dfrac{1}{0.001t + 0.016}$

83. False; $\ln 1 = 0$ **85.** False; $\ln(x - 2) \neq \ln x - \ln 2$

87. False; $u = v^2$

89. The Power Property cannot be used because $\ln e$ is raised to the second power, not just e.

91.

No; $\dfrac{x}{x - 3} > 0$ when $x < 0$.

93. $\log_2 \frac{32}{4} = \log_2 32 - \log_2 4$; Property 2

95. $f(x) = \dfrac{\log x}{\log 2} = \dfrac{\ln x}{\ln 2}$ **97.** $f(x) = \dfrac{\log x}{\log \frac{1}{4}} = \dfrac{\ln x}{\ln \frac{1}{4}}$

99. $\ln 1 = 0$ $\ln 9 \approx 2.1972$
$\ln 2 \approx 0.6931$ $\ln 10 \approx 2.3025$
$\ln 3 \approx 1.0986$ $\ln 12 \approx 2.4848$
$\ln 4 \approx 1.3862$ $\ln 15 \approx 2.7080$
$\ln 5 \approx 1.6094$ $\ln 16 \approx 2.7724$
$\ln 6 \approx 1.7917$ $\ln 18 \approx 2.8903$
$\ln 8 \approx 2.0793$ $\ln 20 \approx 2.9956$

101. $x = -8, x = 3$ **103.** $x = 1, x = 9$ **105.** 4

107. -3 **109.** 0, 5 **111.** $-1, 4$ **113.** -4

115. (a) Yes (b) Yes **117.** (a) Yes (b) No

119. (a) Yes (b) No

Section 5.4 *(page 395)*

1. (a) $x = y$ (b) $x = y$ (c) x (d) x

3. Rewrite 125 as 5^3. By the One-to-One Property, $x = 3$.

5. (a) No (b) Yes **7.** (a) Yes (b) No

9. 2 **11.** 2 **13.** $\ln 2 \approx 0.693$ **15.** $e^{-1} \approx 0.368$

17. 64 **19.** $(3, 8)$ **21.** $2, -1$ **23.** $\dfrac{\ln 5}{\ln 3} \approx 1.465$

25. $\ln 39 \approx 3.664$ **27.** $2 - \dfrac{\ln 400}{\ln 3} \approx -3.454$

29. $\frac{1}{3}\log \frac{3}{2} \approx 0.059$ **31.** $\dfrac{\ln 12}{3} \approx 0.828$

33. 0 **35.** $\dfrac{\ln \frac{8}{3}}{3 \ln 2} + \dfrac{1}{3} \approx 0.805$ **37.** $0, \dfrac{\ln 4}{\ln 5} \approx 0.861$

39. $\ln 5 \approx 1.609$ **41.** $\ln \frac{4}{5} \approx -0.223$

43. $\dfrac{\ln 4}{365 \ln\left(1 + \dfrac{0.065}{365}\right)} \approx 21.330$

45. $e^{-3} \approx 0.050$ **47.** $\dfrac{e^{2.1}}{6} \approx 1.361$ **49.** $e^{-2} \approx 0.135$

51. $2(3^{11/6}) \approx 14.988$ **53.** No solution **55.** No solution

57. 2 **59.** 3.328 **61.** -0.478 **63.** 20.086

65. 1.482 **67.** (a) 27.73 yr (b) 43.94 yr **69.** $-1, 0$

71. 1 **73.** $e^{-1} \approx 0.368$ **75.** $e^{-1/2} \approx 0.607$

77. (a) $y = 100$ and $y = 0$; The range falls between 0% and 100%.
(b) Males: 69.51 in. Females: 64.49 in.

79. 5 years **81.** 2013

83. For $rt < \ln 2$ years, double the amount you invest. For $rt > \ln 2$ years, double your interest rate or double the number of years, because either of these will double the exponent in the exponential function.

85. $\log_b uv = \log_b u + \log_b v$
True by Property 1 in Section 5.3.

87. $\log_b(u - v) = \log_b u - \log_b v$
False.
$1.95 \approx \log(100 - 10) \neq \log 100 - \log 10 = 1$

CHAPTER 5

89. Yes. See Exercise 55.

91. Vertical asymptote: $x = 1$
Horizontal asymptote: $y = 1$

93. $1 - \dfrac{3}{\ln 2}$ **95.** $0, \pm\dfrac{\sqrt{8 + \ln 3}}{4}$ **97.** $\dfrac{100}{x^6}$

Section 5.5 *(page 405)*

1. $y = a + b \ln x;\ y = a + b \log x$

3. An exponential growth model increases over time. An exponential decay model decreases over time.

5. $P = \dfrac{A}{e^{rt}};\ t = \dfrac{\ln\left(\dfrac{A}{P}\right)}{r}$

7. 19.8 yr; $1419.07 **9.** 8.9438%; $1834.37

11. $6376.28; 15.4 yr

13. (a) 7.27 yr (b) 6.96 yr (c) 6.93 yr (d) 6.93 yr

15. (a)

r	2%	4%	6%	8%	10%	12%
t	54.93	27.47	18.31	13.73	10.99	9.16

(b)

r	2%	4%	6%	8%	10%	12%
t	55.48	28.01	18.85	14.27	11.53	9.69

17. The values given by the model are close to the original data values; November

19. $y = e^{0.7675x}$ **21.** $y = 5e^{-0.4024x}$ **23.** 6.48 g

25. 2.26 g

27. (a) Bulgaria: $y = 8.1e^{-0.00741t}$; 6.1 million
Canada: $y = 32.9e^{0.00672t}$; 42.8 million
China: $y = 1359.5e^{0.00115t}$; 1421.9 million
United Kingdom: $y = 60.4e^{0.00419t}$; 71.1 million
United States: $y = 291.1e^{0.00665t}$; 377.3 million
(b) b; The greater the rate of growth, the greater the value of b.

29. About 800 bacteria

31. (a) $V = -150t + 575$ (b) $V = 575e^{-0.3688t}$
(c)

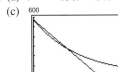

The exponential model depreciates faster.
(d) Linear model: $425; $125
Exponential model: $397.65; $190.18
(e) Answers will vary.

33. About 12,180 yr old

35.

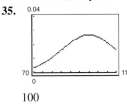

37. (a) 1998: 64,770 sites
2008: 240,797 sites
2015: 312,340 sites

(b)

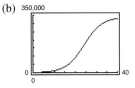

(c) and (d) 2007

39. (a) 203 animals (b) 13 mo
(c)

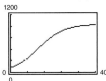

Horizontal asymptotes: $p = 0, p = 1000$. The population size will approach 1000 as time increases.

41. (a) 20 dB (b) 70 dB (c) 40 dB (d) 90 dB

43. 95%

45. (a) $10^{7.6} \approx 39,810,717$ (b) $10^{5.6} \approx 398,107$
(c) $10^{6.6} \approx 3,981,072$

47. 4.64 **49.** 1.58×10^{-6} moles/L

51. $10^{5.1}$ **53.** 3:00 A.M.

55. (a) $t_3 = 0.2729s - 6.0143$
$t_4 = 1.5385e^{0.02913s}$ or $t_4 = 1.5385(1.0296)^s$

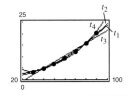

(b)

s	30	40	50	60	70	80	90
t_1	3.6	4.6	6.7	9.4	12.5	15.9	19.6
t_2	3.3	4.9	7.0	9.5	12.5	15.9	19.9
t_3	2.2	4.9	7.6	10.4	13.1	15.8	18.5
t_4	3.7	4.9	6.6	8.8	11.8	15.8	21.2

(c) Model t_1: Sum = 2.0, Model t_2: Sum = 1.1
Model t_3: Sum = 5.6, Model t_4: Sum = 2.7
The quadratic model (t_2) fits best.

57. False. The domain can be the set of real numbers for a logistic growth function.

59. True. The graph of a Gaussian model will never have a zero.

61. Gaussian **63.** Exponential growth

65. $\dfrac{x^2 + 4x - 9}{(x - 3)(x + 1)}$ **67.** $\dfrac{x - 1}{2}, x \neq -1, 0$

69. Quadrant III **71.** Quadrant IV

73.

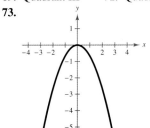

Reflection in the x-axis and a vertical shrink

75.

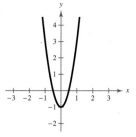

Vertical shift one unit downward and a horizontal shrink
77. All real numbers x
79. All real numbers x such that $x \neq 0$ and $x \neq -3$
81.

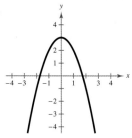

Review Exercises   *(page 412)*

1. 0.164   **3.** 1.587   **5.** 1456.529
7.

x	-1	0	1	2	3
$f(x)$	8	5	4.25	4.063	4.016

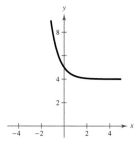

9.

x	-1	0	1	2	3
$f(x)$	4.008	4.04	4.2	5	9

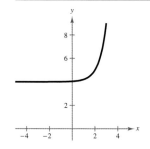

11.

x	-2	-1	0	1	2
$f(x)$	3.25	3.5	4	5	7

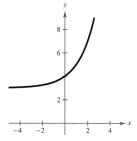

13. 1   **15.** 4   **17.** Shift the graph of f one unit up.
19. Reflect f in the x-axis and shift one unit up.
21. 29.964   **23.** 1.822
25.

x	-2	-1	0	1	2
$h(x)$	2.72	1.65	1	0.61	0.37

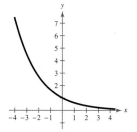

27.

x	-3	-2	-1	0	1
$f(x)$	0.37	1	2.72	7.39	20.09

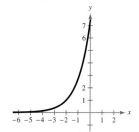

29. (a) 0.283   (b) 0.487   (c) 0.811
31.

n	1	2	4	12
A	\$6719.58	\$6734.28	\$6741.74	\$6746.77

n	365	Continuous
A	\$6749.21	\$6749.29

33. $\log_3 27 = 3$ **35.** $\ln 2.2255\ldots = 0.8$ **37.** 3
39. -2 **41.** 7   **43.** -5

45. Domain: $(0, \infty)$
x-intercept: $(1, 0)$
Vertical asymptote: $x = 0$

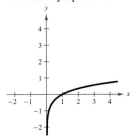

47. Domain: $(-5, \infty)$
x-intercept: $(9995, 0)$
Vertical asymptote: $x = -5$

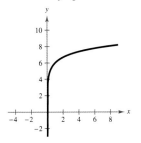

49. 3.118   **51.** 0.25

53. Domain: $(0, \infty)$
x-intercept: $(e^{-6}, 0)$
Vertical asymptote: $x = 0$

55. Domain: $(6, \infty)$
x-intercept: $(7, 0)$
Vertical asymptote: $x = 6$

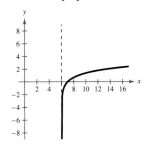

57. About 14.32 parsecs   **59.** (a) and (b) 2.585
61. (a) and (b) -2.322 **63.** $\log_2 5 - \log_2 3$
65. $2 \log_2 3 - \log_2 5$ **67.** $\log 7 + 2 \log x$
69. $2 - \frac{1}{2} \log_3 x$ **71.** $2 \ln x + 2 \ln y + \ln z$

73. $\ln 7x$ **75.** $\log \dfrac{x}{\sqrt{y}}$ **77.** $\log_3 \dfrac{\sqrt{x}}{(y + 8)^2}$

79. (a) $0 \le h < 18,000$
(b)

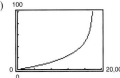

Vertical asymptote: $h = 18,000$
(c) The plane is climbing at a slower rate, so the time required increases.
(d) 5.46 min

81. 3   **83.** $\ln 3 \approx 1.099$ **85.** $e^4 \approx 54.598$ **87.** 1, 3

89. $\dfrac{\ln 32}{\ln 2} = 5$ **91.** $\frac{1}{3} e^{8.2} \approx 1213.650$

93. $\dfrac{3}{2} + \dfrac{\sqrt{9 + 4e}}{2} \approx 3.729$ **95.** No solution

97. 0.900   **99.** 2.447   **101.** 9.182   **103.** 73.2 yr
105. e   **106.** b   **107.** f   **108.** d   **109.** a   **110.** c
111. $y = 2e^{0.1014x}$
113.

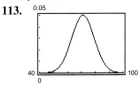

71

115. (a) 10^{-6} W/m^2 (b) $10\sqrt{10}$ W/m^2
(c) 1.259×10^{-12} W/m^2
117. True by the inverse properties.   **119.** Answers will vary.

Chapter Test *(page 415)*

1. 0.410   **2.** 0.032   **3.** 0.497   **4.** 22.198

5.

x	-1	$-\frac{1}{2}$	0	$\frac{1}{2}$	1
$f(x)$	10	3.162	1	0.316	0.1

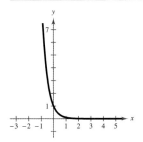

6.

x	-1	0	1	2	3
$f(x)$	-0.005	-0.028	-0.167	-1	-6

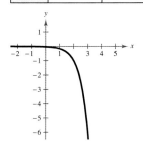

7.

x	-1	$-\frac{1}{2}$	0	$\frac{1}{2}$	1
$f(x)$	0.865	0.632	0	-1.718	-6.389

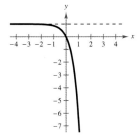

8. (a) -0.89 (b) 9.2
9. Domain: $(0, \infty)$
x-intercept: $(10^{-4}, 0)$
Vertical asymptote: $x = 0$

10. Domain: $(5, \infty)$
x-intercept: $(6, 0)$
Vertical asymptote: $x = 5$

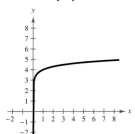

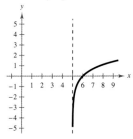

11. Domain: $(-6, \infty)$
x-intercept: $(e^{-1} - 6, 0)$
Vertical asymptote: $x = -6$

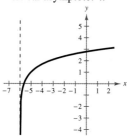

12. 2.209   **13.** -0.167 **14.** -11.047

15. $\log_2 3 + 4 \log_2 a$ **16.** $\frac{1}{2} \ln x - \ln 7$

17. $1 + 2 \log x - 3 \log y$ **18.** $\log_3 13y$ **19.** $\ln \dfrac{x^4}{y^4}$

20. $\ln \dfrac{x^3 y^2}{x + 3}$ **21.** $1, -1$ **22.** $\dfrac{\ln 44}{-5} \approx -0.757$

23. $\dfrac{\ln 197}{4} \approx 1.321$ **24.** $e^{-1/2} \approx 0.607$ **25.** $e^{-11/4} \approx 0.064$

26. 1   **27.** $y = 2745e^{0.1570t}$ **28.** 55%

29. (a)

x	$\frac{1}{4}$	1	2	4	5	6
H	58.720	75.332	86.828	103.43	110.59	117.38

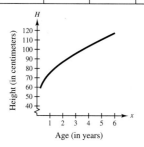

(b) 103 cm; 103.43 cm

Cumulative Test for Chapters 3–5   (page 416)

1. $y = -\frac{3}{4}(x + 8)^2 + 5$

2.

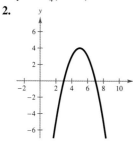

3.

4.

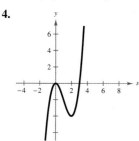

5. $-2, \pm 2i$ **6.** $-7, 0, 3$ **7.** $3x - 2 - \dfrac{3x - 2}{2x^2 + 1}$

8. $3x^3 + 6x^2 + 14x + 23 + \dfrac{49}{x - 2}$ **9.** 1.196

10. $f(x) = x^4 + 3x^3 - 11x^2 + 9x + 70$

11. $y = \frac{4}{3}x$

12. Domain: all real numbers
x except $x = 3$
Intercept: $(0, 0)$
Vertical asymptote: $x = 3$
Horizontal asymptote:
$y = 2$

13. Domain: all real numbers
x except $x = 5$
Intercept: $(0, 0)$
Vertical asymptote: $x = 5$
Slant asymptote:
$y = 4x + 20$

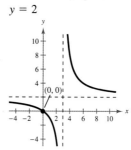

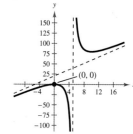

14. Domain: all real numbers
except -3 and 1
Intercept: $(0, 0)$
Vertical asymptotes:
$x = -3, x = 1$
Horizontal asymptote:
$y = 0$

15. Domain: all real numbers
except -2 and 1
y-intercept: $(0, 2)$
x-intercept: $(2, 0)$
Vertical asymptote: $x = 1$
Horizontal asymptote:
$y = 1$
Hole at $\left(-2, \frac{4}{3}\right)$

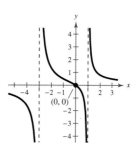

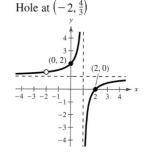

16. Domain: all real numbers except -3 and -1
y-intercept: $(0, 6)$
x-intercepts: $(2, 0), (3, 0)$
Vertical asymptote: $x = -1$
Slant asymptote: $y = x - 6$
Hole at $(-3, -15)$

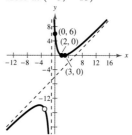

17.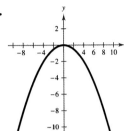

18.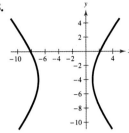

19.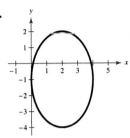

20. $(x - 3)^2 = \frac{3}{2}(y + 2)$

21. $\frac{(y + 2)^2}{9} - \frac{(x + 1)^2}{16} = 1$

22. Reflect f in the x-axis and y-axis, and shift three units to the right.

23. Reflect f in the x-axis, and shift four units up.

24. 1.991   **25.** -0.067 **26.** 1.717   **27.** 0.390

28. 0.906   **29.** -1.733 **30.** -4.087

31. $\ln(x + 5) + \ln(x - 5) - 4 \ln x$

32. $\ln \frac{x^2}{\sqrt{x + 5}}, x > 0$ **33.** $\frac{\ln 12}{2} \approx 1.242$

34. $\frac{\ln 9}{\ln 4} + 5 \approx 6.585$ **35.** $\ln 6 \approx 1.792$ or $\ln 7 \approx 1.946$

36. $\frac{64}{5} = 12.8$ **37.** $\frac{1}{2}e^8 \approx 1490.479$

38. $e^6 - 2 \approx 401.429$ **39.** $20,000

40. (a) $y = \frac{600}{x}$ (b) $(0, \infty)$

(c)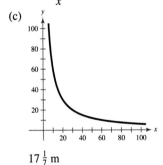

$17\frac{1}{7}$ m

41. $16,302.05   **42.** 6.3 h   **43.** 2026

44. (a) 300   (b) 570   (c) About 9 yr

Problem Solving   (page 419)

1.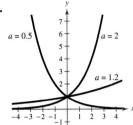

$y = 0.5^x$ and $y = 1.2^x$
$0 < a \le e^{1/e}$

3. As $x \to \infty$, the graph of e^x increases at a greater rate than the graph of x^n.

5. Answers will vary.

7. (a) 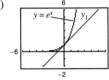   (b)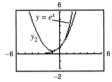

(c)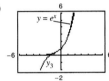

9.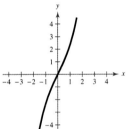

$f^{-1}(x) = \ln\left(\frac{x + \sqrt{x^2 + 4}}{2}\right)$

11. c

13. $t = \dfrac{\ln c_1 - \ln c_2}{\left(\dfrac{1}{k_2} - \dfrac{1}{k_1}\right)\ln \dfrac{1}{2}}$

15. (a) $y_1 = 252,606(1.0310)^t$

(b) $y_2 = 400.88t^2 - 1464.6t + 291,782$

(c)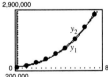

(d) The exponential model is a better fit. No, because the model is rapidly approaching infinity.

17. $1, e^2$

19. $y_4 = (x-1) - \frac{1}{2}(x-1)^2 + \frac{1}{3}(x-1)^3 - \frac{1}{4}(x-1)^4$

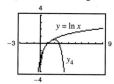

The pattern implies that
$\ln x = (x-1) - \frac{1}{2}(x-1)^2 + \frac{1}{3}(x-1)^3 - \cdots .$

21.

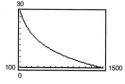

17.7 ft³/min

23. (a)

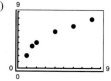

(b)–(e) Answers will vary.

25. (a)

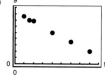

(b)–(e) Answers will vary.

Chapter 6

Section 6.1   *(page 430)*

1. Trigonometry   **3.** degree   **5.** $A = \frac{1}{2}r^2\theta$

7. To find an angle that is coterminal with a $5\pi/4$ angle, add or subtract a multiple of 2π.

9. Multiply by $\dfrac{\pi \text{ rad}}{180°}$; multiply by $\dfrac{180°}{\pi \text{ rad}}$.

11. $210°$ **13.** $-60°$

15. (a)   (b)

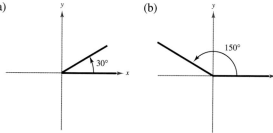

No; Quadrant I   No; Quadrant II

17. (a)   (b)

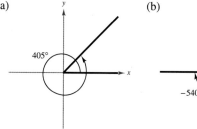

No; Quadrant I   Yes

19. (a) $480°, -240°$ (b) $150°, -570°$

21. (a) $60°, -660°$ (b) $20°, -340°$ **23.** $135.01°$

25. $-408.272°$ **27.** $2° 30'$ **29.** $-345° 7' 12''$

31. Complement: $72°$; Supplement: $162°$

33. Complement: none; Supplement: $87°$

35. Complement: $66°$; Supplement: $156°$

37. 2 rad   **39.** -3 rad

41. (a) Quadrant I   (b) Quadrant II

43. (a) Quadrant IV   (b) Quadrant III

45. (a) Quadrant IV   (b) Quadrant II

47. (a)   (b)

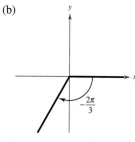

49. (a)   (b)

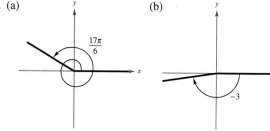

51. (a) $\dfrac{13\pi}{6}, -\dfrac{11\pi}{6}$ (b) $\dfrac{7\pi}{6}, -\dfrac{17\pi}{6}$

53. (a) $\dfrac{8\pi}{3}, -\dfrac{4\pi}{3}$ (b) $\dfrac{23\pi}{12}, -\dfrac{25\pi}{12}$

55. (a) $\dfrac{7\pi}{4}, -\dfrac{\pi}{4}$ (b) $\dfrac{28\pi}{15}, -\dfrac{32\pi}{15}$

57. (a) Complement: $\dfrac{5\pi}{12}$; Supplement: $\dfrac{11\pi}{12}$

   (b) Complement: none; Supplement: $\dfrac{\pi}{12}$

59. (a) Complement: none; Supplement: none

   (b) Complement: $\dfrac{\pi}{4}$; Supplement: $\dfrac{3\pi}{4}$

61. (a) $\dfrac{\pi}{6}$ (b) $\dfrac{\pi}{4}$ **63.** (a) $\dfrac{\pi}{9}$ (b) $-\dfrac{\pi}{3}$

65. (a) $270°$ (b) $210°$ **67.** (a) $75°$ (b) $-420°$

69. 0.785 **71.** 9.298 **73.** -0.014 **75.** $25.714°$

77. $-114.592°$ **79.** 10π in. ≈ 31.42 in.   **81.** 6 m

83. $\frac{9}{2}$ rad   **85.** $\frac{40}{29}$ rad   **87.** $\frac{6}{5}$ rad

89. 6π in.² ≈ 18.85 in.²   **91.** About 12.27 km²

93. About 592 mi   **95.** About 23.87°

97. About 436.97 km/min

99. (a) $10,400\pi$ rad/min $\approx 32,672.56$ rad/min

   (b) $\dfrac{9425\pi}{3}$ ft/min ≈ 9869.84 ft/min

101. (a) $\dfrac{1000\pi}{3}$ rad/sec ≈ 1047.20 rad/sec

   (b) 20π m/sec ≈ 62.83 m/sec

103.

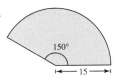

$A = 93.75\pi$ m$^2 \approx 294.52$ m^2

105. (a) $\dfrac{14\pi}{3}$ ft/sec; About 10 mi/h   (b) $d = \dfrac{7\pi}{7920}n$

(c) $d = \dfrac{7\pi}{7920}t$ (d) The functions are both linear.

107. False. $\dfrac{180°}{\pi}$ is in degree measure.

109. False. 1 rad $= \left(\dfrac{180}{\pi}\right)° \approx 57.3°$

111. The speed increases. The linear velocity is proportional to the radius.

113. 25   **115.** 12   **117.** $y^2 = -8x$

119. $\dfrac{x^2}{25} - \dfrac{y^2}{11} = 1$

Section 6.2 *(page 441)*

1. (a) 5   (b) 12   (c) 13   **3.** complementary

5. $\sin \theta = \frac{3}{5}$ $\csc \theta = \frac{5}{3}$

$\cos \theta = \frac{4}{5}$ $\sec \theta = \frac{5}{4}$

$\tan \theta = \frac{3}{4}$ $\cot \theta = \frac{4}{3}$

7. $\sin \theta = \dfrac{\sqrt{2}}{2}$ $\csc \theta = \sqrt{2}$

$\cos \theta = \dfrac{\sqrt{2}}{2}$ $\sec \theta = \sqrt{2}$

$\tan \theta = 1$ $\cot \theta = 1$

9. $\sin \theta = \frac{9}{41}$ $\csc \theta = \frac{41}{9}$

$\cos \theta = \frac{40}{41}$ $\sec \theta = \frac{41}{40}$

$\tan \theta = \frac{9}{40}$ $\cot \theta = \frac{40}{9}$

11. $\sin \theta = \frac{8}{17}$ $\csc \theta = \frac{17}{8}$

$\cos \theta = \frac{15}{17}$ $\sec \theta = \frac{17}{15}$

$\tan \theta = \frac{8}{15}$ $\cot \theta = \frac{15}{8}$

The triangles are similar, and corresponding sides are proportional.

13. $\sin \theta = \dfrac{1}{3}$ $\csc \theta = 3$

$\cos \theta = \dfrac{2\sqrt{2}}{3}$ $\sec \theta = \dfrac{3\sqrt{2}}{4}$

$\tan \theta = \dfrac{\sqrt{2}}{4}$ $\cot \theta = 2\sqrt{2}$

The triangles are similar, and corresponding sides are proportional.

15.

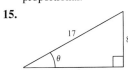

$\sin \theta = \frac{8}{17}$ $\csc \theta = \frac{17}{8}$

$\sec \theta = \frac{17}{15}$

$\tan \theta = \frac{8}{15}$ $\cot \theta = \frac{15}{8}$

17.

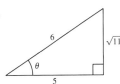

$\sin \theta = \dfrac{\sqrt{11}}{6}$ $\csc \theta = \dfrac{6\sqrt{11}}{11}$

$\cos \theta = \dfrac{5}{6}$

$\tan \theta = \dfrac{\sqrt{11}}{5}$ $\cot \theta = \dfrac{5\sqrt{11}}{11}$

$\csc \theta = 5$

19. $\cos \theta = \dfrac{2\sqrt{6}}{5}$ $\sec \theta = \dfrac{5\sqrt{6}}{12}$

$\tan \theta = \dfrac{\sqrt{6}}{12}$ $\cot \theta = 2\sqrt{6}$

21. $\sin \theta = \dfrac{\sqrt{10}}{10}$ $\csc \theta = \sqrt{10}$

$\cos \theta = \dfrac{3\sqrt{10}}{10}$ $\sec \theta = \dfrac{\sqrt{10}}{3}$

$\tan \theta = \dfrac{1}{3}$

23. $\dfrac{\pi}{6}; \dfrac{\sqrt{3}}{3}$ **25.** $45°; \sqrt{2}$ **27.** (a) 0.3453   (b) 1.0655

29. (a) 5.0273   (b) 0.1989   **31.** (a) 1.1884   (b) 0.5463

33. (a) $\dfrac{2\sqrt{3}}{3}$ (b) $\dfrac{1}{2}$ (c) $\sqrt{3}$ (d) $\dfrac{\sqrt{3}}{3}$

35. (a) $\dfrac{2\sqrt{2}}{3}$ (b) $2\sqrt{2}$ (c) 3   (d) 3

37. (a) $\dfrac{1}{3}$ (b) $\sqrt{10}$ (c) $\dfrac{\sqrt{10}}{3}$ (d) $\dfrac{\sqrt{10}}{10}$

39–47. Answers will vary.

49. (a) $30° = \dfrac{\pi}{6}$ (b) $30° = \dfrac{\pi}{6}$

51. (a) $60° = \dfrac{\pi}{3}$ (b) $45° = \dfrac{\pi}{4}$

53. (a) $60° = \dfrac{\pi}{3}$ (b) $45° = \dfrac{\pi}{4}$ **55.** $x = 9, y = 9\sqrt{3}$

57. $x = \dfrac{32\sqrt{3}}{3}, r = \dfrac{64\sqrt{3}}{3}$

59. About 443.2 m; about 323.3 m

61. $30° = \dfrac{\pi}{6}$ **63.** (a) About 219.9 ft   (b) About 160.9 ft

65. $(x_1, y_1) = \left(28\sqrt{3}, 28\right)$
$(x_2, y_2) = \left(28, 28\sqrt{3}\right)$

67. $\sin 20° \approx 0.34, \cos 20° \approx 0.94, \tan 20° \approx 0.36,$
$\csc 20° \approx 2.92, \sec 20° \approx 1.06, \cot 20° \approx 2.75$

69. (a) About 519.33 ft   (b) About 1174.17 ft

71. True. $\csc x = \dfrac{1}{\sin x}$ **73.** False. $\dfrac{\sqrt{2}}{2} + \dfrac{\sqrt{2}}{2} \neq 1$

75. False. $1.7321 \neq 0.0349$

77. Yes, $\tan \theta$ is equal to opp/adj. You can find the value of the hypotenuse by the Pythagorean Theorem. Then you can find $\sec \theta$, which is equal to hyp/adj.

79. $\cos 60° = \dfrac{\text{adj}}{\text{hyp}} = \dfrac{1}{2}$

81. Quadrant I   **83.** Quadrant II

85. Quadrants II and IV; *Sample answer:* $(1, -1), (-1, 1)$

87. Quadrants I and III; *Sample answer:* $(1, 2), (-1, -2)$

89. (a) $380°, -340°$ (b) $330°, -390°$

91. Odd; origin symmetry   **93.** Even; y-axis symmetry
95. Neither; no symmetry

Section 6.3 *(page 453)*

1. periodic   **3.** Reference angle

5. (a) $\sin \theta = \frac{3}{5}$ $\csc \theta = \frac{5}{3}$
$\cos \theta = \frac{4}{5}$ $\sec \theta = \frac{5}{4}$
$\tan \theta = \frac{3}{4}$ $\cot \theta = \frac{4}{3}$

(b) $\sin \theta = \frac{15}{17}$ $\csc \theta = \frac{17}{15}$
$\cos \theta = -\frac{8}{17}$ $\sec \theta = -\frac{17}{8}$
$\tan \theta = -\frac{15}{8}$ $\cot \theta = -\frac{8}{15}$

7. (a) $\sin \theta = -\frac{1}{2}$ $\csc \theta = -2$

$\cos \theta = -\frac{\sqrt{3}}{2}$ $\sec \theta = -\frac{2\sqrt{3}}{3}$

$\tan \theta = \frac{\sqrt{3}}{3}$ $\cot \theta = \sqrt{3}$

(b) $\sin \theta = -\frac{\sqrt{17}}{17}$ $\csc \theta = -\sqrt{17}$

$\cos \theta = \frac{4\sqrt{17}}{17}$ $\sec \theta = \frac{\sqrt{17}}{4}$

$\tan \theta = -\frac{1}{4}$ $\cot \theta = -4$

9. $\sin \theta = \frac{12}{13}$ $\csc \theta = \frac{13}{12}$
$\cos \theta = \frac{5}{13}$ $\sec \theta = \frac{13}{5}$
$\tan \theta = \frac{12}{5}$ $\cot \theta = \frac{5}{12}$

11. $\sin \theta = -\frac{2\sqrt{29}}{29}$ $\csc \theta = -\frac{\sqrt{29}}{2}$

$\cos \theta = -\frac{5\sqrt{29}}{29}$ $\sec \theta = -\frac{\sqrt{29}}{5}$

$\tan \theta = \frac{2}{5}$ $\cot \theta = \frac{5}{2}$

13. $\sin \theta = \frac{4}{5}$ $\csc \theta = \frac{5}{4}$
$\cos \theta = -\frac{3}{5}$ $\sec \theta = -\frac{5}{3}$
$\tan \theta = -\frac{4}{3}$ $\cot \theta = -\frac{3}{4}$

15. Quadrant I   **17.** Quadrant II

19. $\sin \theta = \frac{15}{17}$ $\csc \theta = \frac{17}{15}$
$\cos \theta = \frac{8}{17}$ $\sec \theta = \frac{17}{8}$
        $\cot \theta = \frac{8}{15}$

21. $\csc \theta = \frac{5}{3}$
$\cos \theta = -\frac{4}{5}$ $\sec \theta = -\frac{5}{4}$
$\tan \theta = -\frac{3}{4}$ $\cot \theta = -\frac{4}{3}$

23. $\sin \theta = -\frac{\sqrt{10}}{10}$ $\csc \theta = -\sqrt{10}$

$\cos \theta = \frac{3\sqrt{10}}{10}$ $\sec \theta = \frac{\sqrt{10}}{3}$

$\tan \theta = -\frac{1}{3}$

25. $\sin \theta = 1$
        $\sec \theta$ is undefined.
$\tan \theta$ is undefined. $\cot \theta = 0$

27. $\sin \theta = 0$ $\csc \theta$ is undefined.
$\cos \theta = -1$ $\sec \theta = -1$
$\tan \theta = 0$ $\cot \theta$ is undefined.

29. $\sin \theta = \frac{\sqrt{2}}{2}$ $\csc \theta = \sqrt{2}$

$\cos \theta = -\frac{\sqrt{2}}{2}$ $\sec \theta = -\sqrt{2}$

$\tan \theta = -1$ $\cot \theta = -1$

31. $\sin \theta = \frac{2\sqrt{5}}{5}$ $\csc \theta = \frac{\sqrt{5}}{2}$

$\cos \theta = \frac{\sqrt{5}}{5}$ $\sec \theta = \sqrt{5}$

$\tan \theta = 2$ $\cot \theta = \frac{1}{2}$

33. 0   **35.** Undefined   **37.** 1   **39.** Undefined
41. 0

43. $\theta' = 20°$ **45.** $\theta' = 55°$

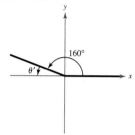

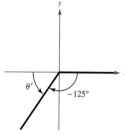

47. $\theta' = \frac{\pi}{3}$ **49.** $\theta' = 2\pi - 4.8$

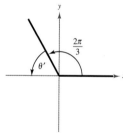

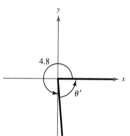

51. $\sin 225° = -\frac{\sqrt{2}}{2}$ **53.** $\sin 750° = \frac{1}{2}$

$\cos 225° = -\frac{\sqrt{2}}{2}$ $\cos 750° = \frac{\sqrt{3}}{2}$

$\tan 225° = 1$ $\cos 750° = \frac{\sqrt{3}}{2}$

55. $\sin(-120°) = -\frac{\sqrt{3}}{2}$ **57.** $\sin \frac{2\pi}{3} = \frac{\sqrt{3}}{2}$

$\cos(-120°) = -\frac{1}{2}$ $\cos \frac{2\pi}{3} = -\frac{1}{2}$

$\tan(-120°) = \sqrt{3}$ $\tan \frac{2\pi}{3} = -\sqrt{3}$

59. $\sin\left(-\frac{\pi}{6}\right) = -\frac{1}{2}$ **61.** $\sin \frac{11\pi}{4} = \frac{\sqrt{2}}{2}$

$\cos\left(-\frac{\pi}{6}\right) = \frac{\sqrt{3}}{2}$ $\cos \frac{11\pi}{4} = -\frac{\sqrt{2}}{2}$

$\tan\left(-\frac{\pi}{6}\right) = -\frac{\sqrt{3}}{3}$ $\tan \frac{11\pi}{4} = -1$

63. $\sin\left(-\dfrac{17\pi}{6}\right) = -\dfrac{1}{2}$

$\cos\left(-\dfrac{17\pi}{6}\right) = -\dfrac{\sqrt{3}}{2}$

$\tan\left(-\dfrac{17\pi}{6}\right) = \dfrac{\sqrt{3}}{3}$

65. $\dfrac{4}{5}$ **67.** $-\dfrac{\sqrt{13}}{12}$ **69.** $\dfrac{8\sqrt{39}}{39}$ **71.** 0.1736

73. -0.3420 **75.** -28.6363 **77.** 1.4142

79. 0.3640   **81.** -2.6131 **83.** -0.6052 **85.** 1.8382

87. (a) $30° = \dfrac{\pi}{6}, 150° = \dfrac{5\pi}{6}$ (b) $210° = \dfrac{7\pi}{6}, 330° = \dfrac{11\pi}{6}$

89. (a) $60° = \dfrac{\pi}{3}, 300° = \dfrac{5\pi}{3}$ (b) $60° = \dfrac{\pi}{3}, 300° = \dfrac{5\pi}{3}$

91. (a) $45° = \dfrac{\pi}{4}, 225° = \dfrac{5\pi}{4}$ (b) $150° = \dfrac{5\pi}{6}, 330° = \dfrac{11\pi}{6}$

93. $\left(\dfrac{\sqrt{2}}{2}, \dfrac{\sqrt{2}}{2}\right)$ **95.** $\left(-\dfrac{\sqrt{3}}{2}, \dfrac{1}{2}\right)$

$\sin\dfrac{\pi}{4} = \dfrac{\sqrt{2}}{2}$ $\sin\dfrac{5\pi}{6} = \dfrac{1}{2}$

$\cos\dfrac{\pi}{4} = \dfrac{\sqrt{2}}{2}$ $\cos\dfrac{5\pi}{6} = -\dfrac{\sqrt{3}}{2}$

$\tan\dfrac{\pi}{4} = 1$ $\tan\dfrac{5\pi}{6} = -\dfrac{\sqrt{3}}{3}$

97. $\left(-\dfrac{1}{2}, -\dfrac{\sqrt{3}}{2}\right)$ **99.** $(0, 1)$

$\sin\dfrac{4\pi}{3} = -\dfrac{\sqrt{3}}{2}$ $\sin\dfrac{\pi}{2} = 1$

$\cos\dfrac{4\pi}{3} = -\dfrac{1}{2}$ $\cos\dfrac{\pi}{2} = 0$

$\tan\dfrac{4\pi}{3} = \sqrt{3}$ $\tan\dfrac{\pi}{2}$ is undefined.

101. (a) -1 (b) -0.4

103. (a) 0.25 or 2.89   (b) 1.82 or 4.46

105. (a) 2 cm   (b) About 0.14 cm   (c) About -1.98 cm

107. (a) $B = 21.865 \sin(0.540t - 2.343) + 60.438$
$F = 61.120 \sin(0.325t - 0.561) + 15.787$

(b) February: $B \approx 39.6°, F \approx 21.2°$
April: $B \approx 56.5°, F \approx 57.0°$
May: $B \approx 68.1°, F \approx 69.2°$
July: $B \approx 82.1°, F \approx 76.3°$
September: $B \approx 73.2°, F \approx 58.7°$
October: $B \approx 62.3°, F \approx 42.5°$
December: $B \approx 42.1°, F \approx 3.8°$

(c)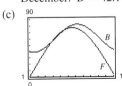

Answers will vary.

109. About 0.79 amp

111. False. In each of the four quadrants, the signs of the secant function and the cosine function are the same because these functions are reciprocals of each other.

113. Odd; $h(-t) = \sin(-t)\cos(-t)$
$= -\sin t \cos t$
$= -h(t)$

115. The calculator was in degree mode instead of radian mode.

117. (a)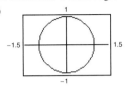

Circle of radius 1 centered at $(0, 0)$

(b) The t-values represent the central angle in radians. The x- and y-values represent the location in the coordinate plane.

(c) $-1 \le x \le 1, -1 \le y \le 1$

119. (a) $f(x) = \sqrt{x}$

(b) Reflection in the x-axis, left shift of 12 units, and an upward shift of nine units

121. Relative maximum: $(-4, 5)$ **123.** $-12, 11$

125. $5 + \ln 2 \approx 5.693$

Section 6.4   *(page 464)*

1. cycle   **3.** Vertical shift of d units

5. Amplitude: 2; Period: $\dfrac{2\pi}{5}$

7. Amplitude: $\dfrac{3}{4}$; Period: 4

9. Amplitude: $\dfrac{1}{2}$; Period: $\dfrac{8\pi}{5}$

11. Amplitude: $\dfrac{5}{3}$; Period: 24

13. The period of g is one-fifth the period of f.

15. g is a reflection of f in the x-axis.

17. g is a shift of f π units to the right.

19. g is a shift of f three units up.

21. The graph of g has twice the amplitude of the graph of f.

23. The graph of g is a horizontal shift of the graph of f π units to the right.

25. 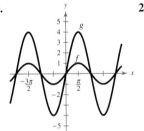   **27.**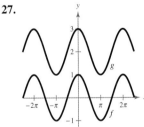

29. 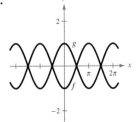   **31.**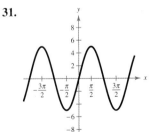

33.

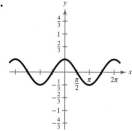

35.

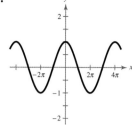

37.

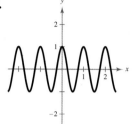

39.

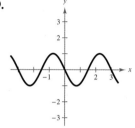

41.

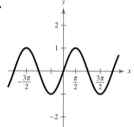

43.

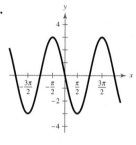

45.

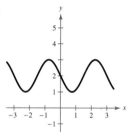

47.

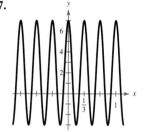

49.

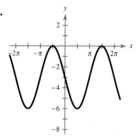

51.

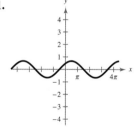

53. (a) Horizontal shrink and a phase shift $\pi/4$ unit right

(b)   (c) $g(x) = f(4x - \pi)$

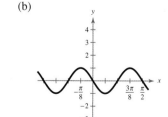

55. (a) Shift two units up and a phase shift $\pi/2$ units right

(b)   (c) $g(x) = f\left(x - \dfrac{\pi}{2}\right) + 2$

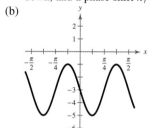

57. (a) Horizontal shrink, a vertical stretch, a shift three units down, and a phase shift $\pi/4$ unit right

(b)   (c) $g(x) = 2f(4x - \pi) - 3$

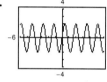

59.

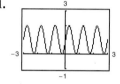

61.

63.

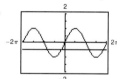

65. $a = 2, d = 1$ **67.** $a = -4, d = 4$

69. $a = -3, b = 2, c = 0$ **71.** $a = 2, b = 1, c = -\dfrac{\pi}{4}$

73.

$$x = -\frac{\pi}{6}, -\frac{5\pi}{6}, \frac{7\pi}{6}, \frac{11\pi}{6}$$

75. $y = 1 + 2\sin(2x - \pi)$ **77.** $y = \cos(2x + 2\pi) - \dfrac{3}{2}$

79. (a) 4 sec

(b) 15 cycles/min

(c)

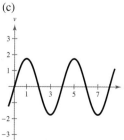

CHAPTER 6

81. (a) $\frac{6}{5}$ sec   (b) 50 heartbeats/min

83. (a) and (c)

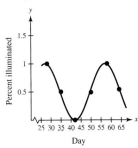

The model fits the data well.

(b) $y = 0.5 \cos\left(\dfrac{\pi}{15}x - \dfrac{9\pi}{5}\right) + 0.5$

(d) 30 days   (e) 100%

85. (a) 20 sec; It takes 20 seconds to complete one revolution on the Ferris wheel.

(b) 50 ft; The diameter of the Ferris wheel is 100 feet.

(c)

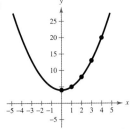

87. False. The function $y = \frac{1}{2}\cos 2x$ has an amplitude that is one-half that of $y = \cos x$. For $y = a \cos bx$, the amplitude is $|a|$.

89. The key points are for the graph of $y = \cos x$.

91.

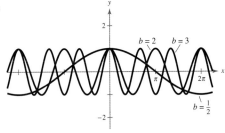

The value of b affects the period of the graph.

$b = \frac{1}{2} \rightarrow \frac{1}{2}$ cycle

$b = 2 \rightarrow 2$ cycles

$b = 3 \rightarrow 3$ cycles

93. (a)

x	0	1	2	3	4
y	4	5	8	13	20

(b) y-axis symmetry

(c)

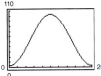

95. (a)

x	0	$\frac{1}{2}$	1	$\frac{3}{2}$	2
y	0	$-\frac{3}{8}$	0	$\frac{15}{8}$	6

(b) origin symmetry

(c)

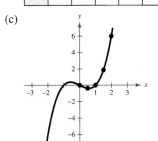

97.

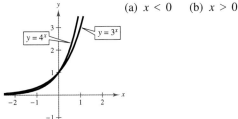

(a) $x < 0$ (b) $x > 0$

Section 6.5   (page 475)

1. odd; origin   **3.** damping   **5.** $(-\infty, -1] \cup [1, \infty)$

7. e, π **8.** c, 2π **9.** a, 1   **10.** d, 2π

11. f, 4   **12.** b, 4

13.

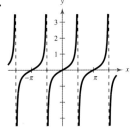

15.

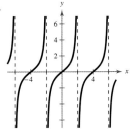

17.

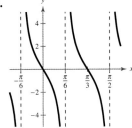

19.

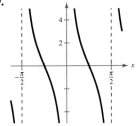

21.

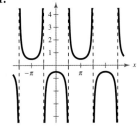

23.

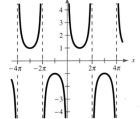

25.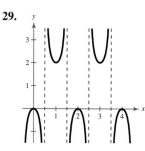

27.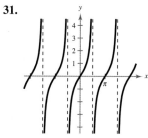

29.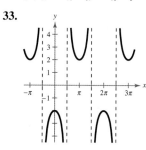

31.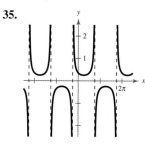

33.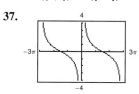

35.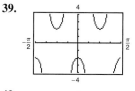

37.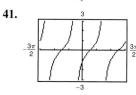

39.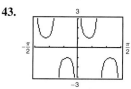

41.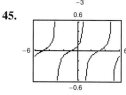

43.

45.

47. $-\dfrac{7\pi}{4}, -\dfrac{3\pi}{4}, \dfrac{\pi}{4}, \dfrac{5\pi}{4}$ **49.** $-\dfrac{7\pi}{6}, -\dfrac{\pi}{6}, \dfrac{5\pi}{6}, \dfrac{11\pi}{6}$

51. $-\dfrac{5\pi}{3}, -\dfrac{\pi}{3}, \dfrac{\pi}{3}, \dfrac{5\pi}{3}$

53. Even   **55.** Odd   **57.** Odd   **59.** Even

61. d, $f \to 0$ as $x \to 0$.   **63.** b, $g \to 0$ as $x \to 0$.

65.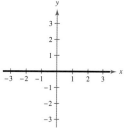

The functions are equal.

67.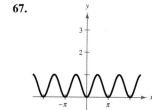

The functions are equal.

69.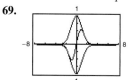

As $x \to \infty$, $g(x) \to 0$.

71.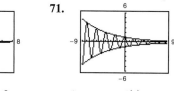

As $x \to \infty$, $f(x) \to 0$.

73.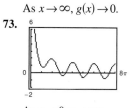

As $x \to 0$, $y \to \infty$.

75.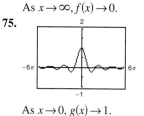

As $x \to 0$, $g(x) \to 1$.

77.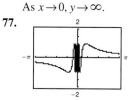

As $x \to 0$, $f(x)$ oscillates between 1 and -1.

79. (a) Period of $H(t)$: 12 mo
      Period of $L(t)$: 12 mo
   (b) Summer; winter   (c) About 0.5 mo

81. $d = 27 \sec x$

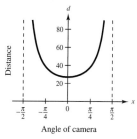

83. True. For a given value of x, the y-coordinate of $\csc x$ is the reciprocal of the y-coordinate of $\sin x$.

85. (a)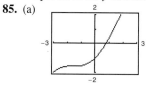

   0.7391

   (b) 1, 0.5403, 0.8576, 0.6543, 0.7935, 0.7014, 0.7640, 0.7221, 0.7504, 0.7314, . . . ; 0.7391

87. (a) $x^{2/3}$ (b) $x^{2/3}$
   Domains of f, g, $f \circ g$, and $g \circ f$: all real numbers x

CHAPTER 6

89. (a) $(x - 1)^2$ (b) $\sqrt{x^4 - 1}$
Domain of f: all real numbers x
Domains of g and $f \circ g$: all real numbers x such that $x \geq 1$
Domain of $g \circ f$: all real numbers x such that $x \leq -1$
or $x \geq 1$

91. No inverse function   **93.** $h^{-1}(x) = \dfrac{1}{\sqrt[3]{x}}$

95. $f^{-1}(x) = \frac{1}{3}(x^2 + 6), x \geq 0$ **97.** $\ln \frac{1}{2} \approx -0.693$

Section 6.6 *(page 484)*

1. $y = \sin^{-1} x; -1 \leq x \leq 1$

3. $y = \tan^{-1} x; -\infty < x < \infty; -\dfrac{\pi}{2} < y < \dfrac{\pi}{2}$

5. $\csc^{-1} x$ or $\operatorname{arccsc} x$ **7.** $\dfrac{\pi}{6}$ **9.** $\dfrac{\pi}{2}$ **11.** $\dfrac{\pi}{6}$

13. Not possible   **15.** $-\dfrac{\pi}{3}$ **17.** $\dfrac{2\pi}{3}$ **19.** $-\dfrac{\pi}{3}$

21.

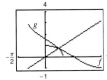

23. 1.19   **25.** -0.85 **27.** -1.25 **29.** Not possible

31. Not possible   **33.** -1.50

35. $-\dfrac{\pi}{3}, -\dfrac{\sqrt{3}}{3}, 1$ **37.** $\theta = \arctan \dfrac{x}{4}$

39. $\theta = \arcsin \dfrac{x + 2}{5}$ **41.** 0.3   **43.** $\dfrac{\pi}{4}$ **45.** $\dfrac{3}{5}$

47. $\dfrac{\sqrt{5}}{5}$ **49.** $\dfrac{13}{12}$ **51.** $-\dfrac{5}{3}$ **53.** $-\dfrac{\sqrt{5}}{2}$ **55.** 2

57. $\sqrt{1 - 4x^2}, 0 \leq x \leq \frac{1}{2}$ **59.** $\dfrac{1}{x}, x > 0$

61. $\sqrt{1 - x^2}, 0 \leq x \leq 1$

63. $\dfrac{\sqrt{9 - x^2}}{x}, 0 < x \leq 3$ **65.** $\dfrac{\sqrt{x^2 + a^2}}{x}, a \geq 0, x > 0$

67.

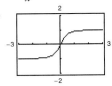

Asymptotes: $y = \pm 1$

69. $\dfrac{9}{\sqrt{x^2 + 81}}$ **71.** $\dfrac{|x - 1|}{\sqrt{x^2 - 2x + 10}}$

73.

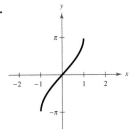

Vertical stretch

75.

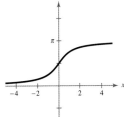

Shift $\dfrac{\pi}{2}$ units up

77.

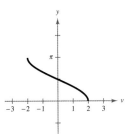

Horizontal stretch

79.

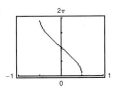

81.

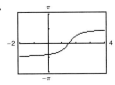

83.

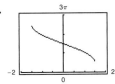

85. $3\sqrt{2} \sin\left(2t + \dfrac{\pi}{4}\right)$

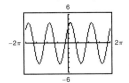

The graph implies that the identity is true.

87. $\dfrac{\pi}{2}$ **89.** $\dfrac{\pi}{2}$ **91.** π

93. (a) $\theta = \arcsin \dfrac{5}{s}$ (b) About 0.13, about 0.25

95. (a) About 32.9°   (b) About 6.5 m

97. (a) $\theta = \arcsin \dfrac{6}{x}$ (b) 30°; about 59.0°

99. False. $\arctan(-1) = -\dfrac{\pi}{4} \neq \dfrac{\arcsin(-1)}{\arccos(-1)} = \dfrac{-\pi/2}{\pi} = -\dfrac{1}{2}$

101. False. $\dfrac{5\pi}{6}$ is not in the range of the arcsine function.

103. $\dfrac{5\pi}{4}$ is not in the range of the arcsine function.

105. Domain: $(-\infty, \infty)$
Range: $(0, \pi)$

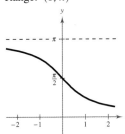

107. Domain: $(-\infty, -1] \cup [1, \infty)$
Range: $[-\pi/2, 0) \cup (0, \pi/2]$

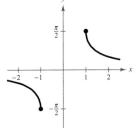

109. $\dfrac{\pi}{4}$ **111.** $\dfrac{3\pi}{4}$ **113.** $\dfrac{\pi}{6}$ **115.** $\dfrac{\pi}{3}$ **117.** $\dfrac{2\pi}{3}$

119. 1.17   **121.** -0.12 **123.** 0.19   **125.** 0.54

127. (a) $f \circ f^{-1}$ $f^{-1} \circ f$

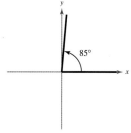
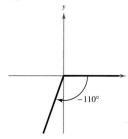

(b) The domains and ranges of the functions are restricted. The graphs of $f \circ f^{-1}$ and $f^{-1} \circ f$ differ because of the domains and ranges of f and f^{-1}.

129. (a) $\dfrac{\pi}{4}$ (b) $\dfrac{\pi}{2}$ (c) About 1.25   (d) About 2.03

131. $4\sqrt{5}$ **133.** $4\sqrt{2}$ **135.** $\cos 40° = \dfrac{c}{5}$

137. $\tan 40° = \dfrac{5}{c}$

Section 6.7   *(page 494)*

1. bearing   **3.** No

5. $a \approx 10.39$ **7.** $b \approx 14.21$
   $b = 6$ $c \approx 14.88$
   $B = 30°$ $A = 17.2°$

9. $c = 5$ **11.** $a \approx 52.88$
   $A \approx 36.87°$ $A \approx 73.46°$
   $B \approx 53.13°$ $B \approx 16.54°$

13. 3.00   **15.** 2.50   **17.** About 214.45 ft

19. About 19.7 ft   **21.** About 20.5 ft   **23.** About 11.8 km

25. (a) About 25.98 ft   (b) $\arctan \dfrac{25.98}{d}$

   (c) 45.0 ft $\le d \le$ 55.7 ft

27. About 56.3°   **29.** About 75.97°   **31.** N 56.31° W

33. (a) About 429.26 mi north, about 2434.44 mi west
   (b) 280°

35. (a) About 4820.7 ft   (b) About 124.5 sec

37. (a) N 58° E   (b) About 68.82 m   **39.** About 29.4 in.

41. About 35.3°   **43.** $d = 4 \sin \pi t$ **45.** $d = 3 \cos \dfrac{4\pi t}{3}$

47. $\omega = 524\pi$ **49.** (a) 9   (b) $\frac{3}{5}$ (c) 9   (d) $\frac{5}{12}$

51. (a) $\frac{1}{4}$ (b) 3   (c) 0   (d) $\frac{1}{6}$

53. (a)

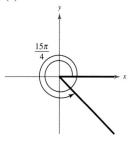

(b) $\dfrac{\pi}{8}$ (c) $\dfrac{\pi}{32}$

55. (a)

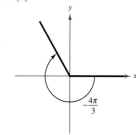

(b) $S = 8 + 6.3 \cos \dfrac{\pi}{6} t$ or $S = 8 + 6.3 \sin\left(\dfrac{\pi}{6}t + \dfrac{\pi}{2}\right)$

   The model is a good fit.

(c) 12. Yes, sales of outerwear are seasonal.

(d) Maximum displacement from average sales of $8 million

57. False. The scenario does not create a right triangle because the tower is not vertical.

59. $2t(3t^2 + 5t - 1)$ **61.** $(5y + 1)(2y - 3)$

63. $(5z^2 + x)(5z^2 - x)$

65. (a) $\dfrac{\sqrt{15}}{4}$ (b) $\sqrt{15}$ (c) 4   (d) $\sqrt{15}$

67. $\log_2 5x$ **69.** $\log \dfrac{x^3}{(x + 2)(x - 2)}$

Review Exercises   *(page 500)*

1. 60°

3. (a)   **5.** (a)

(b) Quadrant I   (b) Quadrant III
(c) 445°, $-275°$ (c) 250°, $-470°$

7. (a)   **9.** (a)

(b) Quadrant IV   (b) Quadrant II
(c) $\dfrac{7\pi}{4}, -\dfrac{\pi}{4}$ (c) $\dfrac{2\pi}{3}, -\dfrac{10\pi}{3}$

11. 7.854   **13.** -0.279 **15.** 0.355   **17.** -0.149

19. 54°   **21.** $-108°$ **23.** 297.938°   **25.** $-123.186°$

27. About 48.17 in.   **29.** $\dfrac{500\pi}{3} \approx 523.6$ in.^{2}

31. $\sin\theta = \dfrac{4\sqrt{41}}{41}$　　　　$\csc\theta = \dfrac{\sqrt{41}}{4}$

$\cos\theta = \dfrac{5\sqrt{41}}{41}$　　　　$\sec\theta = \dfrac{\sqrt{41}}{5}$

$\tan\theta = \dfrac{4}{5}$　　　　　$\cot\theta = \dfrac{5}{4}$

33. (a) 0.7782　　(b) 5.3860

35. (a) 3　(b) $\dfrac{2\sqrt{2}}{3}$　(c) $\dfrac{3\sqrt{2}}{4}$　(d) $\dfrac{\sqrt{2}}{4}$

37. (a) $\dfrac{1}{4}$　(b) $\dfrac{\sqrt{15}}{4}$　(c) $\dfrac{4\sqrt{15}}{15}$　(d) $\dfrac{\sqrt{15}}{15}$

39. About 73.3 m

41. $\sin\theta = \frac{4}{5}$　　　　$\csc\theta = \frac{5}{4}$

$\cos\theta = \frac{3}{5}$　　　　$\sec\theta = \frac{5}{3}$

$\tan\theta = \frac{4}{3}$　　　　$\cot\theta = \frac{3}{4}$

43. $\sin\theta = \dfrac{9\sqrt{82}}{82}$　　　$\csc\theta = \dfrac{\sqrt{82}}{9}$

$\cos\theta = -\dfrac{\sqrt{82}}{82}$　　$\sec\theta = -\sqrt{82}$

$\tan\theta = -9$　　　　$\cot\theta = -\dfrac{1}{9}$

45. $\sin\theta = -\dfrac{\sqrt{11}}{6}$　　$\csc\theta = -\dfrac{6\sqrt{11}}{11}$

$\cos\theta = \dfrac{5}{6}$　　　　$\cot\theta = -\dfrac{5\sqrt{11}}{11}$

$\tan\theta = -\dfrac{\sqrt{11}}{5}$

47. $\sin\theta = -\dfrac{7\sqrt{58}}{58}$　　$\sec\theta = -\dfrac{\sqrt{58}}{3}$

$\cos\theta = -\dfrac{3\sqrt{58}}{58}$　　$\cot\theta = \dfrac{3}{7}$

$\csc\theta = -\dfrac{\sqrt{58}}{7}$

49. $\sin\theta = \frac{40}{41}$　　　　$\sec\theta = -\frac{41}{9}$

$\cos\theta = -\frac{9}{41}$　　　$\cot\theta = -\frac{9}{40}$

$\csc\theta = \frac{41}{40}$

51. $\theta' = 84°$　　　　**53.** $\theta' = \dfrac{\pi}{5}$

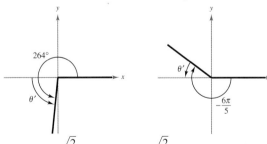

55. $\sin 495° = \dfrac{\sqrt{2}}{2}$; $\cos 495° = -\dfrac{\sqrt{2}}{2}$; $\tan 495° = -1$

57. $\sin(-150°) = -\dfrac{1}{2}$; $\cos(-150°) = -\dfrac{\sqrt{3}}{2}$;

$\tan(-150°) = \dfrac{\sqrt{3}}{3}$

59. $\sin\dfrac{\pi}{3} = \dfrac{\sqrt{3}}{2}$; $\cos\dfrac{\pi}{3} = \dfrac{1}{2}$; $\tan\dfrac{\pi}{3} = \sqrt{3}$

61. $\sin\left(-\dfrac{7\pi}{3}\right) = -\dfrac{\sqrt{3}}{2}$; $\cos\left(-\dfrac{7\pi}{3}\right) = \dfrac{1}{2}$;

$\tan\left(-\dfrac{7\pi}{3}\right) = -\sqrt{3}$

63. 0.9613　　**65.** -0.4452

67. $\left(-\dfrac{1}{2}, \dfrac{\sqrt{3}}{2}\right)$; $\sin t = \dfrac{\sqrt{3}}{2}$, $\cos t = -\dfrac{1}{2}$, $\tan t = -\sqrt{3}$

69. $\left(-\dfrac{\sqrt{3}}{2}, -\dfrac{1}{2}\right)$; $\sin t = -\dfrac{1}{2}$, $\cos t = -\dfrac{\sqrt{3}}{2}$, $\tan t = \dfrac{\sqrt{3}}{3}$

71.　　　　　　　　　　**73.**

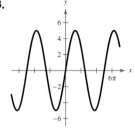

75.　　　　　　　　　　**77.**

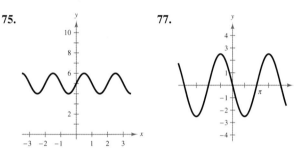

79. (a) $y = 2\sin 528\pi x$　　(b) 264 cycles/sec

81.　　　　　　　　　　**83.**

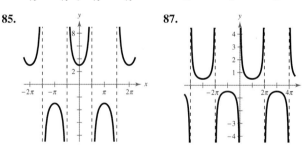

85.　　　　　　　　　　**87.**

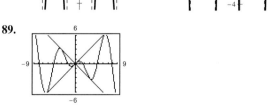

89.

As $x \to \infty$, $f(x)$ oscillates.

91. $-\dfrac{\pi}{6}$　**93.** $\dfrac{3\pi}{4}$　**95.** 0　　**97.** Not possible

99. -0.17 **101.** 1.19 **103.** -0.92

105. 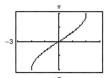   **107.**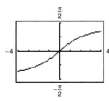

109. $\dfrac{4}{5}$ **111.** $\dfrac{13}{5}$ **113.** $\dfrac{\sqrt{4-x^2}}{x}$

115.

$\theta \approx 66.8°$

117. About 42.43 nm north, about 42.43 nm east

119. False. For each θ there corresponds exactly one value of y.

121. The function is undefined because $\sec \theta = 1/\cos \theta$.

123. The ranges of the other four trigonometric functions are $(-\infty, \infty)$ or $(-\infty, -1] \cup [1, \infty)$.

125. (a) $A = 72(\tan \theta - \theta)$

(b)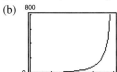

Area increases without bound as θ approaches $\pi/2$.

Chapter Test   *(page 503)*

1. (a) 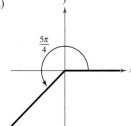   (b) $\dfrac{13\pi}{4}, -\dfrac{3\pi}{4}$

(c) $225°$

2. 3500 rad/min   **3.** About 709.04 ft^2

4. $\sin \theta = \dfrac{3\sqrt{13}}{13}$ $\csc \theta = \dfrac{\sqrt{13}}{3}$

$\cos \theta = \dfrac{2\sqrt{13}}{13}$ $\sec \theta = \dfrac{\sqrt{13}}{2}$

$\cot \theta = \dfrac{2}{3}$

5. $\sin \theta = \dfrac{3\sqrt{10}}{10}$ $\csc \theta = \dfrac{\sqrt{10}}{3}$

$\cos \theta = -\dfrac{\sqrt{10}}{10}$ $\sec \theta = -\sqrt{10}$

$\tan \theta = -3$ $\cot \theta = -\dfrac{1}{3}$

6. $\sin 210° = -\dfrac{1}{2}, \cos 210° = -\dfrac{\sqrt{3}}{2}, \tan 210° = \dfrac{\sqrt{3}}{3}$

7. Quadrant III   **8.** $150°, 210°$

9. $\sin \theta = -\dfrac{4}{5}$ **10.** $\sin \theta = \dfrac{21}{29}$

$\tan \theta = -\dfrac{4}{3}$ $\cos \theta = -\dfrac{20}{29}$

$\csc \theta = -\dfrac{5}{4}$ $\tan \theta = -\dfrac{21}{20}$

$\sec \theta = \dfrac{5}{3}$ $\csc \theta = \dfrac{29}{21}$

$\cot \theta = -\dfrac{3}{4}$ $\cot \theta = -\dfrac{20}{21}$

11. 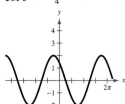   **12.**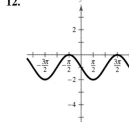

13. 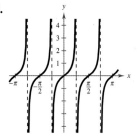   **14.** $a = -2, b = \dfrac{1}{2}, c = -\dfrac{\pi}{4}$

15. 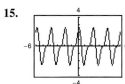   **16.**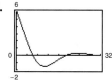

Period: 2   Not periodic;
As $x \to \infty$, $y \to 0$.

17. $\dfrac{\sqrt{55}}{3}$

18. 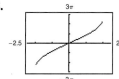   **19.** About $309.3°$

20. $d = -6 \cos \pi t$

Problem Solving   *(page 505)*

1. (a) $\dfrac{11\pi}{2}$ rad or $990°$ (b) About 816.42 ft

3. $h = 51 - 50 \sin\left(8\pi t + \dfrac{\pi}{2}\right)$

5. (a) About 4767 ft   (b) About 3705 ft

(c) $w \approx 2183$ ft, $\tan 63° = \dfrac{w + 3705}{3000}$

7. (a) 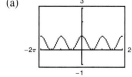   (b)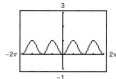

Even   Even

CHAPTER 6

9. (a)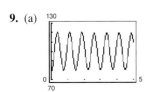

(b) Period $= \dfrac{3}{4}$ sec; This is the time between heartbeats.

(c) 20 mm; The blood pressure ranges between 80 mm and 120 mm.

(d) 80 beats/min   (e) Period $= \dfrac{15}{16}$ sec; $\dfrac{32\pi}{15}$

11. (a)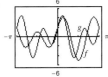

(b) Period of f: 2π
Period of g: π

(c) Yes, because the sine and cosine functions are periodic.

13. (a) About $40.5°$

(b) $x \approx 1.71$ ft; $y \approx 3.46$ ft

(c) About 1.75 ft

(d) As you move closer to the rock, d must get smaller and smaller. The angles θ_1 and θ_2 will decrease along with the distance y, so d will decrease.

Chapter 7

Section 7.1 *(page 513)*

1. $\cot u$ **3.** $\csc^2 u$

5. $\sin x = \dfrac{\sqrt{21}}{5}$ $\csc x = \dfrac{5\sqrt{21}}{21}$

$\cos x = -\dfrac{2}{5}$ $\sec x = -\dfrac{5}{2}$

$\tan x = -\dfrac{\sqrt{21}}{2}$ $\cot x = -\dfrac{2\sqrt{21}}{21}$

7. $\sin \theta = -\dfrac{3}{4}$ $\csc \theta = -\dfrac{4}{3}$

$\cos \theta = \dfrac{\sqrt{7}}{4}$ $\sec \theta = \dfrac{4\sqrt{7}}{7}$

$\tan \theta = -\dfrac{3\sqrt{7}}{7}$ $\cot \theta = -\dfrac{\sqrt{7}}{3}$

9. $\sin x = \dfrac{2\sqrt{13}}{13}$ $\csc x = \dfrac{\sqrt{13}}{2}$

$\cos x = \dfrac{3\sqrt{13}}{13}$ $\sec x = \dfrac{\sqrt{13}}{3}$

$\tan x = \dfrac{2}{3}$ $\cot x = \dfrac{3}{2}$

11. c   **12.** b   **13.** f   **14.** a   **15.** e   **16.** d

17. $\cos \theta$ **19.** $\sin^2 x$ **21.** $\sec x + 1$ **23.** $\sin^4 x$

25. $\csc^2 x(\cot x + 1)$ **27.** $(3 \sin x + 1)(\sin x - 2)$

29. $(\csc x - 1)(\csc x + 2)$ **31.** $\sec \theta$ **33.** $\cos^2 \phi$

35. $\sec \beta$ **37.** $\sin^2 x$ **39.** $2 \csc^2 x$ **41.** $-2 \tan x$

43. $-\cot x$ **45.** $1 + \cos y$ **47.** $\sin x$ **49.** $0 \le \theta \le \pi$

51. $3 \sin \theta$ **53.** $2 \tan \theta$ **55.** $\ln|\cos x|$

57. $\ln|\csc t \sec t|$ **59.** $\mu = \tan \theta$ **61.** True. $\sin \theta = \dfrac{1}{\csc \theta}$

63. True. $\csc x = \dfrac{1}{\sin x}$, $\sec x = \dfrac{1}{\cos x}$, $\tan x = \dfrac{\sin x}{\cos x}$, $\cot x = \dfrac{\cos x}{\sin x}$

65. $\cos \theta = \pm\sqrt{1 - \sin^2 \theta}$

$\tan \theta = \pm\dfrac{\sin \theta}{\sqrt{1 - \sin^2 \theta}}$

$\cot \theta = \pm\dfrac{\sqrt{1 - \sin^2 \theta}}{\sin \theta}$

$\sec \theta = \pm\dfrac{1}{\sqrt{1 - \sin^2 \theta}}$

$\csc \theta = \dfrac{1}{\sin \theta}$

67. $a \sec \theta$ **69.** $\dfrac{x - 4}{2}$, $x \ne -4, 0$

71. $\dfrac{4}{(x - 1)^2}$, $x \ne -1, 0$ **73.** 58   **75.** 32

77. 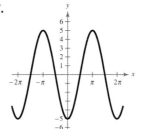   **79.**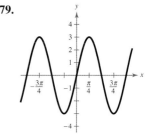

81. True. $\ln(xy) = \ln x + \ln y$

Section 7.2 *(page 520)*

1. identity

3. *Sample answer:* Actual equivalence of two expressions can only be verified algebraically.

5–41. Answers will vary.

43. (a) 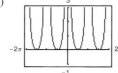   (b)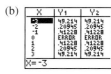

Identity

(c) Answers will vary.

45. (a) 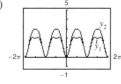   (b)

Not an identity

(c) Answers will vary.

47–49. Answers will vary.

51. (a) Answers will vary.

(b) *Sample answer:*

θ	5°	15°	30°	45°	60°	75°	90°
s	57.15	18.66	8.66	5	2.89	1.34	0

(c) Minimum: 90°   (d) Noon

53. False. $(1 + \csc x)(1 - \csc x) = 1 - \csc x + \csc x - \csc^2 x$
$$= 1 - \csc^2 x = -\cot^2 x$$

55. $\cot(-x) = -\cot x$

57. The equation is not an identity because $\tan \theta = \pm\sqrt{\sec^2 \theta - 1}$.

Sample answer: $\dfrac{3\pi}{4}$

59. $x = -2, 17$ **61.** $x = -9, 7$

63. (a) $\theta = 0° = 0, \theta = 180° = \pi$

(b) $\theta = 45° = \dfrac{\pi}{4}, \theta = 135° = \dfrac{3\pi}{4}$

Section 7.3 *(page 530)*

1. isolate   **3.** quadratic   **5.** $\cos 0 = 1$

7–11. Answers will vary.

13. $\dfrac{\pi}{3} + 2n\pi, \dfrac{2\pi}{3} + 2n\pi$ **15.** $\dfrac{2\pi}{3} + 2n\pi, \dfrac{4\pi}{3} + 2n\pi$

17. $\dfrac{\pi}{6} + n\pi, \dfrac{5\pi}{6} + n\pi$ **19.** $\dfrac{\pi}{3} + n\pi, \dfrac{2\pi}{3} + n\pi$

21. $n\pi, \dfrac{3\pi}{2} + 2n\pi$ **23.** $\dfrac{n\pi}{2}$ **25.** $n\pi, \dfrac{\pi}{6} + n\pi, \dfrac{5\pi}{6} + n\pi$

27. $\dfrac{\pi}{3} + 2n\pi, \pi + 2n\pi, \dfrac{5\pi}{3} + 2n\pi$

29. $\pi + 2n\pi, \dfrac{\pi}{3} + 2n\pi, \dfrac{5\pi}{3} + 2n\pi$ **31.** $\dfrac{\pi}{4}, \dfrac{5\pi}{4}$

33. $\dfrac{\pi}{2}, \dfrac{3\pi}{2}, \dfrac{2\pi}{3}, \dfrac{4\pi}{3}$ **35.** $\dfrac{\pi}{3}, \dfrac{2\pi}{3}, \dfrac{4\pi}{3}, \dfrac{5\pi}{3}$ **37.** No solution

39. $\dfrac{\pi}{2}$ **41.** $\dfrac{\pi}{6} + n\pi, \dfrac{5\pi}{6} + n\pi$ **43.** $\dfrac{\pi}{12} + \dfrac{n\pi}{3}$

45. $\dfrac{\pi}{2} + 4n\pi, \dfrac{7\pi}{2} + 4n\pi$ **47.** $\dfrac{\pi}{3} + 2n\pi$ **49.** $3 + 4n$

51. $3.553, 5.872$ **53.** $1.249, 4.391$ **55.** 0.739

57. $0.955, 2.186, 4.097, 5.328$ **59.** $1.221, 1.921, 4.362, 5.062$

61. $\arctan(-4) + n\pi, \arctan 3 + n\pi$

63. $\dfrac{\pi}{4} + n\pi, \arctan 5 + n\pi$ **65.** $\dfrac{\pi}{3} + 2n\pi, \dfrac{5\pi}{3} + 2n\pi$

67. $\arctan \dfrac{1}{3} + n\pi, \arctan\left(-\dfrac{1}{3}\right) + n\pi$

69. $\arccos \dfrac{1}{4} + 2n\pi, -\arccos \dfrac{1}{4} + 2n\pi$

71. $\dfrac{\pi}{2} + 2n\pi, \arcsin\left(-\dfrac{1}{4}\right) + 2n\pi, \arcsin \dfrac{1}{4} + 2n\pi$

73. $0.3398, 0.8481, 2.2935, 2.8018$

75. $1.9357, 2.7767, 5.0773, 5.9183$

77. $-1.154, 0.534$

79. 1.110

81. (a)

(b) $\dfrac{\pi}{3} \approx 1.0472$

$\dfrac{5\pi}{3} \approx 5.2360$

0

$\pi \approx 3.1416$

Maximum: $(1.0472, 1.25)$
Maximum: $(5.2360, 1.25)$
Minimum: $(0, 1)$
Minimum: $(3.1416, -1)$

83. (a)

(b) $\dfrac{\pi}{4} \approx 0.7854$

$\dfrac{5\pi}{4} \approx 3.9270$

Maximum: $(0.7854, 1.4142)$
Minimum: $(3.9270, -1.4142)$

85. (a)

(b) $\dfrac{\pi}{4} \approx 0.7854$

$\dfrac{5\pi}{4} \approx 3.9270$

$\dfrac{3\pi}{4} \approx 2.3562$

$\dfrac{7\pi}{4} \approx 5.4978$

Maximum: $(0.7854. 0.5)$
Maximum: $(3.9270, 0.5)$
Minimum: $(2.3562, -0.5)$
Minimum: $(5.4978, -0.5)$

87. 1   **89.** 0.04 sec, 0.43 sec, 0.83 sec

91. January, November, December

93. (a) and (c)

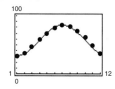

The model fits the data well.

(b) $C = 26.55 \cos\left(\dfrac{\pi t}{6} - \dfrac{7\pi}{6}\right) + 57.55$

(d) Above 72°F: June through September
Below 72°F: October through May

95. (a)

$A \approx 1.12$

(b) $0.6 < x < 1.1$

97. (a)

Graphs intersect when $x = \dfrac{\pi}{2}$ and $x = \pi$.

(b)

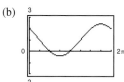

x-intercepts: $\left(\dfrac{\pi}{2}, 0\right), (\pi, 0)$

(c) Yes; Answers will vary.

99. 1

101. True. The first equation has a smaller period than the second equation, so it will have more solutions in the interval $[0, 2\pi)$.

103. The equation becomes $\cos^2 x = 2$; No

105. $x = 2 \pm 2\sqrt{3}$ **107.** $x = -1 \pm \sqrt{22}$

109. $x = 6 \pm 2\sqrt{5}$ **111.** $x = \pm 1$

113. $\dfrac{\sqrt{4 - x^2}}{x}, 0 < x \le 2$ **115.** $\dfrac{x}{\sqrt{x^2 - 9}}, x > 3$

117. $x, 0 \le x < 2$

Section 7.4   (page 538)

1. $\sin u \cos v - \cos u \sin v$ **3.** $\dfrac{\tan u + \tan v}{1 - \tan u \tan v}$

5. $\cos u \cos v + \sin u \sin v$

7. Sample answer: $\sin\left(\dfrac{\pi}{6} + \dfrac{\pi}{4}\right)$

9. (a) $\dfrac{\sqrt{2} - \sqrt{6}}{4}$ (b) $\dfrac{\sqrt{2} + 1}{2}$

11. (a) $\dfrac{\sqrt{6} + \sqrt{2}}{4}$ (b) $\dfrac{\sqrt{2} - \sqrt{3}}{2}$

13. $\sin\dfrac{11\pi}{12} = \dfrac{\sqrt{2}}{4}(\sqrt{3} - 1)$

$\cos\dfrac{11\pi}{12} = -\dfrac{\sqrt{2}}{4}(\sqrt{3} + 1)$

$\tan\dfrac{11\pi}{12} = -2 + \sqrt{3}$

15. $\sin\dfrac{17\pi}{12} = -\dfrac{\sqrt{2}}{4}(\sqrt{3} + 1)$

$\cos\dfrac{17\pi}{12} = \dfrac{\sqrt{2}}{4}(1 - \sqrt{3})$

$\tan\dfrac{17\pi}{12} = 2 + \sqrt{3}$

17. $\sin 105° = \dfrac{\sqrt{2}}{4}(\sqrt{3} + 1)$

$\cos 105° = \dfrac{\sqrt{2}}{4}(1 - \sqrt{3})$

$\tan 105° = -2 - \sqrt{3}$

19. $\sin(-195°) = \dfrac{\sqrt{2}}{4}(\sqrt{3} - 1)$

$\cos(-195°) = -\dfrac{\sqrt{2}}{4}(\sqrt{3} + 1)$

$\tan(-195°) = -2 + \sqrt{3}$

21. $\sin\dfrac{13\pi}{12} = \dfrac{\sqrt{2}}{4}(1 - \sqrt{3})$

$\cos\dfrac{13\pi}{12} = -\dfrac{\sqrt{2}}{4}(1 + \sqrt{3})$

$\tan\dfrac{13\pi}{12} = 2 - \sqrt{3}$

23. $\sin\left(-\dfrac{5\pi}{12}\right) = -\dfrac{\sqrt{2}}{4}(1 + \sqrt{3})$

$\cos\left(-\dfrac{5\pi}{12}\right) = \dfrac{\sqrt{2}}{4}(\sqrt{3} - 1)$

$\tan\left(-\dfrac{5\pi}{12}\right) = -2 - \sqrt{3}$

25. $\sin 285° = -\dfrac{\sqrt{2}}{4}(\sqrt{3} + 1)$

$\cos 285° = \dfrac{\sqrt{2}}{4}(\sqrt{3} - 1)$

$\tan 285° = -(2 + \sqrt{3})$

27. $\sin(-165°) = -\dfrac{\sqrt{2}}{4}(\sqrt{3} - 1)$

$\cos(-165°) = -\dfrac{\sqrt{2}}{4}(1 + \sqrt{3})$

$\tan(-165°) = 2 - \sqrt{3}$

29. $\sin 1.8$ **31.** $\cos 170°$ **33.** $\tan\dfrac{7\pi}{15}$

35. $\cos(3x - 2y)$ **37.** $\dfrac{\sqrt{3}}{2}$ **39.** $-\dfrac{1}{2}$

41. 0   **43.** $-\dfrac{13}{85}$ **45.** $-\dfrac{13}{84}$ **47.** $\dfrac{85}{36}$ **49.** $\dfrac{3}{5}$

51. $-\dfrac{44}{117}$ **53.** $-\dfrac{125}{44}$ **55.** 1   **57.** 0

59–65. Answers will vary.   **67.** $-\sin\theta$ **69.** $-\sec\theta$

71–73. Answers will vary.

75. (a) $\sqrt{2}\sin\left(\theta + \dfrac{\pi}{4}\right)$ (b) $\sqrt{2}\cos\left(\theta - \dfrac{\pi}{4}\right)$

77. (a) $13\sin(3\theta + 0.3948)$ (b) $13\cos(3\theta - 1.1760)$

79. $\sqrt{2}\sin\theta + \sqrt{2}\cos\theta$

81. (a) $y = \dfrac{5}{12}\sin(2t + 0.6435)$ (b) $\dfrac{5}{12}$ ft   (c) $\dfrac{1}{\pi}$ cycle/sec

83. $\dfrac{\pi}{6}, \dfrac{5\pi}{6}$ **85.** $\dfrac{5\pi}{4}, \dfrac{7\pi}{4}$ **87.** $0, \dfrac{\pi}{3}, \pi, \dfrac{5\pi}{3}$

89. $\dfrac{\pi}{4}, \dfrac{7\pi}{4}$ **91.** $\dfrac{\pi}{2}, \pi, \dfrac{3\pi}{2}$

93. True. $\sin(u \pm v) = \sin u \cos v \pm \cos u \sin v$

95. The denominator should be $1 - \tan x \tan\dfrac{\pi}{4} = 1 - \tan x$.

97.

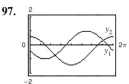

No, $y_1 \ne y_2$ because their graphs are different.

99. Proof

101. (a) All real numbers h except $h = 0$

(b)

h	0.5	0.2	0.1
$f(h)$	0.267	0.410	0.456
$g(h)$	0.267	0.410	0.456

h	0.05	0.02	0.01
$f(h)$	0.478	0.491	0.496
$g(h)$	0.478	0.491	0.496

(c)

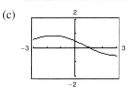

(d) As $h \to 0^+$, $f \to 0.5$ and $g \to 0.5$.

103. $\sin \theta = -\dfrac{12}{13}$

$\tan \theta = -\dfrac{12}{5}$

$\csc \theta = -\dfrac{13}{12}$

$\sec \theta = \dfrac{13}{5}$

$\cot \theta = -\dfrac{5}{12}$

105. $\sin \theta = \dfrac{\sqrt{5}}{5}$

$\cos \theta = -\dfrac{2\sqrt{5}}{5}$

$\csc \theta = \sqrt{5}$

$\sec \theta = -\dfrac{\sqrt{5}}{2}$

$\cot \theta = -2$

107. $x = \dfrac{\pi}{2} + 2\pi$　　**109.** $x = \dfrac{3\pi}{2} + 2n\pi$

111. $x = \dfrac{\pi}{6} + \dfrac{n\pi}{3}$

Section 7.5　(page 548)

1. $\cos 2u$

3. *Sample answer:* Product-to-sum formulas are used in calculus to solve problems involving the products of sines and cosines of two different angles.

5. $n\pi, \dfrac{\pi}{3} + 2n\pi, \dfrac{5\pi}{3} + 2n\pi$　　**7.** $\dfrac{2n\pi}{3}$　　**9.** $\dfrac{n\pi}{2}$

11. $\dfrac{\pi}{6} + n\pi, \dfrac{\pi}{2} + n\pi, \dfrac{5\pi}{6} + n\pi$　　**13.** $3 \sin 2x$

15. $3 \cos 2x$　　**17.** $4 \cos 2x$

19. $\sin 2u = -\dfrac{24}{25}, \cos 2u = \dfrac{7}{25}, \tan 2u = -\dfrac{24}{7}$

21. $\sin 2u = \dfrac{15}{17}, \cos 2u = \dfrac{8}{17}, \tan 2u = \dfrac{15}{8}$

23. $8 \cos^4 x - 8 \cos^2 x + 1$　　**25.** $\dfrac{1}{8}(3 + 4 \cos 2x + \cos 4x)$

27. $\dfrac{1}{8}(3 - 4 \cos 4x + \cos 8x)$　　**29.** $\dfrac{(3 - 4 \cos 4x + \cos 8x)}{(3 + 4 \cos 4x + \cos 8x)}$

31. $\dfrac{1}{8}(1 - \cos 8x)$

33. $\sin 75° = \dfrac{1}{2}\sqrt{2 + \sqrt{3}}$

$\cos 75° = \dfrac{1}{2}\sqrt{2 - \sqrt{3}}$

$\tan 75° = 2 + \sqrt{3}$

35. $\sin 112° \, 30' = \dfrac{1}{2}\sqrt{2 + \sqrt{2}}$

$\cos 112° \, 30' = -\dfrac{1}{2}\sqrt{2 - \sqrt{2}}$

$\tan 112° \, 30' = -1 - \sqrt{2}$

37. $\sin \dfrac{\pi}{8} = \dfrac{1}{2}\sqrt{2 - \sqrt{2}}$

$\cos \dfrac{\pi}{8} = \dfrac{1}{2}\sqrt{2 + \sqrt{2}}$

$\tan \dfrac{\pi}{8} = \sqrt{2} - 1$

39. (a) Quadrant I

(b) $\sin \dfrac{u}{2} = \dfrac{3}{5}, \cos \dfrac{u}{2} = \dfrac{4}{5}, \tan \dfrac{u}{2} = \dfrac{3}{4}$

41. (a) Quadrant II

(b) $\sin \dfrac{u}{2} = \dfrac{\sqrt{26}}{26}, \cos \dfrac{u}{2} = -\dfrac{5\sqrt{26}}{26}, \tan \dfrac{u}{2} = -\dfrac{1}{5}$

43. π　　**45.** $\dfrac{\pi}{3}, \pi, \dfrac{5\pi}{3}$　　**47.** $\dfrac{1}{2}(\cos 2\theta - \cos 8\theta)$

49. $\dfrac{1}{2}(\cos(-2\theta) + \cos 6\theta)$　　**51.** $2 \cos 4\theta \sin \theta$

53. $2 \cos 4x \cos 2x$　　**55.** $\dfrac{\sqrt{6}}{2}$　　**57.** $-\sqrt{2}$

59. $0, \dfrac{\pi}{4}, \dfrac{\pi}{2}, \dfrac{3\pi}{4}, \pi, \dfrac{5\pi}{4}, \dfrac{3\pi}{2}, \dfrac{7\pi}{4}$　　**61.** $\dfrac{\pi}{6}, \dfrac{5\pi}{6}$

63–65. Answers will vary.

67. (a) $\cos \theta = \dfrac{M^2 - 2}{M^2}$　　(b) $\dfrac{\pi}{3}$　　(c) 1520 mi/h

69. Answers will vary.

71. True. $\sin(-2x) = 2 \sin(-x) \cos(-x) = -2 \sin x \cos x$.

73. $B = 63°$　　**75.** $B = C = 45°$　　**77.** 0.8829

79. -1.4945

81. (a) About 104.95 nm south, about 58.18 nm west

(b) S 36.7° W; about 130.9 nm

Review Exercises　(page 552)

1. $\cot x$　　**3.** $\cos x$

5. $\sin \theta = -\dfrac{\sqrt{21}}{5}$　　$\csc \theta = -\dfrac{5\sqrt{21}}{21}$

$\cos \theta = -\dfrac{2}{5}$　　$\sec \theta = -\dfrac{5}{2}$

$\tan \theta = \dfrac{\sqrt{21}}{2}$　　$\cot \theta = \dfrac{2\sqrt{21}}{21}$

7. $\sin^2 x$　　**9.** 1　　**11.** $\tan u \sec u$　　**13.** $\cot^2 x$

15. $-2 \tan^2 \theta$　　**17.** $5 \cos \theta$　　**19–25.** Answers will vary.

27. $\dfrac{\pi}{3} + 2n\pi, \dfrac{2\pi}{3} + 2n\pi$　　**29.** $\dfrac{\pi}{6} + n\pi$

31. $\dfrac{\pi}{3} + n\pi, \dfrac{2\pi}{3} + n\pi$　　**33.** $0, \dfrac{\pi}{2}, \pi, \dfrac{3\pi}{2}$　　**35.** $0, \dfrac{\pi}{2}, \pi$

37. $\dfrac{\pi}{8}, \dfrac{3\pi}{8}, \dfrac{9\pi}{8}, \dfrac{11\pi}{8}$　　**39.** $\dfrac{\pi}{2}$

41. $0, \dfrac{\pi}{8}, \dfrac{3\pi}{8}, \dfrac{5\pi}{8}, \dfrac{7\pi}{8}, \dfrac{9\pi}{8}, \dfrac{11\pi}{8}, \dfrac{13\pi}{8}, \dfrac{15\pi}{8}$

43. $n\pi, \arctan 2 + n\pi$　　**45.** $\arctan(-3) + n\pi, \arctan 2 + n\pi$

47. $\sin 75° = \dfrac{\sqrt{2}}{4}(1 + \sqrt{3})$　　**49.** $\sin \dfrac{25\pi}{12} = \dfrac{\sqrt{2}}{4}(\sqrt{3} - 1)$

$\cos 75° = \dfrac{\sqrt{2}}{4}(\sqrt{3} - 1)$　　$\cos \dfrac{25\pi}{12} = \dfrac{\sqrt{2}}{4}(\sqrt{3} + 1)$

$\tan 75° = 2 + \sqrt{3}$　　$\tan \dfrac{25\pi}{12} = 2 - \sqrt{3}$

51. $\sin\left(\dfrac{-19\pi}{12}\right) = \dfrac{\sqrt{2}}{4}(\sqrt{3} + 1)$

$\cos\left(\dfrac{-19\pi}{12}\right) = \dfrac{\sqrt{2}}{4}(\sqrt{3} - 1)$

$\tan\left(\dfrac{-19\pi}{12}\right) = 2 + \sqrt{3}$

53. $\sin 15°$　　**55.** $-\dfrac{24}{25}$　　**57.** -1

59–61. Answers will vary.　　**63.** $\dfrac{\pi}{4}, \dfrac{7\pi}{4}$

65. *Sample answer:* $4 \sin x \cos x - 8 \sin^3 x \cos x$

67. $8 \cos^3 x \sin x - 4 \cos x \sin x$

69. $\dfrac{1 - \cos 6x}{1 + \cos 6x}$

71. $\sin(-75°) = -\dfrac{1}{2}\sqrt{2 + \sqrt{3}}$

$\cos(-75°) = \dfrac{1}{2}\sqrt{2 - \sqrt{3}}$

$\tan(-75°) = -2 - \sqrt{3}$

73. (a) Quadrant II

(b) $\sin \dfrac{u}{2} = \dfrac{2\sqrt{5}}{5}, \cos \dfrac{u}{2} = -\dfrac{\sqrt{5}}{5}, \tan \dfrac{u}{2} = -2$

75. (a) Quadrant I

(b) $\sin \dfrac{u}{2} = \dfrac{3\sqrt{14}}{14}, \cos \dfrac{u}{2} = \dfrac{\sqrt{70}}{14}, \tan \dfrac{u}{2} = \dfrac{3\sqrt{5}}{5}$

CHAPTER 7

77. $\frac{1}{2}[\sin 10\theta - \sin(-2\theta)]$ **79.** $2 \cos \frac{11\theta}{2} \cos \frac{\theta}{2}$

81. $\theta = 15°$ or $\frac{\pi}{12}$

83. False. If $(\pi/2) < \theta < \pi$, then $\theta/2$ lies in Quadrant I.

85. False. $4 \sin(-x) \cos(-x) = 4(-\sin x) \cos x$
$$= -4 \sin x \cos x$$
$$= -2(2 \sin x \cos x)$$
$$= -2 \sin 2x$$

87. Yes. *Sample answer:* $\sin x = \frac{1}{2}$ has an infinite number of solutions.

Chapter Test *(page 554)*

1. $\sin \theta = \frac{2}{5}$ $\csc \theta = \frac{5}{2}$

$\cos \theta = -\frac{\sqrt{21}}{5}$ $\sec \theta = -\frac{5\sqrt{21}}{21}$

$\tan \theta = -\frac{2\sqrt{21}}{21}$ $\cot \theta = -\frac{\sqrt{21}}{2}$

2. 1   **3.** 1   **4.** $\csc \theta \sec \theta$ **5–10.** Answers will vary.

11. $2(\sin 5\theta + \sin \theta)$ **12.** $-2 \sin \theta$

13. $0, \frac{3\pi}{4}, \pi, \frac{7\pi}{4}$ **14.** $\frac{\pi}{6}, \frac{\pi}{2}, \frac{5\pi}{6}, \frac{3\pi}{2}$ **15.** $\frac{\pi}{6}, \frac{5\pi}{6}, \frac{7\pi}{6}, \frac{11\pi}{6}$

16. $\frac{\pi}{6}, \frac{5\pi}{6}, \frac{3\pi}{2}$ **17.** $0, 2.596$ **18.** $\frac{\sqrt{2} - \sqrt{6}}{4}$

19. $\sin 2u = -\frac{20}{29}, \cos 2u = -\frac{21}{29}, \tan 2u = \frac{20}{21}$

20. Day 30 to day 310

21. 0.26 min, 0.58 min, 0.89 min, 1.20 min, 1.52 min, 1.83 min

Problem Solving *(page 557)*

1. $\sin \theta = \pm\sqrt{1 - \cos^2 \theta}$

$\tan \theta = \pm\frac{\sqrt{1 - \cos^2 \theta}}{\cos \theta}$

$\csc \theta = \pm\frac{1}{\sqrt{1 - \cos^2 \theta}}$

$\sec \theta = \frac{1}{\cos \theta}$

$\cot \theta = \pm\frac{\cos \theta}{\sqrt{1 - \cos^2 \theta}}$

3. Answers will vary.   **5.** $u + v = w$; Proof

7. (a) $A = 100 \sin \frac{\theta}{2} \cos \frac{\theta}{2}$ (b) $A = 50 \sin \theta; \theta = \frac{\pi}{2}$

9. (a) $F = \frac{0.6W \cos \theta}{\sin 12°}$

(b)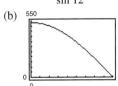

(c) Maximum: $\theta = 0°$
Minimum: $\theta = 90°$

11. (a) High tides: 6:12 A.M., 6:36 P.M.
Low tides: 12:00 A.M., 12:24 P.M.

(b) The water depth never falls below 7 feet.

(c)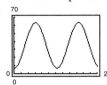

13. (a) $n = \frac{1}{2}\left(\cot \frac{\theta}{2} + \sqrt{3}\right)$ (b) $\theta \approx 76.5°$

15. (a) $\frac{\pi}{6} \leq x \leq \frac{5\pi}{6}$ (b) $\frac{2\pi}{3} \leq x \leq \frac{4\pi}{3}$

(c) $\frac{\pi}{2} < x < \pi, \frac{3\pi}{2} < x < 2\pi$

(d) $0 \leq x \leq \frac{\pi}{4}, \frac{5\pi}{4} \leq x < 2\pi$

Chapter 8

Section 8.1 *(page 566)*

1. oblique   **3.** AAS or ASA, and SSA

5. $A = 30°, a \approx 14.14, c \approx 27.32$

7. $C = 105°, a \approx 5.94, b \approx 6.65$

9. $B = 60.9°, b \approx 19.32, c \approx 6.36$

11. $B = 42° 4', a \approx 22.05, b \approx 14.88$

13. $C = 80°, a \approx 5.82, b \approx 9.20$

15. $B = 83°, a = 0.51, c = 0.62$

17. $B \approx 21.55°, C \approx 122.45°, c \approx 11.49$

19. $B \approx 9.43°, C \approx 25.57°, c \approx 10.53$

21. $A \approx 10° 11', C \approx 154° 19', c \approx 11.03$

23. $B \approx 48.74°, C \approx 21.26°, c \approx 48.23$ **25.** No solution

27. Two solutions:
$B \approx 72.21°, C \approx 49.79°, c \approx 10.27$
$B \approx 107.79°, C \approx 14.21°, c \approx 3.30$

29. No solution

31. $B = 45°, C = 90°, c \approx 1.41$

33. (a) $b \leq 5, b = \frac{5}{\sin 36°}$ (b) $5 < b < \frac{5}{\sin 36°}$

(c) $b > \frac{5}{\sin 36°}$

35. (a) $b < 80$ (b) Not possible   (c) $b \geq 80$

37. 22.1   **39.** 94.4   **41.** 218.0   **43.** 22.3

45. From Pine Knob: about 42.4 km
From Colt Station: about 15.5 km

47. About 77 m

49. (a) $\frac{h}{\sin 30°} = \frac{40}{\sin 56°}$ (b) About 24.1 m

51. (a) 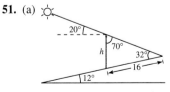   (b) $\frac{16}{\sin 70°} = \frac{h}{\sin 32°}$

(c) About 9 m

53. (a) $\alpha = \arcsin(0.5 \sin \beta)$

(b)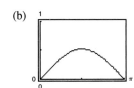

Domain: $0 < \beta < \pi$

Range: $0 < \alpha < \dfrac{\pi}{6}$

(c) $c = \dfrac{18 \sin[\pi - \beta - \arcsin(0.5 \sin \beta)]}{\sin \beta}$

(d)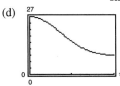

Domain: $0 < \beta < \pi$

Range: $9 < c < 27$

(e)

β	0.4	0.8	1.2	1.6
α	0.1960	0.3669	0.4848	0.5234
c	25.95	23.07	19.19	15.33

β	2.0	2.4	2.8
α	0.4720	0.3445	0.1683
c	12.29	10.31	9.27

As β increases from 0 to π, α increases and then decreases, and c decreases from 27 to 9.

55. True. If an angle of a triangle is obtuse, then the other two angles must be acute.

57. False. When just three angles are known, the triangle cannot be solved.

59. False. The sine of the angle included between the sides is needed.

61. $\dfrac{3}{2}$ **63.** 0

65. (a) 5   (b) Even multiplicity   (c) 1

(d)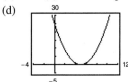

67. (a) $-3, 0, \dfrac{1}{2}$

(b) $-3, \dfrac{1}{2}$, odd multiplicity; 0, even multiplicity   (c) 3

(d)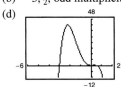

Section 8.2   (page 573)

1. alternative   **3.** $\cos B = \dfrac{a^2 + c^2 - b^2}{2ac}$

5. $A \approx 38.62°, B \approx 48.51°, C \approx 92.87°$

7. $A = 17.61°, B = 28.96°, C = 133.43°$

9. $B \approx 23.79°, C \approx 126.21°, a \approx 18.59$

11. $A \approx 43.31°, B \approx 28.69°, c \approx 13.87$

13. $A \approx 30.11°, B \approx 43.16°, C \approx 106.73°$

15. $A \approx 132.77°, B \approx 31.91°, C \approx 15.32°$

17. $B \approx 27.46°, C \approx 32.54°, a \approx 11.27$

19. $A \approx 141° 45', C \approx 27° 40', b \approx 11.87$

21. $A = 27° 10', C = 27° 10', b \approx 65.84$

23. $A \approx 33.80°, B \approx 103.20°, c \approx 0.54$

25. 12.07; 5.69; 135°   **27.** 13.86; 68.2°; 111.8°

29. 16.96; 77.2°; 102.8°

31. Law of Cosines; $A \approx 102.44°, C \approx 37.56°, b \approx 5.26$

33. Law of Sines; No solution

35. Law of Sines; $C = 103°, a \approx 0.82, b \approx 0.71$

37. 23.53   **39.** 5.35   **41.** 0.24   **43.** 1514.14

45. About 373.3 m   **47.** About 103.9 ft

49. (a) N 59.7° E   (b) N 72.8° E   **51.** 41.2°, 52.9°

53. 72.3°   **55.** About 46,837.5 ft²   **57.** \$83,336.37

59. False. For s to be the average of the lengths of the three sides of the triangle, s would be equal to $(a + b + c)/3$.

61. $c^2 = a^2 + b^2$; The Pythagorean Theorem is a special case of the Law of Cosines.

63. Proof

65. (a) $\dfrac{7}{4}$ (b) $\sqrt{65}$ **67.** (a) $-\dfrac{1}{4}$ (b) $\sqrt{17}$

69. (a) 0   (b) 6   **71.** (a) $\dfrac{5}{6}$ (b) $4\sqrt{61}$

73. (a) $\sin \theta = \dfrac{\sqrt{2}}{2}$ $\csc = \sqrt{2}$

$\cos \theta = -\dfrac{\sqrt{2}}{2}$ $\sec \theta = -\sqrt{2}$

$\tan \theta = -1$ $\cot \theta = -1$

(b) $\sin \theta = -\dfrac{\sqrt{2}}{2}$ $\csc \theta = -\sqrt{2}$

$\cos \theta = \dfrac{\sqrt{2}}{2}$ $\sec = \sqrt{2}$

$\tan \theta = -1$ $\cot \theta = -1$

Section 8.3   (page 585)

1. directed line segment   **3.** multiplication; addition

5. Two vectors are equivalent if they have the same magnitude and direction.

7. Equivalent; $\mathbf{u}$ and $\mathbf{v}$ have the same magnitude and direction.

9. Not equivalent; $\mathbf{u}$ and $\mathbf{v}$ do not have the same direction.

11. Equivalent; $\mathbf{u}$ and $\mathbf{v}$ have the same magnitude and direction.

13. $\mathbf{v} = \langle 1, 3 \rangle, \|\mathbf{v}\| = \sqrt{10}$ **15.** $\mathbf{v} = \langle 0, 5 \rangle; \|\mathbf{v}\| = 5$

17. $\mathbf{v} = \langle -8, 6 \rangle; \|\mathbf{v}\| = 10$ **19.** $\mathbf{v} = \langle -9, -12 \rangle; \|\mathbf{v}\| = 15$

21. $\mathbf{v} = \langle 16, -26 \rangle; \|\mathbf{v}\| = 2\sqrt{233}$

23. 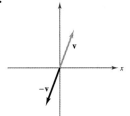   **25.**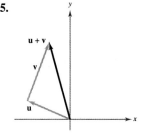

27.

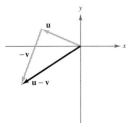

29. (a) $\langle 3, 4 \rangle$ (b) $\langle 1, -2 \rangle$

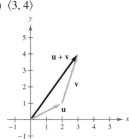 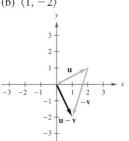

(c) $\langle 1, -7 \rangle$

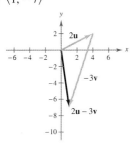

31. (a) $\langle -5, 3 \rangle$ (b) $\langle -5, 3 \rangle$

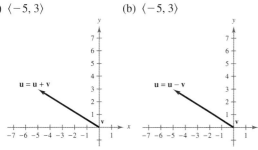

(c) $\langle -10, 6 \rangle$

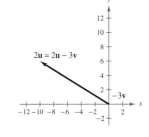

33. (a) $\langle 1, -9 \rangle$ (b) $\langle -1, -5 \rangle$

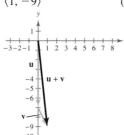 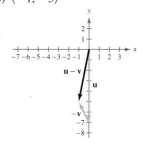

(c) $\langle -3, -8 \rangle$

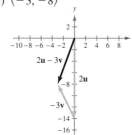

35. 10 **37.** $9\sqrt{5}$ **39.** $\langle 1, 0 \rangle$ **41.** $\left\langle -\dfrac{\sqrt{2}}{2}, \dfrac{\sqrt{2}}{2} \right\rangle$

43. $\left\langle \dfrac{\sqrt{37}}{37}, -\dfrac{6\sqrt{37}}{37} \right\rangle$ **45.** $\mathbf{v} = \langle -6, 8 \rangle$

47. $\mathbf{v} = \left\langle \dfrac{18\sqrt{29}}{29}, \dfrac{45\sqrt{29}}{29} \right\rangle$ **49.** $5\mathbf{i} - 3\mathbf{j}$ **51.** $-6\mathbf{i} + 3\mathbf{j}$

53. $\mathbf{v} = \langle 3, -\tfrac{3}{2} \rangle$ **55.** $\mathbf{v} = \langle 4, 3 \rangle$

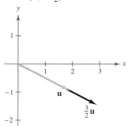 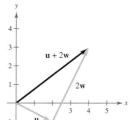

57. $\mathbf{v} = \langle 0, -5 \rangle$

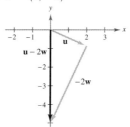

59. $\|\mathbf{v}\| = 6\sqrt{2};\ \theta = 315°$ **61.** $\|\mathbf{v}\| = 3;\ \theta = 60°$

63. $\mathbf{v} = \langle 3, 0 \rangle$ **65.** $\mathbf{v} = \left\langle -\dfrac{7\sqrt{3}}{4}, \dfrac{7}{4} \right\rangle$

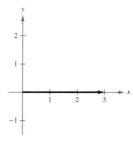 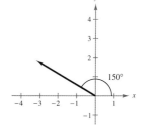

67. $\mathbf{v} = \left\langle \frac{9}{5}, \frac{12}{5} \right\rangle$

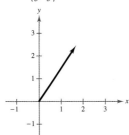

69. $\langle 2, 4 + 2\sqrt{3} \rangle$ **71.** 90°   **73.** About 62.7°

75. Vertical $\approx$ 2.30 ft/sec, horizontal $\approx$ 21.88 ft/sec

77. About 12.8°; about 398.32 N

79. About 71.3°; about 228.5 lb

81. $T_L \approx$ 15,484 lb
     $T_R \approx$ 19,786 lb

83. $\sqrt{2}$ lb; 1 lb   **85.** About 20.8 lb   **87.** About 19.5°

89. N 21.4° E; about 138.7 km/h

91. True. See Example 1.   **93.** True. $a = b = 0$

95. $u_1 = 6 - (-3) = 9$ and $u_2 = -1 - 4 = -5$, so $\mathbf{u} = \langle 9, -5 \rangle$.

97. Proof   **99.** $\langle 1, 3 \rangle$ or $\langle -1, -3 \rangle$

101. (a) $5\sqrt{5 + 4\cos\theta}$

(b)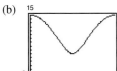

(c) Range: [5, 15]
    Maximum is 15 when $\theta = 0$.
    Minimum is 5 when $\theta = \pi$.

(d) The magnitudes of $\mathbf{F}_1$ and $\mathbf{F}_2$ are not the same.

103. 8.49   **105.** 3.16   **107.** 6.08   **109.** 17.09

111. $\frac{\pi}{2} + n\pi$ **113.** 1

115. $A = 21.00°$, $B = 18.57°$, $C = 140.43°$

117. $A = 37°$, $a = 12.10$, $b = 17.24$

Section 8.4   *(page 595)*

1. *Sample answer:* $\mathbf{u} \cdot (\mathbf{v} + \mathbf{w}) = \mathbf{u} \cdot \mathbf{v} + \mathbf{u} \cdot \mathbf{w}$

3. $\left(\dfrac{\mathbf{u} \cdot \mathbf{v}}{\|\mathbf{v}\|^2} \right)\mathbf{v}$ **5.** -19 **7.** 0   **9.** 6   **11.** 18; scalar

13. $\langle 24, -12 \rangle$; vector   **15.** **0**; vector   **17.** 60; scalar

19. $\sqrt{10} - 1$; scalar   **21.** -12; scalar   **23.** 17

25. $5\sqrt{41}$ **27.** 6   **29.** $\frac{\pi}{2}$ **31.** About 2.50

33. 0   **35.** About 0.93   **37.** $\frac{5\pi}{12}$ **39.** About 91.33°

41. 90°   **43.** 26.57°, 63.43°, 90°

45. 41.63°, 53.13°, 85.24°   **47.** -20 **49.** $12,500\sqrt{3}$

51. Not orthogonal   **53.** Orthogonal   **55.** Not orthogonal

57. $\frac{1}{37}\langle 84, 14 \rangle, \frac{1}{37}\langle -10, 60 \rangle$ **59.** $\langle 0, 0 \rangle, \langle 4, 2 \rangle$

61. $\langle 3, 2 \rangle$ **63.** $\langle 0, 0 \rangle$ **65.** $\langle -5, 3 \rangle, \langle 5, -3 \rangle$

67. $\frac{2}{3}\mathbf{i} + \frac{1}{2}\mathbf{j}, -\frac{2}{3}\mathbf{i} - \frac{1}{2}\mathbf{j}$ **69.** 32

71. (a) \$35,727.50
     This value gives the total amount paid to the employees.
     (b) Multiply **v** by 1.02.

73. (a) Force = 30,000 sin d
     (b)

d	0°	1°	2°	3°	4°	5°
Force	0	523.6	1047.0	1570.1	2092.7	2614.7

d	6°	7°	8°	9°	10°
Force	3135.9	3656.1	4175.2	4693.0	5209.4

(c) About 29,885.8 lb

75. 735 N-m   **77.** About 779.4 ft-lb

79. About 1174.62 ft-lb

81. False. Work is represented by a scalar.   **83.** 4

85. The dot product of two vectors is a scalar, not a vector.
     $\langle 5, 8 \rangle \cdot \langle -2, 7 \rangle = 5(-2) + 8(7) = -10 + 56 = 46$

87. 1; $\mathbf{u} \cdot \mathbf{u} = \|\mathbf{u}\|^2$ **89.** Proofs

91. (a) 5   (b) $\left(\frac{3}{2}, 11 \right)$ **93.** (a) $\sqrt{74}$ (b) $(0, 0)$

95. $-12 + 14i$ **97.** $-6 - 4i$ **99.** $1 - 3i$

101. 4

103. (a) $\langle 1, -1 \rangle$ (b) $\langle -7, 5 \rangle$

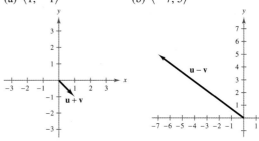

Section 8.5   *(page 603)*

1. real; imaginary

3. Reverse the sign of the imaginary part of the complex number.

5. b   **6.** d   **7.** f   **8.** a   **9.** e   **10.** c

11.

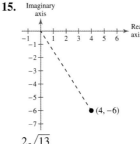

7

13.

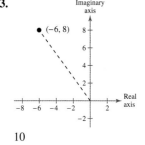

10

15.

$2\sqrt{13}$

17. $5 + 6i$ **19.** $10 + 4i$ **21.** $1 + 7i$ **23.** $-2 - 2i$
25. $10 - 3i$ **27.** $-6i$

29.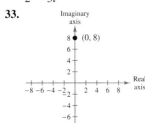

$2 - 3i$

31.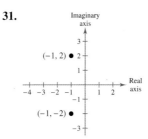

$-1 + 2i$

33.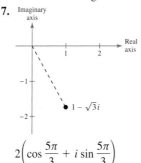

$-8i$

35. $2\sqrt{2} \approx 2.83$ **37.** $\sqrt{109} \approx 10.44$ **39.** $(4, 3)$
41. $\left(\frac{9}{2}, -\frac{3}{2}\right)$
43. (a) Ship A: $3 + 4i$, Ship B: $-5 + 2i$
     (b) *Sample answer:* Find the modulus of the difference of the
        complex numbers.
45. False. The modulus is always real.
47. False. $|1 + i| + |1 - i| = 2\sqrt{2}$ and $|(1 + i) + (1 - i)| = 2$
49. A circle; The modulus represents the distance from the origin.
51. $\pm i, -2$ **53.** $2 \pm 2i, 5$

55. $\sin \theta = \dfrac{4}{5}$

$\tan \theta = -\dfrac{4}{3}$

$\csc \theta = \dfrac{5}{4}$

$\sec \theta = -\dfrac{5}{3}$

$\cot \theta = -\dfrac{3}{4}$

57. $\sin \theta = -\dfrac{5\sqrt{26}}{26}$

$\cos \theta = -\dfrac{\sqrt{26}}{26}$

$\csc \theta = -\dfrac{\sqrt{26}}{5}$

$\sec \theta = -\sqrt{26}$

$\cot \theta = \dfrac{1}{5}$

59. $\dfrac{7\pi}{6} + 2n\pi, \dfrac{11\pi}{6} + 2n\pi$ **61.** $\dfrac{\pi}{6} + n\pi, \dfrac{5\pi}{6} + n\pi$

Section 8.6   *(page 612)*

1. DeMoivre's   **3.** r is the modulus and θ is an argument of z.
5.

$\sqrt{2}\left(\cos \dfrac{\pi}{4} + i \sin \dfrac{\pi}{4}\right)$

7.

$2\left(\cos \dfrac{5\pi}{3} + i \sin \dfrac{5\pi}{3}\right)$

9.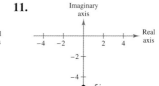

$4\left(\cos \dfrac{4\pi}{3} + i \sin \dfrac{4\pi}{3}\right)$

11.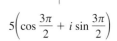

$5\left(\cos \dfrac{3\pi}{2} + i \sin \dfrac{3\pi}{2}\right)$

13.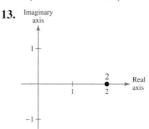

$2(\cos 0 + i \sin 0)$

15.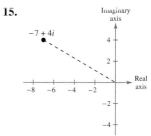

$\sqrt{65}\,(\cos 2.62 + i \sin 2.62)$

17.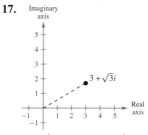

$2\sqrt{3}\left(\cos \dfrac{\pi}{6} + i \sin \dfrac{\pi}{6}\right)$

19.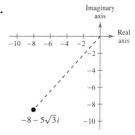

$\sqrt{139}(\cos 3.97 + i \sin 3.97)$

21. $1 + \sqrt{3}i$

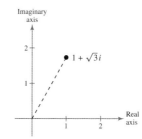

23. $-\dfrac{9\sqrt{2}}{8} + \dfrac{9\sqrt{2}}{8}i$

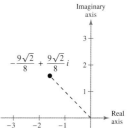

25. $6 - 2\sqrt{3}i$

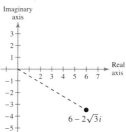

27. $4.6985 + 1.7101i$ **29.** $-1.8126 + 0.8452i$

31. $12\left(\cos \dfrac{\pi}{3} + i \sin \dfrac{\pi}{3}\right)$ **33.** $\dfrac{10}{9}(\cos 150° + i \sin 150°)$

35. $\dfrac{1}{3}(\cos 30° + i \sin 30°)$ **37.** $\cos \dfrac{2\pi}{3} + i \sin \dfrac{2\pi}{3}$

39. -1 **41.** $\dfrac{125}{2} + \dfrac{125\sqrt{3}}{2}i$ **43.** -1

45. $608.0 + 144.7i$ **47.** $\dfrac{81}{2} + \dfrac{81\sqrt{3}}{2}i$ **49.** $-4 - 4i$

51. $8i$ **53.** $1024 - 1024\sqrt{3}i$

55. (a) $\sqrt{5}(\cos 60° + i \sin 60°)$
$\quad\quad \sqrt{5}(\cos 240° + i \sin 240°)$

(b) $\dfrac{\sqrt{5}}{2} + \dfrac{\sqrt{15}}{2}i, \; -\dfrac{\sqrt{5}}{2} - \dfrac{\sqrt{15}}{2}i$

(c)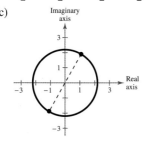

57. (a) $2\left(\cos \dfrac{2\pi}{9} + i \sin \dfrac{2\pi}{9}\right)$ (b) $1.5321 + 1.2856i,$
$\quad\quad 2\left(\cos \dfrac{8\pi}{9} + i \sin \dfrac{8\pi}{9}\right)$ $\quad\quad -1.8794 + 0.6840i,$
$\quad\quad 2\left(\cos \dfrac{14\pi}{9} + i \sin \dfrac{14\pi}{9}\right)$ $\quad\quad 0.3473 - 1.9696i$

(c)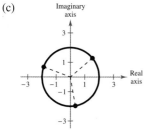

59. (a) $10(\cos 0 + i \sin 0)$ (b) $10, -5 + 5\sqrt{3}i,$
$\quad\quad 10\left(\cos \dfrac{2\pi}{3} + i \sin \dfrac{2\pi}{3}\right)$ $\quad\quad -5 - 5\sqrt{3}i$
$\quad\quad 10\left(\cos \dfrac{4\pi}{3} + i \sin \dfrac{4\pi}{3}\right)$

(c)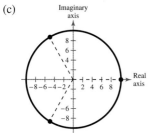

61. (a) $\sqrt{2}\left(\cos \dfrac{\pi}{4} + i \sin \dfrac{\pi}{4}\right)$ (b) $1 + i, -1 + i,$
$\quad\quad \sqrt{2}\left(\cos \dfrac{3\pi}{4} + i \sin \dfrac{3\pi}{4}\right)$ $\quad\quad -1 - i, 1 - i$
$\quad\quad \sqrt{2}\left(\cos \dfrac{5\pi}{4} + i \sin \dfrac{5\pi}{4}\right)$
$\quad\quad \sqrt{2}\left(\cos \dfrac{7\pi}{4} + i \sin \dfrac{7\pi}{4}\right)$

(c)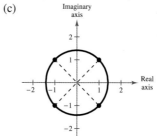

63. (a) $\cos \dfrac{\pi}{8} + i \sin \dfrac{\pi}{8}$ (b) $0.9239 + 0.3827i,$
$\quad\quad \cos \dfrac{5\pi}{8} + i \sin \dfrac{5\pi}{8}$ $\quad\quad -0.3827 + 0.9239i,$
$\quad\quad \cos \dfrac{9\pi}{8} + i \sin \dfrac{9\pi}{8}$ $\quad\quad -0.9239 - 0.3827i,$
$\quad\quad \cos \dfrac{13\pi}{8} + i \sin \dfrac{13\pi}{8}$ $\quad\quad 0.3827 - 0.9239i$

(c)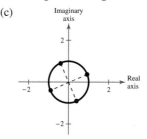

65. (a) $5\left(\cos \dfrac{3\pi}{4} + i \sin \dfrac{3\pi}{4}\right)$ (b) $-\dfrac{5\sqrt{2}}{2} + \dfrac{5\sqrt{2}}{2}i,$
$\quad\quad 5\left(\cos \dfrac{7\pi}{4} + i \sin \dfrac{7\pi}{4}\right)$ $\quad\quad \dfrac{5\sqrt{2}}{2} - \dfrac{5\sqrt{2}}{2}i$

(c)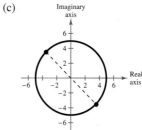

CHAPTER 8

67. (a) $\sqrt{2}\left(\cos\dfrac{7\pi}{20} + i\sin\dfrac{7\pi}{20}\right)$

$\sqrt{2}\left(\cos\dfrac{3\pi}{4} + i\sin\dfrac{3\pi}{4}\right)$

$\sqrt{2}\left(\cos\dfrac{23\pi}{20} + i\sin\dfrac{23\pi}{20}\right)$

$\sqrt{2}\left(\cos\dfrac{31\pi}{20} + i\sin\dfrac{31\pi}{20}\right)$

$\sqrt{2}\left(\cos\dfrac{39\pi}{20} + i\sin\dfrac{39\pi}{20}\right)$

(b) $0.6420 + 1.2601i$,
$-1 + i$,
$-1.2601 - 0.6420i$,
$0.2212 - 1.3968i$,
$1.3968 - 0.2212i$

(c)
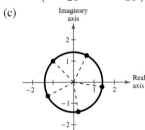

69. (a) $5\left(\cos\dfrac{4\pi}{9} + i\sin\dfrac{4\pi}{9}\right)$

$5\left(\cos\dfrac{10\pi}{9} + i\sin\dfrac{10\pi}{9}\right)$

$5\left(\cos\dfrac{16\pi}{9} + i\sin\dfrac{16\pi}{9}\right)$

(b) $0.8682 + 4.9240i$,
$-4.6985 - 1.7101i$,
$3.8302 - 3.2139i$

(c)
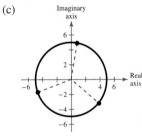

71. $\cos\dfrac{3\pi}{8} + i\sin\dfrac{3\pi}{8}$

$\cos\dfrac{7\pi}{8} + i\sin\dfrac{7\pi}{8}$

$\cos\dfrac{11\pi}{8} + i\sin\dfrac{11\pi}{8}$

$\cos\dfrac{15\pi}{8} + i\sin\dfrac{15\pi}{8}$

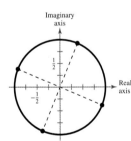

73. $3\left(\cos\dfrac{\pi}{5} + i\sin\dfrac{\pi}{5}\right)$

$3\left(\cos\dfrac{3\pi}{5} + i\sin\dfrac{3\pi}{5}\right)$

$3(\cos\pi + i\sin\pi)$

$3\left(\cos\dfrac{7\pi}{5} + i\sin\dfrac{7\pi}{5}\right)$

$3\left(\cos\dfrac{9\pi}{5} + i\sin\dfrac{9\pi}{5}\right)$

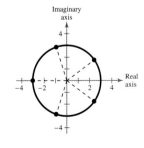

75. $2\left(\cos\dfrac{3\pi}{8} + i\sin\dfrac{3\pi}{8}\right)$

$2\left(\cos\dfrac{7\pi}{8} + i\sin\dfrac{7\pi}{8}\right)$

$2\left(\cos\dfrac{11\pi}{8} + i\sin\dfrac{11\pi}{8}\right)$

$2\left(\cos\dfrac{15\pi}{8} + i\sin\dfrac{15\pi}{8}\right)$

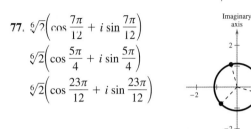

77. $\sqrt[6]{2}\left(\cos\dfrac{7\pi}{12} + i\sin\dfrac{7\pi}{12}\right)$

$\sqrt[6]{2}\left(\cos\dfrac{5\pi}{4} + i\sin\dfrac{5\pi}{4}\right)$

$\sqrt[6]{2}\left(\cos\dfrac{23\pi}{12} + i\sin\dfrac{23\pi}{12}\right)$

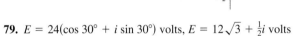

79. $E = 24(\cos 30° + i\sin 30°)$ volts, $E = 12\sqrt{3} + \frac{1}{2}i$ volts

81. False. They are equally spaced around the circle centered at the origin with radius $\sqrt[n]{r}$.

83. Answers will vary.   **85.** The answer is $2\sqrt{3} + 2i$.

87. Yes   **89.** No   **91.** $y = 4 - x$

93. $y = \dfrac{5}{3}x - 2$ **95.** $-2, \dfrac{4}{3}$ **97.** $-\dfrac{1}{2} \pm \dfrac{\sqrt{3}}{2}i$

Review Exercises   (page 616)

1. $C = 72°, b \approx 12.21, c \approx 12.36$

3. $A = 26°, a \approx 24.89, c \approx 56.23$

5. $C = 66°, a \approx 2.53, b \approx 9.11$

7. $B = 108°, a \approx 11.76, c \approx 21.49$

9. $A \approx 20.41°, C \approx 9.59°, a \approx 20.92$

11. $B \approx 39.48°, C \approx 65.52°, c \approx 48.24$

13. 19.1 **15.** 47.2 **17.** About 31.1 m

19. $A \approx 16.99°, B \approx 26.00°, C \approx 137.01°$

21. $A \approx 29.92°, B \approx 86.18°, C \approx 63.90°$

23. $A = 36°, C = 36°, b \approx 17.80$

25. $A \approx 45.76°, B \approx 91.24°, c \approx 21.42$

27. Law of Sines; no solution

29. Law of Cosines; $A \approx 28.62°, B \approx 33.56°, C \approx 117.82°$

31. About 4.3 ft, about 12.6 ft

33. 7.64 **35.** 8.36

37. Not equivalent; $\mathbf{u}$ and $\mathbf{v}$ have a different direction.

39. $\langle 7, -7 \rangle; 7\sqrt{2}$

41. (a) $\langle -4, 3 \rangle$ (b) $\langle 2, -9 \rangle$

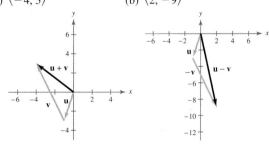

(c) $\langle -4, -12 \rangle$ (d) $\langle -14, 3 \rangle$

47. (a) $3\mathbf{i} + 6\mathbf{j}$ (b) $5\mathbf{i} - 6\mathbf{j}$

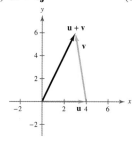

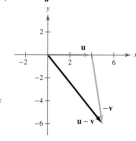

43. (a) $\langle -1, 6 \rangle$ (b) $\langle -9, -2 \rangle$

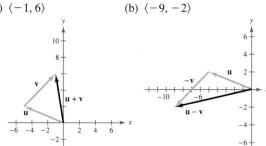

(c) $16\mathbf{i}$ (d) $17\mathbf{i} + 18\mathbf{j}$

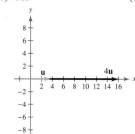

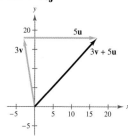

(c) $\langle -20, 8 \rangle$ (d) $\langle -13, 22 \rangle$

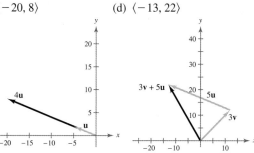

49. $-\mathbf{i} + 5\mathbf{j}$ **51.** $6\mathbf{i} + 4\mathbf{j}$

53. $\langle 30, 9 \rangle$ **55.** $\langle 22, -7 \rangle$

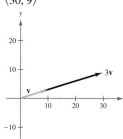

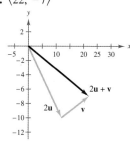

45. (a) $7\mathbf{i} + 2\mathbf{j}$ (b) $-3\mathbf{i} - 4\mathbf{j}$

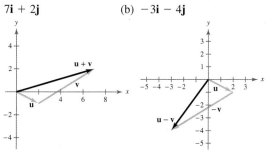

57. $\langle -10, -37 \rangle$

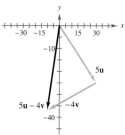

(c) $8\mathbf{i} - 4\mathbf{j}$ (d) $25\mathbf{i} + 4\mathbf{j}$

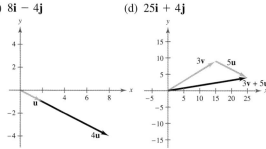

59. $\|\mathbf{v}\| = \sqrt{41};\ \theta \approx 38.7°$ **61.** $\|\mathbf{v}\| = 3\sqrt{2};\ \theta = 225°$

63. $\|\mathbf{v}\| = 7;\ \theta = 60°$

65. $\mathbf{v} = \langle -4, 4\sqrt{3} \rangle$

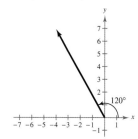

67. About $50.5°$; about 133.92 lb   **69.** 45 **71.** -2

73. 40; scalar   **75.** $4 - 2\sqrt{5}$; scalar

77. $\langle 72, -36 \rangle$; vector   **79.** 38; scalar   **81.** About $160.5°$

83. $165°$ **85.** Orthogonal

CHAPTER 8

87. Not orthogonal   **89.** $-\frac{13}{17}\langle 4, 1\rangle, \frac{16}{17}\langle -1, 4\rangle$

91. $\frac{5}{2}\langle -1, 1\rangle, \frac{9}{2}\langle 1, 1\rangle$ **93.** 48   **95.** 72,000 ft-lb

97.

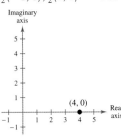

4

99.

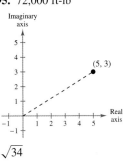

$\sqrt{34}$

101. $3 + i$ **103.** $-2 + i$

105.

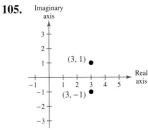

$3 - i$

107. $\sqrt{10}$ **109.** $\left(\frac{5}{2}, 2\right)$

111.

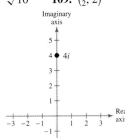

$4\left(\cos\frac{\pi}{2} + i\sin\frac{\pi}{2}\right)$

113.

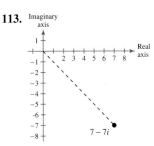

$7\sqrt{2}\left(\cos\frac{7\pi}{4} + i\sin\frac{7\pi}{4}\right)$

115.

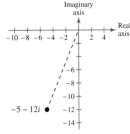

$13(\cos 4.32 + i\sin 4.32)$

117. $4\left(\cos\frac{7\pi}{12} + i\sin\frac{7\pi}{12}\right)$ **119.** $\frac{2}{3}(\cos 45° + i\sin 45°)$

121. $\frac{625}{2} + \frac{625\sqrt{3}}{2}i$ **123.** $2035 - 828i$

125. (a) $3\left(\cos\frac{\pi}{4} + i\sin\frac{\pi}{4}\right)$

$3\left(\cos\frac{7\pi}{12} + i\sin\frac{7\pi}{12}\right)$

$3\left(\cos\frac{11\pi}{12} + i\sin\frac{11\pi}{12}\right)$

$3\left(\cos\frac{5\pi}{4} + i\sin\frac{5\pi}{4}\right)$

$3\left(\cos\frac{19\pi}{12} + i\sin\frac{19\pi}{12}\right)$

$3\left(\cos\frac{23\pi}{12} + i\sin\frac{23\pi}{12}\right)$

(b) $\frac{3\sqrt{2}}{2} + \frac{3\sqrt{2}}{2}i,$

$-0.7765 + 2.8978i,$

$-2.8978 + 0.7765i,$

$-\frac{3\sqrt{2}}{2} - \frac{3\sqrt{2}}{2}i,$

$0.7765 - 2.8978i,$

$2.8978 - 0.7765i$

(c)

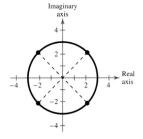

127. (a) $2(\cos 0 + i\sin 0)$

$2\left(\cos\frac{2\pi}{3} + i\sin\frac{2\pi}{3}\right)$

$2\left(\cos\frac{4\pi}{3} + i\sin\frac{4\pi}{3}\right)$

(b) $2, -1 + \sqrt{3}\,i,$

$-1 - \sqrt{3}\,i$

(c)

129. $3\left(\cos\frac{\pi}{4} + i\sin\frac{\pi}{4}\right) = \frac{3\sqrt{2}}{2} + \frac{3\sqrt{2}}{2}i$

$3\left(\cos\frac{3\pi}{4} + i\sin\frac{3\pi}{4}\right) = -\frac{3\sqrt{2}}{2} + \frac{3\sqrt{2}}{2}i$

$3\left(\cos\frac{5\pi}{4} + i\sin\frac{5\pi}{4}\right) = -\frac{3\sqrt{2}}{2} - \frac{3\sqrt{2}}{2}i$

$3\left(\cos\frac{7\pi}{4} + i\sin\frac{7\pi}{4}\right) = \frac{3\sqrt{2}}{2} - \frac{3\sqrt{2}}{2}i$

131. $2\left(\cos\dfrac{\pi}{2} + i\sin\dfrac{\pi}{2}\right) = 2i$

$2\left(\cos\dfrac{7\pi}{6} + i\sin\dfrac{7\pi}{6}\right) = -\sqrt{3} - i$

$2\left(\cos\dfrac{11\pi}{6} + i\sin\dfrac{11\pi}{6}\right) = \sqrt{3} - i$

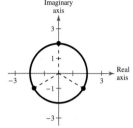

133. True. $\sin 90°$ is defined in the Law of Sines.
135. Direction and magnitude

Chapter Test *(page 619)*

1. Law of Sines; $C = 88°, b \approx 27.81, c \approx 29.98$
2. Law of Sines; $A = 42°, b \approx 21.91, c \approx 10.95$
3. Law of Sines; Two solutions:
  $B \approx 29.12°, C \approx 126.88°, c \approx 22.03$
  $B \approx 150.88°, C \approx 5.12°, c \approx 2.46$
4. Law of Cosines; $A \approx 19.12°, B \approx 23.49°, C \approx 137.39°$
5. Law of Sines; No solution
6. Law of Cosines; $A \approx 21.90°, B \approx 37.10°, c \approx 78.15$
7. 2052.5 m² **8.** 606.3 mi; 29.1° **9.** $\langle 14, -23 \rangle$
10. $\left\langle \dfrac{18\sqrt{34}}{17}, -\dfrac{30\sqrt{34}}{17} \right\rangle$
11. $\langle -4, 12 \rangle$ **12.** $\langle 8, 2 \rangle$

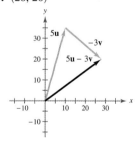 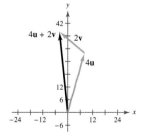

13. $\langle 28, 20 \rangle$ **14.** $\langle -4, 38 \rangle$

15. 5 **16.** About 14.9°; about 250.15 lb **17.** 135°
18. Orthogonal **19.** $\dfrac{37}{26}\langle 5, 1 \rangle; \dfrac{29}{26}\langle -1, 5 \rangle$ **20.** About 104 lb
21. $4\sqrt{2}\left(\cos\dfrac{7\pi}{4} + i\sin\dfrac{7\pi}{4}\right)$ **22.** $-3 + 3\sqrt{3}\,i$
23. $-\dfrac{6561}{2} - \dfrac{6561\sqrt{3}}{2}i$ **24.** 5832i **25.** $4, -4, 4i, -4i$

26. $3\left(\cos\dfrac{\pi}{6} + i\sin\dfrac{\pi}{6}\right)$

$3\left(\cos\dfrac{5\pi}{6} + i\sin\dfrac{5\pi}{6}\right)$

$3\left(\cos\dfrac{3\pi}{2} + i\sin\dfrac{3\pi}{2}\right)$

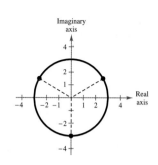

Cumulative Test for Chapters 6–8 *(page 620)*

1. (a)

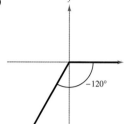

(b) 240°
(c) $-\dfrac{2\pi}{3}$
(d) 60°

(e) $\sin(-120°) = -\dfrac{\sqrt{3}}{2}$ $\csc(-120°) = -\dfrac{2\sqrt{3}}{3}$

$\cos(-120°) = -\dfrac{1}{2}$ $\sec(-120°) = -2$

$\tan(-120°) = \sqrt{3}$ $\cot(-120°) = \dfrac{\sqrt{3}}{3}$

2. $-83.079°$ **3.** $\dfrac{20}{29}$

4.

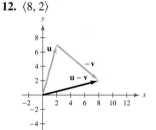

5.

6.

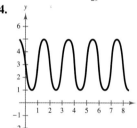

7. $a = -3, b = \pi, c = 0$

8.

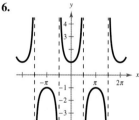

9. 4.9
10. $\dfrac{3}{4}$
11. $\sqrt{1 - 4x^2}$
12. 1
13. $2\tan\theta$
14–16. Answers will vary.
17. $\dfrac{\pi}{3}, \dfrac{\pi}{2}, \dfrac{3\pi}{2}, \dfrac{5\pi}{3}$

18. $\dfrac{\pi}{6}, \dfrac{5\pi}{6}, \dfrac{7\pi}{6}, \dfrac{11\pi}{6}$ **19.** $\dfrac{3\pi}{2}$ **20.** $\dfrac{16}{63}$ **21.** $\dfrac{4}{3}$

CHAPTER 8

22. $\dfrac{\sqrt{5}}{5}$ **23.** $\dfrac{5}{2}\left(\sin\dfrac{5\pi}{2} - \sin\pi\right)$ **24.** $-2\sin 8x\sin x$

25. Law of Sines; $B \approx 26.39°$, $C \approx 123.61°$, $c \approx 14.99$

26. Law of Cosines; $B \approx 52.48°$, $C \approx 97.52°$, $a \approx 5.04$

27. Law of Sines; $B = 60°$, $a \approx 5.77$, $c \approx 11.55$

28. Law of Cosines; $A \approx 26.28°$, $B \approx 49.74°$, $C \approx 103.98°$

29. Law of Sines; $C = 109°$, $a \approx 14.96$, $b \approx 9.27$

30. Law of Cosines; $A \approx 6.88°$, $B \approx 93.12°$, $c \approx 9.86$

31. 41.48 in.^{2} **32.** 599.09 m^2 **33.** $7\mathbf{i} + 8\mathbf{j}$

34. $\left\langle \dfrac{\sqrt{2}}{2}, \dfrac{\sqrt{2}}{2}\right\rangle$ **35.** -5 **36.** $-\dfrac{1}{13}\langle 1, 5\rangle$; $\dfrac{21}{13}\langle 5, -1\rangle$

37.

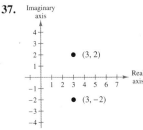

$3 + 2i$

38. $2\sqrt{2}\left(\cos\dfrac{3\pi}{4} + i\sin\dfrac{3\pi}{4}\right)$ **39.** $24(\cos 150° + i\sin 150°)$

40. $\cos 0 + i\sin 0 = 1$

$\cos\dfrac{2\pi}{3} + i\sin\dfrac{2\pi}{3} = -\dfrac{1}{2} + \dfrac{\sqrt{3}}{2}i$

$\cos\dfrac{4\pi}{3} + i\sin\dfrac{4\pi}{3} = -\dfrac{1}{2} - \dfrac{\sqrt{3}}{2}i$

41. $\dfrac{5\sqrt{2}}{2} + \dfrac{5\sqrt{2}}{2}i$

$-\dfrac{5\sqrt{2}}{2} + \dfrac{5\sqrt{2}}{2}i$

$-\dfrac{5\sqrt{2}}{2} - \dfrac{5\sqrt{2}}{2}i$

$\dfrac{5\sqrt{2}}{2} - \dfrac{5\sqrt{2}}{2}i$

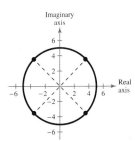

42. About 395.8 rad/min; about 8312.7 in./min

43. 42π yd$^2 \approx 131.95$ yd^2 **44.** 5 ft

45. About $22.6°$ **46.** $d = 4\cos\dfrac{\pi}{4}t$

47. About 543.9 km/h; about $32.6°$

48. 425 ft-lb

Problem Solving   *(page 625)*

1. About 2.01 ft

3. (a)

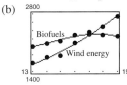

(b) Station A: about 27.45 mi; Station B: about 53.03 mi

(c) About 11.03 mi; S $21.7°$ E

5. (a) (i) $\sqrt{2}$ (ii) $\sqrt{5}$ (iii) 1
    (iv) 1 (v) 1 (vi) 1

(b) (i) 1 (ii) $3\sqrt{2}$ (iii) $\sqrt{13}$
    (iv) 1 (v) 1 (vi) 1

(c) (i) $\dfrac{\sqrt{5}}{2}$ (ii) $\sqrt{13}$ (iii) $\dfrac{\sqrt{85}}{2}$
    (iv) 1 (v) 1 (vi) 1

(d) (i) $2\sqrt{5}$ (ii) $5\sqrt{2}$ (iii) $5\sqrt{2}$
    (iv) 1 (v) 1 (vi) 1

7. Proof

9. (a) $2(\cos 30° + i\sin 30°)$ (b) $3(\cos 45° + i\sin 45°)$
    $2(\cos 150° + i\sin 150°)$ $3(\cos 135° + i\sin 135°)$
    $2(\cos 270° + i\sin 270°)$ $3(\cos 225° + i\sin 225°)$
    $3(\cos 315° + i\sin 315°)$

11. a; The angle between the vectors is acute.

Chapter 9

Section 9.1   *(page 635)*

1. system; equations   **3.** substitution

5. Break-even point

7. (a) No   (b) No   (c) No   (d) Yes

9. $(2, 2)$ **11.** $(2, 6)$, $(-1, 3)$

13. $(0, 0)$, $(2, -4)$ **15.** $(0, 1)$, $(1, -1)$, $(3, 1)$ **17.** $(6, 4)$

19. $(-5, 2)$ **21.** $\left(\dfrac{1}{2}, 3\right)$ **23.** $(1, 1)$ **25.** $(2, 2.5)$

27. $\left(\dfrac{20}{3}, \dfrac{40}{3}\right)$ **29.** No solution   **31.** $\left(\dfrac{22}{5}, \dfrac{14}{5}\right)$

33. $\$5500$ at 2%; $\$6500$ at 6%

35. $\$6000$ at 2.8%; $\$6000$ at 3.8%

37. $(-2, 4)$, $(0, 0)$ **39.** No solution   **41.** $(6, 2)$

43. $\left(-\dfrac{3}{2}, \dfrac{1}{2}\right)$ **45.** $(2, 2)$, $(4, 0)$ **47.** No solution

49. $(4, 3)$, $(-4, 3)$ **51.** $(0, 1)$ **53.** $(5.31, -0.54)$

55. $(1, 2)$ **57.** No solution   **59.** $(0.287, 1.751)$

61. $\left(\dfrac{1}{2}, 2\right)$, $\left(-4, -\dfrac{1}{4}\right)$ **63.** 293 units

65. (a) 344 units   (b) 2495 units

67. (a) Biofuels:

$C_1 = -3.250t^3 + 141.96t^2 - 1978.3t + 10{,}874$

Wind energy: $C_2 = 11.20t^2 - 162.5t + 1797$

(b)

(c) $(17.16, 2306.34)$; The average rate of consumption of biofuels and wind energy were approximately equal around 2017.

(d) Answers will vary.

69. 12 m $\times$ 16 m

71. False. You can solve for either variable in either equation and then back-substitute.

73. *Sample answer:* After substituting, the resulting equation may be a contradiction or have imaginary solutions.

75. Answers will vary. *Sample answer:*

(a) $y = x + 1$

(b) $y = 2x - 1$

(c) $y = x - 1$

77. (a)–(b) Answers will vary.   **79.** $20x + 4y = -4$
81. $6x - 12y = 42$ **83.** $2x - 5y = -38$ **85.** 2
87. $-\frac{1}{2}$ **89.** 3   **91.** -1
93.

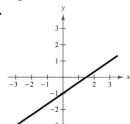

95.

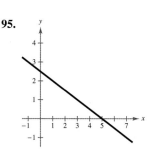

Section 9.2   *(page 646)*

1. elimination   **3.** equilibrium point   **5.** Inconsistent
7. $(1, 5)$ **9.** $(1, -1)$

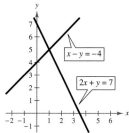

11. No solution   **13.** $\left(a, \frac{3}{2}a - \frac{5}{2}\right)$

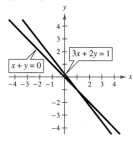

15. $(4, 1)$ **17.** $\left(\frac{3}{2}, -\frac{1}{2}\right)$ **19.** $\left(-3, \frac{5}{3}\right)$ **21.** $(4, -1)$
23. $\left(-\frac{6}{35}, \frac{43}{35}\right)$ **25.** $(101, 96)$ **27.** No solution
29. Infinitely many solutions: $\left(a, -\frac{1}{2} + \frac{5}{6}a\right)$
31. $(5, -2)$ **33.** a; infinitely many solutions; consistent
35. d; no solutions; inconsistent   **37.** $(4, 1)$
39. $(10, 5)$ **41.** $(19, -55)$ **43.** 550 mi/h; 50 mi/h
45. Cheeseburger: 570 calories; fries: 320 calories
47. $(240, 404)$ **49.** \$18,000
51. (a) $y = -44.21x + 89.53$ (b) 12 units
53. $y = 0.97x + 2.1$
55. (a) $y = 14x + 19$ (b) 41.4 bushels/acre
57. False. Two lines that coincide have infinitely many points of intersection.
59. $k = -4$ **61.** Answers will vary.
63. $(39,600, 398)$. It is necessary to change the scale on the axes to see the point of intersection.
65. No. Two lines will intersect only once or will coincide, and if they coincide the system will have infinitely many solutions.
67. $u = 1, v = -\tan x$ **69.** -1 **71.** $-1, \frac{5}{2}$
73. $\pm 1, \pm 3$ **75.** $y = \frac{11}{4}x - \frac{19}{4}$ **77.** $y = -\frac{3}{2}x + 5$

79. $6 \pm 2\sqrt{3}$ **81.** $\pm 1, \frac{7}{6}$

Section 9.3   *(page 658)*

1. row-echelon   **3.** Gaussian
5. Interchange two equations; multiply one of the equations by a nonzero constant, add a multiple of one equation to another equation to replace the latter equation.
7. (a) No   (b) No   (c) No   (d) Yes
9. (a) No   (b) No   (c) Yes   (d) No
11. $(-13, -10, 8)$ **13.** $(3, 10, 2)$ **15.** $\left(\frac{11}{4}, 7, 11\right)$
17. $\begin{cases} x - 2y + 3z = 5 \\ \quad\ y - 2z = 9 \\ 2x \quad\ - 3z = 0 \end{cases}$

First step in putting the system in row-echelon form.
19. $(-2, 2)$ **21.** $(4, 3)$ **23.** $(4, 1, 2)$ **25.** $\left(1, \frac{1}{2}, -3\right)$
27. No solution   **29.** $\left(\frac{1}{8}, -\frac{5}{8}, -\frac{1}{2}\right)$ **31.** $(0, 0, 0)$
33. No solution   **35.** $(-a + 3, a + 1, a)$
37. $(-3a + 10, 5a - 7, a)$ **39.** $(1, 1, 1, 1)$
41. $(2a, 21a - 1, 8a)$ **43.** $\left(-\frac{3}{2}a + \frac{1}{2}, -\frac{2}{3}a + 1, a\right)$
45. $s = -16t^2 + 144$ **47.** $s = -16t^2 - 32t + 400$
49. $y = \frac{1}{2}x^2 - 2x$ **51.** $y = x^2 - 6x + 8$
53. $y = 4x^2 - 2x + 1$ **55.** $x^2 + y^2 - 10x = 0$
57. $x^2 + y^2 + 6x - 8y = 0$
59. Brand X = 4 lb
    Brand Y = 9 lb
    Brand Z = 9 lb
61. $x = 60°, y = 67°, z = 53°$ **63.** $I_1 = 1, I_2 = 2, I_3 = 1$
65. $y = x^2 - x$
67. (a) $y = 0.0514x^2 + 0.8771x + 1.8857$
    (b)

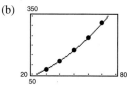

    The model fits the data well.
    (c) About 356 ft
69. $x = \pm\frac{\sqrt{2}}{2}, y = \frac{1}{2}, \lambda = 1$ or $x = 0, y = 0, \lambda = 0$
71. False. See Example 6 on page 655.
73. Answers will vary.   **75–77.** Answers will vary.
79. $-\dfrac{x - 2}{5}, x \neq -4$ **81.** $\dfrac{3(x + 1)(x + 2)}{x^3 - 2x^2 - x - 2}$
83. $-3x + 1, x \neq 6$ **85.** $x^2 + x + 2 + \dfrac{13}{x - 3}$
87.

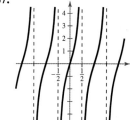

Section 9.4 *(page 668)*

1. improper   **3.** Two

5. $\dfrac{A}{x} + \dfrac{B}{x - 2}$ **7.** $\dfrac{A}{x + 2} + \dfrac{B}{(x + 2)^2} + \dfrac{C}{(x + 2)^3} + \dfrac{D}{(x + 2)^4}$

9. $\dfrac{A}{x} + \dfrac{B}{x^2} + \dfrac{Cx + D}{x^2 + 3} + \dfrac{Ex + F}{(x^2 + 3)^2}$ **11.** $\dfrac{1}{x} - \dfrac{1}{x + 1}$

13. $\dfrac{1}{x - 1} - \dfrac{1}{x + 2}$ **15.** $\dfrac{1}{2}\left(\dfrac{1}{x - 1} - \dfrac{1}{x + 1}\right)$

17. $-\dfrac{3}{x} - \dfrac{1}{x + 2} + \dfrac{5}{x - 2}$ **19.** $\dfrac{3}{x - 3} + \dfrac{9}{(x - 3)^2}$

21. $\dfrac{3}{x} - \dfrac{1}{x^2} + \dfrac{1}{x + 1}$ **23.** $\dfrac{2}{3}\left(\dfrac{1}{x - 1} - \dfrac{x - 1}{x^2 + x + 1}\right)$

25. $-\dfrac{1}{x - 1} + \dfrac{x + 2}{x^2 - 2}$ **27.** $\dfrac{1}{8}\left(\dfrac{1}{2x + 1} + \dfrac{1}{2x - 1} - \dfrac{4x}{4x^2 + 1}\right)$

29. $\dfrac{1}{x + 1} + \dfrac{2}{x^2 - 2x + 3}$ **31.** $\dfrac{2}{x^2 + 4} + \dfrac{x}{(x^2 + 4)^2}$

33. $\dfrac{2}{x} - \dfrac{3}{x^2} - \dfrac{2x - 3}{x^2 + 2} - \dfrac{4x - 6}{(x^2 + 2)^2}$ **35.** $1 - \dfrac{2x + 1}{x^2 + x + 1}$

37. $2x - 7 + \dfrac{17}{x + 2} + \dfrac{1}{x + 1}$

39. $x + 3 + \dfrac{6}{x - 1} + \dfrac{4}{(x - 1)^2} + \dfrac{1}{(x - 1)^3}$

41. $x + \dfrac{2}{x} + \dfrac{1}{x + 1} + \dfrac{3}{(x + 1)^2}$ **43.** $\dfrac{3}{2x - 1} - \dfrac{2}{x + 1}$

45. $\dfrac{2}{x} + \dfrac{4}{x + 1} - \dfrac{3}{x - 1}$ **47.** $\dfrac{1}{x^2 + 2} + \dfrac{x}{(x^2 + 2)^2}$

49. $2x + \dfrac{1}{2}\left(\dfrac{3}{x - 4} - \dfrac{1}{x + 2}\right)$ **51.** $\dfrac{60}{100 - p} - \dfrac{60}{100 + p}$

53. True. The expression is an improper rational expression.

55. The expression is improper, so first divide the denominator

into the numerator to obtain $1 + \dfrac{x + 1}{x^2 - x}$.

57. Answers will vary. *Sample answer:* You can substitute any convenient values of x that will help determine the constants. You can also find the basic equation, expand it, then equate coefficients of like terms.

59. 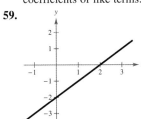   **61.**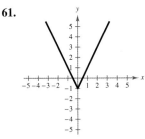

63. 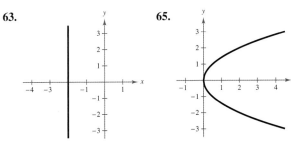   **65.**

67.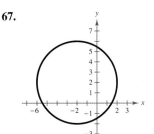

69. (a) Yes   (b) Yes   **71.** (a) Yes   (b) No

Section 9.5 *(page 677)*

1. graph

3. The graph of an inequality with a $<$ sign does not include the points on the graph of the corresponding equation, and the graph of an inequality with a $\leq$ sign does include the points on the graph of the corresponding equation.

5.   **7.**

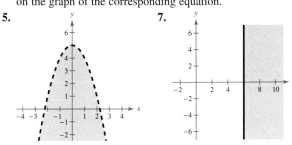

9.   **11.**

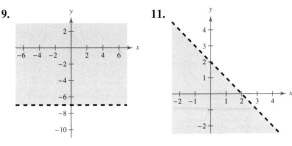

13.   **15.**

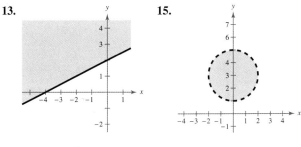

17.   **19.**

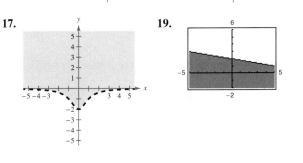

21. 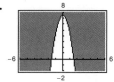   **23.**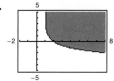

25.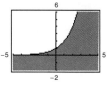

27. $y < 5x + 5$ **29.** $y \geq x^2 - 4$

31. 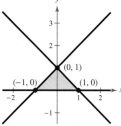   **33.**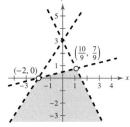

35. 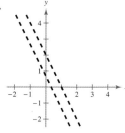   **37.**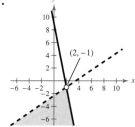

No solution

39. 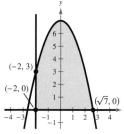   **41.**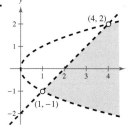

43. 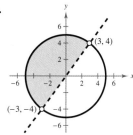   **45.**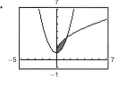

47. 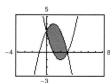   **49.**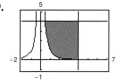

51. $\begin{cases} x \geq 0 \\ y \geq 0 \\ y \leq 6 - x \end{cases}$ **53.** $\begin{cases} x \geq 0 \\ y \geq 0 \\ x^2 + y^2 < 64 \end{cases}$

55. $\begin{cases} x \geq 4 \\ x \leq 9 \\ y \geq 3 \\ y \leq 9 \end{cases}$ **57.** $\begin{cases} y \geq 0 \\ y \leq 5x \\ y \leq -x + 6 \end{cases}$

59. (a) 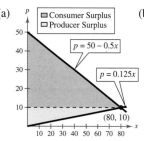   (b) Consumer surplus: $1600
Producer surplus: $400

61. (a) 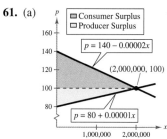   (b) Consumer surplus:
$40,000,000
Producer surplus:
$20,000,000

63. $\begin{cases} x + y \leq 20{,}000 \\ y \geq 2x \\ x \geq 5{,}000 \\ y \geq 5{,}000 \end{cases}$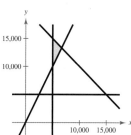

65. $\begin{cases} x + \frac{3}{2}y \leq 12 \\ \frac{4}{3}x + \frac{3}{2}y \leq 15 \\ x \geq 0 \\ y \geq 0 \end{cases}$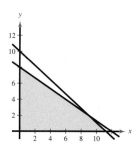

67. (a) $\begin{cases} 180x + 100y \geq 1000 \\ 6x + y \geq 18 \\ 220x + 40y \geq 400 \\ x \geq 0 \\ y \geq 0 \end{cases}$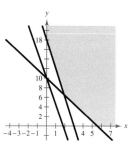

(b) Answers will vary.

69. (a) $\begin{cases} x \geq 50 \\ y \geq 40 \\ 55x + 70y \leq 7500 \end{cases}$

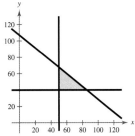

(b) Answers will vary.

71. True. The figure is a rectangle with a length of 9 units and a width of 11 units.

73. False. The solution region lies between the parallel lines $y = x/3 - 8/3$ and $y = x/3 + 5/3$.

75. Test a point on each side of the line.

77. Slope: -3
y-intercept: $(0, 6)$

79. Slope: -4
y-intercept: $\left(0, -\frac{3}{2}\right)$

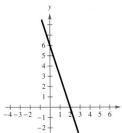 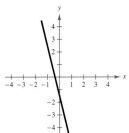

81. Relative minimum: $(-1.25, -3.13)$

83. Relative minimum: $(-0.67, -0.38)$

85. $\sin 15° = \dfrac{\sqrt{2 - \sqrt{3}}}{2}$

$\cos 15° = \dfrac{\sqrt{2 + \sqrt{3}}}{2}$

$\tan 15° = 2 - \sqrt{3}$

87. $\sin\left(-\dfrac{5\pi}{12}\right) = -\dfrac{\sqrt{2} + \sqrt{6}}{4}$

$\cos\left(-\dfrac{5\pi}{12}\right) = \dfrac{\sqrt{6} - \sqrt{2}}{4}$

$\tan\left(-\dfrac{5\pi}{12}\right) = -2 - \sqrt{3}$

89. $\sin(-22° 30') = -\dfrac{\sqrt{2 - \sqrt{2}}}{2}$

$\cos(-22° 30') = \dfrac{\sqrt{2 + \sqrt{2}}}{2}$

$\tan(-22° 30') = -\left(\sqrt{2} - 1\right)$

Section 9.6 *(page 686)*

1. objective

3. Find the points of intersection of the graphs of the equations corresponding to adjacent constraints.

5. Minimum at $(0, 0)$: 0
Maximum at $(5, 0)$: 20

7. Minimum at $(1, 0)$: 2
Maximum at $(3, 4)$: 26

9. Minimum at $(0, 20)$: 140
Maximum at $(60, 20)$: 740

11.

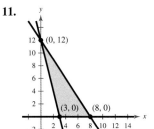

Minimum at $(3, 0)$: 9
Maximum at any point on the line segment connecting $(0, 12)$ and $(8, 0)$: 24

13.

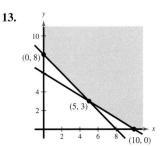

Minimum at $(5, 3)$: 35
No maximum

15.

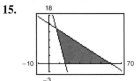

Minimum at $(7.2, 13.2)$: 34.8
Maximum at $(60, 0)$: 180

17.

Minimum at $(7.2, 13.2)$: 7.2
Maximum at $(60, 0)$: 60

19. Minimum at $(0, 0)$: 0
Maximum at $(0, 5)$: 25

21. Minimum at $(0, 0)$: 0
Maximum at $\left(\frac{22}{3}, \frac{19}{6}\right)$: $\frac{271}{6}$

23. Minimum at $(4, 3)$: 10
No maximum

25. No minimum
Maximum at $(12, 5)$: 7

27.

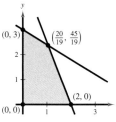

The maximum, 5, occurs at any point on the line segment connecting $(2, 0)$ and $\left(\frac{20}{19}, \frac{45}{19}\right)$. Minimum at $(0, 0)$: 0

29.

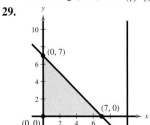

The constraint $x \leq 10$ is extraneous. Minimum at $(7, 0)$: -7; maximum at $(0, 7)$: 14

31.

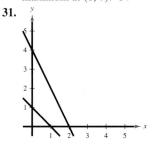

The feasible set is empty.

33.

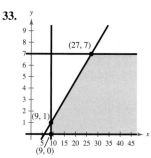

The solution region is unbounded.

Minimum at $(9, 0)$: 9

No maximum

35. 13 audits   **37.** 2 bottles of brand X
0 tax returns   5 bottles of brand Y
Optimal revenue: $20,800

39. 400 units of model X
1200 units of model Y
Optimal profit: $570,000

41. True. The objective function has a maximum value at any point on the line segment connecting the two vertices.

43. The solution region is unbounded and there is no maximum value.

45. $-\frac{3}{5}$ **47.** $(-7, -6, -4)$

49. $\left(\frac{7}{3}, \frac{8}{3}, -\frac{1}{3}\right)$ **51.** $(1, 1, 1, 1)$

Review Exercises *(page 691)*

1. $(1, 1)$ **3.** $\left(\frac{3}{2}, 5\right)$ **5.** $(0.25, 0.625)$ **7.** $(5, 4)$

9. $(0, 0), (2, 8), (-2, 8)$ **11.** $(4, -2)$

13. $(1.41, -0.66), (-1.41, 10.66)$ **15.** $(0, -2)$

17. No solution

19.

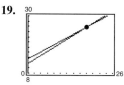

The BMI for males exceeds the BMI for females after age 18.

21. 16 ft × 18 ft   **23.** $\left(\frac{5}{2}, 3\right)$ **25.** $(0, 0)$

27. $\left(\frac{8}{5}a + \frac{14}{5}, a\right)$ **29.** d, one solution, consistent

31. b, no solution, inconsistent

33. $(100{,}000, 23)$ **35.** $(2, -4, -5)$

37. $(-6, 7, 10)$ **39.** $\left(\frac{24}{5}, \frac{22}{5}, -\frac{8}{5}\right)$ **41.** $\left(-\frac{3}{4}, 0, -\frac{5}{4}\right)$

43. $(a - 4, a - 3, a)$ **45.** $y = 2x^2 + x - 5$

47. $x^2 + y^2 - 4x + 4y - 1 = 0$

49. 10 gal of spray X   **51.** $16,000 at 7%
5 gal of spray Y $13,000 at 9%
12 gal of spray Z $11,000 at 11%

53. $s = -16t^2 + 150$ **55.** $\dfrac{A}{x} + \dfrac{B}{x + 20}$

57. $\dfrac{A}{x} + \dfrac{B}{x^2} + \dfrac{C}{x - 5}$ **59.** $\dfrac{3}{x + 2} - \dfrac{4}{x + 4}$

61. $1 - \dfrac{25}{8(x + 5)} + \dfrac{9}{8(x - 3)}$ **63.** $\dfrac{1}{2}\left(\dfrac{3}{x - 1} - \dfrac{x - 3}{x^2 + 1}\right)$

65. $\dfrac{3}{x^2 + 1} + \dfrac{4x - 3}{(x^2 + 1)^2}$

67.

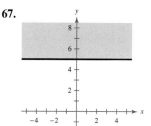

69.

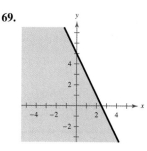

71.

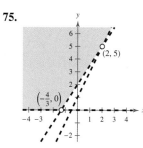

73.

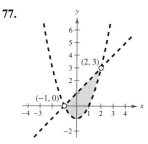

75.

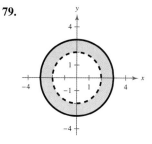

77.

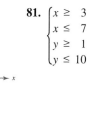

79.

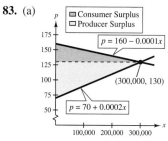

81. $\begin{cases} x \geq 3 \\ x \leq 7 \\ y \geq 1 \\ y \leq 10 \end{cases}$

83. (a)

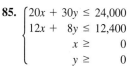

(b) Consumer surplus: $4,500,000
Producer surplus: $9,000,000

85. $\begin{cases} 20x + 30y \leq 24{,}000 \\ 12x + 8y \leq 12{,}400 \\ \quad\ x \geq 0 \\ \quad\ y \geq 0 \end{cases}$

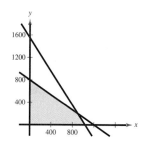

87.

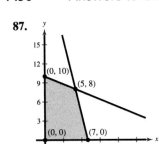

Minimum at $(0, 0)$: 0
Maximum at $(5, 8)$: 47

89.

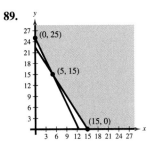

Minimum at $(15, 0)$: 26.25
No maximum

91. 72 haircuts
0 permanents
Optimal revenue: $1800

93. True. The nonparallel sides of the trapezoid are equal in length.

95. $\begin{cases} 4x + y = -22 \\ \frac{1}{2}x + y = 6 \end{cases}$ **97.** $\begin{cases} 3x + y = 7 \\ -6x + 3y = 1 \end{cases}$

99. $\begin{cases} x + y + z = 6 \\ x + y - z = 0 \\ x - y - z = 2 \end{cases}$ **101.** $\begin{cases} 2x + 2y - 3z = 7 \\ x - 2y + z = 4 \\ -x + 4y - z = -1 \end{cases}$

103. An inconsistent system of linear equations has no solution.

Chapter Test   *(page 695)*

1. $(-4, -5)$ **2.** $(0, -1), (1, 0), (2, 1)$ **3.** $(8, 4), (2, -2)$
4. $(4, 2)$ **5.** $(-3, 0), (2, 5)$ **6.** $(1, 4), (0.034, 0.619)$
7. $(-2, -5)$ **8.** $(10, -3)$ **9.** $(2, -3, 1)$

10. No solution   **11.** $-\dfrac{1}{x + 1} + \dfrac{3}{x - 2}$ **12.** $\dfrac{2}{x^2} + \dfrac{3}{2 - x}$

13. $x - \dfrac{5}{x} + \dfrac{3}{x + 1} + \dfrac{3}{x - 1}$ **14.** $-\dfrac{2}{x} + \dfrac{3x}{x^2 + 2}$

15.

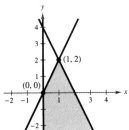

16.

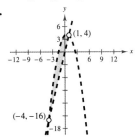

17.

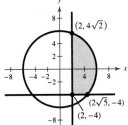

18. Minimum at $(0, 0)$: 0
Maximum at $(12, 0)$: 240
20. $y = -\frac{1}{2}x^2 + x + 6$

19. $24,000 in 4% fund
$26,000 in 5.5% fund
21. 900 units of model I
4400 units of model II
Optimal profit: $203,000

Problem Solving   *(page 697)*

1.

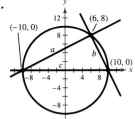

$a = 8\sqrt{5}, b = 4\sqrt{5}, c = 20$
$\left(8\sqrt{5}\right)^2 + \left(4\sqrt{5}\right)^2 = 20^2$
So, the triangle is a right triangle.

3. $ad \neq bc$

5. (a)

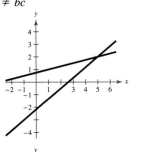 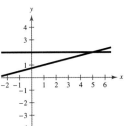

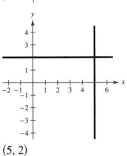

$(5, 2)$
Answers will vary.

(b)

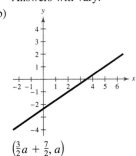

$\left(\frac{3}{2}a + \frac{7}{2}, a\right)$
Answers will vary.

7. 10.1 ft; About 252.7 ft   **9.** $12.00

11. (a) $(3, -4)$ (b) $\left(\dfrac{2}{-a + 5}, \dfrac{1}{4a - 1}, \dfrac{1}{a}\right)$

13. (a) $\left(\dfrac{-5a + 16}{6}, \dfrac{5a - 16}{6}, a\right)$

(b) $\left(\dfrac{-11a + 36}{14}, \dfrac{13a - 40}{14}, a\right)$

(c) $(-a + 3, a - 3, a)$ (d) Infinitely many

15. $\begin{cases} a + t \leq 32 \\ 0.15a \geq 1.9 \\ 193a + 772t \geq 11,000 \end{cases}$

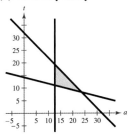

17. (a) $\begin{cases} 0 < y < 130 \\ x \ge 60 \\ x + y \le 200 \end{cases}$ (b)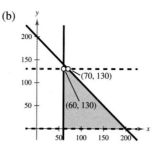

(c) No. The point $(90, 120)$ is not in the solution region.

(d) *Sample answer:* LDL/VLDL: 135 mg/dL; HDL: 65 mg/dL

(e) *Sample answer:* $(75, 90)$; $\frac{165}{75} = 2.2 < 4$

Chapter 10

Section 10.1 *(page 709)*

1. row-equivalent

3. *Sample answer:* 2×2 square matrix
$$\begin{bmatrix} -2 & 3 \\ 6 & 7 \end{bmatrix}$$

5. 1×2 **7.** 3×1 **9.** 2×2 **11.** 3×3

13. $\begin{bmatrix} 2 & -1 & \vdots & 7 \\ 1 & 1 & \vdots & 2 \end{bmatrix}$ **15.** $\begin{bmatrix} 1 & -1 & 2 & \vdots & 2 \\ 4 & -3 & 1 & \vdots & -1 \\ 2 & 1 & 0 & \vdots & 0 \end{bmatrix}$

17. $\begin{bmatrix} 3 & -5 & 2 & \vdots & 12 \\ 12 & 0 & -7 & \vdots & 10 \end{bmatrix}$

19. $\begin{cases} x + y = 3 \\ 5x - 3y = -1 \end{cases}$ **21.** $\begin{cases} 2x & + 5z = -12 \\ y - 2z = 7 \\ 6x + 3y & = 2 \end{cases}$

23. $\begin{cases} 9x + 12y + 3z & = 0 \\ -2x + 18y + 5z + 2w & = 10 \\ x + 7y - 8z & = -4 \\ 3x & + 2z & = -10 \end{cases}$

25. Add 5 times Row 2 to Row 1.

27. Interchange Row 1 and Row 2.
Add 4 times new Row 1 to Row 3.

29. $\begin{bmatrix} 1 & 2 & \frac{8}{3} \\ 4 & -3 & 6 \end{bmatrix}$ **31.** $\begin{bmatrix} 1 & 1 & 1 \\ 0 & -7 & -1 \end{bmatrix}$

33. $\begin{bmatrix} 1 & 0 & 14 & -11 \\ 0 & 1 & -2 & 2 \\ 0 & 0 & 1 & -7 \end{bmatrix}$

35. $\begin{bmatrix} 1 & 1 & 4 & -1 \\ 0 & 5 & -2 & 6 \\ 0 & 3 & 20 & 4 \end{bmatrix}; \begin{bmatrix} 1 & 1 & 4 & -1 \\ 0 & 1 & -\frac{2}{5} & \frac{6}{5} \\ 0 & 3 & 20 & 4 \end{bmatrix}$

37. (a) (i) $\begin{bmatrix} 3 & 0 & \vdots & -6 \\ 6 & -4 & \vdots & -28 \end{bmatrix}$ (b) $\begin{cases} -3x + 4y = 22 \\ 6x - 4y = -28 \end{cases}$

(ii) $\begin{bmatrix} 3 & 0 & \vdots & -6 \\ 0 & -4 & \vdots & -16 \end{bmatrix}$ Solution: $(-2, 4)$

(iii) $\begin{bmatrix} 3 & 0 & \vdots & -6 \\ 0 & 1 & \vdots & 4 \end{bmatrix}$ (c) Answers will vary.

(iv) $\begin{bmatrix} 1 & 0 & \vdots & -2 \\ 0 & 1 & \vdots & 4 \end{bmatrix}$

39. Reduced row-echelon form   **41.** Not in row-echelon form

43–47. *Sample answers:*

43. $\begin{bmatrix} 1 & 1 & 0 & 5 \\ 0 & 1 & 2 & 0 \\ 0 & 0 & 1 & -1 \end{bmatrix}$ **45.** $\begin{bmatrix} 1 & -1 & -1 & 1 \\ 0 & 1 & 6 & 3 \\ 0 & 0 & 0 & 0 \end{bmatrix}$

47. $\begin{bmatrix} 0 & 2 & \frac{11}{4} & -\frac{7}{2} \\ 0 & 1 & 1 & -\frac{8}{3} \\ 0 & 0 & 0 & 1 \\ 0 & 0 & 0 & 0 \end{bmatrix}$ **49.** $\begin{bmatrix} 1 & 0 & 0 \\ 0 & 1 & 0 \\ 0 & 0 & 1 \end{bmatrix}$

51. $\begin{bmatrix} 1 & 0 & 3 & 16 \\ 0 & 1 & 2 & 12 \end{bmatrix}$

53. $\begin{cases} x - 2y = 4 \\ y = -1 \end{cases}$ **55.** $\begin{cases} x - y + 2z = 4 \\ y - z = 2 \\ z = -2 \end{cases}$
$(2, -1)$ $(8, 0, -2)$

57. $(-3, 5)$ **59.** $(-5, 6)$ **61.** $(-4, -3, 6)$

63. No solution   **65.** $(3, -2, 5, 0)$ **67.** $(3, -4)$

69. $(-1, -4)$ **71.** $(4, -3, 2)$ **73.** $(5a + 4, -3a + 2, a)$

75. $(7, -3, 4)$ **77.** $(0, 2 - 4a, a)$ **79.** $(1, 0, 4, -2)$

81. $(-2a, a, a, 0)$ **83.** $f(x) = -x^2 + x + 1$

85. $f(x) = -9x^2 - 5x + 11$

87. \$1,200,000 at 8%
\$200,000 at 9%
\$600,000 at 12%

89. $y = 7.5t + 28$; About 141 cases; Yes, because the data values increase in a linear pattern.

91. False. It is a 2×4 matrix.

93. The matrix is in row-echelon form, not reduced row-echelon form.

95. -5 **97.** $a = -3, b = 8$ **99.** $a = 8, b = 4$

101. $x = \dfrac{b - a}{3}$ **103.** $x = 6b - a$

105. $\mathbf{u} = \langle 8, 6 \rangle$ **107.** $\mathbf{u} = \langle -10, 0 \rangle$

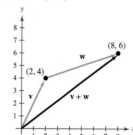   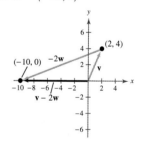

Section 10.2 *(page 724)*

1. scalars   **3.** BA

5. $x = -4, y = 23$ **7.** $x = -1, y = 3$

9. (a) $\begin{bmatrix} 3 & -2 \\ 1 & 7 \end{bmatrix}$ (b) $\begin{bmatrix} -1 & 0 \\ 3 & -9 \end{bmatrix}$ (c) $\begin{bmatrix} 3 & -3 \\ 6 & -3 \end{bmatrix}$

(d) $\begin{bmatrix} -1 & -1 \\ 8 & -19 \end{bmatrix}$

11. (a), (b), and (d) Not possible   (c) $\begin{bmatrix} 18 & 0 & 9 \\ -3 & -12 & 0 \end{bmatrix}$

13. (a) $\begin{bmatrix} 9 & 5 \\ 1 & -2 \\ -3 & 15 \end{bmatrix}$ (b) $\begin{bmatrix} 7 & -7 \\ 3 & 8 \\ -5 & -5 \end{bmatrix}$ (c) $\begin{bmatrix} 24 & -3 \\ 6 & 9 \\ -12 & 15 \end{bmatrix}$

(d) $\begin{bmatrix} 22 & -15 \\ 8 & 19 \\ -14 & -5 \end{bmatrix}$

15. (a) $\begin{bmatrix} 5 & 5 & -2 & 4 & 4 \\ -5 & 10 & 0 & -4 & -7 \end{bmatrix}$

(b) $\begin{bmatrix} 3 & 5 & 0 & 2 & 4 \\ 7 & -6 & -4 & 2 & 7 \end{bmatrix}$

(c) $\begin{bmatrix} 12 & 15 & -3 & 9 & 12 \\ 3 & 6 & -6 & -3 & 0 \end{bmatrix}$

(d) $\begin{bmatrix} 10 & 15 & -1 & 7 & 12 \\ 15 & -10 & -10 & 3 & 14 \end{bmatrix}$

17. $\begin{bmatrix} -8 & -7 \\ 15 & -1 \end{bmatrix}$ 19. $\begin{bmatrix} -24 & -4 & 12 \\ -12 & 32 & 12 \end{bmatrix}$

21. $\begin{bmatrix} 10 & 8 \\ -59 & 9 \end{bmatrix}$ 23. $\begin{bmatrix} -4 & 6 & -2 \\ 4 & 0 & 10 \end{bmatrix}$

25. $\begin{bmatrix} -2 & 0 & 5 \\ -\frac{5}{2} & 0 & \frac{7}{2} \end{bmatrix}$ 27. $\begin{bmatrix} 3 & -\frac{1}{2} & -\frac{13}{2} \\ 3 & 0 & -\frac{11}{2} \end{bmatrix}$

29. $\begin{bmatrix} 2 & -5 & 5 \\ -5 & 0 & -6 \end{bmatrix}$ 31. $\begin{bmatrix} -2 & 51 \\ -8 & 33 \\ 0 & 27 \end{bmatrix}$ 33. Not possible

3×2

35. $\begin{bmatrix} 1 & 0 & 0 \\ 0 & 1 & 0 \\ 0 & 0 & \frac{7}{2} \end{bmatrix}$

3×3

37. (a) $\begin{bmatrix} 0 & 15 \\ 6 & 12 \end{bmatrix}$ (b) $\begin{bmatrix} -2 & 2 \\ 31 & 14 \end{bmatrix}$ (c) $\begin{bmatrix} 9 & 6 \\ 12 & 12 \end{bmatrix}$

39. (a) $\begin{bmatrix} 2 & -2 \\ -3 & 0 \\ 7 & 6 \end{bmatrix}$ (b) and (c) Not possible

41. (a) $\begin{bmatrix} 19 \\ 48 \end{bmatrix}$ (b) Not possible (c) $\begin{bmatrix} 14 & -8 \\ 16 & 142 \end{bmatrix}$

43. (a) $\begin{bmatrix} 7 & 7 & 14 \\ 8 & 8 & 16 \\ -1 & -1 & -2 \end{bmatrix}$ (b) $[13]$ (c) Not possible

45. $\begin{bmatrix} 70 & -17 & 73 \\ 32 & 11 & 6 \\ 16 & -38 & 70 \end{bmatrix}$ 47. $\begin{bmatrix} 5 & 8 \\ -4 & -16 \end{bmatrix}$ 49. $\begin{bmatrix} -4 & 10 \\ 3 & 14 \end{bmatrix}$

51. (a) $\langle 4, 7 \rangle$ (b) $\langle -2, 3 \rangle$ (c) $\langle 8, 1 \rangle$
53. (a) $\langle 3, 6 \rangle$ (b) $\langle -7, -2 \rangle$ (c) $\langle 17, 10 \rangle$
55. $\langle 4, -2 \rangle$; Reflection in the x-axis
57. $\langle 2, 4 \rangle$; Reflection in the line $y = x$
59. $\langle 8, 2 \rangle$; Horizontal stretch

61. (a) $\begin{bmatrix} 2 & 3 \\ 1 & 4 \end{bmatrix}\begin{bmatrix} x_1 \\ x_2 \end{bmatrix} = \begin{bmatrix} 5 \\ 10 \end{bmatrix}$ (b) $\begin{bmatrix} -2 \\ 3 \end{bmatrix}$

63. (a) $\begin{bmatrix} 1 & -2 & 3 \\ -1 & 3 & -1 \\ 2 & -5 & 5 \end{bmatrix}\begin{bmatrix} x_1 \\ x_2 \\ x_3 \end{bmatrix} = \begin{bmatrix} 9 \\ -6 \\ 17 \end{bmatrix}$ (b) $\begin{bmatrix} 1 \\ -1 \\ 2 \end{bmatrix}$

65. (a) $\begin{bmatrix} 1 & -5 & 2 \\ -3 & 1 & -1 \\ 0 & -2 & 5 \end{bmatrix}\begin{bmatrix} x_1 \\ x_2 \\ x_3 \end{bmatrix} = \begin{bmatrix} -20 \\ 8 \\ -16 \end{bmatrix}$ (b) $\begin{bmatrix} -1 \\ 3 \\ -2 \end{bmatrix}$

67. $\begin{bmatrix} 110 & 99 & 77 & 33 \\ 44 & 22 & 66 & 66 \end{bmatrix}$

69. $[\$1037.50 \quad \$1400 \quad \$1012.50]$
The entries represent the profits from both crops at each of the three outlets.

71. $\begin{bmatrix} \$23.20 & \$20.50 \\ \$38.20 & \$33.80 \\ \$76.90 & \$68.50 \end{bmatrix}$
The entries represent the labor costs at each factory for each size of boat.

73. (a) $B = [2 \quad 0.25 \quad 0.5]$
(b) $BA = [1180 \quad 1407.5]$
BA represents the total calories burned by each person.
(c) 130-lb person: 398.25 calories
155-lb person: 475.05 calories

75. False. For most matrices, $AB \neq BA$.

77. $\begin{bmatrix} 1 & 0 \\ 2 & 1 \end{bmatrix} \neq \begin{bmatrix} 0 & 0 \\ 3 & 2 \end{bmatrix}$ 79. $\begin{bmatrix} 3 & -2 \\ 4 & 3 \end{bmatrix} \neq \begin{bmatrix} 2 & -2 \\ 5 & 4 \end{bmatrix}$

81. $AC = BC = \begin{bmatrix} 2 & 3 \\ 2 & 3 \end{bmatrix}$

83. AB is a diagonal matrix whose entries are the products of the corresponding entries of A and B.

85. Answers will vary.

87. (a) $A^2 = \begin{bmatrix} -1 & 0 \\ 0 & -1 \end{bmatrix}$ and $i^2 = -1$

$A^3 = \begin{bmatrix} -i & 0 \\ 0 & -i \end{bmatrix}$ and $i^3 = -i$

$A^4 = \begin{bmatrix} 1 & 0 \\ 0 & 1 \end{bmatrix}$ and $i^4 = 1$

(b) $B^2 = \begin{bmatrix} 1 & 0 \\ 0 & 1 \end{bmatrix} = I$, the identity matrix

89. $1 + \log_3 5 + \log_3 x$ 91. $\ln a - 3 \ln b$
93. $\left(\frac{4}{5}, 1 \right)$ 95. $(-2, 3, 0)$

Section 10.3 (page 734)

1. inverse
3. A square matrix that is singular does not have an inverse.
5–11. $AB = I$ and $BA = I$

13. $\begin{bmatrix} -3 & 2 \\ -2 & 1 \end{bmatrix}$ 15. $\begin{bmatrix} 1 & -\frac{1}{2} \\ -2 & \frac{3}{2} \end{bmatrix}$ 17. $\begin{bmatrix} 1 & 1 & -1 \\ -3 & 2 & -1 \\ 3 & -3 & 2 \end{bmatrix}$

19. Not possible

21. $\begin{bmatrix} -\frac{1}{8} & 0 & 0 & 0 \\ 0 & 1 & 0 & 0 \\ 0 & 0 & \frac{1}{4} & 0 \\ 0 & 0 & 0 & -\frac{1}{5} \end{bmatrix}$ 23. $\begin{bmatrix} -175 & 37 & -13 \\ 95 & -20 & 7 \\ 14 & -3 & 1 \end{bmatrix}$

25. $\begin{bmatrix} -12 & -5 & -9 \\ -4 & -2 & -4 \\ -8 & -4 & -6 \end{bmatrix}$ 27. $\begin{bmatrix} 0 & -1.\overline{81} & 0.\overline{90} \\ -10 & 5 & 5 \\ 10 & -2.\overline{72} & -3.\overline{63} \end{bmatrix}$

29. $\begin{bmatrix} 1 & 0 & 1 & 0 \\ 0 & 1 & 0 & 1 \\ 2 & 0 & 1 & 0 \\ 0 & 1 & 0 & 2 \end{bmatrix}$ **31.** $\begin{bmatrix} \frac{5}{13} & -\frac{3}{13} \\ \frac{1}{13} & \frac{2}{13} \end{bmatrix}$

33. Not possible   **35.** $\begin{bmatrix} -4 & 2 \\ 10 & -\frac{10}{3} \end{bmatrix}$ **37.** $(5, 0)$

39. $(-8, -6)$ **41.** $(3, 8, -11)$ **43.** $(2, 1, 0, 0)$
45. $(-1, 1)$ **47.** No solution   **49.** $(-4, -8)$
51. $(-1, 3, 2)$ **53.** $\left(\frac{13}{16}, \frac{11}{16}, 0\right)$
55. $I_1 = 0.5$ amp   **57.** $I_1 = 4$ amps
    $I_2 = 3$ amps $I_2 = 1$ amp
    $I_3 = 3.5$ amps $I_3 = 5$ amps
59. 100 bags of potting soil for seedlings
    100 bags of potting soil for general potting
    100 bags of potting soil for hardwood plants
61. (a) $\begin{cases} 256a + 16b + c = 82,700 \\ 289a + 17b + c = 86,918 \\ 324a + 18b + c = 89,400 \end{cases}$

(b) $\begin{bmatrix} 0.5 & -1 & 0.5 \\ -17.5 & 34 & -16.5 \\ 153 & -288 & 136 \end{bmatrix}$

(c) $y = -868t^2 + 32,862t - 220,884$

(d)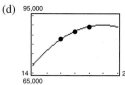

63. True. If B is the inverse of A, then $AB = I = BA$.
65–67. Answers will vary.
69. $k = -\frac{3}{2}$ **71.** 23   **73.** -152 **75.** 8

77. $\begin{bmatrix} 11 & 14 \\ 10 & -9 \end{bmatrix}$ **79.** $\begin{bmatrix} -\frac{7}{2} & 3 \\ 0 & -4 \end{bmatrix}$

Section 10.4   *(page 742)*

1. determinant   **3.** -5 **5.** 4   **7.** 16   **9.** -3
11. 0   **13.** 11   **15.** 0   **17.** $\frac{11}{6}$ **19.** -1924
21. 0.08
23. (a) $M_{11} = -6, M_{12} = 3, M_{21} = 5, M_{22} = 4$
    (b) $C_{11} = -6, C_{12} = -3, C_{21} = -5, C_{22} = 4$
25. (a) $M_{11} = 3, M_{12} = -4, M_{13} = 1, M_{21} = 2, M_{22} = 2,$
    $M_{23} = -4, M_{31} = -4, M_{32} = 10, M_{33} = 8$
    (b) $C_{11} = 3, C_{12} = 4, C_{13} = 1, C_{21} = -2, C_{22} = 2,$
    $C_{23} = 4, C_{31} = -4, C_{32} = -10, C_{33} = 8$
27. (a) $M_{11} = 10, M_{12} = -43, M_{13} = 2, M_{21} = -30, M_{22} = 17,$
    $M_{23} = -6, M_{31} = 54, M_{32} = -53, M_{33} = -34$
    (b) $C_{11} = 10, C_{12} = 43, C_{13} = 2, C_{21} = 30, C_{22} = 17,$
    $C_{23} = 6, C_{31} = 54, C_{32} = 53, C_{33} = -34$
29. (a) and (b) -36 **31.** (a) and (b) 96
33. (a) and (b) -75 **35.** (a) and (b) 0
37. (a) and (b) 225   **39.** -9 **41.** 0   **43.** 0
45. -58 **47.** 72   **49.** 0   **51.** 412   **53.** -126

55. -336 **57.** (a) -3 (b) -2 (c) $\begin{bmatrix} -2 & 0 \\ 0 & -3 \end{bmatrix}$ (d) 6

59. (a) -8 (b) 0   (c) $\begin{bmatrix} -4 & 4 \\ 1 & -1 \end{bmatrix}$ (d) 0

61. (a) 2   (b) -6 (c) $\begin{bmatrix} 1 & 4 & 3 \\ -1 & 0 & 3 \\ 0 & 2 & 0 \end{bmatrix}$ (d) -12

63. $A = \begin{bmatrix} 3 & 3 \\ 1 & 2 \end{bmatrix}$ **65.** $A = \begin{bmatrix} 4 & 2 & -1 \\ 2 & 1 & 0 \\ 1 & 1 & 3 \end{bmatrix}$

67. $A = \begin{bmatrix} 2 & 3 \\ 8 & 12 \end{bmatrix}$

69–73. Answers will vary.   **75.** ± 2 **77.** $-2, 1$
79. $-1, -4$ **81.** $8uv - 1$ **83.** e^{5x} **85.** $1 - \ln x$
87. True. If an entire row is zero, then each cofactor in the expansion is multiplied by zero.
89. Answers will vary.
91. (a) Columns 2 and 3 of A were interchanged.
    $|A| = -115 = -|B|$
    (b) Add -5 times Row 1 to Row 2.
    $|A| = 17 = |B|$
    (c) Multiply Row 1 by 5.
93. (a) 28   (b) -10 (c) -12
    The determinant of a diagonal matrix is the product of the entries on the main diagonal.
95. 3   **97.** 29   **99.** 72   **101.** $(2, -1)$

103. $(2, -3, 1)$ **105.** $\begin{bmatrix} -2 \\ 0 \end{bmatrix}$

Section 10.5   *(page 755)*

1. Cramer's Rule   **3.** -5 **5.** $(1, -1)$
7. Not possible   **9.** $(-1, 3, 2)$ **11.** $(-2, 1, -1)$
13. 7   **15.** 14   **17.** $y = \frac{16}{5}$ or $y = 0$ **19.** 250 mi²
21. Collinear   **23.** Not collinear   **25.** Collinear
27. $y = -3$ **29.** $3x - 5y = 0$ **31.** $x + 3y - 5 = 0$
33. $2x + 3y - 8 = 0$
35. $(0, 0), (0, 3), (6, 0), (6, 3)$

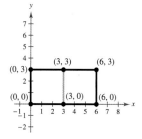

37. $(-4, 3), (-5, 3), (-4, 4), (-5, 4)$

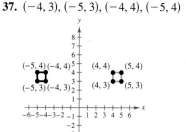

39. 2 square units   **41.** 10 square units

43. (a) Uncoded: $[3 \quad 15], [13 \quad 5], [0 \quad 8], [15 \quad 13], [5 \quad 0],$
$[19 \quad 15], [15 \quad 14]$

(b) Encoded: $48 \quad 81 \quad 28 \quad 51 \quad 24 \quad 40 \quad 54 \quad 95 \quad 5$
$10 \quad 64 \quad 113 \quad 57 \quad 100$

45. (a) Uncoded: $[20 \quad 5 \quad 24], [20 \quad 0 \quad 13], [5 \quad 0 \quad 20],$
$[15 \quad 13 \quad 15], [18 \quad 18 \quad 15], [23 \quad 0 \quad 0]$

(b) Encoded: $-119 \quad 28 \quad 67 \quad -58 \quad 6 \quad 39 \quad -115$
$35 \quad 60 \quad -62 \quad 15 \quad 32 \quad -54 \quad 12 \quad 27 \quad 23 \quad -23 \quad 0$

47. $26 \quad 21 \quad -18 \quad 35 \quad 52 \quad 31 \quad 13 \quad 30 \quad 38 \quad 79 \quad 173 \quad 206$
$-1 \quad -35 \quad -82 \quad 16 \quad -4 \quad -55 \quad 12 \quad 24 \quad 24$

49. $-5 \quad -41 \quad -87 \quad 91 \quad 207 \quad 257 \quad 11 \quad -5 \quad -41 \quad 40$
$80 \quad 84 \quad 76 \quad 177 \quad 227$

51. HAPPY NEW YEAR   **53.** CLASS IS CANCELED

55. MEET ME TONIGHT RON

57. $I_1 = -0.5$ amp
$I_2 = 1$ amp
$I_3 = 0.5$ amp

59. True. The determinant is zero when the points are collinear.

61. The system has either no solution or infinitely many solutions.

65. $2, 4, 2, 4$ **67.** $2, -5, 10, -17$

Review Exercises   (page 760)

1. 1×2 **3.** 2×5

5. $\begin{bmatrix} 3 & -10 & \vdots & 15 \\ 5 & 4 & \vdots & 22 \end{bmatrix}$ **7.** $\begin{cases} x \quad + 2z = -8 \\ 2x - 2y + 3z = 12 \\ 4x + 7y + z = 3 \end{cases}$

9. $\begin{bmatrix} 1 & 2 & 3 \\ 0 & 1 & 1 \\ 0 & 0 & 1 \end{bmatrix}$

11. $\begin{cases} x - 3y = 9 \\ y = -1 \end{cases}$ **13.** $\begin{cases} x + 3y + 4z = 1 \\ y + 2z = 3 \\ z = 4 \end{cases}$

$(6, -1)$ $(0, -5, 4)$

15. $(10, -12)$ **17.** No solution   **19.** $\left(-2a + \frac{3}{2}, 2a + 1, a\right)$

21. $(5, 2, -6)$ **23.** $(1, -2, 2, -4)$ **25.** $(1, 2, 2)$

27. $(2, -3, 3)$ **29.** $(2, 6, -10, -3)$ **31.** $x = 12, y = 11$

33. $x = 1, y = 11$

35. (a) $\begin{bmatrix} -1 & 8 \\ 15 & 13 \end{bmatrix}$ (b) $\begin{bmatrix} 5 & -12 \\ -9 & -3 \end{bmatrix}$

(c) $\begin{bmatrix} 8 & -8 \\ 12 & 20 \end{bmatrix}$ (d) $\begin{bmatrix} -2 & 16 \\ 30 & 26 \end{bmatrix}$

37. (a) $\begin{bmatrix} 5 & 7 \\ -3 & 14 \\ 31 & 42 \end{bmatrix}$ (b) $\begin{bmatrix} 5 & 1 \\ -11 & -10 \\ -9 & -38 \end{bmatrix}$

(c) $\begin{bmatrix} 20 & 16 \\ -28 & 8 \\ 44 & 8 \end{bmatrix}$ (d) $\begin{bmatrix} 10 & 14 \\ -6 & 28 \\ 62 & 84 \end{bmatrix}$

39. $\begin{bmatrix} 22 & -17 \\ 14 & 11 \end{bmatrix}$ **41.** $\begin{bmatrix} -16 & -6 \\ -12 & 4 \\ -14 & -8 \end{bmatrix}$ **43.** $\begin{bmatrix} -11 & -6 \\ 8 & -13 \\ -18 & -8 \end{bmatrix}$

45. $\begin{bmatrix} 3 & \frac{2}{3} \\ -\frac{4}{3} & \frac{11}{3} \\ \frac{10}{3} & 0 \end{bmatrix}$ **47.** $\begin{bmatrix} -30 & 4 \\ 51 & 70 \end{bmatrix}; 2 \times 2$

49. $\begin{bmatrix} 100 & 220 \\ 12 & -4 \\ 84 & 212 \end{bmatrix}; 3 \times 2$

51. (a) $\begin{bmatrix} -1 & -1 \\ 18 & -4 \end{bmatrix}$ (b) $\begin{bmatrix} 1 & 14 \\ -2 & -6 \end{bmatrix}$ (c) $\begin{bmatrix} 13 & 6 \\ 8 & 13 \end{bmatrix}$

53. $\begin{bmatrix} 14 & -22 & 22 \\ 19 & -41 & 80 \\ 42 & -66 & 66 \end{bmatrix}$ **55.** $\begin{bmatrix} -2 & 19 & 21 \\ -2 & 1 & -3 \end{bmatrix}$

57. $\langle 2, -5 \rangle$; Reflection in the x-axis

59. $\langle 1, 5 \rangle$; Horizontal shrink   **61.** $\begin{bmatrix} 76 & 114 & 133 \\ 38 & 95 & 76 \end{bmatrix}$

63–65. $AB = I$ and $BA = I$

67. $\begin{bmatrix} 4 & -5 \\ 5 & -6 \end{bmatrix}$

69. $\begin{bmatrix} \frac{1}{2} & -1 & -\frac{1}{2} \\ \frac{1}{2} & -\frac{2}{3} & -\frac{5}{6} \\ 0 & \frac{2}{3} & \frac{1}{3} \end{bmatrix}$ **71.** $\begin{bmatrix} 13 & 6 & -4 \\ -12 & -5 & 3 \\ 5 & 2 & -1 \end{bmatrix}$

73. $\begin{bmatrix} 1 & -1 \\ 4 & -\frac{7}{2} \end{bmatrix}$ **75.** Not possible   **77.** $(36, 11)$

79. $(-6, -1)$ **81.** No solution   **83.** $(-8, 18)$

85. $(2, -1, -2)$ **87.** $(-65.8, 33.6)$ **89.** 26   **91.** 116

93. (a) $M_{11} = 4, M_{12} = 7, M_{21} = -1, M_{22} = 2$

(b) $C_{11} = 4, C_{12} = -7, C_{21} = 1, C_{22} = 1$

95. (a) $M_{11} = 30, M_{12} = -12, M_{13} = -21, M_{21} = 20,$
$M_{22} = 19, M_{23} = 22, M_{31} = 5, M_{32} = -2, M_{33} = 19$

(b) $C_{11} = 30, C_{12} = 12, C_{13} = -21, C_{21} = -20,$
$C_{22} = 19, C_{23} = -22, C_{31} = 5, C_{32} = 2, C_{33} = 19$

97. -6 **99.** 15   **101.** 130   **103.** $(4, 7)$

105. $(-1, 4, 5)$ **107.** 16   **109.** Collinear

111. $x - 2y + 4 = 0$ **113.** $2x + 6y - 13 = 0$

115. 8 square units   **117.** SEE YOU FRIDAY

119. False. The matrix must be square.

121. If A is a square matrix, then the cofactor C_{ij} of the entry a_{ij} is $(-1)^{i+j}M_{ij}$, where M_{ij} is the determinant obtained by deleting the ith row and jth column of A. The determinant of A is the sum of the entries of any row or column of A multiplied by their respective cofactors.

Chapter Test   (page 764)

1. $\begin{bmatrix} 1 & 0 & 0 \\ 0 & 1 & 0 \\ 0 & 0 & 1 \end{bmatrix}$ **2.** $\begin{bmatrix} 1 & 0 & -1 & 2 \\ 0 & 1 & 0 & -1 \\ 0 & 0 & 0 & 0 \\ 0 & 0 & 0 & 0 \end{bmatrix}$

3. $\begin{bmatrix} 4 & 3 & -2 & \vdots & 14 \\ -1 & -1 & 2 & \vdots & -5 \\ 3 & 1 & -4 & \vdots & 8 \end{bmatrix}, \left(1, 3, -\frac{1}{2}\right)$

4. (a) $\begin{bmatrix} 1 & 5 \\ 0 & -4 \end{bmatrix}$ (b) $\begin{bmatrix} 6 & -3 & 12 \\ 0 & 18 & -9 \end{bmatrix}$

(c) $\begin{bmatrix} 8 & 15 \\ -5 & -13 \end{bmatrix}$ (d) $\begin{bmatrix} 10 & -5 & 20 \\ -10 & -1 & -17 \end{bmatrix}$

(e) Not possible

5. $\langle -3, -2 \rangle$; Reflection in the line $y = -x$

6. $\begin{bmatrix} \frac{2}{7} & \frac{3}{7} \\ \frac{5}{7} & \frac{4}{7} \end{bmatrix}$ **7.** $\begin{bmatrix} -\frac{5}{2} & 4 & -3 \\ 5 & -7 & 6 \\ 4 & -6 & 5 \end{bmatrix}$ **8.** $(12, 18)$

9. -112 **10.** 0 **11.** 43 **12.** $(-3, 5)$

13. $(-2, 4, 6)$ **14.** 7

15. Uncoded: $[11 \ 14 \ 15], [3 \ 11 \ 0], [15 \ 14 \ 0], [23 \ 15 \ 15],$
$[4 \ 0 \ 0]$

Encoded: $115 \ -41 \ -59 \ 14 \ -3 \ -11 \ 29 \ -15$
$-14 \ 128 \ -53 \ -60 \ 4 \ -4 \ 0$

16. 75 L of 60% solution, 25 L of 20% solution

Problem Solving   *(page 767)*

1. (a) $AT = \begin{bmatrix} -1 & -4 & -2 \\ 1 & 2 & 3 \end{bmatrix}$, $AAT = \begin{bmatrix} -1 & -2 & -3 \\ -1 & -4 & -2 \end{bmatrix}$

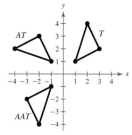

A represents a counterclockwise rotation.

(b) AAT is rotated clockwise 90° to obtain AT. AT is then rotated clockwise 90° to obtain T.

3. (a) Yes   (b) No   (c) No   (d) No   (e) No   (f) No

5. (a) $A^2 - 2A + 5I = \begin{bmatrix} 1 & 2 \\ -2 & 1 \end{bmatrix}^2 - 2\begin{bmatrix} 1 & 2 \\ -2 & 1 \end{bmatrix} + 5\begin{bmatrix} 1 & 0 \\ 0 & 1 \end{bmatrix}$

$= \begin{bmatrix} -3 & 4 \\ -4 & -3 \end{bmatrix} - \begin{bmatrix} 2 & 4 \\ -4 & 2 \end{bmatrix} + \begin{bmatrix} 5 & 0 \\ 0 & 5 \end{bmatrix}$

$= \begin{bmatrix} 0 & 0 \\ 0 & 0 \end{bmatrix}$

(b) $A^{-1} = \frac{1}{5}(2I - A)$

$\begin{bmatrix} 1 & 2 \\ -2 & 1 \end{bmatrix}^{-1} = \frac{1}{5}\left(2\begin{bmatrix} 1 & 0 \\ 0 & 1 \end{bmatrix} - \begin{bmatrix} 1 & 2 \\ -2 & 1 \end{bmatrix}\right)$

$\begin{bmatrix} \frac{1}{5} & -\frac{2}{5} \\ \frac{2}{5} & \frac{1}{5} \end{bmatrix} = \frac{1}{5}\left(\begin{bmatrix} 2 & 0 \\ 0 & 2 \end{bmatrix} - \begin{bmatrix} 1 & 2 \\ -2 & 1 \end{bmatrix}\right)$

$\begin{bmatrix} \frac{1}{5} & -\frac{2}{5} \\ \frac{2}{5} & \frac{1}{5} \end{bmatrix} = \frac{1}{5}\begin{bmatrix} 1 & -2 \\ 2 & 1 \end{bmatrix}$

$\begin{bmatrix} \frac{1}{5} & -\frac{2}{5} \\ \frac{2}{5} & \frac{1}{5} \end{bmatrix} = \begin{bmatrix} \frac{1}{5} & -\frac{2}{5} \\ \frac{2}{5} & \frac{1}{5} \end{bmatrix}$

(c) Answers will vary.

7. $A^T = \begin{bmatrix} -1 & 2 \\ 1 & 0 \\ -2 & 1 \end{bmatrix}$, $B^T = \begin{bmatrix} -3 & 1 & 1 \\ 0 & 2 & -1 \end{bmatrix}$

$(AB)^T = \begin{bmatrix} 2 & -5 \\ 4 & -1 \end{bmatrix} = B^T A^T$

9. $x = 6$ **11.** Answers will vary.

13. $\begin{vmatrix} x & 0 & 0 & d \\ -1 & x & 0 & c \\ 0 & -1 & x & b \\ 0 & 0 & -1 & a \end{vmatrix}$

15. Sulfur: 32 atomic mass units
Nitrogen: 14 atomic mass units
Fluorine: 19 atomic mass units

17. REMEMBER SEPTEMBER THE ELEVENTH

19. $A^{-1} = \begin{bmatrix} 0.0625 & -0.4375 & 0.625 \\ 0.1875 & 0.6875 & -1.125 \\ -0.125 & -0.125 & 0.75 \end{bmatrix}$

$|A^{-1}| = \frac{1}{16}, |A| = 16$

$|A^{-1}| = \frac{1}{|A|}$

Chapter 11

Section 11.1   *(page 777)*

1. recursively   **3.** series   **5.** $6!$ **7.** $-3, 1, 5, 9, 13$

9. $5, 3, 5, 3, 5$ **11.** $\frac{2}{3}, \frac{2}{3}, \frac{2}{3}, \frac{2}{3}, \frac{2}{3}$ **13.** $-2, 4, -8, 16, -32$

15. $\frac{1}{3}, \frac{8}{3}, 9, \frac{64}{3}, \frac{125}{3}$ **17.** $\frac{1}{3}, \frac{1}{2}, \frac{3}{5}, \frac{2}{3}, \frac{5}{7}$ **19.** $0, 0, 6, 24, 60$

21. $-\frac{1}{2}, \frac{2}{3}, -\frac{3}{4}, \frac{4}{5}, -\frac{5}{6}$ **23.** -73 **25.** $\frac{44}{239}$

27.   **29.**

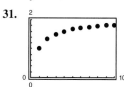

31.

33. $a_n = 4n - 1$ **35.** $a_n = n^3 + 2$ **37.** $a_n = (-1)^{n+1}$

39. $a_n = \dfrac{(-1)^n (n + 1)}{n + 2}$ **41.** $a_n = \dfrac{n + 1}{2n - 1}$ **43.** $a_n = \dfrac{1}{n!}$

45. $a_n = \dfrac{3^{n-1}}{(n - 1)!}$

47. $28, 24, 20, 16, 12$ **49.** $81, 27, 9, 3, 1$ **51.** $1, 2, 2, 3, \frac{7}{2}$

53. $1, 1, 2, 3, 5, 8, 13, 21, 34, 55, 89, 144$

$1, 2, \frac{3}{2}, \frac{5}{3}, \frac{8}{5}, \frac{13}{8}, \frac{21}{13}, \frac{34}{21}, \frac{55}{34}, \frac{89}{55}$

55. $5, 5, \frac{5}{2}, \frac{5}{6}, \frac{5}{24}$ **57.** $6, -24, 60, -120, 210$

59. $\dfrac{1}{30}$ **61.** $n + 1$ **63.** 25 **65.** $\frac{124}{429}$ **67.** 88

69. $\frac{13}{4}$ **71.** $\frac{3}{8}$ **73.** 1.33 **75.** $\displaystyle\sum_{i=1}^{9} \frac{1}{3i}$

77. $\displaystyle\sum_{i=1}^{8} \left[2\left(\frac{i}{8}\right) + 3\right]$ **79.** $\displaystyle\sum_{i=1}^{6} (-1)^{i+1} 3^i$ **81.** $\displaystyle\sum_{i=1}^{7} \frac{i^2}{(i + 1)!}$

83. $\displaystyle\sum_{i=1}^{5} \frac{2^i - 1}{2^{i+1}}$ **85.** (a) $\frac{7}{8}$ (b) $\frac{15}{16}$ (c) $\frac{31}{32}$

87. (a) $-\frac{3}{2}$ (b) $-\frac{5}{4}$ (c) $-\frac{11}{8}$ **89.** $\frac{2}{3}$ **91.** $\frac{7}{9}$

93. (a) $A_1 = \$10,087.50$, $A_2 \approx \$10,175.77$, $A_3 \approx \$10,264.80$,
$A_4 \approx \$10,354.62$, $A_5 \approx \$10,445.22$, $A_6 \approx \$10,536.62$,
$A_7 \approx \$10,628.81$, $A_8 \approx \$10,721.82$
(b) $\$14,169.09$
(c) No. $A_{80} \approx \$20,076.31 \neq 2A_{40} \approx \$28,338.18$

95. True by the Properties of Sums.

97. $\$500.95$ **99.** Proofs

101. $\displaystyle\sum_{k=1}^{4} 3 = 3(4) = 12$

103. $b_1 = \frac{1}{3}$, $b_2 = -\frac{1}{5}$, $b_3 = \frac{1}{7}$, $b_4 = -\frac{1}{9}$; $b_n = -a_n$

105. $y = -\frac{1}{2}x + 3$ **107.** $y = 5x - 4$

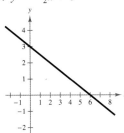

109. Vertices: $(0, \pm 3)$ **111.** Vertices: $(\pm 3, 0)$

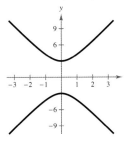
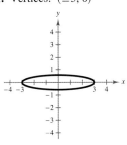

113. $2\mathbf{i} - 4\mathbf{j}$ **115.** $-9\mathbf{i} + 5\mathbf{j}$

Section 11.2   *(page 786)*

1. $a_1 + (n - 1)d$

3. A sequence is arithmetic when the differences between consecutive terms are the same.

5. Not arithmetic   **7.** Arithmetic, $d = -2$

9. Arithmetic, $d = \frac{1}{4}$ **11.** Not arithmetic

13. 8, 11, 14, 17, 20   **15.** $7, 3, -1, -5, -9$
   Arithmetic, $d = 3$ Arithmetic, $d = -4$

17. $-1, 1, -1, 1, -1$ **19.** 2, 8, 24, 64, 160
   Not arithmetic   Not arithmetic

21. $a_n = 3n - 2$ **23.** $a_n = -8n + 108$

25. $a_n = -\frac{5}{2}n + \frac{13}{2}$ **27.** $a_n = \frac{10}{3}n + \frac{5}{3}$

29. $a_n = 3n + 85$ **31.** 5, 11, 17, 23, 29

33. $2, -4, -10, -16, -22$ **35.** $-2, 2, 6, 10, 14$

37. 15, 19, 23, 27, 31   **39.** 15, 13, 11, 9, 7

41. -49 **43.** $\frac{31}{8}$ **45.** 110   **47.** -25 **49.** 10,000

51. 15,100   **53.** -7020 **55.** 1275   **57.** 129,250

59. $-28,300$ **61.** b   **62.** d   **63.** c   **64.** a

65.   **67.**

69. (a) $\$40,000$ (b) $\$217,500$ **71.** 2430 seats

73. (a)

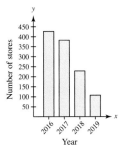

(b) $u_n = 533.33 - 106.33n$

(c)

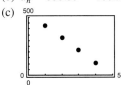

(d) $\displaystyle\sum_{n=1}^{4} (533.33 - 106.33n)$; About 1070 stores

75. 784 ft

77. True. Given a_1 and a_2, $d = a_2 - a_1$ and $a_n = a_1 + (n-1)d$.

79. (a)   (b)

(c) The graph of $y = 3x + 2$ contains all points on the line. The graph of $a_n = 2 + 3n$ contains only points at the positive integers.
(d) The slope of the line and the common difference of the arithmetic sequence are equal.

81. $x, 3x, 5x, 7x, 9x, 11x, 13x, 15x, 17x, 19x$

83. When $n = 50$, $a_n = 2(50) - 1 = 99$.

85. -0.205 **87.** 1.292   **89.** 25   **91.** 9.322

93. $14x + \dfrac{14}{x + 1}$, $x \neq 1$ **95.** $\dfrac{3}{4(x + 2)} + \dfrac{1}{4(x - 2)}$

Section 11.3   *(page 795)*

1. ratio   **3.** Multiply the nth term by the common ratio.

5. Geometric, $r = 2$ **7.** Geometric, $r = 3$

9. Not geometric   **11.** Geometric, $r = -\sqrt{7}$

13. 4, 12, 36, 108, 324   **15.** $1, \frac{1}{2}, \frac{1}{4}, \frac{1}{8}, \frac{1}{16}$

17. $1, e, e^2, e^3, e^4$ **19.** $3, 3\sqrt{5}, 15, 15\sqrt{5}, 75$

21. $2, 6x, 18x^2, 54x^3, 162x^4$ **23.** $a_n = 4\left(\frac{1}{2}\right)^{n-1}$; $\frac{1}{128}$

25. $a_n = 6\left(-\frac{1}{3}\right)^{n-1}$; $-\dfrac{2}{59,049}$ **27.** $a_n = 100e^{x(n-1)}$; $100e^{8x}$

29. $a_n = \left(\sqrt{2}\right)^{n-1}$; $32\sqrt{2}$ **31.** $a_n = 64\left(\frac{1}{2}\right)^{n-1}$

33. $a_n = 9(2)^{n-1}$ **35.** $a_n = 6\left(-\frac{3}{2}\right)^{n-1}$ **37.** 13,122

39. $\frac{1}{768}$ **41.** 9   **43.** -2 **45.** a

46. c   **47.** b   **48.** d

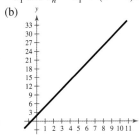

49. 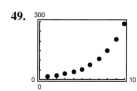   **51.**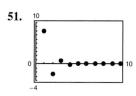

53. 5461   **55.** $-14{,}706$ **57.** 29,921.311   **59.** 1360.383

61. 1.6   **63.** $\displaystyle\sum_{n=1}^{7} 10(3)^{n-1}$ **65.** $\displaystyle\sum_{n=1}^{6} 0.1(4)^{n-1}$ **67.** 2

69. $\dfrac{6}{5}$ **71.** 5   **73.** 32   **75.** Undefined   **77.** $\frac{4}{11}$

79. $\frac{7}{22}$

81.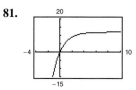

Horizontal asymptote: $y = 12$
Corresponds to the sum of the series

83. \$29,412.25   **85.** Answers will vary.   **87.** $273\frac{8}{9}$ in.^{2}

89. False. A sequence is geometric when the ratios of consecutive terms are the same.

91. (a)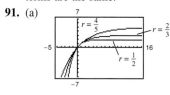

As $x \to \infty$, $y \to \dfrac{1}{1-r}$.

(b)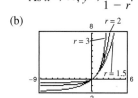

As $x \to \infty$, $y \to \infty$.

93. The common ratio is 2.

95. Prime; 97 has only two positive factors, itself and 1.

97. Prime; 65,537 has only two positive factors, itself and 1.

99–101. Answers will vary.

Section 11.4 (page 806)

1. mathematical induction

3. To find the first differences of a sequence, subtract consecutive terms.

5. $\dfrac{5}{(k+1)(k+2)}$ **7.** $(k+1)^2(k+4)^2$

9–25. Proofs   **27.** $S_n = n(2n-1)$; Proof

29. $S_n = \dfrac{n}{2(n+1)}$; Proof   **31.** 120   **33.** 979

35. 70   **37.** Linear; $a_n = 9n - 4$

39. Quadratic; $a_n = 2n^2 + 2$

41. 0, 3, 6, 9, 12, 15
First differences: 3, 3, 3, 3, 3
Second differences: 0, 0, 0, 0
Linear

43. 4, 10, 19, 31, 46, 64
First differences: 6, 9, 12, 15, 18
Second differences: 3, 3, 3, 3
Quadratic

45. $a_n = n^2 - n + 3$

47. $0, -0.02, -0.04, -0.06, -0.08$; No; The first differences are not equal.

49. False. A sequence with n terms has $n - 2$ second differences.

51. The formula $\dfrac{n(n+1)(2n+1)}{6}$ should be used.

So, $1^2 + 2^2 + 3^2 + \ldots + 9^2 = \dfrac{9(9+1)(18+1)}{6} = 285.$

53. At some point, the pattern may fail.   **55.** $x^2 + 2xy + y^2$

57. $y^4 - 8y^3 + 24y^2 - 32y + 16$ **59.** $x^6 + 12x^4 + 48x^2 + 64$

61. 1   **63.** 120   **65.** 1   **67.** -4

Section 11.5 (page 813)

1. expanding   **3.** Binomial Theorem; Pascal's Triangle

5. 10   **7.** 1   **9.** 210   **11.** 4950   **13.** 20

15. 5   **17.** $x^6 + 6x^5 + 15x^4 + 20x^3 + 15x^2 + 6x + 1$

19. $y^3 - 9y^2 + 27y - 27$ **21.** $r^3 + 9r^2s + 27rs^2 + 27s^3$

23. $243a^5 - 1620a^4b + 4320a^3b^2 - 5760a^2b^3$
$\qquad + 3840ab^4 - 1024b^5$

25. $a^4 + 24a^3 + 216a^2 + 864a + 1296$

27. $y^6 - 6y^5 + 15y^4 - 20y^3 + 15y^2 - 6y + 1$

29. $81 - 216z + 216z^2 - 96z^3 + 16z^4$

31. $x^5 + 10x^4y + 40x^3y^2 + 80x^2y^3 + 80xy^4 + 32y^5$

33. $x^8 + 4x^6y^2 + 6x^4y^4 + 4x^2y^6 + y^8$

35. $\dfrac{1}{x^5} + \dfrac{5y}{x^4} + \dfrac{10y^2}{x^3} + \dfrac{10y^3}{x^2} + \dfrac{5y^4}{x} + y^5$

37. $2x^4 - 24x^3 + 113x^2 - 246x + 207$

39. $120x^7y^3$ **41.** $360x^3y^2$ **43.** $1{,}259{,}712x^2y^7$

45. $-4{,}330{,}260{,}000y^9x^3$ **47.** 160   **49.** 720

51. $-6{,}300{,}000$ **53.** 210

55. $x^{3/2} + 15x + 75x^{1/2} + 125$

57. $x^2 - 3x^{4/3}y^{1/3} + 3x^{2/3}y^{2/3} - y$

59. $81t^2 + 108t^{7/4} + 54t^{3/2} + 12t^{5/4} + t$

61. $3x^2 + 3xh + h^2$, $h \neq 0$

63. $6x^5 + 15x^4h + 20x^3h^2 + 15x^2h^3 + 6xh^4 + h^5$, $h \neq 0$

65. $\dfrac{1}{\sqrt{x+h} + \sqrt{x}}$, $h \neq 0$ **67.** -4 **69.** $512i$

71. 1   **73.** 1.172   **75.** 510,568.785

77. Fibonacci sequence

CHAPTER 11

79.

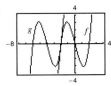

The graph of g is shifted four units to the left of the graph of f.
$g(x) = x^3 + 12x^2 + 44x + 48$

81.

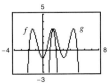

The graph of g is shifted three units to the right of the graph of f.
$g(x) = -x^4 + 12x^3 - 50x^2 + 84x - 46$

83. (a) $g(t) = -5.893t^2 + 67.78t + 141.1, 0 \le t \le 7$

(b)

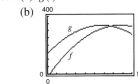

(c) 2015

85. True. The coefficients from the Binomial Theorem can be used to find the numbers in Pascal's Triangle.

87. False. The coefficient of the x^2-term is $_9C_2(4)^7 = -589,824$. The coefficient of the x-term is $_9C_1(4)^8 = 589,824$.

89. The first and last numbers in each row are 1's. Also, each number between the 1's is the sum of the two numbers immediately above the number.

91.

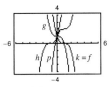

k, f; $k(x)$ is the expansion of $f(x)$.

93. About 0.273   **95–97.** Proofs   **99.** $\frac{1}{3}$ **101.** -9

103. 36   **105.** $\frac{1}{7}$ **107.** $(n + 9)(n + 8)$

109. $\sin 330° = -\dfrac{1}{2}$

$\cos 330° = \dfrac{\sqrt{3}}{2}$

$\tan 330° = -\dfrac{\sqrt{3}}{3}$

111. $\sin\left(-\dfrac{3\pi}{4}\right) = -\dfrac{\sqrt{2}}{2}$

$\cos\left(-\dfrac{3\pi}{4}\right) = -\dfrac{\sqrt{2}}{2}$

$\tan\left(-\dfrac{3\pi}{4}\right) = 1$

113. $\sin\dfrac{11\pi}{3} = -\dfrac{\sqrt{3}}{2}$

$\cos\dfrac{11\pi}{3} = \dfrac{1}{2}$

$\tan\dfrac{11\pi}{3} = -\sqrt{3}$

Section 11.6   *(page 823)*

1. distinguishable permutations

3. A given set of n elements forms exactly one combination, whereas different arrangements of the elements may form numerous permutations.

5. 6   **7.** 5   **9.** 3   **11.** 8   **13.** 30

15. 30   **17.** 64   **19.** 175,760,000

21. (a) 648   (b) 180   (c) 600   **23.** (a) 40,320   (b) 384

25. 120   **27.** 20   **29.** 132   **31.** 2730

33. 5,527,200   **35.** 504   **37.** 1,816,214,400

39. 420   **41.** 2520

43. ABCD, ABDC, ACBD, ACDB, ADBC, ADCB, BACD, BADC, CABD, CADB, DABC, DACB, BCAD, BDAC, CBAD, CDAB, DBAC, DCAB, BCDA, BDCA, CBDA, CDBA, DBCA, DCBA

45. 15   **47.** 1   **49.** 120   **51.** 38,760

53. AB, AC, AD, AE, AF, BC, BD, BE, BF, CD, CE, CF, DE, DF, EF

55. 5,586,853,480   **57.** (a) 7315   (b) 693   (c) 12,628

59. (a) 3744   (b) 24   **61.** 292,600   **63.** 5   **65.** 20

67. $n = 2$ **69.** $n = 3$ **71.** $n = 5$ or $n = 6$

73. $n = 10$ **75.** 36

77. False. It is an example of a combination.

79. $_{10}P_6 > \,_{10}C_6$. Changing the order of any of the six elements selected results in a different permutation but the same combination.

81–83. Proofs   **85.** $\dfrac{1}{74,613}$ **87.** $\dfrac{1}{3,338,335}$

89. $720x^3y^2$ **91.** $28,000,000x^2y^6$

93. $-498,610,169,343x^8y^5$ **95.** $y = -\frac{1}{4}x^2 + \frac{7}{2}x$

Section 11.7   *(page 834)*

1. experiment; outcomes   **3.** event

5. $n(E), n(S)$ **7.** 1

9. $\{(H, 1), (H, 2), (H, 3), (H, 4), (H, 5), (H, 6),$
$(T, 1), (T, 2), (T, 3), (T, 4), (T, 5), (T, 6)\}$

11. $\{ABC, ACB, BAC, BCA, CAB, CBA\}$

13. $\{AB, AC, AD, AE, BC, BD, BE, CD, CE, DE\}$

15. $\frac{3}{8}$ **17.** $\frac{1}{2}$ **19.** $\frac{7}{8}$ **21.** $\frac{3}{13}$ **23.** $\frac{3}{26}$ **25.** $\frac{5}{36}$

27. $\frac{11}{12}$ **29.** $\frac{1}{3}$ **31.** $\frac{1}{5}$ **33.** $\frac{2}{5}$

35. (a) 744,000   (b) $\frac{171}{1000}$ (c) $\frac{267}{500}$ (d) $\frac{171}{1000}$

37. (a) $\frac{13}{16}$ (b) $\frac{3}{16}$ (c) $\frac{1}{32}$ **39.** 19%

41. (a) $\frac{21}{1292}$ (b) $\frac{225}{646}$ (c) $\frac{49}{323}$ **43.** (a) $\frac{1}{120}$ (b) $\frac{1}{24}$

45. (a) $\frac{5}{13}$ (b) $\frac{1}{2}$ (c) $\frac{4}{13}$ **47.** (a) $\frac{14}{55}$ (b) $\frac{12}{55}$ (c) $\frac{54}{55}$

49. (a) $\frac{1}{4}$ (b) $\frac{1}{2}$ (c) $\frac{841}{1600}$ (d) $\frac{1}{40}$

51. 0.27   **53.** $\frac{4}{5}$ **55.** 0.71   **57.** $\frac{11}{25}$

59. (a) 0.9702   (b) 0.0002   (c) 0.9998

61. (a) $\frac{1}{38}$ (b) $\frac{9}{19}$ (c) $\frac{10}{19}$ (d) $\frac{1}{1444}$ (e) $\frac{729}{6859}$

63. True. Two events are independent when the occurrence of one has no effect on the occurrence of the other.

65. The three numbers from 1 to 15 greater than 12 are 13, 14, and 15. So, the probability that both numbers are greater than 12 is $\left(\frac{3}{15}\right)\left(\frac{3}{15}\right) = \frac{9}{225} = \frac{1}{25}$.

67. No; $P(A) + P(B) = 0.76 + 0.58 = 1.34 > 1$.

69. Over an extended period of time with similar weather conditions it will rain 40% of the time.

71. $2\cos\theta = \sqrt{2}$; $\sin\theta = \pm\dfrac{\sqrt{2}}{2}$; $\cos\theta = \dfrac{\sqrt{2}}{2}$

73. $\infty, 0$ **75.** $-1 + 6i$

Review Exercises   *(page 840)*

1. $15, 9, 7, 6, \frac{27}{5}$ **3.** $120, 60, 20, 5, 1$ **5.** $a_n = 2(-1)^n$

7. $a_n = \frac{3^n}{n}$ **9.** $\frac{1}{20}$ **11.** $\frac{1}{n(n+1)}$ **13.** $\frac{205}{24}$

15. $\sum_{k=1}^{20} \frac{1}{2k}$ **17.** $\frac{4}{9}$

19. (a) $A_1 = \$10,018.75$
   $A_2 \approx \$10,037.54$
   $A_3 \approx \$10,056.36$
   $A_4 \approx \$10,075.21$
   $A_5 \approx \$10,094.10$
   $A_6 \approx \$10,113.03$
   $A_7 \approx \$10,131.99$
   $A_8 \approx \$10,150.99$
   $A_9 \approx \$10,170.02$
   $A_{10} \approx \$10,189.09$
   (b) $\$12,520.59$

21. Arithmetic, $d = -6$ **23.** Not arithmetic

25. $a_n = 12n - 5$ **27.** $a_n = -18n + 150$

29. $4, 21, 38, 55, 72$ **31.** $45,450$ **33.** 1640

35. $\frac{11,300}{3}$ **37.** (a) $\$51,600$ (b) $\$238,500$

39. Geometric, $r = 3$ **41.** Geometric, $r = -3$

43. $2, 30, 450, 6750, 101,250$

45. $9, 6, 4, \frac{8}{3}, \frac{16}{9}$ or $9, -6, 4, -\frac{8}{3}, \frac{16}{9}$

47. $a_n = 100(1.05)^{n-1}$;   About 155.133

49. $a_n = 18\left(-\frac{1}{2}\right)^{n-1}$; $-\frac{9}{256}$ **51.** 127 **53.** $\frac{15}{16}$

55. 31 **57.** 23.056 **59.** 8 **61.** $\frac{2}{3}$

63. (a) $a_n = 120,000(0.7)^n$ (b) $\$20,168.40$

65–67. Proofs   **69.** $S_n = n(2n + 7)$; Proof

71. $S_n = \frac{5}{2}\left[1 - \left(\frac{3}{5}\right)^n\right]$; Proof   **73.** 2850

75. $5, 10, 15, 20, 25$
   First differences: $5, 5, 5, 5$
   Second differences: $0, 0, 0$
   Linear

77. 15 **79.** 21 **81.** $x^4 + 16x^3 + 96x^2 + 256x + 256$

83. $64 - 240x + 300x^2 - 125x^3$ **85.** 7 **87.** $10,000$

89. 120 **91.** $225,792,840$ **93.** (a) $\frac{1}{5}$ (b) $\frac{3}{5}$

95. (a) 43% (b) 82% **97.** $\frac{1}{1296}$ **99.** $\frac{3}{4}$

101. False. $\dfrac{(n+2)!}{n!} = \dfrac{(n+2)(n+1)n!}{n!} = (n+2)(n+1)$

103. True by the Properties of Sums.

105. The set of positive integers

107. Each term of the sequence is defined in terms of preceding terms.

Chapter Test   *(page 843)*

1. $-\frac{1}{5}, \frac{1}{8}, -\frac{1}{11}, \frac{1}{14}, -\frac{1}{17}$ **2.** $a_n = \frac{n+2}{n!}$

3. $60, 73, 86$; 329 **4.** $a_n = -3n + 60$ **5.** $a_n = \frac{7}{2}(2)^n$

6. $86,100$ **7.** 477 **8.** 4 **9.** $-\frac{1}{4}$ **10.** Proof

11. $x^4 + 24x^3y + 216x^2y^2 + 864xy^3 + 1296y^4$

12. $3x^5 - 30x^4 + 124x^3 - 264x^2 + 288x - 128$

13. $-22,680$ **14.** (a) 72 (b) $328,440$

15. (a) 330 (b) $720,720$ **16.** $26,000$ **17.** 720

18. $\frac{1}{15}$ **19.** $\frac{1}{27,405}$ **20.** 10%

Cumulative Test for Chapters 9–11   *(page 844)*

1. $(1, 2), \left(-\frac{3}{2}, \frac{3}{4}\right)$ **2.** $(-3, -1)$ **3.** $(5, -2, -2)$

4. $(1, -2, 1)$ **5.** $\$0.75$ mixture: 120 lb; $\$1.25$ mixture: 80 lb

6. $y = \frac{1}{4}x^2 - 2x + 6$ **7.** $-\frac{3}{x} + \frac{5x - 1}{x^2 + 2}$

8. 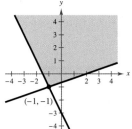   **9.**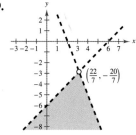

10.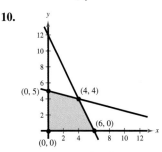

Maximum at $(4, 4)$: 20
Minimum at $(0, 0)$: 0

11. $\begin{bmatrix} -1 & 2 & -1 & \vdots & 9 \\ 2 & -1 & 2 & \vdots & -9 \\ 3 & 3 & -4 & \vdots & 7 \end{bmatrix}$ **12.** $(-2, 3, -1)$

13. $\begin{bmatrix} -3 & 8 \\ 6 & 1 \end{bmatrix}$ **14.** $\begin{bmatrix} 8 & -19 \\ 12 & 9 \end{bmatrix}$ **15.** $\begin{bmatrix} -13 & 6 & -4 \\ 18 & 4 & 4 \end{bmatrix}$

16. Not possible   **17.** $\begin{bmatrix} 19 & 3 \\ 6 & 22 \end{bmatrix}$ **18.** $\begin{bmatrix} 28 & 19 \\ -6 & -3 \end{bmatrix}$

19. $\begin{bmatrix} -175 & 37 & -13 \\ 95 & -20 & 7 \\ 14 & -3 & 1 \end{bmatrix}$ **20.** 203

21. $(0, -2), (3, -5), (0, -5)$ $(3, -2)$

22. Gym shoes: $\$2539$ million
   Jogging shoes: $\$2362$ million
   Walking shoes: $\$4418$ million

23. $(-5, 4)$ **24.** $(-3, 4, 2)$ **25.** 9

26. $\frac{1}{5}, -\frac{1}{7}, \frac{1}{9}, -\frac{1}{11}, \frac{1}{13}$ **27.** $a_n = \frac{(n+1)!}{n+3}$

28. 1536 **29.** (a) 65.4 (b) $a_n = 3.2n + 1.4$

30. $3, 6, 12, 24, 48$ **31.** $\frac{190}{9}$ **32.** Proof

33. $w^4 - 36w^3 + 486w^2 - 2916w + 6561$

34. 2184 **35.** 600 **36.** 70 **37.** 462

38. $453,600$ **39.** $151,200$ **40.** $\frac{1}{45}$ **41.** $\frac{1}{4}$

Problem Solving   *(page 849)*

1. (a)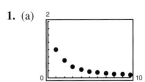

(b) 0

(c)

n	1	10	100	1000	10,000
u_n	1	0.1089	0.0101	0.0010	0.0001

(d) 0

3. $s_d = \dfrac{a_1}{1 - r} = \dfrac{20}{1 - \frac{1}{2}} = 40$

This represents the total distance Achilles ran.

$s_t = \dfrac{a_1}{1 - r} = \dfrac{1}{1 - \frac{1}{2}} = 2$

This represents the total amount of time Achilles ran.

5. (a) Arithmetic sequence, difference $= d$

(b) Arithmetic sequence, difference $= dC$

(c) Not an arithmetic sequence

7. (a) 7, 22, 11, 34, 17, 52, 26, 13, 40, 20, 10, 5, 16, 8, 4, 2, 1, 4, 2, 1

(b) $a_1 = 4$: 4, 2, 1, 4, 2, 1, 4, 2, 1, 4

$a_1 = 5$: 5, 16, 8, 4, 2, 1, 4, 2, 1, 4

$a_1 = 12$: 12, 6, 3, 10, 5, 16, 8, 4, 2, 1

Eventually, the terms repeat: 4, 2, 1.

9. Proof

11. $S_n = \left(\dfrac{1}{2}\right)^n$; $A_n = \dfrac{\sqrt{3}}{4} S_n^2$ **13.** $\dfrac{1}{3}$

15. (a) 3 to 7; 7 to 3   (b) 30 marbles

(c) $P(E) = \dfrac{\text{odds in favor of } E}{\text{odds in favor of } E + 1}$

(d) Odds in favor of event $E = \dfrac{P(E)}{P(E')}$

Appendix A   *(page A6)*

1. numerator

3. $(5 - 4)^2 = 1^2 \neq 9 = 25 - 16 = 5^2 - 4^2$

5. The middle term needs to be included.

$(x + 3)^2 = x^2 + 6x + 9$

7. $\sqrt{x + 9}$ cannot be simplified.

9. Divide out common factors, not common terms.

$\dfrac{2x^2 + 1}{5x}$ cannot be simplified.

11. To add fractions, first find a common denominator.

$\dfrac{3}{x} + \dfrac{4}{y} = \dfrac{3y + 4x}{xy}$

13. $(x + 2)^{-1/2}(3x + 2)$

15. $2x(2x - 1)^{-1/2}[2x^2(2x - 1)^2 - 1]$

17. $5x + 3$ **19.** $2x^2 + x + 15$ **21.** $1 - 5x$

23. $3x - 1$ **25.** $\frac{1}{3}$ **27.** 2

29–31. Answers will vary.

33. $7(x + 3)^{-5}$ **35.** $2x^5(3x + 5)^{-4}$

37. $\frac{4}{3}x^{-1} + 4x^{-4} - 7x(2x)^{-1/3}$

39. $\dfrac{x}{3} + 2 + \dfrac{4}{x}$

41. $4x^{8/3} - 7x^{5/3} + \dfrac{1}{x^{1/3}}$ **43.** $\dfrac{3}{x^{1/2}} - 5x^{3/2} - x^{7/2}$

45. $\dfrac{-7x^2 - 4x + 9}{(x^2 - 3)^3(x + 1)^4}$ **47.** $\dfrac{27x^2 - 24x + 2}{(6x + 1)^4}$

49. $\dfrac{-1}{(x + 3)^{2/3}(x + 2)^{7/4}}$

51. $\dfrac{4x - 3}{(3x - 1)^{4/3}}$ **53.** $\dfrac{x}{x^2 + 4}$

55. $\dfrac{(3x - 2)^{1/2}(15x^2 - 4x + 45)}{2(x^2 + 5)^{1/2}}$

57. (a) Answers will vary.

(b)

x	-2	-1	$-\frac{1}{2}$	0	1	2	$\frac{5}{2}$
y_1	3.67	2.28	1.96	1.85	2.28	3.67	4.80
y_2	3.67	2.28	1.96	1.85	2.28	3.67	4.80

(c) Answers will vary.

59. You cannot move term-by-term from the denominator to the numerator.

61. $y = \log_2 x$, so y is a logarithmic function of x.

63. $72°$ should be converted to radians first.

Technology

Chapter P   *(page 52)*

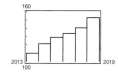

Chapter 2   *(page 201)*

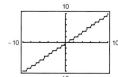 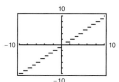

The graph in *dot* mode illustrates that the range is the set of all integers.

Chapter 5   *(page 399)*

$S = 44.73(1.147)^t$

The exponential regression model has the same factor as the model in Example 1. However, the model given in Example 1 contains the natural exponential function.

Chapter 6 *(page 458)*

The graph is so close to the x-axis it is difficult to see. $-\pi \le x \le \pi$ and $-0.5 \le y \le 0.5$ display a good view of the graph.

Chapter 9 *(page 630)*

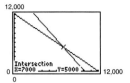

The point of intersection $(7000, 5000)$ agrees with the solution.

(page 632)

$(2, 0)$; $(0, -1)$, $(2, 1)$; None

Chapter 10 *(page 716)*

$$\begin{bmatrix} 1 & 1 \\ 1 & -5 \end{bmatrix}$$

Checkpoints

Chapter P

Section P.1

1. (a) Natural numbers: $\left\{\frac{6}{3}, 8\right\}$ (b) Whole numbers: $\left\{\frac{6}{3}, 8\right\}$
   (c) Integers: $\left\{\frac{6}{3}, -1, 8, -22\right\}$
   (d) Rational numbers: $\left\{-\frac{1}{4}, \frac{6}{3}, -7.5, -1, 8, -22\right\}$
   (e) Irrational numbers: $\left\{-\pi, \frac{1}{2}\sqrt{2}\right\}$

2.
$$-1.6 \quad -\tfrac{3}{4} \quad 0.7 \quad \tfrac{5}{2}$$

3. (a) $1 > -5$ (b) $\frac{3}{2} < 7$ (c) $-\frac{2}{3} > -\frac{3}{4}$
4. (a) The inequality $x > -3$ denotes all real numbers greater than -3.
   (b) The inequality $0 < x \le 4$ denotes all real numbers between 0 and 4, not including 0, but including 4.
5. (a) $[-2, 5)$ denotes the set of all real numbers greater than or equal to -2 and less than 5; $-2 \le x < 5$

   (b) $[-2, 4)$; $-2 \le x < 4$;

6. (a) 1 (b) $-\frac{3}{4}$ (c) $\frac{2}{3}$ (d) -0.7
7. (a) 1 (b) -1
8. (a) $|-3| < |4|$
   (b) $-|-4| = -|4|$
   (c) $|-3| > -|-3|$
9. (a) 58 (b) 12 (c) 12
10. Terms: $-2x, 4$; Coefficients: $-2, 4$ 11. -5
12. (a) Commutative Property of Addition
    (b) Associative Property of Multiplication
    (c) Distributive Property
13. (a) $\dfrac{x}{10}$ (b) $\dfrac{x}{2}$

Section P.2

1. (a) -81 (b) 81 (c) 27 (d) $\frac{1}{27}$
2. (a) $-\frac{1}{16}$ (b) 64
3. (a) $-2x^2y^4$ (b) 1 (c) $-125z^5$ (d) $\dfrac{9x^4}{y^4}$
4. (a) $\dfrac{2}{a^2}$ (b) $\dfrac{b^5}{5a^4}$ (c) $\dfrac{10}{x}$ (d) $-2x^3$
5. 4.585×10^4 6. -0.002718
7. $864,000,000$
8. (a) -12 (b) Not a real number   (c) $\frac{5}{8}$ (d) $-\frac{2}{3}$
9. (a) 5 (b) 25 (c) x (d) $\sqrt[4]{x}$
10. (a) $4\sqrt{2}$ (b) $5\sqrt[3]{2}$ (c) $2a^2\sqrt{6a}$ (d) $-3x\sqrt[3]{5}$
11. (a) $9\sqrt{2}$ (b) $(3x - 2)\sqrt[3]{3x^2}$
12. (a) $\dfrac{5\sqrt{2}}{6}$ (b) $\dfrac{\sqrt[3]{5}}{5}$ 13. $2\left(\sqrt{6} + \sqrt{2}\right)$
14. $\dfrac{2}{3\left(2 + \sqrt{2}\right)}$
15. (a) $27^{1/3}$ (b) $x^{3/2}y^{5/2}z^{1/2}$ (c) $3x^{5/3}$
16. (a) $\dfrac{1}{\sqrt{x^2 - 7}}$ (b) $-3\sqrt[3]{bc^2}$ (c) $\sqrt[4]{a^3}$ (d) $\sqrt[5]{x^4}$
17. (a) $\frac{1}{25}$ (b) $-12x^{5/3}y^{9/10}$, $x \ne 0$, $y \ne 0$
    (c) $\sqrt[4]{3}$ (d) $(3x + 2)^2$, $x \ne -\frac{2}{3}$

Section P.3

1. Standard form: $-7x^3 + 2x + 6$
   Degree: 3; Leading coefficient: -7
2. $2x^3 - x^2 + x + 6$ 3. $3x^2 - 16x + 5$
4. $x^4 + 2x^2 + 9$ 5. $9x^2 - 4$ 6. $x^2 + 20x + 100$
7. $64x^3 - 48x^2 + 12x - 1$ 8. $x^2 - 9y^2 - 4x + 4$
9. Volume $= 4x^3 - 44x^2 + 120x$
   $x = 2$: Volume $= 96$ in.^{3}
   $x = 3$: Volume $= 72$ in.^{3}

Section P.4

1. (a) $5x^2(x - 3)$ (b) $-3(1 - 2x + 4x^3)$
   (c) $(x + 1)(x^2 - 2)$
2. $4(5 + y)(5 - y)$ 3. $(x - 1 + 3y^2)(x - 1 - 3y^2)$
4. $(3x - 5)^2$ 5. $(4x - 1)(16x^2 + 4x + 1)$
6. (a) $(x + 6)(x^2 - 6x + 36)$ (b) $5(y + 3)(y^2 - 3y + 9)$
7. $(x + 3)(x - 2)$ 8. $(2x - 3)(x - 1)$
9. $(x^2 - 5)(x + 1)$ 10. $(2x - 3)(x + 4)$

Section P.5

1. (a) All nonnegative real numbers x
   (b) All real numbers x such that $x \ge -7$
   (c) All real numbers x such that $x \ne 0$
2. $\dfrac{4}{x - 6}$, $x \ne -3$ 3. $-\dfrac{3x + 2}{5 + x}$, $x \ne 1$
4. $\dfrac{5(x - 5)}{(x - 6)(x - 3)}$, $x \ne -3$, $x \ne -\dfrac{1}{3}$, $x \ne 0$
5. $x + 1$, $x \ne \pm 1$ 6. $\dfrac{x^2 + 1}{(2x - 1)(x + 2)}$

7. $\dfrac{7x^2 - 13x - 16}{x(x + 2)(x - 2)}$ **8.** $\dfrac{3(x + 3)}{(x - 3)(x + 2)}$ **9.** $-\dfrac{1}{(x - 1)^{4/3}}$

10. $\dfrac{2(x + 1)(x - 1)}{(x^2 - 2)^{3/2}}$ **11.** $\dfrac{1}{\sqrt{9 + h} + 3}, h \neq 0$

Section P.6

1.

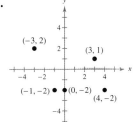

2.

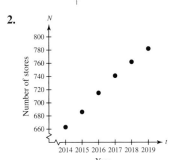

3. $\sqrt{37} \approx 6.08$

4. $d_1 = \sqrt{45}, d_2 = \sqrt{20}, d_3 = \sqrt{65}$
$\left(\sqrt{45}\right)^2 + \left(\sqrt{20}\right)^2 = \left(\sqrt{65}\right)^2$

5. $\sqrt{709} \approx 27$ yd   **6.** $(1, -1)$ **7.** \$66.4 billion

8. $(1, 2), (1, -2), (-1, 0), (-1, -4)$

Chapter 1

Section 1.1

1. (a) No   (b) Yes

2. (a)   (b)

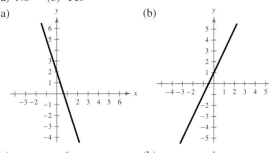

3. (a)   (b)

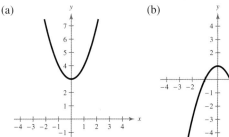

4. (a) x-intercept: $(1, 0)$ (b) x-intercepts: $(0, 0), (3, 0)$
    y-intercept: $(0, -3)$ y-intercept: $(0, 0)$

5. x-axis symmetry

6.

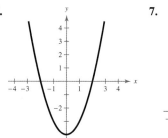

7.

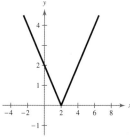

8. $(x + 3)^2 + (y + 5)^2 = 25$ **9.** About 221 lb

Section 1.2

1. (a) -4 (b) 8   **2.** 26   **3.** $-\dfrac{18}{5}$ **4.** No solution

5. (a) x-intercept: $\left(-\dfrac{2}{3}, 0\right)$ (b) x-intercept: $(3, 0)$
    y-intercept: $(0, -2)$ y-intercept: $(0, 5)$

6. (a) $(0, 68{,}676)$; There were about 68,676 male participants in 2010.

(b) 2018

Section 1.3

1. \$1100   **2.** 20%   **3.** \$64,000   **4.** 14 ft by 42 ft

5. 1 hr 24 min   **6.** About 122 ft

7. \$2375 at $2\frac{1}{2}\%$, \$2625 at $3\frac{1}{2}\%$

8. 24-inch television: \$10,000; 50-inch television: \$20,000

9. About 2.97 in.

Section 1.4

1. $-1, \dfrac{5}{2}$ **2.** (a) $\pm 2\sqrt{3}$ (b) $1 \pm \sqrt{10}$ **3.** $2 \pm \sqrt{5}$

4. $1 \pm \dfrac{\sqrt{2}}{2}$ **5.** $\dfrac{5}{3} \pm \dfrac{\sqrt{31}}{3}$ **6.** $-\dfrac{1}{3} \pm \dfrac{\sqrt{31}}{3}$

7. 14 ft by 8 ft   **8.** 3.5 sec   **9.** 2018   **10.** About 27.02 ft

Section 1.5

1. (a) $12 - i$ (b) $-2 + 7i$ (c) i (d) 0

2. (a) $-15 + 10i$ (b) $18 - 6i$ (c) 41   (d) $12 + 16i$

3. (a) 45   (b) 29   **4.** $\dfrac{3}{5} + \dfrac{4}{5}i$ **5.** $-2\sqrt{7}$

6. $-\dfrac{7}{8} \pm \dfrac{\sqrt{23}}{8}i$

Section 1.6

1. $0, \pm\dfrac{2\sqrt{3}}{3}$ **2.** (a) $5, \pm\sqrt{2}$ (b) $0, -\dfrac{3}{2}, 6$

3. (a) $\pm 2, \pm\sqrt{3}$ (b) $\pm\frac{1}{3}, \pm 2$ **4.** -9 **5.** $-59, 69$

6. $-4, -1$ **7.** $-2, 6$ **8.** 32 students   **9.** 7%

Section 1.7

1. (a) Bounded; $-1 \le x \le 3$ (b) Bounded; $-1 < x < 6$

(c) Unbounded; $x < 4$ (d) Unbounded; $x \ge 0$

2. $x \le 2$

3. (a) $x < 3$

(b)
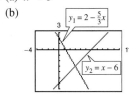

$y_1 > y_2$ for $x < 3$.

4. $(-3, 2)$

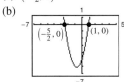

5. $[16, 24]$

6. $0.75x + 10 > 14.50; x > 6$

Section 1.8

1. $(-4, 5)$

2. $\left(-2, \frac{1}{3}\right) \cup (2, \infty)$

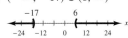

3. (a) $\left(-\frac{5}{2}, 1\right)$

(b)
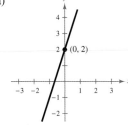

The graph is below the x-axis when x is greater than $-\frac{5}{2}$ and less than 1. So, the solution set is $\left(-\frac{5}{2}, 1\right)$.

4. (a) The solution set is empty.

(b) The solution set consists of the single real number $\{-2\}$.

(c) The solution set consists of all real numbers except $x = 3$.

(d) The solution set is all real numbers.

5. (a) $\left(-\infty, \frac{11}{4}\right] \cup (3, \infty)$ (b) $(-\infty, -17) \cup (6, \infty)$

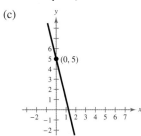

6. $180,000 \leq x \leq 300,000$ **7.** $(-\infty, 2] \cup [5, \infty)$

Chapter 2

Section 2.1

1. (a)

(b)

(c)

2. (a) 2   (b) $-\frac{3}{2}$ (c) Undefined   (d) 0

3. (a) $y = 2x - 13$ (b) $y = -\frac{2}{3}x + \frac{5}{3}$ (c) $y = 1$

4. (a) $y = \frac{5}{3}x + \frac{23}{3}$ (b) $y = -\frac{3}{5}x - \frac{7}{5}$ **5.** Yes

6. The y-intercept, $(0, 1500)$, tells you that the initial value of a copier at the time it is purchased is \$1500. The slope, $m = -300$, tells you that the value of the copier decreases by \$300 each year after it is purchased.

7. $y = -4125x + 24,750$

8. $y = 0.06t + 3.72$; 4.56 million people

Section 2.2

1. (a) Not a function   (b) Function

2. (a) Not a function   (b) Function

3. (a) -2 (b) -38 (c) $-3x^2 + 6x + 7$

4. $f(-2) = 5, f(2) = 1, f(3) = 2$ **5.** ± 4 **6.** $-4, 3$

7. (a) $\{-2, -1, 0, 1, 2\}$

(b) All real numbers x except $x = 3$

(c) All real numbers r such that $r > 0$

(d) All real numbers x such that $x \geq 16$

8. (a) $S(r) = 10\pi r^2$ (b) $S(h) = \frac{5}{8}\pi h^2$ **9.** No

10. 2010: \$0.77 trillion

2013: \$0.641 trillion

2018: \$0.826 trillion

11. $2x + h + 2, h \neq 0$

Section 2.3

1. (a) All real numbers x except $x = -3$

(b) $f(0) = 3; f(3) = -6$ (c) $(-\infty, 3]$

2. Function

3. (a) $x = -8, x = \frac{3}{2}$ (b) $t = 25$ (c) $x = \pm\sqrt{2}$

4.

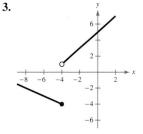

Increasing on $(-\infty, -2)$ and $(0, \infty)$

Decreasing on $(-2, 0)$

5. $(-0.88, 6.06)$ **6.** (a) -3 (b) 0

7. (a) 20 ft/sec   (b) $\frac{140}{3}$ ft/sec

8. (a) Neither; No symmetry   (b) Even; y-axis symmetry

(c) Odd; Origin symmetry

Section 2.4

1. $f(x) = -\frac{5}{2}x + 1$ **2.** $f\left(-\frac{3}{2}\right) = 0, f(1) = 3, f\left(-\frac{5}{2}\right) = -1$

3.

Section 2.5

1. (a) 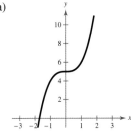   (b)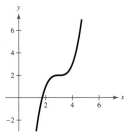

2. $k(x) = x^2 - 1$ **3.** $k(x) = -(x + 2)^2$

4. (a) The graph of g is a reflection in the x-axis of the graph of f.

(b) The graph of h is a reflection in the y-axis of the graph of f.

5. (a) The graph of g is a vertical stretch of the graph of f.

(b) The graph of h is a vertical shrink of the graph of f.

6. (a) The graph of g is a horizontal shrink of the graph of f.

(b) The graph of h is a horizontal stretch of the graph of f.

Section 2.6

1. $x^2 - x + 1; 3$ **2.** $x^2 + x - 1; 11$ **3.** $x^2 - x^3; -18$

4. $\left(\dfrac{f}{g}\right)(x) = \dfrac{\sqrt{x - 3}}{\sqrt{16 - x^2}}$; Domain: $[3, 4)$

5. (a) $8x^2 + 7$ (b) $16x^2 + 80x + 101$ (c) 9

6. All real numbers x **7.** $f(x) = \frac{1}{5}\sqrt[3]{x}, g(x) = 8 - x$

8. (a) $(N \circ T)(t) = 32t^2 + 36t + 204$ (b) About 4.5 h

Section 2.7

1. $f^{-1}(x) = 5x, f(f^{-1}(x)) = \frac{1}{5}(5x) = x, f^{-1}(f(x)) = 5\left(\frac{1}{5}x\right) = x$

2. $g(x) = 7x + 4$

3. 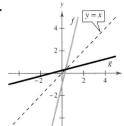   **4.**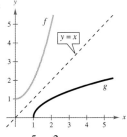

5. (a) Yes   (b) No   **6.** $f^{-1}(x) = \dfrac{5 - 2x}{x + 3}$

7. $f^{-1}(x) = x^3 - 10$

Chapter 3

Section 3.1

1. (a) 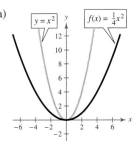   The graph of $f(x) = \frac{1}{4}x^2$ is broader than the graph of $y = x^2$.

(b) 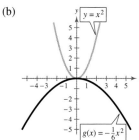   The graph of $g(x) = -\frac{1}{6}x^2$ is a reflection in the x-axis and is broader than the graph of $y = x^2$.

(c) 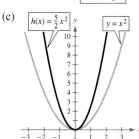   The graph of $h(x) = \frac{5}{2}x^2$ is narrower than the graph of $y = x^2$.

(d) 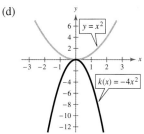   The graph of $k(x) = -4x^2$ is a reflection in the x-axis and is narrower than the graph of $y = x^2$.

2. 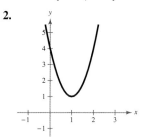   **3.**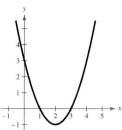

Vertex: $(1, 1)$ Vertex: $(2, -1)$
Axis: $x = 1$ x-intercepts: $(1, 0), (3, 0)$

4. $y = (x + 4)^2 + 11$ **5.** About 39.7 ft

Section 3.2

1. (a) 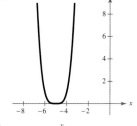   (b)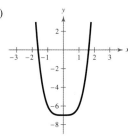

(c) 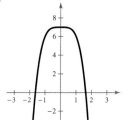   (d)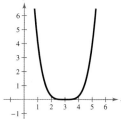

2. (a) Falls to the left, rises to the right
  (b) Rises to the left, falls to the right
3. Real zeros: $x = 0, x = 6$; Turning points: 2
4. 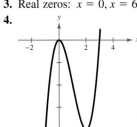   **5.**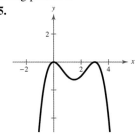
6. $x \approx 3.196$

Section 3.3

1. $(x + 4)(3x + 7)(3x - 7)$ **2.** $x^2 + x + 3, x \neq 3$
3. $2x^3 - x^2 + 3 - \dfrac{6}{3x + 1}$ **4.** $5x^2 - 2x + 3$
5. (a) 1   (b) 396   (c) $-\frac{13}{2}$ (d) -17
6. $f(-3) = 0, f(x) = (x + 3)(x - 5)(x + 2)$

Section 3.4

1. 4   **2.** No rational zeros   **3.** 5   **4.** $-3, \frac{1}{2}, 2$
5. $-1, \dfrac{-3 - 3\sqrt{17}}{4} \approx -3.8423, \dfrac{-3 + 3\sqrt{17}}{4} \approx 2.3423$
6. $f(x) = x^4 + 45x^2 - 196$
7. $f(x) = -x^4 - x^3 - 2x^2 - 4x + 8$ **8.** $\frac{2}{3}, \pm 4i$
9. $f(x) = (x - 1)(x + 1)(x - 3i)(x + 3i); \pm 1, \pm 3i$
10. No positive real zeros, three or no negative real zeros
11. $\frac{1}{2}$ **12.** 7 in. × 7 in. × 9 in.

Section 3.5

1. 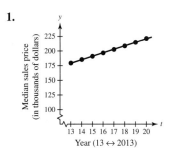   The model is a good fit for the data.

2. 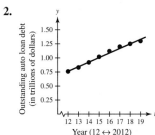 $D = 0.081t - 0.21$; The model is a good fit for the data.

3. $I = 0.075P$ **4.** 576 ft   **5.** 508 units
6. About 1314 ft   **7.** 14,000 joules

Chapter 4

Section 4.1

1. Domain: all real numbers x such that $x \neq 1$
  $f(x)$ decreases without bound as x approaches 1 from the left and increases without bound as x approaches 1 from the right.
2. Vertical asymptotes: $x = \pm 1$
  Horizontal asymptote: $y = 5$
3. Vertical asymptote: $x = -1$
  Horizontal asymptote: $y = 3$
  Hole at $\left(-3, \frac{11}{2}\right)$
4. (a) All real numbers   (b) No vertical asymptote
  (c) $y = 3$
5. (a) \$8.40; \$1.40; \$0.80; \$0.48
  (b) $C = 0.40$; As the number of units increases, the average cost per unit gets closer to \$0.40.

Section 4.2

1. Domain: all real numbers x except $x = -3$ **2.** Domain: all real numbers x except $x = -1$

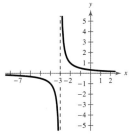   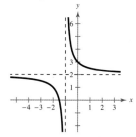

3. Domain: all real numbers x except $x = -2, 1$ **4.** Domain: all real numbers x except $x = -2, 3$

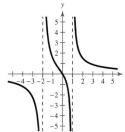   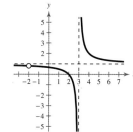

5. Domain: all real numbers x except $x = 0$

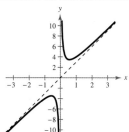

6. 12.9 in. by 6.5 in.

Section 4.3

1. Focus: $(0, 1)$
Directrix: $y = -1$

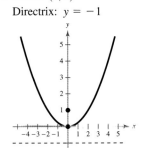

2. $x^2 = \dfrac{3}{2}y$ **3.** $\dfrac{x^2}{16} + \dfrac{y^2}{25} = 1$

4.

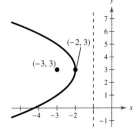

5. $\dfrac{y^2}{9} - \dfrac{x^2}{27} = 1$

Vertices: $(-9, 0), (9, 0)$

6.

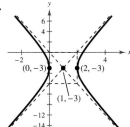

7. $x^2 - \dfrac{y^2}{16} = 1$

Section 4.4

1. (a) Ellipse centered at $(-1, 2)$; horizontal shift one unit to the left and vertical shift two units up
(b) Circle centered at $(-1, 1)$; horizontal shift one unit to the left and vertical shift one unit up
(c) Parabola with vertex at $(-4, 3)$; horizontal shift four units to the left and vertical shift three units up
(d) Hyperbola centered at $(3, 1)$; horizontal shift three units to the right and vertical shift one unit up

2. Vertex: $(-2, 3)$
Focus: $(-3, 3)$
Directrix: $x = -1$

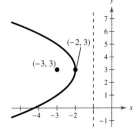

3. $(x + 1)^2 = 4(y - 1)$

4.

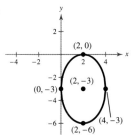

5. $\dfrac{(x - 3)^2}{9} + \dfrac{(y - 5)^2}{25} = 1$

6.

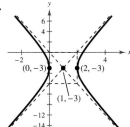

7. $\dfrac{(x - 4)^2}{1} - \dfrac{(y + 1)^2}{8} = 1$

Chapter 5

Section 5.1

1. 0.0528248

2.

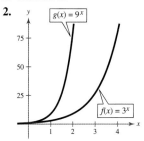

3.

4. (a) 2   (b) 3
5. (a) Shift the graph of f two units to the right.
(b) Shift the graph of f three units up.
(c) Reflect the graph of f in the y-axis and shift three units down.
6. (a) 1.3498588   (b) 0.3011942   (c) 492.7490411
7.

8. (a) \$7927.75   (b) \$7935.08   (c) \$7938.78
9. About 9.970 lb; about 0.275 lb

Section 5.2

1. (a) 0   (b) -3 (c) 3
2. (a) 2.4393327   (b) Error or complex number
(c) -0.3010300
3. (a) 1   (b) 3   (c) 0   (d) 25

4.

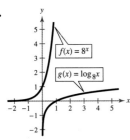

5.

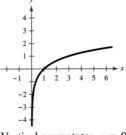

Vertical asymptote: $x = 0$

6. (a)

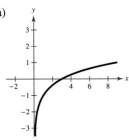

(b)

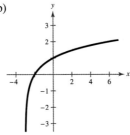

7. (a) -4.6051702 **(b)** 1.3862944 **(c)** 1.3169579
(d) Error or complex number
8. (a) $\frac{1}{3}$ **(b)** 0 **(c)** $\frac{3}{4}$ **(d)** 7 **9.** $(-3, \infty)$
10. (a) 70.84 **(b)** 61.18 **(c)** 59.61

Section 5.3

1. 3.5850 **2.** 3.5850
3. (a) $\log 3 + 2\log 5$ **(b)** $2\log 3 - 3\log 5$ **4.** 4
5. $\log_3 4 + 2\log_3 x - \frac{1}{2}\log_3 y$ **6.** $\log \dfrac{(x+3)^2}{(x-2)^4}$
7. $\ln y = \frac{2}{3}\ln x$

Section 5.4

1. (a) 9 **(b)** 216 **(c)** $\ln 5$ **(d)** $-\frac{1}{2}$
2. $4, -2$ **3. (a)** 1.723 **(b)** 3.401 **4.** -3.415
5. $1.099, 1.386$ **6. (a)** $e^{2/3}$ **(b)** 7 **(c)** 12
7. 0.513 **8.** $\frac{32}{3}$ **9.** 10
10. About 13.2 years; It takes longer for your money to double at a lower interest rate.
11. 6 days

Section 5.5

1. 2021 **2.** 400 bacteria   **3.** About 38,000 yr
4.

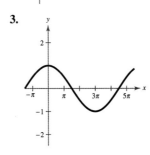

531
5. About 7 days   **6.** 30 decibels

Chapter 6

Section 6.1

1. (a) $415°, -305°$ **(b)** $332°, -388°$
2. (a) Complement: $67°$; Supplement: $157°$
   (b) Complement: none; Supplement: none

3. (a) $\dfrac{2\pi}{3}$ **(b)** $\dfrac{5\pi}{16}$ **(c)** $\dfrac{7\pi}{12}$ **4. (a)** $\dfrac{\pi}{3}$ **(b)** $\dfrac{16\pi}{9}$
5. (a) $30°$ **(b)** $300°$ **6.** 24π in. ≈ 75.40 in.
7. About 0.84 cm/sec
8. (a) 4800π rad/min   **(b)** About 60,319 in./min
9. About 1117 ft²

Section 6.2

1. $\sin\theta = \dfrac{1}{2}$ $\csc\theta = 2$

$\cos\theta = \dfrac{\sqrt{3}}{2}$ $\sec\theta = \dfrac{2\sqrt{3}}{3}$

$\tan\theta = \dfrac{\sqrt{3}}{3}$ $\cot\theta = \sqrt{3}$

2. $\cot 45° = 1$, $\sec 45° = \sqrt{2}$, $\csc 45° = \sqrt{2}$

3. $\tan 60° = \sqrt{3}$, $\tan 30° = \dfrac{\sqrt{3}}{3}$

4. (a) 0.5417082 **(b)** 1.0729164
5. (a) 0.28 **(b)** 0.2917
6. (a) $\frac{1}{2}$ **(b)** $\sqrt{5}$ **7.** Answers will vary.
8. About 40 ft   **9.** $60°$ **10.** About 17.6 ft; about 17.2 ft

Section 6.3

1. $\sin\theta = \dfrac{3\sqrt{13}}{13}$, $\cos\theta = -\dfrac{2\sqrt{13}}{13}$, $\tan\theta = -\dfrac{3}{2}$
2. $\cos\theta = -\frac{3}{5}$, $\tan\theta = -\frac{4}{3}$
3. $-1; 0$ **4. (a)** $33°$ **(b)** $\dfrac{4\pi}{9}$ **(c)** $\dfrac{\pi}{5}$
5. (a) $-\dfrac{\sqrt{2}}{2}$ **(b)** $-\dfrac{1}{2}$ **(c)** $-\dfrac{\sqrt{3}}{3}$ **6. (a)** $-\dfrac{3}{5}$ **(b)** $\dfrac{4}{3}$
7. (a) -1.8040478 **(b)** -1.0428352 **(c)** 0.8090170
8. (a) $-0.4161; 0$ **(b)** $\sqrt{3}$

Section 6.4

1.

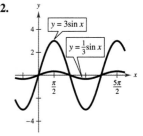

2.

3.

4. Period: 2π

Amplitude: 2

$\left[\dfrac{\pi}{2}, \dfrac{5\pi}{2}\right]$ corresponds to one cycle.

Key points: $\left(\dfrac{\pi}{2}, 2\right)$, $(\pi, 0)$, $\left(\dfrac{3\pi}{2}, -2\right)$, $(2\pi, 0)$, $\left(\dfrac{5\pi}{2}, 2\right)$

5. 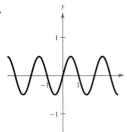   **6.**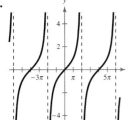

7. $y = 5.6 \sin(0.524t - 0.524) + 5.7$

Section 6.5

1. 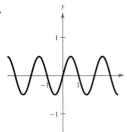   **2.**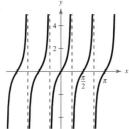

3. 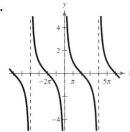   **4.**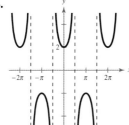

5. 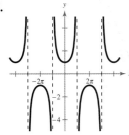   **6.**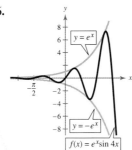

Section 6.6

1. (a) $\dfrac{\pi}{2}$

(b) Not possible

2.

3. (a) $\dfrac{3\pi}{4}$ (b) $\dfrac{\pi}{3}$

4. (a) 1.3670516 (b) Not possible   (c) 1.9273001

5. (a) -14 (b) $-\dfrac{\pi}{4}$ (c) 0.54 **6.** $\dfrac{4}{5}$ **7.** $\sqrt{x^2 + 1}$

Section 6.7

1. $a \approx 5.46$, $c \approx 15.96$, $B = 70°$ **2.** About 16 ft

3. About 15.1 ft   **4.** $3.58°$

5. Bearing: N 53° W, Distance: about 3.2 nm

6. $d = 6 \sin \dfrac{2\pi}{3} t$; $\dfrac{1}{3}$ cycle per second

7. (a) 4 (b) 3 cycles per unit of time   (c) 4 (d) $\dfrac{1}{12}$

Chapter 7
Section 7.1

1. $\sin x = -\dfrac{\sqrt{10}}{10}$, $\cos x = -\dfrac{3\sqrt{10}}{10}$, $\tan x = \dfrac{1}{3}$, $\csc x = -\sqrt{10}$

$\sec x = -\dfrac{\sqrt{10}}{3}$, $\cot x = 3$

2. $-\sin x$

3. (a) $(1 + \cos \theta)(1 - \cos \theta)$ (b) $(2 \csc \theta - 3)(\csc \theta - 2)$

4. $(\tan x + 1)(\tan x + 2)$ **5.** $\sin x$ **6.** $2 \sec^2 \theta$

7. $1 + \sin \theta$ **8.** $3 \cos \theta$ **9.** $\ln|\tan x|$

Section 7.2

1–7. Answers will vary.

Section 7.3

1. $\dfrac{\pi}{4} + 2n\pi$, $\dfrac{3\pi}{4} + 2n\pi$

2. $\dfrac{\pi}{3} + 2n\pi$, $\dfrac{2\pi}{3} + 2n\pi$, $\dfrac{4\pi}{3} + 2n\pi$, $\dfrac{5\pi}{3} + 2n\pi$ **3.** $n\pi$

4. $\dfrac{\pi}{6}, \dfrac{\pi}{2}, \dfrac{5\pi}{6}$ **5.** $\dfrac{\pi}{4} + n\pi$, $\dfrac{3\pi}{4} + n\pi$ **6.** $0, \dfrac{3\pi}{2}$

7. $\dfrac{\pi}{6} + n\pi$, $\dfrac{\pi}{3} + n\pi$ **8.** $\dfrac{\pi}{2} + 2n\pi$

9. $\arctan \dfrac{3}{4} + n\pi$, $\arctan(-2) + n\pi$ **10.** 0.4271, 2.7145

11. $\theta \approx 54.7356°$

Section 7.4

1. $\dfrac{\sqrt{2} + \sqrt{6}}{4}$ **2.** $\dfrac{\sqrt{2} + \sqrt{6}}{4}$ **3.** $-\dfrac{63}{65}$

4. $\dfrac{x + \sqrt{1 - x^2}}{\sqrt{2}}$ **5.** Answers will vary.

6. (a) $-\cos \theta$ (b) $-\tan \theta$ **7.** $\dfrac{\pi}{3}, \dfrac{5\pi}{3}$

8. Answers will vary.

Section 7.5

1. $\dfrac{\pi}{3} + 2n\pi$, $\pi + 2n\pi$, $\dfrac{5\pi}{3} + 2n\pi$

2. $\sin 2\theta = \dfrac{24}{25}$, $\cos 2\theta = \dfrac{7}{25}$, $\tan 2\theta = \dfrac{24}{7}$

3. $4 \cos^3 x - 3 \cos x$ **4.** $\dfrac{\cos 4x - 4 \cos 2x + 3}{\cos 4x + 4 \cos 2x + 3}$

5. $\dfrac{-\sqrt{2 - \sqrt{3}}}{2}$ **6.** $\dfrac{\pi}{3}, \pi, \dfrac{5\pi}{3}$ **7.** $\dfrac{1}{2} \sin 8x + \dfrac{1}{2} \sin 2x$

8. $\dfrac{\sqrt{2}}{2}$ **9.** $\dfrac{\pi}{6} + \dfrac{2n\pi}{3}, \dfrac{\pi}{2} + \dfrac{2n\pi}{3}, n\pi$ **10.** $45°$

Chapter 8

Section 8.1

1. $C = 105°$, $b \approx 45.25$ cm, $c \approx 61.82$ cm   2. 13.40 m
3. $B \approx 12.39°$, $C \approx 136.61°$, $c \approx 16.01$ in.
4. $\sin B \approx 3.0311 > 1$
5. Two solutions:
   $B \approx 70.4°$, $C \approx 51.6°$, $c \approx 4.16$ ft
   $B \approx 109.6°$, $C \approx 12.4°$, $c \approx 1.14$ ft
6. About 213 yd^2 7. About 1856.59 m

Section 8.2

1. $A \approx 26.38°$, $B \approx 36.34°$, $C \approx 117.28°$
2. $B \approx 59.66°$, $C \approx 40.34°$, $a \approx 18.26$ ft   3. About 202 ft
4. N 15.37° E   5. About 19.90 in.^{2}

Section 8.3

1. $\|\overrightarrow{PQ}\| = \|\overrightarrow{RS}\| = \sqrt{10}$, slope$_{\overrightarrow{PQ}}$ = slope$_{\overrightarrow{RS}} = \frac{1}{3}$
   $\overrightarrow{PQ}$ and $\overrightarrow{RS}$ have the same magnitude and direction, so they are equivalent.
2. $\mathbf{v} = \langle -5, 6 \rangle$, $\|\mathbf{v}\| = \sqrt{61}$
3. (a) $\langle 4, 6 \rangle$ (b) $\langle -2, 2 \rangle$ (c) $\langle -7, 2 \rangle$
4. (a) $3\sqrt{17}$ (b) $2\sqrt{13}$ (c) $5\sqrt{13}$
5. $\left\langle \dfrac{6}{\sqrt{37}}, -\dfrac{1}{\sqrt{37}} \right\rangle$ 6. $-6\mathbf{i} - 3\mathbf{j}$ 7. $11\mathbf{i} - 14\mathbf{j}$
8. (a) 135°   (b) About 209.74°   9. $\langle -96.59, -25.88 \rangle$
10. About 2405 lb   11. $\|\mathbf{v}\| \approx 451.8$ mi/h, $\theta \approx 305.1°$

Section 8.4

1. (a) -6 (b) 37   (c) 0
2. (a) $\langle -36, 108 \rangle$ (b) 43   (c) $2\sqrt{10}$ 3. 45°
4. Yes   5. $\frac{1}{17}\langle 64, 16 \rangle$; $\frac{1}{17}\langle -13, 52 \rangle$ 6. About 38.8 lb
7. About 1212 ft-lb

Section 8.5

1. [Imaginary axis graph with point $3 - 4i$]

   5
2. $4 + 3i$ 3. $1 - 5i$
4. [Imaginary axis graph with points $(2, 3)$ and $(2, -3)$]

   $2 + 3i$

5. $\sqrt{82} \approx 9.06$ units   6. $\left(\frac{7}{2}, -2 \right)$

Section 8.6

1. $6\sqrt{2}\left(\cos \dfrac{7\pi}{4} + i \sin \dfrac{7\pi}{4} \right)$ 2. $-4 + 4\sqrt{3}\,i$ 3. 10
4. $12i$ 5. $\dfrac{\sqrt{3}}{2} + \dfrac{1}{2}i$ 6. -8 7. -4

8. $1, i, -1, -i$

9. $\sqrt{2}\sqrt[3]{3}(\cos 45° + i \sin 45°)$
   $\sqrt{2}\sqrt[3]{3}(\cos 165° + i \sin 165°)$
   $\sqrt{2}\sqrt[3]{3}(\cos 285° + i \sin 285°)$

Chapter 9

Section 9.1

1. $(3, 3)$ 2. $6250 at 6.5%, $18,750 at 8.5%
3. $(-3, -1), (2, 9)$ 4. No solution   5. $(1, 3)$
6. About 5172 pairs   7. 5 weeks

Section 9.2

1. $\left(\frac{3}{4}, \frac{5}{2} \right)$ 2. $(4, 3)$ 3. $(3, -1)$ 4. $(9, 12)$
5.

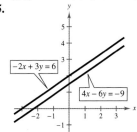

No solution; inconsistent
6. No solution   7. Infinitely many solutions: $(a, 4a + 3)$
8. About 471.18 mi/h; about 16.63 mi/h   9. $(1,500,000, 537)$

Section 9.3

1. $(4, -3, 3)$ 2. $(1, 1)$ 3. $(1, 2, 3)$ 4. No solution
5. Infinitely many solutions: $(-23a + 22, 15a - 13, a)$
6. Infinitely many solutions: $\left(\frac{1}{4}a, \frac{17}{4}a - 3, a \right)$
7. $s = -16t^2 + 20t + 100$; The object was thrown upward at a velocity of 20 feet per second from a height of 100 feet.
8. $y = \frac{1}{3}x^2 - 2x$

Section 9.4

1. $-\dfrac{3}{2x + 1} + \dfrac{2}{x - 1}$ 2. $x - \dfrac{3}{x} + \dfrac{4}{x^2} + \dfrac{3}{x + 1}$
3. $\dfrac{5}{x} + \dfrac{7x}{x^2 + 1}$ 4. $\dfrac{x + 3}{x^2 + 4} - \dfrac{6x + 5}{(x^2 + 4)^2}$
5. $\dfrac{1}{x} - \dfrac{2}{x^2} + \dfrac{2 - x}{x^2 + 2} + \dfrac{4 - 2x}{(x^2 + 2)^2}$

Section 9.5

1.

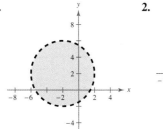

2.

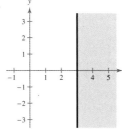

3.

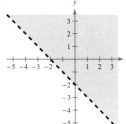

4.

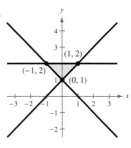

5.

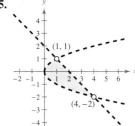

6.

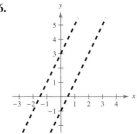

7.

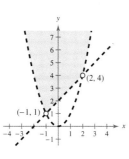

No solution

8. Consumer surplus: $22,500,000
Producer surplus: $33,750,000

9.
$$\begin{cases} 8x + 2y \geq 16 & \text{Nutrient A} \\ x + y \geq 5 & \text{Nutrient B} \\ 2x + 7y \geq 20 & \text{Nutrient C} \\ x \geq 0 \\ y \geq 0 \end{cases}$$

Section 9.6

1. Maximum at $(0, 6)$: 30   **2.** Minimum at $(0, 0)$: 0
3. Maximum at $(60, 20)$: 880   **4.** Minimum at $(10, 0)$: 30
5. $2925; 1050 boxes of chocolate-covered creams, 150 boxes of chocolate-covered nuts
6. 3 bottles of brand X, 2 bottles of brand Y

Chapter 10

Section 10.1

1. 2×3 **2.** $\begin{bmatrix} 1 & 1 & 1 & \vdots & 2 \\ 2 & -1 & 3 & \vdots & -1 \\ -1 & 2 & -1 & \vdots & 4 \end{bmatrix}$; 3×4

3. Add -3 times Row 1 to Row 2.
4. Answers will vary. Solution: $(-1, 0, 1)$
5. Reduced row-echelon form   **6.** $(4, -2, 1)$
7. No solution   **8.** $(7, 4, -3)$ **9.** $(3a + 8, 2a - 5, a)$

Section 10.2

1. $a_{11} = 6, a_{12} = 3, a_{21} = -2, a_{22} = 4$

2. (a) $\begin{bmatrix} 6 & -2 \\ 2 & 3 \end{bmatrix}$ (b) $\begin{bmatrix} 0 & 0 \\ 0 & 0 \\ 0 & 0 \end{bmatrix}$ (c) Not possible   (d) $\begin{bmatrix} 0 \\ 0 \\ 2 \end{bmatrix}$

3. (a) $\begin{bmatrix} 4 & -5 \\ 1 & 1 \\ -4 & 1 \end{bmatrix}$ (b) $\begin{bmatrix} 12 & -3 \\ 0 & 12 \\ -9 & 24 \end{bmatrix}$ (c) $\begin{bmatrix} 12 & -11 \\ 2 & 6 \\ -11 & 10 \end{bmatrix}$

4. $\begin{bmatrix} 1 & 2 \\ 10 & -4 \end{bmatrix}$ **5.** $\begin{bmatrix} -6 & 6 \\ -10 & 6 \end{bmatrix}$ **6.** $\begin{bmatrix} 5 & 0 \\ -1 & 4 \end{bmatrix}$

7. $\begin{bmatrix} -1 & 30 \\ 2 & -4 \\ 1 & 12 \end{bmatrix}$ **8.** $\begin{bmatrix} -3 & -22 \\ 3 & 10 \\ -5 & 10 \end{bmatrix}$

9. (a) $\begin{bmatrix} 3 & -1 \\ -9 & 3 \end{bmatrix}$ (b) $[6]$ (c) Not possible

10. $\begin{bmatrix} 7 & 0 \\ 0 & 7 \end{bmatrix}$ **11.** (a) $\langle -5, 1 \rangle$ (b) $\langle 17, 23 \rangle$

12. $\langle -3, 1 \rangle$; Reflection in the y-axis

13. (a) $\begin{bmatrix} -2 & -3 \\ 6 & 1 \end{bmatrix}\begin{bmatrix} x_1 \\ x_2 \end{bmatrix} = \begin{bmatrix} -4 \\ -36 \end{bmatrix}$ (b) $\begin{bmatrix} -7 \\ 6 \end{bmatrix}$

14. Total cost for women's team: $2310
Total cost for men's team: $2719

Section 10.3

1. $AB = I$ and $BA = I$ **2.** $\begin{bmatrix} 3 & 2 \\ 1 & 1 \end{bmatrix}$

3. $\begin{bmatrix} -4 & -2 & 5 \\ -2 & -1 & 2 \\ -1 & 0 & 1 \end{bmatrix}$ **4.** $\begin{bmatrix} \frac{4}{23} & \frac{1}{23} \\ -\frac{3}{23} & \frac{5}{23} \end{bmatrix}$ **5.** $(2, -1, -2)$

Section 10.4

1. (a) -7 (b) 10   (c) 0
2. $M_{11} = -9, M_{12} = -10, M_{13} = 2, M_{21} = 5, M_{22} = -2,$
$M_{23} = -3, M_{31} = 13, M_{32} = 5, M_{33} = -1$
$C_{11} = -9, C_{12} = 10, C_{13} = 2, C_{21} = -5, C_{22} = -2,$
$C_{23} = 3, C_{31} = 13, C_{32} = -5, C_{33} = -1$
3. -31 **4.** 704

Section 10.5

1. $(3, -2)$ **2.** $(2, -3, 1)$ **3.** 9 square units
4. Collinear   **5.** $x - y + 2 = 0$
6. $(0, 0), (2, 0), (0, 4), (2, 4)$ **7.** 10 square units

8. $[15 \quad 23 \quad 12][19 \quad 0 \quad 1][18 \quad 5 \quad 0][14 \quad 15 \quad 3]$
$[20 \quad 21 \quad 18][14 \quad 1 \quad 12]$

9. $110 \quad -39 \quad -59 \quad 25 \quad -21 \quad -3 \quad 23 \quad -18 \quad -5 \quad 47$
$-20 \quad -24 \quad 149 \quad -56 \quad -75 \quad 87 \quad -38 \quad -37$

10. OWLS ARE NOCTURNAL

Chapter 11

Section 11.1

1. 3, 5, 7, 9   **2.** $1, \frac{3}{2}, \frac{1}{3}, \frac{3}{4}$

3. (a) $a_n = 4n - 3$ (b) $a_n = (-1)^{n+1}(2n)$

4. 6, 7, 8, 9, 10   **5.** 1, 3, 4, 7, 11   **6.** $2, 4, 5, \frac{14}{3}, \frac{41}{12}$

7. $4(n + 1)$ **8.** 44   **9.** (a) 0.5555   (b) $\frac{5}{9}$

10. (a) \$1000, \$1002.50, \$1005.01   (b) \$1127.33

Section 11.2

1. 2, 5, 8, 11; $d = 3$ **2.** $a_n = 5n - 6$

3. $-3, 1, 5, 9, 13, 17, 21, 25, 29, 33, 37$ **4.** 79   **5.** 217

6. (a) 630   (b) $N(1 + 2N)$ **7.** 43,560   **8.** 1470

9. \$2,500,000

Section 11.3

1. $-12, 24, -48, 96; r = -2$

2. 2, 8, 32, 128, 512

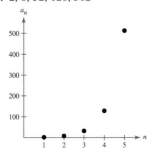

3. 104.02   **4.** $a_n = 4(5)^{n-1}$; 195,312,500   **5.** $\frac{2187}{32}$

6. 2.667   **7.** (a) 10   (b) 6.25   **8.** \$3500.85

Section 11.4

1. (a) $\dfrac{6}{(k + 1)(k + 4)}$ (b) $k + 3 \le 3k^2$

   (c) $2^{4k+2} + 1 > 5k + 5$

2–5. Proofs   **6.** $S_k = k(2k + 1)$; Proof

7. (a) 210   (b) 785   **8.** $a_n = n^2 - n - 2$

Section 11.5

1. (a) 462   (b) 36   (c) 1   (d) 1

2. (a) 21   (b) 21   (c) 14   (d) 14

3. 1, 9, 36, 84, 126, 126, 84, 36, 9, 1

4. $x^4 + 8x^3 + 24x^2 + 32x + 16$

5. (a) $y^4 - 8y^3 + 24y^2 - 32y + 16$
   (b) $32x^5 - 80x^4y + 80x^3y^2 - 40x^2y^3 + 10xy^4 - y^5$

6. $125 + 75y^2 + 15y^4 + y^6$

7. (a) $1120a^4b^4$ (b) $-3,421,440$

Section 11.6

1. Three ways   **2.** Two ways   **3.** 27,000 combinations

4. 2,600,000 numbers   **5.** 24 permutations   **6.** 20 ways

7. 1260 ways   **8.** 21 ways

9. 22,100 three-card poker hands   **10.** 1,051,050 teams

Section 11.7

1. {*HH*1, *HH*2, *HH*3, *HH*4, *HH*5, *HH*6, *HT*1, *HT*2, *HT*3, *HT*4, *HT*5, *HT*6, *TH*1, *TH*2, *TH*3, *TH*4, *TH*5, *TH*6, *TT*1, *TT*2, *TT*3 *TT*4, *TT*5, *TT*6}

2. (a) $\frac{1}{8}$ (b) $\frac{1}{4}$ **3.** $\frac{1}{9}$ **4.** Answers will vary.

5. $\dfrac{606}{4313} \approx 0.141$ **6.** $\dfrac{1}{962,598}$ **7.** $\dfrac{4}{13} \approx 0.308$

8. $\frac{66}{529} \approx 0.125$ **9.** $\frac{121}{900} \approx 0.134$ **10.** About 0.116

11. 0.452

Appendix A

1. Do not apply radicals term-by-term. Leave as $\sqrt{x^2 + 4}$.

2. $(x - 2)^{-1/2}(7x - 12)$ **3.** 3

4. Answers will vary.   **5.** $-6x(1 - 3x^2)^{-2} + x^{-1/3}$

6. (a) $x - 2 + \dfrac{5}{x^3}$ (b) $x^{3/2} - x^{1/2} + 5x^{-1/2}$

Index

Definition of the Six Trigonometric Functions

Right triangle definitions, where $0 < \theta < \pi/2$

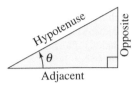

$$\sin \theta = \frac{\text{opp}}{\text{hyp}} \qquad \csc \theta = \frac{\text{hyp}}{\text{opp}}$$

$$\cos \theta = \frac{\text{adj}}{\text{hyp}} \qquad \sec \theta = \frac{\text{hyp}}{\text{adj}}$$

$$\tan \theta = \frac{\text{opp}}{\text{adj}} \qquad \cot \theta = \frac{\text{adj}}{\text{opp}}$$

Circular function definitions, where θ is any angle

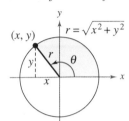

$$\sin \theta = \frac{y}{r} \qquad \csc \theta = \frac{r}{y}$$

$$\cos \theta = \frac{x}{r} \qquad \sec \theta = \frac{r}{x}$$

$$\tan \theta = \frac{y}{x} \qquad \cot \theta = \frac{x}{y}$$

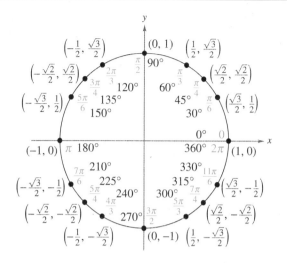

Reciprocal Identities

$$\sin u = \frac{1}{\csc u} \qquad \cos u = \frac{1}{\sec u} \qquad \tan u = \frac{1}{\cot u}$$

$$\csc u = \frac{1}{\sin u} \qquad \sec u = \frac{1}{\cos u} \qquad \cot u = \frac{1}{\tan u}$$

Quotient Identities

$$\tan u = \frac{\sin u}{\cos u} \qquad \cot u = \frac{\cos u}{\sin u}$$

Pythagorean Identities

$$\sin^2 u + \cos^2 u = 1$$

$$1 + \tan^2 u = \sec^2 u \qquad 1 + \cot^2 u = \csc^2 u$$

Cofunction Identities

$$\sin\left(\frac{\pi}{2} - u\right) = \cos u \qquad \cot\left(\frac{\pi}{2} - u\right) = \tan u$$

$$\cos\left(\frac{\pi}{2} - u\right) = \sin u \qquad \sec\left(\frac{\pi}{2} - u\right) = \csc u$$

$$\tan\left(\frac{\pi}{2} - u\right) = \cot u \qquad \csc\left(\frac{\pi}{2} - u\right) = \sec u$$

Even/Odd Identities

$$\sin(-u) = -\sin u \qquad \cot(-u) = -\cot u$$

$$\cos(-u) = \cos u \qquad \sec(-u) = \sec u$$

$$\tan(-u) = -\tan u \qquad \csc(-u) = -\csc u$$

Sum and Difference Formulas

$$\sin(u \pm v) = \sin u \cos v \pm \cos u \sin v$$

$$\cos(u \pm v) = \cos u \cos v \mp \sin u \sin v$$

$$\tan(u \pm v) = \frac{\tan u \pm \tan v}{1 \mp \tan u \tan v}$$

Double-Angle Formulas

$$\sin 2u = 2 \sin u \cos u$$

$$\cos 2u = \cos^2 u - \sin^2 u = 2 \cos^2 u - 1 = 1 - 2 \sin^2 u$$

$$\tan 2u = \frac{2 \tan u}{1 - \tan^2 u}$$

Power-Reducing Formulas

$$\sin^2 u = \frac{1 - \cos 2u}{2}$$

$$\cos^2 u = \frac{1 + \cos 2u}{2}$$

$$\tan^2 u = \frac{1 - \cos 2u}{1 + \cos 2u}$$

Sum-to-Product Formulas

$$\sin u + \sin v = 2 \sin\left(\frac{u + v}{2}\right) \cos\left(\frac{u - v}{2}\right)$$

$$\sin u - \sin v = 2 \cos\left(\frac{u + v}{2}\right) \sin\left(\frac{u - v}{2}\right)$$

$$\cos u + \cos v = 2 \cos\left(\frac{u + v}{2}\right) \cos\left(\frac{u - v}{2}\right)$$

$$\cos u - \cos v = -2 \sin\left(\frac{u + v}{2}\right) \sin\left(\frac{u - v}{2}\right)$$

Product-to-Sum Formulas

$$\sin u \sin v = \frac{1}{2}[\cos(u - v) - \cos(u + v)]$$

$$\cos u \cos v = \frac{1}{2}[\cos(u - v) + \cos(u + v)]$$

$$\sin u \cos v = \frac{1}{2}[\sin(u + v) + \sin(u - v)]$$

$$\cos u \sin v = \frac{1}{2}[\sin(u + v) - \sin(u - v)]$$

FORMULAS FROM GEOMETRY

Triangle:

$h = a \sin \theta$

$\text{Area} = \dfrac{1}{2}bh$

$c^2 = a^2 + b^2 - 2ab \cos \theta$ (Law of Cosines)

Right Triangle:

Pythagorean Theorem
$c^2 = a^2 + b^2$

Equilateral Triangle:

$h = \dfrac{\sqrt{3}s}{2}$

$\text{Area} = \dfrac{\sqrt{3}s^2}{4}$

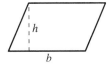

Parallelogram:

$\text{Area} = bh$

Trapezoid:

$\text{Area} = \dfrac{h}{2}(a + b)$

Circle:

$\text{Area} = \pi r^2$

$\text{Circumference} = 2\pi r$

Sector of Circle:

$\text{Area} = \dfrac{\theta r^2}{2}$

$s = r\theta$

θ in radians

Circular Ring:

$\text{Area} = \pi(R^2 - r^2)$

$\qquad = 2\pi pw$

p = average radius,

w = width of ring

Sector of Circular Ring:

$\text{Area} = \theta pw$

p = average radius,

w = width of ring,

θ in radians

Ellipse:

$\text{Area} = \pi ab$

$\text{Circumference} \approx 2\pi \sqrt{\dfrac{a^2 + b^2}{2}}$

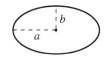

Cone:

$\text{Volume} = \dfrac{Ah}{3}$

A = area of base

Right Circular Cone:

$\text{Volume} = \dfrac{\pi r^2 h}{3}$

$\text{Lateral Surface Area} = \pi r \sqrt{r^2 + h^2}$

Frustum of Right Circular Cone:

$\text{Volume} = \dfrac{\pi(r^2 + rR + R^2)h}{3}$

$\text{Lateral Surface Area} = \pi s(R + r)$

Right Circular Cylinder:

$\text{Volume} = \pi r^2 h$

$\text{Lateral Surface Area} = 2\pi rh$

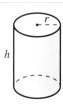

Sphere:

$\text{Volume} = \dfrac{4}{3}\pi r^3$

$\text{Surface Area} = 4\pi r^2$

Wedge:

$A = B \sec \theta$

A = area of upper face,

B = area of base

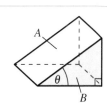